Amateur
Radio
Encyclopedia

Amateur Radio Encyclopedia

Stan Gibilisco, W1GV
Editor-in-Chief

TAB Books

Division of McGraw-Hill, Inc.

New York San Francisco Washington, D.C. Auckland Bogotá
Caracas Lisbon London Madrid Mexico City Milan
Montreal New Delhi San Juan Singapore
Sydney Tokyo Toronto

To Jack and Sherri

© 1994 by **Stan Gibilisco**.
Published by TAB Books.
TAB Books is a division of McGraw-Hill, Inc.

Printed in the United States of America. All rights reserved. The
publisher takes no responsibility for the use of any of the materials
or methods described in this book, nor for the products thereof.

pbk 2 3 4 5 6 7 8 9 10 11 MAL/MAL 9 9 8 7 6 5 4
hc   2 3 4 5 6 7 8 9 10 11 MAL/MAL 9 9 8 7 6 5 4

Library of Congress Cataloging-in-Publication Data

Gibilisco. Stan.
    Amateur radio encyclopedia / by Stan Gibilisco.
        p.   cm.
    Includes index.
    ISBN 0-8306-4095-9   ISBN 0-8306-4096-7 (pbk.)
    1. Amateur radio stations—Encyclopedias. I. Title.
TK9956.G467 1993
621.3841′6′03—dc20   92-35843
                                            CIP
                                            Rev.

Acquisitions Editor: Roland S. Phelps
Editor: Andrew Yoder
Director of Production: Katherine G. Brown
Designer: Jaclyn J. Boone   EE1
Paperbound Cover Design: Carol Stickles   4213

Contents 770973

How to use this book

This encyclopedia is intended as a permanent reference source for radio amateurs of all experience levels, as well as aspiring radio hams, shortwave listeners, and electronics hobbyists.

If you have a subject, device, or specialty in mind, look for it in the main body of articles, just as when using a dictionary. The cross references will provide a nearly unlimited source of reading when you start with a general subject. If your term is not an article title, consult the index. It contains far more terms than the main body of titles, and will guide you to one or more starting points in the book.

Suggestions for future editions are welcome. 73.

Stan Gibilisco, W1GV
Editor-in-Chief

Suggested additional reading

Numerous ham radio books are available. The best listings are in advertisements in ham-radio related magazines. The American Radio Relay League, Inc. has an excellent set of publications covering all aspects of ham radio. At the time of this writing, the major ham magazines are:

CQ, 76 N. Broadway, Hicksville, NY 11801
QST, 225 Main Street, Newington, CT 06111
73, Forest Road, Hancock, NH 03449

The two leading books for ham radio projects are:

American Radio Relay League, Inc., *The ARRL Handbook for Radio Amateurs* (published annually)
Orr, William, *Radio Handbook* (Howard W. Sams & Co., Inc.)

Foreword

Amateur radio, more familiarly known as ham radio, was the forerunner of our electronic age before we even knew we were in one! In an enthralling, hands-on manner, it has managed to keep its enthusiasts interested and abreast of communications technology, lured first by the novelty of talking across town, then across the nation, across the world, and ultimately with hams orbiting the earth in spacecraft.

Since the beginning of this century, radio hams have pioneered communications techniques and protocols, built antennas, learned to operate all kinds of equipment, and communicated with each other in endlessly diverse ways: morse code, voice, radioteletype, television, moonbounce, packet radio. It is an absorbing and fascinating hobby, often starting with shortwave listening, then leading many participants into careers undreamed of, and providing a way for those with physical challenges to reach the world.

Hams measure their operating skills against others in operating competitions and field events. They share experiences without any of the traditional limitations on geography, age, language, religion, ethnic background, or sex. Hams have a well-earned tradition of assisting their communities in times of disaster, when commercial communication is inoperative. It is a lifetime satisfying hobby that one can explore, enjoy, and grow with.

Living in today's technologically advanced world brings just about everyone in contact with the electronic age. For those beginning an association with ham radio, there are even more things to discover! But where does one start? Where do you look when you're not quite sure how much information you want on a given subject, nor how to go about finding it? That initial question in your mind needs a quick and easy-to-find answer. As your background grows, you'll have more questions needing answers. This is what this book is all about. It's an alphabetical, detailed database, starting you on the information track to a specific item — easy-to-find data, understandable and useful.

What an adventure awaits you if you're planning on entering the Amateur Radio service, and what a continuing and interesting challenge faces you as you explore the newer techniques melding computers with transmitters! The more than 500,000 hams in the U.S. alone, with millions more worldwide, bid you welcome with a heartfelt 73: best regards.

Ellen White, W1YL/4

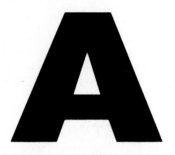

ABSORPTION

When any kind of energy or current is converted to some other form of energy, the transformation is referred to as *absorption* if it is dissipated and not used. Light is be converted to heat, for example, when it strikes a dark surface. Generally, absorption refers to an undesired conversion of energy.

Radio signals encountering the ionosphere undergo absorption as well as refraction. The amount of absorption depends on the wavelength, the layer of the ionosphere, the time of day, the time of year, and the level of sunspot activity. Sometimes most of the radio signal is converted to heat in the ionosphere; sometimes very little is absorbed and most of it is refracted or allowed to continue on into space. The amount of absorption in a given situation is called *absorptance*, and is expressed in *decibels* or *decibels per unit length*.

ABSORPTION WAVEMETER

An *absorption wavemeter* is a device for measuring radio frequencies. It consists of a tuned inductance-capacitance (LC) circuit, which is loosely coupled to the source to be measured. Knowing the resonant frequency of the tuned circuit for various capacitor or inductor settings (by means of a calibrated dial), the LC circuit is adjusted until maximum energy transfer occurs. This condition is indicated by a peaking of RF voltage, as shown by a meter.

When using an absorption wavemeter, it is extremely important that it not be too tightly coupled to the circuit under test. An RF probe, or short-wire pickup, should be connected to the LC circuit of the absorption wavemeter, and the probe brought near the source of RF energy. If too much coupling occurs, reactance can be introduced into the circuit under test, and this might in turn change its frequency, resulting in an inaccurate measurement.

A special kind of absorption wavemeter has its own oscillator built in. Known as a *grid-dip meter*, this device enables easy determination of the resonant frequency or frequencies of LC circuits and antenna systems.

ACCESS TIME

When accessing an electronic memory, such as in a microcomputer calculator, the data is not received instantaneously. Although the process might appear to be instantaneous, there is always a slight delay, called the *access time*. When storing information in an electronic memory, a delay occurs as well between the storage command and actual storage; this, too, is known as *access time*, or it might be called *storage time*. If the computer itself activates a memory circuit, the access time is the delay between arrival of the command pulse at the memory circuit and the arrival of the required information.

In small computers and calculators, access time is so brief that it can hardly be noticed. If much information must be stored or retrieved in a large computer, the access time might be much longer. This is especially likely if the computer time is simultaneously shared among many users. *See also* MEMORY.

ACCURACY

Physical quantities, such as time, distance, temperature, and voltage, can never be determined precisely; there is always some error. Standard instruments or objects are used as the basis for all measurements. The accuracy of a given instrument is expressed as a percentage difference between the reading of that instrument and the reading of a standard instrument. Mathematically, if a standard instrument reads x_s units and the device under test reads x_t units, then the accuracy, A in percent is:

$$A = 100 \, |x_s - X_t| / X_s$$

For meters, accuracy is usually measured at several points on the meter scale, and the accuracy taken as the greatest percentage error. Sometimes only the full-scale reading is used, then the accuracy is stated as a percentage of full scale. For example, a meter might be specified as being accurate to within $\pm$ 10% (plus or minus 10 percent); this guarantees that the meter reading is within 10 percent of the actual value. If full scale is specified, then the only tested point is at full scale. These figures assume that the meter reading is sufficient to cause significant deflection of the needle.

Electronic components, such as resistors and capacitors, are given tolerance factors in percentages, which are an expression of the accuracy of their values. Typical tolerances are 5, 10, and 20 percent. However, certain components are available with much lower tolerances (and therefore greater accuracy).

All units in the United States, and in the engineering world in general, are based on the meter-kilogram-second (MKS) system. The National Bureau of Standards is the ultimate basis in the U.S. for all determinations of accuracy. *See also* CALIBRATION, NATIONAL BUREAU OF STANDARDS, TOLERANCE.

AC GENERATOR

Mechanical energy is converted to electric current, with the aid of a magnetic field, in a device called an *electric generator*. Alternating current is produced by an ac generator.

The ac generator is essentially the same device as an ac motor, except that the conversion of energy occurs in reverse. By rotating the magnet inside the coil of wire, an alternating magnetic field is produced. This change in magnetic flux causes the electrons in the wire to be accelerated, first in one direction and then in the other. The frequency of alternation depends on the speed of rotation of the magnet.

Some ac generators use a rotating coil inside a magnet. The magnet can be operated from electricity itself. Ac generators might provide only a few milliwatts of power, or they might supply many thousands of watts, as is the case with commercial ac power generators.

Any radio-frequency transmitter or audio oscillator is, theoretically, an ac generator because it puts out alternating current. *See also* AUDIO FREQUENCY, and RADIO FREQUENCY.

AC NETWORK

An *alternating-current network* is a circuit that contains resistance and reactance. It differs from a direct-current network, where there is only simple resistance. Resistances are provided by all electrical conductors, resistors, coils, lamps, and the like; when a current passes through a resistance, heat is generated. Pure reactances, though, do not convert electrical current into heat. Instead, they store the energy and release it later. In storing energy, a reactance offers opposition to the flow of alternating current. Inductors and capacitors are the most elementary examples of reactances. Reactance is always either positive (inductive) or negative (capacitive). Certain specialized semiconductor circuits can be made to act like coils or capacitors by showing reactance at a particular ac frequency. Shorted or open lengths of transmission line also behave like reactances at some frequencies.

Although simple resistance is a one-dimensional quantity, and reactance is also one-dimensional (though it might be positive or negative), their combination in an ac network is two-dimensional. Resistance ranges from zero to unlimited values; reactance ranges from zero to unlimited positive or negative values. Their combination in an alternating-current circuit is called *impedance*. Any ac network has a net impedance at a given frequency. The impedance generally changes as the frequency changes, unless it is a pure resistance. Reactance is multiplied by the imaginary number j, called the j *factor*, for mathematical convenience. This quantity is defined as the square root of -1, also sometimes denoted by i. *See also* IMPEDANCE, J OPERATOR, and REACTANCE.

ACOUSTIC FEEDBACK

Positive feedback in a circuit, if it reaches sufficient proportions, will cause oscillation of an amplifier. In a public-address system, feedback might occur not only electrically, between the input and output component wiring, but between the speaker(s) and microphone. The result is a loud audible rumble, tone, or squeal. It might take almost any frequency, and totally disables the public-address system for its intended use. This kind of feedback might also occur between a radio receiver and transmitter, if both are voice modulated and operated in close proximity. This is called *acoustic feedback*.

Another form of acoustic feedback occurs in voice-operated communications systems (*see* VOX). While receiving signals through a speaker, the sound might reach sufficient amplitude to actuate the transmitter switching circuits. This results in intermittent unintended transmissions, and makes reception impossible. Compensating circuits in some radio transceivers equipped with VOX reduce the tendency toward this kind of acoustic feedback.

To prevent acoustic feedback in a public-address system, a directional microphone should be used, and all speakers should be located well outside the microphone-response field. The gain (volume) should be kept as low as possible consistent with the intended operation. The room or environment in which the system is located should have sound-absorbing qualities, if at all possible, to minimize acoustic reflection.

AC RELAY

An *alternating-current (ac) relay* is a device designed for the purpose of power switching from remote points (*see* RELAY). The ac relay differs from the dc relay in that it utilizes alternating current rather than direct current in its electromagnet. This offers convenience because no special power supply is required if the electromagnet is designed to operate from ordinary 117-Vac house outlets.

Ac relays are less likely than dc units to get permanently magnetized. This occurs when the core of the electromagnet becomes a magnet itself. In an ac relay, the magnetic field is reversed every time the direction of current flow changes so that one polarity is not favored over the other. Ac relays must be damped to a certain extent for 60-Hz operation, or the armature will attempt to follow the current alternations and the relay will buzz. The armature and magnetic pole-pieces in ac relays are specially designed to reduce this tendency to buzz.

AC RIPPLE

Alternating-current ripple, usually referred to simply as *ripple*, is the presence of undesired modulation on a signal or power source. The most common form of ripple is 60- or 120-Hz ripple that originates from ac-operated power supplies.

In theory, the output of any power supply will contain some ripple when the supply delivers current. This ripple can, and should, be minimized in practice because it will cause undesirable performance of equipment if not kept under control. Sufficient filtering, if used, will ensure that the ripple will not appear in the output of the circuit.

Ac ripple is virtually eliminated by using large inductors in series with the output of a power supply, and by connecting large capacitors in parallel with the supply output. The more current the supply is required to deliver, the more inductance and capacitance will be required in the filter stage. *See also* POWER SUPPLY.

ACTIVE COMPONENT

In any electronic circuit, certain components are directly responsible for producing gain, oscillation, switching action, or rectification. If such components draw power from an external source, such as a battery or power supply, then they are called *active components*. Active components include integrated circuits, transistors, vacuum tubes, and some diodes. An active component always requires a source of power to perform its function.

A passive component, by contrast, does its job with no outside source of power. Passive components include resistors, capacitors, inductors, and some diodes.

Some devices might act as either passive or active components, depending on the way they are used. One example is the varactor diode (*see* VARACTOR DIODE). When an audio-frequency voltage is applied across this type of diode, its junction capacitance fluctuates without any source of external power, and thus frequency modulation might be produced in an oscillator circuit. A dc voltage from a power supply, however, might be applied across the same diode, in the same circuit, for the purpose of adjusting the carrier frequency of the oscillator. In the first case, the varactor might be considered a passive component because it requires no battery power to produce FM from an audio signal. But in the second case, dc from an external source is necessary, and the varactor becomes an active component.

ACTIVE COMMUNICATIONS SATELLITE

Much of today's communications is done by means of relaying via satellites. This is true in amateur radio, as it is in commercial systems, such as telephone and television.

All modern satellites are of the *active type*. This means that they receive and retransmit the information. The earliest communications satellites, placed in orbit in the 1960s and known as *Echo satellites*, were *passive reflectors* of radio signals. The signals they returned were extremely weak, and they had to be physically large in order to function.

Early Amateur Satellites Active communications satellites can be placed in low, circular orbits. Then they pass close by overhead, and they do not need much transmitter power to be heard on the surface. But the range is limited to whatever horizon exists from the vantage point of the satellite. This will be only a small, and constantly changing, part of the earth if the orbit is low.

The first amateur satellites, called *OSCAR* for *Orbiting Satellite Carrying Amateur Radio*, were placed in low, nearly circular orbits. Contacts had to be carried out within a few minutes via these satellites. It was necessary to keep changing the azimuth and elevation bearings of the antenna in order to keep it pointed at the satellite, or else to use low-gain, omnidirectional, "turnstile" antennas at the surface. It was also necessary to keep constant track of orbital decay effects, because any satellite in low earth orbit is subject to atmospheric drag. Eventually, it will fall into the atmosphere and burn up.

Later Amateur Satellites Later OSCAR satellites have been placed in elliptical orbits. This causes them to swing out far away from the earth, and during this time, they move much more slowly than a satellite in low orbit. It is possible to access the satellite for periods of hours, instead of just minutes, when the satellite is at apogee. Also, a directional antenna can be left alone for some time, not needing constant readjustment. The effective range of the satellite is much greater, because it "sees" a larger part of the earth's surface from its higher altitude, but more gain is generally required for antennas at the surface. This means such antennas must be more complex and more expensive.

Future Amateur Satellites In the future, some amateur satellites will probably be placed in geostationary orbits. This allows for a constant azimuth/elevation setting for surface-station antennas. An antenna can be mounted in a fixed position, aimed at the satellite, and then left alone without the need for tracking, or for azimuth/elevation rotators. The satellite altitude is about 22,300 miles. Such satellites must orbit over the equator. This excludes regions at the very high latitudes, in the Arctic and Antarctic, from coverage. But three geostationary satellites, each spaced equally around the world at angles of 120 degrees, with respect to each other, provide for communications between almost all points in the civilized world. This is the case for most commercial satellite networks today.

Future amateur satellites can also be expected to operate over larger bandwidths, and probably also on higher frequency bands, than current amateur satellites. It will probably be possible to communicate using two satellites, if necessary. In this way, almost every amateur station in the world will have immediate, continuous access to almost every other amateur station, 24 hours a day, every day. The problems (and challenges) associated with ionospheric and tropospheric propagation will be gone. It will be, in essence, a world ham-radio telephone network.

Principles of Operation An amateur satellite *transponder* is actually a sophisticated form of repeater. There are some important differences, however, between a simple repeater, such as the kind you work through on 2-meter FM, and a satellite transponder.

The transponder converts signals from one band to signals in another band. It is, in this way, like a transmitting converter. The bands are portions of the amateur 10-meter, 2-meter, 70-centimeter and 13-centimeter bands. The figure is a simplified block diagram of an amateur active communications satellite, showing the antennas and the main components of the transponder.

The input band and output band are of the same width, but the output is often "upside-down," relative to the input, because of the conversion process. Therefore, if you increase the frequency of your transmitter signal, the output frequency from the satellite might go down, rather than up. In this case, upper-sideband uplink signals will be lower-sideband in the downlink, and vice-versa. In radioteletype, if this mode is allowed through the satellite, the mark and space signals will be reversed when the transponder is of the inverting type. There is no effect on CW signals.

The transponder can handle numerous signals all at once. This makes full duplex operation possible; your QSO can interrupt you while you talk to him/her. You can even listen to yourself. The satellite transponder might use batteries alone, or batteries with solar panels to charge them.

Satellite Use It is a fundamental rule in amateur satellite work that you never use more power than you need. This is a good rule (and actually a law) for all amateur communications, but with satellites, there is an added importance to it. If a single station uses far more power than necessary to access, or to communicate through, the satellite, the transponder will pay a disproportionate amount of attention to that one station. The result will be that the other stations' signals are greatly attenuated while the strong station is transmitting.

Normally, only about 4 watts maximum power is needed for low-orbiting satellites, and about 30 watts with high-altitude satellites. If significantly more power is used than this, the station might "hog" transponder power.

Signals in the downlink are sometimes inverted, relative to those in the uplink. The whole band thus comes out "upside down." In other cases, the band comes out "right-side up." The advantage of inverting transponders is that the Doppler effect is minimized, so that signals in the downlink do not change frequency as rapidly. In future geostationary satellites, Doppler shift will not be a factor.

For additional information, refer to the following terms: APOGEE, ASCENDING NODE, ASCENDING PASS, DESCENDING NODE, DESCENDING PASS, DOWNLINK, GEOSTATIONARY ORBIT, OSCAR, OSCARLOCATOR, PERIGEE, PHASE I SATELLITE, PHASE II SATELLITE, PHASE III SATELLITE, REPEATER, SATELLITE TRANSPONDER MODES, TRANSPONDER, and UPLINK.

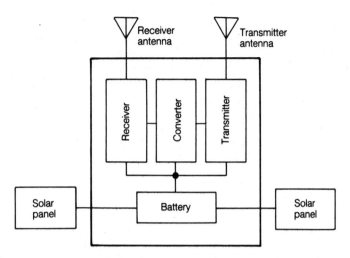

ACTIVE COMMUNICATIONS SATELLITE: Block diagram of the major components of an amateur active communications satellite.

ACTIVE FILTER

An *active filter* is a filter that uses active components to provide selectivity. Generally, active filters are used in the audio range.

Active filters might be designed to have predetermined selective characteristics, and are lightweight and small. Control is easily accomplished by switches and potentiometers. Such filters, being active devices, require a source of dc power, but because the filters consume very little current, a miniature 9-V transistor-radio type battery is usually sufficient to maintain operation for several weeks. Most audio filters use operational amplifiers or "op amps." Active filters are not often seen at radio frequencies.

ACTIVE REGION

In a bipolar transistor, class-A or class-AB amplification occurs when the collector voltage, as measured relative to ground in a common-emitter circuit, is larger than the base voltage. This is called the *active region* of the transistor. This region is between *cutoff* and *saturation*.

Power amplification is possible in the region at or beyond cutoff if there is sufficient drive. No amplification is possible when a transistor is in saturation. *See also* CLASS-A AMPLIFIER, CLASS-AB AMPLIFIER, CLASS-B AMPLIFIER, CLASS-C AMPLIFIER, and CUTOFF.

AC VOLTAGE

There are several different ways of defining voltage in an alternating-current circuit. They are called the *peak, peak-to-peak,* and *root-mean-square (RMS)* methods.

An ac waveform does not necessarily look like a simple sine wave. It might be square, sawtoothed, or irregular in shape. But whatever the shape of an ac waveform, the peak voltage is definable as the largest instantaneous value the waveform reaches. The *peak-to-peak voltage* is the difference between the largest instantaneous values the waveform reaches to either side of zero. Usually, the peak-to-peak voltage is exactly twice the peak voltage. However, if the waveform is not symmetrical, the peak value might be different in the negative direction than in the positive direction, and the peak-to-peak voltage might not be twice the positive peak or negative peak voltage.

The RMS voltage is the most commonly mentioned property of ac voltage. RMS voltage is defined as the dc voltage needed to cause the same amount of heat dissipation, in a simple, nonreactive resistor, as a given alternating-current voltage. For symmetrical, sinusoidal waveforms, the RMS voltage is 0.707 times the peak voltage and 0.354 times the peak-to-peak voltage. *See also* ROOT MEAN SQUARE.

ADAPTOR

Any device that makes two incompatible things work together is an *adaptor*. In electronics, adaptors are most frequently seen in cable connectors because there are so many different kinds of connectors. Such cable adaptors are sometimes called *tweenies* in the popular jargon.

Adaptors should be used as sparingly as possible, especially at radio frequencies, because they sometimes produce impedance irregularities along a section of feed line or cable. However, adaptors are invaluable in engineering and test situations as a convenience. Every test or service shop should have a good supply. *See also* CONNECTOR.

ADDER

In digital electronics, an adder is a circuit that forms the sum of two numbers. An adder is also a circuit that combines two binary digits and produces a carry output. Such a combination is simple in binary arithmetic.

In color television receivers, the circuit that combines the red, green, and blue signals from the receiver is called an adder. *See also* COLOR TELEVISION.

ADDRESS

Computer memory is stored in discrete packages for easy access. Each memory location bears a designator, usually a number, called the *address*. By selecting a particular address by number, the corresponding set of memory data is made available for use.

A calculator or small computer might have several different memory channels for storing numbers. Each channel is itself designated by a number, for example, 1 through 8. By actuating a memory-address function control, followed by the memory address number, the contents of the memory channel are called for use. Some radio receivers and transceivers make use of a memory-address function for convenience in calling, or switching among, frequently used frequencies. The memory-address status is shown by a panel indicator, such as an LCD display. *See also* MEMORY.

ADJACENT-CHANNEL INTERFERENCE

When a receiver is tuned to a particular frequency and interference is received from a signal on a nearby frequency, the effect is referred to as *adjacent-channel interference*.

To a certain extent, adjacent-channel interference is unavoidable. When receiving an extremely weak signal near an extremely strong one, interference is likely—especially if the stronger signal is voice modulated. No transmitter has absolutely clean modulation, and a small amount of off-frequency emission occurs with voice modulation—especially AM and SSB types. *See* AMPLITUDE MODULATION, SINGLE SIDEBAND, and SPLATTER.

Adjacent-channel interference might be reduced by using proper engineering techniques in transmitters and receivers. Transmitter audio amplifiers, modulators, and RF amplifiers should produce as little distortion as the state of the art will permit. Receivers should use selective filters of the proper bandwidth for the signals to be received and the adjacent-channel response should be as low as possible. A flat response in the passband (*see* PASSBAND), and a steep drop-off in sensitivity outside the passband, are characteristics of good receiver design. Modern technology has made great advancements in the area of receiver-passband selectivity.

ADMITTANCE

In some electronic circuit calculations, it is convenient to use a quantity called *admittance*. This is the reciprocal of impedance.

Admittance is a complex quantity, just as is impedance. The components of admittance are *conductance* (the reciprocal of resistance) and *susceptance* (the reciprocal of reactance). Symbolically, the abbreviations for the various quantities are:

$$
\begin{aligned}
Resistance &= R \\
Reactance &= X \\
Impedance &= Z \\
Conductance &= G \\
Susceptance &= B \\
Admittance &= Y
\end{aligned}
$$

Total admittance is the reciprocal of total impedance. That is, in simplified terms, $Y = 1/Z$. Also, $G = 1/R$ and $B = 1/X$.

Knowing the total resistance and reactance in a circuit, the conductance and susceptance can be found by the formulas:

$$
\begin{aligned}
G &= R/(R^2 + X^2) \\
B &= -X/(R^2 + X^2)
\end{aligned}
$$

Knowing the total conductance and the total susceptance, the resistance and reactance can be found by the formulas:

$$
\begin{aligned}
R &= G/(G^2 + B^2) \\
X &= -B/(G^2 + B^2)
\end{aligned}
$$

Total admittance is defined in terms of total conductance and total susceptance, according to the formula:

$$ Y = (G^2 + B^2)^{1/2} $$

Admittance is especially useful when determining the impedance of a network of resistances, capacitances and/or inductances in parallel. This is because admittances add in parallel, just as impedances add in series. Once the total admittance has been found, impedance Z can be determined simply by the reciprocal: $Z = 1/Y$. See also CONDUCTANCE, IMPEDANCE, J OPERATOR, and REACTANCE.

AFC

See AUTOMATIC FREQUENCY CONTROL.

AGC

See AUTOMATIC GAIN CONTROL.

AIR COOLING

Components that generate great amounts of heat, such as vacuum tubes, transistor power amplifiers, and some resistors, must be provided with some means for cooling or damage might result. Such components might be air cooled or conduction cooled (*see* CONDUCTION COOLING). Air cooling might occur as heat radiation, or as convection.

In high-powered vacuum-tube transmitters, a fan is usually provided to force air over the tubes or through special cooling fins (see the illustration). By using such fans, greater heat dissipation is possible than would be the case without them, and this allows higher input and output power levels.

Low-powered transistor amplifiers use small heatsinks to conduct heat away from the body of the transistor (*see* HEATSINK). The heatsink might then radiate the heat into the atmosphere as infrared energy, or the heat might be dissipated into a large, massive object, such as a block of metal. Ultimately, however, some of the heat from conduction-cooled equipment is dissipated in the air as radiant heat.

With the increasing use of solid-state amplifiers in radio and electronic equipment, conduction cooling is becoming more common, replacing the air blowers that are so often seen in tube-type amplifiers. Conduction cooling is quieter and requires no external source of power.

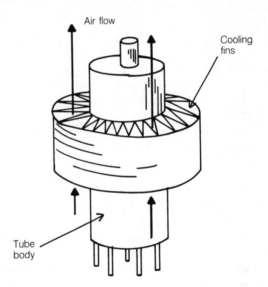

AIR COOLING: The air passes through fins to carry heat away by convection.

AIR CORE

The term *air core* is usually applied in reference to inductors or transformers. At higher radio frequencies, air-core coils are used because the required inductances are small. Powdered-iron and ferrite cores greatly increase the inductance of a coil, as compared to an air core (*see* FERRITE CORE). This occurs because such materials cause a concentration of magnetic flux within the coil. The magnitude of this concentration is referred to as permeability (*see* PERMEABILITY); air is given, by convention, a permeability of 1 at sea level.

Air-core inductors and transformers might be identified in schematic diagrams by the absence of lines near the turns. In a ferrite or powdered-iron core, two parallel straight lines indicate the presence of a permeability-increasing substance in the core (see A and B). Coils are sometimes wound on forms made of dielectric material, such as glass or bakelite. Because these substances have essentially the same permeability as air (with minor differences), they are considered air-core inductors in schematic representations.

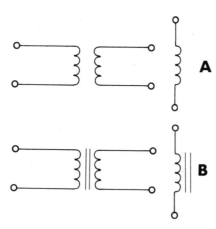

AIR CORE: At A, schematic symbols for air core transformer and inductor. At B, symbols for transformer and coil with ferromagnetic core.

AIR-SPACED COAXIAL CABLE

A coaxial cable might have several kinds of dielectric material. Polyethylene is probably the most common. If loss is to be minimized, air-dielectric coaxial cable is the best.

The principal difficulty in designing air-spaced coaxial cable is the maintenance of proper spacing between the inner conductor and the shield. Usually, disk-shaped pieces of polyethylene or other solid dielectric material are positioned at intervals inside the cable (see illustration). Although sharp bends in an air-dielectric coaxial cable cannot be made without upsetting the spacing and possibly causing a short circuit, the disks keep the center conductor properly positioned while affecting very little the low-loss characteristics of the air dielectric. Each disk is kept in place by adhesive material or bumps or notches in the center conductor.

It is important that moisture be kept from the interior of an air-spaced coaxial cable. If water gets into such a cable, the low-loss properties and characteristic impedance are upset (*see* CHARACTERISTIC IMPEDANCE).

A solid, rigid, coaxial cable, which might have either air or solid dielectric, is called a *hard line* (*see* HARD LINE).

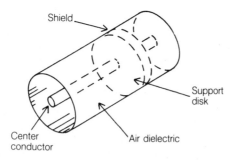

AIR-SPACED COAXIAL CABLE: Supporting disks or beads keep the center conductor spaced away from the shield.

AIR-VARIABLE CAPACITOR

An *air-variable capacitor* is a device whose capacitance is adjustable, usually by means of a rotating shaft. One rotating and one fixed set of metal plates are positioned in meshed fashion with rigidly controlled spacing. Air forms the dielectric material for such capacitors. The capacitance is set to the desired value by rotating one set of plates, called the *rotor plates*, to achieve a cer-

tain amount of overlap with a fixed set of plates, called the *stator plates*. The rotor plates are usually connected electrically to the metal shaft and frame of the unit. The photo shows a common type of air-variable capacitor.

Air-variable capacitors come in many physical sizes and shapes. For receiving, and low-power RF transmitting applications, the plate spacing might be as small as a fraction of a millimeter. At high levels of RF power, the plates might be spaced an inch or more apart. The capacitance range of an air variable has a minimum of a few picofarads (abbreviated pF and equal to 10^{-12} farad or 10^{-6} microfarad) and a maximum that might range from about 10 to 1000 pF. The maximum capacitance depends on the size and number of plates used, and on their spacing.

Because the dielectric material in an air-variable capacitor is air, which has a small amount of loss, air variables are efficient capacitors, as long as they are not subjected to excessive voltages that result in flashover. A special kind of variable capacitor is the *vacuum-variable*, a variable capacitor that is placed in an evacuated enclosure. Such capacitors are even more efficient than air variables.

Air-variable capacitors are frequently found in tuned circuits, RF power-amplifier output networks, and antenna matching systems. *See also* ANTENNA MATCHING, and TRANSMATCH.

AIR-VARIABLE CAPACITOR: A tuning capacitor in a transmatch.

ALC

See AUTOMATIC LEVEL CONTROL.

ALEXANDERSON ANTENNA

An antenna for use at low or very low frequencies, the *Alexanderson antenna* consists of several base-loaded vertical radiators that are connected together at the top and fed at the bottom of one radiator.

At low frequencies, the principal problem with transmitting antenna design is the fact that any radiator of practical height has an exceedingly low radiation resistance (*see* RADIATION RESISTANCE) because the wavelength is so large. This results in severe loss — especially in the earth near the antenna system. By arranging several short, inductively loaded antennas in parallel, and coupling the feed line to only one of the radiators, the effective radiation resistance is greatly increased. This improves the efficiency of the antenna, because more of the energy from the transmitter appears across the larger radiation resistance.

The Alexanderson antenna has not been extensively used at frequencies above the standard AM broadcast band. But where available ground space limits the practical height of an antenna and prohibits the installation of a large system of ground radials, the Alexanderson antenna could be a good choice at frequencies as high as perhaps 5 MHz. The Alexanderson antenna requires a far-less elaborate system of ground radials than a single-radiator vertical antenna worked against ground. The radiation resistance of an Alexanderson array, as compared to a single radiator of a given height, increases according to the square of the number of elements. *See also* INDUCTIVE LOADING.

ALLIGATOR CLIP

For electronic testing and experimentation, where temporary connections are needed, alligator clips (also known as *clip leads*, although there are other kinds of clip leads) are often used. They are easy to use and require no modification to the circuit under test.

Alligator clips come in a variety of sizes, ranging from less than ½ inch long to several inches long. They are clamped to a terminal or a piece of bare wire. Although such clips are convenient for temporary use, they are not good for long-term installations because of their limited current-carrying capacity and the tendency toward corrosion, especially outdoors.

See the drawing, which shows common alligator clips. They derive their somewhat humorous name from their visible resemblance to the mouth of an alligator! *See also* CLIP LEAD.

ALLIGATOR CLIP: These are useful for temporary connections.

ALLOY-DIFFUSED SEMICONDUCTOR

Some semiconductor junctions are formed by a process called *alloy diffusion*. A semiconductor wafer of p-type or n-type material forms the heart of the device. An impurity metal is heated to its melting point and placed onto the semiconductor wafer. As the impurity metal cools, it combines with the semiconductor material to form a region of the opposite type from the semiconductor wafer.

A transistor might be formed in this manner by starting with a wafer of n-type semiconductor. A small amount of metal, such as indium, is melted on each side of this wafer, and the melting process is continued so that the indium diffuses into the n-type wafer. This gives the effect of *doping* (creating an alloy with) the n-type material next to the indium. (*See* DOPING.) Indium is an acceptor impurity (*see* IMPURITY), and thus two p-type regions are formed on either side of the n-type material. The result is a pnp transistor.

Alloy-diffused semiconductor transistors can be made to have extremely thin base regions. This makes it possible to use the transistor at very high frequencies.

ALL-PASS FILTER

An *all-pass filter* is a device or network that is designed to have constant attenuation at all frequencies of alternating current. However, a phase shift might be introduced, and this phase shift is also constant for all alternating-current frequencies. In practice, the attenuation is usually as small as possible (*see* ATTENUATION).

All-pass filters are generally constructed using noninverting operational amplifiers. The amount of phase delay is determined by the values of resistor R and capacitors C. The amount of attenuation is regulated by the values of the other resistors.

ALPHA

In a transistor, the ratio between a change in collector current and a change in emitter current is known as the *alpha* for that particular transistor. Alpha is represented by the first letter of the Greek alphabet, lowercase (α). Alpha is determined in the grounded-base arrangement.

The collector current in a transistor is always smaller than the emitter current. This is because the base draws some current from the emitter-collector path when the transistor is forward biased. Generally, the alpha of a transistor is given as a percentage:

$$\alpha = 100 \ (I_c/I_e),$$

where I_c is the collector current and I_e is the emitter current.

Transistors typically have alpha values from 95 to 99 percent. Alpha must be measured, of course, with the transistor biased for normal operation.

ALPHA-CUTOFF FREQUENCY

As the frequency through a transistor amplifier is increased, the amplification factor of the transistor decreases. The current gain, or beta (*see* BETA) of a transistor is measured at a frequency of 1 kHz with a pure sine-wave input for reference when determining the alpha-cutoff frequency. Then, a test generator must be used, which has a constant output amplitude over a wide range of frequencies. The frequency to the amplifier input is increased until the current gain in the common-base arrangement decreases by 3 dB, with respect to its value at 1 kHz. A decrease in current gain of 3 dB represents a drop to 0.707 of its previous magnitude. The frequency at which the beta is 3 dB below the beta at 1 kHz is called the *alpha-cutoff frequency* for the transistor.

Depending on the type of transistor involved, the alpha-cutoff frequency might be only a few MHz, or perhaps hundreds of MHz. The alpha-cutoff frequency is an important specification in the design of an amplifier. An alpha-cutoff frequency that is too low for a given amplifier requirement will result in poor gain characteristics. If the alpha-cutoff frequency is unnecessarily high, expense becomes a factor; and under such conditions there is a greater tendency toward unwanted VHF parasitic oscillation (*see* PARASITIC OSCILLATION).

As the input frequency is increased past the alpha-cutoff frequency, the gain of the transistor continues to decrease until it reaches unity (zero dB). At still higher frequencies, the gain becomes smaller than unity. *See also* DECIBEL, and GAIN.

ALTERNATE ROUTING

When the primary system for communications between two points breaks down, a backup system must be used to maintain the circuit. Such a system, and its deployment, is called *alternate routing*. Alternate routing might also be used in power transmission, in case of interruption of a major power line, to prevent prolonged and widespread blackouts. Power from other plants is routed to cities that are affected by the failure of one particular generating plant or transmission line.

As an example, the primary communications link for a particular system might be via a geostationary satellite (as shown in A). If the satellite fails, another satellite can be used in its place if one is available (B). This is alternate routing. If the second satellite ceases to function or is not available, further backup systems might be used, such as an HF shortwave link (C) or telephone connection (D). Alternate routing systems should be set up and planned in advance, before the primary system goes down, so that communications might be maintained with a minimum of delay.

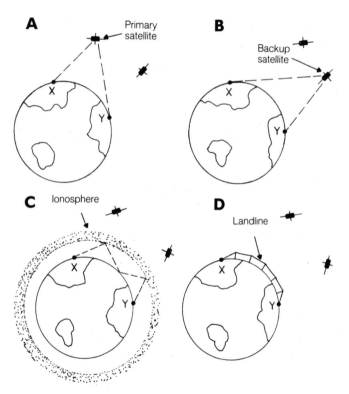

ALTERNATE ROUTING: This is used when primary communication (A) fails. At B, backup satellite; at C, sky wave; at D, wire communication.

ALTERNATOR

See AC GENERATOR.

ALUMINUM

Aluminum is a dull, light, somewhat brittle metal, atomic number 13, atomic weight 27, commonly used as a conductor for electricity. Although it does break rather easily, aluminum is very strong in proportion to its weight, and has replaced much heavier metals, such as steel and copper, in many applications.

Aluminum, like other metallic elements, is found in the earth's crust. It occurs in a rock called *bauxite*. Recent advances

in mining and refining of bauxite have made aluminum one of the most widely used, and inexpensive, industrial metals.

Aluminum is fairly resistant to corrosion, and is an excellent choice in the construction of communications antenna systems. Hard aluminum tubing is available in many sizes and thicknesses. The do-it-yourself electronics hobbyist can build quite sophisticated antennas from aluminum tubing purchased at a hardware store. Most commercially manufactured antennas use aluminum tubing.

Soft aluminum wire is used for grounding systems in communications and utility service. Large-size aluminum wire often proves the best economic choice for such applications. Some municipalities, however, require copper wire for grounding. *See also* LIGHTNING PROTECTION.

AMATEUR RADIO EMERGENCY SERVICE

The American Radio Relay League (ARRL) has numerous public-service organizations. One of these is the *Amateur Radio Emergency Service (ARES)*.

The ARES is organized for each ARRL section. The Section Manager, an elected ARRL official, is in charge of ARES communications for each section. Within sections, local branches operate.

Membership in ARRL is not required for participation in ARES. The primary purpose is to give public assistance; this is an essential aspect of ham radio, and helps to justify the existence of the hobby. Detailed information about ARES can be obtained by writing to the Communications Manager, ARRL Headquarters, Newington, CT 06111. *See also* AMERICAN RADIO RELAY LEAGUE.

AMERICAN RADIO RELAY LEAGUE

Electromagnetic communication has been in existence only since about 1900. Before the Twentieth Century, physics labs were the only places in which such phenomena were even observed, much less put to use. But within just a few years after the discovery of electromagnetic propagation, communicators began to compete for space in the electromagnetic spectrum.

In modern societies, commercial and government interests tend to prevail over amateur interests, because of economic and political factors. One radio amateur, Hiram Percy Maxim, saw the need for an organization to consolidate the power of radio hams, lest amateur privileges eventually be lost. In 1914, the American Radio Relay League (ARRL) was founded by Maxim and some close friends near Hartford, Connecticut. The ARRL is often called simply "the League."

Today, ARRL is a worldwide organization headquartered in Newington, Connecticut. The ARRL works closely with the Federal Communications Commission (FCC), federal legislators, and the International Telecommunication Union (ITU) to ensure that amateur radio continues to exist. The worldwide equivalent of ARRL is the International Amateur Radio Union (IARU). Some other countries have their own "leagues," such as England (Radio Society of Great Britain, RSGB) and Japan (Japan Amateur Radio League, JARL).

The ARRL has done far more than merely keep hostile interests from taking away all radio hams' privileges. Numerous publications are available to help new hams learn about the multiple facets of this hobby. The ARRL maintains a code-practice and bulletin station, bearing Hiram Percy Maxim's original call letters, W1AW. Many other services and activities are carried on by this organization. More than 150,000 American hams belong.

Membership is open to anyone. Even nonhams can be associate members. But to be a full member of ARRL, it is necessary to have at least a novice-class ham license.

For information about ARRL, write to ARRL Headquarters, 225 Main Street, Newington, CT 06111. *See also* INTERNATIONAL AMATEUR RADIO UNION, and W1AW.

AMATEUR RADIO SATELLITE CORPORATION

See RADIO AMATEUR SATELLITE CORPORATION.

AMERICAN MORSE CODE

The *American Morse Code* is a system of dot and dash symbols, first used by Samuel Morse in telegraph communications. The American Morse Code is not widely used today. It has been largely replaced by the International Morse Code. Some telegraph operators still use American Morse.

The American Morse Code differs from the International Morse Code. The American Morse symbols are sometimes called "Railroad Morse." Some letters contain internal spaces. This causes confusion for operators who are familiar with International Morse Code. Some letters are also entirely different between the two codes. *See also* INTERNATIONAL MORSE CODE.

AMATEUR TELEVISION

The term *amateur television (ATV)* refers to either fast-scan or slow-scan television communications by radio hams, with or without accompanying audio. Fast-scan ATV is generally used at UHF and microwave frequencies, because the signals require several megahertz of bandwidth. In fact, ATV signals are just like broadcast signals, except that the power levels are much lower. The signals can be either black-and-white or color. *See* COLOR TELEVISION, and TELEVISION.

Slow-scan TV, abbreviated SSTV, can be used on any ham band. The signals take up only about 3 kHz of spectrum space, the same as a single-sideband (SSB) voice transmission. The pictures can be either black-and-white or in color. *See* COLOR SLOW-SCAN TELEVISION, and SLOW-SCAN TELEVISION.

AMERICAN WIRE GAUGE

Metal wire is available in many different sizes or diameters. Wire is classified according to diameter by giving it a number. The designator for a given wire is known as the *American Wire Gauge (AWG)*. In England, a slightly different system is used (*see* BRITISH STANDARD WIRE GAUGE). The numbers in the American Wire Gauge system range from 1 to 40, although larger and smaller gauges exist. The higher the AWG number, the thinner the wire.

The table shows the diameter vs AWG designator for AWG 1 through 40. The larger the AWG number for a given conductor metal, the smaller the current-carrying capacity becomes. The AWG designator does not include any coatings on the wire such as enamel, rubber, or plastic insulation. Only the metal part of the wire is taken into account.

AMERICAN WIRE GAUGE: AMERICAN WIRE GAUGE EQUIVALENTS IN MILLIMETERS.

AWG	Dia., mm	AWG	Dia., mm
1	7.35	21	0.723
2	6.54	22	0.644
3	5.83	23	0.573
4	5.19	24	0.511
5	4.62	25	0.455
6	4.12	26	0.405
7	3.67	27	0.361
8	3.26	28	0.321
9	2.91	29	0.286
10	2.59	30	0.255
11	2.31	31	0.227
12	2.05	32	0.202
13	1.83	33	0.180
14	1.63	34	0.160
15	1.45	35	0.143
16	1.29	36	0.127
17	1.15	37	0.113
18	1.02	38	0.101
19	0.912	39	0.090
20	0.812	40	0.080

AMMETER

An *ammeter* is a device for measuring electric current. The current passes through a set of coils, which causes rotation of a central armature. An indicator needle attached to this armature shows the amount of deflection against a graduated scale. Ammeters might be designed to have a full-scale deflection as small as a few microamperes (*see also* AMPERE), up to several amperes.

To extend the range of an ammeter, allowing it to register very large currents, a resistor of precisely determined value is placed in parallel with the meter coils. This resistor diverts much, or most, of the current so that the meter actually reads only a fraction of the current.

Ammeters might be used as voltmeters by placing a resistor in series with the meter coils. Then, even a very high voltage will cause a small deflection of the needle (*see* VOLTMETER). The greatest accuracy is obtained when a sensitive ammeter is used with a large-value resistor. This minimizes the current drawn from the circuit. Ammeters should never be connected across a source of voltage without a series resistor because damage to the meter mechanism might result.

Ammeters are available for measuring both ac and dc. Some ammeters register RF current. The devices must be specially designed for each of these applications.

AMPERE

The *ampere* is the unit of electric current. A flow of one coulomb per second, or 6.28×10^{18} electrons per second, past a given fixed point in an electrical conductor, is a current of one ampere.

Various units smaller than the ampere are often used to measure electric current. A milliampere (mA) is one thousandth of an ampere, or a flow of 6.28×10^{15} electrons per second past a given fixed point. A microampere (μA) is one millionth of an ampere, or a flow of 6.28×10^{12} electrons per second. A nanoampere (nA) is a billionth of an ampere; it is the smallest unit of electric current you are likely to use. It represents a flow of 6.28×10^{9} electrons per second past a given fixed point.

A current of one ampere is produced by a voltage of one volt across a resistance of one ohm. This is Ohm's law (*see* OHM'S LAW). The ampere is applicable to measurement of alternating current or direct current.

AMPERE'S LAW

The direction of an electric current is generally regarded by physicists and engineers as the direction of positive charge transfer. This is opposite to the direction of the movement of the electrons themselves because electrons carry a negative charge. By convention, when speaking of the direction of current, it is considered to move from the positive to the negative terminal of a battery or power supply.

According to Ampere's law, the magnetic field or flux lines generated by a current in a wire travel counterclockwise when the current is directed toward the observer. This rule is sometimes also called the *right-handed screw rule for magnetic-flux generation*. A more universal rule for magnetic flux, applying to motors and generators, is called *Fleming's Rule*. As the right hand is held with the thumb pointed outward and the fingers curled, a current in the direction pointed by the thumb will generate a magnetic field in the circular direction pointed by the fingers.

AMPERE TURN

The *ampere turn* is a measure of magnetomotive force. One ampere turn is developed when a current of one ampere flows through a coil of one turn, or, in general, when a current of $1/n$ amperes flows through a coil of n turns.

One ampere turn is equal to 1.26 gilberts. The gilbert is the conventional unit of magnetomotive force. *See also* GILBERT, and MAGNETOMOTIVE FORCE.

AMPLIFICATION

Amplification refers to any increase in the magnitude of a current, voltage, or wattage. Amplification makes it possible to transmit radio signals of tremendous power, sometimes over a million watts. Amplification also makes it possible to receive signals that are extremely weak. It allows the operation of such diverse instruments as light meters, public-address systems, and television receivers.

Usually, amplification involves increasing the magnitude of a change in a certain quantity. For example, a fluctuation from -1 to $+1$ volt, or 2 volts peak-to-peak, might be amplified so that the range becomes 0 to $+10$ volts, or five times greater. Alternating currents, when amplified, produce effective voltage gain (if the impedance is correct) and power gain.

Direct-current amplification is usually done with the intention of increasing the sensitivity of a meter or other measuring instrument. Alternating-current amplification is used primarily in audio-frequency and radio-frequency applications, for the purpose of receiving or transmitting a signal. *See also* AMPLIFIER.

AMPLIFICATION FACTOR

Amplification factor is the ratio of the output amplitude in an amplifier to the input amplitude. The quantity is expressed for current, voltage or power, and is abbreviated by the Greek lowercase letter μ.

Usually, amplification factor is determined from the peak-to-peak voltage or current. If the voltage amplification factor for an amplifier is a certain value, the current amplification factor need not be, and probably will not be, the same. Voltage or current gain is determined from the voltage or current amplification factor by the formula:

$$Gain \text{ (dB)} = 20 \log_{10} \mu$$

Power gain is related to the power-amplification factor by the equation:

$$Gain \text{ (dB)} = 10 \log_{10} \mu$$

See also DECIBEL, and GAIN.

AMPLIFICATION NOISE

All electronic circuits generate some noise. In an amplifier, the transistor, or integrated circuit invariably generates some noise. This is called *amplification noise*. Amplification noise might be categorized as either thermal, electrical, or mechanical.

The molecules of all substances, including the metal and other materials in an electronic circuit, are in constant random motion. The higher the temperature, the more active the molecules. This generates thermal noise in any amplifier.

As the electrons in a circuit hop from atom to atom, or impact against the metal anode of a vacuum tube, electrical noise is generated. The larger the amount of current flowing in a circuit, in general, the more electrical noise there will be.

Mechanical noise is produced by the vibration of the circuit components in an amplifier. Sturdy construction and, if needed, shock-absorbing devices, reduce the problem of mechanical noise.

Although nothing can be done about the thermal noise at a given temperature, some equipment is cooled to extremely low temperatures to minimize thermal noise. Electrical noise is reduced and improved by the use of certain types of amplifying devices, such as the field-effect transistor. It is important that noise is kept as low as possible in the early stages of a multistage circuit because any noise generated in one amplifier will be picked up and amplified, along with the desired signals, in succeeding stages. *See also* NOISE.

AMPLIFIER

An *amplifier* is any circuit that increases the amplitude of a signal. The circuit might amplify voltage or current, or both. Some amplifiers are intended for direct current (dc); many kinds work with alternating current (ac) from extremely low to superhigh frequencies.

The most common devices intended for use as amplifiers are bipolar transistors and field-effect transistors. These might be discrete components, but increasingly, they are fabricated in integrated circuits (ICs). A typical IC amplifier has a gain of many thousands of times.

For further information, *see*: AUDIO AMPLIFIER, CLASS-A AMPLIFIER, CLASS-AB AMPLIFIER, CLASS-B AMPLIFIER, CLASS-C AMPLIFIER, DC AMPLIFIER, DIFFERENTIAL AMPLIFIER, FINAL AMPLIFIER, POWER AMPLIFIER, PUSH-PULL AMPLIFIER, and PUSH-PULL GROUNDED GRID/BASE/GATE AMPLIFIER.

AMPLITUDE

The strength of a signal is called its *amplitude*. Amplitude can be defined in terms of current, voltage, or power for any given signal.

Knowing the root-mean-square (*see* ROOT MEAN SQUARE) current, I, and the root-mean-square voltage, E, for a particular ac signal, the power amplitude in watts is given by:

$$P = IE$$

If we know the circuit impedance, Z, and either the current or voltage, then the power amplitude is:

$$P = I^2 \times Z = E^2/Z$$

Amplitude is usually described in reference to the strength of a radio-frequency signal, either at some intermediate point in a receiver or transmitter circuit, or at the output of a transmitter. Amplitude is measured using a wattmeter or oscilloscope or it can also be measured using a spectrum analyzer (*see* SPECTRUM ANALYZER). On an oscilloscope, signals of increasing amplitude (*see* A, B, and C) appears as waveforms of greater and greater height, but of the same wavelength, assuming the frequency remains constant.

For weak signals at the antenna terminals of a receiver, the term *strength* is usually used. Such signals are measured in microvolts. *See also* SENSITIVITY.

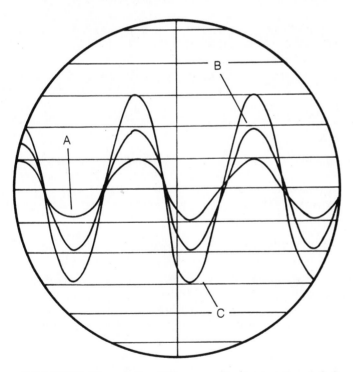

AMPLITUDE: Waves A, B and C have amplitude proportions 1 : 2 : 3.

AMPLITUDE-COMPANDORED SINGLE SIDEBAND

During normal speech, the human voice has a dynamic range of more than 20 dB, or a power ratio of 100 : 1. Peaks tend to be brief and sharp. As a result of this, the average power of a single-sideband (SSB) signal is only a small fraction of the peak power.

By boosting the low-level components of speech, the ratio of average to peak power in an SSB signal is increased. This can be done by means of a *speech-compression* circuit (see SPEECH COMPRESSION). Speech compression differs from *speech clipping*, another method of increasing the ratio of average power to peak power in a voice signal. *See* SPEECH CLIPPING.

At the receiver, the amplitude variations of a speech-compressed signal are spread back out again, so that the voice sounds normal. This mode is called *amplitude-compandored single sideband (ACSSB)*. The term *compandor* derives from the words *compressor* (the circuit at the transmitter) and exp*andor* (the circuit at the receiver).

The main advantage of ACSSB is that it improves the signal-to-noise ratio. Weaker voice components, that might be masked by the noise, are boosted to a level well above the noise. At the receiver, the background noise tends to be reduced in amplitude by the expandor circuit.

A potential problem with ACSSB is that it puts an increased work load on the transmitter final amplifier transistor(s) or tube(s). The average input power, as well as the average output power, is increased. This causes an increase in the collector or plate dissipation. When using ACSSB, it is important to ensure that the final amplifier can tolerate this increased work load. *See also* COMPANDOR.

AMPLITUDE MODULATION

Amplitude modulation is the process of impressing information on a radio-frequency signal by varying its amplitude. The simplest example of amplitude modulation (AM) is probably Morse-code transmissions, where the amplitude switches from zero to maximum.

Generally, amplitude modulation is done for the purpose of relaying messages by voice, television, facsimile, or other modes that are relatively sophisticated. The process is always the same: audio or low frequencies are impressed upon a carrier wave of much higher frequency. A, B, and C illustrate the amplitude modulation of a carrier wave by a sine-wave or sinusoidal audio tone. The amplitude of the carrier is greatest on positive peaks of the sinusoidal tone, and smallest on negative peaks.

The modulation of an AM signal might be considerable, or it might be very small. The degree of modulation is expressed as a percentage. This percentage might vary from zero to more than 100. An unmodulated carrier, as shown in the figure, has zero percent modulation by definition. If the negative peaks drop to zero amplitude, the signal is defined to have 100-percent modulation. At C, we see a signal with modulation of about 75 percent. If the modulation percentage exceeds 100, the negative peaks drop to zero amplitude and remain there for a part of the audio cycle. This is undesirable, because it causes distortion of the information reproduced by the receiver.

When a given radio-frequency signal is amplitude modulated, mixing occurs (*see* MIXER) between the modulation frequencies (f_M) and the carrier frequency (f_C). For sine-wave modulation (such as shown in the figure), this mixing results in new radio-frequency signals, f_{LSB} and f_{USB}, given by:

$$f_{USB} = f_C - f_M$$
$$f_{USB} = f_C + f_M$$

These new frequencies are called *sidebands*. They are referred to as the *upper sideband* (USB) and *lower sideband* (LSB).

A special kind of amplitude modulation, commonly called *single sideband (SSB)*, but properly named *single-sideband, sup-*

pressed-carrier (SSSC), eliminates the carrier frequency (f_C) and also one of the sideband frequencies at the transmitter. Only one sideband is left at the output. This sideband is combined with a local oscillator signal at the receiver, resulting in a perfect reproduction of the modulating signal, provided the receiver frequency is correctly set. *See also* SINGLE SIDEBAND.

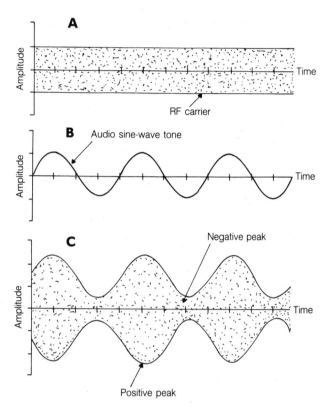

AMPLITUDE MODULATION: At A, the unmodulated carrier; at B, the modulating waveform; at C, the AM signal.

AMSAT
See RADIO AMATEUR SATELLITE CORPORATION.

AMTOR
In recent years, methods have been devised to minimize errors in radioteletype (RTTY) communications. In ham radio, the most commonly used system is *AMTOR*, an acronym that stands for *amateur teleprinting over radio*.

In one type of AMTOR, the sending and receiving stations are in constant contact. The receiving station analyzes the signal, and if there is any doubt about its accuracy, it asks the transmitting station to send a certain piece of data again. In the other mode of AMTOR, as in conventional RTTY, the sending station transmits "blind," but sends each character twice.

On the air, AMTOR signals can be recognized by their "bleep-bleep-bleep" sound. Often the receiving station can also be heard, as its transmitter sends back error checking and correction signals. AMTOR is found on many of the same frequencies as conventional RTTY.

There are two modes in AMTOR. They are called *ARQ*, for *automatic repeat request*, and *FEC*, for *forward error correction*.

ARQ Mode In ARQ mode, characters are sent in groups of three. After every three characters, the receiving station sends a

signal back, either ACK (for acknowledged) or NAK (not acknowledged). If the sending station gets an ACK signal, it sends the next three characters. If it gets a NAK, it repeats the three characters it just sent, until those three characters are acknowledged.

The error-correction feature of ARQ mode only works between the two stations actually in QSO with each other. If a third station listens in, that station can copy either of the two in QSO, but error correction does not take place. The ARQ mode in AMTOR is sometimes called *mode A*.

FEC Mode In this mode, every character is sent twice. The second character starts 0.35 seconds after the first one starts. This is enough time to minimize the chances for a QRN (noise) or QRM (signal) burst to obliterate both characters.

All RTTY characters have a characteristic format. In FEC mode, the receiving station tests the characters to see if they are in the proper format. If one or both of the characters have the right format, then the receiving terminal prints or displays the character, because it is "probably" right. If neither character has the correct format, the receiving terminal rejects that character and prints a blank.

An FEC transmission can be monitored by any number of stations. The transmitter sends at a 100-percent duty cycle, or "blind," just as in ordinary RTTY.

The FEC mode of AMTOR is sometimes called *mode B*.

For Further Information The finer technical details of AMTOR are beyond the scope of this volume. This information can be found in various articles and books on the subject. Operating techniques and regulations are also discussed in these publications. Check the ham magazine ads for the most current available titles. *See also* RADIOTELETYPE.

ANALOG
Quantities or representations that are variable over a continuous range are referred to as being *analog*. In electronics, analog quantities are differentiated from digital quantities by the fact that analog variables can take an infinite number of values, but digital variables are limited to defined states.

Examples of analog quantities include the output of an amplitude-modulated, single-sideband transmitter. The amplitude of such a signal fluctuates over a continuous range from zero to the maximum, or peak, output. An example of a digital amplitude-modulated signal is the output of a cw transmitter sending Morse code. This signal has only two states: *on* and *off*.

Although analog information usually provides more accuracy in reproducing a quantity, variable, or signal, digital information is transferred with greater efficiency. The difference is evident in the example just given. While voice inflections in an AM or SSB signal enable the transmission of emotions, a cw signal travels with better signal-to-noise ratio and hence greater efficiency. When a voice signal becomes unreadable because of interference or poor conditions, a cw signal is often still intelligible. *See also* DIGITAL.

ANALOG CONTROL
A control or adjustment that is variable over a continuous range is an *analog control*. Examples of analog controls are the frequency dials on some communications transmitters and receivers (see illustration), volume controls, and RF gain controls.

Analog controls are sometimes preferred over digital controls because the operator of the equipment can more easily visualize the entire range of a parameter in relation to a specific setting. Analog controls also allow adjustment to the exact desired value, whereas digital controls allow only an approximation.

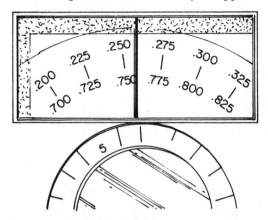

ANALOG CONTROL: An analog frequency readout.

ANALOG METERING

Many of the common panel meters on electronic equipment are analog devices. Current, voltage, and power levels are monitored by devices that show a continuous range of a certain quantity, such as plate current or RF output in a communications transmitter.

In some situations, such as amplifier tuning indicators and signal-strength comparison tests, analog metering allows the operator or technician to visualize a continuous range of possible values in relation to the actual reading. This is important when tuning circuits for optimum performance. It would be difficult to "peak" a signal on a digital meter, but on an analog meter it is easy. *See also* DIGITAL METERING.

ANALOG-TO-DIGITAL CONVERTER

For the transmission of signals, there are advantages to digital mode compared with analog mode. In particular, these advantages include narrower bandwidth, better signal-to-noise ratio, and fewer errors per unit time.

A digital signal has only a few well-defined levels or states, whereas an analog signal, such as a voice or a typical video scene, has infinitely many levels. An analog-to-digital converter (A/D converter or ADC) changes an analog signal into a digital signal at the transmitter, so that the data can be sent in the more efficient digital mode. The data can be converted back into the analog mode at the receiver by means of a digital-to-analog converter (D/A converter or DAC).

A/D conversion works by means of a process called *sampling*, in which the analog signal amplitude is measured periodically, and its level assigned a binary number. If the binary number has just one digit, then there are two possible digital states; if the binary number has two digits, there are four digital states; if there are three digits, there are eight possible digital states. In general, if there are n digits in the binary code, the digital signal will have 2^n possible states. The number of states in the digital signal is known as the *sampling resolution*. The *sampling rate* can also vary; it might be rather slow (a low frequency) or fast (a high frequency).

In general, the greater the sampling resolution of the ADC, the more accurate the digital representation of the analog signal. Also, the greater the ADC sampling rate, the more accurate the representation. The figure illustrates a digital rendition of an analog waveform. It is easy to see that increasing either the sampling resolution or the sampling rate makes the mesh "finer." This means that the digital wave looks more like the analog wave.

There is a tradeoff: increasing either the sampling resolution or the sampling rate will cause the bandwidth to increase. In any case, it is best to use a sampling rate that is at least twice the maximum frequency in the modulating analog signal. The signal-to-noise ratio tends to improve as the sampling resolution increases.

For voice communications, the ADC sampling resolution is usually 8 bits, and the sampling rate is 8000 times per second or 8 kHz. For amateur purposes, the ADC sampling rate could be as low as about 5 to 6 kHz and still provide a reasonably good signal because in amateur work, fidelity is not of great concern as long as intelligibility is optimized with due consideration given to other factors, such as available band space. See also DIGITAL COMMUNICATIONS, DIGITAL MODULATION, DIGITAL-TO-ANALOG CONVERTER, and PULSE MODULATION.

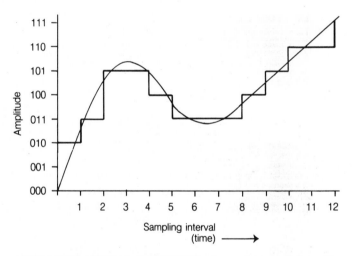

ANALOG-TO-DIGITAL CONVERTER: An analog wave (curved line) and its digital counterpart with eight-level sampling resolution (squared-off line).

AND GATE

An *AND gate* is a circuit that performs the logical operation "AND." It might have two or more inputs.

Logic symbols 1, or high, at all the inputs of an AND gate will produce an output of 1 (high). But if any of the inputs are at logic 0, or low, then the output of the AND gate will be low. *See also* LOGIC GATE.

ANDERSON BRIDGE

An *Anderson bridge* is a device for determining unknown capacitances or inductances. See the schematic diagram for an Anderson bridge designed to measure inductances.

For the proper operation of an Anderson bridge, it is necessary to have a frequency standard, a way to balance this standard with the known reactance and the unknown reactance, and an indicator to show when balance has been achieved. A galvanometer (see GALVANOMETER) is generally used as the indicator. It shows both positive and negative deflections from zero (null).

The Anderson bridge actually measures reactance, from which the inductance or capacitance is easily determined. Inductance bridges are calibrated in millihenrys or microhenrys. Capacitance bridges are calibrated in microfarads or picofarads. Some bridges are capable of measuring either inductance or capacitance, *see also* REACTANCE.

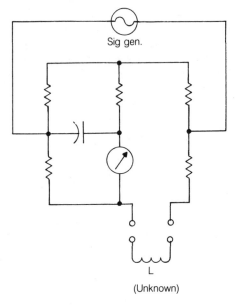

ANDERSON BRIDGE: This circuit is used to measure inductance.

ANGLE OF DEFLECTION

In a cathode-ray tube (*see* CATHODE-RAY TUBE), a narrow beam of electrons is sent through a set of electrically charged deflection plates to obtain the display. The angle of deflection of the beam is the number of degrees the beam is diverted from a straight path. If the beam of electrons continues through the deflection plates in a perfectly straight line, the angle of deflection is zero.

In general, the greater the amplitude of the input signal to an oscilloscope, the greater the angle of deflection of the electron beam. The angle of deflection is directly proportional to the input voltage. Therefore, if an ac voltage of 2 volts peak-to-peak causes an angle of deflection of ± 10 degrees, an ac voltage of 4 volts peak-to-peak will result in an angle of deflection of ± 20 degrees. The angle of deflection in an oscilloscope is always quite small, so the displacement on the screen is essentially proportional to the angle of deflection. Some oscilloscopes are calibrated so that the angle of deflection increases in logarithmic proportion, rather than in direct proportion, to the input signal voltage. *See also* OSCILLOSCOPE.

ANGLE STRUCTURE

An *angle structure* is a method of building a tower for mechanical strength. Braces are placed at angles, with respect to the vertical support rods (see illustration). This provides greater rigidity and resistance to twisting, as compared to a tower without angle braces. This is especially important for towers that must support large antennas because such antennas have a tendency to try to rotate in a high wind.

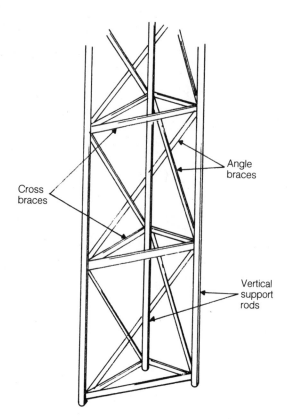

ANGLE STRUCTURE: Angle braces add rigidity to a tower.

ANODE

In a diode, the anode is the electrode toward which the electrons flow. The anode of a vacuum tube is also known as the *plate*. The anode is always positively charged relative to the cathode (*see* CATHODE) under conditions of forward bias, and negatively charged relative to the cathode under conditions of reverse bias. Current therefore flows only when there is forward bias (a small amount of current does flow with reverse bias, but it is usually negligible). If the reverse voltage becomes excessive, there might be a sudden increase in current in the reverse direction.

The term *anode* is sometimes used in reference to the positive terminal or electrode in a cell or battery. *See also* DIODE, and TUBE.

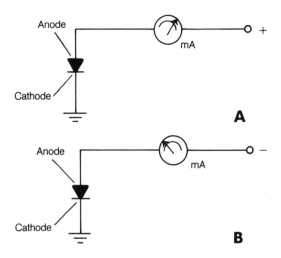

ANODE: When the anode is positive (A), current flows. When the anode is negative (B), current does not flow.

ANTENNA

An *antenna* is a form of *transducer* that converts alternating currents into electromagnetic fields, or vice-versa. Antennas can take an almost limitless variety of forms. Please see the following articles for detailed information: ALEXANDERSON ANTENNA, ANTENNA EFFICIENCY, ANTENNA GROUND SYSTEM, ANTENNA IMPEDANCE, ANTENNA MATCHING, ANTENNA POLARIZATION, ANTENNA POWER GAIN, ANTENNA RESONANT FREQUENCY, ANTENNA TUNING, AUTOMATIC DIRECTION FINDER, BEAM ANTENNA, BEVERAGE ANTENNA, CHARACTERISTIC IMPEDANCE, COAXIAL ANTENNA, COLLINEAR ANTENNA, CONICAL MONOPOLE ANTENNA, DIPOLE ANTENNA, DIRECTIONAL ANTENNA, DISCONE ANTENNA, DISH ANTENNA, EXTENDED DOUBLE ZEPP ANTENNA, END-FIRE ANTENNA, FLAGPOLE ANTENNA, FOLDED DIPOLE ANTENNA, GROUND-PLANE ANTENNA, HALF-WAVE ANTENNA, HALO ANTENNA, HELICAL ANTENNA, HERTZ ANTENNA, HORN ANTENNA, IMPEDANCE, INVERTED-L ANTENNA, INVERTED-V ANTENNA, ISOTROPIC ANTENNA, J ANTENNA, KOOMAN ANTENNA, LOG-PERIODIC ANTENNA, LONGWIRE ANTENNA, LOOP ANTENNA, MARCONI ANTENNA, MULTIBAND ANTENNA, PYRAMID HORN ANTENNA, QUAD ANTENNA, QUAGI ANTENNA, RADIATION RESISTANCE, REACTANCE, RESONANCE, RHOMBIC ANTENNA, STANDING WAVE, STANDING WAVE RATIO, STANDING-WAVE RATIO LOSS, TRANSMATCH, TUNED FEEDERS, TOP-FED VERTICAL ANTENNA, TRAP ANTENNA, TURNSTILE ANTENNA, UMBRELLA ANTENNA, UNIDIRECTIONAL-PATTERN ANTENNA, VERTICAL ANTENNA, VERTICAL DIPOLE ANTENNA, WHIP ANTENNA, WINDOM ANTENNA, WIRE ANTENNA, YAGI ANTENNA, and ZEPPELIN ANTENNA.

ANTENNA EFFECT

A small loop antenna ideally has a sharp null along the line at right angles to the plane of the loop, and running through the loop center. Under some conditions, however, this null is not sharp, and might disappear altogether. This can happen when the loop is inadequately shielded, and/or when it is not electrostatically balanced. Then, the loop acts as a short whip antenna instead of a true loop antenna. This is called *antenna effect*.

The best way to avoid antenna effect is to provide electrostatic shielding for small direction-finding loops. Then the loop responds only to the magnetic component of the radio waves, and will exhibit the desired null pattern. *See also* AUTOMATIC DIRECTION FINDER, and LOOP ANTENNA.

ANTENNA EFFICIENCY

Not all the electromagnetic field received by an antenna from its feed line is ultimately radiated into space. Some power is dissipated in the ground near the antenna in structures such as buildings and trees, the earth itself, and in the conducting material of the antenna. If P represents the total amount of available power at a transmitting antenna, P_R the amount of power eventually radiated into space, and P_L the power lost in surrounding objects and the antenna conductors, then:

$$P_L + P_R = P$$

and:

$$\text{Efficiency (percent)} = 100 \ (P_R/P)$$

The efficiency of a nonresonant antenna is difficult to determine in practice, but at resonance (*see* RESONANCE), when the antenna impedance, Z, is a known pure resistance, the efficiency can be found by the formula:

$$\text{Efficiency (percent)} = 100 \ (R/Z),$$

where R is the theoretical radiation resistance of the antenna at the operating frequency (*see* RADIATION RESISTANCE). The actual resistance, Z, is always larger than the theoretical radiation resistance R. The difference is the loss resistance, and there is always some loss.

Antenna efficiency is optimized by making the loss resistance as small as possible. Some means of reducing the loss resistance are the installation of a good RF-ground system (if the antenna is a type that needs a good RF ground), the use of low-loss components in tuning networks (if they are used), and locating the antenna itself as high above the ground, and as far from obstructions, as possible.

ANTENNA GROUND SYSTEM

Some types of antennas must operate against a ground system, or RF reference potential. Others do not need an RF ground. In general, unbalanced or asymmetrical antennas need a good RF ground, while balanced or symmetrical antennas do not. The ground-plane antenna (*see* GROUND-PLANE ANTENNA) requires an excellent RF ground in order to function efficiently. However, the half-wave dipole antenna (*see* DIPOLE ANTENNA) does not need an RF ground.

When designing an antenna ground system, it is important to realize that a good dc ground does not necessarily constitute a good RF ground. An elevated ground-plane antenna has a very effective RF ground that does not have to be connected physically to the earth in any way. A single thin wire hundreds of feet long might be terminated at a ground rod or the grounded side of a utility outlet and work very well as a dc ground, but it will not work well at radio frequencies.

The earth affects the characteristics of any antenna. The overall radiation resistance and impedance are affected by the height above ground. However, an RF ground system does not necessarily depend on the height of the antenna. Capacitive coupling to ground is usually sufficient, in the form of a counterpoise (*see* COUNTERPOISE).

A ground system for lightning protection must be a direct earth connection. Some antennas are grounded through inductors that do not conduct RF, but serve to discharge static buildup before a lightning strike. *See also* LIGHTNING PROTECTION.

ANTENNA IMPEDANCE

Any antenna displays a defined impedance at its feed point at a particular frequency. Usually, this impedance changes as the frequency changes. Impedance at the feed point of an antenna consists of radiation resistance (*see* RADIATION RESISTANCE) and either capacitive or inductive reactance (*see* REACTANCE). Both the radiation resistance and the reactance are defined in ohms.

An antenna is said to be resonant at a particular frequency when the reactance is zero. Then, the impedance is equal to the radiation resistance in ohms. The frequency at which resonance occurs is called the *resonant frequency* of the antenna. Some antennas have only one resonant frequency. Others have many. The radiation resistance at the resonant frequency of an antenna depends on several factors, including the height above ground and the harmonic order, and whether the antenna is inductively or capacitively tuned. The radiation resistance can be as low as a fraction of an ohm, or as high as several thousand ohms.

When the operating frequency is made higher than the resonant frequency of an antenna, inductive reactance appears at the feed point. When the operating frequency is below the resonant point, capacitive reactance appears. To get a pure resistance, a reactance of the opposite kind from the type present must be connected in series with the antenna. *See also* ANTENNA RESONANT FREQUENCY, and IMPEDANCE.

ANTENNA MATCHING

For optimum operation of an antenna and feed-line combination, the system should be at resonance. This is usually done by eliminating the reactance at the feed point, where the feed line joins the antenna radiator. In other words, the antenna itself is made resonant. The remaining radiation resistance is then transformed to a value that closely matches the characteristic impedance of the feed line.

If the reactance at the dipole antenna feed point is inductive, series capacitors are added to cancel out the inductive reactance. If the reactance is capacitive, series coils are used. The inductances are adjusted until the antenna is resonant. Both coils should have identical inductances to keep the system balanced (because this is a balanced antenna).

Once the antenna is resonant, only resistance remains. This value might not be equal to the characteristic impedance of the line. Generally, coaxial lines are designed to have a characteristic impedance, or Z_o, of 50 to 75 ohms, which closely approximates the radiation resistance of a half-wave dipole in free space. But, if the radiation resistance of the antenna is much different from the Z_o of the line, a transformer should be used to match the two parameters. This results in the greatest efficiency for the feed line. Without the transformer, standing waves on the line will cause some loss of signal. The amount of loss caused by standing waves is sometimes inconsequential, but sometimes it is large (*see* STANDING WAVE, and STANDING-WAVE RATIO).

In some antenna systems, no attempt is made to obtain an impedance match at the feed point. Instead, a matching system (*see* TRANSMATCH) is used between the transmitter or receiver and the feed line. This allows operating convenience when the frequency is changed often. However, it does nothing to reduce the loss on the line caused by standing waves. *See also* TUNED FEEDERS.

ANTENNA NOISE BRIDGE

An *antenna noise bridge* is a device that makes it possible to easily find the complex impedance of an antenna or antenna system. It determines the resistive and reactive components. The device is sometimes called an *R-X noise bridge*. The R stands for resistance; the X stands for reactance.

The photograph shows a typical antenna noise bridge. It is a small unit that is placed in the feed line between a receiver and antenna feed line. A broadband noise generator creates ''hash''

over a wide band of frequencies. Tuned circuits are then used to null out this noise. The values of the tuning components, when the null is found, depend on the resistance and reactance present in the antenna system.

An antenna noise bridge is calibrated by using it with a variety of known complex impedances. The resistance control, usually a potentiometer, and the reactance control, usually a variable capacitor, have pointer knobs with dials calibrated in ohms. The resistance scale goes from some low value, such as 25 ohms, to some high value, such as 250 ohms. The reactance scale is centered at zero and goes from about −70 (capacitive reactance) to +70 ohms (inductive reactance).

There is a limit to the range of resistances and reactances that can be determined accurately with an antenna noise bridge. If extremely high or low resistances, or large reactances must be found, the accuracy of the device is compromised.

Usually, complex impedances of interest lie reasonably close to a match for 50- or 75-ohm coaxial line. Therefore, the reactance range need not be greater than about −70 to +70 ohms, and the resistance range can be from a few ohms to 200 or 300 ohms. Antenna systems with reactances and/or resistances outside these ranges will defy evaluation with most antenna noise bridges. Such impedances almost always require some kind of matching network to be usable with modern amateur equipment. *See also* ANTENNA MATCHING, IMPEDANCE, and STANDING-WAVE RATIO.

ANTENNA NOISE BRIDGE: Allows determination of resonant frequency, feed-point impedance and line characteristics. Palomar Engineers

ANTENNA PATTERN

The directional characteristics of any transmitting or receiving antenna, when graphed on a polar coordinate system, are called the *antenna pattern*. The simplest possible antenna pattern occurs when an isotropic antenna is used (*see* ISOTROPIC ANTENNA), although this is a theoretical ideal. It radiates equally well in all directions in three-dimensional space.

Antenna patterns are represented by diagrams such as those in A and B. The location of the antenna is assumed to be at the center of the coordinate system. The greater the radiation or reception capability of the antenna in a certain direction, the farther from the center the points on the chart are plotted. A dipole antenna, oriented horizontally so that its conductor runs in a north-south direction, has a horizontal-plane (H-plane) pattern similar to that in Fig. A. The elevation-plane (E-plane) pattern depends on the height of the antenna above effective ground at the viewing angle. With the dipole oriented so that its conductor runs perpendicular to the page, and the antenna ¼ wavelength above effective ground, the E-plane antenna pattern will resemble B.

The patterns in A and B are quite simple. Many antennas have patterns that are very complicated. For all antenna pattern graphs, the relative power gain (*see* ANTENNA POWER GAIN) relative to a dipole is plotted on the radial axis. The values thus range from 0 to 1 on a linear scale. Sometimes a logarithmic scale is used. If the antenna has directional gain, the pattern radius will exceed 1 in some directions. Examples of antennas with directional gain are the log periodic, longwire, quad, and the Yagi. Some vertical antennas have gain in all horizontal directions. This occurs at the expense of gain in the E plane.

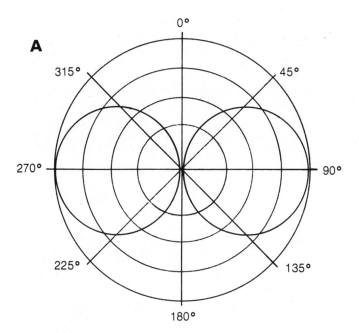

A

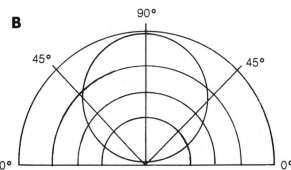

B

ANTENNA PATTERN: Radiation and response for a dipole antenna. At A, as seen from broadside to the wire; at B, as seen from off the end of the wire.

ANTENNA POLARIZATION

The *polarization* of an antenna is determined by the orientation of the electric lines of force in the electromagnetic field radiated or received by the antenna. Polarization might be linear, or it might be rotating (circular). Linear polarization can be vertical, horizontal, or somewhere in between. In circular polarization, the rotation can be either counterclockwise or clockwise (*see* CIRCULAR POLARIZATION).

For antennas with linear polarization, the orientation of the electric lines of flux is parallel with the radiating element. Therefore, a vertical element produces signals with vertical polarization, and a horizontal element produces horizontally polarized fields in directions broadside to the element.

In receiving applications, the polarization of an antenna is determined according to the same factors involved in transmitting. Thus, if an antenna is vertically polarized for transmission of electromagnetic waves, it is also vertically polarized for reception.

In free space, with no nearby reflecting objects to create phase interference, the circuit attenuation between a vertically polarized antenna and a horizontally polarized antenna, or between any two linearly polarized antennas at right angles, is approximately 30 dB compared to the attenuation between two antennas having the same polarization.

Polarization affects the propagation of electromagnetic energy to some extent. A vertical antenna works much better for transmission and reception of surface-wave fields (*see* SURFACE WAVE) than a horizontal antenna. For sky-wave propagation (*see* SKY WAVE), the polarization is not particularly important because the ionosphere causes the polarization to be randomized at the receiving end of a circuit.

ANTENNA POWER GAIN

The *power gain* of an antenna is the ratio of the effective radiated power (*see* EFFECTIVE RADIATED POWER) to the actual RF power applied to the feed point. Power gain might also be expressed in decibels. If the effective radiated power is P_{ERP} watts and the applied power is P watts, then the power gain in decibels is:

$$Power\ Gain\ (dB) = 10\log_{10}(P_{ERP}/P)$$

Power gain is always measured in the favored direction of an antenna. The favored direction is the azimuth direction in which the antenna performs the best. For power gain to be defined, a reference antenna must be chosen with a gain assumed to be unity, or 0 dB. This reference antenna is usually a half-wave dipole in free space (*see* DIPOLE ANTENNA). Power gain figures taken with respect to a dipole are expressed in *dBd*. The reference antenna for power-gain measurements might also be an isotropic radiator (*see* ISOTROPIC ANTENNA), in which case the units of power gain are called *dBi*. For any given antenna, the power gains in dBd and dBi are different by approximately 2.15 dB:

$$Power\ Gain\ (dBi) = 2.15 + Power\ Gain\ (dBd)$$

Directional transmitting antennas can have power gains in excess of 20 dBd. At microwave frequencies, large dish antennas (*see* DISH ANTENNA) can be built with power gains of 30 dBd or more.

Power gain is the same for reception, with a particular antenna, as for transmission of signals. Therefore, when antennas with directional power gain are used at both ends of a communications circuit, the effective power gain over a pair of dipoles is the sum of the individual antenna power gains in dBd.

ANTENNA RESONANT FREQUENCY

An antenna is at resonance whenever the reactance at the feed point (the point where the feed line joins the antenna) is zero. This might occur at just one frequency, or it might occur at several frequencies.

For a half-wave dipole antenna in free space, the resonant frequency is given approximately by:

$$f = 468/s_{ft}$$

or:

$$f = 143/s_m$$

where f is the fundamental resonant frequency in MHz, and s is the antenna length in feet (s_{ft}) or meters (s_m). For a quarter-wave vertical antenna operating against a perfect ground plane:

$$f = 234/h_{ft}$$

or:

$$f = 71/h_m$$

where h is the antenna height in feet (h_{ft}) or meters (h_m).

The dipole antenna and quarter-wave vertical antenna display resonant conditions at all harmonics of their fundamental frequencies. Therefore, if a dipole or quarterwave vertical is resonant at a particular frequency f, it will also be resonant at $2f$, $3f$, $4f$, and so on. The impedance is not necessarily the same, however, at harmonics as it is at the fundamental. At frequencies corresponding to odd harmonics of the fundamental, the impedance is nearly the same as at the fundamental. At even harmonics, the impedance is much higher.

An antenna operating at resonance, where the radiation resistance is almost the same as the characteristic impedance, or Z, of the feed line, will perform with good efficiency, provided the ground system (if a ground system is needed) is efficient. *See also* CHARACTERISTIC IMPEDANCE, and RADIATION RESISTANCE.

ANTENNA TUNING

Antenna tuning is the process of adjusting the resonant frequency of an antenna or antenna system (*see* ANTENNA RESONANT FREQUENCY). This is usually done by means of a tapped or variable inductor at the antenna feed point, or somewhere along the antenna radiator. It is also sometimes done with a transmatch at the transmitter so that the feed line and antenna together form a resonant system (*see* TRANSMATCH, and TUNED FEEDERS). Antenna tuning can be done with any sort of antenna.

An antenna made of telescoping sections of tubing is tuned exactly to the desired frequency by changing the amount of overlap at the tubing joints, thus changing the physical length of the radiating or parasitic elements. In a Yagi array, the director and reflector elements must be precisely tuned to obtain the greatest amount of forward power gain and front-to-back ratio. Phased arrays must be tuned to give the desired directional response. In some antennas, tuning is not critical, while in others it must be done precisely to obtain the rated specifications. In general, the higher the frequency, the more exacting are the tuning requirements.

ANTIHUNT DEVICE

An *antihunt device* is a circuit in an automatic direction-finding system (*see* AUTOMATIC DIRECTION FINDER). In such a device, the circuit sometimes overcorrects itself. The overcorrection in azimuth bearing causes another correction, which also exceeds the needed amount. This can happen over and over, resulting in a back-and-forth oscillation of the antenna. The indicator will read first to the left and then to the right of the desired target. An antihunt device prevents this endless oscillation of an automatic direction-finding system by damping the response. This lessens the extent of correction so that overcorrection does not occur.

APERTURE

The area over which an antenna can effectively intercept an electromagnetic field is called its *aperture*. The true, or effective, aperture can range from an area that is larger than an antenna's physical size, as in the case of an array of wire antennas, to an area that is smaller than the physical size, as with horn or parabolic-reflector antennas. The aperture of an isotropic antenna is large compared to its size (the antenna is a point source, and the aperture is in the shape of a sphere), but the antenna has no gain because it lacks directivity (see A). The aperture of a dipole is not as large as that of an isotropic antenna, but a dipole exhibits some gain because of directivity (see B).

Antenna receiving gain is almost always expressed in decibels with reference to a half-wave dipole (dBd), or with respect to an isotropic antenna (dBi). The term aperture is seldom used (*see* ANTENNA POWER GAIN).

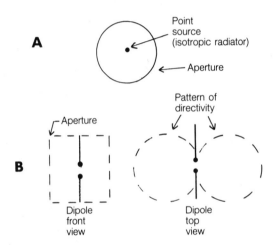

APERTURE: Aperture for an isotropic antenna (A) and a half-wave dipole (B).

APOGEE

Any earth-orbiting satellite follows either of two types of path through space. A *circular* orbit has the center of the earth at the center of the orbit. When a satellite follows an *elliptical* orbit, which is far more common, the center of the earth is at one focus of the ellipse. The extent to which the orbit differs from a circle is called the *eccentricity*.

Whenever a satellite is placed into orbit around the earth, or around any other large celestial body, there is almost always some deviation from a perfectly circular path. This is because the circular orbit represents a very special case, and attaining such an orbit requires precise speeds and launch trajectories. Perfection is difficult to attain.

Whenever a satellite has an elliptical orbit, its altitude varies. The maximum altitude is called the *apogee* of the satellite. It occurs once for every complete orbit. At apogee, the satellite travels more slowly than at any other point in the orbit.

Lunar Apogee The moon's orbit around the earth is elliptical. The distance between the earth and the moon varies between about 225,000 and 253,000 miles. The second of these two numbers represents the moon's apogee.

The eccentricity of the moon's orbit is not very great. But it is enough to affect moonbounce communications, also called *earth-moon-earth* or *EME* (*see* MOONBOUNCE). The moon actually looks a little bit smaller at apogee than at *perigee*, or the point in the orbit where the distance between the earth and moon is the smallest (*see* PERIGEE). Moonbounce is easiest when the moon is at perigee, and is most difficult when it is at apogee.

Apogee of a Communications Satellite   Most ham satellites have orbits that are at least somewhat eccentric. A highly eccentric orbit offers some advantages and some disadvantages. Interestingly, satellite communications is easiest when the satellite is at or near apogee, if it has a very elongated orbit. This is because the satellite moves very slowly then, and its position in the sky does not change much for awhile. The antenna at the ground station does not have to be constantly turned to follow the satellite during this time.

Near perigee, the signals from the satellite are stronger, and less power is needed to reach it. But it moves more rapidly across the sky, and the antennas must be moved more often to track the satellite. *See also* ACTIVE COMMUNICATIONS SATELLITE, and OSCAR.

APPARENT POWER

In an alternating-current circuit containing reactance, the voltage and current reach their peaks at different times. That is, they are not exactly in phase. This complicates the determination of power. In a nonreactive circuit, we might consider:

$$P = EI$$

where P is the power in watts, E is the RMS (*see* ROOT MEAN SQUARE) voltage in volts, and I is the RMS current in amperes. In a circuit with reactance, this expression is referred to as the *apparent power*. It is called *apparent* because it differs from the true power (*see* TRUE POWER) that would be dissipated in a resistor or resistive load. Only when the reactance is zero is the apparent power identical to the true power.

In a nonresonant or improperly matched antenna system, a wattmeter placed in the feed line will give an exaggerated reading. The wattmeter reads apparent power, which is the sum of the true transmitter output power and the reactive or reflected power (*see* REFLECTED POWER). To determine the true power, the reflected reading of a directional wattmeter is subtracted from the forward reading. The more severe the antenna mismatch, the greater the difference between the apparent and true power.

In the extreme, all of the apparent power in a circuit is reactive. This occurs when an alternating-current circuit consists of a pure reactance, such as a coil, capacitor, or short-circuited length of transmission line. *See also* REACTANCE.

ARC

An *arc* occurs when electricity flows through space. Lightning is a good example of an arc. When the potential difference between two objects becomes sufficiently large, the air (or other gas) ionizes between the objects, creating a path of relatively low resistance through which current flows.

An arc might be undesirable and destructive, such as a flashover across the contacts of a wafer switch. Or, an arc can be put to constructive use. A carbon-arc lamp is an extremely bright source of light, and is sometimes seen in large spotlights or searchlights where other kinds of lamps would be too expensive for the illumination needed.

Undesirable or destructive arcing is prevented by keeping the voltage between two points below the value that will cause a flashover. An antenna lightning arrestor allows built-up static potential to discharge across a small gap (*see* LIGHTNING ARRESTOR) before it gets so great that arcing occurs between components of the circuit and ground.

ARITHMETIC SYMMETRY

Arithmetic symmetry refers to the shape of a bandpass or band-rejection filter response (*see* BANDPASS FILTER, and BAND-REJECTION FILTER).

An example of arithmetic symmetry in a bandpass filter curve is shown in the drawing. The frequency scale (horizontal) is linear, so that each unit length represents the same number of Hertz in frequency (in this case, each division represents 1 MHz). The vertical scale might either be linear, calibrated in volts or watts, or logarithmic, calibrated in decibels relative to a certain level. Here, the vertical scale is calibrated in decibels relative to 1 milliwatt, or dBm. *See* DBM.

The curve in the drawing is exactly symmetrical around the center frequency $f_o = 145$ MHz; that is, the left-hand side of the response is a mirror image of the right-hand side. This is arithmetic symmetry.

In practice, arithmetic symmetry represents an ideal condition, and it can only be approximated. Modern technology has developed bandpass and band-rejection filters with almost perfect arithmetic symmetry with a variety of bandwidths and frequency ranges. *See also* CRYSTAL-LATTICE FILTER, and MECHANICAL FILTER.

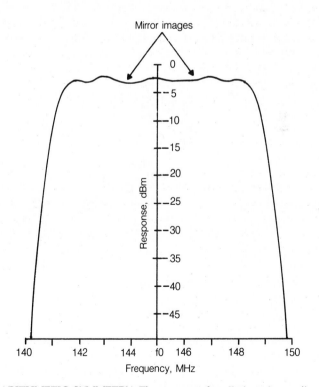

ARITHMETIC SYMMETRY: The response has "mirror images" at either side of center.

ARMSTRONG OSCILLATOR

An *Armstrong oscillator* is a circuit that produces oscillation by means of inductive feedback. See the simple circuit diagram of such an oscillator. A coil called *the tickler* is connected to the collector, and is brought near the coil of the tuned circuit. The tickler is oriented to produce positive feedback. The amount of coupling between the tickler coil and the tuned circuit is adjusted so that stable oscillation takes place. The output is taken from the collector by means of a small capacitance, or by transformer coupling to the tuned circuit.

The frequency of the Armstrong oscillator is determined by the tuned-circuit resonant frequency. Usually, the capacitor is variable and the inductor is fixed, so the amount of feedback remains relatively constant as the frequency is changed. Armstrong oscillators are generally used in regenerative receivers. They are not often seen as variable oscillators in transmitters or superheterodyne receivers. Other types of oscillators are preferred for those applications. *See also* COLPITTS OSCILLATOR.

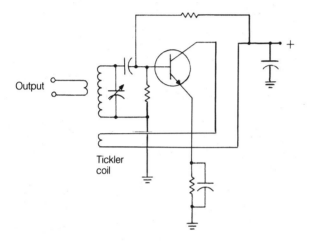

ARMSTRONG OSCILLATOR: This circuit can be used as a regenerative detector.

ARRL
See AMERICAN RADIO RELAY LEAGUE.

ARTICULATION
Articulation is a measure of the quality of a voice-communications circuit. It is given as a percentage of the speech units (syllables or words) understood by the listener. To test for articulation, a set of random words or numbers should be read by the transmitting operator. The words or numbers are chosen at random to avoid possible contextual interpolation by the receiving operator. This gives a true measure of the actual percentage of speech units received.

When plain text or sentences are transmitted, the receiving operator can understand a greater portion of the information because he can figure out some of the missing words or syllables by mental guesswork. The percentage of speech units received with plain-text transmission is called *intelligibility* (*see* INTELLIGIBILITY).

Articulation and intelligibility differ from fidelity. Perfect reproduction of the transmitted voice is not as important in a communications system as the accurate transfer of information. The best articulation generally occurs when the voice frequency components are restricted to approximately the range of 200 to 3000 Hz. Articulation can also be enhanced at times by the use of speech compressors or RF clipping (*see* SPEECH CLIPPING, and SPEECH COMPRESSION).

ARTIFICIAL GROUND
An *artificial ground* is an RF ground not directly connected to the earth. A good example of an artificial ground is the system of quarter-wave radials in an elevated ground-plane antenna (*see* GROUND-PLANE ANTENNA). A counterpoise (*see* COUNTERPOISE) is also a form of artificial ground.

In some situations, it is difficult or impossible to obtain a good earth ground for an antenna system. A piece of wire ¼ wavelength, or any odd multiple of ¼ wavelength, long at the operating frequency can operate as an artificial ground in such a case. This arrangement does not form an ideal ground; the wire will radiate some energy, and thus is actually a part of the antenna. But an artificial ground is much better than no ground at all. Of course, some kind of dc ground should be used in addition to the RF ground to minimize the danger of electrical shock from built-up static on the antenna and from possible short circuits in the transmitting or receiving equipment. *See also* DC GROUND.

ARTWORK
In the construction of an integrated circuit (*see* INTEGRATED CIRCUIT), the pattern is first drawn to scale on a piece of glass or plastic film. If there are several layers to the integrated circuit, all layers are accurately drawn. A special camera then is used to reduce the pattern to its actual size for reproduction in the integrated circuit. In this way, a tremendous number of components can be squeezed into a very small space on a chip of semiconductor material. Transducers, resistors, capacitors, diodes, and wiring are all fabricated in this way.

A printed-circuit board (*see* PRINTED CIRCUIT) is made in a similar manner. Artwork is drawn on a piece of paper or film as shown, usually several times actual size. It is then photographed and reduced and put on a clear plastic film. A photographic process is used to etch the wiring pattern onto a piece of copper-plated phenolic or glass-epoxy material. For this reason, this kind of artwork is called an etching pattern (*see* ETCHING).

Artwork is a tremendous cost saver in the fabrication of integrated circuits, because all wiring errors can be eliminated before the circuit is actually built. Sometimes the artwork for a particularly complicated circuit is drawn by a computer.

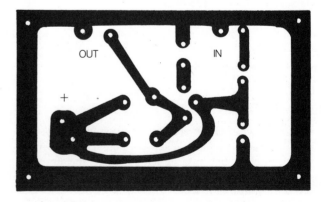

ARTWORK: Etching pattern for a simple printed circuit.

ASCENDING NODE
Most earth-orbiting satellites have a *groundtrack* that crosses the equator twice for each orbit. The only exceptions are geo-

stationary satellites (*see* GEOSTATIONARY ORBIT) and satellites whose orbits are exactly over the equator. The points where, and times when, the groundtrack is exactly over the equator are called *nodes*.

For every orbit, there is one node when the groundtrack moves from the southern hemisphere to the northern, and one node when the groundtrack moves from the northern hemisphere to the southern. The first of these, going south-to-north, is called the *ascending node*.

Ascending nodes are commonly used as reference points for locating satellite positions at future times. The position of the node is given in degrees and minutes of longitude. *See also* DESCENDING NODE.

ASCENDING PASS

Most ham satellites have a *groundtrack* that moves over the surface of the earth, sometimes north of the equator and sometimes south of it. Only geostationary satellites have groundtracks that do not move (*see* GEOSTATIONARY ORBIT). A few satellites orbit directly over the equator.

When a satellite has an orbit that is slanted relative to the equator, its groundtrack is moving generally northwards half of the time. This period starts when the satellite attains the southernmost latitude in its orbit, and lasts until it reaches its northernmost latitude. For any given earthbound location, an *ascending pass* is the time during which the satellite is accessible, while it is moving generally northwards. The pass time depends on the altitude of the satellite, and on how close its groundtrack comes to the earth-based station. *See also* ASCENDING NODE, DESCENDING NODE, and DESCENDING PASS.

ASCII

American National Standard Code for Information Interchange (ASCII) is a seven-unit digital code for the transmission of teleprinter data. Letters, numerals, symbols, and control operations are represented. ASCII is designed primarily for computer applications, but is also used in some teletypewriter systems.

Each unit is either 0 or 1. In the binary number system, there are $2^7 = 128$, possible representations. Table 1 gives the ASCII code symbols for the 128 characters.

The other commonly used teletype code is the Baudot code (*see* BAUDOT). The speed of transmission of ASCII or Baudot is called the baud rate (*see* BAUD RATE). If one unit pulse is *s* seconds in length, then the baud rate is defined as $1/s$. For example, a baud rate of 100 represents a pulse length of 0.01 second, or 10 ms. The speed of ASCII transmission in words per minute, or WPM (*see* WORDS PER MINUTE) is approximately the same as the baud rate. Commonly used ASCII data rates range from 110 to 19,200 baud, as shown in Table 2.

ASPECT RATIO

The *aspect ratio* of a rectangular image is the ratio of its width to its height. For television in the United States, the aspect ratio of a picture frame is 4 to 3. That is, the picture is 1.33 times as wide as it is high. This ratio must be maintained in a television receiver or distortion of the picture will result. *See also* TELEVISION.

ASCII: SYMBOLS FOR ASCII TELEPRINTER CODE.

First four signals	Last three signals							
	000	001	010	011	100	101	110	111
0000	NUL	DLE	SPC	0		P	/	p
0001	SOH	DC1	!	1	A	Q	a	q
0010	STX	DC2	"	2	B	R	b	r
0011	ETX	DC3	#	3	C	S	c	s
0100	EOT	DC4	$	4	D	T	d	t
0101	ENQ	NAK	%	5	E	U	e	u
0110	ACK	SYN	&	6	F	V	f	v
0111	BEL	ETB	'	7	G	W	g	w
1000	BS	CAN	(	8	H	X	h	x
1001	HT	EM	)	9	I	Y	i	y
1010	LF	SUB	*	:	J	Z	j	z
1011	VT	ESC	+	;	K	[	k	{
1100	FF	FS	,	<	L	/	l	/
1101	CR	GS	–	=	M	]	m	}
1110	SO	RS	.	>	N		n	~
1111	SI	US			?	O	-	o	DEL

ACK: Acknowledge
BEL: Bell
BS: Back space
CAN: Cancel
CR: Carriage return
DC1: Device control no. 1

DC2: Device control no. 2
DC3: Device control no. 3
DC4: Device control no. 4
DEL: Delete
DLE: Data link escape
ENQ: Enquiry
EM: End of medium
EOT: End of transmission
ESC: Escape
ETB: End of transmission block
ETX: End of text

FF: Form feed
FS: File separator
GS: Group separator
HT: Horizontal tab
LF: Line feed
NAK: Do not acknowledge
NUL: Null
RS: Record separator
SI: Shift in
SO: Shift out
SOH: Start of heading
SPC: Space
STX: Start of text
SUB: Substitute
SYN: Synchronous idle
US: Unit separator
VT: Vertical tab

Table 1

ASCII: SPEED RATES FOR THE ASCII CODE.

Baud rate	Length of pulse, ms	WPM
110	9.09	110
150	6.67	150
300	3.33	300
600	1.67	600
1200	0.833	1200
1800	0.556	1800
2400	0.417	2400
4800	0.208	4800
9600	0.104	9600
19,200	0.052	19,200

Table 2

ASTABLE MULTIVIBRATOR

An *astable circuit* is a form of oscillator. The word astable means unstable. An *astable multivibrator* consists of two tubes or transistors arranged in such a way that the output of one is fed directly to the input of the other. Two identical resistance-capacitance networks determine the frequency at which oscillation will occur. The amplifying devices are connected in a common-source or common-emitter configuration, as shown.

In the common-source or common-emitter circuit, the output of each transistor is 180 degrees out of phase with the input. An oscillating pulse might begin, for example, at the base of Q1 in the illustration. It is inverted at the collector of Q1, and goes to the base of Q2. It is again inverted at the collector of Q2, and therefore returns to the base of Q1 in its original phase. This produces positive feedback, resulting in sustained oscillation.

The astable multivibrator is frequently used as an audio oscillator, but it is not often seen in RF applications because its output is extremely rich in harmonic products. *See also* OSCILLATOR.

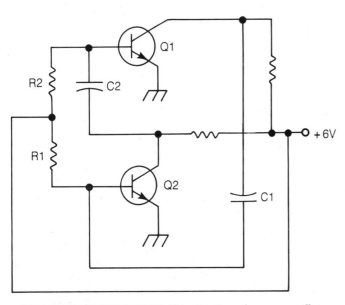

ASTABLE MULTIVIBRATOR: This circuit works as an oscillator.

ASYMMETRICAL DISTORTION

In a binary system of modulation, the high and low conditions (or 1 and 0 states) have defined lengths for each bit of information. The modulation is said to be distorted when these bits are not set to the proper duration. If the output bits of one state are too long or too short, compared with the signal input bits, the distortion is said to be *asymmetrical*.

A simple example of asymmetrical distortion often is found in a Morse-code signal. Morse code is a binary modulation system, with bit lengths corresponding to the duration of one dot. Ideally, a string of dots (such as the letter H) has high and low states of precisely equal length. An electronic keyer can produce signals of this nature. However, because of the shaping network in a CW transmitter (*see* SHAPING), the high state is often effectively prolonged because the decay time is lengthened. While the rise, or attack, time is made slower by the shaping network, the change is much greater on the decay side in most cases. This creates asymmetrical distortion because the dot-to-space ratio, 1-to-1 at the input, is greater than 1-to-1 at the transmitter output.

Asymmetrical distortion makes reception of a binary signal difficult, and less accurate, than would be the case for an undistorted signal. The effect might not be objectionable if it is small; in CW communications, a certain amount of shaping makes a signal more pleasant to the ear. Excessive asymmetrical distortion should be avoided. In teleprinter communications, asymmetrical distortion causes frequent printing errors. An oscilloscope can be used to check for proper adjustment of the high-to-low ratio.

ASYNCHRONOUS DATA

Asynchronous data is information not based on a defined time scale. An example of asynchronous data is manually sent Morse code, or CW. Machine-sent Morse code, in contrast, is synchronous. Synchronous transmission offers a better signal-to-noise ratio in communications systems among machines than asynchronous transmission. Nevertheless, the simple combination of a hand key and human ear for CW — the most primitive communications system — is commonly used when more sophisticated arrangements fail because of poor conditions!

Asynchronous data need not be as simple as hand-sent Morse code. A manually operated teletypewriter station and a voice system are other examples of asynchronous data transfer.

ATMOSPHERE

The *atmosphere* is the shroud of gases that surrounds our planet. Many other planets also have atmospheres; some do not.

Our atmosphere exerts an average pressure of 14.7 pounds per square inch at sea level. As the elevation above sea level increases, the pressure of the atmosphere drops, until it is practically zero at an altitude of 100 miles. Effects of the atmosphere, however, extend to altitudes of several hundred miles.

Scientists define our atmosphere in terms of three layers. The lowest layer, the troposphere, is where all weather disturbances take place. It extends to a height of approximately 8 to 10 miles above sea level. The troposphere affects certain radio-frequency electromagnetic waves (*see* DUCT EFFECT, and TROPOSPHERIC PROPAGATION). The stratosphere begins at the top of the troposphere and extends up to about 40 miles. No weather is ever seen in this layer, although circulation does occur. At altitudes from 40 to about 250 miles above the ground, several ionized layers of low-density gas are found. This region is known as the *ionosphere*. The layers of the ionosphere have a tremendous impact on the propagation of RF energy from dc into the VHF region. Without the ionosphere, radio communication as we know it would be much different. The long-distance shortwave propagation that we take for granted would not exist. *See also* D LAYER, E LAYER, F LAYER, IONOSPHERE, PROPAGATION, and PROPAGATION CHARACTERISTICS.

ATOMIC CHARGE

When an atom contains more or less electrons than normal, it is called an *ion*. The atomic charge of a normal atom is zero. The atomic charge of an ion is positive if there is a shortage of electrons, and negative if there is a surplus of electrons. Some atoms ionize quite easily; others do not readily ionize.

The unit of atomic charge is the amount of electric charge carried by a single electron or proton; they carry equal, but opposite charges. This is called an *electron unit*. One coulomb is a charge of 6.28×10^{18} electron units. Therefore, an electron unit is 1.59×10^{-19} coulomb (*see* COULOMB). An ion might have an atomic charge of $+3$ electron units or -2 electron units. The first case would indicate a deficiency of three electrons; the second case, an excess of two electrons. In general, the atomic charge number is always an integer because the electron unit is the smallest possible quantity of charge.

ATTACK

The rise time for a pulse is sometimes called the *attack* or *attack time*. In music, the attack time of a note is the time required for the note to rise from zero amplitude to full loudness. The attack time for a control system, such as an automatic gain control (*see* AUTOMATIC GAIN CONTROL, AUTOMATIC LEVEL CONTROL), is the time that is needed for that system to fully compensate for a change in input parameters. The graphics show the attack time for a musical tone (A) and a dc pulse (B).

The attack time of a musical note affects its sound quality. A fast rise time sounds "hard" and a slow rise time sounds "soft." With automatic gain or level control, an attack time that is too fast might cause overcompensation, while an attack time that is too slow will cause a loud popping sound at the beginning of each pulse or syllable.

The time required for a note or pulse to drop from full intensity back to zero amplitude is called the *decay* or *release time*. *See also* DECAY.

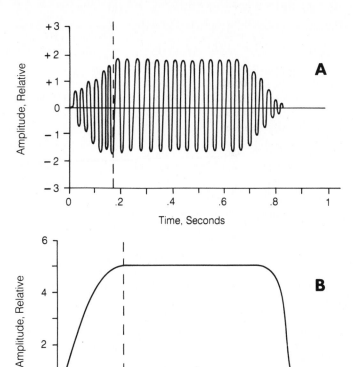

ATTACK: At A, for a musical note; at B, for a dc pulse.

ATTENUATION

Attenuation is the decrease in amplitude of a signal between any two points in a circuit. It is usually expressed in decibels (*see* DECIBEL). Attenuation is the opposite of amplification (*see* AMPLIFICATION), and might be defined for voltage, current, or power. Occasionally the attenuation in a particular circuit is expressed as a ratio. For example, the reduction of a signal in amplitude from ± 5 volts peak to ± 1 volt peak is an attenuation factor of 5.

In general, the attenuation in voltage for an input E_{IN} and an output E_{OUT} is:

$$Attenuation \text{ (dB)} = 20 \log_{10} (E_{IN}/E_{OUT})$$

Similarly, the current attenuation for an input of I_{IN} and an output of I_{OUT} is:

$$Attenuation \text{ (dB)} = 20 \log_{10} (I_{IN}/I_{OUT})$$

For power, given an input of P_{IN} watts and an output of P_{OUT} watts, the attenuation in decibels is given by:

$$Attenuation \text{ (dB)} = 10 \log_{10} (P_{IN}/P_{OUT})$$

If the amplification factor (*see* AMPLIFICATION FACTOR) is X dB, then the attenuation is $-X$ dB. That is, positive attenuation is the same as negative amplification, and negative attenuation is the same as positive amplification. *See also* ATTENUATOR.

ATTENUATION DISTORTION

Attenuation distortion is an undesirable attenuation characteristic over a particular range of frequencies (*see* ATTENUATION VS FREQUENCY CHARACTERISTIC). This can occur in radio-frequency as well as audio-frequency applications. A lowpass response is an advantage for a voice communications circuit because most of the frequencies in the human voice fall below 3 kHz. However, for the transmission of music, such a response would represent a circuit with objectionable attenuation distortion, because music contains audio frequencies as high as 20 kHz or more.

ATTENUATION VS FREQUENCY CHARACTERISTIC

The *attenuation-vs.-frequency characteristic* of a circuit is the amount of loss through the circuit as a function of frequency. This function is generally shown with the amplitude, in decibels relative to a certain reference level, on the vertical scale and the frequency on the horizontal scale. The article ATTENUATION DISTORTION discusses the attenuation-vs-frequency characteristic curve for an audio lowpass filter. This filter is designed for use as an intelligibility enhancer for a voice-communications circuit.

Various devices are used for precise adjustment of the attenuation-vs-frequency characteristic in different situations. In high-fidelity recording equipment, the familiar equalizer or graphic equalizer is often used in place of simple bass and treble controls for obtaining exactly the desired audio response. In some RF amplifiers, it is advantageous to introduce a lowpass, highpass, or bandpass response to prevent oscillation or excessive radiation of unwanted energy. *See also* BANDPASS FILTER, BAND-REJECTION FILTER, HIGHPASS FILTER, and LOWPASS FILTER.

ATTENUATOR

An *attenuator* is a network, usually passive rather than active, designed to cause a reduction in the amplitude of a signal. Such circuits are useful for sensitivity measurements and other calibration purposes. In a well-designed attenuator, the amount of attenuation is constant over the entire range of frequencies in the system to be checked. An attenuator introduces no reactance, and therefore no phase shift. Attenuators are built for a wide variety of input and output impedances. It is important that the impedances be properly matched, or the attenuator will not function properly.

Two simple passive attenuators made from noninductive resistors are shown in A and B. The circuit at A is called a *pi-network attenuator*, and the circuit at B is called a *T-network attenuator*. These circuits will function from the audio-frequency range well into the vhf spectrum. At uhf and above, however, the resistors begin to show inductive reactance because the wavelength is so short that the leads are quite long electrically. Then, the attenuators will no longer perform their intended function.

Attenuators for UHF must be designed especially to suit the short wavelengths at those frequencies. Such circuits must be physically small.

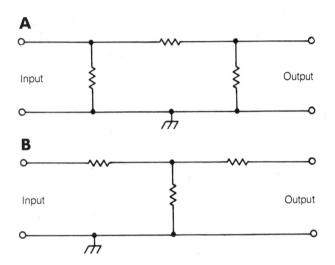

ATTENUATOR: At A, pi network. At B, T network.

AUDIO AMPLIFIER

An *audio amplifier* is an active device that is designed especially for the amplification of signals in the audio-frequency range.

Some audio amplifiers, such as microphone preamplifiers and radio-receiver audio stages, must work with very small signal input. They need not generate much output power. Other audio amplifiers, such as those found in high-fidelity or public-address systems, must develop large amounts of power output, sometimes thousands of watts.

In communications systems, the frequency response of an audio amplifier is restricted to a relatively narrow range. But in a high-fidelity system, a flat response is desired from perhaps 10 Hz to well above the range of human hearing.

Most audio amplifiers use transistors, although some use integrated-circuit operational amplifiers. It is rare to see vacuum tubes in an audio amplifier. *See also* OPERATIONAL AMPLIFIER.

AUDIO FILTERING

In a radio receiver, selectivity is achieved mainly in the intermediate-frequency (IF) stages. The most often-used methods of IF filtering use crystal-lattice filters or mechanical filters. *See* CRYSTAL-LATTICE FILTER, INTERMEDIATE FREQUENCY, MECHANICAL FILTER, and SELECTIVITY. Additional selectivity can be obtained by tailoring the audio response of the receiver after the detection of the signal.

Audio filters can be of the bandpass, band-rejection (notch), or lowpass type. Highpass audio filters are not often used in communications receivers. *See* BANDPASS FILTER, BAND-REJECTION FILTER, LOWPASS FILTER, and NOTCH FILTER.

Voice Passband Audio Filtering A voice signal occupies an audio band ranging from about 300 to 3000 Hz. Although the human voice contains components outside this range, the 2.7-kHz-wide band is enough for good intelligibility under most conditions. Therefore, an audio bandpass filter, with a passband of about 300 to 3000 Hz, can improve the quality of reception with some voice receivers. Sometimes a lowpass filter, with a cutoff frequency of about 3000 Hz, is used instead. The results are about the same for either type of audio filter in voice communications use.

CW Passband Audio Filtering A Morse code, or CW, signal needs only about 100 Hz of bandwidth in order to be clearly read. Most high-frequency communications-receiver IF stages provide a passband of about 2.7 kHz, suitable for single-sideband (SSB). Those with CW filters can narrow this down to 500 Hz or so, and some can provide IF selectivity as narrow as about 200 Hz. A narrower audio bandpass filter can sometimes improve the quality of reception when a band is crowded.

Passbands narrower than 100 Hz are not used often, because *ringing* can occur. Ringing is the tendency for an audio bandpass filter to produce damped oscillations at its center frequency. This effect reduces the speed at which a Morse signal can be clearly received.

Rectangular Response An ideal audio filter has a relatively *rectangular response*. This means that there is very little, or no, attenuation within the passband range, and very great attenuation outside the range. The *skirts* are steep. These are the same features that are desirable in IF filtering stages. There are various reasons why a flat passband response and steep skirts are advantageous in filtering systems. These features are more important in voice communications than they are in CW. Some CW operators prefer a less flat bandpass response for audio filtering. See BANDPASS RESPONSE, and SKIRT SELECTIVITY.

Most CW bandpass audio filters are tunable within an audio range of about 300 to 3000 Hz. Comfortable CW listening frequencies, for most operators, are from about 400 to 1000 Hz.

Notch Filters An audio notch filter is a band-rejection filter with a sharp, narrow response. The idea is that a *heterodyne*, or an interfering carrier that produces a tone of constant frequency in the receiver output, can be nulled out by means of such a filter.

Some receivers have IF notch filters. These work better than any audio notch filter because IF filtering gets rid of AGC (automatic gain control) effects that are caused by interfering heterodynes and audio filters do not. Nonetheless, for receivers that don't have IF notch filters, an audio notch can be useful.

Audio notch filters are usually tunable from about 300 to 3000 Hz. Some tune wider ranges. A few tune automatically: when an interfering heterodyne appears and remains for more than about 0.5 second, the notch is activated and centers itself on the frequency of the heterodyne.

AUDIO FREQUENCY

Alternating current in the range of approximately 10 to 20,000 Hz is called *audio-frequency (AF) current*. When passed through a transducer such as a speaker or headset, these currents produce audible sounds. A young person with excellent hearing can generally detect sine-wave tones from 10 to 20,000 Hz. An older person loses sensitivity to the higher frequencies and to the extremely low frequencies. By age 50 or 60, the limits of the hearing range are usually about 40 Hz to 10,000 Hz.

The range of audio frequencies is called the *AF spectrum*. All sounds, from the simple tone of a sine-wave audio oscillator to the complex noises of speech, are combinations of frequencies in this range.

Radio communications allocations begin at 9 kHz, actually within the audio-frequency range. It is in fact possible at times to hear signals below 20 kHz by simply connecting a sensitive audio amplifier to an antenna.

AUDIO-FREQUENCY TRANSFORMER

An *audio-frequency transformer* is used for the purpose of matching impedances at audio frequencies. The output of an audio amplifier might have an impedance of 200 ohms, and the speaker or headset an impedance of only 8 ohms. The AF transformer provides the proper termination for the amplifier. This assures the most efficient possible transfer of power.

The impedance-matching ratio of a transformer is proportional to the square of the turns ratio. Thus, if the primary winding has N_{PRI} turns and the secondary winding has N_{SEC} turns, the ratio of the primary impedance to the secondary impedance, Z_{PRI} to Z_{SEC} is:

$$Z_{PRI}/Z_{SEC} = (N_{PRI}/N_{SEC})^2$$

Also:

$$N_{PRI}/N_{SEC} = \sqrt{Z_{PRI}/Z_{SEC}}$$

Audio-frequency transformers are available in various power ratings and impedance-matching ratios. There are also different kinds of audio-frequency transformers to meet various frequency-response requirements. Such transformers are physically similar to ordinary alternating-current power transformers (see the illustration). They are wound on laminated or powdered-iron cores. *See also* TRANSFORMER.

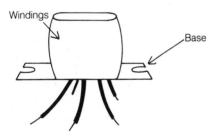

AUDIO-FREQUENCY TRANSFORMER: These devices look like miniature ac power transformers.

AUDIO LIMITER

An *audio limiter* is a device that prevents the amplitude of an audio-frequency signal from exceeding a certain value. All audio-frequency voltages below the limiting value are not affected. A limiter might be either active or passive.

The illustration shows a simple diode limiter. If germanium diodes are used, the limiting voltage is approximately ± 0.3 volts peak. If silicon diodes are used, the limiting voltage is

about ± 0.6 volts peak. Larger limiting voltages are obtained by connecting the diodes in series.

Diode limiters, as well as most other kinds of audio limiters, cause considerable distortion at volume levels above the limiting voltage. If fidelity is important, an automatic gain or level control is preferable to audio limiting (*see* AUTOMATIC GAIN CONTROL, and AUTOMATIC LEVEL CONTROL). In radio communications transmitters, audio limiters are sometimes used in conjunction with amplifiers and lowpass filters to increase the average talk power. The signal is first amplified, and then the limiter is used to prevent overmodulation. Finally, a lowpass filter prevents excessive distortion products from being transmitted over the air. This is called *speech clipping* (*see* SPEECH CLIPPING). Under poor conditions, speech clipping can result in improved intelligibility and articulation (*see* ARTICULATION, and INTELLIGIBILITY).

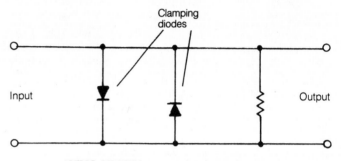

AUDIO LIMITER: A simple two-diode circuit.

AUDIO TAPER

The human ear does not perceive sound intensity in a linear fashion. Instead, the apparent intensity, or audibility, of a sound increases according to the logarithm of the actual amplitude in watts per square centimeter. An increase of approximately 26 percent is the smallest detectable change for the human ear, and is called the *decibel*, abbreviated *dB* (*see* DECIBEL). A threefold increase in sound power is 5 dB; a tenfold increase is 10 dB, and a hundredfold increase represents 20 dB.

The audio-taper volume control compensates for the nonlinearity of the human hearing apparatus. A special kind of potentiometer is used to give the impression of a linear increase in audibility as the volume is raised. Most radio receivers, tape players, and high-fidelity systems use audio-taper potentiometers in their AF amplifier circuits. This kind of potentiometer is sometimes also called a *logarithmic-taper (log-taper) device*.

For linear applications, a linear-taper potentiometer is needed. Use of a linear-taper potentiometer is undesirable in a nonlinear situation. The converse is also true; an audio-taper device does not function well in a linear circuit.

AURORA

A sudden eruption on the sun, called a *solar flare*, causes high-speed atomic particles to be thrown far into space from the surface of the sun. Some of these particles are drawn to the upper atmosphere over the north and south magnetic poles of our planet. This results in dramatic ionization of the thin atmosphere over the Arctic and Antarctic. On a moonless night, the glow of the ionized gas can be seen as the Aurora Borealis (Northern Lights) and Aurora Australis (Southern Lights).

The auroras are important in radio communication because they affect the propagation of electromagnetic waves in the high-frequency and very-high-frequency ranges. Signals might be returned to earth at much higher frequencies than usual when the auroras are active. The unusual solar activity that causes the aurora, however, also can degrade normal ionospheric propagation. Sometimes the disturbance is so pronounced that even wire circuits are disrupted by the fluctuating magnetic fields. *See also* AURORAL PROPAGATION, IONOSPHERE, PROPAGATION, and PROPAGATION CHARACTERISTICS.

AURORAL PROPAGATION

In the presence of unusual solar activity, the auroras (*see* AURORA) often return electromagnetic energy to earth. This is called *auroral propagation*.

The auroras occur in the ionosphere, at altitudes of about 40 to 250 miles above the surface of the earth (*see* IONOSPHERE). Theoretically, auroral propagation is possible, when the auroras are active, between any two points on the ground from which the same part of the aurora can be seen. Auroral propagation seldom occurs when one end of the circuit is at a latitude less than 35 degrees north or south of the equator.

Auroral propagation is characterized by a rapid flutter, which makes voice signals such as AM, FM, or SSB unreadable most of the time. Morse code is generally most effective in auroral propagation, and even this mode might sometimes be heard but not understood because of the extreme distortion. The flutter is the result of severe selective fading and multipath distortion (*see* SELECTIVE FADING, and MULTIPATH FADING). Auroral propagation can occur well into the very-high-frequency region, and is usually accompanied by moderate to severe deterioration in ionospheric conditions of the E and F layers. *See also* E LAYER, F LAYER, PROPAGATION, and PROPAGATION CHARACTERISTICS.

AUTOMATIC BIAS

Automatic bias is a method of obtaining the proper bias for a tube or transistor by means of a resistor, usually in the cathode or emitter circuit. A capacitor might also be used, in parallel with the resistor, to stabilize the bias voltage.

Other means of developing bias include resistive voltage-dividing networks and special batteries or power supplies. *See also* BIAS, and BIAS STABILIZATION.

AUTOMATIC DIRECTION FINDER

An *automatic direction finder (ADF)* is a device that indicates the compass direction from which a radio signal is coming. A loop antenna is used to obtain the bearing. In conjunction with a small whip antenna, the loop gives a unidirectional indication.

A rotating device turns the loop antenna until the signal strength reaches a sharp minimum, or null. At this point the loop, connected to a sensing circuit, stops rotating automatically. The operator can simply read the compass bearing from the azimuth indicator in degrees.

The exact position of a transmitting station is obtained by taking readings with the automatic direction finder at two or more widely separated points. On a detailed map, straight lines are drawn according to the bearings obtained at the observation points. The intersection point of these lines is the location of the transmitter. *See also* DIRECTION FINDER.

AUTOMATIC FREQUENCY CONTROL

Automatic frequency control is a method for keeping a receiver or transmitter on the desired operating frequency. Such devices are commonly used in FM stereo receivers because tuning is critical with this type of signal, and even a small error can noticeably affect sound reproduction.

An automatic-frequency-control device senses any deviation from the correct frequency, and introduces a dc voltage across a varactor diode in the oscillator circuit to compensate for the drift. An increase in oscillator frequency causes the injection of a voltage to lower the frequency; a decrease in the oscillator frequency causes the injection of a voltage to raise the frequency.

A special kind of automatic frequency control is the *phase-locked loop (PLL)*. Such a circuit maintains the frequency of an oscillator within very narrow limits. *See also* PHASE-LOCKED LOOP.

AUTOMATIC GAIN CONTROL

In a communications receiver, it is desirable to keep the output essentially constant regardless of the strength of the incoming signal. This allows reception of strong as well as weak signals, without the need for volume adjustments. It also prevents "blasting," the effect produced by a loud signal as it suddenly begins to come in while the receiver volume is turned up.

Refer to the schematic diagram of a simplified *automatic gain control (AGC)*. An audio or RF amplifier stage can be used for AGC; usually, it is done in the intermediate-frequency stages of a superheterodyne receiver. Part of the amplifier output is rectified and filtered to obtain a dc voltage. This dc voltage is applied to the input circuit of that amplifier stage or a preceding stage. The greater the signal output, the greater the AGC voltage, which changes the amplifier bias to reduce its gain. By properly choosing the component values, the amplifier output can be kept at an almost constant level for a wide range of signal-input amplitudes. Automatic gain control is sometimes also called *automatic level control* or *automatic volume control* (*see* AUTOMATIC LEVEL CONTROL), although the term *automatic level control* is generally used to define a broader scope of applications.

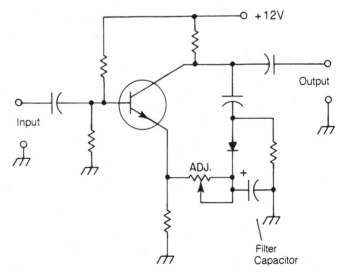

AUTOMATIC GAIN CONTROL: Some of the output is rectified and filtered to reduce the gain when signals are strong.

The attack time (*see* ATTACK) of an AGC system must be rapid, and the decay time (*see* DECAY) slower. Some receivers have AGC with variable decay or release time. The decay time is easily set to the desired value by suitably choosing the value of the filter capacitor in the dc circuit. The time constant of this decay circuit should be slow enough to prevent background noise from appearing between elements of the received signal, but rapid enough to keep up with fading effects. *See also* RECEIVER.

AUTOMATIC LEVEL CONTROL

Automatic level control (ALC) is a generalized form of *automatic gain control (AGC)* (*see* AUTOMATIC GAIN CONTROL). AGC is used in communications receivers, and ALC is found in a wide variety of devices—from tape recorders to AM, SSB, and FM transmitters. Both AGC and ALC function in the same way: Part of the output is rectified and filtered, and the resulting dc voltage is fed back to a preceding stage to control the amplification. This keeps the output essentially constant—even when the input amplitude changes greatly.

Automatic-level control is frequently used in voice transmitters to prevent flat topping of the modulated waveform (*see* FLAT TOPPING). Automatic-level control can also be used to boost the low-level modulation somewhat because the amplifier can be made more sensitive without fear of overmodulation. A circuit designed to substantially increase the talk power in a voice transmitter is called a *speech compressor* or *speech processor* (*see* SPEECH COMPRESSION) and is a form of ALC in conjunction with an audio amplifier. Automatic-level control should not be confused with limiting or clipping (*see* AUDIO LIMITER, CLIPPER, and SPEECH CLIPPING).

In a transmitter modulation circuit, ALC is sometimes called *automatic-modulation control* (*see* AUTOMATIC MODULATION CONTROL).

AUTOMATIC MODULATION CONTROL

Automatic modulation control is a form of automatic-level control (*see* AUTOMATIC LEVEL CONTROL). In an AM, SSB, or FM transmitter, automatic modulation control prevents excessive modulation while allowing ample microphone gain.

In an AM or SSB transmitter, overmodulation causes a phenomenon known as *splatter* (*see* SPLATTER), which greatly increases the signal bandwidth and can cause interference to stations on nearby channels or frequencies. In an FM transmitter, overmodulation results in overdeviation. This makes the signal appear distorted in the receiver, and can also cause interference to stations on adjacent channels. In narrowband FM the proper deviation is $\pm$ 5 kHz. In operation, while observing a deviation monitor, an operator should find it impossible to cause overdeviation, even by shouting into the microphone.

Without automatic modulation control, proper adjustment of a voice transmitter would be exceedingly critical. Therefore, nearly all voice transmitters use some form of automatic modulation control. *See also* AMPLITUDE MODULATION, FREQUENCY MODULATION, and SINGLE SIDEBAND.

AUTOMATIC NOISE LIMITER

An *automatic noise limiter (ANL)* is a circuit that clips impulse and static noise peaks while not affecting the desired signal. An automatic noise limiter sets the level of limiting, or clipping, according to the strength of the incoming signal. The AGC voltage (*see* AUTOMATIC GAIN CONTROL) in the receiver usually forms the reference for the limiter. The stronger the signal, the greater the limiting threshold. A manual noise limiter (*see* NOISE LIMITER) differs from the automatic noise limiter; the manual limiter must be set by the operator, and will cause distortion of received signals if not properly adjusted.

A noise limiter prevents noise from becoming stronger than the signals within the receiver passband, but the signal-to-noise ratio remains marginal in the presence of strong impulse noise. For improved reception with severe impulse noise, the noise blanker (*see* NOISE BLANKER) has been developed. Noise blankers are extremely effective against many kinds of manmade noise, but noise limiters are generally better for atmospheric static.

AUTOMATIC REPEAT REQUEST

See AMTOR.

AUTOPATCH

In FM repeater operation, it is a popular practice to equip the radio transceiver with a telephone keypad (DTMF pad). There are 16 standard pairs of audio tones produced by the keys 0 through 9, *, #, A, B, C and D. These tones are transmitted when the keys are pressed. The 12 tone pairs corresponding to 0 through 9, * and # are exactly the same tone pairs produced by a telephone set.

If a repeater is equipped with a telephone patch, known as an autopatch, the radio operator can make telephone calls from a mobile or handheld unit. A special code is needed to actuate, or "bring up," the autopatch system. This code is a sequence of numbers and/or symbols on the DTMF pad.

There are electronic limitations on the use of an autopatch system. Full duplex is not possible, as it is with a normal telephone. You cannot interrupt the person on the other end. The conversing parties say "over" when they are done speaking. Another limitation is imposed by the time-out system. If you talk too long, the repeater might cut you off.

There are also operating constraints. It is illegal to use an autopatch system for any business-related phone calls. Therefore, autopatch is not like a cellular telephone. Courtesy is important. Private conversations have no place on autopatch because everyone who tunes to the repeater output frequency can overhear both ends of the conversation. It is not good operating practice to carry on long conversations via autopatch; calls should be short and to the point. The ham station must also be sure to identify itself as required by FCC regulations. *See also* REPEATER, and TOUCHTONE.

AUTOSTART

In digital communications, messages can be left at a station even when there is no operator attending the station. *Autostart* is a means of activating a teleprinter or computer when a certain code is received on a certain frequency. This allows the message to be printed out or stored in memory. The station operator can recover the message later.

In the early days of ham radioteletype, autostart devices were simple circuits that started and stopped a mechanical teleprinter. The operator, upon coming to the operating position, would find that paper had "scrolled out" of the teleprinter machine, and printed matter would be there. Sometimes the message made sense; often it did not, because the band conditions might have deteriorated, or there might have been local interference.

Nowadays, with AMTOR and packet radio, autostart is a somewhat archaic term. *See* AMTOR, PACKET RADIO, and RADIOTELETYPE.

AUTOSYN

Autosyn is a trade name for a synchronous motor system developed by Bendix Corporation. The Autosyn is used as a directional indicator for such things as rotatable antenna arrays. When the shaft of the sensor motor is moved, the shaft of the indicator motor follows. The Autosyn might be used as a remote-control device as well as an indicator.

A pair of Autosyn mechanisms can be used as an azimuth-elevation (az-el) indicator for satellite or moonbounce antennas (*see* AZ-EL). The azimuth indicator is geared to a 1-to-1 ratio and calibrated from zero to 360 degrees. The elevation indicator is geared at a 1-to-4 ratio and calibrated from zero to 90 degrees.

A machine that works in a similar way to the Autosyn is called the *selsyn*, used for antenna direction indicators. *See also* SELSYN.

AUTOTRANSFORMER

An *autotransformer* is a special type of step-up or step-down transformer with only one winding. Usually, a transformer has two separate windings, called the *primary* and *secondary*, electrically isolated from each other, as shown in A. The autotransformer, however, uses a single tapped winding.

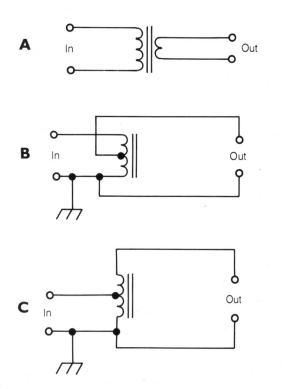

For step-down purposes, the input to the autotransformer is connected across the entire winding, and the output is taken from only part of the winding, as shown in B. For step-up purposes, the situation is reversed: the input is applied across part of the winding, and the output appears across the entire coil (C).

The autotransformer obviously contains less wire than a transformer with two separate windings. If electrical (dc) isolation is needed between the input and output, however, the autotransformer cannot be used.

The step-up or step-down voltage ratio of an autotransformer is determined according to the same formula as for an ordinary transformer (*see* TRANSFORMER). If N_{PRI} is the number of turns in the primary winding, and N_{SEC} is the number of turns in the secondary winding, then the voltage transformation ratio is:

$$E_{PRI}/E_{SEC} = N_{PRI}/N_{SEC}$$

where E_{PRI} and E_{SEC} represent the ac voltages across the primary and secondary windings, respectively. The impedance-transfer ratio is:

$$Z_{PRI}/Z_{SEC} = (N_{PRI}/N_{SEC})^2$$

where Z_{PRI} and Z_{SEC} are the impedances across the primary and secondary windings, respectively.

AVALANCHE BREAKDOWN

When reverse voltage is applied to a p-n semiconductor junction, very little current will flow as long as the voltage is relatively small. But as the voltage is increased the holes and electrons in the p-type and n-type semiconductor materials are accelerated to greater and greater speeds. Finally the electrons collide with atoms, forming a conductive path across the p-n junction. This is called *avalanche breakdown*. It does not permanently damage a semiconductor diode, but does render it useless for rectification or detection because current flows both ways when avalanche breakdown occurs (*see* RECTIFICATION).

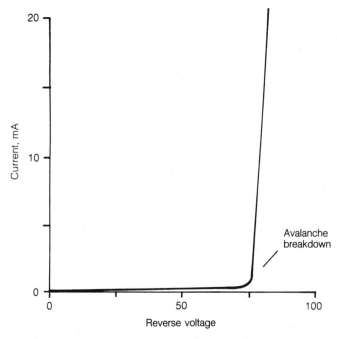

AUTOTRANSFORMER: At A, a conventional transformer. At B, step-down autotransformer; at C, step-up autotransformer.

AVALANCHE BREAKDOWN: The reverse current rises abruptly.

Avalanche breakdown in a particular diode might occur with just a few volts of reverse bias, or it might take hundreds of volts. Rectifier diodes do not undergo avalanche breakdown until the voltage becomes quite large (*see* AVALANCHE VOLTAGE). A special type of diode, called the *zener diode* (*see* ZENER DIODE), is designed to have a fairly low, and very precise, avalanche breakdown voltage. Such diodes are used as voltage regulators. Zener diodes are sometimes called *avalanche diodes* because of their low avalanche voltages.

The illustration shows graphically the current flow through a diode as a function of reverse voltage. The current is so small, for small reverse voltages, that it might be considered to be zero for practical purposes. When the reverse voltage becomes sufficient to cause avalanche breakdown, however, the current rises rapidly.

AVALANCHE IMPEDANCE

When a diode has sufficient reverse bias to cause avalanche breakdown, the device appears to display a finite resistance. This resistance fluctuates with the amount of reverse voltage, and is called the *avalanche resistance* or *avalanche impedance*. It is given by:

$$Z_A = E_R/I_R$$

for direct current, where Z_A is the avalanche impedance, E_R is the reverse voltage, and I_R is the reverse current. Units are ohms, volts, and amperes, respectively.

When E_R is smaller than the avalanche voltage (*see* AVALANCHE VOLTAGE), the value of Z_A is extremely large, because the current is small, as shown in the figure. When E_R exceeds the avalanche voltage, the value of Z_A drops sharply. As E_R is increased further and further, the value of Z_A continues to decrease.

The magnitude of Z_A for alternating current is practically infinite in the reverse direction, as long as the peak ac voltage never exceeds the avalanche voltage. When the peak ac reverse voltage rises to a value greater than the avalanche voltage, the impedance drops. The maximum voltage that a semiconductor diode can tolerate, for rectification purposes, without avalanche breakdown is called the *peak inverse voltage* (*see* PEAK INVERSE VOLTAGE).

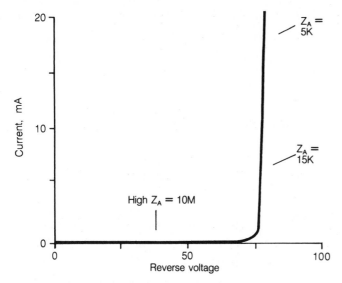

AVALANCHE IMPEDANCE: The avalanche impedance, Z_A, depends on reverse voltage.

The exact value of the avalanche impedance for alternating current is the value of resistor that would be necessary to allow the same flow of reverse current. The average Z_A for alternating current differs from the instantaneous value, which fluctuates as the voltage rises and falls. Avalanche is usually undesirable in alternating-current applications.

AVALANCHE TRANSISTOR

An *avalanche transistor* is an npn or pnp transistor that is designed to operate with a high level of reverse bias at the emitter-base junction. Normally, transistors are forward-biased at the emitter-base junction except when cutoff conditions are desired when there is no signal input.

Avalanche transistors are seen in some switching applications. The emitter-base junction is reverse-biased almost to the point where avalanche breakdown occurs. A small additional reverse voltage, supplied by the input signal, triggers avalanche breakdown of the junction and resultant conduction. Therefore, the entire transistor conducts, switching a large amount of current in a very short time. The extremely sharp "knee" of the reverse-voltage-vs.-current-curve facilitates this switching capability (*see* the illustrations in AVALANCHE BREAKDOWN and AVALANCHE IMPEDANCE). A small amount of input voltage can thus cause the switching of large values of current.

AVALANCHE VOLTAGE

The *avalanche voltage* of a p-n semiconductor junction is the amount of reverse voltage that is required to cause avalanche breakdown (*see* AVALANCHE BREAKDOWN). Normally, the n-type semiconductor is negative with respect to the p-type in forward bias, and the n-type semiconductor is positive, with respect to the p-type in reverse bias.

In some diodes, the avalanche voltage is very low, as small as 6 volts. In other diodes, it might be hundreds of volts. When the avalanche voltage is reached, the current abruptly rises from near zero to a value that depends on the reverse bias (*see* the illustrations in AVALANCHE BREAKDOWN and AVALANCHE IMPEDANCE).

Diodes with high avalanche-voltage ratings are used as rectifiers in dc power supplies. The avalanche voltage for rectifier diodes is called the *peak inverse voltage* or *peak reverse voltage* (*see* PEAK INVERSE VOLTAGE). In the design of a dc power supply, diodes with sufficiently high peak-inverse-voltage ratings must be chosen so that avalanche breakdown does not occur.

Some diodes are deliberately designed so that an effect similar to avalanche breakdown occurs at a relatively low, and well defined, voltage. These diodes are called *Zener diodes* (*see* ZENER DIODE). They are used as voltage-regulating devices in low-voltage dc power supplies. *See also* DC POWER SUPPLY.

AVERAGE CURRENT

When the current flowing through a conductor is not constant, the average current is determined as the mathematical mean value of the instantaneous current at all points during one complete cycle. Most ammeters register average current. Some special devices register peak current.

Consider, for example, a class-B amplifier that has no collector current in the absence of an input signal. Then, under no-signal conditions, the average current at the collector is zero.

When an input signal is applied, the collector current will flow during approximately 50 percent of the cycle as shown. In this particular illustration, the peak collector current is 70 mA. The average current is smaller—about 22 mA.

In the case of alternating current, the average current is usually zero because the polarity is positive during half the cycle and negative during half the cycle, with peak values and waveforms identical in both the positive and negative directions. Of course, the effects of alternating current are very different from the effects of zero current! Average current is given little importance in ac circuits; the *root-mean-square (RMS)* current is more often specified as the effective current in such instances (*see* ROOT MEAN SQUARE). In a sine-wave half cycle, the average current is 0.637 times the peak current.

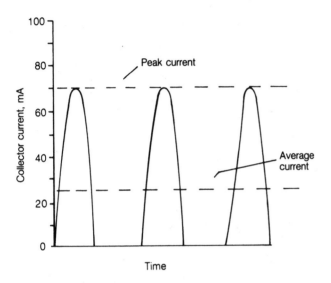

AVERAGE CURRENT: Average and peak current in the output of a class-B amplifier.

AVERAGE POWER

When the level of power fluctuates rapidly, such as in the modulation envelope of a single-sideband transmitter (see the graph), the signal power might be defined in terms of *peak power* or *average power*. An ordinary wattmeter reads average power; special devices are needed to determine peak power.

The average power output of a circuit is never greater than the peak power output. If the power level is constant, as with an unmodulated carrier or frequency-modulated signal, the average and peak power are the same. If the amplitude changes, then the average power is less than the peak power. The output of a single-sideband signal modulated by a voice will have an average power level of about half the peak power level, although the exact ratio depends on the characteristics of the voice. *See also* PEAK ENVELOPE POWER, and PEAK POWER.

AVERAGE VALUE

Any measurable quantity (such as voltage, current, power, temperature, or speed) has an *average value* that is defined for a given period of time. Average values are important in determining the effects of a rapidly fluctuating variable.

To determine the average value of some variable, many instantaneous values are mathematically averaged. The more instantaneous, or sampling, values used, the more accurate the determination of the average value. For a sine-wave half cycle (such as one of the pulses in A or B) having a duration of 10 milliseconds (0.01 second) at an ac frequency of 50 Hz, instantaneous readings might be taken at intervals of 1 millisecond, then 100 microseconds, 10 microseconds, 1 microsecond and so on. This would yield first 10, then 100, then 1000, and finally 10,000 sampling values. The average value that is determined from many sampling values is more accurate than the average value that is determined from just a few sampling values.

There is, of course, no limit to the number of sampling values that can be averaged in this way. The true average value, however, is the arithmetic mean of all the instantaneous values in a given time interval. Although the mathematical construction of this is rather complicated, a true average value always exists for a quantity evaluated over a specified period of time.

An equivalent method of evaluating the average value for a variable quantity involves the construction of a rectangle having a total area equal to the area under the curve for a certain time interval.

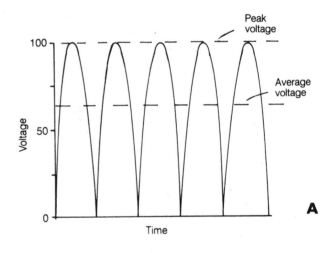

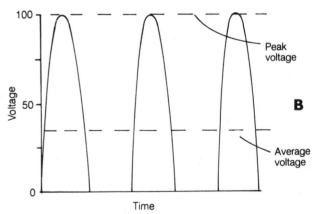

AVERAGE VALUE: At A, average vs peak voltage for a full-wave rectifier. At B, for a half-wave rectifier.

AVERAGE VOLTAGE

When the voltage in a circuit fluctuates, the *average voltage* is the mean value of the instantaneous voltage at all points during one complete cycle. Most voltmeters register average voltage when the frequency is greater than a few Hertz.

The output of a full-wave rectifier circuit is shown in A in AVERAGE VALUE. The peak voltage in this case is 100 volts (*see* PEAK VALUE), but the average voltage is only about 63.7 volts because, in one-half cycle of a sine wave, the average value is 0.637 times the peak value. In the case of a half-wave rectifier, where every other half cycle is cut off, the average voltage is just 31.9 volts when the peak voltage is 100 (B in AVERAGE VALUE).

For a source of alternating current, the average voltage is usually zero over a long period of time because the polarity is positive during half the cycle and negative during the other half, and the peak values are identical in both directions. An ac voltage has, of course, very different effects from zero voltage; therefore, the average voltage is given little importance in ac circuits. The root-mean-square (RMS) voltage is more often specified as the equivalent dc voltage. *See also* ROOT MEAN SQUARE.

AWARDS

In order to make ham radio operating fun and challenging, various organizations give out award certificates to hams. Most of the awards are administered by the American Radio Relay League (ARRL). Smaller groups of hams, and other organizations, give awards also. *See* AMERICAN RADIO RELAY LEAGUE.

The following are some of the awards hams can get.

DX Century Club For contacting 100 or more foreign countries, and for collecting the QSL cards from those contacts, the ARRL will give you a certificate of membership in the *DX Century Club (DXCC)*. Some hams have contacted more than 300 countries. The call signs of these hams are regularly listed in *QST*, the official magazine of the ARRL. Competition is keen among top DXers. *See* DX AND DXING.

Extra Class License Certificate If you obtain the highest grade of ham license, the *Extra Class license*, you can get a certificate from ARRL to hang on your wall. Many professional engineers regard the Extra Class license as a major achievement.

Code Proficiency The ARRL regularly holds on-the-air tests, transmitted by station W1AW, for code proficiency at speeds from 10 to 40 words per minute (wpm). To obtain a certificate for a certain speed, you must get one minute of "solid copy." Once you get a certificate at a certain speed, such as 15 wpm, you can obtain endorsements later for higher speeds in 5-wpm increments. The times and frequencies for qualifying runs are published in *QST*.

WAS and WAC The ARRL will give you a certificate for submitting QSL cards for contacts with each of the 50 states in the U.S. This is called the *Worked All States (WAS) award*.

A certificate is also available for having contacted stations in each of the continents: North America, South America, Europe, Asia, Africa, and Australia. QSL cards are required for verification. This is called the *Worked All Continents (WAC)* award.

A-1 Operators' Club If you show good operating skill and courtesy while in QSO with two or more members of the *A-1 Operators' Club*, you will be invited to join the club. The award is administered by the ARRL.

Contest Awards *See* CONTESTS AND CONTESTING.

Traffic Handling Many hams consider the satisfaction of public service to be an award in itself, whether a certificate is issued, or not. *See* TRAFFIC HANDLING.

Less Serious Awards If you carry on a contact for more than 30 minutes with a ham who is a member of the *Rag Chewers' Club*, you might be invited to join the club. This indicates that you can have a long conversation, as well as a short QSO, on the radio.

If you have been a ham for 20 years or more, and can prove it to ARRL (such as with an old license or *Callbook* listing), you can become a member of the *Old Timers' Club*.

For carrying on a code (CW) conversation at a speed of 80 wpm or more, copying "in your head," but understanding what is said, a group of hams will welcome you into the *Five Star Operators' Club*.

Numerous other awards exist, and they all carry attractive (and sometimes humorous) certificates. You'll learn about these as you gain experience, and make friends, as a radio ham.

AWG

See AMERICAN WIRE GAUGE.

AX.25

AX.25 is a packet-radio communications *protocol*, based on a scheme that was created by the International Standards Organization. The communications scheme is called *OSI-RM*, for *Open Systems Interconnection Reference Model*.

The AX.25 scheme was specifically designed for amateur packet radio. It lets computers talk to each other. It operates at the *link layer* of packet communications, via *digipeaters*.

In recent years, AX.25 has been largely superseded by a more versatile firmware, known as *NET/ROM*. NET/ROM allows all kinds of packets to be sent, such as CQs and bulletin-board messages, along various routes. The routing is done automatically. Node-to-node acknowledgment is used. *See also* DIGIPEATER, NODE, NODE-TO-NODE ACKNOWLEDGMENT, OPEN SYSTEMS INTERCONNECTION REFERENCE MODEL, and PACKET RADIO.

AXIAL LEADS

A component with *axial leads* has the leads protruding along a common, linear axis. Most resistors and semiconductor diodes, as well as many capacitors and inductors, have axial leads.

Axial leads give a component extra mechanical rigidity on a circuit board because the component can be mounted flush with the board. Axial leads have less mutual inductance than parallel leads because of their greater separation and their collinear orientation.

Axial leads are sometimes inconvenient when a component must be mounted within a small space on a printed-circuit board. In such situations, parallel leads are often preferred.

AYRTON SHUNT

The sensitivity of a galvanometer is reduced by means of a shunt resistor. The addition of variable shunt resistors, however, sometimes affects the meter movement, or damping. A special type of shunt, called an *Ayrton shunt,* allows adjustment of the range of a galvanometer without affecting the damping. Such a shunt consists of a series combination of resistors. Several different galvanometer ranges can be selected by connecting the input current across variable portions of the total resistance (*see* illustration).

Because the resistance across the galvanometer coil is the same, regardless of the range selected, the damping of the meter movement does not depend on the range chosen. The effective input resistance to the meter, however, does change. The more sensitive the meter range, the greater the resistance at the meter input terminals. A microammeter, for example, with a range of 0 to 100 microamperes, can be used as a milliammeter or ammeter without changing the damping when a suitable Ayrton shunt system is provided (*see* AMMETER). The com-

mon volt-ohm-milliammeter often utilizes an Ayrton shunt to facilitate operation in its various modes. *See also* VOLT-OHM-MILLIAMMETER.

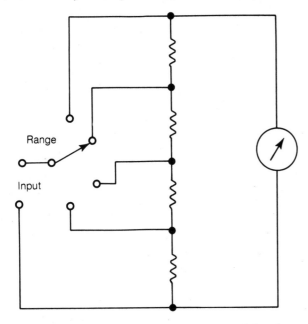

AYRTON SHUNT: This circuit is used to adjust full-scale meter sensitivity.

AZ-EL

The term *az-el* refers to azimuth and elevation. *Azimuth* is the horizontal direction or compass bearing, and *elevation* is the angle relative to the horizon, for a directional device. Azimuth bearings range from zero (true north) clockwise through 90 degrees (east), 180 degrees (south), 270 degrees (west), to 360 degrees (true north). Elevation bearings range from zero (horizontal) to 90 degrees (straight up).

An *az-el mounting* is a two-way bearing in which the azimuth and elevation are independently adjustable. Satellite communication systems, and some telescopes, make use of such mountings. Some moonbounce systems also use this kind of mounting (*see* MOONBOUNCE).

If it is necessary to continuously follow a heavenly body such as the moon, a planet, the sun, or a star, the az-el mounting requires continual adjustment of both the azimuth and elevation bearings. A special kind of mounting, called an equatorial mounting, consists of an az-el system that is oriented with its zenith toward the north celestial pole, or North Star. Then, only the azimuth bearing requires adjustment to compensate for the rotation of the earth. This facilitates prolonged antenna orientation toward one celestial object.

AZIMUTH

The term *azimuth* refers to the horizontal direction or compass bearing of an object or radio signal. True north is azimuth zero. The azimuth bearing is defined in degrees, measured clockwise from true north. Thus, east is azimuth 90 degrees, south is azimuth 180 degrees, and west is azimuth 270 degrees. Provided the elevation is less than 90 degrees, as is usually defined by convention, these azimuth bearings hold for any object in space.

Azimuth bearings are always defined as being at least zero, but less than 360 degrees, avoiding ambiguity. Measures smaller than zero degrees, and greater than or equal to 360 degrees, are not defined.

BACK BIAS

Back bias is a voltage taken from one part of a multistage circuit and applied to a previous stage. This voltage is obtained by rectifying a portion of the signal output from an amplifier circuit. Back bias can be either regenerative or degenerative, or it can be used for control purposes.

A typical application of back bias is the automatic level control (*see* AUTOMATIC GAIN CONTROL, and AUTOMATIC LEVEL CONTROL). A degenerative voltage is applied to either the input of the stage from which it is derived, or to a previous stage. This prevents significant changes in output amplitude under conditions of fluctuating input signal strength.

Back bias can be applied to either the anode or cathode of an active device. A negative voltage, for example, can be applied to the plate of a tube or the collector of an npn transistor to reduce its gain. Alternatively, a positive voltage could be applied to the cathode of a tube or the emitter of an npn transistor, if the cathode or emitter is elevated above dc ground. Still another form of degenerative back bias is the application of a negative voltage to the grid of a tube or the base of an npn transistor.

Sometimes the term *back bias* is used to define reverse bias. This condition occurs when the anode of a diode, transistor, or tube is negative, with respect to the cathode. *See also* REVERSE BIAS.

BACKBONE

The term *backbone* refers to a packet network that works with a *packet-radio bulletin-board system (PBBS)*. The backbone network allows for the automatic routing of packet radio mail. *See* PACKET RADIO, and PACKET-RADIO BULLETIN-BOARD SYSTEM.

BACK DIODE

A *back diode* is a special kind of tunnel diode, operated in the reverse-bias mode. Tunnel diodes are used as oscillators and amplifiers in the UHF and microwave part of the electromagnetic spectrum (*see* TUNNEL DIODE).

The back diode has a negative-resistance region at low levels of reverse bias. This means that, within a certain range of voltage, an increase in reverse voltage will cause a decrease in the flow of current.

BACKLASH

Any rotary, or continuously adjustable, control will have a certain amount of mechanical imprecision. The accuracy of adjustment is limited, of course, by the ability of an operator to interpolate readings (a good example of this is the ordinary receiver or transmitter frequency dial). Some inaccuracy, however, is a consequence of the mechanical shortcomings of such things as gears, planetary drives, slide-contact devices, and the like. This is called *blacklash*.

Often, if a certain setting for a control is desired, that setting will depend on whether the control is turned clockwise or counterclockwise. For example, suppose it is necessary to tune precisely to the frequency of WWV at 15.0000 MHz. This is usually done in receivers by zero beating (*see* ZERO BEAT). When tuning upward from some frequency below 15 MHz, the dial might indicate 15.0002 MHz at zero beat. When tuning downward from some frequency above 15 MHz, the dial might read 14.9997 at zero beat. The difference, 15.0002, 14.9997, or 0.0005 MHz (0.5 kHz) is the dial backlash.

Because backlash is a purely mechanical phenomenon, and not an electrical effect, it can be eliminated by the use of digital displays. The precision of the adjustment is limited only by the resolution of the display; backlash and human error are overcome. *See also* DIGITAL CONTROL, and DIGITAL DISPLAY.

BACKSCATTER

Backscatter is a form of ionospheric propagation via the E and F layers (*see* E LAYER, and F LAYER).

Generally, when a high-frequency electromagnetic field encounters an ionized layer in the upper atmosphere, the angle of return is roughly equal to the angle of incidence. However, a small amount of the field energy is scattered in all directions, as shown in the drawing. Some energy is scattered back toward the transmitting station, and into the skip zone if there is a skip zone (*see* SKIP ZONE). A receiving station within the skip zone, not normally able to hear the transmitting station, might hear this scattered energy if it is sufficiently strong. This is called *backscatter*.

Backscatter signals are often, if not usually, too weak to be heard. Under the right conditions, though, communication is possible at high frequencies via backscatter, over distances that are normally too short for E-layer or F-layer propagation. Backscatter signals are characterized by rapid fluttering and fading. This makes reception of amplitude-modulated or single-sideband signals almost impossible. The multipath nature of backscatter propagation (*see* MULTIPATH FADING) often makes even CW reception very poor. *See also* PROPAGATION, and PROPAGATION CHARACTERISTICS.

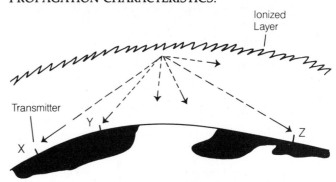

BACKSCATTER: Normally, Y cannot hear X, but Z can. Backscatter allows Y to hear X from within the skip zone.

BALANCE

The term *balance* is used to denote the most desirable set of circuit parameters in a variety of situations. In a radio-frequency transmission line consisting of two parallel conductors, balance is the condition in which the currents in the conductors flow with equal magnitude, but in opposite directions, everywhere along the line. A properly operating parallel-wire transmission line should always be kept in balance (*see* BALANCED TRANSMISSION LINE).

When two or more transistors are operated in push-pull or in parallel, it is desirable to balance the devices so that they both have essentially the same gain characteristics. Supposedly identical transistors often display slightly different amounts of gain under the same conditions. If one transistor draws more current than the other, a phenomenon known as current hogging can take place (*see* CURRENT HOGGING). Differences in input impedances, even if very small, can disrupt the efficiency and linearity of a circuit containing devices in push-pull or in parallel.

In a single-sideband transmitter, carrier balance refers to the most nearly complete suppression of the carrier signal, leaving only the sidebands. An adjustment is usually provided in a balanced modulator to facilitate proper carrier balance (*see* BALANCED MODULATOR).

BALANCED DETECTOR

A *balanced detector* is a special kind of demodulator for reception of frequency-modulated signals (FM). Any FM detector should, if possible, be sensitive only to variations in frequency, and not variations in amplitude, of the received signals. A schematic diagram of a simple circuit that accomplishes this, called a *balanced detector*, is shown in the schematic.

The tuned circuits, consisting of L1/C1 and L2/C2, are set to slightly different frequencies. By adjusting C1 so that the resonant frequency of L1/C1 is slightly above the center frequency of the signal, and by adjusting C2 so that the resonant frequency of L2/C2 is slightly below the center frequency of the signal, the circuit becomes a sort of frequency comparator. Whenever the signal frequency is at the center of the channel, midway between the resonant frequencies of L1/C1 and

L2/C2, the transistors Q1 and Q2 produce equal outputs, and the output of the balanced detector is zero. When the signal frequency rises, the output of Q1 increases and the output of Q2 decreases. When the signal frequency falls, the output of Q2 increases and the output of Q1 decreases. The filtering capacitors, C_F, smooth out the RF components of the signals, and leave only audio frequencies at the output.

An amplitude-modulated signal will produce no output from the balanced detector. The output of Q1 and Q2 are always exactly equal for an AM signal at the center of the channel, although they might both change as the signal amplitude changes. The gain characteristics of Q1 and Q2, if properly chosen, will minimize the response of the balanced detector to AM signals not at the center of the channel. *See also* DISCRIMINATOR, FREQUENCY MODULATION, RATIO DETECTOR, and SLOPE DETECTION.

BALANCED LINE

Any electrical line consisting of two conductors, each of which displays an equal impedance, with respect to ground, is called a *balanced line* if the currents in the conductors are equal in magnitude and opposite in direction at all points along the line.

The advantage of a balanced line is that it effectively prevents external fields from affecting the signals it carries. The equal and opposite nature of the currents in a balanced line results in the cancellation of the electromagnetic fields set up by the currents everywhere in space except in the immediate vicinity of the line conductors.

Precautions must be taken to ensure that balance is maintained in a line that is supposed to be balanced. Otherwise, the line loss increases, and the line becomes susceptible to interference from external fields (in receiving applications) or radiation (in transmitting applications).

Some balanced lines use four, instead of two, parallel conductors arranged at the corners of a geometric square in the transverse plane. Diagonally opposite wires are connected together. This is called a *four-wire line*. It displays more stable balance than a two-wire line. *See also* FOUR-WIRE TRANSMISSION LINE, OPEN-WIRE LINE, TRANSMISSION LINE, and TWIN LEAD.

BALANCED MODULATOR

To generate single-sideband signals, a *balanced modulator* is necessary to suppress the carrier energy (*see* SINGLE SIDEBAND). One such circuit is shown in simplified form (*see* drawing).

Transistors Q1 and Q2 act as ordinary amplitude modulators. While the audio input to Q1 is in phase with the audio input to Q2, however, their RF carrier inputs are out of phase. The collectors are connected in parallel and are thus in phase. The carrier is therefore cancelled in the output circuit, leaving only the sideband energy. The resulting signal is called a *double-sideband, suppressed-carrier signal*.

There are several ways to build balanced modulators, using both active devices (as shown here) and passive devices, such as diodes. All balanced modulators are designed to produce the double-sideband, suppressed-carrier type of signal. To get single-sideband emission, one of the sidebands from the balanced modulator is eliminated, either by filtering or by phase cancellation.

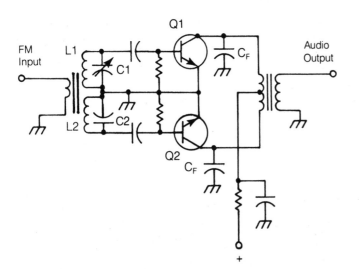

BALANCED DETECTOR: Tuned circuits L1/C1 and L2/C2 are tuned slightly off channel center. Changes in input signal frequency thus cause changes in output amplitude.

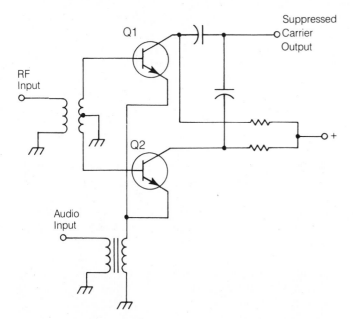

BALANCED MODULATOR: This circuit cancels out the carrier, but leaves the sidebands.

BALANCED OSCILLATOR

Any oscillator with center-tapped, grounded tank-circuit inductors is called a *balanced oscillator*. A simple balanced oscillator might consist of a tuned push-pull amplifier with inductive feedback. Variable capacitors determine the frequency of oscillation; the input circuit is kept in balance by using a split-stator (ganged) capacitor, ensuring identical values on either side of the tank circuit. Both transistors must have the same gain characteristics, as well, for good balance to be maintained.

Of course, the "push-pull oscillator" is not the only kind of balanced oscillator. *See also* OSCILLATOR, PUSH-PULL AMPLIFIER, and PUSH-PULL CONFIGURATION.

BALANCED OUTPUT

A *balanced-output circuit* is designed to be used with a balanced load and a balanced line. There are two output terminals in such a circuit; each terminal displays the same impedance, with respect to ground. The phase of the output is opposite at either terminal, with respect to the other. The peak amplitudes at both terminals are identical.

A balanced output can be used as an unbalanced output by grounding one of the terminals. However, an unbalanced output cannot be used as a balanced output. A special transformer is needed for this purpose. In radio-frequency applications, such a transformer is called a *balun*, short for "balanced/unbalanced." *See also* BALUN.

BALANCED SET

When components, such as transistors, are connected in parallel or push-pull configurations, it is important that they have similar gain characteristics. Two semiconductor transistors might have identical manufacturer numbers, but unless a pair is chosen carefully, the probability is high that minor differences will exist in the actual current gain and amplification factors. Such dissimilarities can result in current hogging (*see* CURRENT HOGGING) or unwanted unbalances in the system.

When two or more components have been chosen on the basis of identical, or nearly identical, gain and load characteristics, the set of components is said to be balanced or matched.

BALLASTED TRANSISTOR

Power transistors often consist of two or more bipolar devices on one substrate. This arrangement amounts to a set of transistors in parallel, increasing the current-handling capacity compared with a single transistor.

With bipolar transistors in parallel, there is a risk that one will take most of the burden, letting the others "loaf." This is known as *current hogging* (*see* CURRENT HOGGING). It is a vicious circle. Excessive current causes heating, which in turn reduces the resistance, increasing the current still more. The more transistors there are in parallel, the greater the likelihood of current hogging.

When discrete transistors are used in parallel, resistors are usually placed in series with the emitters. This is known as *ballast*; it reduces the probability of current hogging. In a power transistor, these resistors, having identical values for each individual emitter, are included as part of the chip. Such a device is called a *ballasted transistor*. *See also* POWER TRANSISTOR.

BALUN

When a balanced load is connected to an unbalanced source of power, a balun is sometimes used. The word *balun* is a contraction of "balanced/unbalanced." A balun has an unbalanced input and a balanced output.

The simplest form of balun is an ordinary transformer. The isolation between the primary and secondary windings allows an unbalanced source of power to be connected to one end, and a balanced load to the other, as shown at A in the illustration on pg. 36. This condition can also be reversed; a balanced source can be connected to the primary and an unbalanced load to the secondary.

At radio frequencies, baluns are sometimes constructed from lengths of coaxial cable. One example of such a device, which provides an impedance step-up ratio of 1 to 4, is shown at B. Coaxial baluns, such as this, are sometimes used for VHF.

Special transformers can be built to act as baluns over a wide range of frequencies. Two such devices are illustrated schematically. The balun at C provides a 1-to-1 impedance-transfer ratio. The balun at D provides a step-up ratio of 1 to 4. The coils are wound adjacent to each other on the same form. Toroidal powdered-iron or ferrite forms are often used for broadband balun transformers.

Although balun coils are quite frequently used when the source of power is unbalanced and the load is balanced, the devices can also be used in the reverse situation—a balanced source and an unbalanced load. An example of the first type of application is a dipole or Yagi fed with coaxial cable. An example of the latter application is a vertical ground-plane antenna fed with balanced line. Baluns help maintain antenna or feedline balance when it is important.

BANANA JACK AND PLUG

A convenient single-lead connector that slips easily in and out of its receptacle is called a *banana connector*. The banana plug looks something like a banana skin, and this is where it gets its name (*see* illustration on pg. 36).

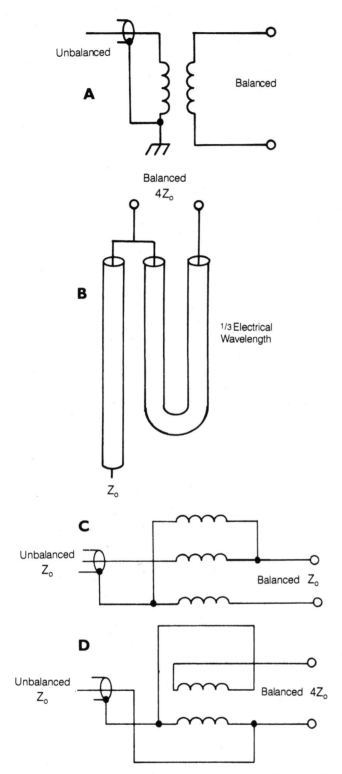

BALUN: At A, a simple transformer; at B, a 1:4 coaxial balun; at C, a 1:1 broadband balun; at D, a 1:4 broadband balun.

Banana jacks are frequently found in screw terminals of low-voltage dc power supplies. The leads can be affixed using the screw terminals for more permanent use. If frequent changing of leads is necessary, the banana jacks allow it with a minimum of trouble. Banana plugs are simply pushed in and pulled out. They lock with a moderate amount of friction. Banana connectors have fairly low dc loss and high current-handling capacity. They are not generally used at high voltages, however, because of the possibility of shock from exposed conductors.

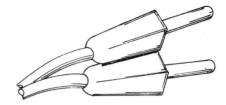

BANANA JACK AND PLUG: These are neat and convenient.

BAND

Any range of electromagnetic frequencies, marked by lower and upper limits for certain purposes, is called a *band*. The AM broadcast band is from 535 to 1605 kHz. The FM broadcast band extends from 88 to 108 MHz. The 40-meter amateur band has a lower limit of 7.000 MHz and an upper limit of 7.300 MHz. There is no limit to how narrow or wide a band can be; it must only cover a certain range.

In radio communication, the electromagnetic spectrum is subdivided into bands according to frequency, as shown in the table. The very-low-frequency (VLF) band is only 20 kHz wide, but it covers a threefold range of frequencies. All of the higher bands cover a tenfold range of frequencies. For a more detailed breakdown of the bands in the radio-frequency spectrum, *see* FREQUENCY ALLOCATIONS, and SUBBAND.

BAND: RADIO-FREQUENCY BANDS.

Designation	Frequency	Wavelength
Very low (VLF)	10 kHz-30 kHz	30 km-10 km
Low (LF)	30 kHz-300 kHz	10 km-1 km
Medium (MF)	300 kHz-3 MHz	1 km-100 m
High (HF)	3 MHz-30 MHz	100 m-10 m
Very high (VHF)	30 MHz-300 MHz	10 m-1 m
Ultra high (UHF)	300 MHz-3 GHz	1 m-100 mm
Super high (SHF)	3 GHz-30 GHz	100 mm-10 mm
Extremely high (EHF)	30 GHz-300 GHz	10 mm-1 mm

BANDPASS FILTER

Any resonant circuit, or combination of resonant circuits, designed to discriminate against all frequencies, except a frequency f_0, or a band of frequencies between two limiting frequencies of f_0 and f_1 is called a *bandpass filter*. In a parallel circuit, a bandpass filter shows a high impedance at the desired frequency or frequencies, and a low impedance at unwanted frequencies. In a series configuration, the filter has a low impedance at the desired frequency or frequencies, and a high impedance at unwanted frequencies. Three types of bandpass filter circuits are illustrated.

Some bandpass filters are built with components other than actual coils and capacitors, but all such filters operate on the same principle. The crystal-lattice filter uses piezo-electric materials, usually quartz, to obtain a bandpass response (*see* CRYSTAL-LATTICE FILTER). A mechanical filter uses vibration resonances of certain substances (*see* MECHANICAL FILTER). A resonant antenna is itself a form of bandpass filter because it allows efficient radiation at one frequency and neighboring frequencies, but discriminates against others. In optics, a simple color filter, discriminating against all light wavelengths, except those within a certain range, is a form of bandpass filter.

Bandpass filters are sometimes designed to have a very sharp, defined, resonant frequency. Sometimes the resonance is spread out over a fairly wide range. *See also* BANDPASS RESPONSE.

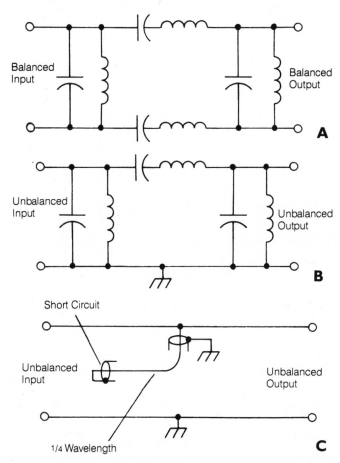

BANDPASS FILTER: Balanced (A), unbalanced (B), and unbalanced coaxial (C).

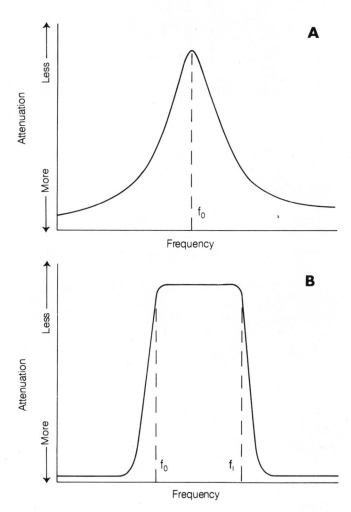

BANDPASS RESPONSE: Single-frequency (A) and wideband (B) curves.

BANDPASS RESPONSE

The attenuation-vs-frequency characteristic of a bandpass filter is called the *bandpass response* (*see* ATTENUATION-VS-FREQUENCY CHARACTERISTIC). A bandpass filter might have a single, well-defined resonant frequency (A in the illustration), denoted by f_0; or, the response might be rectangular, having two well-defined limit frequencies f_0 and f_1, as shown at B. The bandwidth (*see* BANDWIDTH) might be just a few Hertz, such as with an audio filter designed for reception of Morse code. Or, the bandwidth might be several MHz, as in a helical filter designed for the front end of a VHF receiver.

A bandpass response is always characterized by high attenuation at all frequencies, except within a particular range. The actual attenuation at desired frequencies is called the *insertion loss* (*see* INSERTION LOSS).

BAND-REJECTION FILTER

A *band-rejection filter* is a resonant circuit designed to pass energy at all frequencies, except within a certain range. The attenuation is greatest at the resonant frequency, f_0, or between two limiting frequencies, f_0 and f_1. Three examples of band-rejection filters are shown in the illustration on pg. 38. Notice the similarity between the band-rejection and bandpass filters (*see*

BANDPASS FILTER). The fundamental difference is that the band-rejection filter consists of parallel LC circuits connected in series with the signal path, or series LC circuits in parallel with the signal path; in bandpass filters, series-resonant circuits are connected in series, and parallel-resonance circuits in parallel.

Band-rejection filters need not necessarily be made up of actual coils and capacitors, but they usually are. Quartz crystals are sometimes used as band-rejection filters. Lengths of transmission line, either short-circuited or open, act as band-rejection filters for certain frequencies. A common example of a band-rejection filter is the notch filter found in some of the more sophisticated communications receivers. Another example is the antenna trap. Still another is the parasitic suppressor generally seen in the plate lead of a high-power RF amplifier. *See also* NOTCH FILTER, PARASITIC SUPPRESSOR, and TRAP.

BAND-REJECTION RESPONSE

All *band-rejection filters* show an attenuation-vs-frequencies characteristic (*see* ATTENUATION-VS-FREQUENCY CHARACTERISTIC) marked by low loss at all frequencies, except within a prescribed range. The illustration on pg. 38 shows two types of band-rejection response. A sharp response occurs at or near a single resonant frequency f_0. A rectangular response is characterized by low attenuation below a limit f_0 and above a limit f_1, and high attenuation between these limiting frequencies.

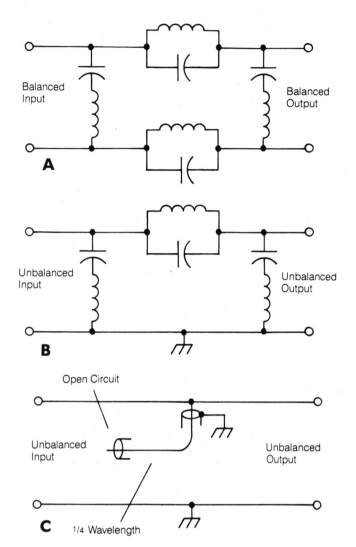

BAND-REJECTION FILTER: Balanced (A), unbalanced (B) and unbalanced coaxial (C).

Most band-rejection filters have a relatively sharp response. This is true of antenna traps and notch filters. Parasitic suppressors, used in high-frequency power amplifiers to prevent VHF parasitic oscillation, have a more broadband response. *See also* NOTCH FILTER, PARASITIC SUPPRESSOR, and TRAP.

BANDWIDTH

Bandwidth is a term used to define the amount of frequency space that is occupied by a signal, and required for effective transfer of the information to be carried by that signal. The term *bandwidth* is also sometimes used in reference to the nature of a bandpass or band-rejection filter response (*see* BANDPASS RESPONSE, and BAND-REJECTION RESPONSE).

The bandwidths of several common types of signals are given by the table. The receivers for such signals must have a bandpass response at least as great as the signal bandwidth. If the receiver bandpass response is too narrow, the signal cannot be readily understood. In the extreme, where the receiver bandpass response is very narrow compared to the signal bandwidth, no information can be conveyed whatsoever.

While an excessively narrow bandpass is obviously undesirable for a signal having a given bandwidth, an unnecessarily wide receiver response is also bad. If the response is much

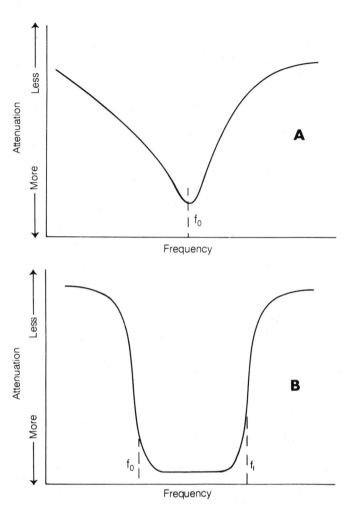

BAND-REJECTION RESPONSE: Single-frequency (A) and wideband (B) curves.

wider than the bandwidth of the signal, a great deal of noise gets into the channel along with the desired signal. Other signals on nearby frequencies might also get into the channel and disrupt communications. So, therefore, the receiver bandpass response should always be tailored to the bandwidth of the signal received.

Bandwidth is sometimes used as a specification for the sharpness of a bandpass filter response. Two frequencies, f_0 and f_1, are found, at which the power attenuation is 3 dB, with respect to the level at the center frequency of the filter. The bandwidth is the difference between these two frequencies. In general, if a signal occupies a certain bandwidth, then the receiver filter bandwidth (as determined from the 3-dB power attenuation points) should be the best for reception.

BANDWIDTH: BANDWIDTHS OF SOME COMMON SIGNALS.

Emission type	Typical bandwidth
Morse code (CW)	10 Hz-250 Hz
Radioteletype (RTTY)	200 Hz-800 Hz
Single sideband (SSB)	3 kHz
Slow-scan television (SSTV)	3 kHz
Amplitude modulation, voice (AM)	6 kHz
Amplitude modulation, music (AM)	10 kHz
Frequency modulation, voice (FM)	10 kHz
Frequency modulation, music (FM)	100 kHz
Television (TV)	3 MHz-10 MHz

BANK

Whenever a group of similar or identical devices is connected together, the composite is called a *bank*. A bank of batteries, for example, is a set of batteries in series or parallel.

Banks are generally used to obtain more power, gain, or power-handling capability than would be possible using just one single component. A series bank of batteries provides greater voltage than one battery. A parallel bank of batteries can deliver more current, or the same amount of current for a longer period of time, as compared to a single battery.

BAR GENERATOR

A *bar generator* is a device that produces regular pulses in such a way that a pattern of bars is produced in a transmitted television picture signal. The bar pattern might be either vertical or horizontal. A bar pattern can have just two bars, or several.

The pattern produced by a bar generator is used to check for vertical or horizontal linearity in a television transmitter and receiver. Vertical bars facilitate alignment of horizontal circuits. Horizontal bars allow adjustment for vertical linearity. In a color television system, colored bars are used to align the circuits for accurate color reproduction. In a black-and-white system, the various shades of gray are used to adjust the relative brightness and contrast. The lines of demarcation between the bars allow adjustment of the focus.

A bar pattern can be used for complete alignment of a television transmitting and receiving system. *See also* TELEVISION.

BARRIER CAPACITANCE

When a semiconductor P-N junction is reverse biased so that it does not conduct, a capacitance exists between the P-type and N-type semiconductor materials. This is called *barrier capacitance*. It is also sometimes called the *depletion-layer* or *junction capacitance*.

Under conditions of reverse bias, a depletion region, or potential barrier, forms between the P-type and N-type materials (*see* DEPLETION LAYER). Positive ions are created in the N-type material, and negative ions are created in the P-type material. The greater the reverse bias, the wider the depletion layer.

Barrier capacitance is a function of the width of the depletion layer in a given P-N junction. The capacitance is generally very small, on the order of a few picofarads. As the reverse-bias voltage increases, the width of the depletion layer increases. The barrier capacitance therefore gets smaller.

The barrier capacitance for a given reverse voltage depends on the cross-sectional area of the P-N junction, and also on the amount of doping and the kind of impurity used for doping (*see* DOPING, and IMPURITY).

BARRIER VOLTAGE

Barrier voltage is the amount of forward bias necessary to cause conduction across a junction of two unlike materials. In a semiconductor P-N junction, the barrier voltage is approximately 0.3 volts for germanium, and 0.6 volts for silicon.

The forward-voltage-vs-current characteristic for a semiconductor diode is such that with small values of forward voltage, very little current flows. But when the voltage reaches the barrier voltage, a sharp rise in the flow of current is observed.

Barrier voltage is not the same thing as avalanche voltage, which takes place under conditions of reverse bias and with much higher voltages. *See also* AVALANCHE VOLTAGE.

BASEBAND

When a signal is amplitude-modulated or frequency-modulated, the range of modulating frequencies is called the *baseband*. For a human voice, intelligible transmissions are possible with baseband frequencies from about 200 to 3000 Hz. Experiments have been conducted to determine the minimum baseband needed for effective voice communication, and with analog methods, efficiency is impaired if the baseband is restricted to a range much smaller than 200 to 3000 Hz. (See NARROW-BAND VOICE MODULATION).

For good reproduction of music, the baseband should cover at least the range of frequencies from 20 to 5000 Hz, and preferably up to 15 kHz or more. The baseband in a standard AM broadcast signal is restricted to the range zero to 5 kHz, since the channel spacing is 10 kHz and a larger baseband would result in interference to stations on adjacent channels. In FM broadcasting, the baseband covers the entire range of normal human hearing. *See also* AUDIO FREQUENCY.

BASE INSULATOR

A *base insulator* is a piece of nonconducting material, such as glass or plastic, used to insulate the lower end of a vertical antenna from RF ground.

A base insulator must be mechanically strong enough to support the weight of an antenna, and to endure stresses caused by wind. The insulator must be able to electrically withstand the voltage that appears between the base of the antenna and the RF ground at the highest level of power used. When the base insulator is exposed to moisture, its dielectric properties should not change.

Some vertical antennas have no base insulator. Instead, they are connected directly to the RF ground, and are fed at some point above the ground. *See also* SHUNT FEED.

BASIC

One of several different higher-order computer languages, *BASIC* is one of the most easily learned. It is easy to understand because the commands and functions are generally in plain English. BASIC is used primarily in mathematical and engineering situations.

For business purposes, the higher-order language, called *COBOL*, is preferred. In some scientific problems, FORTRAN is more efficient than BASIC.

BATTERY

When two or more electrical cells of any kind are connected in series to provide a higher voltage, the result is a battery. A battery might consist of dry cells, such as flashlight cells. Or it might be composed of lead-acid cells; a car battery is a good example. A battery might also be made up of photovoltaic cells, as in a solar panel.

If a single cell has a voltage, E, and there are n cells in the battery, the total voltage is theoretically equal to nE. Thus, you could place 100 flashlight cells, each of 1.5 volts, in series and get a battery that would produce 150 volts. Some older portable radios actually used batteries that supplied in excess of 100 volts. But this is seldom done nowadays; it is not usually necessary in solid-state equipment to provide more than about 24 volts.

When *n* cells are connected in series, the *internal resistance* of the battery is also multiplied by *n*, as compared with that of a single cell. This limits the amount of current that the battery will supply without a serious voltage drop taking place. The current that a battery of *n* identical cells will supply, for a given percentage of voltage drop, is the same as the current that a single cell will supply. For example, if a 1.5-V (no-load) cell can produce 900 mA for a voltage drop to 1.4 V, 10 such cells will have 15 V under no load and 14 V when a current of 900 mA is drawn.

Batteries can be assembled from *series-parallel* combinations of cells to overcome this problem. In the figure, the battery at A can supply just one-third of the current, for a given voltage drop, as compared with the battery at B. Both batteries, however, produce the same voltage, because connecting cells in parallel does not change the no-load output voltage.

Some batteries can be recharged, and some cannot. In general, in order to get a rechargeable battery, all of the individual cells must be rechargeable. Preferably, all the cells will be identical. Series combinations of nickel-cadmium or lead-acid cells are examples of rechargeable batteries.

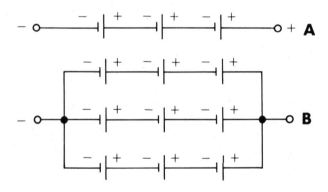

BATTERY: At A, a battery of three cells in series. At B, a battery with nine cells in series-parallel. Both provide the same voltage, but the battery at B can deliver three times the current, assuming that all cells are identical.

Typical battery voltages in amateur electronic work are approximately 6 V (four 1.5-V cells in series), 9 V (six cells), and 12 V (eight cells). Occasionally, batteries of 18 V and even 24 V are used. *See also* CELL.

BATTERY CHARGING

Certain types of batteries, called *storage batteries*, can be recharged after their energy has been used up. A battery charger, consisting of a transformer and rectifier (and sometimes a milliammeter or ammeter to indicate charging current), is used to recharge a storage battery. The positive terminal of the charger is connected to the positive terminal of the battery for charging.

While a battery is charging, it draws the most current initially, and the level drops continuously until, when the battery is fully charged, the current is near zero. The total number of ampere hours indicated by a plot of current vs time would correspond, when charging is complete, to the capacity of the battery. Some chargers operate quickly, within a period of a few minutes, but charging usually takes several hours.

Electronic calculators and some radio equipment contain rechargeable nickel-cadmium batteries. The batteries are charged by means of a small current-regulated power supply that fits standard household electrical outlets. Nickel-cadmium batteries require four to six hours, typically, to reach a fully charged condition. *See also* NICKEL-CADMIUM BATTERY, and STORAGE BATTERY.

BAUDOT

Baudot is a five-unit digital code for the transmission of teleprinter data. Letters, numerals, symbols, and a few control operations are represented. Baudot was one of the first codes used with mechanical printing devices. Sometimes this code is called the Murray code. In recent years, Baudot has been replaced in some applications by the ASCII code (*see* ASCII).

There are $2^5 = 32$, possible combinations of binary pulses in the Baudot code, but this number is doubled by the control operations LTRS (lowercase) and FIGS (uppercase). In the Baudot code, only capital letters are sent. Uppercase characters consist mostly of symbols, numerals, and punctuation. Table 1

BAUDOT: SYMBOLS FOR BAUDOT TELEPRINTER CODE.

Digits	Ltrs	US Figs	CCITT Figs	Digits	Ltrs	US Figs	CCITT Figs
00000	BLANK	BLANK	BLANK	10000	T	5	5
00001	E	3	3	10001	Z	''	+
00010	LF	LF	LF	10010	L	)	)
00011	A	--	--	10011	W	2	2
00100	SPACE	SPACE	SPACE	10100	H	#	£
00101	S	BELL	'	10101	Y	6	6
00110	I	8	8	10110	P	0	0
00111	U	7	7	10111	Q	1	1
01000	CR	CR	CR	11000	O	9	9
01001	D	$	WRU	11001	B	?	?
01010	R	4	4	11010	G	&	&
01011	J	'	BELL	11011	FIGS	FIGS	FIGS
01100	N	,	,	11100	M	•	•
01101	F	;	;	11101	X	/	/
01110	C	:	:	11110	V	;	=
01111	K	(	(	11111	LTRS	LTRS	LTRS

TABLE 1

shows the Baudot code symbols used in the United States and in most foreign countries. In some countries, the uppercase representations differ from those used in the United States. The International Consultative Committee for Telephone and Telegraph (CCITT) version of uppercase Baudot symbols is shown.

The Baudot code is still widely used by amateur radio operators. Baudot equipment is still occasionally seen in commercial systems, but the more efficient ASCII code is becoming the mode of choice, especially in computer applications because ASCII has more symbol representations.

The most common Baudot data speeds range from 45.45 to 100 bauds, or about 60 to 133 words per minute (WPM), *see also* BAUD RATE, and WORDS PER MINUTE.

BAUDOT: SPEED RATES FOR THE BAUDOT CODE.

Baud rate	Length of pulse, ms	WPM
45.45	22.0	60.6
45.45	22.0	61.3
45.45	22.0	65.0
50	20.0	66.7
56.92	17.6	75.9
56.92	17.6	76.7
74.20	13.5	99.0
74.20	13.5	100.0
100	10.0	133.3

TABLE 2

BAUD RATE

Baud rate is a measure of the speed of transmission of a digital code. One baud consists of one element or pulse. The baud rate is simply the number of code elements transmitted per second. Sometimes the baud rate is stated simply as "baud."

Baud rates for teleprinter codes range from 45.45, the slowest Baudot speed, to 19,200 for the fastest ASCII speed. *See also* ASCII, and BAUDOT.

BAYONET BASE AND SOCKET

Some equipment lamps use a special kind of base and socket, called a bayonet (*see* illustration). Such sockets have a spring-loaded contact, and a slot into which the lamp base fits and is held in place. Some bayonet sockets have two contacts for dual-filament lamps, such as automobile tail indicator lights. The lamp can be placed in the socket only one way; it must first be rotated to the proper position.

BAYONET BASE AND SOCKET: Allows quick insertion and removal.

The advantage of bayonet bases and sockets over screw-in type contacts is primarily that the bulb is less likely to come loose, and mechanical jamming is less frequent.

BAZOOKA BALUN

A *bazooka balun* is a means of decoupling a coaxial transmission line from an antenna. When coaxial feed lines are used with antenna radiators, the unbalanced nature of the line sometimes allows RF currents to flow along the outer conductor of the cable. This can create problems such as changes in the antenna directional pattern, and possibly radiation from the line itself. A metal cylinder or braided sleeve, measuring 1/4 wavelength, is placed around the coaxial cable. The end of the sleeve near the antenna is left free, and the other end is connected to the shield of the cable. For a given frequency, f, in MHz, the length of the bazooka in feet is given by

$$Length \text{ (feet)} = 234/f$$

The bazooka is not a transformer, and consequently it cannot correct impedance mismatches. The bazooka, however, is useful for preventing radiation from a coaxial transmission line at high and very high frequencies. *See also* BALUN.

BCD

See BINARY-CODED DECIMAL.

BEACON

A *beacon* is a transmitting station, usually having fairly low RF-power output, designed for the purpose of aiding in the monitoring of propagation conditions. The beacon sends a steady signal with frequent identification by call sign and geographical location. By listening on the beacon frequency, propagation conditions between the beacon and a receiving station are easily checked.

Beacons are used for radio-location purposes, and for easy spotting of certain objects on radar equipment. Beacons are a valuable aid in radar tracking, especially if an object would be difficult to identify. *See also* PROPAGATION.

BEAM-POWER TUBE

A *beam-power tube* is a vacuum tube with a special beam-directing electrode that confines the electrons to certain directions. The result is a concentrated beam, and this is where the tube gets its name. The beam-power tube is more efficient than an ordinary vacuum tube, especially when large amounts of power output are required, such as in RF transmitting equipment. The grids of a beam-power tube are usually aligned, which further aids in efficient operation. A beam-power tube might be a triode, tetrode, or pentode. They are usually pentodes, because a suppressor grid is needed with the high electron speeds. *See also* TUBE.

BEARING

The term *bearing* generally refers to a direction, such as azimuth or elevation, or a combination of directions. The compass bearing is the azimuth based on the north magnetic pole. A signal might be received from great distance, and its bearing found to be 90 degrees. This means the signal arrives from the east. Because of the spherical shape of the earth, however, the station where the signal originates will not usually have the same latitude in such a case. Bearings in terrestrial propagation define great circles (see GREAT CIRCLE). Charts of great circle bearings to various parts of the world, centered on certain locations, are available for communications purposes.

The term *bearing* is also used to mean a support for a rotating shaft. Bearings are found in motors, variable capacitors, potentiometers, and the like.

BEAT

When two waves are combined or superimposed, a beat occurs if the two frequencies are not the same. Waves beat together to create the appearance of either a change in amplitude (if the frequencies differ by just a few Hz) or new frequencies, called *beat frequencies* or *heterodynes* (if the original frequencies are far apart). Beat frequencies are the sum and difference of the original frequencies.

Sound waves often beat together when two musical notes are combined. Two tuning forks, having frequencies of, for example, middle C (261.6 Hz) and D (293.7 Hz) will beat together to create sound waves at 32.1 and 555.3 Hz, the difference and sum frequencies, respectively. Beat frequencies are an important part of the sound of music, even though listeners are not usually aware of them.

Beat frequencies are important in the operation of mixers and superheterodyne receivers and transmitters. Heterodynes are responsible for the undesirable effects of cross modulation and intermodulation distortion. Heterodynes occur in any nonlinear circuit where two or more frequencies exist. *See also* CROSS MODULATION, HETERODYNE, INTERMODULATION, MIXER, and SUPERHETERODYNE RECEIVER.

BELL 103 AND 202

Telephone modem communications standards bear various names, depending on speed and tone frequencies used. The terms are especially pertinent in *packet radio* communications.

Printed data is sent and received using *frequency-shift keying* (FSK). Two audio tones are used to modulate the transmitter. At high frequencies (HF), these tones are inserted into the audio input of a single-sideband (SSB) transmitter, resulting in two transmitted continuous-wave signals that are separated by the difference between the audio frequencies. At very high frequencies (VHF), the audio tones are inserted into the audio input of a frequency-modulated (FM) transmitter, so that the tones are always the same pitch in the receiver.

Bell 103 is used in HF packet. It has frequencies of 2025 and 2225 Hz. This results in 200-Hz shift FSK. The standard speed is 300 bauds.

Bell 202 is used in VHF packet. It has frequencies of 1200 and 2200 Hz. The standard speed is 1200 bauds. *See also* BAUD RATE, FREQUENCY-SHIFT KEYING, and PACKET RADIO.

BENCHMARK

A *benchmark* is a simple program that is often used for testing microcomputers and microprocessors. In a given application, it might be possible to use any of several different chips, but one will perform better than the others. The benchmark program allows an engineer to determine which chip will be the best central processing unit (CPU) for the application.

A benchmark program simulates the actual operating requirements of the microcomputer or microprocessor under test. The program must be simple enough so that the engineer can follow it step by step manually. But the program must be complex enough to provide a reasonable simulation of the operating requirements. If the program is too simple, the wrong chip might be chosen; if it is too complicated, the test procedure will be tedious.

Benchmark programs are often the means by which a microcomputer or microprocessor chip is selected for use in a product, such as a hand calculator or a radio transceiver. The most suitable device, as determined by the benchmark, is used as the CPU. *See also* CENTRAL PROCESSING UNIT, MICROCOMPUTER, and MICROPROCESSOR.

BERYLLIUM OXIDE

Beryllium oxide (BeO) has an exceptional ability to conduct heat. It is an electrical insulator, but has heat-conducting characteristics that are similar to most metals.

Beryllium oxide is used in high-power amplifiers, of both tube and solid-state design, to dissipate the heat and maintain the amplifying device or devices at an acceptable temperature. Beryllium-oxide powder is extremely toxic when inhaled. Therefore, tubes and transistors that contain this substance as part of their heatsink apparatus should be handled with great care. Beryllium oxide is also sometimes called *beryllia*.

BETA

The term *beta* is used to define the current gain of a bipolar transistor when connected in a grounded-emitter circuit. A small change in the base current, I_B, causes a change in the collector current I_C. The ratio of the change in I_C to the change in I_B is called the *beta*, and is abbreviated by the Greek letter β:

$$\beta = \delta I_C / \delta I_B$$

The collector voltage is kept constant when measuring the beta of a transistor.

In a graph of the I_C-vs-I_B curve for a transistor in a grounded-emitter configuration with constant collector voltage. The beta of the transistor is the slope of the curve at the selected operating point. A line drawn tangent to the curve at the no-signal operating point has a definite ratio $(\delta I_C / \delta I_B)$ which is easily determined by inspection. *See also* ALPHA, TRANSISTOR.

BETA-ALPHA RELATION

The *beta* of a transistor is the base-to-collector current gain with the emitter at ground potential (see BETA). The *alpha* is the emitter-to-collector current gain with a grounded-base circuit (see ALPHA). The alpha of a transistor is always less than 1, and the beta is always greater than 1.

The beta of a transistor can be mathematically determined if the alpha is known according to the formula:

$$\beta = \alpha(1 - \alpha)$$

where the Greek letters α and β represent the alpha and beta, respectively. The following ratios also hold:

$$\alpha/\beta = 1 - \alpha = 1/(1 + \beta)$$
$$\beta/\alpha = 1/(1 - \alpha) = 1 + \beta$$

The alpha is mathematically derived from the beta by the equation:

$$\beta = \alpha (1 + \beta)$$

A typical bipolar transistor with an alpha of 0.95 has a beta of 19. As the beta approaches infinity, the alpha approaches 1, and vice-versa.

BETA CIRCUIT

Some amplifiers utilize feedback, either negative (for stabilization) or positive (to enhance gain and selectivity). The part of the amplifier circuit responsible for the feedback is called a beta circuit.

In RF amplifiers, neutralization is a form of feedback, and the neutralizing capacitor forms the beta circuit. *See also* NEUTRALIZATION, and NEUTRALIZING CAPACITOR.

B-H CURVE

A *B-H curve* is a graph that illustrates the magnetic properties of a substance. The magnetic field flux density, B, is plotted on the vertical scale as a function of the magnetizing force, H. Flux density is measured in units, called *gauss*, and magnetizing force is measured in units called *oersteds*. B-H curves are of importance in determining the suitability of a particular magnetic material for use in the cores of inductors and transformers.

When the magnetizing force is increasing, the flux density increases also. While the magnetizing force decreases, so does the flux density. The flux density does not change exactly in step with the magnetizing force in most core materials. This sluggishness of the magnetic response is called *hysteresis*. *See also* HYSTERESIS, and HYSTERESIS LOSS.

BIAS

Bias is a term generally used to refer to a potential difference, applied deliberately between two points for the purpose of controlling a circuit. In a vacuum tube, the bias is the voltage between the cathode and control grid. In a bipolar transistor, the bias is the voltage between the emitter and the base, or the emitter and the collector. In a field-effect transistor, the bias is the voltage between the source and the gate, or the source and the drain.

Certain bias conditions are used for specified purposes. In forward bias, the cathode of a vacuum tube is negative, with respect to the grid or plate. In reverse bias, the opposite is true: the cathode is positive, with respect to the grid or plate. In a semiconductor P-N junction, forward bias occurs when the P-type material is positive, with respect to the N-type materials; in reverse bias, the P-type material is negative, with respect to the N-type material. When two electrodes are at the same potential, they are said to be at zero bias. *See also* FORWARD BIAS, REVERSE BIAS, and ZERO BIAS.

BIAS CURRENT

Bias current is the current between the emitter and the base of a bipolar transistor under conditions of no input signal. When the emitter-base junction is forward-biased, the bias current is normally a few microamperes or milliamperes. When the junction is at zero bias, the bias current is zero. When the junction is reverse-biased, the bias current is negligibly small.

Any particular transistor has an optimum bias-current operating point. When the bias voltage is properly chosen, the bias current is such that the distortion is minimal, but the amplification is sufficient. Bias-current specifications apply to class-A and class-AB amplifiers. In class-B and class-C amplifiers, the bias current is normally zero. *See also* BIAS, CLASS-A AMPLIFIER, CLASS-AB AMPLIFIER, and FORWARD BIAS.

BIAS DISTORTION

When a tube, transistor, or other amplifying device is operated with a bias resulting in nonlinearity, the distortion in the output signal is called *bias distortion*. Bias distortion is sometimes unimportant, such as in RF class-B and class-C amplifiers. In high-fidelity audio equipment, though, the bias must be chosen carefully to avoid this distortion as much as possible.

When the operating point is selected for linear operation, there is essentially no distortion. When the operating point has been improperly selected, the bias current is too large. Therefore, the amplification factor is different on the positive part of the input cycle, as compared with the negative part. At the output, the waveform is lopsided. This is bias distortion.

Bias distortion in an amplifier is not the same as the distortion resulting from excessive input-signal amplitude. Although the two forms of distortion might produce similar output waveforms, excessive input amplitude or overdrive, and cause distortion, no matter what the bias is set at. The drive level must thus be regulated to avoid distortion. *See also* BIAS, and CLASS-A AMPLIFIER.

BIAS STABILIZATION

Bias stabilization is any means of ensuring that the bias in a circuit will remain constant. In a transistor circuit, a common means of bias stabilization is the resistive voltage divider. By choosing the appropriate ratio of resistances and the correct magnitude of resistance values, the base voltage can be maintained within a precise range.

Without bias stabilization, the base of a transistor might become improperly biased because of temperature changes or because of variations in the strength of an input signal. In some cases, changes in bias can lead to thermal runaway, resulting in component overheating, loss of gain and linearity, and possibly even destruction of a transistor. *See also* BIAS, and THERMAL RUNAWAY.

BICONICAL ANTENNA

A *biconical antenna* is a balanced broadband antenna that consists of two metal cones, arranged so that they meet at or near the vertices. The biconical antenna is fed at the point where the vertices meet. The exact feed-point impedance of a biconical antenna depends on the flare angle of the cones and the separation between their vertices.

A biconical antenna displays resonant properties at frequencies above that at which the height h of the cones is ¼ wavelength in free space. The highest operating frequency is several times the lowest operating frequency. A biconical antenna oriented vertically, emits and receives vertically polarized electromagnetic waves.

The bionical antenna is often used at VHF, but its size gets prohibitively large at lower frequencies. However, one of the cones can be replaced by a disk or ground plane for the purpose of reducing the physical dimensions of the antenna while retaining the broadband characteristics. If the top cone is replaced by a disk of a certain radius, the antenna becomes a discone. If the lower cone is replaced by a ground plane, the antenna becomes a conical monopole. Both the discone and the conical monopole are practical for use at frequencies as low as about 2 MHz. *See also* CONICAL MONOPOLE ANTENNA, and DISCONE ANTENNA.

BIFILAR TRANSFORMER

A *bifilar transformer* is a transformer in which the primary and secondary are wound directly adjacent to each other, as shown in the illustration. This provides the maximum possible amount of coupling between the windings. The mutual inductance of the windings of a bifilar transformer is nearly unity (*see* MUTUAL INDUCTANCE).

Bifilar transformers are especially useful when the greatest possible energy transfer is necessary. Capacitive coupling between the primary and secondary of a bifilar transformer, however, is great, and this allows the transfer of harmonic energy with essentially no attenuation. *See also* TRANSFORMER.

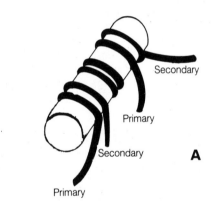

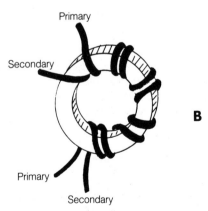

BIFILAR TRANSFORMER: Solenoidal (A) and toroidal (B).

BIFILAR WINDING

A wirewound resistor is made noninductive by a technique called *bifilar winding*. The resistance wire is first bent double. Then the wire is wound on the resistor form, beginning at the joined end of the double wire and continuing until all the wire has been put on the form. This results in cancellation of most of

the inductance because the currents in the adjacent windings flow in opposite directions. The field from one half of the winding thus cancels the field from the other half.

In general, a bifilar winding is any winding that consists of two wires wound adjacent to each other on the same form. A bifilar transformer (*see* BIFILAR TRANSFORMER) uses bifilar windings for maximum energy transfer. Some broadband balun transformers (*see* BALUN) use bifilar, trifilar, or even quadrifilar windings.

BILATERAL NETWORK

Any network or circuit with a voltage-vs-current curve that is symmetrical, with respect to the origin, is called a *bilateral network*. Another way of saying that the voltage-vs-current curve is symmetrical, with respect to the origin, is to say that no matter what the voltage, the current magnitude will remain the same if the voltage polarity is reversed.

Any network of resistors is a bilateral network. Certain combinations of other components are also bilateral circuits. Most tube and transistor circuits are not bilateral because current ordinarily flows through such devices in only one direction.

BILLBOARD ANTENNA

A *billboard antenna* is a form of broadside array (*see* BROADSIDE ARRAY), consisting of a set of dipoles and a reflecting screen.

The reflecting screen is positioned in such a way that the electromagnetic waves at the resonant frequency are reinforced as they return in the favored direction. The more dipoles in a billboard antenna, the greater the forward gain. Because of the reflecting screen, a billboard antenna has an excellent front-to-back ratio (*see* FRONT-TO-BACK RATIO).

Billboard antennas are used at UHF and above. Below about 150 MHz or so, the physical size of a billboard antenna becomes prohibitively large for most installations.

BINARY-CODED DECIMAL

The *binary-coded decimal (BCD)* system of writing numbers assigns a 4-bit binary code to each numeral 0 through 9 in the base-10 system, as shown in the table. The 4-bit BCD code for a given digit is its representation in binary notation. Four places are always assigned.

BINARY-CODED DECIMAL: BINARY-CODED DECIMAL. REPRESENTATIONS OF THE DIGITS 0 THROUGH 9.

Digit in base 10	BCD notation
0	0000
1	0001
2	0010
3	0011
4	0100
5	0101
6	0110
7	0111
8	1000
9	1001

Numbers larger than 9, having two or more digits, are expressed digit by digit in the BCD notation. For example, the number 189 in base-10 form is written as 0001 1000 1001 in BCD notation. The BCD representation of a number is not the same as its binary representation; in binary form, for example, 189 is written 10111101. *See also* BINARY-CODED NUMBER.

BINARY-CODED NUMBER

A *binary-coded number* is a number expressed in the binary system. The binary system of numbers is sometimes called *base 2*. Normally, we write numbers in base 10, called the *decimal system*.

In the base-10 system, the digit farthest to the right is the ones digit, and the digit immediately to its left is the tens digit. In general, if a given digit (m) represents $m \times 10^P$ in the decimal system, then the digit n, to the immediate left of m, represents $n \times 10^{P+1}$.

In base 2, the digit farthest to the right is the ones digit. Instead of 10 possible values for a digit, there are only two, 0 and 1. The digit to the left of the ones digit is the twos digit. Then comes the fours digit, the eights digit, and so on. In general, if a given digit (m) represents $m \times 2^P$ in the binary number system, then the digit (n) to its left represents $n \times 2^{P+1}$.

Consider the decimal and binary notations for the number 189. In base 10:

$$189 = (9 \times 10^0) + (8 \times 10^1) + (1 \times 10^2)$$

In binary notation:

$$10111101 = (1 \times 2^0) + (0 \times 2^1) \\ + (1 \times 2^3) + (1 \times 2^4) + (1 \times 2^5) \\ + (0 \times 2^6) + (1 \times 2^7),$$

which is simply another way of saying that:

$$189 = 1 + 4 + 8 + 16 + 32 + 128.$$

Binary-coded numbers are used by calculators and computers because the 0 and 1 digits are easily handled as "off" and "on" conditions. Although a binary-coded number has more digits than its decimal counterpart, this presents no problem for a computer that is capable of handling thousands or millions of bits. *See also* COMPUTER, and DIGITAL.

BINARY COUNTER

A *binary counter* is a device that counts pulses in the binary-coded number system. Each time a pulse arrives, the binary code stored by the counter increases by 1. A simple algorithm ensures that the counting procedure is correct.

Binary counters form the basis for most frequency counters (*see* FREQUENCY COUNTER). Such circuits actually count the number of cycles in a specified time period, such as 0.1 second or 1 second. Obviously, such counters must work very fast to measure frequencies approaching 1 GHz which is 10^9 Hz.

A simple divide-by-two circuit is sometimes called a *binary counter* or *binary scaler*. Such a circuit produces one output pulse for every two input pulses. A "T" flip-flop (*see* T FLIP-FLOP) is sometimes called a *binary counter*. Several divide-by-two circuits, when placed one after the other, facilitate digital division by any power of 2. *See also* DIGITAL CIRCUITRY.

BINDING POST

A *binding post* is a terminal that is used for a temporary, low-voltage connection. Two thumb screws, made of metal and perhaps coated with plastic, are clamped onto the wire leads by hand turning. Sometimes the screw shafts have holes through which the wire leads can be placed to prevent the wire from slipping off the binding post.

Binding posts offer fairly good current-carrying capacity for low-voltage supplies, up to several amperes. Binding posts are not used with power supplies of more than about 30 volts because of the shock hazard from exposed electrodes. Binding posts are not good for prolonged use outdoors, or for permanent installations. Soldered connections are preferable in such instances. Binding posts are frequently used with spade lugs, which fit neatly around the post.

BIPOLAR TRANSISTOR

A *bipolar transistor* is a semiconductor device consisting of P-type and N-type materials in a sandwich configuration. The pnp bipolar transistor consists of a layer of N-type semiconductor in between two layers of P-type material; the npn transistor is just the reverse of this.

Bipolar transistors display relatively low input impedance, and variable output impedance. They are used as oscillators, amplifiers, and switches. Bipolar transistors are available in a variety of sizes and shapes. Some are built especially for use as weak-signal amplifiers; others are built for high-power amplifiers, both AF and RF.

The collector of a pnp device is always negative, with respect to the emitter; the collector of an npn device is always positive, with respect to the emitter. The base bias for operation of a bipolar transistor depends on the class of operation desired.

BIRMINGHAM WIRE GAUGE

The *Birmingham Wire Gauge* is a standard for measurement of the size, or diameter, of wire. Usually, it is used for iron wire. Wire is classified according to diameter by giving it a number. The Birmingham Wire Gauge designators differ from the American and British Standard designators (*see* AMERICAN WIRE GAUGE, and BRITISH STANDARD WIRE GAUGE), but the sizes are nearly the same. The higher the designator number, the thinner the wire.

The table shows the diameter vs Birmingham Wire-Gauge number for designators 1 through 20. The Birmingham Wire-Gauge designator does not include any coatings that might be on the wire, such as enamel, rubber, or plastic insulation. Only the metal part of the wire is taken into account.

BIRMINGHAM WIRE GAUGE: BIRMINGHAM WIRE GUAGE EQUIVALENTS IN MILLIMETERS.

BWG	Dia., mm	BWG	Dia., mm
1	7.62	11	3.05
2	7.21	12	2.77
3	6.58	13	2.41
4	6.05	14	2.11
5	5.59	15	1.83
6	5.16	16	1.65
7	4.57	17	1.47
8	4.19	18	1.25
9	3.76	19	1.07
10	3.40	20	0.889

BISTABLE CIRCUIT

Any circuit that can attain either of two conditions, and maintain the same condition until a change command is received, is called a *bistable circuit*. The flip-flop (see FLIP-FLOP) is probably the most common type of bistable circuit.

A very simple kind of bistable circuit is the pushbutton switch, used with any device such as an electric light. When the button is pushed once, the lamp or appliance is switched on; pressing the button again turns the circuit off. Such switches can be either mechanical or electronic. The bistable circuit always maintains the same condition indefinitely, unless a change-of-state command is received.

BISTABLE MULTIVIBRATOR

See FLIP-FLOP.

BIT RATE

The bit rate of a binary-coded transmission is the number of bits sent per second. Digital computers process and exchange information at very high rates of speed, up to thousands of bits per second.

Computer data is often sent by means of a teleprinter code, called ASCII (see ASCII). In teleprinter codes, a bit is called a *baud*, which can be either of two binary states. ASCII speeds range up to 19,200 bauds. See also BAUD RATE.

BIT SLICE

A multi-chip microprocessor consists of a control section and one or more separate integrated circuits, in which the register information is stored. Each register chip is called a bit slice. Several bits slices, connected in parallel, complete the microprocessor along with the control chip.

A bit slice can contain any number of bits, usually a power of 2, such as 2, 4, 8, 16, and so on. A control IC with three 8-bit slices forms a 24-bit microprocessor. *See also* MICROPROCESSOR.

BIT STUFFING

Bit stuffing is a process in packet radio transmission, in which a logic 0 (zero) bit is inserted after a series of five 1 (one) bits. This prevents an octet from appearing to indicate the beginning or end of a frame.

If not for bit stuffing, also known as *zero bit insertion*, an octet within a frame might be wrongly interpreted by the receiver as the end of the frame, and/or the beginning of a new frame. *See also* FRAME, OCTET, and PACKET RADIO.

BLACK BOX

Any circuit in which the internal details are unknown, or unimportant, is called a *black box*. Such a circuit becomes, in its intended application, a component itself. An example of a black box is an integrated circuit. Although the internal details of various integrated circuits can be quite complicated, and very different, such a device appears as an empty box in a circuit diagram.

Two black boxes that behave in identical fashion under a certain set of circumstances might be used interchangeably even though they are internally different. An engineer does not have to be concerned with differences that are inconsequential in practice.

BLACK TRANSMISSION

Black transmission is a form of amplitude-modulated facsimile signal (see FACSIMILE), in which the greatest copy density, or darkest shade, corresponds to the maximum amplitude of the signal. Black transmission is the opposite of white transmission, in which the brightest shade corresponds to the maximum signal amplitude.

In a frequency-modulated facsimile system, black transmission means that the darkest copy corresponds to the lowest transmitted frequency.

BLANKING

In television picture transmission, a *blanking signal* is a pulse that cuts off the receiver picture-tube during return traces. The blanking signal prevents the return trace from showing up on the screen, where it would interfere with the picture. Such a pulse is a square wave, with rise and decay times that are very short.

BLEEDER RESISTOR

A *bleeder resistor* is a fairly high-valued, low-wattage resistor that is connected across the output terminals of a power supply. Bleeder resistors serve two purposes: first, to aid in voltage regulation, and second, to reduce the shock hazard in high-voltage systems after power has been shut off. The schematic illustrates the connection of a bleeder resistor in a power supply.

The bleeder resistor prevents excessive voltages from developing across the power-supply filter capacitors when there is no load. Without such a current-draining device, the voltage at the supply output might build up to values substantially larger than the rated value.

In a high-voltage power supply, the filter capacitors will remain charged for some time after the power has been shut off, unless some means of discharging them is provided. Dangerous electrical shocks, from seemingly safe equipment, are made less likely by a bleeder resistor. However, the high-voltage terminal of such a supply should always be shorted to ground before servicing the equipment. This will ensure that no voltage is present. A metal rod with a well-insulated handle should be used to discharge any residual voltages.

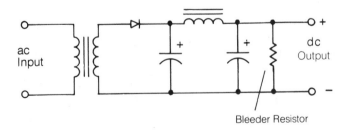

BLEEDER RESISTOR: Connection of a bleeder resistor in a power supply.

BLIND ZONE

In any communications system, a *blind zone* is an area or direction from which signals cannot be received, or to which messages cannot be transmitted. When communicating via the ionosphere, reception is often impossible within a certain distance called the skip zone (*see* SKIP ZONE). At VHF and above, a hill or building can create a blind zone. As the wavelength decreases, smaller and smaller obstructions cause blind zones.

BLOCK DIAGRAM

A circuit diagram that shows the general configuration of a piece of electronic equipment, without showing the individual components, is called a *block diagram*. A simple superheterodyne receiver is block diagrammed in the illustration.

The path of the signal in a block diagram is generally from left to right and bottom to top. There are exceptions to this, but the signal flow, if not evident from the nature of the circuit, is often shown by arrows or heavy lines.

In computer programming, block diagrams are used to indicate logical processes. This aids in developing programs. Such a block diagram is called a *flowchart* (*see* FLOWCHART). Block diagrams can also be used to represent algorithms. Block diagrams are sometimes used to denote the chain of command in a system. In such an illustration, the controller is at the top, and subordinate systems are underneath it.

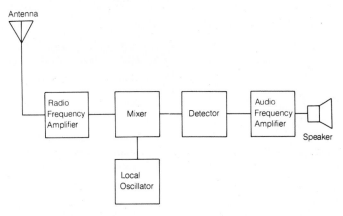

BLOCK DIAGRAM: Stages are shown, but not individual components.

BLOCKING

Blocking is the application of a large negative bias voltage to the base of an npn bipolar transistor, or the gate of an N-channel field-effect transistor, for the purpose of cutting off the device to prevent signal transfer. In a pnp bipolar transistor or a P-channel field-effect transistor, the blocking bias at the base or gate is positive.

Blocking is frequently used as a means of keying a code transmitter. The blocking voltage is applied to an amplifying stage while the key is up. When the key is down, the blocking bias is removed and the signal is transmitted. In transmitters, this is called *base-block* or *gate-block keying*, and is probably the most popular means of keying.

The term *blocking* is sometimes used to refer to the prevention of direct-current flow, while allowing alternating currents to pass. Such an electrical arrangement is necessary in many types of oscillators, and in capacitive coupling networks between amplifier stages. *See also* BLOCKING CAPACITOR.

BLOCKING CAPACITOR

When a capacitor is used for the purpose of blocking the flow of direct current, but allowing the passage of alternating current,

the capacitor is called a *blocking capacitor*. Blocking capacitors facilitate the application of different dc bias voltages to two points in a circuit. Such a situation is common in multi-stage amplifiers. The schematic illustrates the connection of a blocking capacitor between two stages of a transistorized amplifier.

Blocking capacitors are usually fixed, rather than variable, and should be selected so that attenuation does not occur at any point in the operating-frequency range. Generally, the blocking capacitor should have a value that allows ample transfer of signal at the lowest operating frequency, but the value should not be larger than the minimum to accomplish this. At VHF, the value of a blocking capacitor in a high-impedance circuit might be only a fraction of one picofarad. At audio frequencies in low-impedance circuits, values might range up to about 100 microfarads.

Blocking capacitors are used in the feedback circuits of some kinds of oscillators. The value of the feedback capacitor in an oscillator should be the smallest that will allow stable operation. *See also* CAPACITIVE COUPLING.

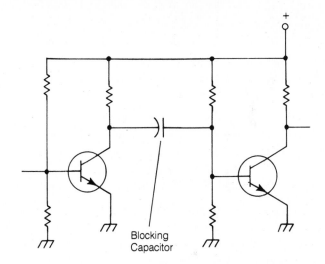

BLOCKING CAPACITOR: This lets ac pass, but blocks dc.

BLOCKING OSCILLATOR

A *blocking oscillator* is a form of relaxation oscillator that uses a feedback capacitor in conjunction with an inductance.

The term *blocking oscillator* is used to refer to an oscillator that is switched on and off periodically by an external source. Such an oscillator is also called a *squegging oscillator*. It transmits a series of pulses at the oscillator frequency. The pulses are used for testing purposes.

BNC CONNECTOR

A small cable connector, used frequently in RF and test applications, is the *BNC connector* (*see* drawing). The BNC connector is designed to provide a constant impedance for 50-ohm coaxial cable connections. Some other types of connectors cause impedance discontinuities, which create losses at VHF and UHF.

The BNC connector has a quick-connect, quick-release feature similar to the bayonet socket (*see* BAYONET BASE AND SOCKET). BNC connectors are not intended for extremely high levels of power, nor for permanent installation outdoors. But they are extremely convenient when impedance matching must be good, or if frequent wiring changes are necessary. *See also* N-TYPE CONNECTOR, and UHF CONNECTOR.

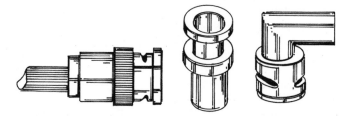

BNC CONNECTOR: These are often used with coaxial cables.

BOBTAIL CURTAIN ANTENNA

The *bobtail curtain antenna* is a vertically polarized, phased array for high-frequency operation. It is usually constructed from wire, although the vertical supports might be metal poles or towers.

The horizontal sections are ½ wavelength electrically, and the three vertical sections, which radiate most of the energy, are ¼ wavelength. The bobtail curtain is a bidirectional antenna, with maximum radiation in line with the vertical radiators and minima in directions broadside to the vertical elements (*see* BIDIRECTIONAL PATTERN).

The bobtail curtain antenna is practical at frequencies as low as about 3 MHz, where two metal supporting poles about 78 feet high can be used for the outer vertical radiators. The bobtail curtain has a low angle of radiation in the elevation plane.

BODY CAPACITANCE

When the human body is brought near an electronic circuit, some capacitance is introduced between the circuit wiring and the ground, unless the circuit is completely shielded. Usually, the effects of this body capacitance are not noticeable because the value of capacitance is never more than a few picofarads. In certain tuned circuits, such as ferrite-rod antennas or loop antennas, however, body capacitance can have a pronounced effect on the tuning—especially when the shunt capacitor is set for a small value. Usually, the hand is brought very near the shunt tuning capacitor as adjustments are made.

In general, the smaller the capacitances in a circuit, the more likely it is that body capacitance will be noticed. Shielding thus becomes very important at high frequencies and above.

BONDING

Bonding is the joining of metals, especially the parts of a shielded enclosure, to prevent unwanted electromagnetic energy from entering or escaping a circuit. Bonding can be accomplished by soldering or welding, but a good electrical bond is usually obtained by simply clamping the two surfaces firmly together.

Examples of electrical bonding include ground clamps, screw-down power-amplifier enclosures, and tubing clamps for antenna construction. Bonding is used in the construction of automobiles to prevent potential differences from developing among the different metal parts. Sometimes, however, additional electrical bonding is necessary in mobile radio installations to prevent interference to the radio receiver from the vehicle electrical system.

Bonds between dissimilar metals are sometimes impossible to obtain by clamping, because the two metals tend to react with each other. In such cases, solder is often used to ensure that the electrical contact does not deteriorate.

BOOK CAPACITOR

A *book capacitor* is a small-value trimmer capacitor consisting of two or more plates. The capacitance is adjusted by changing the angle between the plates. The book capacitor gets this name from its physical resemblance to the covers of a book.

The capacitance is maximum when the two plates are parallel, or closed. The capacitance is minimum when the two plates are open so that they are coplanar. A book capacitor has a maximum value of only a few picofarads, and a minimum value that might be less than 1 pF. *See also* TRIMMER CAPACITOR.

BOOLEAN ALGEBRA

Boolean algebra is a system of mathematical logic, using the functions AND, NOT, and OR. In the Boolean system, AND is represented by multiplication, NOT by complementation, and OR by addition. Thus X AND Y is written XY, NOT X is written X', and X OR Y is written $X + Y$.

Using the Boolean representations for logical functions, some of the mathematical properties of multiplication, addition and complementation can be applied to form equations. The logical combinations on either side of the equation are equivalent. In some cases, the properties of the logical functions are not identical to their mathematical counterparts. As an example, multiplication is distributive, with respect to addition in ordinary arithmetic, and it is true also in Boolean algebra. Thus:

$$X(Y + Z) = XY + XZ,$$

BOOLEAN ALGEBRA: BOOLEAN TRUTH TABLES. AT A, THE AND FUNCTION (MULTIPLICATION); AT B, THE NOT FUNCTION (COMPLEMENTATION); AT C, THE OR FUNCTION (ADDITION).

X	Y	A XY	B X'	C X + Y
0	0	0	1	0
0	1	0	1	1
1	0	0	0	1
1	1	1	0	1

Table 1

BOOLEAN ALGEBRA:
SOME THEOREMS IN BOOLEAN ALGEBRA.

1. $X + 0 = X$ (additive identity)
2. $X1 = X$ (multiplicative identity)
3. $X + 1 = 1$
4. $X0 = 0$
5. $X + X = X$
6. $XX = X$
7. $(X')' = X$ (double negation)
8. $X + X' = 1$
9. $X'X = 0$
10. $X + Y = Y + X$ (commutativity of addition)
11. $XY = YX$ (commutativity of multiplication)
12. $X + XY = X$
13. $XY' + Y = X + Y$
14. $X + Y + Z = (X + Y) + Z = X + (Y + Z)$ (associativity of addition)
15. $XYZ = (XY)Z = X(YZ)$ (associativity of multiplication)
16. $X(Y + Z) = XY + XZ$ (distributivity)
17. $(X + W)(Y + Z) = XY + XZ + WY + WZ$ (distributivity)

Table 2

which means that the logical statements:

$$X \text{ AND } (Y \text{ OR } Z)$$

and:

$$(X \text{ AND } Y) \text{ OR } (X \text{ AND } Z)$$

are equivalent. The statement:

$$XX = X$$

however, is quite alien to the ordinary kind of mathematical algebra.

BOOM

A *boom* is a horizontal support for the elements of a directive antenna such as a quad, log periodic, or Yagi. The boom does not contribute to the radiation from such an antenna; it only supports the elements. The boom might be reinforced by supporting wires if it is long, or if there are many elements in the antenna.

A microphone is sometimes supported by a device called a *boom*, which is simply a long rod or pole that keeps the microphone in position with the base out of the way. Such microphones are often used in radio or television broadcast stations.

BOOST CHARGING

Boost charging is a high-current, short-term method of charging a storage battery. Boost charging is also sometimes called *quick charging*. Generally, a storage battery must be charged for several hours at relatively slow rate in order to obtain a complete charge. However, a partial charge can be realized very quickly. Boost charging offers convenience when the available time is limited. However, slow charging should be done whenever possible.

BOOTSTRAP CIRCUIT

A circuit, in which the output is taken from the source or emitter, is sometimes called a *bootstrap circuit* because the output voltage directly affects the bias. In this type of amplifier, negative output pulses cause an increase in the negative voltage at the input, and positive output pulses cause a reduction in the negative voltage at the input.

The input of a bootstrap circuit is applied between the source and gate of an FET, or between the emitter and base of a transistor. This differentiates the bootstrap circuit from an ordinary grounded drain or grounded-collector amplifier, in which the input is applied between the gate and ground, or between the base and ground.

BOUND ELECTRON

An electron is said to be bound when it is under the influence of the nucleus of an atom, held in place by the electrical attraction between the positive charge of the nucleus and its own negative charge. A bound electron is sometimes stripped away from the nucleus easily; in other cases, the electron is very difficult to remove. The willingness of an electron to leave its orbit around an atomic nucleus depends on the position of the electron, and the number of other electrons in that orbit. Electrons that are stripped away in ordinary reactions among elements are called *valence electrons*.

When electrons are not under the influence of atomic nuclei, they are said to be *free electrons*. A free electron becomes bound when a nucleus captures it. Bound electrons are sometimes shared among more than one nucleus.

BOWTIE ANTENNA

The *bowtie antenna* is a broadbanded antenna, often used at VHF and UHF. It consists of two triangular pieces of stiff wire, or two triangular flat metal plates. The feed point is at the gap between the apexes of the triangles. There might be a reflecting screen to provide unidirectional operation, but this is not always the case.

The bowtie antenna is a two-dimensional form of the biconical antenna (*see* BICONICAL ANTENNA). It obtains its broadband characteristics, according to the same principles as the biconical antenna. The feed-point impedance depends on the apex angles of the two triangles. The polarization of the bowtie antenna is along a line running through the feed point and the centers of the triangle bases.

Bowtie antennas are occasionally used for transmitting and receiving at frequencies as low as about 20 MHz. Below that frequency, the size of the bowtie antenna becomes prohibitive.

B PLUS

The term *B plus* refers to the positive high-voltage dc supply that is used for the operation of a vacuum-tube circuit. Since tubes are becoming less and less common in modern electronics, the expression B plus is heard less often today than a few years ago, but sometimes is applied to the voltage source in transistor circuits.

B-plus voltages are frequently high enough to be dangerous. Some tubes require only a few volts, but most require hundreds of volts. Some tubes must be supplied with thousands of volts to operate properly.

BRAID

Braid is a woven wire, made from many thin conductors. It is commonly used as the outer conductor, or shield, material in prefabricated coaxial cables (*see* COAXIAL CABLE). This allows physical flexibility and little chance of breakage. However, braid allows corrosion to occur more quickly than does solid tubing, since the surface area is not impervious to moisture and chemicals.

Heavy braid is frequently used to electrically ground transmitters and receivers, and to bond station equipment together for safety purposes. Braid has a high current-carrying capacity, and very low inductance at radio frequencies. It is easy to install and adheres well to solder. In fact, a special type of braid called "Solder Wick" is used to remove solder from circuit boards for component replacement.

BRANCH CIRCUIT

A *branch circuit* is a separately protected circuit for the operation of one appliance, or a few appliances. In a home wiring system, each branch circuit has its own separate fuse or breaker. Because each branch has its own fuse or circuit breaker, a short circuit in one branch does not affect the other branches.

Branch circuits are always connected in parallel with the main circuit, since this ensures a constant voltage. Standard house power is supplied at 120 volts, plus or minus approximately 10 volts. Some appliances require twice this voltage. The standard ac frequency is 60 Hz in most of the world.

BREADBOARD TECHNIQUE

The *breadboard technique* is a method of constructing experimental circuits. A circuit board, usually made of phenolic or similar material, is supplied with a grid of holes. Components are mounted in the holes and wired together temporarily, either using hookup wire, or by copper plating on the underside of the board. Connections are easily removed or changed. The term comes from the early radio experimenter's practice of mounting components on a wooden board that was found in most homes and often used to mix bread dough.

Once the circuit has been perfected, a circuit board is designed and printed for permanent use. *See also* CIRCUIT BOARD, and PRINTED CIRCUIT.

BREAK

When a receiving operator takes control of a communications circuit, a break is said to occur. The sending operator might say "break" to give the receiving operator a chance to ask a question or make a comment. The sending operator then pauses, or breaks, awaiting a possible reply.

When a third party with an important message interrupts a contact between two stations, the interruption is called a *break*. The third station indicates its intention to interrupt by saying "break" during a pause in the transmission of one of the other stations.

The term *break* is also used to refer to opening contacts in a circuit that is alternately opened and closed. For example, in Morse-code transmission, the transition between key-down and key-up conditions is called the break. The change from key-up to key-down condition is called *make*. *See also* MAKE.

BREAKDOWN

When the voltage across a space becomes sufficient to cause arcing, a *breakdown* occurs. In a gas, breakdown takes place as the result of ionization, which produces a conductive path. In a solid, breakdown actually damages the material permanently, and the resistance becomes much lower than usual.

In a semiconductor diode, the term *breakdown* is generally used in reference to a reverse-bias condition in which current flow occurs (*see* AVALANCHE BREAKDOWN). Ordinarily, the reverse resistance of a diode is extremely high, and the current flow is essentially zero. When breakdown occurs, the resistance abruptly lowers, and the current flow increases to a fairly large value.

Dielectric materials should be chosen so that breakdown will not occur under ordinary operating conditions. A sufficient margin of safety should be provided, to allow for the possibility of a sudden change in the voltage across the dielectric. Rectifier diodes should have sufficient peak-inverse-voltage ratings so that avalanche breakdown does not occur. Some semiconductor diodes, however, are deliberately designed to take advantage of their breakdown characteristics. *See also* ZENER DIODE.

BREAKDOWN VOLTAGE

The *breakdown voltage* of a dielectric material is the voltage at which the dielectric becomes a conductor because of arcing. In a solid dielectric, this usually results in permanent damage to the material. In a gaseous or liquid dielectric, the arcing causes ionization, but the breakdown damage is not permanent because of the fluid nature of the material.

In a reverse-biased diode, the breakdown voltage is the voltage at which the flow of current begins to rapidly increase. Ordinarily, the current flow is very small in the reverse direction through a semiconductor diode; when the reverse voltage is smaller than the breakdown voltage, virtually no current flows. When the breakdown voltage is reached, a fairly large current flows. This is called *avalanche breakdown*. The transition between low and high current flow is abrupt. *See also* AVALANCHE BREAKDOWN.

BREAKER

A *breaker* is a device deliberately placed in series with an electrical circuit. If the current flow in the circuit becomes greater than a certain prescribed amount, then the breaker opens the circuit, and power is removed until the device is reset. Breakers are commonly used in modern household electrical wiring to prevent the danger of fire.

The term *breaker* is used in radio communication—especially on the Amateur and Citizen's bands. When a conversation between two or more operators is interrupted, the interrupting station is called the *breaker*. Because most CB channels, and many Amateur channels, are occupied, anyone who enters a channel might be loosely called a *breaker*. *See also* CIRCUIT BREAKER.

BREAK-IN OPERATION

Break-in operation is a form of radio communication in which a transmitting operator can hear signals at all times, except during the actual emission of radiation by the transmitter. In break-in operation, when sending code (CW), the receiver is active between the individual dots and dashes—even if the speed is high. In single-sideband (SSB) operation, any pause allows reception when break-in is used. Break-in operation is fairly easy to achieve with low-power equipment, but when the transmitter output power is high, sophisticated techniques are needed to allow true break-in operation. Fast break-in operation is impractical with frequency-shift keying (FSK), amplitude modulation (AM), or frequency modulation (FM). However, semi break-in is sometimes used in these modes (*see* SEMI BREAK-IN OPERATION).

Break-in operation is not the same thing as duplex operation. In duplex, the transmitting operator is able to hear the other operator—even while the transmitter is actually putting out power. This involves the use of two different frequencies. *See also* DUPLEX OPERATION.

BRIDGE CIRCUIT

A *bridge circuit* is a special form of network used in certain alternating-current circuit situations. A voltage is applied at one branch of the bridge circuit, and various components are connected to the other branches.

There are many kinds of bridge circuits. Some are used for signal generation, some for impedance matching, some for rectification or detection, some for mixing of signals, and some in measuring instruments.

An example of a bridge circuit is shown in the drawing. This bridge is used to measure inductance in terms of resistance and frequency. The balance of the circuit, indicated by a zero reading on the galvanometer, depends on the frequency of the applied voltage. Knowing the resistor values and the applied frequency, the unknown inductive reactance can be determined, and the value of inductance easily calculated from this.

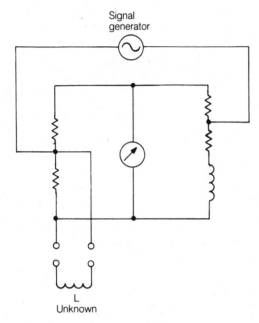

BRIDGE CIRCUIT: Allows determination of unknown inductances.

BRIDGE RECTIFIER

A *bridge rectifier* is a form of full-wave rectifier circuit, consisting of four diodes. The diodes might be either tube type or semiconductor type. Tube type diodes are seldom seen in modern rectifier circuits, however.

Bridge rectifier circuits are commonly used in modern solid-state power supplies. Some integrated circuits are built especially for use as bridge rectifiers; they contain four semiconductor diodes in a bridge configuration, encased in a single package. Sometimes bridge circuits are used as detectors in radio receiving equipment. They might also be used as mixers. *See also* RECTIFIER CIRCUITS.

BRIDGED-T NETWORK

A *bridged-T network* is identical to a T network (*see* T NETWORK), with the exception that the series elements are shunted by an additional impedance. The shunting impedance might be a resistor, capacitor, or an inductor. Generally, the shunting impedance is a variable resistance, used for the purpose of adjusting the sharpness of a filter response. When the value of the resistance is zero, the selectivity of the filter is minimum. When the resistance is greatest, at least several times the impedance of the series elements, the selectivity of the filter is maximum.

BRIDGING CONNECTION

When a high-impedance circuit is connected across a component for the purpose of measurement, the addition of the high-impedance circuit should not noticeably affect the operation of the device under test. Such a connection is called a *bridging connection*. The ordinary voltmeter is used in this way, as are the test oscilloscope and the spectrum analyzer. Any loss caused by a bridging connection is called *bridging loss*. Some bridging loss is unavoidable, but the greater the impedance of the bridging circuit, with respect to the circuit under test, the smaller the bridging loss.

The term *bridging connection* is often used interchangeably with the term shunt connection. *See also* SHUNT.

BRITISH STANDARD WIRE GAUGE

Metal wire is available in many different sizes or diameters. Wire is classified according to diameter by giving it a number. The designator most commonly used in the United States is called the *American Wire Gauge*, abbreviated *AWG* (*see* AMERICAN WIRE GAUGE). In some other countries, the *British Standard Wire Gauge* is used. The higher the number, the thinner the wire. The British Standard Wire Gauge sizes for designators 1 through 40 are shown in the table.

The larger the designator number for a given conductor metal, the smaller the current-carrying capacity becomes. The British Standard Wire Gauge designator does not take into account any coatings on the wire, such as enamel, rubber, or plastic insulation. Only the diameter of the metal itself is included in the measurement.

BRITISH STANDARD WIRE GAUGE:
BRITISH STANDARD WIRE GAUGE EQUIVALENTS IN INCHES.

NBS SWG	Dia., in.	NBS SWG	Dia., in.
1	0.300	21	0.032
2	0.276	22	0.028
3	0.252	23	0.024
4	0.232	24	0.022
5	0.212	25	0.020
6	0.192	26	0.018
7	0.176	27	0.0164
8	0.160	28	0.0148
9	0.144	29	0.0136
10	0.128	30	0.0124
11	0.116	31	0.0116
12	0.104	32	0.0108
13	0.092	33	0.0100
14	0.080	34	0.0092
15	0.072	35	0.0084
16	0.064	36	0.0076
17	0.056	37	0.0068
18	0.048	38	0.0060
19	0.040	39	0.0052
20	0.036	40	0.0048

BROADBAND MODULATION

See SPREAD SPECTRUM.

BROADBAND TRANSFORMER

There are numerous situations where impedance transformation is needed in radio-frequency circuits. *Broadband transformers* are designed to accomplish this over a wide range of frequencies.

In ham radio, most broadband transformers are wound on toroidal cores. This reduces the number of turns, and provides large power-handling capacity in a small physical volume. Depending on the nature of the windings, the impedance transfer ratio can be step-up, step-down, or 1 : 1. In general, for toroidal broadband transformers, the input and output impedances range from a few ohms to about 300 ohms.

The core material most often used is powdered iron. Ferrite will work at low frequencies and low power levels, but its permeability changes when the frequency or power gets too high. The main advantage of ferrite is that it has higher permeability, so that the coils require fewer turns, compared with powdered iron. This appeals to hams who desire the greatest possible amount of miniaturization.

Broadband transformers are used in transmatches, and also for coupling between amplifier stages at all power levels. Some broadband transformers act to couple balanced circuits to unbalanced circuits, or vice-versa. Some work to reverse the phase of the current going through them. Sometimes, two or more broadband transformers are connected in series to obtain isolation between circuits, or to get large impedance transformations. *See also* BALUN, TOROID, and TRANSMATCH.

BROADCASTING

Broadcasting is one-way transmission, intended for reception by the general public. In the early days of radio, broadcasting began as a hobby pursuit. It was soon found to be an effective means of reaching thousands or millions of people simultaneously. Thus it has become a major pastime, and has proven profitable for corporations, as well as for broadcasting stations.

In the United States, broadcasting is done on several different frequency bands, using radio (both AM and FM), and television. Some broadcasting stations are allowed only a few watts of RF power, while others use hundreds of thousands, or even millions, of watts. All broadcasting stations in the United States must be licensed by the Federal Communications Commission.

BROADSIDE ARRAY

A *broadside array* is a phased array of antennas (*see* PHASED ARRAY) that are arranged in such a way that the maximum radiation occurs in directions perpendicular to the plane containing the driven elements. This requires that all of the antennas be fed in phase, and special phasing harnesses are required to accomplish this. The illustration shows the geometric arrangement of a broadside array.

At frequencies as low as about 10 MHz, a broadside array might be constructed from as few as two driven antennas. At VHF and UHF, there might be several antennas. The antennas themselves might have just a single element, as shown in the figure, or they can consist of Yagi antennas, loops, or other systems with individual directive properties. If a reflecting screen is placed behind the array of dipoles in the drawing, the system becomes a billboard antenna (*see* BILLBOARD ANTENNA).

The directional properties of a broadside array depend on the number of elements, whether or not the elements have gain themselves, and on the spacing among the elements. *See also* DIRECTIONAL ANTENNA.

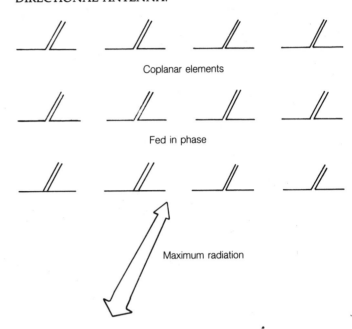

Coplanar elements

Fed in phase

Maximum radiation

BROADSIDE ARRAY: Maximum radiation and response is perpendicular to the plane that contains the elements.

BROWN AND SHARP GAUGE

See AMERICAN WIRE GAUGE.

BUBBLE MEMORY

A *bubble memory* is a magnetic means of storing information. A magnetic field is used to store a binary bit of information this field is maintained until a command is received to change it. The location of a bit in a bubble memory is changed by applying external magnetic fields. *See also* MEMORY.

BUFFER STAGE

A *buffer stage* is a single-tube or single-transistor stage, used for the purpose of providing isolation between two other stages in a radio circuit. Buffer stages are commonly used following oscillators (especially keyed oscillators) to present a constant load impedance to the oscillator. As the amplifiers following an oscillator are tuned, their input impedances change, and this can affect the frequency of the oscillator unless a buffer stage is used. When an oscillator is keyed, the input impedance of the following amplifier stage might change; or, when an amplifier is keyed, its impedance can change. The buffer stage isolates this impedance change from the oscillator, and thereby the oscillator frequency is stabilized.

BUG

The term *bug* is sometimes used in reference to a flaw in an electronic circuit or in a computer program. In a circuit, a "bug" is an imperfection in the general design, resulting in less-than-optimum operation. For example, a radio transmitter might have a problem with frequency stability. The cause of this kind of problem is sometimes difficult to find.

In a computer program, a "bug" results in inaccurate or incomplete output or execution. An example of the effect of this type of "bug" is a search-and-replace word-processing function that does not search an entire file.

A semiautomatic key, used by old-time radiotelegraphers and still used today by some radio amateurs, is often called a *bug*. *See also* SEMIAUTOMATIC KEY.

BULLETIN-BOARD SYSTEM

A *bulletin-board system (BBS)* is a set of stored messages, which are accessible by means of a computer. With a *modem* to connect a personal computer to the telephone line, it is possible for any computer user to contact many different BBSs. Messages can be left for other computer users, or retrieved, in a very short time, with minimal long-distance charges. It's like an instant mail service.

The BBS scheme lends itself well to packet radio. *See also* MODEM, and PACKET-RADIO BULLETIN-BOARD SYSTEM.

BURN-IN

When a piece of equipment is brought into a test lab or service shop to be repaired, the malfunction does not always show up as soon as the device is connected to power. Often, the fault will be seen only occasionally, or the malfunction might not occur until the unit has been warmed or cooled to a certain temperature. Sometimes the humidity is a factor. Under these circumstances, a burn-in must be performed, and environmental parameters varied until the problem is found. The equipment is turned on and left on, sometimes for hours, or even days, until the malfunction is observed and can be diagnosed. Sometimes the problem is never seen. Such a situation is well known as a source of frustration for service technicians and customers in almost all electrical or mechanical fields.

At a factory, the burn-in is the initial testing phase of a unit. This ensures that the unit meets all specifications when it leaves the manufacturing plant. This procedure reduces the number of faulty new units that reach buyers. From a sales point of view, the importance of burn-in is obvious.

BUS

In an electronic circuit, a common ground conductor is sometimes called a *ground bus*, or simply a *bus*. Various pieces of equipment are connected to this common bus; the bus is in turn grounded. This arrangement prevents the formation of ground loops. *See also* GROUND LOOP.

In computer practice, subsystems are connected to the main central processing unit by means of a set of conductors. These conductors, considered as a whole, are called the *bus*. This might take the form of a set of wires, or a set of lines on a circuit board. Often, the bus consists partly of circuit-board traces, and partly of wires and connectors. A single bus might have several dozen conductors. The bus is used with such equipment as disk drives, monitors, printers, and power supplies. It is something like the spinal cord in the human body: a main signal path.

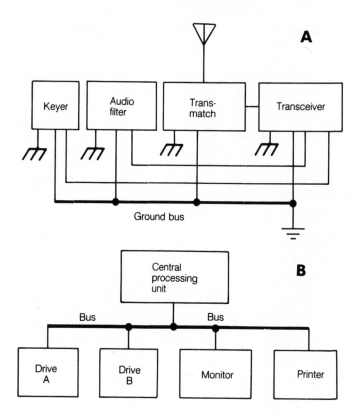

BUS: At A, a ground bus in a typical ham radio station. At B, an example of a computer-system bus.

The size of a bus is sometimes given in terms of *bits*. Typical buses have bit numbers that are powers of 2; for example, an 8-bit bus or a 32-bit bus. *See also* COMPUTER.

BUSHING

A *bushing* is a device used to couple two sections of control shaft together longitudinally. Bushings are used to lengthen the control shafts of rotary switches, variable capacitors, and potentiometers. Some bushings are made of insulating material (such as porcelain or bakelite). Insulating bushings are used when it is necessary to keep a control shaft insulated from chassis ground, or when the effects of body capacitance must be kept at a minimum. *See* BODY CAPACITANCE.

BUTTERFLY CAPACITOR

At very-high and ultra-high frequency ranges, a tuning device called a *butterfly capacitor* is often used in place of the conventional coil-and-capacitor tank circuit. A butterfly capacitor includes inductance as well as capacitance. The device gets its name from the fact that its plates resemble, physically, the opened wings of a butterfly.

The butterfly capacitor has a very high *Q* factor (*see Q* FACTOR). That is, it displays excellent selective characteristics, although the cavity resonator is better at VHF and UHF (*see* CAVITY RESONATOR).

BUTTERWORTH FILTER

A *Butterworth filter* is a special type of selective filter, designed to have a flat response in its passband and a uniform roll-off characteristic. A Butterworth filter might be designed for a lowpass, highpass, bandpass, or band-rejection response.

The figure shows some ideal Butterworth responses for lowpass, highpass, and bandpass filters (A, B, and C). Note the absence of peaks in the passband. The source and load impedances must be correctly chosen for proper operation; the values of the filter inductors and capacitors depend on the source and load impedances.

A similar type of selective filter, called the *Chebyshev filter,* is also commonly used for lowpass, highpass, bandpass, and band-rejection applications. *See also* BANDPASS FILTER, BAND-REJECTION FILTER, CHEBYSHEV FILTER, HIGHPASS FILTER, and LOWPASS FILTER.

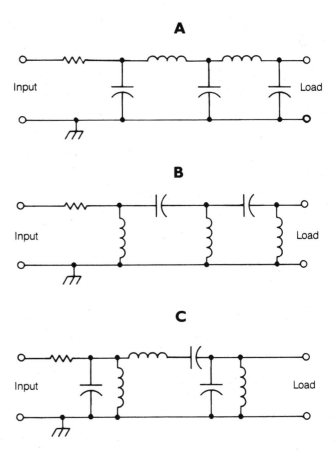

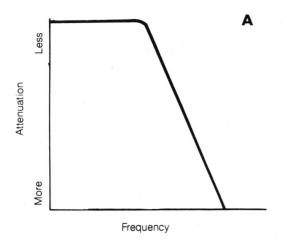

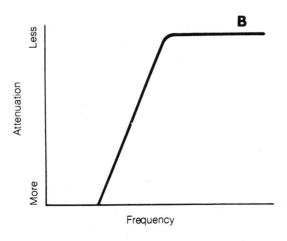

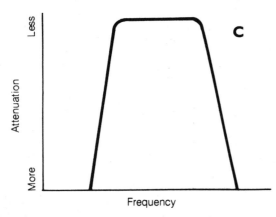

BUTTERWORTH FILTER: Figures 1 and 2 Lowpass (A), highpass (B), and bandpass (C).

BYPASS CAPACITOR

A capacitor that provides a low impedance for a signal, while allowing dc bias, is called a *bypass capacitor.* Bypass capacitors are frequently used in the emitter circuits of transistor amplifiers to provide stabilization. They are also used in the source circuits of field-effect transistors, and in the cathode circuits of tubes. The bypass capacitor places a circuit element at signal ground potential, although the dc potential might be several hundred volts.

The bypass capacitor is so named because the ac signal is provided with a low-impedance path around a high-impedance element. The schematic illustrates a typical bypass capaci-

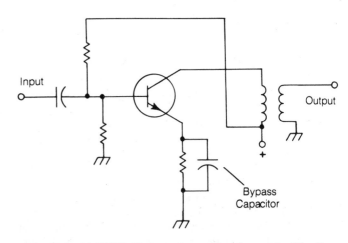

BYPASS CAPACITOR: Keeps emitter at signal ground while allowing for dc bias.

tor in the emitter circuit of a bipolar-transistor amplifier. A bypass capacitor is different from a blocking capacitor (*see* BLOCKING CAPACITOR); a bypass capacitor is usually connected in such a manner that it shorts the signal to ground, while a blocking capacitor is intended to conduct the signal from one part of a circuit to another while isolating the dc potentials.

BYTE

A *byte* is a group of bits that are processed in parallel. A byte might contain any number of bits, but the number is usually a power of 2, such as 2, 4, 8, 16, and so on. A computer processes each byte as one unit.

A byte is the smallest addressable unit of storage in a computer. It generally consists of eight data bits and one parity bit.

C

The letter *C* or *c* can stand for several different things in electronics. The uppercase *C* is used to denote *capacitance*, or to represent a *capacitor*. *See* CAPACITANCE, and CAPACITOR. The uppercase *C* is the name of a versatile computer language, used in some ham-radio software. The uppercase *C* can also represent *degrees Celsius* or *centigrade*.

The lowercase *c* is the standard symbol for the speed of light in a vacuum, about 299,792 kilometers per second or 186,282 miles per second. The lowercase *c* is also a general symbol for a mathematical constant.

In Morse code telegraphy, a letter *c* after the signal report indicates that the transmitter has a chirp. A good, loud, easily readable signal with a chirp might earn a report of 589c. *See* CHIRP, and RST SYSTEM.

CABLE

Any group of electrical conductors, bound together, is a *cable*. A cable might have just two conductors, such as the coaxial cable familiar to RF designers; or it might have many individual conductors.

Cables are usually covered by a tough protective material such as rubber or polyethylene. This protects the conductors against the accumulation of moisture, and also against corrosion. Each individual conductor is, of course, covered with its own layer of insulation.

Cables are used extensively in communications and electronics. Most telephone circuits are completed via cable. In remote areas, cable is sometimes the only available means of obtaining television reception, except for satellite downlinks.

CADMIUM

Cadmium is an element with atomic number 48 and atomic weight 112. It is a metallic substance, often used as a plating material for steel to prevent corrosion. Cadmium adheres well to solder, and it is therefore a good choice for plating of chassis for the construction of electronic equipment.

Cadmium is used in the manufacture of photocells, which are devices that change resistance according to the amount of light that strikes them. Cadmium is also found in certain types of rechargeable cells and batteries, called *nickel-cadmium batteries*. *See also* NICKEL-CADMIUM BATTERY.

CAGE ANTENNA

A *cage antenna* is similar to a dipole antenna (*see* DIPOLE ANTENNA). Several half-wavelength, center-fed conductors are connected together at the feed point. The cage antenna is so named because it often physically resembles a cage. The multiplicity of conductors increases the effective bandwidth of the antenna. It also reduces losses caused by the resistance of the wire conductors.

A cage antenna with wires or rods arranged in the form of two cones, with apexes that meet at the feed point, is called a *biconical antenna*. Such an antenna offers a good impedance match for 50- or 75-ohm feed lines, over a frequency range of several octaves. *See also* BICONICAL ANTENNA.

CALIBRATION

Calibration is the adjustment of a measuring instrument for an accurate indication. The frequency dials of radio receivers and transmitters, as well as the various dials and meters on test instruments, must be calibrated periodically to ensure their accuracy.

The precision of calibration is usually measured as a percentage, plus or minus the actual value. For example, the frequency calibration of a radio receiver can be specified as plus or minus 0.0001 percent, written ± 0.0001 percent. At a frequency of 1 MHz, this is a maximum error of one part in a million (1 Hz). At 10 MHz, it represents a possible error of 10 Hz either side of the indicated frequency.

Calibration requires the use of reference standards. Time and frequency are broadcast by radio stations WWV and WWVH in the United States (*see* WWV/WWVH), with extreme accuracy. Standard values for all units are determined by the National Bureau of Standards (*see* NATIONAL BUREAU OF STANDARDS).

CALLSIGN

Every licensed radio station in the world has a *callsign*. A callsign consists of a sequence of letters or letters and numbers. Commercial radio and television broadcast stations usually have callsigns containing three or four letters. Other radio services assign alphanumeric callsigns to their stations.

Callsigns begin with a letter or number that depends on the country in which the station is licensed. In some cases, the country is subdivided into several zones, each with a distinctive callsign format. The table lists international callsign prefixes currently in use throughout the world. Identification requirements vary from country to country.

The nickname of a broadcast station is not the same as its callsign, and is usually not legal for purposes of identification. The Federal Communications Commission assigns callsigns in the United States.

CAMERA TUBE

A *camera tube* is a device that focuses a visible image onto a small screen, and scans this image with an electron beam to develop a composite video signal. The current in the electron beam varies, depending on the brightness of the image in each part of the screen. The response must be very rapid, since the scanning is done at a high rate of speed. The position of the electron beam, relative to the image, is synchronized between the camera tube and the receiver picture tube.

CALL SIGN: INTERNATIONAL CALL-SIGN PREFIXES, AS ASSIGNED
BY THE INTERNATIONAL TELECOMMUNICATION UNION (ITU).

Prefix	Country	Prefix	Country
AAA-ALZ	United States	SNA-SRZ	Poland
AMA-AOZ	Spain	SSA-SSM	Egypt
APA-ASZ	Pakistan	SSN-STZ	Sudan
ATA-AWZ	India	SUA-SUZ	Egypt
AXA-AXZ	Australia	SVA-SZZ	Greece
AYA-AZZ	Argentina	TAA-TCZ	Turkey
BAA-BZZ	China	TDA-TDZ	Guatemala
CAA-CEZ	Chile	TEA-TEZ	Costa Rica
CFA-CKZ	Canada	TFA-TFZ	Iceland
CLA-CMZ	Cuba	TGA-TGZ	Guatemala
CNA-CNZ	Morocco	THA-THZ	France
COA-COZ	Cuba	TIA-TIZ	Costa Rica
CPA-CPZ	Bolivia	TJA-TJZ	Cameroon
CQA-CUZ	Portugal	TKA-TKZ	France
CVA-CXZ	Uruguay	TLA-TLZ	Central African Republic
CYA-CZZ	Canada	TMA-TMZ	France
DAA-DTZ	Germany	TNA-TNZ	Congo
DUA-DZZ	Philippines	TOZ-TQZ	France
EAA-EHZ	Spain	TRA-TRZ	Gabon
EIA-EJZ	Ireland	TSA-TSZ	Tunisia
EKA-EKZ	Union of Soviet Socialist Republics	TTA-TTZ	Chad
		TUA-TUZ	Ivory Coast
ELA-ELZ	Liberia	TVA-TXZ	France
EMA-EOZ	Union of Soviet Socialist Republics	TYA-TYZ	Benin
		TZA-TZZ	Mali
EPA-EQZ	Iran	UAA-UZZ	Union of Soviet Socialist Republics
ERA-ERZ	Union of Soviet Socialist Republics	VAA-VGZ	Canada
ETA-ETZ	Ethiopia	VHA-VNZ	Australia
EWA-EZZ	Union of Soviet Socialist Republics	VOA-VOZ	Canada
		VPA-VSZ	Great Britain and Northern Ireland
FAA-FZZ	France		
GAA-GZZ	Great Britain and Northern Ireland	VTA-VWZ	India
		VXA-VYZ	Canada
HAA-HAZ	Hungary	VZA-VZZ	Australia
HBA-HBZ	Switzerland	WAA-WZZ	United States
HCA-HDZ	Ecuador	XAA-XIZ	Mexico
HEA-HEZ	Switzerland	XJA-XOZ	Canada
HFA-HFZ	Poland	XPA-XPZ	Denmark
HGA-HGZ	Hungary	XQA-XRZ	Chile
HHA-HHZ	Haiti	XSA-XSZ	China
HIA-HIZ	Dominican Republic	XTA-XTZ	Voltaic Republic
HJA-HKZ	Colombia	XUA-XUZ	Cambodia
HLA-HMZ	South Korea	XVA-XVZ	Vietnam
HNA-HNZ	Iraq	XWA-XWZ	Laos
HOA-HPZ	Panama	XXA-XXZ	Portuguese Overseas Provinces
HQA-HRZ	Honduras	XYA-XZZ	Burma
HSA-HSZ	Thailand	YAA-YAZ	Afghanistan
HTA-HTZ	Nicaragua	YBA-YHZ	Indonesia
HUA-HUZ	El Salvador	YIA-YIZ	Iraq
HVA-HVZ	Vatican City	YJA-YJZ	New Hebrides
HWA-HYZ	France	YKA-YKZ	Syria
HZA-HZZ	Saudi Arabia	YMA-YMZ	Turkey
IAA-IZZ	Italy	YNA-YNZ	Nicaragua
JAA-JSZ	Japan	YOA-YRZ	Romania
JTA-JVZ	Mongolia	YSA-YSZ	El Salvador
JWA-JXZ	Norway	YTA-YUZ	Yugoslavia
JYA-JYZ	Jordan	YVA-YYZ	Venezuela
JZA-JZZ	Indonesia	YZA-YZZ	Yugoslavia
KAA-KZZ	United States	ZAA-ZAZ	Albania
OAA-OCZ	Peru	ZBA-ZJZ	Great Britain and Northern Ireland
ODA-ODZ	Lebanon		
OEA-OEZ	Austria	ZKA-ZMZ	New Zealand
OFA-OJZ	Finland	ZNA-ZOZ	Great Britain and Northern Ireland
OKA-OMZ	Czechoslovakia		
ONA-OTZ	Belgium	ZPA-ZPZ	Paraguay
OUA-OZZ	Denmark	ZQA-ZQZ	Great Britain and Northern Ireland
PAA-PIZ	Netherlands		
PJA-PJZ	Netherlands Antilles	ZRA-ZUZ	South Africa and Namibia
PKA-POZ	Indonesia	ZVA-ZZZ	Brazil

CALL SIGN: INTERNATIONAL CALL-SIGN PREFIXES, AS ASSIGNED
BY THE INTERNATIONAL TELECOMMUNICATION UNION (ITU).

Prefix	Country	Prefix	Country
PPA-PYZ	Brazil	2AA-2ZZ	Great Britain and Northern
PZA-PZZ	Surinam		Ireland
QAA-QZZ	Q signals (no call signs)	3AA-3AZ	Monaco
SAA-SMZ	Sweden	3BA-3BZ	Mauritius
3CA-3CZ	Equatorial Guinea		
3DA-3DM	Swaziland	8PA-8PZ	Barbados
3DN-3DZ	Fiji	8QA-8QZ	Maldive Island
3EA-3FZ	Panama	8RA-8RZ	Guyana
3GA-3GZ	Chile	8SA-8SZ	Sweden
3HA-3UZ	China	8TA-8YZ	India
3VA-3VZ	Tunisia	8ZA-8ZZ	Saudi Arabia
3WA-3WZ	Vietnam	9AA-9ZZ	San Marino
3XA-3XZ	Guinea	9BA-9DZ	Iran
3YA-3YZ	Norway	9EA-9FZ	Ethiopia
3ZA-3ZZ	Poland	9GA-9GZ	Ghana
4AA-4CZ	Mexico	9HA-9HZ	Malta
4DA-4IZ	Philippines	9IA-9JZ	Zambia
4JA-4LZ	Union of Soviet Socialist	9KA-9KZ	Kuwait
	Republics	9LA-9LZ	Sierra Leone
4MA-4MZ	Venezuela	9MA-9MZ	Malaysia
4NA-4OZ	Yugoslavia	9NA-9NZ	Nepal
4PA-4SZ	Sri Lanka	9OA-9TZ	Zaire
4TA-4TZ	Peru	9UA-9UZ	Burundi
4UA-4UZ	United Nations	9VA-9VZ	Singapore
4VA-4VZ	Haiti	9WA-9WZ	Malaysia
4WA-4WZ	North Yemen	9XA-9XZ	Rwanda
4XA-4XZ	Israel	9YA-9ZZ	Trinidad and Tobago
4YA-4YZ	International Civil	A2A-A2Z	Botswana
	Aviation Organization	A3A-A3Z	Tonga
		A4A-A4Z	Oman
4ZA-4ZZ	Israel	A5A-A5Z	Bhutan
5AA-5AZ	Libya	A6A-A6Z	United Arab Emirates
5BA-5BZ	Cyprus	A7A-A7Z	Qatar
5CA-5GZ	Morocco	A8A-A8Z	Liberia
5HA-5IZ	Tanzania	A9A-A9Z	Bahrain
5JA-5KZ	Colombia	C2A-C2Z	Nauru
5LA-5MZ	Liberia	C3A-C3Z	Andorra
5NA-5OZ	Nigeria	C4A-C4Z	Cyprus
5PA-5QZ	Denmark	C5A-C5Z	The Gambia
5RA-5SZ	Malagasy Republic	C6A-C6Z	Bahamas
5TA-5TZ	Mauritania	C7A-C7Z	World Meteorological
5UA-5UZ	Niger		Organization
5VA-5VZ	Togo	C8A-C9Z	Mozambique
5WA-5WZ	Western Samoa	D2A-D3Z	Angola
5XA-5XZ	Uganda	D4A-D4Z	Cape Verde
5YA-5ZZ	Kenya	D5A-D5Z	Liberia
6BA-6BZ	Egypt	D6A-D6Z	State of Comoro
6CA-6CZ	Syria	D7A-D9Z	South Korea
6DA-6JZ	Mexico	H2A-H2Z	Cyprus
6KA-6NZ	Korea	H3A-H3Z	Panama
6OA-6OZ	Somali	H4A-H4Z	Solomon Islands
6PA-6SZ	Pakistan	H5A-H5Z	Bophuthatswana
6TA-6UZ	Sudan	J2A-J2Z	Djibouti
6VA-6WZ	Senegal	J3A-J3Z	Grenada
6XA-6XZ	Malagasy Republic	J4A-J4Z	Greece
6YA-6YZ	Jamaica	J5A-J5Z	Guinea Bissau
6ZA-6ZZ	Liberia	L2A-L9Z	Argentina
7AA-7IZ	Indonesia	P2A-P2Z	Papua New Guinea
7JA-7NZ	Japan	P3A-P3Z	Cyprus
7OA-7OZ	South Yemen	P4A-P4Z	Netherlands Antilles
7PA-7PZ	Lesotho	P5A-P9Z	North Korea
7QA-7QZ	Malawi	S2A-S3Z	Bangladesh
7RA-7RZ	Algeria	S6A-S6Z	Singapore
7SA-7SZ	Sweden	S7A-S7Z	Seychelles
7TA-7YZ	Algeria	S8A-S8Z	Transkei
7ZA-7ZZ	Saudi Arabia	S9A-S9Z	St. Thomas and Principe
8AA-8IZ	Indonesia	T2A-T2Z	Tuvalu
8JA-8NZ	Japan	Y2A-Y9Z	East Germany
8OA-8OZ	Botswana		

The camera tube most often used in television broadcast stations is called the image orthicon. A simpler and more compact camera tube, used primarily in industry and space applications, is called the *vidicon*.

CANNIBALIZATION

When a piece of equipment must be repaired in the field, and the only available components are contained in other equipment that is deemed less essential, those components may be robbed for use in important gear. Such robbing is called *cannibalization*. The equipment from which parts are cannibalized is disabled until replacement parts can be obtained for it.

Cannibalization is a last-resort, emergency procedure, which technicians try to avoid. Cannibalization gets its somewhat humorous name from the fact that it involves a sacrifice of active equipment.

CAPACITANCE

Capacitance is the ability of a device to store an electric charge. Capacitance, represented by the capital letter C in equations, is measured in units called *farads*. One farad is equal to one coulomb of stored electric charge per volt of applied potential. Thus:

$$C \text{ (farads)} = Q/E$$

where Q is the quantity of charge in coulombs, and E is the applied voltage in volts.

The farad is, in practice, an extremely large unit of capacitance. Generally, capacitance is specified in microfarads, abbreviated μF, or in picofarads, abbreviated pF. These units are, respectively, a millionth and a trillionth of a farad:

$$1 \ \mu F = 10^{-6} \ F$$
$$1 \ pF = 10^{-12} \ F$$

Any two electrically isolated conductors will display a certain amount of capacitance. When two conductive materials are deliberately placed near each other to produce capacitance, the device is called a *capacitor* (see CAPACITOR).

Capacitances connected in parallel add together. For n capacitances $C_1, C_2, \ldots, C_n$ in parallel, then, the total capacitance C is:

$$C = C_1 + C_2 + \ldots + C_n$$

In series, capacitances add as follows:

$$C = \frac{1}{1/C_1 + 1/C_2 + \ldots + 1/C_n}$$

See also CAPACITIVE REACTANCE.

CAPACITIVE COUPLING

Capacitive coupling is a means of electrostatic connection among the stages of amplification in a radio circuit. The output of one stage is coupled to the input of the succeeding stage by a capacitor, as illustrated in the schematic diagram. Capacitor C is the coupling capacitor.

Capacitive coupling allows the transfer of an alternating-current signal, but it prevents the short-circuiting of a direct-current potential difference. Thus, the desired bias may be maintained on either side of the capacitor. A coupling capacitor

is sometimes called a blocking capacitor (*see* BLOCKING CAPACITOR).

The chief disadvantage of capacitive coupling is that it is sometimes difficult, or impossible, to achieve a good impedance match between stages. This makes the power transfer inefficient. Also, capacitive coupling allows the transfer of all harmonic energy, as well as the fundamental or desired frequency, from stage to stage. Tuned circuits are necessary to reduce the transfer of this unwanted energy. Sometimes, the use of coupling transformers is better than the use of capacitors for interstage connection. *See also* TRANSFORMER COUPLING.

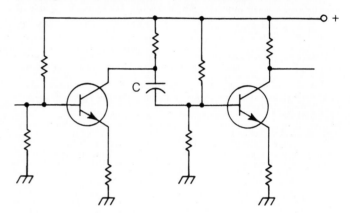

CAPACITIVE COUPLING: The capacitor isolates for dc, but lets the signal pass.

CAPACITIVE FEEDBACK

Capacitive feedback is a means of returning a portion of the output signal of a circuit back to the input, using a capacitor. In most amplifier circuits, the output is 180 degrees out of phase with the input, so connecting a capacitor between the output and the input reduces the gain. If the output is fed back in phase, the gain of an amplifier increases, but the probability of oscillation is much greater.

Capacitive feedback can occur as a result of coupling between the input and output wiring of a circuit. This can produce undesirable effects, such as amplifier instability or parasitic oscillation. This kind of capacitive feedback is also possible within a tube or transistor, because of interelectrode capacitance. It is nullified by a process called *neutralization* (see NEUTRALIZATION).

Capacitive feedback is commonly used in oscillator circuits, such as the multivibrator, Hartley, Pierce, and Colpitts types. The Armstrong oscillator uses inductive feedback, via coupling between coils.

CAPACITIVE LOADING

Capacitive loading can refer to either of two techniques for alternating the resonant frequency of an antenna system. A vertical radiator, operated against ground, is normally resonant when it is $\frac{1}{4}$ wavelength high. However, a taller radiator can be brought to resonance by the installation of a capacitor or capacitors in series. This is commonly called *capacitive loading* or *tuning*.

A vertical radiator shorter than $\frac{1}{4}$ wavelength may be resonated by the use of a device called a capacitance hat. A capacitance hat is often used in conjunction with a loading coil.

The capacitance hat results in an increase in the useful bandwidth of the antenna, as well as a shorter physical length at a given resonant frequency.

Both of the methods of capacitive loading described here are also used with dipole antennas. They can also be used in the driven elements of Yagi antennas, and in various kinds of phased arrangements.

CAPACITIVE REACTANCE

Reactance is the opposition that a component offers to alternating current. Reactance does not behave as simply as dc resistance. The reactance of a component can be either positive or negative, and it varies in magnitude depending on the frequency of the alternating current. Negative reactance is called capacitive reactance, and positive reactance is called inductive reactance. These choices of positive and negative are purely arbitrary, having been chosen by engineers for convenience.

The reactance of a capacitor of C farads, at a frequency of f hertz, is given by:

$$X_C = \frac{-1}{2\pi f C}$$

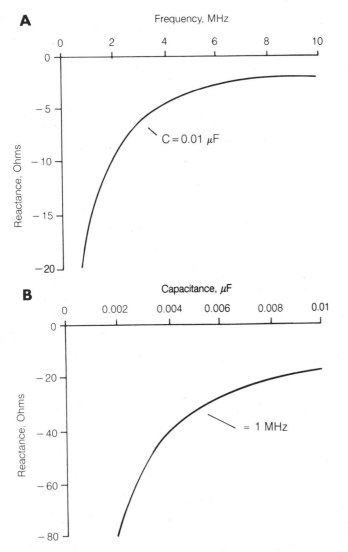

CAPACITIVE REACTANCE: At A, reactance vs frequency for 0.01 μF. At B, reactance vs capacitance at 1 MHz.

This formula also holds for values of C in microfarads, and frequencies in megahertz. When using farads, it is necessary to use hertz; when using microfarads, it is necessary to use megahertz. Units cannot be mixed.

The magnitude of capacitive reactance, for a given component, approaches zero as the frequency is raised. It becomes larger negatively as the frequency is lowered. This effect is illustrated by the graphs. At A, the reactance of a 0.01-μF capacitor is shown as a function of the frequency in MHz. At B, the reactance of various values of capacitance are shown for a frequency of 1 MHz.

Reactances, like dc resistances, are expressed in ohms. In a complex resistance-reactance circuit, the reactance is multiplied by the imaginary number $\sqrt{-1}$, which is mathematically represented by the letter j in engineering documentation. Thus a capacitive reactance of −20 ohms is specified as −20j; a combination of 10 ohms resistance and −20 ohms reactance is an impedance of 10−20j. *See also* IMPEDANCE, and *J* OPERATOR.

CAPACITOR

A *capacitor* is a device deliberately designed to provide a known amount of capacitance in a circuit. Capacitors are available in values ranging from less than 1 picofarad to hundreds of thousands of microfarads. It is rare to see a capacitor with a value of one farad. Capacitors are also specified according to their voltage-handling capability; these ratings must be carefully observed to be certain the potential across a capacitor is not greater than the rated value. If excessive voltage, either from a signal or from a source of direct current, is applied to a capacitor, permanent damage can result.

CAPACITOR: A small disk capacitor.

Capacitors are often very tiny when their voltage ratings and values are small. A small capacitor is shown in the photograph. Some capacitors are very large, such as the electrolytic variety found in high-voltage power supplies, or the wide-spaced air variables that are used in high-power radio-frequency amplifiers.

Capacitors can be constructed in several different ways. Disk ceramic capacitors, often used at radio frequencies, consist of two metal plates separated by a thin layer of porcelain dielectric. Other types of capacitors include air variables, electrolytics, mica-dielectric types, mylar-dielectric capacitors, and many more. The most common kinds of capacitors are listed in the table, with their primary characteristics.

CAPACITOR: COMMON TYPES OF
CAPACITORS AND THEIR APPLICATIONS.

Capacitor type	Approximate frequency range	Voltage range
Air variable	LF, MF, HF, VHF, UHF	Med. to high
Ceramic	LF, MF, HF, VHF	Med. to high
Electrolytic	AF, VLF	Low to med.
Mica	LF, MF, HF, VHF	Low to med.
Mylar	VLF, LF, MF, HF	Low to med.
Paper	VLF, LF, MF, HF	Low to med.
Polystyrene	AF, VLF, LF, HF	Low
Tantalum	AF, VLF	Low
Trimmer	MF, HF, VHF, UHF	Low to med.

Frequency abbreviations

AF: Audio frequency (0 to 20 kHz)
VLF: Very low frequency (10 to 30 kHz)
LF: Low frequency (30 to 300 kHz)
MF: Medium frequency (300 kHz to 3 MHz)
HF: High frequency (3 to 30 MHz)
VHF: Very high frequency (30 to 300 MHz)
UHF: Ultrahigh frequency (300 MHz to 3 GHz)

CAPACITOR TUNING

Capacitor tuning is a means of adjusting the resonant frequency of a tuned circuit by varying the capacitance while leaving the inductance constant. When this is done, the resonant frequency varies according to the inverse of the square root of the capacitance. This means that if the capacitance is doubled, the resonant frequency drops to 0.707 of its previous value. If the capacitance is cut to one-quarter, the resonant frequency is doubled. Mathematically, the resonant frequency of a tuned inductance-capacitance circuit is given by the formula:

$$f = \frac{1}{2 \pi \sqrt{LC}}$$

where *f* is the frequency in Hertz, *L* is the inductance in henrys, and *C* is the capacitance in farads. Alternatively, the units can be given as megahertz, microhenrys, and microfarads.

Capacitor tuning offers convenience when the values of capacitance are small enough to allow the use of air-variable units. If the capacitance required to obtain resonance is larger than about 0.001 μF, however, inductor tuning is generally preferable because it allows a greater tuning range. *See also* PERMEABILITY TUNING.

CAPTURE EFFECT

In a frequency-modulation receiver, a signal is either on or off, present or absent. Fadeouts occur instantly, rather than gradually, as they do in other modes. When an FM signal gets too weak to be received, it comes in intermittently. This is called *breakup*.

If there are two FM stations on the same channel, a receiver will pick up only the stronger signal. This is because of the limiting effect of the FM detector, which sets the receiver gain according to the strongest signal present in the channel. The effect in FM receivers that causes the elimination of weaker signals in favor of stronger ones is called the *capture effect*.

Capture effect does not occur with amplitude-modulated, continuous-wave, or single-sideband receivers. In those modes, both weak and strong signals are audible in proportion, unless the difference is very great.

When two FM signals are approximately equal in strength, a receiver might alternate between one and the other, making both signals unintelligible.

CARBON MICROPHONE

The *carbon microphone* is a device whose conductivity varies with the compression of carbon granules, which changes the influence of passing sound waves.

Vibration of air molecules against the metal diaphragm causes compression of the granules, which changes the resistance through the device. A direct-current potential, applied across the enclosure of carbon granules, is modulated by this resistance change. A transformer can be used at the microphone output to increase the impedance of the device, as seen by the input circuit; the carbon microphone characteristically shows a low impedance.

Carbon microphones are quite susceptible to damage by excessive dc voltages. They have become almost obsolete in modern audio applications. Dynamic microphones are much more often seen today. *See also* DYNAMIC MICROPHONE.

CARCINOTRON

The *carcinotron* is a backward-wave oscillator used to generate signals at extremely high frequencies. Electrons travel through a tube, interacting with electromagnetic fields induced externally. This produces oscillation. The carcinotron can be a linear tube or a circular tube. It operates in a manner similar to the magnetron (*see* MAGNETRON). The electric and magnetic fields cause oscillation in the electron beam as it travels down the tube.

The carcinotron is capable of producing several hundred watts up to frequencies of more than 10 GHz. It will produce several milliwatts at frequencies greater than 300 GHz. The carcinotron tends to produce a large amount of noise along with the desired signal.

CARD

A *card* is an easily-replaced printed-circuit board. Electronic equipment often has modular construction, making in-the-field repairs much more convenient than in past days (*see* MODULAR CONSTRUCTION). The card with the faulty component or components is simply replaced on-site with a properly operating card. Then, the bad card is taken to a central repair facility, where it is repaired and made ready for use as a replacement in another unit.

Some cards can be inserted and pulled out of a piece of equipment as easily as a household appliance is plugged in and unplugged. Others require the use of simple tools such as a screwdriver and needle-nosed pliers for installation and removal. Interchangeable cards have greatly reduced the amount of test equipment needed by field technicians. At the central repair shop, computerized test instruments are sometimes used to diagnose and pinpoint the trouble in seconds.

Modular construction has improved the serviceability of electronic equipment, both in terms of turn-around time and effectiveness of repairs.

CARDIOID PATTERN

A *cardioid pattern* is an azimuth pattern that represents the directional response of certain types of antennas, microphones, and speakers. The cardioid pattern is characterized by a sharp null in one direction, and a symmetrical response about the line running in the direction of the null. The illustration shows a typical cardioid pattern. It gets this name from the fact that it is heart-shaped.

Antennas with a cardioid directional response are frequently used in direction-finding apparatus. The extremely sharp null allows very accurate determination of the bearing.

A microphone with a broad response and a null off the back is called a *cardioid microphone*. Such microphones are useful in public-address systems where the speakers are located around or over the audience. The poor response in that direction minimizes the chances of acoustic feedback.

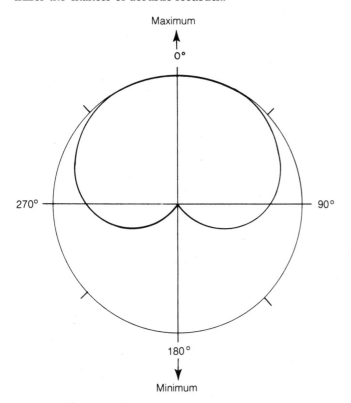

CARDIOID PATTERN: A sharp null exists in one direction.

CARDIOPULMONARY RESUSCITATION

Cardiopulmonary resuscitation (CPR) is a method of maintaining heart and lung function when either or both of these processes has failed in a human victim.

Electric shock can cause heart fibrillation, in which the normal rhythmic beating stops and the heart muscle quivers aimlessly instead. There might also be a failure of respiration.

The heart is made to pump blood, at least to some extent, by forcibly pressing down, at regular intervals, on the victim's breastbone while the victim lies on his/her back. Respiration is done by mouth-to-mouth means. One person can carry out CPR, but it is more effective when done by two people, one for the heart and one for the lungs.

The CPR process must be maintained until paramedics arrive, or until the victim has a regular pulse and regular breathing without assistance. There have been cases in which CPR has been doggedly performed for hours, with seemingly no hope for the victim's recovery, and a life was saved.

A detailed description of the CPR process is beyond the scope of this discussion. This article is *not* intended to teach CPR! The Red Cross offers courses in CPR. To effectively learn CPR, is is necessary to complete an approved course. It is wise for anyone, in general, to know CPR, because it can save lives in many kinds of emergency situations.

CARRIER

A *carrier* is an alternating-current wave of constant frequency, phase, and amplitude. By varying the frequency, phase, or amplitude of a carrier wave, information is transmitted. The term carrier is generally used in reference to the electromagnetic wave in a radio communications system; it can also refer to a current transmitted along a wire. An unmodulated carrier has a theoretical bandwidth of zero; all of the power is concentrated at a single frequency in the electromagnetic spectrum. A modulated carrier has a nonzero, finite bandwidth (*see* BANDWIDTH).

The term carrier is also used for electrons in N-type semiconductor material, or for holes in P-type semiconductor material. In these cases, carriers are the presence or absence of extra electrons in atoms of the substance.

CARRIER FREQUENCY

The *carrier frequency* of a signal is the average frequency of its carrier wave (*see* CARRIER). For continuous-wave (CW) and amplitude-modulated (AM) signals, the carrier frequency should be as constant as the state of the art will permit. The same is true for the suppressed-carrier frequency of a single-sideband (SSB) signal. For frequency-modulated (FM) signals, the carrier frequency is determined under conditions of no modulation. This frequency should always remain as stable as possible.

In frequency-shift keying (FSK), there are two signal frequencies, separated by a certain value, called the *carrier shift*. The shift frequency is usually 170 Hz for amateur communications, and 425 Hz or 850 Hz for commercial circuits. The space-signal frequency is generally considered to be the carrier frequency in a frequency-shift-keyed circuit.

When specifying the frequency of any signal, the carrier frequency is always given.

CARRIER MOBILITY

In a semiconductor material, the electrons or holes are called *carriers* (*see* ELECTRON, and HOLE). The average speed at which the carriers move, per unit of electric-field intensity, is called the *carrier mobility* of the semiconductor material. Electric-field units are specified in volts per centimeter or volts per meter. The mobility of holes is usually not the same as the mobility of electrons in a given material. However, the mobility depends on the amount of doping (*see* DOPING). At very high doping levels, the mobility decreases because of scattering of the carriers. As the electric field is made very intense, the mobility also decreases as the speed of carriers approaches a maximum, or limiting, velocity. This limiting velocity is approximately 10^4 to 10^5 meters per second, or 10^6 to 10^7 centimeters per second, in most materials.

The electron and hole mobility for four different semiconductor materials is given in the table. Units are specified in centimeters per second for an electric-field intensity of 1 volt per centimeter. To determine the drift velocity for an electric-field intensity of E volts per centimeter, the given values must be multiplied by E.

ELECTRON AND HOLE MOBILITY FOR DIFFERENT
SEMICONDUCTOR MATERIALS, IN CENTIMETERS PER
SECOND FOR A FIELD INTENSITY OF
1 VOLT PER CENTIMETER.

	Silicon	Germanium	Gallium arsenide	Indium antimonide
Electron mobility	1350*	3900*	6800**	80,000**
Hole mobility	480*	1900*	680*	4000**

*E. M. Conwell, "Properties of Silicon and Germanium II," *Proc. IRE*, Vol. 46, pp. 1281–1300.
**Bube, "Photoconductivity of Solids," John Wiley & Sons, New York.

CARRIER POWER

Carrier power is a measure of the output power of a transmitter for continuous wave (CW), amplitude modulation (AM), frequency-shift keying (FSK), or frequency modulation (FM). Carrier power is measured by an RF wattmeter under conditions of zero modulation, and with the transmitter connected to a load equivalent to its normal rated operating load.

The output power of a CW transmitter is determined under key-down conditions. With FSK or FM, the output power does not change with modulation, but remains constant at all times. With AM, the output power fluctuates somewhat with modulation.

In single-sideband (SSB) or double-sideband, suppressed-carrier operation, there is no carrier and hence no carrier power. Instead, the output power of the transmitter is measured as the peak-envelope power. *See also* PEAK ENVELOPE POWER.

CARRIER SHIFT

Carrier shift is a method of transmitting a radioteletype signal. The two carrier conditions are called *mark* and *space*. While no information is being sent, a steady carrier is transmitted at the space frequency. When a letter, word, or message is sent, the carrier frequency is shifted in pulses by a certain prescribed number of Hertz. In amateur communications, the standard carrier shift is 170 Hz; in commercial and military communications, values of 425 Hz or 850 Hz are generally used, although nonstandard carrier-shift values are sometimes used.

The transmitter output power remains constant under both mark and space conditions. *See also* FREQUENCY-SHIFT KEYING.

CARRIER SUPPRESSION

In a single-sideband (SSB) transmitter, the carrier is eliminated along with one of the sidebands. The carrier is also eliminated in the double-sideband, suppressed-carrier (DSB) mode. The carrier is usually eliminated by phase cancellation in a special modulator called a *balanced modulator*.

The illustration shows spectral displays of typical amplitude-modulated AM, DSB, and SSB signals. The carrier can never be entirely eliminated, although it is generally at least 60 dB below the level of the peak audio energy. *See also* BALANCED MODULATOR, DOUBLE SIDEBAND, and SINGLE SIDEBAND.

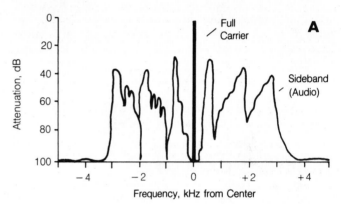

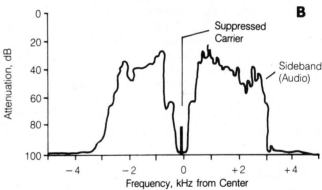

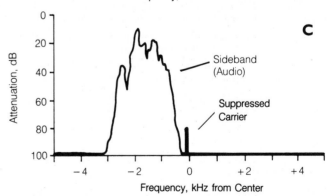

CARRIER SUPPRESSION: A typical AM signal (A); a double-sideband, suppressed-carrier signal (B); and a single-sideband signal (C).

CARRIER SWING

In a frequency-modulation system, *carrier swing* is the total amount of frequency deviation of the signal (*see* DEVIATION).

In a standard FM communications circuit, the maximum deviation is plus or minus 5 kHz with respect to the center of the channel; thus the carrier swing is 10 kHz.

The term carrier *swing* is not often used in FM. More often, the deviation is specified. The modulation index is also occasionally specified (*see* MODULATION INDEX).

CARRIER VOLTAGE

Carrier voltage is the alternating-current voltage of a transmitter radio-frequency carrier. It can be expressed as a peak voltage, a peak-to-peak voltage, or a root-mean-square (RMS) voltage; the magnitude of the carrier voltage depends on the transmitter output power (carrier power) and on the impedance of the feed line and antenna system. It also depends on the standing-wave ratio along the transmission line.

Carrier voltage can be measured with a radio-frequency voltmeter. If the standing-wave ratio is 1 and the feed-line impedance, Z_O is known, the RMS carrier voltage, E, can be calculated from an RF wattmeter reading P watts according to the formula:

$$E = \sqrt{PZ_O}$$

under conditions of zero transmitter modulation. *See also* CARRIER POWER, and ROOT MEAN SQUARE.

CARRYING CAPACITY

Carrying capacity is the ability of a component to handle a given amount of current or power. Generally, the term is used to define the current-handling capability of wire. The maximum amount of current that can be safely carried by a wire depends on the type and diameter of the metal used.

When the carrying capacity of wire is exceeded, there is a possibility of fire because of overheating. If the carrying capacity of a wire is greatly exceeded, the wire becomes soft and subject to stretching or breakage. It can even melt. This sometimes happens when there is a short circuit and no fuse or circuit breaker. It often happens when lightning strikes a conductor.

CARTRIDGE FUSE

A *cartridge fuse* is a tubular fuse with metal end caps. The end caps make electrical contact with the socket. These fuses are often used in radio apparatus, and in the peripheral equipment in a communications installation.

The cartridge fuse fits into a socket for easy replacement. When the fuse blows, it can clearly be seen; the glass appears darkened, or the internal wire obviously appears broken. Cartridge fuses are available in a variety of voltage and current ratings. *See also* FUSE.

CASCADE

When two or more amplifying stages are connected one after another, the arrangement is called a *cascade circuit*. Several amplifying stages can be cascaded to obtain much more signal gain than is possible with only one amplifying stage.

There is a practical limit to the number of amplifying stages that can be cascaded. The noise generated by the first stage, in addition to any noise present at the input of the first stage, will be amplified by succeeding stages along with the desired signal.

When many stages are connected in cascade, the probability of feedback and resulting oscillation increases greatly.

Tuned circuits and stabilizing networks are used to maximize the number of amplifiers that can be connected in cascade. By shielding the circuitry of each stage, and by making sure that the coupling is not too tight, the chances of oscillation are minimized. Proper impedance matching ensures the best transfer of signals, with the least amount of noise. When several stages of amplification are connected in cascade, the entire unit is sometimes called a *cascade amplifier*.

CASCADE VOLTAGE DOUBLER

The *cascade voltage doubler* is a circuit for obtaining high dc voltages with an alternating-current input. It is a form of power supply, used in tube-type circuits with low to medium current requirements.

Cascade voltage-doubler circuits have the advantage of a common input and output terminal, allowing unbalanced operation. *See also* VOLTAGE DOUBLER.

CASCODE

A circuit capable of obtaining high gain using tubes or transistors, with an excellent impedance match between two amplifying stages, is called a *cascode amplifier*.

The input stage is a grounded-cathode or grounded-emitter circuit. The plate or collector of the first stage is fed directly to the cathode or emitter of the second stage, which is a grounded-grid or grounded-base amplifier.

Because of the low-noise characteristics of the cascode circuit, it is often used in preamplifiers for the purpose of improving the sensitivity of HF and VHF receivers. *See also* PREAMPLIFIER.

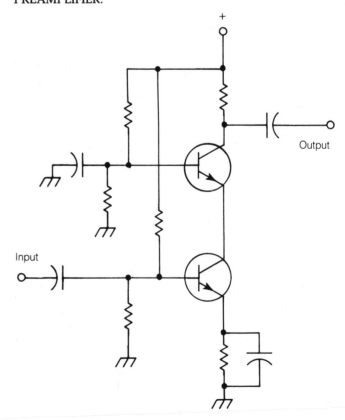

CASCODE: In this case, npn transistors are used.

CASSEGRAIN FEED

The *Cassegrain arrangement* is a method of feeding a parabolic or spherical dish antenna. It completely eliminates the need for mounting the feed apparatus at the focal point of the dish. In the conventional dish feed system, the waveguide terminates at just the right position, so that the RF energy emerges from the dish in a parallel beam (A in the illustration).

In the Cassegrain transmission system, the waveguide terminates at the base of the dish, which is physically more convenient from the standpoint of installation. The waveguide output is beamed to a small convex reflector at the focus of the dish, and the convex reflector directs the energy back to the main dish, shown at B. The beam emerges from the antenna in parallel rays. In reception, the rays of electromagnetic energy follow the same path, but in the opposite direction.

Dish antennas are used at UHF and microwave frequencies, when the diameter of the antenna, and the focal length, are many times the wavelength of the transmitted and received energy. *See also* DISH ANTENNA.

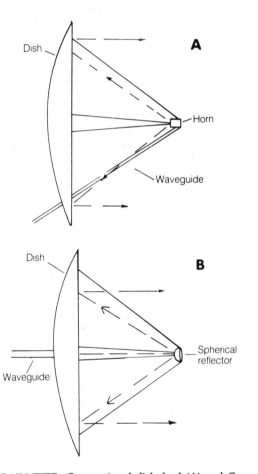

CASSEGRAIN FEED: Conventional dish feed (A) and Cassegrain dish feed (B).

CASSETTE

A device that holds recording tape is called a *cassette*. Cassettes are available in a variety of sizes and shapes for various applications. A typical cassette plays for 30 minutes on each side; longer playing cassettes can allow recording for as much as 60 minutes per side.

Some cassettes, such as the kind used in stereo tape players, have endless loops of tape. Four recording paths, each with two channels for the left and the right half of the sound track, provide a total of eight recording tracks.

CATHODE

The *cathode* is the electron-emitting electrode in a vacuum tube. To enhance its emission properties, the cathode of a tube is heated to a high temperature by a wire, called the *heater* (*see also* FILAMENT).

Two types of cathodes are in common use. The directly-heated cathode consists of a specially treated filament, which itself emits electrons when connected to a potential that is negative with respect to the anode or plate of the tube. The indirectly-heated cathode is a cylinder of metal, separated from the heater but physically close to it.

Directly-heated cathodes require a direct current for heating in amplifier and oscillator applications, since alternating current would produce ripple in the output signal. The indirectly-heated cathode can make use of an alternating-current filament supply; the cathode acts as a shield against the ripple emission from the filament.

In a bipolar transistor, the equivalent of the cathode is the emitter. In a field-effect transistor, the equivalent of the cathode is the source.

CATHODE COUPLING

Cathode coupling is a method of connecting one vacuum-tube circuit to another, or of providing the output from a tube amplifier to an external circuit. The amplifier output is taken from the cathode circuit.

The bipolar-transistor equivalent of cathode coupling is emitter coupling. The field-effect-transistor equivalent is source coupling.

Cathode coupling provides a low output impedance. A cathode follower is used to match a high impedance to a low impedance.

CATHODE-RAY OSCILLOSCOPE

See OSCILLOSCOPE.

CATHODE-RAY TUBE

A *cathode-ray tube* is a device for obtaining a graphic display of an electronic function, such as a waveform or spectral display. A television picture tube is a cathode-ray tube, but in the electronics laboratory, the oscilloscope is the most familiar example of the use of a cathode-ray tube. Cathode-ray tubes are also used as video monitoring devices for computers and word processors.

Oscilloscopes can be used in a variety of different ways. The spectrum-analyzer display uses a cathode-ray tube to obtain a function of signal amplitude vs frequency. Other common oscilloscope functions include amplitude vs time and frequency vs time.

The drawing shows the internal details and operation of a cathode-ray tube. The cathode, or electron gun, emits a stream of electrons. The first anode focuses the electrons into a narrow beam and accelerates them to greater speed. The second anode gives the electrons still more speed. The deflecting plates control the location at which the electron beam strikes the screen. The inside of the viewing screen is coated with phosphor material, which glows when the electrons hit it.

The cathode-ray tube is capable of deflecting an electron beam at an extreme rate of speed. Some cathode-ray tubes can show waveforms at frequencies of hundreds of megahertz. *See also* OSCILLOSCOPE.

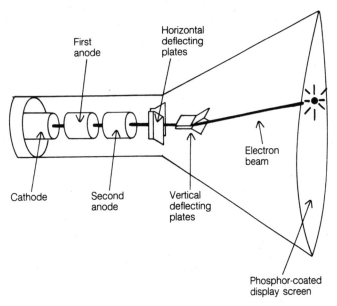

CATHODE-RAY TUBE: Cross-sectional diagram of typical CRT.

CAVITY FREQUENCY METER

A *cavity frequency meter* uses a cavity resonator (*see* CAVITY RESONATOR) to measure wavelengths at very high frequencies and above. An adjustable cavity resonator, and a radio-frequency voltmeter or ammeter, are connected to a pickup wire, or loosely coupled to the circuit under test. The cavity resonator is adjusted until a peak occurs in the meter reading. The free-space wavelength is easily measured and converted to the frequency, according to the formula:

$$F = 300/\lambda$$

where *f* is the frequency in megahertz and λ is the wavelength in meters.

A cavity resonator is generally tuned to a length of ½ wavelength to obtain resonance in the cavity frequency meter. However, any multiple of ½ wavelength, such as 1, 1½, 2, and so on, will allow a resonant reading.

CAVITY RESONATOR

A *cavity resonator* is a metal enclosure that is usually shaped like a cylinder or rectangular prism (see illustration). Cavity resonators operate as tuned circuits, and are practical for use at frequencies above about 200 MHz.

A cavity has an infinite number of resonant frequencies. When the length of the cavity is equal to any integral multiple of ½ wavelength, electromagnetic waves will be reinforced in the enclosure. Thus, a cavity resonator has a fundamental frequency and a theoretically infinite number of harmonic frequencies. Near the resonant frequency or any harmonic frequency, a cavity behaves like a parallel-tuned inductive-capacitive circuit. When the cavity is slightly too long, it shows inductive reactance, and when it is slightly too short, it displays capacitive reactance. At resonance, a cavity has a very high impedance and theoretically zero reactance. Cavity reso-

nators are sometimes used to measure frequency (*see* CAVITY FREQUENCY METER).

The resonant frequency of a cavity is affected by the dielectric constant of the air inside. Temperatures and humidity variations therefore have some effect. If precise tuning is required, the temperature and humidity must be kept constant. Otherwise, a resonant cavity can drift off resonance because of changes in the environment.

A length of coaxial cable, short-circuited at both ends, is sometimes used as a cavity resonator. Such resonators can be used at lower frequencies than can rigid metal cavities, since the cable does not have to be straight. The velocity factor of the cable must be taken into account when designing cavities of this kind. *See also* VELOCITY FACTOR.

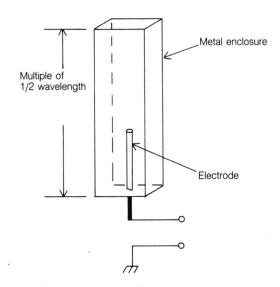

CAVITY RESONATOR: In this case, a piece of waveguide is used.

CCIR

The abbreviation *CCIR* represents the first letters of *International Radio Consultative Committee*. This committee recognizes standards for various forms of radio communications, including some modes and codes used by hams.

CCITT

The abbreviation *CCITT* represents the first letters of *International Telegraph and Telephone Consultative Committee*. This committee recognizes standards for various forms of telecommunications, including some modes and codes used by radio hams.

CELL

A *cell* is a voltage-producing device, usually operating via a chemical reaction. Two electrodes are placed in an electrolyte solution or paste, and a potential difference develops between the electrodes. Cells vary greatly in physical size. Chemical cells usually develop about 1.5 volts. The ampere-hour capacity and current-delivering capacity depend mostly on the physical size of a cell.

When several cells are connected in series to obtain greater voltage, the combination is called a *battery*. Cells are often erroneously called batteries. The two most common types of dry-chemical cell are the alkaline and zinc-carbon cell. Automotive batteries make use of lead-acid cells (*see* LEAD-ACID BAT-TERY). Certain kinds of semiconductor devices generate voltage when light strikes them; these devices are sometimes called *cells* (*see* PHOTOVOLTAIC CELL).

CELLULAR TELEPHONE SYSTEM

A *cellular telephone system*, also sometimes called *cellular radio*, is a special form of mobile telephone that has been developed in recent years.

A cellular system consists of a network of repeaters, all connected to one or more central office switching systems (*see* RE-PEATER). The individual subscribers are provided with radio transceivers operating at very-high or ultra-high frequencies.

The network of repeaters is such that most places are always in range of at least one repeater; ideally, every geographic point in the country would be covered. As a subscriber drives a vehicle, operation is automatically switched from repeater to repeater.

Eventually, most (if not all) telephone communication can be by cellular radio. Worldwide communication of high quality and low cost, using entirely wireless modes, might be achieved by the end of the twentieth century.

CELSIUS TEMPERATURE SCALE

The *Celsius temperature scale* is a scale at which the freezing point of pure water at one atmosphere is assigned the value zero degrees, and the boiling point of pure water at one atmosphere is assigned the value 100 degrees. The Celsius scale was formerly called the *Centigrade scale*. The word *Celsius* is generally abbreviated by the capital letter C.

Temperatures in Celsius and Fahrenheit are related by the equations:

$$C = \frac{5}{9}(F - 32)$$

$$F = \frac{9}{5}C + 32$$

where C represents the Celsius temperature and F represents the Fahrenheit temperature.

Celsius temperature is related to Kelvin, or absolute, temperature by the equation:

$$K = C + 273$$

where K represents the temperature in degrees Kelvin. A temperature of −273 degrees Celsius is called *absolute zero*, the coldest possible temperature.

CENTER FEED

When an antenna element, resonant or nonresonant, is fed at its physical center, the antenna is said to have a *center feed*. Usually, such an antenna is a half-wave dipole, the driven element of a Yagi, or one of the elements of a phased array.

Center feed results in good electrical balance when a two-wire transmission line is used (see drawing) whether the element is ½ wavelength or any other length, provided that the two halves of the antenna are at nearly equal distances from surrounding objects such as trees, utility wires, and the ground.

For antenna elements measuring an odd multiple of ½ wavelength, a center feed results in a purely resistive load impedance between approximately 50 and 100 ohms. For antenna elements measuring an even number of half wavelengths, center feed results in a purely resistive load impedance that ranges from several hundred to several thousand ohms. If the element is not an integral multiple of ½ wavelength, reactance is present at the load in addition to resistance.

Antennas do not necessarily have to be center-fed to work well. *See also* END FEED.

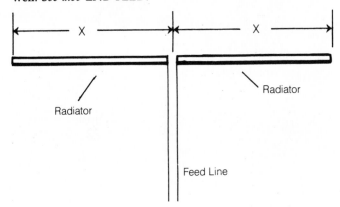

CENTER FEED: The antenna is symmetrical, relative to the line.

CENTER LOADING

Center loading is a method of altering the resonant frequency of an antenna radiator. An inductance or capacitance is placed along the physical length of the radiator, roughly halfway between the feed point and the end. The figure on pg. 68 shows center loading of a vertical radiator fed against ground (A) and a balanced, horizontal radiator (B).

Inductances lower the resonant frequency of a radiator having a given physical length. Generally, for quarter-wave resonant operation with a radiator less than ¼ wavelength in height, some inductive loading is necessary to eliminate the capacitive reactance at the feed point. For quarter-wave resonant operation with a radiator between ¼ and ½ wavelength in height, a capacitor must be used to eliminate the inductive reactance at the feed point.

An 8-foot mobile whip antenna can be brought to quarter-wave resonance by means of inductive center loading at all frequencies below its natural quarter-wave resonant frequency, which is about 29 MHz. While the RF ground in a mobile installation is not anything near perfect, the values given are close enough to be of practical use.

When the inductor or capacitor in an antenna loading scheme is placed at the feed point, the system is called *base loading*. *See also* BASE LOADING.

CENTER TAP

A *center tap* is a terminal connected midway between the ends of a coil or transformer winding. The schematic symbols for center-tapped inductors and transformers are shown in the illustration on pg. 68.

In an inductor, a center tap provides an impedance match. A center-tapped inductor can be used as an autotransformer (*see* AUTOTRANSFORMER) at audio or radio frequencies.

A transformer with a center-tapped secondary winding is often used in power supplies to obtain full-wave operation with

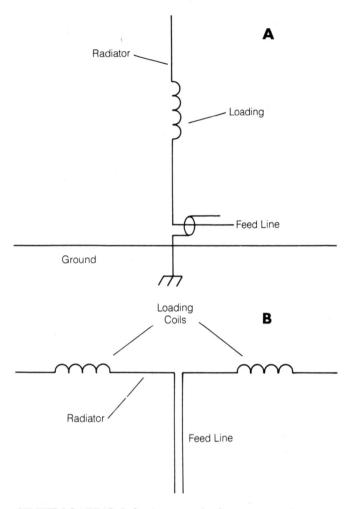

CENTER LOADING: Inductive center loading in a vertical antenna (A) and a dipole antenna (B).

only two rectifier diodes. Audio transformers having center-tapped secondary windings are used to provide a balanced output to speakers. The center tap is grounded, and the ends of the winding are connected to a two-wire line. At radio frequencies, center-tapped output transformers also provide a means of obtaining a balanced feed system.

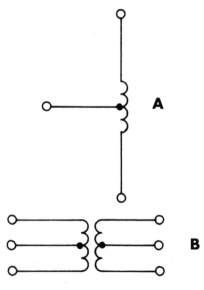

CENTER TAP: At A, in single coil. At B, in transformer windings.

CENTRAL PROCESSING UNIT

The *central processing unit* (also known as a *central processor*), is the part of a computer that coordinates the operation of all systems. The abbreviation, CPU, is often used when referring to the central processing unit.

Any computer-operated device has a CPU. Telephone switching networks, communications equipment, and many other electronic systems are coordinated by a CPU. Sometimes a microcomputer is called a CPU. *See also* COMPUTER, and MICROCOMPUTER.

CERAMIC

Ceramic is a manufactured compound that consists of aluminum oxide, magnesium oxide, and other similar materials. It is a white, fairly lightweight, solid with a dull surface. Materials, such as steatite and barium titanate, are ceramics. A polarized crystal is sometimes called a *ceramic crystal*. The most familiar example of a ceramic material is porcelain.

Ceramics are used in a wide variety of electronic applications. Some kinds of capacitors employ a ceramic material as the dielectric. Certain inductors are wound on ceramic forms, since ceramic is an excellent insulator and is physically strong. Ceramics are used in the manufacture of certain types of microphones and phonographic cartridges. Some bandpass filters, intended for use at radio frequencies, have resonant ceramic crystals or disks. Ceramic materials are used in the manufacture of some kinds of vacuum tubes. *See also* CERAMIC CAPACITOR, CERAMIC FILTER, CERAMIC MICROPHONE, and CERAMIC RESISTOR.

CERAMIC CAPACITOR

A *ceramic capacitor* is a device that consists of two metal plates, usually round in shape, attached to opposite faces of a ceramic disk, as shown in the illustration. The value of capacitance depends on the size of the metal plates and on the thickness of the ceramic dielectric material. Ceramic capacitors are generally available in sizes ranging from about 0.5 picofarad to 0.5 microfarad. Ceramic capacitors have voltage ratings from a few volts to several hundred volts.

The composition of the ceramic material determines the temperature coefficient of the capacitor (*see* TEMPERATURE COEFFICIENT). Ceramic capacitors are used from low frequencies up to several hundred megahertz. At higher frequencies, the ceramic material begins to get lossy, and this results in inefficient operation.

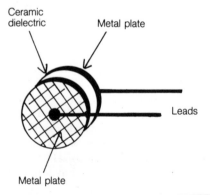

CERAMIC CAPACITOR: The dielectric is sandwiched between plates.

Ceramic capacitors are frequently used in high-frequency communications equipment. They are relatively inexpensive, and have a long operating life provided they are not subjected to excessive voltage. *See also* CERAMIC.

CERAMIC FILTER

A *ceramic filter* is a form of mechanical filter (*see* MECHANICAL FILTER) that makes use of piezoelectric ceramics to obtain a bandpass response. Ceramic disks resonate at the filter frequency. A ceramic filter is essentially the same as a crystal filter in terms of construction; the only difference is the composition of the disk material.

Ceramic filters are used to provide selectivity in the intermediate-frequency sections of transmitters and receivers. When the filters are properly terminated at their input and output sections, the response is nearly rectangular. *See also* BANDPASS RESPONSE, and CRYSTAL-LATTICE FILTER.

CERAMIC MICROPHONE

A *ceramic microphone* uses a ceramic cartridge to transform sound energy into electrical impulses. Its construction is similar to that of a crystal microphone (*see* CRYSTAL MICROPHONE). When subjected to the stresses of mechanical vibration, certain ceramic materials generate electrical impulses.

Ceramic and crystal microphones must be handled with care because they are easily damaged by impact. Ceramic microphones display a high output impedance and excellent audio-frequency response. *See also* CERAMIC.

CERAMIC RESISTOR

A *ceramic resistor* is a device intended to limit the current that flows in a circuit. It is made from carborundum, which is a compound of carbon and silicon. The value of resistance of a ceramic resistor decreases as the voltage across the component increases. Ceramic resistors can be obtained with either positive or negative temperature coefficients. A positive temperature coefficient means that the resistance increases with increasing temperature; a negative temperature coefficient means that the resistance decreases with increasing temperature.

Ceramic resistors generate less electrical noise, as a current passes through them, than ordinary carbon resistors generate. Ceramic resistors are available in fixed or variable form. They have a long service life.

CHANNEL

A *channel* is a particular band of frequencies to be occupied by one signal, or one 2-way conversation in a given mode. The term *channel* is also used to refer to the current path between the source and drain of a field-effect transistor (*see* CHANNEL EFFECT, and FIELD-EFFECT TRANSISTOR).

Some frequency bands, such as the amplitude-modulation (AM) broadcast band and the frequency-modulation (FM) broadcast band, are allocated in channels by legal authority of the Federal Communications Commission in the United States. Some frequency bands, such as the Amateur-Radio FM band at 144 MHz, are allocated into channels by common agreement among the users. Some bands, such as the high-frequency Amateur bands, are not allocated into channels. Operation in such bands is done with variable-frequency oscillators.

Any signal requires a certain amount of bandwidth for efficient transfer of information. This bandwidth is called the *channel*, or *channel width*, of the signal. A typical AM broadcast signal is 10 kHz wide, or 5 kHz above and below the carrier frequency at the center of the channel.

Some signal channels are as narrow as 3 to 5 kHz; some are several megahertz wide, such as fast-scan commercial television channels. *See also* BANDWIDTH.

CHANNEL ANALYSIS

When a signal is checked to ensure that all its components are within the proper assigned channel, the procedure is called *channel analysis*. Channel analysis requires a spectrum analyzer to obtain a visual display of signal amplitude as a function of frequency (*see* SPECTRUM ANALYZER).

The illustration shows an amplitude-modulated (AM) signal as it would appear on a spectrum-analyzer display. The normal bandwidth of an AM broadcast signal is plus or minus 5 kHz, relative to the channel center. (A communications signal is often narrower than this, about plus or minus 3 kHz.) At A, a properly operated AM transmitter produces energy entirely within the channel limits. At B, overmodulation causes excessive bandwidth. At C, an off-frequency signal results in out-of-band emission.

Channel analysis can reveal almost any problem with a modulated signal. But it takes some technical training to learn how different modes should appear on a spectrum analyzer.

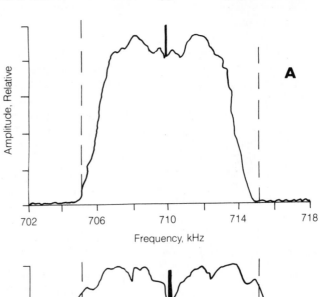

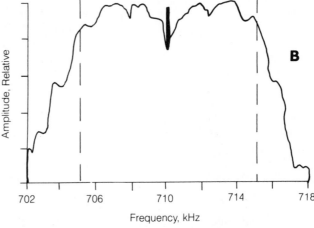

CHANNEL ANALYSIS: At A, proper AM signal at 710 kHz. At B, excessive bandwidth. At C, signal off frequency.

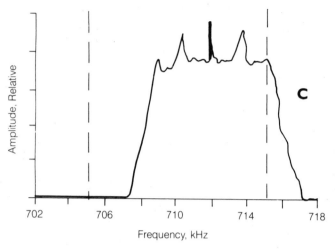

CHANNEL ANALYSIS Continued.

CHANNEL CAPACITY

Any channel is capable of carrying some information at a certain rate of speed. A channel just a few hertz wide can accommodate a code (CW) signal; a channel in the AM broadcast band can be used for the transmission of voices. In general, the greater the bandwidth of the channel (see BANDWIDTH), the more information, measured in characters or words, can be transmitted per unit time. The maximum rate at which information can be reliably sent over a particular channel is called the *channel capacity*.

Reliability is generally defined as an error rate not greater than a certain percentage of characters sent. As the speed of data transmission is increased, the error rate increases gradually at first. Then, as the data rate becomes too fast for the channel, the error percentage rises more and more rapidly. As the data speed is increased without limit, the data reception approaches a condition nearly equivalent to random characters. The precise error percentage, representing the limit of reliability, must be prescribed when defining channel capacity.

The channel capacity depends, to some extent, on the mode of transmission. Digital modes, such as frequency-shift keying, are generally more efficient than analog modes, such as voice transmission.

CHANNEL EFFECT

Channel effect is the tendency for current to leak from the emitter to the collector of a bipolar transistor. This is undesirable in bipolar transistors, but the field-effect transistor (FET) operates on this principle (see FIELD-EFFECT TRANSISTOR).

The illustration shows the channel of a field-effect transistor. A current flows from the source to the drain; its magnitude depends on the voltage at the gate electrodes. In this case, an N-channel field-effect transistor is illustrated. The P-channel unit operates on the same principle, but with reversed polarity.

The channel becomes narrower in the N-channel FET as the gate voltage becomes more and more negative. The channel width also depends on the voltage between the source and the drain.

The narrower the channel, the greater the effective resistance between the source and the drain. The channel effect (as shown), for the ordinary field-effect transistor, is called the *depletion mode*. This is because the channel is normally conduc-

tive, and a charge on the gate cuts it off. Some metal-oxide field-effect transistors, or MOSFETs, operate in the enhancement mode, where there is no channel under conditions of zero gate voltage. In such devices, the gate voltage produces a channel by the same effect as the unwanted channel formation in a bipolar transistor. In the enhancement-mode MOSFET, however, the formation of a channel is desired, and is the basis for its operation. *See also* DEPLETION MODE, ENHANCEMENT MODE, and METAL-OXIDE SEMICONDUCTOR FIELD-EFFECT TRANSISTOR.

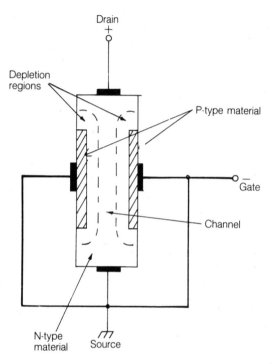

CHANNEL EFFECT: This is the basis for operation of an FET.

CHANNEL SEPARATION

In a channelized band, the frequency difference between adjacent channels is called the *channel separation*. The channel separation must always be at least as great as the channel bandwidth for the signals used; otherwise, interference will result. Occasionally, the channel separation is greater than the signal bandwidth, allowing a small margin of safety against adjacent-channel interference. This is the case in the frequency-modulation amateur band at 144 MHz. Sometimes the channel separation is insufficient to prevent interference between adjacent stations. This is a problem on some of the shortwave broadcast bands, where signals can often be found at a separation of 5 kHz, although their bandwidth is 10 kHz.

CHARACTERISTIC CURVE

A function or relation that defines the interdependence of two quantities is called a *characteristic curve*. In electronics, characteristic curves are generally mentioned in reference to semiconductor or vacuum-tube triode, tetrode, or pentode devices.

A common example of a characteristic curve is illustrated. This curve defines the relation between the gate voltage (E_G) and the drain current (I_D) for an N-channel field-effect transistor, given a drain voltage of 3 volts.

Characteristic curves are used to find the best operating bias for an oscillator or amplifier. Various kinds of amplifiers operate at different points on the characteristic curve for a particular device. A linear amplifier should be biased at a point where the characteristic curve is nearly straight. A class-C amplifier is biased beyond the cutoff or pinchoff point. *See also* FIELD-EFFECT TRANSISTOR, TRANSISTOR, and TUBE.

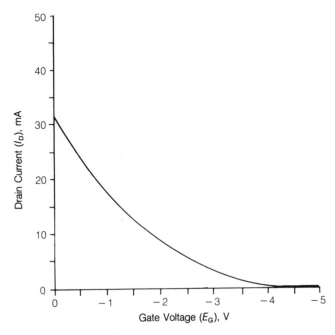

CHARACTERISTIC CURVE: An example of a curve for an FET.

CHARACTERISTIC DISTORTION

Characteristic distortion is a fluctuation in the characteristic curve of a semiconductor device or vacuum tube. The characteristic curve is affected by any change of bias. A change of bias can be caused by the signal itself. Normally, the operation of an amplifier is fairly predictable from the direct-current bias and the characteristic curve of the device. The addition of an alternating-current signal, especially if it is high in voltage, can alter the bias. To get an accurate prediction of the operation of an amplifier in which there is high drive voltage, characteristic distortion must be taken into account. *See also* CHARACTERISTIC CURVE.

CHARACTERISTIC IMPEDANCE

The ratio of the signal voltage (E) to the signal current (I) in a transmission line depends on several things. Ideally, there should be no variations in current and voltage at different places along the line. When the load consists of a noninductive resistor of the correct ohmic value, the voltage-to-current (E/I) ratio Z_O will be constant all along the line so that:

$$Z_O = E/I$$

The value Z_O under these circumstances is called the *characteristic*, or *surge*, *impedance* of the transmission line.

The characteristic impedance of a given line is a function of the diameter and spacing of the conductors used, and is also a function of the type of dielectric material. For air-dielectric coaxial line, in which the inside diameter of the outer conductor

is D, and the outside diameter of the inner conductor is d (see A in the illustration), the characteristic impedance is:

$$Z_O = 138 \log_{10} (D/d)$$

For air-spaced two-wire line, in which the conductor diameter is d and the center-to-center conductor spacing is s (B in the illustration) the formula is:

$$Z_O = 276 \log_{10} (2s/d)$$

The units for D, d, and s must, of course, be uniform in calculation, but any units can be used.

Generally, transmission lines do not have air dielectrics. Solid dielectrics, such as polyethylene, are often used, and this lowers the characteristic impedance for a given conductor size and spacing. For optimum operation of a transmission line, the load impedance should consist of a pure resistance R at the operating frequency, such that $R = Z_O$. *See also* TRANSMISSION LINE.

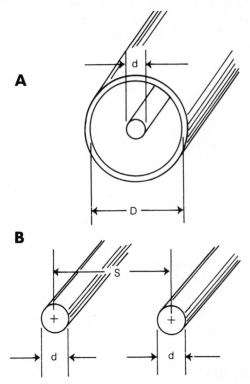

CHARACTERISTIC IMPEDANCE: For coaxial cable (A) and two-wire line (B). See text for formulas.

CHARGE

Charge is an electrostatic quantity, measured as a surplus or deficiency of electrons on a given object. When there is an excess of electrons, the charge is called *negative*. When there is a shortage of electrons, the charge is called *positive*. These choices are purely arbitrary, and do not represent any special qualities of electrons.

Charge is measured in units called *coulombs* (see COULOMB). A coulomb is the charge contained in 6.281×10^{18} electrons. Charge is usually represented by the letter Q in equations. The smallest possible unit of electrostatic charge is the amount of charge contained in one electron.

The quantity of charge per unit length, area, or volume is called the *charge density* on a conductor, surface, or object. Charge density is measured in coulombs per meter, coulombs per square meter, or coulombs per cubic meter. A charge can be carried or retained by an electron, proton, positron, anti-proton, atomic nucleus, or ion. *See also* ELECTRON.

CHARGING

Charging is the process by which an electrostatic charge accumulates. Charging can occur with capacitors, inductors, or storage batteries of various types. In the capacitor, charge is stored in the form of an electric field. In an inductor, charge is stored in the form of a magnetic field. In a battery or cell, charge is stored in chemical form.

The rate of charging is measured in coulombs per second, which represents a certain current in amperes. This charging current can be measured with an ammeter. In a storage battery, charging occurs rapidly at first, and then more and more slowly as the storage capacity is reached. The same is true of capacitors and inductors. The charging current of a battery must be maintained at the proper levels for several hours, generally, in order to ensure optimum charging. *See also* NICKEL-CADMIUM BATTERY, and STORAGE BATTERY.

CHATTER

When the voltage to a relay fluctuates, or is not quite sufficient to keep the contacts closed continuously, the contacts can open and close intermittently. This can occur at a high rate of speed, producing a characteristic sound that is called *chatter*.

Chatter can be deliberately produced by connecting the relay contacts in series with the coil power supply in such a way that oscillation is produced. As soon as the coil receives voltage, the contacts open and the voltage is interrupted. A spring then returns the contacts to the closed condition, and the cycle repeats itself. A resistance-capacitance combination determines the frequency of the oscillation. By varying the oscillation weight and speed, a simple buzzer can be constructed for use with code transmitters (*see* KEYER).

Generally, chatter is an undesirable effect, and can totally disable a piece of electronic equipment. Solid-state switching is preferable to relay switching in many modern applications for this reason. Transistors and diodes do not chatter. *See also* RELAY.

CHEBYSHEV FILTER

(also spelled Tschebyscheff, and Tschebyshev) A Chebyshev filter is a special type of selective filter, having a nearly flat response within its passband, nearly complete attenuation outside the

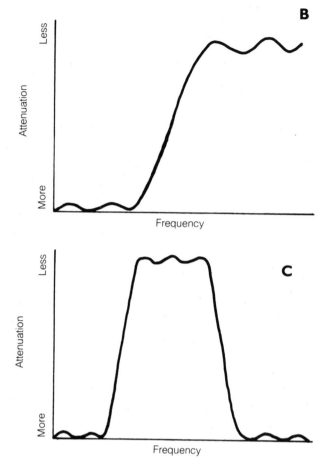

Fig. 1.

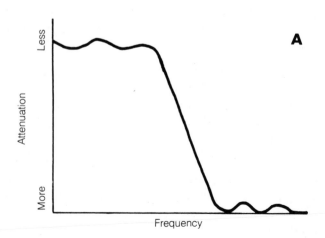

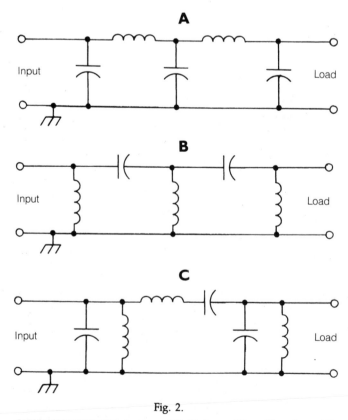

Fig. 2.

CHEBYSHEV FILTER: Lowpass (A), highpass (B), and bandpass (C).

passband, and a sharp cutoff response. The extremely steep skirt selectivity of the Chebyshev filter is the primary advantage of it over other types of filters. The Chebyshev filter design is similar to that of the Butterworth filter (*see* BUTTERWORTH FILTER). Such filters can be designed for lowpass, highpass, and band-rejection applications, as well as for bandpass use.

Figure 1 shows, approximately, the Chebyshev responses for lowpass, highpass, and bandpass filters. Notice the ripple; this ripple is usually of little or no consequence. Figure 2 shows schematic diagrams for simple Chebyshev lowpass, highpass, and bandpass filters. The input and load impedances must be properly chosen for the best filter response. The values of the inductors and capacitors depend on the input and load impedances, as well as the frequency response desired. *See also* BANDPASS FILTER, BAND-REJECTION FILTER, HIGHPASS FILTER, and LOWPASS FILTER.

CHIP

A *chip* is a piece of semiconductor material on which an integrated circuit is fabricated. The word *chip* is often used in place of the term integrated circuit. Chips are sliced from a section of semiconductor material called a *wafer*. Various processes are used in etching individual resistors, capacitors, inductors, diodes, and transistors onto the chip surface. A complete integrated circuit is supplied with external terminals, called *pins*, and is encased in a hard package. *See also* INTEGRATED CIRCUIT.

CHIREIX ANTENNA

A *Chireix antenna*, sometimes also called a *Chireix-Mesny antenna*, is a phased array consisting of two or more square loops connected in series. The loops are coplanar, and have sides measuring ¼ wavelength at the resonant frequency. The radiation pattern is bidirectional, unless a reflecting screen or parasitic element is used.

The gain and radiation pattern of a Chireix antenna, operating at its fundamental frequency, depend on the number of elements. Chireix antennas are used at very-high frequencies and above, where they can be constructed from stiff wire or rigid tubing. *See also* PHASED ARRAY.

CHOKE

An inductor, used for the purpose of passing direct current while blocking an alternating-current signal, is called a *choke*. Typically, an inductor shows no reactance at dc, and increasing reactance at progressively higher ac frequencies. Some chokes are designed for the purpose of cutting off only radio frequencies. Some are designed to cut off audio as well as radio frequencies. Some chokes are designed to cut off essentially all alternating currents—even the 60-Hz utility current, and are used as power-supply filtering components. Together with large capacitances, such chokes make extremely effective power-supply filters.

Chokes can have air cores if they are intended for radio frequencies. Larger chokes use cores of powdered iron or ferrite. This increases the inductance of the winding for a given number of turns. Some chokes are wound on toroidal, or doughnut-shaped, cores. This provides a large increase in the inductance. Some chokes are wound on cores shaped like the core of an ac

power transformer. These chokes can handle large amounts of current.

Chokes are useful when it is necessary to maintain a certain dc bias without short-circuiting the desired signal. *See also* INDUCTOR.

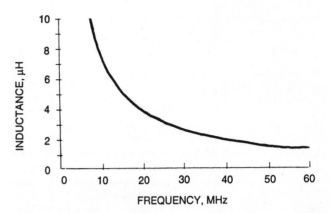

CHOKE: Inductance for 500-ohm choke as function of frequency.

CHRONOGRAPH

A *chronograph* is a plot of any quantity as a function of time. A machine that records a quantity as a function of time is called a *chronograph* or chronograph recorder.

Chronographs are frequently used in electronics. The familiar sine-wave display is an example of a chronograph. Plots of temperature-versus-time, amplitude-versus-time, and other functions are frequently used.

A chronograph is characterized by an accurate, and usually linear, time display along the horizontal axis. This allows precise determination of the period of a function.

CHU

CHU is a time-and-frequency broadcasting station located in Canada. Its primary broadcasting frequency is 7.335 MHz. Time is broadcast each minute in English and French. Eastern Standard Time (EST) is announced; this is five hours behind Coordinated Universal Time (UTC). Brief tones are sent each second. The signal is amplitude-modulated.

In the United States, time and frequency stations are WWV and WWVH, maintained by the National Bureau of Standards. *See also* WWV/WWVH.

CIRCUIT ANALYZER

A device for measuring various electrical quantities, such as current, voltage, and resistance, is called a *circuit analyzer*. Such an instrument is also sometimes called a *multimeter*. Circuit analyzers can have either digital or analog displays.

An analog circuit analyzer generally consists of a sensitive ammeter with a full-scale reading of 30 to 50 microamperes. It has various shunt networks for measuring larger values of current. A battery produces the current needed for measuring resistance, and series resistors allow the measurement of voltage. A range-selector switch is provided for current, voltage, and resistance measurements.

A digital circuit analyzer gives an accurate, numerical reading without the need for visual interpolation, although a range selector is usually supplied for current, voltage, and resistance measurements.

Sometimes an analog device is preferable to a digital device; for example, a maximum or minimum quantity can be required with the adjustment of a certain control. Such an adjustment is more easily done with an analog device than with a digital device. If extreme accuracy is needed, a digital readout is better.

A special kind of circuit analyzer called an *FET voltmeter* is used in high-impedance circuits. This device allows measurement of circuit parameters with almost no effect in the operation. *See also* FET VOLTMETER.

CIRCUIT BREAKER

A *circuit breaker* is a current-sensitive switch. It is placed in series with the power-supply line to a circuit. If the current in the line reaches a certain value, the breaker opens and removes power from the circuit. Breakers are easily reset when they open; this makes them more convenient than fuses, which must be physically removed and replaced when they blow. Circuit breakers are used to protect electronic circuits against damage when a malfunction causes them to draw excessive current. Circuit breakers are also used in utility wiring to minimize the danger of fire in the event of a short circuit.

There are many kinds of circuit breakers, for various voltages and limiting currents. Some breakers will open with just a few milliamperes of current, while others require hundreds of amperes. Some breakers open almost immediately if their limiting currents are reached or exceeded. Some breakers have a built-in delay. *See also* FUSE, and SLOW-BLOW FUSE.

CIRCUIT CAPACITY

In a communications system, the number of channels that can be accommodated simultaneously without overload or mutual interference is called the *circuit capacity*. For example, in a 100-kHz-wide radio-frequency band, using amplitude-modulated signals that occupy 10 kHz of spectrum space apiece, the circuit capacity is 10 channels. It is impossible to increase the number of channels without sacrificing efficiency.

In an electrical system, the circuit capacity is usually specified as the number of amperes that the power supply can deliver without overloading. A typical household branch circuit has a capacity of approximately 10 to 30 amperes. A fuse or circuit breaker prevents overload. *See also* CIRCUIT BREAKER, and FUSE.

CIRCUIT DIAGRAM

See SCHEMATIC DIAGRAM.

CIRCUIT EFFICIENCY

Circuit efficiency is the proportion of the power in a circuit that does the job intended for that circuit. *Efficiency* is usually expressed as a percentage. In an amplifier or oscillator, the efficiency is the ratio of the output power to the input power. For example, if an amplifier has a power input of 100 watts and a power output of 50 watts, its efficiency is 50/100, or 50 percent.

The circuit efficiency of an amplifier depends on the class of operation. Some Class-A amplifiers have a circuit efficiency as low as 20 to 30 percent. Class-B amplifiers usually have an efficiency of about 50 to 60 percent. Class-C amplifiers often have an efficiency rating of more than 80 percent.

An inefficient circuit generates more heat, for the same amount of input power, compared to an efficient one. This heat can, in some instances, destroy the amplifying or oscillating transistor or tube. Regardless of the class of operation, measures should always be taken to maximize the circuit efficiency. *See also* CLASS-A AMPLIFIER, CLASS-AB AMPLIFIER, CLASS-B AMPLIFIER, and CLASS-C AMPLIFIER.

CIRCUIT NOISE

Whenever electrons move in a conductor or semiconductor, some electrical noise is generated. The random movement of molecules in any substance also creates noise. A circuit therefore always generates some noise in addition to the signal it produces or transfers. Noise generated within a piece of electronic equipment is called *circuit noise*.

In a telephone system, circuit noise is the noise at the input of the receiver. This noise comes from the system, and does not include any acoustical noise generated at the transmitter. Such noise is generated within the electrical circuits of the transmitter, and along the transmission lines and switching networks.

Circuit noise limits the sensitivity of any communications system. The less circuit noise produced, the better the signal-to-noise ratio (*see* SIGNAL-TO-NOISE RATIO). Therefore, all possible measures should be used to minimize circuit noise when long-distance communication is desired.

CIRCUIT PROTECTION

Circuit protection is a means of preventing a circuit from drawing excessive current. This is done by means of either a circuit breaker, or by means of a fuse connected in series with the power-supply line (*see* CIRCUIT BREAKER, and FUSE).

The current level at which the circuit breaker or fuse opens is usually about twice the normal operating current drain of the circuit. Circuit protection reduces the chances of extensive component damage in the event of a malfunction. It also reduces the possibility of fire that is caused by overheating of circuit components or wiring.

CIRCULAR ANTENNA

A *circular antenna* is a half-wave dipole bent into a circle. The ends are brought nearly together, but are not physically connected. An insulating brace is often used to add rigidity to the structure. The circular antenna, when oriented in the horizontal plane, produces a nearly omnidirectional radiation pattern in all azimuth directions. Sometimes the circular antenna is called a *halo*.

Circular antennas, constructed from metal tubing, are sometimes used in mobile installations at frequencies above approximately 50 MHz. The horizontal polarization of the halo results in less fading or "picket fencing" than does vertical polarization. Several circular antennas can be stacked at ½-wavelength intervals to produce omnidirectional gain in the horizontal plane.

Circular antennas can be operated at odd multiples of the fundamental frequency, and a reasonably good impedance match will result when 50- or 75-ohm feed lines are used.

CIRCULAR POLARIZATION

The polarization of an electromagnetic wave is the orientation of its electric-field lines of flux. Polarization can be horizontal, vertical, or at a slant (*see* HORIZONTAL POLARIZATION, and VERTICAL POLARIZATION). The polarization can also be rotating, either clockwise or counterclockwise. Uniformly rotating polarization is called circular polarization. The orientation of the electric-field lines of flux completes one rotation for every cycle of the wave, with constant angular speed.

Antennas for circular polarization are not, of course, themselves turned to produce the rotating electromagnetic field; the rotation is easily accomplished by electrical means. The illustration shows a typical antenna for generating waves with circular polarization. The antennas are fed 90 degrees out of phase by making feed-line stub X a quarter wavelength longer than stub Y. The signals from the two antennas thus add vectorially to create a rotating field. The direction, or sense, of the rotation can be reversed by adding ½ wavelength to either stub X or stub Y (but not both).

Circular polarization is compatible, with a 3-dB power loss, with either horizontal or vertical polarization, or with slanted linear polarization. When communicating with another station also using circular polarization, the senses must be in agreement. If a circularly polarized signal arrives with opposite sense from that of the receiving antenna, the attenuation is about 30 dB, compared with matched rotational sense.

In uniform circular polarization, the vertical and horizontal signal components have equal magnitude. But this is not always the case. A more general form of rotating polarization, in which the components can have different magnitude, is called *elliptical polarization. See also* ELLIPTICAL POLARIZATION.

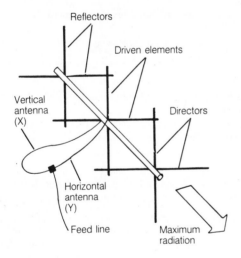

CIRCULAR POLARIZATION: An antenna for transmitting and receiving circularly polarized signals.

CIRCULATING TANK CURRENT

In a resonant circuit consisting of an inductor and capacitor, circulating currents flow back and forth between the two components. As the inductor discharges, the capacitor charges; as the capacitor discharges, the inductor charges.

In class-B or class-C amplifier, only part of the signal cycle is passed through the amplifying device. However, a resonant inductance-capacitance circuit, called a *tank circuit*, completes the cycle in the output by storing and releasing the energy at the resonant frequency. This storage of energy is where the tank circuit gets its name. *See also* CLASS-B AMPLIFIER, CLASS-C AMPLIFIER, and TANK CIRCUIT.

CITIZEN'S BAND

The *Citizen's Radio Service* (*Citizen's Band*), is a public, noncommercial radio service that is available to the general population of the United States. There are four classes of Citizen's Band radio, called *Class A, Class B, Class C,* and *Class D*. Class-A stations are licensed in the band 460 to 470 MHz, and are allowed 60 watts maximum input power. Class-B stations are allowed 5 watts of input power in the same band. Class-C stations are operated in the band 26.96 to 27.23 MHz, and on 27.255 MHz, as well as in the range 72 to 76 MHz, for radio-control purposes. Class-D stations are allowed to use 40 channels in the range 26.965 to 27.405 MHz for general communications. Class-D is by far the most popular of the Citizen's Radio classes.

Citizen's Band or CB, offers a convenient communications system that can be used at home, in a car, in a boat, on an airplane, or even while walking or hiking. The transmitter output power is legally limited on the Class-D band to 4 watts for amplitude modulation and 12 watts peak for single sideband. The communicating range is typically between 10 and 30 miles maximum. However, long-distance propagation is occasionally observed, since the 27-MHz band is susceptible to the effects of sunspot activity.

Citizen's Band enjoyed a great boom in the middle 1970s, when millions of Americans learned how easily they could obtain and operate simple two-way radio equipment. Hundreds of CB clubs exist on the local and national scale. An organization called *Radio Emergency Associated Citizen's Teams* (*REACT*), monitors Channel 9, the officially designated emergency frequency, and provides assistance to motorists in trouble.

CLAMPING

When current passes through a diode in the forward direction, there is always a voltage drop across the diode. For silicon diodes, it is 0.6 to 0.7 V; for germanium diodes it is about 0.3 V. Once there is sufficient current to cause forward conduction, this voltage drop always exists, and its constancy can be put to use in a *clamping circuit*.

The simplest clamping circuit consists of one diode placed across a component. This can regulate bias for an amplifying device such as an FET, and also prevents excessive instantaneous input voltage.

The illustration shows two diodes connected in parallel, with reversed polarity. This serves to limit the signal voltage to 1.4 V peak-to-peak if silicon diodes are used, or 0.6 V peak-to-peak if germanium diodes are used. Such a scheme can be used in the audio stages of a receiver to prevent the volume from exceeding a certain level. But the diodes introduce nonlinearity when clipping occurs, and this will cause signal distortion.

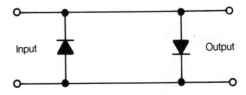

CLAMPING: Back-to-back diodes limit signal amplitude.

CLAPP OSCILLATOR

A specialized form of Colpitts oscillator, using a series-tuned tank circuit, is called a *Clapp oscillator* (*see* COLPITTS OSCIL-

LATOR). The tank circuit is tapped using a capacitive voltage-divider network, but the voltage-divider capacitors are fixed. This is shown in the drawing. The frequency of the oscillator is adjusted either by varying the tank capacitance C, or by varying the tank inductance L. The output can be taken from a secondary winding coupled to L, or by means of capacitive coupling, as shown.

A Clapp oscillator is generally more stable than a Colpitts oscillator because the variable capacitor is independent of the voltage-dividing network.

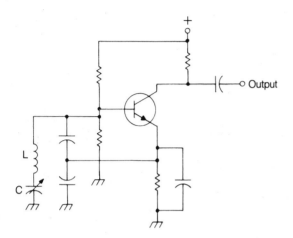

CLAPP OSCILLATOR: Also called *series-tuned Colpitts*.

CLASS-A AMPLIFIER

An amplifier is called *class-A* if the output current flows for the entire input cycle, and if the output waveform is a faithful (linear) reproduction of the input waveform, and if the circuit is single-ended using a certain biasing arrangement.

In a bipolar transistor circuit, class-A operation is accomplished by biasing the base so that the no-signal collector current point is in the middle of the straight-line (linear) part of the characteristic curve. In a field-effect-transistor (FET) circuit, the gate is biased so that the drain current point under no-signal conditions falls in the middle of the linear part of the characteristic curve. This is shown in the graph as solid dots on the curves for bipolar and field-effect transistors.

The advantages of class-A operation are low distortion and good linearity. A class-A amplifier also draws practically no power from the driving source. The disadvantages include relative inefficiency—often less than 30 percent—and a continuous current drain.

Class-A amplifiers are common in audio applications that require good fidelity. They are almost universally used in low-level radio receiving applications. In power amplifiers at radio frequencies, class-A operation is almost never used. *See also* CLASS-AB AMPLIFIER, CLASS-B AMPLIFIER, and CLASS-C AMPLIFIER.

The schematic shows broadband radio-frequency (RF) class-A amplifiers using a bipolar transistor and an FET. Component values are approximate, and will vary depending on the type of device. These amplifiers are low-level amplifiers, such as are found in receiver RF and intermediate-frequency (IF) stages. Tuned circuits are often added at the input and/or output to improve the selectivity, reject unwanted signals, and enhance gain.

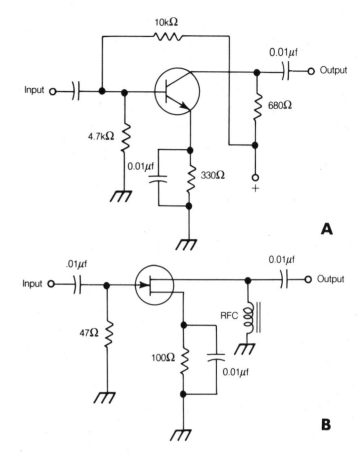

CLASS-A AMPLIFIER: At A, a typical bipolar circuit. At B, a typical FET circuit. Component values are approximate.

CLASS-AB AMPLIFIER

An amplifier is called *class-AB* if the output current flows for more than half, but less than all, of the input cycle. The output waveform is generally not a faithful (linear) reproduction of the input waveform.

In a bipolar transistor circuit, class-AB operation is accomplished by biasing the base so that the collector current point is in the part of the characteristic curve where current flows under no-signal conditions, but it is more nearly at cutoff than is the case with a class-A amplifier. In a field-effect-transistor (FET) circuit, the gate is biased so that the drain current point under no-signal conditions falls closer to the cutoff point than with class-A operation. This is shown in the graph as solid dots with circles around them, on the curves for bipolar and field-effect transistors.

The advantage of class-AB operation over class-A is better efficiency. Sometimes class-AB amplifiers are categorized as class-AB1 and class-AB2. In class-AB1 operation, no power is drawn from the driving source during any part of the cycle; that is, the input signal never drives the base or gate beyond cutoff at any time. In class-AB2 operation, a little power is drawn from the driving source, because the input signal drives the device past cutoff during a small part of the input cycle. Sometimes a class-AB1 amplifier becomes a class-AB2 amplifier because of an increase in drive. Sometimes it happens because of a change in the bias so that the device is closer to cutoff under conditions of no signal.

The disadvantages of class-AB operation include marginal efficiency—around 35 to 45 percent—and a continuous or nearly continuous current drain.

Class-AB amplifiers are often used in power amplifiers at radio frequencies. Sometimes they are used in push-pull. *See also* CLASS-A AMPLIFIER, CLASS-B AMPLIFIER, CLASS-C AMPLIFIER, PUSH-PULL AMPLIFIER, and PUSH-PULL CONFIGURATION.

The schematics for class-A amplifiers also apply for class-AB amplifiers, but the component values are somewhat different, biasing the base or gate more nearly at cutoff. Tuned circuits are often added at the input and/or output to improve the selectivity, reject unwanted signals, and enhance gain.

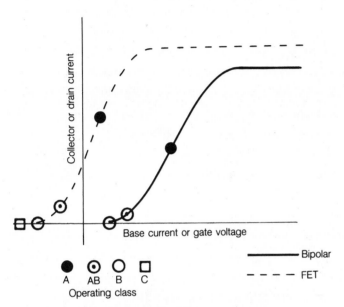

CLASS-A, AB, B, AND C AMPLIFIERS: Characteristic curves for bipolar transistor and FET, showing operating points for amplifier classes A, AB, B, and C.

CLASS-B AMPLIFIER

An amplifier is called *class-B* if the output current flows for exactly half of the input cycle. The output waveform is considerably distorted when this happens, resembling the output of a half-wave rectifier. This can be overcome by the use of high-Q tuned circuits in the output because of the flywheel effect, in which the missing half of the wave is largely replaced. The modulation envelope is generally not distorted in a properly operating class-B RF amplifier. For this reason, class-B RF amplifiers make good linear amplifiers. *See* LINEAR AMPLIFIER.

In a bipolar or field-effect transistor circuit, class-B operation is accomplished by biasing the base or gate at cutoff under conditions of no signal input. In a vacuum-tube circuit, the grid bias is such that cutoff occurs when there is no signal input. This is shown in the graph as open circles on the characteristic curves of the bipolar transistor and FET.

The advantage of class-B amplification over class-A or class-AB is improved efficiency, while still allowing for linearity. Some harmonics are generated in the output, but these can be dealt with by means of tuned circuits in RF power amplifiers. Efficiency is on the order of 50 to 65 percent.

The class-B amplifier needs some driving power from the source. This might be a few watts to obtain 1 kW output with grounded-cathode vacuum tubes. In the case of a grounded-grid configuration, about 100 watts input is needed for 1 kW output.

The disadvantage of the class-B amplifier is that it cannot be used as a single-ended audio circuit. Severe distortion will result. However, class-B push-pull audio amplifiers work quite well. Push-pull is sometimes used at RF also. *See* PUSH-PULL AMPLIFIER, and PUSH-PULL CONFIGURATION.

Class-B circuits are sometimes connected in a sort of full-wave rectifier arrangement, using two devices to double the frequency. This kind of device is called a *push-push circuit*. *See* PUSH-PUSH CONFIGURATION.

The schematics show class-B amplifiers using a bipolar transistor, an FET and a tetrode vacuum tube. These are all tuned-output RF power amplifiers. *See also* CLASS-A AMPLIFIER, CLASS-AB AMPLIFIER, CLASS-C AMPLIFIER, and POWER AMPLIFIER.

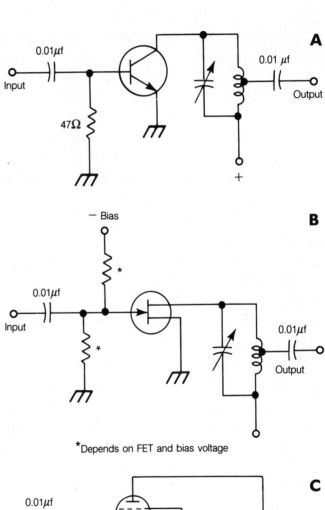

CLASS-B AMPLIFIER: At A, a typical bipolar circuit. At B, a FET circuit. At C, a tetrode vacuum-tube circuit. These are RF power amplifiers.

CLASS-C AMPLIFIER

An amplifier is called *class-C* if the output current flows for less than half of the input cycle. When this occurs, considerable distortion is introduced into the signal waveform. But, as with class-B amplifiers, the flywheel effect of high-*Q* tuned output circuits largely eliminates this.

The modulation envelope will be distorted in class-C amplification if an AM or SSB signal is introduced at the input. Class-C amplifiers are, therefore, normally used only for signals in which the modulation involves *at most* two different levels of amplitude, such as CW, FSK and FM. Class-C circuits do not make good linear amplifiers.

In a bipolar transistor circuit, class-C operation is obtained by reverse-biasing the emitter-base junction under no-signal conditions. In an FET or vacuum tube, the gate or control grid is biased well past cutoff under no-signal conditions. This is shown in the graph for bipolar and field-effect transistors as open squares on the characteristic curves.

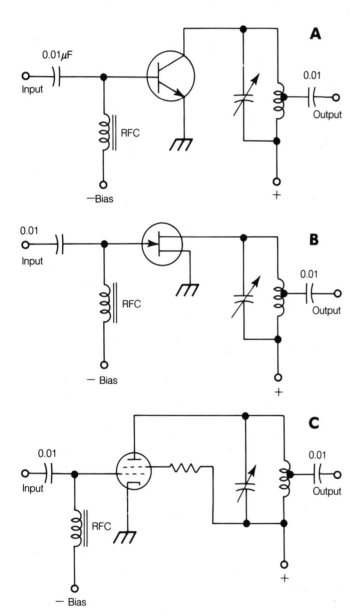

CLASS-C AMPLIFIER: At A, a typical bipolar circuit. At B, a FET circuit. At C, a tetrode vacuum-tube circuit. These are RF power amplifiers.

The main advantage of class-C amplification is high efficiency. In a well-designed, class-C RF power amplifier, the output power might be as much as 75 percent of the input power. Harmonics are generated at the output, but these can be suppressed by means of high-*Q* tuned output circuits. A properly designed transmatch between the transmitter and antenna can also help reduce harmonic emissions.

Class-C amplifiers need considerable driving power. The schematics show class-C amplifiers using a bipolar transistor, an FET and a tetrode vacuum tube. These are all tuned-output RF power amplifiers. *See also* CLASS-A AMPLIFIER, CLASS-AB AMPLIFIER, and CLASS-B AMPLIFIER.

CLEAR

The term *clear* refers to the resetting or reinitialization of a circuit. All active memory contents of a microcomputer are erased by the clear operation. Auxiliary memory is retained when active circuits are cleared.

All electronic calculators have a clear function button. When this button is actuated, the calculation is discontinued and the display reverts to zero. By switching a calculator off and then back on, the clear function is done automatically.

CLICK FILTER

When a switch, relay, or key is opened and closed, a brief pulse of radio-frequency energy is emitted. This is especially true when the device carries a large amount of current. A capacitor connected across the device slows down the decay time from the closed to the open condition, where the click is most likely to occur. These devices are called click filters. Sometimes a choke or resistor is connected in series also.

In a code transmitter, a click filter is used to regulate the rise and decay times of the signal. Without such a filter, the rapid rise and decay of a signal can cause wideband pulses to be radiated at frequencies well above and below that of the carrier itself. This can result in serious interference to other stations. *See also* KEY CLICK, and SHAPING.

CLOCK

A *clock* is a pulse generator that serves as a time-synchronizing standard for digital circuits. The clock sets the speed of operation of a microprocessor, microcomputer, or computer.

The clock produces a stream of electrical pulses with extreme regularity. Some clocks are synchronized with time standards. The speed can also be controlled by a resistance-capacitance network or by a piezoelectric crystal. The clock frequency is generally specified in pulses per second, or hertz.

CLOSED CIRCUIT

Any complete circuit that allows the flow of current is called a *closed circuit*. All operating circuits are closed. A transmission sent over a wire, cable, or fiber-optics medium, and not broadcast for general reception, is called a closed-circuit transmission. A telephone operates via a closed circuit (except, of course, for a radio telephone). Some closed-circuit radio and television systems are used as intercoms or security monitoring devices.

CLOSED LOOP

The gain of an operational amplifier (op amp) depends on the resistance in the feedback circuit. The highest gain occurs when there is no negative feedback.

If a resistor is placed between the output and the inverting input, the gain of the op amp is reduced. This is called the *closed-loop* configuration. The gain depends on the resistance. The smaller the value of the feedback resistor, the greater the amount of negative feedback, and the lower the gain of the op amp. By adjusting the feedback resistance, the gain can be controlled.

Closed-loop op-amp circuits are used more often than *open-loop* circuits, because the closed-loop configuration is more stable. *See also* OPEN LOOP.

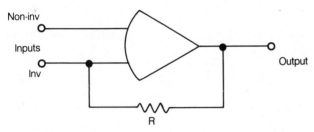

CLOSED LOOP: Negative feedback in an operational amplifier. Resistance R determines gain.

COAXIAL CABLE

Coaxial cable is a two-conductor cable consisting of a single center wire surrounded by a tubular metal shield. Most coaxial cables have a braided shield, insulated from the center conductor by polyethylene. Some coaxial cables have air dielectrics, and the center conductor is insulated from the shield by polyethylene beads or a spiral winding.

Coaxial cable is commercially made in several diameters and characteristic-impedance values. A low-loss, well-shielded type of coaxial cable, with an outer conductor of solid metal tubing, is called *hard line*.

Coaxial cable is convenient to install, and it can be run next to metal objects, and even underground, without adversely affecting the loss performance. However, most coaxial cables have greater loss per unit length than two-conductor lines or waveguides. The characteristic impedance of coaxial lines is generally lower than that of two-wire lines. *See also* TRANSMISSION LINE, and WAVEGUIDE.

COAXIAL SWITCH

A *coaxial switch* is a multi-position switch designed for use with coaxial cable. Coaxial switches must have adequate shielding. This necessitates that the enclosure be made of metal, such as aluminum. At very high frequencies and above, a coaxial switch must be designed to have a characteristic impedance that is identical to that of the transmission line in use; otherwise, impedance discontinuities can contribute to loss in the antenna system.

Some coaxial switches can be operated by remote control. This is especially convenient when there are several different antennas on a single tower, and no two of them have to be used at the same time. A single length of cable can then be used as the main feed line, and each antenna can be connected to a separate branch via the switch.

COAXIAL TANK CIRCUIT

A coaxial cable, cut to any multiple of ¼ electrical wavelength, can be used in place of an inductance-capacitance tuned circuit. If the length of the cable is an even multiple of ¼ wavelength, a low impedance is obtained by short-circuiting the far end, or a high impedance is obtained by opening the far end. If the length of the cable is an odd multiple of ½ wavelength, a high impedance is obtained by short-circuiting the far end and a low impedance is obtained by opening the far end.

Coaxial tank circuits are used mostly at very-high and ultra-high frequencies, where ¼ or ½ wavelength is a short physical length. Coaxial tank circuits have excellent selectivity. Cavity resonators are also used as tuned circuits at very-high and ultra-high frequencies. *See also* CAVITY RESONATOR, and COAXIAL WAVEMETER.

COAXIAL WAVEMETER

For measuring very-high, ultra-high, and microwave frequencies, a *coaxial wavemeter* is sometimes used. This device consists of a rigid metal cylinder with an inner conductor along its central axis, and a sliding disk that shorts the cylinder and the inner conductor. The coaxial wavemeter is thus a variable-frequency coaxial tank circuit.

By adjusting the position of the shorting disk, resonance can be obtained. Resonance is indicated by a dip or peak in an RF voltmeter or ammeter. The length of the resonant section is easily measured; this allows determination of the wavelength of the applied signal. The frequency is determined from the wavelength according to the formula

$$F = 300k/\lambda$$

where f is the frequency in megahertz and λ is the wavelength in meters; k is the velocity factor of the cable tank circuit, typically 0.95 for air dielectric. *See also* COAXIAL TANK CIRCUIT, and VELOCITY FACTOR.

CODE

Any alternative representation of characters, words, or sentences in any language is a *code*. Some codes are binary, consisting of discrete bits in either an "on" or "off" state. The most common binary codes in use today for communications purposes are ASCII, BAUDOT, and the International Morse code (*see* ASCII, BAUDOT, and INTERNATIONAL MORSE CODE). The "Q" and "10" signals, which are abbreviations for various statements, are codes (*see* Q-SIGNAL). Words in computer languages are a form of code.

Binary codes allow accurate and rapid transfer of information, since digital states provide a better signal-to-noise ratio than analog forms of modulation. The oldest telecommunications system, a combination of the Morse code and the human ear, is still used today when all other modes fail.

CODE TRANSMITTER

A *code transmitter* is the simplest kind of radio-frequency transmitter. It consists of an oscillator and one or more stages of amplification. One of the amplifiers is keyed to turn the carrier on and off.

Sophisticated code transmitters use mixers for multiband operation. Many amplitude-modulated, frequency-modulated, or single-sideband transmitters can function as code transmitters. An unmodulated carrier is simply keyed through the amplifying stages.

Ideally, the output of a code transmitter is a pure, unmodulated sine wave at the operating frequency. Changes in amplitude under key-down conditions are undesirable. The rise and decay times of the carrier, as the transmitter is keyed, must be regulated to prevent key clicks. The frequency should be stable to prevent chirp. *See also* CHIRP, and SHAPING.

CODING

The process of formulating a code is called *coding*. When preparing a code language, it is necessary to decide whether the smallest code element will represent a character, a word, or a sentence. The ASCII, BAUDOT, and Morse codes (*see* ASCII, BAUDOT, and INTERNATIONAL MORSE CODE) represent each character by a combination of digital pulses. Computer languages use digital words to perform specific functions. Communications codes use a group of characters, such as QRX or 10-4, to represent an entire thought or sentence (*see* Q SIGNAL).

When a language is translated into code, the process is called *encoding*. When a code is deciphered back into ordinary language, the process is called decoding. These functions can be done either manually or by machine.

COEFFICIENT OF COUPLING

Two circuits can interact to a greater or lesser extent. The degree of interaction, or coupling, between two alternating-current circuits is expressed as a quantity called the *coefficient of coupling*, abbreviated in equations by the letter k. Usually, the coefficient of coupling is used in reference to inductors.

The coefficient of coupling, k, is related to the mutual inductance M and the values of two coils (L_1 and L_2) according to the formula:

$$k = M/\sqrt{L_1 L_2}$$

where the inductances are specified in henrys.

For impedances Z_1 and Z_2 in general, where they are of the same kind (predominantly capacitive or inductive):

$$k = M/\sqrt{Z_1 Z_2}$$

where M is the mutual impedance. *See also* MUTUAL INDUCTANCE.

COHERENT CW

There exists a little-explored mode of continuous-wave (cw) or Morse-code communications, in which the receiver and transmitter are synchronized by means of a primary time standard, such as WWV.

Morse code consists of individual bits, each having a length of one dit (or dot). When the code is broken down this way into its fundamental bits, it becomes a true binary code. A given bit is either *on* or *off*. A dit has a length of one bit, and a dah (or dash) three bits. The space between dits and dahs within any letter is one bit. The space between letters is three bits. The space between words, and after every punctuation mark, is seven bits.

If the receiver "knows" the speed at which the transmitter is sending, the string of bits at the receiver can be synchronized precisely with the string of bits at the transmitter, using the primary time standard, and taking propagation delays into account. When this is done, a sensing circuit at the receiver can consider a bit to be *on* if there is signal for 50 to 100 percent of the time, and *off* if there is signal for 0 to 49 percent of the time. These percentages might be adjusted for further improvement in accuracy; this would require experimentation. This synchronized mode is known as *coherent cw*.

Coherent cw makes it possible to greatly reduce the bandwidth needed by a cw signal at a given speed. This, in turn, provides for a substantial improvement in signal-to-noise ratio. Claims have been made that the signal-to-noise enhancement could be as much as 30 dB over conventional cw.

The main problem with coherent cw is that it is difficult to synchronize the receiver with the transmitter. The trouble is compounded when a station calls CQ, because potential QSO stations might be attuned to the wrong speed, or be in a location where propagation delays are greater or less than anticipated.

Nowadays, packet radio has made coherent cw far less appealing because packet offers error-free communications with greater versatility than coherent cw. Nonetheless, coherent cw can still be of interest to the experimentally inclined ham. *See* PACKET RADIO.

COHERENT LIGHT

Coherent light is light that has a single frequency and phase. Most light, even if it appears to be monochromatic, consists of a certain range of wavelengths, and has random phase combinations. White light is made up of nearly equal radiation intensity at all visible frequencies; red light consists primarily of radiation at long visible wavelengths; green light is composed mostly of light in the middle of the visible frequency range.

The light transmitted by a helium-neon laser appears red, just as sunlight does through a red color filter. However, the laser light is emitted at just one wavelength and all the waves coming from the laser are in perfect phase alignment. Thus, the helium-neon laser emits coherent red light, while the red color filter transmits incoherent light.

Coherent light travels with greater efficiency—that is, lower attenuation per kilometer—than incoherent light. Using coherent light, a nearly parallel beam can be produced, and thus the energy is carried for tremendous distances with very little loss. Modulated-light communications systems generally use lasers, which produce coherent light. *See also* LASER, and MODULATED LIGHT.

COHERENT RADIATION

Coherent radiation is an electromagnetic field with a constant, single, frequency and phase. A continuous-wave radio-frequency signal is an example of coherent radiation. The static, or "sferics," produced by a thunderstorm, is an example of incoherent electromagnetic radio emission.

Energy is transferred more efficiently by coherent radiation than by incoherent radiation. The laser is an example of a visible-light device that produces coherent radiation. *See also* COHERENT LIGHT, and LASER.

COIL

A *coil* is a helical winding of wire, usually intended to provide inductive reactance. The most common form of wire coil is the solenoidal winding. The wire can be wound on an air core, or a core having magnetic permeability to increase the inductance for a given number of turns. Some coils are toroidally wound.

Coils are used in speakers, earphones, microphones, relays, and buzzers to set up or respond to a magnetic field. Coils are used in transformers for the purpose of stepping a voltage up or down, or for the purpose of impedance matching. A coil wound on a ferrite rod can act as a receiving antenna at low, medium, and high frequencies. In electronic circuits, coils are generally used to provide inductance. *See also* COIL WINDING, INDUCTANCE, and INDUCTOR.

COIL WINDING

When winding a coil to obtain a certain value of inductance, the dimensions of the coil, the number of turns, the type of core material, and the shape of the coil all play important roles.

Usually, if a powdered-iron or ferrite core material is used for coil winding, data is furnished with the core as a guide to obtaining the desired value of inductance. For air-core solenoidal coils having only one layer of turns, the inductance L in microhenrys is given by the formula:

$$L = \frac{r^2 N^2}{9r + 10m}$$

where r is the coil radius in inches, N is the number of turns, and m is the length of the winding in inches. The inductance of a single-layer air-core solenoid thus increases with the square of the number of turns, and directly with the coil radius. For a coil with a given radius and number of turns, the greatest inductance is obtained when length m is made as small as possible. *See also* INDUCTANCE, and INDUCTOR.

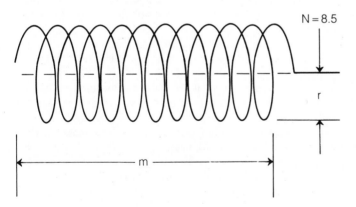

COIL WINDING: See text for discussion.

COINCIDENCE CIRCUIT

A *coincidence circuit* is any digital circuit that requires a certain combination of input pulses in order to generate an output pulse. The input pulses must usually arrive within a designated period of time.

The most common form of coincidence circuit is a combination of AND gates (*see* AND GATE). For an output pulse to occur, all the inputs of an AND gate must be in the high state. Any complex logic circuit can be considered a coincidence circuit, but the term is generally only used with reference to the logical operation AND.

COLLECTOR

The *collector* is the part of a semiconductor bipolar transistor into which carriers flow from the base under normal operating conditions. The base-collector junction is reverse-biased; in a pnp transistor the base is positive, with respect to the collector, and in an npn transistor the base is negative, with respect to the collector.

The output from a transistor oscillator or amplifier is usually taken from the collector. The collector can be placed at ground potential in some situations, but it is usually biased with a direct-current power supply. The amount of power dissipated in the base-collector junction of a transistor must not be allowed to exceed the rated value, or the transistor will be destroyed. Resistors are often used to limit the current through the collector; such resistors are placed in series with either the emitter or collector lead. In some transistors, the collector is bonded to the outer case to facilitate heat conduction away from the base-collector junction.

The collector of a transistor corresponds roughly to the plate of a vacuum tube in circuit engineering applications, although the voltage is much smaller with the transistor than with the tube.

COLLECTOR CURRENT

In a bipolar transistor, the collector current is the average value of the direct current that flows in the collector lead. When there is no signal input, the collector current is a pure, constant direct current, determined by the bias at the base, the series resistance, and the collector voltage. The collector current for proper operation of a transistor varies considerably, depending on the application.

When a signal is applied to the base or emitter circuit of a transistor amplifier, the collector current fluctuates. But its average value, as indicated by an ammeter in the collector circuit, can change only slightly. The collector current is the difference between the emitter current and the base current.

COLLECTOR RESISTANCE

The internal resistance of the base-collector junction of a bipolar transistor is called the *collector resistance*. This resistance can be specified either for direct current or for alternating current.

The direct-current collector resistance, R_{dc}, is given by:

$$R_{dc} = E/I$$

where E is the collector voltage and I is the collector current (see "A" in the illustration). This resistance varies with the supply voltage, the base bias, and any resistances in series with the emitter or collector.

The alternating-current resistance, R_{ac}, is given approximately by:

$$R_{ac} = \Delta E/\Delta I,$$

where ΔE and ΔI are the ranges of maximum-to-minimum collector voltage and current, as the fluctuating output current goes through its cycle. The illustration shows a method of approximately determining this dynamic resistance (B). The value of R_{AC} is affected by the same factors that influence R_{DC}. In addition, the class of operation has an effect, as does the magnitude of the input signal. The alternating-current collector resistance is useful when designing a circuit for optimum impedance matching.

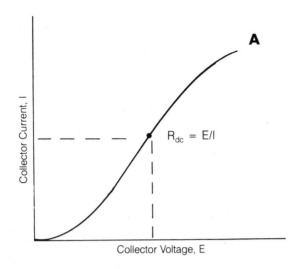

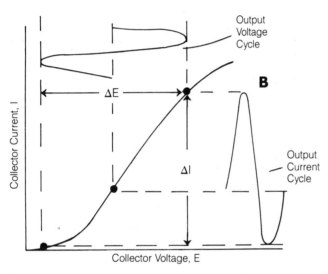

COLLECTOR RESISTANCE: At A, dc collector resistance; at B, ac collector resistance.

COLOR-BAR GENERATOR

A *color-bar generator* is a device used in the testing and adjustment of a color television receiver. It operates in a manner similar to a black-and-white bar generator (*see* BAR GENERATOR).

The bar pattern can be either vertical or horizontal, and in various color combinations. Color reproduction, as well as brightness, contrast, horizontal and vertical linearity, and focus can be adjusted. *See also* COLOR TELEVISION, and TELEVISION.

COLOR CODE

A *color code* is a means of representing component values and characteristics by means of colors. This scheme is used with almost all resistors (*see* RESISTOR COLOR CODE). Color codes are sometimes used on capacitors, inductors, transformers, and transistors.

When a cable has several different conductors, the individual wires are usually color-coded as a means of identification of conductors at opposite ends of the cable. In direct-current power leads, the color black often signifies ground or negative polarity, and red signifies the "hot" or positive lead. In house wiring, color coding can vary.

COLOR PICTURE SIGNAL

A *color picture signal* is a modulated radio-frequency signal that contains all the information needed to accurately reproduce a scene in full color. The channel width of a color-television picture signal is typically 6 MHz.

The horizontal blanking pulse turns off the picture-tube electron beam as it retraces from the end of one line to the beginning of the next line. This pulse is followed by a color-burst signal, which consists of eight or nine cycles at 3.579545 MHz. The phase of this burst provides the color information. The video information is then sent as the electron beam scans from left to right. There are 525 or 625 horizontal lines per frame. *See also* TELEVISION.

COLOR SLOW-SCAN TELEVISION

Slow-scan television (SSTV) signals can be sent and received in color, as well as in black and white. The scan converter generally has three memories instead of the one needed for black and white. One memory is for red, one is for green, and one is for blue.

A common method of sending the color signal requires that three frames be transmitted, using three color filters (red, green, and blue) in front of the black-and-white camera. This amounts to sending three black-and-white pictures, modified by color filters. The images are stored in the three memories of the scan converter. Then, when all three images are complete, they are combined by the scan converter into a color image.

The filter changing can be avoided if a color camera is used. A color monitor is always necessary at the receiving end. *See also* SLOW-SCAN TELEVISION.

COLOR TELEVISION

Amateur television (ATV) signals, like conventional broadcast TV, can be transmitted and received in color as well as in black-and-white. The same kind of receiving set is used; in fact, a standard color TV receiver can be used along with frequency conversion circuits to receive color ATV.

The color picture signal is more complex than the black-and-white signal, simply because an additional "dimension" of information — the color — is conveyed. Color transmission involves a color pulse having precise frequency and phase. The standard color-burst frequency is 3.579545 MHz for TV broadcast. Ironically, this falls within the 80-meter amateur band. You can hear this signal from nearby color TV sets, most of the time in most locations, just by tuning to about 3.580 MHz.

The scan rate and raster in a color TV signal are the same as for a black-and-white signal. In fact, a black-and-white TV receiver will pick up color signals just fine, except that there will be no color observed. Similarly, a color TV receiver will receive a black-and-white transmission.

The color TV camera uses a prism or diffraction grating to separate the incoming light into the three primary colors: red, green and blue. These produce separate signals that are combined by adding and subtracting. The sum signals provide the information for the brightness (the total amount of light, as in a black-and-white picture). The difference signals give the color information. Both the sum and difference signals are transmitted in accordance with strict industry standards. The combinations are such that the individual color signals, as well as the overall brightness signal, can be retrieved at the receiver by a special demodulation process.

The receiving picture tube has three enmeshed sets of phosphor dots, one grid for the red, one for the green, and one for the blue. When observed from a reasonable distance, these dots blend together giving the impression of color. Because all visible colors are combinations of red, green, and blue in various proportions, full and faithful color reproduction is attained. *See also* COLOR PICTURE SIGNAL, COLOR SLOW-SCAN TELEVISION, SLOW-SCAN TELEVISION, and TELEVISION.

COLPITTS OSCILLATOR

A *Colpitts oscillator* is an oscillator, usually of the variable-frequency type, characterized by capacitive feedback and a capacitive voltage-divider network. The illustration shows transistor and field-effect-transistor Colpitts circuits.

The operating frequency of the Colpitts oscillator is determined by the value of the inductance and the series combination of the two capacitors. Generally, the capacitors are variable. Alternatively, the capacitors can be fixed, and the frequency set by means of a variable inductor. The output can be taken from the circuit by inductive coupling, but better stability is usually obtained by capacitive or transformer coupling from the collector or drain circuit.

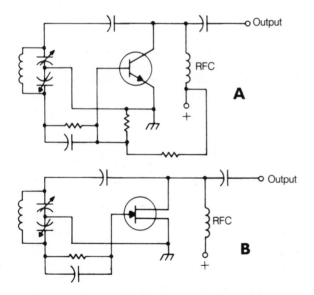

COLPITTS OSCILLATOR: Bipolar (A) and FET (B) circuits.

COMBINATIONAL LOGIC

See BOOLEAN ALGEBRA.

COMMON CATHODE/EMITTER/ SOURCE

The *common-cathode, common-emitter,* and *common-source circuits* are probably the most frequently used amplifier arrangements with tubes, transistors, and field-effect transistors. The cathode, emitter, or source is always operated at ground potential with respect to the signal; it need not necessarily be at ground potential for direct current.

The common-cathode and common-source circuits have high input and output impedances. The common-emitter circuit is characterized by moderately high input impedance and high output impedance. In all three circuits, the input and output signals are 180 degrees out of phase. *See also* FIELD-EFFECT TRANSISTOR, TRANSISTOR, and TUBE.

COMMON GRID/BASE/GATE

The *common-grid, common-base,* and *common-gate circuits* are amplifier or oscillator arrangements using tubes, transistors, and field-effect transistors, respectively. These circuits have excellent stability as amplifiers. They are less likely to break into unwanted oscillation than the common-cathode, common-emitter, and common-source circuits (*see* COMMON CATHODE/EMITTER/SOURCE).

The grid, base, or gate is usually connected directly to ground; occasionally a direct-current bias can be applied and the grid, base, or gate shunted to signal ground with a bypass capacitor. The common-grid, common-base, and common-gate circuits display low input impedance. They require considerable driving power. The output impedance is high. The input and output waveforms are in phase. This kind of amplifier is often used as a power amplifier at radio frequencies. *See also* FIELD-EFFECT TRANSISTOR, TRANSISTOR, and TUBE.

COMMON-MODE HUM

In a direct-conversion radio receiver, the beat-frequency oscillator (BFO) can be modulated by ac from the utility mains. The usual cause of this problem, known as *common-mode hum,* is one or more ground loops in the station arrangement.

Common-mode hum tends to be an increasing annoyance as the operating frequency increases. It is more likely to occur with an end-fed wire that comes right into the shack, as opposed to a center-fed antenna located well away from the station.

Common-mode hum can be reduced or avoided by making sure there are no ground loops at the station, and by using an antenna with a properly balanced feed line, locating the radiating part of the antenna at least a quarter wavelength from the shack. It might also be necessary to install high-value chokes in each lead from the power supply to the radio. *See also* DIRECT-CONVERSION RECEIVER, and GROUND LOOP.

COMMON PLATE/ COLLECTOR/DRAIN

The *common-plate, common-collector,* and *common-drain circuits* are generally used in applications where a high-impedance generator must be matched to a low-impedance load. The input impedances of the common-plate, common-collector, and common-drain circuits are high; the output impedances are low. They are sometimes called cathode-follower, emitter-follower, and source-follower circuits. The gain is always less than unity.

The plate, collector, or drain is sometimes grounded directly. However, this is not always done; biasing can be accomplished in a manner identical to that of the common-cathode, common-emitter, and common-source circuits (*see* COMMON CATHODE/EMITTER/SOURCE); the plate, collector, or drain is then placed at signal ground by means of a bypass capacitor, and the output is taken across a cathode, emitter, or source resistor or transformer. *See also* FIELD-EFFECT TRANSISTOR, TRANSISTOR, and TUBE.

COMMUTATOR

A *commutator* is a mechanical device for obtaining a pulsating direct current from an alternating current. Commutators are used in motors and generators. A high-speed switch that reverses the circuit connections to a transducer, or rapidly exchanges them, is sometimes called a *commutator*.

In the direct-current motor, the commutator acts to reverse the direction of the current every half turn, so that the current in the coils always flows in one direction. As the motor shaft rotates, the commutator, attached to the shaft, connects the power supply to the motor coils.

In a direct-current generator, the commutator inverts every other half cycle of the output to obtain pulsating direct current rather than alternating current. The pulsations can be smoothed out using a capacitor. *See also* DC GENERATOR.

COMPANDOR

A *compandor* is a device that is used for the purpose of improving the efficiency of an analog communications system. The compandor consists of two separate circuits: an amplitude compressor, used at the transmitter, and an amplitude expander, used at the receiver.

The amplitude compressor in a voice transmitter increases the level of the fainter portions of the envelope (see illustration). This increases the average power output of the transmitter, and increases the proportion of signal power that carries the voice information. This kind of compression can be done with amplitude-modulated, frequency-modulated, or single-sideband transmitters.

The amplitude expander follows the detector in the receiver circuit, and returns the voice to its natural dynamic range. Without the amplitude expander, the voice would be understandable, but less intelligible after pauses in speech. Speech expansion is shown at B.

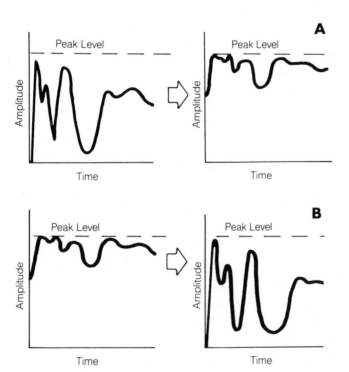

COMPANDOR: At A, amplitude compression at a transmitter; at B, amplitude expansion at a receiver.

Some devices are used to compress the modulating frequencies at the transmitter, and expand them at the receiver. This decreases the bandwidth of the transmitted signal. These devices are sometimes called *frequency compandors*. They are used in an experimental form of transmission called *narrow-band voice modulation. See also* SPEECH COMPRESSION.

COMPARATOR

A comparator is a circuit that evaluates two or more signals, and indicates whether or not the signals are matched in some particular way. Typically, the "yes" output (signals matched) is a high state, and the "no" output (signals different) is a low state. Comparators can test for amplitude, frequency, phase, voltage, current, waveform type, or numerical value. A number comparator has three outputs: greater than, equal to, and less than. A phase or frequency comparator can have an output voltage that varies, depending on which quantity is leading or lagging, or larger or smaller.

In high-fidelity audio testing, it is often desirable to quickly switch back and forth between two systems for the purpose of comparing quality. A device that facilitates convenient switching is called a *comparator.*

COMPENSATION

Compensation is a method of neutralizing some undesirable characteristic of an electronic circuit. For example, a crystal might have a positive temperature coefficient: that is, its frequency might increase with increasing temperature. In an oscillator that must be stable under varying temperature conditions, this characteristic of the crystal can be neutralized by placing, in parallel with the crystal, a capacitor whose value increases with temperature and pulls the crystal frequency lower. If the positive temperature coefficient of the capacitor is just right, the result will be a frequency-stable oscillator.

In an operational-amplifier circuit, the frequency response must sometimes be modified in order to get stable operation. This can be done by means of external components, or it can be done internally. This is called *compensation.*

COMPENSATION THEOREM

Any alternating-current impedance, produced by a combination of resistance, inductance, and capacitance, shows a certain phase relationship between current and voltage at a particular frequency. We usually think of an impedance as being produced by inductors, capacitors, and resistors. However, an equivalent circuit might be a solid-state device, a section of transmission line, or a signal generator.

If the equivalent circuit (black box) is inserted in place of the resistance-inductance-capacitance (RLC) network, the overall operation of the rest of the circuit will remain unchanged. This fact is called the *compensation theorem. See also* BLACK BOX, and IMPEDANCE.

COMPILER

In a digital computer, the *compiler* is the circuit that converts the higher-order language, such as BASIC, COBOL, or FORTRAN, into machine language. The operator understands and uses the higher-order language; the computer operates in machine language. A compiler is thus an electronic translator.

The compiler must itself be programmed to translate a given higher-order language into machine language, and vice versa. This program is called the *assembler*. Assembler programs are written in a language called *assembly language*.

COMPLEMENT

The *complement* is a function of numbers—in particular, binary numbers. In the base-two system, where the only possible digits are 0 and 1, the complement of 0 is 1 and the complement of 1 is 0. In logic, the complement function is the same as negation.

When there are several digits in a binary number, the complement is obtained by simply reversing each digit. For example, the complement of 10101 is 01010. *See also* BINARY-CODED NUMBER.

COMPLEMENTARY CONSTANT-CURRENT LOGIC

Complementary constant-current logic (abbreviated *CCCL* or C^3L, is a form of digital circuitry similar to transistor-transistor logic (TTL). It is a form of large-scale integration that is used in the manufacture of digital integrated circuits. Complementary constant-current logic is characterized by high density.

The switching time in C^3L is very fast—just a few nanoseconds or billionths of a second—allowing more operations per unit time than other logic types. *See also* LARGE-SCALE INTEGRATION.

COMPLEMENTARY METAL-OXIDE SEMICONDUCTOR

Complementary metal-oxide semiconductor, abbreviated *CMOS* and pronounced "seamoss," is the name for a form of digital enhancement-mode switching device. Both N-channel and P-channel field-effect transistors are used. A CMOS integrated circuit is fabricated on a chip of silicon.

The chief advantages of CMOS are its extremely low current consumption and its high speed. The main disadvantage of CMOS is its susceptibility to damage by static electricity. A CMOS integrated circuit should be stored with its pins embedded in a conducting foam material, available especially for this purpose. When building or servicing equipment containing CMOS devices, adequate measures must be taken to prevent static buildup. *See also* ENHANCEMENT MODE, and METAL-OXIDE SEMICONDUCTOR FIELD-EFFECT TRANSISTOR.

COMPLEX NUMBER

A *complex number* is a quantity of the form $a + bi$, where a and b are real numbers, and $i = \sqrt{-1}$. The number a is called the real part of the complex quantity, and the number bi is called the imaginary part. In electronics, the number i is usually called j, and the complex number is written in the form $R \times jX$ (*see* J OPERATOR). Complex numbers are used by electrical engineers to represent impedances: the real part is the resistance and the imaginary part is the reactance. This representation is used because the mathematical properties of complex numbers are well suited to definition of impedance.

Complex numbers are represented on a two-dimensional Cartesian plane. The real part, a, is assigned to the horizontal axis; the imaginary part, bi, is assigned to the vertical axis. The

drawing shows the complex-number plane. Each point on this plane corresponds to one and only one complex number, and each complex number corresponds to one and only one point on the plane.

In electronics applications, only the right-hand side of the complex-number plane is ordinarily used. The top part of the right half of the plane, or first quadrant, represents impedances in which the reactance is inductive. The bottom part of the right half of the plane, or fourth quadrant, represents impedances in which the capacitance is the dominant form of reactance. *See also* IMPEDANCE, and REACTANCE.

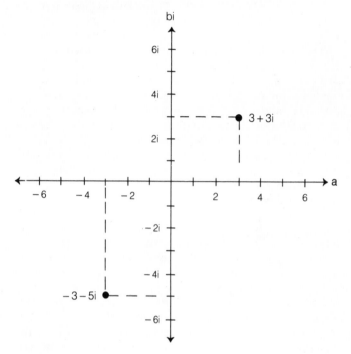

COMPLEX NUMBER: A complex-number plane showing two points.

COMPLEX STEADY-STATE VIBRATION

When several, or many, sine waves are combined, the resulting waveform is called a *complex steady-state vibration*. The sine waves can have any frequency, phase, or amplitude. There can be as few as two sine waves, or infinitely many.

When a sine wave is combined with some of its harmonic frequencies in the proper amplitude relationship, special waveforms result. The sawtooth and square waves are examples of infinite combinations of sine waves with their harmonics.

An audio-frequency, complex, steady-state vibration, when applied to a speaker or headset, produces a sound called a *complex tone*. Essentially all musical instruments produce complex tones. A pure sine-wave tone is unpleasant to the ear. *See also* COMPLEX WAVEFORM, SAWTOOTH WAVE, and SQUARE WAVE.

COMPLEX WAVEFORM

Any nonsinusoidal waveform is a *complex waveform*. The human voice produces a complex waveform. Almost all sounds, in fact, have complex waveforms. The most complex of all waveforms is called white noise, a random combination of sound frequencies and phases. If a complex waveform displays periodic properties, it is called a complex steady-state vibration (*see* COMPLEX STEADY-STATE VIBRATION).

All complex waveforms consist of combinations of sine waves. Even white noise can be resolved to sine-wave tones. A pure sine-wave tone has a rather irritating quality, especially at the midrange and high audio frequencies. *See also* SINE WAVE.

COMPONENT

In any electrical or electronic system, a *component* is a lowest replaceable unit. A resistor, for example, is usually a component, although it can be contained inside an integrated circuit. A tube is a component, but the cathode of a tube is not.

In modular construction, the identity of the components depends on the point of view. To the field technician, a card or circuit board can be the lowest replaceable unit, and therefore it is a component. In the shop, where the cards are repaired individually, the components are the resistors, capacitors, inductors, and other devices that are attached to the board.

A mathematical part of a quantity can sometimes be called a *component*. For example, an electromagnetic wave with slanted polarization has a vertical component and a horizontal component. The collector current in a transistorized amplifier circuit has a direct-current component (usually) and an alternating-current component.

COMPONENT LAYOUT

The arrangement of electronic components on a circuit board is called the *component layout*. The component layout is an important part of the design of a circuit. With modern printed-circuit boards, the placement of foil runs dictates the component layout to a certain extent (two foil runs must be electrically connected if they cross). Poor circuit layout can increase the chances of unwanted feedback, either negative or, positive. It can also cause inconvenience in servicing and adjustment of the equipment.

The component density should be fairly uniform throughout the board; components are not sparse in one area and bunched together in another area. Components are clearly marked for easy identification when adjusting or repairing the equipment. Component layout becomes more and more important as the frequency gets higher. In some very-high-frequency and ultra-high-frequency circuits, the component layout alone can make the difference between a circuit that works and a circuit that doesn't work.

COMPOSITE VIDEO SIGNAL

A *composite video signal* is the modulating waveform of a television signal. It contains video, sync, and blanking information. When the composite video signal is used to modulate a radio-frequency carrier, the video information can be transmitted over long distances by electromagnetic propagation. By itself, the composite video signal is similar to the audio-frequency component of a voice signal, except that the bandwidth is greater.

Most video monitors, used with computers and communications terminals, operate directly from the composite video signal. Some monitors require a modulated very-high-frequency picture signal, usually on one of the standard television broadcast channels. *See also* COLOR PICTURE SIGNAL, and PICTURE SIGNAL.

COMPOUND MODULATION

When a modulated signal is itself used to modulate another carrier, the process is called *compound modulation*. An example of compound modulation is the impression of an amplitude-modulated signal of 100 kHz onto a microwave carrier. The main carrier can have many secondary signals impressed on it. The secondary signals do not all have to use the same kind of modulation; for example, a code signal, an amplitude-modulated signal, a frequency-modulated signal and a single-sideband signal can all be impressed on a microwave carrier at the same time.

With compound modulation, the main signal frequency must be at least several times the frequency of the modulating signals. Otherwise, efficiency is poor and the signal-to-noise ratio is degraded.

Compound modulation is useful in telephone trunk lines, where a large number of conversations must be carried over a single circuit. Compound modulation is used in the transmission of signals via satellites. Compound modulation is used in fiber-optics systems, where many thousands of separate signals can be impressed on a single beam of light. *See also* ACTIVE COMMUNICATIONS SATELLITE, and MODULATED LIGHT.

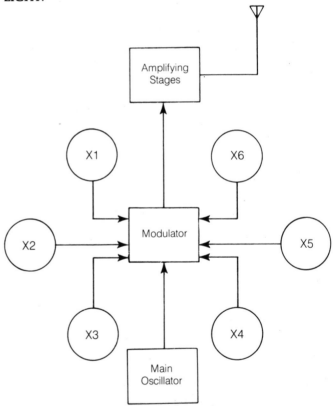

COMPOUND MODULATION: Several signals (shown here as X1 through X6) are impressed on a single carrier.

COMPOUND CONNECTION

When two tubes or transistors are connected in parallel, or in any other configuration, the arrangement is sometimes called a *compound connection*. Examples of compound-connected devices include the Darlington pair, the push-pull configuration, certain rectifier circuits, and many different digital-logic circuits.

Compound connections are used in audio-frequency and radio-frequency circuits for the purpose of improving the efficiency or amplification factor. Compound connections can also be used to provide an impedance match.

COMPRESSION

Compression is the process of modifying the modulation envelope of a signal. This is done in the audio stages, before the modulator. The weaker components of the audio signal are amplified to a greater extent than the stronger components. The peak, or maximum, amplitude of the compressed signal is the same as the amplitude of the normal signal. The weaker the instantaneous amplitude of the signal, the greater the amplification factor.

Compression is often used in communications systems to improve intelligibility under poor conditions. *See also* COMPRESSION CIRCUIT, and SPEECH COMPRESSION.

COMPRESSION CIRCUIT

A *compression circuit* is an amplifier that displays variable gain, depending on the amplitude of the input signal. The lower the input signal level, the greater the amplification factor. A compression circuit operates in a manner similar to an automatic-level-control circuit (*see* AUTOMATIC LEVEL CONTROL).

One means of obtaining compression is to apply a rectified portion of the signal to the input circuit of the amplifying device, changing the bias as the signal amplitude fluctuates. This rectified bias should reduce the gain as the signal level increases. The time constant must be adjusted for the least amount of distortion.

There is a practical limit to the effectiveness of compression. Too much compression will cause the system to emphasize acoustical noise. In a voice communications transmitter, several decibels of effective gain can be realized using audio compression. *See also* SPEECH COMPRESSION.

COMPUTER

A *computer* is an electronic device, usually digital, that processes information. In the developed countries, computers are now in use in private homes, as well as in corporate offices. Many radio hams own personal computers. As the technology becomes less expensive, the proportion of hams owning computers will increase.

A computer can form an important part of an amateur radio station. The most obvious, and probably the most common, application of the personal computer is in packet radio. But computers can also be used for such purposes as aiming directional antennas, calculating satellite position, and figuring out the maximum usable frequency on the HF bands. Computers can be programmed to transmit and receive CW and RTTY. They can be useful in contesting and logging.

Personal computers can be grouped as either *desktop* type or *notebook (laptop)* type. As their names imply, they are intended for fixed and portable operation, respectively. A desktop computer might be found at a home station, and a notebook type used for logging at a field day station.

The figure shows a block diagram of a computer as it might be interconnected in a typical ham shack. A modem (modulator/demodulator) serves to interface the computer with the transceiver for CW, RTTY, AMTOR, and packet operations. The computer is connected to the azimuth/elevation control system for the OSCAR satellite antenna. There is also an inter-

face to the HF beam antenna system rotator, allowing automatic pointing of the antenna in whatever direction is appropriate for a DX region as given by the callsign prefix. The computer has a disk drive to store and retrieve information and programs. It uses a color monitor, and it has a printer for hard copy. It has a second modem for the telephone, allowing remote-control operation via landline, using the laptop and another modem at the operator's condominium in the city.

Related articles in this book include: AMTOR, BYTE, CENTRAL PROCESSING UNIT, COMPUTER PROGRAMMING, DISK DRIVE, HARD DISK, KILOBYTE, MEGABYTE, MICROCOMPUTER, MICROCOMPUTER CONTROL, MICROPROCESSOR, PACKET RADIO, and REMOTE CONTROL.

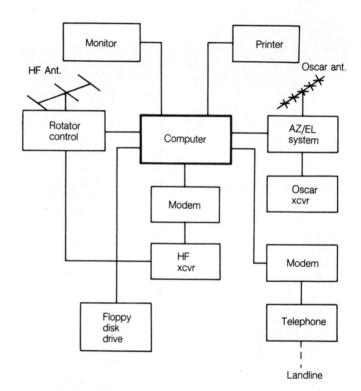

COMPUTER: A block diagram of how a personal computer could be interconnected with other components in a ham radio station.

COMPUTER PROGRAMMING

In order to perform their functions, all computers must be programmed. Generally, a computer requires two different programs. The higher-order program allows the computer to interface, or communicate, with the operator. There are several different higher-order programming languages, such as BASIC, COBOL, FORTRAN, and others. The assembler program converts the higher-order language into binary machine language, and vice versa. The computer actually operates in machine language.

Preparation of assembler and higher-order programs is usually done by people. Some computers can generate their own programs for certain purposes. The program is a logical sequence of statements, telling the computer how to proceed.

CONDENSER

See CAPACITOR.

CONDUCTANCE

Conductance is the mathematical reciprocal of resistance. The unit of conductance is the mho (ohm spelled backwards). A resistance of 1 ohm is equal to a conductance of 1 mho. A resistance of 1 kilohm (1000 ohms) is equal to a conductance of 1 millimho; a resistance of 1 megohm (1,000,000 ohms) is equal to 1 micromho. Mathematically, if R is the resistance in ohms, then the conductance, G, in mhos is:

$$G = 1/R$$

Conductance is occasionally specified instead of resistance when the value of resistance is very low. The word mho is sometimes replaced by an uppercase, but inverted, Greek letter omega, such as 0.3Ω. The mho is also called the *siemens*.

CONDUCTED NOISE

When noise in a communications system comes through the power supply, it is said to be *conducted noise*. This can be a problem with fixed as well as with mobile ham radio stations.

In a fixed station, conducted noise can come from the ac utility mains. Some appliances generate high-frequency noise over a broad spectrum. This is radiated by the power lines in the vicinity of the appliance, and is also conducted along the power lines. Although utility transformers choke off most of this noise, there usually are several households on a single power transformer. Therefore, your neighbor's (as well as your own) hair dryer, vacuum cleaner, electric blanket, light dimmer, or other transient-generating device can interfere with your reception via conducted noise.

Conducted noise in a fixed station can be suppressed by inserting RF chokes in series with both power leads from the station to the wall outlet. These chokes must be rugged enough to carry the current needed by the station. You can wind them on toroidal or solenoidal, powdered-iron cores using No. 12 or No. 14 soft-drawn, insulated copper wire. Inductances of about 1 mH are usually sufficient for the ham bands.

In a mobile station, conducted noise comes from the alternator and from the spark plugs, through the power leads and into the radio.

Conducted noise in a mobile station can be minimized by connecting the radio power leads to the battery, rather than through the cigarette lighter. Filtering the alternator leads using capacitors of about 0.1 μF will help reduce alternator whine. Resistance wiring in the ignition system sometimes helps with spark-plug noise problems.

CONDUCTION COOLING

Conduction cooling is a method of increasing the dissipation capability of a tube or transistor. Heat is carried from the device by a thermally conductive material, such as beryllium oxide or silicone compound, to a large piece of metal. The metal, in turn, uses cooling fins to let the heat escape into space. The illustration shows a typical conduction-cooling arrangement for a transistor.

Conduction cooling is quieter than forced-air cooling, the other most common method of cooling. Blowers are not necessary for conduction cooling. However, care must be exercised to ensure that a good thermal bond is maintained between the cooling device and the tube or transistor. *See also* AIR COOLING, and HEATSINK.

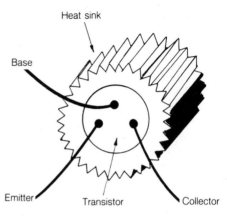

CONDUCTION COOLING: A heatsink conducts heat away from the component.

CONDUCTIVE INTERFERENCE

Interference to electronic equipment can originate in the power lines supplying that equipment. Such interference can be ra-

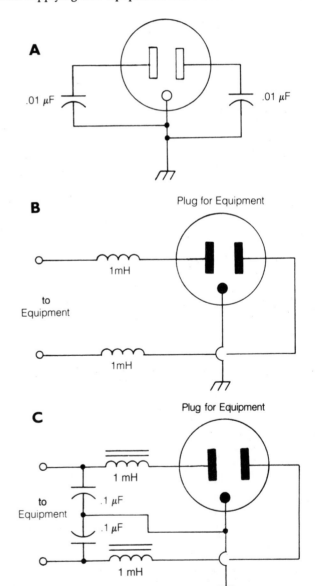

CONDUCTIVE INTERFERENCE: At A, bypassing; at B, chokes; at C, both bypassing and chokes.

diated by the lines and picked up by a radio antenna, or it can be conducted to the equipment and coupled via the power-supply transformer. The latter mode is called *conductive interference.*

Conductive interference originates in many different kinds of electrical appliances, such as razors, fluorescent lights, and electric blankets. Light dimmers are a well-known cause. Faulty power transformers can transmit interference on utility lines. At radio frequencies, these impulses are coupled via the capacitance between the primary and secondary windings of the supply transformer. At audio frequencies, the interference can pass through the transformer inductively.

Conductive interference can be very difficult to eliminate. The illustration shows some common alternating-current line filters for minimizing the effects of conductive interference at audio and radio frequencies.

CONDUCTIVITY

Conductivity is a measure of the ease with which a wire or material carries an electric current. Conductivity is expressed in mhos, millimhos, or micromhos per unit length. Conductivity varies somewhat with temperature. Some materials get more conductive as the temperature rises; most get less conductive as the temperature rises.

The better the conductivity of a material, the more current will flow when a specific voltage is applied. The greater the conductivity, the smaller the voltage required to produce a given current through a length of conductor. Conductivity is the opposite of resistivity.

CONE OF SILENCE

A low-frequency radiolocation beacon radiates energy mainly in the horizontal plane. An inverted, vertical cone over the antenna contains very little signal. This zone is called the *cone of silence.*

Any omnidirectional antenna, designed for the radiation of signals primarily toward the horizon, has a cone of silence. The more horizontal gain, the broader the apex of this cone. In some repeater installations using high-gain collinear vertical antennas, located high above the level of average terrain, a cone of silence (poor repeater response) exists underneath the antenna. This zone can extend outward for several miles in all directions at ground level. This effect often makes a nearby repeater useless. *See also* REPEATER.

CMOS

See COMPLEMENTARY METAL-OXIDE SEMICONDUCTOR.

CONICAL HORN

A *conical horn* is a radiating device used for acoustic or microwave electromagnetic energy. The cross-sectional area, *A*, at a given axial distance, *r*, from the apex of the horn, is given by:

$$A = kr^2$$

where *k* is a constant that depends on the apex angle of the horn.

A conical horn can have a cross section that is circular, square, rectangular, hexagonal, or any other geometric shape. The drawing illustrates a conical horn with a square cross sec-

tion. Conical horns can be visually recognized by the fact that their sides, or faces, are linear. That is, the intersection of a plane passing through the radial axis with the face of the horn is always a straight line. Conical horns have a unidirectional radiation pattern.

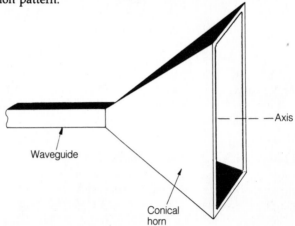

CONICAL HORN: This is often used in microwave communications.

CONICAL MONOPOLE ANTENNA

A *conical monopole antenna* is a form of biconical antenna, in which the lower cone has been replaced by a ground plane. The upper cone is usually bent inward at the top.

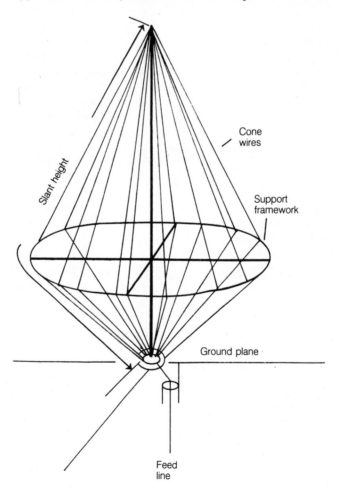

CONICAL MONOPOLE ANTENNA: This is a broadbanded, vertical antenna.

A conical monopole displays a wideband frequency response. The lowest operating frequency is determined by the size of the cone. At all frequencies up to several octaves above that at which the slant height of the cone is $\frac{1}{4}$ wavelength, the conical monopole provides a reasonably good match to a 50-ohm feed line.

Conical monopole antennas have vertical polarization, and are often constructed in the form of a wire cage as shown. This closely approximates a solid metal cone. A good ground system must be furnished for the conical monopole to work well. The conical monopole is an unbalanced antenna, and coaxial cable is best for the feed system. *See also* BICONICAL ANTENNA, and DISCONE ANTENNA.

CONJUGATE IMPEDANCE

An impedance contains resistance and reactance. The resistive component is denoted by a real number *a*, and the reactive component by an imaginary number *jb* (*see* COMPLEX NUMBER, IMPEDANCE, J OPERATOR, REACTANCE, and RESISTANCE). The net impedance is denoted by $R + jX$. If X is positive, the reactance is inductive. If X is negative, the reactance is capacitive.

Given an impedance of $R + jX$, the conjugate impedance is $R - jX$. That is, the conjugate impedance contains the same value of resistance, but an equal and opposite reactance.

Conjugate impedances are important in the design of alternating-current circuits, and for the purpose of impedance matching between a generator and load.

CONNECTION PROTOCOL

In a *packet radio* network, a *connection protocol* is responsible for establishing the pathway for signal transfer. It operates for one communications session.

The pathway or route might be direct between the source and destination station; but more often, there are intermediate *digipeaters*. For example, a station on 2 meters in Florida might wish to send a packet to a station in California. This would involve at least one digipeater so that the packet could be moved to some high-frequency (HF) band, such as 20 meters. More commonly, two or more intermediate digipeaters would be needed. The connection protocol determines the particular digipeaters through which the packet is to be sent. *See also* DIGIPEATER, and PACKET RADIO.

CONNECTIONLESS PROTOCOL

A *connectionless protocol* establishes a pathway for the transfer of one packet in a *packet radio* network. It is similar to a *connection protocol*, except that each individual packet gets its own assigned route independently of other packets. The connection protocol operates for an entire communications session, not just for one packet. *See* CONNECTION PROTOCOL.

CONNECTOR

A *connector* is a device intended for providing, and maintaining, a good electrical connection. Connectors can be either temporary or permanent. The alligator clip (*see* ALLIGATOR CLIP) is an example of a temporary connector. A coaxial-cable connector is suitable for permanent outdoor installation when the contacts are properly soldered to the cable conductors. There are many different kinds of connectors.

Choosing the proper connector for a given application is not difficult. Often, the choice is dictated by convention. In radio-frequency use, especially at very high frequencies and above, special connectors must be used to maintain impedance continuity. The BNC and N-type connectors are usually used in coaxial-cable circuits at the higher frequencies (*see* BNC CONNECTOR, and N-TYPE CONNECTOR). Single-wire connectors are simpler, and consist of such configurations as the binding post, spade lug, and various forms of plugs and clips. Refer to the appropriate connector name for specific information.

CONSONANCE

Consonance is a special manifestation of resonance. If two objects are near each other but not actually in physical contact, and both have identical or harmonically related resonant frequencies, then one can be activated by the other.

An example of acoustic consonance is shown by two tuning forks with identical fundamental frequencies. If one fork is struck and then brought near the other, the second fork will begin vibrating. If the second fork has a fundamental frequency that is a harmonic of the frequency of the first fork, the same thing will happen; the second fork will then vibrate at its resonant frequency.

Electromagnetic consonance is produced when two antennas, both with the same resonant frequency, are brought in close proximity. If one antenna is fed with a radio-frequency signal at its resonant frequency, currents will be induced in the other antenna, and it, too, will radiate. Parasitic arrays operate on this principle. *See also* PARASITIC ARRAY, and RESONANCE.

CONSTANT

In a mathematical equation, a *constant* is a fixed numerical value that does not change as the function is varied over its domain. In an equation with more than one variable, some of the variables can be held constant for certain mathematical evaluations. The familiar equation of Ohm's Law:

$$I = E/R$$

where I represents the current in amperes, E represents the voltage, and R represents the resistance in ohms, can be changed to the more general formula:

$$I^* = kE^*/R^*$$

where I^* is the current in amperes, milliamperes, or microamperes, E^* is the voltage in kilovolts, volts, millivolts, or microvolts, and R^* is the resistance in ohms, kilohms, or megohms. The value of the constant, k, then depends on the particular units chosen. Once the units have been decided upon, k remains constant for all values of the function. For example, I^* can be denoted in milliamperes, E^* in volts, and R^* in megohms; then $k = 0.001$, or 10^{-3}.

Certain constants are accepted as fundamental properties of the physical universe. Perhaps the best known of these is the speed of light, c, which is 3×10^8 meters per second in free space.

In mathematical equations, constants can be denoted by any symbol or letter. Letters from the first half of the alphabet, however, are generally used to represent constants.

CONSTANT-CURRENT MODULATION

Constant-current modulation, also called *Heising modulation*, is a form of plate or collector amplitude modulation. The plates or collectors of the radio-frequency and audio-frequency amplifiers are directly connected through a radio-frequency choke.

The choke allows the audio signal to modulate the supply voltage of the radio-frequency stage. This produces amplitude modulation in the radio-frequency amplifier. The plate or collector current is maintained at a constant level because of the extremely high impedance of the choke at the signal frequency. Constant-current modulation is sometimes called *choke-coupled modulation*, because of the radio-frequency choke. *See also* AMPLITUDE MODULATION.

CONSTANT-CURRENT SOURCE

A *constant-current source* is a device that supplies the same amount of current, no matter what the load resistance. In a constant-current power supply, the voltage across the load varies in direct proportion to the load resistance.

As an illustration, suppose that a constant-current source delivers 100 milliamperes. If the load resistance is 10 ohms, then:

$$E = IR = 0.1 \times 10 = 1 \text{ volt}$$

If the load resistance increases to 100 ohms, then:

$$E = IR = 0.1 \times 100 = 10 \text{ volts}$$

The output voltage rises to its maximum value when there is no load — that is, an open circuit. For extremely high values of load resistance, the supply is not able to deliver its rated current. The load resistance must therefore be within certain limits for a constant-current source to operate properly.

CONSTANT-K FILTER

A *constant-k filter* is a series of L networks (*see* L NETWORK). There might be only one L section, or perhaps several. The L sections can be combined to form pi and T networks (*see* PI NETWORK, and T NETWORK). The filter can be either a highpass or lowpass type (*see* HIGHPASS FILTER, and LOWPASS FILTER). The diagrams illustrate various highpass and lowpass L-section filters.

The primary characteristic of a constant-k filter is the fact that the product of the series and parallel reactances is constant over a certain frequency range. Usually, only one kind of reactance is in series, and only one kind is in parallel. Capacitive reactance is given by the formula:

$$X_C = \frac{-1}{2 \pi fC}$$

where X_C is in ohms, f is in megahertz, and C is in microfarads. Inductive reactance is given by:

$$X_L = 2\pi fL$$

where X_L is in ohms, f is in megahertz, and L is in microhenries. The product of the series and parallel reactances in the illustration is thus:

$$k^2 = X_L X_C = -\frac{2\pi fL}{2\pi fC} = -L/C$$

This value is a constant, and does not depend on the frequency. In practice, the inherent resistance of any circuit limits the frequency range over which it is useful.

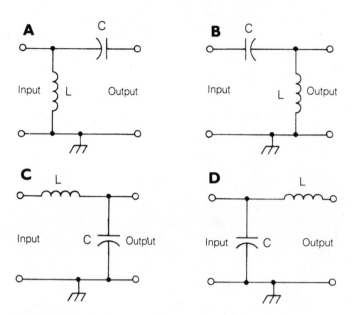

CONSTANT-K FILTER: At A, choke-input highpass; at B, capacitor-input highpass; at C, choke-input lowpass; at D, capacitor-input lowpass.

CONSTANT-VOLTAGE POWER SUPPLY

A *constant-voltage power supply* is a device that provides the same voltage, regardless of the load resistance. The lower the load resistance, the more current the supply delivers, to a certain point. All constant-voltage power supplies have a maximum current rating. As long as the load resistance is high enough to prevent excessive current drain, the voltage at the supply output remains nearly constant. Constant-voltage power supplies are also sometimes called *regulated power supplies*.

Voltage regulation can be accomplished in several different ways (*see* VOLTAGE REGULATION). Regulation is measured in percent. If E_1 is the open-circuit voltage at the output of a power supply and E_2 is the voltage under conditions of maximum rated current drain, then the regulation R in percent is:

$$R = \frac{100 (E_1 - E_2)}{E_1}$$

Most power supplies for radio equipment are intended to deliver a constant voltage for a wide range of load resistances, and are therefore constant-voltage power supplies. *See also* POWER SUPPLY.

CONTACT

A *contact* is a metal disk, prong, or cylinder that is intended to make and break an electric circuit. Contacts are found in all electric switches, relays, and keying devices.

The surface area of the contact determines the amount of current that it can safely carry. Generally, convex contacts cannot carry as much current as flat contacts because the contact area is smaller when the surfaces are curved. Contacts can be made and broken either by a sliding motion or by direct movement. Contacts must be kept clean and free from corrosion in order to maintain their current-handling capacity. Metals that oxidize easily do not make good contacts. Gold and silver are often used where reliability must be high. Some relay contacts are housed in evacuated glass or metal envelopes to keep them from corroding. *See also* REED RELAY, and RELAY.

CONTACT ARC

Contact arc is a spark that occurs immediately following the break between two contacts carrying an electric current. Contact arc occurs to a greater extent when the current is large, as compared to when it is small. Contact arc is undesirable because it speeds up the corrosion of contact faces, eventually resulting in poor reliability and reduced current-carrying capacity.

Contact arc can be minimized by breaking the contacts with the greatest possible speed. Enclosing the contacts in a vacuum chamber prevents dielectric breakdown of the air, and thus eliminates oxidation of the contact metal. *See also* ARC.

CONTACT BOUNCE

Contact bounce is an intermittent make-and-break action of relay or switch contacts as they close. Contact bounce is caused by the elasticity of the metal in the contacts and armature. Contact bounce is undesirable because it often results in contact arc (*see* CONTACT ARC) as the contacts intermittently break. Contact bounce can sometimes cause a circuit to malfunction because of modulation in the flow of current.

The effects of contact bounce can be minimized by placing a small capacitance across the contacts. This helps to smooth out the modulating effects on direct current. The magnetizing current in a relay coil should not be much greater than the minimum needed to reliably close the contacts. Adjustment of the armature spring tension is also helpful in some cases. *See also* RELAY.

CONTACT RATING

The *contact rating* of a switch or relay is the amount of current that the contacts can handle. The contact rating depends on the physical size of the contacts and the contact area. Usually, the contact rating is greatest for direct current, and decreases with increasing alternating-current frequency.

A switch or relay should always be operated well within its contact ratings. Otherwise, arcing will occur, and possibly even contact overheating with resulting damage. Contact ratings are sometimes specified in two ways—continuous and intermittent. The continuous rating is the maximum current that the contacts can handle indefinitely without interruption. The intermittent contact rating is the maximum current that the contacts can handle if they are closed half of the time. *See also* DUTY CYCLE.

CONTESTS AND CONTESTING

One of the most popular ham radio activities is *contesting*. There are many different contests during the calendar year. Most are on weekends. They take place on a variety of different bands, and in any or all of the various modes.

In all contests, the objective is to make as many contacts as possible during the time period. Contests can be recognized by the presence of many stations on a band, carrying on short, rapid-fire, repetitious contacts.

Some hams participate in many contests, and some never participate in any. Contests serve the purpose of keeping operating skill sharp, and this can be invaluable in times of emergency, when large volumes of traffic must be handled accurately and efficiently (*see* TRAFFIC HANDLING). Contests are also great fun for those with a fondness for challenge, and for those who like to place themselves and their stations in competition with others. Most contests have certificates for the winners in various geographical areas.

The following paragraphs are short descriptions of a few contests in which hams can operate. This is by no means a complete list. The ham magazines carry announcements, schedules and rules for contests well in advance of the actual event dates. Contests sponsored by the American Radio Relay League (ARRL) are described in the *ARRL Operating Manual*, available from ARRL Headquarters. *See* AMERICAN RADIO RELAY LEAGUE.

Field Day   On the last full weekend every June, ham operators "go portable" and set up stations using emergency power. The contest lasts 27 hours. Home stations can take part in this contest, but the objective is to simulate a nationwide emergency in which commercial power would not be generally available. Contacts are simple, including only the number of transmitters, the type of power arrangement and the ARRL section. Other activities on Field Day include testing insect repellants, learning new ways in which equipment can fail, and having hamburgers and barbecued chicken late into a summer night.

ARRL November Sweepstakes   Every November, a major contest takes place, known as the *Sweepstakes*. One weekend is for phone (voice), and the other is for CW (Morse code). Most of the activity takes place on the HF bands, or those below 30 MHz.

This is a major event for many hams. The contact exchanges are comprehensive, and the competition is keen, requiring skill and endurance. It is mainly a U.S. contest.

DX Contests   Many hams love the challenge of DX (*see* DX AND DXING). There are several DX contests during the course of the year. Such contests present an excellent opportunity for DXers to get new and/or rare countries.

In a DX contest, a rare station will command a "pileup." High-power and large antennas are an advantage to U.S. hams in these situations. But operating skill can sometimes allow even a modest station to "cut through the heap." There is a special thrill in doing this; it's like making a hole-in-one at golf.

160-meter Contest   Usually held on the first full weekend each December, the 42-hour ARRL 160-meter contest is a special challenge because of the large antennas that are needed for good signals on this band. The band is open for long-distance communications only during the hours of twilight and darkness. Because of the relatively narrow range of frequencies within which contestants operate (about 1.800 to 1.850 MHz), the ARRL contest is CW only.

Another 160-meter contest, the CQ Worldwide, takes place in January. This is part of a much larger contest sponsored by *CQ Magazine*.

QSO Parties The various ARRL sections (usually states, and sometimes parts of states) sponsor local, scaled-down contests. These attract fewer hams than the large contests, but can still be great fun. "QSO parties" are announced in the ham magazines.

CONTINENTAL CODE
See INTERNATIONAL MORSE CODE.

CONTINUOUS-DUTY RATING
The *continuous-duty rating* of a device is the maximum amount of current, voltage, or power that it can handle or deliver with a 100-percent duty cycle (*see* DUTY CYCLE). This means that the device must operate constantly for an indefinite period of time.

A power supply, for example, can be specified as capable of delivering 6 amperes continuous duty. This means that if the necessary load is connected to the supply so that 6 amperes of current is drawn, the supply can be left in operation for an unlimited amount of time, and it will continue to deliver 6 amperes.

Continuous-duty ratings are often given in watts for tubes, transistors, resistors, and other devices. Sometimes such devices can handle or deliver an amount of current, voltage, or power greater than the continuous-rated value for short periods of time.

CONTROL
The term *control* can mean any of several different things in electricity and electronics. Most often, the term *control* refers to an adjustable component, such as a potentiometer, switch, or variable capacitor, that is located for easy access in the operation of a piece of equipment. The channel selector on a television set, for example, is a control.

In computers, control is the maintenance of proper operation. This is accomplished by a central processing unit, CPU (*see* CENTRAL PROCESSING UNIT). In general, control is the maintenance of operation at a certain level, or for a certain purpose, in a piece of electronic apparatus. Control can be either manual or automatic. In some modern communications and electronic devices, control is carried out by means of a microcomputer designed especially for that application. *See also* MICROCOMPUTER, and MICROCOMPUTER CONTROL.

CONTROLLED-CARRIER TRANSMISSION
Controlled-carrier transmission or *modulation* is a special form of amplitude modulation (*see* AMPLITUDE MODULATION). It is sometimes called *quiescent-carrier* or *variable-carrier transmission*. The carrier is present only during modulation. In the absence of modulation, the carrier is suppressed. At intermediate levels of modulation, the carrier is present, but reaches its full amplitude only at modulation peaks.

Controlled-carrier operation is more efficient than ordinary amplitude modulation because the transmitter does not have to dissipate energy during periods of zero modulation. The energy dissipation is also reduced when the modulation is less than 100 percent, which is almost all the time. *See also* CARRIER, and MODULATION.

CONTROLLER
A *controller* is a device or instrument that maintains a certain set of operating parameters for a piece of equipment. Sometimes, the term *controller* refers to a power-regulating device, such as a light dimmer or motor speed adjustment, or a device that shuts off the equipment if it overheats or otherwise exceeds its safe ratings.

Controllers are used in servomechanisms to regulate their movement. As robot devices become more and more commonplace, and are being used to perform increasingly complex tasks, their controllers are getting more sophisticated. Microcomputers are often used in conjunction with controllers for electrical, electronic, and mechanical equipment. *See also* MICROCOMPUTER, and MICROCOMPUTER CONTROL.

CONTROL RECTIFIER
A *control rectifier* is a silicon device used for the purpose of switching large currents. A small control signal, applied to the rectifier, allows power regulation. Control rectifiers are more efficient, and also smaller, than vacuum tubes, relays, or rheostats for the adjustment of power.

Control rectifiers are used in such devices as light dimmers, motor-speed controls, remote switches, and current limiters.

CONTROL REGISTER
A *control register* stores information for the regulation of digital-computer operation. The control register consists of an integrated circuit or a set of integrated circuits. While one program statement is executed, the control register stores the next statement. Once a statement has been carried out by the computer, the statement in the control register is executed, and a new statement enters the register. *See also* COMPUTER.

CONVERGENCE
In a cathode-ray tube with more than one electron beam, *convergence* is the condition in which all of the electron beams intersect in a single point, and this point must lie on the phosphor screen. Convergence is important in a color television picture tube, where three electron guns are used, one for each of the primary colors (red, blue, and green). If convergence is not obtained at all points on a color picture screen, the colors will not line up properly. This results in strange colored borders around certain portions of objects in the picture.

As the electron beams scan across the tube, their point of convergence follows a two-dimensional surface in space, called the convergence surface. In a color television receiver, the three electron beams must not only always converge, but their convergence surface must correspond exactly with the phosphor surface of the picture tube. The controls in a color television receiver that provide precise alignment of the convergence surface are called the convergence and convergence-phase controls. *See also* COLOR TELEVISION, and TELEVISION.

CONVERSION
Conversion is the process of changing frequency, voltage, current, or data-processing mode. A frequency converter changes the frequency of a signal without altering its bandwidth characteristics. A voltage converter changes the voltage of a power supply; such a converter can also change the supply from alter-

nating to direct current or vice versa. A current converter changes the current supplied to a circuit.

In data processing, information can be changed from one form to another. Generally, this consists of altering the transmission mode from serial to parallel, or vice versa. *See also* PARALLEL DATA TRANSFER, and SERIAL DATA TRANSFER.

CONVERSION EFFICIENCY

In a frequency converter, the ratio of the output power, voltage, or current to the input power, voltage, or current is called the *conversion efficiency*. For example, if a frequency converter is designed to change a 21-MHz signal to a 9-MHz signal, the conversion efficiency is the ratio of the 9-MHz signal power, voltage, or current to the 21-MHz signal power, voltage, or current. Conversion efficiency is always less than 100 percent.

In a rectifier, the conversion efficiency is defined as the ratio of the direct-current output power to the alternating-current input power. The conversion efficiency in a rectifier-and-filter power supply is always less than 100 percent because of losses in the rectifier devices and transformers. *See also* CONVERSION, POWER SUPPLY, and RECTIFIER.

CONVERSION GAIN AND LOSS

In a converter, the conversion gain or loss is the increase or decrease, respectively, of signal strength from input to output. This gain or loss can be determined in terms of power, voltage or current. It is specified in decibels.

A passive converter, such as a diode mixer, always has some conversion loss. However, converters with amplifiers can produce conversion gain.

If P_1 is the input power to a converter and P_2 is the output power, then the conversion gain is given in decibels by:

$$\text{Power gain (dB)} = 10 \log_{10} (P_2/P_1)$$

If E_1 is the input voltage and E_2 is the output voltage, then:

$$\text{Voltage gain (dB)} = 20 \log_{10} (E_2/E_1)$$

If I_1 is the input current and I_2 is the output current, then:

$$\text{Current gain (dB)} = 20 \log_{10} (I_2/I_1)$$

When the output amplitude (P_2, E_2, or I_2) is less than the input amplitude (P_1, E_1, or I_1), the conversion gain will be negative. Then we have loss. We can say, for example, that a device has −3-dB conversion gain, or that it has 3-dB conversion loss. *See also* CONVERSION, GAIN, and LOSS.

COOLING

In an electronic circuit, *cooling* is the process of removing undesirable heat to increase the power-handling capability. Cooling also reduces instability, because some component values change with temperature fluctuations. Cooling can occur in three ways: conduction, convection, and radiation. Quite often, a circuit is cooled by a combination of these three modes.

Vacuum tubes are sometimes cooled primarily by convection. However, most modern solid-state equipment uses conduction cooling (*see* CONDUCTION COOLING). The transistor is mounted against a highly heat-conductive surface, which is thermally bonded to a large, finned piece of metal called a heatsink (*see* HEATSINK). The heatsink transfers heat to the environment via convection and radiation.

Cooling becomes especially important in power amplifiers. The tubes or transistors in such amplifiers generate considerable heat because no amplifier is 100-percent efficient. Cooling allows the power input, and hence the output, to be maximized. Without effective cooling, the choice would be between a lower level of operating power and rapid destruction of the tubes or transistors.

COORDINATE SYSTEMS

Coordinate systems are used by scientists in all disciplines, from mathematics to physics, from astronomy to ham radio communications. The purpose of a coordinate system, insofar as hams are likely to use it, is to show relations or functions between or among variables. Coordinate systems can also be used to locate directions for beam headings, or to find the proper orientation for a satellite or moonbounce antenna.

The Cartesian Plane   The simplest two-dimensional coordinate system is the *Cartesian plane*. It consists of two perpendicular axes, each calibrated in numerical units, with graduations of equal size. The horizontal axis usually represents the independent variable, and is known as the *abscissa*. The vertical axis usually represents the dependent variable, and is called the *ordinate*. Illustration A shows a simple Cartesian coordinate plane.

Semilog Graphs   One of the scales on a Cartesian plane can be made logarithmic, rather than linear, to display certain kinds of relations and functions. This is called a *semilog* coordinate system, and is illustrated at B.

Log-log Graphs   Sometimes both of the scales on a Cartesian plane are logarithmic. This type of coordinate system is called *log-log*, and is shown at C.

Polar Coordinates   Rather than having a grid pattern, coordinate systems can have a radial pattern. One set of coordinates consists of lines running outward from *prime center*, like the spokes of a wheel. The other set consists of circles centered at the prime center. One axis is represented by the angle counterclockwise from the *zero line*, and this usually is assigned the independent variable (abscissa). The other axis is represented by the radial distance from prime center, and usually is assigned the dependent variable (ordinate). A simplified polar coordinate plane is shown in at D.

The Smith Chart   *See* SMITH CHART.

Latitude and Longitude   *Latitude and longitude* is the familiar system for locating points on the surface of the earth. *Latitude* is measured north (positive) and south (negative) from the equator, which is assigned zero degrees. Thus, latitude can range from −90 (south pole) to +90 (north pole). *Longitude* is measured east and west from the meridian running through Greenwich, England. Longitude can range from 180 degrees east to 180 degrees west.

Latitude and longitude lines are always circles. The best way to envision them is to use a globe.

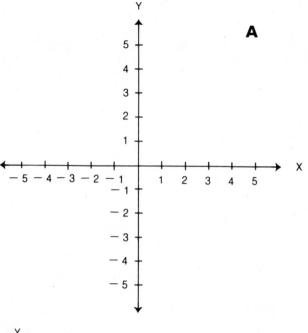

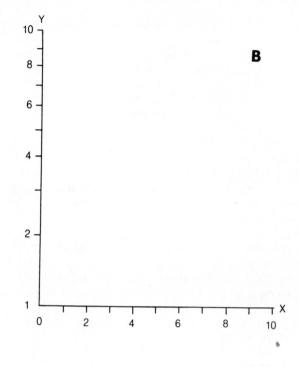

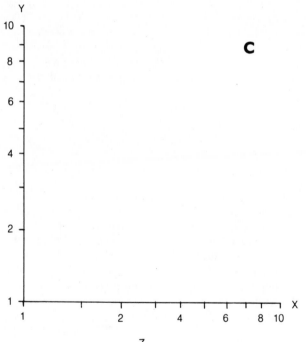

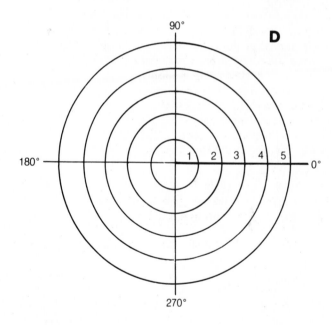

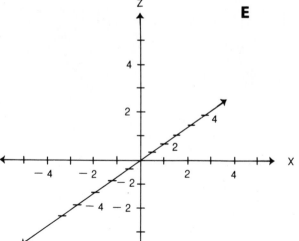

COORDINATE SYSTEMS: Cartesian plane (A), semilog (B), log-log (C), polar (D), Cartesian three-space (E).

Celestial Coordinates Sometimes, latitude and longitude are projected into space to get a system of *celestial coordinates*. A given direction in space might have, for example, the coordinates of 30 degrees north celestial latitude, and 90 degrees west celestial longitude. This would represent a line, running outward from the center of the earth toward the zenith, as seen from 30 degrees north latitude and 90 degrees west longitude on the surface.

The problem with this system of celestial coordinates is that the earth rotates, and this causes any given coordinate pair (except those corresponding to the geographic poles) to move constantly through space. A more common system of celestial coordinates uses *declination* and *right ascension*. *See* DECLINATION, and RIGHT ASCENSION.

Cartesian Three-space *Cartesian coordinates* can be extended into three dimensions by adding a third axis, perpendicular to the two in the Cartesian plane. This third axis is usually called the *z axis*, and represents the dependent variable in a two-variable function of x and y (the independent variables). *Cartesian three-space* is rendered on a flat surface, as shown in the illustration at E.

For Further Information There are other coordinate systems, not covered here, that are used in scientific work. The variety is practically unlimited. Some coordinate systems make use of more than three dimensions. In recent years, computer graphics software has been developed to assist engineers and scientists in working with coordinate systems, especially in three dimensions. Texts on analytic geometry, computer graphics and electrical engineering are recommended if you need in-depth information.

COORDINATED UNIVERSAL TIME

Coordinated Universal Time (UTC), is an astronomical time based on the Greenwich meridian (zero degrees longitude). The UTC day begins at 0000 hours and ends at 2400 hours. Midday is at 1200 hours.

The speed of revolution of the earth varies depending on the time of the year. Coordinated Universal Time is based on the mean, or average, period of the synodic (sun-based) rotation. The earth is slightly behind UTC near June 1, and is slightly ahead near October 1. This variation is caused by tidal effects caused by the gravitational interaction between the earth and the sun.

Coordinated Universal Time is five hours ahead of Eastern Standard Time in the United States. It is six hours ahead of Central Standard Time, seven hours ahead of Mountain Standard Time, and eight hours ahead of Pacific Standard Time. These differences are one hour less during the months of daylight-saving time.

Coordinated Universal Time is broadcast by several time-and-frequency-standard radio stations throughout the world. In the United States, the principal stations are WWV in Colorado and WWVH in Hawaii. These stations are operated by the National Bureau of Standards. *See also* WWV/WWVH/WWVB.

COORDINATION

When two circuit breakers are connected in series, and the breakers have different current ratings, it is advantageous to have the breaker with the lower current rating trip before the one with the higher rating in the event of a short circuit. The same is true of fuses. This is called coordination.

COPPER

Copper is an element with atomic number 29 and atomic weight 64. In its pure form, copper appears as a dark, gold-like metal. It is flexible and malleable. Copper is used extensively in the manufacture of wire since it is an excellent conductor of electric current. Copper is also a good conductor of heat. It is fairly resistant to corrosion.

Oxides of copper are used as rectifiers, especially in the manufacture of photovoltaic cells. Copper oxide exhibits the properties of a semiconductor. *See also* COPPER-CLAD WIRE, and COPPER-OXIDE DIODE.

COPPER-CLAD WIRE

Copper-clad wire is a wire with a core of hard metal, such as iron or steel, and an outer coating of copper. Copper-clad wire is frequently used in radio-frequency antenna systems, where both tensile strength and good electrical conductivity are essential. The central core provides mechanical strength to resist stretching and breakage. The copper coating carries most of the current.

Although copper-clad wire has higher direct-current resistance than pure copper wire of the same diameter, because iron and steel are less effective electrical conductors than copper, the resistance at radio frequencies is almost as low as that of pure copper wire. This is because of the skin effect. At high alternating-current frequencies, most of the electron flow takes place near the outer part of a conductor (*see* SKIN EFFECT).

Copper-clad steel wire is more difficult to work with than soft-drawn or hard-drawn pure copper wire. When installing copper-clad steel wire, the wire must be unwound carefully to avoid kinks. The advantages of greater tensile strength, and less tendency to stretch, however, make copper-clad steel wire preferable to pure copper wire in a variety of applications. *See also* WIRE ANTENNA.

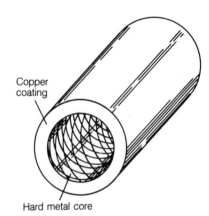

Copper coating

Hard metal core

COPPER-CLAD WIRE: The core gives strength; the cladding carries most of the current.

COPPER-OXIDE DIODE

Copper oxide is a compound that consists of copper and oxygen. It has semiconductor properties, and can be used as a basis for a detector, modulator, rectifier, or photovoltaic cells. A copper-oxide diode is made by placing copper oxide against metallic copper.

Copper-oxide diodes are suitable only for low-voltage applications. In modern circuits, semiconductor diodes are almost always made using germanium, selenium, or silicon. Copper-oxide photovoltaic cells (consisting of copper-oxide diodes covered with clear, protective plastic) are still found in cameras and other devices. Such photovoltaic cells do not require an external source of power. *See also* DIODE, and PHOTOVOLTAIC CELL.

COPY

The term *copy* is used in reference to reception in radiocommunication circuits, and sometimes also in land-line circuits. The received message, whether written by a receiving operator or printed by a machine, is called the *copy*.

The expression *solid copy* means that everything has been received perfectly. It can also indicate to a transmitting operator that the signal is perfectly clear at the receiving end, with no significant interference, either manmade or otherwise. The ease or correctness with which a receiving operator or machine copies a signal is sometimes indicated by a readability number of 1 to 5. The readability designator 1 represents an unreadable signal; the designator 5 represents a perfectly readable signal. Numbers between these extremes represent intermediate levels of readability.

Radio operators use the term copy in a variety of ways. A telegraph operator "copies" the code. A printer "copies" a teletype signal. A voice operator "copies" a message. *See also* RST SYSTEM.

CORE

A *core* is a piece of ferrite or powdered-iron material that is placed inside a coil or transformer to increase its inductance. Cores for solenoidal inductors are cylindrical in shape. Cores for toroidal inductors resemble a doughnut.

The amount by which a ferromagnetic core increases the inductance of a coil depends on the magnetic permeability of the core material. Ferrite has a higher permeability, in most cases, than powdered iron.

The form on which a coil is wound, whether or not it is made of a material with high magnetic permeability, is sometimes called the core. Generally, however, the term *core* refers to a ferrite or powdered-iron substance placed within an inductor. *See also* FERRITE CORE, PERMEABILITY, and TRANS-FORMER.

CORE LOSS

In a powdered-iron or ferrite inductor core, losses occur because of circulating currents and hysteresis effects (*see* EDDY-CURRENT LOSS, and HYSTERESIS LOSS). All ferromagnetic core materials exhibit some loss.

Generally, the loss in a core material increases as the frequency increases. Ferrite cores usually have a lower usable-frequency limit than powdered-iron cores. In general, the higher the magnetic permeability of a core material, the lower the maximum frequency at which it can be used without objectionable loss.

Core losses tend to reduce the Q factor of a coil (*see* Q FACTOR). In ferrite-rod antennas, core losses limit the frequency range at which reception can be realized. In applications that require large currents or voltages through or across an inductor, excessive core losses can result in overheating of the core material. The core can actually break under such conditions. *See also* FERRITE CORE, PERMEABILITY, and TRANSFORMER.

CORE MEMORY

A *core memory* is a means of storing binary information in the form of magnetic fields. A group of small toroidal ferromagnetic cores comprises the memory. A magnetic field of a given polarity represents a digit 1; a field of the opposite polarity represents a digit 0. Wires are run through the cores to facilitate reversing the polarity of the magnetic flux in the cores.

Core memory is retained even when power is removed from the equipment. The cost of core memory is low, but the operating speed is lower than with some other forms of memory. *See also* MEMORY.

CORE SATURATION

As the magnetizing force applied to a ferrite or powdered-iron core is increased, the number of lines of flux through the core is also increased. However, there is a maximum limit to the number of flux lines a given core can accommodate. When the core is as magnetized as it can possibly get, it is said to be saturated. Some cores saturate more easily than others.

In inductors with ferrite or powdered-iron cores, saturation reduces the effective value of inductance for large currents. This is because the magnetic permeability of a saturated core is lessened. The change in inductance can have adverse effects on circuit performance in some situations. In a transformer, core saturation results in reduced efficiency and excessive heating of the core material. The current through an inductor or transformer should be kept below the level at which saturation occurs. *See also* CORE, and FERRITE CORE.

CORNER EFFECT

The response of a bandpass filter should ideally look like a perfect rectangle: zero attenuation within the passband, high attenuation outside the passband, and an instantaneous transition from zero attenuation to high attenuation at the limits of the passband. Lowpass, highpass, and band-rejection filters should have similar instantaneous transitions in the mathematically ideal case.

In practice, the rectangular response is never perfect. The corners of the rectangle—the transitions from zero to high attenuation—are not instantaneous. Immediately next to the passband, the corners appear rounded rather than sharp. This is called the corner effect, and it always occurs with selective filters. *See also* BANDPASS RESPONSE, BAND-REJECTION RESPONSE, HIGHPASS RESPONSE, and LOWPASS RESPONSE.

CORNER REFLECTOR

A *corner reflector* is a device for producing antenna gain at very-high frequencies and above. Two flat metal sheets or screens are attached together. Placed behind the radiating element of an antenna, the corner reflector produces forward gain. The beamwidth is quite broad (*see* BEAMWIDTH). The gain is the same for reception as for transmission at the same frequency.

Corner-reflector antennas are often used for reception of television broadcast signals in the ultra-high-frequency band.

A corner reflector can also consist of three flat metal surfaces or screens, attached together in a manner identical to the way two walls meet the floor or ceiling in a room. Such a device, if it is at least several wavelengths across, will always return electromagnetic energy in exactly the same direction from which it arrives. Because of this effect, corner reflectors of this type make excellent radar dummy targets. Such reflectors are sometimes called *tri-corner reflectors*.

CORONA

When the voltage on an electrical conductor, such as an antenna or high-voltage transmission line, exceeds a certain value, the air around the conductor begins to ionize. The result is a blue or purple glow called *corona*. This glow can sometimes be seen at the end or ends of an antenna at night when a large amount of power is present. The ends of an antenna generally carry the largest voltages. Corona is more likely to occur when the humidity is high than when it is low because moist air has a lower ionization potential than dry air.

Corona can occur inside a cable just before the dielectric material breaks down. Corona is sometimes observed between the plates of air-variable capacitors handling large voltages. A pointed object, such as the end of a whip antenna, is more likely to produce corona than will a flat or blunt surface. Some inductively loaded whip antennas have capacitance hats or large spheres at the ends to minimize corona.

Corona sometimes occurs as a result of high voltages caused by static electricity during thunderstorms. Such a display is occasionally seen at the tip of the mast of a sailing ship. Corona was observed by seafaring explorers hundreds of years ago. They called it *Saint Elmo's fire*.

CORRECTION

In an approximation of the value of a quantity, a *correction* or *correction factor* is a small increment, added to or subtracted from one approximation to obtain a better approximation.

Corrections are generally used when external factors influence the reading or result obtained with an instrument. For example, a frequency counter can be based on a crystal intended for operation at 20 degrees Celsius. If the temperature is above or below 20 degrees Celsius, a correction factor can be necessary, such as 2 Hertz for each degree Celsius.

Sometimes correction is performed by a circuit component especially designed for the purpose. In the previous example, a capacitor with a precisely chosen temperature coefficient can be placed in parallel with the crystal to correct the drift that is caused by temperature changes. This kind of correction is usually called *compensation*. See also COMPENSATION.

CORRELATION

Correlation is a mathematical expression for the relationship between two quantities. Correlation can be positive, zero, or negative. A positive correlation indicates that an increase in one parameter is accompanied by an increase in the other. A negative correlation indicates that an increase in one parameter occurs with a decrease in the other. Zero correlation means that variations in the two parameters are unrelated.

The coefficient of correlation between two variables is usu-

ally expressed as a number in the range between -1 and $+1$. The most negative correlation is given by -1, and the most positive by $+1$. You might say that the failure rate of a certain component (such as a transistor) is correlated with the temperature. The higher the temperature, for example, the more frequently the component fails. A statistical sampling can determine this correlation, and assign it a correlation coefficient.

The closer the correlation coefficient to $+1$, the more nearly the points lie along a straight line with slope of a positive value. When two parameters are not correlated, or have zero correlation, the points are randomly scattered. When the parameters are negatively correlated, the points lie near a line with a negative slope. When the correlation coefficient is exactly -1 or $+1$, all the points lie along a perfectly straight line having either a negative or positive slope, respectively.

Correlation is important in determining the causes of various kinds of circuit malfunctions. Often, two quantities that might intuitively seem correlated are actually not significantly correlated. Sometimes, two quantities that do not intuitively seem correlated actually are.

CORRELATION DETECTION

Correlation detection is a form of detection involving the comparison of an input signal with an internally generated reference signal. The output of a correlation detector varies depending on the similarity of the input signal to the internally generated signal; maximum output occurs when the two signals are identical.

A phase comparator is an example of a correlation detector. In-phase signals produce maximum output; if the signals are not perfectly in phase, the output is reduced. A frequency comparator is another example of correlation detection. Two signals of the same frequency produce zero output. The greater the difference between the input and the internally generated signal frequency, the greater the output frequency. Other types of correlation detection are also possible. *See also* FREQUENCY COMPARATOR, and PHASE COMPARATOR.

CORROSION

Corrosion is a chemical reaction between a metal and the atmosphere. The oxygen in the air is primarily responsible for corrosion. However, salt from the ocean, dissolved in water droplets in the air, can also cause corrosion. Chemicals created by man-made reactions, such as sulfur dioxide, can act to precipitate corrosion of metals.

Corrosion is characterized by deterioration of the surface of a metal. Sometimes, the oxide of a metal can act to protect the metal against further corrosion; an example of this is aluminum oxide. Sometimes, the oxide of a metal increases the vulnerability of the metal to further corrosion by increasing the surface area exposed to the air. An example of this is iron oxide (rust).

Strong electrical currents can sometimes accelerate corrosion. An example of this is the rapid corrosion of the surface of a ground rod placed in acid soil. Conducted currents, through the electrolyte soil, cause rapid disintegration of the metal by electrolytic action. Reactions between certain chemicals that are placed in physical contact can also result in corrosion.

COSINE

The *cosine function* is a trigonometric function. In a right triangle, the cosine is equal to the length of the adjacent side, or base, divided by the length of the hypotenuse. In the unit circle, $x^2 + y^2 = 1$, plotted on the Cartesian (x,y) plane, the cosine of the angle Θ measured counterclockwise from the x axis is equal to x. The cosine function is periodic, and begins with a value of 1 at the point $\Theta = 0$. The shape of the cosine function is identical to that of the sine function except that the cosine function is displaced to the left by 90 degrees.

In mathematical calculations, the cosine function is abbreviated *cos*.

COSINE LAW

The *cosine law* is a rule for diffusion of electromagnetic energy reflected from, or transmitted through, a surface or medium (*see* DIFFUSION).

The energy intensity from a perfectly diffusing surface or medium is the most intense in a direction perpendicular to that surface. As the angle from the normal increases, the intensity drops until it is zero parallel to the surface. The intensity, according to the cosine law, varies with the cosine of angle Θ, relative to the normal. The intensity also varies with the sine of the angle, relative to the surface.

COSMIC NOISE

Cosmic noise is electromagnetic energy arriving from distant planets, stars, galaxies, and other celestial objects. Cosmic noise occurs at all wavelengths from the very-low-frequency radio band to the X-ray band and above. At the lower frequencies, the ionosphere of our planet prevents the noise from reaching the surface. At some higher frequencies, atmospheric absorption prevents the noise from reaching us.

Cosmic noise limits the sensitivity obtainable with receiving equipment, since this noise cannot be eliminated. Radio astronomers deliberately listen to cosmic noise in an effort to gain better understanding of our universe. To them, it is manmade noise rather than cosmic noise that limits the sensitivity of receiving equipment (*see* RADIO ASTRONOMY, and RADIO TELESCOPE).

Cosmic noise is easy to mistake for tropospheric noise, but cosmic noise can be identified by the fact that it correlates with the plane of the galaxy. Perhaps the most intriguing form of cosmic noise, however, arrives with equal strength from all directions. In 1965, Arno Penzias and Robert Wilson of the Bell Laboratories observed faint cosmic noise that seemed to be coming from the entire universe. All other possible sources were ruled out. Astronomers have concluded that the noise originated with the fiery birth of our universe, in an event called the *Big Bang*.

COULOMB

The *coulomb* is the unit of electrical charge quantity. A coulomb of charge is contained in 6.28×10^{18} electrons. One electron thus carries 1.59×10^{-19} coulomb of negative electrical charge.

With a current of 1 ampere flowing in a conductor, exactly 1 coulomb of electrons (or other charge carriers) passes a fixed point in 1 second. The electron flow can occur in the form of actual electron transfer among atoms, or in the form of positive charge carriers called holes. A coulomb of positive charge indicates a deficiency of 6.28×10^{18} electrons on an object; a coulomb of negative charge indicates a surplus of 6.28×10^{18} electrons. *See also* ELECTRON, and HOLE.

COULOMB'S LAW

The properties of electrostatic attraction and repulsion are given by a rule called *Coulomb's law*. Given two charged objects X and Y, separated by a charge-center to charge-center distance d containing charges Q_X and Q_Y, Coulomb's law states that the force (F) between the objects, caused by the electrostatic field is:

$$F = \frac{kQ_XQ_Y}{d^2}$$

where k is a constant that depends on the nature of the medium between the objects. The value of k is given by:

$$k = \frac{1}{4\pi\epsilon}$$

where ϵ is the permittivity of the medium between the objects.

If charges Q_X and Q_Y are opposite, then force F is an attraction. If Q_X and Q_Y are similar charges, force F is a repulsion. If positive charges are given positive values and negative charges are given negative values in the equation, then attraction is indicated by a negative force, and repulsion is indicated by a positive force.

COULOMETER

A *coulometer* is a device that measures a quantity of electric charge. An electrolytic cell, capable of being charged and discharged, makes an excellent coulometer.

When the charge is transferred to the electrolytic cell from an object, a certain amount of chemical action is produced. This chemical action is proportional in magnitude to the amount of charge. Knowing the relation between the charge and the extent of chemical action, the number of coulombs can be accurately determined. *See also* COULOMB.

COUNTER

A digital circuit that keeps track of the number of cycles or pulses entering it is called a *counter*. A counter consists of a set of flip-flops (*see* FLIP-FLOP) or equivalent circuits. Each time a pulse is received, the binary number stored by the counter increases by 1.

Counters can be used to keep track of the number of times a certain event occurs. Some counters measure the number of pulses within a specific interval of time, for the purpose of accurately determining the frequency of a signal. *See also* FREQUENCY COUNTER.

COUNTERMODULATION

Countermodulation is the bypassing of the cathode, emitter, or source resistor of the front end of a receiver, for the purpose of eliminating cross modulation (*see* CROSS MODULATION) in the circuit. The capacitor value is chosen so that the radio-frequency signal is shunted to ground, but the audio frequencies are not. The result is that audio-frequency signals are cancelled, or greatly reduced, by degenerative feedback. The desired radio-frequency signal is, however, easily passed through the amplifier.

The capacitor should have a reactance of less than one-fifth the resistor value at frequencies below 20 kHz, and it should have a reactance of least five times the resistor value at the signal frequency. Therefore, the capacitance depends on the value of the resistor. Countermodulation becomes less effective at low and very low frequencies.

COUNTERPOISE

A *counterpoise* is a means of obtaining a radio-frequency ground or ground plane without a direct earth-ground connection. A grid of wires is placed just above the actual surface to provide capacitive coupling to the ground. This greatly reduces ground loss at radio frequencies.

A simple counterpoise is shown in the illustration. Ideally, the radius of a counterpoise should be at least ¼ wavelength at the lowest operating frequency for a given system. The counterpoise is especially useful at locations where the soil conductivity is poor, rendering a direct ground connection ineffective. A counterpoise can be used in conjunction with a direct ground connection. *See also* GROUND PLANE.

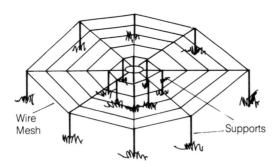

Wire Mesh

Supports

COUNTERPOISE: This is a radio-frequency ground.

COUNTER VOLTAGE

When the current through a conductor is cut off, a reverse voltage, called a *counter voltage*, appears across the coil. This voltage can be very high if the current through the coil is high, and if the coil inductance is large. In some electric appliances containing motors, an interruption in current can present a serious shock hazard because the counter voltage can reach hundreds or even thousands of volts.

A counter voltage is used in every automobile using spark plugs. An inductor called a spark coil stores the electric charge from the battery, which supplies 12 volts direct current, and discharges a pent-up electromotive force of thousands of volts. Counter voltage is also used in some electric fences.

COUPLED IMPEDANCE

When an oscillator or amplifier circuit is followed by another stage, or by a tuning network, the impedance that the oscillator or amplifier "sees" is called the *coupled impedance*. Ideally, the coupled impedance should contain only resistance, and no reactance. A device called a *coupler* or *coupling network* (*see* COUPLER) is sometimes used to eliminate stray reactances—especially in antenna systems.

The actual value of the resistive coupled impedance can range from less than 1 ohm to hundreds of thousands of ohms. Whatever its value, however, it should be matched to the output impedance of the amplifier or oscillator to which it is connected. The coupled impedance should also remain constant at

all times. If it changes, the alteration can be passed back from stage to stage, affecting the gain of the circuit. If a change in impedance is passed all the way back to the oscillator stage, the frequency or phase of the signal will change. This can produce severe distortion of an amplitude-modulated or frequency-modulated signal. It results in chirp on a code signal.

COUPLER

A *coupler* is a device, usually consisting of inductors and/or capacitors, for the purpose of facilitating the optimum transfer of power from an amplifier or oscillator to the next stage. A coupler can also be used between the output of a transmitter and an antenna. Some couplers have fixed components, and some are adjustable. It is intended for impedance matching between a radio-frequency transmitter and an antenna having an unknown impedance.

The resistive component of the impedance is matched by adjusting the tap on an inductor. If the reactive component is capacitive an inductance is switched in series with the antenna to exactly cancel the capacitive reactance. If the reactance is inductive, a capacitance is switched in series with the antenna. It is adjusted until the inductive reactance is balanced. *See also* ANTENNA MATCHING, and ANTENNA TUNING.

COUPLING

Coupling is a means of transferring energy from one stage of a circuit to another. Coupling is also the transfer of energy from the output of a circuit to a load.

Interstage coupling, such as between an oscillator or mixer and an amplifier, can be done in a variety of ways. Four methods of coupling between two bipolar transistor stages are illustrated in the drawing.

In *capacitive coupling*, the signal is transferred through a capacitor. Capacitive coupling isolates the stages for direct current, so that their bias can be independently set.

In *diode coupling*, the diode passes signal energy in one direction, but isolates the stages for direct current in that direction (notice that the second stage uses a pnp transistor, and the first stage uses an npn transistor). The second stage operates in class B or class C.

In *direct coupling*, the voltage at the collector of the first transistor is the same as the voltage at the base of the second. For this method to function, the collector voltage of the second npn transistor must be considerably more positive than the collector voltage of the first transistor. Also, the base voltage of the first transistor must be carefully set to avoid saturation.

Transformer coupling is the most expensive of these techniques; it is preferable because it allows precise impedance matching, and offers good harmonic attenuation. Transformer coupling isolates the two stages for direct current, and allows the use of tuned circuits for improved efficiency. The phase can be reversed if desired. With a well-designed transformer-coupled circuit, electrostatic coupling is kept to a minimum. This improves the stability of the circuit.

These four methods of coupling are only a sampling of the many different arrangements possible. The most common method of interstage coupling is the capacitive method.

Coupling between a radio-frequency transmitter and its antenna is accomplished by means of a circuit called a *coupling network* or *coupler*. *See also* COUPLER.

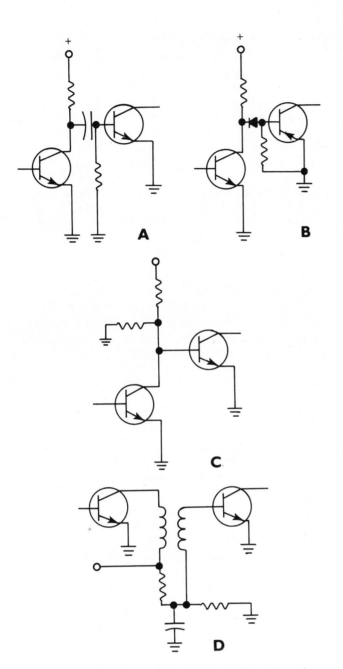

COUPLING: Capacitive (A), diode (B), direct (C), and transformer (D).

COVERAGE

Coverage refers to the frequency range of a receiver or transmitter. The term is also used to define the service area of a communications or broadcast station.

When specifying the frequency coverage of a transmitter or receiver, either the actual frequency range or the approximate wavelength range can be indicated. An amateur-radio receiver might, for example, be specified to cover 80 through 10 meters. This usually means that it operates only on the amateur bands designated in this range. A general-coverage receiver might be specified to work over the range 535 kHz to 30 MHz. This implies continuous coverage.

The coverage area of a broadcast station is determined by the level of output power and the directional characteristics of the antenna system. The Federal Communications Commis-

sion limits the coverage allowed to broadcasting stations in the United States. This prevents mutual interference among different stations on the same frequency. Stations can be designated to have local, regional, or national (clear-channel) coverage.

CQ

In radio communication, the term *CQ* is used to mean "calling anyone." It is used especially by Amateur-Radio operators. A radio operator of station W1GV, for example, can say, in a code transmission, "CQ CQ CQ DE W1GV K," which translates literally to "Calling anyone. Calling anyone. Calling anyone. This is W1GV. Go ahead."

Often, a directional or selective CQ call is used when an operator wants answers from only certain types of stations. For example, CQ DX indicates that the caller wants a station in a country different from his own. CQ MSN might mean "Calling all stations in the Minnesota Section Net."

In some radio services, calling CQ is considered boorish, or the mark of an inexperienced operator. This is the case in the 144-MHz amateur band, when using frequency modulation. It is also true on 27 MHz, in the Citizen's Band.

CREST FACTOR

The ratio of the peak amplitude to the root-mean-square amplitude of an alternating-current or pulsating direct-current waveform is called the *crest factor*. Sometimes it is called the *amplitude factor*. The crest factor depends on the shape of the wave.

In the case of a sine wave, the crest factor is equal to $\sqrt{2}$, or approximately 1.414. In the case of a square wave, the peak and root-mean-square amplitudes are equal, and therefore the crest factor is equal to 1.

In a complicated waveform, the crest factor can vary considerably, and can change with time. It is never less than 1, because the root-mean-square (RMS) voltage, current, or power is never greater than the peak voltage, current, or power. For a voice or music waveform, the crest factor is generally between 2 and 4. *See also* ROOT MEAN SQUARE.

CRITICAL ANGLE

When a beam of light or radio waves passes from one medium to another having a lower index of refraction, the energy can continue on into the second medium, as shown at A in the illustration, or it might be reflected off the boundary and remain within the original medium, as shown at B. Whether refraction or reflection occurs depends on the angle of incidence.

If the angle of incidence is very large, reflection will take place. Then the energy remains in the region having the larger index of refraction. If the angle of incidence is 0 degrees, the energy passes into the medium having the lower index of refraction, generally, and there is no change in its path. At some intermediate angle, called the *critical angle*, reflection just begins to occur as the angle of incidence is made larger and larger. The critical angle depends on the ratio of the indices of refraction of the two media, at the energy wavelength involved.

When radio waves encounter the *E* or *F* layers of the ionosphere (*see* E LAYER, F LAYER), the waves can be returned to the earth, or they might continue on into space (shown at C and D). The smallest angle of incidence, at which energy is returned to the earth is called the critical angle. The critical angle for radio waves depends on the density of the ionosphere, and on the wavelength of the signal. Sometimes, even energy arriving perpendicularly will be returned to the earth; in such a case, the critical angle is considered to be 0 degrees. Sometimes, electromagnetic energy is never returned to the earth by the ionosphere, no matter how great the angle of incidence. Then, the critical angle is undefined. *See also* PROPAGATION CHARACTERISTICS.

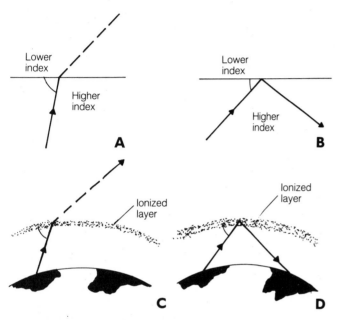

CRITICAL ANGLE: For visible light (A and B), and for radio waves in the ionosphere (C and D).

CRITICAL COUPLING

When two circuits are coupled, the optimum value of coupling (for which the best transfer of power occurs) is called *critical coupling*. If the coupling is made tighter or looser than the critical value, the power transfer becomes less efficient.

The coefficient of coupling, *k*, for critical coupling is given by:

$$k = 1/\sqrt{Q_1 Q_2}$$

where Q_1 is the Q factor of the primary circuit and Q_2 is the Q factor of the secondary circuit. *See also* COEFFICIENT OF COUPLING, and Q FACTOR.

CRITICAL DAMPING

In an analog meter, the *damping* is the rapidity with which the needle reaches the actual current reading. The longer the time required for this, the greater the damping (*see* DAMPING). If the damping is insufficient, the meter needle can overshoot and oscillate back and forth before coming to rest at the actual reading. If the damping is excessive, the meter cannot respond fast enough to be useful for the desired purpose.

Critical damping is the smallest amount of meter damping that can be realized without overshoot. This gives the most accurate and meaningful transient readings. *See also* D'ARSONVAL.

CRITICAL FREQUENCY

At low and very low radio frequencies, all energy is returned to the earth by the ionosphere. This is true even if the angle of incidence of the radio signal with the ionized layer is 90 degrees.

As the frequency of a signal is raised, a point is reached where energy sent directly upward will escape into space. All signals impinging on the ionosphere at an angle of incidence smaller than 90 degrees will, however, still be returned to the earth. The frequency at which this occurs is called the *critical frequency*.

The critical frequency depends on the density of the ionized layers. This density changes with the time of day, the time of year, and the level of sunspot activity. The critical frequency for the ionospheric F layer is typically between about 3 and 5 MHz. *See also* PROPAGATION CHARACTERISTICS.

CROSBY CIRCUIT

The *Crosby circuit* is a method of obtaining frequency modulation. A reactance tube is connected across the tank-circuit coil or capacitor of the oscillator. The modulating signal is applied in series with the control-grid bias supply of the tube. This causes the reactance of the tube to fluctuate in accordance with the modulating signal. In turn, this changes the frequency of the oscillator.

Vacuum tubes are not generally used for obtaining frequency modulation nowadays. A much simpler way to get this kind of modulation is by means of a varactor diode. *See also* VARACTOR DIODE.

CROSS ANTENNA

A *cross antenna* consists of two or more horizontal antennas, connected to the same feed line. The antennas may or may not be fed in phase.

A cross antenna consisting of two horizontal dipoles produces a horizontally polarized signal. The radiation pattern is similar to that of an ordinary dipole, when the two dipoles are fed in phase.

A special form of cross antenna consists of an array of antennas arranged like a cross. They are fed in such a phase combination that a strong gain lobe occurs toward a certain part of the sky. There can be several, or perhaps many, of these antennas. The azimuth and elevation of maximum radiation or reception can be adjusted by varying the phasing among the individual antennas. If its axes measure many wavelengths, this type of antenna can produce an extremely narrow beam. Therefore, such arrays are sometimes used with radio telescopes.

CROSSBAR SWITCH

A *crossbar switch* is a special kind of switch that provides a large number of different connection arrangements. A set of contacts is arranged in a matrix. The matrix can be two-dimensional or three-dimensional. The matrix can be square or rectangular in two dimensions; it can be shaped like a cube or a rectangular prism in three dimensions.

A shorting bar is used to select any adjacent pair of contacts that lie along a common axis. In this way, a large number of different combinations are possible with a relatively small number of switch contacts. The shorting bar is magnetically controlled.

CROSS-CONNECTED NEUTRALIZATION

In a push-pull amplifier (*see* PUSH-PULL AMPLIFIER), instability can cause unwanted oscillation. This is prevented in all radio-frequency amplifiers by means of a procedure called *neutralization* (*see* NEUTRALIZATION). The neutralization method most often used in a push-pull amplifier is called *cross-connected neutralization*, or simply *cross neutralization*.

In cross neutralization, two feedback capacitors are used. The output of one half of the amplifier is connected to the input of the other half. The fed-back signals are out of phase with the input signals for undesired oscillation energy. When the capacitor values are correctly set, the probability of oscillation is greatly reduced. The values of the two neutralizing capacitors must always be identical to maintain circuit balance.

CROSS COUPLING

Cross coupling is a means of obtaining oscillation using two stages of amplification. Normally, an amplifying stage produces a 180-degree phase shift in a signal passing through. Some amplifiers produce no phase shift. In either case, the signal will have its original phase after passing through two amplifying stages. By coupling the output of the second stage to the input of the first stage via a capacitor, oscillation can be obtained.

Cross coupling frequently occurs when it is not wanted. This is especially likely in multistage, high-gain amplifiers. The capacitance between the input and output wiring is often sufficient to produce oscillation because of positive feedback. This oscillation can be very hard to eliminate. The chances of oscillation resulting from unwanted cross coupling can be minimized by keeping all leads as short as possible. The use of coaxial cable is advantageous when lead lengths must be long. Individual shielded enclosures for each stage are sometimes necessary to prevent oscillation in such circuits. Neutralization sometimes works. *See also* NEUTRALIZATION.

CROSSED-NEEDLE METER

A *crossed-needle meter* is a device that consists of two pointer-type analog meters inside a single enclosure. The pointer movements are centered at different positions at the base of the meter. The meters are connected to different circuits, and their point of crossing illustrates some relationship between the two readings.

The illustration shows an example of a crossed-needle meter used for the purpose of showing forward power, reflected power, and the standing-wave ratio on a radio-frequency transmission line. One meter reads the forward power on its scale; the other meter reads reflected power on its scale. A third scale, consisting of lines on the meter face, indicates the standing-wave ratio (SWR), which is a function of forward and reflected power. The point of crossing between the two meter needles is observed against this scale, and the SWR can then be easily determined (*see* STANDING-WAVE RATIO).

Crossed-needle meters are more convenient in some cases than switched or separate meters. They allow the operator to constantly visualize the relationship between two parameters.

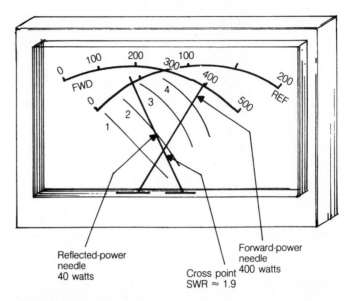

Reflected-power needle 40 watts

Cross point SWR ≈ 1.9

Forward-power needle 400 watts

CROSSED-NEEDLE METER: Shows forward power, reflected power, and standing-wave ratio.

CROSSHATCH

When interference occurs to a television picture, a *crosshatch pattern* sometimes forms on the screen. A weak, unmodulated carrier close to the frequency of the television picture carrier will produce a diagonal set of parallel lines across the screen. If the interfering signal is modulated, sound bars can accompany the diagonal lines. A crosshatch pattern does not necessarily wipe out the desired picture completely, although the contrast often appears reduced, and the overall picture brightness can change.

The term *crosshatch* is also used to refer to a pattern for television testing. The crosshatch pattern consists of a set of horizontal and vertical lines, arranged like a grid, and transmitted into the television receiver. The horizontal lines are used to adjust the vertical linearity of the set. The vertical lines are used to adjust the horizontal linearity.

CROSS MODULATION

Cross modulation is a form of interference to radio and television receivers. Cross modulation is caused by the presence of a strong signal, and also by the existence of a nonlinear component in or near the receiver.

Cross modulation causes all desired signal carriers to appear modulated by the undesired signal. This modulation can usually be heard only if the undesired signal is amplitude-modulated, although a change in receiver gain might occur in the presence of extremely strong unmodulated signals. If the cross modulation is caused entirely by a nonlinearity within the receiver, it is sometimes called *intermodulation* (*see* INTERMODULATION DISTORTION).

Cross modulation can be prevented by attenuating the level of the undesired signal before it reaches nonlinear components, or drives normally linear components into nonlinear operation. If the nonlinearity is outside the receiver and antenna system, the cross modulation can be eliminated only by locating the nonlinear junction and getting rid of it. Marginal electrical bonds between two wires, or between water pipes, or even between parts of a metal fence, can be responsible.

CRYOGENICS

Cryogenics is the science of the behavior of matter and energy at extremely low temperatures. The coldest possible temperature, called *absolute zero*, is the absence of all heat. This temperature is approximately −459.72 degrees Fahrenheit (−273.16 degrees Celsius). The Kelvin temperature scale is based on absolute zero.

When the temperature of a conductor is brought to within a few degrees of absolute zero, the conductivity increases dramatically. If the temperature is cold enough, a current can be made to flow continuously in a closed loop of wire. This is called *superconductivity*.

Supercooling of a receiving antenna allows the use of amplifiers with greater gain than is possible without such cooling. Thermal noise is thereby reduced, and this improves the sensitivity of the receiver.

CRYSTAL

A *crystal* is a piece of piezoelectric material that is used to transform mechanical vibrations into electrical impulses. Crystals are used in some microphones for this purpose (*see* CRYSTAL MICROPHONE).

Quartz crystals are widely used to generate radio-frequency energy. Such devices are typically housed in a metal can. Two wire leads protrude from the base of the can. These leads are internally connected to the faces of the crystal, which consists of a thin wafer of quartz.

The frequency at which a quartz crystal vibrates depends on the manner in which the crystal is cut, and also on its thickness. The thinner the crystal, the higher the natural resonant frequency. The highest fundamental frequency of a common quartz crystal is in the neighborhood of 15 to 20 MHz; above this frequency range, harmonics must be used to obtain radio-frequency energy.

Quartz crystals have excellent frequency stability. This is their main advantage. Crystals are much better than coil-and-capacitor tuned circuits in this respect. A crystal, however, cannot be tuned over a wide range of frequencies. Some crystals are used as selective filters because of their high Q factors. *See also* CRYSTAL CONTROL, CRYSTAL-LATTICE FILTER, CRYSTAL OSCILLATOR, and Q FACTOR.

CRYSTAL CONTROL

Crystal control is a method of determining the frequency of an oscillator by means of a piezoelectric crystal. Such crystals, usually made of quartz (*see* CRYSTAL), have excellent oscillating frequency stability. Crystal control is much more stable than the coil-and-capacitor method.

Crystal oscillators can operate either at the fundamental frequency of the crystal, or at one of the harmonic frequencies. Crystals designed especially for operation at a harmonic frequency are called *overtone crystals*. Overtone crystals are almost always used at frequencies above 20 MHz because a fundamental-frequency crystal would be too thin at these wavelengths, and might easily crack.

The operating frequency of a piezoelectric crystal can be adjusted slightly by placing an inductor or capacitor in parallel with the crystal leads. Generally, the amount of frequency adjustment possible with these methods is very small—approximately ±0.1 percent of the fundamental operating frequency. *See also* CRYSTAL OSCILLATOR.

CRYSTAL-LATTICE FILTER

A *crystal-lattice filter* is a selective filter, usually of the bandpass type. It is similar in construction to a ceramic filter (*see* BANDPASS FILTER, and CERAMIC FILTER) when housed in one container. Some crystal-lattice filters consist of several separate piezoelectric crystals. The crystals are cut to slightly different resonant frequencies to obtain the desired bandwidth and selectivity characteristics.

Crystal-lattice filters are found in the intermediate stages of superheterodyne receivers. They are also used in the filtering stages of single-sideband transmitters. Properly adjusted crystal-lattice filters have an excellent rectangular response, with steep skirts and high adjacent-channel attenuation. A simple crystal-lattice filter uses only two piezoelectric crystals at slightly different frequencies.

CRYSTAL MICROPHONE

A *crystal microphone* is a device that uses a piezoelectric crystal to convert sound vibrations into electrical impulses. The impulses can then be amplified for use in public-address or communications circuits.

In the crystal microphone, which operates in a manner similar to the ceramic microphone (*see* CERAMIC MICROPHONE), vibrating air molecules set a metal diaphragm in motion. The diaphragm, connected physically to the crystal, puts mechanical stress on the piezoelectric substance. This, in turn, results in small currents at the same frequency or frequencies as the sound.

Crystal microphones are rather fragile. If the microphone is dropped, the crystal can break, and the microphone will be ruined. Crystal microphones have excellent fidelity characteristics. Dynamic microphones are also quite commonly used today. *See also* DYNAMIC MICROPHONE.

CRYSTAL OSCILLATOR

A *crystal oscillator* is an oscillator in which the frequency is determined by a piezoelectric crystal. Crystal oscillators can be built using bipolar transistors, field-effect transistors, or vacuum tubes. The circuit generally consists of an amplifier with feedback, with the frequency of feedback governed by the crystal. Oscillation can occur at the fundamental frequency of the crystal, or at one of the harmonic frequencies.

The schematic illustrates two common types of crystal-oscillator circuits. A tuned output circuit provides harmonic attenuation, or it can be used to select one of the harmonic frequencies. If the oscillator is used at the fundamental frequency of the crystal, and if the amount of feedback is properly regulated, a tuned output circuit is not usually needed. However, the oscillator output will contain more energy at unwanted frequencies when a tuned circuit is not used. *See also* OSCILLATOR.

CRYSTAL OVEN

A *crystal oven* is a chamber in which the temperature is kept at an extremely constant level. Most piezoelectric oscillator crystals shift slightly in frequency as the temperature changes. Some crystals get higher in frequency with an increase in the temperature. This is called a *positive temperature coefficient*.

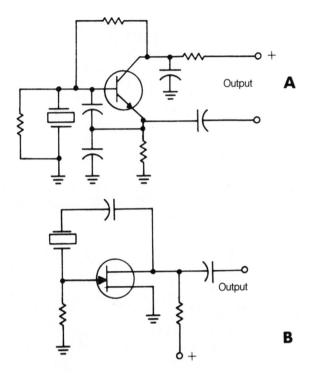

CRYSTAL OSCILLATOR: Bipolar (A) and FET (B) circuits.

Some crystals get lower in frequency when the temperature rises; this is called a *negative temperature coefficient*. A crystal oven is used to house crystals in circuits where extreme frequency accuracy is needed.

Crystal ovens employ thermostat mechanisms and small heating elements, just like ordinary ovens. The temperature is kept at a level just a little above the temperature in the room where the circuit is operated. Sometimes, several ovens are used, one inside the other, to obtain even more precise temperature regulation! Frequency-standard oscillators often use this kind of crystal oven.

CRYSTAL TEST CIRCUIT

A *crystal test circuit* is a device for testing piezoelectric crystals for proper operation. Most crystal test circuits ascertain only that the crystal will oscillate on the correct frequency under the specified conditions. This allows faulty crystals to be easily identified.

More sophisticated crystal testing circuits are used to determine the temperature coefficient, crystal current, and other operating variables. The properties of a crystal or ceramic transducer are checked by a special type of crystal tester.

Semiconductor diodes are sometimes called *crystals*, although this is an antiquated and somewhat inaccurate term. A circuit for testing semiconductor diodes can be called a *crystal tester*. See also CRYSTAL.

CURRENT

Current is a flow of electric charge carriers past a point, or from one point to another. The charge carriers can be electrons, holes, or ions (see ELECTRON, and HOLE). In some cases, atomic nuclei can carry charge.

Electric current is measured in units called amperes. A current of one ampere consists of the transfer of one coulomb of charge per second (see AMPERE, and COULOMB). Current can be either alternating or direct. Current is symbolized by the letter I in most equations involving electrical quantities.

The direction of current flow is theoretically the direction of the positive charge transfer. Thus, in a circuit containing a dry cell and a light bulb, for example, the current flows from the positive terminal of the cell, through the interconnecting wires, the bulb, and finally to the negative terminal. This is a matter of convention. The electron movement is actually in opposition to the current flow. Physicists use this interpretation of current flow purely as a mathematical convenience.

CURRENT AMPLIFICATION

Current amplification is the increase in the flow of current between the input and output of a circuit. It is also sometimes called current gain. In a transistor, the current-amplification characteristic is called the *beta* (see BETA).

Some amplifier circuits are designed to amplify current. Others are designed to amplify voltage. Still others are designed to amplify the power, which is the product of the current and the voltage. A current amplifier requires a certain amount of driving power to operate because such a circuit draws current from its source. Current amplifiers generally have an output impedance that is lower than the input impedance; therefore such circuits are often used in step-down matching applications. That is, they are used as impedance transformers.

Current amplification is measured in decibels. Mathematically, if I_{IN} is the input current and I_{OUT} is the output current, then:

$$Current\ gain\ (dB) = 20\ \log_{10}(I_{OUT}/_{IN})$$

See also DECIBEL, and GAIN.

CURRENT-CARRYING CAPACITY
See CARRYING CAPACITY.

CURRENT DRAIN

The amount of current that a circuit draws from a generator, or other power supply, is called the *current drain*. The amount of current drain determines the size of the power supply needed for proper operation of a circuit.

If the current drain is too great for a power supply, the voltage output of the supply will drop. Ripple can occur in supplies designed to convert alternating current to direct current. With battery power, the battery life is shortened, the voltage drops, and the battery can overheat dangerously.

Current drain is measured in three ways. The peak drain is the largest value of current drawn by a circuit in normal operation. The average current drain is measured over a long period; the total drain in ampere hours is divided by the operating time in hours. Standby current drain is the amount of current used by a circuit during standby periods. Power supplies should always be chosen to handle the peak current drain without malfunctioning. See also POWER SUPPLY.

CURRENT FEED

Current feed is a method of connecting a transmission line to an antenna at a point on the antenna where the current is maximum. Such a point is called a *current loop* (see CURRENT

LOOP). In a half-wavelength radiator, the current maximum occurs at the center, and therefore current feed is the same as center feed. In an antenna longer than ½ wavelength, current maxima exist at odd multiples of ¼ wavelength from either end of the radiator. There can be several different points on an antenna radiator that are suitable for current feed.

The impedance of a current-fed antenna is relatively low. The resistive component varies between about 70 and 200 ohms in most cases. Current feed results in good electrical balance in a two-wire transmission line, provided the current in the antenna is reasonably symmetrical. *See also* VOLTAGE FEED.

CURRENT HOGGING

When two active components are connected together in parallel, or in push-pull configuration, one of them can draw most of the current. This situation, called *current hogging*, occurs because of improper balance between components. Current hogging can sometimes take place with poorly matched tubes or transistors connected in push-pull or parallel amplifiers.

Initially, one of the tubes or transistors exhibits a slightly lower resistance in the circuit. The result is that this component draws more current than its mate. If the resistive temperature coefficient of the device is negative, the tube or transistor carrying the larger current will show a lower and lower resistance as it heats. The result is a vicious circle: The more the component heats up, the lower its resistance becomes, and the more current it draws. Ultimately, one of the tubes or transistors does all the work in the circuit. This can shorten its operating life. It also disturbs the linearity of a push-pull circuit, and upsets the impedance match between the circuit and the load.

Current hogging can be prevented, or at least made unlikely, by placing small resistors in series with the emitter, source, or cathode leads of the amplifying devices. Careful selection of the devices, to ensure the most nearly identical operating characteristics, is also helpful.

CURRENT LIMITING

Current limiting is a process that prevents a circuit from drawing more than a certain predetermined amount of current. Most low-voltage, direct-current power supplies are equipped with current-limiting devices.

A current-limiting component exhibits essentially no resistance until the current, I, reaches the limiting value. When the load resistance R_L becomes smaller than the value at which current I is at its maximum, the limiting component introduces an extra series resistance R_S. If the supply voltage is E volts, then:

$$E = I (R_L + R_S)$$

when the limiting device is active. The resistance R_S increases as R_L decreases so that:

$$R_S = (E/I) - R_L$$

Current-limiting devices help to protect both the supply and the load from damage in the event of a malfunction. Transistors with large current-carrying capacity are used as limiting devices. Current limiting is sometimes called *foldback*.

CURRENT LOOP

In an antenna radiating element, the current in the conductor depends on the location. At any free end, the current is negligi-

ble; the small capacitance allows only a tiny charging current to exist. At a distance of ¼ wavelength from a free end, the current reaches a maximum. This maximum is called a *current loop*. A ½-wavelength radiator has a single current loop at the center. A full-wavelength radiator has two current loops. In general, the number of current loops in a longwire antenna radiator is the same as the number of half wavelengths.

Current loops can occur along a transmission line not terminated in an impedance identical to its characteristic impedance. These loops occur at multiples of ½ wavelength from the resonant antenna feed point when the antenna impedance is smaller than the feed-line characteristic impedance. The loops exist at odd multiples of ¼ wavelength from the feed point when the resonant antenna impedance is larger than the feed-line characteristic impedance. Ideally, the current on a transmission line should be the same everywhere, equal to the voltage divided by the characteristic impedance. *See also* CURRENT NODE, and STANDING WAVE.

CURRENT NODE

A *current node* is a current minimum in an antenna radiator or transmission line. The current in an antenna depends, to some extent, on the location of the radiator. Current nodes occur at free ends of a radiator, and at distances of multiples of ½ wavelength from a free end. The number of current nodes is usually equal to 1 plus the number of half wavelengths in a radiator.

Current nodes can occur along a transmission line not terminated in an impedance equal to its characteristic impedance. These nodes occur at multiples of ½ wavelength from the resonant antenna feed point when the antenna impedance is larger than the feed-line characteristic impedance. They exist at odd multiples of ¼ wavelength from an antenna feed point when a resonant antenna impedance is smaller than the characteristic impedance of the line.

Current nodes are always spaced at intervals of ¼ wavelength from current loops. Ideally, the current on a transmission line is the same everywhere, being equal to the voltage divided by the characteristic impedance. *See also* CURRENT LOOP, and STANDING WAVE.

CURRENT REGULATION

Current regulation is the process of maintaining the current in a load at a constant value. This is done by means of a constant-current power supply (*see* CONSTANT-CURRENT SOURCE). A variable-resistance device is necessary to accomplish current regulation. When such a device is placed in series with the load, and the resistance increases in direct proportion to the supply voltage, the current remains constant.

CURRENT SATURATION

As the bias between the input points of a tube or semiconductor device is varied in such a way that the output current increases, a point will eventually be reached at which the output current no longer increases. This condition is called *current saturation*. In a vacuum tube, the grid bias must usually be positive with respect to the cathode to obtain current saturation. In an npn bipolar transistor, the base must be sufficiently positive with respect to the emitter. In a pnp bipolar transistor, the base must be sufficiently negative, with respect to the emitter. In a field-effect transistor, the parameters for current saturation are affected by the gate-source and drain-source voltages.

Saturation is usually not a desirable condition. It destroys the electrical ability of a tube or semiconductor device to amplify. Saturation sometimes occurs in digital switching transistors, where it can be induced deliberately in the high state to produce maximum conduction.

CURRENT TRANSFORMER

A *current transformer* is a device for stepping current up or down. An ordinary voltage transformer functions as a current transformer, but in the opposite sense (*see* TRANSFORMER). The current step-up ratio of a transformer is the reciprocal of the voltage step-up ratio. If N_{PRI} is the number of turns in the primary winding and N_{SEC} is the number of turns in the secondary winding, then:

$$I_{SEC}/I_{PRI} = N_{PRI}/N_{SEC}$$

where I_{PRI} and I_{SEC} are the currents in the primary and secondary, respectively. The impedance of the primary and secondary, given by Z_{PRI} and Z_{SEC}, are related to the currents by the equation:

$$Z_{PRI}/Z_{SEC} = (I_{SEC}/I_{PRI})^2$$

These formulas assume a transformer efficiency of 100 percent. Although this is an ideal theoretical condition, and it never actually occurs, the equations are usually accurate enough in practice. *See also* TRANSFORMER.

CURVE

A *curve* is a graphical illustration of a relation between two variables. In electronics, two-dimensional graphs are commonly used to show the characteristics of circuits and devices. Generally, the Cartesian coordinate plane is used, although other coordinate systems are sometimes used.

CURVE TRACER

A *curve tracer* is a test circuit used to check the response of a component or circuit under conditions of variable input. A test signal is applied to the input of the component or circuit, and the output is monitored on an oscilloscope.

One common type of curve tracer is used to determine the characteristic curve of a transistor (*see* CHARACTERISTIC CURVE). A predetermined, direct-current voltage is applied between the emitter and the collector. Then, a variable voltage is applied to the base. The variable base voltage is also applied to the horizontal deflecting plates of an oscilloscope. The collector current is measured by sampling the voltage drop across a resistor in the collector circuit; this voltage is supplied to the vertical deflection plates of the oscilloscope. The result is a visual display of the base voltage versus collector-current curve.

Another common type of curve tracer uses a sweep generator and an oscilloscope. This provides a display of attenuation as a function of frequency for a tuned circuit.

Curve tracers allow comparison of actual circuit parameters with theoretical parameters. They are, therefore, invaluable in engineering and test applications.

CUT-IN/CUT-OUT ANGLE

A semiconductor diode requires between 0.3 and 0.6 volt of forward bias in order to conduct. In a rectifier circuit using semi-

conductor diodes, therefore, the conduction period is not quite one-half cycle. Instead, the conduction time is a little less than 180 degrees.

The cut-in angle is the phase angle at which conduction begins. The cut-out angle is the phase angle at which conduction stops. In a 60-Hz rectifier circuit, if a phase angle of 0 degrees is represented by $t = 0$ second and a phase angle of 180 degrees is represented by $t = 1/120$ second or 8.33×10^{-3} second, then:

$$\Theta_1 = 2.16 \times 10^4 \, t_1$$

and:

$$\Theta_2 = 2.16 \times 10^4 \, t_2$$

where Θ_1, and Θ_2 are the cut-in and cut-out angles, respectively, and t_1 and t_2 are the cut-in and cut-out times, respectively.

The cut-in and cut-out angles become closer to 0 and 180 degrees as the voltage of a sine-wave, alternating-current waveform increases. In a square-wave rectifier circuit, the angles Θ_1 and Θ_2 are essentially equal to 0 and 180 degrees. *See also* RECTIFICATION.

CUTOFF

Cutoff is a condition in which the grid or base voltage prevents current from flowing through the device, In a field-effect transistor, the condition of current cutoff is usually called *pinchoff* (*see* PINCHOFF).

In a vacuum tube, cutoff is achieved when the grid voltage is made sufficiently negative with respect to the cathode. In an npn bipolar transistor, the base must generally be at either the same potential as the emitter, or more negative. In a pnp bipolar transistor, the base must usually be at either the same potential as the emitter, or more positive. In a field-effect transistor, pinchoff depends on the bias relationship among the source, gate, and drain.

The cutoff condition of an amplifying device is often used to increase the efficiency when linearity is not important, or when waveform distortion is of no consequence. This is the case in the class-B and class-C amplifier circuits (*see* CLASS-B AMPLIFIER, and CLASS-C AMPLIFIER). A cut-off tube or transistor can also be used as a rectifier or detector.

The term *cutoff* is used also to refer to any point at which a certain parameter is exceeded in a circuit. For example, we can speak of the cutoff frequency of a lowpass filter, or the alpha-cutoff frequency of a transistor.

CUTOFF ATTENUATOR

A waveguide has a certain minimum operating frequency, below which it is not useful as a transmission line because it causes a large attenuation of a signal. The cutoff, or minimum usable, frequency of a waveguide depends on its cross-sectional dimensions (*see* WAVEGUIDE).

When a section of waveguide is deliberately inserted into a circuit, and its cutoff frequency is higher than the operating frequency of the circuit, the waveguide becomes an attenuator. This sort of device, used at very-high and ultra-high frequencies, is called a *cutoff attenuator*. The amount of attenuation depends on the difference between the operating frequency, f_o, and the cutoff frequency, f_c of the waveguide. As $f_c - f_o$ becomes larger, so does the attenuation. The amount of attenuation also depends on the length of the section of waveguide. The longer the lossy section of waveguide, the greater the attenuation. *See also* ATTENUATOR.

CUTOFF FREQUENCY

In any circuit or device, the term *cutoff frequency* can refer to either a maximum usable frequency or a minimum usable frequency. In a transistor, the gain drops as the frequency is increased. The cutoff in such a case is called the *alpha-cutoff frequency* or *beta-cutoff frequency*, depending on the application (*see* ALPHA-CUTOFF FREQUENCY, and BETA-ALPHA RELATION).

As the frequency is raised in a lowpass filter, the frequency at which the voltage attenuation becomes 3 dB, relative to the level in the operating range, is called the *cutoff frequency*. In a highpass filter, as the frequency is lowered, the frequency at which the voltage attenuation becomes 3 dB, relative to the level within the operating range, is called the *cutoff frequency*. A bandpass or band-rejection filter has two cutoff frequencies. (*See* BANDPASS RESPONSE, BAND-REJECTION RESPONSE, HIGHPASS RESPONSE, and LOWPASS RESPONSE.)

Many different kinds of circuits exhibit cutoff frequencies, either at a maximum, minimum, or both. A broad-band antenna has a maximum and a minimum cutoff frequency. All amplifiers have an upper cutoff frequency. Sections of coaxial or two-wire transmission line have an upper cutoff frequency; waveguides have a lower cutoff that is well defined. Usually, the specification for cutoff is 3 dB, representing 70.7 percent of the current or voltage in the normal operating range. However, other attenuation levels are sometimes specified for special purposes. *See also* ATTENUATION.

CUTOFF VOLTAGE

The *cutoff voltage* of a vacuum tube or transistor is the level of grid, base, or gate voltage at which cutoff occurs. In a field-effect transistor, the cutoff voltage is usually called the *pinchoff voltage*.

A bipolar transistor is normally cut-off. That is, when the base voltage is zero with respect to the emitter, the device does not conduct. Until approximately 0.3 to 0.6 volts of base voltage is applied in the forward direction, the transistor remains cut-off. Above +0.3 to +0.6 volts, an npn transistor will begin to conduct; below −0.3 to −0.6 volt, a pnp transistor begins to conduct. This is based on the assumption that the collector is properly biased—positive for an npn device and negative for a pnp device.

In a vacuum tube, cutoff usually requires the presence of a large negative voltage on the control grid. This bias can be anywhere from a few to several hundred volts, depending on the tube type. *See also* TRANSISTOR, and TUBE.

CUTOUT

A *cutout* or *cutout device*, is a circuit-breaking component, such as a common breaker or fuse. When the current exceeds a certain predetermined level, the supply line is broken to protect the circuit from damage. This condition is sometimes called cutout. *See also* CIRCUIT BREAKER, and FUSE.

CUT-OUT ANGLE

See CUT-IN/CUT-OUT ANGLE.

CW

See MORSE CODE.

CW ABBREVIATIONS

See the appendix CW ABBREVIATIONS.

CYCLE

In any periodic wave—that is, a waveform that repeats itself many times—a cycle is the part of a waveform between any point and its repetition. For example, in a sine wave, a cycle can be regarded as that part of the waveform between one positive peak and the next, or between the point of zero, positive-going voltage and the next point of zero, positive-going voltage. It actually does not matter which point of reference is chosen to determine a cycle, as long as the waveform ends at the same place that it begins.

Cycles can be identified for any periodic waveform. Although the waveshape varies slightly from one cycle to the next, a periodic variation is definitely present. In fact, the sunspot fluctuation, which repeats at intervals of approximately 11 years, is often called the *sunspot cycle* because its recurrence is so predictable.

A cycle is routinely divided into 360 small, equal increments, called *degrees*. You can therefore identify small parts of a cycle, such as the 30-degree point and the 130-degree point. The difference between two points is called an *angle of phase*. Hence the angle between the 130-degree point and the 30-degree point is 100 degrees of phase. Engineers sometimes divide a cycle into radians. A radian is roughly equal to 57.3 degrees. There are exactly 2π radians in a complete cycle.

CYCLES PER SECOND

The term *cycles per second* is an obsolete expression for the frequency of a periodic wave. The commonly accepted electronic or electrical term nowadays is *hertz* (*see* HERTZ). In older text and reference books, the frequency is still sometimes expressed in cycles, kilocycles, or megacycles per second, abbreviated cps, kc, and Mc, respectively.

CYCLIC IONOSPHERIC VARIATION

The density of ionization in the upper atmosphere varies periodically with the time of day, the time of year, and the level of sunspot activity. Such fluctuations, which occur on a regular basis, are called cyclic ionospheric variations. Such variations affect the properties of radiocommunication on the medium and high frequencies.

Generally, the density of ions is greater during the daylight hours than during the night. This is because the ultraviolet radiation from the sun causes the atoms in the upper atmosphere to ionize, at heights ranging from about 40 to 250 miles. This produces a daily cycle, which reaches its peak sometime after midday and reaches its minimum shortly before sunrise.

During the summer months, the level of ionization is usually greater, on the average, than during the rest of the year. During the winter months, the level of ionization is the least. Of course, when it is winter in the Northern Hemisphere, it is summer in the Southern Hemisphere and vice versa. The sun remains above the horizon for the longest time in the summer, allowing more atoms to become ionized. Also, the ultraviolet radiation is somewhat more intense in the summer, especially at lower levels in the atmosphere. The daily cycle is impressed upon this annual cycle.

The level of sunspot activity varies over a period of about 11 years. The years of maxima for this era are 1958, 1969, 1980, 1991, and 2002 A.D. The years of minima are 1964, 1975, 1986, 1997, and 2008 A.D. Ionospheric density is, on the average, greatest during the sunspot maxima and least during the minima. The annual and daily cyclic variations are impressed on the 11-year cycle. It is quite possible that even longer periodic sunspot variations occur, resulting in even longer ionospheric variations. We have not been able to measure the solar flux and sunspot numbers for a long enough time, however, to find such a cycle with certainty. *See also* PROPAGATION CHARACTERISTICS, and SUNSPOT CYCLE.

CYCLIC SHIFT

A *cyclic shift* is a transfer of information in a storage register in one direction or the other, usually called the right or the left. In a cyclic shift toward the left, each digit or bit is moved one place toward the left, except for the extreme left-hand digit or bit, which replaces the one originally at the far right. In a cyclic shift toward the right, each digit or bit is moved one position to the right, except for the extreme right-hand digit or bit, which replaces the one on the far left.

In an n-bit register, a succession of n cyclic shifts in the same direction results in the original information. Also, if m cyclic shifts are performed in one direction, followed by or combined with m cyclic shifts in the opposite direction, where m is any positive integer, the initial storage is obtained. *See also* SHIFT REGISTER.

CZOCHRALSKI METHOD

Semiconductor materials, such as silicon and germanium, are often obtained by "crystal growing." A tiny piece of the material, called a *seed*, is placed into a molten bath of the same substance. The seed crystal is slowly rotated in this bath, and the crystal increases in size over a period of time. This produces a large single crystal. This technique for producing large semiconductor crystals is called the *Czochralski method.*

DAISY-WHEEL PRINTER

A *daisy-wheel printer* is a high-speed mechanical printing device that is used in electronic computers and typewriters. A circular type wheel, consisting of several dozen radial spokes, each with one character molded to its end face, rotates rapidly, then stops so that the proper character is at the top. Then, a hammer strikes the spoke from behind, presses it against the back of the printing ribbon, and onto the page. The printing rate of a daisy-wheel device varies, but it is generally from 10 to 15 characters per second.

A daisy-wheel printer has the advantage of high reliability and relatively simple operation. Type wheels are easily interchanged when a different format or character style is desired. *See also* PRINTER.

DAMPED WAVE

A *damped wave* is an oscillation whose amplitude decays with time, as shown in the drawing. The damping might occur rapidly, in a few microseconds, to the point where the wave amplitude is essentially zero; the damping can also occur slowly, over a period of milliseconds or even over several seconds. Damping can occur within the time of one cycle or less, or it can occur over a period of millions of cycles. Generally, the higher the Q factor in a circuit (*see* Q FACTOR), the more cycles occur before the signal amplitude deteriorates to essentially zero.

The decay in a damped wave occurs in the form of a logarithmic function, called a *logarithmic decrement.*

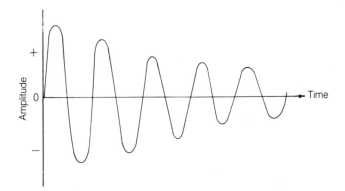

DAMPED WAVE: Decreases in amplitude logarithmically.

DAMPING

Damping is the prevention of overshoot in an analog meter device, or the nature in which the needle comes to rest at a particular reading. Overshoot is generally undesirable, especially when meter readings change often. This is because overshoot creates confusion as to what the actual meter reading is. The greater the damping, the more slowly the meter needle responds to a change in current. The ideal amount of damping in a meter for a given application is called the critical damping (*see*

CRITICAL DAMPING). If the damping is in excess of the critical damping, the meter will not respond fast enough to changes in the current through it.

Any absorption in a circuit that reduces the amount of stored energy is called *damping.* A resistor placed in a tuned circuit to lower the Q factor, for example, constitutes damping. This tends to reduce the chances of oscillation in a high-gain, tuned amplifier circuit. Mechanical resistance can be built into a transducer, such as an earphone or microphone, to limit the frequency response. This also is called *damping. See also* Q FACTOR.

DAMPING FACTOR

In a high-fidelity sound system, the actual output impedance of the amplifier can be much smaller than the impedance of the speaker. The ratio of the speaker impedance, which is usually 4, 8, or 16 ohms, to the amplifier output impedance, which is often less than 1 ohm, is called the *damping factor.*

The effect of this difference in impedance is to minimize the effects of speaker acoustic resonances. The sound output should not be affected by such resonances in a high-fidelity system. The frequency response should be as flat as is practical. The damping factor in a high-fidelity system is somewhat dependent on the frequency of the audio energy. It also is a function of the extent of the negative feedback in the audio amplifier circuit. Damping factors in excess of 60 are quite common.

In a damped oscillation, the quotient of the logarithmic decrement and the oscillation period is sometimes called the *damping factor.* In a damped-wave circuit, where the coil inductance is given by L and the radio-frequency resistance is given by R, the damping factor, a, is defined as:

$$a = \frac{R}{2L}$$

See also DAMPED WAVE.

DAMPING RESISTANCE

If the Q factor in a resonant circuit becomes too great (*see* Q FACTOR), an undesirable effect called *ringing* can occur. Ringing is especially objectionable in audio filters that are used in radioteletype demodulators and in code communications. To reduce the Q factor sufficiently, a resistor, called a *damping resistor,* is placed across a parallel-resonant circuit. Such a resistor can also be placed in series with a series-resonant circuit. In a parallel-resonant circuit, the Q factor decreases as the shunt resistance decreases. In a series-resonant circuit, the Q factor decreases as the series resistance increases. The lower the Q factor, the less the tendency for the resonant circuit to ring.

Damping resistance is an expression sometimes used in reference to a noninductive resistor placed across an analog meter to increase the damping. *See also* CRITICAL DAMPING, and DAMPING.

DARAF

The *daraf* is the reciprocal unit of the farad. The word *daraf* is, in fact, *farad* spelled backwards! A value of 1 daraf is the reciprocal equivalent of 1 farad. The quantity $1/C$, where C is capacitance, is called *elastance*.

Generally, values of capacitance in practical circuits are much less than 1 farad. A capacitance of 1 microfarad (10^{-6} farad), corresponds to an elastance of 1 megadaraf, or 10^6 daraf. A capacitance of 1 picofarad, or 10^{12} farad, is an elastance of 1 teradaraf (10^{12} daraf).

DARLINGTON AMPLIFIER

A *Darlington amplifier*, or *Darlington pair*, is a form of compound connection between two bipolar transistors. In the Darlington amplifier, the collectors of the transistors are connected together. The input is supplied to the base of the first transistor. The emitter of the first transistor is connected directly to the base of the second transistor. The emitter of the second transistor serves as the emitter for the pair. The output is generally taken from both collectors.

The amplification of a Darlington pair is equal to the product of the amplification factors of the individual transistors as connected in the system. This does not necessarily mean that a Darlington amplifier will produce far more gain than a single bipolar transistor in the same circuit. The impedances must be properly matched at the input and output to ensure optimum gain. Some Darlington pairs are available in a single case. Such devices are called *Darlington transistors*. The Darlington amplifier is sometimes called a *double emitter follower* or a *beta multiplier*. See also TRANSISTOR.

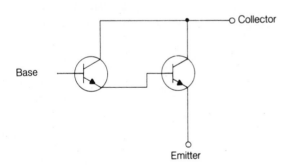

DARLINGTON AMPLIFIER: Two bipolar transistors are directly coupled.

D'ARSONVAL

A *D'Arsonval meter*, or *D'Arsonval movement*, is a device used in analog monitoring. The current to be measured is passed through a coil that is attached to an indicating needle. The coil is operated within the field of a permanent magnet. As a current passes through the coil, a magnetic field is set up around the coil, and a torque appears between the permanent magnet field and the field of the coil. A spring allows the coil, and hence the pointer, to rotate only a certain angular distance; the greater the current, the stronger the torque, and the farther the coil turns.

Generally, the coil in a D'Arsonval meter is mounted on jeweled bearings for maximum accuracy. D'Arsonval meters are widely used as ammeters, milliammeters, and microammeters. With suitable peripheral circuitry, D'Arsonval meters are used as voltmeters and wattmeters as well, both for direct current and alternating current at all frequencies. *See also* AMMETER, and ANALOG METERING.

DATA COMMUNICATION

Data communication is the transfer of data in both directions between two points, or in all possible ways among three or more points. Each station must have a transmitting and receiving device.

Ideally, the data should arrive at a receiving station in exactly the same form as it is transmitted. Interference in the system occasionally changes the data as it is picked up by the receiver. However, some sophisticated data communications systems can tell when something appears to be out of place; the receiving station will then ask the transmitting station to repeat that part of the data in which the apparent error has occurred.

Data communication is generally carried out in binary form because this provides a better signal-to-noise ratio than analog methods. The alphabetic-numeric format of most data lends itself well to digital communications techniques. The impulses can be transferred via radio, optical fibers, lasers, or landline (wire). Sometimes the data is coded, or scrambled, at the transmitting end and decoded at the receiving end. This reduces the chances of interception by unwanted parties. *See also* DATA TRANSMISSION.

DATA CONVERSION

When data is changed from one form to another, the process is called *data conversion*. Data can, for example, be sent in serial form, bit by bit, and then be converted into parallel form for use by a certain circuit (*see* PARALLEL DATA TRANSFER, and SERIAL DATA TRANSFER). Data can initially be in analog form, such as a voice or television-picture signal, and then be converted to digital form for transmission. In the receiving circuit, the digital data can then be converted back into analog data (*see* ANALOG, ANALOG-TO-DIGITAL CONVERTER, DIGITAL, and DIGITAL-TO-ANALOG CONVERTER).

Data conversion is performed for the purpose of improving the efficiency and/or accuracy of data transmission. For example, the digital equivalent of a voice signal is propagated with a better signal-to-noise ratio, in general, than the actual analog signal. It therefore makes sense to convert the signal to digital form when accuracy is of prime importance.

DATA SIGNAL

Data signals are modulated waves, or pulse sequences, used for the purpose of transmitting data from one place to another. A data signal can be either analog or digital in nature (*see* ANALOG, and DIGITAL). Analog data signals consist of a variable parameter, such as amplitude, frequency, or phase, that fluctuates over a continuous range. Theoretically, an analog data signal can have an infinite number of different states. A digital signal has only a few possible states, or levels. A television broadcast signal is an example of an analog data signal. A video-display terminal, showing only alphabetic-numeric characters, is an example of a device that uses a digital data signal. Other examples of digital data signals are Morse code and radioteletype frequency-shift keying.

Data signals can be transmitted in sequential form, that is, one after the other; or they can be sent in bunches. The former method of sending data is called *serial transmission*. The latter method is called *parallel transmission.* There are advantages and disadvantages to both methods. *See also* DATA TRANSMISSION, PARALLEL DATA TRANSFER, and SERIAL DATA TRANSFER.

DATA TRANSMISSION

Data is sent from one place to another by means of a process called *data transmission.* Data can be transmitted via wires, beams of light, radio-frequency energy, sound, or simple direct-current impulses.

Data transmission is generally classified as either analog or digital. Analog data transmission fluctuates over a given range in terms of some parameter, such as amplitude, frequency, or phase. Digital data has a finite number of discrete states, such as on/off or various precisely defined levels. Digital or analog data can be categorized as either serial or parallel. Serial data is transmitted sequentially; parallel data is sent several bits at a time.

There are advantages and disadvantages to analog and digital data-transmission modes. Although analog data transmission allows a greater level of reproduction accuracy, and even helps carry meaning via inflection (such as tone of voice), digital transmission is generally faster and more efficient, especially under marginal conditions. Serial data transmission requires only one line or signal, but the transfer rate is comparatively slow. Parallel data transmission, although it necessitates the use of several lines or channels at once, is much faster. *See also* ANALOG, DATA, DIGITAL, PARALLEL DATA TRANSFER, and SERIAL DATA TRANSFER.

dB

See DECIBEL.

dBa

The abbreviation *dBa* represents *adjusted decibels.* Adjusted decibels are used to express relative levels of noise. First, a reference noise level is chosen, and assigned the value 0 dBa. All noise levels are then compared to this value. Noise levels lower than the reference level have negative values, such as -3 dBa. Noise levels greater than the reference level have positive values, such as $+6$ dBa.

The decibel is a means of expressing a ratio between two currents, power levels, or voltages. A reference level is therefore always necessary for the decibel to have meaning. *See also* DECIBEL.

dBd

The acronym *dBd* refers to the power gain of an antenna, in decibels, with respect to a half-wave dipole antenna. The dBd specification is the most common way of expressing antenna power gain (*see* DECIBEL, and DIPOLE ANTENNA). The reference direction of the antenna under test is considered to be the direction in which it radiates the most power. The reference direction of the dipole is broadside to the antenna conductor.

Power gain in dBd is given by the formula:

$$dBd = 10 \log_{10} (P_a/P_d)$$

where P_a is the effective radiated power from the antenna in question with a transmitter output of P watts, and P_d is the ef-

fective radiated power from the dipole with a transmitter output of P watts.

An alternative method of measuring antenna power gain in dBd is possible using the actual field-strength values. If E_a is the field strength in microvolts per meter at a certain distance from the antenna in question, and E_d is the field strength at the same distance from a half-wave dipole getting the same amount of transmitter power, then:

$$dBd = 20 \log_{10} (E_a/E_d)$$

See also ANTENNA POWER GAIN, and dBi.

dBi

The acronym *dBi* refers to the power gain of an antenna, in decibels, relative to an isotropic antenna (*see* DECIBEL, and ISOTROPIC ANTENNA). The direction is chosen in which the antenna under test radiates the best. An isotropic antenna, in theory, radiates equally well in all directions. The gain of any antenna in dBi is 2.15 dB greater than its gain in dBd; that is:

$$dBi = 2.15 + dBd$$

Power gain in dBi is given by the formula:

$$dBi = 10 \log_{10} (P_a/P_i)$$

where P_a is the effective radiated power from the antenna in question with a transmitter output of P watts, and P_i is the effective radiated power from the isotropic antenna with a transmitter output of P watts.

An alternative method of measuring antenna power gain in dBi is possible using actual field-strength values. If E_a is the field strength in microvolts per meter at a certain distance from the tested antenna, and E_i is the field strength from an isotropic antenna getting the same amount of power from the transmitter, then:

$$dBi = 20 \log_{10} (E_a/E_i)$$

Actually, an isotropic antenna is not seen in practice. It is essentially impossible to construct a true isotropic antenna. Gain figures in dBi are sometimes used instead of dBd for various reasons. *See also* ANTENNA POWER GAIN, and dBd.

dBm

The acronym *dBm* refers to the strength of a signal, in decibels, compared to 1 milliwatt, with a load impedance of 600 ohms. If the signal level is exactly 1 milliwatt, its level is 0 dBm. In general:

$$dBm = 10 \log_{10} P$$

where P is the signal level in milliwatts.

With a 600-ohm load, 0 dBm represents 0.775 volt, or 775 millivolts. With respect to voltage in a 600-ohm system, then:

$$dBm = 20 \log_{10} (E/775)$$

where E is the voltage in millivolts. A level of 0 dBm also represents a current of 1.29×10^{-3} ampere, or 1.29 milliamperes. With respect to current in a 600-ohm system:

$$dBm = 20 \log_{10} (I/1.29)$$

where I is the current in milliamperes. *See also* DECIBEL.

dBW

The abbreviation *dBW* refers to the strength of a signal, in decibels, compared to 1 watt, with a load impedance of 600 ohms. If the signal level is exactly 1 watt, the level is 0 dBW. In general:

$$dbW = 10 \log_{10}P$$

where *P* is the signal level in watts. *See also* DECIBEL.

DC

See DIRECT CURRENT.

DC AMPLIFIER

Any device intended to increase the current, power, or voltage in a direct-current circuit is called a *direct-current (dc) amplifier*. The most common type of dc amplifier is used for the purpose of increasing the sensitivity of a meter or other indicating device. Such an amplifier can be extremely simple, resembling an elementary alternating-current amplifier.

Direct-current amplifiers can be used to amplify the voltage in an automatic-level-control circuit, for the purpose of accomplishing speech compression. In such an application, the time constant of the dc amplifier is critical. The microphone amplifiers are set to a high level of gain when there is no audio input; the greater the audio signal fed to the amplifiers, the more dc is supplied by the dc amplifiers acting on the automatic-level-control voltage. This dc voltage reduces the gain of the amplifying stages as the audio input increases. *See also* AMPLIFICATION, AMPLIFIER, AUTOMATIC LEVEL CONTROL, and SPEECH COMPRESSION.

DC COMPONENT

All waveforms have a *direct-current (dc) component* and an *alternating-current (ac) component*. Sometimes one component is zero. For example, in pure dc, such as the output of a dry cell, the ac component is zero. In a 60-Hz household outlet, the dc component is zero, because the average voltage from such a source is zero.

In a complex waveform, the dc component is the average value of the voltage. This average must be taken over a sufficient period of time. Some waveforms, such as the voltage at the collector of any amplifier circuit, have significant dc components. The dc component does not always change the practical characteristics of the signal, but the dc component must be eliminated to obtain satisfactory circuit operation in some situations.

DC GENERATOR

A direct-current, or dc, generator is a source of direct current. A dc generator can be mechanical in nature, such as an alternating-current generator followed by a rectifier. A dc generator might consist of a chemical battery, a photovoltaic cell or thermocouple. The direct-current amplitude can remain constant or it can fluctuate.

Direct-current generators are commonly used for a variety of purposes. The most common example is the dry cell, which generates electricity from a chemical reaction. Solar cells are another common type of dc generator. *See also* GENERATOR.

DC GROUND

A *direct-current (dc) ground* is a dc short circuit (*see* DC SHORT CIRCUIT) to ground potential. Such a dc short circuit is provided by connecting a circuit point to the chassis of a piece of electronic equipment, either directly or through an inductor.

It is often desirable to place a component lead at dc ground potential, but still apply an alternating-current signal. An example of this is the zero-bias vacuum tube, in which the control grid is at dc ground, but which carries a large driving signal. Another example is the grounding of an antenna system through large inductors. This prevents the buildup of hazardous dc voltages on the antenna system, but does not interfere with the radiation or reception of radio-frequency signals.

DC POWER

Dc power is the rate at which energy is expended in a direct-current circuit. It is equal to the product of the direct-current voltage, *E*, and the current, *I*. This can be determined at one particular instant, or as an average value over a specified period of time. If *R* is the direct-current resistance in a circuit, and *P* is the direct-current power, then:

$$
\begin{aligned}
P &= EI \\
&= E^2/R \\
&= I^2R
\end{aligned}
$$

when units are given in volts, amperes, and watts for voltage, current, and power, respectively.

Direct-current energy is the average direct-current power multiplied by the time period of measurement. The standard unit of energy is the watt hour, although it can also be specified in watt seconds, watt minutes, kilowatt hours, or other variations. If *W* is the amount of energy expended in watt hours, then:

$$
\begin{aligned}
W &= Pt \\
&= EIt \\
&= E^2t/R \\
&= I^2Rt
\end{aligned}
$$

where *t* is the time in hours. *See also* POWER.

DC POWER SUPPLY

A *dc power supply* is either a dc generator (*see* DC GENERATOR), or a device for converting alternating current to direct current for the purpose of operating an electronic circuit. Generally, the term *dc power supply* refers to the latter type of device.

A transformer provides the desired voltage in alternating-current form. A semiconductor diode rectifies this alternating current. A capacitor smooths out the pulsations in the direct-current output of the rectifier.

Many direct-current supplies have voltage-regulation devices, highly sophisticated rectifier circuits, current-limiting circuits, and other features. Some dc power supplies deliver only a few volts at a few milliamperes; others can deliver thousands of volts and/or hundreds of amperes. *See also* CURRENT LIMITING, RECTIFICATION, RECTIFIER CIRCUITS, TRANSFORMER, and VOLTAGE REGULATION.

DC SHORT CIRCUIT

A *dc short circuit* is a path that offers little or no resistance to direct current. A resistance might or might not be present in such a situation for alternating currents.

The simplest example of a direct-current short circuit is, of course, a length of electrical conductor. However, an inductor also provides a path for direct current. But an inductor, unlike a plain length of conductor, offers reactance to alternating-current energy. (*See* INDUCTIVE REACTANCE.) Coils are often used in electronic circuits to provide a dc short circuit while offering a high amount of resistance to alternating-current signals. Such coils are called *chokes*. When it is necessary to place a circuit point at a certain dc potential, without draining off the ac signal, a choke is used. Chokes also can be used to eliminate the alternating-current component of a dc power-supply output. *See also* CHOKE.

DC-TO-AC CONVERTER

A *dc-to-ac (direct-current-to-alternating-current) converter* is a form of power supply, often used for the purpose of obtaining household-type power from a battery or other source of low-voltage direct current. The chopper power supply makes use of a dc-to-ac converter.

A dc-to-ac converter operates by modulating, or interrupting, a source of direct current. A relay or oscillating circuit is used to accomplish this. The resulting modulated direct current is then passed through a transformer to eliminate the direct-current component, and to get the desired alternating-current voltage.

When a dc-to-ac converter is designed especially to produce 120-volt, 60-Hz alternating current for the operation of household appliances, the device is called a *power inverter. See also* INVERTER.

DC-TO-DC CONVERTER

A *dc-to-dc converter* is a circuit that changes the voltage of a direct-current power supply. Such a device consists of a modulator, a transformer, a rectifier, and a filter. This circuit is similar to that of a dc-to-ac converter, except for the addition of the rectifier and filter (*see* DC-TO-AC CONVERTER).

A dc-to-dc converter can be used for either step-up or step-down purposes. Usually, such a circuit is used to obtain a high direct-current voltage from a comparatively low voltage. A fairly common type of dc-to-dc converter is used as a power supply for vacuum-tube equipment when the only available source of power is a 12-volt automotive battery or electrical system. The regulation of such a voltage step-up circuit depends on the ability of the battery or car alternator to handle large changes in the load current. A special regulator circuit is required if the voltage regulation must be precise.

A dc-to-dc converter is sometimes called a *dc transformer*. Low-power dc-to-dc converters can be built into a small integrated-circuit package.

DEAD BAND

When no signals are received within a certain frequency range in the electromagnetic spectrum, that band of frequencies is said to be *dead*. A dead band can result from geomagnetic activity that disrupts the ionosphere of the earth; in fact, the term dead band is used only on those frequency bands that are affected by ionospheric propagation. During a severe geomagnetic storm, propagation becomes virtually impossible via the ionosphere. *See* GEOMAGNETIC STORM, and IONOSPHERE.

A dead band can be caused by the deterioration of the ionosphere with the setting of the sun, with low sunspot activity, and perhaps with coincidences of unknown origin.

A band can sometimes appear dead simply because no one is transmitting on it at a particular time. Amateur radio operators, when communicating for recreation, have sometimes listened to what they thought was a dead band, called CQ (*see* CQ), and found that the band was far from dead!

A band can go dead for just a few seconds or minutes, or it can remain unusable for hours or days. The range of frequencies can be as small as a few kilohertz, or can extend for several megahertz. *See also* PROPAGATION CHARACTERISTICS.

The term *dead band* is also used to describe the lack of response of a servomotor system through a part of its arc or range of operation. This lack of response can be caused by backlash in the gears or rotor of the servo, or by a lack of resolution of the position-sensing potentiometer (or other device) that feeds angular or position information to the servo system. The servo dead band can be expressed as degrees of arc, or as a percentage of total travel.

DEBUGGING

The process of perfecting the operation of an electronic circuit or computer program is called *debugging*. Literally, this means getting the bugs out! Usually, when an electronic circuit is first tested after it has been built exactly according to the plans, or when a computer program is run after it has been meticulously composed, a problem becomes evident. Sometimes, the circuit or program fails to work at all. Only rarely does the circuit or program work perfectly the first time it is tested. Therefore, debugging is almost always necessary.

The process of debugging can be very simple; it might, for instance, involve only a small change in the value of a component. Sometimes, the debugging process requires that the entire design process be started all over. Occasionally, the bugs are hard to find, and do not appear until the device has been put into mass production or the program has been published and extensively used. A debugging test must thus be very thorough so that the chances of production problems are minimized.

DECADE

A range of any parameter, such that the value at one end of the range is 10 times the value at the other end, is called a decade. The radio-frequency bands are arbitrarily designated as decades: 30 to 300 kHz is called the *low-frequency band*, 0.3 to 3 MHz is called the *medium-frequency band*, 3 to 30 MHz is called the *high-frequency band*, and so on.

There are an infinite number of decades between any quantity and the zero value for that parameter. For example, we can speak of frequency decades of 30 to 300 kHz, 3 to 30 kHz, 0.3 to 3 kHz, 30 to 300 Hz, and so on, without end, and we will never actually reach a frequency of zero. Of course, a parameter can be increased indefinitely, too, without end.

The decade method of expressing quantities is used by scientists and engineers quite often, because its logarithmic nature allows the evaluation of a larger range of quantities than is the case with a simple linear system.

DECADE BOX

A *decade box* is a device used for testing a circuit. A set of resistors, capacitors, or inductors is connected together via switches in such a way that values can be selected digit-by-digit in decade fashion.

The illustration shows a schematic diagram of a two-digit decade capacitance box (usually, more digits are provided, but for simplicity, this circuit shows only two). Switch S1 selects any of ten capacitance values in microfarads: 0.00, 0.01, 0.02, 0.03, . . ., 0.09. Switch S2 also selects any of 10 values of capacitance in microfarads, each 10 times the values of capacitance in the circuit containing S1: 0.0, 0.1, 0.2, 0.3, 0.9.

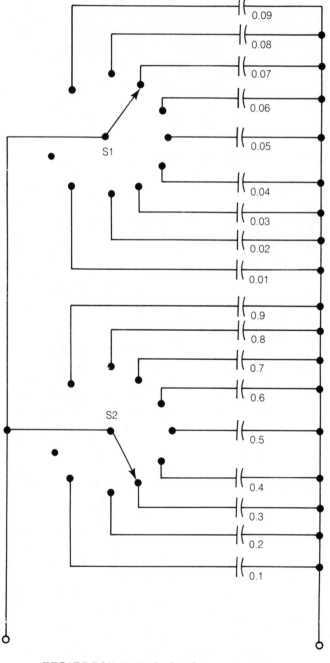

DECADE BOX: A simple decade capacitance box.

Therefore, there are 100 possible values of capacitance that can be selected by this system, ranging from 0.00 to 0.99 μF in increments of 0.01 μF.

Decade boxes are sometimes used to set the frequency of a digital radio receiver or transmitter. A signal generator and monitor can have a range of zero to 999.999999 MHz or higher, with frequencies selectable in increments as small as 1 Hz. This gives as many as 10^9 possible frequencies, with only nine independent selector switches. *See also* DECADE.

DECADE COUNTER

A counter that proceeds in decimal fashion, beginning at zero and going through 9, then to 10 and up to 99, and so on, is called a *decade counter*. A decade counter operates in the familiar base-10 number system. Counting begins with the ones digit (10^0), then the tens digit (10^1), and so on up to the 10^n digit. In a counter going to the 10^n digit, there are $n + 1$ possible digits.

The display on a decade counter often, if not usually, contains a decimal point. Decades can proceed downward toward zero, as well as upward to ever-increasing levels. However, fractions of a pulse or cycle are never actually counted. Rather, the pulses or cycles are counted for a longer time when greater accuracy is needed. For example, a counting time of 10 seconds gives one additional digit of accuracy, as compared to a counting time of 1 second. This allows decimal parts of a cycle to be determined. *See also* DECADE, and FREQUENCY COUNTER.

DECAY

The falling-off in amplitude of a pulse or waveform is called the *decay*. The decay of a pulse or waveform, although appearing to be instantaneous, is never actually so. A certain amount of time is always required for decay.

When you switch off a high-wattage incandescent bulb, for example, you can actually watch its brightness decay. But the decay in brilliance of a neon bulb or light-emitting diode is too rapid to be seen. Nevertheless, even a light-emitting diode has a finite brightness-decay time.

The decay curve is a logarithmic function. *See also* DAMPED WAVE, and DECAY TIME.

DECAY TIME

The *decay time* of a pulse or waveform is the time required for the amplitude to decay to a certain percentage of the maximum amplitude. The time interval begins at the instant the amplitude starts to fall, and ends when the determined percentage has been attained.

The decay of a pulse or waveform proceeds in a logarithmic manner. Therefore, in theory, the amplitude never reaches zero. In practice, of course, a point is always reached at which the pulse or wave amplitude can be considered to be zero. This point can be chosen for the determination of the decay time interval. In a capacitance-resistance circuit, the decay time is considered to be the time required for the charge voltage to drop to 37 percent of its maximum value. In this case, the final amplitude is equal to the initial amplitude divided by e, where e is approximately 2.718. *See also* DECAY, and TIME CONSTANT.

DECIBEL

The *decibel* is a means of measuring relative levels of current, voltage, or power. A reference current I_0, or voltage E_0, or power P_0 must first be established. Then, the ratio of an arbitrary current I to the reference current I_0 is given by:

$$dB = 20 \log_{10} (I/I_0)$$

The ratio of an arbitrary voltage E to the reference voltage E_0 is given by:

$$dB = 20 \log_{10} (E/E_0)$$

A negative decibel figure indicates that I is smaller than I_0, or that E is smaller than E_0. A positive decibel value indicates that I is larger than I_0, or that E is larger than E_0. A tenfold increase in current or voltage, for example, is a change of $+20$ dB:

$$dB = 20 \log_{10} (10E_0/E_0)$$
$$= 20 \log_{10} (10)$$
$$= 20 \times 1 = 20$$

For power, the ratio of an arbitrary wattage P to the reference wattage P_0 is given by:

$$dB = 10 \log_{10} (P/P_0)$$

As with current and voltage, a negative decibel value indicates that P is less than P_0; a positive value indicates that P is greater than P_0. A tenfold increase in power represents an increase of $+10$ dB:

$$dB = 10 \log_{10} (10P_0/P_0)$$
$$= 10 \log_{10} (10)$$
$$= 10 \times 1 = 10$$

The reference value for power is sometimes set at 1 milliwatt (0.001 watt). Decibels measured relative to 1 milliwatt, across a pure resistive load of 600 ohms, are abbreviated dBm. Decibel figures are extensively used in electronics to indicate circuit gain, attenuator losses, and antenna power gain figures. *See also* ANTENNA POWER GAIN, dBa, dBd, dBi, and dBm.

DECIBEL METER

A *decibel meter* is a meter that indicates the level of current, voltage, or power, in decibels, relative to some fixed reference value. The reference value can be arbitrary, or it can be some specific quantity, such as 1 milliwatt or 1 volt. In any case, the reference value corresponds to 0 dB on the meter scale. Levels greater than the reference level are assigned positive decibel values on the scale. Levels lower than the reference are assigned negative values.

The "S" meters on many radio receivers, calibrated in S units and often in decibels as well, are forms of decibel meters. A reading of 20 dB over S9 indicates a signal voltage 10 times as great as the voltage required to produce a reading of S9. An S meter can be helpful in comparing the relative levels of signals received on the air. The reading of S9 corresponds to some reference voltage at the antenna terminals of the receiver, such as 10 microvolts. *See also* S METER.

DECIMAL

The term *decimal* is used to refer to a base-10 number system. In this system, which is commonly used throughout the world, and which is the most familiar to us, numbers are represented by combinations of 10 different digits in various decimal places.

The digit farthest to the right, but to the left of the decimal point, is multiplied by 10^0, or 1; the digit next to the left is multiplied by 10^1, or 10; the digit to the left of this is multiplied by 10^2, or 100. With each move to the left, the base value of the digit increases by a factor of 10 so that the nth digit to the left of the decimal point is multiplied by 10^{n-1}.

The digit first to the right of the decimal point is multiplied by 10^{-1}, or $1/10$; the digit next to the right is multiplied by 10^{-2}, or $1/100$. This process continues so that with each move to the right, the base value of the digit decreases by a factor of 10. Therefore, the nth digit to the right of the decimal point is multiplied by 10^{-n}.

Ultimately, the number represented by a decimal sequence is determined by adding the decimal values of all the digits. For example, the number 27.44 is equal to $2 \times 10^1 + 7 \times 10^0 + 4 \times 10^{-1} + 4 \times 10^{-2}$. We do not normally think of the value of 27.44 in this way, however.

Although we use the decimal system in our everyday lives, digital circuits generally operate in a base-2 number system, where the only digits are 0 and 1, and where the values increment and decrement in powers of 2. *See also* BINARY-CODED NUMBER.

DECLINATION

For satellite communications and moonbounce, it is often necessary to know the *celestial coordinates* toward which the antenna should be aimed. One of these coordinates is called the *declination*.

Declination is a measure of the extent to which an object lies north or south of the equator. It is specified in degrees north or south, like latitude. North is positive, and south is negative.

If the declination of an object in the sky is $+40$ degrees, the object lies 40 degrees north of the *celestial equator*. If a line could be drawn connecting this object and the center of the earth, that line would pass through the earth's surface at 40 degrees north latitude. For an object with declination -66 degrees, such a line would pass through the earth's surface at 66 degrees south latitude. Declination angles can be thought of as latitude circles in the sky. *See also* RIGHT ASCENSION.

DECODING

Decoding is the process of converting a message, received in code, into plain language. This is generally done by a machine, although in the case of the Morse code, a human operator often acts as the decoding medium.

Messages can be coded either for the purpose of efficiency and accuracy, such as with the Morse code or other codes, or for the purpose of keeping a message secret, as with voice scrambling or special abbreviations. Both types of code can be used at the same time. In this case, decoding requires two steps: one to convert the scrambling code to the plain text, and the other to convert the code itself to English or another language.

The conversion of a digital signal to an analog signal is sometimes called decoding. The opposite of the decoding process—the conversion of an analog signal to a digital signal, or the transformation of a plain-language message into coded form — is called encoding. Decoding is always done at the receiving end of a communications circuit. *See also* DIGITAL-TO-ANALOG CONVERTER.

DECOMPOSITION VOLTAGE

See BREAKDOWN VOLTAGE.

DECOUPLING

When undesired coupling effects must be minimized, a technique called *decoupling* is used. For example, a multistage amplifier circuit will often oscillate because of feedback among the stages. This oscillation usually takes place at a frequency different from the operating frequency of the amplifier. In order to reduce this oscillation, or eliminate it entirely, the interstage coupling should be made as loose as possible, consistent with proper operation at the desired frequency.

Another form of decoupling consists of the placement of chokes and/or capacitors in the power-supply leads to each stage of a multistage amplifier. This minimizes the chances of unwanted interstage coupling through the power supply.

When several different loads are connected to a single transmission line, such as in a multiband antenna system, resonant circuits are sometimes used to effectively decouple all undesired loads from the line at the various operating frequencies. The trap antenna decouples a part of the radiator, to obtain resonance on two or more different frequencies. *See also* DECOUPLING STUB, TRAP, and TRAP ANTENNA.

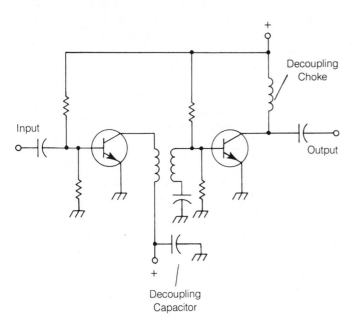

DECOUPLING: Reduces the effects of feedback.

DECOUPLING STUB

A *decoupling stub* is a length of transmission line that acts as a resonant circuit at a particular frequency, and is used in an antenna system in place of a trap (*see* TRAP). Such a stub usually consists of a ¼-wavelength section of transmission line, short-circuited at the far end. This arrangement acts as a parallel-resonant inductance-capacitance circuit.

At the resonant frequency of the stub, the impedance be-

tween the input terminals is extremely high. Therefore, such a stub can be used to decouple a circuit at the resonant frequency (*see* DECOUPLING). Such decoupling might be desired in a multiband antenna system, or to aid in the rejection of an unwanted signal.

A ¼-wavelength section of transmission line, open at the far end, will act as a series-resonant inductance-capacitance circuit. At the resonant frequency, such a device, also called a *stub*, has an extremely low impedance, essentially equivalent to a short circuit. This kind of stub can be extremely effective in rejecting signal energy at unwanted frequencies. By connecting a series-resonant stub across the antenna terminals, spurious responses or emissions are suppressed at the resonant frequency of the stub. The drawing illustrates the use of parallel-equivalent (A) and series-equivalent (B) stubs in antenna and feedline systems.

Some stubs are ½ wavelength long, rather than ¼ wavelength. A short-circuited ½-wavelength stub acts as a series-resonant circuit, and an open-circuited ½-wavelength stub acts as a parallel-resonant circuit. All stubs show the same characteristics at odd harmonics of the fundamental frequency.

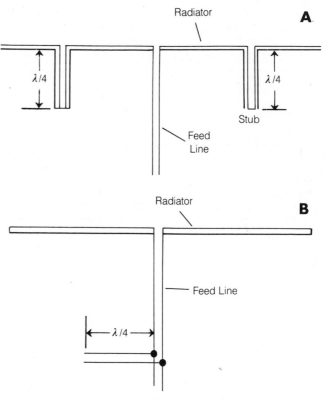

DECOUPLING STUB: At A, stub traps; at B, a stub used as a band-rejection filter.

DEEMPHASIS

Deemphasis is the deliberate introduction of a lowpass type response into the audio-frequency stages of a frequency-modulation receiver. This is done to offset the preemphasis introduced at the transmitter.

By introducing preemphasis at the transmitter and deemphasis at the receiver in a frequency-modulation communications system, the signal-to-noise ratio at the upper end of the audio range is improved. This is because, as the transmitted-signal modulating frequency increases, the amplitude increases (because of preemphasis at the transmitter); when this amplitude is brought back to normal by deemphasis in the receiver, the noise is attenuated as well. *See also* FREQUENCY MODULATION, and SIGNAL-TO-NOISE RATIO.

DEFIBRILLATION

An electric shock can cause the regular rhythm of the heartbeat to stop, and an uncoordinated twitching of the heart muscles occurs instead. This is called *fibrillation* (see ELECTRIC SHOCK). Unless the normal heart action is restored within a few minutes, death results because the blood supply to the body is cut off. A fibrillating heart does not effectively pump blood through the lungs to be oxygenated, nor can the blood get to the rest of the body.

The heart functions by means of electrical nerve impulses. A heart pacemaker actually regulates the heartbeat by transmitting electrical signals to the heart muscle. A device called a defibrillator also works via electrical impulses. The defibrillator produces an electric shock or series of shocks, which often gets the heart beating normally when it is in a state of fibrillation. Two metal electrodes are placed on the chest of the victim, in such a position that the current is sent through the heart.

If a defibrillator is not immediately available when a shock victim goes into fibrillation, the only alternative is to apply cardiopulmonary resuscitation until medical help arrives. *See also* CARDIOPULMONARY RESUSCITATION.

DEFLECTION

Deflection is a deliberately induced change in the direction of an energy beam. The beam can consist of sound waves, radio waves, infrared radiation, visible light, ultraviolet radiation, X rays, or atomic-particle radiation such as a stream of electrons.

In a cathode-ray tube, electron beams are deflected to focus and direct the energy to a certain spot on a phosphor screen. In loudspeaker enclosures, deflecting devices are used to get the best possible fidelity. Deflectors called *baffles* are used for acoustic purposes in concert halls and auditoriums. Deflecting mirrors are used in optical telescopes. Heat deflectors are sometimes used to improve energy efficiency in homes and buildings. *See also* CATHODE-RAY TUBE.

DEGREE

The *degree* is a unit of either temperature or angular measure. There are three common temperature scales. In the Fahrenheit scale, which is most generally used in the United States, the freezing point of pure water at one atmosphere pressure is assigned the value 32 degrees. The boiling point of pure water at one atmosphere is 212 degrees on the Fahrenheit scale. In the Celsius temperature scale, water freezes at 0 degrees and boils at 100 degrees. Thus, a degree in the Celsius scale is representative of a greater change in temperature than a degree in the Fahrenheit scale. In the Kelvin, or absolute, temperature scale, the degrees are the same size as in the Celsius scale, except that 0 degrees Kelvin corresponds to absolute zero, which is the coldest possible temperature in the physical universe. In the respective temperature scales, readings are given in degrees F (Fahrenheit), degrees C (Celsius), or degrees K (Kelvin).

In angular measure, a degree represents $1/360$ of a complete circle. The number 360 was chosen in ancient times, when scientists and astronomers noticed that the solar cycle repeated itself approximately once in 360 days. One day thus corresponds to a degree in the circle of the year (it is fortunate indeed that the ancients were slightly off in their count; otherwise, we might be using degrees of measure of one part in 365.25!).

Phase shifts or differences are usually expressed in degrees, with one complete cycle represented by 360 degrees of phase. *See also* CYCLE.

DEHUMIDIFICATION

The operation of some electronic circuits is affected not only by the temperature, but also by the relative humidity. If the amount of water vapor in the air is too great, corrosion is accelerated, and condensation can occur. In an electrical circuit, condensation can cause unwanted electrical conduction between parts that should be isolated. Condensation can also cause the malfunction of high-speed switches. *Dehumidification* is the process of removing excess moisture from the air.

There are various ways to obtain dehumidification. The simplest method is to raise the temperature; for a given amount of water vapor in the air, the relative humidity decreases as the temperature rises. Dry crystals of calcium chloride or cobalt chloride, packed in a cloth sack, will absorb water vapor from the air, and will help to dehumidify an airtight chamber. Various dehumidifying sprays are also available.

DELAY CIRCUIT

A *delay circuit* is a set of electronic components designed deliberately for the purpose of introducing a time or phase delay. Such a circuit might be a passive combination of resistors, inductors, and/or capacitors. A delay circuit can consist of a simple length of transmission line. Or, the device can be an active set of integrated circuits and peripheral components.

Delay circuits are used in a wide variety of applications. Broadcast stations delay their transmissions by approximately seven seconds. This allows the signal to be cut off, if necessary, before it is transmitted over the air. Phase-delay circuits are extremely common. Certain kinds of switches and circuit breakers have a built-in time delay. Sometimes, delay is undesirable and hinders the performance of a circuit. *See also* DELAY DISTORTION, DELAYED AUTOMATIC GAIN CONTROL, DELAYED MAKE/BREAK, DELAYED REPEATER, DELAY TIME, and DELAY TIMER.

DELAY DISTORTION

In some electronic circuits, the propagation time varies with the signal frequency. When this happens, distortion occurs because signal components having different frequencies arrive at the receiving end of the circuit at different times. This is *delay distortion*, which can happen in a radio communications circuit, in a telephone system, or even within a single piece of electronic equipment.

Generally, higher frequencies are propagated at a lower rate of speed than lower frequencies. If the propagation time is extremely short, the delay distortion will be inconsequential. But the longer the propagation time, the greater the chances of delay distortion. Delay distortion can be minimized by making the percentage difference between the lowest and highest frequencies in a signal as small as possible. A baseband signal, with components as low as 100 Hz and as high as 3000 Hz, for example, is more subject to delay distortion than a single-sideband signal with the same audio characteristics and a suppressed-carrier frequency of 1 MHz. In the former case, the percentage difference between the lowest and highest frequencies is very large, but in the latter case, it is extremely small.

DELAYED AUTOMATIC GAIN CONTROL

A *delayed automatic-gain-control circuit* is a special form of automatic-gain-control, or AGC, circuit (*see* AUTOMATIC GAIN CONTROL). It is used in many communications receivers.

In a delayed AGC circuit, signals below a certain threshold level are passed through the receiver with maximum gain. Only when the signal strength exceeds this threshold amplitude does the AGC become active. Then, as the signal strength continues to increase, the AGC provides greater and greater attenuation.

The delayed AGC circuit allows better weak-signal reception than does an ordinary automatic-gain-control circuit.

DELAYED MAKE/BREAK

When a circuit is closed or opened a short while after the actuating switch or relay is energized or deenergized, the condition is called *delayed make or break*. For example, the contacts of a relay can close several milliseconds, or even several seconds or minutes, following application of current to the circuit. The contacts of a relay or other switching device might not open until some time after current has been removed. The former device is a delayed-make circuit; the latter is a delayed-break circuit.

Delayed-make and delayed-break devices are sometimes used in power supplies. For example, in a tube-type power amplifier, the filament voltage should be applied a few seconds or minutes before the plate voltage. A delayed-make circuit can be used in the plate supply, accomplishing this function automatically. *See also* DELAY TIMER.

DELAYED REPEATER

A *delayed repeater* is a device that receives a signal and retransmits it later. The delay time between reception and retransmission can vary from several milliseconds to seconds or even minutes. Generally, a delayed repeater records the signal

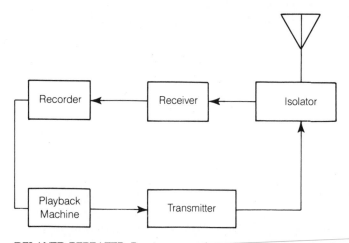

DELAYED REPEATER: Receives, records, and retransmits a signal.

modulation envelope on magnetic tape, and plays the tape back into its transmitter.

A delayed repeater operates in essentially the same way as an ordinary repeater. The signal is received, demodulated, and retransmitted at a different frequency. An isolator circuit prevents interference between the receiver and the transmitter. The block diagram shows a simple delayed-repeater circuit. *See also* REPEATER.

DELAY LINE

A delay line is a circuit, often (but not necessarily) a length of transmission line, that provides a delay for a pulse or signal traveling through it. All transmission lines carry energy with a finite speed. For example, electromagnetic fields travel along a solid-polyethylene-dielectric coaxial cable at approximately 66 percent of the speed of light.

In this type of line, the center conductor is wound into a helix, like a spring, thus greatly increasing its length. This increases the propagation delay per unit length of the line. *See also* DELAY TIME.

DELAY TIME

The *delay time* is the length of time required for a pulse or signal to travel through a delay circuit or line, as compared to its travel over the same distance through free space. (*See* DELAY CIRCUIT, and DELAY LINE.) Delay time is measured in seconds, milliseconds (10^{-3} second), microseconds (10^{-6} second), or nanoseconds (10^{-9} second).

Delay time is ordinarily measured with an oscilloscope. A pulse or signal, supplied by the signal generator, is run through the delay circuit and also directly to the oscilloscope. This results in two pulses or waveforms on the oscilloscope screen. The frequency of the pulses or signal is varied to be certain that the delay time observed is correct (it could be more than one cycle, misleading a technician if only one wavelength is used). The delay time is indicated by the separation of pulses or waves on the oscilloscope display.

In a delay line with length m, in meters, and velocity factor v, in percent, the delay time t, in nanoseconds, is given by:

$$t = 333m/v$$

assuming the transmission and reception points are located negligibly close together for a signal traveling through free space. If the transmission and reception points are located n meters apart, then the delay time in nanoseconds is:

$$t = 333m/v - 3.33n$$

See also VELOCITY FACTOR.

DELAY TIMER

Any device that introduces a variable delay in the switching of a circuit is called a *delay timer*. Such a timer usually has a built-in, resettable clock. After the prescribed amount of time has elapsed, the switching is performed.

An example of the use of a delay timer is the power-up of a large vacuum-tube radio-frequency transmitter. The filament voltage is applied as the delay timer is first set to approximately 2 minutes. The plate voltage, however, does not get switched in until the timer has completed its cycle. Such a timer can be mechanical or electronic.

DELLINGER EFFECT

A sudden solar eruption causes an increase in the ionization density of the upper atmosphere of the earth. At high frequencies, especially those wavelengths known for long-distance daytime propagation, such ionization can cause an abrupt cessation of communications. This is called the *Dellinger effect.*

Normally, signals are propagated via the ionospheric E and F layers. The lower layer, called the *D layer,* is not ordinarily ionized to a sufficient density to affect signals in long-distance propagation. However, a solar eruption causes a dramatic increase in the ionization density of all layers, including the D layer. This results in absorption of electromagnetic energy by the D layer, with a consequent disappearance of signals at distant points. *See also* D LAYER, and PROPAGATION CHARACTERISTICS.

DELTA MATCH

A *delta match* is a method of matching the impedance of an antenna to the characteristic impedance of a transmission line. The delta-matching technique is used with balanced antennas and two-wire transmission lines.

The illustration shows a delta matching system. The length of the network, *m,* and the width or spacing between the connections, *s,* is adjusted until the standing-wave ratio on the feed line is at its lowest value. The length of the radiating element is ½ wavelength.

A variation of the delta match is called the *T match.* When the transmission line is unbalanced, such as is the case with a coaxial cable, a gamma match can be used for matching to a balanced radiating element. *See also* GAMMA MATCH, and T MATCH.

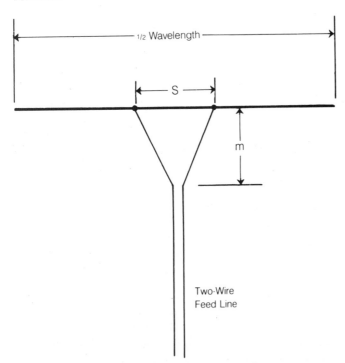

DELTA MATCH: Matches a feed line to an antenna.

DELTA MODULATION

Delta modulation is a form of pulse modulation (*see* PULSE MODULATION). In the delta-modulation scheme, the pulse spacing is constant, generated by a clock circuit. The pulse amplitude is also constant. But the pulse polarity can vary, being either positive or negative.

When the amplitude of the modulating waveform is increasing, positive unit pulses are sent. When the modulating waveform is decreasing in amplitude, negative unit pulses are sent. When the modulating waveform amplitude is not changing, the pulses alternate between positive and negative polarity. Delta modulation gets its name from the fact that it follows the difference, or derivative, of the modulating waveform.

In the delta-modulation detector, the pulses are integrated. This results in a waveform that closely resembles the original modulating waveform. A filter eliminates most of the distortion caused by sampling effects. The integrator circuit can consist of a series resistor and a parallel capacitor; the filter is usually of the lowpass variety. *See also* INTEGRATION.

DEMODULATION
See DETECTION.

DE MORGAN'S THEOREM

De Morgan's theorem, also called the *de Morgan laws,* involves sets of logically equivalent statements. Let the logical operation AND be represented by multiplication and the operation OR be represented by addition, as in Boolean algebra (*see* BOOLEAN ALGEBRA). Let the NOT operation be represented by comple-

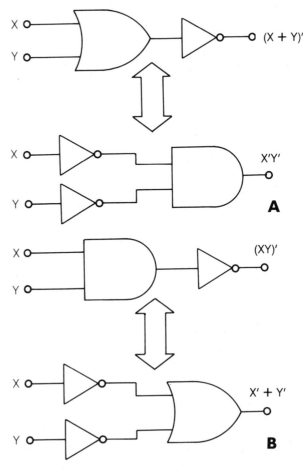

DE MORGAN'S THEOREM: Both pairs of circuits (A and B) are logically equivalent.

mentation, indicated by an apostrophe ('). Then, de Morgan's laws are stated as:

$$(X + Y)' = X'Y'$$
$$(XY)' = X' + Y'$$

The illustration shows these rules in schematic form, showing logic gates, at A and B.

From these rules, it can be proven that the same laws hold for any number of logical statements $X_1, X_2, \ldots, X_n$:

$$(X_1 + X_2 + \ldots + X_n)' = X_1'X_2' \ldots X_n'$$
$$(X_1 X_2 \ldots X_n)' = X_1' + X_2' + \ldots + X_n'$$

De Morgan's theorem is useful in the design of digital circuits because a given logical operation can be easier to obtain in one form than in the other.

DEMULTIPLEXER

A *demultiplexer* is a circuit that separates multiplexed signals (*see* MULTIPLEX). Multiplexing is a method of transmitting several low-frequency signals on a single carrier having a higher frequency. Multiplexing can also be done on a time-sharing basis among many signals. A demultiplexer at the receiving end of a multiplex communications circuit allows interception of only the desired signal, and rejection of all the other signals.

In a demultiplexer, the main carrier is amplified and detected (*see* DETECTION), obtaining the subcarriers. A selective circuit then filters out all subcarriers except the desired one. A second detector demodulates the subcarrier, obtaining the desired signal. The main carrier modulation does not have to be the same as the subcarrier modulation. Thus, different types of detectors can be needed for the two carriers. For example, the main carrier might be amplitude-modulated, while the subcarrier is frequency-modulated. *See also* MODULATION.

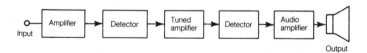

DEMULTIPLEXER: A block diagram of demultiplexer.

DENSITY MODULATION

Density modulation is the modulation of the intensity of a beam of particles, by varying only the number of particles. This type of modulation might be used in the electron stream of a cathode-ray tube. The greater the density of the electrons in the beam, the brighter the phosphorescent screen will glow when the electrons strike it. This is because, with increasing electron-beam density, more electrons strike the screen per unit time. The density is directly proportional to the number of particles passing any given point in a certain span of time.

In density modulation, all the particles of the beam move at constant speed, regardless of the intensity of the beam. When the speed of the electrons changes, the beam is said to be velocity-modulated. Density modulation can be combined with velocity modulation in some situations.

DEPENDENT CURRENT

In a bipolar transistor, the collector current depends on various factors, primarily the base current. In a field-effect transistor, the drain current depends mainly on the gate voltage. In a vacuum tube, the plate current depends mainly on the grid voltage. In a resistor, the current depends on the voltage across the component, and also on the resistance.

Any current that is a predictable function of one or more variables is called a *dependent current*. Dependent current is an important consideration in active amplifying devices.

DEPLETION LAYER

In a depletion-mode field-effect transistor, the region within the channel that does not conduct is called the *depletion layer* (*see* DEPLETION MODE, and FIELD-EFFECT TRANSISTOR). A depletion layer forms at the junction of an N-type semiconductor and a P-type semiconductor whenever the junction is reverse-biased (*see* P-N JUNCTION).

A depletion layer is entirely devoid of charge carriers, and therefore no current can flow through the region. This is why a semiconductor diode will not conduct in the reverse direction. The depletion layer acts as a dielectric, and, in fact, a reverse-biased diode can be used as a capacitor.

The depletion layer reduces the cross-sectional area of the channel in a depletion-mode field-effect transistor when the gate-channel junction is reverse-biased. The greater the reverse-bias voltage, the larger the depletion layer becomes, until finally the channel is totally pinched off, and conduction ceases. *See also* PINCHOFF.

DEPLETION MODE

The *depletion mode* is a form of field-effect-transistor operation (*see* FIELD-EFFECT TRANSISTOR). In the field-effect transistor functioning as a depletion-mode device, the current flows through a region called the channel. Two electrodes on either side of the channel (consisting of the opposite type of semiconductor material, connected together, and called the gate) are used to regulate the current through the device. The construction of a depletion-mode field-effect transistor is shown in the illustration.

In the depletion-mode field-effect transistor, the channel can consist of either N-type or P-type semiconductor material. When the gates are at the same potential as the source, a fairly large current flows through the channel. In an N-channel device, the application of a negative voltage to the gate constricts the channel because of the reverse bias between the gate-channel P-N junction. A depletion layer is formed within the channel (*see* DEPLETION LAYER). In a P-channel field-effect transistor, a positive voltage at the gate has this same effect. As the reverse bias is increased, the device will eventually stop conducting, because the depletion layer gets wider and wider. When the depletion layer extends all the way across the channel, the field-effect transistor is said to be pinched off.

Some field-effect transistors normally do not conduct, and require a forward bias for operation. These devices are called *enhancement-mode field-effect transistors. See also* DRAIN, ENHANCEMENT MODE, GATE, and PINCHOFF.

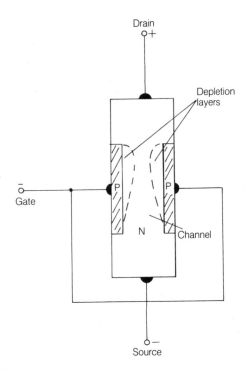

DEPLETION MODE: Conductivity depends on channel width.

DEPOLARIZATION

Any electrolyte substance is subject to the formation of gases or other deposits around the electrodes in the presence of the flow of current. When this occurs, it is called *polarization* (*see* PO-LARIZATION). Polarization can prevent a dry cell or storage battery from operating properly. The deposits can act as electrical insulators, cutting off the flow of current by increasing the internal resistance.

Depolarization is the process of preventing, or at least minimizing, polarization in an electric cell. In dry cells, the compound manganese dioxide is used in the electrolyte material to retard polarization.

DERATING CURVE

A *derating curve* is a function that specifies the amount of dissipation a component can withstand safely. Such a function is plotted with time as the dependent variable, or ordinate, and dissipation or temperature as the independent variable, or abscissa.

DESCENDING NODE

Most earth-orbiting satellites have a *groundtrack* that crosses the equator twice for each orbit. The only exceptions are geostationary satellites (*see* GEOSTATIONARY ORBIT) and satellites whose orbits are exactly over the equator. The points where, and times when, the groundtrack is exactly over the equator are called *nodes.*

For every orbit, there is one node when the groundtrack moves from the southern hemisphere to the northern, and one node when the groundtrack moves from the northern hemisphere to the southern. The second of these, going north-to-south, is called the *descending node.* The position of the node is given in degrees and minutes of longitude. *See also* ASCENDING NODE.

DESCENDING PASS

Most ham satellites have a *groundtrack* that moves over the surface of the earth, sometimes north of the equator and sometimes south of it (*see* GROUNDTRACK). Only geostationary satellites have groundtracks that do not move (*see* GEOSTATIONARY ORBIT). A few satellites orbit directly over the equator.

When a satellite has an orbit that is slanted relative to the equator, its groundtrack is moving generally southwards half of the time. This period starts when the satellite attains the northernmost latitude in its orbit, and lasts until it reaches its southernmost latitude. For any given earthbound location, a *descending pass* is the time during which the satellite is accessible, while it is moving generally southwards. The pass time depends on the altitude of the satellite, and on how close its groundtrack comes to the earth-based station. *See also* ASCENDING NODE, ASCENDING PASS, and DESCENDING NODE.

DESENSITIZATION

Desensitization is a reduction in the gain of an amplifier, particularly when the amplifier is used in the radio-frequency stages of a receiver. Desensitization can be introduced deliberately; the radio-frequency (RF) gain control is used for the purpose of reducing the sensitivity of a receiver. Desensitization can occur when it is not wanted, and this sometimes presents a problem in radio reception.

The most frequent cause of unwanted desensitization of a radio-frequency receiver is an extremely strong signal near the operating frequency. Such a signal, after passing through the front-end tuning network, can drive the first RF-amplifier stage to the point where its gain is radically diminished. When this happens, the desired signal appears to fade out. If the interfering signal is strong enough, reception can become impossible near its frequency.

Some receivers are more susceptible to unwanted desensitization, or "desensing," than others. Excellent front-end selectivity, and the choice of amplifier designs that provide uniform gain for a wide variety of input signal levels, make this problem less likely. Sometimes, a tuned circuit in the receiver antenna feed line is helpful. *See also* FRONT END.

DESOLDERING TECHNIQUE

When replacing a faulty component in an electronic circuit, it is usually necessary to desolder some connections. The method of desoldering used in a given situation is called the *desoldering technique.*

With most printed-circuit boards, desoldering technique consists of the application of heat with a soldering iron, and the conduction of solder away from the connection by means of a braided wire called solder wick. The connection and the wick must both be heated to a temperature sufficient to melt the solder. Excessive heat, however, should be avoided so that only the desired connection is desoldered, and so that the circuit board and nearby components are not damaged.

Many sophisticated desoldering devices are available. One popular device uses an air-suction nozzle, which literally swallows the solder by vacuum action as a soldering iron heats the connection. This sort of apparatus is especially useful when a large number of connections must be desoldered because it works very fast. It is also useful in desoldering tiny connections, where there is little room for error.

For large connections, such as wire splices and solder-welded joints, it is sometimes better to remove the entire connection than to try to desolder it. *See also* SOLDER, SOLDERING GUN, SOLDERING IRON, and SOLDERING TECHNIQUE.

DETECTION

Detection is the extraction of the modulation from a radio-frequency signal. A circuit that performs this function is called a detector. Different forms of modulation require different detector circuits.

The simplest detector consists of a semiconductor diode, which passes current in only one direction. Such a detector is suitable for demodulation of amplitude-modulated (AM) signals. By cutting off one half of the signal, the modulation envelope is obtained. A Class-AB, B, or C amplifier can be used to perform this function, and also provide some gain. This is called envelope detection (*see* CLASS-AB AMPLIFIER, CLASS-B AMPLIFIER, CLASS-C AMPLIFIER, and ENVELOPE DETECTOR).

There are several different ways of detecting a frequency-modulated (FM) signal. These methods also apply to phase modulation. An ordinary AM receiver can sometimes be used to receive FM by means of a technique called *slope detection* (*see* SLOPE DETECTION). Circuits that actually sense the frequency or phase fluctuations of a signal are called the *discriminator* and *ratio detector* (*see* DISCRIMINATOR, and RATIO DETECTOR).

Single-sideband (SSB) and continuous-wave (CW) signals, as well as signals that use frequency-shift keying (FSK), require a product detector for their demodulation (*see* PRODUCT DETECTOR). Such a circuit operates by mixing the received signal with the output of a local oscillator, resulting in an audio-frequency beat note.

Other kinds of detectors, such as digital-to-analog converters, are needed in specialized situations. The detector circuit in a radio receiver is generally placed after the radio-frequency or intermediate-frequency amplifying stages, and ahead of the audio-frequency amplifying stages. *See also* AMPLITUDE MODULATION, FREQUENCY MODULATION, FREQUENCY-SHIFT KEYING, PHASE MODULATION, and SINGLE SIDEBAND.

DETECTION METHODS

Detection is the process of recovering information from a signal. At the transmitter, the information is impressed on the carrier wave by means of a *modulator*. *See* MODULATION METHODS.

The earliest form of detection was via a diode. This device rectifies the signal coming through it, outputting a pulsating direct current that varies in accordance with amplitude changes in the incoming signal. The modulation *envelope* is recovered from the carrier wave of an amplitude-modulated (AM) signal. This kind of detector is still used in AM receivers today. *See* ENVELOPE DETECTOR.

For code (CW), frequency-shift keying (FSK), single-sideband (SSB) and high-frequency slow-scan television (SSTV) signals, a *product detector* is used. *See* PRODUCT DE-

TECTOR. A *direct-coversion* technique will also work (although not as well as the product detector) and is sometimes used in inexpensive, homebrew portable receivers. *See* DIRECT-CONVERSION RECEIVER.

Frequency and phase modulation are detected by means of the *discriminator*, *phase-locked loop* or *ratio detector*. *See* DISCRIMINATOR, PHASE-LOCKED LOOP, and RATIO DETECTOR.

Digital forms of modulation are detected in various different ways. A method of enhancing voice communications, increasingly used in ham radio, is *digital signal processing* (*see* DIGITAL SIGNAL PROCESSING). Personal computers can be used as terminals for communications in "written" digital modes. *See* PACKET RADIO, and RADIOTELETYPE.

DETUNING

Detuning is a procedure in which a resonant circuit is deliberately set to a frequency other than the operating frequency of a piece of equipment. Sometimes detuning is used in the intermediate-frequency stages of a receiver to reduce the chances of unwanted interstage oscillation. This is called *stagger tuning* (*see* STAGGER TUNING).

Sometimes the preselector stage of a receiver is tuned to a higher or lower frequency than the intended one, for the purpose of reducing the gain in the presence of an extremely strong signal. A transmission line is usually made nonresonant, or detuned, to reduce its susceptibility to unwanted coupling with the antenna radiating element. *See also* PRESELECTOR.

DEVIATION

In a frequency-modulated (FM) signal, deviation is the maximum amount by which the carrier frequency changes either side of the center frequency. The greater the amplitude of the modulating signal in an FM transmitter, the greater the deviation, up to a certain maximum. Deviation can be measured on a signal monitor.

Deviation is generally expressed as a plus-or-minus ($\pm$) frequency figure. For example, the standard maximum deviation in a frequency-modulated communications system is ± 5 kHz. When the deviation is greater than the allowed maximum, an FM transmitter is said to be over-deviating. This often causes the received signal to sound distorted. Sometimes, the downward deviation is not the same as the upward deviation. This is the equivalent of a shift in the signal carrier center frequency with modulation, and it, too, can result in a distorted signal at the receiver. *See also* CARRIER SWING, FREQUENCY MODULATION, and MODULATION INDEX.

DIAC

A *diac* is a semiconductor device that physically resembles a pnp bipolar transistor, except that there is no base connection. A diac acts as a bidirectional switch.

The diac can be used as a variable-resistance device for alternating current. The output voltage can be regulated independently of the load resistance. In conjunction with a triac, the diac can be used as a motorspeed control, light dimmer, or other similar device.

DIAL SYSTEM

Dial systems are the mechanical devices used for the purpose of setting the frequency of a radio transmitter or receiver to the desired value. Dial systems are also widely used in such controls as preselectors, amplifier tuning networks, and antenna impedance-matching circuits.

A dial system can be constructed in any of several ways. The simplest method is to mark the face of the equipment in a graduated scale, and to employ a control-shaft pointer knob. More sophisticated mechanisms have drive devices, rotating gears or disks, and a dial face with an adjustable indicator line.

DIAMAGNETIC MATERIAL

Any material with a magnetic permeability less than 1 is called a *diamagnetic material*. The permeability of free space is defined as being equal to 1.

Examples of diamagnetic materials include bismuth, paraffin wax, silver, and wood. In all these cases, the magnetic permeability is only a little bit less than 1. *See also* FERROMAGNETIC MATERIAL, and PERMEABILITY.

DIAPHRAGM

A *diaphragm* is a thin disk, used for the purpose of converting sound waves into mechanical vibrations and vice versa. Diaphragms are used in speakers, headphones, and microphones.

In the speaker or headset, electrical impulses cause a coil to vibrate within a magnetic field. The diaphragm is attached to the coil, and moves along with the coil at the frequency of the electrical impulses. The result is sound waves because the diaphragm imparts vibration to the air molecules immediately adjacent to it.

In a dynamic microphone, the reverse occurs. Vibrating air molecules set the diaphragm in motion at the same frequency as the impinging sound. This causes a coil, attached to the diaphragm, to move back and forth within a magnetic field supplied by a permanent magnet or electromagnet. In a crystal or ceramic microphone, vibration is transferred to a piezoelectric material, which generates electrical impulses. In any microphone, the electrical currents have the same frequency characteristics as the sound hitting the diaphragm.

Some devices can operate as either a microphone or a speaker/headphone. *See also* CERAMIC MICROPHONE, CRYSTAL MICROPHONE, DYNAMIC LOUDSPEAKER, DYNAMIC MICROPHONE, and HEADPHONE.

DICHOTOMIZING SEARCH

In a digital computer, a *dichotomizing search* is a method of locating an item, by number key, in a large set of items. Each item in the set is given a key. If there are 16 items, for example, they might be numbered 1 through 16.

The desired number key is first compared with a number halfway down the list. If the desired key is smaller than the halfway number, then the first half of the table is accepted, and the second half is rejected. If the desired key is larger than the halfway number, then the second half of the table is accepted, and the first half is rejected. The desired key is then compared with a number in the middle of the accepted portion of the table. On this basis, one half of this portion is accepted and the other half is rejected, just as in the first case. This process is repeated, selecting smaller and smaller parts of the table, until only one item remains. That key is then the desired key.

The dichotomizing search is a form of algorithm often used in data processing.

DIELECTRIC

A *dielectric* is an electrical insulator, generally used for the purpose of manufacturing cables, capacitors, and coil forms. Dry, pure air is an excellent dielectric. Other examples of dielectrics include wood, paper, glass, and various rubbers and plastics. Distilled water is also a fairly good dielectric; water is known as a conductor only because it so often contains impurities that enhance its conductivity.

Dielectric materials are classified according to their ability to withstand electrical stress, and according to their ability to cause a charge to be retained when they are used in a capacitor. Dielectric materials are also classified according to their loss characteristics. Generally, dielectric materials become lossier and less able to withstand electrical stress as the operating frequency is increased. *See also* DIELECTRIC ABSORPTION, DIELECTRIC AMPLIFIER, DIELECTRIC BREAKDOWN, DIELECTRIC CONSTANT, DIELECTRIC CURRENT, DIELECTRIC HEATING, DIELECTRIC LENS, DIELECTRIC LOSS, and DIELECTRIC POLARIZATION.

DIELECTRIC ABSORPTION

After a dielectric material has been discharged, it can retain some of the electric charge originally placed across it. This effect is called *dielectric absorption*, because the material literally seems to absorb some electric charge. In capacitors, the dielectric absorption can make it necessary to discharge the component several times before the open-circuit voltage remains at zero.

Some dielectric materials absorb an electric charge to a greater extent than other substances. If a capacitor seems to have been completely discharged, it should still be checked for residual voltage before the assumption is made that there is no voltage across the component. Partially charged capacitors can present a dangerous shock hazard. *See also* CAPACITOR, and DIELECTRIC.

DIELECTRIC AMPLIFIER

A *dielectric amplifier* is a voltage amplifier that functions because of variations in the dielectric constant of a special kind of capacitor (*see* DIELECTRIC CONSTANT). The input signal, applied to the capacitor, causes the dielectric constant of the device to fluctuate. This, in turn, changes its value. An alternating current, supplied by a local oscillator, is modulated by this fluctuation in the capacitance of the component. When this modulated current flows through a resistance of the correct value, a fluctuating voltage having a greater magnitude than the input voltage develops across the resistor. The output from the amplifier is taken at this point.

DIELECTRIC BREAKDOWN

When the voltage across a dielectric material becomes sufficiently high, the dielectric, normally an insulator, will begin to conduct. When this occurs in an air-dielectric cable, capacitor, or feed line, it is called *arcing* (*see* ARC). The breakdown voltage of a dielectric substance is measured in volts or kilovolts per unit length. Some dielectric materials can withstand much greater electrical stress than others. The amount of voltage, per unit length, that a dielectric material can withstand is called the *dielectric strength*.

In most materials, dielectric breakdown is the result of ionization. At a pressure of one atmosphere, air breaks down at a potential of about 2 to 4 kilovolts per millimeter. The value depends on the relative humidity and the amount of dust and other matter in the air; the greater the humidity or the amount of dust, the lower the breakdown voltage. In a solid dielectric material such as polyethylene, which has a dielectric strength of about 1.4 kilovolts per millimeter, permanent damage can result from excessive voltage. *See also* DIELECTRIC, DIELECTRIC HEATING, and DIELECTRIC RATING.

DIELECTRIC CONSTANT

The *dielectric constant* of a material, usually abbreviated by the lowercase letter k, is a measure of the ability of a dielectric material to hold a charge. The dielectric constant is generally defined in terms of the capability of a material to increase capacitance. If an air-dielectric capacitor has a value of C, then the same capacitor, with a dielectric substance of dielectric-constant value k, will have a capacitance of kC. The dielectric constant of air is thus defined as 1.

Various insulating materials have different dielectric constants. Some materials show a dielectric constant that changes considerably with frequency; usually the constant is smaller as the frequency increases. The table shows the dielectric constants of several materials at frequencies of 1 kHz, 1 MHz, and 100 MHz, at approximately room temperature. *See also* DIELECTRIC.

DIELECTRIC CURRENT

Dielectric materials are generally regarded as electrical insulators, but they are not perfect insulators. A small current flows through even the best quality dielectric substances. This current results in some heating, and therefore some loss (*see* DIELECTRIC HEATING, and DIELECTRIC LOSS).

For direct current, the resistivity of a dielectric material is specified in ohm-millimeters or ohm-centimeters. This resistivity is always extremely high, assuming the dielectric material is not contaminated; it ranges from about 10^{12} to 10^{20} ohm-centimeters at room temperature. The table shows the direct-current resistivity of several types of dielectric materials. The greater the resistivity, the lower the dielectric current for a given voltage, provided the voltage is not so great that breakdown occurs. In the event of dielectric breakdown, a conductive path forms through the material, and the current increases dramatically. The substance then loses its dielectric properties. *See also* DIELECTRIC BREAKDOWN.

DIELECTRIC HEATING

Dielectric heating is the result of losses that occur in a dielectric material when it is subjected to an electric field (*see* DIELECTRIC LOSS). Dielectric materials heat up in direct proportion to the intensity of the electric field. The lossier a dielectric material, the hotter it will get in the presence of an electric field of a certain intensity.

Dielectric heating is sometimes deliberately used in the forming process of certain plastics. By subjecting the plastic to an intense radio-frequency electric field, the material becomes soft and pliable. In a prefabricated transmission line, dielectric heating can cause permanent damage. If the dielectric material in such a line, usually polyethylene, becomes soft, the conductors can move and cause a short circuit. Such dielectric heating might be the result of excessive transmitter power for the line in use. The large voltages caused by an excessive standing-wave ratio can also cause dielectric heating with consequent damage. *See also* STANDING-WAVE RATIO.

DIELECTRIC LENS

A *dielectric lens* is a device that is used for the purpose of focusing or collimating electromagnetic energy in the microwave frequency range. Such a lens operates in essentially the same manner as an optical lens. Any dielectric material can be used, although the dielectric loss should be as low as possible for maximum transmission of energy.

The focal length of a dielectric lens depends on the degree of curvature in the lens, and on the dielectric constant of the material from which the lens is made. The greater the curvature and/or the larger the dielectric constant, the shorter the focal length.

DIELECTRIC CONSTANT, DIELECTRIC CURRENT, AND DIELECTRIC RATING: DIELECTRIC CHARACTERISTICS OF VARIOUS MATERIALS AT ROOM TEMPERATURE (APPROXIMATELY 25 DEGREES CELSIUS).

| Material | Dielectric constant | | | dc Resistivity, ohm-cm | Rating, kV/mm |
	1 kHz	1 MHz	100 MHz		
Bakelite	4.7	4.4	4.0	10^{11}	0.1
Balsa wood	1.4	1.4	1.3	—	—
Epoxy resin	3.7	3.6	3.4	4×10^7	0.13
Fused quartz	3.8	3.8	3.8	10^{19}	0.1
Paper	3.3	3.0	2.8	—	0.07
Polyethylene	2.3	2.3	2.3	10^{17}	1.4
Polystyrene	2.6	2.6	2.6	10^{18}	0.2
Porcelain	5.4	5.1	5.0	—	—
Teflon	2.1	2.1	2.1	10^{17}	6
Water (pure)	78	78	78	10^6	—

DIELECTRIC LOSS

Some dielectric materials transform very little of an electric field into heat. Some dielectrics transform a considerable amount of the field into heat. The *dielectric loss* in a particular material is an expression of the tendency of the material to become hot in the presence of a fluctuating electric field. Dielectric loss is usually expressed in terms of a quantity called the *dissipation factor* (*see* DISSIPATION FACTOR).

The dielectric loss of an insulating substance usually increases as the frequency of the fluctuating electric field is raised. Some of the best dielectric materials, such as polystyrene, have very low loss levels. Some materials, such as nylon, show generally lower losses as the frequency increases.

In transmission lines that carry radio-frequency energy, it is extremely important that the loss in the dielectric material be kept as low as possible. For this reason, many types of prefabricated transmission lines employ foamed dielectric material, rather than solid material. From the standpoint of low loss, air is one of the best dielectrics. Some transmission lines are sealed and filled with an inert gas, such as helium, to minimize the dielectric loss; spacers keep the inner conductor of such a coaxial line from short-circuiting to the braid. In some lines, such as the kind used in automobile radio antenna systems, the braid has a thin layer of polyethylene inside it, but most of the dielectric is air. *See also* DIELECTRIC.

DIELECTRIC POLARIZATION

When an electric field is placed across a dielectric material, the location of the positive-charge center in each atom is slightly displaced relative to the negative-charge center. The greater the intensity of the surrounding electric field, the greater the charge displacement. This results in what is known as *dielectric polarization*.

Dielectric polarization is caused by the forces of attraction between opposite charges, and repulsion between like charges. If the intensity of the electric field is sufficiently great, electrons are stripped from the atomic nuclei. This results in conduction because the atoms then easily flow from one atom to the next, throughout the material. Sometimes this ionization causes permanent damage, such as in a solid dielectric. In the case of a gaseous dielectric material, such as air, the ionization does not cause permanent damage. *See also* DIELECTRIC, and DIELECTRIC BREAKDOWN.

DIELECTRIC RATING

The *dielectric rating* of an insulating material refers to its ability to withstand electric fields without breaking down (*see* DIELECTRIC BREAKDOWN). The term can also be used to refer to the general characteristics of a dielectric material, such as the dielectric constant, the resistivity, and the loss.

In a device that employs a dielectric material, such as a capacitor, the dielectric rating is usually the breakdown voltage of the entire device. This rating depends, in the case of a capacitor, on the thickness of the dielectric layer, as well as on the particular material used.

DIETZHOLD NETWORK

A *Dietzhold network* is a shunt *m*-derived circuit, used for the purpose of tailoring the frequency response of a wideband radio-frequency power amplifier. The values of capacitance and inductance are chosen for optimum amplifier performance within a certain frequency range. *See also* M-DERIVED FILTER.

DIFFERENTIAL AMPLIFIER

A circuit that responds to the difference in the amplitude between two signals, and produces gain, is called a *differential amplifier*. Such an amplifier has two input terminals and two output terminals. Differential amplifiers are often found in integrated-circuit packages. The diagram illustrates a simple differential amplifier.

When two identical signals are applied to the input terminals of a differential amplifier, the output is zero. The greater the difference in the amplitudes of the signals, the greater the output amplitude. Differential amplifiers can be used as linear amplifiers, and have a broad operating-frequency range. Differential amplifiers can also be used as mixers, detectors, modulators, and frequency multipliers. *See also* DIFFERENTIAL VOLTAGE GAIN.

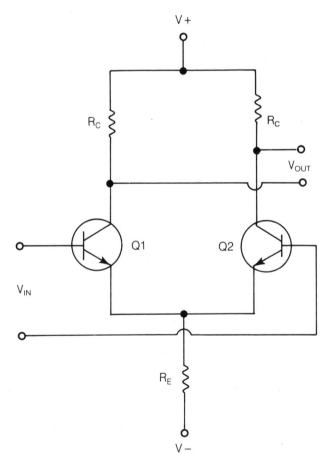

DIFFERENTIAL AMPLIFIER: Amplifies the difference between two signals.

DIFFERENTIAL CAPACITOR

A form of variable capacitor having two sets of stator plates, and a single rotor-plate set, is called a *differential capacitor. See also* AIR-VARIABLE CAPACITOR, and VARIABLE CAPACITOR.

DIFFERENTIAL INSTRUMENT

A *differential instrument* or indicating device is a meter that shows the difference between two input signals. Two identical coils are connected to two sets of input terminals. The coils carry currents in opposite directions. When the currents are equal, their effects in the meter mechanism balance, and there is no deflection of the needle. When one current is greater in magnitude than the other, the net current flow is in one direction or the other.

Differential instruments can use amplifiers for better sensitivity. *See also* DIFFERENTIAL AMPLIFIER.

DIFFERENTIAL KEYING

Differential keying is a form of amplifier/oscillator keying in a code transmitter. Although oscillator keying has the advantage of allowing full break-in operation because the transmitter is off during the key-up intervals, chirp often results because of loading effects of the amplifier stages immediately following the keyed oscillator (*see* BREAK-IN OPERATION, and CODE TRANSMITTER).

An ingenious method of eliminating the chirp (that almost always accompanies oscillator keying) was devised several decades ago. In this technique, called *differential keying*, the amplifier is keyed along with the oscillator, but in a delayed fashion. When the key is pressed, the oscillator comes on first. Then, a few milliseconds later, the amplifier is switched on. This gives the oscillator time to chirp before the signal is actually transmitted over the air! When the key is lifted, both the oscillator and the amplifier are switched off together; alternatively, the amplifier can be switched off a few milliseconds before the oscillator. *See also* KEYING.

DIFFERENTIAL TRANSFORMER

A *differential transformer* is a special kind of transformer that has one or two primary windings and two secondary windings, and an adjustable powdered-iron or ferrite core. The windings are generally placed on a solenoidal form. The primaries, if dual, are connected in series. The secondary windings are connected in phase-opposing series fashion (*see* TRANSFORMER).

As the core is moved in and out of the solenoidal form, the coupling ratio between winding pair X and winding pair Y varies. This affects the amplitude and phase of the transformer output. When the core is at the center, so that the coupling between the two winding pairs is equal, the output of the transformer is zero. The farther off center the core is positioned, the greater the output amplitude. However, the output phase is reversed with the core nearer winding pair X, as compared to when it is closer to pair Y.

DIFFERENTIAL VOLTAGE GAIN

The gain of a differential amplifier, in terms of voltage, is known as *differential voltage gain*. Any differential device has a differential voltage-gain figure. It is measured in decibels (*see* DECIBEL). However, a differential amplifier usually displays a signal gain of significant proportions; passive devices generally show a loss.

Differential voltage gain is the ratio, in decibels, between a change in output voltage and a change in input voltage applied to either input terminal. *See also* DIFFERENTIAL AMPLIFIER.

DIFFERENTIATION

The mathematical determination of the derivative of a function is called *differentiation*. Differentiation can be performed either at a single point on a function, or as a general operation involving the entire function.

Certain electronic circuits act to differentiate the waveform supplied to their inputs. The output waveform of such a circuit is the derivative of the input waveform, representing the rate of change of the input amplitude. If the amplitude is becoming more positive, then the output of the differentiator circuit is positive. If the amplitude is becoming more negative, then the output is negative. If the input amplitude to a differentiating circuit is constant, the output is zero.

Differentiation is, both electrically and mathematically, the opposite of integration. When an integrator and a differentiator circuit are connected in cascade, the output waveform is usually identical to the input waveform. *See also* DIFFERENTIATOR CIRCUIT, INTEGRATION, and INTEGRATOR CIRCUIT.

DIFFERENTIATOR CIRCUIT

An electronic circuit that generates the derivative with respect to time, of a waveform is called a *differentiator circuit*. When the input amplitude to a differentiator is constant—that is, a direct current—the output is thus zero. When the input amplitude increases at a constant rate, the output is a direct-current voltage. When the input is a fluctuating wave, such as a sine wave, the output varies according to the instantaneous derivative of the input. A perfectly sinusoidal waveform is shifted 90 degrees by a differentiator.

The differentiator circuit generally acts in an exactly opposite manner to an integrator circuit. This does not mean, however, that cascading an integrator and a differentiator will always result in the same signal at the output and input. Such is usually the case, but certain waveforms cannot be duplicated in this way. *See also* INTEGRATION, and INTEGRATOR CIRCUIT.

DIFFRACTION

Any effects that show wavelike properties, such as sound, radio-frequency energy, and visible light, have the ability to turn sharp corners and to pass around small obstructions. This property is called *wave diffraction*. The wavelike nature of visible light was discovered by observing interference patterns caused by diffraction.

Diffraction allows you to hear a friend speaking from around the corner of an obstruction, such as a building, even when there are no nearby objects to reflect the sound. This is called *razor-edge diffraction*. The corner of the obstruction acts as a second source of wave action. Electromagnetic fields having extremely long wavelength, comparable to or greater than the diameter of an obstruction, are propagated around that obstruction with very little attenuation. For example, low-frequency radio waves are easily transmitted around a concrete-and-steel building that is small compared to a wavelength. As the wavelength of the energy becomes shorter, however, the obstruction causes more and more attenuation. At very high frequencies, a concrete-and-steel building casts a definite shadow in the wave train of electromagnetic signals.

A piece of clear plastic, having thousands of dark lines drawn on it, is called a diffraction grating. Light passing through such a grating is split into its constituent wavelengths, in very much the same way as light is transmitted through a prism. This effect is the result of interference caused by diffraction through the many tiny slits.

DIFFUSION

When one material spreads into another, the process is called *diffusion.* This process is used in the manufacture of certain semiconductor devices, particularly integrated circuits. *See* INTEGRATED CIRCUIT.

DIGIPEATER

The term *digipeater* is a contraction of the words *digital* and *repeater*. It refers to a digital repeater, commonly used in packet radio. A digipeater works somewhat differently than a common voice repeater.

An ordinary repeater receives and retransmits all the signals it picks up, at exactly the same time (no delay) but on a different frequency, although sometimes a subaudible tone is needed to access it.

A digipeater stores transmissions for a short time, and will accept only signals that are intended specifically for it, and not for other digipeaters. The *routing frame* in the packet transmission contains information concerning which digipeater(s) the signal is to go through, and in what order. As the signal is routed along, *secondary station identifier (SSID)* codes tell each succeeding digipeater that the signal is intended specifically for it.

Any *AX.25 terminal node controller* can operate as a digipeater. A packet can be routed through as many as eight digipeaters along its way from the sender to the receiver. *See also* AX.25, REPEATER, PACKET RADIO, SECONDARY STATION IDENTIFIER, and TERMINAL NODE CONTROLLER.

DIGITAL

The term *digital* is used to define a quantity that exists at discrete states, or levels, rather than over a continuously variable range. Digital signals can have two or more different states. Usually, the total possible number of levels is a power of 2, such as 2, 4, 8, 16, and so on.

Probably the simplest kind of digital signal is Morse code. The length of a bit in this code is called a *dot length*. More complex digital signals include frequency-shift keying, pulse modulation, etc. *See* FREQUENCY-SHIFT KEYING, MORSE CODE, and PULSE MODULATION.

Circuits that operate in a digital fashion are becoming the rule in all electronic applications. The digital mode is generally more efficient, more precise, and more straightforward than older analog techniques. Digital displays of current, voltage, time, frequency, and other parameters are extensively used because of easy readability and precision. Digital operations can be carried out with extreme speed. *See also* DIGITAL CIRCUITRY, and DIGITAL DEVICE.

DIGITAL CIRCUITRY

Any circuit that operates in the digital mode is called *digital circuitry* (*see* DIGITAL). Examples of digital circuits include the flip-flop and the various forms of logic gates. Complicated digital devices are built up from simple fundamental units such as

the inverter, AND gate, and OR gate. There is no limit, in theory, to the level of complexity attainable in digital circuit design. Physical space presents the only constraint. With increasingly sophisticated methods of miniaturization, more and more complex digital circuits are being packed into smaller and smaller packages. Entire computers can now be installed on a desktop for home or business use.

Digital circuitry is generally more efficient than older analog circuitry (*see* ANALOG), because of the finite number of possible digital states, compared to the infinite number of levels in an analog circuit. Many digital circuits can operate at greater speed than similar analog devices. *See also* AND GATE, FLIP-FLOP, INVERTER, and OR GATE.

DIGITAL COMMUNICATIONS

Digital communications refers to any mode in which signals attain a finite number of levels or states. Digital modes differ from analog modes; the latter have infinitely many levels or states. Analog signals range continuously over a certain range of frequencies, phases or amplitudes. Digital signals change in discrete steps. *See* ANALOG, and DIGITAL.

Presently, the common digital modes of amateur radio communications include Morse code (CW), radioteletype (RTTY), AMTOR and packet radio. All of these modes are "written" or "printed." *See* AMTOR, MORSE CODE, PACKET RADIO, and RADIOTELETYPE.

Voice and video signals, analog by nature, can be processed into digital form for transmission, and then back to analog form for listening or display. The technology for this has been around for awhile, but in the past few years research and development has accelerated. Digital communications provides a better signal-to-noise ratio than is possible with analog signals. *See also* DIGITAL SIGNAL PROCESSING.

DIGITAL CONTROL

Digital control is a method of adjusting a piece of equipment by digital means. Parameters frequently controlled by digital methods include the frequency of a radio receiver or transmitter, the programming of information into a memory channel, and the selection of a memory address. There are, of course, many other applications of digital control.

Digital control involves the selection of discrete states or levels, rather than the adjustment of a parameter over a continuous range. When the adjustment is continuous, the device is analog controlled (*see* ANALOG CONTROL). Digital controls are becoming increasingly common in all sorts of electronic devices.

Some modern electronic equipment is controlled almost entirely by digital circuits. But some functions are still regulated by analog means in most cases. The volume and tone controls in a high-fidelity stereo amplifier or receiver, for example, are often of analog design, although it is possible to make them digital. Undoubtedly, there are some aspects of machine control that will never be digitized. An example of this is the steering mechanism in an automobile. *See also* DIGITAL.

DIGITAL DEVICE

A *digital device* is a circuit, or a set of circuits, that operates in the digital mode (*see* DIGITAL, and DIGITAL CIRCUITRY). Digital devices include most computers and calculators, and various kinds of test instruments. Some radio receivers and transmitters use digital devices in their frequency-control systems.

Digital devices can be extremely simple, or highly complex. The simplest of all digital devices is the ordinary single-pole, single-throw switch. It is either on or off! Large networks of interconnected computers form extremely complicated digital devices. They can be made up of millions of individual switches.

DIGITAL DISPLAY

A *digital display* shows a quantity (such as current, voltage, frequency, or resistance) in terms of numerals. This eliminates the reading errors that can occur with analog displays; there is no doubt about the indication. The photograph shows a typical digital frequency display. In this case, the display happens to be the readout of a clock radio. Digital displays can use light-emitting diodes or liquid crystals.

Although a digital display has obvious advantages over an analog display, there are some drawbacks in certain applications. For example, the operator of a piece of equipment that uses an analog display can get an intuitive feeling for how the indicated reading compares with other possible values of the parameter. It is visually apparent, for example, whether a receiver frequency is near the bottom, in the middle, or near the top of a frequency band, when an analog display is used. With a digital display, there is no such spatial reinforcement. *See also* ANALOG METERING, DIGITAL METERING, LIGHT-EMITTING DIODE, LIQUID-CRYSTAL DISPLAY, and SEVEN-SEGMENT DISPLAY.

DIGITAL INTEGRATED CIRCUIT

A digital-logic circuit, built onto a chip or wafer of semiconductor material, is a *digital integrated circuit*. There are many different kinds of commercially manufactured digital integrated circuits. Sometimes, different manufacturers make digital integrated circuits that can be directly substituted for each other.

Some of the more sophisticated digital integrated circuits are the single-chip calculator, the microcomputer, the microprocessor, and the digital clock. Most digital integrated circuits were first developed using bipolar techniques. Now, metal-oxide-semiconductor design is also common (*see* METAL-OXIDE SEMICONDUCTOR).

A special kind of digital integrated circuit is the memory integrated circuit. This device has helped to make the handheld calculator and the desktop computer a reality. *See also* MEMORY.

DIGITAL METERING

An electrical or physical quantity can be monitored by means of a digital display of its value. This is called *digital metering*. In recent years, digital metering has become increasingly common, because of the proliferation, and the drop in the cost, of digital-circuit devices. The photograph shows one extremely common type of digital meter—a clock.

Digital methods are used in many metering applications, such as ammeters, ohmmeters, voltmeters, power meters, and frequency meters. Almost any analog meter can be replaced with a digital meter. Digital meters have the advantage of easy, quick readability. There is no margin for error because of inaccurate interpolation on the part of the person reading the meter. Digital meters also have the advantage of no moving parts.

Digital meters are less preferable than analog meters in certain situations. This is especially true when it is necessary to adjust a circuit for a minimum or maximum reading. Such a "dip" or "peak" is much more easily seen on an analog meter than on a digital meter. *See also* ANALOG METERING, and DIGITAL DISPLAY.

DIGITAL METERING: A clock-radio display.

DIGITAL MODULATION

Whenever a specific characteristic of a signal is varied for the purpose of conveying information, the process is called *modulation* (*see* MODULATION). If this is done by restricting the signal parameter to two or more discrete levels, or states, the process is known as *digital modulation*. Digital modulation differs from analog modulation (*see* ANALOG), in which a parameter varies over a continuous range, and therefore has a theoretically infinite number of possible levels. The parameter in a digital-modulation system can be any signal characteristic, such as amplitude, frequency, or phase. In addition, pulses can be digitally modulated (*see* PULSE MODULATION).

The drawing illustrates a hypothetical digital pulse-amplitude modulation system. The analog waveform is shown by the solid line. A series of pulses, transmitted at uniform time intervals, approximates this waveform. The pulses can achieve any of eight different amplitude levels, as shown on the vertical scale. In this particular example, the amplitude of a given pulse corresponds to the instantaneous level nearest the amplitude of the analog signal. There are, of course, many other possible ways of getting a digital modulation output from an analog signal.

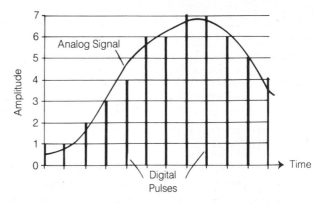

DIGITAL MODULATION: Digital pulses approximate the signal amplitude.

The simplest form of digital modulation is Morse code. It has two possible states: on and off. These are generally called the *key-down* and *key-up conditions*, respectively. The common emission called *frequency-shift keying (FSK)*, is another simple form of digital modulation. It is used extensively in teletype. *See also* DIGITAL SIGNAL PROCESSING, FREQUENCY-SHIFT KEYING, and MORSE CODE.

DIGITAL SIGNAL PROCESSING

A comparatively new, and rapidly advancing, ham radio communications technique, *digital signal processing (DSP)* promises to revolutionize voice, digital, and image communications. In some ham radio transceivers, DSP chips are included as standard or optional equipment. The DSP integrated circuit (IC) is a form of *microprocessor*.

In analog modes, the DSP chip works by converting the voice or video signal input into digital data by means of an *analog-to-digital (A/D) converter* (*see* ANALOG-TO-DIGITAL CONVERTER). The digital signal is then processed, and is reconverted back to the original voice or video via a *digital-to-analog (D/A) converter* (*see* DIGITAL-TO-ANALOG CONVERTER).

In digital modes, A/D and D/A conversion is not necessary, but processing can still be used to advantage to "clean up" the signal. This reduces the number of errors.

It is in the digital part of the DSP chip that the signal enhancement occurs. Digital signals have a finite number of discrete, well-defined states. It is easier to process a signal of this kind than to process an analog signal, which has a theoretically infinite number of possible states. *See* ANALOG, and DIGITAL. The microprocessor acts to get rid of any confusion between different digital states. The result is an output that is essentially free from interference.

Digital signal processors are available from several commercial sources. The photograph shows a multimode DSP data controller useful for digital communications modes.

The benefits of digital signal processing are improved signal-to-noise ratio, superior intelligibility and enhanced fidelity. In single sideband (SSB), static (QRN) and interference from stations on adjacent channels (QRM) are greatly reduced or eliminated. In slow-scan television (SSTV), "snow," modulation bars and cross-hatching are minimized.

DIGITAL SIGNAL PROCESSING: A DSP multimode data controller. It interfaces the radio with a computer for digital communications. Advanced Electronic Applications, Inc.

DIGITAL STORAGE OSCILLOSCOPE

Oscilloscopes can be used to evaluate digital signals as well as analog signals. Before the advent of digital communications modes, oscilloscopes were used primarily to check analog signals for distortion, to view modulation envelopes of analog signals, or to check for presence or absence of signals. Calibration

was approximate. The instruments were crude by today's standards.

Nowadays, the *digital storage oscilloscope (DSO)* is commonly used for laboratory testing, design and maintenance of electronic equipment of all kinds (both analog and digital), from audio frequencies well into the ultra-high range. *See also* OSCILLOSCOPE.

The DSO cannot only give a clear rendition of a waveform, but can store it for later comparison with other waveforms. In recent years, DSO quality has been increasing, while prices have been going down. This can be attributed to a general improvement in digital technology at all levels.

Digital pulses show up clearly on a DSO. Two or more pulse trains can be displayed, one above the other, and evaluated for timing, frequency, shaping and duration.

In digital equipment maintenance manuals, test points are specified and the appropriate pulse shapes and durations are diagrammed. The repair technician needs only to check each test point, in the prescribed order from one integrated circuit (IC) to the next, for the presence of the proper digital pulses. If the DSO display differs substantially from the standard at some test point, the technician can identify and replace the faulty IC immediately.

DIGITAL-TO-ANALOG CONVERTER

Digital signal transmission offers advantages over analog transmission. These improvements include narrower bandwidth, better signal-to-noise ratio, and fewer errors per unit time. The digital signal differs from the analog signal, in that the digital signal has only a few discrete levels or states, while the analog signal has, in theory, infinitely many. The human voice and a typical picture signal have amplitudes that vary in an analog manner. But they can be digitized, and the benefits of digital transmission can be realized.

When a digitized signal arrives at the receiving end of a communications circuit, it can be converted back to the original analog form, if needed, by means of a digital-to-analog converter. This is also sometimes called a *D/A converter (DAC)*.

A DAC is also used for another, entirely different purpose: the generation of artificial analog signals. A voice synthesizer is a good example. In recent years, it has become possible to digitally encode voice information in memories, such as integrated-circuit chips. This information can be recalled, and a DAC used to produce a natural-sounding voice.

The output of a DAC is a synthesized, quantized waveform. This causes the signal to sound "coarse" or "rough" unless filtering is used to smooth out the abrupt transitions. This can be done using an operational-amplifier (op amp) circuit with the proper resistance and capacitance values.

There are numerous integrated circuits that are commercially supplied to perform complete DAC functions on a single chip. *See also* ANALOG-TO-DIGITAL CONVERTER, DIGITAL COMMUNICATIONS, DIGITAL MODULATION, DIGITAL TRANSMISSION SYSTEM, and PULSE MODULATION.

DIGITAL TRANSMISSION SYSTEM

Any system that transfers information by digital means is a *digital transmission system*. The simplest digital transmission system is a Morse code transmitter and receiver, along with the attendant operators. A teletype system uses digital transmission methods. Computers communicate by digital transmission.

Analog signals, such as voice and picture waveforms, can be transmitted by digital methods. At the transmitting station, a circuit called an *analog-to-digital converter* changes the signal to digital form. This signal is then transmitted, and the receiver uses a digital-to-analog converter to get the original analog signal back.

Digital transmission often provides a better signal-to-noise ratio over a given communications link than analog transmission. This results in better efficiency. *See also* ANALOG-TO-DIGITAL CONVERTER, DIGITAL MODULATION, DIGITAL SIGNAL PROCESSING, and DIGITAL-TO-ANALOG CONVERTER.

DIODE

A *diode* is a tube or semiconductor device that is intended to pass current in only one direction. Diodes can be used for a wide variety of different purposes. The semiconductor diode is far more common than the tube diode in modern electronic circuits. The drawing at A shows the construction of a typical semiconductor diode; it consists of N-type semiconductor material, usually germanium or silicon, and P-type material. Electrons flow into the N-type material and out of the P terminal. The schematic symbol for a semiconductor diode is shown at B. Positive current flows in the direction of the arrow. Electron movement is contrary to the arrow. The positive terminal of a diode is called the anode, and the negative terminal is called the cathode, under conditions of forward bias (conduction).

Semiconductor diodes can be very small, and still handle hundreds or even thousands of volts at several amperes. The older tube type diodes are much bulkier and less efficient than the semiconductor diodes. Some of the tube type diodes require a separate power supply for the purpose of heating a filament. Semiconductor diodes are used for many different purposes in electronics. They can be used as amplifiers, frequency controllers, oscillators, voltage regulators, switches, mixers, and in many other types of circuits. *See also* DIODE ACTION, DIODE CAPACITANCE, DIODE CLIPPING, DIODE DETECTOR, DIODE FEEDBACK RECTIFIER, DIODE FIELD-STRENGTH METER, DIODE MIXER, DIODE OSCILLATOR, DIODE-TRANSISTOR LOGIC, and DIODE TYPES.

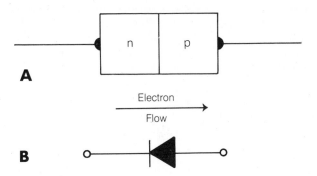

DIODE: At A, a P-N junction. At B, a schematic symbol.

DIODE ACTION

Diode action is the property of an electronic component to pass current in only one direction. In a tube or semiconductor diode, the electrons can flow from the cathode to the anode, but not

vice versa. Diode action occurs in all tubes and bipolar transistors, as well as in semiconductor diodes.

A voltage that allows current to flow through a diode is called *forward bias*. This occurs when the cathode is negative, with respect to the anode. A voltage of the opposite polarity is called *reverse bias*. With most tubes and transistors, as well as with semiconductor diodes, a certain amount of forward bias voltage is necessary in order for current to flow. In a germanium diode, this bias is about 0.3 volt; in a silicon diode, it is about 0.6 volt. In mercury-vapor rectifier tubes, it is about 15 volts. *See also* DIODE, and P-N JUNCTION.

DIODE CAPACITANCE

When a diode is reverse-biased, so that the anode is negative, with respect to the cathode, the device will not conduct. Under these conditions in a semiconductor diode, a depletion layer forms at the P-N junction (*see* DEPLETION LAYER, and P-N JUNCTION). The greater the reverse-bias voltage, the wider the depletion region.

The depletion region in a semiconductor diode has such a high resistance that it acts as a dielectric material (*see* DIELECTRIC). Because the P and N materials both conduct, the reverse-biased diode acts as a capacitor, assuming the bias remains reversed during all parts of the cycle.

Some diodes are deliberately used as variable capacitors. These devices are called *varactors* (*see* VARACTOR DIODE). The capacitance of a reverse-biased diode limits the frequency at which it can effectively be used as a detector because at sufficiently high frequencies the diode capacitance allows significant signal transfer in the reverse direction. Diode capacitance, when undesirable, is minimized by making the P-N junction area as small as possible.

A tube type diode also displays capacitance when reverse bias is applied to it. This is because of interelectrode effects. The capacitance of a reverse-biased tube diode does not depend to a great extent on the value of the voltage.

DIODE CHECKER

A *diode checker* is a device that is intended for the purpose of testing semiconductor diodes. The simplest kind of diode checker consists of a battery or power supply, a resistor, and a milliammeter. This device can easily be used to determine if current will flow in the forward direction (anode positive) and not in the reverse direction (cathode positive).

An ohmmeter makes a good diode checker, However, before using an ohmmeter for this purpose, check the polarity of the voltage present at the leads. In some volt-ohm-milliammeters, the red lead presents a negative voltage, with respect to the black lead when the instrument is in the ohmmeter mode.

More sophisticated diode checkers show whether or not a particular diode is within its rated specifications. However, for most purposes, the circuit covered here is adequate. When a diode fails, the failure is generally catastrophic, such as an open or short circuit. *See also* DIODE.

DIODE CLIPPING

A *diode clipper*, or *diode limiter*, is a device that uses diodes for the purpose of limiting the amplitude of a signal. Generally, such a device consists of two semiconductor diodes connected in reverse parallel.

A silicon semiconductor diode has a forward voltage drop of about 0.6 volt. Thus, when two such diodes are placed in reverse parallel, the signal is limited to an amplitude of ±0.6 volt, or 1.2 volts peak-to-peak. If the signal amplitude is smaller than this value, the diodes have no effect, except for the small amount of parallel capacitance they present in the circuit. When the signal without the diodes would exceed 1.2 volts peak-to-peak, however, the diodes flatten the tops of the waveforms at +0.6 and −0.6 volts. This results in severe distortion. Thus, diode limiters are not generally useful in applications where complex waveforms are present, or in situations where fidelity is important. *See also* DIODE.

DIODE DETECTOR

A *diode detector* is an envelope-detector circuit (*see* ENVELOPE DETECTOR). Such a detector is generally used for the demodulation of an amplitude-modulated signal.

When the alternating-current signal is passed through a semiconductor diode, half of the wave cycle is cut off. This results in a pulsating direct-current signal of variable peak amplitude. The rate of pulsation corresponds to the signal carrier frequency, and the amplitude fluctuations are the result of the effects of the modulating information. A capacitor is used to filter out the carrier pulsations, in a manner that is similar to the filter capacitor in a power supply. The remaining waveform is the audio-frequency modulation envelope of the signal. This waveform contains a fluctuating direct-current component, the result of the rectified and filtered carrier. A transformer or series capacitor can be used to eliminate this. The resulting output is then identical to the original audio waveform at the transmitter. *See also* DETECTION.

DIODE FEEDBACK RECTIFIER

A *diode feedback rectifier* is a device for obtaining a fluctuating voltage suitable for use in an automatic-gain-control circuit (*see* AUTOMATIC GAIN CONTROL, AUTOMATIC LEVEL CONTROL).

In a series of amplifying stages, a portion of the output from one of the later stages is rectified and filtered. This provides a direct-current voltage that varies in proportion to the strength of the signal. The voltage can be either positive or negative, depending on the direction in which the diode is connected. The polarity should be chosen so that the gain of a preceding stage is reduced when the voltage is applied to the base, gate, or grid of the earlier stage.

The effect of the diode feedback rectifier is to keep the output level constant, or almost constant, for a wide variety of signal amplitudes at the input. This maximizes sensitivity for weak signals, and reduces it for strong signals.

DIODE FIELD-STRENGTH METER

A field-strength meter that uses a semiconductor diode, for the purpose of obtaining a direct current to drive a microammeter, is called a *diode field-strength meter.* Such a field-strength meter is the simplest kind of device possible for measuring relative levels of electromagnetic field strength. This kind of field-strength meter is not very sensitive. More sophisticated field-strength meters often have amplifiers and tuned circuits built in, and more nearly resemble radio receivers than simple meters.

The diode field-strength meter is easy to carry, and can be used to check a transmission line for proper shielding or balance. A handheld diode field-strength meter should show very little electromagnetic energy near a transmission line, but in the vicinity of the antenna radiator, a large reading is normally obtained. Some commercially made SWR (standing-wave-ratio) meters have a built-in diode field-strength meter, so that a small whip antenna can be used to monitor relative field strength. *See also* FIELD-STRENGTH METER.

DIODE IMPEDANCE

The impedance of a diode is the vector sum of the resistance and reactance of the device in a particular circuit (*see* IMPEDANCE). Both the resistance and the capacitive reactance of a diode depend on the voltage across the device. The inductive reactance of a diode is essentially constant, and is primarily the result of inductance in the wire leads.

Generally, the resistance of a heavily forward-biased diode is extremely low, and the device in this case acts as a nearly perfect short circuit. When the forward bias is not strong, the resistance is higher, and the capacitive reactance is small. A reverse-biased diode has extremely high resistance. The capacitive reactance of a reverse-biased semiconductor diode is greater as the reverse-bias voltage rises. This is because the depletion region gets wider and wider (*see* DEPLETION LAYER, and DIODE CAPACITANCE). The smaller the capacitance, the larger the capacitive reactance.

DIODE MATRIX

A *diode matrix* is a form of high-speed, digital switching circuit, using semiconductor diodes in a large array. Two sets of wires, one shown horizontally on a circuit diagram and the other shown vertically, are interconnected at various points by semiconductor diodes. Diode matrices can be fairly small, such as in a simple counter. Or they can be huge, as in a digital computer.

Diode matrices are used as decoders, memory circuits, and rotary switching circuits. *See also* DECODING, MEMORY, and SWITCHING.

DIODE MIXER

A *diode mixer* is a circuit that uses the nonlinear characteristics of a diode for the purpose of mixing signals (*see* MIXER). Whenever two signals that have different frequencies are fed into a nonlinear circuit, the sum and difference frequencies are obtained at the output, in addition to the original frequencies.

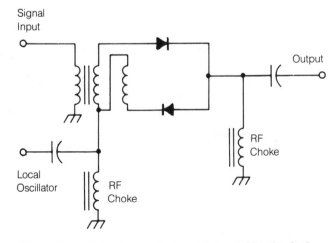

DIODE MIXER: Mixing products are generated by the diodes.

The illustration is a schematic diagram of a typical, simple diode mixer circuit. Such a circuit has no gain because it is passive. In fact, there is a certain amount of insertion loss. However, amplification circuits can be used to boost the output to the desired level. Selective circuits reject all unwanted mixing products and harmonics, and allow only the desired frequency to pass. Diode mixers are often found in superheterodyne receivers, and also in transmitters. Frequency converters, used with receivers to provide operation on frequencies far above or below their normal range, sometimes use diode mixers. These circuits can function well into the microwave spectrum. *See also* FREQUENCY CONVERSION, and MIXING PRODUCT.

DIODE OSCILLATOR

Under the right conditions, certain types of semiconductor diodes will oscillate at ultra-high or microwave frequencies. A circuit that is deliberately designed to produce oscillation at these frequencies, using a diode as the active component, is called a *diode oscillator*. The drawing is a simple schematic diagram of a microwave oscillator that uses a semiconductor device called a *Gunn diode*. The Gunn diode has largely replaced the Klystron tube for ultra-high and microwave frequency oscillator applications.

The Gunn diode is mounted inside a resonant cavity. A direct-current bias is supplied to cause oscillation. The efficiency of the Gunn-diode oscillator is low, only a few percent. The frequency stability tends to be rather poor because the slightest change in the bias voltage or temperature can cause a radical change in the oscillating frequency. The bias voltage must therefore be carefully regulated, and the temperature maintained at a level that is as nearly constant as possible. Automatic frequency control is sometimes used to improve the stability of the Gunn-diode oscillator (*see* AUTOMATIC FREQUENCY CONTROL). A phase-locked-loop device is also sometimes used (*see* PHASE-LOCKED LOOP).

The diode oscillator provides a maximum output of considerably less than one watt. Gunn-diode oscillators can be frequency-modulated by varying the bias voltage. A device called a *tunnel diode*, now essentially obsolete because of the superior characteristics of the Gunn diode, can also be used in a diode-oscillator circuit. *See also* GUNN DIODE, KLYSTRON, and TUNNEL DIODE.

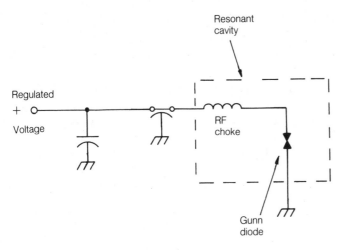

DIODE OSCILLATOR: A Gunn diode generates microwave energy.

DIODE RATING

The rating of a diode refers to its ability to handle current, power, or voltage. Some semiconductor diodes are intended strictly for small-signal applications, and can handle only a few microamperes or milliamperes of current. Other semiconductor diodes are capable of handling peak-inverse voltages of hundreds or even thousands of volts, and currents of several amperes. These rugged diodes are found in power-supply rectifier circuits.

Diode ratings are generally specified in terms of the peak inverse voltage (PIV) and the maximum forward current. Zener diodes, used mostly for voltage regulation, are rated in terms of the breakdown or avalanche voltage and the power-handling capacity. Other characteristics of diodes, that can be called ratings or specifications, are temperature effects, capacitance, voltage drop, and the current-voltage curve. *See also* AVALANCHE VOLTAGE, DIODE CAPACITANCE, DIODE IMPEDANCE, PEAK INVERSE VOLTAGE, and ZENER DIODE.

DIODE-TRANSISTOR LOGIC

Diode-transistor logic (DTL), is a form of digital-logic design in which a diode and transistor act to amplify and invert a pulse. Diode-transistor logic has a somewhat less rapid switching rate than most other bipolar logic families. The power- dissipation rating is medium to low. Diode-transistor logic is sometimes mixed with transistor-transistor logic (TTL) in a single circuit.

Diode-transistor logic gates are generally fabricated into an integrated-circuit package. *See also* DIRECT-COUPLED TRANSISTOR LOGIC, HIGH-THRESHOLD LOGIC, METAL-OXIDE SEMICONDUCTOR LOGIC FAMILIES, NAND GATE, NEGATIVE LOGIC, and POSITIVE LOGIC.

DIODE TYPES

There are several different types of semiconductor diodes, each intended for a different purpose. The most obvious application of a diode is the conversion of alternating current to direct current. Detection and rectification use the ability of a diode to pass current in only one direction. But there are many other uses for these semiconductor devices.

Light-emitting diodes (LEDs) produce visible light when forward-biased. Solar-electric diodes do just the opposite, and generate direct current from visible light. Zener diodes are used as voltage regulators and limiters. Gunn diodes and tunnel diodes can be used as oscillators at ultra-high and microwave frequencies. Varactor diodes are used for amplifier tuning. A device called a *PIN diode*, which exhibits very low capacitance, is used as a high-speed switch at radio frequencies. Hot-carrier diodes are used as mixers and frequency multipliers. Frequency multiplication is also accomplished effectively using a step-recovery diode. The *impact-avalanche-transit-time diode (IMPATT diode)*, can act as an amplifying device.

Details of various diodes types and uses are discussed under the following headings: DIODE, DIODE ACTION, DIODE DETECTOR, DIODE FEEDBACK RECTIFIER, DIODE FIELD-STRENGTH METER, DIODE MATRIX, DIODE-TRANSISTOR LOGIC, GUNN DIODE, HOT-CARRIER DIODE, IMPATT DIODE, LIGHT-EMITTING DIODE, P-N JUNCTION, TUNNEL DIODE, VARACTOR DIODE, and ZENER DIODE.

DIP

In electronics, the term *dip* usually refers to the adjustment of a certain parameter for a minimum value. A common example is the dipping of the plate current in a tube type radio-frequency amplifier. The dip indicates that the output circuit is tuned to resonance, or optimum condition. Antenna tuning networks are adjusted for a dip in the standing-wave ratio. A dip is also sometimes called a *null*.

The dual-inline package, a familiar form of integrated circuit, is sometimes called a *DIP* for short. *See also* DUAL IN-LINE PACKAGE.

DIPLEX

When more than one receiver or transmitter are connected to a single antenna, the system is called a *diplex* or *multiplex circuit*. The diplexer allows two transmitters or receivers to be operated with the same antenna at the same time.

The most familiar example of a diplexer is a television feed-line splitter, which allows two television receivers to be operated simultaneously using the same antenna. Such a device must have impedance-matching circuits to equalize the load for each receiver. Simply connecting two or more receivers together by splicing the feed lines will result in ghosting because of reflected electromagnetic waves along the lines. Diplexers for transmitters operate in a similar manner to those for receivers.

Multiplex transmission is sometimes called *diplex transmission* when two signals are sent over a single carrier. Each of the two signals in a diplex transmission consists of a low-frequency, modulated carrier called a *subcarrier*. The main carrier, much higher in frequency than the subcarriers, is modulated by the subcarriers. *See also* MULTIPLEX.

DIPLEXER

A *diplexer* is a device that allows a single antenna to be used with two receivers or transmitters. Diplexers usually incorporate impedance-matching circuitry so that the radios are both properly matched to the antenna. Simply connecting two radios in "parallel" will not always work, because a 2:1 mismatch is introduced. This mismatch can sometimes be tolerated, but in certain applications, especially in the ultra-high-frequency (UHF) and microwave range, it cannot.

A device that allows more than one signal to be sent on a single carrier wave is sometimes called a *diplexer*. *See* DIPLEX.

DIPOLE ANTENNA

The term dipole, or dipole antenna, is often used to describe a half-wavelength radiator fed at the center with a two-wire or coaxial transmission line. Such an antenna can be oriented horizontally or vertically, or at a slant. The radiating element is usually straight; variations of the dipole go by other nicknames.

A half-wavelength conductor displays resonant properties for electromagnetic energy. In free space—that is, when there are no objects near the radiator—the impedance at the center of a dipole is about 73 ohms, purely resistive. The impedance is a pure resistance at all harmonic frequencies. At odd harmonics, the value is about the same as at the fundamental frequency; but at even harmonics, it is very high. The radiation pattern of a dipole in free space is rather doughnut-shaped, as

shown in the illustration. Maximum radiation occurs in directions perpendicular to the conductor.

Because of their relative simplicity, dipole antennas are quite popular among shortwave listeners and radio amateurs. This is especially true at frequencies below about 10 MHz, where more complicated antennas are often impractical. At higher frequencies, parasitic elements are often added to the dipole, creating power gain. Dipoles can also be fed in various multiple configurations to obtain power gain. *See also* PARASITIC ARRAY, PHASED ARRAY, VERTICAL DIPOLE ANTENNA, and YAGI ANTENNA.

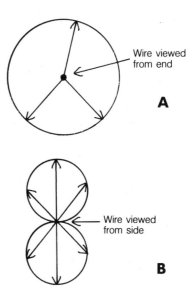

DIPOLE ANTENNA: Directional pattern as seen from the end of the radiator (A) and from the side (B).

DIP SOLDERING TECHNIQUE

Dip soldering is a method of soldering an electrical connection, or coating a terminal or lead with solder, by dipping the entire connection, terminal, or lead into a container of molten solder. Excess solder is then removed. Solder melts at a fairly low temperature, so components can often be dipped without damage.

Printed-circuit boards are sometimes tinned, or coated with solder, by dipping the entire board into molten solder, or the face of the board to be tinned may be placed against the surface of the molten solder bath. This results in a coating of solder on all foil runs, making assembly easier and helping to protect the board against corrosion. Excess solder is easily removed from the non-foil parts of the board. The solder is an excellent conductor of electricity. *See also* SOLDER, and SOLDERING TECHNIQUE.

DIRECT-CONVERSION RECEIVER

A *direct-conversion receiver* is a receiver whose intermediate frequency is actually the audio signal heard in the speaker or headset. The received signal is fed into a mixer, along with the output of a variable-frequency local oscillator. As the oscillator is tuned across the frequency of an unmodulated carrier, a high-pitched audio beat note is heard, which becomes lower until it vanishes at the zero-beat point. Then, it rises in pitch again as the oscillator frequency gets farther and farther away from the signal frequency. The drawing shows a simple block diagram of a direct-conversion receiver.

For reception of code signals, the local oscillator is set slightly above or below the signal frequency. The audio tone will have a frequency equal to the difference between the oscillator and signal frequencies. For reception of amplitude-modulated or single-sideband signals, the oscillator should be set to zero beat with the carrier frequency of the incoming signal.

The direct-conversion receiver normally cannot provide the selectivity of a superheterodyne because single-signal reception with the direct-conversion receiver is impossible. Audio filters are often used to provide some measure of selectivity. *See also* INTERMEDIATE FREQUENCY, SINGLE-SIGNAL RECEPTION, and SUPERHETERODYNE RECEIVER.

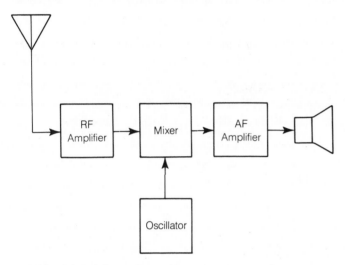

DIRECT-CONVERSION RECEIVER: A block diagram of a direct-conversion receiver.

DIRECT-COUPLED TRANSISTOR LOGIC

Direct-coupled transistor logic (DCTL) is a bipolar logic family. It was the earliest logic design used in commercially manufactured integrated circuits. The DCTL logic scheme is fairly simple.

Direct-coupled transistor logic has rather poor noise rejection characteristics. Current hogging can also cause some problems. Newer forms of direct-coupled transistor logic are available, and these designs have better operating characteristics than the original form. The signal voltages in DCTL are low. The switching speed and power-handling capabilities are about average, compared with other bipolar logic families. *See also* DIODE-TRANSISTOR LOGIC, HIGH-THRESHOLD LOGIC, METAL-OXIDE-SEMICONDUCTOR LOGIC FAMILIES.

DIRECT COUPLING

Direct coupling is a form of circuit coupling, usually used between stages of an amplifier. In direct coupling, the output from one stage is connected, by a direct wire circuit, to the input of the next stage. This increases the gain of the pair over the gain of a single component, provided the bias voltages are correct. The Darlington amplifier is an example of direct coupling (*see* DARLINGTON AMPLIFIER).

Direct coupling is characterized by a wideband frequency response. This is because there are no intervening reactive components, such as capacitors or inductors, to act as tuned circuits.

Therefore, direct-coupled amplifiers are more subject to noise than tuned amplifiers. Direct coupling transmits the alternating-current and direct-current components of a signal. Current hogging can sometimes be a problem, especially if adequate attention is not given to the maintenance of proper bias. *See also* CURRENT HOGGING.

DIRECT CURRENT

A *direct current* is a current that always flows in the same direction. That is, the polarity never reverses. If the direction of the current ever changes, it is considered to be an alternating current.

Physicists consider the current in a circuit to flow from the positive pole to the negative pole. This is purely a convention. The movement of electrons in a direct-current circuit is therefore contrary to the theoretical direction of the current. In a P-type semiconductor material, however, the motion of the positive charge carriers (holes) is the same as the direction of the current.

Typical sources of direct current include most electronic power supplies, as well as batteries and cells. The intensity, or amplitude, of a direct current can fluctuate with time, and this fluctuation might be periodic. In this case, the current can be considered to have an alternating as well as a direct component, but the current itself is direct.

DIRECT-DRIVE TUNING

When the tuning knob of a radio receiver or transmitter is mounted directly on the shaft of a variable capacitor, the tuning is said to be *directly driven*. A half turn of the tuning knob thus covers the entire range. Direct-drive tuning is found in small portable transistor radio receivers for the standard broadcast band. It is difficult to obtain precise tuning with a direct-drive control.

Most controls for the adjustment of frequency are indirectly driven. Thus, several turns of the control knob are needed to cover the entire range. This method of control is called *vernier drive*, and it can be used in various other situations besides radio tuning. Cable-driven controls are also sometimes used to spread out the tuning range of a radio receiver or transmitter.

Most circuit-adjustment controls, such as volume and tone, are directly driven. However, vernier or cable drives may be used with any control to obtain precise adjustment.

DIRECTIONAL ANTENNA

A *directional antenna* is a receiving or transmitting antenna that is deliberately designed to be more effective in some directions than in others. For most radio-communications purposes, antenna directionality is considered to be important only in the azimuth, or horizontal, plane if communication is terrestrial. But for satellite applications, both the azimuth and altitude directional characteristics are important (*see* AZIMUTH, and ELEVATION). Directional antennas are usually either bidirectional or unidirectional; that is, their maximum gain is either in two opposite directions or in one single direction. Some antennas have a large number of high-gain directions.

A vertical radiator, by itself, is omnidirectional in the azimuth plane. In the elevation plane, it shows maximum gain parallel to the ground, and minimum gain directly upward. A single horizontal radiator, such as a dipole antenna, produces more gain off the sides than off the ends, and therefore it is directional in the azimuth plane, as shown in the illustration. A dipole is considered a directional antenna, since it shows a bidirectional pattern.

Sophisticated types of directional antennas provide large amounts of signal gain in their favored directions. This gain and directionality can be obtained in a variety of ways. *See also* ANTENNA POWER GAIN, DIPOLE ANTENNA, DISH ANTENNA, PHASED ARRAY, QUAD ANTENNA, and YAGI ANTENNA.

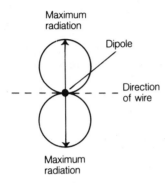

DIRECTIONAL ANTENNA: In this case, the pattern is bidirectional.

DIRECTIONAL GAIN

Directional gain is a means of expressing the sound-radiating characteristics of a speaker. The term can also be applied to the acoustic-pickup characteristics of a microphone.

Generally, speakers do not radiate sound equally well in all directions. Instead, most of the sound energy is concentrated in a narrow cone, with its axis perpendicular to the speaker face. If a speaker could be designed that would radiate equally well in all directions, then its directional gain would theoretically be zero.

Suppose a hypothetical speaker, supplied with P watts of audio-frequency power, produces a sound-intensity level of p_x dynes per square centimeter, at a fixed distance of m meters, in all directions. The directional gain of this speaker is zero because its sound radiation is the same in all directions. Imagine that this theoretical speaker is replaced by a real speaker, which radiates more effectively in some directions than in others. Suppose the real speaker has the same efficiency as the hypothetical one, and receives the same amount of power. If the sound pressure at a distance of m meters from this speaker, along the axis of its maximum sound output, is p dynes per square centimeter, then the directional gain of the real speaker, in decibels (dB), is:

$$\text{Directional gain (dB)} = 10 \log_{10}(p/p_x)$$

All real speakers have positive directional gain. When sound radiation is enhanced in one direction, it must be sacrificed in other directions. The converse of this is also true: a reduction in sound in some directions results in an increase in other directions, assuming the same total power output from the speaker.

Directional gain is expressed in decibels for microphones, too. But for a microphone, the directional gain is given in terms of pickup sensitivity, or the sound pressure required to produce a certain electrical output at the microphone terminals. For transducers in general, the directional gain is expressed as the directivity index (*see* DIRECTIVITY INDEX).

An omnidirectional microphone has zero directional gain. All other microphones have a positive directional gain. *See also* DIRECTIONAL MICROPHONE.

DIRECTIONAL MICROPHONE

A *directional microphone* is a microphone designed to be more sensitive in some directions than in others. Usually, a directional microphone is unidirectional—that is, its maximum sensitivity occurs in only one direction.

Directional microphones are commonly used in communications systems to reduce the level of background noise that is picked up. This is important for intelligibility, especially in industrial environments where the ambient noise level is high. Directional microphones are also useful in public-address systems because they minimize the amount of feedback from the speakers. Directional microphones can usually be recognized by their physical asymmetry.

Most directional microphones have what is called a *cardioid response* (*see* CARDIOID PATTERN). The maximum sensitivity of a cardioid microphone exists in a very broad lobe. The minimum sensitivity is sharply defined, and usually occurs directly opposite the direction of greatest sensitivity.

A microphone without directional properties is called an *omnidirectional microphone*.

DIRECTION FINDER

A *direction finder* is a radio receiver with a precise signal-strength indicator, connected to a rotatable loop antenna having a defined directional response. Direction-finding equipment is often used for the purpose of locating a transmitter.

The receiver in a direction-finding system need not be especially sophisticated. It is the antenna design that is critical. A loop antenna is generally used, and it is often shielded against the electric component of the radio-wave front so that it picks up only the magnetic field. The circumference of the loop (*see* LOOP ANTENNA) should be less than about 0.1 wavelength at the operating frequency. Such an antenna displays a sharp null along a line passing through its center and perpendicular to the plane of the loop.

By rotating the loop antenna until a null is seen in the receiver, the line toward the transmitter can be found. This line provides an ambiguous bearing however because it is not possible to tell whether the transmitter is at a certain bearing or 180 degrees opposite. Some direction finders have special antennas that eliminate this ambiguity. However, by taking readings from two different locations, the transmitter can be pinpointed by finding the intersection point of the two lines on a map.

Some direction finders work automatically. The antenna is rotated by a mechanical device until it points to the transmitter. *See also* AUTOMATIC DIRECTION FINDER.

DIRECTIVITY INDEX

The *directivity index* of a transducer is a measure of its directional properties. The directional index is similar to the directional gain of a speaker or microphone (*see* DIRECTIONAL GAIN); however, the mathematical definition is slightly different in terms of its expression.

For a sound-emitting transducer, let p_{av} be the average sound intensity in all directions from the device, at a constant radius m, assuming that the transducer receives an input power P. Then if p is the sound intensity on the acoustic axis, or favored direction, at a distance m from the device, the directivity index in decibels is given by:

$$Directivity\ index\ (\text{dB}) = 10\ \log_{10}(p/p_{av})$$

For a pickup transducer, the same mathematical concept applies. However, it is in the reverse sense. If a given sound source, at a distance m from the transducer and having a certain intensity, produces an average voltage E_{av} at the transducer terminals as its orientation is varied in all possible directions, and the same source provides a voltage (E) when at a distance m from the transducer on the acoustic axis, then:

$$Directivity\ index\ (\text{dB}) = 20\ \log_{10}\ (E/E_{av})$$

See also DECIBEL.

DIRECT MEMORY ACCESS

In a digital computer, *direct memory access (DMA)* is a means of obtaining information from the memory circuits without having to go through the central processing unit (CPU). The CPU is disabled while the memory circuits are being accessed.

Direct memory access saves time. It is much more efficient than getting memory information by routing through the central processing unit. There are several different methods of obtaining direct memory access. The process varies among different computer models. Direct memory access is used for the purpose of transferring memory data from one location to another, when it is not necessary to actually perform any operations on it. See also MEMORY.

DIRECTOR

A *director* is a form of parasitic element in an antenna system, designed for the purpose of generating power gain in certain directions. This increases the efficiency of a communications system, both by maximizing the effective radiated power of a transmitter, and by reducing the interference from unwanted directions in a receiving system. See ANTENNA POWER GAIN, and PARASITIC ARRAY.

An example of the operation of a director is found in all Yagi- or quad-type antennas. A half-wave dipole antenna has a gain of 0 dBd (see dBd) in free space. This means, literally, that its gain, with respect to a dipole, is zero.

If a length of conductor measuring approximately 1/2 wavelength, and not physically attached to anything, is brought near the half-wave dipole and parallel to it, the directivity pattern of the antenna changes radically. This was noticed by a Japanese engineer named *Yagi*, and the Yagi antenna is thus named after him. When the free element is a certain distance from the dipole, gain is produced in the direction of the free element. The free element is then called a director. At certain other separations, the free element causes the gain to occur in the opposite direction, and then it is called a *reflector*.

The most power gain that can be obtained using a dipole antenna and a single director is, theoretically, about 6 dBd. In practice, it is closer to 5 dBd, because of ohmic losses in the antenna conductors. When two full-wavelength loops are brought in close proximity parallel to each other, with one loop driven (connected to the transmission line) and the other loop free, the same effect is observed. An antenna using loops in this manner is called a *quad antenna*.

The design of parasitic arrays is a very sophisticated art, and beyond the scope of this book. However, many excellent sources are available that discuss the operation and design of parasitic arrays. See also QUAD ANTENNA, and YAGI ANTENNA.

DIRECT SEQUENCE

Direct sequence is a method of *spread-spectrum* modulation and reception. In direct-sequence spread spectrum, a binary sequence is generated by a circuit according to a specific program. This program is identical, and is precisely synchronized, at the transmitter and receiver. The sequence shifts the phase of the transmitted carrier at a rapid rate. It repeats at regular intervals.

The sequence appears to be random within each repetition. For this reason, it is called a *pseudorandom* sequence. The transmitter frequency jumps around according to this rapid, pseudorandom pattern; the receiver, if synchronized and using the same pattern, follows. Receivers not programmed to follow exactly along with the transmitter sequence will not receive the signal. See also SPREAD SPECTRUM.

DIRECT WAVE

In radio communications, the *direct wave* is the electromagnetic field that travels from the transmitting antenna to the receiving antenna along a straight line through space. Direct waves are responsible for part, but not all, of the signal propagation between two antennas when a line connecting the antennas lies entirely above the ground. The surface wave and the reflected wave also contribute to the overall signal at the receiving antenna in such a case. The combination of the direct wave, the reflected wave, and the surface wave is sometimes called the *ground wave* (see GROUND WAVE). Depending on the relative phases of the direct, reflected, and surface waves, the received signal over a line-of-sight path may be very strong or practically nonexistent.

The range of communication via direct waves is, of course, limited to the line of sight. The higher the transmitting antenna, the larger the area covered by the direct wave. In mountainous areas, or in places where there are many obstructions, such as buildings, it is advantageous to locate the antenna in the highest possible place. Direct waves are of little importance at low frequencies, medium frequencies, and high frequencies. But at very high frequencies and above, the direct wave is very important in propagation.

DISCHARGE

When an electronic component that holds an electric charge (such as a storage battery, capacitor, or inductor) loses its charge, the process is called *discharge*. Charge is measured in coulombs, or units of 6.218×10^{18} electrons (see CHARGE).

Discharging can occur rapidly, or it can occur gradually. The discharge rate is the time required for a component to go from a fully charged state to a completely discharged condition. In a storage battery, the discharge rate is the amount of current the battery can provide for a specified length of time. The discharging process occurs exponentially. With a given load resistance, the current is greatest at the beginning of the discharge process, and grows smaller and smaller with time.

Components, such as capacitors and inductors, can build up a charge over a long period of time, and then release the charge quickly. When this happens, large values of current or voltage can be produced. An automobile spark coil works on this principle. This property of charging and discharging can create a shock hazard, and precautions should be taken to ensure that a component has been completely discharged before any service work is performed. This is especially important with high-voltage power supplies.

DISCONE ANTENNA

A *discone antenna* is a wideband antenna, resembling a biconical antenna, except that the upper conical section is replaced by a flat, round disk. A discone antenna is very similar to a conical monopole antenna as well (*see* BICONICAL ANTENNA, and CONICAL MONOPOLE ANTENNA).

The discone, often used at very high frequencies, is fed at the point where the vertex of the cone joins the center of the disk, as shown in the illustration. The lowest operating frequency is determined by the height of the cone, h, and the radius of the disk, r. The value of h should be at least $1/4$ wavelength in free space, and the value of r should be at least $1/10$ wavelength in free space. The discone presents a nearly constant, nonreactive load at all frequencies above the lower-limit frequency, for at least a range of several octaves. The exact value of the resistive impedance depends on the flare angle Θ of the cone. Typical values of Θ range between 25 and 40 degrees, resulting in impedances that present a good match for coaxial transmission lines.

A discone antenna is usually oriented so that the disk is horizontal, and on top of the cone. This produces a vertically polarized wave. The disk and the cone are made of sheet metal or fine wire mesh. The dimensions of a discone make it a practical choice at frequencies above 30 MHz; occasionally, it is used at frequencies as low as 3 MHz. The maximum radiation occurs approximately in the plane of the disk, or slightly below. The discone is omnidirectional in the azimuth plane.

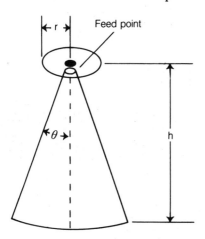

DISCONE ANTENNA: Frequency range depends on disk radius, r, and on height of cone, h.

DISCONTINUITY

A *discontinuity* in an electrical circuit is a break, or open circuit, that prevents current from flowing. Discontinuity can occur in the power-supply line to a piece of equipment, resulting in failure. Sometimes a discontinuity occurs because of a faulty solder connection, or a break within a component. Such circuit breaks can be extremely difficult to find. Sometimes the discontinuity

is intermittent, or affects the operation of a circuit only in certain modes or under certain conditions. This kind of problem is a well-known thing to experienced service technicians.

In a transmission line, a discontinuity is an abrupt change in the characteristic impedance (*see* CHARACTERISTIC IMPEDANCE). This can occur because of damage or deterioration in the line, a short or open circuit, or a poor splice. A transmission-line discontinuity can be introduced deliberately, by splicing two sections of line having different characteristic impedances. This technique is used for certain impedance-matching applications, and for distributing the power uniformly among two or more separate antennas.

DISCRETE COMPONENT

An electronic component (such as a resistor, capacitor, inductor, or transistor) is called a *discrete component* if it has been manufactured before its installation. In contrast to this, the resistors, capacitors, diodes, and transistors of an integrated circuit are not discrete; they are manufactured with the whole package, which can contain thousands of individual components.

In the early days of electronics, all circuits were built from discrete components. Only after the advent of solid-state technology, and especially miniaturization, did other designs emerge. At first, discrete components were assembled and sealed in a package called a *compound circuit*. But modern technology has provided the means for fabricating thousands of individual components on the surface of a semiconductor wafer.

Although discrete components are used less often than they were only a few years ago, there will always be a place for them. Such devices as fuses, circuit breakers, and switches must remain discrete. However, we will probably see fewer and fewer discrete components in electronic circuits in the coming years, as digital-control techniques become more refined. *See also* DIGITAL CONTROL, and INTEGRATED CIRCUIT.

DISCRIMINATOR

A detector often used in frequency-modulation receivers is called a *discriminator* (*see* FREQUENCY MODULATION). The discriminator circuit produces an output voltage that depends on the frequency of the incoming signal. In this way, the circuit detects the frequency-modulated waveform.

When a signal is at the center of the passband of the discriminator, the voltage at the output of the circuit is zero. If the signal frequency drops below the channel center, the output voltage becomes positive. The greater the deviation of the signal frequency below the channel center, the greater the positive voltage at the output of the discriminator. If the signal frequency rises above the channel center, the discriminator output voltage becomes negative; and, again, the voltage is proportional to the deviation of the signal frequency. The amplitude of the voltage at the output of the discriminator is linear, in proportion to the frequency of the signal. This ensures that the output is not distorted.

The illustration is a schematic diagram of a simple discriminator circuit suitable for use in a frequency-modulation receiver. A shift in the input signal frequency causes a phase shift in the voltages on either side of the transformer. When the signal is at the center of the channel, the voltages are equal and opposite, so that the net output is zero.

A discriminator circuit is somewhat sensitive to amplitude variations in the signal, as well as to changes in the frequency. Therefore, a limiter circuit is usually necessary when the discriminator is used in a frequency-modulation receiver. A circuit called a *ratio detector*, developed by RCA, is not sensitive to amplitude variations in the incoming signal. Thus, it acts as its own limiter. Immunity to amplitude variations is important in frequency-modulation reception, because it enhances the signal-to-noise ratio. *See also* LIMITER, and RATIO DETECTOR.

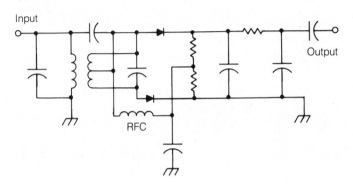

DISCRIMINATOR: This is an FM detector.

DISH ANTENNA

A *dish antenna* is a high-gain antenna that is used for transmission and reception of ultra-high-frequency and microwave signals. The dish antenna consists of a driven element or other form of radiating device, and a large spherical or parabolic reflector, as shown in the illustration. The driven element is placed at the focal point of the reflector.

Signals arriving from a great distance, in parallel wavefronts, are reflected off the dish and brought together at the focus. Energy radiated by the driven element is reflected by the dish and sent out as parallel waves. The principle is exactly the same as that of a flashlight or lantern reflector, except that radio waves are involved instead of visible light.

A dish antenna must be at least several wavelengths in diameter for proper operation. Otherwise, the waves tend to be

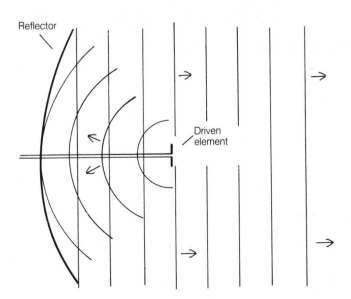

DISH ANTENNA: A parabolic or spherical reflector focuses received waves and collimates transmitted waves.

diffracted around the edges of the dish reflector. The dish is thus an impractical choice of antenna, in most cases, for frequencies below the ultra-high range. The reflecting element of a dish antenna can be made of sheet metal, or it can be fabricated from a screen or wire mesh. In the latter case, the spacing between screen or mesh conductors must be a very small fraction of a wavelength in free space.

Dish antennas typically show very high gain. The larger a dish with respect to a wavelength, the greater the gain of the antenna. It is essential that a dish antenna be correctly shaped, and that the driven element be located at the focal point. Dish antennas are used in radar, and in satellite communications systems. Some television receiving antennas use this configuration as well. *See* ANTENNA POWER GAIN.

DISK CAPACITOR

A capacitor consisting of two round metal plates with leads attached, and with a layer of dielectric sandwiched between the metal plates, is called a *disk capacitor*, or sometimes disc capacitor. Disk capacitors are very common in radio-frequency electronic circuits, and are also sometimes seen in high-impedance audio applications. The leads of the capacitor protrude from the disk-shaped body, and are parallel to each other. The entire body of the capacitor is coated with a sealant to protect it against moisture and contamination.

The physical dimensions of a disk capacitor vary greatly. The capacitance depends on the size of the disks, the spacing between them, and the kind of dielectric material used. The working voltage depends mostly on the thickness of the dielectric. The photograph shows a typical disk capacitor, which is quite small in size, both physically and electrically.

Capacitance values of disk capacitors generally range from less than 1 pF to about 1 μF. Voltage ratings are usually between about 10 and 1000 volts. The dielectric material is often a ceramic substance, which has very low loss. *See also* CERAMIC CAPACITOR.

DISK CAPACITOR: The shape is characteristic.

DISK DRIVE

In computers, digital information can be stored on magnetic disks, in much the same way that sound is stored on recording tape. There are two main types of disk in personal computers (PCs): the *hard disk*, capable of storing up to about 1 gigabyte (1,000,000,000 bytes or 1,000 megabytes) of data, and the *floppy disk,* usually capable of storing about 1.1 or 1.5 megabytes. The hard disk is usually within the PC housing, and forms the core of the data storage for the computer. The floppy, measuring either 3.5 or 5.25 inches in diameter, can be easily changed by the operator. Floppies are kept in cases, like a miniature library.

Disk drives are used to store and retrieve the information from hard disks and floppy disks. When speaking simply of a "disk drive," it is assumed that one is speaking of a floppy disk drive. In the case of a hard disk, it is common to call it a "hard drive."

Hard drives and floppy drives get their commands from a controller circuit. This circuit interfaces between the PC central processing unit (CPU) and the disk drive mechanics. When you give certain commands via the PC keyboard, the disk drive(s) react accordingly when it is necessary to store or recover data from either the hard disk or the floppy disk. You can sometimes hear the disk drives working; they make soft humming and clicking sounds.

A disk drive has a motor and a magnetic head. The head moves along *tracks* on the disk, recording digital highs and lows (ones and zeroes) in the form of magnetization regions on the disk. These regions are microscopic in size. The disk itself has a coating of magnetic material consisting of an extremely fine powder. This is why it is possible to get so much information on a single disk. A disk is also arranged in *sectors*, or wedge-shaped regions resembling pieces of a pie.

A hard disk drive commonly has two or more disks, stacked in a manner similar to the way plastic disks were arranged in old phonograph players. These disks are called *platters*.

A floppy disk drive usually can accommodate just one 3.5-inch or 5.25-inch disk at a time; to change the disk, the operator must physically pull one out and then place another one in the drive.

A disk drive makes it possible to store and recover data quite fast. No two bits of information are ever farther apart than the diameter of the disk. In practice, as the disk turns within the drive, the maximum separation distance is about half the circumference of the disk, or a little more than eight inches. Compare this with a magnetic tape, in which two data bits might be hundreds of feet apart.

Another advantage of the disk over the older ribbon tape is that a disk won't stretch, and if the drive malfunctions, the disk is almost never ruined. Contrast this with tape, that can stretch, causing misreading of data, which can jam up in its drive, and cause a catastrophe.

Sophisticated hard drives can store or read data on more than one individual disk simultaneously. The data can be accessed in a fraction of a second. A 40-megabyte hard drive provides enough space for various kinds of software, along with plenty of room for documentation and other semipermanent data that suits the needs of the individual operator. *See also* BYTE, COMPUTER, FLOPPY DISK, HARD DISK, KILOBYTE, and MEGABYTE.

DISKETTE

A *diskette* is a small magnetic disk used for data storage in personal computers. Common diskettes are either 3.5 inches or 5.25 inches in diameter.

A 3.5-inch diskette is encased in a rigid housing and is sometimes called a *microdisk*. A 5.25-inch diskette has a flexible case and is therefore "floppy." *See* DISK DRIVE, FLOPPY DISK, and HARD DISK.

DISK OPERATING SYSTEM

The *disk operating system (DOS)* is a set of computer programs that governs the operation of a computer using disk drives. Most personal computers (PCs), of the type used by radio hams, have one or more disk drives. The acronym *DOS* is a trademark of Microsoft Corporation.

DOS acts as an *interface* between you, the operator, and the computer hardware. There are other operating systems, but DOS is common in personal computers today. DOS can be installed on practically any computer. It can be purchased on a set of floppy disks.

DISPLACEMENT CURRENT

When a voltage is applied to a capacitor, the capacitor begins to charge. A current flows into the capacitor as soon as the voltage is applied. At first, this *displacement current* is quite large. But as time passes, with the continued application of the voltage, it grows smaller and smaller, approaching zero in an exponential manner.

The rate of the decline in the displacement current depends on the amount of capacitance in the circuit, and also on the amount of resistance. The larger the product of the capacitance and resistance, the slower the rate of decline of the displacement current, and the longer the time necessary for the capacitor to become fully charged. The magnitude of the displacement current at the initial moment the voltage is applied depends on the quotient of the capacitance and resistance. The larger the capacitance for a given resistance, the greater the initial displacement current. The larger the resistance for a given capacitance, the smaller the initial displacement current. Of course, the larger the charging voltage, the greater the initial displacement current.

In electromagnetic propagation, a change in the electric flux causes an effective flow of current. This current is called displacement current. The more rapid the change in the intensity of the electric field, the greater the value of the displacement current. The displacement current is 90 degrees out of phase with the electric-field cycle. The displacement current is perpendicular to the direction of wave propagation. *See also* ELECTROMAGNETIC FIELD.

DISPLAY

A *display* is a visual indication of the status of a piece of electronic equipment. Displays are also used in all metering devices. A display can be as simple as the frequency readout in a communications receiver or transmitter. Or, a display can be as complicated as the video monitors used with computers. The physical layout of a display is important from the standpoint of operating efficiency and convenience.

The cathode-ray-tube screen of an oscilloscope or spectrum analyzer is a form of display. So is the face of a digital watch or timer, or the speedometer of an automobile. Displays can be either electronic or mechanical. Electronic displays can use tubes, light-emitting diodes, or liquid crystals. An analog display shows a range of values in a continuous manner. A digital display shows either a set of numerals, or a bar indication. *See also* ANALOG, ANALOG METERING, DIGITAL, DIGITAL ME-

TERING, LIGHT-EMITTING DIODE, LIQUID-CRYSTAL DISPLAY.

DISPLAY LOSS

Display loss is a term that is generally used with regard to receiver output monitoring devices, such as a spectrum monitor. A human operator, listening to the output of a receiver, will always be able to hear faint signals that do not show up on a spectrum monitor or other instrument (*see* SPECTRUM ANALYZER, and SPECTRUM MONITOR). The ratio, expressed in decibels, between the minimum signal input power P_1 detected by an ideal instrument, and the minimum signal input power P_2 detected by a human operator using the same receiver, is called the display loss or visibility factor. Mathematically:

$$\text{Display loss (dB)} = 10 \log_{10}(P_1/P_2)$$

This is always a positive value. That is, the human operator is always better than the instrument. This is exemplified by the fact that a Morse-code copying machine will have difficulty in a marginal situation in which a human operator can get the message adequately. Although code readers can "copy" signals at extremely high speed, given good propagation conditions and a good signal-to-noise ratio, a faint signal is often imperceptible to the machine even at a slow speed which the human operator can "copy" fairly well.

DISPOSABLE COMPONENT

Some circuit components are repairable, and others are not. Components that are not repairable, or are so inexpensive that it is cheaper to just throw them out and replace them, are called *disposable components*. Capacitors, diodes, integrated circuits, resistors, and transistors are all disposable. However, printed-circuit boards are often repairable, as are interconnecting cables, inductors, and the like. *See also* COMPONENT.

DISSECTOR TUBE

A *dissector tube*, also known as an *image dissector*, is a form of photomultiplier television camera tube. Light is focused, by means of a lens, onto a translucent surface called a *photocathode*. This surface emits electrons in proportion to the light intensity. The electrons from the photocathode are directed to a barrier containing a small aperture. The vertical and horizontal deflection plates, supplied with synchronized scanning voltages, move the beam from the photocathode across the aperture. Thus, the aperture scans the entire image. The electron stream passing through the aperture is thus modulated depending on the light and dark nature of the image.

After the electrons have passed through the aperture, they strike a dynode or series of dynodes. Each dynode emits several secondary electrons for each electron that strikes it (*see* DYNODE). In this way, the electron stream is intensified. Several dynodes in cascade can provide an extremely large amount of gain.

The resolving power, or image sharpness, of the dissector tube depends on the size of the aperture. The smaller the aperture, to a point, the sharper the image. However, there is a limit to how small the aperture can be, while still allowing enough electrons to pass, and avoiding diffraction interference patterns.

The image dissector tube, unlike other types of television camera tubes, produces very little dark noise. That is, there is essentially no output when the image is dark. This results in an excellent signal-to-noise ratio. *See also* TELEVISION.

DISSIPATION

Dissipation is an expression of power consumption, and is measured in watts. Power is defined as the rate of expenditure of energy; dissipation always occurs in a particular physical location.

When energy is dissipated, it can be converted into other forms, such as heat, light, sound, or electromagnetic fields. It never just disappears. The term *dissipation* is used especially with respect to consumption of power resulting in heat. A resistor, for example, dissipates power in this way. A tube or transistor converts some of its input power into heat; that power is said to be dissipated.

Generally, dissipated power is an undesired waste of power. Dissipated power does not contribute to the function of the circuit. Its effect can be detrimental; excessive dissipated power in a tube or transistor can destroy the device. Engineers use the term dissipation to refer to any form of power consumption. *See* also POWER, and WATT.

DISSIPATION FACTOR

The *dissipation factor* of an insulating, or dielectric, material is the ratio of energy dissipated to energy stored in each cycle of an alternating electromagnetic field. This quantity, expressed as a number, is used as an indicator of the amount of loss in a dielectric material (*see* DIELECTRIC LOSS). The larger the dissipation factor, the more lossy the dielectric substance.

When the dissipation factor of a dielectric material is smaller than about 0.1, the dissipation factor is very nearly equal to the power factor, and the two can be considered the same for all practical purposes. The power factor is defined as the cosine of the angle by which the current leads the voltage. In general, the dissipation factor (D) is the tangent of the loss angle (Θ), which is the complement of the phase angle ϕ. Thus:

$$D = \tan(90° - \phi)$$

and, when $D < 0.1$,

$$D = \cos\phi$$

In a perfect, or lossless, dielectric material, the loss angle Θ is zero. This means that the current leads the voltage by 90 degrees, as in a perfect capacitive reactance. In a conductor, the loss angle is 90 degrees. This means that the current and voltage are in phase. *See also* PHASE.

DISSIPATION RATING

The *dissipation rating* of a component is a specification of the amount of power it can safely consume, usually as heat loss. The dissipation rating is given in watts. Resistors are rated in this way; the most common values are ⅛, ¼, ½, and 1 watt, although much larger ratings are available. Zener diodes are also rated in terms of their power-dissipation capacity.

Transistors and vacuum tubes are rated according to their maximum safe collector, drain, or plate dissipation. In an amplifier or oscillator, the dissipated power P_D is the difference between the input power P_I and the output power P_O. That is:

$$P_D = P_I - P_O$$

The lower the efficiency, for a given amount of input power, the greater the amount of dissipation. It is extremely important that tubes and transistors be operated well within their dissipation ratings.

The dissipation rating for a tube or transistor is sometimes specified in two forms: the continuous-duty rating and the intermittent-duty rating. A device can often (but not always) handle more dissipation in intermittent service. For example, a tube can usually handle a higher amount of dissipation when used in a code transmitter, as compared to in an amplitude-modulated or frequency-modulated system in which the carrier is always present. *See also* DUTY CYCLE, and POWER.

DISTORTION

Distortion is a change in the shape of a waveform. Generally, the term *distortion* is used to refer to an undesired change in a wave, but distortion is sometimes introduced into a circuit deliberately.

The drawing illustrates a nearly perfect sine waveform, A, typical of radio-frequency transmitters. All of the energy is concentrated at a single wavelength. There is no harmonic energy in a perfect wave, but in practice there is always a little bit of distortion. Distortion results in the presence of the harmonics in various proportions (*see* HARMONIC), in addition to the fundamental frequency. Harmonics can be suppressed more than 80 dB in radio-frequency practice. This means that a harmonic is just one part in 100 million in amplitude, compared with the fundamental frequency. The sine wave in such a case is almost completely distortion-free.

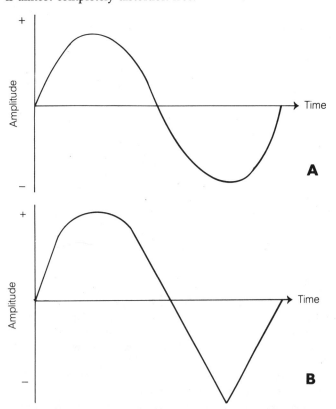

DISTORTION: At A, a pure sine wave; at B, a distorted wave.

A severely distorted waveform is shown at B. On a spectrum analyzer, distortion is visible as pips at harmonic frequencies in addition to the fundamental frequency (*see* SPECTRUM ANALYZER). The harmonics present, and their amplitudes, depend on the type and extent of the distortion.

In audio applications, distortion can occur with complex as well as simple waveforms. Sometimes the distortion can be so severe that a voice is unreadable or unrecognizable. Sound quality is measured in terms of fidelity, or accuracy of reproduction.

Distortion (such as that illustrated) is sometimes introduced into a circuit on purpose. Mixers and frequency multipliers require distortion in order to work properly. *See also* DISTORTION ANALYZER, FREQUENCY MULTIPLIER, and MIXER.

DISTORTION ANALYZER

A *distortion analyzer* is a device, usually a meter, for measuring total harmonic distortion in an audio circuit. Such a meter has a special notch filter set for 1000 Hz. This filter can be used to remove the fundamental frequency of a standard audio test tone at the output of the amplifier. A voltmeter measures the audio signal remaining after notching out the fundamental frequency; this remaining energy consists entirely of harmonics, or distortion products. Total harmonic distortion is expressed as a percentage, measured with respect to the level of the output without the filter.

A typical distortion analyzer can also measure the SINAD, or signal-to-noise-and-distortion, sensitivity of a communications receiver. The meter is extremely easy to use. *See also* SINAD.

DISTRESS FREQUENCY

In radio communication, a *distress frequency* is a frequency or channel specified for use in distress calling. Other calls may or may not be permitted.

Radiotelephone distress frequencies are 2.182 MHz and 156.8 MHz. Survival craft use 243 MHz. The distress frequencies are the same in the maritime mobile service.

In the fixed and land mobile radio services, the disaster band is at 1.75 to 1.8 MHz. In the Citizen's-Band service, Channel 9, at 27.065 MHz, is the distress frequency. There is no specifically assigned distress frequency in the Amateur Radio Service. *See also* DISTRESS SIGNAL.

DISTRESS SIGNAL

A *distress signal* is a transmission that indicates a life-threatening situation. The most familiar distress signals are mayday and SOS. The word *mayday* is used in radiotelephony. The SOS signal is used in radiotelegraphy, and in the Morse codes consists of three dots, three dashes, and three dots, usually sent in a continuous string.

In government radio services, standard frequencies are used for distress calling. *See also* DISTRESS FREQUENCY.

DISTRIBUTED AMPLIFIER

A *distributed amplifier* is a broadbanded amplifier sometimes used as a preamplifier at very-high and ultra-high frequencies. Several tubes or transistors are connected in cascade, with delay lines between them (*see* DELAY LINE).

The more tubes or transistors in a distributed amplifier, the greater the gain. However, there is a practical limit to the number of amplifying stages that can be effectively cascaded in this way. Too many tubes or transistors will result in instability, possible oscillation, and an excessive noise level at the output of the circuit.

Because a distributed amplifier is not tuned, no adjustment is required within the operating range for which the circuit is designed. Distributed amplifiers are useful as preamplifiers for television receivers.

DISTRIBUTED ELEMENT

A constant parameter (such as resistance, capacitance, or inductance) is usually thought to exist in a discrete form, in a component especially designed to have certain electrical properties (*see* DISCRETE COMPONENT). However, there is always some resistance, capacitance, and inductance in even the most simple circuits. The equivalent resistance, capacitance, and inductance in the wiring of a circuit are called *distributed elements* or *distributed constants*.

The effects of distributed elements in circuit design and operation are not usually significant at low frequencies. However, in the very-high-frequency range and above, distributed elements are an important consideration. The lead length of a discrete component, at a sufficiently high frequency, becomes an appreciable fraction of the wavelength, and then the lead inductance and capacitance will affect the operation of a circuit. An inductor has distributed resistance, on account of the ohmic loss in the conductor, and distributed capacitance, because of interaction between the windings.

In a transmission line, the inductance, resistance, and capacitance are distributed uniformly. The inductance and resistance appear in series with the conductors. The capacitance appears across the conductors.

When discrete components are used in a circuit, or when the inductances, capacitances, or resistances are in defined locations, the elements are said to be lumped.

DIVERGENCE LOSS

A radiated energy beam generally diverges from the point of transmission. Sound, visible light, and all electromagnetic fields diverge in this manner. The exception is the laser beam (*see* LASER), which puts out a parallel beam of light.

Divergence results in a decrease in the intensity of the energy reaching a given amount of surface area, as a receiver is moved farther and farther from a transmitter. With most energy effects, the decrease follows the well-known inverse-square law.

Given an energy source with a constant output, such as a speaker, the amount of energy striking a given area is inversely proportional to the square of the distance. Thus, if the distance doubles, the sound intensity is cut to one-fourth its previous level. If the distance becomes 10 times as great, the sound intensity drops to 1/100, or 0.01, of its previous value. Electromagnetic field strength, measured in microvolts per meter, decreases in proportion to the distance from the antenna.

DIVERSITY RECEPTION

Diversity reception is a technique of reception that reduces the effects of fading in ionospheric communication. Two receivers

are used, and they are tuned to the same signal. Their antennas are spaced several wavelengths apart. The outputs of the receivers are fed into a common audio amplifier, as shown in the illustration.

Signal fading usually occurs over very small areas, and at different rates. Even at a separation of a few wavelengths in the receiving location, a signal can be very weak in one place and very strong in the other. Thus, when two antennas are used for receiving, and they are positioned sufficiently far apart, and they are connected to independent receivers, the chances are small that a fade will occur at both antenna locations simultaneously. At least one antenna almost always gets a good signal. The result of this, when the receiver outputs are combined, is the best of both situations! The fading is much less pronounced.

Although diversity reception provides some measure of immunity to fading, the tuning procedure is critical and the equipment is expensive. More than two antennas and receivers can be used. In order to tune the receivers to exactly the same frequency, a common local oscillator should be used. Otherwise, the audio outputs of the different receivers will not be in the proper phase, and the result can be a garbled signal, totally unreadable.

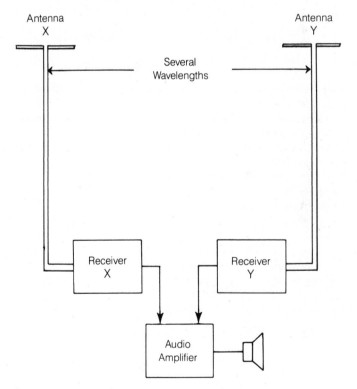

DIVERSITY RECEPTION: Two separate receivers are used.

DIVIDE-BY-N CIRCUIT

A *divide-by-n* circuit is a digital logic circuit, usually an *integrated circuit (IC)*. The output frequency of a divide-by-*n* IC is $1/n$ of the input frequency. For example, if a pulse train has a frequency of 10 MHz, a divide-by-10 circuit will produce an output of 1 MHz.

Divide-by-*n* circuits operate by counting the pulses in the train in sets of *n*. An output pulse occurs every time there are *n* input pulses. The value of *n* must always be a whole number (positive integer).

DIVIDER

A *divider* is a circuit that is used for the purpose of splitting voltages or currents. A common example of a divider is the resistive network often found in the biasing circuits of transistor oscillators and amplifiers.

In a calculator or computer, a divider is a circuit that performs mathematical division. A divider circuit is often used in a frequency counter to regulate the gate time. Divider circuits are used in digital systems to produce various pulse rates for frequency-control and measurement purposes. For example, a crystal calibrator having an oscillator frequency of 1 MHz can be used in conjunction with a divide-by-10 circuit so that 100-kHz markers can be obtained. The divider can be switched in and out of the circuit so that the operator can select the markers he wants. A divide-by-100 circuit will allow the generation of 10-kHz markers from a 1-MHz signal. *See also* FREQUENCY DIVIDER.

D LAYER

The *D layer* is the lowest region of ionization in the upper atmosphere of our planet. The D layer generally exists only while the sun is above the visual horizon. During unusual solar activity, such as a solar flare, the D layer can become ionized during the hours of darkness. The D layer is about 35 to 60 miles in altitude, or 55 to 95 kilometers.

At very low frequencies, the D layer and the ground combine to act as a huge waveguide, making worldwide communication possible with large antennas and high-power transmitters. At the low and medium frequencies, the D layer becomes highly absorptive, limiting the effective daytime communication range to about 200 miles. At frequencies above about 7 to 10 MHz, the D layer begins to lose its absorptive qualities, and long-distance daytime communication is typical at frequencies as high as 30 MHz.

After sunset because the D layer disappears, low-frequency and medium-frequency propagation changes, and long-distance communication becomes possible. This is why, for example, the standard amplitude-modulation (AM) broadcast band behaves so differently at night than during the day. Signals passing through the D layer, or through the region normally occupied by the D layer, are propagated by the higher E and F layers. *See also* E LAYER, F LAYER, IONOSPHERE, and PROPAGATION CHARACTERISTICS.

DMA

See DIRECT MEMORY ACCESS.

DOHERTY AMPLIFIER

A *Doherty amplifier* is an amplitude-modulation amplifier consisting of two transistors used for different functions. One of the devices is biased to cutoff during unmodulated-signal periods, while the other acts as an ordinary amplifier. Enhancement of the modulation peaks is achieved by an impedance inverting line between the output circuits of the two transistors. An FET-type Doherty amplifier is also sometimes called a *Terman-Woodyard modulated amplifier*.

Under conditions of zero modulation, Q1, called the *carrier FET*, acts as an ordinary radio-frequency amplifier. But Q2, called the *peak FET*, contributes no power during unmodulated periods. During negative modulation, when the instantaneous input power is less than the carrier power under conditions of zero modulation, the output of the carrier FET Q1 drops in proportion to the change in the input amplitude. At the negative peak, where the instantaneous amplitude can be as small as zero, the output of the FET is minimum.

When a positive modulation peak occurs, the peak FET contributes some of the power to the output of the amplifier. The carrier-FET output power doubles at a 100-percent peak, compared to the output under conditions of no modulation. The peak FET also provides twice the output power to the carrier FET with zero modulation. Consequently, the output power of the amplifier is quadrupled over the unmodulated-carrier level, when a 100-percent peak comes along. This is normal for amplitude modulation at 100 percent. *See also* AMPLITUDE MODULATION.

DON'T-CARE STATE

In a binary logic function or operation, some states do not matter; they are unimportant insofar as the outcome is concerned. This occurs when the function is defined for some, but not all, of the logical states. In such a case, those states that do not affect conditions, one way or the other, are called *don't-care states*.

As an example, suppose you wish to represent the decimal digits 0 through 9 in binary form. Then, you need four binary places, and we obtain:

0 = 0000	5 = 0101
1 = 0001	6 = 0110
2 = 0010	7 = 0111
3 = 0011	8 = 1000
4 = 0100	9 = 1001

Actually, then, there are six binary numbers that are not used: 1010, 1011, 1100, 1101, 1110, and 1111. These, of course, correspond to the decimal numbers 10 through 15. If the left-hand binary digit is a 1, then you know the decimal representation must be either 8 or 9. You might say that the binary numbers 1000, 1010, 1100, and 1110 correspond to the decimal value 8, and the binary numbers 1001, 1011, 1101, and 1111 correspond to the decimal number 9. It doesn't matter, then, what the two middle digits are when the far left digit is 1. In that situation, the two middle binary digits are don't-care states. *See also* BINARY-CODED NUMBER.

DOPING

Doping is the addition of impurity materials to semiconductor substances. Doping alters the manner in which such substances conduct currents. This makes the semiconductor, such as germanium or silicon, into an N-type or P-type substance.

When a doping impurity containing an excess of electrons is added to a semiconductor material, the impurity is called a *donor impurity*. This results in an N-type semiconductor. The conducting particles in such a substance are the *excess electrons*, passed along from atom to atom.

When an impurity containing a shortage of electrons is added to a semiconductor material, it is called an *acceptor impurity*. The addition of acceptor impurities results in a P-type semiconductor material. The charge carriers in this kind of substance are called *holes*, which are atoms lacking electrons. The flow of holes is from positive to negative in a P-type material (*see* HOLE).

Typical donor elements, used in the manufacture of N-type semiconductors, include antimony, arsenic, phosphorus, and bismuth. Acceptor elements, for P-type materials, include boron, aluminum, gallium, and indium. The N-type and P-type materials resulting from the addition of these elements to silicon, germanium, and other semiconductor materials, are the basis for all of solid-state electronics. *See also* SEMICONDUCTOR.

DOPPLER EFFECT

An emitter of wave energy, such as sound or electromagnetic fields, shows a frequency that depends on the radial speed of the source, with respect to an observer. If the source is moving closer to the observer, the apparent frequency increases. If the source moves farther from the observer with time, the apparent frequency decreases. For a given source of wave energy, different observers can thus measure different emission frequencies, depending on the motion of each observer, with respect to the source. The illustration shows the Doppler effect for a moving wave source. Only the radial component of the motion, with respect to the observer, results in a Doppler shift.

For sound, where the speed of propagation in air at sea level is about 1100 feet per second, the apparent frequency f^* of a source having an actual frequency f is:

$$f^* = f(1 + v/1100)$$

where v is the approach velocity in feet per second. The approach velocity is considered negative if the observer is moving away from the source.

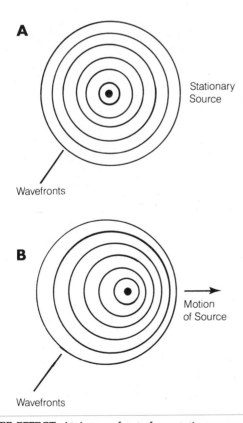

A

Stationary
Source

Wavefronts

B

Motion
of Source

Wavefronts

DOPPLER EFFECT: At A, wavefronts from stationary source; at B, from moving source.

For electromagnetic radiation, where the speed of propagation is c, the apparent frequency f^* of the emission from a source having an actual frequency f is:

$$f^* = f(1 + v/c)$$

where v is the approach velocity in the same units as c. Again, the approach velocity is considered negative if the source and the observer are moving farther apart.

The above formula for electromagnetic Doppler shifts holds only for velocities up to about $\pm 0.1c$, or 10 percent of the speed of light. For greater speeds, the relativistic correction factor must be added, giving:

$$f^* = f(1 + v/c) \sqrt{1 - v^2/c^2}$$

The wavelength becomes shorter as the frequency gets higher, and longer as the frequency gets lower. Thus, if the actual wavelength of the source is λ, the apparent wavelength λ^* is given by:

$$\lambda^* = \frac{\lambda}{(1 + v/c)\sqrt{1 - v^2/c^2}}$$

If there is no radial change in the separation of two objects, there will be no Doppler effect observed between them, although, if the tangential velocity is great enough, there can be a lowering of the apparent frequency because of relativistic effects.

DOPPLER SHIFT

When a source of electromagnetic energy moves toward or away from an observer, the observer will see an increase or decrease in the frequency emanating from that source. This is known as *Doppler shift*. In ham radio, Doppler shift is important in satellite and EME (moonbounce) work. It can also occur with auroral propagation. *See also* DOPPLER EFFECT.

The downlink signals from a satellite will appear to change frequency during a satellite pass, as the satellite first approaches your station and then recedes. This frequency change must be compensated for by tuning the receiver at the ground station. This effect does not occur with geostationary satellites. *See* ACTIVE COMMUNICATIONS SATELLITE.

In moonbounce work, Doppler shift occurs mainly because the rotation of the earth causes your station to move toward the moon when it is in the eastern sky, and away from the moon when it is in the western sky. *See* MOONBOUNCE.

In auroral propagation, the positions of the ionized regions often move so fast that the reflected energy is raised and lowered in frequency at a rapid rate. This causes the received signal to appear "smeared out" over a few hundred Hertz of spectrum. *See* AURORAL PROPAGATION.

DOS

See DISK OPERATING SYSTEM.

DOT GENERATOR

In a color television receiver, it is important that the three color electron beams converge at all points on the phosphor screen. Red, green, and blue beams meet to form white. If the alignment is not correct, the picture will appear distorted in color. A *dot generator* is a device that is used by television technicians to aid in the adjustment of the beam convergence. The pattern produced by the dot generator contains a number of white circular regions distributed over the picture frame.

If the convergence alignment of a color television receiver is perfect, all of the dots will appear white and in good focus. But if the convergence alignment is faulty, the red, green, and blue electron beams will not land together at all points of the screen. Some of the white dots will then have colored borders. If the convergence alignment is very bad, the red, green, and blue color beams might land so far away from each other that a picture is hardly recognizable in color. Alignment with the pattern produced by the dot generator is comparatively easy. With an actual picture, it would be very difficult. *See also* CONVERGENCE, PICTURE TUBE, and TELEVISION.

DOT-MATRIX PRINTER

In personal computer (PC) systems, it is usually necessary to have some way of keeping a "hard copy" of data. This is done by means of a printer. There are several different kinds of printers available.

The *dot matrix printer* offers high speed at low cost. It is ideal for plain text when it isn't necessary to have publication-quality, camera-ready copy. Dot-matrix printers cost from about $175 up.

The cheapest dot-matrix printers provide a rather coarse, but readable text, using nine pins. The copy is characterized by its appearance, best described as "dotty." More sophisticated dot-matrix printers can produce copy that is comparable to that of a typewriter using a cloth ribbon. The best dot-matrix printers offer a variety of type styles and sizes. They have 24 pins. The quality of the printed matter is enhanced by making more than one "pass" per printed line, filling in the gaps between the dots and also allowing for boldface or serifs.

The most common type sizes are 10 characters per inch (10 char/in) and 12 char/in. These are known to typewriter users as *pica* and *elite*. The usual paper size is 8.5 inches wide by 11 inches long, but larger paper sizes are sometimes used. Fan-folding the paper allows for continuous feeding. The paper feeding process is stabilized by means of detachable strips with holes that accommodate sprocket feed.

Dot-matrix printer options include type sizes larger than pica or smaller than elite, boldface, italics, underlining, double spacing, and other functions. Many dot-matrix printers can produce graphics of a fair quality in black-and-white format. *See also* DAISY WHEEL PRINTER, and LASER PRINTER.

DOUBLE BALANCED MIXER

A *double balanced mixer* is a mixer circuit that operates in a manner similar to a balanced modulator (*see* BALANCED MODULATOR, MIXER, and MODULATOR). The input energy in a double balanced mixer does not appear at the output. All of the output ports and input ports are completely isolated so that the coupling between the input oscillators is negligible, and the coupling between the input and output is also negligible. The double balanced mixer is different from the single balanced mixer; in the latter circuit, there is some degree of interaction between the input signal ports (*see* SINGLE BALANCED MIXER).

Hot-carrier diodes are generally used in modern balanced-mixer circuits. They can handle large signal amplitudes without distortion, and they generate very little noise. They are also effective at frequencies up to several gigahertz. Because these cir-

cuits, such as the one shown in the diagram, are passive rather than active, they show some conversion loss. This loss is typically about 6 dB. However, the loss is easily overcome by the use of amplifiers following the mixer. *See also* HOT-CARRIER DIODE.

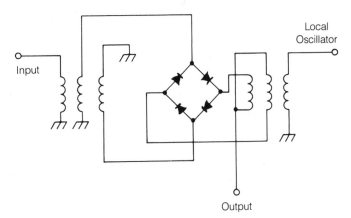

DOUBLE BALANCED MIXER: Isolates input, output, and local oscillator ports.

DOUBLE-DIFFUSED MOSFET

The most common method of manufacture for a power metal-oxide-semiconductor field-effect transistor (MOSFET) is the *double-diffusion method*. A device that is manufactured in this way is called a *double-diffused MOSFET* or *DMOS transistor*.

Double diffusion is an improvement of the *vertical MOS (VMOS)*, method that was popular for radio-frequency power FETs about two decades ago. These devices are excellent for medium-power transmitting amplifiers at high and very-high frequencies (HF and VHF). There are several variations on the basic DMOS scheme; all have similar characteristics.

DMOS technology offers high input impedance and high gain. Amplifiers using DMOS devices are simple to design and easy to build.

DOUBLE CIRCUIT TUNING

When the input circuit and the output circuit of an amplifier are independently tunable, the system is said to have *double circuit tuning*. In an oscillator, double tuning involves separate tank circuits in the base and collector portions for a bipolar transistor, the gate and plate portions for a vacuum tube. In a coupling transformer, double circuit tuning refers to the presence of a tuning capacitor across the secondary winding, as well as across the primary winding.

Double circuit tuning provides, as you should expect, a greater degree of selectivity and harmonic suppression than the use of only one tuned circuit per stage. In a multistage amplifier, however, double circuit tuning increases the possibility of oscillation at or near the operating frequency. This tendency can be reduced by tuning each circuit to a slightly different frequency, thereby reducing the Q factor of the entire amplifier chain. In such a situation, the circuit is said to be stagger-tuned.

DOUBLE-CONVERSION RECEIVER

A *double-conversion receiver*, also known as a *dual-conversion receiver*, is a superheterodyne receiver that has two different intermediate frequencies. The incoming signal is first heterodyned to a fixed frequency, called the *first intermediate frequency*. Selective circuits are used at this point. In a single-conversion receiver (*see* SINGLE-CONVERSION RECEIVER), the first intermediate-frequency signal is fed to the detector and subsequent audio amplifiers. However, in a double-conversion receiver, the first intermediate frequency is heterodyned to a second, much lower frequency, called the *second intermediate frequency*. The illustration is a block diagram of a double-conversion receiver. The second intermediate frequency can be as low as 50 to 60 kHz.

The low-frequency signal from the second mixer stage is very easy to work with from the standpoint of obtaining high gain and selectivity. Such low frequencies are not practical in a single-conversion receiver because the local-oscillator frequency would be too close to the actual signal frequency; this would cause very poor image rejection. A double-conversion receiver, while offering superior selectivity, usually has more "birdies" or local-oscillator heterodynes. But, judicious selection of the intermediate frequencies will minimize this problem. *See also* INTERMEDIATE FREQUENCY, MIXER, and SUPERHETERODYNE RECEIVER.

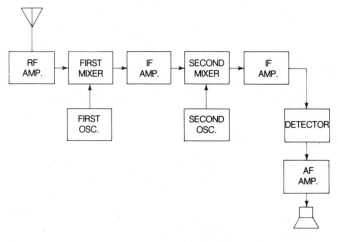

DOUBLE-CONVERSION RECEIVER: The signal frequency is heterodyned twice.

DOUBLE POLE

In a switching arrangement, the term *double pole* refers to the simultaneous switching of more than one circuit. Double-pole switches are commonly used in many electronics and radio circuits. They are especially useful when two circuits must be switched on or off at once. For example, the plate voltages to the driver and final amplifier stages of a vacuum-tube transmitter are generally of different values, and must be switched separately. A single switch, having two poles, is much more practical than two single-pole switches.

In band-switching applications in radio receivers and transmitters, several poles of switching are sometimes used. There can be three, four, or more poles in the same switch. Such multiple-pole switches are usually ganged rotary switches. *See also* MULTIPLE POLE.

DOUBLER

A *doubler* is an amplifier circuit designed to produce an output at the second harmonic of the input frequency. Such a circuit is therefore a *frequency multiplier* (*see* FREQUENCY MULTI-

PLIER). The input circuit can be untuned, or it can be tuned to the fundamental frequency. The output circuit is tuned to the second-harmonic frequency. The active element in the circuit — that is, the transistor or tube — is deliberately biased to result in nonlinear operation. This means it must be biased for class AB, B, C. Alternatively, a nonlinear passive element, such as a diode, can be used. The nonlinearity of the device produces a signal rich in harmonic energy.

A special type of circuit that lends itself well to application as a doubler is the push-push circuit. In this arrangement, which resembles an active full-wave rectifier, two transistors or tubes are connected with their inputs in phase opposition, and their outputs in parallel. This tends to cancel the fundamental frequency and all odd harmonics in the output circuit, but the even harmonics are reinforced. *See also* PUSH-PUSH CONFIGURATION.

DOUBLE SIDEBAND

Whenever a carrier is amplitude-modulated, sidebands are produced above and below the carrier frequency. These sidebands represent the sum and difference frequencies of the carrier and the modulating signal (*see* SIDEBAND). A typical amplitude-modulated signal, as it might appear on a spectrum

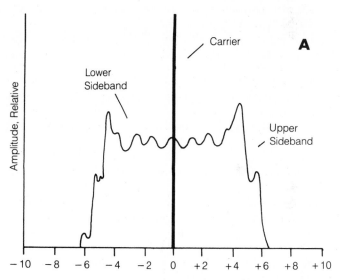

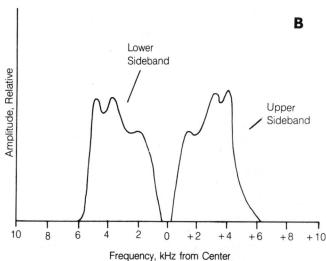

DOUBLE SIDEBAND: At A, a carrier not suppressed; at B, a carrier suppressed.

analyzer, is shown at A in the illustration. This is called a *double-sideband, full-carrier signal*; both sidebands and the carrier are plainly visible on the display.

If a balanced modulator (*see* BALANCED MODULATOR) is used for the purpose of obtaining the amplitude-modulated signal, the output lacks a carrier, as shown at B. The sidebands are left intact. Actually, of course, the carrier is not completely eliminated, but it is greatly attenuated, typically by 40 to 60 dB or more. Such a signal is called a *double-sideband, suppressed-carrier signal*. To obtain a normal amplitude-modulated signal at the receiver, a local oscillator is used to reinsert the missing carrier. The double-sideband, suppressed-carrier signal makes more efficient use of the audio energy than the full-carrier signal. This is because, in the suppressed-carrier signal, all of the transmitter output power goes into the sidebands, which carry the intelligence. In the full-carrier signal, no less than two-thirds of the transmitter power is used by the carrier, which alone carries no information at all.

A double-sideband, suppressed-carrier signal is difficult to receive, because the local oscillator must be on exactly the same frequency as the suppressed carrier. The slightest error will cause beating between the two sidebands, and the result will be a totally unreadable signal. Therefore, this kind of modulation is not often found in radiocommunication situations; single-sideband emission is much more common. *See also* AMPLITUDE MODULATION, and SINGLE SIDEBAND.

DOUBLE THROW

A *double-throw switch* is used to connect one line to either of two other lines. For example, it might be desired to provide two different plate voltages to a tube-type amplifier for operation at two different power-input levels. In such a case, the common point of a double-throw switch can be connected to the plate circuit, and the two terminals to the different power supplies.

Some switches have several different throw positions. These are called *multiple-throw switches*; a rotary switch is the most common configuration. There can be 15 or even 20 different positions in such a rotary switch. Multiple-throw switches are used in band-switching applications for radio receivers and transmitters. They are also used in multifunction controls, digital frequency controls, and in a wide variety of other situations. *See also* MULTIPLE THROW.

DOUBLET ANTENNA

See DIPOLE ANTENNA.

DOUBLE-V ANTENNA

A *double-V (fan) antenna* is a form of dipole antenna. In any antenna, the bandwidth increases when the diameter of the radiating element is made larger. A double-V takes advantage of this by using two elements rather than one, and geometrically separating them so that they behave as a single, very broad radiator. Two dipoles are connected in parallel and positioned at an angle.

The maximum radiation from a double-V antenna, and the maximum gain for receiving, occur in all directions perpendicular to a line bisecting the angle between the dipoles. In the horizontal plane, this is broadside to the plane containing the elements. The polarization is parallel to the line bisecting the apex angles of each pair of jointed conductors. The impedance

at the feed point is somewhat higher than that of an ordinary dipole antenna; it is generally on the order of 200 to 300 ohms. This provides a good match for the common types of prefabricated twin-lead line.

Double-V antennas can be used for either transmitting or receiving, and are effective at any frequency on which a dipole antenna is useful. They are quite often used as receiving antennas for the standard frequency-modulation (FM) broadcast band at 88 to 108 MHz, because of their broad frequency-response characteristics. *See also* DIPOLE ANTENNA.

DOUBLE-ZEPP ANTENNA

A full-wavelength, straight conductor, fed at the center, is sometimes called a *double-zepp antenna*. Such an antenna is actually a two-element collinear antenna, and shows a small power gain over a half-wave dipole in free space. The directional pattern of the double-zepp antenna is similar to that of the dipole antenna. Maximum radiation occurs perpendicular to the element, but the lobes are somewhat more oblong than for a dipole. The polarization is in directions parallel to the wire.

The impedance at the feed point of a double-zepp antenna is a pure resistance, and the value is quite high. If wire is used in the construction of the antenna, the feedpoint resistance can be as high as several thousand ohms. If tubing is used or the antenna is placed close to the ground or other obstructions, the value is lower, ranging from perhaps 300 to 1000 ohms.

Double-zepp antennas are used by radio amateurs, especially at the lower frequency bands (such as 1.8 and 3.5 MHz). The double zepp can, however, be used to advantage at any frequency. The versatility of this antenna is enhanced by the fact that it is resonant on all harmonic frequencies, with about the same feed-point impedance as on the fundamental frequency. It will function as a halfwave dipole at an operating frequency of half the fundamental. Generally, the double-zepp antenna is fed with a low-loss transmission line such as open wire so that the loss caused by standing waves will not be significant. *See also* DIPOLE ANTENNA, EXTENDED DOUBLE-ZEPP ANTENNA, and ZEPPELIN ANTENNA.

DOWN CONDUCTOR

In an antenna installation, a down conductor is a length of heavy wire or tubing, run from the top of a supporting mast to a ground rod. The purpose of the down conductor is to provide a low-impedance path for the discharge of lightning strokes. Down conductors are especially important for wooden masts because a lightning stroke can set such a structure on fire. Down conductors are sometimes used to protect trees against lightning.

If an antenna mast or lightning rod is located on the roof of a house or other building, or in some location where it is not possible to obtain a straight run to the ground rod, the down conductor should be positioned in the shortest possible path between the mast or rod and the ground. Ground rods should be driven at least eight feet into the soil, and at least a few feet away from underground objects (such as sewer pipes, gas mains, and building foundations). *See also* LIGHTNING, LIGHTNING PROTECTION, and LIGHTNING ROD.

DOWN CONVERSION

Down conversion refers to the heterodyning of an input signal with the output of a local oscillator, resulting in an intermediate frequency that is lower than the incoming signal frequency. Technically, any situation in which this occurs is down conversion. But usually, the term is used with reference to converters designed especially for reception of very-high, ultra-high, or microwave signals with communications equipment designed for much lower frequencies than that contemplated.

The available amount of spectrum space at ultra-high and microwave frequencies is extremely great, covering many hundreds of megahertz, and the operator of a receiver using a down converter can find the selectivity too great! For example, with communications receivers that are designed to tune the high-frequency band in 1-MHz ranges, the 10,000- to 10,500-MHz band would appear, following down-conversion, to occupy 500 such ranges. Each of these 500 ranges would require the receiver to be tuned across its entire bandspread. One signal might then require a long time to find!

This same spreading-out property, however, promises no lack of available spectrum space in the future, especially with the advent of microwave satellites. *See also* MIXER, and UP CONVERSION.

DOWNLINK

The term *downlink* is used to refer to the band on which an active communications satellite transmits its signals back to the earth. The downlink frequency is much different from the uplink frequency, on which the signal is sent up from the earth to the satellite. The different uplink and downlink frequencies allow the satellite to act as a repeater, retransmitting the signal immediately.

Satellite uplink and downlink frequencies are usually in the very-high or ultra-high frequency range. They can even be in the microwave range. The downlink transmitting antenna on a satellite must produce a fairly wide-angle beam so that all of the desired receiving stations get some of the signal. *See also* ACTIVE COMMUNICATIONS SATELLITE, REPEATER, and UPLINK.

DOWNTIME

Downtime is the time during which an electronic circuit or system is not operational. This can be because of routine servicing or because of a catastrophe, such as a power failure. The term *downtime* is used especially with respect to computers and other highly sophisticated electronic systems.

Obviously, the objective of any user of any system is to minimize the amount of downtime. Periodic shutdowns for maintenance are generally preferable, and result in less overall downtime, than waiting for a malfunction and then repairing it. The problem might not occur when it is convenient!

DOWNWARD MODULATION

If the average power of an amplitude-modulated transmitter decreases when the operator speaks into the microphone, the modulation is said to be downward. Generally, in downward modulation, the instantaneous carrier amplitude is never greater than the amplitude under zero-modulation conditions. That is, the positive modulation peaks never exceed the level of the unmodulated carrier. Negative modulation peaks can go as low as the zero amplitude level, just as is the case with ordinary amplitude modulation, but clipping should not occur. The drawing illustrates a typical amplitude-versus-time function for a downward-modulated signal.

In theory, downward modulation is less efficient than ordinary amplitude modulation because more transmitter power is required in proportion to the intelligence transmitted. That is, less effective use is made of the energy in the transmitter. Downward modulation is not often used in communications systems because the large amplitude of the unmodulated carrier is a waste of power. *See also* AMPLITUDE MODULATION.

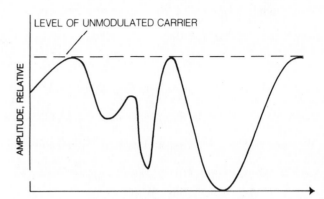

DOWNWARD MODULATION: Maximum amplitude is with zero modulation.

DRAIN

The term *drain* is used to refer to the current or power drain from a source. For example, the current drain of a 12-ohm resistor connected to a 12-volt power supply is 1 ampere.

In a field-effect transistor, the drain is the electrode from which the output is generally taken. The drain of a field-effect transistor is at one end of the channel, and is the equivalent of the plate of a vacuum tube or the collector of a bipolar transistor. In an N-channel field-effect transistor, the drain is biased positive with respect to the gate and the source. In the P-channel device, the drain is negative, with respect to the gate and source. *See also* FIELD-EFFECT TRANSISTOR.

DRAIN-COUPLED MULTIVIBRATOR

A *drain-coupled multivibrator* is a form of oscillator that uses two field-effect transistors. The circuit is analogous to a tube-type multivibrator (*see* MULTIVIBRATOR). The drain of each field-effect transistor is coupled to the gate of the other through a blocking capacitor.

The drain-coupled multivibrator is, in effect, a pair of field-effect-transistor amplifiers, connected in cascade with positive feedback from the output to the input. Because each stage inverts the phase of the signal, the output of the two-stage amplifier is in phase with the input, and oscillation results. The values of the resistors and capacitors determine the oscillating frequency of the drain-coupled multivibrator. Alternatively, tuned circuits can be used in place of, or in series with, the drain resistors. This will create a resonant situation in the feedback circuit, confining the feedback to a single frequency.

DRIFT

In a conductor or semiconductor, *drift* is the movement of charge carriers. The term is especially used in conjunction with the flow of current in a semiconductor substance. In an N-type semiconductor material, the charge carriers are extra electrons among the atoms of the substance. In a P-type material, the charge carriers are atoms deficient in electrons; these are called holes (*see* ELECTRON, and HOLE).

Drift velocity is often mentioned with respect to a semiconductor device. Drift velocity is given in centimeters per second or meters per second. In most materials, an increase in the applied voltage results in an increase in the drift velocity, but only up to a certain maximum. The greater the drift velocity for a given material, the greater the current per unit cross-sectional area.

The ease with which the charge carriers in a semiconductor are set in motion is called the *drift mobility*. Generally, the drift mobility for holes is less than that for electrons. The drift mobility determines, in a transistor or field-effect transistor, the maximum frequency at which the device will produce gain in an amplifier circuit. This frequency also depends on the thickness of the base or channel region.

The term *drift* is sometimes used in electronics to refer to an unwanted change in a parameter. This is especially true of frequency in an oscillator. The oscillator frequency of a radio transmitter can change, in either an upward or a downward direction, because of temperature variations among the components. Frequency drift is not nearly as serious a problem today as it was before the advent of solid-state technology. Transistors and integrated circuits generate almost no heat in an oscillator circuit; vacuum tubes in previous years were notorious for this. The phase-locked-loop oscillator is an extremely drift-free form of oscillator, which uses a crystal and various frequency-dividing integrated circuits. *See also* PHASE-LOCKED LOOP.

DRIFT-FIELD TRANSISTOR

A *drift-field transistor* is a special form of bipolar transistor, intended as an amplifier or oscillator at very high frequencies. The base region of the device is made very thin by an alloy-diffusion process. The thinner the base region, the shorter the time required for charge carriers to get from one side of the base to the other, and therefore the higher the frequency at which the device can be effectively used.

In the base region of the drift-field transistor, the impurity concentration is greater toward the emitter side than toward the collector side. This causes the drift mobility of the charge carriers to increase toward the collector side (*see* DRIFT). The charge carriers are thus accelerated as they move from the emitter-base junction to the base-collector junction. The result is a very high maximum-usable frequency. *See also* TRANSISTOR.

DRIVE

Drive is the application of power to a circuit for amplification, dissipation, or radiation. In particular, the term *drive* refers to the current, voltage, or power applied to the input of the final amplifier of a radio transmitter. The greater the drive, the greater the amplifier output. But excessive drive can cause undesirable effects (such as harmonic generation and signal distortion).

Some amplifiers, such as the class-A and class-AB amplifier, require very little driving power. The class-A amplifier theoretically needs no driving power at all; it runs solely off of the voltage supplied to its input. Some amplifiers, such as the class-C radio-frequency amplifier, must have a large amount of driving power to function properly. *See* CLASS-A AMPLIFIER,

CLASS-AB AMPLIFIER, CLASS-B AMPLIFIER, and CLASS-C AMPLIFIER.

In an antenna, the element that is connected directly to the feed line, and therefore receives power from the transmitter directly, is called the *driven element*. An antenna can have one or more such elements (*see* DRIVEN ELEMENT).

The term *drive* is sometimes used to refer to the mechanical device in a disk recorder. The device used to operate a rotating control, such as a tuning dial, can also be called a *drive*. *See also* DIRECT-DRIVE TUNING, and DISK DRIVE.

DRIVEN ELEMENT

In an antenna with parasitic elements (*see* PARASITIC ARRAY), those elements connected directly to the transmission line are called driven elements. In most parasitic arrays, there is one driven element, one reflector, and one or more directors.

When several parasitic antennas are operated together, such as in a collinear or stacked array, there are several driven elements. Each driven element receives a portion of the output power of the transmitter. Generally, the power is divided equally among all of the driven elements. In some phased arrays, all of the elements are driven.

The driven element in a parasitic array is always resonant at the operating frequency. The parasitic elements are usually (but not always) slightly off resonance; the directors are generally tuned to a higher frequency than that of the driven element, and the reflector is generally set to a lower frequency. The impedance of the driven element, at the feed point, is a pure resistance when the antenna is operated at its resonant frequency. When parasitic elements are near the driven element, the impedance of the driven element is low compared to that of a dipole in free space.

For the purpose of providing an impedance match between a driven element and a transmission line, the driven element can be folded or bent into various configurations. Among the most common matching systems are the delta, gamma, and T networks. Sometimes the driven element is a folded dipole rather than a single conductor. *See also* DELTA MATCH, FOLDED DIPOLE ANTENNA, GAMMA MATCH, QUAD ANTENNA, T MATCH, and YAGI ANTENNA.

DRIVER

A *driver* is an amplifier in a radio transmitter that is designed for the purpose of providing power to the final amplifier. The driver stage must be designed to provide the right amount of power into a certain load impedance. Too little drive will result in reduced output from the final amplifier, and reduced efficiency of the final amplifier. Excessive drive can cause harmonic radiation and distortion of the signal modulation envelope.

The driver stage receives a signal at the operating frequency and provides a regulated output with a minimum of tuning. Most or all of the transmitter tuning adjustments are performed in the final amplifier stage. *See also* DRIVE, and FINAL AMPLIFIER.

DROOP

Droop is a term used to define the shape of an electric pulse. Droop is generally specified as a percentage of the maximum amplitude of a pulse.

With an electric pulse on an oscilloscope display screen, the maximum amplitude of the pulse is indicated by E_1. After the initial rise to maximum amplitude, the pulse should ideally remain at that amplitude for its duration, and then rapidly fall back to the zero or minimum level. However, in practice, this ideal is seldom attained. The amplitude of the pulse decreases until it drops to zero, but the drop is often followed by overshoot, or a reversal of the polarity. The pulse duration at maximum amplitude is t. At time t, the amplitude of the pulse is E_2, somewhat less than E_1. The droop, in percent, is given by:

$$D = 100(E_1 - E_2)/E_1$$

and represents the amount by which the maximum pulse amplitude drops during the high state. In a perfectly rectangular pulse, there is no droop, and therefore it is 0 percent.

DROOPING RADIAL

The radial system in a ground-plane antenna (see GROUND-PLANE ANTENNA) is usually horizontal, and the radiating element is vertical. A quarter-wave vertical radiator in this configuration gives a resistive impedance of about 37 ohms. This represents a fairly good match to 50-ohm coaxial-cable transmission line.

A perfect match to 50-ohm cable, using a ground-plane antenna, can be obtained by slanting the radials downward at an angle of about 45 degrees. The greater the slant, with respect to the horizontal plane, the larger the feed-point impedance becomes. When the radials are oriented straight down, the antenna impedance is about 73 ohms, identical with that of a dipole in free space. In fact, such an antenna is called a *vertical dipole*.

Ground-plane antennas with drooping radials are often used at frequencies above approximately 27 MHz, or the Class-D Citizen's Band. At these frequencies, rigid tubing can be used for the radials, as well as for the vertical radiator. In some amateur-radio installations, groundplane antennas are used at frequencies as low as 1.8 MHz, and the radials serve the function of guy wires. *See also* VERTICAL DIPOLE ANTENNA.

DROPPING RESISTOR

A *dropping resistor* is a resistor installed in a circuit for the purpose of lowering the voltage supplied to a load. The amount by which the voltage is dropped depends on the power-supply output voltage, and on the load resistance.

Dropping resistors should be used only when the load resistance remains constant. Otherwise, the voltage regulation will be extremely poor. For example, a dropping resistor can be used to operate a 12-volt incandescent bulb, which normally draws a current of 100 mA from 120-volt alternating-current utility mains. In such a case, the normal operating resistance of the bulb is:

$$R_{BULB} = 12/0.1 = 120 \text{ ohms}$$

In a 120-volt circuit, the total resistance R_{LOAD} necessary to provide a current of 100 mA is:

$$R_{LOAD} = 120/0.1 = 1200 \text{ ohms}$$

Thus, the resistor R should have a value of $1200 - 120$ ohms, or 1080 ohms. Under these conditions the bulb will light nor-

mally, provided the dropping resistor can handle a power level of at least:

$$P = (0.1)^2 \times 1080 = 10.8 \text{ watts}$$

To illustrate the poor regulation of this arrangement under conditions of varying load resistance, suppose a second 12-volt bulb, identical to the first, is connected in parallel with the first. The result would be a decrease in the current through both bulbs, because of the lowered resistance of the bulb combination. In fact, the current would drop to about half its original value in either bulb. To provide the bulbs with 12 volts, the value of the dropping resistor would have to be cut in half.

Dropping resistors are used only in circuits having a constant load resistance. They are also quite wasteful of power. In the above example, with the single 12-volt bulb and the 1080-ohm resistor, the bulb consumes just 1.2 watts, but the resistor takes 10.8 watts!

DRY CELL

A *dry cell* is a voltage-generating dry chemical device. It consists of two electrodes separated by a thick electrolyte paste. The cells commonly used in flashlights and portable radio receivers are dry cells. They are available in hardware, grocery, and drug stores everywhere.

Dry cells exist in a variety of different physical sizes, ranging from the tiny camera cells, smaller than a dime, to large cylindrical cells measuring more than a foot high. In general, the milliampere-hour capacity of a dry cell is proportional to its physical size. The amount of current that a cell can deliver at a given time is also proportional to its physical size.

Dry cells are commercially made primarily in two different forms: the alkaline cell and the zinc cell. There are other forms of dry cells.

DRY-REED SWITCH

The *dry-reed switch* is a form of relay. Two metal strips, called *reeds*, are sealed in a cylindrical glass tube. Some dry-reed switches contain several poles, and can be either single-throw or double-throw.

One of the strips in the dry-reed switch is made of a magnetic material. Therefore, in the presence of a magnetic field, it is moved from one position to the other. Some single-throw dry-reed switches are normally open, and some are normally closed.

Dry-reed switches are noteworthy for their high speed of operation. However, their current-carrying capacity is rather limited because of their generally small physical size. They cannot be subjected to very high voltages, or arcing will occur across the contacts. Dry-reed switches are favored for use in radio-frequency switching applications because of the low capacitance between the contacts, resulting from the small physical dimensions of the devices.

DTL

See DIODE-TRANSISTOR LOGIC.

DTMF

See TOUCHTONE.

DUAL-BEAM OSCILLOSCOPE

Some oscilloscopes have the ability to display two signals at the same time, independently of each other. This is accomplished by means of two electron guns in the cathode-ray tube, producing two beams. Each beam produces its own trace on the screen. One trace can be "zeroed" several divisions above the other, making it easy to tell them apart. This type of scope is called a *dual-beam* or *dual-trace oscilloscope.*

In the scope cathode-ray tube, there are two sets of vertical deflection plates, one for each beam. One set of deflection plates is connected to one input (for example, the A input), and the other is connected to the B input. The horizontal sweep rate is generally the same for both the A and B traces. However, the waveforms applied to the inputs can be entirely different; the vertical deflection plates operate independently of each other.

Dual-beam oscilloscopes are useful for such purposes as comparing the input and output of a circuit, or for comparing the outputs of two circuits having a common input signal.

Suppose you want to check a differentiator circuit to see if it really is doing its job. You connect the A scope probe to the differentiator input, and the B scope probe to the differentiator output. If the circuit is working properly, the B trace will appear on the screen as the derivative of the A trace. That is, the B trace will show the rate at which the A trace is changing. The figure shows the resulting oscilloscope display. In this example, the A trace is a sine wave, and the B trace appears as another sine wave, shifted by 90 degrees, or $\frac{1}{4}$ cycle, to the left (a cosine wave). Because the cosine is the derivative of the sine, the circuit is indeed differentiating the signal.

Some oscilloscopes have four or even eight traces. These are used by professional engineers who design sophisticated equipment, especially digital devices. *See also* OSCILLOSCOPE.

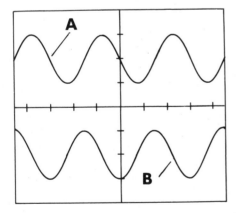

DUAL-BEAM OSCILLOSCOPE: Example of display showing a sine wave (A) and its derivative (B).

DUAL COMPONENT

When two individual components are combined into a single package, the total package is called a *dual component.* The components can be identical, or they can be substantially different.

The most common dual components are dual diodes and dual capacitors, often used in full-wave power supplies. Dual transistors are sometimes seen as well; one such component consists of two transistors that are connected in a Darlington pair and housed in a single case.

Sometimes several components are contained in a single package. An example of this is the bridge rectifier circuit, which contains four semiconductor diodes in a single housing. In the case of the integrated circuit, of course, hundreds or even thousands of components are fabricated on a semiconductor wafer and placed into a single package. *See also* INTEGRATED CIRCUIT.

DUAL IN-LINE PACKAGE

The *dual in-line package* is a common type of housing for integrated circuits. A flat, rectangular box that contains the chip is fitted with lugs along either side, as shown in the photo. There can be just a few pins on each side, or there can be 15, 20, or even 30. The number of lugs is generally the same on either side of the device.

Dual in-line packages are convenient for installation in sockets. Sometimes they are soldered directly to printed-circuit boards. The short length of the lugs, and their small physical diameter, result in very low interelectrode capacitances. A dual in-line package device is sometimes called a *DIP. See also* INTEGRATED CIRCUIT, and SINGLE IN-LINE PACKAGE.

DUAL IN-LINE PACKAGE: The pins are in two rows.

DUCT EFFECT

The *duct effect* is a form of propagation that occurs with very-high-frequency and ultra-high-frequency electromagnetic waves. Also called *ducting,* this form of propagation is entirely confined to the lower atmosphere of our planet.

A duct forms in the troposphere when a layer of cool air becomes trapped underneath a layer of warmer air, or when a layer of cool air becomes sandwiched between two layers of warmer air. These kinds of atmospheric phenomena are fairly common along and near weather fronts in the temperate latitudes. They also occur frequently above water surfaces during the daylight hours, and over land surfaces at night. Cool air, being more dense than warmer air at the same humidity level, exhibits a higher index of refraction for radio waves of certain frequencies. Total internal reflection therefore occurs inside the region of cooler air, in much the same way as light waves are trapped inside an optical fiber or under the surface of a body of water.

For the duct effect to provide communications, both the transmitting and receiving antennas must be located within the same duct, and this duct must be present continuously between the two locations. Sometimes, a duct is only a few feet wide. Over cool water, a duct can extend just a short distance above the surface, although it can persist for hundreds or even thousands of miles in a horizontal direction. Ducting allows over-the-horizon communication of exceptional quality on the very-high and ultra-high frequencies. Sometimes, ranges of over 1,000 miles can be obtained by this mode of propagation. *See also* TROPOSPHERIC PROPAGATION.

DUMMY ANTENNA

A *dummy antenna* is a resistor, having a value equal to the characteristic impedance of the feed line, and a power-dissipation rating at least as great as the transmitter output power. There is no inductance in the resistor. A dummy antenna is used for the purpose of testing transmitters off the air, so that no interference will be caused on the frequencies at which testing is carried out. A dummy antenna can also be useful for testing communications receivers for internal noise generation, when no outside noise sources can be allowed to enter the receiver antenna terminals.

The resistor in a dummy antenna is always noninductive. This ensures that the feed line can be terminated in a perfect impedance match. The sizes of dummy loads can range from very small to very large. Some dummy loads consist of large resistors immersed in oil-coolant containers. They can handle a kilowatt or more of power on a continuous basis.

Dummy antennas must be well-shielded to prevent accidental signal radiation. This is especially important at the higher power levels.

DUMP

The term *dump* is used in computer practice to refer to the transfer or loss of memory. Memory bits can be moved from one storage location to another to make room in the first memory for more data. In the second memory, sometimes called a *hard memory*, the data is not affected by programming changes or power failures.

Some computer programs use dump points in their execution at periodic intervals. This transfers the data from the active, or soft, memory into a permanent, or hard, memory. During a pause in the execution of a program, the computer can backtrack to the most recent dump point, and use the data there to continue the program. This process is called *dump-and-restart*. *See also* MEMORY.

DUPLEXER

A *duplexer* is a device in a communications system that allows duplex operation. This means that the two operators can interrupt each other at any time — even while the other operator is actually transmitting. Duplex operation is usually carried out using two different frequencies. Notch filters between the transmitter and receiver at each station prevent overloading of the receiver front end. A repeater uses a duplexer to allow the simultaneous retransmission of received signals on a different frequency (*see* NOTCH FILTER, and REPEATER). The illustration shows a block diagram of a duplex communications system, including the duplexer.

In a radar installation, a duplexer is a device that automatically switches the antenna from the receiver to the transmitter whenever the transmitter puts out a pulse. This prevents the transmitter from damaging or overloading the receiver. In this application, the duplexer acts as a high-speed, radio-frequency-actuated transmit-receive switch. The receiver is not

actually able to operate while the transmitter sends out the pulse. *See also* TRANSMIT-RECEIVE SWITCH.

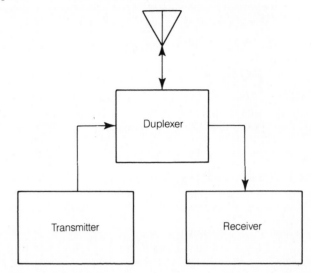

DUPLEXER: Allows simultaneous transmission and reception on different frequencies.

DUPLEX OPERATION

Duplex operation is a form of radiocommunication in which the transmitters and receivers of both stations are operated simultaneously and continuously. Each station operator can interrupt the other at any time, even while the other is actually transmitting. Therefore, a normal conversation is possible, similar to that over a telephone. There are no "blind" transmissions, and therefore there is no need for exchange signals such as "over."

In order to achieve full duplex operation, it is necessary that the transmitting and receiving frequencies be different. Two stations in duplex communication (for example, station X and station Y) must have opposite transmitting and receiving frequencies. For example, station X can transmit on 146.01 MHz and receive on 146.61 MHz; then, station Y must transmit on 146.61 MHz and receive on 146.01 MHz. A duplexer (*see* DUPLEXER) is usually required to prevent the station receiver from being desensitized or overloaded by the transmitter.

In radiotelegraphy, a form of nearly full duplex operation is sometimes called *break-in operation*. A fast transmit-receive switch blocks the receiver during a transmitted dot or dash, but allows extremely rapid recovery. If the receiving operator desires to interrupt the transmitting operator, a tap on the key will be heard by the sending operator within a few milliseconds. Break-in operation is not true duplex, however, because the sending operator cannot actually hear the receiving operator interrupt during a transmitted pulse. *See also* BREAK-IN OPERATION.

DURATION TIME

Duration time is the length of an electric pulse, measured from the turn-on time to the turn-off time. If the pulse is rectangular, then the duration time is quite easy to determine. But electric pulses are usually not perfectly rectangular. They often have rounded corners, droop, or other irregularities. Then, the duration of the pulse can be expressed in several different ways. Generally, the duration time of a nonrectangular pulse is considered to be the theoretical duration time of an equivalent rectangular pulse. The equivalent pulse can be generated by area redistribution, or by amplitude averaging.

The average amplitude of a pulse train depends on the duration and the frequency of the pulses. *See also* DUTY FACTOR.

DUTY CYCLE

The *duty cycle* is a measure of the proportion of time during which a circuit is operated. Suppose, for example, that during a given time period t_0, a circuit is in operation for a total length of time t, measured in the same units as t_0. Of course, t cannot be larger than t_0; if $t = t_0$, then the duty cycle is 100 percent. In general, the duty cycle is given by:

$$Duty\ cycle\ (percent) = 100t/t_0$$

When determining the duty cycle, it is important that the evaluation time t_0 be long enough. The evaluation time must be sufficiently lengthy so that an accurate representation is obtained. For example, a circuit can be on for one minute, and then off for one minute, and then on again for one minute, in a repeating cycle. In such a case, an evaluation time of 10 seconds is obviously too short; t_0 must be at least several minutes for a reasonably accurate estimate of the duty cycle.

Power-dissipating components (such as resistors, transistors, or vacuum tubes), can often handle more power as the duty cycle decreases. A transmitter transistor can have a dissipation rating of 100 watts for continuous duty; but at a 50-percent duty cycle, the rating can increase to 150 or 200 watts. The duty cycle is assumed to be reasonable under such circumstances, such as a maximum of 15 seconds on and 15 seconds off. *See also* DISSIPATION RATING.

DUTY FACTOR

The *duty factor* of a system is the ratio of average power to peak power, expressed as a percentage. The duty factor is similar to the duty cycle, except that the system does not necessarily have to shut down completely during the low period (*see* DUTY CYCLE). In a continuous-wave transmitter, the duty factor and the duty cycle are the same. Letting P_{max} be the peak power and P_{av} be the average power, then:

$$Duty\ factor\ (percent) = 100P_{av}/P_{max}$$

In a pulse train, or string of pulses, the duty factor is the product of the pulse duration and the frequency. As a percentage, letting the duration time be t and the frequency in Hertz be f, then:

$$Duty\ factor\ (percent) = 100tf$$

In a string of square or rectangular pulses, the duty factor is the same as the duty cycle. *See also* DURATION TIME.

DX AND DXING

One of the most interesting, challenging and exciting kinds of ham radio operation involves making contacts with stations in foreign lands. Ham radio signals know no geographic bounds, and ham-radio friendships recognize no political lines. A contact with a station in another country is called *DX*. The pursuit of contacts in many foreign countries is *DXing*. A ham who likes to work DX, or primarily DX, is a *DXer*.

Station Design for DX It is not necessary to have a huge antenna or high power in order to work DX, even "rare" DX. A simple wire antenna, strung between two trees and fed with coaxial cable or open-wire line, can produce plenty of DX contacts, even with a low-power transmitter. This is particularly true on the 10-, 12- and 15-meter amateur bands.

The single most important piece of hardware for working DX is a good receiver. It should be sensitive, stable and selective, and should have good dynamic range. In recent years, receiver design has been improved dramatically. Most commercially manufactured ham receivers and transceivers are satisfactory for working DX. In older receivers, a preamplifier at the antenna input can sometimes make a big difference in sensitivity, especially on the bands above 14 MHz. *See* PREAMPLIFIER.

The second most important thing for a DX station is a good antenna. A quad or yagi antenna (*see* QUAD ANTENNA, and YAGI ANTENNA) is preferred for 10, 12, 15, 17, and 20 meters, and also for 30 and 40 meters if space and money are available. Such antennas should be at a height of at least ½ wavelength, and preferably a full wavelength or more, above the ground and surrounding obstructions. For the lower bands —40, 80, and 160 meters — vertical antennas work well if they are provided with a good RF ground (*see* GROUND-PLANE ANTENNA, and VERTICAL ANTENNA). Dipole antennas are satisfactory if they are at a height of at least ¼ wavelength, but preferably ½ wavelength or more. Inverted-V antennas are noted for their DX performance on 80 and 160 meters if the center is at least ¼ wavelength above ground. *See* DIPOLE ANTENNA, and INVERTED-V ANTENNA.

The third, and optional, ingredient for working DX is a linear amplifier that can deliver the maximum legal power, or close to it.

Operating DX Not mentioned as station equipment, but as important as the hardware, is a good, patient, courteous operator. DX stations are often easy to find on the ham bands, not because their signals sound unique, but because of the presence of "pileups," or groups of stations calling. You might hear one of these and wonder how you can ever hope to be heard by the DX station, when you are just one of dozens or even hundreds of stations calling. It's easy to imagine that the combined power of all these stations will drown you out. It's not hard to envision gigantic yagis and quads, and big linears run by hardened DXers, making it impossible for you to get through. This writer has faced all of that with a wire thrown over a sapling and a mere 85 watts output, and snared rare DX on the very first call. This experience is not too unusual.

One method of being heard is to call the DX station on a frequency no one else appears to be using. This is easier on CW than on phone. But sometimes, especially in a thick pileup or when the DX operator is experienced, the DX station will give specific instructions concerning the calling frequency. You should always follow these instructions, making sure, of course, that your license class allows you to use that frequency.

Another method of being heard is to try to stand out because of your operating courtesy. Keep calls short. Never call while the DX station is in QSO with someone else.

QSLs Some DXers like to collect QSL cards from their DX contacts. This is necessary in order to demonstrate that you have had QSOs with the stations; you will need them for the DX

Century Club, for example. The easiest way to send, receive and keep track of QSL cards is via a *QSL bureau. See* QSL BUREAU, and QSL CARD. Some hams don't care about QSLing, but you should send one if you receive one from a DX station and the operator specifically requests it by checking the box "PSE QSL."

For More Information There are books available concerning ham operating, including DXing. Any of these make good reading. They are advertised in all the ham-radio-oriented magazines. You can also learn about operating DX from a local ham who is "into" this aspect of amateur radio. Most DXers like to show off their stations and their accomplishments.

DYNAMIC CHARACTERISTIC

Electronic components that carry alternating currents behave differently from those that carry only pure direct currents. Transistors and tubes are especially significant in this respect. The characteristic curves of such devices (*see* CHARACTERISTIC CURVE) are generally determined by testing with direct currents only; such curves are therefore static characteristics, meaning that they have been determined with steady signals. In actual operation of most amplifiers and oscillators, of course, the signal is not constant, but fluctuating. This can cause the characteristics to be slightly different. Real-operation conditions are reflected in the dynamic characteristics of a component or circuit.

In general, the dynamic characteristics are the actual operating parameters of circuits and components. These parameters are not always the same as the static characteristics, which are usually theoretical or are derived under rigid test conditions.

DYNAMIC CONTACT RESISTANCE

In a relay, the resistance caused by the contacts depends on several factors. Such factors include the pressure with which the contacts press against each other, as well as their surface area and the amount of corrosion. The greater the pressure, the lower the contact resistance. The effective resistance of a current-carrying pair of relay contacts can vary because of mechanical vibration between the contacts. The actual contact resistance, measured with the relay in an actual circuit in normal operating circumstances, is the dynamic contact resistance.

Normally, the dynamic resistance of any pair of relay contacts should be as low as possible. This can be ensured by keeping the contacts clean, and by making certain that the relay coil gets the right amount of current, so that the contacts will open and close properly. Too much coil current will cause contact bounce; too little coil current will result in insufficient pressure between the contacts. *See also* RELAY.

DYNAMIC INPUT

In a digital-logic circuit, a *dynamic input* is sampled whenever, but only when, the clock changes levels. For this reason, the dynamic input is sometimes called an *edge-triggered input* because it is actuated by the edge (beginning and/or ending) of a digital pulse.

Some dynamic inputs are actuated only when the state changes from 0 to 1. This is called *positive edge triggering*. If the input is actuated by a change from 1 to 0, it is called *negative edge triggering*. *See also* STATIC INPUT.

DYNAMIC LOUDSPEAKER

A *dynamic loudspeaker* is a speaker that uses magnetic fields and electric currents to produce sound vibrations. In the dynamic loudspeaker, a permanent magnet supplies the necessary magnetic field. A solenoidal coil carries the audio-frequency alternating or pulsating direct current. As the current in the coil fluctuates, the changing magnetic field, created by the current, interacts with the field around the permanent magnet. The resulting forces cause sound to be propagated.

The coil in the dynamic loudspeaker, called the voice coil, is mounted in such a way that it can move back and forth as the magnetically induced forces act on it. A large paper cone is attached to the coil, and moves along with it. The moving cone imparts some of the energy to the surrounding air molecules, resulting in acoustical waves. Most speakers today are of this type.

A device similar to the dynamic loudspeaker can be used to convert sound energy into audio-frequency electric currents. *See also* DYNAMIC MICROPHONE.

DYNAMIC MICROPHONE

A *dynamic microphone* is a device that utilizes magnetic fields and mechanical movements to create alternating electric currents. The principle of operation is exactly the same as that of the dynamic loudspeaker, except that it operates in the reverse sense (*see* DYNAMIC LOUDSPEAKER).

Vibrating air molecules cause mechanical movement of a diaphragm. A coil, attached rigidly to the diaphragm, moves along with it in the field of a permanent magnet. As a result of this movement, currents are induced in the coil. These currents are in exact synchronization with the sound vibrations impinging against the diaphragm. Amplifiers can be used to boost the currents to any desired level.

Dynamic microphones are one of the most common types of microphones used today. They are physically and electrically rugged, and last a long time. The magnet should not be subjected to extreme heat or severe physical blows. Other types of microphones are sometimes used, such as the ceramic and crystal microphones. *See also* CERAMIC MICROPHONE, CRYSTAL MICROPHONE, and MICROPHONE.

DYNAMIC RANGE

In an audio sound system or a communications receiver, the dynamic range is often specified as an indicator of the signal variations that the system can accept, and reproduce, without objectionable distortion.

In high-fidelity sound systems, there is a weakest signal level that can be reproduced without being masked by the internal noise of the amplifiers. There is also a strongest signal level that can be reproduced without exceeding a certain amount of total harmonic distortion in the output. Usually, the maximum allowable level of total harmonic distortion is 10 percent (*see* TOTAL HARMONIC DISTORTION). If P_{max} represents the maximum level of input power that can be reproduced without appreciable distortion, and P_{min} represents the smallest audible input power, then the dynamic range in decibels is given by:

$$Dynamic\ range\ (dB) = 10\ \log_{10}(P_{max}/P_{min})$$

In a communications receiver, the dynamic range refers to the amplitude ratio between the strongest signal that can be handled without appreciable distortion in the front end, and the weakest detectable signal. The formula for the dynamic range in a communications receiver is exactly analogous to that for audio equipment. Most receivers today have adequate sensitivity, so this is not a primary limiting factor in the dynamic range of a receiver. Much more important is the ability of a receiver to tolerate strong input signals without developing nonlinearity. Such nonlinearity in the front end, or initial radio-frequency amplification stages, can result in problems with intermodulation and desensitization. *See also* DESENSITIZATION, and FRONT END.

DYNAMIC REGULATION

In a voltage-regulated power supply, *dynamic regulation* refers to the ability of the supply to handle large, sudden changes in the input voltage with a minimum of change in the output voltage or current. The most common kind of sudden change in the input voltage to a power supply is called a *transient*. A transient can be the result of a lightning stroke or arc in a power transmission line (*see* TRANSIENT).

Poor dynamic regulation in a power supply can result in sudden changes in the output voltage. This can cause damage to modern solid-state equipment. *See also* POWER SUPPLY, and VOLTAGE REGULATION.

DYNAMIC RESISTANCE

According to Ohm's law, we think of the resistance between two points as the quotient of the voltage and the current (*see* OHM'S LAW). This quantity is the static resistance in a circuit. In operation, the effective resistance or impedance of a component might not be the same as the static resistance or impedance. The actual, or effective, operating resistance is called the *dynamic resistance*.

DYNODE

A *dynode* is an electrode often used in a photomultiplier or dissector tube to amplify a stream of electrons. When the dynode is hit by an electron, it emits several secondary electrons. Approximately five to seven secondary electrons are given off by the dynode for each electron striking it.

Dynodes in photomultiplier tubes are slanted in such a way that an electron must strike a series of the dynodes in succession. The electron hits first one dynode, where several secondary electrons are emitted along with it. The electrons then hit a second dynode, where on incident electron again produces several secondary electrons. It does not take many dynodes to get very large gain figures in this way. For example, four dynodes, each of which emits five secondary electrons in addition to the one that strikes it, will result in a gain of $6^4 = 1296$. This is 62 dB in terms of current!

Dynodes in a photomultiplier or dissector tube are placed around the inner circumference, and are oriented at the proper angles so that an impinging electron strikes them one after the other. With the electrons in the tube, as with light against a mirror, the angle of incidence is equal to the angle of reflection. *See also* DISSECTOR TUBE, and PHOTOMULTIPLIER.

EARPHONE

A small speaker, designed to be worn in or over the ear, is called an *earphone*. Earphones are sometimes used in communications for privacy or for quiet reception. With an earphone worn on one ear, the other ear is free, allowing the operator to hear external sounds.

A pair of earphones is often mounted on a headband and worn over both ears. The combination is then called a headphone or headset. Such a combination of earphones essentially obstructs the listener's sensitivity to external noises.

Earphones are available in impedances as low as 4 ohms and as high as several thousand ohms. An earphone consumes far less power than a speaker, and so the earphone can be useful when it is necessary to conserve power, as in battery operation. Some earphones operate on the dynamic principle; others are crystal or ceramic transducers. *See also* DYNAMIC LOUD-SPEAKER, and HEADPHONE.

EARTH CONDUCTIVITY

The earth is a fair conductor of electric current. The conductivity of the soil is, however, highly variable depending on the geographic location. In some places, the soil can be considered an excellent conductor; in other locations it is a very poor conductor. In general, the better the soil for farming, the better it will be in terms of electrical conductivity. Wet, black soil is a very good conductor, but rocky or sandy dry soil is poor. A brief rain shower can drastically alter the soil conductivity in some locations.

The conductivity of the earth is generally measured in millimhos per meter. A *mho* (also known as a *Siemens*) is the unit of conductance, and is expressed as the reciprocal of the resistance. A millimho is equivalent to a resistance of 1,000 ohms. A conductance of 1,000 millimhos is equivalent to a resistance of 1 ohm. In the United States, the earth conductivity ranges from approximately 1 millimho per meter to more than 30 millimhos per meter. All earth is, by comparison to seawater, a very poor conductor; saltwater in the ocean has an average conductivity of about 5,000 millimhos per meter.

In radiocommunications, the earth conductivity is important from the standpoint of obtaining a good electrical ground. The earth conductivity is also important for surface-wave propagation at frequencies below about 20 to 30 MHz. The better the conductivity of the earth, the better the radio-station ground system will be for a given installation, and the better the surface-wave propagation. Good ground conductivity is also advantageous in electrical installations, and for the grounding of equipment for protection against lightning. *See also* GROUND CONNECTION, SURFACE WAVE.

EARTH CURRENT

Electric wires under or near the ground, carrying alternating currents, cause currents to be induced in the ground. This occurs because the ground is a fairly good conductor, and the electromagnetic field around the wire causes movement of the charge carriers in any nearby conductor. High-tension lines can induce considerable current in the ground near the wires. Radio broadcast stations cause radio-frequency currents to flow in the ground in the vicinity of their antennas.

Currents flow in the earth as a natural part of the environment. Two ground rods, driven several feet apart, will often have a small potential difference, and this can be measured with an ordinary alternating-current voltmeter. The intensity of the ground current in the vicinity of a conductor carrying alternating current depends on the conductivity of the ground in the area, and on the current intensity in the wire and the distance of the wire above the ground.

The earth conductivity is quite variable in different geographic locations. In rocky or sandy soil, the conductivity is poor and the earth currents can be expected to be small. In black soil, especially if wet, ground currents can be very large. *See also* EARTH CONDUCTIVITY.

EARTH OPERATION

In amateur-radio satellite communications, *earth operation* refers to the station working from the surface of the earth. This is contrasted with *space operation*, referring to the station (usually a transponder) working from earth orbit. Earth operation can be carried out with an operator in attendance, or via remote control. *See also* ACTIVE COMMUNICATIONS SATELLITE.

E BEND

In a waveguide, an *E bend* is a change in the direction of the waveguide, made parallel to the plane of electric-field lines of flux. It is called an *E bend* for exactly this reason: It follows the E-field component of the electromagnetic wave.

The bending of radio waves by the ionospheric E layer is sometimes called *E bend* or *E skip*, although both of these terms are somewhat inaccurate from a technical standpoint. Signal propagation via the E layer is fairly common at frequencies below about 50 MHz, and may sometimes occur well into the very-high-frequency spectrum. *See also* E LAYER, and WAVEGUIDE.

ECCENTRICITY OF ORBIT

Almost any satellite has an orbit that deviates somewhat from a perfect circle. When an orbit differs from a circle, it follows an elliptical path relative to the earth. The degree to which this ellipse is elongated is called the *eccentricity of orbit*.

A perfectly circular orbit, in which the satellite is always at the same altitude, has eccentricity zero. Orbits that are nearly circular have small, or low, eccentricity. Orbits in which the apogee of the satellite is much greater than the perigee, and which are therefore very elongated, have large, or high, eccentricity. *See also* ACTIVE COMMUNICATIONS SATELLITE, APOGEE, ELLIPTICAL ORBIT, and PERIGEE.

ECHO

Whenever energy is sent outward from a source, is reflected off of a distant object, and then returns in detectable magnitude, the return signal is called an *echo*. The most familiar echoes are, of course, the sound echoes that we all hear when we shout in the direction of a distant, acoustically reflective object. In radar, the term echo is used to refer to any return of the transmitted pulses, creating a blip in the screen (*see* RADAR).

Echoes can sometimes be heard on radio signals in the high-frequency part of the electromagnetic spectrum, where ionospheric propagation is prevalent. A few milliseconds after receiving a strong signal pulse, a weaker echo may be heard. This occurs because the signal has propagated via two paths: a short path and a long path. The echo delay time is the difference in the time required for the signal to propagate by means of the two paths. *See* LONG-PATH PROPAGATION, and SHORT-PATH PROPAGATION.

ECHO INTERFERENCE

When a high-frequency radio signal is propagated via the ionosphere, an echo sometimes occurs (*see* ECHO). This can be the result of long-path propagation in addition to the normal short-path propagation. Or, echo interference can be the result of propagation of the signal over several more-or-less direct paths of variable distance.

When echo interference is severe, it can render a signal almost unintelligible, because phase modulation is superimposed on the envelope of the signal. This phenomenon is especially noticeable with auroral propagation, when the signal path can change at a very rapid rate.

In a long-distance telephone circuit, the echo is sometimes of sufficient amplitude to create interference at the transmitting station. This can distract the person speaking because he hears his own words with a slight delay in the earphones, and it can actually be difficult to maintain a train of thought! Echo eliminators are commonly used in telephone systems to prevent this kind of interference, which results from reflection of signals from impedance discontinuities along the line.

In a radar system, echo interference is caused by the presence of unwanted ''false'' targets such as buildings and thunderstorms. During the second world war, it was discovered that thundershowers and rain squalls produced echoes on radar sets; since then, the radar has been a valuable tool in the determination of short-term weather forecasts and warnings. *See also* LONG-PATH PROPAGATION, RADAR, and TELEPHONE.

ECLIPSE EFFECT

The conditions of the ionospheric D, E, and F layers vary greatly with the time of day. During a solar eclipse, when the moon passes between the earth and the sun, nighttime conditions can occur for a brief time within the shadow of the moon. This effect on the ionized layers of the upper atmosphere is called the *eclipse effect*.

The D layer, the lowest layer of the ionosphere, seems to be affected to a greater extent during a solar eclipse than the E or F layers, which are at greater altitudes, and ionize more slowly. The ionization of the D layer depends on the immediate level of ultraviolet radiation from the sun, but the E and F layers have some ''lag'' time. An eclipse in the sun causes a dramatic decrease in the ionization level of the D layer, but the E and F layers are generally not affected to a great extent. The reduction in the ionization density of the D layer causes improved propagation at some frequencies, because the absorption created by the D layer is not as great during an eclipse, as under normal circumstances during daylight hours.

Investigations are continuing into the effects of solar eclipses on the behavior of the ionosphere. This can help us learn more about the nature of the ionosphere, and how it propagates radio signals. Of course, because solar eclipses are rare at any given location, they are not of great practical concern in ionospheric propagation. *See also* D LAYER, E LAYER, F LAYER, and PROPAGATION CHARACTERISTICS.

E CORE

An *E core* is a form of transformer core, made of ferromagnetic material (such as powdered iron) and shaped like a capital letter E. Wire coils are wound on the horizontal bars of the letter E; up to three different windings are possible so that a transformer can have two secondary windings giving different voltages and impedances, with respect to a single primary winding.

The interaction among the windings on an E core are, of course, enhanced by the action of the ferromagnetic material, which has a high permeability for magnetic fields. Most of the magnetic flux is contained within the E-shaped core, and very little exists in the surrounding air. *See also* TRANSFORMER.

EDDY-CURRENT LOSS

In any ferromagnetic material, the presence of an alternating magnetic field produces currents that flow in circular or elliptical paths, around and around. These currents, because of their resemblance to vortex motions, are called *eddy currents*.

Eddy currents are responsible for some degradation in the efficiency of any transformer that uses an iron core. To minimize the effects of eddy currents, transformers sometimes have laminated cores, consisting of a number of thin, flat pieces of ferromagnetic material that are glued together, but insulated from each other. This chokes off the flow of eddy currents to a large extent, because circular paths are cut by the insulating layers. *See also* TRANSFORMER.

EDGE CONNECTOR

An *edge connector* is a rigid plug-and-socket device used for multiple-conductor interconnections. Edge connectors provide a reliable means of contact when many conductors must be used between circuits. Some edge connectors contain dozens of pins.

The male edge connector is usually fabricated from printed-circuit material, such as phenolic, and can actually be part of a printed circuit containing other components. The female part of the connector contains spring pins, which make contact with the thin foil strips on the male edge connector. Several connections can be made on either side of the connector board. A typical edge-connector arrangement is shown in the photograph.

In modern, highly sophisticated solid-state equipment, edge connectors are commonly used. They are sometimes provided on circuit boards, in digital devices such as communications monitors, for the purpose of attaching a troubleshooting computer system. *See also* PRINTED CIRCUIT.

EDGE CONNECTOR: Allows easy exchange of PC boards.

EDGE EFFECT

In a capacitor, most of the electric field is confined between the two plates. However, near the edges of the surfaces, some of the electric lines of flux extend outward from the space between the plates. This is called the *edge effect.*

The edge effect is most pronounced with air-dielectric capacitors because the surrounding air has the same dielectric constant as the medium between the plates. However, the edge effect also occurs to some extent in other types of capacitors. It is least pronounced in electrolytic capacitors, which usually have shielded enclosures. The edge effect can result in unwanted capacitive coupling among components in a circuit, and for this reason, the component layout of a circuit is critical at frequencies where the smallest intercomponent capacitance can cause feedback or other undesirable effects. *See also* CAPACITOR.

EFFECTIVE BANDWIDTH

All bandpass filters allow some signal to pass, and reject energy at frequencies outside the passband range (*see* BANDPASS RESPONSE). The bandwidth of a filter of this kind is usually

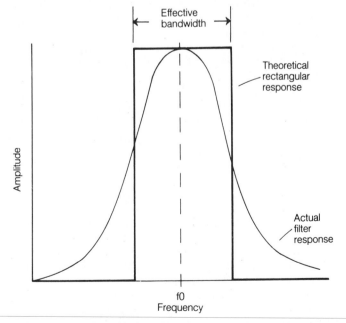

EFFECTIVE BANDWIDTH: Theoretical rectangular response passes the same energy as an actual filter.

determined according to the 3-dB power-attenuation points. However, there is another method of specifying the bandpass characteristic of a bandpass filter. This is the *effective-bandwidth method.*

Suppose a hypothetical bandpass filter, having a perfectly rectangular response (*see* RECTANGULAR RESPONSE), is centered at the same frequency f_0 as an actual bandpass filter (see the drawing). Suppose that the bandwidth of the rectangular filter can be varied at will, all the way from zero to a value much greater than that of the real filter, but always remaining centered at f_0. At some value of bandwidth of the rectangular filter, the characteristics will be identical to that of the actual filter, with respect to the amount of energy transferred. This rectangular bandwidth is the effective bandwidth of the real filter.

The effective bandwidth of a filter depends on the steepness of the response, and also on the nature of the response within the passband. *See also* BANDPASS FILTER.

EFFECTIVE GROUND

In radiocommunications practice, below a frequency of about 100 MHz, the ground plays a significant role in the behavior of antenna systems, both for receiving and for transmitting. This is true whether or not an antenna is physically connected to an earth ground. The earth, because it is a fairly good conductor at most locations, absorbs some radio-frequency energy and reflects some. The lower the frequency, in general, the better the conductivity of the soil in a particular location (*see* EARTH CONDUCTIVITY). Some soils are good conductors at virtually all frequencies below 100 MHz; some soils are poor conductors even for direct current.

Because reflected electromagnetic waves combine with the direct wave for any antenna system, the height of the antenna above the ground is an important consideration because this height determines the elevation angles at which the direct and reflected waves will act together, and also the elevation angles at which they will oppose each other. The earth surface as we see it is not the surface from which the electromagnetic waves are reflected. The effective ground plane, in most cases, lies several feet below the actual earth surface. The poorer the earth conductivity, the farther below the surface lies the effective ground plane at a certain electromagnetic frequency.

Irregularity in the terrain near an antenna, whether natural or man-made, affects the level of the effective ground. In areas with numerous buildings, utility wires, fences, and other obstructions, the determination of the effective ground plane is practically impossible from a theoretical standpoint. Antenna installations in such environments usually proceed on a trial-and-error basis, insofar as the optimum height and position are concerned. *See also* DIRECT WAVE, and REFLECTED WAVE.

EFFECTIVE ISOTROPIC RADIATED POWER

In space communications, the most common expression for radiated power is the *effective isotropic radiated power (EIRP).* It is measured in decibels relative to one watt (dBW). An EIRP of 0 dBW is the equivalent radiation that occurs if 1 W of radio-frequency energy is fed into an *isotropic antenna.*

A gain of 3 dBW is the equivalent of 2 W fed to an isotropic antenna; a gain of 20 dBW would result if 100 W was fed to it. But 3 dBW would also result from 1 W at an antenna with 3-dBi gain; 20 dBW EIRP would occur if 1 W was fed to an antenna with a gain of 20 dBi. In theory, there exist infinitely many different ways in which a particular EIRP can be produced.

The EIRP from an installation depends on the transmitter output power, the loss in the line between the transmitter and the antenna, and the gain of the antenna itself. *See also* dBi, dBW, DECIBEL, and ISOTROPIC ANTENNA.

EFFECTIVE RADIATED POWER

When a transmitting antenna has power gain, the effect in the favored direction is equivalent to an increase in the transmitter power, compared to the situation in which an antenna with no gain is used. To obtain a doubling of the effective radiated power in a certain direction, for example, either of two things can be done: The transmitter output power can actually be doubled, or the antenna power gain can be increased in that direction by 3 dB. Effective radiated power takes into account both the antenna gain and the transmitter output power. It is also affected by losses in the transmission line and impedance matching devices.

With an antenna that shows zero gain, with respect to a dipole (0 dBd), the effective radiated power (*ERP*) is the same as the power (*P*) reaching the antenna feed point. If the antenna has a power gain of A_p dB, then:

$$A_p = 10 \log_{10}(ERP/P)$$

and the ERP is thus given by:

$$ERP = P\text{antilog}_{10}\left(\frac{A_p}{10}\right)$$

The greater the forward gain of the antenna, the greater the effective radiated power.

Effective radiated power is usually determined by using a half-wave dipole as the reference antenna. However, occasionally the effective radiated power can be determined, with respect to an isotropic radiator (*see* DIPOLE ANTENNA, and ISOTROPIC ANTENNA).

Effective radiated power is an important criterion in the determining the effectiveness of a transmitting station. The ERP can be many times the transmitter output power. The ERP specification is more often used at very-high frequencies and above, than at high frequencies and below. *See also* ANTENNA POWER GAIN, dBd, dBi, and DECIBEL.

EFFECTIVE VALUE

A component, pulse, or other electrical quantity can have, in practice, a value that appears different from the actual value. This actual value is called the *effective value of the quantity.*

An example of the difference between an actual value and an effective value can be found when a resistor is used at different alternating-current frequencies. This is especially true of wirewound resistors. The actual value, at direct-current and very low alternating-current frequencies, might be 10 ohms. As the frequency is raised, however, a point is reached at which the inductance of the windings introduces an additional resistance in the circuit. At a frequency of, for example, 10 MHz, the effective resistance might be 20 ohms, consisting of 10 ohms pure resistance and 10 ohms of inductive reactance.

In alternating-current theory, the root-mean-square value of a waveform is sometimes called the *effective value. See also* ROOT MEAN SQUARE.

EFFICIENCY

Efficiency is the ratio, usually expressed as a percentage, between the output power from a device and the input power to that device. All kinds of electronic apparatus (such as speakers, amplifiers, transformers, antennas, and light bulbs) can be evaluated in terms of efficiency. Generally, only the output in the desired form is taken as the actual output power; dissipated power in unwanted forms is considered to have been wasted. The efficiency of any device is less than 100 percent; although some circuits or devices approach 100-percent efficiency, this ideal is never actually achieved.

Given that the input power to a device is P_{IN} watts and the useful output power is P_{OUT} watts, then the efficiency in percent is given by:

$$\text{Efficiency (percent)} = 100P_{OUT}/P_{IN}$$

Some devices, such as air-core transformers, are very efficient. Others, such as speakers and incandescent lamps, are very inefficient, and deliver only a small proportion of their input power in the desired output form. Amplifiers are often evaluated in terms of circuit efficiency; the input power is the product of the collector, drain, or plate current and voltage, and the output power is the actual wattage of audio-frequency or radio-frequency signal available. *See also* CIRCUIT EFFICIENCY.

EFFICIENCY MODULATION

Amplitude modulation of a carrier signal can be obtained in many different ways (*see* AMPLITUDE MODULATION). One rarely used method is called *efficiency modulation.* In this modulation scheme, the efficiency of a radio-frequency amplifier is varied by an audio input signal, and this causes changes in the level of the carrier output.

One technique for obtaining efficiency modulation is described as follows. An unmodulated carrier is supplied to the input of a radio-frequency amplifier stage. The output circuit is tuned so that it is slightly off resonance. The applied audio signal, acting via a varactor diode, affects the resonant frequency of the output tank circuit, alternately bringing it closer and farther away, with respect to the resonant frequency of operation. As the circuit approaches resonance, the output of the amplifier increases because of improved efficiency; as the circuit deviates from resonance, the output drops because of poorer efficiency. The resulting amplitude variations are in step with the audio signal, and amplitude modulation is the result.

Efficiency modulation is not often used in amplitude-modulated transmitters. More practical, and efficient, methods are available. *See also* MODULATION METHODS.

EIA

See ELECTRONIC INDUSTRIES ASSOCIATION.

EIA-232-D

In packet radio, *EIA-232-D* refers to a standard for data transfer at the physical level. This standard specifies the characteristics for serial binary data. The EIA-232-D standard replaces the old RS-232-C. *See also* OPEN SYSTEMS INTERCONNECTION REFERENCE MODEL, and PACKET RADIO.

ELASTANCE

Elastance is a measure of the opposition a capacitor offers to being charged or discharged. The unit of elastance is called the *daraf*, which is *farad* spelled backwards. Elastance is the reciprocal of capacitance. That is, if C is the capacitance in farads, then:

$$\text{Elastance (darafs)} = 1/C$$

A value of capacitance of 1 picofarad is equivalent to an elastance of 1 teradaraf (10^{12} darafs). A capacitance of 1 nanofarad is an elastance of 1 gigadaraf (10^9 darafs); a value of capacitance of 1 microfarad is 1 megadaraf.

The daraf is an extremely small unit of elastance, seldom observed in actual capacitors. The elastance of 1 daraf is the value at which 1 volt of applied electric charge will produce 1 coulomb of displacement (*see* COULOMB). In most actual capacitors, 1 volt of charge produces no more than a few thousandths of a coulomb of charge displacement. *See also* DARAF.

E LAYER

The *E layer* of the atmosphere is an ionized region that lies at an altitude of about 60 to 70 miles (100 to 115 kilometers) above sea level. The E layer is frequently responsible for medium-range communication on the low-frequency through very-high-frequency radio bands. The ionized atoms of the atmosphere at that altitude affect electromagnetic waves at wavelengths as short as about 2 meters.

The ionization density of the E layer is affected greatly by the height of the sun above the visual horizon. Ultraviolet radiation and X rays from the sun are primarily responsible for the existence of the E layer. The maximum ionization density generally occurs around midday, or slightly after. The minimum ionization density occurs during the early morning hours, after the longest period of darkness. The E layer almost disappears shortly after sundown under normal conditions.

Occasionally, a solar flare will cause a geomagnetic disturbance, and the E layer will ionize at night in small areas or "clouds." The resulting propagation is called *sporadic-E propagation,* and is well known among communications people who use the upper portion of the high-frequency band and the lower portion of the very-high-frequency band. Often, the range of communication in sporadic-E mode can extend for over 1,000 miles. However, the range is not as great as with conventional F-layer propagation (*see* SPORADIC-E PROPAGATION).

At frequencies above about 150 MHz (a wavelength of 2 meters), the E layer becomes essentially transparent to radio waves. Then, instead of being returned to earth, they continue on into space. *See also* D LAYER, F LAYER, IONOSPHERE, and PROPAGATION CHARACTERISTICS.

ELECTRET

An *electret* is the electrostatic equivalent of a permanent magnet. Although a magnet has permanently aligned magnetic dipoles, the electret has permanently aligned electric dipoles. An electret is generally made of an insulating material (dielectric). Some varieties of waxes, ceramic substances, and plastics are used to make electrets. The material is heated and then cooled while in a strong, constant electric field.

The electret can be used in a microphone, in much the same way that a permanent magnet is used in the dynamic microphone. The impinging sound waves cause a diaphragm, attached to an electret, to vibrate. This produces a fluctuating electric field, which produces an audio-frequency voltage at the output terminals. *See also* DYNAMIC MICROPHONE.

ELECTRICAL ANGLE

A sine-wave alternating current or voltage can be represented by a counterclockwise rotating vector in a Cartesian coordinate plane. The origin of the vector is at the point (0,0), or the center of the plane, and the end of the vector is at a point having a constant distance from the origin.

In such a representation, the amplitude of the current or voltage is indicated by the length of the vector. The greater the amplitude, the longer the vector, and the farther its end point is from the point (0,0). The frequency of the sine wave is represented by the rotational speed of the vector; one cycle is given by a full-circle, or 360-degree, rotation. This model is illustrated in the drawing. The y axis, or ordinate, represents the amplitude scale, and the instantaneous amplitude of the wave is therefore given by the y value of the end point of the vector.

The electrical angle θ is the angle, in degrees or radians, that the vector subtends relative to the positive x axis at any given instant. The wave cycle begins at 0 degrees, with the vector pointing along the x axis in the positive direction. The amplitude, or y component, at this time is zero. At an electrical angle of 90 degrees, the vector points along the y axis in a positive direction. This is the maximum positive point of the wave cycle. At 180 degrees, the vector points along the x axis in a negative direction, and the amplitude is zero once again. At 270 degrees, the vector points along the y axis in a negative direction, and this is the maximum negative amplitude of the wave.

While the y component of the electrical-angle vector gives the instantaneous amplitude, the x component gives the instantaneous rate of change of the signal amplitude. The speed of vector rotation, in radians per second, is called the angular frequency of the wave, and is equal to 2π (about 6.28) times the frequency in Hertz. *See also* PHASE ANGLE, and SINE WAVE.

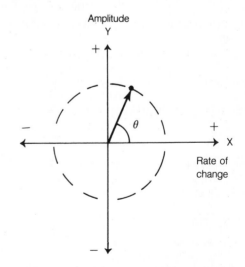

ELECTRICAL ANGLE: Measured counterclockwise from +x axis.

ELECTRICAL BANDSPREAD

In a radio receiver, *electrical bandspread* refers to bandspread obtained by adjustment of a separate inductor or capacitor, in parallel with the main-tuning inductor or capacitor. The main-tuning control in a shortwave receiver can cover several megahertz in one revolution. Bandspread is considered electrical if it is derived using a separate component; simply gearing down the main-tuning control constitutes mechanical bandspread.

An electrical-bandspread control can be made to have any desired degree of resolution, consistent with the stability of the frequency of the receiver. Generally, one turn of a bandspread knob covers between 10 and 100 kHz of spectrum in a high-frequency receiver, with 25 kHz probably being about average.

ELECTRICAL DISTANCE

Electrical distance is the distance, measured in wavelengths, that is required for an electromagnetic wave to travel between two points. The electrical distance depends on the frequency of the electromagnetic wave, and on the velocity factor of the medium in which the wave travels.

In free space, an electrical wavelength at 1 MHz is about 300 meters. In general, in free space:

$$\lambda = 300/f$$

where λ is the length of a wavelength in meters and f is the frequency in megahertz. If v is the velocity factor, given as a fraction, rather than a percent, then:

$$\lambda = 300 \, v/f$$

Electrical distances are often specified for antenna radiators and transmission-line stubs. Such distances are always given in wavelengths.

ELECTRICAL INTERLOCK

An *electrical interlock* is a switch that is designed to remove power from a circuit when the enclosing cabinet is opened. Interlocks are especially important in high-voltage circuits because they protect service personnel from the possibility of electric shock.

An interlock switch is usually attached to the cabinet door or other opening, in such a way that access cannot be gained without removing the high voltage from the interior circuits.

Interlock switches can fail, and therefore they should not be blindly relied upon to protect the lives of service personnel. Interlock switches should, of course, never be defeated, unless it is absolutely necessary. Even then, this should be done only by technicians familiar with the equipment. *See also* ELECTRIC SHOCK.

ELECTRICAL WAVELENGTH

In an antenna or transmission line carrying radio-frequency signals, the *electrical wavelength* is the distance between one part of the cycle and the next identical part. The electrical wavelength depends on the velocity factor of the antenna or transmission line, and also on the frequency of the signal.

In free space, where a signal has a frequency f in megahertz, the wavelength λ is given by the equations:

$$\lambda \text{ (meters)} = 300/f$$
$$\lambda \text{ (feet)} = 984/f$$

Along a thin conductor, the electrical wavelength at a given frequency is somewhat shorter than in free space. The velocity factor (*see* VELOCITY FACTOR) in such a situation is about 95 percent, or 0.95, therefore:

$$\lambda \text{ (meters)} = 285/f$$
$$\lambda \text{ (feet)} = 935/f$$

along a wire conductor.

In general, in a conductor or transmission line with a velocity factor v, given as a fraction:

$$\lambda \text{ (meters)} = 300v/f$$
$$\lambda \text{ (feet)} = 984v/f$$

The electrical wavelength of a signal in a transmission line is always less than the wavelength in free space. When designing impedance-matching networks or resonant circuits using lengths of transmission line, it is imperative that the line be cut to the proper length in terms of the electrical wavelength, taking the velocity factor into account. *See also* ELECTRICAL DISTANCE.

ELECTRIC CONSTANT

The electrical permittivity of free space is sometimes called the *electric constant*, represented by ϵ_0. This value is about 8.854 picofarads per meter. Essentially all dielectric materials have an electrical permittivity greater than that of a vacuum.

The electric constant is the basis for the determination of dielectric constants for all insulating materials. *See also* DIELECTRIC CONSTANT, and PERMITTIVITY.

ELECTRIC COUPLING

Electric coupling exists between or among any objects that show mutual capacitance. Electric coupling can be desirable or undesirable.

When two objects are electrically coupled, the charged particles on both objects exert a mutual attractive force or a mutual repulsive force. Like charges repel, and opposite charges attract. The plates of a capacitor provide a good illustration of the effects of electric coupling. A negative charge on one plate produces a positive charge on the other plate, by repelling the electrons in the other plate and literally pushing them from it. A positive charge on one plate produces a negative charge on the other, by attracting extra electrons to that plate. *See also* CHARGE, and MAGNETIC COUPLING.

ELECTRIC FIELD

Any electrically charged body sets up an electric field in its vicinity. The electric field produces demonstrable effects on other charged objects. The electric field around a charged object is represented, by the physicist and the engineer, as lines of flux. A single charged object produces radial lines of flux. The direction of the field is considered to be from the positive pole outward, or inward toward the negative pole.

When two charged objects are brought near each other, their electric fields interact. If the charges are both positive or both negative, a repulsion occurs between their electric fields. If the charges are opposite, attraction occurs.

The intensity of an electric field in space is measured in volts per meter. Two opposite charges with a potential difference of 1 volt, separated by 1 meter, produce a field of 1 volt per meter. The greater the electric charge on an object, the greater the intensity of the electric field surrounding the object, and the greater the force that is exerted on other charged objects in the vicinity. *See also* CHARGE, ELECTRIC COUPLING, and ELECTRIC FLUX.

ELECTRIC FLUX

Electric flux is the presence of electric lines of flux in space or in a dielectric substance. The greater the intensity of the electric field, the more lines of flux present in a given space, and therefore the greater the electric flux density.

Of course, the "lines" of an electric field are not actual lines that can be seen. They are simply imaginary, with each "line" representing a certain amount of electric charge. Electric flux density is measured in coulombs per square meter.

The electric flux density falls off, or decreases, with increasing distance from a charged object, according to the law of inverse squares (*see* INVERSE-SQUARE LAW). If the distance from a charged object is doubled, the electric flux density is reduced to one-quarter its previous value. This can be envisioned by imagining spheres centered on the charged particle. No matter what the radius of the sphere, all of the electric flux around the charged particle must pass through the surface of the sphere. As the radius of the sphere is increased, the surface area of the sphere grows according to the square of the radius. Therefore, a region of the sphere with a certain area, such as 1 square meter, becomes smaller in proportion to the square of the radius of the sphere; fewer electric lines of flux will cut through the region. *See also* CHARGE, and ELECTRIC FIELD.

ELECTRIC SHOCK

Electric shock is the harmful flow of electric current through living tissue. The effects of such current can be very damaging, and can even cause death. All living cells contain some electrolyte matter, and therefore they are capable of carrying electric current. Harmful currents are produced by varying amounts of voltage, depending on the resistance of the circuit through the living tissue.

In the human body, the most susceptible organ, as far as life-threatening effects are concerned, is the heart. A current of 100 to 300 mA for a short time can cause the heart muscle to stop its rhythmic beating and begin to twitch in an uncoordinated way. This is called heart *fibrillation*. An electric shock is much more dangerous when the heart is part of the circuit, as compared to when the heart is not in the circuit.

Under certain conditions, even a low voltage can produce a lethal electric shock. When working on equipment carrying more than perhaps 12 volts, therefore, strict precautions must be taken to minimize the chances of electric shock. The 120-volt utility lines produce an extreme hazard because this voltage is more than sufficient to cause death.

If electric shock occurs, the power source should be cut off immediately. No attempt should be made to rescue a person who is in contact with live wires. Cardiopulmonary resuscitation (CPR) or mouth-to-mouth resuscitation might be necessary for a victim of electric shock. A medical professional should be called immediately. *See also* CARDIOPULMONARY RESUSCITATION, HEART FIBRILLATION, and MOUTH-TO-MOUTH RESUSCITATION.

ELECTROCHEMICAL DIFFUSED-COLLECTOR TRANSISTOR

All transistors generate heat in operation, especially at the base-collector junction. In transistors that are designed to act as power amplifiers at audio or radio frequencies, heatsinks are often used to prevent the base-collector junction from heating to destructive temperatures. The electrochemical diffused-collector transistor is a device designed especially for the effective removal of heat from the base-collector region. *See also* DIFFUSION, HEATSINK, and TRANSISTOR.

ELECTROCUTION

See ELECTRIC SHOCK.

ELECTRODE

An *electrode* is any object to which a voltage is deliberately applied. Electrodes are used in such things as vacuum tubes, electrolysis devices, cells, and other applications. An electrode is generally made of a conducting material, such as metal or carbon.

In a device that has both a positive and a negative electrode, the positive electrode is called the *anode*, and the negative electrode is called the *cathode*. The cathode in a vacuum tube emits electrons, and the anode, sometimes called the *plate*, collects them.

All electrodes display properties of capacitance, conductance, and impedance. These characteristics, in individual cases, depend on many things, such as the size and shape of the electrodes, their mutual separation, and the medium between them. *See also* ELECTRODE CAPACITANCE, ELECTRODE CONDUCTANCE, ELECTRODE IMPEDANCE, and TUBE.

ELECTRODE CAPACITANCE

Electrode capacitance is the effective capacitance that an electrode shows relative to the ground, the surrounding environment, or other electrodes in a device. All electrodes have a certain amount of capacitance, which depends on the surface area of the electrode, the distances and surface areas of nearby objects, and the medium separating the electrode from the nearby objects.

In a vacuum tube, the interelectrode capacitance, or capacitance among the cathode, grids, and plate, is an important parameter in the design of circuits at very high frequencies and above. The interelectrode capacitance in a tube can, at such frequencies, affect the resonant frequency of associated tank circuits. The interelectrode capacitance also places an upper limit on the frequency at which the tube can be effectively used to produce gain. Generally, the interelectrode capacitance in a vacuum tube is on the order of just a few picofarads (trillionths of a farad). *See also* ELECTRODE, and TUBE.

ELECTRODE CONDUCTANCE

The conductance of an electrode is the ease with which it emits or picks up electrons. *Electrode conductance* is measured in units called *mhos* or *Siemens*, which are the standard units of conductance, and are equivalent to the reciprocal of the ohmic resistance.

The conductance of an electrode or group of electrodes depends on many factors, including the physical size of the electrodes, the material from which they are made, the coating (if any) on electron-emitting electrodes, and the medium among the electrodes. In a vacuum tube, the interelectrode conductance is the ease with which electrons flow from the cathode to the plate. This depends, of course, on the grid bias, as well as on the characteristics of the tube itself. *See also* ELECTRODE, and TUBE.

ELECTRODE DARK CURRENT

In a photomultiplier or television camera tube, the electrode dark current is the level of current that flows in the absence of light input to the tube. Dark current is important as a measure of the noise generated by a television camera tube; the smaller the dark current, the better.

The electrode dark current in most photomultiplier tubes is very small. The anode dark current originates partially from the photocathode, and partly from the other components within the tube. Some of the dark current also originates because of thermal disturbances. The electrode dark current results in what is called dark noise, as the electrons strike the anode. Dark noise limits the sensitivity of a television picture camera tube. *See also* PHOTOMULTIPLIER.

ELECTRODE IMPEDANCE

Electrode impedance is the resistance encountered by a current as it flows through an electrode. In the case of direct current, electrode impedance is simply the reciprocal of the electrode conductance (*see* ELECTRODE CONDUCTANCE). It is measured, therefore, in ohms.

For alternating current, the electrode impedance consists of both resistive and reactive components. The electrode impedance is determined in the same manner as ordinary impedance, by adding, vectorially, the resistance and the reactance (*see* IMPEDANCE, REACTANCE, and RESISTANCE).

Electrode impedance in a particular situation depends on several factors, including the surface area of the electrode, the material from which it is made, the surrounding medium, and, in a vacuum tube, the bias and the grid voltage. The frequency of the applied alternating-current signal also affects the electrode impedance in some cases because the capacitive and inductive reactances change as the frequency changes. *See also* ELECTRODE CAPACITANCE, and TUBE.

ELECTRODYNAMOMETER

An *electrodynamometer*, also sometimes called a *dynamometer*, is a form of analog metering device. The principle of operation is similar to that of the D'Arsonval movement, but the magnetic field is supplied by an electromagnet, rather than by a permanent magnet. The current being measured supplies the magnetic-field current for the electromagnet.

Two stationary coils are connected in series with a moving coil. The moving coil is attached to the indicating needle, and is mounted on bearings. The moving coil is held at the meter zero position, under conditions of no current, by a set of springs.

When a current is applied to the coils, magnetic fields are produced around both the stationary coils and the movable coil. The resulting magnetic forces cause the movable coil to rotate on its bearings. The larger the current, the more the coil is allowed to rotate by the springs.

Electrodynamometers can be used in any situation where a D'Arsonval meter is used, although the D'Arsonval meter is generally more sensitive when it is necessary to measure very small currents. *See also* D'ARSONVAL.

ELECTROLYTIC CAPACITOR

An *electrolytic capacitor* is a device that makes use of an electrochemical reaction to generate a large capacitance within a small volume. An electrolyte paste is contained in a cylindrical aluminum can. One of the leads is connected to the aluminum can, and the other to the electrolyte paste. The photograph shows a typical electrolytic capacitor, and the drawing illustrates a cutaway view inside such a capacitor.

The electrolysis process within the electrolytic capacitor causes a thin, nonconducting layer of aluminum oxide to form on the surface of the aluminum-foil sheet. This dielectric coating provides a very small physical separation between the conductive electrolyte and the aluminum. Therefore, the capacitance between the two substances is very large per unit surface area. Electrolytic capacitors can have values as high as hundreds of thousands of microfarads. Other types of capacitors are generally much smaller in value. A disk-ceramic capacitor, for example, rarely has a value of more than 1 μF.

Unlike other kinds of capacitors, electrolytic devices are sensitive to polarity. A positive and a negative lead are clearly marked on all commercially manufactured electrolytic capacitors. The capacitor must be connected so that the voltage across the capacitor has the polarity shown by the markings. *See also* CAPACITANCE, and CAPACITOR.

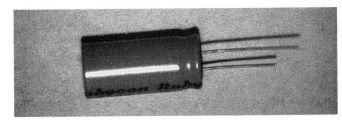

ELECTROLYTIC CAPACITOR: A photo of a low-voltage electrolytic.

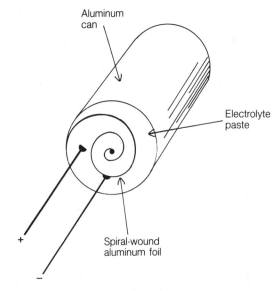

ELECTROLYTIC CAPACITOR: A cross-sectional drawing that shows the construction.

ELECTROMAGNETIC CONSTANT

In a vacuum, the speed of propagation of electric, electromagnetic, and magnetic fields is a constant of about 186,282 miles per second (299,792 kilometers per second). This constant is sometimes abbreviated by the lowercase letter c.

The electromagnetic constant is always the same in a vacuum, no matter from what viewpoint it is measured. A passenger on a rapidly moving space ship would observe the same value as a person standing on the earth, or anywhere else. This fact was established by the physicist Albert Einstein in the early part of the twentieth century, and became part of his famous theory of special relativity. Before that time, scientists thought that the speed of light, and the speed of other electromagnetic effects, might depend on the motion of the observer. The special theory of relativity postulated that the speed of light is always the same, no matter from what point of view it is measured. This dramatically affected the course of physics. *See also* ELECTRIC FIELD, ELECTROMAGNETIC FIELD, and MAGNETIC FIELD.

ELECTROMAGNETIC DEFLECTION

Electromagnetic deflection is the tendency of a beam of charged particles to follow a curved path in the presence of a magnetic field. This occurs because any moving particle itself produces a magnetic field, and the two fields interact to produce forces on the particles.

In a television picture tube, as well as in some cathode-ray oscilloscope tubes, electromagnetic deflection is used to guide the electron beam, emitted by the electron gun or guns, across the phosphor screen. Electric fields can also be used for this purpose.

Electromagnetic deflection can be demonstrated in a simple home experiment. Bring a powerful magnet, either a permanent magnet or an electromagnet, near the phosphor screen of a television receiver picture tube. Tune the receiver to a channel that comes in clearly. The magnet will produce a noticeable distortion in the picture. *See also* ELECTRIC FIELD, and MAGNETIC FIELD.

ELECTROMAGNETIC FIELD

An *electromagnetic field* is a combination of electric and magnetic fields. The electric lines of force are perpendicular to the magnetic lines of force at all points in space. An electromagnetic field can be either static or alternating. An alternating electromagnetic field propagates in a direction that is perpendicular to both the electric and magnetic components, as shown in the diagram.

Propagating electromagnetic fields are produced whenever charged particles are subjected to acceleration. The most common example of acceleration of charged particles is the movement of electrons in a conductor carrying an alternating current. The constantly changing velocity of the electrons causes the generation of a fluctuating magnetic field, which results in an electromagnetic field that propagates outward from the conductor.

The frequency of an alternating electromagnetic field is the same as the frequency of the alternating current in the conductor. The frequency can be as low as a fraction of a cycle per second (hertz); it can be as high as many trillions or even quadrillions of hertz. Visible light is a form of electromagnetic-field

disturbance; so are infrared rays, ultraviolet rays, X rays, and gamma rays. Of course, radio waves are an electromagnetic phenomenon.

All propagating electromagnetic fields display properties of wavelength λ as well as frequency f. In a vacuum, where the speed of electromagnetic propagation is about 300 million (3×10^8) meters per second, the wavelength in meters is given by:

$$\lambda = 300/f$$

where f is specified in megahertz. The higher the frequency, the shorter the wavelength.

Electromagnetic fields were first discovered by physicists during the nineteenth century. They observed that the fluctuating fields had a peculiar way of exerting their effects over great distances. Today, the results of this discovery are all around us, in the form of a vast and complex network of wireless communication systems, having frequencies ranging over several orders of magnitude. *See also* ELECTROMAGNETIC RADIATION, and ELECTROMAGNETIC SPECTRUM.

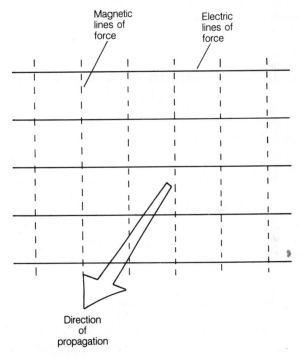

ELECTROMAGNETIC FIELD: Propagation is at right angles to both the electric and the magnetic lines of flux.

ELECTROMAGNETIC FOCUSING

Electromagnetic focusing is a form of electromagnetic deflection, used in cathode-ray tubes (*see* ELECTROMAGNETIC DEFLECTION). A stream of electrons is brought to a sharp focus at the phosphor surface of the tube, by means of a coil or set of coils that carry electric currents. The coils produce magnetic fields, which direct the electrons so that they all land on the phosphor screen at the same point.

Electromagnetic focusing provides optimum resolution in the picture of a television receiver. It also operates in any device using a cathode-ray tube, such as an oscilloscope, electron microscope, or electron telescope. *See also* CATHODE-RAY TUBE.

ELECTROMAGNETIC INDUCTION

When an alternating current flows in a conductor, a nearby conductor that is not physically connected to the current-carrying wire will show some electron movement at the same frequency. This phenomenon is known as *electromagnetic induction*, and is a result of the effects of the electromagnetic field set up by a conductor having an alternating current. The varying electric and magnetic fields produced by acceleration of charged particles cause forces to be exerted on nearby charged particles.

All radio communication is possible because of electromagnetic induction acting over long distances. The effect, originally observed by the discoverers of the propagation as happening over small distances, actually occurs over an unlimited distance. The effect travels through space with the speed of light (about 186,282 miles per second). Such a disturbance takes the form of mutually perpendicular electric and magnetic fields. *See also* ELECTROMAGNETIC FIELD.

ELECTROMAGNETIC INTERFERENCE

Electromagnetic interference (EMI) is a term used to describe a wide range of phenomena in which different electronic devices upset each other's proper operation. The kinds of EMI most affecting amateurs are outlined here.

From Ham Transmitters to Consumer Devices
Nothing hurts the reputation of amateur radio, and of radio hams, more than the type of EMI in which ham transmitters cause malfuntions in consumer devices. In recent years, this problem has become more widespread because there are more consumer electronic devices, and these devices have become more susceptible to EMI.

This kind of EMI, also often called *radio-frequency interference (RFI)*, can be the fault of the ham radio transmitter and/or operator, as well as the fault of consumer equipment design. Are you using the minimum amount of power necessary for the contacts you want to make? Is your transmitter working properly? These are things that you can control, and they can often reduce or eliminate RFI in mild cases.

Unfortunately, a large proportion (perhaps most) of RFI cases result from inadequate protection of the consumer equipment. The ham can still sometimes solve the problem at his/her station by reducing power. It is neither your fault nor the consumer's fault when equipment is poorly engineered. Diplomacy can go a long way toward securing the cooperation of a neighbor who is experiencing RFI from your station.

It is unwise to make any modifications to a neighbor's equipment yourself. The safest course of action is to contact the manufacturer of the equipment, or recommend that the neighbor contact a professional engineer. *See* FOR MORE INFORMATION below.

From Consumer Devices to Ham Receivers   It is common also for EMI to happen the opposite way, from nonham apparatus to your ham station. Vacuum cleaners, hair dryers and television sets are well known for this. There is very little regulation concerning the emissions from consumer apparatus, and manufacturers do not worry about whether or not such devices interfere with ham radio operation.

A good receiver, with a well-designed noise blanker, can reduce the effects of this type of EMI, especially when caused by devices that create electric sparks, such as gasoline-powered engines. Directional antennas for receiving, such as the small loop, can also work. Sometimes it is necessary to find the source of the interference and to try to get rid of it. *See* FOR MORE INFORMATION below.

From Power Lines to Ham Equipment   Power lines can cause EMI that is especially hard to eradicate. Such interference is almost always caused by arcing somewhere near the ham station. A malfunctioning transformer, or a bad streetlight, or a salt-encrusted insulator can all be responsible.

Techniques for dealing with these problems are beyond the scope of this discussion. Often, help can be obtained by calling the utility company. Perhaps the power line is interfering not only with your ham radio operation, but with your television or FM stereo radio reception. In these cases, your neighbors might be affected as well. This can give additional clout to your requests for help from utility companies. *See* FOR MORE INFORMATION below.

Within the Shack   Personal computers (PCs) have become increasingly common in households throughout the U.S., and many hams now own them. It is possible for the circuitry of the computer to generate signals that interfere with your reception. Your transmissions can also sometimes upset the working of the computer.

Surge suppressors in the power lines to the computer are essential for any PC. Bypassing the power lines to each individual piece of equipment, and/or installing RF chokes in series with all power-carrying wires, might help. Keep the computer away from antenna tuners and feed lines. Be sure the station equipment is well grounded and that there are no ground loops (*see* GROUND LOOP). You might write or call the manufacturer of the PC.

Sometimes, the antenna feeders will radiate enough so that RF is present "in the shack," and this can cause keyers and other peripheral equipment to malfunction. This can be an especially persistent and tricky problem. Good antenna balance, proper grounding and shielding, and sometimes even proper choice of feed-line length are necessary to minimize this. Reducing power can often prove a temporary solution until an electronic means is found to correct this type of EMI.

Other kinds of EMI   As the number and variety of consumer electronic devices proliferates, EMI problems will probably become more frequent and more severe.

Telephone Interference   This can be especially bothersome if you have a high-power single-sideband (SSB) station. Filtering devices are available for telephone sets, but they do not always work. This is one instance in which it is to the consumer's advantage to purchase a well-engineered telephone set. A bad phone line can sometimes cause an EMI problem. The telephone company is responsible for the upkeep of their lines, but not usually for telephone sets.

Medical Equipment   Electromagnetic fields can upset the workings of certain medical electronic devices. If ham radio operation causes such problems in any situation, the amateur operation must cease until a resolution is found.

Cross Modulation   Unwanted mixing of radio signals can occur in the most unsuspected places. This produces a kind of EMI that can be difficult to track down and correct. *See* CROSS MODULATION.

Intermodulation This problem is particularly likely in the downtown areas of large cities, where many radio transmitters are in operation simultaneously. It causes false signals in radio receivers, often sounding "hashy" or broken-up. *See* INTERMODULATION.

Harmonics Television interference (TVI) and hi-fi FM stereo interference is sometimes caused by excessive harmonic emission from a ham rig. You should always be certain that your transmitter's harmonic emissions are within legal limits, but this is not always sufficient to prevent TVI and FM stereo interference. Lowpass filters and antenna tuners can provide additional attenuation of harmonics. Linear amplifiers should not be overdriven or underloaded. See HARMONIC, HARMONIC SUPPRESSION, and LOWPASS FILTER.

Spurious Emissions Ham radio equipment sometimes malfunctions in such a way that signals are transmitted on unintended (and often unknown) frequencies. If you suspect such a malfunction, write or call the manufacturer of your ham equipment. Technicians are usually cooperative concerning these types of problems. *See* SPURIOUS EMISSIONS.

For More Information The American Radio Relay League (ARRL), 225 Main Street, Newington, CT publishes books about EMI, its causes, and ways to deal with it. You can also check the various ham magazine ads for new publications that might come out concerning this problem.

Useful information appears in many articles herein. EMI is a continuing problem. It is a good idea to write to legislators to express concern.

Devices that react unfavorably to ham radio transmissions are also frequently affected by commercial and government transmissions. As more and more transmitters are built, manufacturers will feel increasing pressure to provide EMI protection in electronic equipment.

ELECTROMAGNETIC PULSE

An *electromagnetic pulse* is a sudden burst of electromagnetic energy, caused by a single, abrupt change in the speed or position of a group of charged particles. An electromagnetic pulse does not generally have a well-defined frequency or wavelength, but instead it exists over practically the entire electromagnetic spectrum. This can include the radio wavelengths, infrared, visible light, ultraviolet, X rays, and even gamma rays. Electromagnetic pulses can be generated by arcing, and on a radio receiver they sound like popping or static bursts. Lightning produces an electromagnetic pulse of bandwidth extending from the very-low-frequency range to the ultraviolet range.

An electromagnetic pulse can contain a fantastic amount of power for a short time. Lightning discharges have been known to induce current and voltage spikes in nearby electrical conductors, of such magnitude that equipment is destroyed and fires are started.

The detonation of an atomic bomb creates a strong electromagnetic pulse. The explosion of a multi-megaton bomb at a very high altitude, while not creating a devastating shock wave or heat blast at the surface of the earth, could cause the generation of a damaging electromagnetic pulse over an area of many thousands of square miles (see illustration). It is possible that a major nuclear attack might be preceded by the explosion of bombs aboard satellites. The resulting electromagnetic pulses could induce damaging voltages and currents in radio antennas, telephone wires, and power transmission lines over a vast geographic area. For this reason, all communications systems should have some form of protection against the effects of such an electromagnetic pulse. One possible means of defense is provided by equipment for direct-hit lightning protection. *See also* ELECTROMAGNETIC FIELD, and LIGHTNING PROTECTION.

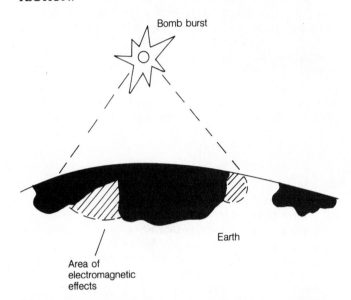

ELECTROMAGNETIC PULSE: A nuclear explosion will cause a damaging burst of electromagnetic energy.

ELECTROMAGNETIC RADIATION

Electromagnetic radiation is the propagation of an electromagnetic field through space (*see* ELECTROMAGNETIC FIELD). All forms of radiant energy, such as infrared, visible light, ultraviolet, and X rays, are forms of electromagnetic radiation.

The wavelength is the property of electromagnetic radiation that determines the form in which it is manifested. Radio waves can be as long as thousands of meters, or as short as a few millimeters, or anywhere in between these two extremes. Other forms of electromagnetic radiation have very short wavelengths. The shortest known gamma rays measure about 10^{-15} meter (10^{-5} Angstrom unit). This is so small that even the most powerful microscope could not resolve it. Theoretically, no limits exist for how long or how short an electromagnetic wave can be.

In a vacuum, electromagnetic radiation travels at the speed of light, or about 186,282 miles per second or 299,792 kilometers per second. In media other than a perfect vacuum, the speed of propagation is slower. The precise velocity depends on the substance, and also on the wavelength of the electromagnetic radiation. Some materials are opaque to electromagnetic radiation at certain wavelengths. *See also* ELECTROMAGNETIC CONSTANT, ELECTROMAGNETIC FIELD, and ELECTROMAGNETIC SPECTRUM.

ELECTROMAGNETIC SHIELDING

Electromagnetic shielding is a means of keeping an electromagnetic field from entering or leaving a certain area. The most common form of electromagnetic shielding consists of a grounded enclosure, made of sheet metal or perforated metal. A screen can also be used.

An electromagnetic shield is important in many different kinds of electronic systems. Shielding prevents unwanted electromagnetic coupling between circuits. If a shielded enclosure is not made of solid metal, the openings should be very small compared with the wavelength of the electromagnetic field. Perforated shields allow ventilation for cooling.

Coaxial cable provides an example of an electromagnetic shield that is used for the purpose of preventing radiation from a transmission line. By keeping the electromagnetic field inside the line, the shield guides the wave along the line. The shield continuity or effectiveness is a measure of the degree to which such a cable confines the electromagnetic field. This parameter depends on the frequency of the field. The higher the frequency for a given cable, the less the effective shielding. *See also* ELECTROMAGNETIC FIELD, and SHIELDING EFFECTIVENESS.

ELECTROMAGNETIC SPECTRUM

Electromagnetic wave disturbances can take place at any frequency, in theory, without limits as to how slow or how fast the oscillations occur. It is possible for a wave to require millions of years to complete a cycle; or an octillion cycles can occur every second. Some of the slowest cycles exist in the magnetic fields of stars. The fastest cycles result from bombardment of matter by heavy subatomic particles at near-light speeds.

The *electromagnetic spectrum* is the limitless wavelength range of all possible electromagnetic disturbances. It is a continuum, often displayed in logarithmic form according to wavelength or frequency.

In general, the higher the frequency, the shorter the wavelength. If frequency f is given in hertz and wavelength w is given in meters, then in free space:

$$w = 300,000,000/f$$

If frequency f is in kilohertz, then:

$$w = 300,000/f$$

If frequency f is in megahertz, then:

$$w = 300/f$$

If frequency f is in gigahertz, then:

$$w = 0.3/f$$

If frequency f is in gigahertz and the wavelength w is in centimeters, then

$$w = 30/f$$

The figure is a logarithmic diagram of portions of the electromagnetic spectrum. At A is the range in which most electromagnetic fields occur in the universe. The heavy portion is expanded at B, to show the part of the radio spectrum in which some of the more commonly used ham bands are found. There are some ham bands above ¾ meters, or the 432-MHz band, extending well into the microwave spectrum.

Infrared, visible, ultraviolet and shorter wavelengths are all, in theory, available for amateur use. An example of infrared or visible-light amateur communication is the modulated laser or maser.

See also ELECTROMAGNETIC RADIATION, and FREQUENCY ALLOCATIONS.

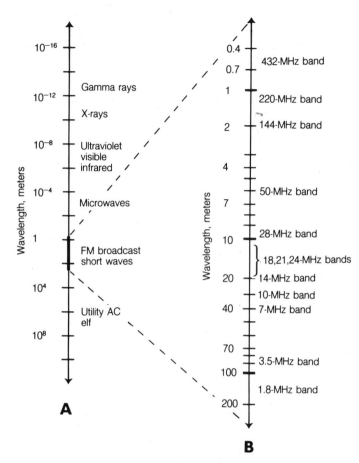

ELECTROMAGNETIC SPECTRUM: At A, ELF to gamma radiation. At B, the spectrum containing ham bands at 75 cm and longer.

ELECTROMAGNETIC SWITCH

An *electromagnetic switch* is a device that allows remote-control switching. A current, either direct or alternating, is supplied to a coil of wire with a ferromagnetic, solenoidal core, creating an electromagnet. The field attracts an armature toward or away from a fixed set of contacts, opening and closing various circuits.

Electromagnetic switches, also often called relays, are used in a variety of applications for switching at different speeds, and with frequencies that range from direct current up to several hundred megahertz. *See also* RELAY.

ELECTROMAGNETISM

When a charged particle, or a stream of charged particles, is set in motion, a magnetic field is produced. The lines of magnetic force occur in directions perpendicular to the motion of the charged particles. This effect is called *electromagnetism*.

Electromagnetism is well illustrated by the flow of current in a straight wire conductor. The magnetic lines of flux lie in a plane or planes orthogonal to the wire. They appear as concentric circles around the wire axis. The intensity of the magnetic field, or the number of flux lines in a given amount of area, is proportional to the intensity of the current in the wire.

If an alternating current is flowing in the wire, the intensity of the magnetic field varies, and these fluctuations result in a spatial charge differential or electric field. A changing, or oscillating, magnetic field therefore contains an electric-field component as well. The electric lines of flux occur parallel to the wire, or perpendicular to the magnetic lines of force. The combination of the two fields has an ability to propagate long distances through space, and is called an electromagnetic field. *See also* ELECTROMAGNETIC FIELD, and ELECTROMAGNETIC RADIATION.

ELECTROMECHANICAL RECTIFIER

An *electromechanical rectifier* is a fast-acting relay device, which alternately opens and closes in the presence of an alternating current. During one half of the cycle, the contacts are open, and that half of the cycle is cut off. During the other half of the cycle, the contacts are closed, and that half is passed. For example, if the contacts are open during the negative half of the cycle, the resulting current is positive. This is a pulsating direct current, exactly the same as would be produced by a half-wave diode rectifier circuit.

Another form of electromechanical rectifier is the *commutator*. This device operates as a shaft rotates, alternating the polarity in step with the frequency of the current. This results in full-wave rectification. Commutators are used in motors to allow them to operate from direct current; they are also used to produce direct current from generators. *See also* COMMUTATOR, GENERATOR, and MOTOR.

ELECTROMETER

An *electrometer* is a highly sensitive device, incorporating a meter and an amplifier, that is used for measuring small voltages. The input resistance is extremely high, and can be as large as several quadrillion ohms (a quadrillion is 10^{15}, or a million billion). The electrometer, because of its high input resistance, draws essentially no current from the source under measurement. This allows the device to be used for measuring electrostatic charges.

An electrometer uses a special vacuum tube called an electrometer amplifier or electrometer tube (*see* TUBE). The interelectrode resistance is very high, and the amplifier circuit is designed in a manner similar to that of a class-A audio or radio-frequency amplifier. The noise generation is minimal; this is important from the standpoint of obtaining the maximum possible sensitivity. A tiny change in the input current, in some cases as small as 1 picoampere (a millionth of a millionth of an ampere), can be detected.

The electrometer is used whenever it is necessary that the current drain be negligible. An electrometer can also be used to measure electrostatic voltages in which the amount of charge (quantity) is actually small. *See also* FET VOLTMETER.

ELECTROMOTIVE FORCE

Electromotive force is the force that causes movement of electrons in a conductor. The greater the electromotive force, the greater the tendency of electrons to move. Other charge carriers can also be moved by electromotive force; in some types of semiconductor, the deficiency of an electron in an atom can be responsible for conduction. Such a deficiency is called a *hole*; electromotive force can move holes as well as electrons. *See also* ELECTRON, HOLE, and VOLTAGE.

ELECTRON

An *electron* is a subatomic particle that carries a unit negative electric charge. Electrons can be free in space, or they can be under the influence of the nuclei of atoms. An electron is tiny indeed; it is hard to imagine its size. A single electron is to a particle of dust as the particle of dust is to the whole planet earth! The mass of an electron is 9.11×10^{-31} kilogram; therefore, a gram of them would have a count of 1.1×10^{27}, or about 1 trillion quadrillion. A single electron carries a charge of 1.59×10^{-19} coulomb; thus a coulomb of charge represents 6.28×10^{18} electron charges (*see* COULOMB).

Electrons are believed to move around the nuclei of atoms, in such a way that their average position is represented by a sphere at a certain distance from the nucleus. The force of electrostatic repulsion prevents gravity from causing atoms to fall into one another. It is primarily the electrons that are responsible for this repulsion.

In an electric conductor, the flow of electrons occurs as a passing-on of the particles from one atom to another. It does not occur as a simple flow, like water in a hose. Some atoms have a tendency to pick up extra electrons, and some have a tendency to lose electrons. Sometimes a flow of current occurs because of a deficiency of electrons, and sometimes it occurs because of an excess of electrons among the constituent atoms of a particular substance. The deficiency of an electron in an atom is called a *hole*. Hole conduction is important in the operation of semiconductor diodes and transistors. *See also* HOLE.

ELECTRON-BEAM GENERATOR

An *electron-beam generator*, also sometimes called an *electron gun*, is an electrode designed for the purpose of producing a stream of electrons in a vacuum tube. An electron-beam generator is used in cathode-ray tubes, such as those used in television receivers or oscilloscopes.

The electron-beam generator is made up of a heated cathode, a control electrode, accelerating electrodes, and focusing electrodes. The cathode emits electrons because of the negative voltage applied to it. The control electrode regulates the intensity of the electron emission from the cathode, and is used to control the brightness of the image. The accelerating electrodes give the electron beam more speed. The focusing electrodes direct the electrons into a narrow defined beam.

After being emitted from an electron-beam generator, a beam of electrons is generally passed through a set of deflecting electrodes. These control the position at which the beam lands on the phosphor screen. Voltages applied to the deflecting electrodes cause the beam to scan across the screen; the intensity or position of the beam can be further modified by signal information. *See also* CATHODE-RAY TUBE.

ELECTRON-BEAM TUBE

An *electron-beam tube* is a vacuum tube that uses an electron-beam generator rather than a typical cathode. The beam-power tube, the klystron tube, the magnetron tube, the cathode-ray tube, and the photomultiplier tube are examples of electron-beam tubes.

The electron-beam tube, in contrast to the conventional vacuum tube, operates using a focused and directed beam of electrons. This can create amplification or oscillation in the same way as with conventional tubes, but according to special parameters not available with other kinds of tubes. Electric and magnetic fields are used in electron-beam tubes to control the paths of the beams. In an ordinary vacuum tube, the intensity of the electron beam is regulated but its directional properties are not. *See also* BEAM-POWER TUBE, CATHODE-RAY TUBE, MAGNETRON, PHOTOMULTIPLIER, and TUBE.

ELECTRON EMISSION

Electron emission occurs whenever an object gives off electrons to the surrounding medium. This occurs in all vacuum tubes, and also in incandescent and fluorescent light bulbs. It also occurs in a variety of other devices. Some materials, such as barium oxide or strontium oxide, are especially well-suited to electron emission. Electrodes are sometimes coated with such because this enhances their capability to give off electrons when heated and subjected to a negative voltage.

Electron emission can occur because of thermionic effects, where the temperature of a negatively charged electrode is raised to the point that electrons break free from the forces that normally hold them to their constituent atoms.

Electron emission can be secondary in nature, resulting from the impact of high-speed electrons against a metal surface. This sometimes takes place at the plate of a vacuum tube, and in that case it can be detrimental. The suppressor grid of a pentode tube keeps the secondary electrons from escaping from the plate. A dynode operates on the principle of secondary emission to amplify an electron beam (*see* DYNODE).

If an electric charge is applied to the surface of an object, and the voltage is great enough, electrons will be thrown off. Also, if a nearby positive voltage is sufficiently strong, electrons will be pulled from an object.

In a photoelectric tube, light striking a barrier called a *photocathode* results in the emission of electrons. This is called *photoemission*. The intensity of photoemission depends on the brightness of the light, and also on its wavelength (*see* PHOTOMULTIPLIER).

The electrons emitted from an electrode are always the same elementary particles, carrying a unit charge and having the same mass, no matter what the cause of their emission. *See also* ELECTRON, and TUBE.

ELECTRON-HOLE PAIR

In an atom of a semiconductor material, the conduction band and the valence band are separated by an energy gap. An electron in the conduction band is free to move to another atom or escape entirely. But an electron in the valence band is held to the atom by the positive charge of the protons in the nucleus (*see* VALENCE BAND). An electron can move from the valence band to the conduction band of an atom if it receives a certain amount of energy. This leaves an electron vacancy, called a *hole*, in the valence band of the atom, as shown in the illustration. The electron and the hole form what is called an *electron-hole pair*.

The current in certain semiconductor materials, called *N-type semiconductors*, flows because of the extra electrons that are passed from atom to atom. In other substances, called *P-type semiconductors*, the current flows as the result of a transfer of electron deficiencies, or holes, from atom to atom. The electron carries a negative charge, and the deficiency, or hole, car-

ries a positive charge. Therefore, the flow of electrons is from the negative to the positive, but the flow of holes is in the other direction, from the positive to the negative. *See also* ELECTRON, HOLE, N-TYPE SEMICONDUCTOR, and P-TYPE SEMICONDUCTOR.

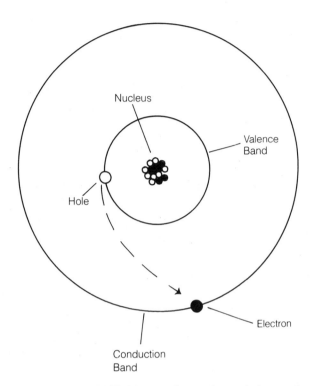

ELECTRON-HOLE PAIR: Moving electron leaves hole in valence band.

ELECTRONIC INDUSTRIES ASSOCIATION

The *Electronic Industries Association (EIA)*, is an agency in the United States that sets standards for electronic components and test procedures. The EIA is also responsible for the setting of performance standards for various types of electronic equipment. Standardization is advantageous both to the manufacturers (because it broadens their markets) and to the users of electronic equipment (because it makes it easier for them to get replacement parts and peripheral devices).

The Electronic Industries Association works with other national and international standards agencies. The American National Standard Institute (ANSI) is one such association. The International Electrotechnical Commission (IEC), headquartered in Geneva, Switzerland, is responsible for electronic standards worldwide.

ELECTRON-MULTIPLIER TUBE

An *electron-multiplier tube* is a device that amplifies a beam of electrons. Such tubes operate on the principle of secondary emission (*see* SECONDARY EMISSION). The electron beam is reflected off of a series of plates, each of which throws off several electrons for each one that strikes it. With several such plates, called *dynodes*, in succession, very large current-gain factors can be realized. The gain can exceed 60 dB.

Electron-multiplier tubes are commonly used in a device called a *photomultiplier*, which is in turn often used in sensitive television cameras and infrared detectors. *See also* DYNODE, and PHOTOMULTIPLIER.

ELECTRON ORBIT

In an atom, the electrons orbit the nucleus in defined ways. The positions of the electrons are sometimes called *orbits* or *shells*, but actually, the orbit or shell of an electron represents the average path that it follows around the nucleus. Electrons orbit only at precise distances from the nucleus, in terms of their average positions. The greater the radius of an orbit, the greater the energy contained in the individual electron.

The defined orbits or shells for the electrons in an atom are called, beginning with the innermost shell, the *K, L, M, N, O, P,* and *Q shells*. The K shell is the least energetic, and can contain at most two electrons. The L shell, slightly more energetic, can have up to eight electrons. The M shell can have as many as 18 electrons; the N shell, 32; the O shell, 50; the P shell, 72; and the Q shell, 98. The outermost shells of an atom are never filled. The inner ones tend to fill up first, and the number of electrons in an atom is limited.

An electron can change its orbit from one shell to another. If an electron absorbs energy, it will move into a more distant shell, provided it gets enough extra energy. By moving inward to a shell closer to the nucleus, an electron gives off energy. When an electron absorbs energy, it leaves a definite trace in a spectral dispersion, called an *absorption line*. When it gives off energy, it creates an emission line. Astronomers have used absorption and emission lines to detect the presence of various elements on the planets, in the sun, in other stars, and in other galaxies. *See also* ELECTRON.

ELECTRON VOLT

The *electron volt* is a unit of electrical energy. An electron, or any object having a unit charge, moving through a potential difference of 1 volt, has an energy of 1 electron volt (abbreviated eV). An electron volt is a very small amount of energy. It is only about 1.6×10^{-19} joule (*see* JOULE). Common units of energy in this system are the kiloelectron volt (keV), mega-electron volt (MeV), and giga-electron volt (GeV or BeV); these units represent, respectively, a thousand, a million, and a billion electron volts.

The electron volt is generally specified to represent the energy contained in fast-moving atomic particles. This can include electrons, protons, alpha particles, or heavier nuclei. Photons can also be assigned an energy in electron volts; the amount of energy in a photon of 1.24×10^{-6} meters, or 12,400 Angstrom units in the infrared spectrum, is equal to 1 eV. *See also* ELECTRON, and ENERGY.

ELECTROPHORESIS

Electrophoresis is a method of coating a metal with an insulating material. A suspension is prepared, consisting of dielectric particles in a liquid. Two electrodes are immersed into the suspension, and a voltage is applied between the electrodes.

When the current flows in the suspension, particles of the dielectric material are attracted to, and deposited on, the anode or positively charged electrode. The resulting coating of insulation adheres very well. Electrophoresis is a superior means of applying insulation to certain metals.

Electrophoresis effects are useful in the manufacture of certain kinds of electronic displays. Charged particles, suspended in a colored liquid, will migrate toward the top or the bottom of a flat container, depending on the polarity of an applied electric field. The container is extremely thin, so that the particles can move from one face to the other in a short time. The liquid in which the particles are suspended is colored or cloudy so that the particles will be seen only when they are at the top of the container. Such an arrangement is called an *electrophoretic display*. The familiar liquid-crystal display operates according to a principle much like this. *See also* LIQUID-CRYSTAL DISPLAY.

ELECTROPLATING

Electroplating is a method of coating one substance with another substance by means of electrolysis. This is quite often done with metals, to reduce the rate at which they corrode; a less corrosive metal plating is applied to the surface of a metal that tends to corrode easily. This is especially important in marine environments because the chlorine ions from salt-water vapor can cause extreme problems with corrosion in metals. Tropical climates, because of their high humidity and temperature, are also very hard on some metals.

Electroplating allows two dissimilar metals, which would normally react with each other, to be put in electrical contact. Both metals are electroplated with the same coating metal. For example, copper and iron can both be electroplated with tin. This results in a good electrical contact that is immune to interaction between dissimilar metals.

Aluminum, steel, and zinc should not be used in an outdoor environment, especially where corrosion is a problem, without first being electroplated with a less reactive metal.

ELECTROSTATIC DEFLECTION

When a beam of electrons or other charged particles is passed through an electric field, the direction of the beam is altered. Because a beam of electrons is composed of particles having negative charge, the beam will be accelerated away from the negative pole and toward the positive pole of an electric field, as shown in the illustration.

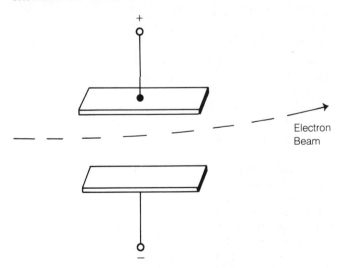

ELECTROSTATIC DEFLECTION: Causes a change in the direction of an electron beam.

The phenomenon of electrostatic deflection is used in the oscilloscope cathode-ray tube. A pair of deflecting plates causes the electron beam in such a tube to be deflected; if an alternating field is applied, the electron beam swings back and forth or up and down. This causes movement of the spot of the waveforms applied to the deflecting plates.

Electrostatic deflection is important in the operation of many different kinds of devices. Some television picture tubes make use of this effect, although electromagnetism is more often used in this kind of device. See also CATHODE-RAY TUBE, ELECTROMAGNETIC DEFLECTION, and OSCILLOSCOPE.

ELECTROSTATIC FORCE

Electrostatic force is the attraction between opposite electric charges and the repulsion between like electric charges. The electrostatic force depends on the amount of charge on an object or objects, and the amount of distance separating the objects. It also depends on the medium between the objects.

Electrostatic forces cause movement of the electrons in substances subjected to electric fields. Electrostatic forces are also responsible for the action of the deflecting plates in a cathode-ray oscilloscope, and in some other devices. The flow of current in a conductor or semiconductor is made possible by electrostatic forces. The force is generally measured in newtons per coulomb of electric charge. See also CHARGE.

ELECTROSTATIC HYSTERESIS

In a dielectric substance, the polarization of the electric field within the material does not always follow the polarization of the external field under alternating-current conditions. A small amount of delay, or lag, occurs, because any dielectric substance requires some time to charge and discharge. This effect is called electrostatic hysteresis.

Electrostatic hysteresis is not usually significant, in most dielectrics, at the lower radio frequencies. As the frequency is increased into the very high and ultra high ranges, however, the effects of electrostatic hysteresis become noticeable. Different dielectric materials exhibit different electrostatic hysteresis properties. Ferroelectric substances are especially noted for their large amount of electrostatic hysteresis. See also DIELECTRIC ABSORPTION, and HYSTERESIS.

ELECTROSTATIC INDUCTION

When any object is placed in an electric field, a separation of charge occurs on that object. Electrons move toward the positive pole of the electric field, and away from the negative pole, within the object. This creates a potential difference between different regions in the object.

An example of electrostatic induction is provided by bringing a charged object near the ball of an electroscope. If the object is negatively charged, the gold-foil leaves of the electroscope will receive a surplus of electrons. This is because the electrons will be driven away from the ball. If the object brought near the ball is positively charged, electrons will be attracted to the ball, and away from the gold-foil leaves. In either case, the leaves will stand apart, indicating that they have a charge, even though no actual contact has been made between the charged object and the electroscope. See also CHARGE.

ELECTROSTATIC INSTRUMENT

An electrostatic instrument is a form of metering device that operates via electrostatic forces only. Generally, a stationary, fixed metal plate is mounted near a rotatable plate, creating a sort of air-variable capacitor. The indicating needle is attached to the rotatable plate.

When a voltage is applied between the two metal plates, they are attracted to each other because of their opposite polarity (see ELECTROSTATIC FORCE). A spring, attached to the rotatable plate, works against this force to regulate the amount of rotation. The greater the voltage between the plates, the greater the force of attraction, and the farther the rotatable plate will turn against the tension of the spring.

An electrostatic voltmeter draws very little current. Only the tiny leakage current through the air between the plates contributes to current drain. Thus, the electrostatic voltmeter has an extremely high impedance. When the appropriate peripheral circuitry is added, the electrostatic voltmeter can be used as an ammeter or wattmeter. See also CHARGE.

ELECTROSTATIC MICROPHONE

An electrostatic microphone is a microphone that produces a variable capacitance as sound waves strike its diaphragm. For this reason, such a microphone is sometimes called a capacitor microphone or condenser microphone.

In an electrostatic microphone, the capacitor has one fixed plate, which is rigid. The other plate comprises the diaphragm of the device. A direct-current voltage is applied between the plates. When an acoustic disturbance causes the diaphragm to vibrate, the instantaneous separation between the plates is affected. This causes the capacitance between the plates to fluctuate, exactly in step with the sound waves impinging on the diaphragm.

A variable capacitance, applied across a source of direct-current voltage, results in a certain amount of charging and discharging. In the electrostatic microphone, the changing capacitance between the diaphragm and the fixed plates causes current to flow because of the changing amount of charge. This current can be amplified, and the result is an excellent reproduction of the sound waves in the form of electrical impulses. See also MICROPHONE.

ELECTROSTATIC POTENTIAL

Any electric field produces a voltage gradient, measured in volts per meter of distance. Two points in space, separated by a given distance within the influence of an electric field, will have a voltage difference. This is called an electrostatic potential. The voltage between the two electrodes, from which the field results, is the largest possible electrostatic potential within the field.

An electrostatic potential between two objects produces an electrostatic force, and this force is directly proportional to the potential difference for a given constant separation distance. For a given amount of electrostatic potential between two objects, the force drops off as the distance increases. Electrostatic potential is measured by a device that draws very little current. This is necessary, because although the voltages can be quite large, the total amount of electric charge can be rather small. See also ELECTROSTATIC FORCE, and ELECTROSTATIC INSTRUMENT.

ELECTROSTATIC SHIELDING

Electrostatic shielding, also commonly known as Faraday shielding, is a means of blocking the effects of an electric field, while allowing the passage of magnetic fields. Electrostatic shielding reduces the amount of capacitive coupling between two objects to practically zero. However, inductive coupling is not affected.

An electrostatic shield consists of a wire mesh, a screen, or sometimes a plate of nonmagnetic metal, such as aluminum or copper, grounded at a single point. This causes the mesh, screen, or plate to acquire a constant electric charge, but little or no current can flow in it. Such a shield, placed between the primary and secondary windings of an air-core transformer as shown at A in the illustration, practically eliminates capacitive interaction between the windings. Electrostatic shields are often used in the tuned circuits of radio-frequency equipment, when transformer coupling is used. The shield, by eliminating the electrostatic coupling between stages, reduces the amount of unwanted harmonic and spurious-signal energy that is transferred from stage to stage (*see* TRANSFORMER COUPLING).

An electrostatic shield is sometimes used in the construction of a direction-finding loop antenna, as shown at B. A section of nonmagnetic tubing or braid is placed around, and insulated from, the loop itself. The shield is grounded at the feed point, but is broken at the top, preventing the circulation of current, but maintaining a fixed electric charge. This allows the magnetic component of an electromagnetic wave to get to the loop, but the electric component is kept out. This enhances the directional characteristics of the antenna. It also can improve the signal-to-noise ratio. *See also* LOOP ANTENNA.

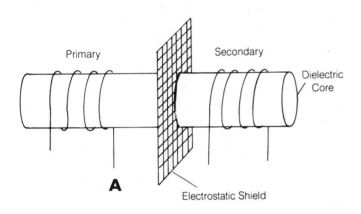

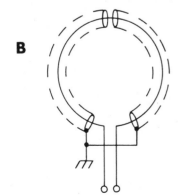

ELECTROSTATIC SHIELDING: In an RF transformer (A) and around a loop antenna (B).

ELECTROSTRICTION

When an electric field is applied to a dielectric material, an inward force is created on the material. This is called *electrostriction*. The charge within a dielectric causes attraction between one face of the dielectric and the other. The electrodes, having opposite polarity, are also attracted to each other. If the intensity of the electric field becomes too great, the dielectric can physically crack under the stress of electrostriction. This sometimes happens in disk capacitors and other ceramic devices when they are subjected to excessive voltages. *See also* ELECTROSTATIC FORCE.

ELEMENT

An *element* is one of the fundamental building blocks of matter. According to contemporary theories of matter, all elements are made up of atoms. The atoms consist of a nucleus, having some positively charged particles and some neutral particles, and the electrons, which are negatively charged and orbit around the nucleus. The positively charged particles in the nucleus are called *protons*. The number of protons in an atom is normally the same, or about the same, as the number of electrons so that the atom has no net charge. The neutral particles are called *neutrons*.

An atom can have only one proton and one electron; this is the simplest atom, hydrogen. Hydrogen is the most abundant element in the universe. Some atoms have more than 100 protons and electrons. The elements with many protons and electrons are called *heavy elements*, and they tend to be unstable, breaking up into elements with fewer protons and electrons.

The term *element* is sometimes used to define a component in an electrical or electronic system. This is true especially of antennas. The antenna element that receives the energy directly from the feed line is called the *driven element*. Some antennas have elements that are not connected to the feed line directly; these are called *parasitic elements. See also* DRIVEN ELEMENT, ELECTRON, and PARASITIC ELEMENT.

ELEMENT SPACING

In an antenna having more than one element, such as a parasitic or phased array (*see* PARASITIC ARRAY, and PHASED ARRAY), the element spacing is the free-space distance, in wavelengths, between two specified antenna elements. This might be the director spacing, with respect to the driven element or another director; it might be the reflector spacing, with respect to the driven element; it might be the spacing between two driven elements. In an antenna with several elements, the spacing between adjacent elements can differ depending on which elements are specified.

For a given frequency (f) in megahertz, the spacing (s) in wavelengths is given by the simple formula:

$$s = df/300$$

if d is the spacing in meters. If d is in feet, then:

$$s = df/984$$

At shorter wavelengths, d can be given in centimeters or inches. If d is in centimeters, then:

$$s = df/30,000$$

and if d is in inches, then:

$$s = df/11,800$$

See also DIRECTOR, DRIVEN ELEMENT, QUAD ANTENNA, and YAGI ANTENNA.

ELEVATION

Elevation is the angle, in degrees, that an object in the sky sub-tends with respect to the horizon. It is also specified as the vertical deviation of the major lobe of an antenna from the horizontal. The smallest possible angle of elevation is 0 degrees, which represents the horizontal; the largest possible angle of elevation is 90 degrees, which represents the zenith.

Elevation is also sometimes called *altitude*. It is one of two coordinates needed for uniquely determining the direction of an object in the sky. The other coordinate, called *azimuth*, represents the angle measured clockwise around the horizon from true north (*see* AZIMUTH).

Astronomers use a variation of the azimuth/elevation system. The coordinates, rather than being fixed with respect to the horizon of the earth at a particular location, are fixed with respect to the heavens. The equivalent of azimuth is called *right ascension*; the equivalent of elevation is called *declination*. The declination in this system can vary between +90 degrees (toward the north celestial pole) and −90 degrees (toward the south celestial pole).

ELLIPTICAL ORBIT

Almost all satellites have orbits that deviate from perfect circles. The main exception to this rule is the *geostationary orbit*. If the orbit of a satellite is not a circle, then it is always an ellipse. Such orbits have the center of the earth at one focus of the ellipse. The extent to which the ellipse is elongated, or different from a circle, is known as the *eccentricity* (*see* ECCENTRICITY OF ORBIT).

The speed, as well as the altitude, of a satellite varies when it has an elliptical orbit. When it is at, or near, its lowest altitude in an elliptical orbit, the satellite moves fastest. When it is at, or near, its highest altitude, it moves slowest. *See also* ACTIVE COMMUNICATIONS SATELLITE, APOGEE, and PERIGEE.

ELLIPTICAL POLARIZATION

The *polarization* of an electromagnetic wave is the orientation of the electric lines of flux in the wave. Often, this orientation remains constant; but sometimes it is deliberately made to rotate as the wave propagates through space. If the orientation of the electric lines of flux changes as the signal is propagated from the transmitting antenna, the signal is said to have elliptical polarization (*see* HORIZONTAL POLARIZATION, POLARIZATION, and VERTICAL POLARIZATION).

An elliptically polarized electromagnetic field can rotate either clockwise or counterclockwise as it moves through space (see the illustration). The intensity of the signal can or cannot remain constant as the wave rotates. If the intensity does remain constant as the wave rotates, the polarization is said to be *circular* (*see* CIRCULAR POLARIZATION).

Elliptical polarization is useful because it allows the reception of signals having unpredictable or changing polarization, with a minimum of fading and loss. Ideally, the transmitting and receiving antennas should both have elliptical polarization, although signals with linear polarization can be received with an elliptically polarized antenna. If the transmitted signal has opposite elliptical polarization from the receiving antenna, however, there is substantial loss.

Elliptical polarization is generally used at ground stations for satellite communication. In receiving, the use of an elliptically polarized antenna reduces the fading that is caused by changing satellite orientation. In transmitting, the use of elliptical polarization ensures that the satellite will always receive a good signal for retransmission. *See also* LINEAR POLARIZATION.

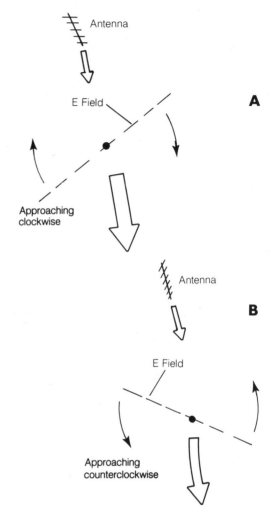

ELLIPTICAL POLARIZATION: At A, clockwise; at B, counterclockwise as wavefronts approach.

EMERGENCY BROADCAST SYSTEM

In the event of a national emergency, especially the threat of a large-scale nuclear war, all normal broadcasting would cease. An attention tone would first be transmitted over all stations. This tone would be followed by specific instructions concerning the frequencies to which to tune in a given area. This system is called the *Emergency Broadcast System (EBS)*. All broadcasting stations in the United States are required by the federal government to conduct periodic tests of the Emergency Broadcast System.

The reason for changing the frequencies and locations of broadcasting transmitters, in case of nuclear attack, is that enemy missiles might use the signals from broadcast stations to zero in on large cities. The Emergency Broadcast System is an updated version of the older system called *Control of Electromagnetic Radiation (CONELRAD)*.

EMERGENCY COMMUNICATIONS

Emergency communications is the highest-priority form of communications. Messages having emergency status always take precedence over other kinds of messages. An emergency or distress signal implies a threat to life. The emergency designation is to be used only in such a situation. In some communications services, certain frequencies are set aside for use in emergency communications (*see* DISTRESS FREQUENCY, and DISTRESS SIGNAL).

An emergency communications system should be periodically tested, to be sure that it is working perfectly. Such a communications system must have an independent source of power (*see* EMERGENCY POWER SUPPLY). Portable and mobile equipment can be valuable in a large-scale emergency, but fixed stations are generally capable of more radio-frequency output power.

Distress signals in radiotelegraphy are easily recognizable; the standard signal is the SOS (... --- ...). In radiotelephony, the word "mayday" is used. An emergency message should be sent slowly and clearly. If interference is encountered, the continuous-wave mode is best. The type A1 emission (continuous-wave Morse code) is still the best choice of emission for adverse conditions. Code transmitters are simple and can sometimes be built from receivers. *See also* CODE TRANSMITTER.

EMERGENCY COORDINATOR

An *Emergency Coordinator* is a ham who volunteers his or her time for the *Amateur Radio Emergency Service (ARES)*. This is the person you should contact if there is an emergency in your area, and if you want to assist in emergency communications.

To find out the name and address of your Emergency Coordinator, you can talk to local hams, and/or contact the American Radio Relay League. *See* AMATEUR RADIO EMERGENCY SERVICE, and AMERICAN RADIO RELAY LEAGUE.

EMERGENCY POWER SUPPLY

An *emergency power supply* is an independent, self-contained power source that can be used in the event of a failure in the normal utility power. Such an emergency supply is usually either a set of rechargeable batteries, or a fossil-fuel generator. The primary requirements for an emergency power supply are:

- It must supply the necessary voltages for the operation of the equipment to be used.
- It must be capable of delivering sufficient current to run the equipment to be used.
- It must be capable of operating for an extended period of time.
- It must be accessible immediately when needed.

A gasoline generator is an excellent emergency power supply. Such a generator can supply 120 volts at about the standard line frequency, up to power levels of several kilowatts. Department stores often carry small gasoline-powered generators for portable use. A supply of gasoline is needed for the operation of these generators, and the gasoline should be stored safely to prevent explosions.

A set of storage batteries, such as 12-volt automotive batteries, can also be used as a source of emergency power. Many radio receivers and transmitters, as well as some specialized appliances, will run directly from 12 volts direct current. Power inverters can be used to get 120 volts alternating current from such a source. A solar recharging system can be used to keep the batteries operational. *See also* GENERATOR, INVERTER, SOLAR POWER, and STORAGE BATTERY.

EMISSION CLASS

In amateur radio, communications can be carried out in many different modes. The main modes are International Morse code or continuous wave (CW), single sideband (SSB), frequency modulation (FM), phase modulation, frequency-shift keying (FSK), digital communications, facsimile (FAX), slow-scan television (SSTV) and fast-scan amateur television (ATV). Other modes, less often used, include such modulation techniques as *pulse modulation* and *spread spectrum*.

Emissions are formally classified by a five-symbol designator system. The first symbol, always a letter of the alphabet, indicates the type of modulation. The second symbol, always a numeral, designates the nature of the modulating signal. The third symbol, always a letter, indicates the type of data sent. The fourth and fifth symbols are not often used in amateur work, and depict the details of the signal code and the methods of multiplexing. This new system was devised in 1979 to replace the older, more limited system.

Table A shows the first, second and third symbols in the emission classification code developed at the World Administrative Radio Conference 1979 (WARC-79). Table B gives the modes most often used by ham radio operators, followed by the formal emission designators.

EMISSION CLASS: TABLE A.
DESIGNATORS AND THEIR MEANINGS.

	First symbol
N	Steady, unmodulated, pure carrier wave
A	Double-sideband amplitude modulation, full carrier (AM)
R	Double-sideband amplitude modulation, reduced carrier (DSB)
J	Single-sideband modulation, reduced carrier (SSB)
C	Vestigial-sideband amplitude modulation
F	Frequency modulation (FM)
G	Phase modulation
P	Pulse modulation
K	Alternative method of pulse modulation
L	"
M	"
Q	"
V	"
W	"
X	"

	Second symbol
0	Unmodulated carrier
1	One channel of digital modulation
2	One channel of digital modulation with subcarrier
3	One channel of analog modulation
7	More than one channel of digital modulation
8	More than one channel of analog modulation
9	One or more channels of digital modulation, plus one or more channels of analog modulation
X	Forms other than the above

	Third symbol
N	No data transmitted
A	Morse code signals (CW or MCW) for manual reception
B	Morse code signals (CW or MCW) for machine reception
C	Facsimile (FAX)
D	Telemetry or command signals
E	Voice signals (telephony)
F	Television signals (SSTV, ATV)

EMISSION CLASS: TABLE B. COMMONLY USED HAM
MODES AND THEIR FORMAL EMISSION DESIGNATORS.

Mode description and abbreviation	Symbol
Morse code by keying carrier (CW)	A1A, A1B
Morse code by keying a tone, AM (MCW)	A2A, A2B
Morse code by keying a tone, FM (MCW)	F2A, F2B
Single sideband voice (SSB)	J3E
Frequency modulated voice (FM)	F3E
Teletype by frequency-shift keying (FSK)	F1B
Teletype by audio tones in an FM transmitter (AFSK)	F2B
Slow-scan television, SSB (SSTV)	J3F
Slow-scan television, FM (SSTV)	F3F
Fast-scan television, AM (ATV)	A3F
Fast-scan television, FM (ATV)	F3F
Packet radio	F1D

See also the articles AMPLITUDE MODULATION, DIGI-
TAL COMMUNICATIONS, FACSIMILE, FREQUENCY
MODULATION, FREQUENCY-SHIFT KEYING, PHASE
MODULATION, PULSE MODULATION, SINGLE SIDE-
BAND, SLOW-SCAN TELEVISION, SPREAD SPECTRUM,
and TELEVISION.

EMISSIVITY

Emissivity is the ease with which a substance emits or absorbs
heat energy. This property is important in the choice of materials for, and in the design of, heatsinks (*see* HEATSINK) because such devices give up much of their heat by means of radiation. Emissivity is a measure of the brightness of an object in
the infrared part of the electromagnetic spectrum. The darker
the object appears in the infrared, the better the emissivity.
Simply painting an object black does not guarantee that the
emissivity will be excellent because visually black paint might
not be as black in the infrared region.

The capability of a material to dissipate power depends on
the temperature; the higher the temperature, in general, the
more power can be dissipated by a given material. Emissivity
depends on the shape of an object; irregular surfaces are better
for heat radiation than smooth surfaces. A sphere is the poorest
possible choice for a heat emitter; in heatsink design, finned
surfaces are used to maximize the ratio of the surface area to the
volume. Different substances have different emissivity characteristics under identical conditions. Graphite is one of the most
heat-emissive known substances.

EMITTER

In a semiconductor bipolar transistor, the *emitter* is the region
from which the current carriers are injected. The emitter of a
transistor is somewhat analogous to the cathode of a vacuum
tube, or the source of a field-effect transistor. But the actual
operating principles of the bipolar-transistor emitter are different from those of the cathode or source of a tube or field-effect
device.

The emitter of a bipolar transistor can be made from either
an N-type semiconductor wafer (in an npn device) or a P-type
semiconductor wafer (in a pnp device). The drawing shows the
schematic symbols for both of these kinds of bipolar transistors,
illustrating the position of the emitter. The emitter lead is indicated by the presence of an arrowhead (*see* N-TYPE SEMI-
CONDUCTOR, and P-TYPE SEMICONDUCTOR).

The input of a transistor amplifier can be supplied to the
emitter circuit or to the base circuit. If it is supplied to the emitter
circuit, the impedance is low. The output of a transistor amplifier can be taken from either the collector or the emitter circuit;
if the output is taken from the emitter circuit, the impedance is
low. In either the emitter-input or emitter output case, an amplifier is said to be emitter-coupled. Generally, the input of a
transistor amplifier is supplied to the base, and the output is
taken from the collector. The emitter is usually kept at signal
ground potential. *See also* COLLECTOR, EMITTER-BASE
JUNCTION, EMITTER-COUPLED LOGIC, EMITTER COUP-
LING, EMITTER CURRENT, EMITTER DEGENERATION,
EMITTER FOLLOWER, EMITTER KEYING, EMITTER MOD-
ULATION, EMITTER RESISTANCE, EMITTER STABILIZA-
TION, EMITTER VOLTAGE, and TRANSISTOR.

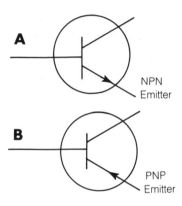

EMITTER: At A, for an npn transistor; at B, for a pnp transistor.

EMITTER-BASE JUNCTION

The *emitter-base junction* of a bipolar transistor is the boundary
between the emitter semiconductor and the base semiconductor. It is always a P-N junction (*see* P-N JUNCTION). In an npn
transistor, the emitter is made from N-type material and the
base is made from P-type; in the pnp transistor it is the other
way around.

In most transistor oscillators and low-level amplifiers, the
emitter-base junction is forward-biased. In the npn device,
therefore, the base is usually somewhat positive with respect to
the emitter, and in the pnp device, somewhat negative. The
emitter-base junction of a transistor will conduct only when it is
forward-biased. In this respect, it behaves exactly like a diode.
For the current to pass to the collector circuit, the emitter-base
junction must be forward-biased. But a small change in this bias
will result in a large change in the collector current. The input
signal to a bipolar-transistor circuit is always applied at the
emitter-base junction.

In class-B and class-C amplifiers, the emitter-base junction
does not conduct under conditions of zero input signal. Thus, it
conducts for only part of the signal cycle. Sometimes, a reverse
bias is deliberately applied, and the collector current flows for
just a tiny part of input cycle. When no current flows in the
emitter-base junction because of a lack of forward bias, the
transistor is said to be cut off. *See also* CLASS-B AMPLIFIER,
CLASS-C AMPLIFIER, and TRANSISTOR.

EMITTER-COUPLED LOGIC

Emitter-coupled logic (ECL), is a bipolar form of logic. Most digital switching circuits in the bipolar family utilize saturated transistors; they are either fully conductive or completely cut off.

But in emitter-coupled logic, this is not the case. The emitter-coupled logic circuit acts as a comparator between two different current levels.

The input impedance with ECL is very high. The switching speed is very rapid. The output impedance is relatively low. The principal disadvantages of emitter-coupled logic are the large number of components required, and the susceptibility of the circuit to noise. The noise susceptibility of ECL arises from the fact that the transistors, not operated at saturation, tend to act as amplifiers of analog signals. *See also* DIODE-TRANSISTOR LOGIC, DIRECT-COUPLED TRANSISTOR LOGIC, HIGH-THRESHOLD LOGIC, INTEGRATED INJECTION LOGIC, METAL-OXIDE SEMICONDUCTOR LOGIC FAMILIES, RESISTOR-CAPACITOR-TRANSISTOR LOGIC, RESISTOR-TRANSISTOR LOGIC, TRANSISTOR-TRANSISTOR LOGIC, and TRIPLE-DIFFUSED EMITTER-FOLLOWER LOGIC.

EMITTER COUPLING

When the output of a bipolar-transistor amplifier is taken from the emitter circuit, or when the input to a transistor amplifier is applied in series with the emitter, the device is said to have *emitter coupling*. Emitter coupling can be capacitive, or it can make use of transformers.

Emitter coupling results in a low input or output impedance, depending on whether the coupling is in the input or output of the amplifier stage. Emitter input coupling is used with grounded-base amplifiers. Such circuits are occasionally used as radio-frequency power amplifiers. Emitter output coupling is used when the following stage requires a low driving impedance. A circuit utilizing emitter output coupling is sometimes called an emitter follower. *See also* EMITTER FOLLOWER.

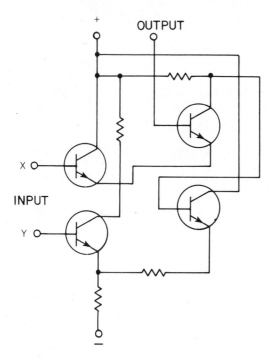

EMITTER-COUPLED LOGIC: Four bipolar transistors form one logic gate.

EMITTER CURRENT

In a bipolar-transistor circuit, *emitter current* is the rate of charge-carrier transfer through the emitter. Emitter current is generally measured by means of an ammeter connected in series with the emitter lead of a bipolar-transistor amplifier circuit.

Most of the emitter current flows through a transistor and appears at the collector circuit. A small amount of the emitter current flows in the base circuit. When the emitter-base junction is reverse-biased, no emitter current flows, and the transistor is cut off. As the emitter-base junction is forward-biased to larger and larger voltages, the emitter current increases, as does the base current and the collector current. The emitter current fluctuates along with the signal input at the emitter-base junction of a transistor amplifier. *See also* EMITTER-BASE JUNCTION.

EMITTER DEGENERATION

Emitter degeneration is a simple means of obtaining degenerative, or negative, feedback in a common-emitter transistor amplifier (*see* COMMON CATHODE/EMITTER/SOURCE). A series resistor is installed in the emitter lead of a transistor amplifier stage; no capacitor is connected across it.

When an alternating-current input signal is applied to the base of an amplifier having emitter degeneration, the emitter voltage fluctuates along with the input signal. This happens in such a way that the gain of the stage is reduced. The distortion level is also greatly reduced. The amplifier is able to tolerate larger fluctuations in the amplitude of the input signal when emitter degeneration is used; the output is virtually distortion-free by comparison with an amplifier that does not use emitter degeneration.

Emitter degeneration is often used in high-fidelity audio amplifiers. Distortion must be kept to a minimum in such circuits. Emitter degeneration acts to stabilize an amplifier circuit, protecting the transistor against destruction by such effects as current hogging or thermal runaway. *See also* EMITTER CURRENT, EMITTER STABILIZATION, and EMITTER VOLTAGE.

EMITTER FOLLOWER

An *emitter follower* is an amplifier circuit in which the output is taken from the emitter circuit of a bipolar transistor. The output impedance of the emitter follower is low. The voltage gain is always less than 1; that is to say, the output signal voltage is smaller than the input signal voltage. The emitter-follower circuit is generally used for the purpose of impedance matching. The amplifier components are often less expensive, and offer greater bandwidth, than transformers.

The output of the emitter follower is in phase with the input. The impedance of the output circuit depends on the particular characteristics of the transistor used, and also on the value of the emitter resistor. Sometimes, two series-connected emitter resistors are used. The output is taken, in such a case, from between the two resistors. Alternatively, a transformer output can be used. Emitter-follower circuits are useful because they offer wideband impedance matching. *See also* EMITTER COUPLING.

EMITTER KEYING

In an oscillator or amplifier, *emitter keying* is the interruption of the emitter circuit of a bipolar-transistor stage for the purpose of obtaining code transmission. Emitter keying completely shuts off an oscillator in the key-up condition. In an amplifier, only a negligible amount of signal leakage occurs when the key is up. Emitter keying is an entirely satisfactory method of keying in bipolar circuits although, in high-power amplifiers, the switched current can be very large. Usually, emitter keying is done at low-level amplifier circuits. In a code transmitter, emitter keying is often performed on two or more amplifier stages simultaneously. A shaping circuit provides reduction of key clicks at the instants of make and break in code transmission. The series resistor, in conjunction with the parallel capacitor, slows down the emitter voltage drop when the key is closed. This allows a controlled signal rise time. When the key is released, the capacitor discharges through the resistor and the emitter-base junction, slowing down the decay time of the signal. *See also* KEY CLICK, and SHAPING.

EMITTER MODULATION

Emitter modulation is a method of obtaining amplitude modulation in a radio-frequency amplifier. The radio-frequency input signal is applied to the base of a bipolar transistor in most cases, although it can be applied in series with the emitter. The audio signal is always applied in series with the emitter. Emitter modulation offers the advantage of requiring very little audio input power for full modulation. *See also* AMPLITUDE MODULATION, and MODULATION.

EMITTER RESISTANCE

The *emitter resistance* of a bipolar transistor is the effective resistance of the emitter in a given circuit. The emitter resistance depends on the bias voltages at the base and collector of a transistor. It also depends on the amplitude of the input signal, and on the characteristics of the particular transistor used. The emitter resistance can be controlled, to a certain extent, by inserting a resistor in series with the emitter lead in a transistor circuit.

The external resistor in the emitter lead of a bipolar transistor oscillator or amplifier is, itself, sometimes called the *emitter resistance*. Such a resistor is used for impedance-matching purposes, for biasing, for current limiting, or for stabilization. The emitter resistance can have a capacitor connected in shunt. *See also* EMITTER CURRENT, EMITTER STABILIZATION, and EMITTER VOLTAGE.

EMITTER STABILIZATION

Emitter stabilization is a means of preventing undesirable effects that sometimes occur in transistor amplifier circuits as the temperature varies. Emitter stabilization is used in common-emitter transistor amplifiers to prevent the phenomenon known as thermal runaway (*see* THERMAL RUNAWAY). A resistor, connected in series with the emitter lead of the transistor, accomplishes the objective of emitter stabilization.

If the collector current increases because of a temperature rise in the transistor, and there is no resistor in series with the emitter lead, the increased current will cause further heating of the transistor, which can in turn cause more current to be drawn in the collector circuit. This results in a vicious circle. It can end with the destruction of the transistor base-collector junction. If an emitter-stabilization resistor, having the proper value, is installed in the circuit, the collector-current runaway will not take

place. Any increase in the collector current will then cause an increase in the voltage drop across the resistor; this will cause the bias at the emitter-base junction to change in such a way as to decrease the current through the transistor. The emitter-stabilization resistor thus regulates the current in the collector circuit. A decrease in the current through the transistor will, conversely, cause a decrease in the voltage drop across the emitter resistor, and a tendency for the collector current to rise.

Emitter stabilization is especially important in transistor amplifiers using two or more bipolar transistors in a parallel or push-pull common-emitter configuration. *See also* COMMON CATHODE/EMITTER/SOURCE, EMITTER CURRENT, and EMITTER VOLTAGE.

EMITTER VOLTAGE

In bipolar transistor circuits, the *emitter voltage* is the direct-current potential difference between the emitter and ground. If the emitter is connected directly to the chassis ground, then the emitter voltage is zero. But if a series resistor is used, then the emitter voltage will not be zero. In such a case, if an npn transistor is used, the emitter voltage will be positive; if a pnp transistor is used, it will be negative. The emitter voltage increases as the current through the transistor increases, and also as the value of the series resistance is made larger. In the common-collector configuration, when the collector is directly connected to chassis ground, the emitter voltage is negative in the case of an npn circuit, and positive in the case of a pnp circuit. Letting I be the emitter current in a transistor circuit, and R the value of the series resistor in the emitter circuit, specified in amperes and ohms respectively, then the emitter voltage V is given in volts by:

$$V = IR$$

A capacitor, placed across the emitter resistor, keeps the emitter voltage constant under conditions of variable input signal. If no such capacitor is used, the emitter voltage will fluctuate along with the input cycle. *See also* EMITTER CURRENT, EMITTER DEGENERATION, and EMITTER STABILIZATION.

EMPIRICAL DESIGN

Empirical design is a method of circuit engineering in which experience and intuition are used more than theoretical techniques. The empirical-design process can be described as a trial-and-error method of arriving at an optimum circuit design.

Empirical design is used by the professional as well as by the hobbyist. Engineers are all familiar with the phenomenon in which a circuit performs well "on paper," but inadequately in practice!

Sophisticated circuits are generally designed theoretically before they are put together for testing and debugging. But more simple circuits, such as oscillators and amplifiers, can often be thrown together and then tailored to perfection by empirical methods. Such an approach often saves considerable time in these situations.

ENABLE

The term *enable* means the initialization of a circuit. To enable a device is to switch it on, or make it ready for operation. This can be done by means of a simple switch, or by a triggering pulse. The initializing command or pulse is itself called the *enable* command or *pulse*. The term *enable* **is used** mainly in computer and digital-circuit applications.

When a magnetic core is induced to change polarity by an electronic signal, the signal is called an *enable pulse*. This signal allows a binary bit to be written into, or removed from, the magnetic memory. *See also* MAGNETIC CORE, and MEMORY.

ENAMELED WIRE

When a wire is insulated by a thin coating of paint, it is called *enameled wire*. Enameled wire is available in sizes ranging from smaller than AWG 40 to larger than AWG 10 (*see* AMERICAN WIRE GAUGE). Enameled wire is often preferable to ordinary insulated wire, because the enamel does not add significantly to the wire diameter.

Enameled wire is commonly used in coil windings, when many turns must be put in a limited space. Examples of such situations include audio-frequency chokes, transformers, and coupling or tuning coils. Some radio-frequency chokes also use enameled wire. Whenever miniaturization is important, enameled wire will probably be found. Because the enamel is thin, however, this type of wire is not suitable for high-voltage use. *See also* WIRE.

ENCAPSULATION

When a group of discrete components is enclosed or embedded in a rigid material, such as wax or plastic, the circuit is said to be *encapsulated*. Encapsulation protects the components against physical damage. Encapsulation also reduces the effects of mechanical vibration on the operation of a circuit. *See also* MICROPHONICS.

ENCODER

An *encoder* is a device that translates a common language into a code. Such a device often has a keyboard resembling that of a typewriter. When the operator presses a particular key, the machine produces the code element corresponding to that character. The code can have any of many different forms. Some of the most common are the *ASCII, BAUDOT,* and *morse codes* (*see* ASCII, BAUDOT, CODE, and MORSE CODE).

An analog-to-digital converter can be called an *encoder* if a communications system uses digital transmission. In general, any device that interfaces a human operator with a transmission medium, for the purpose of sending signals, can be called an *encoder*. At the receiving end of the circuit, a decoder performs exactly the opposite function, interfacing the transmission medium with another human operator or machine. *See also* ANALOG-TO-DIGITAL CONVERTER, DIGITAL-TO-ANALOG CONVERTER.

ENCRYPTION

Encryption is the sending of a message in cryptic form so that only certain people can tell what it says. Encryption differs from encoding, in which a signal is sent using commonly agreed-upon combinations of pulses. The ASCII code and morse code are examples of encoding. *See* ENCODER.

An example of encryption might be to send every letter in a morse code transmission backwards. Then A would become N, B would become V, C would become a character currently not used (dit-dah-dit-dah), D would become U, and so on.

Encryption is used for sending commands to satellites. But this is the only purpose for which the Federal Communications Commission allows hams to use cryptic signals.

END EFFECT

When a solenoidal inductor carries an alternating current, some of the magnetic lines of flux extend outside the immediate vicinity of the coil. Although the greatest concentration of magnetic flux lines is inside the coil itself, especially if the coil has a ferromagnetic core (such as powdered iron), the external flux can result in coupling with other components in a circuit. Most of the external magnetic flux appears off the ends of a solenoidal coil; hence, it is sometimes called the *end effect*. In recent years, toroidal inductor cores have become more commonly used, because toroidal coils do not exhibit the end effect (*see* TOROID).

In an antenna radiator, the impedance at any free end is theoretically infinite; that is, no current should flow. In practice, a small charging current exists at a free end of any antenna, because of capacitive coupling to the environment. This is called the *end effect*. The amount of end capacitance, and therefore the value of the charging current, depends on many factors, including the conductor diameter, the presence of nearby objects, and even the relative humidity of the air surrounding the antenna. The end effect reduces the impedance at the feed point of an end-fed antenna or a center-fed antenna measuring a multiple of one wavelength. *See also* END FEED.

END FEED

When the feed point of an antenna radiator is at one end of the radiator, the arrangement is called *end feed*. End feed always constitutes voltage feed because the end of an antenna radiating element is always at a current node (*see* VOLTAGE FEED). End feed is usually accomplished by means of a two-wire balanced transmission line. One of the feed-line conductors is connected to the radiator, and the other end is left free, as shown in the illustration.

In an end-fed antenna system, the radiator must have an electrical length that is an integral multiple of ½ wavelength. If *f* is the frequency in megahertz, then ½ wavelength, denoted by *L*, is given approximately by the formulas:

$$L \text{ (meters)} = 143/f$$
$$L \text{ (feet)} = 468/f$$

The length of the radiating element is quite critical when end feed is used. If the antenna is slightly too long or too short, the current nodes in the transmission line will not occur in the same place. This will result in unbalance between the two conductors of the line, with consequent radiation from the line. Even if the antenna radiator is exactly a multiple of ½ electrical wavelength, the end effect will cause the terminated end of the line to have a somewhat different impedance than the unterminated end (*see* END EFFECT). This, again, results in unbalance between the currents in the two line conductors. The zeppelin, or zepp, antenna is the most common example of an end-fed system. End feed is often much more convenient, from an installation standpoint, than other methods of antenna feed. This is particularly true at the lower frequencies. The slight amount of radiation from the feed line in a zepp antenna can be tolerated under most conditions. *See also* ZEPPELIN ANTENNA.

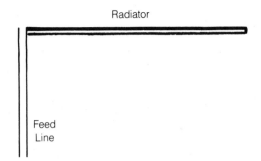

END FEED: The radiator must be an integral multiple of 1/2 wavelength.

END-PLATE MAGNETRON

An *end-plate magnetron* is a special kind of magnetron tube (*see* MAGNETRON). In the ordinary magnetron, the electrons are accelerated into spiral paths by means of a magnetic field. The greater the intensity of the magnetic field, the more energy is imparted to the electrons, up to a certain practical maximum.

With the addition of an electric field between two plates at the ends of the cylindrical chamber, the electrons are accelerated longitudinally as well as in an outward spiral. The radial-velocity component remains the same, and is the result of the magnetic field. The longitudinal component, imparted by the electric field between the two end plates, thus gives the electrons extra speed. With the end plates, greater electron speed can be realized than without the plates, and this increases the amplitude of the tube output. *See also* ELECTRIC FIELD, ELECTRON, and ELECTROSTATIC FORCE.

END-TO-END ACKNOWLEDGMENT

In packet radio, *end-to-end acknowledgment* refers to a message sent back from the destination station to the originating station, which tells the originating station that the message was received. The acknowledgment signal goes back through all of the digipeaters along the way, via which the signal was sent.

When the originating station gets an end-to-end acknowledgement, the originating operator can be 100-percent certain that the destination station got the packet. *See also* DIGIPEATER, NODE-TO-NODE ACKNOWLEDGMENT, and PACKET RADIO.

ENERGY

Energy is the capacity for doing work. Energy is manifested in many forms, including chemical, electrical, thermal, electromagnetic, mechanical, and nuclear. Energy is expended at a rate that can be specified in various ways; the standard unit of energy is the joule (*see* JOULE). The rate of energy expenditure is known as power (*see* POWER); the unit of power is the watt, which is a rate of energy expenditure of 1 joule per second. The units specified for energy depend on the application. The British thermal unit is a measure of thermal energy; the foot pound, of mechanical energy, and so on (*see* ENERGY UNITS).

Energy is becoming an increasingly familiar term in the modern world. The supply of energy, in readily usable form, is not unlimited, and in some places, there have been spot shortages of this indispensable commodity. Raw energy from the universe is, for all practical purposes, infinite in supply, but people must learn how to harness it, and this will require a great research effort. When controlled, energy provides us with light, transportation, and (indirectly) food. Unharnessed energy leaves us at its mercy. It can even be destructive.

ENERGY CONVERSION

Energy occurs in many different forms. *Energy conversion* is the process of changing energy from one form to another. For example, mechanical energy can be changed into electrical energy; this is done by a generator system. Visible-light energy can be converted to electrical energy by means of a solar cell. Chemical energy can be converted to electrical energy by means of a battery. The number of possible conversion modes is practically unlimited.

Energy conversion always occurs with an efficiency of less than 100 percent. No device has ever been invented that will produce the same output energy, in desired form, as it receives in another form. Some devices are almost 100 percent efficient; others are much less than perfect. An incandescent light bulb, which converts electrical energy into visible-light energy, has a very poor efficiency; such a bulb is better for generating heat than light! The fluorescent light bulb is a much more efficient energy converter.

In general, if P_{IN} is the input power to an energy converter, and P_{OUT} is the output from the converter in the desired form, then the conversion efficiency in percent is given by:

$$\text{Efficiency (percent)} = 100 P_{OUT}/P_{IN}$$

Energy is measured in different units, depending on the application, and on the form in which the energy occurs. The universal unit of energy is the joule. *See also* ENERGY UNITS.

ENERGY DENSITY

In an energy-producing cell, the *energy density* is the ability of the cell to produce a given amount of power for a given length of time, for a given amount of cell mass. Theoretically, the energy density is expressed as the total energy in joules divided by the cell mass in kilograms. Obviously, it is advantageous to have an energy density that is as high as possible.

Energy density in a cell can also be represented as the total energy in joules divided by the cell volume in cubic meters. Again, it is desirable to maximize the energy density, as expressed in terms of cell volume.

Technology has greatly improved the energy density of the common chemical cell over the past several decades. Ordinary dry cells, available in grocery, drug, and hardware stores, have excellent energy density. Some such cells are smaller than a dime, and yet will run a digital watch for months. *See also* ENERGY.

ENERGY EFFICIENCY

The *energy efficiency* of a device is the ratio between the energy output of the device, in the desired form, and the amount of energy it receives. This value is always smaller than 100 percent, but it can vary greatly. The exact value depends on many things, including the type of device and the form of energy that is desired in the output.

In recent years, with ever-growing concern about the availability of energy supplies, energy efficiency has become a more and more important consideration in the manufacture of consumer and industrial appliances and machines. The greater the energy efficiency of a device, the less energy it requires to perform a given task. This results in more work for a fixed amount of energy. As an example of the application of energy efficiency, incandescent bulbs are being replaced by fluorescent bulbs in homes and businesses. The incandescent bulb wastes much energy producing heat, when the desired form of energy is visible light. Fluorescent bulbs produce more light, and less heat, for a given amount of input energy. In automotive engineering, the concept of energy efficiency is well known, and cars with greater gasoline mileage per gallon are constantly being developed. Energy efficiency depends, to some extent, on the proper adjustment and operation of a device. Therefore, apparatus that is out of alignment should be realigned for maximum energy efficiency, and operation should be conducted in an energy-efficient manner. The maximization of energy efficiency today will give civilization the greatest possible amount of time to perfect new methods of energy utilization. *See also* ENERGY.

ENERGY UNITS

Energy is manifested in many different forms, and used for different purposes. Energy can be measured in any of several different systems of units, depending on the application. The fundamental unit of energy is the joule, or watt second (*see* JOULE). The other units are used as a matter of convention, and they each bear a relationship to the joule that can be given in terms of a numerical constant, or conversion factor. The most common energy units are: the British thermal unit (BTU), the electron volt, the erg, the foot pound, the kilowatt hour, and the watt hour.

The unit chosen for a given application depends mostly on the consensus of the users. Although joules represent a perfectly satisfactory way to express energy in any situation the quantities are sometimes rather unwieldy and can appear ridiculous. For example, an average residential dwelling might consume 1,000 kilowatt hours of electricity in a month; this is 3.6 billion joules. It makes more sense to use a kilowatt-hour meter in this application, and not a joule meter!

ENHANCEMENT MODE

The *enhancement mode* is an operating characteristic of certain kinds of semiconductor devices, notably the metal-oxide-semiconductor field-effect transistor (MOSFET). Some of the MOSFET devices have no channel under conditions of zero gate bias. In other words, they are normally pinched off. These MOSFET devices are called *enhancement-mode devices*, because a voltage must be applied in order for conduction to take place. The channel is enhanced, rather than depleted, with increasing bias. The other type of field-effect transistor is normally conducting, and the application of gate bias reduces the conduction. This is the depletion mode (*see* DEPLETION MODE).

In an N-channel enhancement-mode device, a negative voltage at the gate electrode causes a channel to be formed. In the P-channel enhancement-mode device, a positive voltage at the gate is necessary for a channel to be produced. The larger the voltage, assuming the polarity is correct, the wider the channel gets, and the better the conductivity from the source to the drain. This occurs, however, only up to a certain maximum. The enhancement-mode effect is shown schematically in the drawing.

The enhancement-mode MOSFET has an extremely low capacitance between the gate and the channel. This is also true of the depletion-mode MOSFET. This characteristic allows the MOSFET to be used at the very-high-frequency range and above, with good efficiency. The enhancement-mode device because it is normally cut off, works well as a class-B or class-C amplifier, although it can also be used in the class-A and class-AB configurations. *See also* METAL-OXIDE SEMICONDUCTOR FIELD-EFFECT TRANSISTOR.

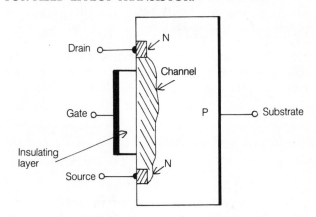

ENHANCEMENT MODE: A channel forms when proper bias is applied.

ENVELOPE

The *envelope* of a modulated signal is the modulated waveform. Generally, the term *envelope* is used for amplitude-modulated or single-sideband signals. An imaginary line, connecting the peaks of the radio-frequency carrier wave, illustrates the envelope of a signal.

The appearance of a signal envelope is an indicator of how well a transmitter is functioning. Different kinds of distortion appear as different abnormalities in the modulation envelope of a transmitted signal. An experienced engineer can immediately tell, by looking at the radio-frequency modulation envelope of a transmitter supplied with a test-modulation signal, whether or not severe distortion is present. The most serious kind of distortion in an amplitude-modulated wave is called *peak clipping* or *flat topping*. Such a condition causes the generation of sidebands at frequencies far removed from that of the carrier. This can result in interference to stations on frequencies outside the normal channel required by such a transmitter (*see* AMPLITUDE MODULATION, and SINGLE SIDEBAND).

The glass enclosure in which a vacuum tube is housed is sometimes called the *tube envelope*. With the decreasing use of vacuum tubes in electronic circuits, however, this use of the term is getting less and less common. *See also* TUBE.

ENVELOPE DETECTOR

An *envelope detector* is a form of demodulator in which changes of signal amplitude are converted into audio-frequency impulses. The most simple form of envelope detector is the simple half-wave rectifier circuit for radio frequencies. A semiconductor diode is adequate to perform the function of an envelope detector.

Although an envelope detector is suitable for demodulating an amplitude-modulated signal having a full carrier, it is not satisfactory for the demodulation of a single-sideband, suppressed-carrier signal. When an envelope detector is used in an attempt to demodulate a suppressed-carrier signal, the resulting sound at the receiver output is unintelligible, and has been called *monkey chatter*. This sound is familiar to those involved in communications. In order to properly demodulate a single-sideband, suppressed-carrier signal, a beat-frequency oscillator is necessary. This replaces the missing carrier wave in the receiver.

An envelope detector can be used to demodulate a frequency-modulated signal by means of a technique called *slope detection*. The receiver is tuned slightly away from the carrier frequency of the signal so that the frequency variations bring the signal in and out of the receiver passband. This produces amplitude fluctuations in the receiver; when these are demodulated by the envelope detector, the result is amplitude variations identical to those of an amplitude-modulated signal. *See also* AMPLITUDE MODULATION, and DETECTION.

EPITAXIAL TRANSISTOR

An *epitaxial transistor* is a transistor that is made by "growing" semiconductor layers on top of each other. The process of semiconductor-crystal growth is called *epitaxy*.

In the epitaxy process, a silicon substrate is used to begin the growth. This substrate, which eventually serves as the collector of the transistor, is immersed in a prepared gas or liquid solution. The base is later diffused into this epitaxial layer (*see* DIFFUSION). The emitter is alloyed to the top of the epitaxial layer.

The epitaxial transistor is characterized by a very thin base region, which is important for effective operation at high frequencies. It is also possible to get a relatively large surface area at the base-collector junction, and this is important for the effective dissipation of heat. Sometimes it is called an *epitaxial mesa transistor*. *See also* TRANSISTOR.

E PLANE

The *E plane (electric-field plane)*, is a theoretical plane in space surrounding an antenna. There are an infinite number of E planes surrounding any antenna radiator; they all contain the radiating element itself. The principal E plane is the plane containing both the radiating element and the axis of maximum radiation. Thus, for a Yagi antenna, the E plane is oriented horizontally; for a quarter-wave vertical element without parasitics, there are infinitely many E planes, all oriented vertically.

The magnetic lines of flux in an electromagnetic field cut through the E plane at right angles at every point in space. The direction of field propagation always lies within the E plane, and points directly away from the radiating element. In a parallel-wire transmission line, the E plane is that plane containing both of the feed-line conductors in a given location; the E plane follows all bends that are made in the line. In a coaxial transmission line, the E planes all contain the line of the center conductor. In waveguides, the orientation of the E plane depends on the manner in which the waveguide is fed. *See also* ELECTRIC FIELD, and ELECTROMAGNETIC FIELD.

EPROM

The term *EPROM*, pronounced "e-prom," is an abbreviation for *electrically programmable read-only memory*. This type of memory is erased and reprogrammed by an electric signal. *See also* PROGRAMMABLE READ-ONLY MEMORY.

EQUATOR CROSSING

See ASCENDING NODE, and DESCENDING NODE.

EQUIVALENT CIRCUITS

When two different circuits behave the same way under the same conditions, the circuits are said to be equivalent. There are many examples of, and uses for, equivalent circuits. One good example is the dummy antenna (*see* DUMMY ANTENNA). Such a device consists of a pure resistance, and when connected to a transmission line having a characteristic impedance equal to this resistance, the dummy load behaves exactly like an antenna with a purely resistive impedance of the same value.

Equivalent circuits are often used in the test laboratory and on the technician's bench, for the purpose of simulating actual operating conditions when designing and repairing electronic devices. The equivalent circuit is often much simpler than the actual circuit. The equivalent must have the same current, voltage, impedance, and phase relationship as the actual circuit. The equivalent circuit must also be capable of handling the same amount of power as the actual circuit; in the dummy-antenna example given above, a ⅛-watt resistor would obviously not suffice for use with a transmitter delivering 25 watts of RF output!

An important aspect of technological advancement is the development of simpler and simpler equivalent circuits. This has led, for example, to the handheld calculator, which, for a price of a few dollars, replaces the large tube-type devices of just a few decades ago. It has also led to the development of more and more complex computers, as a larger number of components can be fit into the same space. *See also* MINIATURIZATION.

ERASE

The *erase process* is the removal of information or data from a certain medium. Some media are erasable while others are not. Examples of erasable media include magnetic tape and disks, and various integrated-circuit memories.

In some devices, it is necessary to erase the stored data before new data can be recorded. In other instances, the new data is automatically recorded, without the need to erase or re-initialize the medium.

Erasing can be done accidentally as well as intentionally, and care should be exercised, especially with magnetic storage, to prevent this. Magnetic tapes and disks should be kept far away from magnetic fields. They should not be placed near current-carrying wires, coils, or devices containing permanent magnets, such as speakers!

ERROR

Whenever a quantity is measured, there is always a difference between the instrument indication and the actual value. This difference, expressed as a percentage, is called the *error* or *instrument error*. Mathematically, if x is the actual value of a parameter and y is the value indicated by a measuring instrument, then the error is given by:

$$Error \text{ (percent)} = 100(y - x)/x$$

if y is larger than x, and:

$$Error \text{ (percent)} = 100(x - y)/x$$

if x is larger than y. Not all of the error in instrumentation is the fault of the instrument. There is a limit to how well the person reading an instrument can interpolate values. For example, the needle of a meter might appear to be 0.3 of the way from one division to the next, as far as the person can tell, but in reality it is 0.38 of the way from one division to the next. Digital meters, of course, eliminate this problem, but the resolution of a digital device is limited by the number of digits in the readout. The term *error* is often used in reference to an improper command to a computer. This will result in an undesired response from the computer, or will cause the computer to inform the operator of the error. *See also* INTERPOLATION.

ERROR ACCUMULATION

Errors in measurement (*see* ERROR) tend to add together when several measurements are made in succession. This is called *error accumulation*. If the maximum possible error in a given measurement is x units, for example, and the measurement is repeated n times, then the maximum possible error becomes nx units. The percentage of maximum error, however, does not change.

An example of error accumulation is the measurement of a long distance, such as the length of a piece of antenna wire, with a short measuring stick such as a ruler. It is not convenient, of course, to measure long distances with rulers, but sometimes this is all that is available! Suppose the actual length of a piece of wire is 100 feet, and this length is measured with a ruler, resulting in a maximum error of 0.01 foot. The ruler must be placed against the wire 100 times, and this results in a maximum possible error of 100×0.01 foot = 1 foot. The maximum possible error per measurement is 1 percent, or $0.01/1$; the maximum possible overall error is also 1 percent $(1/100)$.

When making measurements, it is best, if possible, to avoid iterating the procedure, such as in the above example, because this multiplies the total error. *See also* ERROR.

ERROR CORRECTION

Error correction is a computer operating system in which certain types of errors are automatically corrected. An example of such a routine is a program that maintains a large dictionary of common English words. The operator of a word-processing machine can make mistakes in typing or spelling, but when the error-correction program is run, the computer automatically corrects all typographical errors, except in words not contained in its dictionary. With modern computers, extremely large dictionary files are realizable, and a program using such a file can perform almost all of the editing functions. This saves the wordprocessor operator the trouble of having to peruse the text for minor errors.

In measurement of parameters such as voltage, current, power, or frequency, error correction is used when an instrument is known to be inaccurate. For example, the operator of a radio transmitter can know that the station wattmeter reads 3 percent too high; this can have been determined by using a standard wattmeter borrowed from a laboratory. In such a case, if the transmitter output is supposed to be 100 watts, the operator will know that this will be shown on his instrument by a reading of 103 watts. *See also* ERROR.

ERROR-SENSING CIRCUIT

An *error-sensing circuit* is a means of regulating the output current, power, or voltage of a device. In a power supply, for example, an error-sensing circuit might be used to keep the output voltage at a constant value; such a circuit requires a standard reference with which to compare the output of the supply. If the power-supply voltage increases, the error-sensing circuit produces a signal to reduce the voltage. If the power-supply output voltage decreases, the error-sensing circuit produces a signal to raise it. An error-sensing circuit is therefore a form of amplitude comparator.

An error circuit does not necessarily have to provide regulation for the circuit to which it is connected; some error-sensing circuits merely alert the equipment personnel that something is wrong by causing a bell to ring or a warning light to flash. *See also* ERROR SIGNAL.

ERROR SIGNAL

An *error signal* is the signal put out by an error-sensing circuit (*see* ERROR-SENSING CIRCUIT). The error signal is actuated whenever the output parameter of a circuit or device differs from a standard reference value. An error signal can be used simply for the purpose of informing personnel of the discrepancy, or the signal can be used to initiate corrective measures.

An example of an error signal is the rectified voltage in an automatic-level-control circuit (*see* AUTOMATIC LEVEL CONTROL). The output of the amplifier chain in such a circuit is held to a constant value. If the signal tends to increase in amplitude, the error signal reduces the gain of the system.

Error signals are used to regulate the operation of servo systems. Each mechanical movement is monitored by an error-sensing circuit; error signals are generated whenever the system deviates from the prescribed operating conditions, and the signals cause corrective action. *See also* SERVOMECHANISM, and SERVO SYSTEM.

ESNAULT-PELTERIE FORMULA

The *Esnault-Pelterie formula* is a formula for determining the inductance of a single-layer, solenoidal, air-core coil based on its physical dimensions. The inductance increases roughly in proportion to the coil radius for a given number of turns. The inductance increases with the square of the number of turns for a given coil radius. If the number of turns and the radius are held constant, the inductance decreases as the solenoid is made longer.

Letting r be the coil radius in inches, N the number of turns, and m the length of the coil in inches, then inductance L in microhenrys is given by:

$$L = r^2 N^2 / (9r + 10m)$$

See also INDUCTANCE.

ETCHING

Etching is a process by which printed-circuit boards are made. Initially, all circuit boards are fabricated of phenolic or glass-epoxy material, plated on one or both sides with copper.

The circuit layout is drawn on a piece of paper, and perfected for optimum use of the available space. This is called the *etching pattern*. An example of such a pattern is shown in the illustration. The conductive paths are dark, and the portions of the board to be etched are white. The etching pattern is photographed and reduced or enlarged to the size of the circuit board. The photograph is printed on clear mylar.

The mylar sheet is placed over the copper plating of the board, and then the board and mylar are immersed in an etching solution. Ammonium persulfate crystals are added to water, in a ratio of about one part ammonium persulfate to two parts pure water. To this, a small amount of mercuric chloride is added. Alternatively, the etching solution can consist of one part ferric chloride and two parts pure water. When the circuit board, immersed in the etching solution, is exposed to bright light, the etching solution reacts with the copper. The etching pattern on the mylar prevents the copper from disappearing underneath the black areas. The circuit board should be continuously moved to speed up the etching process, and it should be kept at a temperature somewhat above normal room temperature. The etchant solution, ready-made, is commercially available. The skin should never be allowed to come into contact with etching solutions. *See also* PRINTED CIRCUIT.

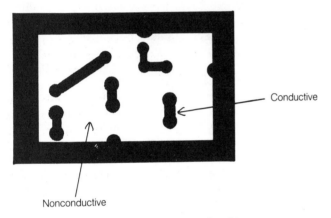

ETCHING: A simple circuit-board etching pattern.

EVEN-ORDER HARMONIC

An *even-order harmonic* is any even multiple of the fundamental frequency of a signal. For example, if the fundamental frequency is 1 MHz, then the even-order harmonics occur at frequencies of 2 MHz, 4 MHz, 6 MHz, and so on.

Certain conditions favor the generation of even-order harmonics in a circuit, and other circumstances tend to cancel out such harmonics. The push-push circuit accentuates even-order harmonics (*see* PUSH-PUSH CONFIGURATION), and is often used as a frequency doubler. The push-pull circuit (*see* PUSH-PULL CONFIGURATION) tends to cancel out the even-order harmonics.

A half-wave dipole antenna generally discriminates against the even-order harmonics. The impedance at the feed point of a half-wave dipole is very high at the even-order harmonic frequencies; although it can range from 50 to 150 ohms at the fundamental and odd harmonics, it can be as large as several thousand ohms at the even harmonics. *See also* ODD-ORDER HARMONIC.

EXCITATION

Excitation is the driving power, current, or voltage to an amplifier circuit. The term is generally used in reference to radio-fre-

quency power amplifiers. Excitation is sometimes called *drive* or *driving power*.

Class-A amplifiers theoretically require only a driving voltage, but no excitation power. In practice, a small amount of excitation power is required in the class-A amplifier. In the class-AB and class-B amplifiers, some excitation is required to obtain proper operation. In the class-C amplifier, a large amount of excitation power is needed in order to obtain satisfactory operation.

The circuit that supplies the excitation to a power amplifier is called the driver or exciter. When a transmitter is used in conjunction with an external power amplifier, the transmitter itself is called an exciter. *See also* CLASS-A AMPLIFIER, CLASS-AB AMPLIFIER, CLASS-B AMPLIFIER, CLASS-C AMPLIFIER, DRIVE, DRIVER, and EXCITER.

EXCITER

An *exciter* is a circuit that supplies the driving power to an amplifier. Usually, the term *exciter* is applied to radio-frequency equipment. The exciter must be capable of delivering sufficient power to operate the amplifier circuit properly. With class-A amplifiers, this power level can be almost negligibly small, although considerable peak-to-peak voltage can be necessary. With class-AB and class-B amplifiers, the power level varies depending on the exact bias level. With class-C amplifiers, a large amount of exciter power is needed for satisfactory operation.

The exciter generally has a tuned output so that optimum impedance matching can be obtained between it and the power amplifier. The exciter can be run by itself without the power amplifier, provided the exciter output tuning circuits are capable of providing a good direct match to the antenna system. When a transmitter is used in conjunction with an external power amplifier, the transmitter is sometimes called the exciter. The exciter can also be called the driver. *See also* CLASS-A AMPLIFIER, CLASS-AB AMPLIFIER, CLASS-B AMPLIFIER, CLASS-C AMPLIFIER, DRIVER, and EXCITATION.

EXCLUSIVE-OR GATE

The OR function can be either inclusive or exclusive; normally, unless otherwise specified, it is considered inclusive. The normal OR function is true if either or both of the inputs are true, and false only if both of the inputs are false. The exclusive OR function is true only if the inputs are opposite; if both inputs are true or both are false, the exclusive-OR function is false.

The illustration shows the schematic symbol of the exclusive-OR logic gate, along with a truth table of the function. The logic symbol 1, or high, indicates true, and 0, or low, indicates false. *See also* LOGIC GATE, and OR GATE.

EXPANDER

An *expander* is a circuit that increases the amplitude variations of a signal. Expansion is the opposite, electrically, of compression (*see* COMPRESSION, and COMPRESSION CIRCUIT). At low signal levels, the expander has little effect; the amplification factor increases as the input-signal amplitude increases.

An amplitude expander is used at the receiving end of a circuit in which compression is used at the transmitter. By compressing the amplitude variations at the transmitting end of a circuit, and expanding the amplitude at the receiver, the signal-to-noise ratio is improved, and communications efficiency

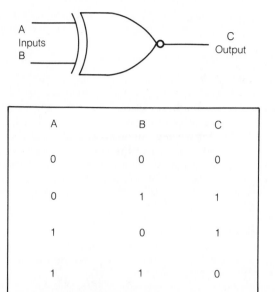

A	B	C
0	0	0
0	1	1
1	0	1
1	1	0

EXCLUSIVE-OR GATE: A schematic symbol and truth table.

thereby is made better. This is because the weaker parts of the signal are boosted before propagation via the airwaves; when the weak components are attenuated at the receiver, the noise is attenuated also. A system that makes use of compression at the transmitter and expansion at the receiver is sometimes called an *amplitude compandor*. *See also* COMPANDOR.

EXPLORING COIL

An *exploring coil*, sometimes called a *sniffer*, is a form of radio-frequency or magnetic-field detector. The device consists basically of a small inductor at the end of a probe, and a shielded or balanced cable connection to an amplifier. The amplifier output is connected to an indicating device such as a meter, light-emitting diode, or an oscilloscope.

When a magnetic field is present in the vicinity of the exploring coil, an indication is obtained whenever the coil is moved across the magnetic lines of flux. If a radio-frequency field is present, an indication will be obtained even if the coil is held still. An exploring coil can be used to determine if the shielding is adequate in a particular circuit. The exploring coil can also be used to obtain a small signal for monitoring purposes, without affecting the load impedance of the circuit under observation. *See also* EXPLORING ELECTRODE, and RADIO-FREQUENCY PROBE.

EXPLORING ELECTRODE

An *exploring electrode* is a device similar to an exploring coil. The device consists of a small pickup electrode, connected to an amplifier circuit. The output of the amplifier circuit is connected to an indicating meter, light-emitting diode, or an oscilloscope.

Either an exploring coil or an exploring electrode can be used to detect radio-frequency fields. However, for some applications, the electrode is preferable to the coil, and in other situations, the coil is more effective. An exploring electrode, like the exploring coil, causes little or no change in the load impedance of a circuit under test. The coupling mode of the exploring electrode is capacitive (electrostatic); that of the exploring coil is inductive (magnetic).

An exploring electrode generally presents an extremely high impedance to the input of the amplifier. Shielding the cable from the probe to the amplifier is an absolute necessity, and the radio-frequency ground must be excellent. Otherwise, stray electromagnetic fields will be picked up by the cable and amplified, resulting in severe interference. *See also* EXPLORING COIL, and RADIO-FREQUENCY PROBE.

EXTENDED DOUBLE-ZEPP ANTENNA

The *extended double-zepp antenna* is similar to the double-zepp antenna, except that it is slightly longer. The optimum length of the extended-double zepp antenna is about 1.3 wavelengths, compared with 1 wavelength for the double zepp. This extension of the antenna provides approximately 1 dB additional gain; although the power gain of the double zepp is 2 dB over a dipole, the extended version provides 3 dB power gain over a dipole.

The radiation pattern of the extended double zepp is very much like that of the double zepp, except that the lobes are slightly broader, and there are four minor lobes in addition to the two major lobes. The drawing illustrates the directional pattern of the extended double zepp, compared with that of a half-wave dipole, as viewed from directly above. The maximum radiation occurs in directions perpendicular to the axis of the antenna conductor. The minor lobes are at angles of about 35 degrees, with respect to the conductor.

The extended double-zepp antenna is popular among amateur radio operators, especially on the lower-frequency bands such as 1.8 and 3.5 MHz. For proper operation, the entire antenna should be run in a single straight line, and should be reasonably clear of obstructions. The feed-point impedance of the antenna contains some reactance as well as resistance, because it is not fed at a current or voltage loop. Low-loss, parallel-wire transmission line should be used to feed the antenna. *See also* DOUBLE-ZEPP ANTENNA.

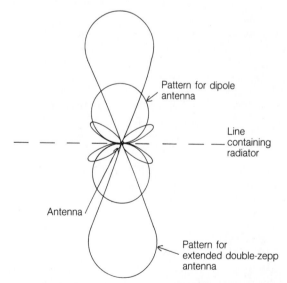

EXTENDED DOUBLE-ZEPP ANTENNA: Power gain in major lobes is about 3 dB over a half-wave dipole.

EXTERNAL FEEDBACK

When feedback, either positive or negative, occurs because of the external connections to a circuit, the phenomenon is called *external feedback*. External feedback can occur because of improper shielding of the input and/or output leads of a device. For example, in an audio-amplifier system, poor shielding of the microphone and speaker leads can result in coupling between the input and output of the circuit. This can, in turn, cause oscillation.

External feedback is generally not desired. If feedback is wanted, it is usually incorporated into a circuit, so that it can be regulated for optimum operation. If external feedback becomes a problem in a circuit, it can usually be eliminated by grounding the system, shielding all input, output, and power-supply leads, and bypassing the power-supply leads. Under certain circumstances, in radio-frequency apparatus, accidental resonances in external leads can cause feedback that is difficult to eliminate. In such cases, it is often necessary to change the lengths of one or more external leads. *See also* FEEDBACK.

EXTRACTOR

An *extractor* is a device that is used for the purpose of removing a signal from a circuit for monitoring purposes, or for sampling a constituent of a signal. For example, a radio-frequency probe (*see* RADIO-FREQUENCY PROBE) can be used to extract a small amount of voltage from the tank circuit of a radio-frequency amplifier, for monitoring on an oscilloscope. The modulation can then be monitored, in addition to other signal parameters.

A device for removing integrated circuits from a printed-circuit board is sometimes called an *extractor*. This device generally consists of a multi-pronged soldering iron, which heats up all of the pins of the device simultaneously. An absorbing braid or suction device removes the solder from the connections. Then, the integrated circuit can be easily pulled from the board. *See also* SOLDER WICK.

EXTRAPOLATION

When data is available within a certain range, an estimate of values outside that range is sometimes made by a technique called *extrapolation*. Extrapolation can be done by "educated guesswork," although computers provide the best results. Some functions can be extrapolated, while others cannot.

Random-number functions are impossible to extrapolate. Other very complicated functions can be difficult or impossible to extrapolate with reasonable accuracy. The accuracy of extrapolation diminishes as the independent variable gets farther from the domain of values given. Thus, while we can get a very good idea of the attenuation of a filter near its bandpass, we cannot readily ascertain the attenuation at any frequency far removed from the bandpass range, on the basis of the data given. *See also* INTERPOLATION.

EXTREMELY HIGH FREQUENCY

The *extremely-high-frequency range* of the radio spectrum is the range from 30 to 300 GHz. The wavelengths that correspond to these limit frequencies are 10 to 1 mm, and thus the extremely-high-frequency waves are called *millimetric waves*. Most of the frequencies in the millimetric range are allocated by the International Telecommunication Union, headquartered in Geneva, Switzerland.

Electromagnetic waves in the extremely-high-frequency range are not affected by the ionosphere of the earth. Signals in this range pass through the ionosphere as if it were not there. Millimetric waves behave very much like infrared rays, visible light, and ultraviolet. They can be focused by parabolic reflectors of modest size. Because of their tendency to propagate in straight lines unaffected by the atmosphere of the earth, millimetric waves are used largely in satellite communications. The frequencies allow modulation by wideband signals. *See also* ELECTROMAGNETIC SPECTRUM, and FREQUENCY ALLOCATIONS.

EXTREMELY LOW FREQUENCY

The *extremely-low-frequency band* is the range of the electromagnetic spectrum from 30 to 300 Hz. The wavelengths corresponding to these frequencies are 10,000 to 1,000 km (10 million meters to 1 million meters). For this reason, the waves are called *megametric*.

The ionosphere is an almost perfect reflector for megametric waves. Signals in this range cannot reach us from space; they are all reflected back by the ionized layers. The wavelength is many times the height of the ionosphere above the ground, and for this reason, severe attenuation occurs when megametric waves are propagated over the surface of the earth. Antennas for the extremely-low-frequency range are ridiculously long; for example, at 30 Hz, a half-wave dipole must be approximately 3,100 miles in length!

Megametric waves can be used to generate ground currents, in which the entire planet can become resonant. Research is continuing in this area. Megametric waves penetrate underwater, and therefore are useful in communicating with submarines. However, only very narrowband modulation is possible. *See also* ELECTROMAGNETIC SPECTRUM, and FREQUENCY ALLOCATIONS.

EXTRINSIC SEMICONDUCTOR

An *extrinsic semiconductor* is a semiconductor to which an impurity has been deliberately added. The pure form of the semiconductor is called *intrinsic*. Impurities must generally be added to semiconductor materials to make them suitable for use in solid-state devices such as transistors and diodes. The process of adding impurities to semiconductors is called *doping* (*see* DOPING).

Generally, the greater the amount of impurity added to a material such as silicon or germanium, the better the conductivity becomes. Certain impurities result in charge transfer via electrons; this kind of substance is called an *N-type semiconductor*. Other impurity elements form a substance in which holes carry the charge; these materials are called *P-type semiconductors*. *See also* ELECTRON, HOLE, INTRINSIC SEMICONDUCTOR, N-TYPE SEMICONDUCTOR, and P-TYPE SEMICONDUCTOR.

FACSIMILE

It is possible to send *facsimile (fax)*, images by means of narrow-band ham transmissions, in a manner very much like the way it is done over telephone channels.

Amateur fax transmissions have resolution comparable to that of any commercial transmission. The only problems occur because ham fax is via radio, rather than by wire, and the image can be degraded because of fading (QSB), noise (QRN) or interference from other stations (QRM). These problems are most likely on the HF bands. They are less severe or frequent at VHF and UHF.

A ham fax transmission requires from about three to 15 minutes to send. The longer the time for sending the image, the better the resolution, in general. The image is sent in lines, similar to the way in which a television raster is scanned. In ham work, 120 or 240 lines per minute (two or four lines per second) are scanned.

The modulation method can be either *positive* or *negative*. In positive modulation, the instantaneous signal level is directly proportional to the brightness at any given point on the image. In negative modulation, the instantaneous signal level is inversely proportional to the brightness at any given point on the image.

Hams can use surplus commercial fax apparatus to build their fax stations. Often the equipment must be modified. This is a challenge to the ham who likes to tinker and perfect station equipment. Most commercial units are designed to work with amplitude modulation (AM), but they can be converted to work with frequency modulation (FM). Some hams like to build all of their fax equipment from scratch.

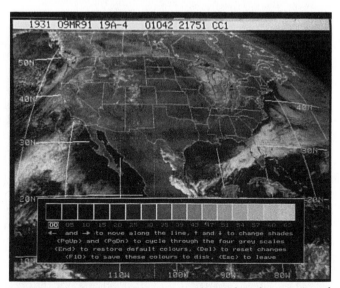

FACSIMILE: A high-resolution, gray-scale weather photo as viewed on video monitor, using a AEA-FAX converter. ADVANCED ELECTRONIC APPLICATIONS, INC.

In recent years, a several manufacturers have come out with interface units that allow personal computers (PCs) to be used to receive, and also to send, fax transmissions. The images can be displayed on the monitor screen with excellent resolution (see photo). It is even possible to use these interfaces for color fax. There are numerous functions that enhance the versatility of the system, such as "zoom" for viewing a portion of the image close-up.

The most popular use of ham fax is in the receipt and re-transmission of weather-satellite data. These satellites send their pictures back to earth via high-resolution fax. Two-way contacts (QSOs) are sometimes also carried out on fax; this is somewhat like slow-scan television (SSTV), except the pictures take longer to send, and show more detail.

For more information on fax, and for sources of supplies and equipment, see current issues of any of the ham magazines. *See also* SLOW-SCAN TELEVISION.

FADING

Fading is a phenomenon that occurs with radio-frequency signals propagated via the ionosphere. Fading occurs because of changes in the conditions of the ionosphere between a transmitting and receiving station, at frequencies where ionospheric propagation is common.

Fading can occur in two different ways. One way in which fading occurs is the combination of waves of variable phase at the receiving station in a communications circuit. The ionosphere is somewhat unstable, and as its height above the earth fluctuates, the phase of a signal reflected from it also changes. Several signals can propagate over different paths simultaneously, and the resulting phase combination at a receiver can fluctuate greatly within a short time interval.

Fading can also occur because of the ionosphere between the transmitting and receiving station. This kind of fading is apt to occur near the maximum usable frequency for the ionosphere at a given time (*see* MAXIMUM USABLE FREQUENCY). A signal can be perfectly readable and then suddenly disappear.

Fading sometimes occurs because of multipath propagation, in which a signal is reflected from the ionosphere at several different points en route to the receiving station (*see* MULTIPATH FADING). When the ionosphere changes orientation rapidly, the fading can occur first at one frequency, and then at another, progressing across the passband of a signal from top to bottom or vice versa. This is called selective fading. The narrower the bandwidth of a signal, the less susceptible it is to selective fading. Since selective fading is common, narrowband signals are preferred over wideband signals in the frequency range where ionospheric propagation is predominant. *See also* SELECTIVE FADING.

FAILSAFE

Failsafe is a term that refers to a form of failure protection in electronic circuits and systems. A failsafe circuit is so designed that, in the event of malfunction of one or several components, the circuit will remain safe for operation; no further damage will occur, and no hazard will be presented to personnel using the equipment. A failsafe system is a system that places a backup circuit in operation if a circuit fails. This allows continued operation. A warning signal alerts the personnel that a circuit has failed, and this expedites repair.

In a computer, failsafe is sometimes called *failsoft*. A malfunction in a failsoft computer will result in some degradation of efficiency, but will not cause a total shutdown. The computer alerts the user that a problem is present, giving its location in the system.

No system is 100-percent reliable, but failsafe circuits and systems reduce the amount of downtime in the event of component failures. *See also* DOWNTIME, and FAILURE RATE.

FAILURE RATE

Failure rate is an expression of the frequency with which a component, circuit, or system fails. Generally, failure rate is specified in terms of the average number of failures per unit time. Of course, the longer the time interval chosen for the specification, the greater the probability that a given component, circuit, or system will malfunction.

An example of component failure rate is as follows. Suppose that, out of 100 diodes, three of them malfunction within a period of 1 year. We assume that all of the diodes are operational at the start of the 1-year period. Then the failure rate is 3 percent per year for that particular application of the diode. The probability is 3 percent that a given diode will fail within 1 year.

When determining the failure rate for a component, circuit or system, it is customary to test many units at the same time. This increases the number of observed failures, and allows a better probability estimate than would be possible with just one test unit.

As the time interval approaches zero in the failure-rate specification, the failure rate itself does not approach zero. Some components, circuits, or systems fail the instant they are switched on. Any failure occurs at a defined instant. The probability that a component will fail at any given moment is called the *hazard rate*. *See also* HAZARD RATE, QUALITY CONTROL, and RELIABILITY.

FAMILY

Any set of similar things, such as components, devices, or mathematical functions, is called a *family*. For example, in digital logic, there are several families of gate design, including the bipolar and complementary-metal-oxide semiconductor families.

The constituents of a family are always related in some way. Totally different items cannot be members of the same family. However, there are various ways of specifying the nature of the relationship among members of a family. A bipolar transistor and a field-effect transistor are not both members of the family of bipolar devices, but they are both members of the family of solid-state devices. A synonym for "family" might be the word "category."

FARAD

A *farad* is the unit of capacitance. When the voltage across a capacitor changes at a rate of 1 volt per second, and a current flow of 1 ampere results, the capacitor has a value of 1 farad. A capacitance of 1 farad also results in 1 volt of potential difference for a charge of 1 coulomb.

The farad is, in practice, an extremely large unit of capacitance. Rarely does a capacitor have a value of 1 farad. Capacitance is generally measured in microfarads, or millionths of a farad (abbreviated μF); for small capacitors, the picofarad, or trillionth of a farad (abbreviated pF) is often specified. Sometimes the nanofarad, or billionth of a farad (abbreviated nF) is indicated.

At radio frequencies, typical capacitances range from 1 pF to perhaps 1,000 pF in tuned circuits, and from 0.001 to about 1 μF for bypassing purposes. At audio frequencies, typical capacitances are in the range 0.1 to 100 μF. In power-supply filtering applications, capacitances can range up to 10,000 μF or more. *See also* CAPACITANCE, and CAPACITOR.

FARADAY

The *faraday* is a unit of electrical quantity or charge. A charge of 1 faraday represents the amount of electrical quantity required to free 1 gram atomic weight of a univalent element in the electrolysis process. A charge of 1 faraday represents about 96,500 coulombs, or 6.06×10^{23} electrons.

Generally, the faraday is not specified in electronics applications. The coulomb is much more often used to represent charge quantities. *See also* COULOMB.

FARADAY CAGE

A *Faraday cage* is an enclosure that prevents electric fields from entering or leaving. However, magnetic fields can pass through. The Faraday cage consists of a wire screen or mesh, or a solid non-magnetic metal, broken so that there is not a complete path for current flow. The Faraday cage is similar to the electrostatic shield used in some radio-frequency transformers to reduce the transfer of harmonic energy (*see* ELECTROSTATIC SHIELDING).

Faraday cages are used in experimentation and testing, when the presence of electric fields is not desired. An entire room can be shielded. Sometimes a complete electromagnetic shield is called a *Faraday cage*, although this is technically a misnomer. *See also* ELECTROMAGNETIC SHIELDING.

FARADAY EFFECT

When radio waves are propagated through the ionosphere, their polarization changes because of the effects of the magnetic field of the earth. This is called the *Faraday effect*. Because of the Faraday effect, sky-wave signals arrive with random, and fluctuating, polarization. It makes very little difference what the polarization of the receiving antenna can be; signals originating in a vertically polarized antenna can arrive with horizontal polarization some of the time, vertical polarization at other times, and slanted polarization at still other times.

The Faraday effect results in signal fading over ionospheric circuits. There are other causes of fading, as well. The fading caused by changes in signal polarization can be reduced by the use of a circularly polarized receiving antenna (*see* CIRCULAR POLARIZATION).

When light is passed through certain substances in the presence of a strong magnetic field, the plane of polarization of the light is made to rotate. This is a form of Faraday effect. In order for this to occur, the magnetic lines of force must be parallel to the direction of propagation of the light. *See also* POLARIZATION.

FARADAY ROTATION

When a radio wave passes through the earth's ionosphere, its polarization can be changed. In fact, at frequencies below 1 GHz (1000 MHz), some change almost always occurs. The effect, known as *Faraday rotation*, becomes more pronounced as the wavelength increases.

Faraday rotation tends to occur in a random and fluctuating manner. A wave might undergo several complete rotations on its way through the ionosphere. Because of this, the receiving station will "see" a wave whose polarization constantly changes, and that can arrive at any angle.

Faraday rotation practically necessitates the use of a circularly polarized antenna, either at the transmitter, the receiver, or both, for space communications (such as satellite work or moonbounce). *See* CIRCULAR POLARIZATION, and POLARIZATION.

FARADAY SHIELDING

See ELECTROSTATIC SHIELDING, and FARADAY CAGE.

FARADAY'S LAWS

Faraday's laws are concerned with two different phenomena, and are therefore placed into two categories. The law of electromagnetic induction is sometimes called *Faraday's law*. This is the familiar principle of the generation of a current in a wire that moves in a magnetic field. Actually, the wire can be stationary and the field can be moving, but as long as the wire cuts across magnetic lines of flux, a current is induced. The current is directly proportional to the rate at which the wire crosses the lines of force of the magnetic field. The faster the motion, the greater the current; the more intense the field, the greater the current. It is this effect, in part, that makes radio communication possible (*see* ELECTROMAGNETIC INDUCTION).

The other principle form of Faraday's law is involved with electrolytic cells. In any electrolytic cell, the greater the amount of charge passed through, the greater the mass of substance deposited on the electrodes. The actual mass deposited for a given amount of electric charge depends on the electrochemical equivalent of the substance.

FAR FIELD

The *far field* of an antenna is the electromagnetic field at a great distance from the antenna. The far field has essentially straight lines of electric and magnetic flux, and the lines of electric flux are perpendicular to the lines of magnetic flux. The wavefronts are essentially flat planes.

The far field of an antenna has a polarization that depends on several factors. The polarization of the transmitting antenna dictates the polarization of the far field under conditions in which there is no Faraday effect (*see* FARADAY EFFECT). The power density of the far field diminishes with the square of the distance from the antenna. The field strength, in microvolts per meter, diminishes in direct proportion to the distance from the antenna.

The far field of an antenna is sometimes called the *Fraunhofer region*. The far field begins at a distance that depends on many factors, including the wavelength and the size of the antenna. The signal normally picked up by a receiving antenna is the far field, except at extremely long wavelengths. *See also* ELECTROMAGNETIC FIELD, NEAR FIELD, and TRANSITION ZONE.

FATIGUE

As a circuit or component gets older, it becomes less reliable. This aging process occurs rapidly with some components or circuits, and slowly with others. When a component or circuit needs to be replaced or repaired because of the simple effects of aging, the condition is called *fatigue*.

Fatigue usually occurs only in power-dissipating components, such as vacuum tubes and transistors. To a lesser extent, fatigue occurs in all active devices. Passive components (such as coils and capacitors) generally do not suffer from fatigue unless they are forced to handle currents or voltages that put them under strain. That should not happen in a well-designed circuit.

Equipment should be checked periodically for possible fatigue. A drop in the power output of an amplifier, for example, can go unnoticed if it is gradual and no periodic tests are conducted. A change in the standing-wave ratio on a transmission line, especially a gradual drop, often indicates feed-line aging, with a resulting loss increase.

FAULT

Any discontinuity or problem in a circuit is sometimes called a *fault*. The fault is always localized in nature, such as a short-circuited capacitor or a burned-out diode. With proper tools and equipment, the fault can always be found and corrected.

A fault will often result in a leakage current, known as a *fault current*, that is not present under proper operating conditions. This current can, in some cases, cause destruction of additional components. For example, a partial short circuit in a coupling capacitor can change the bias at the next stage sufficiently to cause excessive current in that stage, and consequent damage to its components.

Troubleshooting is sometimes called *fault finding*, and an instrument used by a technician is called a *fault finder*. *See also* TROUBLESHOOTING.

FEDERAL COMMUNICATIONS COMMISSION

The *Federal Communications Commission (FCC)*, is an agency of the United States Government. The FCC is responsible for the allocation of frequencies for radiocommunications and broadcasting within the United States. The FCC is also responsible for the enforcement of the laws concerning telecommunications.

The FCC issues various kinds of licenses for communications and broadcasting personnel. Some licenses can be obtained only by passing an examination. The FCC composes and administers these examinations. Other licenses require only the filing of an application.

When the FCC deems it necessary to create a new rule, it first publishes a notice of proposed rulemaking. Then, within a certain specified time period, interested or concerned parties are allowed to make comments on the subject. The FCC, after considering the comments, makes its decision. The power of the FCC is limited, however, by the Congress.

Each country determines its own frequency allocations and system of radiocommunications laws. The International Telecommunication Union provides coherence among the many nations of the world, for the purpose of optimum utilization of the limited space in the spectrum. *See also* INTERNATIONAL TELECOMMUNICATION UNION.

FEED

Feed is the application of current, power, or voltage to a circuit. The term feed is used particularly with reference to antenna systems. There are basically three electrical methods of antenna feed: current feed, voltage feed, and reactive feed.

A current-fed antenna has its transmission line connected where the current in the radiating element is greatest. This occurs at odd multiples of ¼ wavelength from free ends of the radiator. A voltage-fed antenna has the transmission line connected at points where the current is minimum; such points occur at even multiples of ¼ wavelength from free ends, and also at the free ends. Both current feed and voltage feed are characterized by the absence of reactance, only resistance is present. In current feed, the resistance is relatively low; in voltage feed, it is high.

When an antenna is not fed at a current or voltage loop, the feed is considered reactive. This can be the case when an antenna is not resonant at the operating frequency, or when it is fed at a point not located at an integral multiple of ¼ wavelength from a free end.

Feed can be classified in other ways, such as according to the geometric position of the point where the transmission line joins the radiating element. In end feed, the line is connected to the end of the element; in center feed, to the center; in off-center feed, somewhat to either side of the center. A vertical radiator can be fed at the base, or part of the way up from the base to the top. A special method of feed for a vertical radiator is called *shunt feed*. *See also* CENTER FEED, CURRENT FEED, END FEED, OFF-CENTER FEED, SHUNT FEED, TRANSMISSION LINE, and VOLTAGE FEED.

FEEDBACK

When part of the output from a circuit is returned to the input, the situation is known as *feedback*. Sometimes feedback is deliberately introduced into a circuit; sometimes it is not wanted. Feedback is called *positive* when the signal arriving back at the input is in phase with the original input signal. Feedback is called *negative (or inverse)* when the signal arriving back at the input is 180 degrees out of phase, with respect to the original input signal. Positive feedback often results in oscillation, although it can enhance the gain and selectivity of an amplifier if it is not excessive. Negative feedback reduces the gain of an amplifier stage, makes oscillation less likely, and enhances linearity.

Feedback can result from a deliberate arrangement of components. All oscillators use positive feedback. Unwanted feedback can result from coupling among external wires or equipment. This external coupling can be capacitive, inductive, or perhaps acoustic. We have all heard the acoustic feedback in an improperly operated public-address system.

The number of amplifying stages that can be cascaded, for the purpose of obtaining high gain, is limited partly by the effects of feedback. The more stages connected in cascade, the higher the probability that some positive feedback will occur, with consequent oscillation. *See also* NEGATIVE FEEDBACK, and OSCILLATION.

FEEDBACK AMPLIFIER

A *feedback amplifier* is a circuit, placed in the feedback path of another circuit, to increase the amplitude of the feedback signal. The phase can also be inverted. A feedback amplifier is used when the feedback signal would not otherwise be strong enough to obtain the desired operation.

The term *feedback amplifier* is sometimes used to describe an amplifier that uses feedback to obtain a certain desired characteristic. Negative feedback, for example, is often used to reduce the possibility of unwanted oscillations in a radio-frequency amplifier. Combinations of negative and positive feedback can be used to modify the frequency response of an amplifier. This is especially true of operational-amplifier circuits. *See also* FEEDBACK.

FEEDBACK CONTROL

Whenever feedback is deliberately introduced into a circuit for any reason, some means must be provided to control the amount of feedback. Otherwise, the desired circuit operation cannot be obtained. A feedback control can consist of a simple potentiometer, used to adjust the amount of feedback signal arriving at the input of the circuit. More sophisticated methods of feedback control consist of self-regulating circuits, such as the feedback amplifier used in the amplified automatic-level-control circuit (*see* AUTOMATIC LEVEL CONTROL, and FEEDBACK AMPLIFIER).

The elimination of undesired feedback in a circuit is sometimes called feedback control. The output of a circuit can be partially coupled back to the input by the capacitance or inductance between the input and output peripheral wiring. In a public-address system, acoustic feedback can sometimes occur as the microphone picks up sound from the speakers. Coupling between the input and output should be minimized. This might be done by means of bypass capacitors, series chokes, or phase-inverting circuits. *See also* FEEDBACK.

FEEDBACK RATIO

The *feedback ratio*, usually represented by the Greek letter beta (β), is a measure of the amount of feedback in a feedback amplifier. If e_o is the output voltage of an amplifier with no load, and e_f is the feedback voltage, then the feedback ratio is given simply as:

$$\beta = e_f/e_o$$

When the feedback ratio is to be expressed as a percentage, it is generally represented by n, and:

$$n = 100e_f/e_o$$

This quantity is sometimes called the *feedback percentage*. The feedback ratio is never greater than 1, and the feedback percentage is never greater than 100.

If A is the open-loop gain of an amplifier, and β is the feedback ratio, then the feedback factor m is the quantity given by:

$$m = 1 - \beta A$$

See also FEEDBACK.

FEED LINE

A *feed line* is a transmission line used for the purpose of transferring an electromagnetic field from a transmitter to an antenna. The point where the feed line joins the antenna is called the *feed point*.

Feed lines can take a variety of different forms. The most common type of feed line in use today is the coaxial line, which is prefabricated and usually has a solid or foamed polyethylene dielectric material. Two-wire line, also called *open-wire* or *parallel-wire line*, is often used in the feed system of a television receiving antenna. Homemade open-wire line is sometimes used by radio amateurs. Such a feed line has very low loss, and can tolerate high power levels and large impedance mismatches.

A feed line should be chosen on the basis of low loss, cost effectiveness, and power-handling capacity. Other important factors include convenience of installation, the antenna impedance, and antenna-system balance. The best feed line for a particular antenna depends on all these factors.

FEEDTHROUGH CAPACITOR

A *feedthrough capacitor* is a device that is used for passing a lead through a chassis, while simultaneously bypassing the lead to chassis ground. The center (axial) part of the feedthrough capacitor is connected to the lead; the outer part of the component is connected to chassis ground.

Feedthrough capacitors are useful in power-supply leads, when it is desirable to prevent radio-frequency energy from being transferred into or out of the equipment along these leads. Feedthrough capacitors are also sometimes used in the speaker or microphone leads of an audio amplifier circuit.

Feedthrough capacitors are available in many different physical and electrical sizes. The proper value of capacitance should be chosen for the application desired. *See also* BYPASS CAPACITOR, and CAPACITOR.

FEEDTHROUGH INSULATOR

A *feedthrough insulator* is used for the purpose of passing a lead through a metal chassis while maintaining complete electrical insulation from the chassis. Feedthrough insulators are also sometimes used when it is necessary to pass leads through a partial conductor or nonconductor with a minimum amount of dielectric loss. Feedthrough insulators generally consist of a threaded shaft with two conical sections of porcelain or glass held in place by nuts on the shaft.

Feedthrough insulators are available in a variety of different sizes. The voltage across a feedthrough insulator can, in some applications, reach substantial values. Therefore, a sufficiently large insulator should always be used. In an antenna-tuning network, the feedthrough insulators can be subjected to several thousand volts under certain conditions. The feedthrough insulators should have the lowest possible loss at the highest frequency of the equipment in which they are used. The insulators should also have the smallest possible amount of capacitance. *See also* INSULATOR.

FELICI MUTUAL-INDUCTANCE BALANCE

The *Felici mutual-inductance balance* is a device used for the purpose of determining the mutual inductance, or degree of coupling, between two coils. The Felici balance is used primarily at lower frequencies. At higher frequencies, the capacitance among the windings of the inductors results in difficulty of measurement. *See also* MUTUAL INDUCTANCE.

FEMALE

Female is an electrical term used to describe a jack. A female plug is recessed, and the conductors are not exposed. An example of a female plug is the common utility wall outlet.

The female jack is usually, but not always, mounted in a fixed position on the panel of a piece of electronic apparatus. The male plug (*see* MALE) is generally attached to the cord or cable and fits into the female plug. Connectors can have only one conductor, or they can have dozens of conductors.

FEMTO

See PREFIX MULTIPLIERS.

FERRITE

Ferrite is a substance designed for the purpose of increasing the magnetic permeability in an inductor core. Ferrite has a higher degree of permeability than ordinary powdered iron. Ferrite materials are classified as either soft or permanent. Ferrite acts as as electrical insulator, resembling a ceramic substance, with respect to current conductivity. The magnetic conductivity, however, is excellent, and the eddy-current losses are very small.

Ferrite materials are available in a great variety of shapes and permeability values. The tiny receiving antennas used in small transistor radios consist of a coil wound on a solenoidal ferrite core. Such antennas have excellent sensitivity at frequencies below the shortwave range. Ferrite is used in toroidal inductor cores to obtain large values of inductance with a relatively small amount of wire. Ferrite is also used to make pot cores, which allows even larger inductances to be obtained. Typical permeability values for ferrites range from about 40 to more than 2,000. Ordinary powdered-iron cores generally have much smaller permeability values. Ferrite is well suited to low-frequency and medium-frequency applications in which loss must be kept to a minimum. *See also* FERRITE BEAD, FERRITE CORE, FERRITE-ROD ANTENNA, PERMEABILITY, POT CORE, POWDERED-IRON CORE, and TOROID.

FERRITE BEAD

A *ferrite bead* is a small, toroidal piece of ferrite. Ferrite beads are often used for the purpose of choking off radio-frequency currents on wire leads and cables. The bead is simply slipped around the wire or cable, as illustrated. This introduces an inductance for high-frequency alternating currents without affecting the direct currents and low-frequency alternating currents.

Ferrite beads are especially useful for choking off antenna currents on coaxial transmission lines in the very-high-frequency range and above. Antenna currents, or radio-frequency currents flowing on the outside of a coaxial feed line, can cause problems with transmitting equipment. One or more ferrite beads along the feed line are effective in choking off these unwanted currents. But the desired transmission of electromagnetic fields inside the cable is not affected.

Ferrite beads are used in certain digital memory systems to store information by means of magnetic fields. The polarity of the magnetic field can be either clockwise or counterclockwise. The polarity will remain the same until a current is sent through a conductor passing through the bead. If the current is such as to reverse the polarity of the magnetic field stored in the ferrite, then the polarity will remain reversed until another surge of current, in the opposite direction, changes the field orientation again. *See also* FERRITE, MAGNETIC CORE, and TOROID.

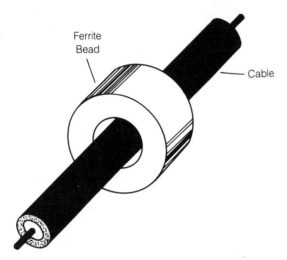

FERRITE BEAD: Chokes off RF currents in the outer conductor of a coaxial cable.

FERRITE CORE

A *ferrite core* is an inductor core made from ferrite material. Ferrite cores are extensively used in electronic applications at low, medium, and high frequencies, for the purpose of maximizing the inductance of a coil while minimizing the number of turns. This results in low-loss coils.

The most common ferrite-core configurations are the solenoidal type and the toroidal type. The toroid core has gained popularity in recent years because the entire magnetic flux is contained within the core material. This eliminates coupling to external components (*see* TOROID).

At audio frequencies and very low frequencies, the pot core is often used. This type of core actually surrounds the coil, whereas in other inductors the coil surrounds the core. Pot cores allow inductance values in excess of 1 henry to be realized using a moderate amount of wire. Such coils, because of the low loss in the ferrite material, exhibit extremely high Q factors, and are thus useful in applications, where extreme selectivity is needed (*see* POT CORE).

At higher frequencies, ferrite becomes lossy, and powdered-iron cores are preferred. *See also* FERRITE, and POWDERED-IRON CORE.

FERRITE-ROD ANTENNA

For receiving applications at low, medium, and high frequencies, up to approximately 20 MHz, a ferrite-rod antenna is sometimes used. This type of antenna consists of a coil wound on a solenoidal ferrite core. A series or parallel capacitor, in conjunction with the coil forms a tuned circuit. The operating frequency is determined by the resonant frequency of the inductance-capacitance combination.

Ferrite-rod antennas display directional characteristics similar to the dipole antenna (*see* DIPOLE ANTENNA). The sensitivity is maximum off the sides of the coil, and a sharp null occurs off the ends. This has little effect with sky-wave signals, which tend to arrive from varying directions. However, the null can be used to advantage to eliminate interference from local signals and from man-made sources of noise. The ferrite-rod antenna is physically very small. This makes it easy to orient the rod in any direction. Some ferrite-rod antennas are affixed to az-el mountings (*see* AZ-EL).

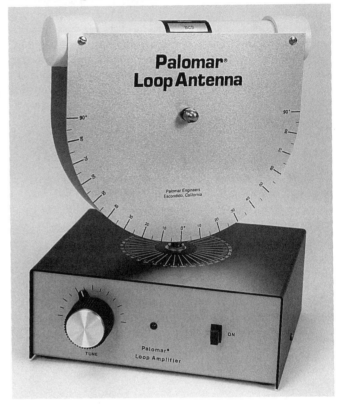

FERRITE-ROD ANTENNA: This antenna rotates and tilts to eliminate interference. Plug-in loops cover 10 kHz to 16 MHz. PALOMAR ENGINEERS

Ferrite-rod antennas are not generally used at VHF and UHF, because the required inductance is so small that the coil would have very few turns, and capacitive effects would predominate. A further limitation is imposed by the fact that ferrite materials are seldom suitable for use above about 50 MHz. They become lossy because of hysteresis effects. *See also* FERRITE.

FERROELECTRIC CAPACITOR

The *ferroelectric capacitor* is a form of ceramic capacitor in which the dielectric is a ferroelectric substance. Ferroelectric materials show excellent dielectric properties and have very large dielectric constants. Ferroelectric capacitors therefore have large values of capacitance for a relatively small demand on physical space.

Because of hysteresis effects, the value of a ferroelectric capacitor depends on the voltage across it. In general, the greater the voltage across a ferroelectric capacitor, the smaller its value. Ferroelectric capacitors tend to drop somewhat in value as they age, because the polarization of the ferroelectric material deteriorates, reducing the dielectric constant.

Ferroelectric capacitors tend to be self-resonant at frequencies ranging from a few megahertz to the very-high frequency region. This is because of inherent inductance in the leads and plates. Such capacitors are used mostly for coupling and by-passing purposes, where a constant value of capacitance is not important. They are generally not suitable for use much above the high-frequency region. *See also* CAPACITOR, and DIELECTRIC.

FERROMAGNETIC MATERIAL

A *ferromagnetic material* is a substance that has a high magnetic permeability. The most common ferromagnetic materials used in electronics are ferrite and powdered iron. Such substances greatly increase the inductance of a coil when used as the core material.

Ferromagnetic materials can consist of various combinations of iron fragments and other substances. Such materials as ferrite and powdered iron are highly resistive to the direct application of electric current, but will carry magnetic fields with very little loss. The permeability of a ferromagnetic material is a measure of the degree to which the magnetic lines of force are concentrated within the substance (*see* PERMEABILITY).

Ferromagnetic materials are formed into a variety of different shapes for different applications. *See also* FERRITE, FERRITE-ROD ANTENNA, POT CORE, POWDERED-IRON CORE, and TOROID.

FET

See FIELD-EFFECT TRANSISTOR.

FET VOLTMETER

An *FET voltmeter* is a device similar to a vacuum-tube voltmeter (*see* VACUUM-TUBE VOLTMETER). The input impedance is extremely high, so the device does not draw much current from the source. An FET voltmeter uses a field-effect-transistor amplifier circuit to achieve the high input impedance.

The illustration shows a simplified schematic diagram of an FET voltmeter. The ranges are selected by varying the gain of the amplifying stage, or by connecting (by means of a switch) various resistances in parallel with the microammeter.

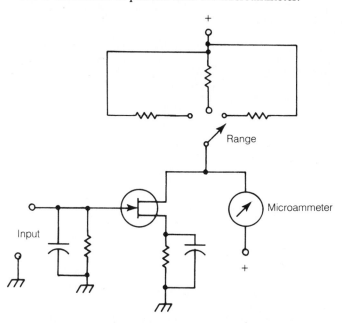

FET VOLTMETER: Draws essentially no current.

The FET voltmeter can also be used to measure current and resistance; this can be done simply by using the meter and power supply in conjunction with various resistors, or the field-effect transistor can be used to act as an amplifier to increase the sensitivity. Such a combination meter is called an *FET volt ohm-milliammeter (FET VOM)*. *See also* FIELD-EFFECT TRANSISTOR.

FIBER

A *fiber* is a thin, transparent strand of material, usually glass or plastic, that is used to carry light beams. A fiber confines the light energy inside its walls because of high internal reflection. The index of refraction of the fiber substance is, for visible light, much higher than the index of refraction for air; therefore, light rays are reflected from the inner walls as they propagate lengthwise along the fiber.

A bundle of optical fibers can be used to transfer many different light beams at the same time. A single light beam can be simultaneously modulated by hundreds, or even thousands, of independent signals. Fiber cables of current design can carry millions of different conversations at once!

A large bundle of optical filaments can carry a picture. The image is focused onto one end of the bundle of fibers, and is re-magnified at the other end. This principle makes it possible to look deep inside the human body. A device called a *fiberscope* can be swallowed by a patient and slowly fed through the entire gastrointestinal system. *See also* FIBER OPTICS.

FIBER OPTICS

Fiber optics is a modern method of conveying information. Light beams are transferred from one place to another by an optical fiber or fibers (*see* FIBER). Light beams can be modulated at frequencies as high as hundreds of megahertz, and can be modulated by many different signals at the same time. Optical fibers are inexpensive and efficient. They allow the transmission of data at extremely high speeds. Optical fibers are compact and lightweight, and are essentially immune to electromagnetic interference. Modern optical fibers are replacing ordinary wire conductors in long-distance, high-volume communication cables.

A fiber-optics communications system requires a light source that can be easily modulated, a light receptor that can follow the modulation of the light, and a system of amplifiers. Certain kinds of diodes are excellent sources of light for this purpose; some such diodes emit coherent light, which suffers less attenuation than ordinary light. Gallium-arsenide (GaAs) electroluminescent diodes are widely used in fiber-optics communications systems. Phototransistors, silicon solar cells, and similar devices can be used to intercept the light.

Modulation of a light beam for fiber-optics communications can be accomplished electrically or externally. The electrical method essentially resembles amplitude modulation, in which the intensity of the light source is varied periodically with the signal information. External modulation is accomplished by passing the light through, or reflecting it from, a device with variable transmittivity. Magnetic fields, if sufficiently strong and properly oriented, can be used to modulate the polarization of a light beam. *See also* MODULATED LIGHT, and OPTICAL COMMUNICATION.

FIBRILLATION

See HEART FIBRILLATION.

FIDELITY

Fidelity is the faithfulness with which a sound, or other signal, is reproduced. In general, perfect fidelity represents a complete lack of distortion in the signal waveform at all frequencies of modulation. Fidelity is particularly important in the reproduction of music, and in modern stereo high-fidelity equipment, great attention is given to minimizing the distortion.

Fidelity is not very important in communications practice. Intelligibility is of greater concern in this application. Maximum fidelity does not always occur along with maximum intelligibility. Although the human voice contains components of very low and very high audio frequencies, the range from about 300 Hz to 3 kHz is sufficient for communications purposes, and thus the audio-frequency response in a voice communications circuit is usually limited to this range.

For the best possible fidelity, an audio amplifier must usually be operated in class A or class B push-pull. The amplifier must be linear over the range of amplitudes to be reproduced. This requires careful engineering. *See also* CLASS-A AMPLIFIER.

FIELD

Many effects occur between or among objects separated by empty space. Examples of this are electricity, magnetism, and gravitation. The means by which an effect is propagated with no apparent medium to carry it is called a *field*. In electronics, the most significant fields are the electric field, the electromagnetic field, and the magnetic field.

The effects of fields travel through empty space at a speed of approximately 186,282 miles per second, or 299,792 kilometers per second. This is the familiar speed of light. In material substances, the speed is slower than that in a vacuum, and depends on several factors.

Radio waves, infrared, visible light, ultraviolet, X rays, and gamma rays are all electromagnetic-field effects. The whole range of electromagnetic-field effects is called the electromagnetic spectrum (*see* ELECTROMAGNETIC SPECTRUM).

In television, half of the image, or 262.2 lines, is called a field (*see* TELEVISION). Computer-record subdivisions are also sometimes called *fields*. Generally, the term field is used in reference to the effects of electricity and magnetism. *See also* ELECTRIC FIELD, ELECTROMAGNETIC FIELD, and MAGNETIC FIELD.

FIELD-EFFECT TRANSISTOR

A *field-effect transistor (FET)*, is a semiconductor device used as an amplifier, oscillator, or switch. The field-effect transistor operates on the basis of the interaction of an electric field with semiconductor material.

The drawing illustrates a cross-sectional representation of a field-effect transistor. The source is at ground potential, and the drain, at the opposite end of the N-type wafer, is connected to a positive voltage. The path from the source to the drain is called the *channel*. In this case, the channel consists of an N-type semiconductor wafer, and thus the device is called an *N-channel FET*. Electrons flow from the source to the drain, through the channel.

The gate electrode consists of two sections of P-type semiconductor material in the N-channel FET. Alternatively, the P-type semiconductor can be wrapped around the channel. When a negative voltage is applied to the gate electrode, a de-

pletion region, or area of nonconduction, forms in the channel, as shown. The larger the negative voltage, the bigger the depletion region becomes. This narrows the channel, and increases the resistance from the source to the drain. In this way, the FET acts as a sort of current valve. An alternating-current signal can be applied to the gate, and the current through the FET will fluctuate with much greater amplitude than the original signal. The result is amplification.

If the negative voltage at the gate electrode becomes large enough, the channel will be entirely severed by the depletion region. This condition is called *pinchoff* and is somewhat analogous to cutoff in a tube or bipolar transistor.

If the N-type and P-type semiconductors are reversed, the FET will operate in much the same way, except that the channel conducts via holes, rather than electrons. Such a device is called a *P-channel FET*. The voltage at the gate electrode must be positive in order to narrow the channel. The drain is normally made negative, with respect to the grounded source in this type of FET.

There are numerous other kinds of field-effect transistors. The type shown here is sometimes called a *junction field-effect transistor (JFET)*. The other common type of field-effect transistor is the *metal-oxide semiconductor FET (MOSFET)*.

Some field-effect transistors operate by enhancement, rather than depletion, of the channel. These are called *enhancement-mode devices*; the example shown here is of a depletion-mode FET (*see* DEPLETION MODE, and ENHANCEMENT MODE).

All field-effect transistors are characterized by high input impedance. The gate draws almost no current, and this makes the FET useful in weak-signal amplifier applications. Some field-effect transistors are useful as power amplifiers. *See also* METAL-OXIDE SEMICONDUCTOR FIELD-EFFECT TRANSISTOR, and VERTICAL METAL-OXIDE SEMICONDUCTOR FIELD-EFFECT TRANSISTOR.

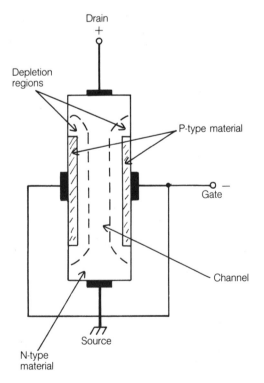

FIELD-EFFECT TRANSISTOR: A simplified cross-sectional diagram of a junction FET.

FIELD STRENGTH

Field strength is a measure of the intensity of an electric, electromagnetic, or magnetic field. The strength of an electric field is measured in volts per meter. The strength of a magnetic field is measured in gauss (*see* GAUSS). The intensity of an electromagnetic field is generally measured in volts per meter, as registered by a field-strength device. The intensity of an electromagnetic field can also be measured in watts per square meter.

The electromagnetic field strength from a transmitting antenna, as measured in volts, millivolts, or microvolts per meter, is proportional to the current in the antenna and to the effective length of the antenna. The field strength is inversely proportional to the wavelength and distance from the antenna. The field strength in watts, milliwatts, or microwatts per square meter varies with the power applied to the antenna and inversely with the square of the distance. The field strength always depends on the direction from the antenna as well, and is influenced by such things as phasing systems and parasitic elements. *See also* ELECTROMAGNETIC FIELD.

FIELD-STRENGTH METER

A *field-strength meter* is a device that is designed to measure the intensity of an electromagnetic field. Such a meter can be extremely simple and broadbanded, or highly complex, incorporating amplifiers and tuned circuits. The field-strength meter usually provides an indication in volts, millivolts, or microvolts per meter.

The simplest type of field-strength meter consists of a microammeter, a semiconductor diode, and a short length of wire that serves as the pickup. Such a device is uncalibrated, but is useful for getting an idea of the level of radio-frequency energy in a particular location. This kind of field-strength meter might, for example, be used to test a transmission line in an amateur-radio station.

A more sophisticated field-strength meter can use an amplification circuit to measure weak fields. Generally, the amplifier uses a tuned circuit to avoid confusion resulting from signals on frequencies other than that desired. The drawing illustrates a simplified schematic diagram of such a field-strength meter.

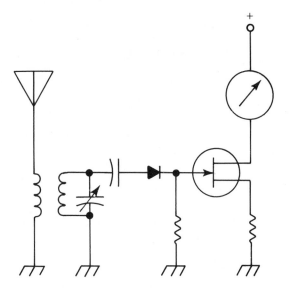

FIELD-STRENGTH METER: A simple circuit that uses a diode and an FET.

The most complex field-strength meters are built into radio receivers. Accurately calibrated S meters can serve as field-strength meters in advanced receivers. Most S meters in common receivers are calibrated, but are not precise enough for actual measurements of electromagnetic field strength. The precision field-strength meter is used in antenna testing and design, primarily for measurements of gain and efficiency. *See also* S METER.

FIFO

See FIRST-IN/FIRST-OUT

FILAMENT

A *filament* is a thin piece of wire that emits heat and/or light when an electric current is passed through it. In an incandescent bulb, the filament glows white hot, and is intended mainly to produce light. In a vacuum tube, the filament is intended to generate heat, which drives electrons off the cathode.

Filaments in vacuum tubes take two basic forms. The directly heated cathode consists of a filament, through which a current is passed, and this filament also serves as the cathode, connected to a negative source of voltage for tube operation. The directly heated tube has a thin, cylindrical cathode around the filament; the filament heats the cathode, but is not directly connected to it.

Tube filaments can require as little as 1 volt for their operation. Some filaments require hundreds of volts. In small tubes, the filaments can hardly be seen to glow while the device is operating. In large power-amplifier tubes, the filaments glow as brightly as small incandescent bulbs. *See also* CATHODE, and TUBE.

FILAMENT CHOKE

A *filament choke* is a heavy-duty, radio-frequency choke, used in the filament leads of vacuum-tube power amplifiers. Such chokes are necessary in the directly heated, grounded-grid type amplifier, which receives its drive at the cathode circuit. Without the filament chokes in such a circuit, the RF driving power would be short-circuited to ground.

Filament chokes must be capable of carrying large currents because they are connected in series with the filaments of large tubes. A filament-choke pair can be wound on a single core, which is made of powdered iron or ferrite. The reactance of the filament choke must be large enough to prevent significant drive power from being short-circuited to ground. *See also* FILAMENT, and COMMON GRID/BASE/GATE.

FILAMENT SATURATION

In a vacuum tube, given a constant level of plate voltage and a constant grid bias, the plate current increases as the filament voltage increases. This is because the higher the filament voltage gets, the more heat is developed, and the better the filament acts to drive electrons from the cathode.

There is a limit, however, to the plate current as the filament voltage is increased. Beyond a certain point, further increases in the filament voltage will not result in any increase in the plate current. When the filament is operated at this point, the condition is called *filament saturation*. It is pointless to have a filament voltage larger than the saturation value; this will unnecessarily heat the filament and shorten its life.

The filament-saturation voltage depends, to a certain extent, on the grid bias and the plate voltage. When a tube is cut off because of large negative grid bias, no amount of filament voltage will result in plate current. The same is true, of course, if the plate voltage is zero. If the plate voltage becomes large enough, the tube will begin to conduct even in the absence of voltage at the filament. *See also* FILAMENT, and TUBE.

FILAMENT TRANSFORMER

In a vacuum-tube circuit, the *filament transformer* provides the filament or filaments with the needed voltage. Most tubes use either 6 or 12 volts for the operation of their filaments; this voltage need not usually be rectified or filtered. Some power-supply transformers have several secondary windings, some of which are designed for the filaments.

The secondary winding or windings of a filament transformer must be capable of handling the current drawn by all of the tubes in a circuit. The tube filaments are connected in parallel; consequently, the current requirement is proportional to the number of tubes used with a given winding. If too many tubes are connected across a given filament transformer, the transformer will overheat. This will result in reduced voltage and current for all of the filaments, and can cause damage to the transformer itself.

Filament transformers and power supplies are not needed in solid-state devices, and thus such transformers are rarely seen today, except in large power-amplifier circuits. *See also* FILAMENT, and TUBE.

FILE

A *file* is a data store. All of the elements in a file are generally related, according to format or application. Data files can be modified easily; it is a simple matter to add, delete, or change a file.

An example of a file is a section of text in a word-processing device. The file is stored on a magnetic disk. There is a limit to how large the file can be; it can hold information only up to a certain maximum. Anything consisting of the standard alphabetic, numeric, and punctuation symbols, along with spaces in any arrangement, can be placed in the file by the operator of the system. The file can be deleted, changed, or appended at any time. The file is given a name, and this name is used to retrieve the file and to distinguish it from others on the same disk or tape.

In a larger sense, a file set is a collection of data, consisting of smaller, individual files. For example, a file can be kept of tax-deductible expenditures for a given year; the combination of all the annual files, over a long period of time, is the file set. *See also* WORD PROCESSING.

FILE TRANSFER PROTOCOL

In packet radio, an operator might want to work with the files in a computer at some distant node. In order to do this, a *file transfer protocol (FTP)* is used.

There are three different modes, or security levels, in FTP. The lowest level is called *read only*. This allows you to see the contents of a distant file, but not to change the file or create any new files. The second level is *read-only, write-new*. In this mode, you cannot change existing files, but you can add new ones at the distant node. The highest level is *read-only, write-new,*

overwrite-old. This allows full manipulation of files at the distant node. *See also* PACKET RADIO.

FILTER

A *filter* is a passive or active circuit designed primarily for the purpose of modifying a signal or source of power. A *passive filter* requires no power for its operation, and always has a certain amount of loss. An *active filter* requires its own power supply, but can provide gain.

In a power supply, the filter is the network of capacitors, resistors, and inductors that eliminates the fluctuations in the direct current from the rectifier. A good power supply filter will produce a smooth direct-current output under variable load conditions, with good regulation.

Filters are used in communications practice to eliminate energy at some frequencies, while allowing energy at other frequencies to pass with little or no attenuation. Such filters, if passive, are constructed using capacitors, inductors, and sometimes resistors. Active filters generally utilize operational amplifiers (*see* OPERATIONAL AMPLIFIER). Mechanical filters use the resonant properties of certain substances. Crystal and ceramic filters use piezoelectric materials to obtain the desired frequency response.

Filters can be categorized, in terms of their frequency-response characteristics, into four groups. The *bandpass filter* allows signals between two predetermined frequencies to pass, but attenuates all other frequencies. The *band-rejection (band-stop) filter* eliminates all energy between two frequencies, and allows signals outside the limit frequencies to pass. The *high-pass filter* allows only the signals above a certain frequency to pass. The *lowpass filter* allows only the signals below a certain frequency to pass. The actual response characteristics of all four types of filter are highly variable, and depend on many different factors. *Selective filters* are used in both transmitting and receiving applications. Some are designed for use at radio frequencies, while others operate at audio frequencies. *See also* ACTIVE FILTER, BANDPASS FILTER, BAND-REJECTION FILTER, CERAMIC FILTER, CRYSTAL-LATTICE FILTER, FILTER ATTENUATION, FILTER CAPACITOR, FILTER CUTOFF, FILTER PASSBAND, FILTER STOPBAND, HIGH-PASS FILTER, LOWPASS FILTER, MECHANICAL FILTER, and POWER SUPPLY.

FILTER ATTENUATION

Filter attenuation is the loss caused by a selective filter at a certain frequency. If the specified frequency is within the passband of the filter, the attenuation is known as *insertion loss*.

Outside the passband of a selective filter, the attenuation depends on many things, including the distance of the frequency from the passband and the sharpness of the filter response. The ultimate attenuation of the filter is the greatest amount of attenuation obtained at any frequency. This normally occurs well outside the passband of the filter.

Filter attenuation is specified in decibels. If E is the root-mean-square signal voltage at a given frequency with the filter, and e_o is the root-mean-square signal voltage with the filter short-circuited, then the attenuation of the filter at that frequency is:

$$Attenuation \ (dB) = 20 \ \log_{10} \ (e_o/E)$$

When an active filter is used, the filter can produce gain. In that instance, the attenuation is negative. The filter gain is given by:

$$Gain \text{ (dB)} = 20 \log_{10} (E/e_o)$$

See also FILTER.

FILTER CAPACITOR

A *filter capacitor* is used in a power supply to smooth out the ripples in the direct-current output of the rectifier circuit. Such a capacitor is usually quite large in value, ranging from a few microfarads in high-voltage, low-current power supplies to several thousand microfarads in low-voltage, high-current power supplies. Filter capacitors are often used in conjunction with other components such as inductors and resistors.

The filter capacitor operates because it holds the charge from the output-voltage peaks of the power supply. The smaller the load resistance, the greater the amount of capacitance required in order to make this happen to a sufficient extent.

Filter capacitors in high-voltage power supplies can hold their charge even after the equipment has been shut off. Resistors of a fairly large value should be placed in parallel with the filter capacitors in such a supply so that the shock hazard is reduced. Before servicing any high-voltage equipment, the filter capacitors should be discharged with a shorting stick several times. *See also* POWER SUPPLY.

FILTER CUTOFF

In a selective filter, the *cutoff frequency (filter cutoff)* is that frequency or frequencies at which the signal output voltage is 6 dB below the level in the passband. A bandpass or band-rejection filter normally has two cutoff frequencies. A highpass or lowpass filter has just one cutoff point. The filter cutoff frequency is an important characteristic in choosing a filter for a particular application, but it is not the only significant factor. The sharpness of the response is also important, as is the general shape of the response within the passband. *See also* BANDPASS FILTER, BAND-REJECTION FILTER, FILTER, HIGHPASS FILTER, and LOWPASS FILTER.

FILTER PASSBAND

The passband of a selective filter is the range of frequencies over which the attenuation is less than a certain value. Usually, this value is specified as 6 dB for voltage or current, and 3 dB for power.

The passband of a filter depends on the kind of filter. If L represents the lower filter cutoff (*see* FILTER CUTOFF) and U represents the upper cutoff, then the passband of a bandpass filter is given according to:

$$L < x < U$$

where x represents frequencies in the passband. For a band-rejection filter:

$$x < L \text{ or } x > U$$

That is, the passband consists of all frequencies outside the limit frequencies. For a highpass filter, if L is the cutoff, then:

$$x > L$$

and for a lowpass filter, if U represents the cutoff, then:

$$x < U$$

This means that, for the highpass filter, the passband consists of all frequencies above the cutoff; for a lowpass filter, the passband consists of all frequencies below the cutoff. *See also* BANDPASS FILTER, BAND-REJECTION FILTER, FILTER, HIGHPASS FILTER, and LOWPASS FILTER.

FILTER STOPBAND

The stopband of a selective filter consists of those frequencies not inside the filter passband. Generally, this means those frequencies for which the filter causes a voltage or current attenuation of 6 dB or more, or a power attenuation of 3 dB or more (*see* FILTER PASSBAND).

For a bandpass filter, the stopband consists of two groups of frequencies, one below or equal to the lower cutoff (L), and the other above or equal to the upper cutoff (U). For a band-rejection filter, the stopband consists of all frequencies between and including L and U. For a highpass filter, the stopband is that range of frequencies less than or equal to the cutoff. For a lowpass filter, the stopband is the range of frequencies higher than or equal to the cutoff. *See also* BANDPASS FILTER, BAND-REJECTION FILTER, FILTER, HIGHPASS FILTER, and LOWPASS FILTER.

FINAL AMPLIFIER

A *final amplifier* is the last stage of amplification in any system containing several amplifying circuits. The final amplifier drives the load. Generally, the term *final amplifier* is used in reference to radio-frequency transmitting equipment.

In a transmitter, the final amplifier is often tuned in the output circuit to allow precise adjustment of the resonant frequency and impedance-matching characteristics. However, in recent years, untuned or broadbanded final amplifiers have become increasingly common. These amplifiers are simple to use because they require little or no adjustment for proper operation; however, they cannot compensate for wide fluctuations in the load impedance.

An amplifier attached to the output of a transmitter for the purpose of increasing the power level is sometimes called a *final amplifier*. Such an amplifier usually has its own tuned circuits, and if the transmitter has a tuned output, this means that two sets of adjustments must be made.

The final-amplifier tube or transistor is itself often called the *final*. It is important that this device be operated according to its specifications. Otherwise, excessive harmonic output can occur, and distortion can be introduced into the modulation envelope. Finals are often operated quite close to their maximum power-dissipation limits, and in some cases it is necessary to observe a duty cycle of less than 100 percent. *See also* TRANSMITTER.

FINE TUNING

Fine tuning is the precise adjustment of the frequency of a radio transmitter or receiver. Fine tuning can be accomplished in many different ways, both electrical and mechanical. In a general-coverage communications receiver, the bandspread control is used for fine tuning. In some communications transceivers, a "clarifier" control accomplishes fine tuning.

Mechanical fine-tuning controls are not often seen in modern equipment, but can occasionally be encountered in older receivers. Such a fine-tuning control uses knobs, connected to the main-tuning shaft through gears, to obtain a spread-out control.

Electrical fine tuning can be obtained using small variable capacitors, inductors, or potentiometers. The fine-tuning control is normally connected in parallel with the main-tuning control, and has a much smaller minimum-to-maximum range.

In modern digital transmitters and receivers, fine-tuning controls are essentially obsolete. Each individual digit can be selected independently, and this obviates the need for a fine-tuning control. For example, to obtain a frequency of 14.2311 MHz, each of the six digits can be selected independently in some digital communications equipment. *See also* ELECTRICAL BANDSPREAD, and VERNIER.

FINNED SURFACE

In the design of heatsinks (*see* HEATSINK), it is desirable to maximize the surface area, from which heat can be carried away by convection and radiation. This has resulted in the use of finned surfaces.

The finned surface of a well-designed heatsink allows vertical air flow to enhance convection. Air warmed by the heatsink can rise upward through the spaces between the fins. Cool air can flow inward from the bottom.

The finned surface has a much larger surface area than a flat surface, and this increases the radiation of heat. Typical finned surfaces can have 10 times the surface area of a flat surface that occupies roughly the same amount of space.

FIRE PROTECTION

Wherever electrical or electronic equipment is used, the danger of fire exists. Electrical fires can start because the overheating of wires or components. Fires can also begin as the result of sparks or electrical arcing.

Prevention of fire is far preferable to dealing with a fire already started. Care should be exercised in the installation of electrical wiring, to be certain that the conductors can safely handle the current they will be required to carry (*see* CARRYING CAPACITY). Connections should be made so that the chances of sparking or short circuits are minimized. Wall outlets should not be overloaded; electrical circuits should be protected with conservatively rated fuses or breakers.

In the construction of electronic equipment, again, the wiring should be capable of handling the current. Components should be rated such that they will not overheat and catch fire; this is especially true of resistors. Electronic circuits should always be fused to protect them from damage in the event of a short circuit.

No matter what precautions are taken against fire, it is always possible that a fire will start. For electrical fires, special extinguishers are necessary. Baking-soda or carbon-dioxide types are best. Water extinguishers are not sufficient because the water can result in further short circuiting and can increase the chances of electric shock to personnel.

The proper fire-protection measures can save thousands or millions of dollars, and make electrical and electronic equipment safe to operate. Local fire departments are excellent sources of further information regarding fire protection. Standards for fire protection are established by the American National Standards Institute, Inc.

FIRING ANGLE

The *firing angle* is an expression of the phase angle at which a silicon-controlled rectifier fires (*see* SILICON-CONTROLLED RECTIFIER). The firing angle is measured in degrees or radians; it represents the point on the control-voltage cycle at which the device is activated.

In a magnetic amplifier, the firing angle is denoted by the lowercase Greek letter phi (ϕ). As the input-voltage vector rotates, the core of the magnetic amplifier is driven into saturation at a certain point (*see* MAGNETIC AMPLIFIER). This point, measured as a phase angle in degrees or radians, is called the *firing angle*. See also PHASE ANGLE.

FIRMWARE

Firmware is a form of microcomputer or computer programming, or software (*see* SOFTWARE). Firmware is programmed into a circuit permanently; however, to change the programming, it is necessary to replace one or more components in the system. The read-only memory (ROM), is an example of firmware.

In many different kinds of consumer and industrial devices, microcomputers are found in which firmware programming is used. Simple control systems are especially well suited to firmware programming. For example, a radio transceiver can have certain control functions performed by a microcomputer which has been programmed in a certain way. It is impossible to change the programming of the device because it uses firmware. *See also* COMPUTER, and MICROCOMPUTER.

FIRST AID

When a person has been injured or becomes suddenly ill, the immediate attention he or she needs is called *first aid*. Medical assistance is not always available soon enough. Proper administration of first aid can make the difference, in some situations, between life and death, or between short-term disability and long-term disability. Proper first aid can also reduce the duration of a hospital stay.

Personnel who work with electronic equipment run the risk of electric shock. Even comparatively low voltages can sometimes cause a serious shock; many low-voltage power supplies are operated from 120-volt alternating current sources. First aid for electric-shock victims must be directed at the possible stoppage of breathing and/or failure of the heart.

In the event of injury requiring first aid, proper training is invaluable. The American National Red Cross publishes a booklet about first aid; courses are widely available for training purposes. *See also* CARDIOPULMONARY RESUSCITATION, ELECTRIC SHOCK, and MOUTH-TO-MOUTH RESUSCITATION.

FIRST HARMONIC

Whenever frequency multipliers are used in a circuit, harmonics are produced. The input signal to a frequency multiplier is called the *first harmonic*. Higher-order harmonics appear at integral multiples of the first harmonic.

A crystal calibrator, often used in communications receivers, provides a good example of deliberate harmonic generation. A 1-MHz crystal can be used in the oscillator circuit. A nonlinear device follows, resulting in harmonic output; the second harmonic is 2 MHz, the third harmonic is 3 MHz, and so on. The 1-MHz output signal is the first harmonic.

Some crystal calibrators use frequency dividers. For example, a divide-by-10 circuit can be switched in between the oscillator and the nonlinear device in the above example, producing markers at multiples of 100 kHz. The first harmonic of the divider then becomes 100 kHz, the second harmonic becomes 200 kHz, and so on. The first harmonic can be identified because all of the other signals are integral frequency multiples of it. This is true only of the first harmonic.

The first harmonic is sometimes called the *fundamental frequency*. In circuits where harmonic output is not desired, the term *first harmonic* is generally not used. *See also* FREQUENCY MULTIPLIER, FUNDAMENTAL FREQUENCY, and HARMONIC.

FIRST-IN/FIRST-OUT

A *first-in/first-out circuit (FIFO)* is a form of read-write memory. The buffer circuits used in such devices as electronic typewriters, word processors, and computer terminals are examples of first-in/first-out memory stores.

The operation of a first-in/first-out buffer is evident from its name. The illustration shows an FIFO circuit with eight characters of storage. If certain characters are fed into the input at an irregular rate of speed, the buffer eliminates some of the irregularity without changing the order in which the characters are transmitted. Suppose, for example, that the operator of a teletype machine can type 60 words per minute, but not at a smooth, constant rate of speed. If the system speed is 60 words per minute, a FIFO buffer of sufficient capacity can be used to produce continuous, regular output characters. The FIFO circuit transmits the characters in exactly the same order in which they are received.

Not all buffers operate on the FIFO principle. Sometimes it is desirable to have a first-in/last-out buffer; as its name implies, this type of buffer inverts the order of the characters it receives. *See also* MEMORY.

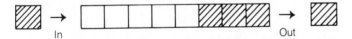

FIRST-IN/FIRST-OUT: Data output is in the same sequence as data input.

FISHBONE ANTENNA

A *fishbone antenna* is a form of wideband antenna that has a unidirectional end-fire response. The antenna gets its name from its physical appearance. The fishbone antenna is used primarily for receiving applications at low, medium, and high frequencies.

Several collector antennas are coupled by means of capacitors to a transmission line. The spacing of the antennas is not critical, although they are normally separated by about 0.1 to 1 wavelength. The terminating resistor provides a unidirectional response; if the resistor is removed, the response becomes bidirectional, along the direction in which the feed line runs.

The fishbone antenna is not rotatable, but it can be installed fairly close to the ground. The antenna is also sometimes called a *herringbone antenna*. Such an antenna can provide greatly improved reception, when compared with a dipole or ground-plane antenna, under conditions of severe interference.

FIXED BIAS

When the bias at the base, gate, or grid of an amplifying or oscillating transistor, field-effect transistor, or tube is unchanging with variations in the input signal, the bias is called *fixed bias*. Fixed bias can be supplied by resistive voltage-divider networks, or by an independent power supply. Fixed bias is often used in amplifying circuits for both audio and radio frequencies.

The schematics illustrate two ways of getting fixed bias with a bipolar-transistor amplifier. At A, a voltage divider is shown. The values of the resistors determine the voltage at the base of the transistor. At B, an independent power supply is used. The circuit at A is a class-A amplifier because the base is biased to draw current even in the absence of signal; the resistors should be chosen for operation in the middle of the linear part of the collector-current curve. At B, a class-C amplifier is shown; the independent power supply is used to bias the transistor beyond the cutoff point.

When the bias is not fixed, the characteristics of an amplifier will change, along with whatever parameter is responsible for the changes in bias. An automatic-level-control circuit, for example, uses variable bias to change the gain of an amplifier. *See also* AUTOMATIC BIAS, and AUTOMATIC LEVEL CONTROL.

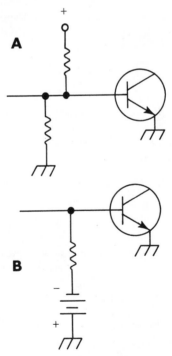

FIXED BIAS: A voltage divider (A) and a separate battery (B).

FIXED FREQUENCY

A *fixed-frequency device* is an oscillator, receiver, or transmitter that is designed to operate on only one frequency. Such devices are usually crystal-controlled (*see* CRYSTAL CONTROL).

Fixed-frequency communication offers the advantage of instant contact; no search is necessary at the receiving station in order to locate the frequency of the transmitter. However, if this advantage is to be realized, the frequencies must be accurately matched. The slightest error can cause reduction in communications efficiency, or even total system failure. To maintain the operating frequency, a standard source must be used, such as WWV or WWVH. Phase-locked-loop circuits can, in conjunction with such frequency standards, keep the transmitter and receiver frequencies matched to a high degree of precision. *See also* PHASE-LOCKED LOOP, VARIABLE FREQUENCY, and WWV/WWVH/WWVB.

FIXED RADIO SERVICE

The *Fixed Radio Service* is a communications network operated by the U.S. Government. Similar services exist in many other countries. The Fixed Service is intended, as its name implies, for communication between fixed locations. Such links include circuits at all frequencies, from the very-low to the ultra-high and microwave ranges.

The Fixed Service is assigned many different frequency bands on a world-wide basis. Some bands are assigned differently in various countries. Power limitations vary, depending on the frequency band and the country. In the United States, the frequency allocations are administered by the Federal Communications Commission. *See also* FEDERAL COMMUNICATIONS COMMISSION.

FLAG FIELD

In packet radio, a *flag field* is used at the start and end of a frame. Its sole purpose is to indicate beginnings and endings of frames. Sometimes, a single flag field is used to specify the end of one frame and the start of the next.

A flag field is always one octet in length, and has the code 01111110. *See also* FRAME, OCTET, and PACKET RADIO.

FLAGPOLE ANTENNA

A *flagpole antenna* is a vertical antenna in which the ground system is either underground or oriented directly downward. Such an antenna gets its name because it has the appearance of a flagpole; it is simply a vertical rod without appendages. A coaxial antenna is an example of a flagpole antenna (*see* COAXIAL ANTENNA). Flagpole antennas are characterized by their unobtrusiveness. The polarization is vertical, and the radiation pattern is omnidirectional in the horizontal plane. Flagpole antennas can be inductively loaded for use at medium and high frequencies, while maintaining a height of only a few feet.

Some inventive radio amateurs and CB operators have used actual flagpoles as vertical antennas. There are many different ways of feeding a flagpole to make it work as a radiating element in the medium-frequency and high-frequency ranges. Probably the most common feed method is *shunt feed*. The feed point is connected part of the way from the base of the vertical mast to the top. Another method is to insulate the base of the flagpole from the ground, and feed it at that point. *See also* SHUNT FEED.

FLASHOVER

Flashover is a term used to describe the arcing that occurs when the voltage across an insulator becomes excessive. Flashover is undesirable because it results in conduction over the ionized path of the arc, where conduction is not wanted. Flashover can cause fire. It can also cause permanent damage to the insulating material.

Flashover can occur in radio-frequency circuits when a severe impedance mismatch exists. If the mismatch is of enough magnitude, extreme voltages can appear at various points along a transmission line, or across feed insulators. High voltages can also appear across switch contacts. The resulting flashover causes a change in the load impedance as seen by the transmitter; this can cause damage to the final amplifier of the transmitter. *See also* ARC, and FLASHOVER VOLTAGE.

FLASHOVER VOLTAGE

The *flashover voltage* is the minimum voltage, under certain conditions, at which flashover or arcing occurs. The flashover voltage in a given situation depends on several things, including the space between the electrodes that carry opposite polarity, the dielectric substance separating the electrodes, the temperature and, in some instances, the humidity.

Dielectric materials should not be subjected to voltages too near the flashover voltage. Flashover can result in permanent damage to some materials. In other cases, the flashover voltage becomes much lower once an ionized path has formed in the insulating material. *See also* ARC, and FLASHOVER.

FLAT PACK

A *flat pack* is a form of housing for integrated circuits. The flat pack is characterized by physical thinness; the pins protrude straight outward from the main package, in the plane of the housing. The photograph shows a typical example of a flat-pack integrated circuit. The entire package is smaller than a penny.

Flat-pack integrated circuits are used in situations where space is restricted. The devices are normally soldered directly to the circuit boards. Extreme care is necessary in soldering and desoldering flat-pack devices because they are easily damaged, and solder bridges can form between adjacent pins because of the small spacing. The integrated-circuit flat pack shown contains a microcomputer. *See also* INTEGRATED CIRCUIT.

FLAT PACK: Used in microminiature circuits.

FLAT RESPONSE

When a transducer or filter displays uniform gain or attenuation over a wide range of frequencies, the device is said to have a *flat response*. Normally, the response is considered flat if the gain or attenuation is more or less constant throughout the operating range. At frequencies above or below the operating range, the response is not important.

A flat response is desirable for speakers, headphones, and microphones in high-fidelity equipment. Such devices should have uniform response throughout the audio-frequency spectrum, or from about 20 Hz to 20 kHz. In communications practice, the response is normally flat over a much smaller range of frequencies. The illustration shows a flat response, such as is found in high-fidelity transducers, and a narrower response, such as is typical of communications equipment.

Controls in a high-fidelity recording or reproducing system can be used to tailor the response to the liking of a particular listener. In the recording process, the amplifier response is not always flat; the gain can be greater at some frequencies than at others. This is corrected in the playback process. *See also* FREQUENCY RESPONSE.

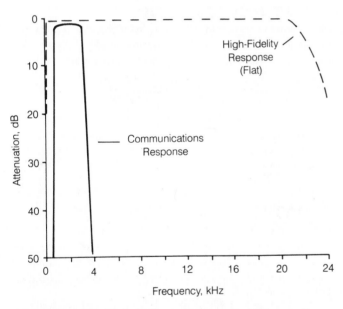

FLAT RESPONSE: High-fidelity versus communications audio responses.

FLAT-TOP ANTENNA

Any antenna system in which the radiating element is horizontal can be called a *flat-top antenna*. Examples of flat-top antennas are the dipole antenna, the double-zepp antenna, and the extended double-zepp antenna. *See* DIPOLE ANTENNA, DOUBLE-ZEPP ANTENNA, and EXTENDED DOUBLE-ZEPP ANTENNA.

A vertical antenna can be connected to a horizontal radiating wire for the purpose of lowering the resonant frequency. The resulting antenna is sometimes called a *flat-top antenna*. Most of the radiation in such an antenna occurs from the vertical portion. The top section provides additional capacitance, and carries a relatively high voltage. The flat-top antenna radiates a signal that has predominantly vertical polarization.

FLAT TOPPING

Flat topping is a form of distortion that sometimes occurs on an audio-frequency waveform or a modulation envelope. Flat topping occurs because of a severe nonlinearity in an amplifying circuit. Flat topping is undesirable because it results in the generation of harmonic energy, and degrades the quality of a signal.

Flat topping in a modulation envelope of a single-sideband transmitter is shown in the illustration. The peaks of the signal are cut off, and appear flat. Flat topping on a single-sideband or amplitude-modulated signal results in excessive bandwidth because of the harmonic distortion. This can cause interference to stations on frequencies near the distorted signal. This type of distortion is sometimes called *splatter*.

Flat topping is usually caused by improper bias in an amplifying stage, or by excessive drive. If the bias is incorrect, a tube or transistor can saturate easily. Saturation can also be caused by too much driving voltage or power. *See also* DISTORTION.

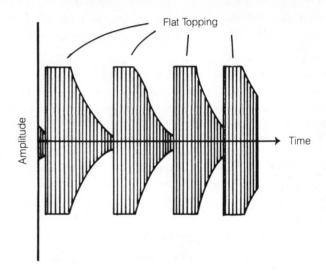

FLAT TOPPING: Modulation peaks are clipped.

F LAYER

The *F layer* is a region of ionization in the upper atmosphere of the earth. The altitude of the F layer ranges from about 100 to 260 miles, or 160 to 420 kilometers. The F layer of the ionosphere is responsible for most long-distance radio propagation at medium and high frequencies.

The F layer actually consists of two separate ionized regions during the daylight hours. The lower layer is called the *F1 layer* and the upper layer is the *F2 region*. During the night, the F1 layer usually disappears. The F1 layer rarely affects radio signals, although under certain conditions it can intercept and return them.

The F layer attains its maximum ionization during the afternoon hours. But the effect of the daily cycle is not as pronounced in the F layer as in the D and E layers. Atoms in the F layer remain ionized for a longer time after sunset. During times of maximum sunspot activity, the F layer often remains ionized all night long.

Because the F layer is the highest of the ionized regions in the atmosphere, the propagation distance is longer via F-layer circuits than via E-layer circuits. The single-hop distance for signals returned by the F2 layer is about 500 miles for a radiation angle of 45 degrees. For a radiation angle of 10 degrees, the single-hop distance can be as great as 2,000 miles. For horizontal waves, the single-hop F2-layer distance is from 2,500 to 3,000 miles. For signals to propagate over longer distances, two or more hops are necessary.

The maximum frequency at which the F layer will return signals depends on the level of sunspot activity. During sunspot maxima, the F layer can occasionally return signals at frequencies as high as about 80 to 100 MHz. During the sunspot minimum, the maximum usable frequency can drop to less than 10 MHz. *See also* D LAYER, E LAYER, IONOSPHERE, and PROPAGATION CHARACTERISTICS.

FLEMING'S RULES

Fleming's rules are simple means of remembering the relationship among electric fields, magnetic fields, voltages, currents, and forces. If the fingers of the right hand are curled and the thumb is pointed outward, then a current flowing in the direction of the thumb will cause a magnetic field to flow in a circle, the sense of which is indicated by the fingers. This is called the *right-hand rule* for the magnetic flux generated by an electric current.

If the thumb, first finger, and second finger are oriented at right angles to each other, then the right hand will show the relationship among the direction of wire motion, the magnetic field, and the current in an electric generator, and the left hand will indicate the relationship among the direction of wire motion, the magnetic field, and the current in an electric motor. The thumb shows the direction of wire motion. The index finger shows the direction of the magnetic field from north to south. The middle finger shows the direction of electric current from positive to negative. *See also* GENERATOR, and MOTOR.

FLEXIBLE COUPLING

A *flexible coupling* is a mechanical coupling between two rotary shafts. The flexibility allows the shafts to be positioned so that they are not perfectly aligned with each other, while still transferring rotation from one shaft to the other. Flexible couplings are used in a variety of different situations.

Flexible couplings can be made of metal springs or rubber or plastic tubing. Sometimes, special gear-drive systems are used. *See also* BUSHING.

FLEXIBLE WAVEGUIDE

A *flexible waveguide* is a section of waveguide used to join rigid waveguides that are not in precise orientation with respect to each other. Flexible-waveguide sections make the installation of a waveguide transmission line much easier because exact positioning is not required. Flexible waveguides also allow for the expansion and contraction of rigid waveguides with changes in temperature.

Flexible-waveguide sections are made in a variety of ways. Metal ribbons can be joined together. Metal foil or tubing can be used. Some flexible waveguides can be twisted as well as bent or stretched. If a section of flexible waveguide is installed between two sections of rigid waveguide, the impedances should be matched. That is, the flexible section should show the same characteristic impedance as the rigid waveguide. The joints themselves must be made properly, as well, to prevent the formation of impedance "bumps." *See also* WAVEGUIDE.

FLIP-FLOP

A *flip-flop* is a simple electronic circuit with two stable states. The circuit is changed from one state to the other by a pulse or other signal. The flip-flop maintains its state indefinitely unless a change signal is received. There are several different kinds of flip-flop circuits.

The D-type flip-flop operates in a delayed manner, from the pulse immediately preceding the current pulse. The J-K flip-flop has two inputs, commonly called the *J* and *K* inputs. If the J input receives a high pulse, the output is set to the high state; if the K input receives a high pulse, the output is set low. If both inputs receive high pulses, the output changes its state either from low to high or vice versa. The R-S flip-flop has two inputs, called the *R and S inputs*. A high pulse at the R input sets the output low; a high pulse on the S input sets the output high. The circuit is not affected by high pulses at both inputs. The R-S-T flip-flop has three inputs called *R, S,* and *T*. The R-S-T flip-flop operates exactly as the R-S flip-flop works, except that a high pulse at the T input causes the circuit to change states.

A T flip-flop has only one input. Each time a high pulse appears at the T input, the output state is reversed. Flip-flop circuits are interconnected to form the familiar logic gates, which in turn comprise all digital apparatus. *See also* LOGIC GATE.

FLOATING CONTROL

Most control components, such as switches, variable capacitors, and potentiometers, have grounded shafts. This is convenient from an installation standpoint because the shafts of controls are normally fed through a metal front panel. However, there are certain instances in which it is not possible to ground the shaft. The control is then said to be floating because it is grounded at no point.

Floating controls are more subject to the effects of body capacitance than grounded controls. Therefore, floating controls should be used only when there is no alternative. *See also* BODY CAPACITANCE.

FLOATING PARAPHASE INVERTER

A *floating paraphase inverter* is a circuit devised for inverting the phase of a signal. Two transistors, field-effect transistors, or vacuum tubes are used in the circuit. The schematic diagram shows a floating paraphase inverter.

A signal applied to the input is inverted in phase at output X, because of the phase-reversing effect of the bipolar transistor amplifier Q1. A portion of the signal from output X is fed to the base of transistor Q2. This signal is obtained by means of a resistive network, so that it is equal in amplitude to the input signal. Amplifier Q2 inverts the phase of this signal, delivering it to output Y. Consequently, the waves at the two outputs are out of phase, and equally elevated above the common terminal. Either output can be used to obtain signals in phase opposition, or they can be combined to form a floating pair of terminals. *See also* PARAPHASE INVERTER.

FLOATING POINT

In an electronic circuit, a point is called *floating* if it is ungrounded and not directly connected to the power supply terminal. In the illustration for FLOATING PARAPHASE INVERTER, for example, the output terminals X and Y are floating.

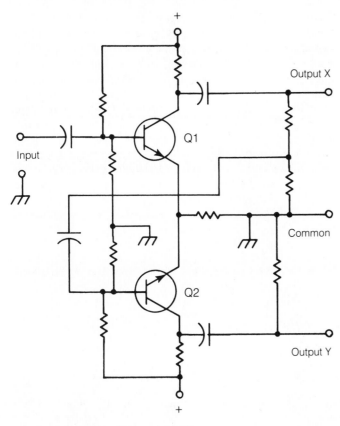

FLOATING PARAPHASE INVERTER: Outputs X and Y are inverted, with respect to each other.

FLOOD GUN

A *flood gun* is an electron gun that is used in a storage cathode-ray tube. The flood gun operates in conjunction with another electron gun, called the *writing gun*, to store a display for a period of time (*see* ELECTRON-BEAM GENERATOR). The writing gun emits a thin beam of electrons, which strikes the phosphor screen and forms the image. The beam is modulated by means of deflecting plates, exactly as in the conventional oscilloscope. A storage mesh holds the image. The flood gun is used to illuminate the phosphor screen to view the stored image. Low-energy electrons from the flood gun can pass through the mesh only where the image has been stored. When the image is to be erased, the flood gun is used to clear the mesh. *See also* OSCILLOSCOPE, STORAGE OSCILLOSCOPE, and WRITING GUN.

FLOPPY DISK

Practically all personal computers (PCs), of the kind often found in ham radio stations, use one or more magnetic data storage disks. Some PCs have an internal set of rigid disks called *hard disks*. But a PC always makes use of at least one *floppy disk*.

"Floppies" or "diskettes" are most commonly found in two sizes. The 5.25-inch floppy is encased in a flexible package, and the 3.5-inch floppy has a rigid case. Either type of disk can usually store a little more than one megabyte of data. Some store less (several hundred kilobytes) and a few can contain several megabytes. A typical floppy disk has enough text space for a medium-sized novel. A dozen 5.25-inch floppy disks, stacked on top of each other, approximate the physical size of such a book.

The disk itself is made of a flexible, durable synthetic resin, covered by a layer of fine-grained magnetic powder. It is similar to magnetic tape, except for its shape and the geometry of the way data is stored on it.

Floppy disks might have data stored on both sides; then they are called *double-sided*. Some disks have the capacity for storage on only one side, and are known as *single-sided floppies*. Some disks have an extra fine grain of magnetic material and can store more data per unit surface area than others. The finer-grained floppies are called *double-density diskettes*.

Information is written onto, and retrieved from, a floppy disk in a pattern similar to that of a phonograph disk. There are circular *tracks* and pie-shaped *sectors* (see the illustration). Generally, there are 96 tracks per inch in a floppy disk.

To read from, and/or to write on, a floppy disk, the disk is placed in a *disk drive*. A personal computer usually has at least one floppy drive. More expensive PCs have an internal hard drive and a floppy drive. The most sophisticated and versatile PCs have two floppy drives in conjunction with an internal hard drive.

Floppy disks have largely replaced magnetic tape for storing computer data. One advantage of the diskette is that no two bits of memory are ever farther apart than the diameter of the diskette. On tape, two data bits could be separated by the whole length of the tape. Data can always be accessed from, or written onto, a floppy within a few seconds. But on tape, it can take a long time to get at certain parts of the stored information. Magnetic tape is still sometimes used for storage of data when it is not necessary to change it very often. *See also* DISK DRIVE, MAGNETIC RECORDING, and MAGNETIC TAPE.

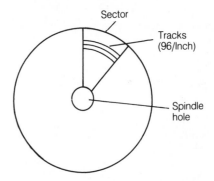

FLOPPY DISK: A view of a 5-1/4-inch diskette with the case removed.

FLOWCHART

A *flowchart* is a diagram that depicts a logical algorithm or sequence of steps. The flowchart looks very much like a block diagram. Boxes indicate conditions, and arrows show procedural steps. Flowcharts are often used to develop computer programs. Such charts can also be used to document trouble-shooting processes for all kinds of electronic equipment.

The symbology used in flowcharts is not well standardized. An example of a simple troubleshooting flowchart is shown in the illustration. Decision steps are indicated by diamonds. Various conditions are indicated by boxes.

A flowchart must represent a complete logical process. No matter what the combination of conditions and decisions, a conclusion must always be reached. Sometimes the conclusion consists simply of an instruction to return to an earlier stage in the algorithm, but infinite loops should not exist.

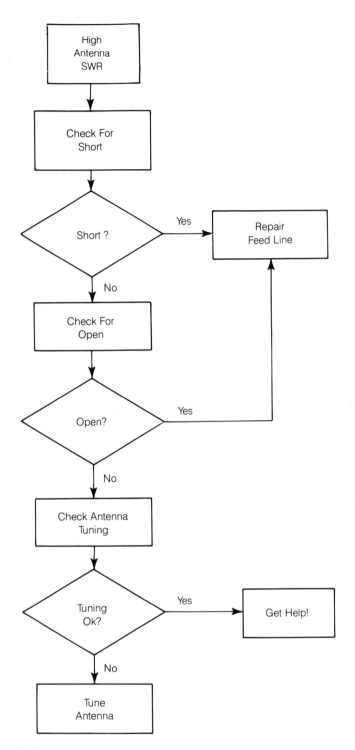

FLOWCHART: This example depicts repair procedure for an antenna with excessive SWR.

FLUCTUATING LOAD

The resistance or impedance of a circuit load might not remain constant. Some loads maintain an essentially constant impedance as a function of time; an example of such a load is the antenna system in a radio-frequency transmitting station. Some loads show variable resistance or impedance. An example of this is an ordinary household electrical system. The load impedance of such a circuit depends on the number and kinds of appliances that are used.

A generator subjected to a fluctuating load must be capable of delivering the proper voltages and currents to the load, no matter what the state of the load at a given instant. The generator must be able to respond rapidly to changes in the load impedance. Under certain extreme conditions, the generator might not be capable of dealing with a change in load resistance; this sometimes happens, for example, when all the residents of a large city operate air conditioners at the same time.

The load impedance that a transmitter oscillator circuit "sees," at the input of the following amplifier stage, often fluctuates with the keying of the oscillator. A buffer stage between the oscillator and the first amplifier is used to reduce the magnitude of the load fluctuations. Oscillators tend to change frequency when the load impedance changes; the buffer stage therefore improves the stability of the oscillator. *See also* BUFFER STAGE, and IMPEDANCE.

FLUTTER

Flutter is the term used to describe a warbling, or change in pitch, of the sound in an audio recording/reproducing system. Slow flutter is sometimes called *wow*, because of the way it sounds. Flutter can occur in any recording or reproducing process in which the vibrations are stored in a fixed medium.

When the local oscillator in a superheterodyne receiver is unstable, the received signal appears to fluctuate in frequency. Sometimes the oscillator is affected by mechanical vibration, resulting in rapid and dramatic changes in frequency. The sound of the beat note resulting from this instability is called *flutter*.

The flutter in a recording/reproducing system, as well as that in a superheterodyne receiver, can result from engineering deficiencies, or they can take place because a recording and reproducing device is used for an application other than that intended. Flutter problems can be corrected, if necessary, by replacing the faulty component or device with one having greater precision.

FLUX

Flux is a measure of the intensity of an electric, electromagnetic, or magnetic field. Any field has an orientation, or direction, which can be denoted by imaginary "lines of flux." In an electric field, the lines of flux extend from one charge center to the other in various paths through space. In a magnetic field, the lines extend between the two poles. In an electromagnetic field, the electric lines of flux are generally denoted. *See* ELECTRIC FIELD, ELECTRIC FLUX, ELECTROMAGNETIC FIELD, MAGNETIC FIELD, and MAGNETIC FLUX.

The flux lines of any field have a certain concentration per unit of surface area through which they pass at a right angle. This field intensity varies with the strength of the charges and/or magnetic poles, and with the distance from the poles. The greatest concentration of flux is near either pole, and at points lying on a line connecting the two poles. For radiant energy (such as infrared, visible light, and ultraviolet), the flux is considered to be the number of photons that strike an orthogonal surface per unit time. Alternatively, the flux can be expressed in power per unit area, such as watts per square meter. *See also* FLUX DENSITY, and FLUX DISTRIBUTION.

FLUX DENSITY

The *flux density* of a field is an expression of its intensity. Generally, flux density is specified for magnetic fields. The standard unit of magnetic flux density is the tesla, which is equivalent to 1 volt second (weber) per square meter. Magnetic flux density is abbreviated by the letter B.

Another unit commonly used to express magnetic flux density is the gauss. A flux density of 1 gauss is considered to represent a line of force per square centimeter. The "lines" are imaginary, and represent an arbitrary concentration of magnetic field. There are 10,000 gauss per tesla. *See also* FLUX, GAUSS, MAGNETIC FIELD, MAGNETIC FLUX. TESLA, and WEBER.

FLUX DISTRIBUTION

Flux distribution is the general shape of an electric, electromagnetic, or magnetic field. The flux distribution can be graphically shown, for a particular field, by lines of force. The flux tends to be the most concentrated near the poles of an electric or magnetic dipole.

The flux distribution around a bar magnet can be vividly illustrated by the use of iron filings and a sheet of paper. The paper is placed over the bar magnet, and the filings are sprinkled onto the paper. The filings align themselves in such a way as to render the field lines visible. The general distribution of the magnetic field is characteristic.

The greatest electric or magnetic flux is always near the poles, and along a line connecting two poles. *See also* ELECTRIC FLUX, FLUX, and MAGNETIC FLUX.

FLUXMETER

A *fluxmeter* is a device for measuring the intensity of a magnetic field. Such a meter is also called a *gaussmeter*.

The fluxmeter generally consists of a wire coil, which can be moved back and forth across the flux lines of the magnetic field. This causes alternating currents to flow in the coil, and the resulting current is indicated by a meter. The greater the intensity of the magnetic field for a given coil speed, the greater the amplitude of the alternating currents in the coil. Knowing the rate of acceleration of the coil, and the amplitude of the currents induced in it, the flux density can be determined.

Fluxmeters are usually calibrated in gauss. A flux density of 1 gauss is the equivalent of a line of flux per square centimeter. *See also* FLUX, GAUSS, and MAGNETIC FLUX.

FLYBACK

Flyback is a term that denotes the rapid fall of a current or voltage that has been increasing for a period of time. The flyback time is practically instantaneous under ideal conditions; in practice, however, it is finite.

Sawtooth waves are used in such devices as oscilloscopes and television receivers to sweep the electron beam rapidly back after a line has been scanned. The flyback sweep is much more rapid than the forward sweep; the display is blanked during the flyback to prevent any possible interference with the image created by the forward sweep. In an oscilloscope or television picture tube, the flyback is from right to left, while the visible sweep is from left to right.

When a coil or capacitor is rapidly discharged, the voltage or current can be very large for a short time. This can cause electric shock, arcing, and possible damage to components of a circuit. This kind of flyback is called *kickback*. The spark generator in an automobile engine, and in any other internal combustion engine (except a diesel engine), operates on this principle. *See also* DISCHARGE, FLYBACK POWER SUPPLY, PICTURE SIGNAL, and PICTURE TUBE.

FLYBACK POWER SUPPLY

In a television picture receiver, the horizontal scanning is controlled by a device called a *flyback power supply*. It is also often called a *kickback power supply*. The flyback supply provides a high-voltage sawtooth wave, which is fed to the horizontal deflecting coils in the picture tube.

The output waveform of the flyback power supply is such that the deflection of the electron beam occurs rather slowly from left to right, but rapidly from right to left. The picture modulation is impressed on the electron beam during the left-to-right sweep. The beam is blanked out during the return sweep, to prevent interference with the picture. A similar power supply is used in cathode-ray oscilloscope, but the left-to-right (forward) scanning speed is variable, rather than constant. *See also* FLYBACK, PICTURE SIGNAL, and PICTURE TUBE.

FLYWHEEL EFFECT

In any tuned circuit containing inductance and capacitance, oscillations tend to continue at the resonant frequency of the circuit, even after the energy has been removed. The higher the Q factor, or selectivity, of the circuit, the longer the decay period. The Q factor increases as the amount of resistance in the tuned circuit decreases. This phenomenon, wherein resonant circuits tend to "ring," is called *flywheel effect*, and it occurs because the inductor and capacitor in a tuned circuit store energy.

The flywheel effect makes it possible to operate a class-AB, class-B, or class-C radio-frequency amplifier, with minimal distortion in the shape of the signal wave. The output waveforms of such amplifiers not having tuned output circuits would be nonsinusoidal. The class-B and class-C output waveforms would be greatly distorted, and would contain great amounts of harmonic energy. But, because of the flywheel effect, the tuned output circuit requires only brief pulses, occurring at the resonant frequency, to produce a nearly pure sine-wave output (see illustration).

The flywheel effect does not make an inherently nonlinear amplifier, such as a class-C amplifier, operate in linear fashion.

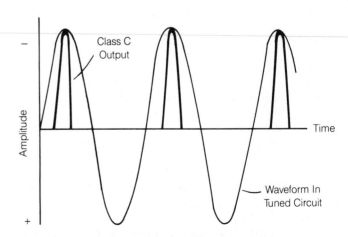

FLYWHEEL EFFECT: Completes the ac cycle in an amplifier output.

The modulation envelope of a single-sideband signal, for example, will be badly distorted if the signal is passed through a class-C amplifier circuit. The shape of the carrier wave is, however, maintained (even in a class-C amplifier) because of the flywheel effect of the tuned output circuit. *See also* CLASS-AB AMPLIFIER, CLASS-B AMPLIFIER, CLASS-C AMPLIFIER, *Q* FACTOR, and TUNED CIRCUIT.

FLYWHEEL TUNING

The tuning controls of some radio communications receivers use heavy weights to obtain a large amount of angular momentum. This weight can take the form of a thick metal disk attached to the shaft of the control. This is known as *flywheel tuning*.

Flywheel tuning offers some advantages. It creates a smooth "feel" to the tuning dial of a receiver. By spinning a dial with flywheel tuning, several revolutions of the knob can be rapidly executed; the dial just keeps on going because of the momentum. Many operators like flywheel tuning. However, in some situations, flywheel tuning is not desirable. In mobile or portable operation, the flywheel can cause misadjustment of the tuning control in the event of mechanical vibration. And, just as some operators like the "feel" of flywheel tuning, some operators dislike it. Flywheel tuning is most often found in older communications and short-wave broadcast receivers, especially those with slide-rule tuning dials.

FM

See FREQUENCY MODULATION.

FOCAL LENGTH

The *focal length* of a lens or reflector is the distance from its center to its focal point. Parallel rays of energy arriving at the lens or reflector from a great distance converge, and form an image, at the focal point. Conversely, a point source of energy at the focal point of a lens or reflector will produce parallel rays.

The focal length of a lens or reflector determines the size of the image resulting from impinging parallel rays. The greater the focal length, the larger the image. For a given source of energy, the greater the focal length, the more nearly parallel the transmitted rays. *See also* PARABOLOID ANTENNA.

FOCUS CONTROL

In a television receiver or cathode-ray oscilloscope, it is desirable to have the electron beam as narrow as possible. This is accomplished by means of focusing electrodes. The voltage applied to these electrodes controls the deflection of electrons so that they are brought to a point as they land on the phosphor screen.

Most television receivers have internal focusing potentiometers, and the controls are preset. Once properly adjusted, they seldom require further attention. In the oscilloscope, the focus control is usually located on the front panel.

Focusing can be accomplished by electromagnetic, as well as electrostatic deflection. Some cathode-ray tubes use coils, which produce magnetic fields, to direct and focus the electron beam. *See also* ELECTROMAGNETIC DEFLECTION, ELECTROMAGNETIC FOCUSING, ELECTROSTATIC DEFLECTION, and PICTURE TUBE.

FOLDBACK

See CURRENT LIMITING.

FOLDED DIPOLE ANTENNA

A *folded dipole antenna* is a half-wavelength, center-fed antenna constructed of parallel wires in which the outer ends are connected together. The folded-dipole antenna can be thought of as a "squashed" full-wave loop (see illustration).

The folded dipole has exactly the same gain and radiation pattern, in free space, as a dipole antenna. However, the feed-point impedance of the folded dipole is four times that of the ordinary dipole. Instead of approximately 73 ohms, the folded dipole presents a resistive impedance of almost 300 ohms. This makes the folded dipole desirable for use with high-impedance, parallel-wire transmission lines. It also can be used to obtain a good match with 75-ohm coaxial cable when four antennas are connected in phase, or with 50-ohm coaxial cable when six antennas are connected in phase.

Folded dipoles are often found in vertical collinear antennas, such as are used in repeaters at the very high frequencies. Folded dipoles have somewhat greater bandwidth than ordinary dipoles, and this makes them useful for reception in the frequency-modulation (FM) broadcast band, between 88 and 108 MHz. *See also* DIPOLE ANTENNA.

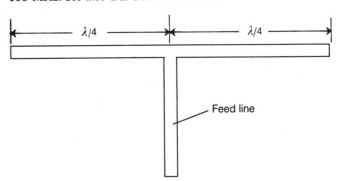

FOLDED DIPOLE ANTENNA: Feed impedance is four times that of a single-wire dipole.

FOLLOWER

A *follower* is a circuit in which the output signal is in phase with, or follows, the input signal. Follower circuits always have a voltage gain of less than 1. The output impedance is lower than the input impedance. Follower circuits are typically used as broadbanded impedance-matching circuits; they are often cheaper and more efficient than transformers.

The follower circuit is characterized by a grounded collector if a bipolar transistor is used or a grounded drain if a field-effect transistor is used. The bipolar circuit is called an *emitter follower*; the field-effect-transistor circuit is called a *source follower*. *See also* EMITTER FOLLOWER, and SOURCE FOLLOWER.

FORCED-AIR COOLING

Forced-air cooling is a method of maintaining the temperature of electronic equipment by passing air through an enclosure. Forced-air cooling is accomplished by means of electric fans. Generally, the term *forced-air cooling* is used in reference to localized cooling of a particular component or components, especially vacuum tubes in high-power amplifiers.

Certain vacuum tubes are provided with heat-dissipating fins, positioned radially inside a cylinder surrounding the plate. The air is guided through the fins by means of a hose. Large glass-envelope vacuum tubes are sometimes placed inside cylindrical chambers, called *chimneys*. The air is forced up the chimney, and as the air passes over the surface of the tube, heat is removed. The forced-air cooling allows the tubes to be operated at a higher level of input power. If the cooling system fails, some means must be provided to shut down the amplifier, or the tubes will be damaged. *See also* AIR COOLING.

FORM FACTOR

The *form factor* is a function of the diameter-to-length ratio of a solenoidal coil. It is used for the purpose of calculating the inductance. Given the value of the form factor F, the number of turns N, and the diameter of the coil d in inches, the coil inductance L in microhenrys is given by the formula:

$$L = FN^2d$$

In a tuned-circuit resonant response, the shape factor is sometimes called the *form factor* (*see* SHAPE FACTOR).

In an alternating-current wave, the ratio of the root-mean-square (RMS), value to the average value of a half cycle is sometimes called the *form factor* (*see* ROOT MEAN SQUARE).

FORWARD BIAS

In any diode or semiconductor junction, forward bias is a voltage applied so that current flows. A negative voltage is applied to the cathode, and a positive voltage to the anode. In some diode or semiconductor devices, a certain amount of forward bias is necessary before current will flow.

The emitter-base junction of a bipolar transistor is normally forward-biased. In the npn device, the voltage applied to the base is usually positive, with respect to the voltage at the emitter; in the pnp device, the base voltage is negative, with respect to the emitter voltage. This causes a constant current to flow through the junction. The exception is the case of the class-C amplifier, where the junction is reverse-biased to put it at cutoff.

Any semiconductor diode will conduct, for all practical purposes, only when it is forward-biased. Under conditions of reverse bias, there is no current flow. This effect makes rectification and detection possible. *See also* REVERSE BIAS.

FORWARD BREAKOVER VOLTAGE

In a silicon-controlled rectifier, a certain amount of voltage must be applied to the control, or gate, terminal in order for the device to conduct in the forward direction. However, if the forward bias is great enough (*see* FORWARD BIAS), the device will conduct no matter what the voltage on the control electrode. The minimum forward voltage at which this happens is called the *forward breakover voltage*. This voltage is extremely high in the absence of a signal at the control electrode; the greater the signal voltage, the lower the forward breakover voltage.

After the control signal has been removed, the silicon-controlled rectifier continues to conduct. However, if forward bias is removed and then reapplied, the device will not conduct

again until the control signal returns. *See also* SILICON-CONTROLLED RECTIFIER.

FORWARD CURRENT

In a semiconductor junction, the *forward current* is the value of the current that flows in the presence of a forward bias (*see* FORWARD BIAS). Generally, the greater the forward bias, the greater the forward current, although there are exceptions.

In any diode, a certain amount of forward bias is necessary to cause the flow of current in the forward direction. This minimum forward voltage varies from about 0.3 volt in the case of a germanium diode to 15 volts for a mercury-vapor device. The function of forward current versus forward bias voltage is called the *forward characteristic* of a diode, transistor, or tube. *See also* REVERSE BIAS.

FORWARD RESISTANCE

The *forward resistance* of a diode device, or any device that typically conducts in only one direction, is the value of the forward voltage divided by the forward current. If E is the forward voltage in volts and I is the forward current in amperes, then the forward resistance R, in ohms, is:

$$R = E/I$$

This is simply an expression of Ohm's law (*see* OHM'S LAW).

The forward resistance of a P-N junction or other unidirectionally conducting device depends on the value of the forward voltage. Until the forward voltage reaches the minimum value that is required for current to flow, the resistance is extremely high, and in some cases can be considered practically infinite. In some devices, the resistance increases with increasing forward voltage; in other situations it decreases. In most semiconductor diode devices, the forward resistance is very high until a forward voltage of 0.3 to 0.6 volt is obtained. Then, the resistance abruptly drops to a low value, and with further increases in the forward voltage, the resistance approaches a minimum limit. *See also* FORWARD BIAS, and FORWARD CURRENT.

FORWARD SCATTER

When a radio wave strikes the ionospheric E and F layers, it can be returned to earth. This kind of propagation most often occurs at frequencies below about 30 MHz, although it can occasionally be observed at frequencies well into the very-high part of the spectrum. When the electromagnetic field encounters the ionosphere, most of the energy is returned at an angle equal to the angle of incidence, much like the reflection of light from a mirror. However, some scattering does occur. *Forward scatter* is the scattering of electromagnetic waves in directions away from the transmitter.

Forward scatter results in the propagation of signals over several paths at once. This contributes to fading, because the signal reaches the receiving antenna in varying phase combinations. Some scattering occurs in the backward direction, toward the transmitter; this is called backscatter. *See also* BACKSCATTER, IONOSPHERE, and PROPAGATION CHARACTERISTICS.

FORWARD VOLTAGE DROP

In a rectifying device, such as a semiconductor diode or vacuum tube, the forward voltage drop is the voltage, under conditions of forward bias, that appears between the cathode and the anode. The forward voltage drop of most rectifying devices remains constant at all values of forward bias. For a germanium semiconductor P-N junction, the forward voltage drop is about 0.3 volt. For silicon semiconductor devices, it is about 0.6 volt. For vacuum-tube rectifiers, the forward voltage drop varies, depending on the type of tube. Mercury-vapor tube rectifiers have a voltage drop of about 15 volts with forward bias.

If the forward bias is less than the forward voltage drop for a particular device, the component will normally not conduct. When the voltage reaches or exceeds the forward voltage drop, the component abruptly begins to conduct. *See also* FORWARD BIAS, FORWARD CURRENT, and FORWARD RESISTANCE.

FOSTER-SEELEY DISCRIMINATOR

The *Foster-Seeley discriminator* is a form of detector circuit for reception of frequency-modulated signals. A center-tapped transformer, with tuned primary and secondary windings, is connected to a pair of diodes. The center tap of the secondary winding is coupled to the primary. The circuit is recognized by the single, center-tapped secondary winding.

As the frequency of the input signal varies, the voltage across the secondary fluctuates in phase. The diodes produce audio-frequency variations from these changes in phase. The Foster-Seeley discriminator circuit is sensitive to changes in the signal amplitude, as well as phase and frequency. Therefore, the circuit must have a limiter stage preceding it for best results. *See also* DISCRIMINATOR, FREQUENCY MODULATION, and RATIO DETECTOR.

FOUR-LAYER SEMICONDUCTOR

A *four-layer semiconductor*, also called a *four-layer diode*, is a special form of semiconductor device that contains four semiconductor wafers with three P-N junctions. The four-layer diode is sometimes called a *Shockley diode* or *pnpn device*. Four-layer diodes are commonly used as switching devices.

In the reverse-bias condition, a four-layer diode behaves in the manner that is similar to an ordinary diode. The resistance is high, and very little current flows. In the forward direction, as the voltage increases, the current remains small until a certain point is reached. Then, the current becomes large abruptly. Once this critical point has been reached, the forward voltage can be reduced almost to zero, and the current will continue to rise. Four-layer diodes are switched on by means of a brief forward pulse called a *trigger pulse*. To be switched off again, the voltage must be removed or reversed. The switching time is quite rapid: in some cases, just a few nanoseconds (billionths of a second). The time required for the device to switch on is shorter than that required for it to switch off. Other pnpn or four-layer devices include the silicon-controlled rectifier and the triac. *See also* SILICON-CONTROLLED RECTIFIER, and TRIAC.

FOUR-LAYER TRANSISTOR

A *four-layer transistor* is a form of four-layer semiconductor of the pnpn form that has three terminals. Such devices are generally used for switching purposes. Examples include the silicon-controlled rectifier and the thyristor.

Four-layer transistors have rapid switching times, sometimes as short as a few nanoseconds (billionths of a sec-

ond). The devices are frequently used in such circuits as light dimmers and motor-speed controls. *See also* FOUR-LAYER SEMICONDUCTOR, SILICON-CONTROLLED RECTIFIER, THYRISTOR, and TRIAC.

FOUR-WIRE TRANSMISSION LINE

A *four-wire transmission line* is a special form of balanced line, used to carry radio-frequency energy from a transmitter to an antenna, or from an antenna to a receiver. Four conductors are run parallel to each other, in such a way that they form the edges of a long square prism (see illustration). Each pair of diagonally opposite conductors is connected together at either end of the line.

Four-wire transmission line can be used in place of parallel-wire line in almost any installation. For a given conductor size and spacing, four-wire line has a somewhat lower characteristic impedance because the four-wire line is essentially two two-wire lines in parallel. The four-wire line is not as susceptible to the influences of nearby objects as is the two-wire line; more of the electromagnetic-field energy is contained within the line. Thus, the four-wire line has inherently better balance, and is less likely to radiate or pick up unwanted noise. Four-wire line is somewhat more difficult to install, and is more expensive, than two-wire lines. Four-wire lines are not commonly used.

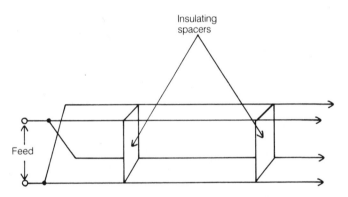

FOUR-WIRE TRANSMISSION LINE: This is a balanced line.

FOX MESSAGE

The *fox message* is a familiar test signal to operators of teletype equipment, especially radioteletype. The short message contains every letter of the alphabet and all 10 numeric digits. The most common fox message is:

THE QUICK BROWN FOX JUMPS OVER THE LAZY DOG
0123456789

There are variations. *See also* TEST MESSAGE.

FRAME

A *frame* is a single, complete television picture. The picture is generally repeated 25 times per second. The electron beam scans 625 horizontal lines per frame, with a horizontal-to-vertical size ratio of 4 to 3. Scanning occurs from left to right along each horizontal line, and from top to bottom for the lines in a complete frame.

There are some variations, in different parts of the world, in the number of scanning lines in a frame and in the overall bandwidth of the picture signal. However, each frame is always a still picture. The human eye can perceive only about 15 to 25 still pictures per second; if the pictures are repeated more often than this, they combine to form an apparent moving picture. Television picture frames are repeated so often that they combine to form a moving image. *See also* TELEVISION.

In packet radio, a *frame* is a fundamental unit of data. There are three types of frame, called *information (I) frame, supervisory (S) frame* and *unnumbered (U) frame*. Frames consist of smaller units, called *fields*.

The communications link is controlled by an S frame or a U frame. There are several variations for either kind of frame, depending on the particular function. One type of U frame is for unconnected transmissions, such as calling CQ.

The aspects and interactions of packet frames are rather complex. There are several good books on packet radio that go into detail concerning frame functions. These books, advertised in the ham magazines, are updated periodically to reflect improvements and refinements. *See also* PACKET RADIO.

FRANKLINE

Frankline is a name that has been suggested as a unit of electric charge. A charge of 1 frankline is that amount of electric potential that exerts a force of 1 dyne on an equal charge of 1 centimeter in empty space.

The frankline is not commonly used as a unit of charge—the standard unit is the coulomb. *See also* COULOMB.

FRANKLIN OSCILLATOR

A *Franklin oscillator* is a form of variable-frequency oscillator circuit, that uses transistors, field-effect transistors, or tubes. A resonant circuit is connected in the base, gate, or grid circuit of one device. Feedback is provided by a second stage, which amplies the signal and inverts its phase. The output of the amplifying stage is coupled back to the base, gate, or grid of the first device.

In the common-emitter, common-source, or common-cathode configurations, active devices show a 180-degree phase difference between their input and output circuits. In the Franklin oscillator, because there are two devices, the output is in phase with the input. Coupling between the two devices is usually accomplished by means of capacitors. However, tuned circuits can be used. The output of the oscillator can be taken from any point in the circuit. *See also* OSCILLATION, and OSCILLATOR.

FREE-RUNNING MULTIVIBRATOR

See ASTABLE MULTIVIBRATOR.

FREE SPACE

Free space is a term used to denote an environment in which there are essentially no objects or substances that affect the propagation of an electromagnetic field. Perfect free space exists nowhere in the universe; the closest thing to true free space is probably found in interstellar space. Free space has a magnetic permeability of 1 and a dielectric constant of 1. Electromagnetic fields travel through free space at approximately 186,282 miles per second.

When designing and testing antenna systems, the freespace performance is evaluated for the purpose of determining the directional characteristics, impedance, and power gain. A special testing environment provides conditions approaching those of true free space. An antenna installation always deviates somewhat from the theoretical free-space ideal. However, antennas placed several wavelengths above the ground, and away from objects such as trees, utility wires, and buildings, operate in nearly the same fashion as if they were in free space. *See also* FREE-SPACE LOSS, and FREE-SPACE PATTERN.

FREE-SPACE LOSS

As an electromagnetic field is propagated away from its source, the field becomes less intense as the distance increases. In free space, this happens in a predictable and constant manner, regardless of the wavelength.

The field strength of an electromagnetic field, in volts per meter, varies inversely with distance. If F is the field strength expressed in volts per meter and d is the distance from the source, then:

$$F = k/d$$

where k is a constant that depends on the intensity of the source.

The field strength as measured in watts per square meter varies inversely with the square of the distance from the source. If G is the field strength in watts per square meter and d is the distance from the source, then:

$$G = m/d^2$$

where m is, again, a constant that depends on the source intensity (though not the same constant as k).

In free space, the field strength drops by 6 dB when the distance from the source is doubled. This is true no matter how the field strength is measured in the far field. *See also* FAR FIELD, FIELD STRENGTH, FREE SPACE, and INVERSE-SQUARE LAW.

FREE-SPACE PATTERN

The *free-space pattern* is the radiation pattern exhibited by an antenna in free space. Free-space patterns are usually specified when referring to the directional characteristics and power gain of an antenna. Under actual operating conditions, in which an antenna is surrounded by other objects, the free-space pattern is modified. If the antenna is close to the ground, or is operated in the midst of trees, buildings, and utility wires within a few wavelengths of the radiator, the antenna pattern can differ substantially from the free-space pattern.

An antenna that radiates equally well in all possible directions is called an isotropic antenna (*see* ISOTROPIC ANTENNA). The free-space pattern of the isotropic antenna is a sphere, with the antenna at the center. A dipole antenna (*see* DIPOLE ANTENNA) radiates equally well in all directions perpendicular to the axis of the radiator, but the field strength drops as the direction approaches the axis of the wire. Along the axis of the wire, there is no radiation from the dipole. The free-space pattern of a dipole is torus-shaped. The dipole and the isotropic antenna are used as references for antenna power-gain measurements. *See also* ANTENNA PATTERN, ANTENNA POWER GAIN, dBd, and dBi.

FREQUENCY

For any periodic disturbance, the *frequency* is the rate at which the cycle repeats. Frequency is generally measured in cycles per second (hertz, abbreviated Hz). Rapid oscillation frequencies are specified in kilohertz (kHz), megahertz (MHz), gigahertz (GHz), and terahertz (THz). A frequency of 1 kHz is 1,000 Hz; 1 MHz = 1,000 kHz; 1 GHz = 1,000 MHz; and 1 THz = 1,000 GHz. If the time required for 1 cycle is denoted by p, in seconds, then frequency f is equal to $1/p$.

Wave disturbances can have frequencies ranging from less than 1 Hz to many trillions of terahertz. The audio-frequency range, at which the human ear can detect acoustic energy, ranges from about 20 Hz to 20 kHz. The radio-frequency spectrum of electromagnetic energy ranges up to hundreds or thousands of gigahertz. The frequency of electromagnetic radiation is related to the wavelength according to the formula:

$$\lambda = 300v/f$$

where λ is the wavelength in meters, v is the velocity factor of the medium in which the wave travels, and f is the frequency in megahertz. *See also* ELECTROMAGNETIC SPECTRUM, VELOCITY FACTOR, and WAVELENGTH.

FREQUENCY ALLOCATIONS

Radio amateurs are granted the privilege to communicate via electromagnetic waves, within certain frequency ranges called *bands*. The amateur bands are specified according to the approximate wavelength in meters, centimeters or millimeters. The ham bands can also be indicated by the lower-limit frequency in megahertz or gigahertz.

For example, the allocation from 7.000 to 7.300 MHz is sometimes called "40 meters" or "7 MHz." The allocation at 1.240 to 1.300 GHz is known as "23 centimeters," and also as "1240 MHz." You might hear hams talking about narrow spectrum regions around certain frequencies, used for specialized communications. Examples are "432" (MHz) and "1296" (MHz) for moonbounce.

Table A gives the amateur bands according to wavelength and frequency, as of December 1, 1991. Some of the bands are broken down according to license class. Table B shows frequency allocations for hams on bands that are so partitioned.

In bands not included in Table B, the entire band is open to hams of license classes Extra, Advanced and General. In addition, Technician class operators are allowed all privileges at amateur frequencies above 50 MHz. If a Technician licensee passes a code test of five words per minute, that person can use the Novice-class segments of 80, 40, 15, and 10 meters. *See* LICENSE REQUIREMENTS.

On some bands, special power and/or geographic restrictions apply. Some bands are not exclusively allocated to radio hams, but are shared with other services. These other services often have priority over ham communications. Current detailed information about all restrictions is available from the American Radio Relay League (ARRL), 225 Main Street, Newington, CT 06111. It is the responsibility of every radio amateur to have a timely copy of these specifics for reference.

Amateur radio operation is a privilege that has been won and kept only by community effort. World Radio Conferences are held periodically, in which other interests compete with ham radio for spectrum space. Hams must always be ready to meet challenges from other services, whose users might want to obtain some ham bands. Hams can also convince the proper authorities that new amateur bands are justifiable.

Changes in the ham frequency allocations are certain to occur in the future. It's a good idea to keep track of them according to the most recent available data from the Federal Communications Commission (FCC), the ARRL, or other current sources. In this way, you can be sure that you are aware of all the privileges you have, and are also informed of limitations on those privileges. *See also* AMERICAN RADIO RELAY LEAGUE, and FEDERAL COMMUNICATIONS COMMISSION.

FREQUENCY CALIBRATOR

A *frequency calibrator* is a device that generates unmodulated-carrier markers at precise, known frequencies. Frequency calibrators generally consist of crystal-controlled oscillators. The crystal can be housed in a special heat-controlled chamber for maximum stability; phase-locked-loop circuits can also be used in conjunction with standard broadcast frequencies.

Frequency calibrators are a virtual necessity in all receivers. Without a properly adjusted calibrator, the operator of a receiver cannot be certain of the accuracy of the frequency dial or readout. Calibrators are especially important in transmitting stations because out-of-band transmission can cause harmful interference to other stations.

The frequency calibrator in a receiver should be checked periodically against a standard frequency source, such as WWV

FREQUENCY ALLOCATIONS: TABLE A. AMATEUR BANDS BY WAVELENGTH AND FREQUENCY.

Wavelength	Frequency range
160 m	1.800–2.000 MHz
80 m, 75 m	3.500–4.000 MHz
40 m	7.000–7.300 MHz
30 m	10.100–10.150 MHz
	(200 W PEP output maximum)
20 m	14.000–14.350 MHz
17 m	18.068–18.168 MHz
15 m	21.000–21.450 MHz
12 m	24.890–24.990 MHz
10 m	28.000–29.700 MHz
6 m	50.000–54.000 MHz
2 m	144.00–148.00 MHz
1-1/4 m	222.00–225.00 MHz
3/4 m or 70 cm	420.00–450.00 MHz
1/3 m or 33 cm	902.00–928.00 MHz
1/4 m or 23 cm	1.240–1.300 GHz
13 cm	2.300–2.310 GHz
12.5 cm	2.390–2.450 GHz
8.8 cm	3.300–3.500 GHz
5.2 cm	5.650–5.925 GHz
30 mm	10.00–10.50 GHz
12.5 mm	24.00–24.25 GHz
6.4 mm	47.00–47.20 GHz
4 mm	75.50–81.00 GHz
2.5 mm	119.98–120.02 GHz
2.1 mm	142.0–149.0 GHz
1 mm and less	300 GHz and above

FREQUENCY ALLOCATIONS: TABLE B:
HAM SUBALLOCATIONS BY LICENSE CLASS.

Wavelength	Class: mode(s)	Frequency range
80 m	N/TC: CW	3.675–3.725 MHz
80 m	G/A: CW, FSK	3.525–3.750 MHz
80 m	E: CW, FSK	3.500–3.750 MHz
75 m	G: CW, SSB, image	3.850–4.000 MHz
75 m	A: CW, SSB, image	3.775–4.000 MHz
75 m	E: CW, SSB, image	3.750–4.000 MHz
40 m	N/TC: CW	7.100–7.150 MHz
40 m	G/A: CW, FSK	7.025–7.150 MHz
40 m	E: CW, FSK	7.000–7.150 MHz
40 m	G: CW, SSB, image	7.225–7.300 MHz
40 m	A/E: CW, SSB, image	7.150–7.300 MHz
20 m	G/A: CW, FSK	14.025–14.150 MHz
20 m	E: CW, FSK	14.000–14.150 MHz
20 m	G: CW, SSB, image	14.225–14.350 MHz
20 m	A: CW, SSB, image	14.175–14.350 MHz
20 m	E: CW, SSB, image	14.150–14.350 MHz
15 m	N/TC: CW	21.100–21.200 MHz
15 m	G/A: CW, FSK	21.025–21.200 MHz
15 m	E: CW, FSK	21.000–21.200 MHz
15 m	G: CW, SSB, image	21.300–21.450 MHz
15 m	A: CW, SSB, image	21.225–21.450 MHz
15 m	E: CW, SSB, image	21.200–21.450 MHz
10 m	N/TC: CW, FSK	28.100–28.300 MHz
10 m	N/TC: CW, SSB	28.300–28.500 MHz
10 m	G/A/E: CW, FSK	28.00–28.300 MHz
10 m	G/A/E: CW, SSB, image	28.300–29.700 MHz
1-1/4 m	N: CW, FSK, phone, image, test	222.10–223.91 MHz
1-1/4 m	G/A/E: CW, FSK, phone, image, test	222.00–225.00 MHz
23 cm	N: CW, FSK, phone, image, test	1.270–1.295 GHz
23 cm	G/A/E: CW, FSK, phone, image, test	1.240–1.300 GHz

Notes:

1. License class designators: N = Novice; T = Technician; TC = Technician + code; G = General; A = Advanced; E = Extra. See text for information about bands not listed here.

2. Power output must not exceed 200 W PEP for all operators in N/TC subbands below 10 m, and for Novices at 10 m. Otherwise it is 1500 W PEP maximum.

3. Power output must not exceed 200 W PEP at 30 meters (not shown in table).

4. Novice power output limit is 25 W PEP at 1-1/4 m, and 5 W PEP at 23 cm.

5. Consult current ARRL publications for additional details and changes.

or WWVH. The calibrator should be adjusted, if it produces markers at even-megahertz points, to zero beat with the carrier frequency of the standard source. *See also* CRYSTAL CONTROL, CRYSTAL OVEN, PHASE-LOCKED LOOP, and WWV/WWVH/WWVB.

FREQUENCY COMPARATOR

A *frequency comparator* is a circuit that detects the difference between the frequencies of two input signals. The output of a frequency comparator can be in video or audio form. Basically, all frequency comparators operate on the principle of beating. The beat frequency (difference between the two input frequencies) determines the output.

A simple frequency comparator circuit is illustrated in the schematic diagram. When the frequencies of the two input sig-

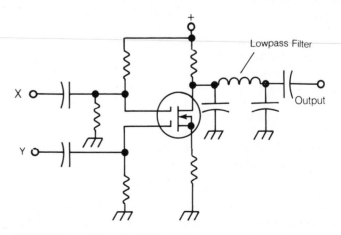

FREQUENCY COMPARATOR: The output is proportional to the difference between input signal frequencies.

nals are identical, there is no output. When the frequencies are different, a beat note is produced, in proportion to the difference between the input-signal frequencies. This output signal is fed to an audio amplifier. The audio beat note appears at the output of the amplifier, where it can be fed to a speaker, oscilloscope, or frequency counter.

Frequency comparators are used in such devices as phase-locked-loop circuits, where two signals must be precisely matched in frequency. The output of the frequency comparator is used to keep the frequencies identical, by means of controlling circuits. *See also* PHASE-LOCKED LOOP.

FREQUENCY COMPENSATION

When the frequency response of a circuit must be tailored to have a certain characteristic, frequency-compensating circuits are often used. The simplest form of frequency-compensating circuit is a capacitor or inductor, in conjunction with a resistor. More sophisticated frequency-compensating circuits use operational amplifiers or tuned circuits.

The bass and treble controls in a stereo hi-fi system are a form of frequency compensation. The desired response can be obtained by adjusting these controls. In frequency-modulation communications systems, a form of frequency compensation called preemphasis is often used at the transmitter, with deemphasis at the receiver, for improved intelligibility. Some communications receivers have an audio tone control, to compensate for the speaker response. *See also* DEEMPHASIS, and PREEMPHASIS.

FREQUENCY CONTROL

A *frequency control* is a knob, switch, button, or other device that is used to set the frequency of a receiver, transmitter, transceiver, or oscillator. The most common method of frequency control is the tuning knob; the frequency can be adjusted over a continuous range by rotating the knob. This is *analog frequency control*. The knob is connected either to a variable capacitor or a variable inductor, which in turn sets the frequency of an oscillator-tuned circuit.

In recent years, *digital frequency control* has become increasingly common. There are several different configurations of digital frequency control, but this method of frequency adjustment is always characterized by discrete frequency intervals instead of a continuous range. One method of digital frequency control uses a rotary knob, very similar in appearance to the analog tuning knob. However, when the knob is turned, it produces a series of discrete changes in frequency, at defined intervals such as 10 Hz. Another form of digital control uses spring-loaded toggle switches or buttons; by holding down the switch lever or button, the frequency increments at a certain rate in discrete steps, such as 1 kHz per second in 10-Hz jumps. The third common form of digital frequency control uses several buttons, each of which sets one digit of the desired frequency. Some communications devices have more than one of these digital tuning methods so that the operator can choose the particular frequency-control mode he wants. The electrical control of the frequency is usually accomplished by means of varactor diodes in the oscillator tuned circuit, and/or synthesized oscillators.

Both analog and digital frequency-control methods have advantages and disadvantages. Often, the optimum method is

determined largely by the preference of the individual operator. *See also* ANALOG CONTROL, and DIGITAL CONTROL.

FREQUENCY CONVERSION

In some situations, it is necessary to change the carrier frequency of a signal, without changing the modulation characteristics. When this is deliberately done, it is called *frequency conversion*. Frequency conversion is an integral part of the superheterodyne receiver, and it is found in many communications transmitters as well.

Frequency conversion is accomplished by means of *heterodyning*. The signal is combined with an unmodulated carrier in a circuit called a mixer (*see* MIXER). If the input signal has a frequency f, and the desired output frequency is h, then the frequency of the unmodulated oscillator g must be such that either $f - g = h$, $g - f = h$, or $f + g = h$. The output frequency must, in other words, be equal to either the sum or the difference of the input frequencies.

Frequency conversion is not the same thing as frequency multiplication. With certain types of modulation, the bandwidth changes when the frequency of the carrier is multiplied. But with true frequency conversion, the signal bandwidth is never affected. Frequency conversion is desirable in many applications, because it simplifies the design of the selective circuits in a receiver or transmitter. *See also* FREQUENCY CONVERTER, FREQUENCY MULTIPLIER, and SUPERHETERODYNE RECEIVER.

FREQUENCY CONVERTER

A *frequency converter* is a mixer, designed for use with a communications receiver, to allow operation on frequencies not within the normal range of the receiver. If the operating frequency is above the receiver range, the converter is called a *down converter*. If the operating frequency is below the receiver range, the converter is called an *up converter*. Converters are often used with high-frequency, general-coverage receivers to allow reception at very-low, low, very-high, and ultra-high frequencies.

Converters are relatively simple circuits. With a good communications receiver, capable of operating in the range of 3 to 30 MHz, frequency converters can be used to obtain excellent sensitivity and selectivity throughout the radio spectrum. The use of a converter with an existing receiver is often not only superior to a separate receiver, but far less expensive.

Some converters, designed to operate with transceivers, allow two-way communication on frequencies outside the range of the transceiver. This kind of device has two separate converters, and the entire circuit is called a *transverter*. *See also* DOWN CONVERSION, FREQUENCY CONVERSION, MIXER, and UP CONVERSION.

FREQUENCY COUNTER

A *frequency counter* is a device that measures the frequency of a periodic wave by actually counting the pulses in a given interval of time. The usual frequency-counter circuit consists of a gate, which begins and ends each counting cycle at defined intervals. The accuracy of the frequency measurement is a direct function of the length of the gate time; the longer the time base, the better the accuracy. At higher frequencies, the accuracy is limited by the number of digits in the display. Of course, an accurate reference frequency is necessary in order to have a precision counter; crystals are used for this purpose. The reference oscillator frequency can be synchronized with a time standard by means of a phase-locked loop.

Frequency counters are available that allow accurate measurements of signal frequencies into the gigahertz range (see photo). Frequency counters are invaluable in the test laboratory because they are easy to use and read. Typical frequency counters have readouts that show six to ten significant digits.

FREQUENCY COUNTER: A handheld frequency counter allows maximum versatility in the ham lab and around the station. OPTO-ELECTRONICS, INC. MODEL 2810

FREQUENCY DEVIATION

See DEVIATION.

FREQUENCY DIVIDER

A *frequency divider* is a circuit whose output frequency is some fraction of the input frequency. All frequency dividers are digital circuits. Although frequency multiplication can be accomplished by means of simple nonlinear analog devices, frequency division requires counting devices. Fractional components of a given frequency are sometimes called subharmonics; a wave does not naturally contain subharmonics.

Frequency dividers can be built up from bistable multivibrators. Dividers, in conjunction with frequency multipliers, can be combined to produce almost any rational-number multiple of a given frequency. For example, four divide-by-two circuits can be combined with a tripler, resulting in multiplication of the input frequency by 3/16. *See also* FREQUENCY MULTIPLIER.

FREQUENCY-DIVISION MULTIPLEX

One method of sending several signals simultaneously over a single channel is called *frequency-division multiplex*. An available channel is subdivided into smaller segments, all of equal size, and these segments are called *subchannels*. Each subchannel carries a separate signal. Frequency-division multiplexing can be used in wire transmission circuits or radio-frequency links.

One method of obtaining frequency-division multiplexing is shown in the illustration. The available channel is the frequency band from 10,000 to 10.009 MHz, or a space of 9 kHz. This space is divided into nine subchannels, each 1 kHz wide. Nine different medium-speed carriers are impressed on the main carrier of 10.000 MHz. This is accomplished by using audio tones of 500, 1500, 2500, 3500, 4500, 5500, 6500, 7500, and 8500 Hz, all of which modulate the main carrier to produce a single-sideband signal. The main carrier is suppressed by means of a balanced modulator. Each audio tone can be keyed independently in a digital code, such as morse, BAUDOT, or ASCII. The tones must be fed into the modulator circuit using appropriate isolation methods to avoid the generation of mixing products. The resulting output signals appear at 1-kHz intervals from 10.0005 to 10.0085 MHz.

This example is given only for the purpose of illustrating the concept of frequency-division multiplex. There are many different ways of obtaining this form of emission. Frequency-division multiplex necessitates a sacrifice of data-transmission speed in each subchannel, as compared with the speed obtainable if the entire channel were used. For example, each signal in the previous illustration must be no wider than 1 kHz; this limits the data-transmission speed to 1/9 the value possible if the whole 9-kHz channel was used. Frequency-division multiplex is a form of parallel data transfer. *See also* MULTIPLEX, and PARALLEL DATA TRANSFER.

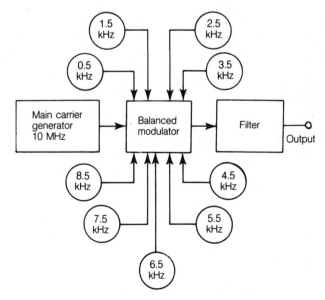

FREQUENCY-DIVISION MULTIPLEX: Several signals are impressed on one carrier.

FREQUENCY DOMAIN

When amplitude is displayed as a function of frequency, such as on a spectrum analyzer, the display is said to be in the *frequency domain*. This is because the domain of the function, shown on the horizontal axis, is frequency. Another way of saying this is that frequency is the independent variable in the display function. *See also* SPECTRUM ANALYZER, and TIME DOMAIN.

FREQUENCY HOPPING

One common method of achieving *spread-spectrum modulation* is to make the transmitter frequency "jump around," according to a sequence programmed into a microcomputer. The sequence is repeated over and over at a rapid rate. This is called *frequency hopping*.

Frequency hopping is done along with whatever other form of modulation (CW, RTTY, SSB, SSTV, FM, or packet) might be used.

The receiver frequency must be programmed to follow the same sequence as the transmitter frequency, at precisely the same rate. The receiver must also be synchronized to begin each cycle at the same instant as the transmitter, taking propagation delay into account. When this is done, the receiver "thinks" that the transmitter frequency is constant, but in reality, the signal is spread over a large bandwidth. *See also* SPREAD SPECTRUM.

FREQUENCY METER

A *frequency meter* is a device that is used for measuring the frequency of a periodic occurrence. There are several different kinds of frequency meters. An increasingly common method of frequency measurement is the direct counting of the cycles within a given interval of time; this is done with a frequency counter. Frequency counters provide extremely accurate indications of the frequency of a periodic wave. In some cases, the resolution is better than 10 significant digits.

The absorption wavemeter uses a tuned circuit in conjunction with an indicating meter. The capacitor or inductor in the tuned circuit is adjustable, and has a calibrated scale showing the resonant frequency over the tuning range. When the device is placed near the signal source, the meter shows the energy transferred to the tuned circuit; the coupling is maximum, resulting in a peak reading of the meter, when the tuned circuit is set to the same frequency as the signal source.

The gate-dip meter operates on a principle similar to the absorption wavemeter, except that the gate-dip meter contains a variable-frequency oscillator. An indicating meter shows when the oscillator is set to the same frequency as a tuned circuit under test. Gate-dip meters are commonly used to determine unknown resonant frequencies in tuned amplifiers and antenna systems.

The heterodyne frequency meter contains a variable-frequency oscillator, a mixer, and an indicator (such as a meter). The oscillator frequency is adjusted until zero beat is reached with the signal source. This zero beat is shown by a dip in the meter indication.

The cavity frequency meter is often used to determine resonant frequencies at the very-high, ultra-high, and microwave parts of the spectrum. The principle of operation is similar to that of the absorption wavemeter, except that an adjustable resonant cavity is used rather than a tuned circuit.

The lecher-wire system of frequency measurement uses a tunable section of transmission line to determine the wavelength of an electromagnetic signal in the very-high, ultra-high, and microwave regions. The system consists of two parallel wires with a movable shorting bar. The shorting bar is set until the system reaches resonance, as shown by an indicating meter.

Spectrum analyzers are sometimes used for the purpose of frequency measurement. General-coverage receivers can be used for determining the frequency of a signal. An oscilloscope can be used for very approximate measurements. *See also* ABSORPTION WAVEMETER, CAVITY FREQUENCY METER, FREQUENCY COUNTER, HETERODYNE FREQUENCY METER, LECHER WIRE, OSCILLOSCOPE, and SPECTRUM ANALYZER.

FREQUENCY MODULATION

Information can be put onto a carrier wave in many different ways. One of the most popular methods is to vary the instantaneous frequency of the wave, in step with the waveform of the data to be transmitted. This is known as *frequency modulation* (FM).

Methods of Obtaining FM   The most direct approach to frequency-modulating a carrier wave is to introduce the desired audio (or other type) signal into the transmitter oscillator, in such a way that the resonant frequency of the oscillator varies according to the instantaneous amplitude of the data. A *varactor diode* accomplishes this purpose very well (*see* VARACTOR DIODE). This component acts like a capacitor, the reactance of which fluctuates as the data voltage is applied across it. A frequency modulator of this kind is called a *reactance modulator* (*see* REACTANCE MODULATOR).

Another way to get frequency modulation is to alter the phase of the carrier wave. An instantaneous change in phase is always attended by an instantaneous change in frequency, and vice-versa. But the audio response produced by phase modulation is different than that resulting from direct frequency modulation. This discrepancy is corrected by deemphasis and/or preemphasis of the audio. *See* DEEMPHASIS, PHASE MODULATION, and PREEMPHASIS.

Sidebands   When a carrier is frequency modulated over just a narrow range, sidebands are produced in much the same way as they are for amplitude modulation. As the frequency swing is increased, additional sidebands appear, farther away from the carrier frequency. The number and relative amplitudes of these sidebands is a rather complex function of the deviation, or extent to which the carrier frequency swings away from its center (unmodulated) frequency, and also of the bandwidth of the modulating data. *See* DEVIATION, and MODULATION INDEX.

An FM signal is called *narrowband* if its bandwidth is no greater than that of a standard amplitude-modulated (AM) signal having the same waveform. For voice communications, this is about 6 kHz, or 3 kHz above and below the carrier.

If an FM signal has a bandwidth greater than that of a standard AM signal with the same modulating waveform, the signal is said to be *wideband FM*. Stereo FM broadcasting is done via this mode.

Reception of FM   Frequency-modulated signals are received by circuits sensitive to fluctuations in frequency or phase. One common type of FM detector is the *discriminator*; another is the *phase-locked loop*; still another is the *ratio detector*. See DISCRIMINATOR, PHASE-LOCKED LOOP, and RATIO DETECTOR. Most FM receivers use limiter circuits to minimize the response to changes in amplitude. Sometimes the limiter is a separate stage, and in other receivers it is done in the detector.

Narrowband FM can be received with an AM receiver that has a bandpass response several kilohertz wide. The AM receiver is tuned slightly off frequency, and the FM signal moves in and out of the AM receiver passband so that its strength varies right along with the modulating waveform. This is called *slope detection*. It usually causes some distortion of the FM signal, because the shape of the receiver passband (the "slope" of the attenuation-vs-frequency curve) isn't intended for the reception of FM.

Advantages and Disadvantages of FM

Frequency modulation offers a certain degree of immunity to static and impulse noise. This is because these types of noise are amplitude-modulated. Receivers for FM can be made unresponsive to changes in signal amplitude. In practice, weak FM signals are usually affected by static or impulse noise because the signals aren't strong enough to let the receiver limiting circuit function optimally.

The fidelity of FM is better than that of single sideband (SSB) when signals are reasonably strong. However, SSB offers better communications when conditions are bad and signals are weak or fading.

Frequency modulation does not work very well for signals propagated via the ionosphere. This is because ionospheric bending causes phase modulation of signals that come back to the earth. This phase modulation is usually not noticed on code (CW), frequency-shift-keyed (FSK) or SSB signals. But it causes severe distortion of FM signals because phase modulation is so much like frequency modulation. Because of this effect, FM voice is done mainly at very-high frequencies (VHF) and above. *See also* EMISSION CLASS, and MODULATION METHODS.

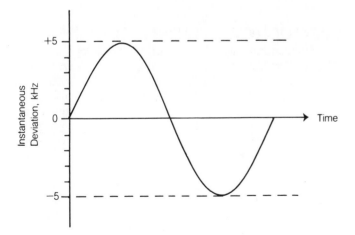

FREQUENCY MODULATION: Deviation versus time for a hypothetical FM signal.

FREQUENCY MULTIPLIER

A *frequency multiplier* is a device that produces an integral multiple, or many integral multiples, of a given input signal. A frequency multiplier circuit can also be called a *harmonic generator*.

The design of a frequency multiplier is straightforward. In order to cause the generation of harmonic energy, a nonlinear element must be placed in the path of the signal. A semiconductor diode is ideal for this purpose. However, the type of diode must be chosen so that the capacitance is not too great. Then, the signal waveform is distorted, and harmonics are produced. Following the nonlinear element, a tuned circuit can be used to select the particular harmonic desired. The tuned circuit is set to resonate at the harmonic frequency. The illustration is a simple schematic diagram of a frequency multiplier. There are other methods of obtaining frequency multiplication; odd multiples can be produced by means of a push-pull circuit, and even multiples by a push-push circuit (*see* PUSH-PULL CONFIGURATION, and PUSH-PUSH CONFIGURATION).

Frequency multiplication does not change the characteristics of an amplitude-modulated or continuous-wave signal. However, the bandwidth of a single-sideband, frequency-shift-keyed, or frequency-modulated signal is multiplied by the same factor as the frequency. With frequency-shift keying or frequency modulation, it is possible to compensate for this effect, but with a single-sideband signal, severe distortion is introduced. Frequency multiplication increases the tuning range and the tuning rate of a variable-frequency oscillator by the factor of multiplication. Any signal instability, such as chirp or drift, is also multiplied. For this reason, frequency multiplication is generally limited to a factor of about four or less. *See also* DOUBLER, and TRIPLER.

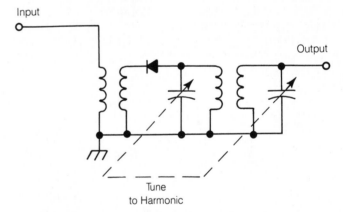

FREQUENCY MULTIPLIER: This circuit uses a single diode.

FREQUENCY OFFSET

In a communications transceiver, the *frequency offset* is the difference between the transmitter frequency and the receiver frequency. If the transmitter and the receiver are set to the same frequency, then the offset is zero, and the mode of operation is called *simplex*.

In continuous-wave (code) communications, the two stations usually set their transmitters to exactly the same frequency. This necessitates that their receivers be offset by several hundred hertz so that audible tones appear at the speakers. The receiver offset is obtained by a control called a *clarifier* or a *receiver-incremental-tuning (RIT)* control.

In transceivers designed for operation with repeaters, the transmitter frequency is significantly different from the receiver frequency. The offset can be several megahertz at the ultra-high-frequency range. This offset is needed because the repeater cannot transmit on the same frequency at which it receives. If the transmitter frequency of the transceiver is higher than the receiver frequency, the offset is called *positive*; if the transmitter frequency is lower, the offset is considered *negative*. For example, a typical amateur transceiver operating on the 2-meter band, might be set to receive on 146.94 MHz and transmit on 146.34 MHz. The offset is thus −600 kHz. The repeater, of course, has an input or receiving frequency of 146.34 MHz, and an output or transmitting frequency of 146.94 MHz. *See also* RECEIVER INCREMENTAL TUNING, and REPEATER.

FREQUENCY RESPONSE

Frequency response is a term used to define the performance or behavior of a filter, antenna system, microphone, speaker, or headphone. There are basically five kinds of frequency response: flat, bandpass, band-rejection, highpass, and lowpass.

The frequency responses of some devices are more complicated than the elementary functions. There can be several "peaks" and "valleys" in the attenuation-versus-frequency functions of some systems. An example of this is the audio high-fidelity system using a graphic equalizer. The equalizer allows precise adjustment of the response.

Devices with various kinds of frequency response are useful in particular electronics applications. Bandpass filters are used in radio-frequency transmitters and receivers. Lowpass filters are used to minimize the harmonic output from a transmitter. Highpass filters can be used to reduce the susceptibility of television receivers to interference by lower-frequency signals. Band-rejection filters are often used to attenuate a strong, undesired signal in the front end of a receiver. *See also* BAND-PASS RESPONSE, BAND-REJECTION RESPONSE, FLAT RESPONSE, HIGHPASS RESPONSE, and LOWPASS RESPONSE.

FREQUENCY SHIFT

A *frequency shift* is a change in the frequency of a signal, either intentional or unintentional. The term *shift* implies a sudden, or rapid, change in frequency.

The frequency of a transmitter or receiver can suddenly shift because of a change in the capacitance or inductance of the oscillator tuned circuit. The amount of the shift is given by the difference between the signal frequencies before and after the change. An increase in frequency represents a positive shift; a decrease is a negative shift.

Frequency shift is deliberately introduced into a carrier wave in all forms of frequency-modulation transmission. *See also* FREQUENCY MODULATION, and FREQUENCY-SHIFT KEYING.

FREQUENCY-SHIFT KEYING

Frequency-shift keying is a digital mode of transmission, commonly used in radioteletype applications. In a continuous-wave or code transmission, the carrier is present only while the key is down, and the transmitter is turned off while the key is up. In frequency-shift keying, however, the carrier is present all the time. When the key is down, the carrier frequency changes by a predetermined amount. The carrier frequency under key-up conditions is called the *space frequency*. The carrier frequency under key-down conditions is called the *mark frequency*.

In high-frequency radioteletype communications, the space frequency is usually higher than the mark frequency by a value ranging from 100 Hz to 900 Hz. The most common values of frequency shift are 170 Hz, 425 Hz, and 850 Hz. In recent years, because of the improvement in selective-filter technology, the narrower shift values have become increasingly common.

Frequency-shift keying is considered a form of frequency modulation, and is designated type F1 emission. Frequency-shift keying can be obtained in two ways. The first, and obvious, method is the introduction of reactance into the tuned circuit of the oscillator in a transmitter. The other method uses an audio tone generator with variable frequency, the output of which is fed into the microphone input of a single-sideband transmitter. If this method is used, precautions must be taken to ensure that there is no noise in the injected signal, for such noise will appear on the transmitted signal.

In some radioteletype links, especially at the very-high frequencies and above, audio frequency-shift keying is used. This is called *F2 emission*. An audio tone generator with variable frequency is coupled to the microphone input of an amplitude-modulated or frequency-modulated transmitter. The tone frequencies are between about 1000 Hz and 3000 Hz. This mode of frequency-shift keying requires considerably greater bandwidth than F1 emission. Its primary advantage is that the receiver frequency setting is not critical, whereas with F1 emission, even a slight error in the receiver setting will seriously degrade the reception.

Frequency-shift keying provides a greater degree of transmission accuracy than is possible with an ordinary continuous-wave system. This is because the key-up condition, as well as the key-down condition, is positively indicated. The frequency-shift-keyed system is therefore less susceptible to interference from atmospheric or manmade sources. *See also* RADIOTELETYPE.

FREQUENCY SWING

See CARRIER SWING.

FREQUENCY SYNTHESIZER

In older ham radio equipment, variable frequency oscillators (VFOs) were prone to *drift*—to change frequency with time. Fixed-frequency crystal oscillators were much more stable than those using tuned inductance-capacitance (LC) circuits. But in order to change frequency with this kind of transmitter, it was necessary to change crystals. This usually meant unplugging one crystal, searching for another with the desired frequency, and then plugging the new crystal into a socket similar to a utility outlet.

The design of VFOs gradually improved so that drift became less and less of a problem. But LC-tuned VFOs never achieved the stability of crystal-controlled circuits until the advent of the *frequency synthesizer*.

A synthesizer uses a crystal frequency standard, and derives thousands or even millions of selectable, precise working frequencies from the crystal output. The crystal oscillator is known as the *reference oscillator*. Digital circuits divide and multiply this frequency by integral amounts. In this way, any rational number multiplier (ratio of integer values) can be obtained. In theory, there are infinitely many possible ratios, and a synthesizer can be tuned to any desired frequency via the proper ratio.

Actually, the synthesizer circuit is a little more complicated than this. A *voltage-controlled oscillator (VCO)*, whose stability by itself is not that good, is governed by phase-detection circuitry, matching the VCO frequency to the desired rational-number multiple of the crystal frequency. This makes the VCO work like a VFO, but it's as stable as the crystal *master oscillator (MO)*. Such a circuit is a *phase-locked loop (PLL)* and is widely used in modern ham equipment.

State-of-the-art digital transceivers are so stable that the frequency can be read to better than 10 Hz at 10 MHz (one part per million). But the digital readout will not show if there is anything wrong with the PLL. It is good ham operating practice to check the transmit and receive frequencies periodically against some primary standard, such as WWV/WWVH. *See also* PHASE-LOCKED LOOP, and VOLTAGE-CONTROLLED OSCILLATOR.

FREQUENCY TOLERANCE

Frequency tolerance is a means of expressing the accuracy of a signal generator. The more stable an oscillator, the narrower the frequency tolerance. Tolerances are always expressed as a percentage of the fundamental quantity under consideration; in this case, it is the frequency of the oscillator.

Suppose, for example, that a 1-MHz oscillator has a frequency tolerance of plus or minus 1 percent. Because 1 percent of 1 MHz is 10 kHz, this means that the oscillator can be expected to be within 10 kHz of its specified frequency of 1 MHz. This is a range of 0.990 to 1.010 MHz.

Frequency-tolerance specifications are often given for communications receivers and transmitters. The tolerances can be given as a percentage, or as an actual maximum expected frequency error. The frequency tolerance required for one communications service can be much different from that needed for another service. *See also* TOLERANCE.

FRONT END

Front end is another name for the first radio-frequency amplifier stage in a receiver. The front end is one of the most important parts of any receiver, because the sensitivity of the front end dictates the sensitivity of the entire receiver. A low-noise, high-gain front end provides a good signal for the rest of the receiver circuitry. A poor front end results in a high amount of noise or distortion.

The front end of a receiver becomes more important as the frequency gets higher. At low and medium frequencies, there is a great deal of atmospheric noise, and the design of a front-end circuit is rather simple. But at frequencies above 20 or 30 MHz, the amount of atmospheric noise becomes small, and the main factor that limits the sensitivity of a receiver is the noise generated within the receiver itself. Noise generated in the front end of a receiver is amplified by all of the succeeding stages, so it is most important that the front end generate very little noise.

In recent years, improved solid-state devices, especially field-effect transistors, have paved the way for excellent receiver front-end designs. As a result of this, receiver sensitivity today is vastly superior to that of just a few years ago.

Another major consideration in receiver front-end design is low distortion. The front-end amplifier must be as linear as pos-

sible; the greater the degree of nonlinearity, the more susceptible the front end is to the generation of mixing products. Mixing products result from the heterodyning, or beating, of two or more signals in the circuits of a receiver; this causes false signals to appear in the output of the receiver.

The front end of a receiver should be capable of handling very strong as well as very weak signals, without nonlinear operation. The ability of a front end to maintain its linear characteristics in the presence of strong signals is called the *dynamic range* of the circuit. If a receiver has a front end that cannot handle strong signals, desensitization will occur when a strong signal is present at the antenna terminals. Mixing products can also appear because the front end is driven into nonlinear operation. *See also* DESENSITIZATION, INTERMODULATION, NOISE FIGURE, and SENSITIVITY.

FRONT-TO-BACK RATIO

Front-to-back ratio is one expression of the directivity of an antenna system. The term applies only to antennas that are unidirectional, or directive in predominantly one direction. Such antennas include (among others) the Yagi, the quad, and the parabolic-dish antenna.

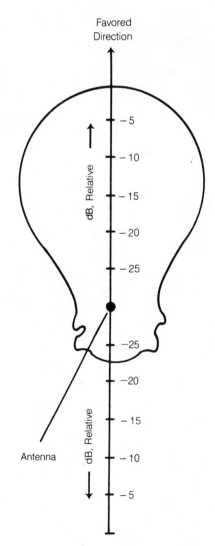

FRONT-TO-BACK RATIO: A measure of antenna directivity.

Mathematically, the front-to-back ratio is the ratio, in decibels, of the field strength in the favored direction to the field strength exactly opposite the favored direction, given an equal distance (see illustration). In terms of power, if P is the power gain of an antenna in the forward direction and Q is the power gain in the reverse direction, both expressed in decibels with respect to a dipole or isotropic radiator, then the front-to-back ratio is given simply by $P - Q$.

The other primary consideration in antenna directive behavior is the forward gain, or power gain. Power gain is more important in terms of transmitting; the front-to-back ratio is of more interest in terms of receiving directivity. *See also* ANTENNA POWER GAIN, dBd, and dBi.

FUCHS ANTENNA

A *Fuchs antenna* is the simplest possible form of antenna. It consists simply of a length of wire, one end of which is attached to the transmitter output terminals, and the other end of which is left free. An antenna coupler is used if the antenna is not an odd multiple of ¼ wavelength at the operating frequency.

The Fuchs antenna is also called a *random-wire antenna*, and is used primarily in portable or emergency installations in the high and very-high frequency ranges. The main advantage of the Fuchs antenna is that it is simple to install, and works relatively well. There is no transmission line, and thus there is no loss in the line. However because the radiating part of the antenna is immediately adjacent to the transmitter, electromagnetic interference can sometimes be a problem when this kind of antenna is used. A good ground system is essential when the Fuchs antenna is used. If possible, the antenna feed point should be removed at least a few feet from the transmitter, to minimize the possibility of electromagnetic-interference effects. *See also* END FEED.

FULL-SCALE ERROR

The *full-scale error* is a means of determining the accuracy of an analog metering device. Often, the accuracy specification of such a meter is given in terms of a percentage of the full-scale reading.

As an example, suppose that a milliammeter, with a range of 0 to 1 mA, is stated as having a maximum error of plus or minus 10 percent of full scale. This means that, if the meter reads 1 mA (or full scale), the actual current might be as small as 0.9 mA or as large as 1.1 mA. If the meter reads half scale, or 0.5 mA, the actual current might be as small as 0.4 mA or as large as 0.6 mA. In other words, the maximum possible error is always plus or minus 0.1 mA, regardless of the meter reading.

Meters are specified for full-scale readings because that is one point where all errors can conveniently be summed and compensated. Errors introduced by resistive components in the circuit are constant percentages, regardless of the current flowing through them. Errors introduced by the meter movement are not necessarily linear; the springs that return the pointer to its zero position produce nonlinear forces, which can cause an error at midscale to be significantly different from the error at full scale. Calibration of all the errors in a meter at many points on the scale would be costly and time-consuming. *See also* ANALOG METERING.

FULL-WAVE RECTIFIER

A *full-wave rectifier* is a circuit that converts alternating current into pulsating direct current. The full-wave rectifier operates on both halves of the cycle. One half of the wave cycle is inverted, while the other is left as it is. The input and output waveforms of a full-wave rectifier in the case of a sine-wave alternating-current power source are illustrated at A.

A typical full-wave rectifier circuit is shown at B. The transformer secondary winding is center-tapped, and the center tap is connected to ground. The opposite sides of the secondary winding are out of phase. Two diodes, usually semiconductor types, allow the current to flow either into or out of either end of the winding, depending on whether the rectifier is designed to supply a positive or a negative voltage. A form of full-wave rectifier that does not require a center-tapped secondary winding, but necessitates the use of four diodes, rather than two, is called a *bridge rectifier* (*see* BRIDGE RECTIFIER).

The primary advantage of a full-wave rectifier circuit over a half-wave rectifier circuit is that the output of a full-wave circuit is easier to filter. The voltage regulation also tends to be better. The ripple frequency at the output of a full-wave rectifier circuit is twice the frequency of the alternating-current input. *See also* HALF-WAVE RECTIFIER, and RECTIFICATION.

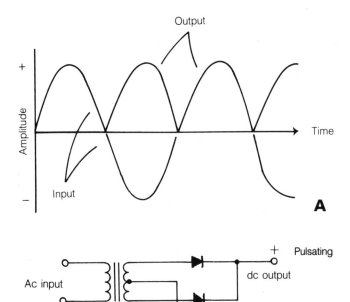

FULL-WAVE RECTIFIER: At A, input and output waveforms; at B, a typical circuit.

FULL-WAVE VOLTAGE DOUBLER

A *full-wave voltage doubler* is a form of full-wave power supply. The direct-current output voltage is approximately twice the alternating-current input voltage. Voltage-doubler power supplies are useful because smaller transformers can be used to obtain a given voltage, as compared with ordinary half-wave or full-wave supplies. Medium-power and low-power vacuum-tube circuits sometimes have voltage-doubler supplies.

See the schematic diagram of a full-wave voltage doubler. During one half of the cycle, the top capacitor charges to the peak value of the alternating-current input voltage; during the other half of the cycle, the other capacitor charges to the peak of the alternating-current input voltage, but in the opposite direction. Both capacitors hold their charge throughout the input cycle, resulting in the effective connection of two half-wave, direct-current power supplies in series, with twice the voltage of either supply.

Full-wave voltage doublers are not generally suitable for applications in which large amounts of current are drawn. The regulation and filtering tends to be rather poor in such cases. *See also* HALF-WAVE RECTIFIER, and RECTIFICATION.

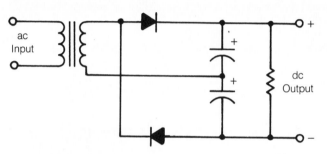

FULL-WAVE VOLTAGE DOUBLER: This is basically two half-wave rectifiers connected in series.

FUNCTION GENERATOR

A *function generator* is a signal generator that can produce various different waveforms. All electrical waveforms can be expressed as mathematical functions of time; for example, the instantaneous amplitude of a sine wave can be expressed in the form $f(t) = a \sin(bt)$, where a is a constant that determines the peak amplitude, and b is a constant that determines the frequency. Square waves, sawtooth waves, and all other periodic disturbances can be expressed as mathematical functions of time, although the functions are quite complicated in some cases.

Most function generators can produce sine waves, sawtooth waves, and square waves. Some can also produce sequences of pulses. More sophisticated function generators can create a large variety of different waveforms that are used for testing purposes in the design, troubleshooting, and alignment of electronic apparatus, such as audio amplifiers and digital circuits. *See also* SAWTOOTH WAVE, SINE WAVE, and SQUARE WAVE.

FUNDAMENTAL FREQUENCY

The *fundamental frequency* of an oscillator, antenna, or tuned circuit is the primary frequency at which the device is resonant. All waveforms or resonant circuits have a fundamental frequency, and integral multiples of the fundamental frequency, called *harmonics*. A theoretically pure sine wave contains energy at only its fundamental frequency. But in practice, sine waves are always slightly impure or distorted, and contain some harmonic energy. Complex waves, such as the sawtooth

or square wave, often contain large amounts of harmonic energy in addition to the fundamental.

The amplitude of the fundamental frequency is used as a reference standard for the determination of harmonic suppression. The degree of suppression of a particular harmonic is expressed in decibels, with respect to the level of the fundamental. The fundamental frequency need not necessarily be the frequency at which the most energy is concentrated; a harmonic can have greater amplitude. The fundamental also might not be the desired frequency in a particular application; harmonics are often deliberately generated and amplified. *See also* HARMONIC, and HARMONIC SUPPRESSION.

FUNDAMENTAL SUPPRESSION

In the measurement of the total harmonic distortion in an amplifier circuit, the fundamental frequency must be eliminated so that only the harmonic energy remains. This is called *fundamental suppression*. Fundamental suppression is generally done by means of a trap circuit, or band-rejection filter, centered at the fundamental frequency (*see* TOTAL HARMONIC DISTORTION).

In a frequency multiplier, the fundamental frequency is suppressed at the output while one or more harmonics are amplified. In such circuits, the fundamental frequency is not wanted. Band-rejection filters can be used in the output of the frequency multiplier to attenuate the fundamental signal. Highpass filters can also be used. Bandpass filters, centered at the frequency of the desired harmonic, also provide fundamental suppression. *See also* FREQUENCY MULTIPLIER, and HARMONIC.

FUSE: Miniature fuses for a signal monitor/analyzer.

FUSE

A *fuse* is a device used in power supplies and power sources for electronic equipment. The fuse protects the components of the power supply, as well as the components of the equipment, against possible damage in the event of a malfunction of either the supply or the equipment. A fuse is a simple device that consists of a thin piece of wire in an enclosure. The thickness of the wire determines the current level at which the fuse will open when the fuse is connected in series with the power-supply leads.

Fuses are commonly found in household electrical systems. Electronic devices such as communications receivers and transmitters have their own independent fuses. A fuse can be as tiny as those shown in the photograph — about the size of a pencil eraser — or they can be extremely large and bulky.

Fuses can be obtained with a variety of time constants. Some fuses blow out almost instantly when the current exceeds the rated value. Others have a built-in delay, which allows a brief period of excessive current without blowing the fuse. The latter type of fuse is known as a *slow-blow fuse*.

The main disadvantage of fuses is that they must be replaced when they blow out. This is not always convenient. Extra fuses must be kept on hand in case of an accidental short circuit or other temporary malfunction that causes a fuse to blow. Circuit breakers, which can be reset an indefinite number of times without replacement, are becoming more and more common. But there are some situations where fuses are still used because they offer better circuit protection. *See* CIRCUIT BREAKER.

GAIN

Gain is a term that describes the extent of an increase in current, voltage, or power. The gain of an amplifier circuit is an expression of the ratio between the amplitudes of the input signal and the output signal. In an antenna system, gain is a measure of the effective radiated power compared to the effective radiated power of some reference antenna (*see* ANTENNA POWER GAIN). Gain is almost always given in decibels.

In terms of voltage, if E_{IN} is the input voltage and E_{OUT} is the output voltage in an amplifier circuit, then the gain G, in decibels, is given by:

$$G = 20 \log_{10} (E_{OUT}/E_{IN})$$

The current gain of an amplifier having an input current of I_{IN} and an output current of I_{OUT} is given in decibels by:

$$G = 20 \log_{10} (I_{OUT}/I_{IN})$$

In a power amplifier, where P_{IN} is the input power and P_{OUT} is the output power, the power gain G, in decibels, is:

$$G = 10 \log_{10} (P_{OUT}/P_{IN})$$

When the gain of a particular circuit is negative, there is insertion loss. Some types of circuits, such as the follower, exhibit insertion loss. For typical radio-frequency amplifiers, the gain ranges from about 15 to 25 dB per stage of amplification. Sometimes, the gain is as high as 30 dB per stage, but this is rare. *See also* AMPLIFICATION FACTOR, DECIBEL, and INSERTION LOSS.

GAIN CONTROL

A *gain control* is an adjustable control, such as a potentiometer, that changes the gain of an amplifier circuit. The volume control in every radio receiver or audio device is a form of gain control. Communications equipment generally has an audio-frequency gain control and a radio-frequency gain control.

A gain control is fairly simple to incorporate into an amplifier circuit. A potentiometer may be used to provide varying amounts of drive to an amplifier circuit; this should be done at a low power level.

A gain control should not affect the linearity of the amplifier circuit in which it is placed. Therefore, changing the bias on the transistor or tube in an amplifier is not an acceptable way of obtaining gain control. The gain control should be chosen so that succeeding stages cannot be overdriven if the gain is set at maximum. The gain control should have sufficient range so that, when the gain is set at minimum, the output of the amplifier chain is reduced essentially to zero. *See also* GAIN.

GALACTIC NOISE

All celestial objects radiate a certain amount of energy at radio wavelengths. This is especially true of stars and galaxies, and the diffuse matter among the stars.

Galactic noise was first observed by Karl Jansky, a physicist working for the Bell Telephone Laboratories, in the 1930s. It was an accidental discovery. Jansky was investigating the nature of atmospheric noise at a wavelength of about 15 m (20 MHz). Jansky's antenna consisted of a rotatable array, not much larger than the Yagi antennas commonly used by radio amateurs at the wavelength of 15 m.

Most of the noise from our galaxy comes from the direction of the galactic center. This is not surprising because that is where most of the stars are concentrated. Galactic noise contributes, along with noise from the sun, the planet Jupiter, and a few other celestial objects, to most of the cosmic radio noise arriving at the surface of the earth. Other galaxies radiate noise, but because the external galaxies are much farther away from us than the center of our own galaxy, sophisticated equipment is needed to detect the noise from them. *See also* RADIO ASTRONOMY, and RADIO TELESCOPE.

GALLIUM-ARSENIDE FIELD-EFFECT TRANSISTOR

The field-effect transistor is well known as a low-noise amplifying device. Field-effect transistors are available in many different forms. One type of field effect transistor is fabricated from gallium-arsenide (GaAs) semiconductor material, and is called a *gallium-arsenide field-effect transistor* (*GaAsFET*, pronounced just as it looks).

The GaAsFET is ideal for use in high-gain, low-noise amplifier circuits, especially at ultra-high and microwave frequencies. Gallium arsenide (an N-type material) has higher carrier mobility than the germanium or silicon that is usually found in semiconductor devices. Carrier mobility is the speed with which the charge carriers—either electrons or holes—move within a substance with a given electric-field intensity (*see* CARRIER MOBILITY). The GaAsFET is a depletion-mode field-effect transistor. That is, it normally conducts, and the application of bias to the gate electrode reduces the conductivity of the channel.

The GaAsFET can be used in receiver preamplifiers, converters, and in power amplifiers up to several watts. They are generally N-channel devices. *See also* DEPLETION MODE, and FIELD-EFFECT TRANSISTOR.

GALVANISM

The production of electric current by chemical action is called *galvanism*. The word comes from the name of an eighteenth-century scientist, Luigi Galvani. The principles of galvanism are used in common energy cells, such as the dry cell and the storage battery. Galvanism also causes certain metals to corrode, or react with other substances to form compounds.

When two dissimilar metals are brought into contact with each other in the presence of an electrolyte, they will act like a small cell, and a voltage difference will be produced between the metals. Salt in the air, or even slightly impure water vapor, can act as the electrolyte. The metals gradually corrode because of the chemical reaction; this is called *galvanic corrosion*.

Electrolytic action can be used to coat iron or steel with a thin layer of zinc. This is called *galvanizing*. Galvanized metal is more resistant to corrosion than bare metal, although in a marine or tropical environment, even galvanized iron or steel will eventually corrode.

GALVANOMETER

A *galvanometer* is a sensitive device for detecting the presence of, and measuring, electric currents. The galvanometer is similar to an ammeter, but the needle of the galvanometer normally rests in the center position. Current flowing in one direction results in a deflection of the needle to the right; current in the other direction causes the needle to move to the left.

The galvanometer is especially useful as a null indicator in the operation of various bridge circuits because currents in both directions can be detected. An ordinary meter in such a situation will read zero when a reverse current is present, thus making an accurate null indication impossible to obtain. *See also* AMMETER.

GAMMA MATCH

A *gamma match* is a device for coupling a coaxial transmission line to a balanced antenna element, such as a halfwave dipole. The radiating element consists of a single conductor. The shield part of the coaxial cable is connected to the radiating element at the center. The center conductor is attached to a rod or wire running parallel to the radiator, toward one end. The illustration shows the gamma match. The device gets its name from its resemblance in shape to the upper-case Greek letter gamma.

The impedance-matching ratio of the gamma match depends on the physical dimensions: the spacing between the matching rod and the radiator, the length of the matching rod, and the relative diameters of the radiator and the matching rod. The gamma match allows the low feed-point impedance of a Yagi antenna driven element to be precisely matched with the 50-ohm or 75-ohm characteristic impedance of a coaxial feed line. The gamma match also acts as a balun, which makes the balanced radiator compatible with the unbalanced feed line. Gamma matching is often seen in multielement parasitic arrays, especially rotatable arrays. *See also* BALUN, IMPEDANCE MATCHING, and YAGI ANTENNA.

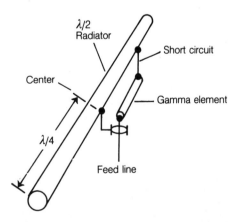

GAMMA MATCH: Used with coaxial lines and balanced antennas.

GANG CAPACITOR

Variable capacitors are sometimes connected together in parallel, on a common shaft. Such a combination capacitor is called a *gang capacitor*. The individual capacitors rotate together. The sections might or might not have the same values of capacitance.

Gang capacitors are used in circuits having more than one adjustable resonant stage, and a requirement for tuning each resonant stage along with the others. This is often the case in communications equipment.

The rotor plates of the gang capacitor are usually all connected to the frame of the unit, and the frame is in turn put at common ground. Gang capacitors with electrically separate sets of rotor plates are rare. *See also* AIR-VARIABLE CAPACITOR, and GANGED CONTROLS.

GANGED CONTROLS

When several adjustable components, such as potentiometers, switches, variable capacitors, or variable inductors are connected in tandem, the controls are said to be ganged. Ganged controls are frequently found in electronic devices, because it is often necessary to adjust several circuits at the same time.

An example of the use of a ganged control is the volume adjustment in a stereo high-fidelity system. There are two channels, and usually a single volume control. This requires a two-gang potentiometer (some systems have a separate volume control for each channel). Ganged rotary switches are almost always found in multiband communications receivers and transmitters because of the need to switch several stages of amplification simultaneously. When several tuned circuits must be adjusted together, a gang of variable capacitors is employed. *See also* GANG CAPACITOR.

GATE

The *gate* of a field-effect transistor is the electrode that allows control of the current through the channel of the device. The gate is analogous to the base of a bipolar transistor, or the grid of a vacuum tube. In an N-channel junction field-effect transistor (JFET), the gate voltage is usually negative with respect to the source voltage. In the P-channel device, the gate is biased positively.

In the JFET, the gate consists of two sections of P-type semiconductor material in the N-channel device, placed on either side of the channel. Alternatively, the gate may have the configuration of a cylinder of P-type material around the channel. In the P-channel device, of course, the gate is fabricated from N-type semiconductor material.

The greater the reverse bias applied to the gate, the less the channel of the JFET conducts, until, at a certain value of reverse bias, the channel does not conduct at all. This condition is called *pinchoff*. The pinchoff voltage depends on the voltage between the source and the drain of the field-effect transistor.

In the insulated-gate (metal-oxide-semiconductor) field-effect transistor (MOSFET), the gate operates differently. There are two basic modes of operation, known as depletion mode and enhancement mode (*see* DEPLETION MODE, ENHANCEMENT MODE, and METAL-OXIDE-SEMICONDUCTOR FIELD-EFFECT TRANSISTOR). A MOSFET can have more than one gate.

Small fluctuations in the gate voltage result in large changes in the current through the field-effect transistor. This is the reason that amplification is possible with such devices. *See also* FIELD-EFFECT TRANSISTOR.

In digital logic, the term gate refers to a switching device with one or more inputs, and one output. The gate performs a logical function, depending on the states of the inputs. The three basic logic gates are the inverter, or NOT gate, the AND gate, and the OR gate. *See also* LOGIC GATE.

GATE CIRCUIT

In a field-effect-transistor amplifier, the *gate circuit* is the combination of components associated with the gate of the device. Usually, the input to a field-effect-transistor amplifier is applied to the gate circuit.

In digital electronics, a gate circuit is a combination of logic gates, designed for the purpose of performing a certain function. Many different logical operations can be carried out by combining the simple NOT, AND, and OR gates. In fact, all logical switching arrangements can theoretically be realized by various combinations of logic gates. *See also* GATE, and LOGIC GATE.

GATE-DIP METER

A *gate-dip meter* is a field-effect-transistor device which operates on the same principle as a grid-dip meter. The gate-dip and grid-dip meters are used to determine the resonant frequency of a tuned circuit or antenna.

The gate-dip meter has the obvious advantage of requiring less power than a grid-dip meter. There is no filament, and the necessary operating voltage is much smaller. The characteristics of the gate-dip meter are, however, identical to those of the grid-dip meter in practical application.

GATE-PROTECTED MOSFET

A *gate-protected MOSFET* is a type of metal-oxide semiconductor field-effect transistor (MOSFET). The gate electrode in a MOSFET is susceptible to breakdown in the presence of static electricity. This can be such a problem that handling the devices can be risky. The gate-protected MOSFET has built-in Zener diodes connected in reverse parallel, which prevent the accumulation of charges of over a few volts. *See also* METAL-OXIDE SEMICONDUCTOR FIELD-EFFECT TRANSISTOR.

GAUGE

A *gauge* is a device that measures a certain quantity, such as current, voltage, power, or frequency. All meters are examples of gauges. Gauges usually are found in the form of a meter-type device.

The diameter of an electric wire is sometimes specified in terms of the gauge, which is a number assigned to indicate the approximate size of a conductor. In the United States, the American Wire Gauge (AWG) designator is most commonly used for this purpose. There are numerous other wire gauges. The term *gauge* is also sometimes used in reference to the thickness of a piece of sheet metal.

GAUSS

The *gauss* is a unit of magnetic flux density. A field strength of 1 gauss is considered to be 1 line per square centimeter of surface perpendicular to the flux lines in a magnetic field. The tesla is the other major unit of magnetic flux density; 1 tesla is equal to 10,000 gauss. *See also* MAGNETIC FLUX, and TESLA.

GAUSS'S THEOREM

Gauss's Theorem is an expression for determining the intensity of an electric field, depending on the quantity of charge. For any closed surface in the presence of an electric field, the electric flux passing through that surface is directly proportional to the enclosed quantity of charge. *See also* ELECTRIC FIELD.

GENERATOR

A *generator* is a source of signal in an electronic circuit. Such a device can be an oscillator or it can be an electromechanical circuit. Signal generators are widely used in electronic design, testing, and troubleshooting (*see* SIGNAL GENERATOR).

A device for producing alternating-current electricity, by means of a rotating coil and magnetic field, is called a *generator*. Whenever a conductor moves within a magnetic field so that the conductor cuts across magnetic lines of force, current is induced in the conductor. A generator can consist of either a rotating magnet inside a coil of wire, or a rotating coil of wire inside a magnet. The illustration shows the principle of an alternating-current generator. The shaft is driven by a motor powered by gasoline or some other fossil fuel; alternatively, steam turbines can be used.

Small portable gasoline-powered generators, capable of delivering from about 1 to 5 kW, can be purchased in department stores. Larger generators can supply enough electricity to run an entire house or building. The largest electric generators, found in power plants, are as large as a small building, produce millions of watts, and can provide sufficient electricity for a community.

A generator, like a motor, is an electromechanical transducer. The construction of an electric generator is, in fact, almost identical to that of a motor; some motors can operate as generators when their shafts are turned by an external force. *See also* MOTOR.

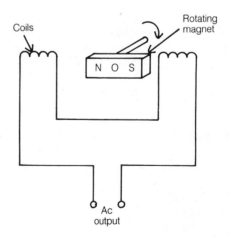

GENERATOR: A magnet rotates within a coil of wire.

GEOMAGNETIC FIELD

The *geomagnetic field* is a magnetic field surrounding our planet. The earth is magnetized like a huge bar magnet. The north magnetic pole is located near the north geographic pole, and the south magnetic pole near the south geographic pole. The magnetic lines of force extend far out into space. The earth is not the only planet that has a magnetic field. Other planets, especially Jupiter, as well as stars, such as the sun, are known to have magnetic fields.

The magnetic field of the earth is responsible for the behavior of the magnetic compass. The needle of a compass is actually a small bar magnet, suspended on a bearing that allows free rotation. One pole of the bar magnet points south and the other pole points north in most locations on the surface of the earth. There is a slight error because of the difference in location between the magnetic and geographic poles.

The geomagnetic field attracts charged particles emitted by the sun, and causes them to be concentrated near the magnetic poles. When there is a solar flare, resulting in large amounts of charged-particle emission from the sun, the upper atmosphere glows in the vicinity of the magnetic poles. This is known as the *Aurora*. The appearance of the Aurora heralds a disturbance in the magnetic environment of the earth. This is called a *geomagnetic storm. See also* AURORA, GEOMAGNETIC STORM, and MAGNETIC FIELD.

GEOMAGNETIC STORM

A *geomagnetic storm* is a disturbance or fluctuation in the magnetic field of the earth. Such a ''storm'' is caused by charged particles from the sun. A solar flare causes the emission of high-speed protons, electrons, and alpha particles, all of which carry an electric charge. Because these charges are in motion, they have an effective electric current, which is influenced by the magnetic field of the earth. These particles also create their own magnetic field as they move.

A geomagnetic storm changes the character of the ionosphere of the earth, with a profound impact on radio communications at the low, medium, and high frequencies. Often, communications circuits will be completely severed because of changes in the ionization in the upper atmosphere. Instead of returning signals to the earth, the ionosphere can absorb electromagnetic energy, or can allow it to pass into space.

To a certain extent, we can predict a geomagnetic storm. A solar flare can be seen with optical telescopes, and it is from one hour to several hours before the charged particles arrive at the earth. During the night, the Aurora can sometimes be seen during a geomagnetic storm. If a disturbance is especially severe, telephone circuits can be disrupted, as well as other wire services. *See also* AURORA, GEOMAGNETIC FIELD, IONOSPHERE, and SOLAR FLARE.

GERMANIUM

Germanium is an element, with atomic number 32 and atomic weight 73. Germanium is a semiconductor material. In its pure form, it is an insulator; its conductivity depends on the amount of impurity elements added. It is used in the manufacture of some diodes, photocells, and transistors.

Germanium has been largely replaced by silicon in modern semiconductor devices. This is because germanium is sensitive to heat, and germanium devices can be destroyed by the soldering process that is used to manufacture electronic circuitry.

Germanium diodes are still occasionally found in electronic devices. The germanium diode has a forward voltage drop of approximately 0.3, compared with 0.6 for the silicon diode.

Germanium diodes are used primarily in radio-frequency applications. *See also* DIODE.

GHOST

A *ghost* is a false image in a television receiver, caused by reflection of a signal from some object, just prior to its arrival at the antenna. The ghost appears in a position different from the actual picture because of the delay in propagation of the ghost signal, with respect to the actual received signal.

Ghosts usually appear as slightly displaced images on a television screen. In especially severe cases, the ghost can be almost as prominent as the actual picture, which makes it difficult or impossible to view the television signal. Ghosting can often be minimized simply by reorienting an indoor television antenna, or rotating an outdoor antenna. In some cases, ghosts can be difficult to eliminate. The object responsible for the signal reflection must then be moved far away from the antenna. In certain cases, this is not practical.

With modern cable television, ghosting is seldom a problem. The signals are entirely confined to the cable, and propagation effects are unimportant. Ghosting can occur in a cable system, however, if impedance ''bumps'' are present in the line near the receiver. *See also* TELEVISION.

GIGA

Giga is a prefix meaning 1 billion (1,000,000,000). This prefix is attached to quantities to indicate the multiplication factor of 1 billion. For example, 1 gigohm is 1 billion ohms; 1 gigaelectron-volt is 1 billion electron volts; 1 gigahertz is 1 billion Hertz; and so on.

Giga is one of many different prefix multipliers commonly used in electronics. *See also* PREFIX MULTIPLIERS.

GILBERT

The *gilbert* is a unit of magnetomotive force, named after the scientist William Gilbert. The standard symbol for the gilbert is the upper-case Greek letter sigma (Σ).

In a coil, a magnetomotive force of 1 gilbert is equivalent to 0.796 ampere turns. In general, if N is the number of turns in a coil and I is the current, in amperes, that flows through the coil, then the magnetomotive force Σ in gilberts is equal to:

$$\Sigma = 1.26 \, NI$$

See also MAGNETOMOTIVE FORCE.

GOLD

Gold is a metallic element, with atomic number 79 and atomic weight 197. We recognize this yellowish metal as an international money standard. It is also useful in some electronic applications.

Gold plating is occasionally used to reduce the electrical resistance of a set of switch or relay contacts. Gold is highly malleable, and can withstand the physical stress of repeated contact openings and closings. Also, gold is highly resistant to corrosion. Gold-plated contacts maintain their reliability over a long period of time without the need for frequent cleaning. Gold is a good conductor of electricity.

Gold is used in the manufacture of certain semiconductor diodes. A fine gold wire is bonded to a tiny wafer of germanium. This device is called a *gold-bonded diode*. Gold can be used as an impurity substance in the doping of semiconductor transistors. *See also* DOPING.

GONIOMETER

A *goniometer* is a direction-finding device commonly used in radiolocation and radionavigation. Many direction finders have rotatable antennas, with certain characteristic directional patterns. The goniometer uses a mechanically fixed antenna, and the directional response is varied electrically.

A simple pair of phased vertical antennas can serve as a goniometer. The two antennas can be spaced at any distance between ¼ and ½ wavelength. Depending on the phase in which the antennas are fed, the null can occur in any compass direction. The null can be unidirectional in some situations, and bidirectional in other cases. The precise direction is determined by knowing the phase relationship of the two feed systems. *See also* DIRECTION FINDER, RADIOLOCATION, and RADIONAVIGATION.

GRACEFUL DEGRADATION

When a portion of a computer system malfunctions, it is desirable to have the computer continue operating, even though the efficiency is impaired. The failure of a single diode should not result in the shutdown of an entire system. This property is called *graceful degradation*.

In the event of a subsystem malfunction, a sophisticated computer can use other circuits to temporarily accomplish the tasks of the failed portions of the system. The attendant personnel are notified by the computer that a malfunction has occurred so that it can be corrected as soon as possible. Graceful degradation is sometimes called *failsafe* or *failsoft*. *See also* COMPUTER, and FAILSAFE.

GRADED FILTER

A *graded filter* is a form of power-supply output filter that provides various degrees of ripple elimination. Some circuits can tolerate more ripple than others. Those circuits that can operate with higher ripple levels are connected to earlier points in the filter sequence. Some circuits need less power-supply filtering because they draw very little current; these circuits are also connected to the earlier points in the graded filter.

The graded filter offers the advantage of superior ripple elimination and voltage regulation for those circuits that require the purest direct current. The current drain is reduced at the farthest point in the graded filter. *See also* POWER SUPPLY.

GRADED JUNCTION

A *graded semiconductor junction* is a P-N junction that is grown, by means of epitaxial methods, in a carefully controlled manner. Graded-junction devices have reverse-bias characteristics that differ from ordinary semiconductor devices. The capacitance across the junction of a reverse-biased, graded-junction diode decreases more rapidly, with increasing reverse voltage, than the capacitance across an ordinary P-N junction. *See also* P-N JUNCTION.

GRAPHITE

Graphite is a form of carbon, atomic number 12. Graphite is a soft, grayish-black substance. It is used in many different kinds of electronic components, especially resistors and the plates of certain vacuum tubes. Graphite is used in the manufacture of common pencils.

Graphite, because of its high tolerance for heat, is often combined with nonconducting material, such as clay, to make resistors. Most small noninductive resistors are of the graphite variety.

In power-amplifier vacuum tubes, the plates dissipate large amounts of heat. Graphite can withstand heat better than the metals normally used in vacuum-tube plates; graphite plates are found in thicknesses as great as several millimeters. The use of graphite plates increases the tube life. *See also* CARBON RESISTOR, and TUBE.

GREAT CIRCLE

A *great circle* is a path between two points over the surface of a sphere, such that the path forms part of a circle whose center is the same as the center of the sphere. A great-circle path lies in a plane that passes through the geometric center of the sphere (see illustration). On the earth, great-circle paths represent the shortest distance between two points. Great-circle paths are sometimes called *geodesic paths*. Examples of great circles on the earth include all longitude lines and the equator; but the latitude lines are not great circles.

When radio waves propagate around the world, they usually follow paths that are nearly great circles. For this reason, in high-frequency radio communications, it is often desirable to know the great-circle bearing toward a particular place. This is especially important if a directional antenna system is in use. The great-circle bearing can be determined by means of a great-circle map of the world. Such a map shows the geography of the earth, as seen from a viewpoint that is centered on your location.

Great-circle bearings are often surprising. For example, South Africa lies almost directly east, along a great-circle path, from the center of the United States. India lies north, over the pole. Radio waves occasionally travel the long way around the world, rather than the short way; in such cases, the antenna should be pointed west for South Africa and south for India, from a receiving location in the center of the United States. *See also* LONG-PATH PROPAGATION, and SHORT-PATH PROPAGATION.

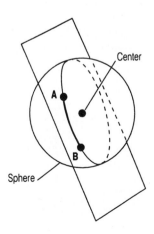

GREAT CIRCLE: The sphere and circle have the same center point.

GREENWICH MEAN TIME

Greenwich Mean Time (GMT), is the time in the zone along the meridian of zero degrees, passing through Greenwich, England. Greenwich Mean Time is based on the position of this meridian, with respect to the sun. At 1200 GMT, the Greenwich meridian lines up precisely with the sun. At 0600 and 1800 GMT, the Greenwich meridian is 90 degrees (¼ rotation) from the sun. At 0000 GMT, the Greenwich meridian lies exactly opposite the sun. The 24-hour time scale is always specified for GMT.

Greenwich mean time used to be the standard for time throughout the world. In recent years, however, GMT has been replaced with coordinated universal time (UTC). The two are the same for all practical purposes. *See also* COORDINATED UNIVERSAL TIME.

GREMLIN

Gremlin is a term that is sometimes used by technicians to refer to a continuing problem that is difficult or impossible to track down. The expression originated among aircraft personnel. A gremlin is a little devil, wandering around the apparatus, on which the problem can be blamed!

Gremlins are a frustratingly common phenomenon in electronics. Every technician or engineer knows how gremlins behave. *See also* TROUBLESHOOTING.

GRID

A *grid* is an element in a vacuum tube, placed between the *cathode*, which emits electrons, and the *plate*, which collects electrons. The grid resembles a screen, and surrounds the cathode in the shape of a cylinder or oblate cylinder. A vacuum tube can have two, three, or more concentric grids.

The function of the grid is to control the current through the tube. When the grid is at the same potential as the cathode, a certain amount of current will flow from the cathode to the plate. As the grid voltage is made more negative, less current flows through the tube. If the grid is made positive, it draws some current away from the path between the cathode and the plate. When a rapidly varying signal voltage is applied to the grid of a tube, the current in the plate circuit fluctuates along with the signal. When this current is put through a load, the resulting voltage fluctuations are often much larger than those of the original input signal; this is how amplification is obtained.

The innermost grid in a tube is called the *control grid*. The second grid is called the screen; the third grid is called the *suppressor*. All have different functions. The input signal is usually applied to the control grid. *See also* TUBE.

GRID BIAS

In a vacuum tube, the *grid bias* is the direct-current voltage that is applied to the control grid. The grid bias is usually measured with respect to the cathode. The grid bias can be obtained in a variety of different ways.

Most vacuum tubes require a negative grid bias, the exact value depending on the characteristics of the tube and the application for which the tube is to be used. Some tubes operate with zero bias.

As the grid bias is made progressively more negative, the plate current decreases until, at a certain point, the tube stops conducting altogether. This value of grid bias is called the *cutoff*

bias, and it depends on the plate voltage as well as the particular type of tube used. If the grid bias is made positive, the grid will draw current away from the plate circuit. An input signal can drive the control grid positive for part of the cycle, but deliberate positive bias is almost never placed on the control grid of a tube.

The level of grid bias must be properly set for a tube to function according to its specifications in a given application. The required grid bias is different among class-A, class-AB, class-B, and class-C amplifiers. *See also* CLASS-A AMPLIFIER, CLASS-AB AMPLIFIER, CLASS-B AMPLIFIER, CLASS-C AMPLIFIER, GRID, and TUBE.

GROUND

Ground is the term generally used to describe the common connection in an electrical or electronic circuit. The common connection is usually at the same potential for all circuits in a system.

The common connection for electronic circuits is almost always ultimately routed to the earth. The earth is a fair to good conductor of electricity, depending on the characteristics of the soil. The earth itself provides an excellent common connection where the conductivity is good.

For an earth ground, the best connection is found in areas where salt comprises part of the soil. Such places include beach areas. Dark, moist soil also provides a good ground. *See also* EARTH CONDUCTIVITY.

GROUND ABSORPTION

Electromagnetic energy propagates, to a certain extent, along the surface of the earth. This occurs mostly at the very low, low, and medium frequencies. It can also take place at the high frequencies, but to a lesser degree. In this situation, the earth actually forms part of the circuit by which the wave travels; this mode is *called surface-wave propagation* (*see* SURFACE WAVE).

The ground is not a perfect conductor, and therefore a surface-wave circuit tends to be rather lossy. The better the earth conductivity in a given location, the better the surface-wave propagation, and the lower the ground absorption. Salt water forms the best surface, over which electromagnetic waves can travel. Dark, moist soil is also good. Rocky, sandy, or dry soil is relatively poor for the propagation of electromagnetic waves, and the absorption is high.

No matter what the earth conductivity, the ground absorption always increases as the frequency increases. At very high frequencies and above, the ground absorption is so high that signals do not propagate along the surface, but instead travel directly through space. *See also* EARTH CONDUCTIVITY.

GROUND BUS

A *ground bus* is a thick metal conductor, such as a strap, braid, or section of tubing, to which all of the common connections in a system are made. The ground bus is routed to an earth ground by the most direct path possible.

The advantage of using a ground bus in an electronic system is that it avoids the formation of ground loops (*see* GROUND LOOP). All of the individual components of the system are separately connected to the bus, as shown in the schematic diagram.

The size of the conductor used for a ground bus depends on the number of pieces of equipment in the system, and on the total current that the system carries. Typical ground-bus conductors are made from copper tubing that have a diameter of ¼ to ½-inch. For radio-frequency applications, the larger conductor sizes are preferable; also, the distance to the ground connection should be made as short as possible in such cases. *See also* GROUND.

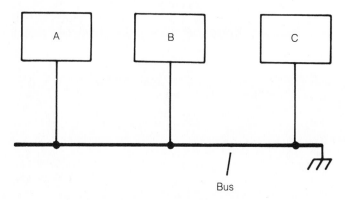

GROUND BUS: This scheme avoids ground loops.

GROUND CONNECTION

A *ground connection* is the electrical contact between the common point of an electrical or electronic system and the earth. A ground connection should have the least possible resistance. The most effective ground connection consists of one or more ground rods driven into the soil to a depth of at least 6 to 8 feet. Alternatively, a ground connection can be made to a cold-water pipe, although the continuity of the pipe should be checked before assuming the connection will be good. In modern homes, pipes often have plastic joints; the entire pipe can even be made of plastic.

The ground connection is of prime importance to the operation of unbalanced radio-frequency systems. In balanced radio-frequency systems, the ground connection is less critical, although from a safety standpoint, it should always be maintained. All components of a system should be connected together, preferably via a ground bus leading to an earth ground, to prevent possible potential differences that can cause electric shock. *See also* GROUND, and GROUND BUS.

GROUND EFFECT

The directional pattern of an antenna system, especially at the very low, low, medium, and high frequencies, is modified by the presence of the surface of the earth underneath the antenna. This is called *ground effect*. It is more pronounced in the vertical, or elevation, plane than in the horizontal plane.

The effective surface of the earth usually lies somewhat below the actual surface. The difference depends on the earth conductivity, and on the presence of conducting objects (such as buildings, trees, and utility wires). The effective ground surface reflects radio waves to a certain extent. At a great distance, the reflected wave and the direct wave add together in variable phase.

Ground effects have a large influence on the takeoff angle of electromagnetic waves from an antenna. This, in turn, affects the optimum distance at which communications is realized. In general, the higher a horizontally polarized antenna is positioned above the ground, the better the long-distance communication will be. For a vertically polarized antenna with a good electrical ground system, the height of the radiator above the ground is of lesser importance.

Ground effect occurs in the same manner for reception, with a given antenna system, as for transmission at the same frequency. *See also* EARTH CONDUCTIVITY, EFFECTIVE GROUND, and GROUND WAVE.

GROUND FAULT

A *ground fault* is an unwanted interruption in the ground connection in an electrical system. This can result in loss of power to electrical equipment because there is no circuit via which the current can return. Excessive current can flow in the ground conductor of an ac electrical system, and dangerous voltages can be present at points that should be at ground potential.

A ground fault is usually not difficult to locate, but it might not immediately manifest itself. Only one piece of equipment can be involved, and the ground fault will not be noticed until that piece of equipment is operated. Ground faults can sometimes result in dangerous electrical shock.

Ground connections should, if possible, be soldered into place so that corrosion will not result in ground faults. The number of joints in a ground circuit should be kept to a minimum. *See also* GROUND, and GROUND CONNECTION.

GROUND LOOP

A *ground loop* occurs when two pieces of equipment are connected to a common ground bus, and are also connected together by means of separate wires or cables. Ground loops are quite common in all kinds of electrical and electronic apparatus, and normally they do not present a problem. However, in certain situations, they can result in susceptibility to electromagnetic interference, such as hum pickup or interaction with nearby transmitters.

In the event of undesirable hum pickup or other electromagnetic interference that defies attempts at troubleshooting, ground loops should be suspected. Ground loops can be extremely difficult to locate. However, removing interconnecting ground wires one at a time, and checking for interference each time, will eventually pinpoint the source of the trouble. *See also* ELECTROMAGNETIC INTERFERENCE, GROUND, and GROUND BUS.

GROUND PLANE

A *ground plane* is an artificial radio-frequency ground, constructed from wires or other conductors, and intended to approximate a perfectly conducting ground. Ground planes are often used with vertical antennas to reduce the losses caused by ground currents. All amplitude-modulation broadcast stations use vertical antennas with ground-plane systems.

A ground plane can be constructed by running radial wires outward from the base of a quarter-wave vertical antenna. The conductors can be buried under the soil or placed just above the surface of the earth. They can be insulated or bare. The more radials that are used, the more effective the ground plane. Also, the quality of the ground plane improves with increasing radial length. An optimum system consists of at least 100 radials of ½-wavelength or longer, arranged at equal angular intervals. However, smaller radial systems provide considerable improvement in antenna performance, compared to no radials at all. The radials are connected to a driven ground rod at the base of the antenna.

In some high-frequency antenna systems, the ground plane is elevated above the actual surface of the earth. When the ground plane is ¼ wavelength or more above the surface, only three or four quarter-wave radials are necessary in order to obtain a nearly perfect ground plane. The ground-plane antenna operates on this principle. *See also* GROUND-PLANE ANTENNA.

GROUND-PLANE ANTENNA

A *ground-plane antenna* is a vertical radiator operated against a system of quarter-wave radials and elevated at least a quarter wavelength above the effective ground. The radiator itself can be any length, but should be tuned to resonance at the desired operating frequency.

When a ground plane is elevated at least 90 electrical degrees above the effective ground surface, only three or four radials are necessary in order to obtain an almost lossless system. The radials are usually run outward from the base of the antenna at an angle that can vary from 0 degrees to 45 degrees, with respect to the horizon. The drawing illustrates a typical ground-plane antenna.

A ground-plane antenna is an unbalanced system, and should be fed with coaxial cable. A balun can be used, however, to allow the use of a balanced feed line. The feed-point impedance of a ground-plane antenna having a quarter-wave radiator is about 37 ohms if the radials are horizontal; this impedance increases as the radials are drooped, reaching about 50 ohms at an angle of 45 degrees (*see* DROOPING RADIAL). The radials can be run directly downward, in the form of a quarter-wave tube concentric with the feed line. Then, the feed-point impedance is approximately 73 ohms. This configuration is known as a *coaxial antenna. See also* EFFECTIVE GROUND, GROUND PLANE, and RADIAL.

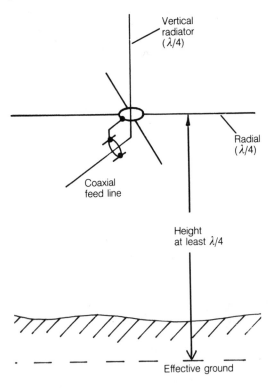

Vertical radiator (λ/4)

Radial (λ/4)

Coaxial feed line

Height at least λ/4

Effective ground

GROUND-PLANE ANTENNA: Radials form an artificial ground.

GROUND RETURN

In direct-current transmission, as well as some alternating-current circuits, one leg of the connection is provided by the earth ground. This part of the circuit is called the *ground return*. The ground return allows the use of only one conductor, rather than two conductors, in a transmission system. A good ground connection is essential at each end of the circuit if the ground return is to be effective.

In radar, objects on the ground cause false echoes. This effect is especially prevalent in populated areas, where the numerous manmade structures "fool" the radar. This is called the *ground-return effect*; it is also sometimes called *ground clutter*. Ground-return effects make it difficult to track a target at a low altitude. This phenomenon can also confuse weather forecasters attempting to track thunderstorms by radar. *See also* RADAR.

GROUND ROD

A *ground rod* is a solid metal rod, usually made of copper-plated steel, used for the purpose of obtaining a good ground connection for electrical or electronic apparatus.

Ground rods are available in a wide variety of diameters and lengths. The most effective ground is obtained by driving one or more rods at least 8 feet into the soil, well away from the foundations of buildings. Smaller ground rods are sometimes adequate for very low-current or low-power installations in locations where the soil conductivity is excellent. Ground rods can be obtained at most electrical equipment stores. *See also* GROUND, and GROUND CONNECTION.

GROUND WAVE

In radio communication, the *ground wave* is that part of the electromagnetic field that is propagated parallel to the surface of the earth. The ground wave actually consists of three distinct components: the direct (line-of-sight) wave, the reflected wave, and the surface wave. Each of these three components contributes to the received groundwave signal.

The direct wave travels in a straight line from the transmitting antenna to the receiving antenna. At most radio frequencies, the electromagnetic fields pass through objects, such as trees and frame houses, with little attenuation. Concrete-and-steel structures cause some loss in the direct wave at higher frequencies. Obstructions, such as hills, mountains, or the curvature of the earth, cut off the direct wave completely.

A radio signal can be reflected from the earth or from certain structures, such as concrete-and-steel buildings. The reflected wave combines with the direct wave (if any) at the receiving antenna. Sometimes, the two are exactly out of phase, in which case the received signal is extremely weak. This effect occurs mostly in the very-high-frequency range and above.

The surface wave travels along the earth, and occurs only with vertically polarized energy at the very low, low, medium, and high frequencies. Above 30 MHz, there is essentially no surface wave. At the very low and low frequencies, the surface wave propagates for hundreds or even thousands of miles. Sometimes, the surface wave is called the *ground wave*, in ignorance of the fact that the direct and reflected waves can also contribute to the ground wave in very short-range communications. The actual ground-wave signal is the phase combination of all three components at the receiving antenna. *See also* DIRECT WAVE, REFLECTED WAVE, and SURFACE WAVE.

GUARD BAND

In a channelized communications system, a *guard band* or *guard zone* is a small part of the spectrum allocated for the purpose of minimizing interference between stations on adjacent channels. It is desirable to have some unused frequency space between channels so that the sidebands of one signal will not cause interference in the passband of a receiver tuned to the next channel.

The width of the guard band should be sufficient to allow for the imperfections in the bandpass filters of the transmitter and receiver circuits. However, excessive guard-band space is wasteful of the spectrum. It is sometimes a matter of trial- and-error to determine the optimum guard-band width in a communications system. An example of this is the 2-meter frequency-modulation amateur band. The original band plan used channels spaced at 15-kHz intervals between 146 and 148 MHz. The deviation of the signals in this band is plus or minus 5 kHz, or a total theoretical bandwidth of 10 kHz. However, adjacent-channel interference was often observed in this band. Later, when the band was extended below 146 MHz, the channels were established at 20-kHz increments. This has reduced adjacent-channel interference.

In some bands, notably the international shortwave broadcast band, there is often no guard zone at all. Stations in this part of the spectrum usually have a bandwidth of 10 kHz, but can be spaced only 5 kHz apart, resulting in severe adjacent-channel interference. *See also* ADJACENT-CHANNEL INTERFERENCE.

GUNN DIODE

A *Gunn diode* is a form of semiconductor device that operates as an oscillator in the ultra-high-frequency and microwave parts of the electromagnetic spectrum. The Gunn diode has replaced the Klystron tube in many situations.

The Gunn diode is mounted in a resonant enclosure, as shown at A in the illustration. A direct-current voltage is applied to the device and, if this voltage is large enough, the Gunn diode will oscillate. A schematic diagram of an oscillator using a Gunn diode is shown at B.

Gunn diodes are not very efficient. Only a small fraction of the consumed power actually results in radio-frequency power. Gunn diodes tend to be highly sensitive to changes in temperature and bias voltage. The frequency can vary considerably— even with a small change in the ambient temperature; for this reason, the temperature must be carefully regulated. The change in frequency with voltage can be useful for frequency-modulation purposes, but voltage regulation of some sort is essential. Most Gunn diodes require about 12 volts for proper operation. Oscillation can occur at frequencies in excess of 20 GHz. The output power from a Gunn-diode oscillator can, with proper bias and circuit design, be more than 0.1 W. *See also* DIODE OSCILLATOR.

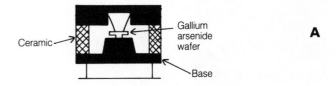

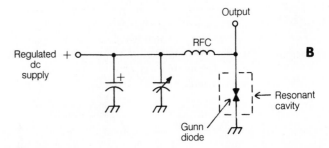

GUNN DIODE: At A, a cutaway view; at B, a typical connection.

GUYING

Guying is the installation of supporting wires for tall structures. Some antenna supports cannot stand up without guy wires. Others need guying only in the event of high winds.

Guy wires are usually installed in sets of three or four, and can be attached to the antenna support structure at one or more levels. If three sets of guy wires are used, they should be brought out from the support structure at 120-degree angles (as viewed from above); if sets of four guys are used, they should be positioned at 90-degree angles.

The number of sets of guy wires needed for a particular support structure depends on its height, its rigidity, and the expected maximum wind loading. Wind loading depends on the diameter of the structure, the kind of antenna array it supports, and the wind velocity. The maximum expected wind velocity varies considerably depending on geographic location. In general, the worst areas are those exposed to hurricanes arriving from the ocean.

Guy wires should generally not be brought down from a tower at an angle of less than 30 degrees, with respect to the tower in the vertical plane. The steeper the angle at which a guy wire is installed, the greater the strain on the wire and on the guy anchor.

Guy anchors should be installed so that they are not easily pulled from the ground. With large towers, a separate guy anchor should be used for each guy wire. In smaller towers having multiple guy sets, a wire from each set can be attached to a single, common guy anchor.

The guying of antenna support structures should not be undertaken by an amateur. It is advisable that a professional be consulted so that the optimum system can be installed in terms of safety, cost, and longevity. Guesswork can result in disaster. *See also* TOWER, and WIND LOADING.

HAIRPIN MATCH

A *hairpin match* is a means of matching a half-wave, center-fed radiator to a transmission line. The hairpin match is especially useful in the driven element of a Yagi antenna, where the feed-point impedance is lowered by the proximity of parasitic elements.

The hairpin match requires that the radiating element be split at the center. It also requires the use of a balanced feed system; if a coaxial line is used, a balun is needed at the feed point. The hairpin match consists of a section of parallel-wire transmission line, short-circuited at the far end. The length of the section is somewhat less than ¼ wavelength, and thus it appears as an inductance.

The adjustment of the hairpin match is quite simple; a sliding bar can be used to vary the length of the section. The section itself should be perpendicular to the driven element; this generally necessitates mounting it along the boom of the Yagi antenna. The use of a hairpin match causes a slight lowering of the resonant frequency of a radiating element. Therefore, the radiator must be shortened by a few percent to maintain operation at the desired frequency. When the length of the hairpin section and the length of the radiating element are just right, a nearly perfect match can be obtained with either 50- or 75-ohm coaxial feed lines. *See also* IMPEDANCE MATCHING, and YAGI ANTENNA.

HALF-ADDER

A *half-adder* is a digital logic circuit with two input terminals and two output terminals. The output terminals are called the *sum and carry outputs*. The sum output of a half-adder circuit is the exclusive-OR function of the two inputs. That is, the sum output is 0 when the inputs are the same and 1 when they are different. The carry output is the AND function of the two inputs: It is 1 only when both inputs are 1 (see the truth table).

The half-adder circuit differs from the adder in that the half-adder will not consider carry bits from previous stages. There are several different combinations of logic gates that can function as half adders. *See also* ADDER, and BINARY-CODED NUMBER.

HALF-ADDER: SUM AND CARRY OUTPUTS
AS A FUNCTION OF LOGIC INPUT.

Inputs X Y	Sum output	Carry output
0 0	0	0
0 1	1	0
1 0	1	0
1 1	0	1

HALF-BRIDGE

A *half-bridge* is a form of rectifier circuit, similar to the bridge rectifier, except that two of the diodes are replaced by resistors. The two resistors have equal values. The voltages at either end of the series combination are always equal and opposite. Thus, at the center point, the voltage is zero, and this point is generally grounded. By reversing the diodes, a negative-voltage supply is obtained.

The efficiency of the half-bridge circuit is less than that of the conventional bridge rectifier. This is because the resistors tend to dissipate some power as heat as the current flows through them. The half-bridge operates over the entire alternating-current input cycle, and is therefore a form of full-wave rectifier. *See also* BRIDGE RECTIFIER.

HALF CYCLE

In any alternating-current system, a complete cycle occurs between any two identical points on the waveform. These points can be chosen arbitrarily; they may be positive peaks, zero-voltage points, negative peaks, or any other point. The complete cycle requires a certain length of time, *P*, which is called the *period of the wave*.

A half cycle is simply any part of the waveform that occurs during a time interval of *P*/2, if the period is *P*. Usually, the term *half-cycle* is used in reference to either the negative or positive portion of a sine-wave alternating-current waveform. A half-cycle represents 180 electrical degrees. *See also* CYCLE, and ELECTRICAL ANGLE.

HALF-POWER POINTS

The sharpness of an antenna directive pattern, or of the selective response of a bandpass filter, is often specified in terms of the half-power points. In the case of a directive antenna, the variable parameter is compass direction. In the case of a bandpass filter, the variable parameter is frequency.

In antenna systems, the reference power level is the effective radiated power in the favored direction. This can occur in more than one direction. In most parasitic arrays, the favored direction is a single compass point; in the case of the half-wave dipole, the favored direction is two compass points; an unterminated longwire generally has favored directions at four compass points. The effective radiated power at the half-power points is 3 dB below the level in the favored direction. The field strength, in volts per meter, is 0.707 times the field strength in the favored direction. The angle, in degrees, between the half-power points of a single lobe is called *the beamwidth*. *See also* FIELD STRENGTH.

For a bandpass filter, the half-power points are the frequencies at which the power from the filter drops 3 dB below the power output at the center of the passband. The bandwidth of the filter is sometimes specified in terms of the half-power points, but more often it is given in terms of the 6-dB attenuation points and the 60-dB attenuation points. This combined figure gives an indication of the skirt selectivity, as well as the actual bandwidth. *See also* BANDPASS FILTER, BANDPASS RESPONSE, SHAPE FACTOR, and SKIRT SELECTIVITY.

HALF-WAVE ANTENNA

A *half-wave antenna* is a radiating element that measures an electrical half wavelength in free space. Such an antenna can have a physical length anywhere from practically zero to almost the physical dimensions of a half wavelength in free space. A dipole antenna is the simplest example of a half-wave antenna.

In theory, a half wavelength in free space is given in feet according to the equation:

$$L = 492/f$$

where L is the linear distance and f is the frequency in megahertz. A half wavelength in meters is given by:

$$L = 150/f$$

In practice, an additional factor must be added to the above equations, because electromagnetic fields travel somewhat more slowly along the conductors of an antenna than they do in free space. For ordinary wire, the results (as obtained above) are multiplied by about 0.95. For tubing or large-diameter conductors, the factor is slightly smaller, and can range down to about 0.90 (*see* VELOCITY FACTOR).

A half-wave antenna can be made much shorter than a physical half wavelength. This is accomplished by inserting inductances in series with the radiator. The antenna can be made much longer than the physical half wavelength by inserting capacitances in series with the radiator. *See also* CAPACITIVE LOADING, ELECTRICAL WAVELENGTH, and INDUCTIVE LOADING.

HALF-WAVE RECTIFIER

A *half-wave rectifier* is the simplest form of rectifier circuit. It consists of nothing more than a diode in series with one line of an alternating-current power source, and a transformer (if necessary) to obtain the desired voltage. A is a schematic diagram of a half-wave rectifier circuit that is designed to supply a positive output voltage. The diode can be reversed to provide a negative output voltage.

The output of the half-wave rectifier contains only one half of the alternating-current input cycle, as shown at B. The other half is simply blocked. The half-wave rectifier gets its name from the fact that it operates on only one half of the input cycle.

Half-wave rectifiers have the advantage of being extremely simple in terms of design. An unbalanced alternating-current input source can be used; in some instances, no transformer is needed and the direct-current voltage can be derived straight from a wall outlet. However, the half-wave rectifier has rather poor voltage-regulation characteristics. The pulsating output is more difficult to filter than that of the full-wave rectifier. Half-wave rectifier circuits are often used in situations where the current drain is low and the voltage regulation need not be especially precise. *See also* BRIDGE RECTIFIER, and FULL-WAVE RECTIFIER.

HALF-WAVE TRANSMISSION LINE

A *half-wave transmission line* is a section of electromagnetic feed line that measures an electrical half wavelength. The physical length (L) of such a transmission-line section, for a frequency (f) in megahertz, is given in feet by:

$$L = 492v/f$$

where v is the velocity factor of the line (*see* VELOCITY FACTOR). The length (L) in meters is:

$$L = 150v/f$$

A half-wave section of transmission line has certain properties that make it useful as a tuned circuit. The impedance at one end of such a line is exactly the same, neglecting line loss, as the impedance at the other end. This is true, however, only within a narrow range of frequencies, centered at the half-wave resonant frequency. If the far end of a half-wave transmission line is an open circuit, the line behaves as a parallel-resonant tuned circuit. If the far end is short-circuited, the section behaves as a series-resonant tuned circuit.

At frequencies above approximately 20 MHz, where half-wave sections of transmission line have reasonable length, these sections are often used in place of coils and capacitors as resonant circuits. If the line loss is reasonably low, the Q factor of the half-wave transmission line is very high. The half-wave transmission line may be either balanced or unbalanced, depending on the circuit in which it is used. *See also* Q FACTOR.

HALO ANTENNA

A *halo antenna* is a special form of horizontal half-wave antenna. Basically, the halo consists of a dipole whose elements have been bent into a circle so that the circumference of the circle is ½ electrical wavelength. The ends of the dipole are insulated from each other at the opposite side of the circle from the feed point (see illustration).

The halo antenna is used mostly in the very-high-frequency part of the radio spectrum. At 30 MHz, for example, the circumference of a halo antenna is only about 15 feet so that the diameter is less than 5 feet. Copper or aluminum tubing is practical for the construction of a halo in the very-high-frequency range.

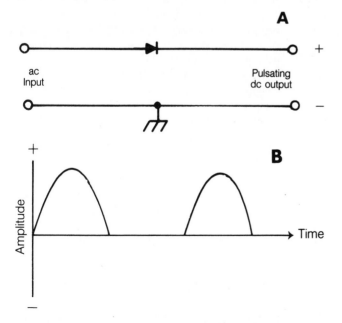

HALF-WAVE RECTIFIER: Circuit (A) and an output waveform (B).

Halo antennas exhibit a nearly omnidirectional radiation pattern in the horizontal plane. The polarization is horizontal. Halo antennas are often vertically stacked to obtain omnidirectional gain in the horizontal plane. *See also* DIPOLE ANTENNA.

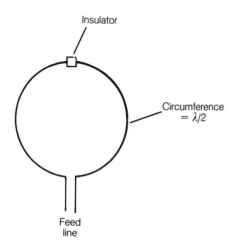

HALO ANTENNA: This is a half-wave dipole that is bent into a circle.

HAM

Amateur radio operators are sometimes called *hams*, and amateur radio itself is sometimes called *ham radio*. The exact origin of the term *ham* is uncertain, although in the old days of telegraphy, the expression was used to refer to poor operators! It is believed that the amateurs embraced the name in a light-hearted way, and the term has stuck ever since.

HANDSHAKING

In a digital communications system, accuracy can be improved by synchronizing the transmitter and receiver precisely before the beginning of data transfer. This is called *handshaking*. The process can be repeated at intervals to maintain synchronization.

Handshaking is becoming more and more universal in electronic communications because it dramatically improves the signal-to-noise ratio. In general, the more frequently the handshaking operation is done in the process of signal transmission, the better the signal-to-noise ratio, although there is a point of diminishing returns.

In some digital systems, an independent reference standard, such as a time and frequency station, is used to synchronize the digital signals between the transmitter and receiver. This is called *coherent* or *synchronized digital communications*. *See also* SYNCHRONIZED COMMUNICATIONS.

HARD LINE

Hard line is a form of coaxial cable with 100 percent shielding continuity. The perfect shield is obtained because the outer conductor of the hard line is made from metal tubing. Aluminum tubing is commonly used for the manufacture of hard line; copper or galvanized metal can also occasionally be used.

Hard line is not as flexible as coaxial cable with a braided-wire outer conductor. Hard line is therefore somewhat more difficult to install. The dielectric material can be solid, or it can consist of beaded sections; it is often made from polyethylene, although sometimes Teflon is used.

Hard lines are known for their durability and relatively low loss. They are available in a number of sizes and characteristic-impedance ratings. The smallest common hard line is about 3/8 inch in diameter; the largest commonly available is about 1 inch across.

HARDWARE

In computer engineering, *hardware* is a term used to describe the actual circuitry that makes up the device. The wiring, circuit boards, integrated circuits, diodes, transistors, and resistors comprise the hardware. So do the operator interface devices, such as display terminals, printers, and keyboards. The computer programs, in contrast, are called *software* or *firmware*, depending on whether or not they are easily modified (*see* FIRMWARE, and SOFTWARE).

A person who designs computer hardware is called a *hardware engineer*. The hardware engineer is an electrical engineer.

HARMONIC

Any signal contains energy at multiples of its frequency, in addition to energy at the desired frequency. The lowest frequency component of a signal is called the *fundamental frequency*; all integral multiples are called *harmonic frequencies*, or simply *harmonics*.

In theory, a pure sine wave contains energy at only one frequency, and has no harmonic energy. In practice, this ideal is never achieved. All signals contain some energy at harmonic frequencies, in addition to the energy at the fundamental frequency. The signal having a frequency of twice the fundamental is called the *second harmonic*. The signal having a frequency of three times the fundamental is called the *third harmonic*, and so on.

Wave distortion always results in the generation of harmonic energy. Although the nearly perfect sine wave has very little harmonic energy, the sawtooth wave, square wave, and other distorted periodic oscillations contain large amounts of energy at the harmonic frequencies. Whenever a sine wave is passed through a nonlinear circuit, harmonic energy is produced. A circuit designed to deliberately create harmonics is called a *harmonic generator* or *frequency multiplier*.

Harmonic output from radio transmitters is undesirable, and the designers and operators of such equipment often go to great lengths to minimize this energy as much as possible. *See also* FREQUENCY MULTIPLIER, FUNDAMENTAL FREQUENCY, and HARMONIC SUPPRESSION.

HARMONIC SUPPRESSION

Harmonic suppression is an expression of the degree to which harmonic energy is attenuated, with respect to the fundamental frequency, in the output of a radio transmitter. *Harmonic suppression* is also used to denote the process of minimizing harmonic energy in the output of a signal generator, especially a radio transmitter. Harmonics are undesirable in the output of such equipment because they cause interference to other services and operations.

If a given harmonic signal has a power level of Q watts in the output circuit of a transmitter, and the fundamental-frequency output is P watts, then the harmonic suppression S in decibels is given by:

$$S = 10 \log_{10} (P/Q)$$

Harmonic suppression can be accomplished in three ways. The most frequently used method is the insertion of one or more tuned bandpass filters in the output of the transmitter. The filter frequency is centered at the operating frequency of the transmitter. This provides additional harmonic attenuation in the output of the final amplifier.

The second method of obtaining harmonic suppression is the insertion of a lowpass filter in the transmitter output. The cutoff frequency of the filter should be the lowest that will result in negligible attenuation at the fundamental frequency.

The third method of obtaining harmonic suppression is the use of band-rejection filters, also sometimes called *traps*. The trap offers harmonic attenuation at only one frequency; the bandpass and lowpass filters provide rejection of all harmonics above the fundamental. However, the trap circuit often gives better results at the design frequency.

Of course, if an amplifier is intended to operate as a linear amplifier, the bias and drive should be maintained at the proper levels to ensure minimal generation of harmonic energy. The class-C amplifier causes more harmonics to be produced than other types of amplifiers. *See also* BANDPASS FILTER, BAND-REJECTION FILTER, HARMONIC, and LOWPASS FILTER.

HARP ANTENNA

A *harp antenna* is a special form of broadband unbalanced ground-plane antenna. The radiating portion of the antenna is vertically polarized, and consists of a number of different vertical conductors of various lengths. The conductors extend upward from a common feed line, giving the antenna the appearance of a harp.

Generally, the lowest operating frequency of the harp antenna is determined by the length of the longest radiator. The highest operating frequency is determined by the length of the shortest radiator. The radiating elements are an electrical quarter wavelength at resonance. Harp antennas are most practical at the very-high frequencies, where the element lengths are easily manageable, allowing construction from metal rods or tubing. The harp antenna requires a good ground plane. The polarization is vertical, and the radiation pattern is essentially omnidirectional in the azimuth plane. *See also* GROUNDPLANE ANTENNA.

HARTLEY OSCILLATOR

The *Hartley oscillator* is a form of variable-frequency oscillator. The operating frequency is determined by a parallel combination of inductance and capacitance. The feedback system is provided by a tap in the coil of the tank circuit. The illustration shows schematic diagrams for Hartley oscillators using a bipolar transistor and a field-effect transistor.

The emitter or source of the amplifying device is always connected to the coil tap in the Hartley configuration. This is how it can be recognized. The output is usually taken from the collector or drain circuit of the oscillator. However, the output can also be obtained by means of a loosely coupled coil near the tank coil. *See also* OSCILLATOR.

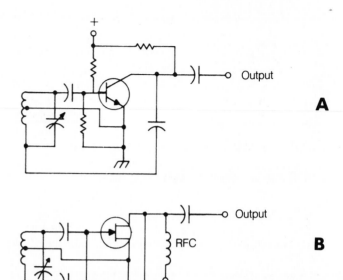

HARTLEY OSCILLATOR: Bipolar (A) and FET (B) circuits.

HAY BRIDGE

The Hay bridge is a circuit that is designed for the purpose of measuring the value of an unknown inductance. The Hay bridge also gives an indication of the Q factor of the inductor under test.

Two balance controls, consisting of potentiometers, are used in the bridge. One control is calibrated in terms of inductance, in millihenrys or microhenrys. The other is calibrated in terms of the Q factor, giving a measure of the reactance-to-resistance ratio of the tested coil. Correct adjustment is indicated by a null meter or headset.

The Hay bridge is generally used for the measurement of relatively large inductances. The signal generator is thus an audio device in most cases. At very-high frequencies, the inductance of the component leads affects the accuracy of the bridge.

HAZARD RATE

The *hazard rate* of an electronic circuit or component is the instantaneous failure rate. It is an expression of the probability that a device will fail immediately when it is put to use.

As the interval of time approaches zero, the failure rate of a component or circuit approaches the hazard rate. Some components do fail immediately when put to use, and therefore the hazard rate is never zero. *See also* FAILURE RATE.

HEADPHONE

A *headphone* is a pair of earphones, designed to be worn over the head so that both ears are covered. Headphones provide excellent intelligibility in communications because external sounds are minimized.

Some headphones operate with both earphones connected in parallel or in series. Such devices are called *monaural headphones*. Some headphones have two independent earphones, and are used for listening to high-fidelity stereo.

Headphones are available in impedances ranging from 4 ohms to more than 2,000 ohms. Some headphones have a wide-range frequency response; these are ideal for high-fidelity applications. Other types of headphones have peaked responses, intended for communications. The proper impedance and frequency response should be chosen for the desired situation. Most headphones are of the dynamic type, and resemble pairs of miniature speakers. *See also* EARPHONE, and SPEAKER.

HEART FIBRILLATION

When a current of about 100 to 200 milliamperes passes through the heart muscle, the normal beating sometimes stops, and the muscles contract in a random manner. This condition is called *heart fibrillation*. Low blood potassium and chronic heart disease can also cause fibrillation.

Heart fibrillation is an extremely dangerous condition. Unless normal heart activity is restored, the body cells cannot receive the oxygen they need to survive. Brain damage occurs within a few minutes. Death follows shortly thereafter.

Cardiopulmonary resuscitation is sometimes effective in keeping a fibrillation victim alive until medical help arrives. An ambulance or physician should be called immediately if heart fibrillation is suspected. Medical professionals can often get the heart beating normally again by means of a device called a *defibrillator*. *See also* CARDIOPULMONARY RESUSCITATION, and ELECTRIC SHOCK.

HEATSINK

A device that helps to remove excess heat from electronic devices, by means of conduction, convection, and radiation, is called a *heatsink*. The use of a heatsink increases the power-dissipating capability of a transistor, tube, integrated circuit, or other active device. Some heatsinks are small, and fit around a single miniature component (see photograph). Other heatsinks are large, and can be used with several different components at once.

The heatsink is made of metal, usually aluminum or iron, and is an excellent conductor of heat. The heat is thus carried away from the component by means of thermal conduction. The surface of the heatsink is deliberately designed to facilitate radiation and convection. This is usually accomplished by

HEAT SINK: Facilitates conduction cooling.

means of a finned surface. Heatsinks are often painted black to enhance the radiation.

When a heatsink is used, it is necessary that the component be thermally bonded to the device. Otherwise, the benefits will be lost. Silicone compound is generally used for thermal bonding. If it is necessary to electrically insulate the component from the heatsink, a flat piece of mica can be placed between the component and the heatsink.

HEIGHT ABOVE AVERAGE TERRAIN

At very-high frequencies and above, antenna performance is dependent on the height. The higher the antenna, the better the performance in most cases. In locations where the terrain is irregular, it is often difficult to determine the effective height of an antenna for communications purposes. Thus, the *height-above-average-terrain (HAAT)* figure has been devised.

In order to accurately determine the HAAT of an antenna, a topographical map is necessary. The location of the antenna must be found on the map. A circle is then drawn around the point corresponding to the location of the antenna. This circle should have a radius large enough to encompass at least several hills and valleys in the vicinity of the antenna. Within the circle, the average elevation of the terrain is calculated. This process involves choosing points within the circle in a grid pattern, adding up all of the elevation figures, and then dividing by the number of points in the pattern.

Once the average elevation, A, is found in the vicinity of the antenna, the elevation B of the antenna must be found by adding the height of the supporting structure to the ground-elevation figure at the antenna location. The HAAT is then equal to $B - A$.

The Federal Communications Commission imposes certain limitations, in some services, on antenna HAAT. This is especially true for repeater antennas. *See also* REPEATER.

HELICAL ANTENNA

A *helical antenna* is a form of circularly polarized, high-gain antenna that is used mostly in the ultra-high-frequency part of the radio spectrum. Circular polarization offers certain advantages at these frequencies (*see* CIRCULAR POLARIZATION).

The drawing illustrates a typical helical antenna. The reflecting device can consist of sheet metal or screen, in a disk configuration with a diameter d of at least 0.8 wavelength at the lowest operating frequency. The radius r of the helix should be approximately 0.15 wavelength at the center of the intended operating frequency range. The longitudinal spacing between turns of the helix, given by s, should be approximately ¼ wavelength in the center of the operating frequency range. The overall length of the helix, shown by L, can vary, but should be at least 1 wavelength at the lowest operating frequency. The longer the helix, the greater the forward power gain. Gain figures in excess of 15 dB can be realized with a single, moderate-sized helical antenna; when several such antennas are phased, the gain increases accordingly. Bays of two or four helical antennas are quite common.

The helical antenna illustrated will show a useful operating bandwidth equal to about half the value of the center frequency. An antenna centered at 400 MHz will function between approximately 300 and 500 MHz, for example. The helical antenna is normally fed with coaxial cable. The outer conductor should be connected to the reflecting screen or sheet, and the center conductor should be connected to the helix. The feed-point impedance is about 100 to 150 ohms throughout the useful operating frequency range.

Helical antennas are ideally suited to satellite communications because the circular polarization of the transmitted and received signals reduces the amount of fading as the satellite orientation changes. The sense of the circular polarization can be made either clockwise or counterclockwise, depending on the sense of the helix.

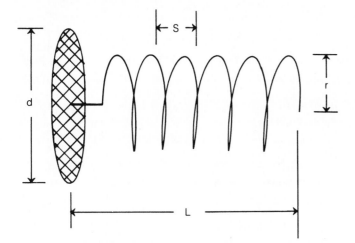

HELICAL ANTENNA: Dimensions are discussed in text.

HELICAL FILTER

A *helical filter (helical resonator)*, is a quarter-wave device often used at very-high and ultra-high frequencies as a bandpass filter. A shielded enclosure contains a coil, one end of which is connected to the enclosure and the other end of which is either left free or connected to a variable capacitor.

The filter sections are tuned to slightly different frequencies, resulting in very little attenuation throughout the passband, but steep skirts and high attenuation outside the passband. Helical filters provide a high Q factor, which is important in the reduction of out-of-band interference. *See also* BANDPASS FILTER, and Q FACTOR.

HELMHOLTZ COIL

A *Helmholtz coil* is an inductive device that provides continuously variable phase shift for an alternating-current signal. Two primary windings are oriented at right angles and split into two sections. The currents in the two primary coils differ by 90 degrees; this phase shift is provided by a resistor and capacitor. The secondary coil is mounted on rotatable bearings. As the secondary coil is turned through one complete rotation, the phase of the signal at the output terminals changes continuously from 0 to 360 degrees. Any desired signal phase can be chosen by setting the coil to the proper position.

The Helmholtz coil works because the fields from the primary windings add together as vectors. The magnitudes of the two component vectors change in the secondary coil as the secondary coil is turned. The Helmholtz coil is frequency sensitive; that is, it will work at only one frequency. To change the operating frequency, the values of the resistor and capacitor must be changed to provide a 90-degree phase difference between the two primary coils. *See also* PHASE ANGLE.

HENRY

The *henry (H)* is the standard unit of inductance. In a circuit in which the current is changing at a constant rate of 1 ampere per second, an inductance of 1 henry results in the generation of 1 volt of potential difference across an inductor.

The henry is an extremely large unit of inductance. It is rare to find a coil with a value of 1 H. Therefore, inductance values are generally given in millihenrys (mH), microhenrys (μH), or nanohenrys (nH). An inductance of 1 mH is equal to 0.001 H; 1 μH is 0.001 mH; 1 nH is 0.001 μH. *See also* INDUCTANCE.

HERMAPHRODITIC CONNECTOR

A *hermaphroditic connector* is an electrical plug that mates with another plug exactly like itself. Such a connector has an equal number of male and female contacts.

Hermaphroditic connectors can be put together in only one way. This makes them useful in polarized circuits, such as direct-current power supplies. *See also* CONNECTOR.

HERMETIC SEAL

Some electronic components are susceptible to damage or malfunction from moisture in the air. This is especially true of piezoelectric crystals and certain semiconductor devices. Such components are often enclosed in hermetically sealed cases. A hermetic seal is simply an airtight, durable seal. The common oscillator crystal, housed in a metal can, is hermetically sealed.

A hermetic seal must be long-lasting and physically rugged. Many different kinds of glues and cements are acceptable for hermetic sealing. Sometimes the interior of a hermetically sealed enclosure is filled with an inert gas, such as helium, to further retard the deterioration of the component or components inside.

HERTZ

Hertz (Hz) is the standard unit of frequency. A frequency of 1 complete cycle per second is a frequency of 1 Hz. The term *hertz* became widespread in the late 1960s, and is now used instead of the term *cycles per second*.

In radiocommunication, signals are typically thousands, millions, or billions of hertz in frequency. A frequency of 1,000 Hz is called *1 kilohertz* (kHz); a frequency of 1,000 kHz is 1 megahertz (MHz); a frequency of 1,000 MHz is 1 gigahertz (GHz). Sometimes the terahertz (THz) is used as a measure of frequency; 1 THz = 1,000 GHz.

The angular frequency in radians per second is equal to approximately 6.3 times the frequency in hertz.

HERTZ ANTENNA

A *Hertz antenna* is any horizontal, half-wavelength antenna. The feed point can be at the center, at either end, or at some intermediate point. The Hertz antenna operates independently of the ground, and is therefore a balanced antenna.

Examples of the Hertz antenna include the dipole and the zeppelin antenna. The driven element of a Yagi antenna is often of the Hertz configuration. *See also* DIPOLE ANTENNA, YAGI ANTENNA, and ZEPPELIN ANTENNA.

HETERODYNE

The term *heterodyne* can refer to either of two things. A heterodyne is a mixing product resulting from the combination of two signals in a nonlinear component or circuit. Mixing is sometimes called *heterodyning*.

When two waves, having frequencies f and g, are combined in a nonlinear component or circuit, the original frequencies appear at the output along with energy at two new frequencies. These frequencies are the sum and difference frequencies, $f + g$ and $f - g$. They are sometimes called *heterodyne frequencies*. A tuned circuit can be used to choose either the sum or the difference frequency. In the mixer, $f + g$ and $f - g$ are usually much different. In the heterodyne or product detector, they can be very close together.

Heterodyning occurs to a certain extent whenever two signals are present in the same medium. This is because no circuit is perfectly linear, and some distortion always takes place. A diode, capable of effectively handling frequencies f and g, or a transistor biased to cutoff, are ideal for heterodyning purposes. *See also* FREQUENCY CONVERSION, FREQUENCY CONVERTER, HETERODYNE DETECTOR, and MIXER.

HETERODYNE DETECTOR

A *heterodyne detector* is a detector that operates by beating the signal from a local oscillator against the received signal. This form of detector is more often called a *product detector*, and is required for the reception of continuous-wave, frequency-shift-keyed, and single-sideband signals.

The incoming signal information is extracted by heterodyning, resulting in audible difference frequencies. In the case of a continuous-wave signal, the difference can be as small as about 100 Hz or as large as about 3 kHz; the same is true with frequency-shift keying. For single-sideband reception, the local oscillator frequency should correspond to the frequency of the suppressed carrier.

A heterodyne detector is used in many superheterodyne receivers and all direct-conversion receivers. *See also* PRODUCT DETECTOR.

HETERODYNE FREQUENCY METER

A *heterodyne frequency meter* is a device similar to a direct-conversion receiver, used for the purpose of measuring unknown frequencies. A calibrated local oscillator is used, along with a mixer and amplifier. The signal of unknown frequency is fed into the mixer along with the output of the local oscillator. It is helpful to have some idea of the frequency of the signal beforehand, because harmonics of the local oscillator will produce false readings. As the local oscillator is tuned, or heterodynes occur in the output. These heterodynes occur at frequencies f, $f/2, f/3, f/4$, and so on, as read on the local-oscillator calibrated scale. The correct reading is f, the highest frequency. Readings can also be obtained at $2f$, $3f$, $4f$, and so on, corresponding to harmonics of the input signal; but these components are actually present, and the lower-frequency indications are false signals. Some heterodyne frequency meters have tuned input circuits, which track along with the local oscillator fundamental frequency, reducing harmonic effects. The correct reading usually gives the loudest heterodyne. The local oscillator should be adjusted for zero beat before the final reading is taken.

All receivers with product detectors can be used as heterodyne frequency meters within their operating ranges, provided that the dial is calibrated with reasonable accuracy. It is helpful to have a crystal calibrator in such a receiver, and this calibrator should be adjusted against a frequency standard, such as WWV or WWVH. *See also* FREQUENCY METER, and HETERODYNE.

HETERODYNE REPEATER

When a signal is amplified and retransmitted, the frequency of the signal must usually be changed. Otherwise, feedback is likely to result in oscillation between the receiver and transmitter circuits. Virtually all repeaters use frequency converters, and they are therefore called *heterodyne repeaters*.

The input signal to the heterodyne repeater might have a frequency, f. This signal is heterodyned with the output of a local oscillator at frequency g, producing mixing products at frequencies $f + g$ and $f - g$. Generally, the difference frequency is chosen. A tuned bandpass filter, centered at frequency $f - g$, serves to eliminate other signals. An amplifier tuned to frequency $f - g$ provides the output. At the input of the repeater, a notch or band-rejection filter can be used, centered at frequency $f - g$, to prevent desensitization of the receiver by the transmitter.

In general, the greater the difference between the input and output frequencies of a heterodyne repeater, the more easily the circuit is designed, and the less difficulty is encountered with receiver desensitization. *See also* FREQUENCY CONVERSION, and REPEATER.

HEXADECIMAL NUMBER SYSTEM

The *hexadecimal number system* is a base-16 system. Hexadecimal notation is commonly used by computers, because 16 is a power of 2. The numbers 0 through 9 are the same in hexadecimal notation as they are in the familiar decimal number system. However, the values 10 through 15 are represented by single "digits," usually the letters A through F.

Addition, subtraction, multiplication, and division are somewhat different in hexadecimal notation. For example, in decimal language, we say that $5 + 7 = 12$; but in hexadecimal, $5 + 7 = C$. The carry operation does not occur until a sum is greater than F. Thus, $8 + 7 = F$, and $8 + 8 = 10$.

HIGHER-ORDER LANGUAGE

A *higher-order* or *high-level computer language* is the interface between the machine and the human operator. The computer itself "thinks" in binary or machine language, and the construction of programs in this language is tedious. The higher-order language is often quite similar to ordinary language. A translation program, called an *assembler*, converts the machine language back into terms the operator can readily understand.

Examples of higher-order languages are BASIC (used for mathematical and scientific calculations), COBOL (used in business applications), FORTRAN (used mostly by scientists and engineers), and the languages of computer games.

HIGH FREQUENCY

High frequency is the designator applied to the range of frequencies from 3 to 30 MHz. This corresponds to a wavelength between 100 and 10 meters. The high frequencies include all of the so-called "shortwave bands." Ionospheric propagation is of great importance at the high frequencies.

At the lower end of the high-frequency part of the electromagnetic spectrum, the ionosphere almost always reflects energy back to the earth. Conditions are highly variable within the high-frequency range, and depend largely on the sunspot cycle, the time of day, and the time of year. At the upper end of the high-frequency range, near 30 MHz, the ionosphere is often transparent to radio signals. The F layer is the primary ionospheric layer responsible for long-distance communication in the high-frequency spectrum. *See also* F LAYER, IONOSPHERE, and PROPAGATION CHARACTERISTICS.

HIGHPASS FILTER

A *highpass filter* is a combination of capacitance, inductance, and/or resistance, intended to produce large amounts of attenuation below a certain frequency and little or no attenuation above that frequency. The frequency at which the transition occurs is called the cutoff frequency (*see* CUTOFF FREQUENCY). At the cutoff frequency, the voltage attenuation is 3 dB, with respect to the minimum attenuation. Above the cutoff frequency, the voltage attenuation is less than 3 dB. Below the cutoff, the voltage attenuation is more than 3 dB.

The simplest highpass filters consist of a parallel inductor or a series capacitor. More sophisticated highpass filters have a combination of parallel inductors and series capacitors, such as the filters shown in the illustration. The filter at A is called an *L-section filter*; that at B is called a *T-section filter*. These names are derived from the geometric shapes of the filters as they appear in the schematic diagram.

Resistors are sometimes substituted for the inductors in a highpass filter. This is especially true if active devices are used, in which case many filter sections can be cascaded.

Highpass filters are used in a wide variety of situations in electronic apparatus. One common use for the highpass filter is at the input of a television receiver. The cutoff frequency of such a filter is about 40 MHz. The installation of such a filter reduces the susceptibility of the television receiver to interference from sources at lower frequencies. *See also* HIGHPASS RESPONSE.

HIGHPASS RESPONSE

A *highpass response* is an attenuation-versus-frequency curve that shows greater attenuation at lower frequencies than at higher frequencies. The sharpness of the response can vary considerably. Usually, a highpass response is characterized by a high degree of attenuation up to a certain frequency, where the attenuation rapidly decreases. Finally, the attenuation levels off at near zero insertion loss. The highpass response is typical of highpass filters.

The cutoff frequency of a highpass response is that frequency at which the insertion loss is 3 dB, with respect to the minimum loss. The ultimate attenuation is the level of attenuation well below the cutoff frequency, where the signal is virtually blocked. The ideal highpass response should look like the attenuation-versus-frequency curve in the illustration. The curve is smooth, and the insertion loss is essentially zero everywhere well above the cutoff frequency. *See also* CUTOFF FREQUENCY, HIGHPASS FILTER, and INSERTION LOSS.

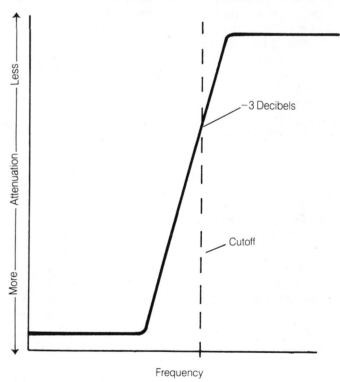

HIGHPASS RESPONSE: Passes all frequencies above the cutoff.

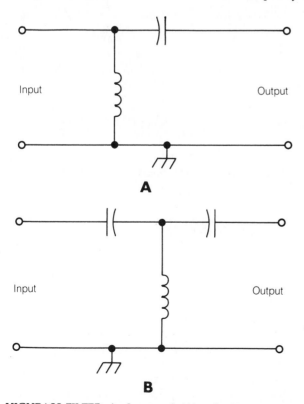

HIGHPASS FILTER: An L network (A) and a T network (B).

HIGH Q

High Q is a term used to describe a filter circuit with a great deal of selectivity. The term is usually applied to bandpass and band-rejection filters.

High-*Q* filters are desirable in situations where the response must be confined to a single signal, or to a very narrow range of frequencies. However, this characteristic is not always wanted. In general, high-*Q* circuits require low-loss inductors and capacitors, and the resistance should be as low as possible. *See also* Q FACTOR.

HIGH-THRESHOLD LOGIC

High-threshold logic (HTL) is a form of bipolar digital logic. High-threshold logic is exactly the same as diode-transistor logic (DTL), except that Zener diodes are added in series with each input line. The Zener diodes greatly reduce the noise susceptibility of DTL.

High-threshold logic devices require rather large operating voltages, and the power dissipation is therefore higher than with DTL. But, in situations where noise must be minimized, HTL gives superior performance. *See also* DIODE-TRANSISTOR LOGIC.

HISS

A form of audio noise, in which the amplitude is peaked near the midrange or treble regions, is often called *hiss*. Hiss is always heard in active audio-frequency circuits. The sound of hiss is familiar to anyone who has worked with high-fidelity equipment, public-address systems, or radio receivers.

In audio equipment, the hiss is generated as the result of random electron or hole movement in components. The hiss generated in the early stages is amplified by subsequent stages. In a radio receiver, hiss can be generated in the intermediate-frequency stages and in the front end, mixers, and oscillators. Some hiss even originates outside the receiver, in the form of thermal noise in the antenna conductors and atmosphere. *See also* WHITE NOISE.

H NETWORK

An *H network* is a form of filter section, sometimes seen in balanced circuits. The H network gets its name from the fact that the schematic-diagram component arrangement looks like the capital letter H turned sideways.

The H configuration can be used in the construction of bandpass, band-rejection, highpass, and lowpass filters. The H configuration is a popular arrangement for attenuator design; noninductive resistors are used. Several H networks can be cascaded to obtain better filter or attenuator performance.

The H network is not the only configuration for filters and attenuators. *See also* L NETWORK, PI NETWORK, and T NETWORK.

HOLDING CURRENT

Holding current is the minimum amount of current that will keep a switching device actuated. The term applies to relays, and also to various tube and solid-state devices, especially the silicon-controlled rectifier and thyristor.

In any switching device, a certain amount of current must flow before a change of state occurs. This turn-on current is always larger than the holding current. In relays, the difference between the turn-on and holding currents is usually rather small; the initial current must therefore be nearly maintained. But in the silicon-controlled rectifier and thyristor, the current can be reduced considerably following turn-on, and the device will remain actuated. Relays, silicon-controlled rectifiers, and thyristors with various ratios of turn-on current to holding current are available.

Silicon-controlled rectifiers and thyristors with low holding-current levels are preferable when the load impedance is high. This keeps the device in the actuated state under conditions of variable current. A larger holding-current value can be wanted when the load impedance is low, so that the device can be switched off easily. *See also* RELAY, SILICON-CONTROLLED RECTIFIER, and THYRISTOR.

HOLE

A *hole* is a carrier of electric charge. In certain semiconductor materials, the charge carriers are predominantly holes rather than electrons. A hole is an atom with one electron missing. Holes have positive charge, and electrons have negative charge.

An electric current in a P-type semiconductor material flows as the result of hole movement. The electron-deficient atoms do not themselves move; the electrons migrate from atom to atom, creating a chain of vacancies which moves from the positive to the negative (see illustration). The holes in a P-type material behave in much the same way as the electrons in an N-type substance or ordinary wire conductor.

Although an electron is an identifiable particle of matter, a hole is not. The concept of the hole makes it easier to explain the operation of a semiconductor device in some cases. *See also* ELECTRON, ELECTRON-HOLE PAIR, N-TYPE SEMICONDUCTOR, P-N JUNCTION, and P-TYPE SEMICONDUCTOR.

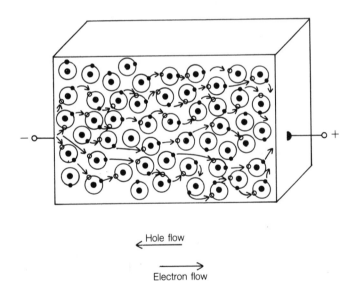

Hole flow
←

Electron flow
→

HOLE: Solid large dots represent nuclei. Solid small dots represent electrons. Small open circles represent holes. Large circles are electron orbits.

HOOK TRANSISTOR

A *hook transistor* is a four-layer semiconductor device, also called a *PNPN transistor*. The outer P-type and N-type semiconductor layers serve as the emitter and collector; the inner N-type layer serves as the base.

The hook transistor contains an extra P-type layer between the base and the collector. This extra layer results in increased gain at the higher frequencies. The extra P-N junction enhances the flow of charge carriers. It also tends to reduce the capacitance in the base-collector junction, allowing the device to operate effectively at higher frequencies than is possible with a conventional pnp or npn bipolar transistor. *See also* TRANSISTOR.

HORIZON

The *horizon* is the apparent edge of the earth, as viewed from a particular height above the surface at a particular electromagnetic frequency. The radio horizon is generally farther away than the visual horizon. The radio horizon is of considerable importance in communication at very-high frequencies and above, because most communication in this part of the spectrum takes place via so-called direct or line-of-sight propagation.

The distance to the horizon depends on several factors. In general, the higher the viewing point, the greater the distance to the horizon, regardless of frequency. The distance is also dependent on the frequency; at some wavelengths the atmosphere displays refractive effects to a greater extent than at other wavelengths. In calculating the effective distance to the radio horizon, it must be realized that the values obtained are theoretical, and are based on smooth terrain. In areas having many large hills or mountains, the actual distance to the radio horizon can vary greatly from the theoretical values.

Let h be the height of the viewpoint in feet over smooth earth, and let d be the distance to the horizon in miles. Then, at visual wavelengths:

$$d = \sqrt{1.53\,h}$$

The infrared horizon is essentially the same as the visual horizon. The ultraviolet and gamma-ray horizons are also the same, for all practical purposes, as the visual horizon. However, as the wavelength becomes much longer than the infrared, the horizon begins to lengthen. This is equivalent to an effective increase in the radius of the earth. For radio waves in the very-high and ultra-high frequency bands, a good approximation for the distance to the horizon is given by:

$$d = \sqrt{2h}$$

where d is again specified in miles and h is specified in feet.

For a complete radio circuit, with a transmitting antenna at height g feet and a receiving antenna at height h feet, the effective line-of-sight path can be calculated by:

$$d = \sqrt{2g} + \sqrt{2h}$$

See also DIRECT WAVE, and LINE-OF-SIGHT COMMUNICATION.

HORIZONTAL LINEARITY

Horizontal linearity is one expression of the degree to which a television picture is an accurate reproduction of a scene. Most television receivers have an internal adjustment for calibrating the horizontal linearity. In a few receivers, the control is external.

The horizontal linearity of a television receiver should be adjusted only by a technician with the proper test equipment. Improper horizontal linearity results in a distorted picture. If the horizontal linearity is not properly set, objects that appear to have a certain width in one part of the screen will seem to be wider or narrower at another location on the screen.

All television picture signals are broadcast with a linear horizontal scan. Improper horizontal linearity is usually the fault of the receiver. *See also* TELEVISION.

HORIZONTAL POLARIZATION

When the electric lines of force of an electromagnetic wave are oriented horizontally, the field is said to be *horizontally polar-* *ized*. In communications, horizontal polarization has certain advantages and disadvantages at various wavelengths.

At the low and very-low frequencies, horizontal polarization is not often used. This is because the surface wave, an important factor in propagation at these frequencies, is more effectively transferred when the electric field is oriented vertically. Most standard AM broadcast stations, operating in the medium-frequency range, also use vertical, rather than horizontal polarization.

In the high-frequency part of the electromagnetic spectrum, horizontal polarization becomes practical. The polarization is always parallel to the orientation of the radiating antenna element; horizontal wire antennas are simple to install above about 3 MHz. The surface wave is of lesser importance at high frequencies than at low and very-low frequencies; the sky wave is the primary mode of propagation above 3 MHz. This becomes increasingly true as the wavelength gets shorter. Horizontal polarization is just as effective as vertical polarization in the sky-wave mode.

In the very-high and ultra-high frequency range, either vertical or horizontal polarization can be used. Horizontal polarization generally provides better noise immunity and less fading than vertical polarization in this part of the spectrum. *See also* CIRCULAR POLARIZATION, POLARIZATION, and VERTICAL POLARIZATION.

HORIZONTAL SYNCHRONIZATION

In television communications, the picture signals must be synchronized at the transmitter and receiver. The electron beam in the television picture tube scans from left to right and top to bottom, in just the same way as you read the page of a book. At any given instant of time, in a properly operating television system, the electron beam in the receiver picture tube is in exactly the same relative position as the scanning beam in the camera tube. This requires synchronization of the horizontal as well as the vertical position of the beams.

When the horizontal synchronization in a television system is lost, the picture becomes totally unrecognizable. This is illustrated by misadjustment of the horizontal-hold control in any television set. Even a small synchronization error results in severe "tearing" of the picture. The transmitted television signal contains horizontal-synchronization pulses at the end of every line. This tells the receiver to move the electron beam from the end of one line to the beginning of the next line. Vertical synchronization pulses tell the receiver that one complete picture, called a frame, is complete, and that it is time to begin the next frame. *See also* PICTURE SIGNAL, TELEVISION, and VERTICAL SYNCHRONIZATION.

HORN ANTENNA

A *horn antenna* is a device used for transmission and reception of signals at ultra-high and microwave frequencies. There are several different configurations of the horn antenna, but they all look similar. The illustration on pg. 240 is a pictorial drawing of a commonly used horn antenna.

The horn antenna provides a undirectional radiation pattern, with the favored direction coincident with the opening of the horn. Horn antennas are characterized by a lowest usable frequency. The feed system generally consists of a waveguide, which joins the horn at its narrowest point.

Horn antennas are often used in feed systems of large dish antennas. The horn is pointed toward the center of the dish, in the opposite direction from the favored direction of the dish. When the horn is positioned at the focal point of the dish, extremely high gain and narrow-beam radiation are realized. *See also* DISH ANTENNA.

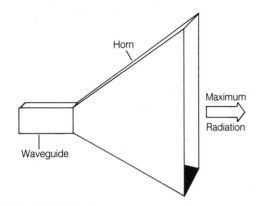

HORN ANTENNA: Used in microwave communications.

HOT-CARRIER DIODE

The *hot-carrier diode (HCD)* is a special kind of semiconductor diode. The HCD is used in various circuits, especially mixers and detectors at very-high and ultra-high frequencies. The HCD will work at much higher frequencies than most other semiconductor diodes.

The hot-carrier diode generates relatively little noise, and exhibits a high breakdown or avalanche voltage. The reverse current is minimal. The internal capacitance is low because the diode consists of a tiny point contact (see illustration). The wire is gold-plated to minimize corrosion effects. An N-type silicon wafer is used as the semiconductor. *See also* DIODE.

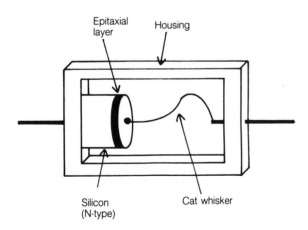

HOT-CARRIER DIODE: A point-contact semiconductor device.

HOT SPOT

In a communications system, a *hot spot* is a region in which reception is much better than in the immediate surroundings. This effect can be caused by phase addition of several components of a signal as a result of geography in the area. It can also be caused by unusually good earth conductivity. Sometimes the cause is unknown. A receiving hot spot is not necessarily a good transmitting location. *See also* EARTH CONDUCTIVITY, and PROPAGATION CHARACTERISTICS.

HOT-WIRE METER

A *hot-wire meter* is a device that makes use of the thermal expansion characteristics of a metal wire for the purpose of measuring current. When a current flows through the tightly stretched wire, the wire expands. This causes a pointer, attached to the wire, to move across a graduated scale. Hot-wire current-measuring devices can be used as ammeters, voltmeters, or wattmeters. Hot-wire ammeters can be attached to a variety of different devices to indicate such parameters as wind speed, motor speed, or rate of fluid flow.

The hot-wire meter is not particularly sensitive, but can register large amounts of current without damage. The damping, or rate at which the meter responds to changes in the current passing through it, is slow. This is an advantage in situations where rapid fluctuations are of little interest, but is not desirable when precise indications are required for rapidly changing parameters. The hot-wire meter can measure alternating current just as well as it can measure direct current. For this reason, hot-wire ammeters are often used for the determination of radio-frequency current in antenna transmission lines. *See also* AMMETER.

HUM

Hum is the presence of 60- or 120-Hz modulation in an electronic circuit. Hum can occur in the carrier of a radio transmitter, in a radio-frequency receiving system, or in an audio system.

When the filtering is inadequate in the output of a power supply, hum is often introduced into the circuits operating from the supply. With half-wave power supplies, the hum has a frequency of 60 Hz; with full-wave supplies, the frequency is 120 Hz. Both of these frequencies are within the range of human hearing, and produce objectionable modulation.

Hum can be picked up by means of inductive or capacitive coupling to nearby utility wires. This hum is always at a frequency of 60 Hz. Poorly shielded amplifier-input wiring is a major cause of hum in audio circuits. Hum can also be picked up by the magnetic heads of a tape recorder. Improper shielding or balance in the output leads of an audio amplifier can also cause hum modulation to be introduced into the circuit.

Power-supply hum can be remedied by the installation of additional filtering chokes and/or capacitors. Hum from utility wiring can be minimized by the use of excellent shielding in the input leads to an amplifier, and by proper shielding or balance in the output. A good ground system, without ground loops, can be helpful as well. *See also* ELECTROMAGNETIC SHIELDING, and POWER SUPPLY.

HUMAN ENGINEERING

The art of designing electronic equipment to suit the needs of the operator is called *human engineering*. Even if a circuit is well-designed electrically, it will be difficult to use if proper attention has not been given to human engineering.

Human engineering involves judicious positioning of controls and indicators. Meters and displays should be easy to read, and controls convenient to adjust. The number of controls should be sufficient to accomplish the desired functions, but too many adjustments make it difficult to use the apparatus.

The art of human engineering also involves the proper choice of electrical characteristics. A receiver with a product detector, but no envelope detector, can be used to receive amplitude-modulated signals, but tuning is difficult. Many telegraph operators prefer a receiver in which the automatic level control can be switched off. The tuning rate should be slow enough so that signals are easy to pinpoint, but it should be fast enough so that the operator does not have to spend a lot of time spinning the dial to get from one frequency to another. The above examples refer to radio receivers, but human engineering is important in the design of computer terminals, test equipment, transmitters, and all other electronic devices.

As the field of electronics gets more and more advanced, human engineering should play a greater role.

HUMIDITY

Humidity is the presence of water vapor in the air. The humidity is usually specified in terms of a ratio, called the *relative humidity*. This is the amount of water vapor actually in the air, compared to the amount of water vapor the air is capable of holding without condensation. The higher the temperature, the more water vapor can be contained in the air.

Relative humidity is determined by the use of two thermometers, one with a dry bulb and the other with its bulb surrounded by a wet cloth or wick. Evaporation from the wick causes the wet-bulb reading to be lower than the dry-bulb reading. For a given temperature, the water from the wet bulb evaporates more and more rapidly as the humidity gets lower. The two readings are found on a table, and the relative humidity is indicated at the point corresponding to both readings.

In northern latitudes during winter, the indoor relative humidity can become very low. This is because the cold outside air cannot hold much moisture, and when the cold air is heated, its capacity for holding moisture goes up while the actual amount of moisture stays the same.

In some mountainous, cold locations, the relative humidity can drop to less than 5 percent. The resulting static electricity can cause damage to electronic circuits when they are handled. This is especially true of metal-oxide semiconductor (MOS) devices. In marine or tropical regions, the humidity can rise to the point where condensation occurs on circuit boards and components. This can degrade performance and accelerate corrosion.

The specifications for a piece of electronic equipment often contain humidity limitations. The primary reason for this is the danger to MOS devices when it is very dry, and the possibility of condensation, with resultant component damage, when it is wet. *See also* SPECIFICATIONS.

HUNTING

Hunting is the result of overcompensation in an electronic circuit. Hunting is particularly common in direction-finding apparatus and improperly adjusted phase-locked-loop devices.

Any circuit that is designed to lock on some signal is subject to hunting if it is set in such a way that overcompensation occurs. The circuit will then oscillate back and forth on either side of the desired direction, frequency, or other parameter. The oscillation can be fast or slow. It can eventually stop, and the circuit achieve the desired condition. But if the misadjustment is especially severe, the hunting can continue indefinitely.

Hunting can be eliminated by proper alignment of a circuit. Additional damping usually gets rid of this problem. *See also* DIRECTION FINDER, and PHASE-LOCKED LOOP.

HYBRID DEVICE

A component or circuit that uses two or more different technological forms of design is sometimes called a *hybrid device*. For example, a radio transmitter can utilize bipolar transistors and field-effect transistors in every stage except the final amplifier, which uses a vacuum tube. Such a transmitter would be considered to have hybrid design because it uses both solid-state and vacuum-tube technology.

Whether or not a circuit is to be called a *hybrid device* depends, in part, on the distinction that is drawn among various technological methods. An integrated circuit that uses both microminiature discrete components and integrated components is often called a *hybrid integrated circuit*, even though all of the components are solid-state. Computers that use both analog and digital techniques are often called *hybrid computers*. A switch that makes use of relays and transistors might be called a *hybrid switching circuit*.

Hybrid devices often contain the best of various different forms of design, resulting in a circuit that is more efficient than would be possible if only one form of design is used.

HYSTERESIS

Hysteresis is the tendency for certain electronic devices to act "sluggish." This effect is especially important in the cores of transformers, because it limits the rate at which the core can be magnetized and demagnetized. This, in turn, limits the alternating-current frequency at which a transformer will operate efficiently.

Hysteresis effects occur in many different devices, not just in transformer cores. For example, the thermostat in a heater or air conditioner must have a certain amount of hysteresis, or "sluggishness." If there was no hysteresis in such a device, the heater or air conditioner would cycle on and off at a rapid rate, once the temperature reached the thermostat setting. With too much hysteresis, the temperature would fluctuate far above and below the thermostat setting. The correct amount of hysteresis in a thermostat results in a nearly constant temperature without excessive cycling of the heater or air conditioner.

Hysteresis is an important consideration in the design of certain electronic controls, such as a receiver squelch. Too little hysteresis results in squelch popping. Too much hysteresis makes it difficult to obtain a precise squelch setting. Certain digital control devices have built-in hysteresis. *See also* B-H CURVE, HYSTERESIS LOOP, and HYSTERESIS LOSS.

HYSTERESIS LOOP

A *hysteresis loop* is a graphical representation of the effects of hysteresis. Such a graph can also be called a *hysteresis curve* or a *box-shaped curve*. The B-H curve is the graphical representation of magnetization versus magnetic force (*see* B-H CURVE), and thus is a special form of hysteresis loop.

The hysteresis loop is useful for evaluating the performance of any device that exhibits hysteretic properties. In general, the wider the hysteresis loop, the greater the effect. *See also* HYSTERESIS, and HYSTERESIS LOSS.

HYSTERESIS LOSS

The hysteretic properties of magnetic materials cause losses in transformers having ferromagnetic cores. *Hysteresis loss* occurs as a result of the tendency of ferromagnetic materials to resist rapid changes in magnetization. Hysteresis and eddy currents are the two major causes of loss in transformer core materials.

The amount of hysteresis loss in a core material depends on two factors. In general, the higher the alternating-current frequency of the magnetizing force, the greater the hysteresis loss for any given material. The way in which the material is manufactured affects the amount of hysteresis loss at a specific frequency.

Solid or laminated iron displays considerable hysteresis loss. For this reason, transformers having such cores are useful at frequencies up to only about 15 kHz. Ferrite has much less hysteresis loss, and can be used at frequencies as high as 20 to 40 MHz, depending on the particular mix. Powdered-iron cores have the least hysteresis loss of any ferromagnetic material, and are used well into the VHF range. Of course, air has practically no hysteresis loss, and is preferred in many radio-frequency applications for this reason. *See also* EDDY-CURRENT LOSS, FERRITE, FERRITE CORE, FERROMAGNETIC MATERIAL, HYSTERESIS, LAMINATED CORE, and POWDERED-IRON CORE.

IC

See INTEGRATED CIRCUIT.

ICE LOADING

Ice loading is the additional stress placed on an antenna or power line, and the associated supporting structures, by accumulation of ice. When a structure is designed and installed in an area subject to ice storms, allowances must be made for ice loading. Otherwise, the antennas, power lines, or towers can actually collapse under the stress.

The most susceptible areas of the United States, in terms of the probability and frequency of ice storms, are the northeast and midwest regions. The northwest and southeast are somewhat less likely to have ice storms. The desert southwest and the extreme southern United States seldom have such storms.

The presence of ice on a structure increases the wind loading dramatically. This effect, too, must be taken into account when power lines or antenna systems are designed or installed in a particular geographic region. *See also* WIND LOADING.

ICONOSCOPE

An *iconoscope* is a form of television camera tube. The operation of the iconoscope is similar to that of the vidicon.

The visible image is focused onto a flat photocathode by means of a convex lens. The photocathode consists of a fine grid of light-sensitive capacitors, called a *mosaic*. Each capacitor is supplied with a charging voltage. The light from the image causes each capacitor to discharge as the electron beam, supplied by the electron gun, scans the mosaic. The amount of discharging depends on the intensity of the light at that particular location: with no light, the capacitors remain almost fully charged, and under bright light, they almost completely discharge. As the electron beam strikes each capacitor in turn, an output pulse occurs. The intensity of the pulse is proportional to the brightness of the light striking the capacitor.

The iconoscope, like the vidicon, is a bit less sensitive, and has a slightly longer image lag time, than the image orthicon. *See also* CAMERA TUBE, IMAGE ORTHICON, and VIDICON.

IDEAL CIRCUIT OR COMPONENT

An ideal circuit is hypothetical, not real. It is perfection, for which engineers strive; it is the mathematical example they use to represent the best possible performance for a device.

An example of an ideal circuit is a lowpass filter with zero attenuation below its cutoff frequency, infinite attenuation above the cutoff, and an infinitely steep cutoff slope. This is illustrated by the accompanying diagram of an *ideal lowpass response curve* (illustration A).

An ideal receiver front end would generate no noise of its own. An ideal antenna would radiate every last microwatt of power fed into it. An ideal ground system would have zero ohmic resistance from dc through the microwave spectrum. You can probably think of numerous other examples of ideal circuits.

An *ideal component* is a hypothetical component that behaves perfectly. An example is a capacitor with zero dielectric loss and a value that does not change with temperature (unless it is designed to). Another example is an inductor with no loss in the winding, no loss in the core, constant inductance regardless of temperature and frequency changes, and immunity to the effects of surrounding objects.

Sometimes a mathematically ideal circuit or component for one purpose would be far from ideal in another application. An example is a diode. For envelope detection or for rectification, the ideal behavior would be low resistance in one direction, and infinite resistance in the opposite direction, no matter what the voltage or current (illustration B). But for audio clamping (see CLAMPING), a diode should have a constant, nonzero *forward breakover voltage*, so that two of them can be connected back-to-back to limit the amplitude. If the forward voltage is less than the breakover value, the resistance should ideally be infinite in such a diode; if the forward voltage exceeds breakover, the resistance should abruptly drop (illustration C).

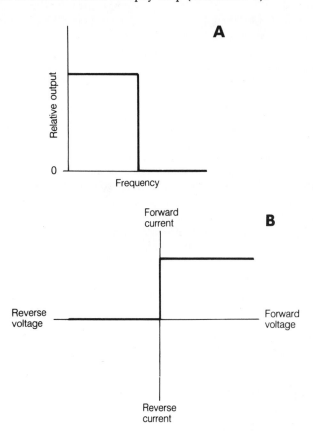

IDEAL CIRCUIT OR COMPONENT: Ideal lowpass response (A), ideal rectifier diode response (B), ideal clamping diode response (C).

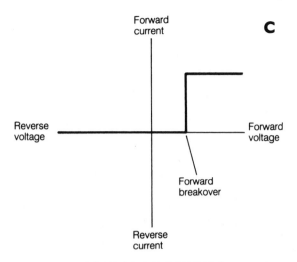

IDEAL CIRCUIT OR COMPONENT Continued

IEEE

See INSTITUTE OF ELECTRICAL AND ELECTRONICS ENGINEERS.

IF

See INTERMEDIATE FREQUENCY.

IGNITION NOISE

Ignition noise is a wideband form of impulse noise, generated by the electric arc in the spark plugs of an internal combustion engine. Ignition noise is radiated from many different kinds of devices, such as automobiles and trucks, lawn mowers, and gasoline-engine-driven generators.

Ignition noise can usually be reduced or eliminated in a receiver by means of a noise blanker. The pulses of ignition noise are of very short duration, although their peak intensity can be considerable. Ignition noise is a common problem for mobile radio operators, especially if communication in the high-frequency range is contemplated. Ignition noise can be worsened by radiation from the distributor to wiring in a truck or automobile. Sometimes, special spark plugs, called *resistance plugs*, can be installed in place of ordinary spark plugs, and the ignition noise will be reduced. An automotive specialist will know whether or not this is feasible in a given case. An excellent ground connection is imperative in the mobile installation. Mobile receivers for high-frequency work should have effective, built-in noise blankers.

Ignition noise is not the only source of trouble for the mobile radio operator. Noise can be generated by the friction of the tires against a pavement. High-tension power lines often radiate noise at radio frequencies. *See also* IMPULSE NOISE, NOISE, NOISE LIMITER, and POWER-LINE NOISE.

IGNITRON

The *ignitron* is a form of high-voltage, high-current rectifier. The device consists of a mercury-pool cathode and a single anode. Forward conduction occurs as a result of arcing during the part of the cycle in which the anode is positive with respect to the cathode.

During the forward-voltage part of the alternating-current cycle, an igniting electrode dips into the mercury and initiates the arc. The ignitron generates large amounts of heat, and therefore it must be cooled in some manner. Water, pumped around the outside of the tube, serves this purpose well.

The ignitron is characterized by the ability to withstand temporary currents or voltages far in excess of its normal ratings. *See also* RECTIFICATION.

IMAGE COMMUNICATIONS

Any mode that involves the transmission and reception of pictures, either still or moving, is *image communications*. These modes include facsimile (FAX), slow-scan television (SSTV) and amateur television (ATV or TV). See COLOR PICTURE SIGNAL, COLOR SLOW-SCAN TELEVISION, COLOR TELEVISION, COMPOSITE VIDEO SIGNAL, FACSIMILE, SLOW-SCAN TELEVISION, TELEVISION.

In practice, images can be sent and received either in *analog* form or in *digital* form. See ANALOG, DIGITAL. The amplitude, frequency or phase can be modulated, or the data can be encoded as a modulated pulse train. See AMPLITUDE MODULATION, FREQUENCY MODULATION, PHASE MODULATION, PULSE MODULATION.

IMAGE FREQUENCY

In a superheterodyne receiver, frequency conversion results in a constant intermediate frequency. This simplifies the problem of obtaining good selectivity and gain over a wide range of input-signal frequencies. However, it is possible for input signals on two different frequencies to result in output at the intermediate frequency. One of these signal frequencies is the desired one, and the other is undesired. The undesired response frequency is called the *image frequency*.

The illustration is a simple block diagram of a superheterodyne receiver having an intermediate frequency of 455 kHz. This is a common value for the intermediate frequency of a superheterodyne circuit. The local oscillator is tuned to a frequency 455 kHz higher than the desired signal; for example, if reception is wanted at 10.000 MHz, the local oscillator must be set to 10.455 MHz. However, an input signal 455 kHz higher than the local-oscillator frequency, at 10.910 MHz, will also mix with the local oscillator to produce an output at 455 kHz. If signals are simultaneously present at 10.000 and 10.910 MHz, then, they can interfere with each other. We call 10.910 MHz, in this example, the image frequency.

The image frequency in a superheterodyne receiver always differs from the signal frequency by twice the value of the intermediate frequency. A selective circuit in the front end of the receiver must be used to attenuate signals arriving at the image frequency, while passing those arriving at the desired frequency. The lower the intermediate frequency in a superheterodyne receiver, the closer the image frequency is to the desired frequency, and the more difficult it becomes to attenuate the image signal. For this reason, modern superheterodyne receivers use a high intermediate frequency, such as 9 MHz so that the image and desired frequencies are far apart. The high intermediate frequency can later be heterodyned to a low frequency, making it easier to obtain sharp selectivity. *See also* DOUBLE-CONVERSION RECEIVER, IMAGE REJECTION, SINGLE-CONVERSION RECEIVER, and SUPERHETERODYNE RECEIVER.

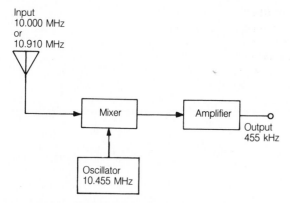

Input
10.000 MHz
or
10.910 MHz

Mixer

Amplifier

Output
455 kHz

Oscillator
10.455 MHz

IMAGE FREQUENCY: The image frequency is separated from the desired frequency by twice the intermediate frequency.

IMAGE IMPEDANCE

An *image impedance* is an impedance that is "seen" at one end of a network, when the other end is connected to a load or generator having a defined impedance. The image impedance can be the same as the terminating impedance at the other end of the network, but this is not always the case.

As an example, suppose that a pure resistive impedance Z_1 is connected to one end of a quarter-wave transmission like having a characteristic impedance Z_0. Then the image impedance Z_2 at the opposite end of the line is related to the other impedances according to the formula:

$$Z_0 = \sqrt{Z_1 Z_2}$$

If the transmission line measures a half electrical wavelength, however, Z_1 and Z_2 will always be equal.

The relationships become more complicated when reactance is present in the terminating impedance. For a quarter-wave line, capacitive reactance at one end is transformed into an inductive reactance at the opposite end, and vice versa. For a half-wave line, the terminating and image impedances are equal even when reactance is present.

There are many different kinds of networks that produce various relationships between terminating and image impedances. *See also* IMPEDANCE, and REACTANCE.

IMAGE INTENSIFIER

An *image intensifier* is a device that makes a visual or electronic image brighter. Such a device can be used to literally see in the dark. Image intensifiers also can be used to help people with visual handicaps to see better in dim light.

The image intensifier generally consists of a television camera and picture tube, thus forming a self-contained closed-circuit television system. An amplifier increases the brightness of the picture signal from the camera tube. *See also* CAMERA TUBE, and PICTURE TUBE.

IMAGE ORTHICON

The camera tube most often used for live commercial television broadcasting is called the *image orthicon*. The image orthicon is a highly sensitive camera tube. It responds rapidly to motion and changes in light intensity.

Light is focused by means of a convex lens onto a translucent plate called the photocathode. The photocathode emits

electrons, according to the intensity of the light in various locations. A target electrode attracts these electrons, and accelerator grids give the electrons additional speed. When the electrons from the photocathode, called photoelectrons, strike this target electrode, secondary electrons are produced. Several secondary electrons are emitted from the target for each impinging photoelectron. This is part of the reason for the high sensitivity of the image orthicon.

A fine electron beam, generated by an electron gun, scans the target electrode (the scanning rate and pattern correspond with the scanning in television receivers). Some of these electrons are reflected back toward the electron gun by the secondary emissions from the target. Areas of the target having greater secondary-electron emission result in an intense return beam; areas of the target that emit few secondary electrons result in lower returned-beam intensity. Therefore, the returned beam is modulated as it scans the target electrode. A receptor electrode picks up the returned beam.

The main disadvantage of the image orthicon is its rather high noise output. However, in situations where there is not much light, or where a fast response time is required, the image orthicon is superior to other types of camera tubes. *See also* CAMERA TUBE, ICONOSCOPE, and VIDICON.

IMAGE PROCESSING

Any means of improving an image signal at the transmitter or receiver, or both, is *image processing*. Image processing is similar to schemes for enhancing single-sideband voice modulation. Methods include compandoring and digital signal processing. See AMPLITUDE-COMPANDORED SINGLE SIDEBAND, COMPANDOR, DIGITAL SIGNAL PROCESSING.

The noticeable effects of image processing are just what you would expect. Pictures that contain interference, such as "snow," cross hatching or splatter lines, have less of these nuisance factors after processing.

There is a limit to what image processing can do. It cannot make a "buried" signal look clear. An excellent signal can't be made much better; if image processing isn't needed, it won't do any good. And a poorly sent signal, such as might result from improper video camera focus, can't be fixed up by a process that does not address the cause of the defect.

See also COLOR SLOW-SCAN TELEVISION, COLOR TELEVISION, FACSIMILE, SLOW-SCAN TELEVISION, TELEVISION.

IMAGE REJECTION

In a superheterodyne receiver, *image rejection* is the amount by which the image signal is attenuated, with respect to a signal on the desired frequency. Image rejection is always specified in decibels.

In general, the higher the intermediate frequency of a receiver, the better the image rejection. *See also* IMAGE FREQUENCY, and SUPERHETERODYNE RECEIVER.

IMPATT DIODE

The *IMPATT (impact-avalanche-transit-time)* diode is a form of semiconductor device that is useful as an amplifier of microwave radio-frequency energy. Some IMPATT diodes can provide output power in excess of 1 watt, providing considerable power gain in transmitters using Gunn-diode oscillators. In future years, improved devices of this kind can be expected. *See also* GUNN DIODE.

IMPEDANCE

Broadly speaking, *impedance* is the opposition that a component or circuit offers to alternating current (ac). Impedance is a two-dimensional quantity. That is, it consists of two independent components.

Resistance and Reactance   In a direct-current (dc) circuit, the opposition to current is known as *resistance*. The resistance is measured in ohms, and is always zero or more. The larger the value of resistance, the greater the opposition to dc. In an ac circuit, dc resistance also opposes the flow of current, and the action is just the same. *See* OHM'S LAW, and RESISTANCE.

In ac circuits, there is another phenomenon, *reactance*, which also puts up opposition to the flow of current. Reactance can be either positive (inductive) or negative (capacitive). *See* CAPACITIVE REACTANCE, and INDUCTIVE REACTANCE.

Resistance does not change with the frequency of the ac signal. But reactance does, if the circuit contains any reactive components (inductances and/or capacitances). In the simplest case, as the frequency goes up, so does the reactance. At very low frequencies, in a given ac circuit, the reactance is negative, or capacitive. As the frequency is raised, the reactance passes through zero (*resonance*) and then becomes positive, or inductive. In practice, many circuits have reactance that rises and falls, over and over, as the frequency is raised, passing through zero (resonance) many times. *See* RESONANCE.

The Complex Impedance Plane   Engineers use a special notation to indicate reactance, and to differentiate it from resistance. Although resistance can be delineated along a half-line corresponding to the non-negative real numbers (see A in the figure), reactance needs a whole real-number line, negative as well as positive and zero, to be depicted (see B in the figure).

Reactance can vary independently of resistance, and vice-versa. In order to fully and unambiguously show impedance, therefore, it is necessary to place the resistance and reactance lines together into a *complex impedance plane* (see C in the figure).

The resistive component of impedance is abbreviated R, and the reactance is indicated by jX. Although R can be any non-negative real number, X can be any real number whatsoever. The operator j is equal to the square root of −1. This might seem esoteric, but it is used because it creates a mathematical model that perfectly explains the behavior of complex impedances. A detailed discussion of this can be found in any good electrical engineering textbook. *See* J OPERATOR.

Impedance Points and Plots   In dc circuits, resistances are indicated by points on the non-negative real number line. In ac circuits, a two-dimensional plane (actually a half plane) is needed. Any impedance, denoted R + jX, corresponds to a unique point on this plane. Any point on the plane matches a unique complex impedance value.

Sometimes impedance is given as a single number value Z, corresponding to the distance of the complex point R + jX from the origin 0 + j0. This expression of impedance is ambiguous, because it can correspond to any of an infinite number of points on a half circle with radius Z. Nonetheless, this expression is

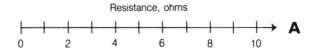

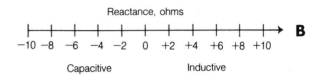

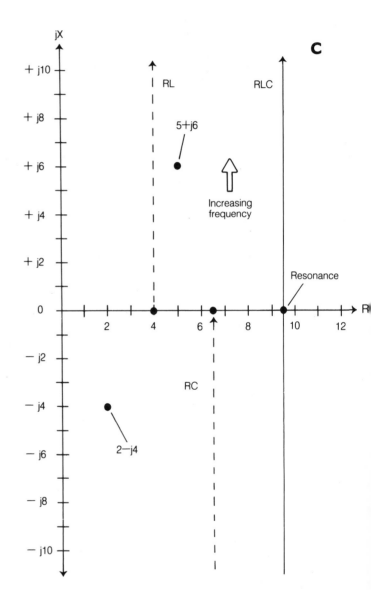

IMPEDANCE: At A, resistance corresponds to non-negative real numbers. At B, reactance corresponds to all real numbers. At C, complex impedance plane. Details are in the text.

convenient in some cases when reactance is either zero or of no concern.

Also shown are three curves. The dotted-line curves show the impedances for resistance-capacitance (RC) and resistance-inductance (RL) circuits, as the frequency varies. The solid-line curve shows the impedance of a series-resonant inductance-capacitance circuit in which there is also some resistance (RLC). Changes in frequency are denoted by moving along the curves.

Plots like this are of value to the engineer in designing resistance/inductance/capacitance circuits.

Related articles in this encyclopedia, not already referenced include: ADMITTANCE, ANTENNA IMPEDANCE, ANTENNA RESONANT FREQUENCY, CHARACTERISTIC IMPEDANCE, IMPEDANCE BRIDGE, IMPEDANCE MATCHING, IMPEDANCE TRANSFORMER, and REACTANCE.

IMPEDANCE BRIDGE

An *impedance bridge* is a device used for the purpose of determining the resistive and reactive components of an unknown impedance. The impedance of any reactive circuit changes with the frequency, so the bridge can function only at a specific frequency. Impedance bridges are often used by radio-frequency engineers to determine the impedances of tuned circuits and antenna systems.

The impedance bridge generally has a null indicator and two adjustments. One adjustment indicates the resistance, and the other indicates the reactance. *See also* IMPEDANCE.

IMPEDANCE MATCHING

An alternating-current circuit always functions best when the impedance of a power source is the same as the impedance of the load to which power is delivered. This does not happen by chance; often, the two impedances must be made the same by means of special transformers or networks. This process is called impedance matching. Impedance matching is important in audio-frequency, as well as radio-frequency applications.

In a high-fidelity system, audio transformers ensure that the output impedance of an amplifier is the same as that of the speakers; this value is generally standardized at 8 ohms. Radio-frequency transmitting equipment is usually designed to operate into a 50-ohm, nonreactive load, although some systems have output tuning circuits that allow for small resistance fluctuations and/or small amounts of reactance in the load.

Radio-frequency engineers probably face the most difficult impedance-matching tasks. At very-high, ultra-high, and microwave frequencies, poorly matched impedances can result in large amounts of signal loss. If the generator and load impedances are not identical, some of the electromagnetic field is reflected from the load back toward the source. In any alternating-current system, the load will accept all of the power only when the impedances of the load and source are identical. *See also* IMPEDANCE, REACTANCE, and RESISTANCE.

IMPEDANCE TRANSFORMER

An *impedance transformer* is used to change one pure resistive impedance to another. Impedance transformers are sometimes used at audio frequencies to match the output impedance of an amplifier to the input impedance of a set of speakers. In radio-frequency work, impedance transformers comprise part of an impedance-matching system, generally used for optimizing the antenna-system for a transmitter.

Suppose that the impedance of a purely resistive (that is, nonreactive) load is given by Z_S, and this load is connected to the secondary winding or a transformer having a primary-to-secondary turns ratio of T. Then the impedance Z_P appearing across the primary winding, neglecting transformer losses, is given by:

$$Z_P = Z_S T^2$$

An impedance transformer must be designed for the proper range of frequencies. This generally means that the reactances of the windings must be comparable to the source and load reactances at the frequency in use. *See also* IMPEDANCE, IMPEDANCE MATCHING, and TRANSFORMER.

IMPULSE

An *impulse* is a sudden surge in voltage or current. An impulse that appears at a utility outlet is sometimes called a *transient* or *transient spike*. Because impulses are of very short duration, they contain high-frequency components. This can cause radio-frequency emissions from electrical wiring and appliances, resulting in interference to nearby radio receivers. *See also* IMPULSE NOISE.

IMPULSE NOISE

Any sudden, high-amplitude voltage pulse will cause radio-frequency energy to be generated. In electrical systems without shielded wiring, the result can be electromagnetic interference. Impulse noise is generated by internal combustion engines because of the spark-producing voltage pulses. This form of noise is called *ignition noise* (*see* IGNITION NOISE). Impulse noise can also be produced by household appliances (such as vacuum cleaners, hair dryers, electric blankets, thermostat mechanisms, and fluorescent-light starters).

Impulse noise is usually the most severe at very-low and low radio frequencies. However, serious interference can often occur in the medium and high frequency ranges. Impulse noise is seldom a problem above about 30 MHz, except in extreme cases. Impulse noise can be picked up by high-fidelity audio systems, resulting in interference to phonograph devices and tape-deck playback equipment.

Impulse noise in a radio receiver can be reduced by the use of a good ground system. Ground loops should be avoided. A noise blanker or noise limiter can be helpful. A receiver should be set for the narrowest response bandwidth consistent with the mode of reception. Impulse noise in a high-fidelity sound system can be more difficult to eliminate. An excellent ground connection, without ground loops, is imperative. It might be necessary to shield all speaker leads and interconnecting wiring. *See also* ELECTROMAGNETIC INTERFERENCE.

IMPURITY

An *impurity* is a substance that is added to a semiconductor material. The type and amount of impurity determines whether the semiconductor will conduit by means of electron transfer or hole transfer. The process of adding an impurity is called doping (*see* DOPING).

An impurity that results in the creation of an N-type material is called a donor impurity. Donor substances include arsenic, phosphorus, and bismuth. The N-type material conducts by passing excess electrons from atom to atom. The electron mobility depends on the amount and kind of donor impurity that is added to the silicon or germanium semiconductor. This, in turn, affects the characteristics of the diode, transistor, or field-effect transistor that is made from the material. *See also* N-TYPE SEMICONDUCTOR.

An impurity that results in the formation of a P-type semiconductor is called an acceptor impurity. Acceptors include boron, aluminum, gallium, and indium. Such substances are electron-deficient. The P-type material conducts by passing electron deficiencies, or holes, from atom to atom. Hole mobility is generally lower than electron mobility. However, the hole mobility is determined, as is the electron speed in the N-type material, by the kind and amount of impurity. The velocity of electrons or holes in a semiconductor, for a given electric-field intensity, is called the *carrier mobility. See also* CARRIER MOBILITY, GERMANIUM, HOLE, P-TYPE SEMICONDUCTOR, and SILICON.

INCOHERENT RADIATION

Incoherent electromagnetic radiation occurs when there are many different frequency and/or phase components in the wave. An example of incoherent radiation is the light from any ordinary bulb, such as an incandescent lamp or fluorescent tube.

Most radiation, including radio-frequency noise, infrared, visible light, ultraviolet rays, X rays, and gamma rays, is incoherent. This is because such emissions result from a random sort of disturbance. When radiation is concentrated at a single wavelength, and all of the wavefronts are lined up in phase, the radiation is said to be coherent. A radio transmitter generates coherent radiation. So does a maser or laser. *See also* COHERENT RADIATION.

INCREMENT

An *increment* is a small (or unit) change in a quantity. In a high-frequency (HF) transceiver with digital tuning, for example, the frequency might be variable in increments of 10 Hz. In a 2-meter FM transceiver, the frequency might be varied in increments of 5, 10, 15, or 20 kHz (selectable).

In satellite communications, the increment refers to a factor important in locating the satellite. Unless the satellite is geostationary, its groundtrack moves over the earth nearly in great circles. But the earth rotates eastward underneath the satellite,

so the groundtrack for each orbit is to the west of that for the preceding orbit. For each orbit there is an *ascending node*, at which the groundtrack crosses the equator going north. The node lies west, by a certain longitude angle, of the node for the preceding orbit. This angle is the increment for the satellite. See the illustration. *See also* ACTIVE COMMUNICATIONS SATELLITE, ASCENDING NODE, DESCENDING NODE, GREAT CIRCLE, OSCAR, and OSCARLOCATOR.

INDEPENDENT SIDEBAND

Single-sideband (SSB) transmission can use either the upper sideband (USB) or the lower sideband (LSB). This saves spectrum space over the older amplitude-modulation (AM) mode of emission. It also saves considerable power, because the carrier is suppressed. *See* AMPLITUDE MODULATION, LOWER SIDEBAND, SINGLE SIDEBAND, and UPPER SIDEBAND.

It is possible to transmit different data on the upper and lower sidebands of a modulated signal. This can be done using double-sideband, suppressed-carrier emission (DSB). *See* DOUBLE SIDEBAND. Such a transmission is obviously different than the signal that would result from simply suppressing the carrier of an AM transmission. That is conventional DSB emission, with both sidebands identical.

Obtaining *independent sideband (ISB) emission (DSB)*, in which one sideband contains information completely independent of the contents of the other sideband, is rather complicated. One scheme is to "slave" two SSB transmitters to a single suppressed-carrier frequency, using a phase-locked loop. One transmitter operates on USB, and the other on LSB. The outputs of the transmitters are fed to a single antenna or a pair of antennas adjacent to each other.

The receiving setup needs two separate receivers, tuned to the same frequency, with one set for USB and the other set for LSB. An ISB signal has the same bandwidth as a conventional AM transmission, or about 6 kHz.

Hams have used ISB to send and receive narrated slow-scan television (SSTV). The image data is contained in one sideband, say the USB, while the voice is transmitted over the other sideband, say the LSB. A transmission like this, if properly received, "plays" like a well-presented slide show. *See also* SLOW-SCAN TELEVISION.

INDICATOR

An *indicator* is a device that displays the status of a circuit function, such as current, voltage, power, on/off condition, or frequency. Indicators usually take the form of analog or digital meters (*see* ANALOG METERING, and DIGITAL METERING). However, light-emitting devices are also sometimes used as indicators.

Indicators are characterized by a direct reading of the quantity to be measured. No calculations are necessary. If the measured parameter changes value or state, the indicator follows the change. *See also* METER.

INDIUM

Indium is an element with atomic number 49 and atomic weight 115. In the manufacture of semiconductor materials, indium is used as an impurity or dopant. Indium is an acceptor impurity. This means that it is electron-deficient. When indium is added to a germanium or silicon semiconductor, a P-type substance is

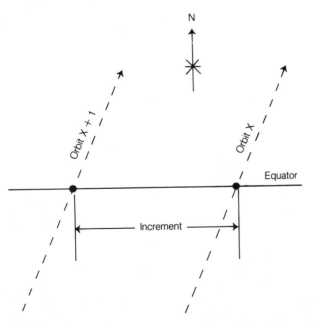

INCREMENT: The increment of a satellite orbit is the longitude angle between successive ascending nodes.

the result. P-type semiconductors conduct via holes. *See also* DOPING, HOLE, IMPURITY, and P-TYPE SEMICONDUCTOR.

INDOOR ANTENNA

For radio and television reception or transmission, an outdoor antenna is always preferable to an indoor antenna. However, in certain situations it is not possible to install an outdoor antenna.

Indoor antennas are always a compromise, especially at the very-low, low, medium, and high frequencies. Electrical wiring interferes with wave propagation at these frequencies. In a concrete-and-steel structure, the shielding effect is even more pronounced than in a frame building. Indoor antennas are generally more subject to manmade interference from electrical appliances. If an indoor antenna is used for transmitting purposes, the chances of electromagnetic interference are greatly increased, compared with the use of an outdoor antenna. The efficiency of an indoor transmitting antenna is generally lower than that of an identical outdoor antenna.

Indoor antennas have some advantages over outdoor antennas. An indoor antenna is less susceptible to induced voltages resulting from nearby lightning strikes. Indoor antennas do not corrode as rapidly as outdoor antennas, and maintenance is simpler. At very-high and ultra-high frequencies, indoor antennas can perform very well if they are located high above the ground.

In general, if an indoor antenna must be used, the antenna should be kept reasonably clear of metallic obstructions. It should be placed as high as possible. At frequencies below about 30 MHz, the antenna should be cut to resonance or made as long as possible.

INDUCED EFFECTS

Voltages and currents can be induced in a material in a variety of ways. When a conductor is moved through a magnetic field, current flows in the conductor. An object in an electric field exhibits a potential difference, in certain instances, between one region and another. Alternating currents in one conductor can induce similar currents in nearby conductors.

Induced effects are extremely important in many applications. The transformer works because a current in the primary winding induces a current in the secondary. Radio communication is possible because of the induced effects of alternating currents at high frequencies. *See also* ELECTRIC FIELD, ELECTROMAGNETIC FIELD, and MAGNETIC FIELD.

INDUCTANCE

Inductance is the ability of a device to store energy in the form of a magnetic field. Inductance is represented by the capital letter L in mathematical equations. The unit of inductance is the henry. One henry is the amount of inductance necessary to generate 1 volt with a current that changes at the rate of 1 ampere per second. Mathematically:

$$L \text{ (henrys)} = E/(dI/dt)$$

where E is the induced voltage and dI/dt is the rate of current change in amperes per second.

In practice, the henry is an extremely large unit of inductance. Usually, inductance is specified in millihenrys (mH),

microhenrys (uH), or nanohenrys (nH). These units are, respectively, a thousandth, a millionth, and a billionth of 1 henry:

$$1 \text{ mH} = 10^{-3} \text{ H}$$
$$1 \,\mu\text{H} = 10^{-3} \text{ mH} = 10^{-6} \text{ H}$$
$$1 \text{ nH} = 10^{-3} \,\mu\text{H} = 10^{-9} \text{ H}$$

Any length of electrical conductor displays a certain amount of inductance. When a length of conductor is deliberately coiled to produce inductance, the device is called an *inductor* (*see* INDUCTOR).

Inductances in series add together. For n inductances L_1, L_2, . . ., L_n in series, then, the total inductance L is:

$$L = L_1 + L_2 + \ldots + L_n$$

In parallel, inductances add according to the equation:

$$L = 1/(1/L_1 + 1/L2 + \ldots + 1/L_n)$$

See also INDUCTIVE REACTANCE.

INDUCTANCE MEASUREMENT

Inductance is usually measured by means of a device incorporating known capacitances that cause resonance effects at measurable frequencies. In a tuned circuit, an unknown inductance can be combined with a known capacitance, and the resonant frequency determined by means of a signal generator and indicating device.

Once the resonant frequency (f) has been found for a known capacitance (C) and an unknown inductance (L), the value of L can be determined according to the following equation:

$$L = 1/(39.5f^2C)$$

where L is given in microhenrys, f in megahertz, and C in microfarads. Alternatively, L can be found in henrys if f is given in hertz and C is given in farads.

A more sophisticated device for determining unknown inductances is called the *Hay bridge*. This circuit allows not only the inductance, but also the Q factor, to be found for a particular inductor. *See also* HAY BRIDGE, INDUCTANCE, Q FACTOR, and RESONANCE.

INDUCTION

See INDUCED EFFECTS.

INDUCTION LOSS

When a conductor carries radio-frequency energy, it is important that the conductor either be shielded, or that it be kept away from metallic objects. This is because the inductive coupling to nearby metallic objects can cause losses in the conductor.

Induction loss can occur when an antenna radiator is placed too close to utility wiring, a metal roof, or other conducting medium. Circulating currents occur, resulting in heating of the nearby material. A parallel-wire transmission line should be kept at least several inches away from metal objects such as wires, towers, or siding so that the possibility of induction loss is minimized. Induction losses increase the overall loss in an antenna system.

INDUCTIVE CAPACITOR

An *inductive capacitor* is a capacitor with built-in inductance. Such capacitors are useful as self-contained tuned circuits. Inductive capacitors are generally wound in a spiral pattern, with two concentric plates separated by a rigid dielectric material. The amount of capacitance depends on the area and separation of the metal plates, and also on the nature of the dielectric. The amount of inductance depends largely on the number of turns in the spiral structure.

In the very-high-frequency range and above, all capacitors begin to show significant inductive reactance, simply because of the finite length of the component leads. In the ultra-high and microwave parts of the spectrum, the wiring inductance of all components must be taken into account when circuits are designed and constructed. *See also* LEAD INDUCTANCE.

INDUCTIVE COUPLING

See TRANSFORMER COUPLING.

INDUCTIVE FEEDBACK

Inductive feedback results from inductive coupling between the input and output circuits of an amplifier. The feedback can be intentional or unintentional; it can be positive (regenerative) or negative (degenerative), depending on the number of stages and the circuit configuration. Inductive feedback can occur at audio or radio frequencies.

Some oscillators use inductive feedback for the purpose of signal generation. An example of such an oscillator is the Armstrong oscillator, in which a feedback coil from the output circuit is deliberately placed near the tank coil (*see* ARMSTRONG OSCILLATOR).

Unwanted inductive feedback is generally more of a problem in the very-high-frequency range and above, as compared to its likelihood at lower frequencies. This is because the inductive reactance of intercomponent wiring becomes greater as the frequency increases, and this in turn increases the inductive coupling among different parts of a circuit. Inductive feedback of a regenerative nature can result in spurious oscillations in a receiver or transmitter. The spurious signals in a transmitter can be radiated along with the desired signal, causing interference to other stations. In audio-frequency equipment, inductive feedback can sometimes occur if unshielded speaker leads are placed near the input cables of the amplifier. The result, if the feedback is regenerative, will be hum or an audio-frequency tone. *See also* FEEDBACK.

INDUCTIVE LOADING

In radio-frequency transmitting installations, it is often necessary to lengthen an antenna electrically, without making it physically longer. This is especially true at the very-low, low, and medium frequencies. Inductive loading is the most common method of accomplishing this.

In an unbalanced antenna, such as a quarter-wave resonant vertical working against ground, the inductor can be placed at any location less than approximately 90 percent of the way to the top of the radiator. The most common positions for the inductor are at the base and near the center. With a given physical radiator length, larger and larger inductances result in lower and lower resonant frequencies. There is no theoretical limit to how low the resonant frequency of an inductively loaded

quarter-wave vertical can be made. In practice, however, the losses become prohibitive when the physical length of the radiator is less than about 0.05 wavelength.

In a balanced antenna system, such as a half-wave resonant dipole, identical inductors are placed in each half of the antenna, in a symmetrical arrangement, with respect to the feed line (see illustration). For a given physical antenna length, larger and larger inductances result in lower and lower resonant frequencies. As with the quarter-wave vertical antenna, there is no theoretical lower limit to the resonant frequency that can be realized with a given physical antenna length. However, the losses become prohibitive when the physical length becomes too short. If the dipole is placed well above the ground and away from obstructions, and if the conductors and inductors are made from material with the least possible ohmic loss, good performance can be had with antennas as short as 0.05 free-space wavelength.

Inductively loaded antennas always show a low radiation resistance, because the free-space physical length of the radiator is so short. As the length of an antenna becomes shorter than 0.1 wavelength, the radiation resistance drops rapidly. This makes it difficult to achieve good transmitting efficiency with very short radiators. *See also* CENTER LOADING, and RADIATION RESISTANCE.

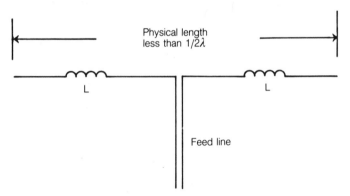

INDUCTIVE LOADING: Coils L lower the resonant frequency of an antenna.

INDUCTIVE REACTANCE

Reactance is the opposition that a component offers to alternating current. Reactance does not behave as simply as resistance. The reactance of a component can be either positive or negative, and it varies in magnitude, depending on the frequency of the alternating current. Positive reactance is called *inductive reactance*; negative reactance is called *capacitive reactance*. These choices of positive and negative are purely arbitrary, and are used as a matter of mathematical convenience.

The reactance of an inductor of *L* henrys, at a frequency of *f* hertz, is given by:

$$X_L = 2\pi f L,$$

where X_L is specified in ohms. If the frequency (*f*) is given in kilohertz and the inductance (*L*) in millihenrys, the formula also applies. The frequency can be given in megahertz and the inductance in microhenrys, as well (units should not be mixed).

The magnitude of the inductive reactance, for a given component, approaches zero as the frequency is lowered. The inductive reactance gets larger without limit as the frequency is raised. This effect is illustrated in the graph. At A, the reactance of a 100-μH inductor is shown as a function of the frequency in megahertz. At B, the reactance of various values of inductance are shown for a constant frequency of 1 MHz.

Reactances, like direct-current resistances, are always expressed in ohms. In a complex circuit containing resistance and reactance, the reactance is multiplied by $\sqrt{-1}$, which is mathematically represented by the letter j in engineering documentation. Thus, an inductive reactance of +20 ohms is specified as $+j20$; a combination of 10 ohms resistance and +20 ohms inductive reactance is a complex impedance of $10 + j20$. *See also* IMPEDANCE, and *J* OPERATOR.

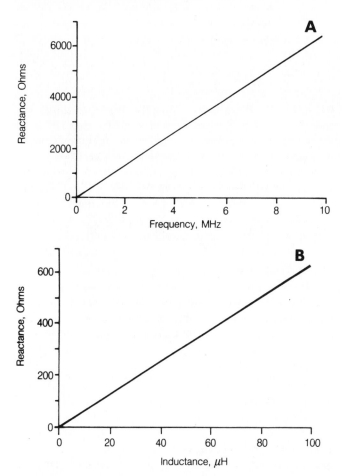

INDUCTIVE REACTANCE: At A, reactance versus frequency for 100-μH inductor. At B, reactance versus inductance at 1 MHz.

INDUCTOR TUNING

Inductor tuning is a means of adjusting the resonant frequency of a tuned circuit by varying the inductance while leaving the capacitance constant. When this is done, the resonant frequency varies according to the inverse square root of the inductance. This means that if the inductance is doubled, the resonant frequency drops to 0.707 of its previous value. If the inductance is cut to one-quarter, the resonant frequency is doubled. Mathematically, the resonant frequency of a tuned inductance-capacitance circuit is given by the formula:

$$f = 1/(2\pi \sqrt{LC})$$

where f is the frequency in hertz, L is the inductance in henrys, and C is the capacitance in farads. Alternatively, the units can be given as megahertz, microhenrys, and microfarads.

Inductor tuning is somewhat more difficult to obtain than capacitor tuning, from a mechanical standpoint. However, inductor tuning usually offers a more linear frequency readout than capacitor tuning. If the capacitance required to obtain resonance is larger than the values normally provided for variable capacitors, inductor tuning is preferable. For the purpose of inductor tuning, a rotary device can be used. A movable core, consisting of powdered iron or ferrite, can also be used. The latter form of inductor tuning is sometimes called permeability tuning. *See also* CAPACITOR TUNING.

INFRADYNE RECEIVER

An *infradyne receiver* is a form of superheterodyne receiver. Most superheterodyne receivers produce an intermediate-frequency signal that has a frequency equal to the difference between the input and local-oscillator frequencies. The infradyne receiver, however, produces an intermediate frequency that is equal to the sum of the input and local-oscillator frequencies. The term *infradyne* arises from the fact that both the input and local oscillator signals are lower than the intermediate frequency.

If f is the input-signal frequency to an infradyne receiver and g is the frequency of the local oscillator, heterodynes are produced at frequencies corresponding to $f - g$ (if f is larger than g) or $g - f$ (if f is less than g); a signal is also produced at $f + g$. In the infradyne, $f + g$ is the intermediate frequency.

The infradyne receiver is characterized by the fact that the signal always appears right side up. That is, the frequency components are not inverted. In a superheterodyne receiver using the difference frequencies, the signal sometimes appears upside down. The main disadvantage of the infradyne is that harmonics of the local-oscillator signal can fall within the input-tuning range of the receiver. *See also* HETERODYNE, and SUPERHETERODYNE RECEIVER.

INHIBIT

In a digital system such as a computer, an *inhibit command* is a signal or pulse that prevents or delays a certain operation. If the inhibit line is in the low or zero-voltage state, certain operations do not occur, but they will occur if the inhibit line is high. In negative logic, the situation is reversed; a low inhibit line allows the operation to proceed, but the high state prevents it.

Inhibit gates are used in memory circuits for the purpose of preventing a change of state. This keeps the contents of the memory intact and inalterable. Until the inhibit command is removed, the memory cannot be changed. *See also* MEMORY.

INITIALIZATION

Initialization is a process in which all of the lines in a microcomputer are set to the low (zero-voltage) condition prior to operation. Most microcomputers are initialized every time power is removed and reapplied. In some cases, certain memory stores are retained by memory-backup power supplies. Removal of the power source results in initialization.

It is sometimes necessary to initialize a microcomputer device during the course of normal operation. This is especially true of the complementary metal-oxide semiconductor (CMOS) integrated circuit. Such devices are characterized by extremely low leakage current, and "false signals" can be maintained for extended periods, causing what appears to be a malfunction.

In some microcomputer devices, initialization consists of a resident routine for the purpose of aiding in the programming of the system. Such a routine is called the *initial instruction. See also* MICROCOMPUTER.

INITIAL SURGE

When power is first applied to a device that draws a large amount of current, an initial surge can occur in the circuit current. The surge is often considerably higher than the normal operating current load of the device. This effect is exemplified by the momentary dimming of the lights in a house when the refrigerator, air conditioner, or electric heater first starts. The current surge is normally not dangerous, but it can occasionally result in a blown fuse or tripped circuit breaker. The same effect occurs when power is applied to radio equipment, although it is not as dramatic. Slow-blow fuses are employed in devices that cause an initial surge (*see* SLOW-BLOW FUSE).

In the event of an electric power failure, an initial voltage surge can occur when electricity is restored. This surge can, in some instances, cause damage to appliances. This kind of initial surge is most likely to occur if the power failure covers a widespread area. In the event of a power failure, it is wise to shut off devices that draw large amounts of current. The same precautions should be taken with appliances that might be damaged by voltages slightly in excess of the normal 110 to 120 volts.

The initial surge is not the same thing, and is not caused by the same conditions, as a transient. Transients reach much higher peak amplitudes. *See also* TRANSIENT.

INJECTION LASER

The *injection laser* is a form of light-emitting diode (LED), with a relatively large and flat P-N junction. The injection laser emits coherent light, provided the applied current is sufficient. If the current is below a certain level, the injection laser behaves much like an ordinary LED, but when the so-called threshold current is reached, the charge carriers recombine in such a manner that laser action occurs.

Most injection-laser devices are fabricated from gallium arsenide (GaAs), and have a primary emission wavelength of about 905 nanometers (nm). This is in the infrared range. Other types of injection lasers are available for different wavelengths. Among these, the GaAsP and GaP injection lasers produce outputs at approximately 660 nm and 550 nm, respectively; these wavelengths are in the visible-light range. The maximum peak power, in short pulses, can be as great as 100 watts, but the pulse duration must be very short. Injection lasers produce emission at exceedingly narrow bandwidth. This is characteristic of laser devices.

Injection lasers are used mainly in optical communications systems. The coherent light output from the injection laser suffers less attenuation than incoherent light as it passes through the atmosphere. *See also* OPTICAL COMMUNICATIONS.

INPUT

Input is the application of a signal to a processing circuit. An amplifier circuit, for example, receives its input in the form of a small signal; the output is a larger signal. A logic gate can have several signal inputs. The input terminals are themselves sometimes called *inputs*.

The input signal in a given situation must conform to certain requirements. The amplitude must be within certain limits, and the impedance of the signal source must match the impedance presented by the input terminals of the circuit. *See also* INPUT CAPACITANCE, INPUT DEVICE, INPUT IMPEDANCE, and INPUT RESISTANCE.

INPUT CAPACITANCE

The input terminals of any amplifier, logic gate, or other circuit always contain a certain amount of parallel capacitance. The capacitance appears between the two terminals in a balanced circuit, and between the active terminal and ground in an unbalanced circuit.

The amount of input capacitance must, in most circuits, be smaller than a certain value for proper circuit operation. In general, the higher the frequency, the smaller the maximum tolerable input capacitance. Also, the higher the input impedance of a circuit (*see* INPUT IMPEDANCE), the smaller the maximum tolerable input capacitance. Generally, the input capacitance should not exceed 10 percent of the resistive input impedance. When the input capacitance becomes too high, the input impedance changes so much that some of the incident signal is reflected back toward the source.

In some circuits, the input capacitance can be very large. This is the case, for example, in the output filtering network of a power supply when a capacitor is used at the input. This condition is normal. *See also* INPUT RESISTANCE.

INPUT DEVICE

In a computer system, an *input device* consists of any peripheral equipment that serves the function of providing information to the computer. Examples of input devices include the keyboard in a terminal unit, a tape recorder or disk drive, and a telephone or radio interface.

In any electronic system, a circuit through which an input is applied is called an input device. Examples are the power transformer, a signal transformer, or a coupling network. *See also* INPUT.

INPUT IMPEDANCE

The *input impedance* of a circuit is the complex impedance that appears at the input terminals. Input impedance normally contains resistance, but no reactance, under ideal operating conditions. However, in certain situations, large amounts of reactance can be present. This occurs as either an inductance or a capacitance.

Input impedances are often matched to the output impedance of the driving device. For example, an audio amplifier with a 600-ohm resistive input impedance should be used with a microphone having an impedance of close to 600 ohms. A half-wave dipole antenna with a pure resistive input impedance of 73 ohms should be fed with a transmission line having a characteristic impedance of nearly this value; the transmitter output circuits should also have an impedance of nearly this value.

If reactance is present in the input impedance of a device, some of the electromagnetic energy is reflected back toward the source. This can happen in some circuits, to a limited extent, with little or no adverse effect. In other circuits, such a condition must be avoided. *See also* IMPEDANCE, IMPEDANCE MATCHING, OUTPUT IMPEDANCE, and REACTANCE.

INPUT-OUTPUT PORT

In computer systems, an *input-output port* or *device* is a peripheral unit that serves two purposes: to provide data to the computer, and to transfer data from the computer. Most input devices, such as a tape recorder, disk drive, and terminal unit, can also serve as output ports. Punched cards, in some computers, serve as input-output ports. The modem, which provides an interface with a telephone or radio circuit, is a common form of input-output device (*see* MODEM).

An input-output port can serve any of three different functions. It can operate as an interface between the computer and a human operator; a video-display terminal and keyboard function in this manner. The input-output device can serve as a read-write memory medium. The disk drive is an example of this. The input-output device can also act as an interface between one computer and another. The modem is generally used for this purpose.

INPUT RESISTANCE

In a direct-current circuit, a certain resistance is always present at the input terminals. This is called the *input resistance,* and is equivalent to the input voltage divided by the input current.

The input resistance of a direct-current circuit can remain constant, regardless of the applied voltage. This is generally the case with passive direct-current networks that contain only resistors. Often, however, the input resistance changes with the amount of applied voltage. This is generally true with passive circuits containing nonlinear elements, such as semiconductor diodes. Active devices, such as audio amplifiers and radio equipment, usually have an input resistance that changes, not only with the level of applied voltage, but with the level of the input signal.

In an alternating-current circuit, the input resistance is the resistive component of the input impedance. *See also* INPUT IMPEDANCE.

INPUT TUNING

When the input circuit of an amplifier contains a tuned inductance-capacitance circuit, the amplifier is said to have *input tuning.* Not all amplifiers require input tuning. The illustration shows three radio-frequency amplifier circuits. The amplifier at

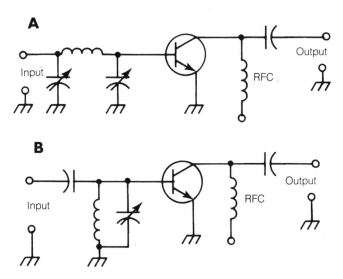

A

B

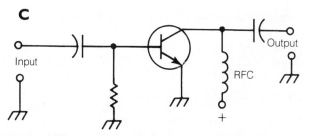

C

INPUT TUNING: At A, a pi network; at B, a parallel-resonant; at C, an untuned input.

A has a pi-network input-tuning circuit. The amplifier shown at B uses a resonant inductance-capacitance network at its input. The two configurations are used for different purposes. At C, an amplifier is illustrated that does not make use of input tuning.

Input tuning has certain advantages and disadvantages. The primary advantage is realized in situations where the amplifier input impedance differs from the output impedance of the driving stage, or when additional selectivity is desired. The pi-network circuit (A) allows unequal impedances to be matched. The parallel-resonant network (B) provides high input selectivity. The main disadvantage of input tuning is that it increases the complexity of adjustment of the amplifier circuit. A change in the frequency of the driving signal usually necessitates retuning of the input networks. *See also* IMPEDANCE MATCHING, and PI NETWORK.

INSERTION GAIN

Insertion gain is an expression of signal amplitude with and without the use of some intermediate circuit. Insertion gain is expressed in decibels.

If the voltage amplitude at the output of a device is E_1 without the intermediary circuit and E_2 with the intermediary circuit, then the insertion gain (G) is given in decibels by:

$$G \text{ (dB)} = 20 \log_{10}(E_2/E_1)$$

For current, the formula is similar; if I_1 represents the output current without the intermediary circuit and I_2 is the output current with the circuit in place, the gain G is:

$$G \text{ (dB)} = 20 \log_{10}(I_2/I_1)$$

For power, if P_1 is the power without the intermediary circuit and P_2 is the power with the circuit in place, then the insertion gain G is:

$$G \text{ (dB)} = 10 \log_{10}(P_2/P_1)$$

It is evident from these equations that if the insertion of the intermediary circuit causes a reduction in voltage, current, or power, then the insertion gain for that parameter is negative. In such cases, the intermediary circuit is said to have insertion loss (*see* INSERTION LOSS). Passive circuits always have negative insertion gain. Active circuits can display positive or negative insertion gain, depending on whether voltage, current, or power is evaluated.

Amplifier circuits can show insertion gain for some parameters, but a loss for other parameters. It is necessary to define insertion gain in terms of current, voltage, or power, if the expression is to have meaning. *See also* DECIBEL, and GAIN.

INSERTION LOSS

Insertion loss is a reduction in the output voltage, current, or power of a system, resulting from the addition of an intermediary network, such as a filter or attenuator. Insertion loss is generally specified in decibels.

For a selective filter, the insertion loss is normally specified for the frequency, or band of frequencies, at which the attenuation is the least. Most selective filters have low insertion loss at the frequencies to be passed, and high loss at other frequencies.

If the output voltage of a circuit is E_1 before the insertion of an intermediary network and E_2 after the insertion of the network, the insertion loss L is given by:

$$L \text{ (dB)} = 20 \log_{10}(E_1/E_2)$$

For current, if I_1 is the output current before insertion of the device and I_2 is the current with the device installed, then the loss in decibels is given by:

$$L \text{ (dB)} = 20 \log_{10}(I_1/I_2)$$

For power, if P_1 is the output wattage before the insertion of the device and P_2 is the wattage afterward, then:

$$L \text{ (dB)} = 10 \log_{10}(P_1/P_2)$$

Passive networks always have a certain amount of insertion loss, but an efficient device normally exhibits less than 1 dB of loss at the frequencies to be passed. Active networks can have negative insertion loss. In general, if the insertion loss in decibels is $-x$, then the insertion gain in decibels is x. That is, negative loss is the same as positive gain, and vice-versa. *See also* DECIBEL, INSERTION GAIN, and LOSS.

INSTANTANEOUS EFFECT

Whenever a parameter fluctuates, rather than maintaining a constant magnitude or value, instantaneous effects become important. Amplitude and frequency are the most common parameters, in electronic circuits, that change value. However, resistance and impedance also change rapidly under some conditions.

The instantaneous magnitude of a parameter is the value at a given instant. This is, of course, a function of time. The familiar sine-wave alternating-current waveform provides a good example of the difference between an instantaneous value and the average and effective values. The instantaneous voltage can attain any value between − 165 volts and + 165 volts. The precise voltage changes with time, completing one cycle every 16.67 milliseconds. The average voltage is actually zero, but this is obviously not the effective voltage! The root-mean-square (RMS) voltage, which is usually considered the effective voltage, is approximately 117 volts.

Instantaneous effects are sometimes of little concern, but in other cases, they can be very important. A sudden surge in the voltage from a household utility line might last for only a few millionths of a second, and thus not affect the average or RMS voltages significantly. However, the instantaneous voltage can reach several hundred volts during that brief time, and this can be damaging to some kinds of electronic equipment. Instantaneous effects can be measured only with devices capable of displaying the details of a waveform. The oscilloscope is the most common such instrument. *See also* AVERAGE VALUE, EFFECTIVE VALUE, OSCILLOSCOPE, PEAK VALUE, and ROOT MEAN SQUARE.

INSTITUTE OF ELECTRICAL AND ELECTRONICS ENGINEERS

The *Institute of Electrical and Electronic Engineers (IEEE)* is a professional organization in the United States and some other countries. The purpose of the association is to advance the state of the art in electricity and electronics. Originally, the IEEE consisted of two separate groups: *the American Institute of Electrical Engineers (AIEE)* and *the Institute of Radio Engineers (IRE)*.

The IEEE sets standards for construction and performance of many kinds of electrical and electronic equipment. Members work in conjunction with other organizations, such as the Electronic Industries Association. *See also* ELECTRONIC INDUSTRIES ASSOCIATION.

INSTRUCTION

An *instruction* is a set of data bits that directs a computer to follow certain procedures. A computer program, for example, is a string of instructions. The execution of the instructions results in the computer performing the complete operations desired. A program can contain in excess of 1,000,000 individual instructions.

A computer is capable of performing only a finite number of different kinds of instruction; the complete set of instructions that the machine can execute is called the *instruction repertoire*. The total number of individual instructions that a machine can handle, in a specific program, depends on the memory capacity. Most modern computers have an instruction repertoire of fewer than 100 different identifiable commands, although the number of possible combinations in a program is enormous.

An instruction normally consists of a code signal and a memory address. The code signal denotes the particular instruction, such as "multiply" or "divide." The memory address tells the machine where to get and store the information. Instructions can be fed into a computer by means of a disk drive, a magnetic tape, a set of punched cards, a modem, or by a human operator via a keyboard and video display unit.

INSTRUMENT AMPLIFIER

An *instrument amplifier* is an input amplifier for an electronic instrument. The amplifier is used for the purpose of increasing the sensitivity of the instrument. Such amplifiers can be used with direct-current or alternating-current instruments. Some devices that employ amplifiers include the active field-strength meter, the oscilloscope, and the FET voltmeter.

There is a limit to the sensitivity that can be obtained with any instrument, no matter how high the gain of the amplifier. This is because all amplifiers generate some noise, and a small amount of background noise occurs as a result of currents and thermal effects. Quantities of magnitude smaller than the level of the noise, therefore, cannot be reliably measured even when instrument amplifiers are used.

Instrument amplifiers should be housed in shielded enclosures, and bypass capacitors or appropriate chokes should be used so that the instrument will not be affected by signals other than those to be measured. The power supply must be checked periodically. Calibration should also be regularly checked against a known standard when an instrument amplifier is used.

INSTRUMENT ERROR

All indicating devices have some degree of inaccuracy; none display the actual value of a parameter. The difference between the actual value and the instrument reading is called the *instrument error*.

In an analog instrument, the amount of error depends on the electrical and mechanical calibration of the device, and also on the ease with which the operator can interpolate among the divisions on the scale. The error is usually specified, for a given instrument, as a maximum percentage at full scale (*see* FULL-SCALE ERROR).

In a digital instrument, operator error is not a factor because the parameter is displayed as a numeral. The limitations on the accuracy of a digital device are imposed by the electrical precision of the indicator circuit, and by the resolution of the display. A digital meter that has a resolution of three decimal places for a given range (such as 0.00 to 9.99 volts) cannot provide as much accuracy as a meter with a resolution of four decimal places (such as 0.000 to 9.999 volts). The instrument error becomes just a tenth as great when one digit is added to the display, assuming the electrical precision is sufficient.

If the value of a parameter changes rapidly, the instrument error can increase dramatically. This occurs, for example, in the collector-current metering of a single-sideband amplifier. The meter cannot follow the rapid modulation of the voice signal. The needle appears to jump up during voice peaks, but it does not accurately register the actual instantaneous current. A digital meter would be entirely useless in this kind of application; the display would present a constantly changing array of digits. For optimum indication of the instantaneous current values in such a situation, an oscilloscope would be necessary. *See also* ANALOG METERING, and DIGITAL METERING.

INSTRUMENT TRANSFORMER

In an instrument intended for measuring alternating currents, an *instrument transformer* is sometimes used to provide a variety of metering ranges. Transformers can be used for alternating-current ammeters or voltmeters. Such transformers usually have several tap positions, connected to a rotary switch for choosing the desired range.

In a voltmeter, the range is inversely proportional to the number of turns in the secondary winding of the transformer. For example, if the secondary has n turns and the maximum voltage on the scale is x, then switching to $10n$ turns will change the full-scale reading to $0.1x$. In an ammeter, the situation is reversed; the full-scale range is directly proportional to the number of turns. Thus, if n turns provide a reading of x amperes at full scale, $10n$ turns will result in a full-scale reading of $10x$. These relations are based on the assumptions that the input of the meter is provided at the primary winding of the transformer, and that the number of turns in the primary winding remains the same under all conditions. *See also* TRANSFORMER.

INSULATED CONDUCTOR

An *insulated conductor* is a wire, length of tubing, or other electrically conducting substance that is covered with a layer of non-conducting material. Insulated conductors are used in electrical wiring to prevent short circuits, and to reduce the danger of shock or fire. Insulated conductors are also sometimes used in radio-frequency antenna systems.

The insulation covering a conductor must be thick enough to prevent accidental arcing. Wire is available with a variety of insulating materials and thicknesses. Enamel is the thinnest insulation. Cloth is sometimes used because of its durability, although it tends to lose some of its insulating properties if it gets wet. Rubber and plastic are commonly used in insulation on household electrical wires. *See also* ENAMELED WIRE, and INSULATING MATERIAL.

INSULATING MATERIAL

Electrical insulating materials are used for a variety of purposes. In general, insulating substances are used in situations where conduction must be prevented. Insulating materials are classified according to their ability to withstand heat. In order of ascending temperature rating, these categories are designed as follows: O (90 degrees Celsius), A (105 degrees Celsius), B (130 degrees Celsius), F (155 degrees Celsius), H (180 degrees Celsius), C (220 degrees Celsius), and over Class C (more than 220 degrees Celsius).

Class-O materials include paper and cloth. Class-A insulating substances include impregnated paper and cloth. Class-B, class-F, class-H, and class-C materials are substances such as glass fiber, asbestos, and mica. All insulators over class C consist of inorganic materials, such as specially treated glass, porcelain, and quartz. Teflon is able to withstand temperatures of more than 250 degrees Celsius, and it is therefore rated over class C. Enamel is usually a class-O material. Vinyl, often used for insulating electrical wiring, is generally a class-A substance, although silicone enamel can rate as high as class H.

The class of insulating material needed for a given application depends on the amount of current expected to flow in the conductors, and on the conductor size. If the class of insulation is too low, fire can occur. If the class is unnecessarily high, the expense is unwarranted.

Insulating materials are also rated according to dielectric constant, dielectric strength, tensile (breaking) strength, and leakage resistance. *See also* DIELECTRIC, DIELECTRIC CONSTANT, DIELECTRIC RATING, and INSULATION RESISTANCE.

INSULATION RESISTANCE

The resistance of an insulating material is a measure of the ability of the substance to prevent the undesired flow of electric currents. In general, the higher the insulation resistance, the more effective the material for purposes of preventing current flow.

Insulation resistance is generally measured for direct currents. This value, given in ohm-centimeters (ohm-cm), can change for radio-frequency alternating currents. Among the substances with the highest known insulation resistance are polystyrene, at 10^{18} ohm-cm, fused quartz, at 10^{19} ohm-cm, and teflon and polyethylene, which have direct-current resistivity of 10^{17} ohm-cm. Vinyl substances have insulation resistances ranging from 10^{14} to 10^{16} ohm-cm. Rubbers rate from about 10^{12} to 10^{15} ohm-cm. *See also* DIELECTRIC, and DIELECTRIC RATING.

INSULATOR

An insulating material is sometimes called an *insulator* (*see* INSULATING MATERIAL). However, the term *insulator* generally refers to a device specifically designed to prevent current from flowing between or among nearby electrical conductors. An insulator is a rigid object, usually made from porcelain or glass. Insulators are commonly used in radio-frequency antenna systems and in utility lines.

Insulators are found in a tremendous range of sizes and shapes. Two common types are shown (A and B). At A is a small porcelain insulator, which is intended for use in receiving antenna systems and in transmitting systems at power levels up to about 1 kW. At B are large insulators, which are typical of high-tension power-line structures. The small insulator is about 2 inches long, while the large ones are from 2 feet to 6 feet in length. The extremely high-voltage cross-country power lines, with voltages in the hundreds of thousands, have insulators up to several meters long.

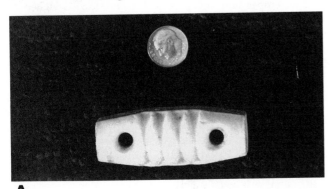

A

B

INSULATOR: At A, a small porcelain insulator. At B, power-line insulators.

INTEGRATED CIRCUIT

Much of today's electronic equipment relies on *integrated circuit* (IC) technology. A typical IC can contain amplifiers, timers, counters, memory, and almost any kind of electronic circuit.

ICs are generally categorized as either *linear* or *digital*, depending on whether the output varies as a smooth function, or in discrete states.

All ICs share certain characteristics. First, all the components are impressed on a single "chip" of semiconducting material. ICs are often called *chips* for this reason.

Second, all ICs occupy far less space than an equivalent circuit using discrete components. A typical IC package is about the size of a paper clip. The same circuitry using individual ca-

pacitors, transistors and other components might take up a container as big as a shoebox.

Third, ICs work like "black boxes." This simplifies repair of complex electronic equipment. All the technician needs to do is find the bad IC and replace it.

Linear ICs   All linear ICs have a continuously variable output that depends on the input signal level. One of the most common linear ICs is the *operational amplifier*. Other types include power amplifiers for audio or radio frequencies, specialized function ICs, and arrays of diodes and/or transistors.

Linear ICs are used in amplifiers, oscillators, signal filters, test instruments, control systems, and other applications requiring a device that can have infinitely many different states, and/or that require high gain.

Digital ICs   Digital ICs differ from analog ICs in the way they handle signals. A digital circuit works in defined levels or states, rather than in a continuous, smooth manner. This makes digital ICs simpler, in some ways, than analog ICs.

Digital ICs are used extensively in calculators, computers, digital communications equipment, control systems, and any other apparatus in which large digital networks are used.

In digital IC technology, engineers sometimes speak of *generations*. It has become possible to get more and more logic gates onto a single chip. Nowadays, *large-scale integration (LSI)* and *very-large-scale integration (VLSI)* are common. In LSI, there are from 100 to 1000 logic gates on a single chip; in VLSI, 1000 to 10,000 gates exist on one semiconductor chip.

Engineers and technicians are concerned with other aspects of digital IC operation. There are different ways of obtaining logic gates on a chip, and these manufacturing methods provide various advantages and disadvantages in terms of speed, power consumption, impedance, and the number of inputs that can work from a single IC output.

The Future of IC Technology   There is a physical limit to the density of gate circuitry that can be achieved with semiconductor materials. This limit is imposed by the crystalline structure of the silicon or metal oxide used to make the IC. This limit has not yet been reached, and it is quite probable that engineers will soon speak often of *ULSI (ultra-large-scale integration)*: ICs with 10,000 to 100,000 logic gates on a single chip.

There is no doubt that all possible methods of microminiaturization will be explored. The day might come when it is possible to build a machine with memory capacity equal to, or greater than, that of a human being. But there are things people can do that no IC will ever be able to achieve—for example, to enjoy ham radio!

For Further Information   A detailed overview of current ICs, available from companies worldwide, can be found in the most recent edition of *International Encyclopedia of Integrated Circuits*, published by TAB Books. For those interested mainly in the hobby applications of ICs, a good reference is *Hot ICs for the Electronics Hobbyist*, also published by TAB Books.

Related terms in this encyclopedia include DIGITAL INTEGRATED CIRCUIT, LARGE-SCALE INTEGRATION, LINEAR INTEGRATED CIRCUIT, LOGIC, MEDIUM-SCALE INTEGRATION, METAL-OXIDE SEMICONDUCTOR, METAL-OXIDE-SEMICONDUCTOR LOGIC FAMILIES, THICK-FILM AND THIN-FILM INTEGRATED CIRCUITS, ULTRA-LARGE-SCALE INTEGRATION, and VERY-LARGE-SCALE INTEGRATION.

INTEGRATED CIRCUIT: Typical ICs next to a dime for a size comparison.

INTEGRATED INJECTION LOGIC

A bipolar form of logic, using both npn and pnp transistors, is called integrated injection logic (abbreviated I^2L). This type of logic circuit allows the smallest physical size of any member of the bipolar logic family.

The manufacture of I^2L is simple. No resistors are needed; a single pair of transistors completes the circuit for one gate. The operating speed is very rapid, approaching the speed of transistor-transistor logic. The current requirement of I^2L is quite small, resulting in low power consumption. *See also* DIODE-TRANSISTOR LOGIC, DIRECT-COUPLED TRANSISTOR LOGIC, EMITTER-COUPLED LOGIC, HIGH-THRESHOLD LOGIC, METAL-OXIDE-SEMICONDUCTOR LOGIC FAMILIES, RESISTOR-CAPACITOR-TRANSISTOR LOGIC, RESISTOR-TRANSISTOR LOGIC, TRANSISTOR-TRANSISTOR LOGIC, and TRIPLE-DIFFUSED EMITTER-FOLLOWER LOGIC.

INTEGRATION

Integration can refer to either of two different things in electronics, or one in mathematics. In the manufacture of integrated circuits, integration is the process of forming the diodes, transistors, capacitors and resistors onto the semiconductor material. *See* INTEGRATED CIRCUIT, LARGE-SCALE INTEGRATION, MEDIUM-SCALE INTEGRATION, ULTRA-LARGE-SCALE INTEGRATION, and VERY-LARGE-SCALE INTEGRATION.

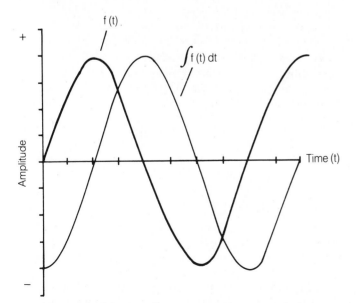

INTEGRATION: An example of integration for a sine wave.

Electrically, integration is the reverse of differentiation: cumulatively adding up the area under the curve of a function. This can be done only with certain functions. Some, such as a continuous dc voltage, in theory require that the integral function increase without limit over time. Usually, it is done with periodic waveforms such as sine waves, square waves and ramp waves. *See* INTEGRATOR CIRCUIT.

In mathematics, integration is a process in which the area under the curve is determined, either in general (indefinite integration) or between specific limits in the domain of the function (definite integration).

INTEGRATOR CIRCUIT

An electronic circuit that generates the integral, with respect to time, of a waveform is called an *integrator circuit*. When the input amplitude to an integrator is constant—that is, a direct current—the output constantly increases. If the input is zero, the output is zero. When the input amplitude fluctuates alternately toward the positive and the negative, the output reflects the integral of the waveform.

An integrator circuit can be extremely simple. The diagrams show two examples of circuits designed for electrical integration. At A, a resistance-capacitance network is shown. At B, an operational amplifier is used.

The integrator circuit generally acts in a manner exactly opposite to a differentiator. This does not mean, however, that cascading the two forms of circuit will always result in an output that is the same as the input. This is usually the case, but there are certain waveforms that are not duplicated when passed through an integrator and differentiator in cascade. *See also* DIFFERENTIATION, DIFFERENTIATOR CIRCUIT, and INTEGRATION.

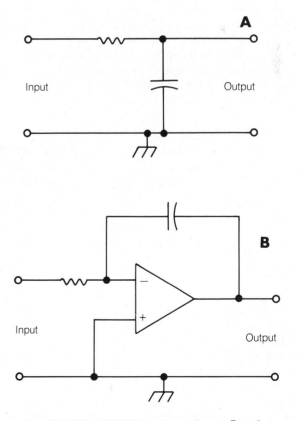

INTEGRATOR CIRCUIT: At A, passive; at B, active.

INTELLIGIBILITY

Intelligibility is a measure of the ease with which a signal is accurately received. It is measured as a percentage of correctly received single syllables in a plain-text transmission. The intelligibility can be affected by the context of the transmission because the receiving operator can sometimes guess what is coming.

The percentage of correctly received syllables in a random transmission, in which the receiving operator cannot guess what is coming, is called *articulation* (*see* ARTICULATION).

Good intelligibility is imperative in a communications circuit. Normally, a band of audio frequencies ranging from about 300 to 3000 Hz is sufficient to convey a voice message with good intelligibility. Intelligibility is not highly correlated with fidelity; excellent fidelity requires a much wider band of audio frequencies, and this can actually reduce the intelligibility. Intelligibility is directly related to the signal-to-noise ratio in a voice circuit. *See also* FIDELITY, and SIGNAL-TO-NOISE RATIO.

INTERACTION

When a change in a parameter affects the value of some other parameters, *interaction* is said to occur. Interaction occurs in many different forms in electrical and electronic devices. For example, the setting of the loading control in a transmitter affects the optimum setting of the output tuning control; the two controls therefore interact. In a shortwave general-coverage radio receiver, the main-tuning control and the bandspread control often interact. Interaction can be desired, or it can be unwanted.

When one effect (such as the presence of a fluctuating magnetic field) produces another (such as an alternating current), interaction is said to take place. This form of interaction is very common, and occurs in a variety of different situations.

Two or more circuits that operate in a state of mutual dependence, such as a group of interconnected computers, comprise an interactive system. Information is transmitted from one point to another in such a system; the functioning of one member affects the functioning of the other members. This mode of operation is often called the *interactive* or *conversational mode*, in which two-way data exchanges occur.

INTERCARRIER RECEIVER

An *intercarrier receiver* is a form of television receiver. All of the signal components, including the picture information, sound signal, and synchronization pulses, are heterodyned to the intermediate frequency. The intermediate-frequency stages amplify the signal. The signal is then fed to the detector.

In the detector stage, the picture signal, sound signal, and synchronizing pulses are separated and fed to the picture tube modulator, speaker, and picture tube deflection plates. *See also* TELEVISION.

INTERCHANGEABILITY

Interchangeability is the degree of ease with which one component type can be replaced by another type, while providing acceptable operation for the whole system. Standardization of electrical and electronic components makes interchangeability feasible in many cases. The American National Standards Institute and the Electronic Industries Association set standards for component interchangeability in commercial electronics in the United States. International standardization is carried out by a number of organizations run by the International Electrotechnical Commission.

In circuits made up of discrete components, interchangeability is often possible among many items. This is especially true of vacuum tubes and transistors. Other components, too, are interchangeable; for example, a mylar capacitor can sometimes be used in place of a ceramic capacitor. A 2-watt, 47-ohm resistor can often, of course, replace a 1-watt, 47-ohm resistor.

In modern electronic systems, the trend is toward integrated-circuit construction, in which complex circuits are housed in discrete packages. Interchangeability is often impossible in such systems, but the troubleshooting and repair process is greatly simplified. *See also* INTEGRATED CIRCUIT.

INTERELECTRODE CAPACITANCE

When electrodes are placed in close proximity, they show a certain amount of mutual capacitance. For any two electrodes, this capacitance varies inversely with the separation, and directly according to the size of either electrode. *Interelectrode capacitance* is not very important at low and medium frequencies, but as the frequency increases, the interelectrode capacitance in a system becomes more significant.

Interelectrode capacitance is specified for tubes and transistors. In a vacuum tube, the grid and plate electrodes display mutual capacitance, and this is the primary limiting factor on the frequency range in which the device can produce gain. The reverse-biased base-collector junction of a transistor behaves in a similar manner. In the field-effect transistor, capacitance exists between the channel and the gate or gates. The interelectrode capacitance in a typical tube, transistor, or field-effect transistor is a few picofarads. *See also* ELECTRODE CAPACITANCE.

INTERFACE

In an electronic system, an *interface* is a point at which data is transferred from one component to another. An interface generally involves conversion of data so that compatibility is achieved for both components. Thus, we can speak of a computer-to-operator interface (a terminal) or a radio-to-telephone interface (autopatch). The circuit that performs the data conversion is often called an *interface*.

An interface can perform a relatively simple task, such as the wiring of audio-frequency energy into a transmitter. The data conversion can be exceedingly complicated, such as the translation of visual data and keyboard commands into machine language for the operation of a computer. In computer systems, there are many interface points; a home computer, for example, requires a video display unit, a tape recorder or disk drive, and often a printer and telephone modem. *See also* DATA CONVERSION, and MODEM.

INTERFERENCE

Interference is the presence of unwanted signals or noise that increases the difficulty of radio reception. There are three kinds of interference: natural noise, manmade noise, and manmade signals.

Natural electromagnetic noise originates in outer space and in the atmosphere of the earth. There is nothing man can do to reduce the actual level of this noise. Atmospheric noise tends to be greatest at the lower frequencies and in the vicinity of thundershowers. Noise from outer space is prevented from reaching the surface of the earth in the very-low, low, and medium-frequency ranges because of the shielding effect of the ionosphere. However, at higher frequencies, such noise can sometimes be significant in communications applications. In general, the narrower the signal bandwidth in a communications link, the greater the immunity to natural noise (see NOISE).

Manmade noise is caused mostly by electrical wiring and appliances, and by internal-combustion engines and motors. Manmade noise is most intense at the very-low frequencies. As the electromagnetic frequency increases, the level of manmade noise goes down. As with natural noise, the immunity of a system depends on the bandwidth. In some cases, special circuits can reduce the level of manmade noise (see IGNITION NOISE, IMPULSE NOISE, and NOISE LIMITER).

Interference from manmade communications transmitters can be further categorized as either unintentional or intentional. The radio-frequency spectrum is heavily used in modern industrialized nations, and a certain amount of accidental interference is to be expected. The narrower the average signal bandwidth, in a given band of frequencies, the lower the probability of accidental interference. Intentional interference is called *jamming*, and is often used in wartime for the purpose of cutting enemy communications links. See INTERFERENCE FILTER, and INTERFERENCE REDUCTION.

When two or more electromagnetic waves combine in variable phase, producing areas of stronger and weaker field intensity, the interaction is sometimes called interference. See also INTERFERENCE PATTERN, and INTERFEROMETER.

INTERFERENCE FILTER

An *interference filter* is a device, used in a radio receiver, for the purpose of reducing or eliminating interference (see INTERFERENCE). There are several different types of interference filters.

For natural noise, such as the noise originating in the atmosphere, a noise limiter can provide some reduction in the level of interference (see NOISE LIMITER). Narrow-band transmission and reception is preferable to wide-band communications in the presence of high levels of natural noise. The noise limiter and narrow-band selective filter can thus act as noise filters.

Manmade noise can be reduced or eliminated by the same means as natural noise. However, a noise blanker is often effective against manmade ignition or impulse noise, whereas it is usually not effective against natural noise.

Interference from manmade signals is best reduced by means of highly selective bandpass filters, in conjunction with narrow-band transmission. A device called a notch filter can be used to attenuate fixed-frequency, unmodulated carriers within the communications passband (see NOTCH FILTER). Broadband, deliberate interference, or jamming, can be almost impossible to overcome by means of filtering. In such cases, a directional antenna system is usually needed. See also DIRECTIONAL ANTENNA.

INTERFERENCE PATTERN

When two electromagnetic fields interact, regions of phase reinforcement and phase opposition are created. The result is a pattern of more and less intense emission. This pattern is known as an *interference pattern*.

The simplest interference patterns are the directional radiation and response patterns of antenna systems. The configuration of the pattern depends on the spacing of the antennas, measured in wavelengths, and on the relative phase of the signals at the different antennas (see PHASED ARRAY).

INTERFERENCE REDUCTION

The reduction of interference is important in a communications system. The lower the interference level, the better the signal-to-noise ratio in a communications link. Interference reduction can be effected at the transmitting station, by the use of the narrowest possible emission bandwidth. With this single exception, however, all interference reduction must be accomplished at the receiving end of a circuit.

Interference can be reduced in a receiver by means of various filtering devices (see INTERFERENCE FILTER). Directional antenna systems (see DIRECTIONAL ANTENNA) can also be used. If the interference originates at a single point in space, and this point remains fixed or moves slowly, a loop antenna (see LOOP ANTENNA) can be used to "null out" the interference. If the interference comes from a broad general direction different from the direction of the desired signal, a unidirectional antenna such as a quad, Yagi, or dish can be used.

In certain situations, interference reduction can be extremely difficult. This is especially true in cases of deliberate interference. A new technique of communication, called *spread-spectrum operation*, can sometimes be used to advantage to reduce interference. See also SPREAD-SPECTRUM.

INTERFEROMETER

The *interferometer* is a form of radio telescope in which two antennas are used in a phased configuration. The interferometer provides much greater resolving power, using two small antennas, than is possible with a single radio antenna. The technique was pioneered by two radio astronomers, Martin Ryle of England and J. L. Pawsey of Australia.

A single dish antenna, having a diameter of many wavelengths, can yield good resolution in radio astronomy. But two such antennas, separated by a great distance, allow far greater resolution. The sensitivity of the interferometer is not very much higher than that of a single-antenna system, but in radio astronomy, resolving power is often more important than sensitivity.

The greater the spacing between the antennas in the interferometer, the better the resolution. If the spacing in meters between the aerials is L and the wavelength in meters is λ, then the angular separation a, in degrees, between the lobes of the directional pattern is given by:

$$a = 57.3\lambda/L$$

Of course, the shorter the wavelength, the greater the resolving power for a given antenna separation. Some interferometer systems have resolution of a few seconds of arc at wavelengths of several centimeters. See also RADIO ASTRONOMY, and RADIO TELESCOPE.

INTERMEDIATE FREQUENCY

In a superheterodyne receiver, the *intermediate frequency (IF)*, is the output frequency from the mixer stage (*see* MIXER, and SUPERHETERODYNE RECEIVER). The intermediate frequency of the superheterodyne is usually a fixed frequency. This makes it easy to obtain high gain and excellent selectivity because the IF amplifier stages can be tuned precisely for optimum performance at a single frequency (*see* INTERMEDIATE-FREQUENCY AMPLIFIER).

Some superheterodyne receivers use one intermediate frequency; these are called *single-conversion receivers*. Other receivers have two intermediate frequencies. The incoming signal is heterodyned to a fixed frequency called the *first IF*, which is in turn heterodyned in a later stage to the *second IF*. The second IF is generally a low frequency, which facilitates high selectivity. This type of receiver is called a *double-conversion* or *dual-conversion receiver*. Some receivers even have three intermediate frequencies; these are called *triple-conversion receivers*.

A high intermediate frequency, such as several megahertz, is preferable to a low IF for purposes of image rejection. However, a low IF is better for obtaining sharp selectivity. This is why double-conversion receivers are common: They provide the advantages of both a high first IF and a low second IF. *See also* DOUBLE-CONVERSION RECEIVER, IMAGE REJECTION, and SINGLE-CONVERSION RECEIVER.

INTERMEDIATE-FREQUENCY AMPLIFIER

An *intermediate-frequency amplifier* is a fixed radio-frequency amplifier commonly used in superheterodyne receivers. Such amplifiers generally are cascaded two or more in a row, with tuned-transformer coupling (see A). The intermediate-frequency (IF) amplifiers follow the mixer stage, and precede the detector stage. Double-conversion receivers (*see* DOUBLE-CONVERSION RECEIVER) have two sets of IF amplifiers; the first set follows the first mixer and precedes the second mixer, and the second set follows the second mixer and precedes the detector, as at B.

The intermediate-frequency amplifier chain serves two main purposes: to provide high gain, and to provide excellent selectivity. Gain and selectivity are much easier to obtain with amplifiers that operate at a single frequency, as compared with tuned radio-frequency amplifiers. *See also* INTERMEDIATE FREQUENCY, and SUPERHETERODYNE RECEIVER.

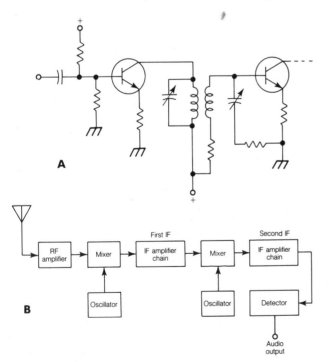

INTERMEDIATE-FREQUENCY AMPLIFIER: At A, coupling between IF amplifiers. At B, placement of an IF chain in a receiver.

INTERMITTENT-DUTY RATING

The *intermittent-duty rating* of a component is its ability to handle current, voltage, or power with a duty cycle of less than 100 percent (*see* DUTY CYCLE). Usually, intermittent duty is specified as a duty cycle of 50 percent. However, the length of a single active period must also be specified. The intermittent-duty rating of a device is almost always greater than the continuous-duty rating (*see* CONTINUOUS DUTY RATING).

Some electronic devices are intended primarily for intermittent service, while others are designed for continuous service. Sometimes, specifications will be given for both continuous and intermittent service.

INTERMITTENT FAILURE

When a system, circuit, or component malfunctions only occasionally, the malfunction is called an *intermittent failure*, or simply an *intermittent*.

Intermittents are well known to every experienced service technician. Troubleshooting is difficult if the problem is not always evident. In some cases, the failure refuses to manifest itself under test conditions, but only occurs in actual service, where repair is inconvenient or impossible. The failure can occur for only a fraction of 1 percent of the operating time, or it can occur for more than 99 percent of the time.

Intermittent failures can be induced by mechanical vibration, thermal shock, excessive humidity, or operation at excessive voltage, current, or power levels. Sometimes there is no apparent cause for the failure. The repair process consists of locating and replacing the defective components, or improving the operating environment to conform with equipment specifications. Components must be wiggled around, or the unit bumped and shaken. It can be necessary to use cooling compounds, such as "freeze mist," localized heating devices, and the like. The troubleshooting process can be very time-consuming. When it appears as though the problem has been corrected, extensive testing is needed to ensure that this is actually the case. The most harrowing intermittent problems are sometimes called *gremlins*—meaning little devils—by the technician.

INTERMODULATION

In a receiver, an undesired signal sometimes interacts with the desired signal. This usually, but not always, occurs in the first radio-frequency amplifier stages, or front end. The desired signal appears to be modulated by the undesired signal, as well as having its own modulation. This effect is known as *intermodulation*. Intermodulation is sometimes mistaken for interference from another signal; occasionally, interference is mistaken for intermodulation. Intermodulation can be comparatively mild, or it can be so severe that reception becomes impossible.

Intermodulation can occur for any or all of several reasons. Inadequate selectivity in the receiver input can result in unwanted signals passing into the front-end transistor. Nonlinear operation of the front-end transistors increases the chances of intermodulation. An excessively strong signal can drive the front end into nonlinear operation, even though it is properly designed. A poor electrical connection in the antenna system can cause intermodulation in a receiver used with that antenna system. In the extreme, a poor electrical connection near the antenna, but physically external to it, can produce intermodulation effects. The poor connection literally forms a diode mixer.

Intermodulation is, of course, never desirable, and modern receivers are designed to keep this form of distortion to a minimum. Manufacturers and designers speak of a specification called *intermodulation spurious-response attenuation*. This is usually expressed in decibels. Two signals are applied simultaneously to the antenna terminals of the receiver under test. The amplitude ratio is prescribed according to preset specifications. The strength of the unwanted signal required to produce a given amount of intermodulation is then compared with the strength of the desired signal, and the ratio is calculated in decibels.

Alternatively, the strength of the unwanted signal can be held constant, and the extent of the distortion measured as a modulation percentage in the desired signal. This percentage is called the *intermodulation percentage*.

Standards for intermodulation spurious-response attenuation are set in the United States by the Electronic Industries Association. When testing a receiver for intermodulation response, the procedures and standards of this association should be consulted. The susceptibility or immunity of a receiver to intermodulation is primarily dependent on the design of its front end. *See also* FRONT END.

INTERMODULATION DISTORTION

When two or more strong signals mix with each other in a receiver and produce other, "false" signals, the effect is called *intermodulation distortion (IMD)*. This is sometimes simply called *intermod*. It is similar to *cross modulation* (see CROSS MODULATION).

The effects of intermod are most likely to be a problem when a radio is operated in an environment with many transmitters running high power. Poor front-end design in a radio receiver increases the chances of IMD occurring in any given environment. If IMD does exist, it will be worse in a poorly designed receiver, as compared with a well-designed one. *See* FRONT END.

Perhaps you've ridden in a taxi through the center of a large city, and heard intermittent popping and bursts of voice on the driver's radio that were not being sent by the taxi dispatcher. These bursts are the result of IMD in the front end of the receiver.

Another example of IMD often is seen when using a very long wire antenna that is resonant in the AM broadcast band. You might be tuned to 80 meters, but the mixing of strong local AM broadcast signals creates a "goulash" in the receiver.

The best safeguard against IMD is to use a well-designed radio receiver. The antenna system should have good electrical connections everywhere; corrosion in the antenna and/or feed line can introduce IMD into a system—even if the receiver is good.

Sometimes, filters can be placed between a receiver and the antenna to attenuate signals causing intermod. This scheme will work only if the offending signals are well away from the operating frequency. The filters might be of the bandpass, band-rejection, highpass or lowpass type, depending on the number of offending signals, and on their relationship to the operating frequency. *See* BANDPASS FILTER, BAND-REJECTION FILTER, HIGHPASS FILTER, and LOWPASS FILTER.

INTERNAL IMPEDANCE

The *internal impedance* of a component or device is the impedance between its terminals, at a given frequency, without external components. Internal impedance consists of direct-current or ohmic resistance and either positive (inductive) or negative (capacitive) reactance. Internal impedance is an important consideration when using test instruments, such as meters and oscilloscopes.

The internal impedance of current-measuring devices should be as low as possible. When voltage is to be measured, the internal impedance should be high. In general, the internal impedance of a test instrument should be such that the performance of the circuit is not altered when the instrument is used. *See also* IMPEDANCE.

INTERNAL NOISE

In a radio receiver, noise is an important consideration because it limits the sensitivity of the circuit. Noise can originate outside the receiver, in the form of atmospheric and cosmic noise (*see* COSMIC NOISE, and SFERICS). Noise is also produced by the movement of atoms as a result of temperature (*see* THERMAL NOISE). Some noise is produced by the active components within the receiver itself. This is known as *internal noise*.

At very-low, low, and medium frequencies, external noise is quite intense, and internal receiver noise normally does not play a significant part in limiting the sensitivity. In the high-frequency spectrum, external noise becomes less intense, and in the very-high, ultra-high, and microwave parts of the spectrum, internal noise is much more important than external noise. Low-noise receiver design is of the greatest concern, therefore, at frequencies above about 30 MHz.

Modern semiconductor technology allows the design and manufacture of low-noise receivers having sensitivity unheard-of just a few years ago. In some scientific applications, noise is further reduced by supercooling, or operation of equipment at extremely low temperatures. *See also* NOISE, SENSITIVITY, SIGNAL-TO-NOISE RATIO, and SUPERCONDUCTIVITY.

INTERNATIONAL AMATEUR RADIO UNION

There are amateur radio organizations in many countries throughout the world. In the United States, the ham radio union is the American Radio Relay League, ARRL (see AMERICAN RADIO RELAY LEAGUE).

All of the ARRL type ham radio unions in the world are combined to form members of the *International Amateur Radio Union (IARU)*. The IARU has successfully represented ham radio in the world for decades. Noteworthy among its contributions has been the addition of several new ham bands during the World Administrative Radio Conference in 1979.

INTERNATIONAL MORSE CODE

International Morse Code is a system of dot and dash symbols commonly used by radiotelegraph operators throughout the world. The International Morse Code is much more often used than the American Morse Code, although a few telegraph operators still use the American Morse.

The International Morse Code differs from the American Morse Code; the table shows the International Morse symbols. The dot represents one unit length. The dash represents three units and the spaces between dots and dashes is one unit. The space between letters is three units, and between words, seven units. The International Morse Code is less confusing than the American Morse, because the latter code contains odd spaces and unit lengths in some instances. *See also* AMERICAN MORSE CODE.

INTERNATIONAL MORSE CODE:
INTERNATIONAL MORSE CODE SYMBOLS.

Character	Symbol	Character	Symbol
A	·—	U	··—
B	—···	V	···—
C	—·—·	W	·——
D	—··	X	—··—
E	·	Y	—·——
F	··—·	Z	——··
G	——·	1	·————
H	····	2	··———
I	··	3	···——
J	·———	4	····—
K	—·—	5	·····
L	·—··	6	—····
M	——	7	——···
N	—·	8	———··
O	———	9	————·
P	·——·	0	—————
Q	——·—	Period	·—·—·—
R	·—·	Comma	——··——
S	···	Question	
T	—	Mark	··——··

INTERNATIONAL SYSTEM OF UNITS

The *International System of Units (SI)* is based on the meter, kilogram, second, ampere, degree Kelvin, candela, and mole (Avogadro constant). The system is sometimes called the *meter-kilogram-second (MKS)* or *meter-kilogram-second-ampere (MKSA)* system of units. All physical units can be derived in terms of these standard units. The SI, or Systeme International d'Unites, was agreed upon in the Treaty of the Meter in 1960.

Prior to 1948, the international system of units was based on the centimeter, gram, second, and degree Kelvin. The older international-unit system was discarded by international treaty on January 1, 1948. The most recent agreement was made in 1960. Sometimes a distinction is made between "absolute" and "international" electrical quantities. The difference is primarily a matter of the standard quantities from which the units are derived.

INTERNATIONAL TELECOMMUNICATION UNION

The *International Telecommunication Union (ITU)* is an international organization that sets worldwide standards for electro-magnetic communication. The ITU governs the frequency allocations and callsign prefixes for all regions of the world. This helps maximize the efficiency with which the electromagnetic spectrum is utilized. The ITU has established a set of Radio Regulations. These regulations can be obtained by contacting the Secretary General, International Telecommunication Union, Place des Nations, CH-1211, Geneva 20, Switzerland.

In the United States, telecommunications regulations are determined by the Federal Communications Commission, although their rules are based on those given by the ITU. *See also* FEDERAL COMMUNICATIONS COMMISSION.

INTERPOLATION

When reading an analog dial or instrument, the indication is often between two divisions of the scale. It is usually possible to make a good guess, down to the tenth of a division, as to what the dial or meter says. An example is shown in illustration A. This guesswork is called *interpolation*.

Digital readouts and meters do not allow for interpolation. For this reason, some operators prefer analog dials and readouts over digital ones, even when digital ones might be better suited for the application. And also for this reason, some operators like digital displays better. Digital displays save the operator from having to scrutinize the reading, and they practically eliminate visual/mental errors. *See* ANALOG METERING, and DIGITAL METERING.

In a graphical representation of a quantity, the data is sometimes given only in the form of a set of points. The complete graph must be drawn in using a process of interpolation. An example is shown in illustration B. The easiest method is to draw straight lines between the points, as shown by the dotted lines. This is the type of graph you might see in corporate profit reports. A more accurate method is to draw in the graph as a curve, based on the "feel" of the contour of the set of points

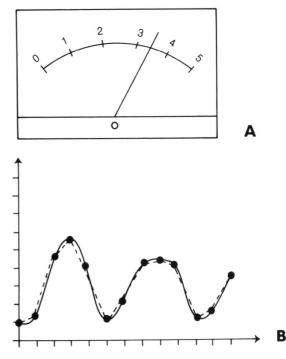

INTERPOLATION: At A, the analog meter reads about 3.3 units. At B, a graph is drawn connecting a set of points by straight lines (dotted) or by guessing at the most likely curve (solid line).

(solid line). This mental guesswork is a form of interpolation familiar to physical scientists. Sometimes a computer can be used to obtain a more accurate curve than a person can draw in by "feel."

INTERRUPTER

An *interrupter* is a device that intermittently opens a circuit. The interrupter can be used for a variety of purposes, such as generation of pulsating direct current from pure direct current, half-wave rectification of alternating current, or keying of a continuous-wave transmitter.

Interrupters generally fall into two categories: electrical and mechanical. Electrical interrupters are such devices as switching diodes and transistors and silicon-controlled rectifiers. Mechanical interrupters include switches and relays. In a direct-current to alternating-current power inverter, the interrupter is sometimes called a *chopper*.

INTERSTAGE COUPLING

Interstage coupling is the means by which a signal is transferred from one circuit to another. The circuits can be oscillators, amplifiers, mixers, or detectors. There are several different methods of interstage coupling.

Capacitive coupling is commonly used at all frequencies, from the low audio range into the ultra-high-frequency radio spectrum. A series capacitor passes the signal while allowing different direct-current voltages to be applied to the stages (*see* CAPACITIVE COUPLING).

Direct coupling consists of a short circuit between two stages. This method of coupling does not allow for different direct-current voltages between stages. Direct coupling exhibits a constant attenuation level over a wide range of frequencies (*see* DIRECT COUPLING).

Transformer coupling makes use of a two-winding audio-frequency or radio-frequency transformer for the purpose of passing a signal. Either or both of the windings can contain a parallel capacitor for tuning purposes. An electrostatic shield can be placed between the windings to reduce the stray capacitance. Transformer coupling results in a narrow range of operating frequencies when tuning is used. Direct-current isolation is excellent (*see* TRANSFORMER COUPLING).

INTRINSIC SEMICONDUCTOR

A pure semiconductor material, containing no added impurities, is called an *intrinsic semiconductor*. Such substances are almost insulators; they do not conduct electricity very well. Usually, impurities must be added for the purpose of manufacturing semiconductor devices. Then, the semiconductor material becomes extrinsic. *See also* DOPING, ELECTRON, EXTRINSIC SEMICONDUCTOR, HOLE, N-TYPE SEMICONDUCTOR, and P-TYPE SEMICONDUCTOR.

INVERSE-SQUARE LAW

The *inverse-square law* is a physical principle that defines the manner in which three-dimensional radiation is propagated. This principle applies to radio waves, infrared radiation, visible light, ultraviolet, X rays, and gamma rays. It applies to barrages of particles, such as high-speed protons or neutrons. It also applies to sound waves. The inverse-square law is used for the purpose of determining, or predicting, the power intensity, per unit area, of a diverging effect at a given distance from a point source. The inverse-square law does not apply to parallel energy beams such as the maser or laser. The rule is also invalid when a parameter is measured in terms other than intensity per unit area.

To illustrate the derivation of the inverse-square principle, suppose we have a light bulb that emits 100 watts of power. If you completely surround this bulb by a sphere you know that 100 watts of power is striking the inside surface of the sphere. This is true no matter what the radius of the sphere because any photon from the bulb must eventually encounter the sphere. Recall the formula for the surface area of the sphere:

$$A = 4\pi r^2$$

where A is the area, r is the radius, and π is approximately equal to 3.14. The energy intensity per unit area depends on the size of the sphere. If you double the radius of the sphere, you quadruple its surface area, and this results in just $1/4$ as much light for each square meter, square inch, or square foot of the sphere. If you triple the radius, the surface area increases by a factor of 9, resulting in $1/9$ the energy density. In general, given a power level of P watts per square centimeter at a distance d from a source of energy, the power Q at distance xd will be:

$$Q = P/x^2$$

where x can be any positive real number.

The inverse-square law only applies to quantities that are measured over a specific surface area. The rule does not apply to parameters that are measured along a line. Therefore, the field strength from an antenna in watts per square meter does obey the inverse-square law, but the field strength in microvolts per meter does not obey the principle. Some effects, such as magnetic-field intensity surrounding an inductor, also do not obey the inverse-square law because flux density is determined partly in terms of three-dimensional volume, and is not a simple radiant effect.

INVERSE VOLTAGE

The *inverse voltage* across a rectifier is the extent to which the cathode becomes positive, with respect to the anode. Inverse voltage results in nonconduction of a rectifier device, unless the potential becomes so great that avalanche or flashover occurs. In a semiconductor device, inverse voltage is sometimes called *reverse voltage*.

All rectifier devices are rated according to the maximum peak value of inverse voltage they can withstand. If this rating is exceeded, avalanche can occur. *See also* AVALANCHE BREAKDOWN, and PEAK INVERSE VOLTAGE.

INVERSION

Normally, the temperature of the lower atmosphere drops with increasing altitude. This is primarily because the surface of the earth is warmed by radiation from the sun, and heat is transferred less and less effectively into higher and higher parts of the atmosphere. The increased density of the air near the surface results in a greater index of refraction for radio waves near the earth; the index gradually decreases with altitude. This effect results is a form of propagation called *tropospheric bending* (*see* TROPOSPHERIC PROPAGATION).

Occasionally, a layer of cold air exists near the surface, and the index of refraction greatly increases. At low levels, the air temperature increases with increasing altitude under such conditions. This is called a *temperature inversion*, or simply *inversion*. The cold air is more dense than the warm air, and this results in a more pronounced difference in the index of refraction at various altitudes. At the boundary between the cold air and the overlaid warm air, the radio-wave index of refraction drops abruptly.

Temperature inversions often occur during the early morning hours, after the land has cooled. Over water, an inversion can persist for 24 hours a day, although this depends on the water temperature. Weather fronts are commonly associated with inversions. Inversions produce a form of tropospheric propagation known as ducting, in which a radio wave is actually trapped within a cold layer of air. Inversions are also associated with severe air pollution in large cities. *See also* DUCT EFFECT.

INVERTED-L ANTENNA

An *inverted-L antenna* is a form of receiving and transmitting antenna that is sometimes used at medium and high frequencies. The antenna gets its name from the fact that it is shaped like an upside-down letter L.

The inverted-L antenna can be either end-fed or center-fed. The illustration at A shows a quarter-wave, end-fed inverted-L

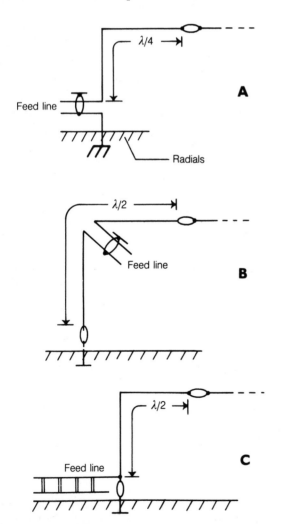

INVERTED-L ANTENNA: At A, quarter-wave end-fed; at B, half-wave center-fed; at C, half-wave end-fed.

antenna. This antenna requires an excellent ground system. The height of the vertical section is normally about $1/10$ to $1/8$ wavelength, with the rest of the length made up in the horizontal part. This antenna is often used in situations where a full-size quarter-wave vertical structure is impractical. The feed-point impedance is approximately 30 to 40 ohms at resonance.

The arrangement at B is a center-fed, half-wave inverted-L antenna. This arrangement is nearly the electrical equivalent of a dipole antenna (*see* DIPOLE ANTENNA). A good ground is not essential for proper functioning of this antenna. The feed point is elevated, and the feed line runs down at about a 45-degree angle. The physical lengths of the antenna legs, each $1/4$ electrical wavelength, can be reduced by inductive loading (*see* INDUCTIVE LOADING). The feed-point impedance is about 50 ohms, but lower if inductive loading is used.

At C, an end-fed, half-wave inverted-L antenna is illustrated. This antenna is fed in the same manner as a zeppelin antenna (*see* ZEPPELIN ANTENNA). It does not require a particularly good ground system for proper operation, although a radial network enhances performance. The feed-point impedance of this antenna is quite high, but in the other arrangements it is relatively low. For this reason, open-wire line or an impedance-matching network must be used.

The inverted-L antenna produces both vertical and horizontal polarization. At medium and high frequencies, this allows good performance both for DX (distance) and local work. *See also* HORIZONTAL POLARIZATION, and VERTICAL POLARIZATION.

INVERTED-V ANTENNA

A half-wave dipole antenna is usually strung horizontally. This requires two supports, one for each end. A third support is necessary in the middle if the feed line is heavy. *See* DIPOLE ANTENNA.

A similar antenna can be made with one support in the middle, running two quarter-wave wires down at 45-degree angles, as shown in the illustration. The support is $1/4$ wavelength high, or higher. This is called an *inverted-V antenna*.

The inverted-V antenna has a somewhat lower feed-point impedance than a half-wave dipole at resonance. Typically, the value is 40 to 50 ohms. This provides a good match for 50-ohm coaxial cable.

The directional pattern for an inverted-V antenna differs from that of a dipole. The inverted-V will radiate well at low

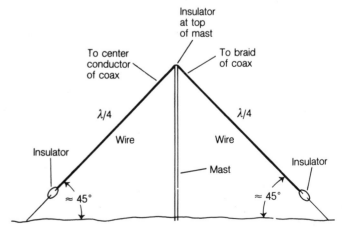

INVERTED-V ANTENNA: Geometry of a typical inverted-V antenna.

angles, in all compass directions. The polarization is horizontal broadside to the plane containing the antenna wires, and vertical in the plane of the wires.

INVERTER

The term *inverter* is used to denote either of two different kinds of electronic circuits. A logical inverter is a NOT gate, and a power inverter converts low-voltage direct current into 120-volt alternating current.

A logical inverter has one input and one output. If the input is high, the output is low; if the input is low, the output is high. The illustration shows the symbol for the logical inverter, along with a truth table showing its characteristics.

A power inverter is essentially a chopper power supply. An interrupting device produces a pulsating direct-current output from the power source, which is usually a 12-volt battery. This pulsating direct current is regulated in frequency so that it is as close as possible to 60 Hz. A step-up transformer then provides the 120-volt output needed to operate low-power household appliances.

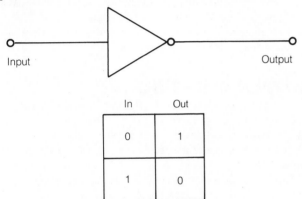

In	Out
0	1
1	0

INVERTER: A schematic symbol and truth table.

INVERTING AMPLIFIER

An *inverting amplifier* is any amplifier that produces a 180-degree phase shift in the process of amplification. Virtually all common-cathode, common-emitter, and common-source amplifiers are of this variety (*see* COMMON CATHODE/EMITTER/SOURCE). In general, the grounded-grid, grounded-base, and grounded-gate configurations are not inverting amplifiers (*see* COMMON GRID/BASE/GATE).

A pushpull amplifier can be inverting or noninverting, depending on the transformer output arrangement. Inverting amplifiers are less susceptible to oscillation than noninverting amplifiers because any stray coupling between input and output results in degenerative, rather than regenerative, feedback.

ION

An *ion* is an atom from which one or more electrons have been removed or added. Normally, atoms are electrically neutral; this is the state in which they "prefer" to be. But some atoms give up or accept electrons readily. Ions are created by a process called *ionization*. See also IONIZATION.

IONIZATION

An atom normally has the same number of negatively charged *electrons* and positively charged *protons*. Because both particles

have equal charge quantities, most atoms are electrically neutral. Sometimes, however, electrons are lost or captured by an atom so that the charge is not neutral. The nonneutral atom is called an *ion*, and any process or event that creates ions is known as *ionization*.

Ionization occurs when substances are heated to high temperatures, or when large voltages are impressed across them. Lightning is the result of ionization of the air. An electric spark is caused by a large buildup of charges, resulting in forces on the electrons in the intervening medium. These forces pull the electrons away from individual atoms. Ionized atoms generally conduct electric currents with greater ease than electrically neutral atoms.

Ionized gases are prevalent in the upper atmosphere of our planet. These layers are known as the *ionosphere*. The ionosphere has refracting effects on electromagnetic waves at certain frequencies. This makes long-distance communication possible on the shortwave bands. *See also* IONOSPHERE.

IONOSPHERE

The atmosphere of our planet becomes less and less dense with increasing altitude. Because of this, the amount of ultraviolet and X-ray energy received from the sun gets greater at increased altitude. At certain levels, the gases in the atmosphere become ionized by solar radiation. These regions comprise the ionosphere of the earth.

Ionization in the upper atmosphere occurs mainly at three levels, called *layers*. The lowest region is called the *D layer*. The D layer exists at altitudes ranging from 35 to 60 miles, and is ordinarily present only on the daylight side of the planet. The E layer, at about 60 to 70 miles above the surface, also exists mainly during the day, although night-time ionization is sometimes observed. The upper layer is called the *F layer*. This region sometimes splits into two regions, known as the F1 (lower) and F2 (upper) *zones*. The F layer can be found at altitudes as low as 100 miles and as high as 260 miles.

The ionosphere has a significant effect on the propagation of electromagnetic waves in the very-low, low, medium, and high-frequency bands. Some effect is observed in the very-high-frequency part of the spectrum. At frequencies above about 150 MHz, the ionosphere has essentially no effect on radio waves. Ionized layers cause absorption and refraction of the waves. This makes long-distance communication possible on some frequencies. At the longer wavelengths, energy from space cannot reach the surface of the earth because of the ionosphere. *See also* D LAYER, E LAYER, and F LAYER.

IONOSPHERIC PROPAGATION

See PROPAGATION CHARACTERISTICS.

I/O PORT

See INPUT-OUTPUT PORT.

IRON

Iron is an element with atomic number of 26 and atomic weight 56. Iron is best known for its magnetic properties. Iron is found in great abundance on our planet, especially deep under the surface. The core of the earth is believed to consist largely of iron. This is the reason for the existence of a magnetic field surrounding the planet.

Iron is used in many applications in electricity and electronics. The atoms of iron tend to align themselves in the presence of a magnetic field. If the field is strong enough, the alignment persists — even after the field is removed. In this case, the iron is said to be permanently magnetized. Permanent magnets are used in dynamic microphones, speakers, earphones, and other transducers. Permanent magnets are also used in analog metering devices.

Because of its high magnetic permeability, iron makes an excellent core material for transformers. Various forms of iron are used at different frequencies. Laminated iron is found in transformers for frequencies in the audio range. Powdered iron is used at radio frequencies up to the very high region. Ferrite transformer cores are used at radio frequencies in the very-low, low, medium, and high ranges. *See also* FERRITE, FERROMAGNETIC MATERIAL, LAMINATED CORE, PERMEABILITY, and POWDERED-IRON CORE.

IRON-VANE METER

An *iron-vane meter* is a device used for measuring alternating currents. The ordinary direct-current meter will not respond to alternating current because such a meter produces directional readings, according to the current polarity.

The iron-vane instrument contains a movable iron vane, attached to a pointer. The pointer is held at the zero position on a calibrated scale by a spring under conditions of zero current. A stationary iron vane is positioned next to the movable vane. Both vanes are placed inside a coil of wire, the ends of which form the meter terminals.

When current is supplied to the coil, a magnetic field is produced in the vicinity of the iron vanes. The field causes the vanes to become temporarily magnetized. The vanes are positioned in such a way that like magnetic poles form near each other on each vane end; the strength of the magnetism is proportional to the current in the coil. Because similar magnetic poles repel, the movable vane is deflected away from the stationary vane. The spring allows the pointer to move a specific distance, depending on the force of repulsion. This effect occurs for alternating or direct currents in the coil. The meter deflection is therefore proportional to the current.

The iron-vane instrument is, by itself, an ammeter. However, it can be used as a voltmeter or wattmeter with the addition of appropriate peripheral components. The scale of the iron-vane meter is usually nonlinear. The meter is accurate only at power-line frequencies because of the coil inductance and losses in the iron vanes. Thus, the iron-vane meter is used only for measurement of currents and voltages in ordinary utility devices. It is not suitable for use at audio or radio frequencies.

ISOLATION

Isolation is any means of electrical, acoustical, or electromagnetic separation between components, circuits, or systems. Isolation is important from the standpoint of circuit independence in certain instances.

In direct-current and low-frequency alternating-current applications, isolation is often called *insulation*. In radio-frequency applications, isolation is called *shielding*.

ISOLATION TRANSFORMER

When a transformer is used for the purpose of reducing or minimizing the interaction between stages of a circuit, the device is sometimes called an *isolation transformer*. Such a transformer does not completely eliminate the effects of the two stages or circuits on each other, but does provide more isolation than direct coupling or capacitor/resistor coupling. See TRANSFORMER, and TRANSFORMER COUPLING.

In an antenna system, a balun is sometimes used as an *isolation transformer*. See BALUN.

ISOTROPIC ANTENNA

An *isotropic antenna* is a device that radiates electromagnetic energy equally well in all directions. Such an antenna is a theoretical construct, and does not actually exist. However, the isotropic-antenna concept is occasionally used for antenna-gain comparisons.

The power gain of an isotropic antenna is about −2.15 dB, with respect to a half-wave dipole in free space. That is, the field strength from a half-wave dipole antenna, in its favored direction, is approximately 2.15 dB greater than the field strength from an isotropic antenna at the same distance and at the same frequency.

The radiation pattern of the isotropic antenna, in three dimensions, appears as a perfect sphere because the device works equally well in all directions. In any given plane, the radiation pattern of the isotropic antenna is a perfect circle, centered at the antenna. *See also* ANTENNA PATTERN, ANTENNA POWER GAIN, dBd, and dBi.

ITU

See INTERNATIONAL TELECOMMUNICATION UNION.

JACK

A *jack* is a receptacle for a plug. A jack is sometimes called a *female connector*. Usually, the jack is mounted on the equipment panel in a fixed position, and the plug is mounted on the end of a cord. However, in some cases this is reversed. *See also* PLUG.

JACKET

A *jacket* is the coating on a multi-conductor cable. The term is used especially with coaxial cables. The coating on an electronic component, such as a capacitor, is also sometimes called a *jacket*.

The jacket is generally made of insulating material, rather than conducting material. This isolates the component from surrounding objects. It also protects the component or cable against corrosion.

JAM ENTRY

When a logic-gate input is deliberately supplied with a high or low voltage by force, the entry is called a *jam entry* or *jam input*. The jam entry overrides whatever logic state is present on the line initially.

Jam entry is often used for the purpose of initializing, or presetting, a logic circuit or microcomputer. Once the desired state has been achieved, the circuit is free to operate normally. *See also* INITIALIZATION.

J ANTENNA

A *J antenna* is a form of end-fed antenna. It is electrically similar to the zeppelin antenna because it is an end-fed, half-wave antenna (*see* ZEPPELIN ANTENNA). The J antenna is sometimes known as a *J pole*. The antenna gets these names from its physical shape.

The J antenna consists of a half-wave, vertical radiator, with a quarter-wave section of transmission line that serves as a matching stub (see illustration). If the radiator is made from a length of metal rod or tubing, the matching section can be incorporated into the physical construction of the antenna.

The matching section is short-circuited at the bottom. The antenna is fed with coaxial cable at a point between the shorted end of the stub and the antenna end. The feed-point impedance is a pure resistance at all points on the matching stub; the value of this pure resistance is zero at the shorted end and very high —several hundred ohms or more—at the antenna end. If 50-ohm coaxial cable is used, the best impedance match is obtained when the feed point is about ⅙ of the way from the shorted end to the antenna end. It actually does not matter which coaxial conductor goes to the longer element, but in most cases the center conductor is placed there.

The J antenna has an omnidirectional radiation pattern in the horizontal plane. The J antenna is used primarily at very-high and ultra-high frequencies. The primary advantage of this antenna is that it does not require a ground system.

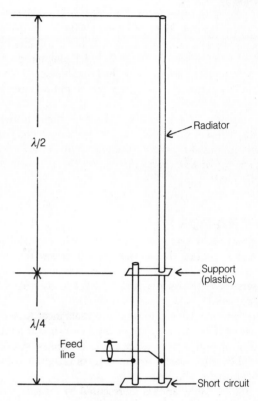

J ANTENNA: A half-wave radiator and a quarter-wave stub.

JITTER

A *jitter* is a rapid fluctuation in the amplitude, frequency, or other characteristic of a signal. Usually, the term *jitter* is used in video systems to refer to instability in the displayed picture on a cathode-ray tube.

Jitter in a cathode-ray-tube display can result from changes in the voltages on the deflecting plates. It can also occur because of variations in the current through the deflecting coils, or because of changes in the speed of the electrons passing through the tube. Mechanical vibration, an interfering signal, or internal noise can also cause jitter. Sometimes, the synchronizing circuits fail to "lock" onto the signal in the display, causing vertical or horizontal jitter. *See also* CATHODE-RAY TUBE, OSCILLOSCOPE, and TELEVISION.

JOINT COMMUNICATIONS

When a communications medium or facility is used by more than one service, the arrangement is called *joint communications.* Many frequency bands in the electromagnetic spectrum are shared by more than one service. In such situations, one service generally has priority over the others. The service having the top priority is called the primary service, and the others are called *secondary services.*

The International Telecommunication Union is responsible for assignment of frequency bands throughout the world. In the United States, the Federal Communications Commission makes frequency allocations. *See also* FEDERAL COMMUNICATIONS COMMISSION, and INTERNATIONAL TELECOMMUNICATION UNION.

JONES PLUG

A *Jones plug* is a special form of connector often used for providing power to electronic equipment. The female Jones plug has two or more recessed contacts, oriented in a defined pattern of vertical and horizontal. The male Jones plug has protruding contacts that mate with those of the female plug. The contacts are arranged so that the two plugs can be put together in only one way.

Jones plugs are available in numerous configurations and sizes, for different power-supply applications. Jones plugs are normally intended for indoor use only, unless they are sealed with a coating of waterproof material, such as silicone. *See also* CONNECTOR.

J OPERATOR

The *J operator* is an imaginary number, mathematically defined as the square root of −1. Mathematicians denote this quantity by the lowercase letter *i*; however, electrical and electronics engineers prefer the lower case letter *j* because *i* is used to represent current.

The *j* operator is used to represent reactance. An inductive reactance of X ohms, where X is any positive real number, is written jX. A capacitive reactance of $-X$ ohms is written $-jX$. The set of possible reactances is the set of imaginary numbers, or real-number multiples of *j*.

A complex impedance is represented by the sum of a pure resistance and an inductive (positive) or capacitive (negative) reactance. If the resistive component is R, then the impedance (Z) takes the form $R + jX$ or $R - jX$.

Complex impedances can be added together, or even multiplied, in this form. The important thing to remember is that when *j* is multiplied by itself, the result is −1. *See also* CAPACITIVE REACTANCE, IMPEDANCE, and INDUCTIVE REACTANCE.

JOULE

The *joule* is the standard unit of energy or work. When a force of 1 newton is applied over a distance of 1 meter, the amount of work done is 1 joule. The joule is equivalent in energy to 0.24 calories; that is, 1 joule of energy will raise the temperature of 1 gram of water by 0.24 degrees Celsius. A kilowatt hour is equivalent to 3.6 million joules. A power level of 1 watt is a rate of expenditure of 1 joule per second. *See also* ENERGY, and ENERGY UNITS.

JOULE'S LAW

When a current flows through a resistance, heat is produced. This heat is called *joule heat* or *joule effect*. The amount of heat produced is proportional to the power dissipated. The power P, in watts, dissipated in a circuit carrying I amperes, and having resistance of R ohms, is given by the formula:

$$P = I^2 R$$

Joule's law recognizes this, by stating that the amount of heat generated in a constant-resistance circuit is proportional to the square of the current.

JUNCTION

A *junction* is the point, line, or plane at which two or more different components or substances are joined. A simple electrical connection may be called a junction. In a waveguide, a junction is a fitting used to join one section to another. In a transmission line, a junction is the splice between two sections of the feed system.

A semiconductor junction is a plane surface at which P-type and N-type materials meet. This is called a *P-N junction*. Such a junction forms a one-way barrier for current; that is, it has diode action. *See also* P-N JUNCTION.

JUNCTION CAPACITANCE

When a semiconductor P-N junction is reverse-biased (*see* P-N JUNCTION), the resistance is high. However, there is some capacitance, known as *junction capacitance*.

The junction capacitance of a given diode, under conditions of reverse bias, depends on several factors. The surface area of the junction is important; in general, the larger the area, the greater the junction capacitance. The value of the reverse voltage affects the junction capacitance in some cases. This is especially true in the varactor diode, which acts as a voltage-controlled, variable capacitor (*see* VARACTOR DIODE).

The capacitance of a reverse-biased semiconductor junction determines the maximum frequency at which a device can be used. The smaller the junction capacitance, the higher the limiting frequency. At very-high and ultra-high frequencies, therefore, it is necessary that semiconductor devices have small values of junction capacitance. Rectifier diodes have large junction capacitance. Gallium-arsenide devices usually have small values of junction capacitance; the same is true of metal-oxide semiconductor components. The point-contact junction exhibits a smaller capacitance than the P-N junction. *See also* DIODE CAPACITANCE, GALLIUM-ARSENIDE FIELD-EFFECT TRANSISTOR, METAL-OXIDE SEMICONDUCTOR, METAL-OXIDE SEMICONDUCTOR FIELD-EFFECT TRANSISTOR, and POINT-CONTACT JUNCTION.

JUNCTION DIODE

A *junction diode* is a common type of semiconductor device, manufactured by diffusing P-type material into N-type material, thereby obtaining a P-N junction (*see* P-N JUNCTION). Junction diodes are usually made from germanium or silicon, with added dopants.

The junction diode conducts when the N-type material is negatively charged, with respect to the P-type material. This negative charge must be at least 0.3 volts if germanium is used as the semiconductor, or 0.6 volts if silicon is used. If the potential difference is less than this threshold value, or if the P-type material is negative, with respect to the N-type material, the junction diode will exhibit a very large direct-current resistance, infinite for all practical purposes.

Junction diodes are commonly used in rectifier and detector circuits. They can also be used as mixers, modulators, converters, and frequency-control devices.

A silicon junction diode can withstand higher temperatures than a germanium device. However, germanium has certain advantages because of its lower threshold. *See also* DIODE, DIODE CAPACITANCE, JUNCTION CAPACITANCE, and POINT-CONTACT JUNCTION.

JUNCTION TRANSISTOR

A *junction transistor* is a form of bipolar transistor. The junction transistor is formed by joining N-type and P-type semiconductor materials. The emitter-base and base-collector junctions are thus semiconductor P-N junctions. The junction transistor is characterized by relatively high junction capacitance. This limits the frequency at which the device can be used. *See also* JUNCTION CAPACITANCE, POINT-CONTACT JUNC-TION, P-N JUNCTION, and POINT-CONTACT TRANSIS-TOR.

JUSTIFICATION

Justification is a term sometimes used in electronic word-processing applications. Justification literally means the vertical alignment of lines. Most word-processing systems use left-margin justification. This means that the lines are vertically aligned on the left side. Some word processors have the option of right-margin justification. This function requires that extra spaces be inserted between words in some lines so that each line is exactly the same physical length. Typewriters produce copy without right-margin justification; books are usually typeset in a format with right-margin justification. *See also* WORD PRO-CESSING.

KELVIN ABSOLUTE ELECTROMETER

The *Kelvin absolute electrometer* is a device for measuring electrostatic voltages. The meter consists of two stationary metal plates, with a movable plate between them. The movable plate is attached to a pointer, and is held at the zero position by a set of return springs.

When an alternating-current or direct-current voltage is applied to the input terminals of the electrometer, the movable plate is deflected away from the stationary plates. The extent of the movement is determined by the voltage; the greater the input voltage, the farther the plate moves before equilibrium is reached with the return-spring tension. The pointer moves along a graduated scale.

The Kelvin absolute electrometer has extremely high input impedance, and therefore draws essentially no current from the circuit under test. The device does not use an amplifying device; this is why it is called an absolute electrometer. *See also* ELECTROMETER.

KELVIN DOUBLE BRIDGE

The measurement of small resistance requires special techniques. One device, specifically designed for the measurement of low resistance, is the *Kelvin double bridge*.

The unknown resistance, R, is connected in the circuit with known resistances. A precisely calibrated, low-resistance potentiometer is adjusted for a zero reading on the meter. The value *R* is calculated from the other values. The Kelvin double bridge is sometimes called the *Thomson bridge.*

KELVIN TEMPERATURE SCALE

The *Kelvin temperature scale*, also called the *absolute temperature scale*, is based on the coldest possible temperature. All readings on the Kelvin scale are positive readings; 0 degrees Kelvin is known as *absolute zero*, and it represents the total absence of thermal energy.

The Kelvin degree is the same size as the Celsius degree. If C represents the temperature in degrees Celsius and K represents the temperature in degrees Kelvin, then the readings are approximately related according to the simple formula:

$$K = C + 273$$

For conversion from Fahrenheit (F) to Kelvin, the following formulas apply:

$$K = 0.555F + 255$$
$$F = 1.8K - 459$$

KEY CLICK

A *key click* is a spurious signal from a radio transmitter, resulting from excessively rapid rise time or fall time. In a properly ad-

justed continuous-wave radio transmitter, the rise and fall time should be finite so that the modulation envelope appears similar to the illustration at A. If the rise time and/or fall time are too rapid, as shown at B or C, sidebands are generated at frequencies far removed from the signal frequency. These sidebands have short duration, and occur at the "make" or "break" instants. Such sidebands sound like clicks to an operator listening away from the signal frequency. These clicks can cause objectionable interference to other communications.

In order to ensure that a transmitter will not produce key clicks, it is necessary to provide a shaping network. Such a network is usually comprised of series resistors and parallel capacitors in the keying circuit. *See also* CODE TRANSMITTER, and SHAPING.

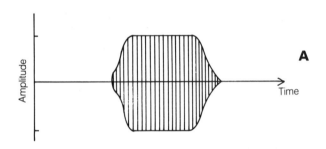

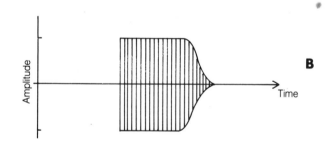

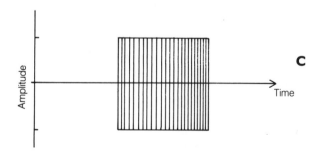

KEY CLICK: A properly shaped signal (A); key clicks on "make" (B); key clicks on both "make" and "break" (C).

KEYER

A *keyer* is a device for switching a continuous-wave (code) transmitter on and off, or for shifting its frequency, to produce morse code. Keyers are available in several different basic configurations. Commercial units (see photo) can be found in an almost countless selection of styles, some with sophisticated memory circuits and operating aids.

The basic electronic keyer is actuated by a single-pole, double-throw, center-off, lever switch that can be moved from side to side. This lever switch, called a *paddle*, resembles the semiautomatic key or "bug" (*see* SEMIAUTOMATIC KEY). The paddle is normally at center, or off. When the lever is pushed to the left, the keyer produces a string of dashes. When the lever is pressed to the right, the keyer produces a string of dots. The dots and dashes are perfectly timed and formed by the keyer circuit. The actual duration of the dot is ⅓ of the duration of the dash. The keyer inserts a space of one dot length after each dot or dash. When these spaces are included in the element length, the dash is twice as long as the dot. The dot-to-space ratio is usually 1 to 1, but it can be adjusted in some keyers, via a weight control (*see* WEIGHT).

Some keyer devices contain two single-pole, single-throw, momentary-contact switches arranged in such a way that either or both can be actuated. Pressing one switch produces dashes, and pressing the other results in the generation of dots. Pressing both switches together produces an alternating string of dots and dashes. This is called *squeeze keying* because of the physical arrangement of the paddles.

The most sophisticated keyers are actuated by means of a keyboard. With the keyboard keyer, the speed of transmission is limited only by the speed at which the operator can type. With a conventional paddle keyer or squeeze keyer, the most proficient operators can send code at speeds of 60 to 70 words per minute. But with a keyboard keyer, speeds of more than 100 words per minute are possible. *See also* KEYING.

KEYER: A sophisticated electronic keyer. ADVANCED ELECTRONIC APPLICATIONS, INC.

KEYING

Keying is the means by which a code transmitter is modulated. Keying is accomplished either by switching the carrier on and off, or by changing its frequency. The latter method is called *frequency-shift keying* (*see* FREQUENCY-SHIFT KEYING).

The code transmitter can be keyed at the oscillator, or at any of the amplifiers following the oscillator. If a heterodyne circuit is used, any of the local oscillators can be keyed. Oscillator keying makes it possible for the operator to listen in between dots and dashes. This is called *break-in operation* (*see* BREAK-IN OPERATION). Amplifier keying usually results in the transmission of a local signal when the key is up; this disables the receiver unless excellent isolation is provided in the transmitter circuit.

The most common method of keying in a transmitter is called *gate block keying*. In the bipolar circuit, the equivalent is called base-block keying. *See also* CODE TRANSMITTER, KEY CLICK, and SHAPING.

KEYSTONING

Keystoning is a form of television picture distortion in which the horizontal gain is not uniform at various vertical picture levels. This form of distortion occurs when the picture tube circuitry is out of alignment.

Keystoning can be detected by directing the camera toward a square object. The distortion will cause the object to appear trapezoidal — that is, narrower at the top than at the bottom, or vice-versa. *See also* TELEVISION.

KICKBACK

When the current through an inductor is suddenly interrupted, a voltage surge, called *kickback*, occurs. The polarity of this voltage is opposite to the polarity of the original current. The surge can be quite large; the peak voltage depends on the coil inductance and the original current.

Kickback voltage can, in some instances, reach lethal values. Whenever large inductances carry large amounts of current, the components of the circuit must be handled carefully because of the kickback shock hazard. The electric-fence generator makes use of the kickback phenomenon, as does the spark generator in an automobile.

KILO

Kilo is a prefix multiplier. The addition of "kilo-" before a designator denotes a quantity 1,000 (one thousand) times as large as the designator without the prefix. *See also* PREFIX MULTIPLIERS.

KILOBYTE

The *kilobyte* is a commonly used unit by which information quantity is expressed. A kilobyte is equivalent to 1024 (2^{10}) bytes. Most personal computers have a memory capacity of several tens of kilobytes. The term *kilobyte* is often abbreviated K, so people often speak of a computer having 32K or 64K bytes of memory. *See also* BYTE, and MEGABYTE.

KILOCYCLE

See KILOHERTZ.

KILOHERTZ

The *kilohertz* (*kHz*) is a unit of frequency, equal to 1,000 hertz (1,000 cycles per second). Audio frequencies and radio frequencies in the very low, low, and medium ranges, are often specified in kilohertz. The human ear can detect sounds at frequencies as high as 16 to 20 kHz.

KILOWATT

The *kilowatt (kW)*, is a unit of power, or rate of energy expenditure. A power level of 1 kW represents 1,000 watts. In terms of mechanical power, 1 kW is about 1.34 horsepower.

The kilowatt is a fairly large unit of power. An average incandescent bulb uses only about 0.1 kW. Radio broadcast transmitters produce radio-frequency signals ranging in power from less than 1 kW to over 100 kW. An average residence uses from about 1 kW to 10 kW. *See also* POWER, and WATT.

KILOWATT HOUR

The *kilowatt hour (kWh)* is a unit of energy. An energy expenditure of 1 kWh represents an average power dissipation of P kilowatts, drawn for t hours, such that $Pt = 1$.

The kilowatt hour is the standard unit in which electric energy consumption is measured in businesses and residences. An average household consumes about 500 to 1,500 kWh each month. *See also* ENERGY.

KINESCOPE

See ICONOSCOPE, IMAGE ORTHICON, PICTURE TUBE, and VIDICON.

KIRCHHOFF'S LAWS

The physicist Gustav R. Kirchhoff is credited with two fundamental rules for behavior of currents in networks. These rules are called *Kirchhoff's First Law* and *Kirchhoff's Second Law*.

According to Kirchhoff's First Law, the total current flowing into any point in a direct-current circuit is the same as the total current flowing out of that point. This is true no matter what the number of branches intersecting at the point (see A).

Kirchhoff's Second Law states that the sum of all the voltage drops around a circuit is equal to zero (see B). This sum must include the voltage of the generator, and polarity must be taken into account: equal negative and positive voltages total zero.

Using Kirchhoff's Laws, the behavior of networks, from the simplest to the most complex, can be evaluated.

KNEE

A sharp bend in a response curve is sometimes called a *knee*. This is especially true of the current-versus-voltage curve for a semiconductor P-N junction.

As the forward voltage across such a junction increases, the current remains small until the voltage reaches 0.3 V to 0.6 V. At this point, the current abruptly rises; this point is the knee. Under conditions of reverse voltage, the response is similar, but the knee occurs at a much greater voltage than in the forward condition. *See also* DIODE, P-N JUNCTION.

KNIFE-EDGE DIFFRACTION

When an electromagnetic wave encounters a barrier with a sharp edge, the energy propagates around the edge to a certain extent. The sharper the edge, with respect to the size of the wavelength, the more pronounced this effect. It is called *knife-edge diffraction*.

In general, the lower the electromagnetic frequency, the less the knife-edge diffraction loss in any given case. For this reason, knife-edge diffraction is common at the very low, low, and medium frequencies; it becomes less prevalent at high and very-high frequencies. In the presence of natural obstacles such as hills and mountains, knife-edge diffraction is not common at ultra-high and microwave frequencies. *See also* DIFFRACTION.

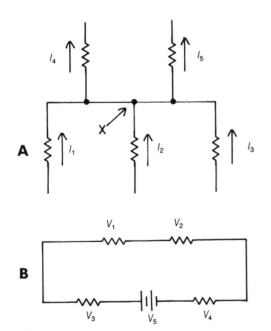

KIRCHHOFF'S LAWS: At A, current into point X equals current out of X. At B, sum of voltages around loop is equal to zero.

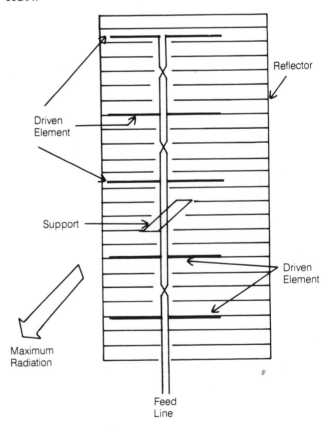

KOOMAN ANTENNA: A stacked array with plane reflector.

KNIFE SWITCH

A *knife switch* is a mechanical switch capable of handling very large voltages and currents. The switch gets its name from its resemblance to a knife. A movable blade slides in and out of a pair of contacts. Knife switches range in size from about 1 inch to over 1 foot in length. There can be just one blade or there can be two or more connected in parallel. The switch can have one or two throw positions.

The principal advantage of the knife switch is its ability to withstand extreme voltages without arcing. This makes it valuable for lightning protection, and in high-voltage applications. However, the knife switch usually has exposed contacts, and this presents a shock hazard.

KOOMAN ANTENNA

A *Kooman antenna* is a high-gain, undirectional antenna that is often used at ultra-high and microwave frequencies. The antenna uses a reflector in conjunction with collinear and broadside characteristics (see illustration).

Several full-wave, center-fed conductors are oriented horizontally and stacked above each other. They are separated by intervals of ½ electrical wavelength (as measured along the phasing line). The phasing line consists of two parallel conductors. The conductors are transposed at each succeeding driven element. Thus, all of the driven elements are fed in the same phase. The reflecting network can consist of a wire mesh or another set of conductors. *See also* BROADSIDE ARRAY.

LADDER ATTENUATOR

A *ladder attenuator* is a network of resistors, connected in such a way that the input and output impedances remain constant as the attenuation is varied. Some ladder attenuators have selector switches in banks, facilitating operation at any desired decibel value.

The ladder attenuator is often used in the test laboratory for comparing the amplitudes of different signals when calibrated oscilloscopes or spectrum analyzers are not sufficiently accurate. Ladder attenuators are also used in situations where the test signal is too strong for the input circuit of the test equipment. Ladder attenuators are characterized by negligible reactance at frequencies well into the ultra-high range, provided that the resistor leads are very short. *See also* ATTENUATION, and ATTENUATOR.

LADDER NETWORK

A *ladder network* is, in general, any cascaded series of L networks or H networks (*see* H NETWORK, and L NETWORK). The L-network sequence is used in unbalanced systems; a series of H networks is used in balanced systems. The ladder attenuator is a network of resistors (*see* LADDER ATTENUATOR). However, capacitors and inductors can also be used in ladder networks.

Ladder networks, consisting of inductances and capacitances, are used as lowpass and highpass filters in radio-frequency applications. By cascading several L-section or H-section filters, greater attenuation is obtained outside the passband, and a sharper cutoff frequency is achieved. Some power-supply filter sections consist of ladder networks, using series inductors and/or resistors, and parallel capacitors. *See also* HIGHPASS FILTER, and LOWPASS FILTER.

LAGGING PHASE

When two alternating-current waveforms, having identical frequency, do not precisely coincide, the waveforms are said to be out of phase. The difference in phase can be as great as ½-cycle.

In a display of amplitude versus time, one wave occurs less than one half cycle later than the other. One wave occurs less than ½-cycle later than the other. The later wave is called the *lagging wave*. If the waveforms are exactly ½-cycle apart in phase, the two signals are said to be in phase opposition. If they are precisely aligned in phase, they are reinforcing. *See also* PHASE ANGLE.

LAMINATED CORE

A *laminated core* is a form of transformer core, generally used at power-line and audio frequencies. The transformer core, rather than consisting of a solid piece of iron, is made up of several thin cross-sectional sheets, glued together with an insulating adhesive (see illustration).

The laminated core reduces losses caused by eddy currents in the transformer core. If the core is solid iron, circulating currents develop in the material. These currents do not contribute to the operation of the transformer, but they simply heat up the core and result in loss of efficiency. The laminated core chokes off these eddy currents (*see* EDDY-CURRENT LOSS). At radio frequencies, laminated cores become too lossy for practical transformer use. For this reason, ferrite and powdered iron are preferred at these frequencies. *See also* FERRITE CORE, and POWDERED-IRON CORE.

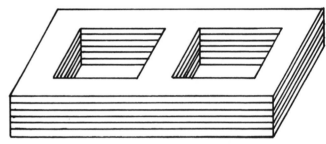

LAMINATED CORE: Several layers of iron are glued together.

LAND-MOBILE RADIO SERVICE

A two-way radio station in a traveling land vehicle, such as a car or bus, engaged in nonamateur communication, is called a *land-mobile station*. Any fixed station, used for the purpose of nonamateur communication with the operator of a land vehicle, is also a land-mobile station.

The land-mobile radio service is allocated a number of frequency bands in the radio spectrum. Many of these bands are in the very-high and ultra-high ranges. The allocations vary somewhat from country to country.

LANGUAGE

In computer applications, a *language* is a system of representing data by means of digital bits, or by characters, such as letters and numerals. Languages fall into three basic categories. Machine language is the actual digital information that is processed by the computer. This language appears almost meaningless to the casual observer. The operator of the computer must converse in a higher-order language, such as BASIC, COBOL, or FORTRAN. These languages use actual words and phrases, as well as standard numerals and symbols. The higher-order language is converted into the machine language by the compiler program. This program must be written in an intermediate language, known as *assembly language*.

Different computer languages are best suited to different applications. For example, BASIC and FORTRAN are intended mainly for mathematical and scientific applications, and COBOL is used primarily by business. *See also* HIGHER-ORDER LANGUAGE, and MACHINE LANGUAGE.

L ANTENNA

Any antenna that is bent at a right angle can be called an *L antenna*. The term is used mainly in reference to medium-frequency and high-frequency wire antennas. The antenna can have any orientation. If the L appears upside-down, a common configuration, the radiator is called an *inverted-L antenna* (*see* INVERTED-L ANTENNA).

The various forms of L antenna are commonly used in locations where space is limited. Such situations often arise for the shortwave listener and amateur-radio operator. The L antenna is characterized by a single bend, at approximately the center of the radiating element. The bend subtends an angle of about 90 degrees.

LARGE-SCALE INTEGRATION

Large-scale integration is the process by which integrated-circuits, containing more than 100 logic elements per chip, are fabricated. Large-scale integration, or LSI, has advanced solid-state electronics to the point where complex systems can be housed in tiny packages.

Both bipolar and metal-oxide-semiconductor (MOS) technology have been adapted to LSI. The electronic wrist watch, single-chip calculator, and microcomputer are some of the most recent developments.

There is evidently a physical limit to the amount of miniaturization that is possible in semiconductor technology. The emphasis in future years will probably be directed toward lower cost and improved efficiency and reliability, as well as toward miniaturization. *See also* INTEGRATED CIRCUIT, and SOLID-STATE ELECTRONICS.

LASER

Most of us are familiar with the word *laser*, an acronym for *l*ight *a*mplification by *s*timulated *e*mission of *r*adiation. The common notion seems to be that a laser is some kind of ray gun. This is true in theory, although it leads people to exaggerated fantasies. Many people get the idea that lasers can knock satellites out of orbit, or that lasers routinely melt holes through steel plates. These things are possible, but in ham radio, the importance of the laser is in less spectacular applications.

Modulated Light   Modulated-light communications is not new. The earliest voice-modulated light beams made use of mirrors that were vibrated while they reflected sunlight. The pickup device was a circuit consisting of a phototube, a battery, and a headset.

A primitive modulated-light transmitter can be made using an audio amplifier connected to a light bulb. A receiver can use a photovoltaic cell or phototransistor along with an audio amplifier. *See* MODULATED LIGHT, and OPTICAL COMMUNICATIONS.

Lasers offer superior performance over long distances for modulated-light communications. This is because the beam from a typical laser, even a rather inexpensive one, diverges gradually. The light from a point source decreases in intensity according to the inverse-square law (*see* INVERSE-SQUARE LAW). A laser beam is so narrow and so intense that this divergence can be essentially ignored within a certain distance of the laser source.

A helium-neon hobby laser has a beam about the diameter of a pencil at close range; this enlarges to a spot about the size of a half dollar at a distance of several hundred feet. High-gain, low-noise amplifiers can be used for reception, and therefore it is easily within the means of any radio ham or electronics hobbyist to make a system that will work for several miles, or even tens of miles, given clear air and an unobstructed line of sight.

Amateur experimentation with laser communications has been done, but the field is still largely unexplored. Many possibilities exist for communications over paths not connected by a single, direct line of sight. For example, the beam from a laser might be aimed at a cloud, or at a mountainside, or even straight up into the atmosphere, and the scatter detected via highly sensitive receivers. Large parabolic mirrors, sensitive photovoltaic cells and FET preamplifiers in the receiving apparatus should make these modes practicable and worthy of future experimentation by radio hams.

Basic Principles   All lasers work according to the principle of *resonance*. A cavity is designed so that the light waves bounce back and forth inside it, reinforcing each other in phase, in much the same way as the electromagnetic field reinforces itself in a VHF or UHF radio-frequency cavity resonator.

The heart of a laser is the cavity in which resonance occurs. This cavity might contain gaseous matter a liquid or a solid. The elements and/or compounds chosen determine the wavelength of the output. Mirrors are placed at each end of the cavity so that energy bounces back and forth many times between them. One mirror reflects 100 percent of the energy that strikes it, and the other mirror is about 95 percent reflective (see illustrations A and B on pg. 276).

Energy is supplied to the cavity by means of pumping so that the intensity of the beam increases, and it finally emerges from the partially silvered mirror in the form of coherent radiation. Some lasers emit a continuous output beam, and others produce a series of pulses. *See* COHERENT LIGHT, and COHERENT RADIATION.

Some Types of Lasers   The *helium-neon laser* is the most often-used in hobby work because of its low cost and versatility. It gets its name from the two gases in its resonant cavity. The light from a helium-neon laser is bright red.

Other gases can be used. Argon produces a laser with blue light. A mixture of nitrogen, carbon dioxide, and helium provides an infrared beam. Hydrogen, xenon, oxygen, and chlorine have all been used to make gas lasers.

The *ruby laser* makes use of solid aluminum oxide in the form of a rod, with a trace of chromium mixed in. The reflectors are silvered onto the ends of the rod. The output is in pulses that last about half a millisecond (0.0005 s). Pumping is done by means of a helical flash tube wrapped around the rod. The output is in the red visible range. Slow-speed digital modulation is possible.

Semiconductor lasers work differently than the types of lasers mentioned previously. Instead of a resonant cavity, semiconductor lasers use P-N junctions that emit coherent light when they are forward biased. Gallium arsenide works well for this purpose. This device, the *GaAs laser diode*, is widely available and lends itself well to home experimentation. Short-range modulated-light communications systems are easy to build using these devices.

Another type of semiconductor laser is the *injection laser*. It works by principles similar to those of the laser diode. A GaAs injection laser produces infrared coherent emissions (see illustration C on pg. 276).

Semiconductor lasers do not usually have the narrow beam typical of larger, pumped lasers. But the beam can be focused or collimated, if desired, using lenses or mirrors. The output of a semiconductor laser is very low, so it is not well suited to long-distance communications. But semiconductor lasers work very well in fiber-optic control systems and optoisolators. *See also* FIBER OPTICS, and OPTICAL COUPLING.

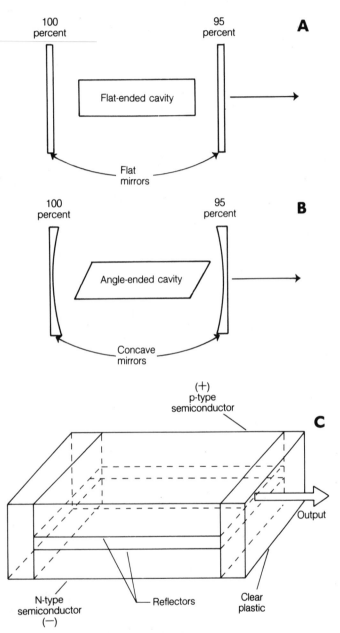

LASER: At A and B, schemes for typical pumped cavity-type lasers. At C, a semiconductor injection laser.

LASER DIODE
See INJECTION LASER.

LASER PRINTER

In recent years, a printing technique has been developed, that works via the same principle by which some photocopying machines work. This is called a *laser printer* or *laser-beam printer* because a semiconductor laser is used to generate the images.

A rotating mirror causes the laser beam to scan rapidly across the page in fine lines as the paper moves through. The laser beam is modulated so that it forms a pattern of light on the paper corresponding to a thin horizontal "slice" of the image to be printed. A full line of text is comprised of several horizontal scans of the laser beam, in a manner similar to the way a video monitor produces its images of text.

The paper moves through the printer, and the laser beam activates a toner compound that causes the paper to darken in regions, according to the modulation of the laser beam. Some laser printers produce copy so crisp that it looks as if it were run on a printing press.

For amateur communications work, dot-matrix printers are more commonly used than laser printers. The main reason is cost; laser printers tend to be expensive. Another reason is that the quality of the copy from a laser printer is often better than a ham radio operator really needs.

LATCH

A *latch* is a digital circuit intended for the purpose of maintaining a particular condition. A latch consists of a feedback loop. The latch prevents changes from high to low, or vice versa, resulting from external causes.

A simple circuit for storing a logic element is called a *latch*. A flip-flop can be used for this purpose; a cross-coupled pair of logic gates can also be used. *See also* LOGIC GATE.

LATCHUP

Latchup is an undesirable and abnormal operating condition in which a transistor or logic circuit becomes disabled because of the application of an excessive voltage or current. Latchup results in a circuit malfunction, even though there is nothing physically wrong with any of the circuit components.

In a switching circuit, a transistor will sometimes fall into the avalanche region at the base-collector junction, because of excessive reverse bias, and it will not return to normal until the voltage is entirely removed. This is called *P-N-junction latchup* or *transistor latchup*. It sometimes occurs in regulated power supplies.

In a digital circuit, latchup can occur in numerous ways. The application of an excessive or improper voltage, at some point in the circuit, is usually responsible. Microcomputer circuits can sometimes operate improperly because of a stray voltage at some point. In such cases, reinitialization is usually needed to bring the device back to normal (*see* REINITIALIZATION).

In an amplifier circuit, unwanted oscillation is sometimes called *latchup* because it disables the amplifier. This can occur in improperly shielded audio-frequency or radio-frequency systems. *See also* FEEDBACK, and PARASTIC OSCILLATION.

LATENCY

In a digital computer system, the response to a command can be very rapid, but it is not instantaneous. The delay time is called the *latent time*, and the condition of the computer during this time is called *latency*. Of course, the latent time should be as short as possible when high operating speed is needed. In a serial storage system, *latency* is defined as the difference between access time and word time. *See also* ACCESS TIME.

LATTICE

In crystalline substances, the atoms are arranged in an orderly pattern. This pattern is the same wherever the substance is found. The piezoelectric crystal is an example of a substance with such a structure, called a *lattice structure*. The term *lattice* is used to describe the arrangement of components in certain electronic circuits. *See also* CRYSTAL-LATTICE FILTER, and LATTICE FILTER.

LATTICE FILTER

A *lattice filter* is a selective filter in which the components are arranged in a lattice configuration. The details of the impedance can vary greatly; in the simplest form it can be simply an inductor or capacitor. However, the impedances in a lattice filter are usually bandpass, or resonant, circuits. The impedances can also be piezoelectric crystals.

Lattice filters provide a better bandpass response than is obtainable with a single selective circuit. *See also* BANDPASS RESPONSE, and CRYSTAL-LATTICE FILTER.

LCD

See LIQUID-CRYSTAL DISPLAY.

LEAD-ACID BATTERY

A *lead-acid battery* is a combination of cells, forming a storage battery. The negative electrodes in each cell are made from spongy lead. The positive electrodes are made of lead peroxide. The electrodes are immersed in a dilute solution of sulfuric acid, which acts as an electrolyte.

The lead-acid battery is rechargeable, so it can be used over and over. The most common example of a lead-acid battery is the common automobile storage battery, which produces between 12 and 14 volts, depending on its state of charge and the load imposed on it. *See also* BATTERY, CELL, and STORAGE BATTERY.

LEAD CAPACITANCE

Component leads have a certain amount of capacitance, with respect to the surrounding environment, because the leads have a measurable, finite length. This capacitance is, in most cases, extremely small—a tiny fraction of 1 pF. This is not usually of concern at frequencies below the ultra-high range. However, above about 300 MHz, even this small amount of capacitance can have a significant effect on the performance of a circuit.

In the design of ultra-high-frequency and microwave circuits, component leads should be as short as possible to minimize the capacitance. Lead inductance can also influence circuit operation at these frequencies.

Long leads, such as test leads for voltmeters and oscilloscopes, have much larger capacitance than component leads. Typical coaxial cable has a capacitance of 15 to 30 pF per foot. Parallel-wire leads exhibit somewhat lower lead capacitance than coaxial cable, although it can still be significant at high and very-high frequencies. *See also* LEAD INDUCTANCE.

LEAD-IN

An antenna feed line is sometimes called a *lead-in*. Generally, the term *lead-in* is used to refer to a single-wire feed line in a shortwave receiving antenna. A single-wire line displays a characteristic impedance of between 600 and 1,000 ohms under ordinary circumstances (*see* CHARACTERISTIC IMPEDANCE). Little or no attempt is made to match the characteristic impedance of a single-wire lead-in to the impedance of the antenna itself.

The lead-in, if it consists of a single wire, contributes somewhat to the received signal. However, most of the signal is picked up by the main antenna. If the lead-in consists of a coaxial cable or a parallel-wire line, and the lead-in is properly terminated, the lead-in does not contribute to reception or transmission of signals. *See also* FEED LINE.

LEAD INDUCTANCE

Component leads, because they consist of wire and have measurable length, invariably display a certain amount of inductance. This inductance is normally very small—on the order of a few nanohenrys (billionths of a henry) for the average resistor or capacitor with 1-inch leads.

The lead inductance of electronic components is not generally of great concern at frequencies below the ultra-high range. However, above 300 MHz, the effects can become significant because at these frequencies the lead lengths become a substantial fraction of a wavelength. At ultra-high and microwave frequencies, therefore, leads must be cut very short to minimize the effects of lead inductance.

Long leads, such as test leads for voltmeters and oscilloscopes, have much larger inductance than component leads. A typical cable or parallel-wire test lead has sufficient inductance to affect circuit operation at very-high frequencies. In conjunction with the lead capacitance, resonance can occur at various frequencies in the very-high or ultra-high range, and this will have an effect on measurements at those frequencies. *See also* LEAD CAPACITANCE.

LEADING PHASE

When two alternating-current waveforms, having the same frequency, do not precisely coincide, the waveforms are said to be out of phase. The difference in phase can be as great as $\frac{1}{2}$-cycle.

In a display of amplitude versus time, two waveforms with different phase, but identical frequency, might appear with one wave less than $\frac{1}{2}$-cycle earlier than the other. The earlier wave is called the *leading wave*. If the waveforms are exactly $\frac{1}{2}$-cycle apart in phase, the two signals are said to be in phase opposition. If they are precisely aligned in phase, they are reinforcing. *See also* PHASE ANGLE.

LEAKAGE CURRENT

In a reverse-biased semiconductor junction, the current flow is ideally, or theoretically, zero. However, some current flows in practice, even when the reverse voltage is well below the avalanche value. This small current is called the leakage current. *See* P-N JUNCTION.

All dielectric materials conduct to a certain extent, although the resistance is high. Even a dry, inert gas allows a tiny current to flow when a potential difference exists between two points. This current is called the *dielectric leakage current*. *See also* DIELECTRIC CURRENT.

LEAKAGE FLUX

In a transformer, the coupling between the primary and the secondary windings is the result of a magnetic field that passes through both windings. In some transformers, not all of the magnetic lines of flux pass through both windings. Some of the flux generated by the current in the primary does not pass through the secondary. This is called the *leakage flux*.

In general, the closer the primary and secondary windings are located to each other, the smaller the leakage flux. An iron-core transformer has a smaller leakage flux than an air-core transformer of the same configuration. The toroidal transformer has the least leakage flux of any transformer. The leakage flux is inversely related to the coefficient of coupling between two inductive windings. *See also* LEAKAGE INDUCTANCE, MAGNETIC FLUX, and TRANSFORMER.

LEAKAGE INDUCTANCE

In a transformer, the mutual inductance between the primary and secondary windings is often less than 1 because of the existence of leakage flux (*see* LEAKAGE FLUX). As a result of this, the primary and secondary windings contain inductance that does not contribute to the transformer action. This inductance, known as *leakage inductance*, is effectively in series with the actual primary and secondary windings.

A transformer can be considered to consist of a perfect component, having primary and secondary windings with no leakage flux; there is also a series self-inductive component in both the primary and secondary circuits. These series inductances should be as small as possible so that the transformer will have negligible losses in the windings. *See also* TRANSFORMER.

LEAKAGE RESISTANCE

In theory, a capacitor should not conduct any current when a direct-current voltage is placed across it, except for the initial charging current. However, once the capacitor has become fully charged, a small current continues to flow. The leakage resistance of the capacitor is defined as the voltage divided by the current.

Air-dielectric capacitors generally have the highest leakage resistance. Ceramic and mylar capacitors also have very high leakage resistance. Electrolytic capacitors have somewhat lower leakage resistance. Typically, capacitors have leakage resistances of billions or trillions of ohms or more. *See also* DIELECTRIC CURRENT, and LEAKAGE CURRENT.

LECHER WIRES

Lecher wires are a form of selective circuit used at ultra-high frequencies for the purpose of measuring the frequency of an electromagnetic wave. The assembly consists of a pair of parallel wires or rods, mounted on an insulating framework. A movable bar allows adjustment of the length of the section of parallel conductors (see illustration).

The electromagnetic energy is coupled into the system by means of an inductor at the shorted end of the parallel-conductor section. An indicating device, such as a meter or neon lamp, is inductively coupled to the other end. Both the input and output couplings must be loose so that the resonant properties of the parallel-conductor system are not affected.

The position of the movable shorting bar is varied until a peak is indicated on the meter or lamp. This shows that the system of wires is resonant at the input frequency, or at some multiple of the input frequency. The shortest resonant length indicates the fundamental frequency of the input signal. A calibrated scale, adjacent to the movable bar, shows the approximate frequency or wavelength of the input signal. The fundamental frequency also corresponds to the distance between adjacent resonant positions of the bar.

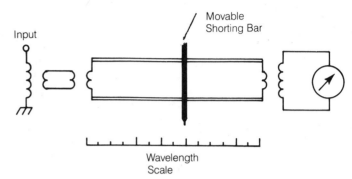

LECHER WIRES: Used to measure wavelength.

LED

See LIGHT-EMITTING DIODE.

LENZ'S LAW

When a current is induced by a changing magnetic field, or by the motion of a conductor across the lines of flux of a magnetic field, that current itself generates a magnetic field. According to Lenz's Law, the induced current causes a magnetic force that acts against the motion.

An example of Lenz's Law can be observed in the operation of an electric generator. When the output of the generator is connected to a load, so that current flows in the generator coils, considerable turning force (torque) is necessary to make the generator work. The lower the load resistance, the greater the current in the coils, and the greater the magnetic force that opposes the coil rotation. This is why a powerful engine or turbine is needed to operate a large generator. *See also* MAGNETIC FIELD.

LIBRATION FADING

In moonbounce communications, also called *earth-moon-earth (EME)*, a peculiar and severe type of fading occurs. It makes the signal "flutter," often by as much as plus-or-minus 10 dB. It sounds somewhat like auroral propagation or backscatter fading (*see* AURORAL PROPAGATION, and BACKSCATTER). But the cause is entirely different.

When a radio wave is reflected from the surface of the moon, it comes back from innumerable hills, valleys, cliffs, and crater walls. Millions of different waves, in random phase, arrive at the receiving antenna of a moonbounce station. These component waves all add up to a composite wave, whose amplitude depends on how much the component waves cancel or add in phase.

The moon keeps the same side always facing the earth, more or less. But there is some back-and-forth "wobbling." This is called *libration*. The fading it causes is known as *libration fading*, and is the result of the constant, and unavoidable, fluctuation in the phases of the component waves.

If the moon were exactly still, keeping in precise alignment as it orbited the earth, the component waves would always arrive in the same relative phase. But the slightest wobble — even if very slow, and even if only a few inches — causes a tremendous jumble in the phases of the waves.

If the moon were as smooth as a cue ball, libration would cause no fading. But, of course, the moon's terrain is far from featureless, and even small bumps give rise to severe phase-induced fading. *See* MOONBOUNCE.

LICENSE REQUIREMENTS

There are several classes of license in ham radio. The simplest, for many, is the Novice-Class license, with a slow-speed morse (code) test and elementary electronic theory test. For some, the code-free Technician-Class license is the easier way to begin, with an intermediate-level electronic theory test but no code requirement.

Novice For this license, a 5-word-per-minute (5 wpm) code test is given. This is called *Element 1A*. An elementary theory test, along with some rules and regulations, must also be passed; this is called *Element 2*. Privileges include code on small portions of the 3.5, 7, and 21-MHz bands, and some teletype and voice on the 28, 222, and 1250-MHz bands. Power output is limited to 200 W through the 28-MHz band, 25 W on 222 MHz, and 5 W on 1250 MHz.

Code-Free Technician This license requires passing the Novice theory test (Element 2) and an intermediate-level theory test, called *Element 3A*. If you have a Code-Free Technician Class license, you get all amateur privileges above 50 MHz.

Technician Plus Code If a 5-wpm code test is passed in addition to the above — that is, Elements 1A, 2 and 3A — then Novice privileges are conveyed, in addition to those for the Code-Free Technician Class license.

General To obtain a General Class license, you need to pass *Element 1B*, a 13-wpm code test. You also need to pass Elements 2 and 3A. In addition, an intermediate-level test on regulations is given. This is *Element 3B*. The General-Class licensee can operate on all ham bands. There are segments reserved for Advanced- and Extra-class licensees on some bands.

Advanced To get an Advanced-Class license, you need to pass all the elements for the General Class, plus *Element 4A*, the advanced theory test. Then, you get all the privileges available to General-Class operators, and additional voice privileges on certain bands.

Extra For the Extra-Class license, the highest class you can get, you need to pass everything for the Advanced, plus *Element 1C*, a 20-wpm code test, and *Element 4B*, a rather sophisticated theory test.

If you were to start right out with the Extra-Class license, you wouldn't need to take the 5-wpm or 13-wpm test for code. Some people have actually done this! Usually, they are engineers or shipboard radiotelegraph operators. The Extra Class license conveys all ham privileges.

LIGHT

Light is visible electromagnetic radiation, in the wavelength range of about 750 nanometers (nm) to 390 nm (a nanometer is equal to a billionth of a meter). The longest wavelengths appear red to the human eye, and the colors change as the wavelength gets shorter, progressing through orange, yellow, green, blue, indigo, and violet.

The earliest theory of light held that it is a barrage of particles. This is known as the *corpuscular theory of light* and it is still accepted today, although we now know that light also has electromagnetic-wave properties. The light particle is called a *photon*. The shorter the wavelength of light, the more energy is contained in a single photon.

In a vacuum, light travels at a speed of about 186,282 miles per second (299,792 kilometers per second). This is the same speed with which all electromagnetic fields propagate. In some materials, light travels more slowly than this.

Light can be amplitude-modulated and polarization-modulated, for the purpose of line-of-sight communications. This is commonly done in optical-fiber systems (*see* FIBER, FIBER OPTICS, and OPTICAL COMMUNICATIONS).

LIGHT-ACTIVATED SILICON-CONTROLLED RECTIFIER

A *light-activated silicon-controlled rectifier (LASCR)* is a switching device activated by visible light. Impinging light waves perform the same function as the gate current in the ordinary silicon-controlled rectifier (SCR).

Ordinarily, the LASCR does not conduct. However, when the incident light reaches a certain intensity, the device conducts. The amount of light necessary for conduction is determined by the characteristics of the device, and also by the value of an externally applied bias. *See also* SILICON-CONTROLLED RECTIFIER.

LIGHT-ACTIVATED SWITCH

A *light-activated switch* is any device that opens or closes a circuit in the presence of irradiation by visible light. Such a device can consist of a phototransistor, or a solar cell and switching circuit.

Light-actuated switches are used for a variety of purposes. Some types of light-actuated switches close a circuit when light impinges on them; some open a normally closed circuit. An example of the latter type of light-actuated switch is a device often found in homes for the purpose of deterring burglars. When it gets dark, the switch turns on one or more lights. The same sort of light-actuated switch is used by some municipalities to turn on street lights at dusk. *See also* LIGHT-ACTIVATED SILICON-CONTROLLED RECTIFIER, PHOTOTRANSISTOR, and PHOTOVOLTAIC CELL.

LIGHT-EMITTING DIODE

A *light-emitting diode (LED)*, technically called an *electroluminescent diode*, is a device that emits infrared, visible light, or ultraviolet radiation when it is supplied with a certain amount of forward voltage. Light-emitting diodes are used in many different kinds of electronic circuits, especially indicators and digital displays. Many different colors are available; the most common are red, yellow, and green.

The LED is generally fabricated from a direct band-gap semiconductor material, such as gallium arsenide (GaAs). Many different kinds of commercially manufactured LEDs are available. The typical LED is mounted inside an epoxy material that allows maximum transmission of light at the emission frequency. Sometimes a lens-type enclosure is used to enhance the radiation or give it desired directional characteristics. The photograph shows a typical light-emitting diode next to a dime, for size comparison.

An LED with a flat and uniform junction, capable of emitting coherent light, is the injection laser. Such a device requires a certain minimum forward current in order to produce coherent-light output. These devices are sometimes called laser diodes or laser LEDs. *See also* INJECTION LASER.

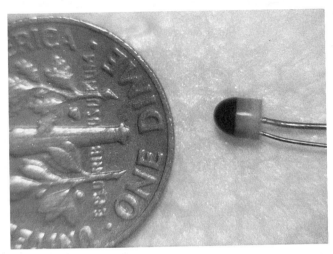

LIGHT-EMITTING DIODE: Some LEDs are almost microscopic; others are large.

LIGHTNING

There is a constant potential difference between the earth's surface and the ionosphere. The lower atmosphere acts as a dielectric, so that the earth and ionosphere form a giant capacitor. The charge in this atmospheric capacitor is stupendous, and the dielectric frequently breaks down with discharges of lightning. These occur dozens of times per second. The discharges are almost always concentrated in areas of precipitation, particularly in and near storms.

The Lightning Stroke Lightning always accompanies thundershowers; this is how these storms get their name. Lightning is also common in tropical showers, disturbances, and hurricanes. But lightning can, and occasionally does, occur in snowstorms (especially mountain storms), light rain showers, and even in places where no precipitation is falling.

A lightning surge is called a *stroke*. It lasts only for a fraction of a second, but in this time it can start fires, cause explosions, destroy electrical and electronic equipment, and electrocute people and animals.

A stroke begins when the charge between a cloud and the ground, or between a cloud and some other part of the cloud, gets so large that electrons begin to advance out from the negative charge pole. A small current, called a *stepped leader*, finds the path of least resistance through the atmosphere by a kind of "trial and error" process. This takes a few milliseconds up to perhaps several tenths of a second. Once the stepped leader has established the circuit, a large return stroke occurs. This carries hundreds of thousands of amperes. It is responsible for the damaging effects of lightning. There can be several return strokes in rapid succession.

The Danger to Hams Lightning is more deadly, in terms of annual deaths, than hurricanes, tornadoes or blizzards. The risk to ham radio operators is exaggerated because most hams have outdoor antennas, and these are susceptible to induced surges from nearby lightning strokes, as well as to direct hits.

Every lightning stroke produces a strong electromagnetic field. You hear this as static (QRN) on the radio, especially on the low bands, such as 160, 80, and 40 meters. This field is so intense, within a short distance of the stroke itself, that it can induce a pulse of current in electrical conductors like power lines and ham radio antennas. This *electromagnetic pulse* can, and often does, damage ham equipment, as well as household appliances (especially television sets and hi-fi equipment).

If you happen to be operating your station and lightning strikes nearby, you might be seriously injured or even killed. And, of course, a direct hit is much worse than the induced charge from a nearby stroke. Fortunately, direct hits don't occur very often — but they can be devastating.

The way to avoid or minimize the personal danger to yourself is to stay away from all electrical and electronic equipment, especially antenna systems, whenever there is a thundershower in the area, or whenever there is any risk of lightning. Ham radio is fun, but it is not worth getting maimed or killed for.

The Danger to Equipment Lightning causes damage from electromagnetic pulses, which induce large currents in antennas, feed lines and power lines, even if there is no direct hit. Electrostatic-sensitive components are most easily destroyed. These include MOS FETs and MOS ICs. The front end circuitry of a receiver is extremely vulnerable. Semiconductor components in general, especially those near the antenna terminals or in the power supply, are also susceptible to damage from the electromagnetic pulse.

A direct hit to an antenna or feed line, if these are not grounded and disconnected from the station equipment, can ruin a ham rig and accessories beyond repair. An ungrounded antenna also presents a fire hazard to the household in general. *See also* ELECTROMAGNETIC PULSE, and LIGHTNING PROTECTION.

LIGHTNING ARRESTOR

A *lightning arrestor* is a device that allows discharge of high voltages on an antenna feed line. Lightning arrestors generally consist of one or more spark gaps, adjusted so that a voltage in excess of a certain value will be routed to ground. The illustration at A shows a cutaway view of a coaxial lightning arrester. A lightning arrestor for balanced line can be home-constructed, as shown at B.

The gap is adjusted so that the maximum transmitter power does not produce a discharge at any operating frequency. This is a trial-and-error process. If the gap is located at a current node on a transmission line having a high standing-wave ratio, the voltage can be sufficient to cause arcing even if the contacts are fairly far apart. The closest possible spacing should be used.

Lightning arrestors provide some reduction in the chances of a direct lightning strike. This is because the charge on an antenna is neutralized, via arcing to ground, before it can build up to large values. However, a lightning arrestor does not guarantee that a direct hit will not occur. Radio equipment should not be used during thunderstorms, unless it is absolutely necessary. The safest precaution is to disconnect the feed line from the radio equipment, and connect the line to a well-grounded point, whenever the radio equipment is not in use. *See also* LIGHTNING, and LIGHTNING PROTECTION.

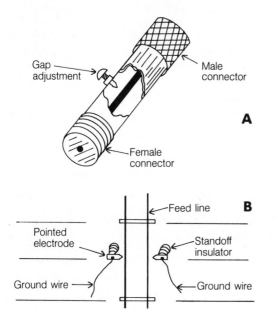

LIGHTNING ARRESTOR: A cutaway view of coaxial unit (A); a homebrew device for open-wire line (B).

LIGHTNING PROTECTION

Lightning is common in many parts of the world. It is also dangerous, killing more people every year in the United States than hurricanes or tornadoes. The main danger to people is from electrocution; in the case of property, damage can result from fires, induced currents, and sometimes even explosions. *See* LIGHTNING.

This article mentions precautions to protect people and ham equipment from harm by lightning. But this article cannot possibly cover every situation. Sometimes people are injured, or electronic equipment is damaged, despite precautions. The best you can do is statistically minimize the danger.

Protecting Yourself   Lightning can occur at any time. But it is most common in or near areas of precipitation. If thunder can be heard, lightning is occurring. Do not assume that the noise is coming from a jet aircraft. Heavy precipitation at or near a place (including snow and sleet), dark clouds, a sand storm, a dust storm, or even an erupting or smoldering volcano can be a source of lightning. The only necessary ingredient is a large potential difference between two regions of the atmosphere that are close enough together, or between some parcel of air and the ground.

The following precautions are recommended if lightning is taking place near you.

1. Stay indoors, or inside a metal enclosure such as a car, bus or train. Stay away from windows.
2. If it is not possible to get indoors, find a low-lying spot on the ground, such as a ditch or ravine, and squat down with feet close together, until the threat has passed.
3. Avoid lone trees or other isolated, tall objects, such as utility poles or flagpoles.
4. Avoid electric appliances or electronic equipment that makes use of the utility power lines, or that has an outdoor antenna.

5. Stay out of the shower or bathtub.
6. Avoid swimming pools, either indoors or outdoors.
7. Do not use the telephone.

Equipment Protection   The antenna is the most dangerous part of a ham radio station when lightning is occurring nearby. A direct hit is not necessary for destructive and dangerous currents to be induced in the antenna element(s) and feed line. The power lines can also get high-voltage surges that can damage equipment plugged into wall outlets. Precautions to minimize the risk are as follows.

1. Never operate, or experiment with, a ham station when lightning is occurring anywhere near your location.
2. When the station is not in use, disconnect all antennas and ground all feed-line conductors to a good dc ground other than the utility power-line ground. Preferably, the lines should be left entirely outside the building and connected to an earth ground at least several feet from the building.
3. When the station is not in use, unplug all equipment from the wall outlets.
4. When the station is not in use, disconnect and ground all rotator cables and other wiring that leads outdoors.
5. Lightning arrestors provide some protection from charge buildup, but they cannot offer complete safety, and should not be relied upon for routine protection of ham equipment or people. *See* LIGHTNING ARRESTOR.
6. Lightning rods reduce the chance of a direct hit, but should not be used as an excuse to neglect the other precautions. *See* LIGHTNING ROD.
7. Power-line surge suppressors reduce computer "glitches" and can sometimes protect sensitive components in a power supply, but these should not be used as an excuse to neglect the other precautions. *See* SURGE PROTECTION, and SURGE SUPPRESSOR.
8. The antenna mast or tower should be connected to an excellent dc earth ground using heavy-gauge wire or braid. Several parallel lengths of AWG No. 8 aluminum ground wire, run in a straight line from the mast or tower to ground, form an adequate conductor. Remember that this conductor might be called upon to carry a momentary current of thousands of amperes. The conductor must be able to survive in case of another strike.
9. Other secondary protection devices are available and can be found in electronic-related and ham-radio-related magazines. None of these should be relied upon for complete protection; the primary methods (1 through 4) are mandatory to minimize the risk to equipment.

For Further Information   Refer to *Lightning Protection Code*, published by the National Fire Protection Association, Batterymarch Park, Quincy, MA 02269. Ham magazines often carry articles on the topic of lightning protection. Various ham-related books also have information, but it is important to be sure it is recent because new, improved protection methods are occasionally found.

LIGHTNING ROD

A *lightning rod* is a grounded metal electrode, placed on the roof of a building for the purpose of lightning protection. Lightning rods are commercially available. They can also be home-constructed.

Generally, a lightning rod gives protection within a cone-shaped region under the rod. The apex of the cone is located at the top of the rod. The apex angle of the cone, with respect to the vertical axis of the rod, is about 45 degrees. If the ground is level, a lightning rod can therefore be expected to provide safety within a circular region whose radius is equal to the height of the rod.

According to Benjamin Franklin, who is given partial credit for inventing the lightning rod, the tip of the rod should be tapered to a sharp point. This is, in fact, the way most lightning rods are built today. King George III of England believed that lightning rods should have blunt or rounded tips. Recently, some evidence has been gathered to suggest that the blunt tip is more effective than the pointed tip. Discharges occur more readily from pointed objects than from blunt objects. Lightning is therefore more likely to strike a blunt rod than a pointed one. There is some disagreement in the scientific community as to which kind of rod is better.

A lightning rod must be properly grounded. Heavy wire should be used, preferably AWG #4 or larger. The ground rod should be driven at least 8 feet into the ground, and it should be located several feet from the foundation of the building. Multiple rods can be used. The ground wire should take the shortest possible path from the base of the lightning rod to the ground. The electrical bonds between the lightning rod and the wire, and between the wire and the ground rod, must be as substantial as possible. The ground rod must be located high enough so that the cone of protection covers the whole building. *See also* LIGHTNING, and LIGHTNING PROTECTION.

LIGHT PEN

A *light pen* is a device used with video-display units for the purpose of creating graphic images. The light pen makes it possible to literally draw pictures on a television screen.

The image on a television screen is created by an electron beam that scans rapidly from left to right, in horizontal lines beginning at the top of the screen and moving downward. It is similar to the path your eyes follow as you read a book. When the light-sensitive tip of the light pen is placed over a certain spot on the screen, a pulse is produced as the electron beam scans past that point. This pulse is fed to a microcomputer, which causes the spot to change state. A letter X can, for example, be converted to a blank (space) or to a solid square. The light pen can also be used to produce certain functions or effects, as in a video game.

As the tip of the light pen is moved around on the screen, successive character blocks are changed from one state to the other, and/or the designated function is applied at the points scanned. The resolution is limited by the number of character blocks in the entire screen.

LIMITER

A *limiter* is a device that prevents a signal voltage from exceeding a certain peak value. When the peak signal voltage is less than the limiting value, the limiter has no effect. But if the peak input-signal voltage exceeds the limiting value, the limiter clips off the tops of the waveform at the peak-limiting value (see illustration).

Limiters are used in frequency-modulation receivers for the purpose of reducing the response to variations in signal amplitude. The limiting threshold is set very low, so even a rather

weak signal will exceed the limiting voltage. Then, changes in amplitude are eliminated. This is why frequency-modulation receivers are less susceptible than amplitude-modulation receivers to impulse noise and atmospheric static. The limiter stage is placed immediately before the discriminator stage. *See also* DISCRIMINATOR, and FREQUENCY MODULATION.

In low-frequency, medium-frequency, and high-frequency communications receivers, audio-peak limiters are sometimes used to improve the signal-to-noise ratio under adverse conditions. *See also* NOISE LIMITER.

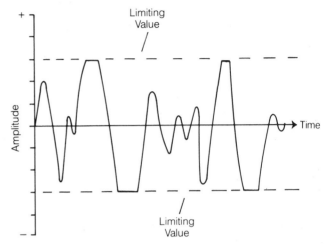

LIMITER: Prevents signal peaks from exceeding a certain level.

LIMIT SWITCH

A *limit switch* is a device that opens or closes a circuit when a certain parameter (such as current, power, voltage, or illumination) reaches a specified value, either from below or above. A common circuit breaker is a form of limit switch that is actuated by excessive current in a circuit. A voice-operated relay is another form of limit switch. There are many other examples.

Limit switches can be electromechanical or electrical. The circuit breaker and voice-actuated relay are electromechanical devices. Solid-state limit switches can be fabricated using diodes, transistors, and silicon-controlled rectifiers. Some limit switches are actuated when the parameter exceeds a certain value. Others are actuated when the parameter falls below a specified cutoff level.

LINE

A *line* is a wire or set of wires over which currents or electromagnetic fields are propagated. A power line, telephone line, or radio transmission line can each be simply called a *line*. *See also* FEED LINE, POWER LINE, and TRANSMISSION LINE.

Electric and magnetic fields are defined in terms of flux lines. The lines of flux represent the theoretical direction of the field. Each line represents a certain quantity of electric or magnetic flux. *See* ELECTRIC FLUX, FLUX, and MAGNETIC FLUX.

LINEAR AMPLIFIER

In amateur radio, a *linear amplifier* is a radio-frequency power amplifier that faithfully reproduces the modulation envelope of a single-sideband (SSB) signal. Such an amplifier is sometimes called a *linear* for short. These amplifiers can be used with any type of emission, amplitude-modulated or frequency-modulated. They will not, when operated properly, introduce any significant distortion into the signal.

Description and Purpose A linear is generally used when the power output from a transceiver is not enough to provide reliable communications. Linears should not be used when they aren't needed. Some hams never own one; others consider them standard station equipment.

Linears are popular among DXers and contesters. They are most valuable in traffic handling, especially when the traffic is of a priority or emergency nature. *See* CONTESTS AND CONTESTING, DX AND DXING, and TRAFFIC HANDLING.

A linear is usually a self-contained unit, sometimes with a built-in power supply and sometimes with an outboard supply. The most rugged ham linears can provide 1500 W continuous (100-percent duty cycle) RF output, the maximum legal limit for amateur service. Some linears are designed to supply this output for only a 50-percent duty cycle, typical of code (CW) and SSB. Other linears cannot provide the maximum legal limit, but something less, such as 1000 W output or 1000 W input.

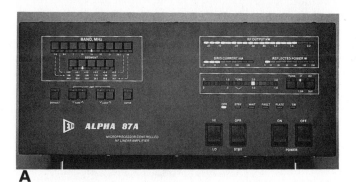

A

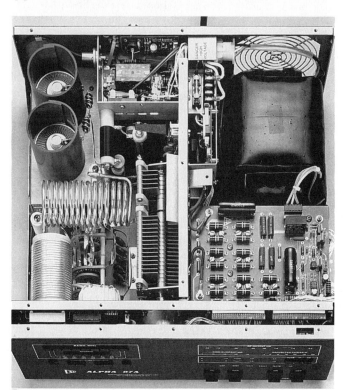

B

LINEAR AMPLIFIER: At A, the front panel of a high-technology, microprocessor-controlled, 1.5-kW linear amplifier. At B, the interior of the same amplifier. Ehrhorn Technological Operations, Inc.

Electrical Characteristics The term *linear* means that the output envelope is the exact same shape as the input envelope. In engineering terms, the amplification curve is a straight line. The instantaneous output is a linear function of the instantaneous input.

Class-A amplifiers are always linear. But this type of circuit is not often used at RF because it is comparatively inefficient. Instead, linear amplifiers are usually operated in class AB or B. *See* CLASS-A AMPLIFIER, CLASS-AB AMPLIFIER, and CLASS-B AMPLIFIER.

Push-pull class-B circuits work very well as linear amplifiers, offering attenuation of the even harmonics, as well as reasonable efficiency. But these circuits have the disadvantage of requiring center-tapped transformers, and the balance between the two amplifying transistors or tubes is rather critical. *See* PUSH-PULL AMPLIFIER, and PUSH-PULL CONFIGURATION.

Class-C amplifiers are never linear. They have high efficiency, and are useful for frequency modulation (FM) and for digital signals, such as CW and frequency-shift keying (FSK), but they will introduce distortion into amplitude-modulated SSB or slow-scan television (SSTV). *See* CLASS-C AMPLIFIER.

A typical amateur linear needs about 100 W of drive power to produce the maximum output. Some linears can work off of much less drive, about 10 W. Linears today have numerous features that make them versatile. These features can include built-in wattmeters, built-in SWR meters, full break-in option, automatic shutdown in case of overheating, automatic gain control, and even automatic tune-up or broadbanded operation.

LINEAR INTEGRATED CIRCUIT

A *linear integrated circuit* is a solid-state analog device. Linear integrated circuits are characterized by a theoretically infinite number of possible operating states; in contrast, the digital integrated circuit usually has just two possible states. Linear integrated circuits are used as amplifiers, oscillators, and regulators.

Within a certain input range, the input and output voltages of the linear integrated circuit are directly proportional to each other. If the instantaneous output is plotted on the Cartesian plane as a function of the instantaneous input, the graph appears as a straight line. *See also* DIGITAL INTEGRATED CIRCUIT, INTEGRATED CIRCUIT, and LINEAR AMPLIFIER.

LINEARITY

Linearity is an expression of the resemblance between the input and output signals of a circuit. In general, the better the linearity, the less distortion is generated in a device.

In a strict mathematical sense, linearity is the condition in which the instantaneous input and output signal amplitudes are related by a constant factor. In audio-frequency applications, the input and output waveforms must be such that all points are related by this constant. In radio-frequency work, a circuit can be considered linear, as long as the input and output envelope amplitudes are related by a constant. The actual radio-frequency waveform can be distorted. *See also* LINEAR AMPLIFIER.

LINEAR POLARIZATION

An electromagnetic wave is said to have *linear polarization* when the electric field maintains a constant orientation. Most radio-frequency signals have linear polarization as they leave an antenna. Any antenna consisting of a single, fixed radiator, with or without parasitic elements, produces a signal with linear polarization. A linearly polarized wave can be oriented horizontally, diagonally, or vertically.

When a radio signal is returned to the earth from the ionosphere, the direction of the electric-field lines can no longer be constant because of phasing effects in the ionospheric layers. The polarization of a radio signal can be deliberately made to rotate as it is emitted from the antenna. This is called *elliptical* or *circular polarization. See also* CIRCULAR POLARIZATION, ELLIPTICAL POLARIZATION, HORIZONTAL POLARIZATION, and VERTICAL POLARIZATION.

LINEAR RESPONSE

A transducer is said to have a *linear response* if the input and output current or voltage are related in a certain way. Specifically, the response is linear if the graph of the output versus input appears as a straight line passing through the origin point (0,0).

A linear response can be defined in algebraic terms. If an input current or voltage of x_1 results in an output current or voltage of y_1, and if an input current or voltage of x_2 results in an output current or voltage of y_2, then the response is linear if and only if $y_1 = kx_1$, $y_2 = kx_2$ and $y_1 + y_2 = k(x_1 + x_2)$.

A transducer can exhibit a linear response over a certain range of input current or voltage. Often though, a transducer is linear only within a certain range; outside of that range it becomes nonlinear. *See also* LINEARITY.

LINEAR TAPER

Some potentiometers have a resistance that varies in linear proportion to the rotation. Such potentiometers are said to have a *linear taper*. If a linear-taper potentiometer is rotated through a certain angle in any part of its range, the change in resistance is always the same.

Linear-taper potentiometers are used in many electronic circuits for alignment or adjustment purposes. For volume control, however, the linear-taper potentiometer is not generally used because people perceive sound in a logarithmic manner. For volume control, audio-taper potentiometers are preferred. *See also* AUDIO TAPER.

LINEAR TRANSFORMER

A *linear transformer* is a radio-frequency transformer, sometimes used at very-high frequencies and above for the purpose of obtaining an impedance match. The linear transformer consists of a quarter-wave section of transmission line, short-circuited at one end and open at the other end (see illustration). The line can be of the parallel-conductor type or the coaxial type. Power is applied via a small link at the short-circuited end of the line. The output is taken from some point along the length of the transmission line.

The output impedance is near zero at the short-circuited end of the linear transformer, and is extremely large (theoretically infinite) at the open end. At intermediate points, the impedance is a pure resistance whose value depends on the distance from the short-circuited end.

The linear transformer is used, in various configurations, in many kinds of antenna systems for radio-frequency reception and transmission. *See also* IMPEDANCE MATCHING.

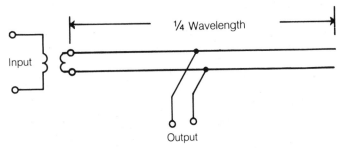

LINEAR TRANSFORMER: A tapped, ¼-wave transmission line.

LINE BALANCE

A parallel-conductor transmission line must be balanced in order to function properly. *Line balance* is achieved when the currents in the two conductors are equal in magnitude, but opposite in direction, at every point along the line.

In an antenna system, line balance can be difficult to obtain because of interaction between the line and the radiating part of the antenna. In a center-fed dipole antenna, the open-wire or twin-lead feed line should be oriented at a right angle to the radiating element. The two halves of the radiating element must be exactly the same electrical length, and they must present the same impedance at the feed point. In an antenna system, poor line balance results in radiation from the feed line.

LINE FAULT

A *line fault* is an open or short circuit in a transmission line, resulting in partial or complete loss of signal or power at the output end of the line. An open circuit in a line can be located by shorting the terminating, or output, end of the line, and measuring the circuit continuity at various points with an ohmmeter or other device. Measurements should be taken first at the input end of the line; subsequent checks should be made at points closer and closer to the output end. The resistance will appear infinite, or very large, until the testing apparatus is moved past the line fault. Then, the resistance will abruptly drop.

A short circuit in a line can be found without breaking the circuit if a high-frequency signal is applied at the input. An indicating device, which can consist of a meter with a radio-frequency diode in series, is moved back and forth along the line. The meter produces readings that fluctuate, depending on the distance from the input, until the short is passed. Beyond the short, the meter reading drops to nearly zero and remains there as the meter is moved further.

LINE FILTER

A *line filter* is a device that can be placed in the alternating-current power-supply cord of an electronic device. Line filters generally consist of series inductors and/or parallel capacitors (see illustration).

A line filter is useful for eliminating transient spikes in utility-line voltage. Such spikes can cause an electronic device to malfunction; in some cases a spike can result in permanent damage. Line filters are also helpful in reducing electromagnetic noise from alternating-current power lines. *See also* POWER-LINE NOISE, and TRANSIENT.

LINE LOSS

Line loss is the dissipation of power in, or radiation of power from, a transmission line. Given a line-input power of P_1 watts,

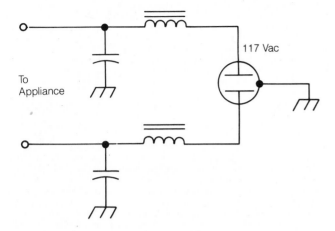

LINE FILTER: A typical utility-line filter.

and an output power of P_2 watts, the loss power is the difference $P_1 - P_2$.

Line loss can result from any or all of several different factors. The conductors of any transmission line have some resistance, known as *ohmic loss*. The dielectric material between the conductors of a transmission line produces some loss. Conductor losses and dielectric losses are manifested in the form of heat. If the line balance is poor, power is lost by radiation from the line.

Line loss is generally specified in decibels per unit length. For a particular transmission line, the loss usually increases as the frequency increases. The decibel value, or proportion, of line loss does not depend on the power level of the applied signal, as long as the power does not exceed the maximum rating for the line. In a radio-frequency feed line, the loss increases as the standing-wave ratio increases. *See also* DECIBEL, and STANDING-WAVE-RATIO LOSS.

LINE-OF-SIGHT COMMUNICATION

Radio communication by means of the direct wave is sometimes called *line-of-sight communication*. The range of line-of-sight communication depends on the height of the transmitting and receiving antennas above the ground, and on the nature of the terrain between the two antennas.

Line-of-sight communication is the primary mode at microwave frequencies. Although the range is obviously limited in this mode, propagation is virtually unaffected by external parameters, such as ionospheric or tropospheric disturbances. Line-of-sight communication range is limited to the radio horizon. *See also* DIRECT WAVE, and HORIZON.

LINE PRINTER

A *line printer* is a device that prints the output of a computer on paper, line by line. The normal line lengths are 80 and 132 characters. Line printers are similar to teletype receiving devices. Such printers can operate on the dot-matrix principle, or they can use ball or daisy-wheel devices. The dot-matrix printer is fastest. *See also* DAISY-WHEEL PRINTER, DOT-MATRIX PRINTER, and PRINTER.

LINES OF FLUX

Electric or magnetic fields are sometimes theoretically described as consisting of *lines of flux*. The lines of flux in a field run in the general direction of the field effect. The field is considered to originate at one end of each line of force, and to terminate at the other end of the same line. The lines of flux converge at the electric charge centers or magnetic poles.

Lines of flux do not physically exist. They simply represent a certain quantity of electric or magnetic field flux. *See also* ELECTRIC FLUX, FLUX, and MAGNETIC FLUX.

LINE TRAP

Any band-rejection filter, used in a transmission line for the purpose of notching out signals at a certain frequency or frequencies, is called a *line trap*. Line traps can be installed in various different ways.

A line trap can be used to suppress the harmonic output of a radio transmitter. The trap is simply tuned to the harmonic frequency. In a receiving antenna system, a line trap is sometimes used to reduce the level of a strong local signal that would otherwise overload the front end of the receiver.

A quarter-wavelength section of transmission line, either coaxial or parallel-wire, can be used as a series-resonant trap. One end of the quarter-wavelength section is connected to the receiver or transmitter antenna terminals along with the feed line, and the other end is simply left open. This kind of trap is sometimes called a *line trap* because it is actually constructed from transmission line. At the resonant frequency, this device appears as a short circuit across the input terminals. *See also* TRAP.

LINE TUNING

Line tuning is a method of tank-circuit construction that is sometimes used at ultra-high and microwave frequencies. A pair of parallel conductors, or a length of coaxial line, is used as a parallel-resonant circuit. This is accomplished by short circuiting one end of a quarter-wavelength section of line (see illustration). The other end of the line then attains the characteristics of a parallel-resonant, inductance-capacitance circuit.

The resonant line can be tuned by means of a movable shorting bar, or the frequency can be fixed. A tuned line has better frequency stability than an inductor-capacitor arrangement at ultra-high and microwave frequencies. However, the quarter-wave tuned line is resonant at odd harmonic frequencies as well as at the fundamental frequency.

A quarter-wavelength section of line can be left open at the far end, resulting in the equivalent of a series-resonant inductance-capacitance circuit. Half-wavelength sections of line are also sometimes used for tuning purposes. A shorted half-wavelength line acts as a series-resonant circuit; if the far end is open, the half-wavelength line acts as a parallel-resonant tank.

A tuned line can be used to eliminate undesired signals in an antenna system. Such a device is called a *line trap*. *See also* LINE TRAP.

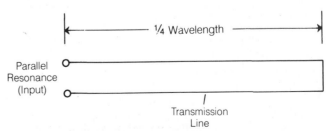

LINE TUNING: A resonant length of transmission line.

LINK

A *link* is a connection, via wire, radio, light beams, or other medium, between two circuits within a larger system. A radio or television broadcasting station can use a link between the studio and the transmitter. A computer can be linked to other computers for the purpose of transferring data or to obtain more memory. In general, any one-way or two-way communications path can be called a *link*.

When two circuits are coupled loosely by means of a pair of inductive transformers, the arrangement is called a *link* or *link coupling*. *See also* LINK COUPLING.

LINK COUPLING

Link coupling is a means of inductive coupling. The output of one circuit is connected to the primary winding of a step-down transformer. The secondary winding of the transformer is directly connected to the primary winding of a step-up transformer. The secondary of the step-up transformer provides the input signal for the next stage or circuit. The illustration is a schematic diagram of link coupling.

The smaller, intermediate windings in a link-coupling arrangement usually have just one or two turns. As a result, there is very little capacitance between the two stages. This minimizes the transfer of unwanted harmonic energy. Link coupling also minimizes loading effects in the output of the first stage.

Test instruments are sometimes connected to a circuit by means of link coupling. Because the capacitance is so small, the presence of the test instrument does not significantly affect the circuit under test. This is important in radio-frequency work, especially at the very-high frequencies and above.

Link coupling can be used to generate feedback in an amplifier system. The feedback can be positive, resulting in oscillation; negative feedback can be used for neutralization. *See also* LINK FEEDBACK.

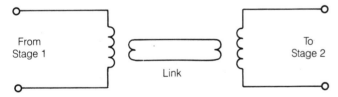

LINK COUPLING: Minimizes capacitance between stages.

LINK FEEDBACK

Link feedback is the use of link coupling between the output and input of an amplifier circuit. The collector, drain, or plate transformer winding is inductively coupled, by means of a link, to the base, gate, or grid winding.

Link feedback can be either positive (regenerative) or negative (degenerative). Positive feedback results in oscillation if the coupling is tight enough. Positive link feedback can be used to make a regenerative receiver. Negative link feedback is sometimes used in radio-frequency power amplifiers for the purpose of neutralization. *See also* LINK COUPLING, NEUTRALIZATION, and REGENERATIVE DETECTOR.

LINK LAYER

See OPEN SYSTEMS INTERCONNECTION REFERENCE MODEL.

LIQUID-CRYSTAL DISPLAY

A *liquid-crystal display (LCD)* is a solid-state, flat-screen device that can show geometric shapes. The simplest LCDs are used as alphanumeric displays in electronic calculators, digital meters, wristwatches, radio receivers, transmitters and receivers. More sophisticated LCDs are used in personal computers (particularly the laptop/notebook variety), and in portable television receivers.

Principles of Operation   All LCDs consist, as their name indicates, of fluid that changes its light-transmitting and light-reflecting properties in regions. The fluid is confined between transparent charged plates. When a voltage is applied to these electrodes, the resulting electric field causes a change in the molecules of the liquid. This changes the way in which light passes through the display, within the region containing the electric field.

Modern LCDs can change not only from light to dark (so-called "black on gray"), but can also exhibit some colors. There is always a limit to how fast the liquid can change state. In recent years, LCDs have been developed that are fast enough, at room temperature, to function as picture displays in miniature fast-scan television sets. The speed of the LCD is affected by the temperature; extreme cold causes the LCD to change state more slowly.

Advantages of the LCD   One of the most significant advantages of the LCD is that it draws very little current. This makes it ideal for digital watches, hand-held radio transceivers and laptop computers, where battery life must be maximized. The display is essentially an electrostatic device, consuming minimal power to change state, and practically zero power to maintain a given state.

Another advantage of the LCD is that it is easy to read in bright sunlight. The older light-emitting-diode (LED) or Nixie-tube displays were plagued by visibility problems under conditions of varying illumination. The LCD can be backlit to allow it to be read in total darkness. Therefore, a good LCD can be easily read in light of any intensity, direct or indirect.

Still another advantage, especially important in recent years, is fineness of detail, or resolution, comparable to that of cathode-ray tubes. This has made it possible to do graphics on laptop computers, and has even resulted in the evolution of the "wrist TV."

Problems with LCDs   One of the main problems of the LCD is that it needs to be backlit if the operating environment is dark. This cancels out the benefit of low current drain, because the lighting itself takes current. This can be minimized by shutting the feature off when there is sufficient illumination to read the display without backlighting.

Another problem is that, from certain angles, the LCD can be hard to read. This problem has been reduced in recent years, but in equipment using early type LCDs, the display is illegible when observed from some directions.

Another disadvantage of the LCD is that it is susceptible to cold. The speed can be affected to such an extent that equipment is hard or impossible to operate. This can be an annoyance in mobile and portable ham radio equipment during the wintertime.

LISSAJOUS FIGURE

A *Lissajous figure* is a pattern that appears on the screen of an oscilloscope when the horizontal and vertical signal frequencies are integral multiples of some base frequency.

When the horizontal and vertical signals are equal in amplitude and frequency and differ in phase by 90 degrees, a circle appears on the display. When the vertical signal has twice the frequency of the horizontal signal, a sideways figure-8 pattern is observed. If the vertical signal has half the frequency of the horizontal signal, an upright figure-8 pattern is traced. Different frequency combinations produce different patterns. The shapes of the ellipses vary, depending on relative amplitude and/or phase angle.

Lissajous figures are useful for precise adjustment of variable-frequency oscillators against a reference oscillator with a known audio frequency. Using an oscilloscope and a reference oscillator, Lissajous figures simplify such procedures as the adjustment of Private-Line® (PL) oscillators and Touchtone® encoders. A fixed pattern indicates a harmonic zero-beat condition. *See also* OSCILLOSCOPE, and ZERO BEAT.

LITHIUM CELL

A *lithium cell* is an electrochemical cell, in which the positive electrode is fabricated from lithium. The lithium cell produces approximately 2.8 volts. Lithium cells are noted for their excellent shelf life, and are ideally suited for low-current applications over long periods of time. Lithium cells are becoming more and more common in electronic clocks and watches, and as memory-backup power supplies for microcomputer devices. *See also* CELL.

LITZ WIRE

At radio frequencies, current tends to flow mostly near the outside surface of a conductor. This is called *skin effect* (*see* SKIN EFFECT). For direct current and low-frequency alternating current, the conductivity of a wire is proportional to the cross-sectional area, the square of the wire diameter. However, at high frequencies, the conductivity is directly proportional to the diameter of the wire. At high frequencies, therefore, it makes better sense to maximize the conductor surface area, rather than to simply maximize the cross-sectional area of the wire. Litz wire is designed with this principle in mind.

Litz wire consists of several individual, enameled conductors, interwoven in a special way. The wire is fabricated so that all inner strands come to the outside at regular intervals, and all outer strands come to the center at equal intervals. Litz wire is basically a stranded wire in which the conductors are insulated from each other.

Litz wire exhibits low losses at radio frequencies because the conducting surface area is much greater than that of an ordinary solid wire of the same diameter.

L NETWORK

An *L network* is a form of filter section that is used in unbalanced circuits. The L network gets its name from the fact that the schematic-diagram component arrangement resembles the capital letter L (see illustration).

The L configuration can be used in the construction of highpass or lowpass filters. The series component can appear either before or after the parallel component. In the illustration, a typ-

ical highpass L network is shown at A; a lowpass network is shown at B. Several L networks can be cascaded to obtain a sharper cutoff response, as shown at C.

The L network is one of several different configurations for selective filters. *See also* H NETWORK, PI NETWORK, and T NETWORK.

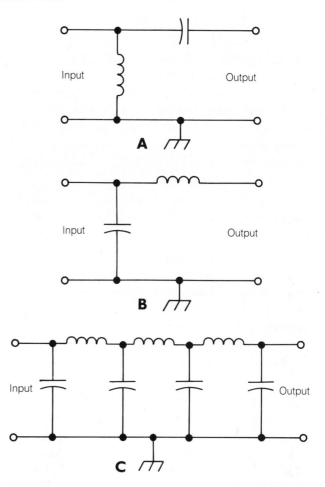

L NETWORK: Highpass (A), lowpass (B), and cascaded lowpass (C).

LOAD

A *load* is any circuit that dissipates, radiates, or otherwise makes use of power. A direct-current load exhibits a definite resistance. This resistance can vary with the amount of power applied. An alternating-current load exhibits resistance, and can also show capacitive or inductive reactance. Examples of loads include audio speakers, electrical appliances, and radio-frequency transmitting antennas.

For optimum transfer of power from a circuit to its load, the load impedance should be the same as the output impedance of the circuit. *See also* IMPEDANCE MATCHING, LOAD IMPEDANCE, and LOADING.

LOAD CURRENT

The amount of current flowing in a load is called the *load current*. For a direct-current load, the current (I) is given by the formula:

$$I = E/R$$

where E is the voltage across the load, and R is the resistance of the load.

For an alternating-current load, the current depends on several different factors. In general, given a load impedance (Z), and a root-mean-square voltage (E) across the load (see ROOT MEAN SQUARE), the current (I) is given by:

$$I = E/Z$$

The load current depends on the power supplied to the load and on the load impedance. *See also* LOAD, and LOAD IMPEDANCE.

LOADED ANTENNA

The natural resonant frequency of an antenna can be changed by placing reactance in series with the radiating element. An antenna that has such a reactance in its radiating element is called a *loaded antenna*.

The most common type of loaded antenna has a resonant frequency that has been lowered by the series installation of an inductor. This type of loaded antenna is often used for mobile operation at medium and high frequencies. A capacitance hat (loading disk) can be placed at the end or ends of such an antenna to increase the bandwidth.

The series connection of capacitors raises the resonant frequency of an antenna element. This kind of loaded antenna is sometimes used at very-high and ultra-high frequencies. *See also* CAPACITIVE LOADING, and INDUCTIVE LOADING.

LOADED LINE

The characteristic impedance of a transmission line depends on the distributed capacitance and inductance. Any transmission line can be theoretically represented as a set of series inductors and parallel capacitors, although in reality these reactances are not discrete components.

Sometimes, it is necessary to change the characteristics of a transmission line at a certain location. This is done by the installation of capacitors and/or inductors in the line. The result of adding reactance to a transmission line is a change in the resonant frequency of a line. A transmission line in which reactance is deliberately added is called a *loaded line*. A line can be loaded for the purpose of matching the characteristic impedance to the output impedance of a radio transmitter. Line loading can also be used to match the characteristic impedance to the impedance of an antenna or other load. *See also* CHARACTERISTIC IMPEDANCE, FEED LINE, IMPEDANCE MATCHING, and LOADING.

LOADED Q

When a tuned circuit, or reactive component (such as an inductor or capacitor), has no other component connected, the circuit is *unloaded*. But when a resistive load is connected to the circuit, the Q factor is lowered. This is the *loaded Q*.

In practice, connecting a load to a tuned circuit causes the resonance curve to broaden out. The resonance curve of an unloaded, tuned circuit is extremely sharp. There is always a little resistance, in the form of loss, even in an unloaded circuit. Otherwise the resonance curve would be infinitely sharp (zero bandwidth).

An example of loaded Q, as compared with unloaded Q, is the effect of connecting an antenna to the output of a transmatch. Without any antenna, the transmatch tunes sharply (if at all), because the Q factor is theoretically infinite. But when an

antenna, containing resistance and perhaps also reactance, is hooked up, the tuning becomes more broad, because of the effect of the resistive part of the load. *See also* Q FACTOR.

LOAD IMPEDANCE

Any power-dissipating or power-radiating load has an impedance. Part of the impedance is a pure resistance, and is given in ohms. Reactance can also be present; it is also specified in ohms. A direct-current load, or a resonant alternating-current load, has resistance but no reactance. A nonresonant alternating-current load can have capacitive or inductive reactance.

The *load impedance* should always be matched to the impedance of the power-delivering circuit. Normally, this means that the load contains no reactance (that is, the load is resonant), and the value of the load resistance is equal to the characteristic impedance of the feed line. If the load impedance is not the same as the characteristic impedance of the line, special circuits can be installed at the feed point to change the effective load impedance.

If the load impedance differs from the feed-system impedance, the load will not absorb all of the delivered power. Some of the electromagnetic field will be reflected back toward the source. This results in an increased amount of power loss in the system. The magnitude of this loss increase might or might not be significant. *See also* IMPEDANCE, IMPEDANCE MATCHING, and REFLECTED POWER.

LOADING

The deliberate insertion of reactance into a circuit is called *loading*. Generally, this is done for the purpose of obtaining resonance. If a load contains capacitive reactance, then an equal amount of inductive reactance must be added to obtain resonance. Conversely, if a load contains inductive reactance, an equal amount of capacitive reactance must be added.

A common example of loading is the insertion of an inductor in series with a physically short antenna. A short antenna has capacitive reactance, and the loading inductor provides an equal and opposite reactance. The larger the value of the loading inductance, the lower the resonant frequency becomes. A capacitor can be inserted in series with a physically long antenna to obtain resonance. *See also* CAPACITIVE LOADING, INDUCTIVE LOADING, LOADED ANTENNA, LOADED LINE, and RESONANCE.

LOAD LOSS

When power is applied to a load, the load ideally dissipates all of the power in the manner intended. For example, a light bulb should produce visible light, but no heat; a radio antenna should radiate all of the energy it receives, and waste none as heat. However, in practice, no load is perfect. If P watts are delivered to a load, and the load dissipates Q watts in the manner intended, then the value $P - Q$ is called the *load loss*. Load loss can be expressed in decibels by the formula:

$$Loss \text{ (dB)} = 10 \log_{10} (P/Q)$$

Some loads exhibit large losses. The incandescent bulb is an example of a lossy load. Other loads have very little loss; a half-wave dipole antenna in free space is an example of an efficient load. *See also* LOAD.

LOAD POWER

Load power can be defined in two ways. The load input power is the amount of power delivered to a load. The load output power is the amount of power dissipated, in the desired form, by the load. The load output power is always less than the load input power; their difference is called the *load loss* (*see* LOAD LOSS).

If the load impedance is a pure resistance, then the load input power (*P*) can be determined in terms of the root-mean-square load current (*I*) and the root-mean-square load voltage (*E*) by the formula:

$$P = EI$$

If the purely resistive load impedance is *R* ohms, then:

$$P = E^2/R = I^2R$$

When reactance is present in a load, the load input power differs from the incident power, as determined by the current and voltage. This is because some power appears in imaginary form across the reactance. *See also* APPARENT POWER, REACTIVE POWER, and TRUE POWER.

LOAD VOLTAGE

The voltage applied across a load is called the *load voltage*. For a direct-current load, the load voltage (*E*) can be determined according to the formula:

$$E = IR$$

where *I* is the current through the load and *R* is the resistance of the load.

For an alternating-current load, the voltage depends on several different things. In general, given a load impedance (*Z*) and a root-mean-square current (*I*) through the load (*see* ROOT MEAN SQUARE), the root-mean-square voltage (*E*) is:

$$E = IZ$$

The load voltage depends on the power supplied to the load, and also on the load impedance. *See also* LOAD, and LOAD IMPEDANCE.

LOBE

In the radiation pattern of an antenna, a *lobe* is a local angular maximum. An antenna pattern can have just one lobe, or it can have several lobes. Different lobes can have different magnitudes. The strongest lobe is called the main or major lobe. The weaker lobes are called *secondary* or *minor lobes*. *See also* ANTENNA PATTERN.

Lobes occur in the directional responses of various kinds of transducers, such as microphones and speakers. *See also* DIRECTIONAL MICROPHONE.

LOCAL LOOP

In a teletype system, the *local loop* is an arrangement that allows paper tapes to be run through the machine for checking. In modern, computerized systems, the information is usually stored in a solid-state memory bank, rather than on paper tape. By running the data through the machine locally, the station operator can correct errors in the text before they are transmitted. The local loop does not involve connection to an external line or circuit; it is contained entirely at one place. *See also* RADIOTELETYPE, and TELETYPE.

LOCAL OSCILLATOR

A *local oscillator* is a radio-frequency oscillator, the output of which is not intended for direct transmission over the air. All of the oscillators in a superheterodyne receiver are local oscillators. In transmitters, local-oscillator signals can be combined in mixers, amplified, perhaps multiplied, and then sent over the air.

Local-oscillator design requires excellent shielding and isolation so that the output will not cause interference to nearby radio receivers. *See also* OSCILLATOR, and SUPERHETERODYNE RECEIVER.

LOCKUP TIME

Lockup time is a general term, used with circuits or devices in which one part, or component, automatically follows another. The term is common in robotics, but also has some applications in amateur radio. As electromechanical and master-slave systems become more common in ham radio, lockup time will be an increasingly important consideration.

Lockup time is significant in all servo systems. In a servo system, it takes a certain amount of time for the slave device to catch on and "lock into" the master device. An example of this is a selsyn, such as might be used as an antenna-direction indicator or in a remote-control antenna tuner. *See* SELSYN, and SERVO SYSTEM.

In most electrical devices, a fast lockup time is wanted, so there will not be unnecessary and inconvenient delays in the functioning of the equipment. But in some electromechanical systems, such as automatic direction finders, there needs to be some delay. Otherwise hunting can occur (*see* AUTOMATIC DIRECTION FINDER, and HUNTING).

The time required for a *phase-locked loop* to stabilize is sometimes called the *lockup time*. Phase-locked loops are extensively used for frequency control in ham radio equipment. *See* PHASE-LOCKED LOOP.

LOFTIN-WHITE AMPLIFIER

A two-transistor audio amplifier circuit, in which the output of the first stage is directly coupled to the input of the second stage, is called a *Loftin-White amplifier*. Either bipolar or field-effect transistors can be used.

The Loftin-White circuit is characterized by a multielement resistive voltage divider that provides bias for the emitters or sources of both active devices and for the collector or drain of the first device.

LOG

A *log* is a record of the operation of a radio station. Logs generally include such information as the date and time of a transmission, the nature of the transmission, the mode of emission, the power input or output of the transmitter, the frequency of transmission, and the message traffic handled. In some situations, logs must be kept by law. Logging requirements are determined, in the United States, by the Federal Communications Commission. The mathematical logarithm operation is sometimes abbreviated as "log."

LOGARITHMIC SCALE

When a coordinate scale is calibrated according to the logarithm of the actual scale displacement, the scale is a *logarithmic scale*.

Such scaling is universally done, according to the base-10 logarithm of the distance.

Logarithmic scales can be used in a Cartesian coordinate system for one axis or both axes. Logarithmic scales are especially useful in plotting amplitudes when the range is expected to vary over many orders of magnitude. Any decibel scale is a logarithmic scale; some meters and oscilloscope displays are calibrated in this way. *See also* DECIBEL.

LOGARITHMIC TAPER

Some controls have a value or function that varies with the logarithm of the angular displacement, rather than directly with the angular displacement. Such controls are said to have a *logarithmic taper*. Logarithmic-taper potentiometers are fairly common in audio-frequency applications and in brightness controls.

The human senses perceive disturbances approximately according to the logarithm of their intensity. Therefore, a linear-taper control is often not suitable for adjustment of variables perceived by the senses, such as loudness and brightness. A logarithmic-taper device compensates for human perception, making the control "seem" linear. *See also* AUDIO TAPER, and LINEAR TAPER.

LOGGING

The Federal Communications Commission once required all ham radio stations to keep detailed logs of all transmissions and contacts. This included fixed, mobile, and portable operations. These requirements no longer apply, but it is still a good idea to keep a station log whenever possible. There are several reasons for this.

First, you might need such logs for verification of contacts, for QSL checking, and in contesting and traffic handling. *See* CONTESTS AND CONTESTING, DX AND DXING, QSL CARD, and TRAFFIC HANDLING.

Second, if you have problems with *electromagnetic interference (EMI)*, you will find a log invaluable for determining the frequencies (if any!) on which your transmissions cause problems. *See* ELECTROMAGNETIC INTERFERENCE.

Third, a station log gives you concrete, and lasting, records of your ham "travels." For many hams, this is a matter of pride. There's nothing quite like looking at the old, yellowed pages of a log you wrote a quarter of a century before.

Station logs generally should include the date and time, the frequency or band used, the emission type, and the power output of the transmitter. It's also good to include the callsigns, locations and operator names for all stations contacted. Some hams include one-way transmissions (CQs), and calls not resulting in contacts, in their logs for completeness.

Optional information, often of interest to you later, might include the antenna type ("new dipole running north/south up 50 feet"), propagation conditions ("band wide open"), and other variables.

Standard log books and log sheets are available from the American Radio Relay League, 225 Main Street, Newington, CT 06111. You can also use a computer to keep your log. Various companies provide software for this purpose. Computerized logs are especially convenient for contesters and DXers. Consult the most recent issues of ham magazines for ads.

LOGIC

Digital computers, and many other types of digital electronic devices, work using data whose bits can have either of two states. These states are called *high* and *low*. The operations that are done with this data are known collectively as *logic* or *digital logic*.

The high state is usually called *logic 1*, or *truth*, and the low state is called *logic 0*, or *falsity*. This scheme is *positive logic*. If logic 1 is low and logic 0 is high, the scheme is known as *negative logic*. Either system will work just as well as the other; it's mostly a matter of semantics.

The highs and lows of digital logic are combined via various operations. These include NOT, inclusive OR (usually called OR), exclusive OR (abbreviated XOR), NOT OR (abbreviated NOR), AND, and NOT AND (abbreviated NAND). The truth table illustrates these operations for positive logic 1's and 0's. The inputs are X and Y, and the output is Z.

By examining the table, and by substituting "true" for "1" and "false" for "0," you can see that the operations are so named because the outputs are derived from the well-known operations of mathematical logic. For example, X AND Y can be true only if both X and Y are true; otherwise, the statement X AND Y must be false.

Combining logic operations lets sophisticated processes be done with data that has, ultimately, bits that are either 1 or 0. Related articles in this encyclopedia include AND GATE, BOOLEAN ALGEBRA, DE MORGAN'S THEOREM, DIGITAL, DIGITAL CIRCUITRY, EXCLUSIVE-OR GATE, INVERTER, LOGIC CIRCUIT, LOGIC DIAGRAM, LOGIC EQUATION, LOGIC FUNCTION, LOGIC GATE, LOGIC PROBE, NAND GATE, NOR GATE, OR GATE, and TRUTH TABLE.

LOGIC CIRCUIT

A combination of logic gates is called a *logic circuit*. Logic circuits vary from the simplest (a single inverter, or NOT gate) to extremely complex arrays.

A logic circuit is designed to perform a certain logic function. That is, each combination of inputs always produces the same output or outputs. Logic circuits are often fabricated onto integrated circuits; sometimes a single integrated circuit contains several different logic circuits. *See also* LOGIC FUNCTION, and LOGIC GATE.

LOGIC: TRUTH TABLE FOR VARIOUS LOGIC OPERATIONS.

X	Y	NOT X	X OR Y	X XOR Y	X NOR Y	X AND Y	X NAND Y
0	0	1	0	0	1	0	1
0	1	1	1	1	0	0	1
1	0	0	1	1	0	0	1
1	1	0	1	0	0	1	0

LOGIC DIAGRAM

A *logic diagram* is a schematic diagram of a logic circuit. The complete logic diagram shows the interconnection of individual gates. The complexity of the logic diagram depends on the function that the circuit performs.

The illustration is an example of a logic diagram. This circuit performs the logic function shown by the accompanying truth table. There are other possible logic circuits that will accomplish this same logic function, and the diagrams of those circuits are much different from the one illustrated. *See also* LOGIC CIRCUIT, LOGIC FUNCTION, and LOGIC GATE.

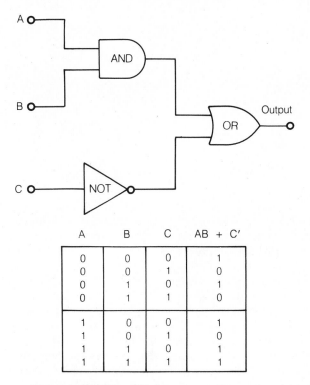

A	B	C	AB + C'
0	0	0	1
0	0	1	0
0	1	0	1
0	1	1	0
1	0	0	1
1	0	1	0
1	1	0	1
1	1	1	1

LOGIC DIAGRAM: A logic circuit and truth table.

LOGIC EQUATION

A *logic equation* is a symbolic statement showing two logically equivalent Boolean expressions on either side of an equal sign (*see* BOOLEAN ALGEBRA). The logic equation $A = B$, where A and B are Boolean expressions, means literally "A is true if B is true, and B is true if A is true." Mathematicians would say, "A if and only if B."

Logic equations are important in electronics. A particular logic circuit, intended to accomplish a given logic function, can often be assembled from logic gates in several different ways. One of these arrangements is usually simpler than all of the others, and is therefore most desirable from an engineering standpoint.

Many logic equations are known to be true, and are called *theorems of Boolean algebra*. By applying these theorems, the digital engineer can find the simplest possible logic circuit for a given function. *See also* LOGIC CIRCUIT, and LOGIC FUNCTION.

LOGIC FUNCTION

A *logic function* is an operation, consisting of one or more input variables and one output variable. The logic function is a simple form of mathematical function in which the input variables can achieve either of two states in various combinations. A logic function can be written in the form:

$$F(x_1, x_2, x_3, \ldots, x_n) = y$$

where x_1 through x_n represent the input variables and y represents the output. The values of the variables can be either 0 (representing falsity) or 1 (representing truth). If there are n input variables, then there are 2^n possible input combinations.

Every logic circuit performs a particular logic function. Logic functions can be written in Boolean form. Most logic functions can be represented in several different Boolean forms. Logic functions can also be written as a truth table, or a listing of all possible input combinations along with the corresponding output states. Logic functions can also be written in the form of a schematic diagram of the logic circuit, showing the interconnection of gates. *See also* BOOLEAN ALGEBRA, LOGIC CIRCUIT, LOGIC DIAGRAM, LOGIC EQUATION, LOGIC GATE, and TRUTH TABLE.

LOGIC GATE

A *logic gate* is a simple logic circuit. The gate performs a single AND, NOT, or OR operation, perhaps preceded by a NOT operation. Logic gates are shown schematically in the illustration. The AND gate, at A, produces a high output only if all of

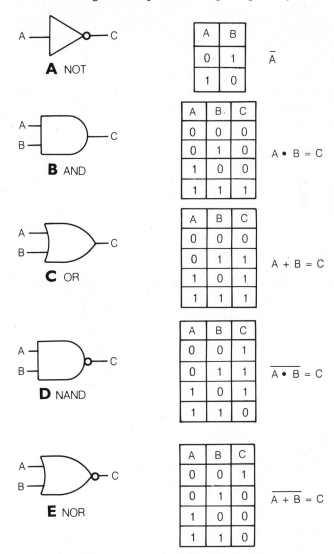

LOGIC GATE: A: NOT. B: AND. C: OR. D: NAND. E: NOR.

the inputs are high. The NOT gate or inverter (B) simply changes the state of the input. The OR gate, at C, produces a high output if at least one of the inputs is high. The exclusive-OR gate, shown at D, produces a high output if and only if the two inputs are different. The AND gate can be followed by an inverter, resulting in a device called a *NAND gate* (E). The OR gate can be followed by an inverter to form a NOR gate (F).

Logic gates are combined to perform complicated logical operations. This is the basis for all of digital technology. *See also* AND GATE, EXCLUSIVE-OR GATE, INVERTER, LOGIC CIRCUIT, LOGIC FUNCTION, NAND GATE, NOR GATE, and OR GATE.

LOGIC PROBE

Troubleshooting and testing of logic circuits is, is some ways, easier than working with analog circuits. This is because digital logic devices generally have only two possible states: *high (logic 1)*, and *low (logic 0)*.

In analog circuits, a meter is needed to read the level of voltage or current at a given point. But in a digital circuit, a two-state meter, known as a *logic probe*, shows whether the point under test is high or low.

If a point in a digital circuit changes between high and low rapidly and repeatedly, a logic probe will not work. If the signal at such a point needs checking, an oscilloscope is the best instrument to use. *See also* DIGITAL, DIGITAL DEVICE, LOGIC, LOGIC CIRCUIT, LOGIC FUNCTION, and LOGIC GATE.

LOG-PERIODIC ANTENNA

A *log-periodic antenna*, also known as a *log-periodic dipole array (LPDA)*, is a special form of unidirectional, broadband antenna. The log-periodic antenna is sometimes used in the high-frequency and very-high-frequency parts of the radio spectrum, for transmitting or receiving.

The log-periodic antenna consists of a special arrangement of driven dipoles, connected to a common transmission line. The illustration is a schematic diagram of the general form of a log-periodic antenna. The design parameters are beyond the scope of this discussion, since they vary depending on the gain and bandwidth desired. In general, the elements become shorter nearer the feed point (forward direction) and longer toward the back of the antenna. Notice that the element-interconnecting line is twisted 180 degrees between any two adjacent elements.

The log-periodic antenna exhibits a fairly constant input impedance over a wide range of frequencies. Typically, the antenna is useful over a frequency spread of about 2 to 1. The forward gain of the log-periodic antenna is comparable to that of a two-element or three-element Yagi. The gain can be increased slightly by slanting the elements forward. The gain is also proportional to the number of elements in the antenna.

The log-periodic antenna is especially useful in situations where a continuous range of frequencies must be covered. This is the case, for example, in television receiving. Therefore, many (if not most) commercially manufactured very-high-frequency television receiving antennas are variations of the log-periodic design.

LONG-PATH PROPAGATION

At some radio frequencies, worldwide communication is possible because of the effects of the ionosphere. Radio waves travel

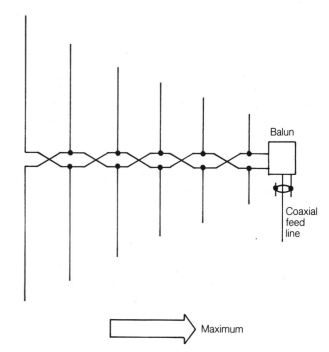

LOG-PERIODIC ANTENNA: General configuration of LPDA.

over, or near, great-circle paths across the globe (*see* GREAT CIRCLE). There are two possible paths over which the radio waves can propagate between two points on the surface of the earth. Both paths lie along the same great circle, but one path is generally much longer than the other.

When the radio waves travel the long way around the world to get from one point to the other, the effect is called *long-path propagation*. Long-path propagation is more common between widely separated points than between nearby points. Long-path propagation always occurs via the F layer, or highest layer, of the ionosphere. Whether the signal will take the long path or the short path depends on the ionospheric conditions at the time. In some cases, a signal will propagate via both paths. For two points located exactly opposite each other on the globe, all paths are the same length, and the signal can travel over many different routes at the same time.

Under some conditions, long-path propagation occurs all the way around the planet. When this happens, the signal from a nearby transmitter can seem to echo: The signal is heard first directly, and then a fraction of a second later as the long-path wave arrives. *See also* F LAYER, IONOSPHERE, PROPAGATION CHARACTERISTICS, and SHORT-PATH PROPAGATION.

LONGWIRE ANTENNA

A wire antenna that measures 1 wavelength or more, and is fed at a current loop or at one end, is called a *longwire antenna*. Longwire antennas are sometimes used for receiving and transmitting at medium and high frequencies.

Longwire antennas offer some power gain over the half-wave dipole antenna. The longer the wire, the greater the power gain. For an unterminated longwire antenna that measures several wavelengths, the directional pattern resembles A in the illustration. As the wire is made longer, the main lobes get more nearly in line with the antenna, and their amplitudes increase. As the wire is made shorter, the main lobes get farther from the axis of the antenna, and their amplitudes decrease. If the longwire is terminated at the far end (opposite the feed point), half of the pattern disappears, as at B.

Longwire antennas have certain advantages. They offer considerable gain and low-angle radiation, provided that they are long enough. The graph (C) shows the theoretical power gain, with respect to a dipole, that can be realized with a longwire antenna. The gain is a function of the length of the antenna. Longwire antennas are inexpensive and easy to install, provided that there is sufficient real estate. The longwire antenna must be as straight as possible for proper operation.

There are two main disadvantages to the longwire antenna. First, it cannot conveniently be rotated to change the direction in which maximum gain occurs. Second, a great deal of space is needed, especially in the medium-frequency spectrum and the longer high-frequency wavelengths. For example, a 10-wavelength longwire antenna measures about 1,340 feet at 7 MHz. This is more than a quarter of a mile. *See also* WIRE ANTENNA.

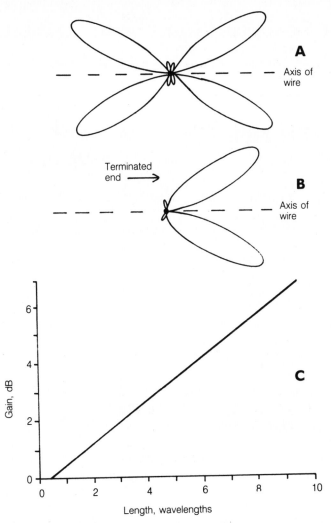

LONGWIRE ANTENNA: At A, directional pattern for unterminated longwire. At B, directional pattern for terminated longwire. At C, gain versus length.

LOOP

In electronics, the term *loop* can have any of several different meanings. A local current or voltage maximum on an antenna or transmission line is called a *loop* (see CURRENT LOOP, and VOLTAGE LOOP).

A closed signal path in an amplifier, oscillator, or other cir-

cuit can be called a loop. Examples are the feedback loop in an oscillator, and the loop in an operational-amplifier circuit (*see* FEEDBACK).

In a computer program, a loop is a portion of the program through which the computer cycles many times. The loop ends when some condition is satisfied.

A single-turn coil of wire is called a *loop*. A large coil, having one or more turns and used for the purpose of receiving or transmitting radio signals, is called a *loop* or *loop antenna*. *See also* LOOP ANTENNA.

LOOP ANTENNA

Any receiving or transmitting antenna, consisting of one or more turns of wire forming a direct-current short circuit, is called a *loop antenna*. Loop antennas can be categorized as either small or large.

Small loops have a circumference of less than 0.1 wavelength at the highest operating frequency. Such antennas are suitable for receiving, and exhibit a sharp null along the loop axis. The small loop can contain just one turn of wire, or it can contain many turns. The loop can be electrostatically shielded to improve the directional characteristics. The ferrite-rod antenna is a form of small loop in which a ferromagnetic core is used to enhance the signal pickup (*see* FERRITE-ROD ANTENNA). Small loops are most often used for direction finding and for eliminating manmade noise or strong local interfering signals. Small loops can be used to transmit if they are expertly engineered (see photo).

Large loops have a circumference of either 0.5 wavelength or 1 wavelength at the operating frequency. The half-wavelength loop presents a high impedance at the feed point, and the maximum radiation occurs in the plane of the loop. The full-wavelength loop presents an impedance of about 50 ohms at the feed point, and the maximum radiation occurs perpendicular to the plane of the loop. A large loop can be used either for transmitting or receiving. The half-wavelength loop exhibits a

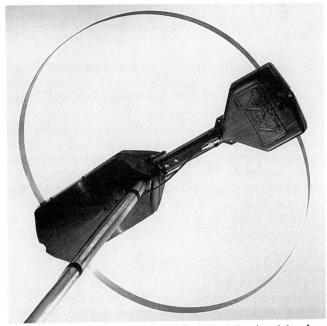

LOOP ANTENNA: A loaded, tuned transmitting/receiving loop covering 14 to 30 MHz continuous, rated at 150 W. This antenna is useful for limited-space and mobile applications. Advanced Electronic Applications, Inc.

slight power loss relative to a dipole, but the full-wavelength loop shows a gain of about 2 dBd. The full-wavelength loop forms the driven element for the popular quad antenna. *See also* QUAD ANTENNA.

LOOPSTICK ANTENNA
See FERRITE-ROD ANTENNA.

LOOSE COUPLING

When two inductors have a small amount of mutual inductance, they are said to be *loosely coupled*. Loose coupling provides relatively little signal transfer. Loose coupling is desirable when it is necessary to pick off a small amount of signal energy for monitoring purposes, without disturbing the operation of the circuit under test. *See also* COEFFICIENT OF COUPLING, and MUTUAL INDUCTANCE.

LORAN

The term *loran* is a contraction of "long-range navigation." Loran is used by ships and aircraft for determination of position, and for navigation. There is one system of loran in common use today. The system, called *loran C*, uses pulse transmission at a rate of about 20 cycles per second. Two transmitters are used, and they are spaced at a distance of 300 miles. The operating frequency is 100 kHz, which corresponds to a wavelength of 3 km. Location is found by comparing the time delay in the arrival of the signal from the more distant transmitter, relative to the arrival time of the signal from the nearer transmitter. The lines of constant time difference are hyperbolic.

An older system of loran, which has now been discontinued, was known as *loran A*. This system operated in the range 1.85 to 1.95 MHz, or a wavelength of about 160 m. Loran C is favored because the ground-wave propagation range is greater, resulting in better reliability. *See also* RADIOLOCATION, and RADIONAVIGATION.

LOSS

Loss is a term that describes the extent of a decrease in current, voltage, or power. Most passive circuits exhibit some loss. The loss is an expression of the ratio between the input and output signal amplitudes in a circuit. Loss is usually specified in decibels.

Loss in a circuit often results in the generation of heat. The more efficient a circuit, the lower the loss, and the smaller the amount of generated heat. Sometimes, loss is deliberately inserted into a circuit or transmission line. A device designed for such a purpose is called an *attenuator* (*see* ATTENUATOR).

If a circuit has a gain of x dB, then the loss is $-x$ dB. Conversely, if a circuit has a loss of x dB, then the gain is $-x$ dB. Loss is just the opposite of gain. *See also* DECIBEL, GAIN, and INSERTION LOSS.

LOSS ANGLE

In a dielectric material, the *loss angle* is the complement of the phase angle. If the phase angle is given by ϕ and the loss angle is given by θ, then:

$$\theta = 90 - \phi$$

where both angles are specified in degrees.

In a perfect, or lossless, dielectric material, the current leads the voltage by 90 degrees, and thus $\phi = 90$. The loss angle is therefore zero. As the dielectric becomes lossy, the phase angle gets smaller, and the loss angle gets larger. This is because resistance is present in a lossy dielectric, in addition to capacitance. The tangent of the loss angle is called the *dissipation factor*. *See also* DISSIPATION FACTOR, and PHASE ANGLE.

LOSSLESS LINE

A theoretically perfect transmission line is called a *lossless line*. Such a line does not exist in practice, but it is useful, for the purposes of some mathematical calculations, to assume that a transmission line has no loss.

In a lossless line, no power is dissipated as heat. The available power at the line output is the same as the input power. If the line is terminated in a pure resistance exactly equal to its characteristic impedance, the current and voltage along the lossless line are uniform at all points.

In a lossless line, an impedance mismatch at the terminating end would result in no additional power loss. But in a real feed line, a mismatch results in some additional loss because of the standing waves. *See also* FEED LINE, and STANDING-WAVE-RATIO LOSS.

LOUDNESS

Loudness is an expression of relative apparent intensity of sound. Loudness is sometimes called *volume*. The human ear perceives loudness approximately in terms of the logarithm of the actual intensity of the disturbance. Loudness is often expressed in decibels (*see* DECIBEL). An increase in loudness of 2 dB represents the smallest detectable change in sound level that a listener can detect if he is expecting the change. The threshold of hearing is assigned the value 0 dB.

In an audio circuit, such as a radio receiver output or high-fidelity system, the control that affects the sound intensity is often called the *loudness control*.

LOUDSPEAKER
See SPEAKER.

LOW-ANGLE RADIATION AND RESPONSE

Often, it is desirable for an antenna system to radiate, and respond to, radio waves nearly parallel to the horizon. The ability of an antenna to concentrate energy near the horizon is called *low-angle radiation*. If an antenna transmits well at low angles, it will usually receive well from low angles. That is, it will exhibit a *low-angle response*.

Radiation Pattern The amount of low-angle radiation or response in an antenna system is evident from looking at the *vertical-plane radiation pattern*. The illustration at A shows an antenna with good low-angle radiation and response. The illustration at B shows one with poor low-angle radiation and response.

Low-angle radiation is specified in terms of the number of degrees, above the horizon, at which the most radiation occurs. In the illustration at A, the low-angle radiation is 0 degrees. At

B, it is about 45 degrees. Sometimes, the low-angle radiation/ response of an antenna is simply called *good*, *fair*, or *poor*.

When is it Wanted?

Low-angle radiation/response is essential for working DX on the high-frequency (HF) bands because it minimizes the number of hops in ionosopheric propagation, and it also keeps the maximum usable frequency as high as possible.

Low-angle radiation and response are almost always wanted for terrestrial communications at very-high frequencies (VHF) and above. This is because such communications on these bands occurs via signals that travel horizontally through the atmosphere.

When is it Not Needed or Wanted?

For regional work at HF, especially on the 160-, 80-, and 40-meter bands, low-angle radiation can be undesirable. This is because, over moderate distances, ionospheric propagation requires that the signals make steep hops. If a wave reaches the ionosphere too far away, it will skip over the desired location. *See* SKIP, and SKIP ZONE.

Low-angle radiation might not be wanted in satellite or moonbounce work. This is because the satellite or moon are often not near the horizon. *See* ACTIVE COMMUNICATIONS SATELLITE, and MOONBOUNCE.

Antennas that Have It

In order to have good low-angle radiation/response, an antenna must usually be at least a wavelength above radio-frequency (RF) ground. The exception is a vertical antenna in which a ground plane is provided. This can be done by means of radials on, or buried just under, the earth. Or, if the antenna is at least ¼ wavelength up, it can be done via a few radials, each of ¼ wavelength. *See* GROUND-PLANE ANTENNA, RADIAL, VERTICAL ANTENNA, and VERTICAL DIPOLE ANTENNA.

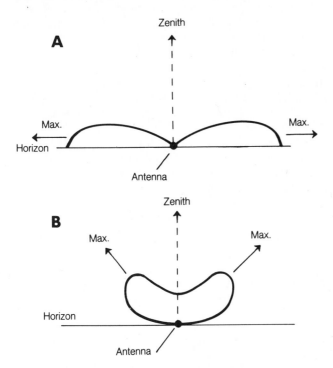

LOW-ANGLE RADIATION AND RESPONSE: At A, a vertical-plane radiation pattern of an antenna with good low-angle radiation and response. At B, an example of a poor pattern.

Collinear vertical arrays, popular with repeater installations, have excellent low-angle radiation/response. In multi-element phased arrays, the phasing should be such that optimum phase addition occurs parallel with the horizon. In any directional system, the array should be pointed horizontally if low-angle radiation/response is wanted.

Antennas that Don't Have It

A horizontally polarized antenna that is less than a wavelength above RF ground will not usually have good low-angle radiation. The low-angle characteristics become especially poor if the antenna is less than ½ wavelength up.

Any directional antenna, such as a Yagi, quad, dish, or helical array, will not have good low-angle radiation/response if it is pointed up into the sky. Of course, the operator might want radiation to go skyward, or be received from the sky, if satellite or moonbounce work is being done.

For further information, *see also* IONOSPHERE, PROPAGATION CHARACTERISTICS, and TROPOSPHERIC PROPAGATION.

LOWER SIDEBAND

An amplitude-modulated signal carries the information in the form of sidebands, or energy at frequencies just above and below the carrier frequency (*see* AMPLITUDE MODULATION, and SIDEBAND). The *lower sideband (LSB)* is the group of frequencies immediately below the carrier frequency.

The LSB frequencies result from difference mixing between the carrier signal and the modulating signal. For a typical voice amplitude-modulated signal, the lower sideband occupies about 3 kHz of spectrum space (see illustration). For reproduction of television video information, the sideband can be several megahertz wide.

The lower sideband contains all of the modulating-signal intelligence. For this reason, the rest of the signal (carrier and upper sideband) can be eliminated. This results in a single-sideband (SSB) signal on the lower sideband. *See also* SINGLE SIDEBAND, and UPPER SIDEBAND.

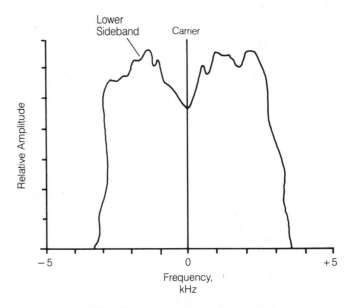

LOWER SIDEBAND: Energy is below the carrier frequency.

LOWEST USABLE FREQUENCY

For any two points separated by more than a few miles, electromagnetic communication is possible only at certain frequencies. At very low frequencies, communication is almost always possible between any two locations in the world, provided the power level is high enough. Usually, however, communication is also possible at some much higher band of frequencies, in the medium or high range.

Given some arbitrary medium or high frequency f, at which communication is possible between two specific points, suppose this frequency is made lower until communication is no longer possible. This cutoff frequency might be called f_L. It is known as the lowest usable high frequency, or simply the *lowest usable frequency (LUF)*. The LUF changes as the locations of the transmitting or receiving stations are changed. The LUF is a function of the ionospheric conditions, which fluctuate with the time of day, the season of the year, and the environment in the sun. *See also* MAXIMUM USABLE FREQUENCY, MAXIMUM USABLE LOW FREQUENCY, and PROPAGATION CHARACTERISTICS.

LOW FREQUENCY

The range of electromagnetic frequencies extending from 30 kHz to 300 kHz is sometimes called the *low-frequency band*. The wavelengths are between 1 and 10 km. For this reason, low-frequency waves are sometimes called *kilometric waves*.

The ionospheric E and F layers return all low-frequency signals to the earth, even if they are sent straight upward. Low-frequency signals from space cannot reach the surface of our planet because the ionosphere blocks them.

Low-frequency waves, if vertically polarized, propagate well along the surface of the earth. This becomes more and more true as the frequency gets lower. Low-frequency signals can, therefore, be used by space travelers on planets that have no ionosphere. Even on the moon, which has no ionosphere to return radio waves back to the ground, low-frequency waves might be useful for over-the-horizon communication via the surface wave. *See also* SURFACE WAVE.

LOWPASS FILTER

A *lowpass filter* is a combination of capacitance, inductance, and/or resistance, intended to produce large amounts of attenuation above a certain frequency and little or no attenuation below that frequency. The frequency at which the transition occurs is called the *cutoff frequency* (*see* CUTOFF FREQUENCY). At the cutoff frequency, the attenuation is 3 dB, with respect to the minimum attenuation. Below the cutoff frequency, the attenuation is less than 3 dB. Above the cutoff, the attenuation is more than 3 dB.

The simplest lowpass filters consist of a series inductor or a parallel capacitor. More sophisticated lowpass filters have a combination of series inductors and parallel capacitors, such as the examples shown in the illustration. The filter at A is called an *L-section lowpass filter*; the circuit at B is a *pi-section lowpass filter*. These names are derived from the geometric arrangement of the components as they appear in the diagram.

Resistors are sometimes substituted for the inductors in a lowpass filter. This is especially true when active devices are used, in which case many filter stages can be cascaded.

Lowpass filters are used in many different applications in radio-frequency electronics. One common use of a lowpass

filter is at the output of a high-frequency transmitter. The cutoff frequency is about 40 MHz. When such a lowpass filter is installed in the transmission line between a transmitter and antenna, very-high-frequency harmonics are greatly attenuated. This reduces the probability of television interference. In a narrow-band transmitter, lowpass filters are often built-in for reduction of harmonic output. *See also* LOWPASS RESPONSE.

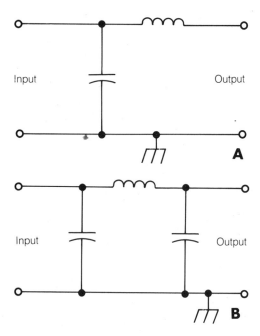

LOWPASS FILTER: At A, L-section. At B, pi-section.

LOWPASS RESPONSE

A *lowpass response* is an attenuation-versus-frequency curve that shows greater attenuation at higher frequencies than at lower frequencies. The sharpness of the response can vary considerably. Usually, a lowpass response is characterized by a low degree of attenuation up to a certain frequency; above that point, the attenuation rapidly increases. Finally, the attenuation levels off at a large value. Below the cutoff frequency, the attenuation is practically zero. The lowpass response is typical of lowpass filters.

The cutoff frequency of a lowpass response is that frequency at which the insertion loss is 3 dB with respect to the minimum loss. The ultimate attenuation is the level of attenuation well above the cutoff frequency, where the signal is virtually blocked. The ideal lowpass response should look like the attenuation-versus-frequency curve in the illustration. The curve is smooth, and the insertion loss is essentially zero everywhere well below the cutoff frequency. *See also* CUTOFF FREQUENCY, INSERTION LOSS, and LOWPASS FILTER.

LSB

See LOWER SIDEBAND.

LSI

See LARGE-SCALE INTEGRATION.

LUF

See LOWEST USABLE FREQUENCY.

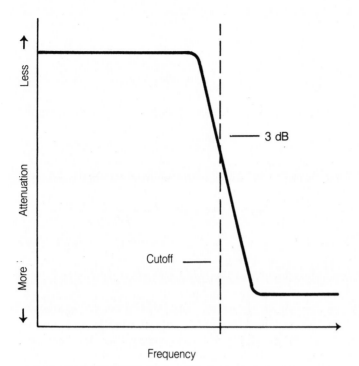

LOWPASS RESPONSE: Passes all frequencies below the cutoff.

LUG

A *lug* is a terminal attached to the end of a wire for convenient connection to a screw or binding post. Lugs provide a better electrical connection than simply bending the wire around the screw or binding post. Some lugs can be soldered to the end of a wire; others are clamped on without soldering.

The terminals on a tie strip, to which wires are soldered in point-to-point electronic assemblies, are sometimes called *lugs*. *See also* TIE STRIP.

LUMISTOR

A *lumistor* is an optical coupling circuit. The signal to be transferred is applied to a light-emitting device. This device can be a light bulb for signals at very low frequencies. A light-emitting diode produces better results at higher frequencies. A laser can also be used. A photocell, photovoltaic cell, or other light-sensitive device picks up the modulated light and delivers it to the load.

The lumistor makes it possible to transfer a signal with essentially no "pulling," or effects on the output impedance of the signal generator. A change in the load impedance has no effect on the operating characteristics of the signal generator when optical coupling is used. Optical coupling also facilitates the transfer of signals between circuits at substantially different potentials. *See also* OPTICAL COUPLING.

LUMPED ELEMENT

When a reactance, resistance, or voltage source appears in a discrete location, the component is called a *lumped element*. All ordinary capacitors, inductors, and resistors are lumped elements. However, a transmission line is not a lumped element because it exhibits inductance and capacitance that are distributed over a region. *See also* DISCRETE COMPONENT, and DISTRIBUTED ELEMENT.

MACHINE LANGUAGE

Machine language is the mode in which a computer actually operates. Small groups of binary digits, or bits, comprise instructions for the computer. These groups of bits are called *microinstructions*. One or more microinstructions together form a macroinstruction (*see* MACROINSTRUCTION, and MICROINSTRUCTION).

Computers are usually not operated or programmed in machine language. Instead an assembler or compiler determines the machine language from a higher-order language more easily understood by people. There are several different higher-order languages for various computer applications.

When a new computer is checked, or if a new higher-order language is to be developed, it might be necessary to work directly with machine language. In some program debugging situations, and in computer maintenance and repair work, machine language must be understood by the programmer or technician. *See also* HIGHER-ORDER LANGUAGE.

MACROINSTRUCTION

A *macroinstruction* is a machine-language computer instruction made up of a group of microinstructions (*see* MICROINSTRUCTION). A macroinstruction corresponds to a specific, unique command in a higher-order language. The translation from command to macroinstruction is done by the compiler. *See also* MACHINE LANGUAGE, and MICROINSTRUCTION.

MACROPROGRAM

A *macroprogram* is a computer program written in the language used by the operator. When we think of a computer program, we generally think of the macroprogram, which takes the form of a series of statements in higher-order language.

A macroprogram can be written in assembly language, for the purpose of translating between machine language and the user language. Such a program is sometimes called a *macroassembler* or a *macroassembly program*. *See also* HIGHER-ORDER LANGUAGE.

MAGNET

A *magnet* is a device that produces a flux field because of the effects of molecular alignment, or because of electric currents. The molecules of iron or nickel, when aligned uniformly, produce a continuous magnetic field, and these substances can be used as permanent magnets. A coil of wire with an iron core will produce a strong magnetic field when a current passes through the coil, but the field disappears when the current stops flowing. Such a device is called an *electromagnet* or a *temporary magnet*.

Magnets are used for many purposes in electronics. The dynamic microphone or speaker uses a magnet to convert mechanical forces into electrical impulses, or vice-versa.

Generators and motors use magnets. Indicating devices, such as meters, often use magnets to obtain needle deflection. The list of devices that use magnets is almost endless. *See also* MAGNET COIL, MAGNETIC FIELD, MAGNETIC POLARIZATION, MAGNETIC POLE, and PERMANENT MAGNET.

MAGNET COIL

An electric current always produces a magnetic field. The magnetic lines of flux occur in directions perpendicular to the flow of current. Thus, if a wire is wound into a helical or circular coil, a strong magnetic field is produced, with a configuration similar to that of the magnetic field surrounding a bar magnet. A coil, deliberately wound around an iron rod to create a magnetic field, is called a *magnet coil* or *magnetic coil*.

All electromagnets contain magnet coils. When a magnet coil carries a large amount of current, the magnetic field becomes extremely strong, and the electromagnet can pick up large iron or steel objects. If the magnet coil is supplied with an alternating current, the magnetic field collapses and reverses polarity in step with the change in the direction of the current.

Some meters, motors, generators, and other devices use magnet coils, usually for the purpose of converting electrical energy to mechanical energy, or vice-versa.

MAGNETIC AMPLIFIER

A *magnetic amplifier* is a device that is used for the purpose of modulating or changing the voltage across the load in an alternating-current circuit. The magnetic amplifier consists of an iron-core transformer, with an extra winding to which a control signal can be applied.

Two windings appear in series with the alternating-current power supply and the load. These windings are connected in opposing phase, and together they are called the *output winding*.

A third coil is wound around the center column of the transformer core. This coil is called the *input coil*. When a direct current is applied to the input coil, the impedance of the output coil changes. This occurs because of the saturable-reactor principle (*see* SATURABLE REACTOR). A small change in the current through the input coil results in a large fluctuation in the output-coil impedance. Therefore, amplification occurs. Magnetic amplifiers can, if properly designed, produce considerable power gain.

MAGNETIC BEARING

The *magnetic bearing*, also known as *magnetic heading*, refers to the azimuth as determined by a magnetic compass. Magnetic bearings are measured in degrees clockwise from magnetic north.

The magnetic bearing is usually a little different from the geographic bearing because the geographic and geomagnetic poles are not in exactly the same places. *See also* GEOMAGNETIC FIELD.

The shafts of some meters, gyroscopes, and other rotating devices can employ bearings using magnetic fields to minimize friction. Such bearings are called *magnetic bearings.* The shaft is literally suspended in the air because of the repulsive magnetic force between like poles of permanent magnets or electromagnets. The friction of such a bearing, when operated in a vacuum, is very nearly zero.

MAGNETIC CIRCUIT

A *magnetic circuit* is a complete path for magnetic flux lines (*see* MAGNETIC FLUX). The lines of flux are considered to originate at the north pole of a magnetic dipole, and to terminate at the south pole. A magnetic circuit has certain similarities to an electric circuit.

The magnetic equivalent of voltage is known as magnetomotive force, and is measured in gilberts. The magnetic equivalent of current is flux density. The magnetic equivalent of resistance is reluctance. These three parameters behave, in a magnetic circuit, in the same way as voltage, current, and resistance behave in an electric circuit. *See also* FLUX DENSITY, MAGNETOMOTIVE FORCE, and RELUCTANCE.

MAGNETIC CORE

A *magnetic core* is a ferromagnetic object, such as an iron rod or powdered-iron toroid core, used for the purpose of concentrating or storing a magnetic field. Magnetic cores are used in many different electrical and electronic devices, such as the electromagnet, the transformer, and the relay.

A special form of magnetic core is used to store digital information in some computers. A toroidal ferromagnetic bead surrounds a wire or set or wires, as shown in the illustration. When a current pulse is sent through the wire, a magnetic field is produced, and the toroid becomes magnetized in either the clockwise or counterclockwise sense (depending on the direction of the current pulse). The toroid remains magnetized in that sense indefinitely, until another current pulse is sent through the wire in the opposite direction. One sense represents the binary digit 0 (false); the other sense represents 1 (true). Magnetic-core storage is retained even in the total absence of power in a computer circuit. Several magnetic cores, when combined, allow the storage of large binary numbers.

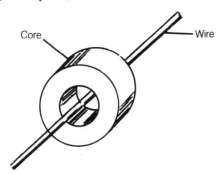

MAGNETIC CORE: A toroid surrounds a current-carrying wire.

MAGNETIC COUPLING

When a magnetic field influences a nearby object or objects, the effect is known as *magnetic coupling.* Examples of magnetic coupling include the deflection of a compass needle by the earth's magnetic field, the interaction among the windings of a motor or generator, and the operation of a transformer. There are many other examples. Magnetic coupling can occur as a direct force, such as attraction or repulsion, between magnetic fields. Magnetic coupling can also occur because of motion among magnetic objects, or because of changes in the intensity of a magnetic field.

A fluctuating current produces a changing magnetic field, which in turn produces changing currents in nearby objects. This is especially true when coils of wire are placed along a common axis. This form of magnetic coupling is called electromagnetic induction (*see* ELECTROMAGNETIC INDUCTION).

Magnetic coupling is enhanced by the presence of a ferromagnetic substance between two objects. At extremely low frequencies, solid iron is an effective magnetic medium. At very low frequencies, laminated iron or ferrite is preferable. At higher frequencies, ferrite and powdered iron are used.

The degree of magnetic coupling between two wires or coils is measured as mutual inductance. *See also* MUTUAL INDUCTANCE.

MAGNETIC DEFLECTION

Magnetic force, like any force, can cause objects to be accelerated. This acceleration can occur as a change in the speed of an object, a change in the direction of its motion, or both.

Any moving charged object generates a magnetic field. The most common example of this is the magnetic field that surrounds a wire carrying an electric current. However, charged particles traveling through space also produce magnetic fields. When such moving charged particles encounter an external magnetic field, the particles are accelerated or deflected.

Charged particles from the sun are deflected by the earth's magnetic field. During an intense solar storm, when many charged particles are emitted from the sun, the deflection causes such a concentration of particles near the poles that a visible glow (aurora) occurs.

The principle of magnetic deflection is used in the common indicating meter, the cathode-ray tube, and many other electronic devices.

MAGNETIC EQUATOR

The *magnetic equator* is a line around the earth, midway between the north and south magnetic poles. Every point on the magnetic equator is equidistant from either magnetic pole. At the magnetic equator, a compass needle experiences no dip; the geomagnetic lines of flux are parallel to the surface of the earth.

The magnetic equator does not exactly coincide with the geographic equator because the magnetic poles do not lie exactly at the geographic poles. *See also* GEOMAGNETIC FIELD.

MAGNETIC FIELD

A *magnetic field* is a region in which magnetic forces can occur. A magnetic field is produced by a magnetic dipole or by the motion of electrically charged particles. Physicists consider magnetic fields to be made up of flux lines. The intensity of the magnetic field is determined according to the number of flux lines per unit surface area (*see* FLUX DENSITY). Although the magnetic lines of flux do not physically exist in any tangible form, they do have a direction that can be determined at any point in space. Magnetic fields can have various shapes.

A magnetic field is considered to originate at a north magnetic pole, and to terminate at a south magnetic pole. A pair of magnetic poles, surrounded by a magnetic field, is called a *magnetic dipole*. *See also* MAGNETIC FLUX, and MAGNETIC POLE.

MAGNETIC FLUX

Magnetic lines of flux are sometimes referred to as *magnetic flux*. Magnetic flux exists wherever an electrically charged particle is moving. Magnetic flux surrounds certain materials with aligned or polarized molecules. Magnetic flux lines are considered to emerge from a magnetic north pole and to enter a magnetic south pole. Every flux line is, however, a continuous, closed loop. This is true no matter what is responsible for the existence of the magnetic field.

The more intense a magnetic field, the greater the number of flux lines crossing a given two-dimensional region in space. A line of flux is actually a certain quantity of magnetic flux, usually 1 weber or 1 maxwell. The strength of a field can be measured in webers per square meter. Another commonly used unit of flux density is the line of flux per square centimeter; this unit is known as the *gauss*. *See also* FLUX, FLUX DENSITY, FLUX DISTRIBUTION, GAUSS, MAGNETIC FIELD, MAXWELL, and WEBER.

MAGNETIC POLARIZATION

In a ferrous substance, such as iron, nickel, or steel, the molecules produce magnetic fields. Each molecule forms a magnetic dipole, containing a north and south magnetic pole (*see* MAGNETIC POLE).

The molecules in a ferrous substance are normally aligned in random fashion, so that the net effect of their magnetic fields is zero. However, in the presence of an external magnetic field, the molecules become more nearly aligned. If the field is very strong, the molecules line up in nearly the same orientation. This greatly increases the strength of the magnetic field within, and near, the substance. When the field is removed, the molecules of some ferrous materials remain aligned, forming a permanent magnet.

Magnetic memory devices operate on the principle of magnetic polarization. A current through a wire produces a magnetic field. This field influences the properties of some ferrous substance, such as powdered iron, causing the substance to become polarized. The magnetic memory retains its polarization until reverse current pulse is sent through the wire (*see* MAGNETIC CORE, and MAGNETIC RECORDING).

The core of the earth is believed to be composed mostly of iron. The molecules in the interior of our planet are aligned in such a way that a permanent magnetic field exists around the earth. The axis of this magnetic field nearly coincides with the axis of rotation of the earth. Some other planets, such as Jupiter, also exhibit this form of magnetic polarization. *See also* GEOMAGNETIC FIELD, and PERMANENT MAGNET.

MAGNETIC POLE

A magnetic pole is a point, or localized region, at which magnetic flux lines converge. Magnetic poles are called either *north* or *south*. This is basically a matter of convention. In the mag-

netic compass, one end of the magnetized needle points toward the north magnetic pole; this is called the north pole of the magnet. Similarly, the end of a magnet that orients itself toward the south is called the south pole. Magnetic flux lines, according to theory, leave north poles and enter south poles.

It is believed that magnetic poles must always exist in pairs. This conclusion arises from one of the fundamental laws of Maxwell. Magnetic flux lines, according to Maxwell, always take the form of closed loops. Thus the magnetic lines of flux must arise as a result of dipoles — pairs of north and south magnetic poles. Magnetic monopoles might be possible, and much interesting theoretical research has been done in an attempt to find an isolated magnetic north or south pole. However, such a phenomenon would appear contrary to the laws of Maxwell. Some scientists think they might have found magnetic monopoles. *See also* MAGNETIC FIELD.

MAGNETIC RECORDING

Magnetic recording is a method of storing and retrieving data by means of magnetic tape or disks. The magnetic medium consists of powdered iron oxide on a flexible plastic base. A magnetic transducer, called the *head*, generates a magnetic field that causes polarization of the molecules in the iron oxide. *See* DISKETTE, FLOPPY DISK, and MAGNETIC TAPE.

In the magnetic-disk recording system, the head follows a spiral path on the disk. The main advantage of magnetic-disk recording is rapid access to any part of the information stored on the disk. Magnetic disks are used for storage of computer data, but they are becoming more common for video recording as well.

Magnetic-tape recording makes use of a long strip of magnetic material, usually 0.5 mil, 1.0 mil, or 1.5 mil thick and $\frac{1}{8}$ inch, $\frac{1}{4}$ inch, or 2 inches wide. In magnetic sound recording, the information is placed along straight paths called *tracks*. A tape can have one, two, four, or eight tracks, each with a different recording impressed on it. For sound applications, the standard tape sizes are $\frac{1}{8}$ inch and $\frac{1}{4}$ inch wide, and the most common speeds are $1\frac{7}{8}$, $3\frac{3}{4}$, and $7\frac{1}{2}$-inches per second. In video recording, the tape speed is 15 inches per second; the video information is recorded on oblique tracks with a frequency-modulated carrier, and the sound is recorded on one or two straight tracks.

Magnetic recording techniques have improved since they were first introduced decades ago. Today, magnetic recordings make it possible to store vast amounts of information.

MAGNETIC TAPE

Magnetic tape is a form of information-storage medium, commonly used for storing sound, video, and digital data. Magnetic tape is available in several different thicknesses and widths for different applications.

Magnetic tape consists of millions of fine particles of iron oxide, attached to a plastic or mylar base. A magnetic field, produced by the recording head, causes polarization of the iron-oxide particles. As the field fluctuates in intensity, the polarization of the iron-oxide particles also varies (see illustration). When the tape is played back, the magnetic fields surrounding the individual iron-oxide particles produce current fluctuations in the playback or pickup head.

For sound and computer-data recording, magnetic tape is available either in cassette form or wound on reels. The thickness can be 0.5 mil, 1.0 mil, or 1.5 mil. (A mil is 0.001 inch.) The thicker tapes have better resistance to stretching, although the recording time, for a given length of tape, is proportionately shorter than with thin tape. In the cassette, the tape can be either ⅛ inch or ¼ inch wide. Reel-to-reel tape is generally ¼-inch wide. Wider tape is often used, however, in recording studios and in video applications. The tape can have up to 24 individual recording tracks. The most common speeds are 1-⅛, 3-¾, and 7-½ inches per second. For voice recording, the slower speeds are adequate, but for music, higher speeds (such as 15 inches per second) are preferable for enhanced sensitivity to high-frequency sound.

Video tape ranges from ½ inch to 2 inches wide, and the standard recording speed is 15 inches per second. Even this speed is not directly sufficient for the reproduction of the high frequencies encountered in video, but the recording is done in a slanted or oblique direction, resulting in much higher effective speeds.

Magnetic tape provides a convenient and compact medium for long-term storage of information. However, certain precautions should be observed. The tape must be kept clean, and free from grease. Magnetic tapes should be kept at a reasonable temperature and humidity and they should not be subjected to magnetic fields. *See also* MAGNETIC RECORDING.

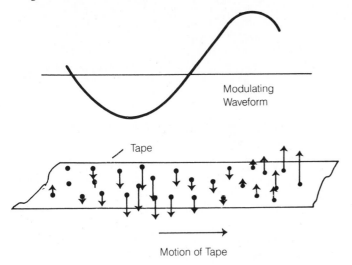

MAGNETIC TAPE: Signal causes magnetization of atoms in tape.

MAGNETISM

Magnetism is a term that is used to describe any magnetic effects. Magnetic attraction and repulsion, interaction with moving charged particles, and polarization of molecules can all be loosely categorized as magnetism. *See also* MAGNETIC DEFLECTION, MAGNETIC FIELD, and MAGNETIC POLARIZATION.

MAGNETOMOTIVE FORCE

Magnetomotive force is sometimes called *magnetic potential* or *magnetic pressure*. Magnetomotive force, in a magnetic circuit, is analogous to electromotive force in an electric circuit.

The most common unit of magnetomotive force is the gilbert. Another often-used unit is the ampere turn. A current of 1 ampere, passing through a one-turn coil of wire, sets up a magnetomotive force of 1 ampere turn or 1.26 gilberts. In general, the magnetomotive force (M) generated by a coil of N turns,

carrying a current of 1 amperes, is given in ampere turns by the formula:

$$M = NI$$

The larger the magnetomotive force in a region of space, the greater the magnetic effects, all other factors being constant. *See also* AMPERE TURN, and GILBERT.

MAGNETORESISTANCE

Certain semiconductor substances, such as bismuth and indium antimonide, exhibit variable resistance depending on the intensity of a surrounding magnetic field. Even ordinary wire conductors can be influenced somewhat, if the magnetic field is sufficiently strong.

Magnetoresistance occurs because of magnetic deflection of the paths of charge carriers. If no magnetic field is present, charge carriers move in paths that, on the average, are parallel to a line connecting the centers of opposite charge poles. If a magnetic field is introduced, however, the paths of the charge carriers are deflected, and their paths tend to curve toward one side of the conducting medium. This increases the effective distance through which the charge carriers must travel to get from one charge pole to the other. The effect is very much like a current in a river, which makes it necessary for a swimmer to work harder to reach a specified point on the opposite shore. *See also* MAGNETIC DEFLECTION.

MAGNETOSPHERE

The magnetic field surrounding the earth produces dramatic effects on charged particles in space. The geomagnetic lines of flux extend far above the surface of our planet, and high-speed atomic particles constantly arrive from the sun and the distant stars. When these charged particles enter the earth's magnetic field, they are deflected toward the poles.

The *magnetosphere* is the boundary, roughly spherical in shape and surrounding the earth, that represents the maximum extent of the magnetic-deflection effects of the geomagnetic field on charged particles from space. *See also* GEOMAGNETIC FIELD.

MAGNETOSTRICTION

Magnetostriction is the tendency for certain substances to expand or contract under the influence of a magnetic field. Magnetostriction, in a magnetic field, is analogous to electrostriction in an electric field (*see* ELECTROSTRICTION).

Substances that exhibit magnetostrictive properties include Alfer (an alloy of aluminum and iron), nickel, and Permalloy (an alloy of iron and nickel). Nickel exhibits negative magnetostriction; that is, it is compressed by a magnetic field. Alfer and Permalloy exhibit positive magnetostriction; they expand as the surrounding magnetic field becomes stronger. Other magnetostrictive substances include various forms of iron oxide, nichrome, powdered iron, and ferrite.

If a magnetostrictive substance is subjected to a sufficiently strong magnetic field, the material can actually break under the strain. This occasionally happens with ferrite and powdered-iron inductor cores when the surrounding coils carry current far in excess of rated values.

Magnetostrictive materials can be employed in some of the same applications as piezoelectric crystals. For example, a magnetostriction oscillator works by means of resonance in a rod of magnetostrictive material. Transducers can also be made with magnetostrictive materials. *See also* MAGNETOSTRICTION OSCILLATOR, and PIEZOELECTRIC EFFECT.

MAGNETOSTRICTION OSCILLATOR

A *magnetostriction oscillator* is similar to a crystal oscillator. Instead of a piezoelectric crystal, the magnetostriction oscillator makes use of a rod of magnetostrictive material. The rod serves two purposes: It vibrates at the frequency of oscillation, and it acts as the core for the feedback transformer.

The fundamental oscillating frequency is determined by the physical length and diameter of the rod, whether it exhibits positive or negative magnetostriction, and the extent of the magnetostrictive effects. *See also* MAGNETOSTRICTION.

MAGNETRON

The *magnetron* is a form of traveling-wave tube used as an oscillator at ultra-high and microwave frequencies. Most magnetrons contain a central cathode and a surrounding plate, as shown in the illustration. The plate is usually divided into two or more sections by radial barriers called cavities. The RF output is taken from a waveguide opening in the anode.

The cathode is connected to the negative terminal of a high-voltage source, and the anode is connected to the positive terminal. This causes electrons to flow radially outward from the cathode to the anode. A magnetic field is applied in a longitudinal direction; this causes the electrons to travel outward in spiral, rather than straight, paths. The electrons tend to travel in bunches because of the interaction between the electric and magnetic fields. This results in oscillation with the frequency being somewhat stabilized by the cavities.

Magnetrons can produce continuous power outputs of more than 1 kW at a frequency of 1 GHz. The output drops as the frequency increases; at 10 GHz, a magnetron can produce about 10 to 20 watts of continuous radio-frequency output. For pulse modulation, the peak-power figures are much higher. *See also* TRAVELING-WAVE TUBE.

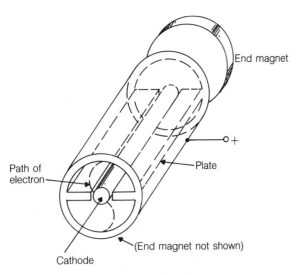

MAGNETRON: Oscillation results from interaction between magnetic field and moving electrons.

MAILBOX

In packet radio, a *mailbox* is a sending, receiving and storage system for messages, kept at a single ham station. The mailbox for your station might contain messages for several other hams. They can retrieve the messages by checking the contents of your mailbox. Another station might leave a message for you in your mailbox.

Generally, other stations cannot store messages for each other in your mailbox. That is the function of a *packet-radio bulletin-board system*. *See also* PACKET RADIO, and PACKET-RADIO BULLETIN-BOARD SYSTEM.

MAIL FORWARDING

In packet radio, *mail forwarding* is a process in which packets, or messages, are transferred between and among packet-radio bulletin-board stations (PBBSs). A mail-forwarding PBBS has an up-to-date list of all the other mail-forwarding PBBSs in the United States or in the world.

Many mail-forwarding routes are on the high-frequency (HF) bands, for long-distance transfer of messages. Using HF keeps the number of relay stations to a minimum, as a given message goes from its source to its destination. Each HF band has a standard frequency for mail forwarding. It is important that these frequencies be kept clear, to keep the packet radio network working efficiently.

If the distance between the source and destination is short, very-high frequencies (VHF) can be used for mail forwarding. *See also* PACKET RADIO, and PACKET-RADIO BULLETIN-BOARD SYSTEM.

MAINFRAME

See CENTRAL PROCESSING UNIT.

MAJORITY CARRIER

All semiconductors carry current in two ways. Electrons transfer negative charge from the negative pole to the positive pole; holes carry positive charge from the positive pole to the negative pole (*see* ELECTRON, HOLE). Electrons and holes are known as *charge carriers*.

In an N-type semiconductor material, electrons are the dominant form of charge carrier. In a P-type material, holes are the more common charge carrier. The dominant charge carrier is called the *majority carrier*. *See also* MINORITY CARRIER, N-TYPE SEMICONDUCTOR, and P-TYPE SEMICONDUCTOR.

MAKE

The action of closing a circuit is often called *make*. The term is generally used with relays, or with keying circuits in a code transmitter. In the code transmitter, make represents the length of time required, after the key is closed, for the output signal to rise from zero to maximum amplitude. This is a rapid process, normally taking a few milliseconds. However, it is not instantaneous. The rate at which the carrier amplitude rises is extremely important; if it is too fast, key clicks will occur, but if it is too slow, the signal will sound "soft" and can even be difficult to read. *See also* CODE TRANSMITTER, KEY CLICK, and SHAPING.

MALE

Male is the term that denotes any connector designed for insertion into a receptacle. Examples of male connectors include the common lamp plug, the phone plug, and the edge connectors found on some printed-circuit boards.

A male connector can be recognized by the fact that, when the plug is pulled, the conductors are exposed. For this reason, male connectors are usually used for the load in a circuit, rather than for the power supply, to reduce the chances for short circuits or electric shock. *See also* FEMALE.

MARCONI ANTENNA

Any antenna measuring ¼ electrical wavelength at the operating frequency, and fed at one end, is called a *Marconi antenna*. One of the earliest communications antennas was of the Marconi type, used at low frequencies, and consisting of an end-fed wire with a short vertical section and a long horizontal span. The most common modern form of Marconi antenna is the ground-mounted, quarter-wave vertical antenna. However, a simple length of wire, connected directly to a transmitter, is also sometimes used by radio amateurs and shortwave listeners today, much as it was in the first days of wireless communication.

The Marconi antenna is simple to install, but it must be operated against a good radio-frequency ground. When this is done, the Marconi antenna can be extremely effective. The grounding system can consist of a set of radial wires, a counterpoise, or a simple ground-rod or waterpipe connection. The feed-point impedance of the Marconi antenna is a pure resistance, and ranges from about 20 to 50 ohms, depending on the surrounding environment and the precise shape of the radiating element. The Marconi antenna can be operated at any odd harmonic of the fundamental frequency. *See also* GROUND-PLANE ANTENNA, MARCONI EFFECT, and VERTICAL ANTENNA.

MARCONI EFFECT

A center-fed or end-fed antenna can show resonance at a frequency determined by the length of the feed line and radiator combined. Such resonance usually occurs at a frequency at which the combined length of the feed line and radiator is ¼ wavelength, or some odd multiple thereof. *Marconi effect* can result in unwanted radiation from an antenna feed line, as the whole system behaves like a Marconi antenna. This can occur with antenna systems using coaxial feeders, as well as with systems using openwire line. The currents flow on the outside of the coaxial outer conductor, or in phase along both conductors of open-wire line.

At low, medium, and high frequencies, antenna systems should generally not be operated at frequencies where Marconi effect might occur. For a particular frequency, it is wise to choose the feed line length to avoid Marconi resonances. *See also* MARCONI ANTENNA.

MARITIME-MOBILE RADIO SERVICE

A two-way radio station in an ocean-going vessel, such as a small boat or a large ship, engaged in nonamateur communication, is called a *maritime-mobile station*. Any fixed station, used for the purpose of nonamateur communication with the operator of an ocean-going vessel, is also part of the maritime-mobile service.

The maritime-mobile radio service is allocated a number of frequency bands in the radio spectrum. The bands range from the very low frequencies through the ultra-high frequencies; precise allocations vary somewhat from country to country.

MARKER

A signal used for receiver calibration purposes is called a *marker*. Markers are usually at intervals of 25 kHz, 100 kHz, or 1 MHz. Markers are especially useful in older radio receivers that do not have digital readouts, and that do not employ phase-locked loops for precise calibration. *See also* MARKER GENERATOR.

MARKER GENERATOR

In a receiver, a calibrating signal can be produced using a circuit called a *marker generator*. This is a crystal oscillator that has a harmonic-generating component, and can also employ a frequency divider.

A typical marker generator uses a crystal with a frequency of 1 MHz. This signal might be phase-locked into WWV, by means of a receiver that picks up WWV at 10 MHz and divides this signal by 10. The harmonic generator then produces markers at all frequencies that are multiples of 1 MHz, up to several tens or hundreds of megahertz. *See* MARKER.

The 1-MHz signal can be divided down to 100 kHz, and a harmonic generator employed to create markers at whole-number multiples of 100 kHz. The 1-MHz signal might even be divided down to 25 kHz, producing markers at whole-number multiples of this frequency. In each case, the markers are produced throughout the high-frequency (HF) range, and often well beyond this, providing calibration for shortwave receivers and ham rigs up through 10, 6, or even 2 meters.

MARK/SPACE

In a digital communications system in which there are two possible states, such as a morse-code system or radioteletype system, the ON condition is usually called *mark*, and the OFF condition is called *space*. Frequency-shift keying is often used in radioteletype; the lower frequency represents mark and the higher frequency represents space. *See also* FREQUENCY-SHIFT KEYING.

MASER

The *maser* is a special form of amplifier for microwave energy. The term itself is an acronym for *Microwave Amplification by Stimulated Emission of Radiation*. The maser output is the result of quantum resonances in various substances.

When an electron moves from a high-energy orbit to an orbit having less energy, a photon is emitted (*see* ELECTRON ORBIT). For a particular electron transition or quantum jump, the emitted photon always has the same amount of energy, and therefore the same frequency. By stimulating a substance to produce many quantum jumps from a given high-energy level to a given low-energy level, an extremely stable signal is produced. This is the principle of the maser oscillator. Ammonia and hydrogen gas are used in maser oscillators as frequency standards. Rubidium gas can also be used. The output-frequency accuracy is within a few billionths of 1 percent in the gas maser.

Solid materials, such as ruby, can be used to obtain maser resonances. This kind of maser is called a *solid-state maser*, and it can be used as an amplifier or oscillator. When an external signal is applied at one of the quantum resonance frequencies, amplification is produced. The solid-state maser must be cooled to very low temperatures for proper operation; the optimum temperature is nearly absolute zero. Although satisfactory maser operation can sometimes be had at temperatures as high as that of dry ice, the most common method of cooling is the use of liquid helium or liquid nitrogen. These liquid gases bring the temperature down to just a few degrees above absolute zero.

The traveling-wave maser is a common form of device for use at ultra-high and microwave frequencies. The cavity maser is another fairly simple device. Both the traveling-wave and cavity masers use the resonant properties of materials cut to precise dimensions. The traveling-wave and cavity masers are solid-state devices, and the temperature must be reduced to a few degrees Kelvin for operation. The gain of a properly operating traveling-wave or cavity type maser amplifier can be as great as 30 to 40 dB.

Maser amplifiers are invaluable in radio astronomy, communications, and radar. Some masers operate in the infrared, visible-light, and even ultraviolet parts of the electromagnetic spectrum. Visible-light masers are called *optical masers* or, more commonly, *lasers. See also* LASER.

MAST

A *mast* is a supporting structure for an antenna. A mast is not part of the radiating system, but serves only to hold the antenna in place. Generally, a mast consists of a vertical pole, from 1 inch to several inches in diameter. A mast can be self-supporting or it can require guying. A short mast is often attached to a rotator atop a tower when directional antenna systems are used.

When using a mast to support an antenna, it is important that the mast have sufficient diameter to withstand twisting forces in high winds. It is also important to be sure that unsupported mast sections are strong enough to stand up against the wind resistance. *See also* TOWER.

MASTER OSCILLATOR

The main frequency-control oscillator for a radio is often called the *master oscillator (MO)*. It is usually a variable-frequency oscillator (VFO) of some kind. Occasionally, the MO for a radio transmitter might be crystal-controlled, operating at a fixed frequency.

Older radios generally have tuned oscillators that use inductors and capacitors to determine their frequencies. But in recent years, the phase-locked loop (PLL) has become one of the most common types of MO. *See also* PHASE-LOCKED LOOP, and VARIABLE-FREQUENCY OSCILLATOR.

MATCHED LINE

A *matched line* is a feed line or transmission line, terminated by a nonreactive load whose impedance is identical to the characteristic impedance of the line. In a matched line, there is no reflected power, and there are no standing waves. A matched

condition is desirable for any line, because the line loss is minimized.

Given a feed-line characteristic impedance of Z_o and a purely resistive load impedance (R), where $Z_o = R$, the current and voltage appear in the same proportion at all points along the line. If E is the voltage at any point on the line, and I is the current at that same point, the following holds:

$$E/I = Z_o = R$$

This uniformity of voltage and current exists only in a matched line.

Because any feed line has some loss, the voltage and current are somewhat higher near the transmitter than near the load, as shown in the illustration. However, the ratio E/I is always the same. *See also* CHARACTERISTIC IMPEDANCE, IMPEDANCE MATCHING, MISMATCHED LINE, REFLECTED POWER, STANDING WAVE, and STANDING-WAVE RATIO.

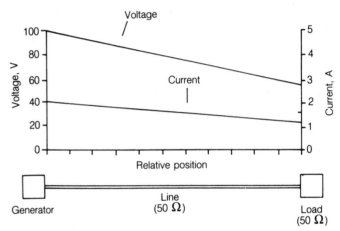

MATCHED LINE: Voltage-to-current ratio is uniform all along the line.

MATCHING

When two or more active devices, such as field-effect transistors, bipolar transistors, or vacuum tubes, are operated in parallel, push-pull, or push-push, it is desirable to ensure that they have operating characteristics that are identical. For a given type of transistor or tube, there are often considerable differences in operating parameters among different sample units. If the components are chosen at random, current hogging or other undesirable effects can occur. Component matching eliminates such potential problems. *See also* CURRENT HOGGING, PUSH-PULL CONFIGURATION, and PUSH-PUSH CONFIGURATION. The term *matching* is sometimes used to describe the process or condition of matched impedances. *See also* IMPEDANCE MATCHING.

MATCHTONE

A *matchtone* is a device that is used for the purpose of monitoring the transmission of code signals. The matchtone device is connected to the output of a transmitter to make the signals audible.

The matchtone operates via the radio-frequency energy from a strong signal. The radio-frequency energy is rectified to produce direct-current pulses. These pulses cause a changing bias in an audio-frequency oscillator so that the key-down condition results in an audible-tone output.

MATRIX

A *matrix* is a high-speed switching array, consisting of wires and diodes in a certain orderly arrangement. The matrix itself is an array of wires; some of the wires are interconnected via diodes (*see* DIODE MATRIX).

MAXIMUM POWER TRANSFER

When the power from a source is optimally transferred to a load, *maximum power transfer* is said to occur. Maximum power transfer requires that the load impedance be matched to the source impedance.

When the load and source impedances differ, some of the electromagnetic field is reflected back to the source. This causes loss in the system. *See also* IMPEDANCE MATCHING, and STANDING-WAVE-RATIO LOSS.

MAXIMUM USABLE FREQUENCY

For any two points separated by more than a few miles, electromagnetic communication is possible only at certain frequencies. At very low frequencies, using high transmitter power, worldwide communication is almost always possible. In general, there is a band of higher frequencies at which communication is also possible, and the power requirements are moderate.

Given some arbitrary medium or high frequency f, at which communication is possible between two specific points, suppose the frequency is raised until communication is no longer possible. The cutoff frequency might be called f_u. This cutoff point is called the *maximum usable frequency (MUF)* for the two locations in question.

The MUF depends on the locations and separation distance of the transmitter and receiver. The MUF also varies with the time of day, the season of the year, and the sunspot activity. All of these factors affect the ionosphere. For a given signal path, the propagation generally improves as the frequency is increased toward the MUF; above the MUF, the communication abruptly deteriorates. When a communications circuit is operated just below the MUF, and the MUF suddenly drops, a rapid and total fadeout occurs. This effect is familiar to users of the high-frequency radio spectrum. *See also* LOWEST USABLE FREQUENCY, MAXIMUM USABLE LOW FREQUENCY, and PROPAGATION CHARACTERISTICS.

MAXIMUM USABLE LOW FREQUENCY

In the very-low-frequency radio band, worldwide communication is almost always possible with high-power transmitters. The ionosphere returns all signals to the earth at these frequencies. Surface-wave propagation alone can provide long-range communications at very-low frequencies.

Given a circuit at some very-low frequency, f, between two defined points on the earth, suppose the frequency is gradually raised. If the transmitter and receiver are more than a few miles apart, the path loss generally increases as the frequency is increased. This happens for two reasons: The ionospheric D layer becomes less reflective and more absorptive, and the surface-wave loss increases (*see* D LAYER, and SURFACE WAVE).

Under some conditions, a frequency f_u exists at which communication deteriorates to the point of uselessness. This frequency can be in the low or medium range, and is known as the maximum usable low frequency (MULF).

The MULF is not the same as the maximum usable frequency. If the frequency is raised above the MULF, communication will often become possible again at a frequency called the *lowest usable high frequency (LUF)*. As the frequency is raised still more, propagation ultimately deteriorates at the maximum usable frequency (MUF). *See also* LOWEST USABLE FREQUENCY, MAXIMUM USABLE FREQUENCY, and PROPAGATION CHARACTERISTICS.

MAXWELL

The *maxwell* is a unit of magnetic flux, representing one line of flux in the centimeter-gram-second system of units. Magnetic flux density can be measured in maxwells per square centimeter.

The weber is the more common, standard international unit of magnetic flux. The weber is equivalent to 10^8 maxwells. *See also* MAGNETIC FLUX, and WEBER.

MAXWELL BRIDGE

The *Maxwell bridge* is a device used for measuring unknown inductances. The internal resistance of a coil can also be determined. The unknown inductance, L, and the unknown series resistance, R, are determined by manipulating two variable resistors. Balance is indicated by a null reading on the indicator meter.

MAYDAY

See DISTRESS SIGNAL.

M-DERIVED FILTER

An *m-derived filter* is a variation of a constant-k inductance-capacitance filter (*see* CONSTANT-K FILTER). The m-derived filter gets its name from the fact that the values of inductance and capacitance are multiplied by a common factor m. The illustration at A shows a typical T-section, constant-k, lowpass filter for unbalanced line. The illustration at B shows an m-derived filter with component values altered by the factor m. An additional inductor is placed in series with the capacitor in the m-derived filter.

The value of the factor m is always between 0 and 1. The optimum value for m in a given situation depends on the type of

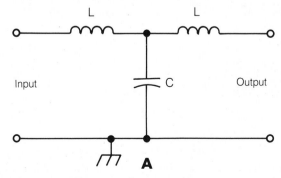

M-DERIVED FILTER: T-section lowpass filters. At A, typical constant-k; at B, typical m-derived.

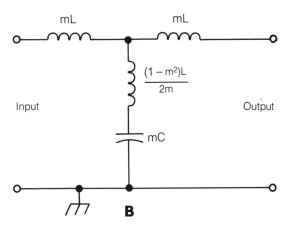

M-DERIVED FILTER Continued.

response and the cutoff frequency desired. A properly designed *m*-derived filter has a sharper cutoff than a constant-k filter for a given frequency. *See also* HIGHPASS FILTER, HIGHPASS RESPONSE, LOWPASS FILTER, and LOWPASS RESPONSE.

MECHANICAL FILTER

A *mechanical filter* is a form of bandpass filter. It is sometimes called an *ultrasonic filter*. The mechanical filter operates on a principle similar to the ceramic or crystal filter. Two transducers are employed, one for the input and the other for the output. The input signal is converted into electromechanical vibration by the input transducer. The resulting vibrations travel through a set of resonant disks to the output transducer. The output transducer converts the vibrations back into an electrical signal by means of magnetostriction.

The mechanical filter offers some advantages over electrical filters at low and medium frequencies. Mechanical elements are of reasonable size above about 75 kHz, allowing a filter to be enclosed in a small package. Above perhaps 750 kHz, mechanical filters become difficult to construct because the elements become too small. No adjustment is needed because the resonant disks are solid and of fixed size. The resonant characteristics are extremely pronounced, often far superior to electrical filters of the same physical size. Mechanical filters can be designed to have a nearly flat response within the passband, and very steep skirts with excellent ultimate attenuation (*see* SKIRT SELECTIVITY). The mechanical filter is a good choice for bandpass applications at intermediate frequencies from 75 to 750 kHz.

The elements of a mechanical filter have more than one resonant frequency. Therefore, it is necessary to provide some external means of attenuation at spurious frequencies. Mechanical filters are somewhat sensitive to physical shock, and care must be exercised to ensure that such filters are not subjected to excessive vibration.

The construction of a mechanical filter is very similar to the construction of a ceramic filter. Instead of ceramic disks, metal disks, usually made of nickel, are used. *See also* CERAMIC FILTER.

MEDIUM-SCALE INTEGRATION

Medium-scale integration is the process by which integrated circuits, containing up to 100 individual gates per chip, are fabricated. Medium-scale integration (MSI), allows considerable miniaturization of electronic circuits, but not to the extent of large-scale integration (LSI).

Both bipolar and metal-oxide semiconductor (MOS) technology have been adapted to MSI. Various linear and digital circuits use MSI circuitry. When extreme miniaturization is necessary, LSI is more often used. *See also* INTEGRATED CIRCUIT, and SOLID-STATE ELECTRONICS.

MEGA-

A prefix multiplier, attached to a quantity to indicate 1 million (10^6), is *mega-*. For example, 1 megahertz is 1 million hertz, and 1 megavolt is 1 million volts. *See also* MEGAHERTZ, MEGAWATT, MEGOHM, and PREFIX MULTIPLIERS.

MEGABYTE

The *megabyte* is a unit of information equal to $2^{20} = 1,048,576$ bytes. The megabyte is a common unit of memory used in minicomputers and larger digital computers.

Personal computers generally have memory of less than one megabyte. With the term *megabyte (M)*, we might speak of a memory capable of holding 10M, for example. *See also* BYTE, and KILOBYTE.

MEGAHERTZ

The *megahertz (MHz)* is a unit of frequency, equal to 1 million hertz (1 million cycles per second). Radio frequencies in the medium, high, very-high, and ultra-high ranges are often specified in megahertz. A frequency of 1 MHz is at approximately the middle of the standard AM broadcast band in the United States.

MEGAWATT

The *megawatt (MW)* is a unit of power, or rate of energy expenditure. A power level of 1 MW represents 1 million watts. In terms of mechanical power, 1 MW is about 1,340 horsepower.

The megawatt is a very large unit of power. A small town of 5,000 residents uses approximately 1 MW of power on an average day. Some radio broadcast transmitters have an input power level approaching 1 MW. *See also* POWER, and WATT.

MEGGER

A *megger* is a device that is used to measure extremely high resistances. The resistances of electrical insulators range from several megohms (millions of ohms) up to billions or trillions of ohms, or even higher. For the measurement of such resistances, a high-voltage source is needed.

The megger can be used for continuity testing and short-circuit location, as well as for the measurement of high direct-current resistances. *See also* OHMMETER.

MEGOHM

A *megohm* is a unit of resistance equivalent to 1 million ohms. With a potential difference of 1 volt, a current of 1 microampere (1 millionth of an ampere) flows through a resistance of 1 megohm.

Some common resistors have values that range as high as a few hundred megohms. Typical insulating materials have resistances of millions of megohms or more. *See also* OHM, and RESISTANCE.

MEMORY

The process of storing data is called *memory*. This especially applies to digital data. The circuit or medium that contains the data, once it has been stored, is also known as *memory*.

Memory can exist on magnetic tapes, magnetic disks, or in various electronic media, such as magnetic cores or digital integrated circuits (ICs or chips). *See* CORE MEMORY, DIGITAL INTEGRATED CIRCUIT, FLOPPY DISK, HARD DISK, and INTEGRATED CIRCUIT.

Memory chips are usually categorized as random-access (RAM), read-only (ROM) or read-write (scratchpad). *See* RANDOM-ACCESS MEMORY, READ-ONLY MEMORY, and READ-WRITE MEMORY.

MEMORY-BACKUP BATTERY

A *memory backup battery* is a set of electrochemical cells used for the purpose of retaining electrical data in integrated circuits. The memory backup battery can be rechargeable, and connected to the main power source, or it can be independent of the main power source.

Memory backup batteries are used in microcomputer-controlled devices in which the main source of power is removed for long periods of time. For example, a transceiver can contain programmed operating frequencies; the memory backup battery keeps these frequencies in storage even if the unit is not in use. The memory backup battery in a microcomputer should be capable of supplying sufficient energy to retain the data for at least a month. *See also* MEMORY.

MERCURY CELL

A *mercury cell* is a form of electrochemical cell that generates about 1.4 volts. The mercury cell is enclosed in a steel housing. The cathode is made from mercuric oxide; this is how the cell gets its name. The anode is made of zinc. The electrolyte is made up of potassium hydroxide and zinc oxide; hence, the mercury cell is an alkaline cell.

The mercury cell is characterized by a relatively long life for its size and mass. Also, the mercury cell tends to produce a very constant voltage throughout its life. A series combination of mercury cells is called a *mercury battery*. *See also* CELL.

MERCURY-VAPOR RECTIFIER

The *mercury-vapor rectifier* is a cold-cathode, tube-type rectifier used in high-voltage power supplies. Although solid-state rectifier diodes have largely replaced tube-type rectifiers in modern electronic power supplies, mercury-vapor tubes are still sometimes found.

The mercury-vapor tube operates because of ionization of a small amount of mercury at a high temperature and a low pressure. The voltage drop across the mercury-vapor tube electrodes is about 15 volts, regardless of the supply voltage. Mercury-vapor rectifiers are usually paired in a full-wave, center-tap configuration. Four tubes can be used as a bridge rectifier. *See also* RECTIFICATION.

MESH

A *mesh* is a collection of circuit branches with the following three properties: (1) The collection forms a closed circuit; (2) Every branch point in the circuit is incident to exactly two branches; and (3) No other branches are enclosed by the collection. Although a circuit loop can contain smaller loops, such as in the example at A in the illustration, a mesh is a "smallest possible loop," as at B.

The mesh provides a means of evaluating the currents and voltages in various parts of a complex circuit. Kirchhoffs laws can be applied to all of the individual meshes in a circuit. *See also* KIRCHHOFF'S LAWS, and MESH ANALYSIS.

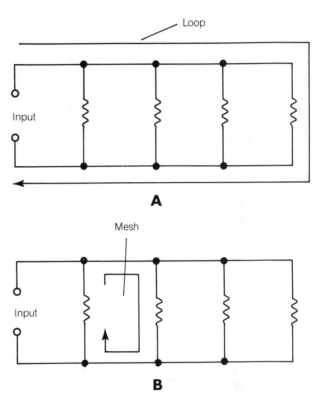

MESH: At A, a loop. At B, a mesh.

MESH ANALYSIS

Mesh analysis is a method of evaluating the operation or characteristics of an electronic circuit containing one or more closed loops. The current, voltage, and resistance through or between various parts of the circuit can be determined according to Kirchhoff's laws. Kirchhoff's laws form the basis for mesh analysis.

Mesh analysis is relatively simple for direct-current circuits, although the equations become rather cumbersome for circuits having many meshes. Mesh analysis is mathematically sophisticated in the alternating-current case, where reactance must be considered, as well as resistance.

A full description of the techniques of mesh analysis is beyond the scope of this book. For details, a circuit-theory text should be consulted. *See also* KIRCHHOFF'S LAWS, and MESH.

MESNY CIRCUIT

A *Mesny circuit* or *Mesny oscillator* is a tuned-input, tuned-output, push-pull circuit used to produce signals at ultra-high and microwave frequencies. A pair of transistors is connected together. A feedback loop causes oscillation.

The Mesny circuit uses quarter-wave matching sections to obtain resonance. A pair of parallel wires, with a movable shorting bar, is connected in the input circuit. A similar pair of wires is connected in the output. The positions of the shorting bars determine the oscillating frequency of the circuit. Normally, the input and output tuned circuits are set for resonance at the same frequency. The output is generally taken from the source or emitter circuit to minimize loading effects.

MESSAGE HEADER

In packet radio, a message is preceded by a preamble or heading, called the *message header*. The header gives essentials concerning the "letter" to follow. A message header generally contains the date, time, callsign of the originating station, callsign of the destination station, callsigns of the packet-radio bulletin-board stations (PBBSs) through which the message has passed, and the nature of the subject matter (in a few words). The text of the message comes after the header. The resulting format resembles that of an office memo. *See also* PACKET RADIO.

METAL-OXIDE SEMICONDUCTOR

The oxides of certain metals exhibit insulating properties. In recent years, so-called metal-oxide-semiconductor (MOS) devices have come into widespread use. MOS materials include such compounds as aluminum oxide and silicon dioxide. Metal-oxide semiconductor devices are noted for their low power requirements. MOS integrated circuits have high component density and high operating speed. Metal oxides are also used in the manufacture of certain field-effect transistors.

All MOS devices are subject to damage by the discharge of static electricity. Therefore, care must be exercised when working with MOS components. MOS integrated circuits and transistors should be stored with the leads inserted into a conducting foam so that large potential differences cannot develop. When building, testing, and servicing electronic equipment in which MOS devices are present, the body and all test equipment should be kept at direct-current ground potential. *See also* METAL-OXIDE-SEMICONDUCTOR FIELD-EFFECT TRANSISTOR, and METAL-OXIDE-SEMICONDUCTOR LOGIC FAMILIES.

METAL-OXIDE-SEMICONDUCTOR FIELD-EFFECT TRANSISTOR

An active component commonly used today is the *metal-oxide-semiconductor field-effect transistor (MOSFET)*. A typical MOSFET cross section is illustrated at A. The schematic symbol for a MOSFET is shown at B. The MOSFET can be recognized because the gate electrode is insulated from the channel by a thin layer of metal oxide. Because the gate is insulated from the channel, the MOSFET is sometimes called an *insulated-gate field-effect transistor (IGFET)*.

The MOSFET device has extremely high input impedance. It is typically billions or trillions of ohms. Thus, the MOSFET requires essentially no driving power. MOSFETs are commonly used in high-gain receiver amplifier circuits. A single-gate MOSFET amplifier stage can produce better than 15 dB gain at

a frequency of 100 MHz. Some MOSFETs have two gates, and are known as *dual-gate MOSFETS*. They are useful as mixers and modulators, as well as high-gain amplifiers. The schematic symbol for a dual-gate MOSFET is shown at C.

Some MOSFETs are normally conducting, or saturated, under conditions of zero gate bias. They are therefore depletion-mode devices. However, enhancement-mode MOSFETs are common. MOSFETs can have a channel consisting of either N-type material or P-type material. *See also* DEPLETION MODE, ENHANCEMENT MODE, and FIELD-EFFECT TRANSISTOR.

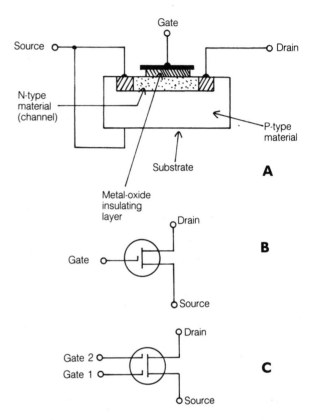

METAL-OXIDE-SEMICONDUCTOR FIELD-EFFECT TRANSISTOR: At A, simplified cutaway drawing. At B, symbol for single-gate MOSFET. At C, symbol for dual-gate MOSFET.

METAL-OXIDE-SEMICONDUCTOR LOGIC FAMILIES

Metal-oxide semiconductor technology lends itself well to the fabrication of digital integrated circuits. In recent years, several metal-oxide-semiconductor (MOS) logic families have been developed. Miniaturization is more easily obtained with MOS technology than with bipolar technology, and the power requirements are lower.

Metal-oxide-semiconductor logic families include complementary (CMOS), N-channel (NMOS), and silicon-on-sapphire (SOS). CMOS utilizes both N-type and P-type materials on the same substrate. CMOS is characterized by extremely low power requirements and relative insensitivity to external noise (*see* COMPLEMENTARY METAL-OXIDE SEMICONDUCTOR). The operating speed is relatively high.

The NMOS technology is commonly used today. It has high operating speed and the advantage of simplicity. P-channel (PMOS) metal-oxide-semiconductor logic is occasionally used, although it operates at a slower speed than NMOS or bipolar logic devices.

SOS provides the fastest operation of currently available MOS logic families. The operating speed of SOS approaches that of transistor-transistor-logic bipolar circuits. SOS processing is more expensive than CMOS, NMOS, or PMOS.

Metal-oxide-semiconductor logic families are especially useful in high-density memory applications. Many microcomputer chips make use of MOS technology today. *See also* METAL-OXIDE SEMICONDUCTOR.

METEOR SCATTER

When a meteor from space enters the upper part of the atmosphere, an ionized trail is produced because of the heat of friction. Such an ionized region reflects electromagnetic waves at certain wavelengths. This phenomenon, known as *meteor scatter*, can result in over-the-horizon propagation of radio signals (see illustration).

A single meteor generally produces a trail that persists from about 1 second to several seconds, depending on the size of the meteor, its speed, and the angle at which it enters the atmosphere. This amount of time is not sufficient for the transmission of very much information, but during a meteor shower, ionization is often almost continuous. Meteor-scatter propagation has been observed well into the very-high-frequency range of the radio spectrum.

Meteor-scatter propagation is mainly of interest to experimenters and radio amateurs; it is used in packet radio communication. Meteor-scatter propagation occurs over distances ranging from just beyond the horizon up to about 1,500 miles, depending on the height of the ionized trail and the relative locations of the trail, the transmitting station, and the receiving station.

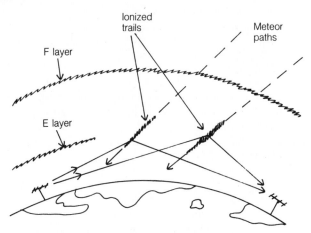

METEOR SCATTER: Usually occurs in or near the E layer.

METEOR SHOWER

The space between the planets is not empty, not a perfect vacuum. There is gas, mostly hydrogen, and dust. And there are chunks of matter ranging in size from sand grains to huge rocks. These chunks are called *meteoroids*. When one of these enters the atmosphere of the earth, it becomes a *meteor*.

Meteoroids exist in clusters, and are the remnants of broken-up, ancient comets. When the earth passes through one of these clusters, many meteors fall. Such an event is known as a *meteor shower*. There are several meteor showers annually.

They occur at the same times every year, as the earth's orbit repeatedly passes through the orbits of the comet remnants.

Intense meteor showers interest hams who work *meteor scatter* because, when meteors are falling at a rapid rate, it is possible to communicate almost continuously. Four annual showers produce meteors that fall often enough to sustain reliable communications. These occur on or about January 3, June 7, August 12, and December 12. *See also* METEOR SCATTER.

METER

A *meter* is a device, either electrical or electromechanical, that is used for the purpose of measuring an electrical quantity. The most familiar kind of meter is the moving-needle device. However, digital meters are becoming increasingly common. *See also* ANALOG METERING, and DIGITAL METERING.

The basic unit of displacement in the metric system is the meter. Originally, the meter was defined as 1 ten-millionth of the distance from the north pole to the equator of the earth. That is, there were supposed to be 10,000,000 meters in that distance. Now, the meter is more precisely defined as the standard international unit of length. A distance of 1 meter represents 1,650,763.73 wavelengths, in a vacuum, of the radiation corresponding to the transition of electrons between the levels $2p_{10}$ (in the L shell) and $5d_5$ (in the 0 shell) of the atom of krypton 86. The meter is approximately 39.37 inches, or a little more than an English yard. *See also* METRIC SYSTEM.

METRIC SYSTEM

The *metric system* is a decimal system for measurement of length, area, and volume. Because it is a decimal system, the metric system has gained widespread use in many countries throughout the world. Scientists routinely use the metric system. The principal unit of length is the meter (*see* METER). The decimeter is 0.1 meter; the centimeter is 0.01 meter; the millimeter is 0.001 meter; the kilometer is 1,000 meters.

The decimal nature of the metric system makes it much easier to use than the older English system. However, the English system is still used for non-scientific applications in the United States.

MHO

The *mho* is a unit of electrical conductance. In recent years, the mho has been called the siemens in the standard international system of units. Electrical conductance is the mathematical reciprocal of resistance. Given a resistance of R ohms, the conductance (S) in siemens is simply:

$$S = I/R$$

A resistance of 1 ohm represents a conductance of 1 siemens. When the resistance is 1,000 ohms, the conductance is 0.001 siemens or 1 millisiemens. When the resistance is 1,000,000 ohms, the conductance is 1 microsiemens. The siemens is also used as the unit of admittance in alternating-current circuits. *See also* ADMITTANCE, and CONDUCTANCE.

MICA

Mica is a silicate material that occurs naturally in the crust of the earth. Physically, mica appears as a transparent, sheet-like substance that is similar to plastic or cellophane. Mica has excellent dielectric properties.

Mica conducts heat quite well, and it is therefore useful in transistor heatsink applications when it is necessary to electrically insulate the case of the transistor from the heatsink. Mica is commonly used in the manufacture of low-value to medium-value capacitors for use at moderately high voltages. *See also* MICA CAPACITOR.

MICA CAPACITOR

Mica capacitors are characterized by low loss, high dielectric strength, and good stability under variable voltage and temperature conditions. Mica capacitors are widely used in radio-frequency circuits, especially in the construction of selective inductance-capacitance filters. Mica capacitors are available in a wide range of values, from a few picofarads up to a few thousand picofarads. Voltage ratings range from about 100 volts to 35 kilovolts. Mica capacitors are nonpolarized.

Several different methods of construction are used in the manufacture of mica capacitors. The simplest method is to clamp a pair of foil electrodes to either side of a thin sheet of mica. Other methods include the bonding of metals (especially silver) to the mica sheet, the attachment of concentric ring-shaped electrodes to the mica, and the stacking of plated mica sheets. Mica capacitors are encased in resin to protect them from the environment. *See also* CAPACITOR.

MICRO

Micro is a prefix multiplier that means $0.000001 = 10^{-6}$. For example, 1 microfarad is 10^{-6} farad, and 1 microvolt is 10^{-6} volt. *See also* PREFIX MULTIPLIERS.

The term *micro* is sometimes used to refer to something very small. For example, a microcircuit is a small, or miniaturized, electronic circuit; a microcomputer is a miniature computer. More examples can be found in the following several terms.

MICROALLOY TRANSISTOR

A *microalloy transistor* is a form of bipolar semiconductor transistor. A thin wafer of semiconductor material forms the base of the transistor. The emitter and collector are formed by alloying a small amount of impurity material at two points on opposite faces of the wafer.

The microalloy transistor is characterized by relatively low-junction capacitance. Microalloy transistors are available in either npn or pnp types. They are not commonly used today. *See also* TRANSISTOR.

MICROCIRCUIT

A *microcircuit* is a highly miniaturized electronic circuit. In modern electronics, the microcircuit is found in the form of densely packed components, all etched onto a single semiconductor wafer. The microcircuit is usually called an *integrated circuit*. *See also* INTEGRATED CIRCUIT.

MICROCOMPUTER

A *microcomputer* is a small computer that has the central processing unit (CPU) enclosed in a single integrated-circuit package. Today, microcomputer CPU chips are available in sizes ranging down to $\frac{1}{4}$ inch on a side. The microcomputer CPU is sometimes called a *microprocessor*.

Microcomputers vary in sophistication and memory storage capacity, depending on the intended use. Some personal microcomputers are available for less than $100 as of this writing. Such devices use liquid-crystal displays and have typewriter-style keyboards. Larger microcomputers are used by more serious computer hobbyists and by small businesses. Such microcomputers typically cost from several hundred to several thousand dollars. *See also* COMPUTER, MICROCOMPUTER CONTROL, and MICROPROCESSOR.

MICROCOMPUTER CONTROL

Microcomputers are often used for the purpose of regulating the operation of electrical and electromechanical devices. This is known as *microcomputer control*. Microcomputer control makes it possible to perform complex tasks with a minimum of difficulty.

An example of a microcomputer-controlled device is the programmable scanning radio receiver. Microcomputer control is also widely used in such devices as robots, automobiles, and aircraft. For example, a microcomputer can be programmed to switch on an oven, heat the food to a prescribed temperature for a certain length of time, and then switch the oven off again. Microcomputers can be used to control automobile engines to enhance efficiency and gasoline mileage. Microcomputers can navigate and fly airplanes. The list of potential applications is almost uncountable.

One of the most recent, and exciting, applications of microcomputer control is in the field of medical electronics. Microcomputers can be programmed to provide electrical impulses to control erratically functioning body organs, to move the muscles of paralyzed persons, and for various other purposes.

MICROELECTRONICS

Microelectronics is the application of miniaturization techniques to the design and construction of electronic equipment. Microelectronics has progressed to the point that a circuit once requiring an entire room, and several thousand watts of power, can now be housed in a package smaller than a dime and requiring only a few milliwatts or microwatts. The biggest breakthrough in microelectronics has been the integrated circuit. *See also* INTEGRATED CIRCUIT.

MICROINSTRUCTION

In computer machine language, a bit pattern comprising an elementary command is called a *microinstruction*. The microinstruction is the smallest form of machine-language instruction. Several microinstructions can be combined to form a macroinstruction, which is a common and identifiable computer instruction. *See also* MACHINE LANGUAGE, and MACROINSTRUCTION.

MICROMINIATURIZATION

The techniques of fabricating electronic components by etching of semiconductors is known as *microminiaturization*. Modern methods of microminiaturization have, within a few decades, reduced the physical size of electronic equipment by a factor of several million. Power requirements have decreased by almost the same proportion. A small computer can now be fit into a case no larger than a wallet.

The most advanced form of microminiaturization is found in integrated-circuit chips. The most compact chips contain hundreds or even thousands of individual logic gates on a semiconductor wafer that has a surface area of a fraction of a square inch. Very-large-scale integration (VLSI) chips contain from 100 to 1,000 gates on a single wafer. There is some research being done in the field of ultra-large-scale integration (ULSI), in which up to 10,000 gates might be placed on a single chip.

Although there must be a limit to how small an electronic circuit can be (electrons are tiny, but are nevertheless bigger than geometric points), we have not yet reached this limit. Some scientists think we are still nowhere near it. *See also* LARGE-SCALE INTEGRATION, VERY-LARGE-SCALE INTEGRATION, and ULTRA-LARGE-SCALE INTEGRATION.

MICROPHONE

A *microphone* is an electroacoustic transducer that is designed to produce alternating-current electrical impulses from sound waves. This can be done in many ways, and thus there are many different types of microphones.

One of the earliest microphones was the carbon-granule microphone. This device required an external source of direct current. The sound waves caused changes in the resistance of a carbon-granule container, resulting in modulation of the current (*see* CARBON MICROPHONE). This kind of microphone is seldom used today.

More recently, the piezoelectric effect has been used to generate electric currents from passing sound waves. Various substances, when subjected to vibration, produce weak electric impulses that can be amplified. The ceramic microphone operates on this principle (see CERAMIC MICROPHONE).

A more rugged type of microphone operates by means of electromagnetic effects. A diaphragm, set in motion by passing sound waves, causes a coil and magnet to move with respect to each other. The fluctuating magnetic field thus results in alternating currents through the coil. This device is called a *dynamic microphone* (*see* DYNAMIC MICROPHONE).

Other, less common, forms of microphones are used for special purposes. Electrostatic and optical devices, for example, can be used as microphones in certain situations.

Microphones are available in a wide variety of sizes, input impedances, and with various frequency-response characteristics. Some microphones are omnidirectional, and others have a cardioid or unidirectional response. The optimum choice of a microphone is important in any audio-frequency system. Specialized microphones are made for communications, high-fidelity, and public-address applications.

MICROPHONICS

Mechanical vibration can cause unwanted modulation of a radio-frequency oscillator circuit, or in an audio- or radio-frequency amplifier circuit. This can occur in a transmitter, resulting in over-the-air noise in addition to the desired modulation; it can occur in a receiver, causing apparent noise on a signal. Such unwanted modulation is known as *microphonics*.

In fixed-station equipment, microphonics are not usually a problem because mechanical vibration is not severe. However, in mobile applications, equipment must be designed for minimum susceptibility to microphonics. Microphonics can, if severe, result in out-of-band modulation in a transmitter. It can also make a signal almost unintelligible.

For immunity to microphonics, equipment circuit boards must be firmly anchored to the chassis. Component leads must be kept short. Especially sensitive circuits, such as oscillators, might have to be encased in wax or some other shock-absorbing substance.

MICROPROCESSOR

The central processing unit of a microcomputer is sometimes called a *microprocessor*. There is some confusion between the terms microcomputer and microprocessor, and the two are often used in place of each other (*see* MICROCOMPUTER).

Technically, a microcomputer consists of the microprocessor integrated circuit and perhaps one or more peripheral integrated circuits. The peripheral circuits can be integrated onto the same chip as the central processing unit, or they can be separate. The peripheral integrated circuits contain memory and programming instructions. *See also* CENTRAL PROCESSING UNIT.

MICROPROGRAM

A *microprogram* is a set of microinstructions, at the machine-language level, that results in the execution of a specific function, independent of the functions of the program being run. The microprograms implement routine operating functions in a computer. Microprograms can be permanently placed in a computer or microcomputer in the form of firmware. *See also* FIRMWARE.

MICROSAT

In 1990, six Phase II amateur satellites were launched into low, sun-synchronous orbits around the earth. The satellites are physically small, approximately $9 \times 9 \times 9$ inches cubic. For this reason they are called *Microsats*. The Microsats are part of the OSCAR program. Amateur satellite organizations from several countries provided support for this set of satellites.

The Microsats are used largely for packet radio communications. *See also* ACTIVE COMMUNICATIONS SATELLITE, OSCAR, PACKET RADIO, and PHASE II SATELLITE.

MICROSTRIP

Microstrip is a form of unbalanced transmission line. A flat conductor is bonded to a ground-plane strip by means of a dielectric material (see illustration on pg. 312). This results in an image conductor, parallel to the flat wire conductor, but on the other side of the ground-plane strip. The effective current in the image conductor is equal in magnitude to the current in the actual conductor, but flows in the opposite direction. Thus, very little radiation occurs (in theory) from a microstrip transmission line.

Microstrip lines are sometimes used at ultra-high and microwave frequencies. Microstrip lines have somewhat lower loss than coaxial lines, but radiate less than most open-wire lines at the short wavelengths. The characteristic impedance of the microstrip line depends on the width of the flat wire conductor, the spacing between the flat wire conductor and the groundplane, and on the type of dielectric material used.

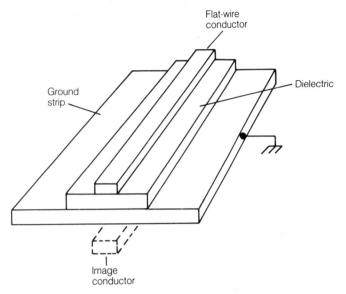

Flat-wire conductor

Ground strip

Dielectric

Image conductor

MICROSTRIP: An unbalanced transmission line for microwaves.

MICROWAVE

Microwaves are defined as that part of the electromagnetic spectrum at which the wavelength falls between about 1 millimeter and 30 centimeters. The microwave frequencies range from approximately 1 to 300 GHz.

Microwaves are very short electromagnetic radio waves, but they have a longer wavelength than infrared energy. Microwaves travel in essentially straight lines through the atmosphere, and are not affected by the ionized layers.

Microwave frequencies are useful for short-range, high-reliability radio and television links. In a radio or television broadcasting system, the studio is usually at a different location than the transmitter; a microwave link connects them. Satellite communication and control is generally accomplished at microwave frequencies. The microwave region contains a vast amount of spectrum space, and can therefore hold many wideband signals.

Microwave radiation can cause heating of certain materials. This heating can be dangerous to human beings when the microwave radiation is intense. When working with microwave equipment, care must be exercised to avoid exposure to the rays. The heating of organic tissue by microwaves can be put to constructive use, however; this idea led to the invention of the microwave oven.

MICROWAVE REPEATER

A *microwave repeater* is a receiver/transmitter combination used for relaying signals at microwave frequencies. Basically, the microwave repeater works in the same way as a repeater at any lower frequency. The signal is intercepted by a horn or dish antenna, amplified, converted to another frequency, and retransmitted.

Microwave repeaters are used in long-distance overland communications links. With the aid of such repeaters, microwave links supplant wire-transmission systems. *See also* REPEATER.

MILLER EFFECT

A transistor, field-effect transistor, or tube exhibits variable input capacitance under conditions of changing direct-current input bias. In general, an active amplifying device shows smaller input capacitance when it is biased at or beyond cutoff, as compared with bias below the cutoff point. This change in capacitance is called the *Miller effect*.

The Miller effect can result in rapidly fluctuating input impedance to a stage driven into the cutoff region during the alternating-current input cycle. For example, in a class-B amplifier circuit, the input capacitance normally increases with increasing drive. This can result in nonlinearity unless the circuit is designed with the Miller effect in mind. *See also* INPUT IMPEDANCE.

MILLER OSCILLATOR

A *Miller oscillator* is a special form of crystal oscillator. A transistor, field-effect transistor, or tube can be used as the active element in the circuit. The Miller oscillator can be recognized by the presence of the crystal between the base, gate, or grid and ground. The output contains a tuned circuit. The internal capacitance of the active device provides the necessary feedback for oscillation. The Miller oscillator is sometimes called a *conventional crystal oscillator. See also* CRYSTAL OSCILLATOR, and OSCILLATION.

MILLI-

Milli- is a prefix multiplier meaning 0.001 or 1/1,000. When this prefix is placed in front of a quantity, the magnitude is decreased by a factor of 1,000. Thus, for example, 1 millimeter is equal to 0.001 meter, 1 milliampere is equal to 0.001 ampere, and so on. *See also* PREFIX MULTIPLIERS.

MILLS CROSS

The *Mills cross* is a form of antenna used mostly by radio astronomers. Two phased arrays are positioned at an angle. The antenna is named after the radio astronomer B.Y. Mills, who originally called the antenna the "super cross."

Mills perfected his design at the University of Sydney in Australia. His antenna consists of two cylindrical parabolic reflectors, each 1 mile long, and oriented at right angles to each other. By changing the phase relationship between the two linear arrays, it is possible to steer the main lobe to a certain extent.

By combining two linear arrays, it is possible to obtain much greater resolution than with either linear array by itself. In fact, the Mills cross, with its two mile-long arrays, has resolution almost equivalent to that of a single dish 1 mile in diameter. At a wavelength of 2.7 meters (about 112 MHz), the main-lobe beamwidth is just 10 minutes of arc, or 0.167 degree. At higher frequencies, the beamwidth is even smaller. *See also* RADIO ASTRONOMY, and RADIO TELESCOPE.

MINIATURIZATION

A few years ago, it would have been absurd to think of building a pocket-sized microcomputer; now such devices are commonplace. Radio receivers and transmitters that once required a large portion of a room can now be combined in a desktop unit. Miniaturization of electronic equipment accomplishes two things: first, it reduces the size of a given piece of apparatus, and second, it makes it possible to put more into a given amount of physical space.

Solid-state technology, especially the integrated circuit, has been the main reason for the rapid advances that have been made in miniaturization. Printed circuit boards and modular construction, along with better use of space, have also been responsible. Miniaturization has not only resulted in greater component and circuit density, but it has also greatly reduced the power requirements and improved electrical efficiency. *See also* INTEGRATED CIRCUIT, and SOLID-STATE ELECTRONICS.

MINICOMPUTER

A *minicomputer* is a small digital computer. It is similar to a microcomputer, but has larger memory capacity and higher speed of operation. The demarcation between a minicomputer and a microcomputer is not precisely defined. You might call a 128-kilobyte computer a *micro* and a 32-megabyte computer a *mini*. As the technology advances, however, we might someday speak of a 32-megabyte computer as a *micro* and a 512-megabyte computer as a *mini*!

Minicomputers are used by medium-size and large businesses. For small-business and personal applications, microcomputers are more often used. *See also* MICROCOMPUTER.

MINORITY CARRIER

Semiconductor materials carry electric currents in two ways. Electrons transfer negative charge from the negative pole to the positive pole; holes carry positive charge from the positive pole to the negative pole (*see* ELECTRON, and HOLE).

Electrons and holes are known as *charge carriers*. In an N-type semiconductor material, most of the charge is carried by electrons, and relatively little by holes. Thus, holes are called the *minority carrier* in N-type material. Conversely, in P-type material, the electrons are the minority carriers. *See also* MAJORITY CARRIER, N-TYPE SEMICONDUCTOR, and P-TYPE SEMICONDUCTOR.

MISMATCHED LINE

When the terminating impedance, or load impedance, of a transmission line is not a pure resistance with a value equal to the characteristic impedance of the line, the line is said to be *mismatched*. A line is always mismatched if the load contains reactance; a purely resistive load of the wrong value can also present a mismatch to a transmission line.

A mismatched line can degrade the performance of a system significantly, but this is not always the case. In a mismatched line, the voltage and current do not exist in uniform proportion. The current and voltage, instead, occur in a pattern of maxima and minima along the line (see illustration). Current minima occur at the same points as voltage maxima, and current maxima correspond to voltage minima. This pattern of current and voltage is known as a *standing-wave pattern*. Standing waves result in increased conductor and dielectric heating in a transmission line, and this causes more power to be dissipated as heat, reducing the efficiency of the system.

Small mismatches in a transmission line can often be tolerated with little or no adverse effects. However, large mismatches can cause problems, such as reduced generator efficiency, increased line loss, and overheating of the line. The degree of the mismatch is called the *standing-wave ratio*. *See also* STANDING-WAVE RATIO, and STANDING-WAVE-RATIO LOSS.

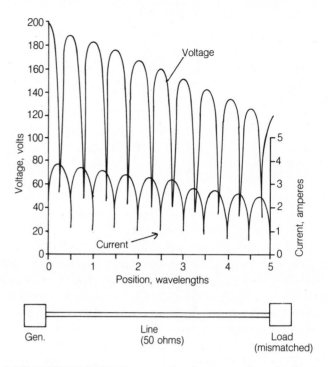

MISMATCHED LINE: Current and voltage vary along the line.

MIXER

A *mixer* is a device that combines two signals of different frequencies to produce a third signal, the frequency of which is either the sum or difference of the input frequencies. Usually, the difference frequency is used as the mixer output frequency. Mixers are widely used in superheterodyne receivers, and in radio transmitters (*see* SUPERHETERODYNE RECEIVER).

A mixer requires some kind of nonlinear circuit element in order to function properly. The nonlinear element can be a diode or combination of diodes; such a circuit is shown in the illustration at A. The diode mixer is a passive circuit because it does not require an external source of power. However, there is some insertion loss when such a mixer is used.

Mixers can have gain; at B, a bipolar-transistor mixer is shown. At C, a dual-gate metal-oxide-semiconductor field-effect transistor (MOSFET) is used.

The mixer operates on a principle similar to that of an amplitude modulator. However, in the mixer, both signals are usually in the radio-frequency spectrum, while in a modulator, one signal is much lower in frequency than the other (*see* MODULATOR). The output of the mixer circuit is tuned to the sum or difference frequency, as desired.

Mixing can occur in a circuit even when it is not desired. This is especially likely in high-gain receiver front-end ampli-

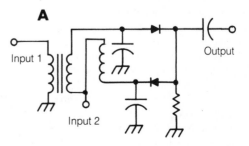

MIXER: At A, passive; at B, active with bipolar transistor; at C, active with dual-gate MOSFET.

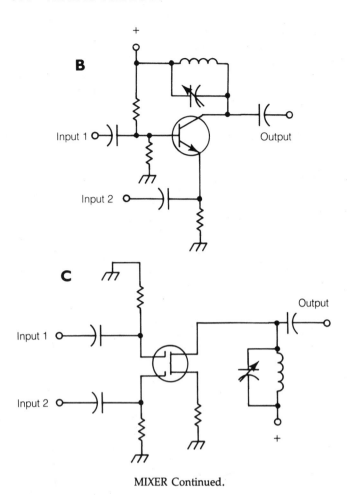

MIXER Continued.

fiers. Mixing can also occur between a desired signal and a parasitic oscillation in an amplifier or oscillator. Mixing sometimes takes place in semiconductor devices when two or more strong signals are present. Such mixing is called *intermodulation*, and the resulting unwanted signals are called *mixing products*.

A device used for combining two or more audio signals in a broadcast or recording studio is known as a *mixer*. In this type of device, nonlinear operation, with the consequent generation of harmonics and distortion products, is not desired. The circuit must therefore be as linear as possible.

MIXING PRODUCT

A *mixing product* is a signal that results from mixing of two other signals. The output of a mixer, for example, is a mixing product (*see* MIXER). Mixing products can occur either as a result of intentional mixing, as in the mixer, or as a result of unintentional mixing, in amplifiers, oscillators, or filters.

Whenever two signals with frequencies f and g are combined in a nonlinear component, mixing products occur at frequencies $f - g$ and $f + g$. If there are three or more signals, many different mixing products will exist. Mixing products can themselves combine with the original signals or other mixing products, producing further signals known as *higher-order mixing products*. It is not difficult to see how, when numerous signals are combined in a nonlinear environment, the situation can get very complex!

Mixing products can result in interference to radio receivers. Two external signals, at frequencies f and g, can mix in the receiver front end and result in interference at frequencies $f - g$ and $f + g$. This is called *intermodulation*. Mixing products generated within a receiver, from signals originating in the local oscillators, are sometimes called *birdies*. *See also* INTERMODULATION.

MOBILE ANTENNA

A *mobile antenna* is any antenna that is primarily intended for use in a mobile installation. Mobile antennas are almost always vertical antennas, and they usually use the metal body of the vehicle as the RF ground. They generally use 50-ohm coaxial feed lines.

Shortened Whips For the high-frequency (HF) bands, and especially below 28 MHz, mobile antennas must be inductively loaded if they are to be kept at a manageable length. In practice, any height over about 10 feet is impractical for a whip antenna because the tip of the whip will strike barriers (such as bridges, trees, and highway overpasses).

In order to accomplish inductive loading, a coil can be placed either near the center of the whip, or at the base. These schemes are known as *center loading* and *base loading*, respectively.

In a center-loaded mobile antenna, interchangeable coil-and-whip combinations are supplied, with different size coils for each band (160, 80/75, 40, 30, 20, 17, 15, 12, and 10 meters). These units can be screwed into place for operation on a given band, and then unscrewed and replaced with another unit for use on another band.

In a base-loaded HF mobile antenna, a tapped coil is placed at the bottom of a whip about 8 feet long. The tap is moved to different points for operation on various bands. Tap points are marked, and a clip lead is generally used.

Quarter-wave Whips At 10 meters and above, full-size, quarter-wave whips are practical. A 28-MHz quarter-wave antenna is about 8 feet high. At 50 MHz, a quarter wave is about 4 feet 8 inches, and at 144 MHz it is about 19 inches. For the 2-meter band and above, magnetic mounts are feasible, and a quarter-wave antenna can be placed in the center of the car roof for good omnidirectional performance.

Longer Whips At 144 MHz and higher frequencies, $1/2$-wave and $5/8$-wave whips are fairly common. This type of antenna usually has a magnetic mount, with an impedance-matching transformer at the base. Such an antenna provides 2 to 3 dB omnidirectional gain over a quarter-wave whip.

Some mobile antennas for ultra-high frequencies (432 MHz and higher ham bands) use two or even three half-wave collinear elements, with coils in between to ensure proper phasing. These are similar to the antennas used at repeater installations. They provide omnidirectional gain that is even greater than that of a $1/2$-wave or $5/8$-wave whip.

Coils or traps can be used to provide multiband performance for mobile VHF/UHF antennas. For example, a trap, placed at just the right point in a whip antenna, can allow dual-band operation at 144 and 432 MHz. It is uncommon to see three-band trapped mobile antennas, although the design would be straightforward and quite practical at VHF/UHF.

Other Mobile Antennas Rarely, you might see horizontally polarized mobile antennas. These are usually small (half-wave) loops or dipoles. Some hams actually use yagi antennas at 2 meters and above in mobile installations. But this is not too practical with a vehicle on the road, because the direction of travel changes often, necessitating constant adjustment of the antenna rotator. More often, these installations are actually *portable*, intended for use when the car or truck is not moving.

Yagis and quads can work well for VHF operation on boats when the bearing is fairly constant. On large vessels, beams, quads, and dipoles can be used—even on the HF bands.

MOBILE EQUIPMENT

Mobile electronic equipment is any apparatus that is operated in a moving vehicle such as a car, truck, train, boat, or airplane. Mobile equipment includes all attendant devices, including antennas, microphones, power supplies, and so on.

The design and operation of mobile electronic equipment is much simpler today than it was just a few years ago. The main reason for the improvement is the development of solid-state technology. Most mobile power supplies provide low-voltage direct-current power; 13.8 volts is by far the most common. A tube type circuit requires a complex power inverter for operation from such a supply, but solid-state equipment can usually be operated directly from the mobile supply.

Mobile equipment is generally more compact than fixed-station apparatus. In addition, mobile equipment must be designed to withstand larger changes in temperature and humidity, as well as severe mechanical vibration.

Most two-way mobile communications installations consist of a vertically polarized omnidirectional antenna and transceiver. The transceiver is a transmitter-receiver combination housed in a single cabinet (*see* TRANSCEIVER). This saves significant space and cost. Most mobile operation is done at very-high and ultra-high radio frequencies, where the antennas are manageable in size. However, some mobile operation is carried out at low, medium, and high frequencies, using inductively loaded antennas.

Equipment designed for portable operation can also be used in mobile applications. Generally, however, portable radio transmitters and receivers are less powerful and versatile than apparatus designed particularly for mobile use. *See also* PORTABLE EQUIPMENT.

MOBILE EQUIPMENT: A compact mobile radio transceiver. Radio Shack, a division of Tandy Corporation

MODE A, B, J, K, L, S, T

See SATELLITE TRANSPONDER MODES.

MODEM

The term *modem* is a contraction of the words *mo*dulator and *dem*odulator. A modem is a two-way interfacing device that can perform either modulation or demodulation. Modems are extensively used in computer communications, for interfacing the digital signals of the computer with a transmission medium such as a radio or telephone circuit.

The illustration is a block diagram of a modem suitable for interfacing a home or business computer with an ordinary telephone. The modulator converts the digital signals from the computer into audio tones. The output is similar to that of an ordinary audio-frequency-shift keyer. The demodulator converts the incoming tones back into digital signals. The audio tones fall within the band of approximately 300 Hz to 3 kHz, so they can be efficiently transmitted over a telephone circuit or narrow-band radio transmitter.

There are many different kinds of devices that perform modulation and demodulation, and can therefore be called *modems*. In general, any signal converter that consists of modulator and demodulator is a modem. *See also* DETECTION, and MODULATION.

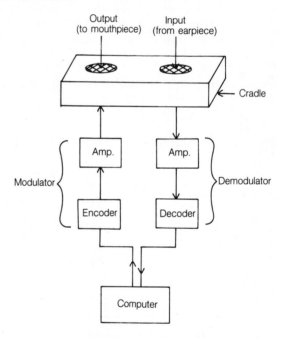

MODEM: Basic construction of telephone modem.

MODULAR CONSTRUCTION

A few decades ago, electronic equipment was constructed in a much different way than it is today. Components were mounted on tie strips, and wiring was done in point-to-point fashion (*see* POINT-TO-POINT WIRING). This kind of wiring is still used in some high-power radio transmitters, but in recent years, modular construction has become the rule. Solid-state technology has been largely responsible for this change; circuits are far more compact, and there is much less shock hazard, than was the case in the vacuum-tube days.

In the modular method of construction, individual circuit boards are used. Each circuit board contains the components for a certain part or parts of the system. The circuit boards are entirely removable, usually with a simple tool that resembles pliers. Edge connectors facilitate easy replacement. The edge connectors are wired together for interconnection between circuit boards.

Modular construction has greatly simplified the maintenance and servicing of complicated apparatus. In-the-field repair consists of nothing more than the identification, removal, and replacement of the faulty module. The faulty circuit board is then sent to a central facility, where it can be fixed using highly sophisticated equipment. Once the faulty board has been repaired, it is ready to serve as a replacement module in another system when the need arises.

MODULATED LIGHT

Electromagnetic energy of any frequency can theoretically be modulated for the purpose of transmitting intelligence. The only constraint is that the frequency of the carrier be at least several times the highest modulating frequency. Modulated light has recently become a significant method of transmitting information. The frequency of visible light radiation is exceedingly high, and therefore modulated light allows the transfer of a great amount of information on a single beam.

A simple modulated-light communications system can be built for a few dollars, using components available in hardware and electronics retail stores. The schematic diagram shows such a system. It is capable of operating over a range of several feet. More sophisticated modulated-light systems use laser beams for greater range. Optical fibers can also be used (see FIBER, and FIBER OPTICS).

The system shown is an amplitude-modulation system. There are other ways of modulating a light beam. Polarization modulation is one such alternative. Lasers make pulse modulation a viable method of transmitting information over light beams. Position modulation is also possible, especially with narrow-beam sources, such as lasers. Modulation of the actual frequency of a light beam is difficult to obtain directly, but phase modulation can be achieved with coherent-light sources.

Modulated-light transmission through the atmosphere is, of course, limited by weather conditions. However, cloudless weather is frequent enough in some parts of the world to make modulated-light satellite communications and control feasible. *See also* OPTICAL COMMUNICATIONS.

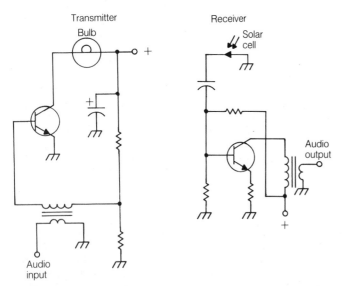

MODULATED LIGHT: Simple transmitter and receiver for visible-light voice communications.

MODULATION

When some characteristic of an electromagnetic wave is deliberately changed or manipulated for the purpose of transmitting information, the energy is said to be *modulated*. Modulation can be accomplished with any form of electromagnetic energy. There are many different forms of modulation.

Amplitude modulation was the first method of transmitting complex information via electromagnetic waves. The simplest form of amplitude modulation is morse-code transmission. Voices and other analog signals can be easily impressed onto a carrier wave. Analog amplitude modulation can take several forms. *See* AMPLITUDE MODULATION, DOUBLE SIDEBAND, and SINGLE SIDEBAND.

Frequency modulation is another common method of conveying intelligence via electromagnetic waves. Phase modulation works in a similar way. *See* FREQUENCY MODULATION, and PHASE MODULATION.

A more recent development has been the technique of pulse modulation. There are several ways of modulating signal pulses; the amplitude, frequency, duration, or position of a pulse can be modified to achieve modulation. *See* PULSE MODULATION.

To recover the intelligence in a modulated signal, some means is necessary to separate the modulating signal from the carrier wave at the receiving end of a communications circuit. This process is called *detection*. Different forms of modulation require different processes for signal detection. *See* DETECTION, DISCRIMINATOR, ENVELOPE DETECTOR, HETERODYNE DETECTOR, and PRODUCT DETECTOR.

In general, if a characteristic of an electromagnetic wave can be made to fluctuate rapidly enough to convey intelligence at the desired rate, then that parameter offers a means of modulation. The more data that must be transmitted in a given amount of time, however, the more difficult it becomes to efficiently modulate an electromagnetic wave in a particular manner. *See also* MODULATION METHODS, and MODULATOR.

MODULATION COEFFICIENT

The *modulation coefficient* is a specification of the extent of amplitude modulation of an electromagnetic wave. The modulation coefficient is abbreviated by the small letter m, and it can range from a value of $m = 0$ (for an unmodulated carrier) to $m = 1$ (for 100-percent or maximum distortion-free modulation).

Let E_c be the peak-to-peak voltage of the unmodulated carrier wave. Let E_m be the maximum peak-to-peak voltage of the modulated carrier. Then:

$$m = (E_m - E_c)/E_c$$

Theoretically, m can be larger than 1, but this represents over-modulation, and distortion inevitably occurs under such conditions. *See also* AMPLITUDE MODULATION, and MODULATION PERCENTAGE.

MODULATION INDEX

In a frequency-modulated transmitter, the modulation index is a specification that indicates the extent of modulation. Frequency modulation differs from all forms of amplitude modulation in terms of the way it must be measured. There is a limit to the extent to which a signal can be usefully amplitude-modulated, but the only limitation on frequency deviation is imposed by the fact that the frequency cannot be lower than zero at any given instant. For practical purposes, such a limitation is nearly meaningless.

Suppose that the maximum instantaneous frequency deviation (*see* DEVIATION) of a carrier is *f* kHz. Suppose the instantaneous audio modulation frequency is *g* kHz. Let us call the modulation index by the letter *m*. Then:

$$m = f/g$$

For example, if the frequency deviation is plus or minus 5 kHz and the modulating frequency is 3 kHz, then $f = 5$ and $g = 3$; therefore:

$$m = \frac{5}{3} = 1.67$$

The modulation index is a useful means of measuring frequency modulation when a pure sine-wave tone and constant deviation are used. However, with a voice signal, this is not the case, and a more general form of the modulation index must be used. This specification is called the *deviation ratio*. The deviation ratio (*d*) is given by:

$$d = f/g$$

where *f* is the maximum instantaneous deviation and *g* is the highest audio modulating frequency, both specified in kilohertz. *See also* FREQUENCY MODULATION.

MODULATION METHODS

There are numerous ways in which data can be conveyed by means of radio waves. If a characteristic of the wave can be varied fast enough, and if the changes can be detected at the receiver, modulation is possible by this means. The circuit in a transmitter that impresses the data onto the carrier wave is known as a *modulator*. Each type of modulation has its own special modulator design. *See* MODULATION, and MODULATOR.

The most common ways to convey data are by varying the amplitude, the frequency or the phase of the carrier wave. *See* AMPLITUDE MODULATION, FREQUENCY MODULATION, and PHASE MODULATION. These methods are usually used to transmit voice information. *Also see* DIGITAL MODULATION, PULSE MODULATION, and SINGLE SIDEBAND.

For the transmission of alphanumeric matter, either printed or "by ear," only two states are required: *on* and *off*. These are also called *1* and *0*, or *high* and *low*, or *mark* and *space*. The simplest such modulation scheme is morse-code keying (*see* CONTINUOUS WAVE). For machine printing (teleprinting), the frequency is shifted, but the carrier is always at full power (*see* FREQUENCY-SHIFT KEYING).

For the transmission of images, facsimile and television are employed. These signals can be sent using forms of amplitude modulation, frequency modulation and sometimes even digital or pulse modulation. *See* FACSIMILE, SLOW-SCAN TELEVISION, and TELEVISION.

There are more exotic ways of getting modulation on a signal. For example, the polarization of a signal can be varied by driving two antennas at right angles to each other, and by varying the relative signal strengths at these antennas. *See* POLARIZATION MODULATION. Each type of modulation requires a specialized detector at the receiving end in order to retrieve the data.

Related articles in this encyclopedia, not mentioned above,

include: COMPOUND MODULATION, CONSTANT-CURRENT MODULATION, DELTA MODULATION, DENSITY MODULATION, DETECTION METHODS, DOWNWARD MODULATION, EFFICIENCY MODULATION, EMISSION CLASS, MODULATED LIGHT, NARROW-BAND VOICE MODULATION, SPREAD SPECTRUM, UPWARD MODULATION, and VELOCITY MODULATION.

MODULATION PERCENTAGE

In an amplitude-modulated signal, the modulation percentage is a measure of the extent to which the carrier wave is modulated. A percentage of zero refers to the absence of amplitude modulation. A percentage of 100 represents the maximum modulating-signal amplitude that can be accommodated without envelope distortion.

Suppose the unmodulated-carrier voltage of a signal is E_c volts, and that the peak amplitude (with maximum modulation) is E_m volts. Then, the percentage of modulation, *m*, is given by:

$$m = 100 \, (E_m - E_c)/E_c$$

Under conditions of 100-percent amplitude modulation, the peak power is four times the unmodulated power, regardless of the waveform of the modulating signal. The extent to which the average power increases over the unmodulated-carrier power, for a given modulation percentage, depends on the waveform. The average-power increase with 100-percent sine-wave modulation is approximately 50 percent. For a human voice, the average power increases by only about half that amount. Generally, amplitude modulation levels of more than 100 percent are not used. This is because negative-peak clipping occurs under such conditions, resulting in distortion of the signal and unnecessary bandwidth. *See also* AMPLITUDE MODULATION.

MODULATOR

A *modulator* is a circuit that combines information with a radio-frequency carrier for the purpose of transmission over the airwaves. Different forms of modulation require different kinds of modulator circuits.

The simplest modulator is the telegraph key for producing code transmissions. Code is a form of amplitude modulation. More complex amplitude modulation is produced by a circuit, such as that shown in the illustration on pg. 318. This circuit is essentially an amplifier with variable gain. The gain is controlled by the input signal, in this case an audio signal from the microphone and audio amplifier. All amplitude modulators work according to this basic principle, although the details vary. *See* AMPLITUDE MODULATION.

A special sort of amplitude modulator is designed to eliminate the carrier wave, leaving only the sideband energy. Such a circuit is called a *balanced modulator. See* BALANCED MODULATOR, DOUBLE SIDEBAND, and SINGLE SIDEBAND.

Frequency or phase modulation requires the introduction of a variable reactance into an oscillator circuit. This can be done in various ways. The variable reactance is controlled by the modulating signal, and causes the phase and/or resonant frequency of the oscillator to fluctuate. *See* FREQUENCY MODULATION, PHASE MODULATION, and REACTANCE MODULATOR.

Pulse modulation is obtained by circuits that cause changes

in the amplitude, timing, or duration of high-powered radio-frequency pulses. The circuit details depend on the type of pulse modulation used. *See* PULSE MODULATION.

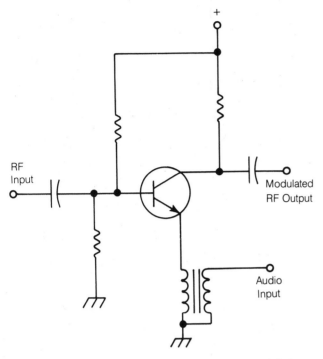

MODULATOR: A simple circuit for amplitude modulation.

MODULATOR-DEMODULATOR

See MODEM.

MONITOR

A *monitor* is an instrument that allows a signal to be analyzed. The oscilloscope is a common type of signal monitor (*see* OSCILLOSCOPE). However, more specialized types of monitors are used for such purposes as analyzing the modulating waveform in an amplitude-modulated or frequency-modulated signal or evaluating the spectral output of a radio transmitter.

In a video display terminal, a monitor is used to view the text or data in alphanumeric form. Such a monitor consists of a cathode-ray tube and associated circuitry, and appears like a television set. *See also* VIDEO DISPLAY TERMINAL.

MONKEY CHATTER

When a single-sideband signal is not tuned in properly, the voice is impossible to understand. This can occur because the receiver and transmitter are not on precisely the same frequency. It can also happen when a receiver is set for one sideband and the transmitted signal is on the opposite sideband. The sound of the signal under such conditions is called *monkey chatter*. In some cases, the signal really sounds like a monkey. *See also* SINGLE SIDEBAND.

MONOPOLE ANTENNA

Any unbalanced antenna, measuring ¼ physical wavelength or less in free space, is known as a *monopole antenna*. The monopole antenna is always quarter-wave resonant. The monopole

is always fed at one end, and requires a ground plane for proper operation. The drawing illustrates the basic concept of the monopole antenna.

A monopole antenna is normally fed with unbalanced transmission line, such as coaxial cable. The feed-point impedance varies, depending on the design. For a straight quarter-wave monopole over perfectly conducting ground, the feed-point impedance is approximately 37 ohms.

There are many variations of the monopole antenna. *See also* CONICAL MONOPOLE ANTENNA, DISCONE ANTENNA, MARCONI ANTENNA, and VERTICAL ANTENNA.

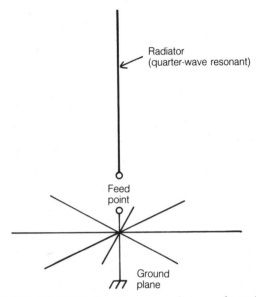

MONOPOLE ANTENNA: Quarter-wave radiator works against RF ground.

MONOSTABLE MULTIVIBRATOR

A *monostable multivibrator* is a circuit with only one stable condition. The circuit can be removed from this condition temporarily, but it always returns to that condition after a certain period of time. The monostable multivibrator is sometimes called a *one-shot multivibrator*.

The illustration on pg. 319 is a simple schematic diagram of a monostable multivibrator. Normally, the output is high, at the level of the supply voltage (+5 V). When a positive triggering pulse is applied to the input, the output goes low (0 V) for a length of time that depends on the values of the timing resistor (R) and the timing capacitor (C). If R is given in ohms and C is given in microfarads, then the pulse duration (T), in microseconds, can be found by the equation:

$$T = 0.69RC$$

After the pulse duration time (T) has elapsed, the monostable multivibrator returns to the high state.

Monostable multivibrators are used as pulse generators, timing-wave generators, and sweep generators for cathode-ray-tube devices. *See also* MULTIVIBRATOR.

MOONBOUNCE

If communication is possible by any means, no matter how esoteric, there are hams who will attempt it. It is well known that radio signals, in multihop ionospheric propagation, are re-

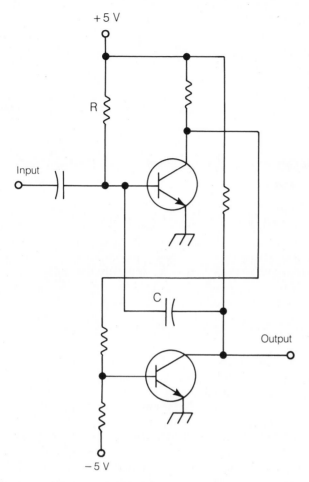

MONOSTABLE MULTIVIBRATOR: Also called a *one-shot circuit*.

flected one or more times from the surface of the earth. This led hams to contemplate bouncing signals off the moon.

The earth's natural satellite is large enough, and near enough, to return signals using ham equipment. *Moonbounce*, also known as *earth-moon-earth (EME)*, was first successfully done in the high-frequency (HF) range at 21 and 28 MHz, where signals often can pass through the ionosphere almost unaffected. Later, 144 MHz and 432 MHz became the main frequencies for this mode. Moonbounce has also been done regularly by some hams at 1296 MHz.

Equipment Needed A sensitive receiver, using a low-noise preamplifier and having narrow bandwidth, is mandatory if EME is contemplated. Some hams use their HF rigs along with converters. In recent years, good VHF and UHF gear has become available. These rigs will receive moonbounce if used along with GaAs FET preamplifiers. *See* PREAMPLIFIER.

A high-gain antenna is always needed. At 144 and 432 MHz, stacked Yagis, or even matrixes (bays) of four or more Yagis, are common. Dish antennas become practical at 1296 MHz and above. *See* DISH ANTENNA, and YAGI ANTENNA. It is considered necessary to use the maximum legal power for transmitting: 1500 W PEP output.

Finally, code (CW) is the only mode that offers any chance of a solid QSO via moonbounce. The signal-to-noise ratio is not usually good enough to allow communications by any other means. Moreover, severe and rapid fading is common, and this would introduce distortion into a single-sideband or frequency-modulated signal, even if it was strong.

It is possible that synchronized communications might offer an improvement in moonbounce mode in future years, when and if such equipment becomes commonly available. *See* SYNCHRONIZED COMMUNICATIONS.

Technical Considerations The primary difficulty with moonbounce is the signal path loss. Any signal received via EME will be very weak. Another problem is that the high-gain antennas must always be kept aimed at the moon. This can be done using a clock drive and an equatorial mount for the array or dish, but the moon does not exactly follow the stars in the sky, so correction is needed to compensate for the moon's orbital motion around the earth.

Still another problem is *solar noise*. Moonbounce is more difficult near the time of the new moon, or when the moon is near the sun in the sky. The sun is a broadband generator of electromagnetic energy, and plenty of this falls within the ham bands. Problems can also occur when the moon passes near other noisy regions in the radio sky.

Moonbounce is subject to fading that results from the random phase interference of signals reflected from the irregular, constantly moving surface of the moon. *See* LIBRATION FADING.

Path loss increases with increasing frequency. But this effect is offset by the more manageable size of high-gain antennas as the wavelength decreases. The bands above 1300 MHz should provide promise in future years, as improved designs are developed for receivers and power amplifiers.

MORSE CODE

Morse code is a binary means of sending and receiving messages. It is a binary code because it has just two possible states: on (key-down) and off (key-up). There are two different morse codes in use by English-speaking operators today. The more commonly used code is called the *international* or *continental* code. A few telegraph operators use the *American morse code*. (*See* AMERICAN MORSE CODE, and INTERNATIONAL MORSE CODE.) The code characters vary somewhat in other languages.

Modern communications devices today can function under weak-signal conditions that would frustrate a human operator. But when human operators are involved, the morse code has always been the most reliable means of getting a message through severe interference. This is because the bandwidth of a morse-code radio signal is extremely narrow, and it is comparatively easy for the human ear to distinguish between the background noise and a code signal.

MOS

See METAL-OXIDE SEMICONDUCTOR.

MOSFET

See METAL-OXIDE-SEMICONDUCTOR FIELD-EFFECT TRANSISTOR.

MOTOR

A *motor* is a device that converts electrical energy into mechanical energy. Motors can operate from alternating or direct current, and can run at almost any speed. Motors range in size from the tiny devices in a wristwatch to huge, powerful machines that can pull a train at over 100 miles per hour.

All motors operate by means of electromagnetic effects. Electric current flows through a set of coils, producing powerful magnetic fields. The attraction of opposite magnetic poles, and the repulsion of similar magnetic poles, results in a rotating force. The greater the current flowing in the coils, the greater the rotating force. When the motor is connected to a load, the amount of force required to turn the motor shaft increases. The more the force, the greater the current flow becomes, and the greater the amount of energy drawn from the power source.

In a typical electric motor, one set of coils rotates with the motor shaft. This is called the *armature coil*. The other set of coils is fixed, and is called the field coil. The commutator reverses the current with each half-rotation of the motor so that the force between the coils maintains the same angular direction.

The electric motor operates on the same principle as an electric generator. In fact, some motors can actually be used as generators. *See also* GENERATOR.

MOUTH-TO-MOUTH RESUSCITATION

When a person has stopped breathing because of electric shock or other injury, resuscitation must be applied. The most commonly used method of artificial respiration is the mouth-to-mouth technique.

If an injured person is not breathing, the heart can also have stopped. The pulse should be checked immediately. If the heart has stopped, cardiopulmonary resuscitation (CPR) must be used. If the heart is beating, mouth-to-mouth resuscitation alone is sufficient.

The person's mouth should be checked for debris, and then the head should be tilted back. The nose should be pinched shut. To perform the resuscitation, air is exhaled from the rescuer, by mouth, into the mouth of the victim. The victim is then allowed to exhale. This process should be repeated about 12 times per minute for an adult victim, and 20 times per minute for a child.

A full and comprehensive discussion of resuscitation techniques is beyond the scope of this book. This essay should not, therefore, be considered as instruction. The American Red Cross publishes books and provides courses in cardiopulmonary resuscitation and mouth-to-mouth resuscitation. These books and courses should be used for training purposes. Everyone should be trained in cardiopulmonary resuscitation and mouth-to-mouth resuscitation. Information can be obtained from local chapters of the American Red Cross. *See also* CARDIOPULMONARY RESUSCITATION.

MOUTH-TO-NOSE RESUSCITATION

Mouth-to-nose resuscitation is basically the same as the mouth-to-mouth method. The only difference is that the rescuer places his or her mouth over the nose of the victim. Mouth-to-nose resuscitation is preferred by some people in the event the victim has vomited. *See* MOUTH-TO-MOUTH RESUSCITATION.

MSI

See MEDIUM-SCALE INTEGRATION.

MU

Mu is a letter of the Greek alphabet. Its symbol is written like a small English letter *u* with a "tail" (μ). The English u is often used in place of the actual symbol μ.

Mu is used as a prefix multiplier meaning micro (*see* MICRO). The symbol μ is also used to indicate amplification factor, permeability, inductivity, magnetic moment, and molecular conductivity.

MUF

See MAXIMUM USABLE FREQUENCY.

MULTIBAND ANTENNA

A *multiband antenna* is an antenna that is designed for operation on more than one frequency. A half-wave dipole antenna is a multiband antenna; it is resonant at all odd multiples of the fundamental resonant frequency. An end-fed half-wave antenna can be operated at any multiple of the fundamental frequency. All antennas have a theoretically infinite number of resonant frequencies; not all of these frequencies, however, are useful.

Multiband antennas can be designed deliberately for operation on specific frequencies. Traps are commonly used for achieving multiband operation (*see* TRAP, and TRAP ANTENNA). A variable inductor can be used for changing the resonant frequency of an antenna, as shown at A in the illustration (*see* INDUCTIVE LOADING). A tuning network can be used to adjust the resonant frequency of an antenna/feeder system (*see* TUNED FEEDERS). Several different antennas can be connected in parallel to a single feed system to achieve multiband operation as at B.

Multiband antennas offer convenience; it is simple to switch from one frequency band to another. However, harmonic-resonant multiband antennas can radiate unwanted

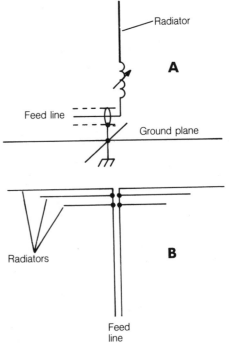

MULTIBAND ANTENNA: At A, a vertical with adjustable base loading; at B, dipoles in parallel.

harmonic signals. For this reason, when a harmonic-resonant antenna is used, care must be exercised to ensure that the transmitter harmonic output is sufficiently attenuated in the final-amplifier and output circuits. *See also* HARMONIC, and RESONANCE.

MULTIELEMENT ANTENNA

Some antennas use multiple elements for the purpose of obtaining a directional response. An antenna element consists of a length of conductor. A conductor can be directly connected to the feed line; then it is called an *active* or *driven element*. A conductor can be physically separate from the feed line; then it is called a *passive* or *parasitic element*.

Multielement antennas can be broadly classified as either parasitic arrays or phased arrays. In the parasitic array, passive elements are placed near a single driven element (*see* DIRECTOR, DRIVEN ELEMENT, PARASITIC ARRAY, and REFLECTOR). In the phased array, two or more elements are driven together (*see* PHASED ARRAY).

Multielement antennas are used mostly at high frequencies and above, although some large broadcast installations make use of phased vertical arrays at low and medium frequencies. The main advantages of a multielement antenna are power gain and directivity, which enhance both the transmitting and receiving capability of a communications station. The design and construction of a multi-element antenna is, however, more critical than for a single-element antenna.

MULTILEVEL TRANSMISSION

Multilevel transmission is a form of digital transmission, in which some signal parameter has three or more discrete values. The number of possible levels must, however, be finite. Therefore, ordinary analog modulation is not considered to be multilevel transmission.

Multilevel transmission can be used to digitally transmit a complex waveform, provided the element (bit) duration is short enough, and provided the number of levels is large enough. The complex waveform shown at A in the illustration is converted to a coarse three-level amplitude-modulated signal at B. Finer multilevel conversion signals are illustrated at C and D.

The amplitude is not the only parameter that can be varied in a multilevel signal. In a slow-scan television signal, for example, the frequency can attain any of several discrete values, resulting in various shades of brightness. Multilevel transmission can be used with any form of modulation. The primary advantage of multilevel transmission is its narrow bandwidth, compared with ordinary analog modulation.

MULTIPATH FADING

Multipath fading is a form of fading that occurs primarily at medium and high frequencies, and results from the random phase combination of received ionospheric signals arriving along more than one path at the same time.

The ionosphere is not a smooth reflector of electromagnetic energy. Instead, the ionization occurs in irregular patches and with variable density. The result is that signals can be transmitted between two points via several different paths simultaneously. These several paths do not all have the same overall length, and therefore the signals combine in random phase at the receiver. As the ionosphere undulates, the various path

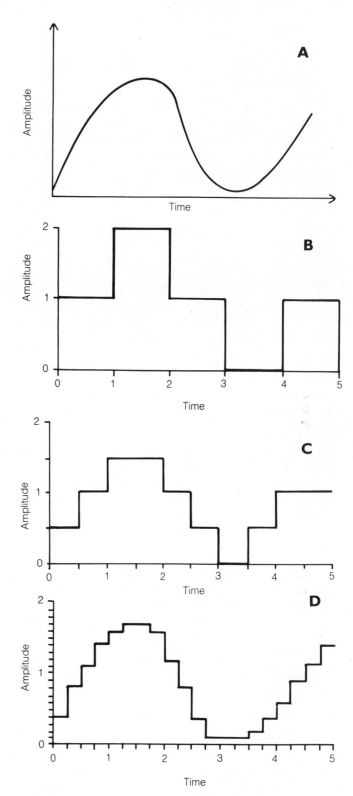

MULTILEVEL TRANSMISSION: At A, analog waveform. At B, a coarse digital representation; at C and D, finer digital representations.

lengths constantly vary, and so does the overall phase combination at the receiver.

Multipath fading effects can be reduced by using two receivers for the same signal, with antennas located at least several wavelengths apart. The probability that a severe fade will occur at both receivers, simultaneously, is less than the probability that a fade will occur at a single receiver. *See also* DIVERSITY RECEPTION, and FADING.

MULTIPLE POLE

A *multiple-pole relay* or *switch* is designed to switch two or more circuits at once. Multiple-pole relays and switches can be of the single-throw or multiple-throw variety (*see* MULTIPLE THROW).

Schematic symbols for multiple-pole switches or relays generally show the same circuit designator for each pole, followed by A, B, C, and so on. The various poles can be connected in the diagram by a dotted line. When one pole changes state, the others all change state.

MULTIPLE THROW

A *multiple-throw relay* or *switch* is a device that can connect a given conductor to two or more other conductors. A relay normally has at most two throw positions because of mechanical constraints, but a switch can have several throw positions. Multiple-throw relays and switches can be ganged to form a multiple-pole, multiple-throw device (*see* MULTIPLE POLE).

Generally, a switch with three or more throw positions is a wafer switch. *See also* WAFER SWITCH.

MULTIPLEX

Multiplex refers to the simultaneous transmission of two or more messages over the same medium or channel at the same time. Multiplex transmission can be achieved in various different ways, but the most common methods are frequency-division multiplex and time-division multiplex (*see* FREQUENCY-DIVISION MULTIPLEX, and TIME-DIVISION MULTIPLEX).

Multiplex transmission requires a special encoder at the transmitting end and a special decoder at the receiving end. The medium must be such that interference does not occur among the various channels transmitted.

A form of digital signal transmission, used especially with display devices, is called *multiplex*. When it is necessary to provide a voltage to many different display components at the same time, the data can be cumbersome to transmit in parallel form. The data can, instead, be sent in serial form, using time-division multiplex. For example, a four-digit light-emitting-diode display can be illuminated by "rotating" the applied voltage from left to right. If this is done rapidly enough, the eye cannot tell that each digit is illuminated for only 25 percent of the time. *See also* PARALLEL DATA TRANSFER, and SERIAL DATA TRANSFER.

MULTIVIBRATOR

A *multivibrator* is a form of relaxation oscillator. A multivibrator circuit consists of a pair of inverting amplifiers. The amplifiers are connected in series, and a direct feedback loop runs from the output of the second stage to the input of the first. A block diagram of the basic scheme is shown in the illustration. Multivibrators are extensively used in digital circuit design.

There are three main types of multivibrator. The astable multivibrator is a free-running oscillator. The bistable multivibrator, often called a *flip-flop*, can attain either of two stable conditions. The monostable multivibrator maintains a single condition except when a triggering pulse is applied; then the state changes for a predetermined length of time. *See also* ASTABLE MULTIVIBRATOR, FLIP-FLOP, and MONOSTABLE MULTIVIBRATOR.

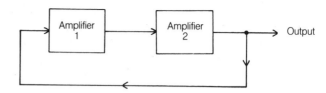

MULTIVIBRATOR: Two amplifiers in cascade with positive feedback.

MUMETAL

Mumetal is a ferromagnetic substance commonly used as a magnetic shield in cathode-ray tubes. Mumetal consists of a mixture of iron, nickel, copper, and chromium. The magnetic permeability is very high. The direct-current resistivity is moderately low, resulting in good magnetic shielding characteristics. *See also* FERROMAGNETIC MATERIAL.

MUTING CIRCUIT

A *muting circuit* is a device that cuts off a television or radio receiver while the power is on. This prevents signals from being heard, but is allows instantaneous reactivation of the circuit when desired by the operator.

A muting circuit generally consists of a switching transistor or relay in series with one of the stages of the receiver. The muted stage is usually one of the radio-frequency or intermediate-frequency amplifiers in the receiver chain. However, an audio-frequency circuit can be muted if an extremely fast recovery is not needed. The schematic diagram shows a simple muting system using a transistor switch. A negative voltage at the muting terminal opens the emitter circuit of the amplifier. Other types of muting circuits also exist; some are actuated by a positive voltage, others by an open circuit at the input terminals, and still others by a short circuit.

Muting circuits are used for break-in operation for morse-code communication. Muting circuits are also used to prevent receiver overload or acoustic feedback in two-way voice radio systems. They can also be found in television channel switching circuits. *See also* BREAK-IN OPERATION.

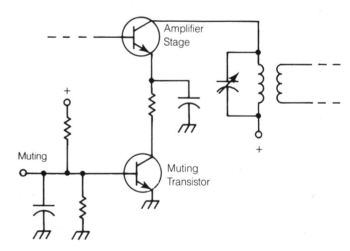

MUTING CIRCUIT: A muting signal cuts off the amplifier.

MUTUAL CAPACITANCE

Any two conductors, no matter where they are located or how they are oriented, have a certain amount of capacitance, with respect to each other. That is, given a device capable of measuring a sufficiently tiny value of capacitance, any set of two conductors can be shown to act as a capacitor. The existence of capacitance among electrical conductors is called *mutual capacitance*. The mutual capacitance between two conductors increases as the surface area of either conductor is made larger; the mutual capacitance decreases as the distance between conductors becomes greater.

Mutual capacitance is often too small to be measured, and is of no consequence. However, in radio-frequency circuit wiring, mutual capacitance is sometimes sufficient to cause feedback and other problems. This is especially true at very-high and ultra-high frequencies. Radio circuits at these frequencies are designed to minimize the mutual capacitance among wires and components.

Mutual capacitance exists among the elements within certain electronic components, particularly diodes, relays, switches, transistors, and vacuum tubes. In radio-frequency circuit design, this capacitance is often significant, and must be taken into account in the design process. *See also* ELECTRODE CAPACITANCE, and INTERELECTRODE CAPACITANCE.

MUTUAL IMPEDANCE

Two electrical conductors invariably display a certain amount of mutual capacitance and mutual inductance (*see* MUTUAL CAPACITANCE, and MUTUAL INDUCTANCE). The combination of mutual capacitance and inductance presents a complex impedance, known as *mutual impedance*, that varies with frequency. Because of this, a resonant circuit is formed. The resonant frequency depends on the amount of mutual capacitance and inductance. Generally, the natural resonant frequency of two conductors or electrodes is very high—on the order of hundreds or even thousands of megahertz.

The impedance between or among the electrodes of certain components can cause oscillations. These oscillations are harnessed in certain types of vacuum tubes for the purpose of generating microwave signals. However, the oscillations can occur in a vacuum tube—even when they are not wanted. Then they are called *parasitic oscillations*. *See also* PARASITIC OSCILLATION.

MUTUAL INDUCTANCE

Any two conductors have a certain amount of inductance, with respect to each other. A fluctuating magnetic field around one conductor will invariably cause an alternating current in the other. The degree of inductive coupling between conductors or inductors is called *mutual inductance*.

Mutual inductance, represented by the letter M and expressed in henrys, is an expression of the degree of coupling between two coils. The mutual inductance is mathematically related to the coefficient of coupling, which can range from 0 (no coupling) to 1 (maximum possible coupling). If the coefficient of coupling is given by k and the inductances of the two coils, in henrys, are given by L_1 and L_2, then:

$$M = k\sqrt{L_1 L_2}$$

For example, suppose that $k = 0.5$, and the coils have inductances of 0.5 henry and 2 henrys. Then:

$$M = 0.5 \sqrt{0.5 \times 2} = 0.5$$

In practice, because most coils have inductances that are a small fraction of 1 henry, the value of M is quite small.

For two inductors having values L_1 and L_2 (given in henrys) and connected in series with the flux linkages in the same direction, the total inductance L, in henrys, is:

$$L = L_1 + L_2 + 2M$$

If the flux linkages are in opposition, then:

$$L = L_1 + L_2 - 2M$$

See also COEFFICIENT OF COUPLING, and INDUCTANCE.

MYLAR CAPACITOR

Mylar, a form of plastic, has excellent dielectric properties, and it is therefore well suited for making capacitors. Mylar capacitors are common in audio-frequency and radio-frequency circuits.

A Mylar capacitor is relatively impervious to the effects of moisture because plastic does not absorb water. Therefore, Mylar capacitors are an excellent choice in marine or tropical environments. Mylar capacitors are relatively inexpensive. Typical values range from several picofarads to several microfarads. The voltage rating is low to moderate. Mylar capacitors are suitable for use at high frequencies. But at very-high frequencies and above, the dielectric loss is somewhat increased. The maximum frequency at which a Mylar capacitor can efficiently operate depends on the particular type of plastic used. *See also* CAPACITOR.

NAND GATE

A NAND gate is a logical AND gate, followed by an inverter (*see* AND GATE, INVERTER). The term *NAND* is a contraction of NOT AND. The illustration shows the schematic symbol for the NAND gate, along with a truth table indicating the output as a function of the input. When both or all of the inputs of the NAND gate are high, the output is low. Otherwise, the output is high.

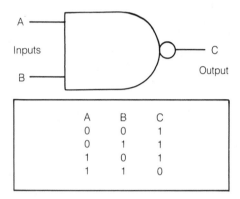

A	B	C
0	0	1
0	1	1
1	0	1
1	1	0

NAND GATE: A schematic symbol and truth table.

NANO

Nano is a prefix multiplier, representing 1 billionth (10^{-9}). For example, 1 nanofarad is equal to 10^{-9} farad, or 0.001 microfarad; 1 nanowatt is equal to 10^{-9} watt. The abbreviation for nano is the small letter n. *See also* PREFIX MULTIPLIERS.

NARROW-BAND VOICE MODULATION

Narrow-band voice modulation (NBVM) is a technique for reducing the bandwidth necessary for the transmission of a human voice. Normally, this bandwidth is about 2.5 to 3 kHz; a band-pass of approximately 300 Hz to 3 kHz is required. Using NBVM, the bandwidth can be substantially reduced.

The sounds of the human voice do not occupy the entire frequency range between 300 Hz and 3 kHz; rather, the sound tends to be concentrated within certain bands called *formants*. For a voice to be clear and understandable, three formants must be passed. The lower formant occurs at a range of about 300 Hz to 600 Hz. The upper two formants exist within the range 1.5 to 3 kHz. Between 600 Hz and 1.5 kHz, very little sound occurs. In NBVM, this unused range is eliminated, reducing the overall signal bandwidth by about 900 Hz. Narrow-band voice modulation is thus a frequency-companding process.

Narrow-band voice modulation is achieved by mixing at the baseband, or audio-frequency, level. The resulting audio signal sounds "scrambled." The audio output of the NBVM compressor can be fed to the microphone input of any voice transmitter. At the receiver, the signal is restored to its original form by frequency expansion. *See also* COMPANDOR.

NATIONAL BUREAU OF STANDARDS

The *National Bureau of Standards (NBS)* is an agency in the United States that maintains values for physical constants in the standard international system of units.

Among shortwave radio listeners and radio amateurs, the National Bureau of Standards is best known for its continuous standard time and frequency broadcasts. These signals are sent by WWV in Fort Collins, Colorado, and WWVH on the island of Kauai, Hawaii. *See also* WWV/WWVH/WWVB.

NATIONAL ELECTRIC CODE

The *National Electric Code (NEC)* is a set of recommendations for safety in electric wiring. The NEC is prepared by the National Fire Protection Association and the Institute of Electrical and Electronic Engineers, and is purely advisory. However, the NEC is enforced by some local governments.

NEC is primarily concerned with types of insulation, and suitable wire sizes and conductor materials, and for various levels of voltage and current. Maximum allowable current values are given for different kinds and sizes of wire and cable. Temperature derating curves are given.

The NEC also makes recommendations regarding the wiring of electronic equipment, especially if high voltage and/or current levels are used. Detailed information can be obtained from the American National Standards Institute, New York, NY.

NATIONAL TRAFFIC SYSTEM

In the United States, nonemergency amateur radio message traffic is coordinated via the *National Traffic System (NTS)*. Emergency traffic is passed mainly through the *Amateur Radio Emergency Service (ARES)*. *See* AMATEUR RADIO EMERGENCY SERVICE, and TRAFFIC HANDLING. For details concerning the NTS, contact the headquarters of the American Radio Relay League, 225 Main Street, Newington, CT 06111.

NATURAL FREQUENCY

The *natural frequency* of a circuit, antenna, or set of electrical conductors is the lowest frequency at which the system is resonant for electromagnetic energy. In acoustics, physical objects have a lowest resonant sound frequency; this is sometimes called the *natural frequency*.

Certain systems have a tendency to oscillate at the natural frequency. For example, a radio-frequency power amplifier, if not neutralized, can develop parasitic oscillations at the natural frequency of interelectrode impedance.

In an antenna system, the radiator and feed line together have a natural frequency that differs from the resonant frequency of the antenna by itself. When tuned feeders are used to obtain broadband operation, the feed line may radiate if the system is used at its natural frequency. *See also* PARASITIC OSCILLATION, RESONANCE, and TUNED FEEDERS.

NEAR FIELD

The *near field* of radiation from an antenna is the electromagnetic field in the immediate vicinity of the antenna. The electric and magnetic lines of flux in the near field are curved because of the proximity of the radiating element (A in the illustration). The wavefronts are not flat, but instead are convex.

For a paraboloidal or dish antenna, the near field is the region within which the radiation occurs mostly within a cylindrical region having the same diameter as the dish, as at B. The near field of radiation from a dish antenna is sometimes called the *Fresnel zone.* The distance to which the near field extends depends on the diameter of the antenna, the antenna aperture, and the wavelength.

In radio communications, the near field is generally of little importance. The signal normally picked up by the receiving antenna results not from the near field, but from the far field or Fraunhofer region. *See also* FAR FIELD.

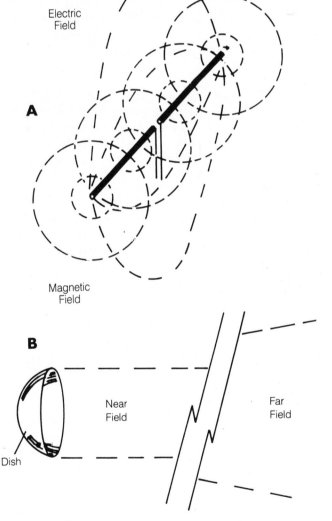

NEAR FIELD: At A, around a dipole; at B, from dish antenna.

NECESSARY BANDWIDTH

The *necessary bandwidth* of a signal is the minimum bandwidth required for transmission of the data with reasonable accuracy. The degree of accuracy is somewhat arbitrary; it is impossible to obtain perfection. In general, the greater the signal bandwidth, the better the accuracy. As the bandwidth is increased beyond a certain point, however, the improvement is small and spectrum space is wasted. Typical values of necessary bandwidth for various emission types are:

- For a morse-code signal, the necessary bandwidth, in hertz, is generally considered to be about 2.4 times the speed in words per minute. Thus, for example, a morse transmission at 20 words per minute (a typical speed) is 48 Hz. The receiver bandpass should be at least 48 Hz for accurate reception of a morse signal at 20 words per minute.
- For BAUDOT and ASCII codes, the necessary bandwidth also depends on the speed. Generally, the necessary bandwidth, in hertz, for such signals is about 2.4 times the speed in words per minute, or 3.2 times the speed in bauds. As with morse transmission, this is a somewhat subjective value, and does not represent an absolute standard.
- For a single-sideband transmission, the necessary bandwidth is about 2.5 kHz. However, with narrowband modulation techniques, this value can be reduced considerably (*see* NARROW-BAND VOICE MODULATION). A slow-scan television signal requires about 2.5 kHz.
- A normal amplitude-modulated or frequency-modulated voice signal requires approximately 5 to 6 kHz of spectrum space for reasonable intelligibility. Some frequency-modulated signals, however, are spread over a much larger space to improve the fidelity. The transmission of high-fidelity music requires a minimum of 40 kHz of spectrum space with amplitude or frequency modulation.
- Video signals and high-speed data transmissions have large necessary bandwidth. The typical television broadcast channel is 6 MHz wide.

See also BANDWIDTH, BAUD RATE, and SPEED OF TRANSMISSION.

NEGATION

Negation is a logical NOT operation. In Boolean algebra, negation is called *complementation.* Electronically, negation is performed by a NOT gate or inverter. *See also* BOOLEAN ALGEBRA, and INVERTER.

NEGATIVE CHARGE

Negative charge (negative electrification) is the result of an excess of electrons on a body. Friction between objects can result in an accumulation of electrons on one object (a negative charge) at the expense of electrons on the other object. When an atom has more electrons than protons, the atom is considered to be negatively charged. The smallest unit of negative charge is carried by a single electron. Conversely, the smallest unit of positive charge is carried by the proton. The terms *negative* and *positive* are arbitrary. *See also* CHARGE, and POSITIVE CHARGE.

NEGATIVE FEEDBACK

When the output of an amplifier circuit is fed back to the input in phase opposition, the feedback is said to be negative. *Negative feedback* is used for a variety of purposes in electronic circuits.

Some amplifiers tend to break into oscillation easily. This can be prevented by a neutralizing circuit, which is a form of negative-feedback arrangement (*see* NEUTRALIZATION).

Negative feedback is used in some audio amplifiers to improve stability and fidelity. Generally, negative feedback can be provided by simply installing a resistor of the proper value in series with the cathode, emitter, or source of the active amplifying device. Negative feedback may also be obtained by directly applying some of the output signal to the input of the amplifier, 180 degrees out of phase.

Negative feedback is routinely used in operational-amplifier circuits to control the gain and improve the bandwidth. When no negative feedback is used, the operational amplifier is said to be operating in the open-loop condition (*see* OPEN LOOP). Negative feedback is provided by placing a resistor in the feedback path, as shown in the illustration. The amount of negative feedback depends on the value of the resistor between the output and the input, denoted by *R* in the illustration. The smaller the value of *R*, the greater the negative feedback, and the smaller the gain of the circuit. *See also* FEEDBACK.

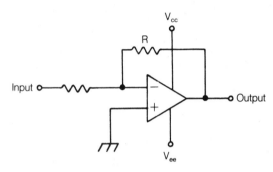

NEGATIVE FEEDBACK: Degeneration is controlled by *R*.

NEGATIVE LOGIC

The way in which logical signals are defined is a matter of convention. Normally, the logic 1 is the more positive of the voltage levels and the logic 0 is the more negative of the voltage levels. That is, logic 1 is high and logic 0 is low. When the voltages are reversed, the logic is said to be negative or inverted. Either positive or negative logic will provide satisfactory operation of a digital device. *See also* LOGIC.

NEGATIVE RESISTANCE

Normally, the current through a device increases as the applied voltage is made larger. This is true of most active and passive components. However, certain components exhibit a different characteristic within a certain range: The current decreases as the applied voltage is made larger. This is called a *negative-resistance characteristic*. The illustration shows an example of a negative-resistance characteristic.

Negative resistance occurs in certain diodes. Some transistors and vacuum tubes also exhibit this property. Negative resistance causes oscillation when the applied voltage is within a certain range. This oscillation usually occurs at ultra-high or microwave frequencies. The magnetron and the tunnel diode operate on this principle. *See also* MAGNETRON, NEGATIVE TRANSCONDUCTANCE, and TUNNEL DIODE.

NEGATIVE TRANSCONDUCTANCE

In a pentode vacuum tube, the transconductance between the suppressor grid and the screen grid is normally positive. How-

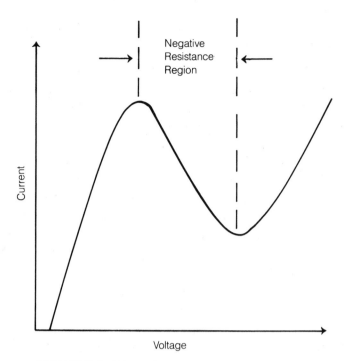

NEGATIVE RESISTANCE: Current drops as voltage rises.

ever, when the screen grid is supplied with a more positive voltage than the plate, the suppressor-screen transconductance becomes negative within a certain range. That is, if the negative voltage of the suppressor is increased, the screen current increases. This occurs because of the repulsion of electrons by the suppressor. The more negative the suppressor, the more electrons it repels back toward the screen, and thus the more electrons are picked up by the screen.

Negative transconductance often results in oscillation at ultra-high or microwave frequencies. The plate voltage should always be higher than the screen voltage in a tetrode or pentode radio-frequency amplifier. Otherwise, parasitics are likely. Negative transconductance can, however, be introduced into a tetrode or pentode deliberately, for the purpose of generating signals. The dynatron and transitron oscillators are examples of circuits in which negative transconductance is used to generate signals. *See also* PARASITIC OSCILLATION, TRANSCONDUCTANCE, and TUBE.

NEMATIC CRYSTAL

A *nematic crystal* or *nematic fluid* is an organic liquid with long molecules. Nematic fluid is widely used in digital displays. The liquid is normally transparent. When an electric field of sufficient intensity is introduced, the liquid becomes opaque until the field is removed. Then, the liquid immediately becomes transparent again.

Displays using nematic crystals have become increasingly popular in calculators, watches, and other electronic devices. The nematic crystal draws practically no current because it is the electric field alone that causes the change of state. This is an advantage in battery-powered devices, where the current drain must be minimized. *See also* LIQUID CRYSTAL DISPLAY.

NEON LAMP

A *neon lamp* is a device consisting of two electrodes in a sealed glass envelope containing neon gas. When a voltage is introduced between the electrodes, the neon gas becomes ionized, and the lamp glows. Neon lamps have a characteristic red-or-ange color.

Neon lamps are used in indicating devices, displays, oscillators, and voltage regulators. *See also* NEON-LAMP OSCILLATOR.

NEON-LAMP OSCILLATOR

A neon lamp, a capacitor, and a resistor can be interconnected to form an audio-frequency oscillator. Such an oscillator produces a sawtooth wave. The frequency is determined by the values of the parallel capacitance and the series resistance.

When power is applied to the neon-lamp oscillator, the capacitor charges at a rate determined by the value of the resistor. When the capacitor is fully charged, the bulb ionizes and depletes the charge in the capacitor. The voltage across the bulb then falls below the minimum level necessary to cause ionization; the bulb is extinguished, and a new cycle begins. The process repeats at an audio-frequency rate.

The neon-lamp oscillator is simple and inexpensive, and is commonly used as an audio-frequency oscillator in applications where the waveform and frequency are not critical. The deionization time of the neon lamp prevents oscillation at frequencies above about 5 kHz.

NEPER

The *neper (Np)* is a unit that is sometimes used for expressing a ratio of currents, voltages, or wattages. The neper can also be called the *napier*. The neper is similar to the decibel. In fact, nepers and decibels are related by a simple constant:

$$1 \text{ Np} = 8.686 \text{ dB}$$
$$1 \text{ dB} = 0.1151 \text{ Np}$$

A gain or loss is determined in nepers according to the formula:

$$\text{Np} = \log_e \sqrt{x_1/x_2},$$

where x_1 and x_2 are the quantities to be compared and $\log_e$ represents the natural (base-e) logarithm. *See also* DECIBEL.

NET

A communications network, intended for the relaying and delivering of information or messages by radio, is called a *net*. Nets are especially popular among amateur radio operators. However, commercial nets also exist.

Most amateur-radio nets are "traffic" nets, and their operators handle a wide variety of messages. Other amateur-radio nets are "conversational" nets, intended for special-interest discussion, such as antenna theory and construction. Still other nets operate for specific purposes, such as the relaying of information in a severe-weather situation. The term *net* is widely used as a synonym for the term network (*see* NETWORK).

A resultant or effective quantity is called a *net quantity*. For example, a circuit can have currents flowing in opposite directions at the same time; the net current is the effective current is one direction. A circuit can have two or more voltage sources, some of which buck each other; the net voltage is the sum of the individual voltages, taking polarity into account.

NETWORK

A communications system, with several or many stations that can contact any of the others, is a *network*. The set of all subscribers in a telephone system, for example, along with the wiring and switching apparatus, forms a network.

In ham radio, the term is shortened to net. Nets are used for sending and receiving messages. The originators and recipients might be hams, but often they are not; traffic is routinely passed on behalf of third parties.

When sending or receiving messages on behalf of third parties, certain restrictions apply. It is illegal to do this to or from most foreign countries. Those countries that have signed third-party agreements with the United States are listed in publications of the American Radio Relay League. *See* AMERICAN RADIO RELAY LEAGUE.

In recent years, nets have become much more sophisticated and efficient than they used to be, as a result of the development of packet radio. Computers take the place of human operators in many capacities. It will someday be possible for hams to send and receive messages via such modes as FAX, with speed and efficiency rivaling that of the telephone. But the law forbids hams to send messages of a business nature. Technically, it is illegal even to order pizza via ham radio. *See* NET, PACKET RADIO, and TRAFFIC HANDLING.

A system of electronic components, in general, is often called a *network. See*, for example, H NETWORK, L NETWORK, PI NETWORK, and T NETWORK.

NETWORKING

The organization of a communications system, especially among computers, is known as *networking*. In ham radio, networking is mainly of interest to packet radio operators.

The ham radio packet network has matured over the years, to the point that a message can now be sent without the operators having to worry about the exact details of the path taken by that message. Computers take care of the routing. *See* PACKET RADIO.

NETWORK LAYER

See OPEN SYSTEMS INTERCONNECTION REFERENCE MODEL.

NETWORK NODE CONTROLLER

In a packet radio network, a *terminal node controller (TNC)* can be easily converted into a device that provides coverage over a larger region. A TNC, thus converted, is called a *network node controller (NNC)*. The conversion is accomplished via computer firmware known as *NET/ROM* and is incorporated in medium-coverage and wide-coverage digipeaters. *See* DIGIPEATER, PACKET RADIO, and TERMINAL NODE CONTROLLER.

NEUTRAL CHARGE

When the total charge on an object is zero—the positive balances the negative—the object is said to be electrically neutral. A neutral charge results in a zero net electric field.

Atoms normally have neutral charge because the number of protons is the same as the number of orbiting electrons. However, atoms can gain or lose orbiting electrons, becoming charged. *See also* NEGATIVE CHARGE, and POSITIVE CHARGE.

NEUTRALIZATION

Radio-frequency amplifiers have a tendency to oscillate. This is especially true of power amplifiers. This unwanted oscillation is called *parasitic oscillation* (*see* PARASITIC OSCILLATION), and can occur at any frequency, regardless of the frequency at which the amplifier is operating. Neutralization is a method of reducing the likelihood of such oscillation.

The neutralizing circuit is a negative-feedback circuit. A small amount of the output signal is fed back, in phase opposition, to the input. This can be done in a variety of different ways; the most common method is the insertion of a small, variable capacitor in the circuit. The drawing illustrates two methods of neutralization in a simple bipolar-transistor power amplifier. The method at A is called *collector neutralization* because the capacitor is connected to the bottom of the collector tank circuit. The method at B is called *base neutralization*, because the capacitor is connected to the bottom of the base tank circuit. The negative feedback occurs through the variable capacitor. The capacitance required for optimum neutralization depends on the operating frequency and impedance of the amplifier. Normally, the neutralizing capacitance is a few picofarads.

The neutralizing capacitor is first adjusted by removing the driving power from the input of the amplifier. A sensitive broadband wattmeter, or other output-sensing indicator, is connected to the amplifier output. The capacitor is adjusted for minimum output indication. *See also* NEGATIVE FEEDBACK, NEUTRALIZING CAPACITOR, and POWER AMPLIFIER.

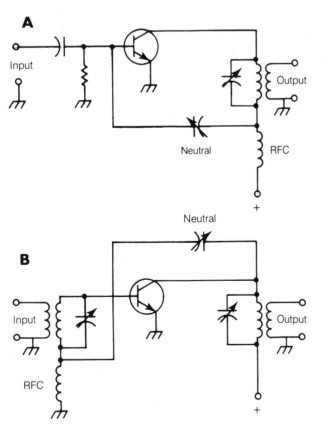

NEUTRALIZATION: At A, collector neutralization. At B, base neutralization.

NEUTRALIZING CAPACITOR

A *neutralizing capacitor* is a small-valued, variable capacitor used for providing negative feedback in a radio-frequency power amplifier. The negative feedback can be varied by adjusting the value of the capacitance.

The capacitor method is the most common means of neutralizing a power amplifier. However, other negative-feedback circuits can also be used. *See also* NEGATIVE FEEDBACK, NEUTRALIZATION, and POWER AMPLIFIER.

NICHROME

Nichrome is an alloy of two common metals, nickel and chromium. Nichrome has high resistivity and a high melting point. It is used for electrical heating elements, and also in the fabrication of thin-film and wirewound resistors.

Nichrome is sometimes also called *nickel-chromium*. The word *Nichrome* was originally coined by Driver-Harris Company. *See also* WIREWOUND RESISTOR.

NICKEL

Nickel is an element with atomic number 28 and atomic weight 59. Nickel is used for many different electrical and electronic purposes. An alloy of nickel and chromium exhibits a high resistance and a high melting temperature. This alloy, called Nichrome, is used in the manufacture of heating elements and resistors (*see* NICHROME).

Nickel is sometimes used as a plating material for steel. Nickel-plated steel has a dull, grayish-white appearance when unpolished; when polished it appears dark gray. Nickel plating retards the corrosion of steel, but does not entirely prevent it (*see* ELECTROPLATING).

Nickel is used in the manufacture of electrodes in some vacuum tubes. Pressed or impregnated nickel is found in the cathodes of such tubes. Nickel cathodes exhibit high electron emission at relatively low temperatures. However, the tungsten cathode can withstand higher temperature levels (*see* CATHODE, and TUBE).

Nickel is used in the manufacture of some ferromagnetic transformer-core and inductor-core materials. The characteristics of such cores vary greatly, depending on the method of manufacture (*see* FERROMAGNETIC MATERIAL).

A compound of nickel, hydrogen, and oxygen is used in the manufacture of a form of rechargeable cell, the nickel-cadmium cell. Such cells, and batteries made from them, are widely used in electronic devices (*see* NICKEL-CADMIUM BATTERY).

Nickel is a magnetostrictive material. In the presence of an increasing magnetic field, nickel contracts. For this reason, nickel is often used in the manufacture of magnetostrictive transducers (*see* MAGNETOSTRICTION, and MAGNETOSTRICTION OSCILLATOR).

NICKEL-CADMIUM BATTERY

A *nickel-cadmium battery* (Ni-Cd or NICAD) is a battery made with cadmium anodes, nickel-hydroxide cathodes, and an electrolyte of potassium hydroxide. A single nickel-cadmium cell is shown in the diagram. A single Ni-Cd cell provides approximately 1.2 volts direct current.

Nickel-cadmium cells and batteries are rechargeable, and thus they can be used repeatedly. For this reason, they have become popular in such items as calculators, radio receivers, and low-powered portable radio transceivers.

A typical, 500-milliampere-hour Ni-Cd battery is about the size of eight AA cells. Such a battery requires about three to five hours to be fully charged. When the battery is used, it should not be allowed to become totally discharged, or its polarity can be reversed and the battery will be ruined. However, Ni-Cd batteries should not be repeatedly charged under partial-discharge conditions either. Most electronic devices operating from these batteries have a warning indicator that shows when it is time to recharge. *See also* BATTERY, CELL, CHARGING, and RECHARGEABLE CELL.

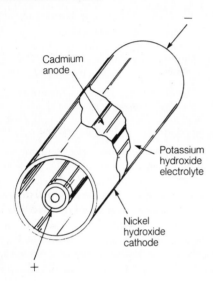

NICKEL-CADMIUM BATTERY: A cutaway view of a NICAD cell.

NIGHT EFFECT

The ionospheric D layer, the lowest ionized layer, generally exists only during daylight hours. As soon as the sun goes down, the D layer deteriorates. When the sun rises again, the D layer rapidly reforms.

Because the D layer absorbs electromagnetic energy over almost all of the low-frequency and medium-frequency spectrum, and also over much of the high-frequency spectrum, the nighttime absence of the D layer is a significant factor in wave propagation at frequencies from about 50 kHz to 10 MHz. Long-distance propagation at these frequencies is better at night than during the day. This is especially true from 300 kHz to 6 MHz. Daytime propagation is better than nighttime propagation over short and moderate distances at these frequencies.

The change in propagation caused by the presence or absence of the D layer is known as the *night effect*. It can be a factor in radio direction finding as well as communications. *See also* D LAYER, IONOSPHERE, and PROPAGATION CHARACTERISTICS.

NITROGEN

Nitrogen is an element with atomic number 7 and atomic weight 14. At room temperature, nitrogen is a gas. The atmosphere of our planet is made up of 78 percent nitrogen.

Nitrogen is a relatively nonreactive gas, although it can form certain compounds under the right conditions. Because nitrogen is so abundant, it is relatively inexpensive to obtain in pure form. One simple method of getting a 99-percent pure sample of nitrogen is to burn the oxygen out of the air in an enclosed chamber.

Nitrogen is used in the fabrication of certain sealed transmission lines because it is a noncorrosive gas. An air-dielectric coaxial cable, for example, can be evacuated and then refilled with pure, dry nitrogen gas. This improves the dielectric qualities, and prolongs the life of the transmission line.

NMOS

See METAL-OXIDE-SEMICONDUCTOR LOGIC FAMILIES.

NO-CODE LICENSE

Recently, the Federal Communications Commission made it possible for people to obtain ham licenses without having to know the morse code. The "code," as it is known, was a psychological barrier to some would-be hams.

The no-code license is known as the *Code-Free Technician Class license*. It's fairly easy for the average person to learn the electronic theory needed to pass the test to obtain this license. Many "no-code" hams, once they are exposed to the digital modes (such as code and teletype), go on to learn the code and get a license to operate these modes. *See* LICENSE REQUIREMENTS.

NODE

A *node* is a local minimum in a variable quantity. In a transmission line or antenna radiator, for example, we can speak of current nodes or voltage nodes. A current node is a point along a transmission line or antenna at which the current reaches a local minimum; at such a point, the voltage is usually at a local maximum. A voltage node is a point at which the voltage reaches a local minimum, and the current is usually at a local maximum. Nodes of a given kind are separated by electrical multiples of $1/2$ wavelength. The opposite of a node is a loop (*see* CURRENT NODE, LOOP, and VOLTAGE NODE).

In a circuit with two or more branches, a node is any circuit point that is common to at least two different branches. At a node, the current inflow is always the same as the current outflow, according to Kirchhoff's laws (*see* KIRCHHOFF'S LAWS).

In a packet-radio network, a node is a station, or location, at which messages are received and retransmitted. Any ham radio station, properly equipped, can serve as a node. *See* PACKET RADIO.

In satellite communications, a node is a longitude point where the satellite groundtrack crosses the equator. Each orbit has two nodes, one going north (ascending) and the other going south (descending), unless the satellite happens to be orbiting directly over the equator. Nodes are measured in degrees of longitude east or west. *See* ASCENDING NODE, and DESCENDING NODE.

NODE-TO-NODE ACKNOWLEDGMENT

In packet radio, *node-to-node acknowledgment* refers to a message sent back from one node in the communications route to the previous node. This backward-going signal informs the sending node that the receiving node got the message. The signal only goes back to the digipeater or node immediately preceding, not to any nodes that might have passed the message before that.

Node-to-node acknowledgment is used in a NET/ROM network. This is a more versatile, updated form of packet firmware that expands the capabilities of the old AX.25.

In node-to-node acknowledgment, if a packet is lost somewhere along the route, it is repeated only by the node that last sent it. The packet does not have to go all the way through the entire route again, as is the case if *end-to-end acknowledgment* is used. *See also* AX.25, DIGIPEATER, END-TO-END ACKNOWLEDGMENT, and PACKET RADIO.

NOISE

Noise is a broadbanded electromagnetic field, generated by various environmental effects and artificial devices. Noise can be categorized as either natural or manmade.

Natural noise can be either thermal or electrical in origin. All objects radiate noise as a result of their thermal energy content. The higher the temperature, the shorter the average wavelength of the noise.

Cosmic disturbances, solar flares, and the movement of ions in the upper atmosphere all contribute to the natural electromagnetic noise present at the surface of the earth. Sferics, or noise generated by lightning, is a source of natural noise (*see* SFERICS).

Electromagnetic noise is produced by many different manmade devices. In general, any circuit or appliance that produces electric arcing will produce noise. Such devices include fluorescent lights, heating devices, automobiles, electric motors, thermostats, and many other appliances.

The level of electromagnetic noise affects the ease with which radio communications can be carried out. The higher the noise level, the stronger a signal must be if it is to be received. The signal-to-noise ratio (*see* SIGNAL-TO-NOISE RATIO) can be maximized in a variety of different ways. The narrower the bandwidth of the transmitted signal, and the narrower the passband of the receiver, the better the signal-to-noise ratio at a given frequency. This improvement occurs, however, at the expense of data-transmission speed capability. Circuits such as noise blankers and limiters are sometimes helpful in improving the signal-to-noise ratio. Noise-reducing antennas can also be used to advantage in some cases. But there is a limit to how much the noise level can be reduced; a certain amount of noise will always exist. *See also* NOISE FILTER, NOISE FLOOR, NOISE LIMITER, and NOISE-REDUCING ANTENNA.

NOISE-EQUIVALENT BANDWIDTH

Any noise source has a spectral distribution. Some noise is very broadbanded, and some noise occurs with a fairly well-defined peak in the spectral distribution. The total noise can be expressed in terms of the area under the curve of amplitude versus frequency.

Let f be the frequency at which the power density of the noise is greatest in the spectral distribution. Let a represent the power density of the noise at the frequency f. We can construct a rectangle with height a, centered at f, that has the same enclosed area P as the curve encloses. The width of this rectangle, N, is a frequency span $f_2 - f_1$, such that $f_2 - f = f - f_1$. The value N is the noise equivalent bandwidth at the frequency f.

Noise equivalent bandwidth is often specified for devices such as filters, bolometers, and thermistors. The noise equiva-

lent bandwidth is an expression of the frequency characteristics of the devices. *See also* NOISE.

NOISE FIGURE

The *noise figure* is a specification of the performance of an amplifier or receiver. Noise figure, expressed in decibels, is an indication of the degree to which a circuit deviates from the theoretical ideal.

Suppose that the noise output of an ideal network results in a signal-to-noise ratio:

$$R_1 = P/Q_1$$

where P is the signal power and Q_1 is the noise power. Let the noise power in the actual circuit be given by Q_2, resulting in an actual signal-to-noise ratio:

$$R_2 = P/Q_2$$

Then, the noise figure, N, is given in decibels by the equation:

$$N \text{ (dB)} = 10 \log_{10} (R_1/R_2)$$
$$= 10 \log_{10} (Q_2/Q_1)$$

A perfect circuit would have a noise figure of 0 dB. This value is never achieved in practice. Values of a few decibels are typical. The noise figure is important at very-high frequencies and above, where relatively little noise occurs in the external environment, and the internal noise is the primary factor that limits sensitivity. *See also* SENSITIVITY, and SIGNAL-TO-NOISE RATIO.

NOISE FILTER

A *noise filter* is a passive circuit, usually consisting of capacitance and/or inductance, that is inserted in series with the alternating-current power cord of an electronic device. The noise filter allows the 60-Hz alternating current to pass with essentially no attenuation, but higher-frequency noise components are suppressed.

A typical noise filter is simply a lowpass filter, consisting of a capacitor or capacitors in parallel with the power leads, and an inductor or inductors in series (see the illustration). The values are chosen for a cutoff frequency just above 60 Hz.

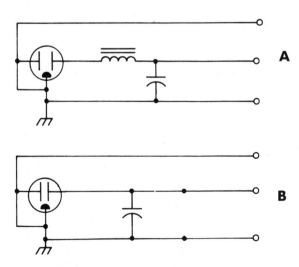

NOISE FILTER: A line-noise attenuator.

NOISE FLOOR

The level of background noise, relative to some reference signal, is called the *noise floor*. Signals normally can be detected if their levels are above the noise floor; signals below the noise floor cannot be detected. In a radio receiver, the level of the noise floor can be expressed as the *noise figure* (*see* NOISE FIGURE).

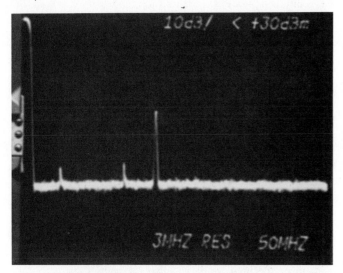

NOISE FLOOR: The bright horizontal line on this spectrum-analyzer display is the noise floor.

In a spectrum-analyzer display, the noise-floor level determines the sensitivity and the dynamic range of the instrument. The noise floor is generally specified in decibels with respect to the local-oscillator signal (see photograph). A typical spectrum analyzer has a noise floor of −50 dB to −70 dB. *See also* SPECTRUM ANALYZER.

NOISE GENERATOR

Any electronic circuit that is designed to produce electromagnetic noise is called a *noise generator*. Noise generators are used in a variety of testing and alignment applications, especially with radio receivers.

Noise can be generated in many different ways. A diode tube, operated at saturation (full conduction), produces broadband noise. A semiconductor diode will also produce broadband noise when operated in the fully conducting condition. Some diodes generate noise when they are reverse-biased. A current-carrying resistor produces thermal noise. *See also* NOISE.

NOISE LIMITER

A *noise limiter* is a circuit, often used in radio receivers, that prevents externally generated noise from exceeding a certain amplitude. Noise limiters are sometimes called *noise clippers*.

A noise limiter can consist of a pair of clipping diodes with variable bias for control of the clipping level (see illustration). The bias is adjusted until clipping occurs at the signal amplitude. Noise pulses then cannot exceed the signal amplitude. This makes it possible to receive a signal that would otherwise be drowned out by the noise. The noise limiter is generally installed between two intermediate-frequency stages of a superheterodyne receiver. In a direct-conversion receiver, the best place for the noise limiter is just prior to the detector stage.

A noise limiter can use a circuit that sets the clipping level automatically, according to the strength of an incoming signal. This is a useful feature because it relieves the receiver operator of the necessity to continually readjust the clipping level as the signal fades. Such a circuit is called an *automatic noise limiter* (*see* AUTOMATIC NOISE LIMITER).

Noise limiters are effective against all types of natural and manmade noise, including noise not affected by noise-blanking circuits. However, the noise limiter cannot totally eliminate the noise.

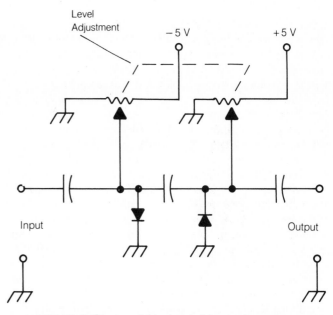

NOISE LIMITER: A variable-threshold noise-limiter circuit.

NOISE PULSE

A *noise pulse* is a short burst of electromagnetic energy. Often, the instantaneous amplitude of a noise pulse rises to a much higher level than the amplitude of a received signal. However, because the duration of a noise pulse is short, the total electromagnetic energy is usually small.

Noise pulses are produced especially by electric arcing. The noise pulse can produce energy over a wide band of frequencies, from the very-low to the ultra-high, at wavelengths from kilometers to millimeters. *See also* IMPULSE NOISE, and NOISE.

NOISE QUIETING

Noise quieting is a decrease in the level of internal noise in a frequency-modulation receiver, as a result of an incoming signal. With the squelch open (receiver unsquelched) and no signal, a frequency-modulated receiver emits a loud hissing noise. This is internally generated noise. When a weak signal is received, the noise level decreases. As the signal gets stronger, the level of the noise continues to decrease until, when the signal is very strong, there is almost no hiss from the receiver.

The noise quieting phenomenon provides a means of measuring the sensitivity of a frequency-modulation receiver. The level of the noise at the speaker terminals is first measured under no-signal conditions. Then, a signal is introduced, by means of a calibrated signal generator. The signal level is increased, without modulation, until the noise voltage drops by

20 dB at the speaker terminals. The signal level, in microvolts at the antenna terminals, is then determined. Typical unmodulated signal levels for 20-dB noise quieting are in the range of 1 μV or less at very-high and ultra-high frequencies, with receivers using modern solid-state amplifiers.

The noise-quieting method is one of two common ways of determining the sensitivity of a frequency-modulation receiver. The other method is called the *signal-to-noise-and-distortion (SINAD)* method. *See also* SINAD, SINAD METER, and SQUELCH.

NOISE-REDUCING ANTENNA

Electromagnetic-noise pickup can be minimized by the use of various specialized types of receiving antennas. The most effective method of reducing noise in an antenna system is the phase-cancellation or nulling method. With antennas designed for transmission as well as reception, noise reduction is generally more difficult, and the results less spectacular.

Any receiving antenna with a sharp directional null can be used to reduce the level of received manmade noise. A ferrite-rod or loop antenna is especially useful for this purpose (*see* FERRITE-ROD ANTENNA, and LOOP ANTENNA). Both of these antennas have sharp directional nulls along their axes. When the null is oriented in the focal direction of a noise source, the noise level drops. When tuned to resonance by means of a variable capacitor, the bandwidth of a ferrite-rod or loop antenna is very narrow, and this serves to further increase the immunity of the antenna system to broadband noise. The feed line must be properly balanced or shielded, however, if the benefits of such antennas are to be realized.

The noise rejection of a small loop antenna can be enhanced by the addition of a Faraday shield surrounding the loop element. The Faraday shield discriminates against the electrostatic component of a signal, while allowing the magnetic component to penetrate unaffected. Many noise signals are locally propagated by predominantly electrostatic coupling.

The overall noise susceptibility of a receiving/transmitting antenna system can be minimized by careful balancing or shielding of the feed line, maintaining an adequate radio-frequency ground, and maximizing the antenna Q factor (that is, minimizing the bandwidth). In general, at the medium and high frequencies, a horizontally polarized antenna is less susceptible to manmade noise than is a vertically polarized antenna.

NO-LOAD CURRENT

When an amplifier is disconnected from its load, some current still flows in the collector, drain, or plate circuit. This current is usually smaller than the current under normal load conditions. The no-load current depends on the class of amplification and the full-load output impedance for which the circuit is designed.

When the secondary winding of an alternating-current power transformer is disconnected from its load, a small amount of current still flows. This current occurs because of the inductance of the secondary winding; it is called *no-load current*. The value of the no-load current depends on the voltage supplied to the primary winding of the transformer, and also on the primary-to-secondary turns ratio. *See also* TRANSFORMER.

NO-LOAD VOLTAGE

When the load is removed from a power supply, the voltage rises at the output terminals. The voltage at the output terminals of a power supply, with the load disconnected, is called the *no-load voltage*.

The difference between the no-load and full-load voltages of a power supply depends on the amount of current drawn by the load. The greater the current demanded by the load, the greater the difference between the no-load voltage and the full-load voltage.

The percentage difference between the no-load and full-load voltages of a power supply is sometimes called the *regulation* or the *regulation factor* of the supply. In some instances, it is important that the output voltage change very little with large fluctuations in the load current. In other cases, a large voltage change can be tolerated. *See also* POWER SUPPLY, REGULATED POWER SUPPLY, and VOLTAGE REGULATION.

NOMINAL VALUE

All components have named, or specified, values. For example, a capacitor can have a specified value of 0.1 μF; a microphone can have a named impedance of 600 ohms. An amplifier can have a specified power-output rating of 25 watts. A piezoelectric crystal can have a frequency rating of 3.650 MHz. The named, or specified, value is called the *nominal value*.

The nominal value for a component or circuit is the average for a large sample of manufactured items, but individual units will vary somewhat either way. Nominal values are given without reference to the tolerance, or deviation that an individual sample can exhibit from the nominal value. *See also* TOLERANCE.

NONLINEAR CIRCUIT

A *nonlinear circuit* is a circuit having a nonlinear relationship between its input and output. That is, if the output is graphed against the input, the function is not a straight line.

Sometimes a circuit will behave in a linear manner within a certain range of input current or voltage, but will be nonlinear outside that range. An example is a radio-frequency power amplifier designed to act as a linear amplifier. It will behave in a linear manner only as long as the drive (input power) is not excessive. But if the amplifier is overdriven, it will no longer act as a linear amplifier. A similar effect is observed in other linear circuits.

Nonlinearity results in increased harmonic and intermodulation distortion in a circuit. In fact, harmonic generators and mixers make use of this characteristic.

Nonlinear circuits are extensively used as radio-frequency power amplifiers. Class-AB, class-B, and class-C amplifiers all cause some distortion of the input waveform. Class-AB and class-B amplifiers do not, however, they cause envelope distortion of an amplitude-modulated signal. *See also* CLASS-A AMPLIFIER, CLASS-AB AMPLIFIER, CLASS-B AMPLIFIER, CLASS-C AMPLIFIER, LINEAR AMPLIFIER, LINEARITY, and NONLINEARITY.

NONLINEARITY

Any departure of a circuit from linear operation is called a *nonlinearity*. Nonlinearity always results in distortion of a signal. This can be deliberately introduced or it can be undesirable.

Nonlinearity can be caused by improper bias voltage in a transistor, field-effect transistor, or tube. Nonlinearity can also be caused by excessive drive at the input and/or underloading at the output of an amplifier. In a push-pull or parallel amplifier configuration, nonlinearity can result from improper balance between the two transistors, field-effect transistors, or tubes.

In a linear radio-frequency amplifier, nonlinearity of the modulation envelope is undesirable, although nonlinearity of the carrier waveform is often produced. Nonlinearity of the modulation envelope results in increased bandwidth because of harmonic and intermodulation products. See also LINEAR AMPLIFIER, LINEARITY, and NONLINEAR CIRCUIT.

NONPOLARIZED COMPONENT

A component is considered *nonpolarized* if it can be inserted either way in a circuit with the same results. Examples of nonpolarized components are resistors, inductors, and many types of capacitors.

If the leads of a component cannot be interchanged or reversed without adversely affecting circuit performance, the component considered is polarized. See also POLARIZED COMPONENT.

NONPOLARIZED ELECTROLYTIC CAPACITOR

Most electrolytic capacitors are polarized components; that is, they must be installed in the correct direction with regard to the dc polarity of the circuit. However, certain types of electrolytic capacitors are deliberately designed so that they can be installed without concern for polarity. Such capacitors are called *nonpolarized electrolytic capacitors*. They are useful in applications where a large amount of capacitance is needed, but there is no direct-current bias to form the oxide layer in a normal electrolytic capacitor.

Both terminals are called *anodes.* The electrolyte is placed between the anodes. An oxide film forms around each anode, insulating it from the electrolyte material. In an alternating-current circuit, each anode receives its bias during half of the cycle. See also ELECTROLYTIC CAPACITOR.

NONREACTIVE CIRCUIT

In theory, all circuits contain at least a tiny amount of inductive or capacitive reactance because of effects among the various leads and electrodes. However, this reactance can be so small that it can be neglected in some instances. A circuit is nonreactive if it contains no effective inductive or capacitive reactance, as seen from its input terminals.

A circuit can be nonreactive at just one frequency. An example of such a circuit is the common resonant circuit. At the resonant frequency, the inductive and capacitive reactances cancel, and only resistance is left (see RESONANCE).

A circuit can be nonreactive at one frequency and an infinite number of harmonic frequencies. For example, a quarter-wave section of transmission line, having a fundamental frequency f, is also resonant at frequencies $2f$, $3f$, $4f$, and so on. At all of these frequencies, the line measures an integral multiple of a quarter wavelength. A quarterwave radiating element exhibits the same property.

Some circuits are essentially nonreactive over a wide range of frequencies. This type of circuit generally contains no reactive components, but only resistances. At all frequencies below a certain maximum frequency, the circuit can be considered nonreactive; but above this subjective maximum, component leads introduce significant reactance. See also REACTANCE, and REACTIVE CIRCUIT.

NONRESONANT CIRCUIT

A circuit is *nonresonant* if either of the following is true: the circuit is nonreactive over a continuous range of frequencies, or the circuit contains capacitive and inductive reactances that do not cancel each other.

Resistive networks do not contain significant reactance over a wide range of frequencies, but such circuits do not exhibit resonance. Thus, resistive networks are nonresonant circuits.

Bandpass filters, band-rejection filters, highpass filters, lowpass filters, piezoelectric crystals, transmission lines, and antennas are all examples of circuits that are nonresonant at most frequencies. This is because either the capacitive reactance or the inductive reactance predominates. The reactances cancel only at certain discrete frequencies or bands of frequencies. See also RESONANCE, and RESONANT CIRCUIT.

NONSATURATED LOGIC

Most digital-logic circuits use semiconductors that are either saturated (fully conducting) or cut off (fully open). However, some types of logic circuits operate under conditions of partial conduction in one or both states. Examples of such logic designs include the emitter-coupled, integrated injection, and triple-diffused emitter-follower varieties (see EMITTER-COUPLED LOGIC, INTEGRATED INJECTION LOGIC, and TRIPLE-DIFFUSED EMITTER-FOLLOWER LOGIC).

Nonsaturated logic has certain advantages. The most notable advantage is its higher switching speed, compared with most saturated logic forms. However, a nonsaturated transistor is somewhat more susceptible to noise than is a saturated transistor. Recent technological improvements have reduced the noise susceptibility of nonsaturated bipolar logic devices. See also SATURATED LOGIC.

NONSINUSOIDAL WAVEFORM

A waveform is nonsinusoidal if its shape cannot be precisely represented by the sine function. That is, a nonsinusoidal waveform is any waveform that is not a sine wave (see SINE WAVE).

Most naturally occurring waveforms are nonsinusoidal. All nonsinusoidal waveforms contain energy at more than one frequency, although a definite fundamental period is usually ascertainable. Most musical instruments, as well as the human voice, produce nonsinusoidal waves.

Certain forms of periodic nonsinusoidal waveforms are used for test purposes. Of these, the most common are the sawtooth and square waves. See also SAWTOOTH WAVE, and SQUARE WAVE.

NOR GATE

A *NOR gate* is an inclusive-OR gate followed by an inverter. The expression NOR derives from NOT-OR. The NOR gate outputs are exactly reversed from those of the OR gate. That is, when both or all inputs are 0, the output is 1; otherwise, the output is 0. See also OR GATE.

NORTON'S THEOREM

Norton's theorem is a principle involving the properties of circuits with more than one generator and/or more than one constant impedance. Given a circuit with two or more constant-current, constant-frequency generators and/or two or more constant impedances, there exists an equivalent circuit, consisting of exactly one constant-current, constant-frequency generator and exactly one constant impedance. If the more complicated circuit is replaced by the simpler circuit, identical conditions will prevail. *See also* COMPENSATION THEOREM, RECIPROCITY THEOREM, SUPERPOSITION THEOREM, and THEVENIN'S THEOREM.

NOTCH FILTER

A *notch filter* is a narrowband-rejection filter. Notch filters are found in many superheterodyne receivers. The notch filter is extremely convenient for reducing interference caused by strong, unmodulated carriers within the passband of a receiver.

Notch-filter circuits are generally inserted in one of the intermediate-frequency stages of a superheterodyne receiver, where the bandpass frequency is constant. There are several different kinds of notch-filter circuit. One of the simplest is a trap configuration, inserted in series with the signal path, as shown at A in the illustration. The notch frequency is adjustable so that the deep null (B) can be tuned to any frequency within the receiver passband.

A properly designed notch filter can produce attenuation well in excess of 40 dB. *See also* BAND-REJECTION FILTER.

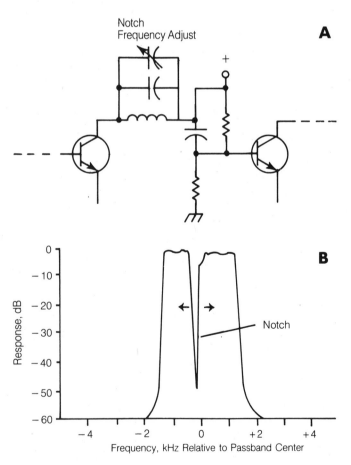

NOTCH FILTER: At A, a simple circuit; at B, typical frequency response.

NOT GATE

See INVERTER.

NPN TRANSISTOR

An npn transistor is a three-layer, bipolar semiconductor device. A layer of P-type material is sandwiched in between two layers of N-type material (see illustration A). One N-type layer is thinner than the other; this layer is the emitter. The P-type layer in the center is the base; the thicker N-type layer is the collector.

The npn transistor is normally biased with the base positive, with respect to the emitter. The collector is positive, with respect to the base. This biasing configuration is shown at B.

For class-B or class-C amplifier operation, the base can be biased at the same voltage as the emitter, or at a more negative voltage (*see* CLASS-B AMPLIFIER, and CLASS-C AMPLIFIER).

The npn transistor uses a biasing polarity that is opposite to that of the pnp transistor. *See also* PNP TRANSISTOR.

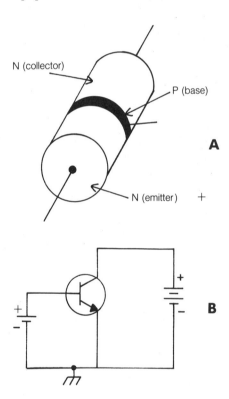

NPN TRANSISTOR: At A, a pictorial representation; at B, typical biasing.

N-TYPE CONNECTOR

The *N-type connector* is a connector similar to the PL-259, used with coaxial cables in radio-frequency applications. The N-type connector is known for its low loss and constant impedance. These features are important at very-high and ultra-high frequencies, and therefore the N-type connector is preferred at these frequencies.

N-type connectors are available in either male or female form (see photograph) and in other configurations such as male-to-male and female-to-female. Adaptors are available for coupling an N-type connector to a cable or chassis having a different type of connector.

N-TYPE CONNECTOR: Male (left) and female (right).

N-TYPE SEMICONDUCTOR

An *N-type material* is a semiconductor substance that conducts mostly by electron transfer. N-type germanium or silicon is fabricated by adding certain impurity elements to the semiconductor. These impurity elements are called *donors* because they contain an excess of electrons. The N-type semiconductor gets its name from the fact that it conducts via negative charge carriers (electrons).

The carrier mobility in an N-type semiconductor is greater than the carrier mobility in a P-type material because electrons are transferred more rapidly, from atom to atom, than are holes, for a given applied electric field.

N-type material, combined with P-type material, is the basis for manufacture of most forms of semiconductor diodes, transistors, field-effect transistors, integrated circuits, and other devices. *See also* CARRIER MOBILITY, DOPING, ELECTRON, HOLE, P-N JUNCTION, and P-TYPE SEMICONDUCTOR.

NULL

A *null* is a condition of zero output from an alternating-current circuit, resulting from phase cancellation or signal balance. An alternating-current bridge circuit is adjusted for a zero-output, or null, reading in order to determine the values of unknown capacitances, inductances, or impedances. The radiation pattern from an antenna system can exhibit zero field strength in certain directions; these directions are called *nulls*. *See also* PHASE BALANCE.

NYQUIST DIAGRAM

A *Nyquist diagram*, also called a *Nyquist plot*, is a complex-number graph used for evaluating the performance of a feedback amplifier. The Nyquist-diagram plane is a Cartesian system. The real-number axis (x axis) is horizontal, and the imaginary-number axis (y axis) is vertical.

In the Nyquist diagram, open-loop gain is represented by circles centered at the origin. The circle with radius 1 is the *unity-gain circle*. Points inside this circle represent amplifier gain less than 1; points outside this circle correspond to gain figures greater than 1. The output phase is represented by radial lines passing through the origin. The Nyquist diagram is made by plotting the real and imaginary components of the amplifier output in the complex plane, for all possible input frequencies.

The Nyquist diagram is used to determine whether or not a feedback amplifier is stable. *See* NYQUIST STABILITY CRITERION.

NYQUIST STABILITY CRITERION

To determine if a single-loop feedback amplifier is stable, the Nyquist diagram can be used. This is known as the *Nyquist stability criterion*.

The Nyquist plot is obtained for all possible input frequencies. If the Nyquist plot contains a closed loop surrounding the point $-1 + j0$, then the system is unstable. If the plot contains no such loop and the open-loop configuration is stable, then the amplifier is stable. *See also* NYQUIST DIAGRAM.

OCTAL NUMBER SYSTEM

The octal number system is a base-8 system. Only the digits 0 through 7 are used. When counting in base 8, the number following 7 is 10; the number following 77 is 100, and so on. Moving toward the left, the digits are multiplied by 1, then by 8, then by 64, and so on, upward in successively larger integral powers of 8.

The octal number system is sometimes used as shorthand for long binary numbers. The binary number is split into groups of three digits, beginning at the extreme right, and then the octal representation is given for each three-digit binary number. For example, suppose we are given the binary number 11000111. This is an eight-digit binary number. We first add a zero to the left-hand end of this number, making it nine digits long (a multiple of three), obtaining 011000111, Breaking up this number gives us 011 000 111. These three binary numbers correspond to 3, 0, and 7 respectively. The octal representation, then, of 11000111 is 307.

The octal number system is used by some computers, since its base, 8, is a power of 2, making it easier to work with than the base-10 system. A base-16 number system is also sometimes used by computers. *See also* HEXADECIMAL NUMBER SYSTEM.

OCTAVE

An *octave* is a 2-to-1 range of frequencies. If the lower limit of an octave is represented in hertz by f, then the upper limit is represented by $2f$, or the second harmonic of f. If the upper limit of an octave is given in hertz by f, then the lower limit is $f/2$.

The term *octave* is sometimes used when specifying the gain-versus-frequency characteristics of an operational amplifier. A circuit can, for example, have a rolloff characteristic (decrease in gain) of 6 dB per octave. This means that if the gain is x dB at frequency f, then the gain is $x - 6$ dB at frequency $2f$, $x - 12$ dB at frequency $4f$, $x - 18$ dB at frequency $8f$, and so on.

For audio frequencies, the octave can be recognized by ear. This phenomenon is familiar to the musician; from a given tone frequency f, it is easy to imagine the notes having frequencies $2f$ and $4f$. It is also easy to imagine the notes having frequencies $f/2$ and $f/4$. *See also* HARMONIC.

OCTET

In packet radio, the term *octet* is used in reference to a unit of digital information that is eight bits long. This is the equivalent of a *byte* in computer language. An octet is a string of eight ones and/or zeros, such as 01111110. *See also* BYTE, and PACKET RADIO.

ODD-ORDER HARMONIC

An *odd-order harmonic* is any odd multiple of the fundamental frequency of a signal. For example, if the fundamental frequency is 1 MHz, then the odd-order harmonics occur at frequencies of 3 MHz, 5 MHz, 7 MHz, and soon.

Certain conditions favor the generation of odd-order harmonics in a circuit, while other circumstances tend to cancel out the odd harmonics. The push-pull circuit accentuates the odd harmonics (*see* PUSH-PULL CONFIGURATION). Such circuits are often used as frequency triplers. The push-push circuit (*see* PUSH-PUSH CONFIGURATION) tends to cancel out the odd harmonics in the output.

A half-wave dipole antenna will operate at all odd-harmonic frequencies, assuming it is a full-size antenna (not inductively loaded). The impedance at the center of such an antenna is purely resistive at all odd-harmonic frequencies. At the odd harmonics, the dipole antenna presents a slightly higher feed-point resistance than at the fundamental frequency, but the difference is not usually large enough to cause an appreciable mismatch when the antenna is operated at the odd harmonics. *See also* EVEN-ORDER HARMONIC.

OERSTED

The *oersted* is a unit of magnetic-field intensity in the centimeter-gram-second (CGS) system, expressed in terms of magnetomotive force per unit length. A field of 1 oersted is equivalent to a field of 1 gilbert, or 0.796 ampere turns, per centimeter. *See also* AMPERE TURN, GILBERT, and MAGNETOMOTIVE FORCE.

OFF-CENTER FEED

Off-center feed is a method of applying power to an antenna at a current or voltage loop not at the center or end of the radiating element. Off-center feed is sometimes called *windom feed*, but the windom is only one form of off-center-fed antenna (*see* WINDOM ANTENNA).

For off-center feed to be possible, an antenna must be at least 1 wavelength long. A 1-wavelength antenna has current loops at distances of ¼ wavelength from either end. A 1-wavelength radiator can thus be off-center-fed with coaxial cable.

In general, for a longwire antenna measuring an integral multiple of ½ wavelength, a coaxial feed line can be connected ¼ wavelength from one end, and a reasonably good impedance match will be obtained. If the longwire is terminated, the feed point should be ¼ wavelength from the unterminated end. A low-loss, high-impedance, balanced feed line can be connected at any multiple of ¼ wavelength from the unterminated end of such an antenna. *See also* CENTER FEED, END FEED, and LONGWIRE ANTENNA.

OFFSET VOLTAGE

In an operational amplifier or "op amp," the output depends on the voltage difference between the two inputs. At a certain voltage difference, the output voltage is zero in the closed-loop configuration. This potential difference is known as the *offset voltage*. It is given as a specification for op amps.

Ideally, the offset voltage would be zero. In practice, it is a few microvolts or millivolts. In high-grade op amps, the offset voltage might be less than a microvolt.

OHM

The *ohm* is the standard unit of resistance, reactance, and impedance. The symbol is the Greek capital letter omega (Ω). A resistance of 1 ohm will conduct 1 ampere of current when a voltage of 1 volt is placed across it. *See also* IMPEDANCE, REACTANCE, and RESISTANCE.

OHMIC LOSS

Ohmic loss is the loss that occurs in a conductor because of resistance. Ohmic loss is determined by the type of material used to carry a current, and by the size of the conductor. In general, the larger the conductor diameter for a given material, the lower the ohmic loss. Ohmic loss results in heating of a conductor.

For alternating currents, the ohmic loss of a conductor generally increases as the frequency increases. This occurs because of a phenomenon known as *skin effect*, in which high-frequency alternating currents tend to flow mostly near the surface of a conductor (*see* SKIN EFFECT).

In a transmission line, ohmic loss is just one form of loss. Additional losses can occur in the dielectric material. Loss can also be caused by unwanted radiation of electromagnetic energy, and by impedance mismatches. *See also* DIELECTRIC LOSS, RADIATION LOSS, and STANDING-WAVE-RATIO LOSS.

OHMMETER

An *ohmmeter* is a simple device used for measuring direct-current resistance. A source of known voltage, usually a battery, is connected in series with a switchable known resistance, a milliammeter, and a pair of test leads (see illustration).

Most ohmmeters have several ranges, labeled according to the magnitude of the resistances in terms of the scale indication. The scale is calibrated from 0 to "infinity" in a nonlinear manner. The range switch, by inserting various resistances, allows measurement of fairly large or fairly small resistances.

Most ohmmeters can accurately measure any resistance between about 1 ohm and several hundred megohms. Specialized ohmmeters are needed for measuring extremely small or large resistances. *See also* MEGGER, and WHEATSTONE BRIDGE.

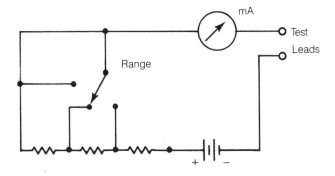

OHMMETER: A schematic diagram of a simple ohmmeter.

OHM'S LAW

Ohm's law is a simple relation between the current, voltage, and resistance in a circuit. The current, voltage, and resistance in a direct-current situation are interdependent. If two of the quantities are known, the third can be found by a simple equation.

Letting I represent the current in amperes, E represent the voltage in volts, and R represent the resistance in ohms, the following relation holds:

$$I = E/R$$

The formula can also be restated as:

$$E = IR, \text{ or}$$
$$R = E/I$$

In an alternating-current circuit, pure reactance can be substituted for the resistance in Ohm's law. If both reactance and resistance are present, however, the situation becomes more complex. *See also* CURRENT, RESISTANCE, and VOLTAGE.

OHMS PER VOLT

A voltmeter is sometimes rated for sensitivity according to a specification known as *ohms-per-volt*. The ohms-per-volt rating is an indication of the extent to which the voltmeter will affect the operation of a high-impedance circuit.

To determine the ohms-per-volt rating of a voltmeter, the resistance through the meter must be measured at a certain voltage. Normally, this is the full-scale voltage. This resistance can be found by connecting the voltmeter in series with a milliammeter or microammeter, and then placing a voltage across the combination such that the meter reads full scale. The ohms-per-volt (R/E) rating is then equal to 1 divided by the milliammeter or microammeter reading expressed in amperes.

For example, suppose that a voltmeter with a full-scale reading of 100 V is connected in series with a microammeter. A source of voltage is connected across this combination, and the voltage adjusted until the voltmeter reads full scale. The meter reads 10 μA. Then, the ohms-per-volt rating of the meter is $1/0.00001 = 100,000$ ohms per volt.

It is desirable that a voltmeter have a very high ohms-per-volt rating. The higher this figure, the smaller the current drain produced by the voltmeter when it is connected into a circuit. In a high-impedance circuit, this is extremely important. Some voltmeters are designed so that they exhibit extremely high ohms-per-volt figures. These meters use field-effect transistors that have extremely high input impedances. Older meters of this type use vacuum tubes, although this technology is now obsolete. *See also* FET VOLTMETER, VACUUM-TUBE VOLTMETER, and VOLTMETER.

OMNIDIRECTIONAL ANTENNA

An *omnidirectional antenna* is an antenna that radiates with equal field strength in all directions. Such an antenna does not exist in reality. However, some antennas do radiate with equal field strength within a given plane. Most vertical antennas are omnidirectional in the horizontal plane (see VERTICAL ANTENNA). The vertical-plane directional pattern of such an antenna exhibits maximum field strength at or near the horizontal plane.

An isotropic antenna is a true omnidirectional antenna, but it exists only in theory. An isotropic antenna radiates with equal field strength in all directions in three dimensions. *See also* ISO-TROPIC ANTENNA.

ONE-SHOT MULTIVIBRATOR
See MONOSTABLE MULTIVIBRATOR.

OPEN LOOP
In an amplifier, the condition of zero negative feedback is called the *open-loop condition.* With an open loop, an amplifier is at maximum gain. In an operational-amplifier circuit, the open-loop gain is sometimes specified as a measure of the amplification factor of a particular integrated circuit. The input and output impedances are generally specified for the open-loop configuration.

OPEN SYSTEMS INTERCONNECTION REFERENCE MODEL (OSI-RM)
There are many different kinds of computer systems. In a network, it is important that they all be able to communicate effectively with each other. In order to ensure this, the computers' protocols must all have certain characteristics in common (*see* PROTOCOL). A standard set of protocols, used in packet radio, is called the *open systems interconnection reference model (OSI-RM).*

There are seven levels, or *layers,* in OSI-RM. These are as follows, in order from lowest to highest:

- Physical
- Link
- Network
- Transport
- Session
- Presentation
- Application

As of this writing, only the first, or lowest, four levels play important roles in amateur packet radio. The higher levels remain available for more sophisticated future systems.

The *physical layer* is responsible for actually moving the messages from place to place. For amateur packet radio, the standard is EIA-232-D (*see* EIA-232-D). The *link layer* puts bits of data into *frames,* and transmits the data in this form from node to node (see FRAME, NODE). The *network layer* finds the routes, or paths, for the messages through the packet-radio network. The first packet establishes this route; the following packets automatically follow the same route until all the information has been sent. The *transport layer* maintains the connection between the originating station and the destination station.

The above mentioned four layers are in current use in ham packet radio systems. The following three layers have yet to be utilized in amateur work as of this writing.

The *session layer* synchronizes the flow of data in transmissions from the source to destination station, through intervening nodes. This optimizes system performance. The *presentation layer* acts as a translator between different forms of data. In other words, this layer governs the way in which the data is presented. The *application layer* interfaces the packet messages with the user application programs used by the computers in the system. This might allow, for example, a ham radio operator in Minneapolis to directly and fully use the computer of another ham in Miami, via packet radio.

With all seven OSI-RM layers in place and fully operational, hams could link groups of personal computers together to form supercomputers — entirely via amateur radio, independent of commercial media. *See also* PACKET RADIO.

OPERATING ANGLE
In an amplifier, the *operating angle* is the number of degrees, for each cycle, during which current flows in the collector, drain, or plate circuit. The operating angle varies, depending on the class of amplifier operation.

In a class-A amplifier, the output current flows during the entire cycle. Therefore, in such an amplifier, the operating angle is 360 degrees. In a class-AB amplifier, the current flows for less than the entire cycle, but for more than half of the cycle; hence, the operating angle is larger than 180 degrees, but less than 360 degrees. In a single-ended class-B amplifier, the operating angle is about 180 degrees. In a class-C amplifier, the current flows for much less than half of the cycle, and the operating angle is thus smaller than 180 degrees.

The operating angle varies with the relative base/collector, gate/drain, or grid/plate bias of an amplifier. The operating angle is also affected by the driving voltage. *See also* CLASS-A AMPLIFIER, CLASS-AB AMPLIFIER, CLASS-B AMPLIFIER, CLASS-C AMPLIFIER, and OPERATING POINT.

OPERATING POINT
The *operating point* of an amplifier circuit is the point at which the direct-current bias is applied along the curve depicting the collector current versus base voltage (for a transistor), the drain current versus gate voltage (for a field-effect transistor), or the plate current versus grid voltage (for a tube). The choice of operating point determines the class of operation for the amplifier. *See* CLASS-A AMPLIFIER, CLASS-AB AMPLIFIER, CLASS-B AMPLIFIER, CLASS-C AMPLIFIER, FIELD-EFFECT TRANSISTOR, TRANSISTOR, and TUBE.

Certain transducers and semiconductor devices require direct-current bias for proper operation. When the signal is received or applied, the instantaneous current fluctuates above and below the no-signal current. The no-signal current is called the *operating current* or *operating point.* The correct choice of operating point ensures that the device will function according to its ratings.

OPTICAL COMMUNICATION
Optical communication is the application of light beams for sending and receiving messages. In recent decades, optical communication has become increasingly important. Early sailors actually used a form of optical communication — flags and flashing lights — but modern systems are capable of transferring millions of times more information than were the seamen who originally used visible light for sending and receiving signals.

Modern optical communications systems use modulated-light sources and receivers sensitive to rapidly varying light intensity (*see* MODULATED LIGHT). Lasers and light-emitting diodes are the favored method of generating modulated light because they have rapid response time and emit light within a narrow range of wavelengths (*see* LASER, and LIGHT-EMITTING DIODE).

Communication can be obtained in the infrared and ultraviolet regions of the spectrum, as well as in the visible range, provided that the wavelength is chosen for low atmospheric attenuation. Infrared and ultraviolet communications can be considered, for practical purposes, as optical communication.

Optical communication can be carried out directly through the atmosphere, in much the same way as microwave links are maintained. However, optical propagation is susceptible to the weather. Rain, snow, and fog can obscure the light and prevent the transfer of data. The transmittivity of clear air varies with the wavelength, as well. The atmosphere is fairly transparent at visible wavelengths. Blue light tends to be scattered more than red light. Considerable attenuation occurs over significant portions of the infrared and ultraviolet spectra.

Glass or plastic fibers can transmit light for long distances, in much the same way that wires carry electricity. A single beam of light can carry far more information than can an electric current, however, and so-called optical fibers are becoming more common for use in place of wire systems (*see* FIBER, and FIBER OPTICS).

OPTICAL COUPLING

Optical coupling is a method of transferring a signal from one circuit to another, by converting it to modulated light and then back to its original electrical form.

The signal is first applied to a light-emitting diode or other light-producing device. The resulting modulated-light signal propagates across a small gap, where it is intercepted by a photocell, phototransistor, or other light-sensitive device and converted back into electrical impulses. The basic principle is shown in the illustration. Some optical-coupling devices contain the light transmitter and the light receptor in a single package with a glass or plastic transmission medium.

Optical coupling is used when it is necessary to prevent impedance interaction between two successive stages in a multistage circuit. Normally, changes in the input signal to an amplifier cause the input impedance to vary, and this in turn affects the preceding stage. Optical coupling avoids this problem by providing essentially complete isolation. Optical coupling also facilitates coupling between circuits at substantially different potentials.

Optical-coupling devices can be assembled in various different forms for different purposes. One common optical coupler is called a *lumistor*. *See also* LIGHT-EMITTING DIODE, LUMISTOR, MODULATED LIGHT, PHOTOCELL, PHOTODIODE, PHOTOFET, PHOTORESISTOR, PHOTOTRANSISTOR, and PHOTOVOLTAIC CELL.

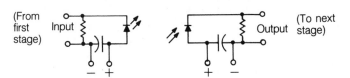

OPTICAL COUPLING: A typical optoisolator.

OPTICAL FIBER

See FIBER, FIBER OPTICS, and MODULATED LIGHT.

OPTICAL SHAFT ENCODER

In digital radios, frequency adjustment is in discrete steps. In high-frequency (HF) gear, the usual increment is 10 Hz. At 6

meters and above, 10- or 100-Hz increments are used for code (CW), radioteletype (RTTY), and single sideband (SSB); 5-kHz increments are common for frequency modulation (FM).

For tuning, an alternative to mechanical switches (which corrode or wear out with time) is the *optical shaft encoder*. This consists of a light-emitting diode (LED), a photodetector, and a chopping wheel.

The LED shines on the photodetector through the chopping wheel. The wheel has radial bands, alternately transparent and opaque. The wheel is attached to the tuning knob. As the knob is turned, the light beam is interrupted. Each interruption causes the frequency to change by the specified amount, such as 10 Hz. The difference between FREQUENCY UP and FREQUENCY DOWN is determined by using two LEDs and two photodetectors side-by-side, to sense the direction in which the wheel turns.

A well-designed optical shaft encoder provides a "feel" just like that of a mechanical dial, but without most of the moving parts and sliding contacts that cause maintenance trouble.

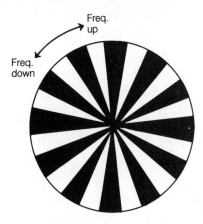

OPTICAL SHAFT ENCODER: A chopping wheel interrupts the light beam from a set of lasers to increment the frequency.

OPTIMIZATION

Optimization is a process or operation, by which a set of parameters is adjusted for the most efficient, the maximum, or the minimum result. In electronics, the process of optimization is a part of all engineering practice. Once a circuit design has been made to work, it is usually possible to improve the performance considerably. This can involve the changing of component values or the changing of component types. Optimization is partly determinable by calculation, but often it is a trial-and-error procedure.

OPTIMUM ANGLE OF RADIATION

When a radio signal is returned to earth by the ionosphere, the angle of incidence is approximately equal to the angle of reflection. Although there exist some irregularities in the ionosphere, this rule can be considered valid on the average. For a given distance between the transmitting and receiving station, and a given ionized-layer altitude, the optimum angle of radiation for single-hop propagation can be determined. The greater the distance between the stations, the lower the optimum angle; the higher the ionized-layer altitude, the higher the optimum angle.

At the high frequencies, ionospheric propagation usually occurs from the F1 layer, at an altitude of about 100 miles. For single-hop propagation via the F1 layer, the optimum angle varies with distance. Propagation from the F2 layer occurs over greater distances because of the greater ionization altitude of about 200 miles. The F1 layer is responsible for most high-frequency propagation during the daylight hours. The F2 layer usually returns the signals at night (*see* F LAYER).

Under certain conditions at high frequencies, but more commonly at very-high frequencies, the *E* layer, at an altitude of about 60 miles, returns signals to the earth (*see* E LAYER, and SPORADIC-E PROPAGATION).

The angle of departure from an antenna can be controlled to a certain extent at the high frequencies. The higher the antenna above the ground, the lower the angle at which maximum radiation occurs.

When the distance between the transmitting and receiving station is greater than the maximum single-hop distance, multi-hop propagation occurs. In such cases, the optimum angle of radiation is less well defined. In general, the best results for multi-hop paths are obtained when the angle of departure is as low as possible. This means that, for long-distance communication, the transmitting antenna should be placed as far above the ground as can be managed. *See also* PROPAGATION CHARACTERISTICS.

OPTIMUM WORKING FREQUENCY

For any high-frequency communications circuit, there exists an *optimum working frequency*, at which the most reliable communications are obtained. The optimum working frequency lies between the lowest usable frequency and the maximum usable frequency (*see* LOWEST USABLE FREQUENCY, and MAXIMUM USABLE FREQUENCY).

The optimum working frequency depends on the distance between the transmitting and receiving stations, the time of day, the time of year, and the level of sunspot activity. In general, the optimum working frequency is increased as the station separation becomes greater. The optimum working frequency is usually greater during the day than at night; it is higher in summer than in winter, and is highest during periods of peak sunspot activity. *See also* PROPAGATION CHARACTERISTICS.

OPTOISOLATOR

See LUMISTOR, and OPTICAL COUPLING.

OR GATE

An *OR gate* is a form of digital logic gate, with two or more inputs and one output. The OR gate performs the Boolean inclusive-OR function. That is, if all inputs are 0 (false), the output is 0; if any or all inputs are 1 (true), the output is 1.

OSCAR

Several ham radio satellites have been launched and operated since the Space Age began in the 1960s. Usually, ham radio transponders have "hitched rides" aboard commercial satellites, rather than occupied satellites of their own. The ham satellites put into space by the United States have been dubbed *OSCAR*, an acronym for *Orbiting Satellite Carrying Amateur Radio*.

The first OSCAR was known as OSCAR 1. It just sent a simple morse code message over and over; it was only an orbiting beacon. Modern OSCAR satellites have transponders that work various crossband modes. All ham satellites operate at 21 MHz or higher frequencies; the majority of the work is done on bands at or above 144 MHz.

Communications through a modern amateur satellite is a unique and fascinating experience. The satellite must first be located, and then the antennas pointed into space in just the right direction. Signals are always rather faint, but are clear when a good antenna is used. You can listen to your own signal as it is intercepted and retransmitted by the satellite: "Calling CQ, CQ, CQ . . ." Many hams report that their first QSO, or contact, via satellite is as exciting as their very first ham radio contact was.

For further information on ham satellites and their operation, *see* the following articles: ACTIVE COMMUNICATIONS SATELLITE, APOGEE, ASCENDING NODE, ASCENDING PASS, DESCENDING NODE, DESCENDING PASS, DOWNLINK, ECCENTRICITY OF ORBIT, ELLIPTICAL ORBIT, INCREMENT, OSCARLOCATOR, PERIGEE, PHASE I SATELLITE, PHASE II SATELLITE, PHASE III SATELLITE, RADIO AMATEUR SATELLITE CORPORATION, SATELLITE TRANSPONDER MODES, TRANSPONDER, and UPLINK.

OSCARLOCATOR

In order to use OSCAR or any other amateur satellite, it is necessary to know where the satellite will be at any given time. It is also important to know the nature of the course the satellite will take across the sky for a given pass, so that the antenna system can be aimed in the right direction, and so the operator can turn the antennas to follow the satellite. One handy device for locating OSCAR is called an *OSCARLOCATOR*.

Information about the OSCARLOCATOR can be obtained from the Radio Amateur Satellite Corporation. Write: AMSAT, P.O. Box 27, Washington, DC 20044. Or contact the American Radio Relay League, 225 Main Street, Newington, CT 06111. *See also* OSCAR.

OSCILLATION

Oscillation is any repeating, periodic effect, such as the swinging of a pendulum, the back-and-forth motion of a spring, or the continual back-and-forth movement of electrons in an electrical conductor. Oscillation can be short-lived, dying down with time, or it can continue indefinitely, if an outside source of energy is provided to keep it going.

In electronics, oscillation generally refers to a phenomenon caused by positive feedback (*see* FEEDBACK). Circuits can be made to oscillate deliberately at a precise frequency. Sometimes, oscillation occurs when it is not desired. This is likely to happen when an amplifier circuit has too much gain or is improperly designed for the application intended.

Oscillation is always the result of periodic storage and release of energy. In the pendulum, energy is stored as the weight rises, and is released as the weight falls. In the spring, energy is stored with compression and released with expansion. In an inductance-capacitance circuit, energy is alternately transferred between the inductive and capacitive reactances. *See also* OSCILLATOR.

OSCILLATOR

An *oscillator* is a circuit that generates a signal for a specific purpose. In ham radio applications, oscillators are used in transmitters and receivers, and they almost always generate radio-frequency energy.

The most stable oscillators use crystal frequency standards. Synthesizers can derive many different frequencies from this standard, creating, in effect, a variable-frequency oscillator (VFO).

Tuned inductance-capacitance (LC), resistance-inductance (RL) or resistance-capacitance (RC) circuits can also be used to determine the frequency of an oscillator. But these circuits are not as stable as crystal-controlled oscillators.

For information about specific types of oscillators, and for related material, please refer to the following articles: BALANCED OSCILLATOR, BLOCKING OSCILLATOR, CLAPP OSCILLATOR, COLPITTS OSCILLATOR, CRYSTAL OSCILLATOR, DIODE OSCILLATOR, FRANKLIN OSCILLATOR, FREQUENCY SYNTHESIZER, HARTLEY OSCILLATOR, LOCAL OSCILLATOR, MAGNETOSTRICTION OSCILLATOR, MILLER OSCILLATOR, NEON-LAMP OSCILLATOR, PHASE-LOCKED LOOP, PIERCE OSCILLATOR, REINARTZ CRYSTAL OSCILLATOR, SIDETONE OSCILLATOR, SWEEP-FREQUENCY FILTER OSCILLATOR, VARIABLE CRYSTAL OSCILLATOR, VARIABLE-FREQUENCY OSCILLATOR, VOLTAGE-CONTROLLED OSCILLATOR, and WIEN-BRIDGE OSCILLATOR.

OSCILLATOR KEYING

Oscillator keying is a method of obtaining code transmission by interrupting the action of an oscillator circuit. Oscillator keying can be accomplished in various ways; the most common method is simply to break the cathode, emitter, or source circuit of the oscillator (see the schematic diagram).

Oscillator keying offers the advantage of completely shutting down the transmitter when the key is opened. This allows reception between code elements. However, oscillator keying results in chirp unless the circuit is carefully designed. *See also* KEYING.

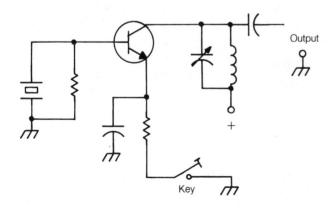

OSCILLATOR KEYING: In this case, the key is in the emitter circuit.

OSCILLOSCOPE

An *oscilloscope* is a test instrument that makes use of amplifiers, oscillators and a cathode-ray tube to produce a display of a signal waveform. An oscilloscope is an important tool in designing and servicing all types of electronic communications equipment. Oscilloscopes can be tailored to serve many different functions.

The display of an oscilloscope appears as a graph. The screen is calibrated in a grid pattern for reference. Deflection plates in the cathode-ray tube work independently, one set moving the electron beam up and down, and the other set moving it from side to side. *See* CATHODE-RAY TUBE.

Operation   The oscilloscope works by causing a narrow beam of electrons to move horizontally and vertically, forming patterns on a phosphor screen. When there is no signal present at the input, the beam sweeps, or traces, horizontally across the screen in a straight line. In the most common oscilloscope configuration, known as *time domain*, the horizontal axis shows time.

The rate of the sweep can be adjusted, and is measured in seconds per division (s/div), milliseconds per division (ms/div) or microseconds per division (μs/div). A division is about 1 centimeter on the screen.

When a signal is provided at the input of the oscilloscope, the trace is deflected vertically. The extent of deflection up or down depends on the instantaneous positive or negative amplitude of the signal. The sensitivity of the display can be adjusted, and is measured in volts per division (V/div), millivolts per division (mV/div) or microvolts per division (μV/div).

The Display   The illustration shows a typical oscilloscope screen, and an alternating-current (ac) waveform of 1 kHz and a peak-to-peak amplitude of 0.4 V. The sweep rate is set to 250 μs/div, and the sensitivity is at 100 mV/div.

An oscilloscope can provide an indication of whether a waveform is the right shape, and whether it is the right amplitude, in a given circuit. It can also be used as a general test probe to see whether the output of an oscillator, amplifier or mixer is what it should be.

An oscilloscope can be used to measure frequency, but not very accurately; a counter is more often used for this purpose. *See* FREQUENCY COUNTER.

Some oscilloscopes have signal inputs for both the horizontal and vertical deflections. This allows comparison of two signals for frequency and phase. The resulting patterns are called *Lissajous figures*. *See* LISSAJOUS FIGURE.

Other Common Features   One of the most important features of an oscilloscope is that the display be stable. For this to be possible, the sweep must be exactly synchronized with the frequency, so that the trace goes over the same places on the screen with each pass. This is ensured by *triggering*, in which the sweep rate locks into the frequency of the input signal. Most scopes can be triggered either via the input signal, or by some external source. *See* TRIGGERING.

Most oscilloscopes will reveal waveform details up to several megahertz. There are special wideband oscilloscopes that will resolve signal details at much higher frequencies. Sometimes it is necessary to compare two waveforms that have the same frequency, or that are harmonically related. A *dual-beam* or *dual-trace* scope does this by displaying a separate trace for each signal, one above the other. *See* DUAL-BEAM OSCILLOSCOPE.

A *persistence trace* oscilloscope holds a display for several seconds or minutes after the signal has been removed. *See* PERSISTENCE TRACE. A *storage oscilloscope* can store and recall waveforms, in much the same way that a memory calculator works. *See* STORAGE OSCILLOSCOPE.

Some oscilloscopes display frequency, rather than time, on the horizontal axis. These are called *frequency-domain oscilloscopes*. The most common type of frequency-domain scope is a *spectrum analyzer*. *See* SPECTRUM ANALYZER.

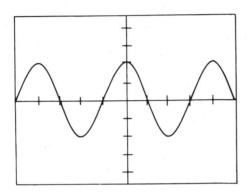

OSCILLOSCOPE: A sine-wave trace on an oscilloscope screen.

OSCILLOSCOPE CAMERA

An *oscilloscope camera* is a special camera designed for taking photographs of oscilloscope displays. This provides a permanent record of the display. Oscilloscope cameras are often used for comparing data obtained at different times in the design of electronic equipment.

An ordinary camera, mounted on a tripod, can be used as an oscilloscope camera. It is necessary that the shutter speed be slow enough to allow at least one complete trace across the oscilloscope screen. Otherwise, only part of the waveform will be shown in the resulting photograph.

Specially designed cameras are made for the purpose of photographing oscilloscope displays. Such a camera is fitted with an opaque hood for keeping out stray light. The hood is attached to the screen frame of the oscilloscope. Many oscilloscopes are provided with mounting screws or flanges around the screen, for the attachment of an oscilloscope camera. *See also* OSCILLOSCOPE.

OSI-RM

See OPEN SYSTEMS INTERCONNECTION REFERENCE MODEL.

OUTDOOR ANTENNA

An *outdoor antenna* is any antenna that is placed outside a house or building. In general, outdoor antennas are preferable to indoor antennas because there are fewer obstructions for the transmission or reception of signals in most outdoor locations as compared with most indoor locations.

An outdoor antenna is almost always much better, in terms of transmitting performance, than an indoor antenna. However, indoor antennas can sometimes be reasonably effective, particularly when the structure is a frame house without metal reinforcements or metal siding. Narrowband antennas, equipped with low-noise preamplifiers, can be successfully used for certain receiving purposes, even when they are not located outdoors.

Indoor antennas usually perform poorly in steel-frame because the frame acts as a shield against radio-frequency energy. *See also* INDOOR ANTENNA and antenna articles, according to specific name.

OUT OF PHASE

Two signals are considered to be *out of phase* when their phase difference is 180 degrees. That is, they must be ½ cycle different in phase. For certain waveforms, such as the sine wave and the square wave, two out-of-phase signals of equal amplitude totally cancel (see illustration). When this happens, the net signal is zero because the instantaneous signal amplitudes are equal and opposite. *See also* PHASE, and PHASE ANGLE.

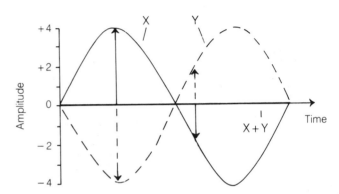

OUT OF PHASE: Sine waves of equal amplitude and opposite phase.

OUTPUT

Output is the signal produced by a circuit, or processed by a circuit. An oscillator generates its own output. An amplifier circuit produces its output from a signal of lesser intensity, applied to the input. A circuit can have just one set of output terminals, or it can have several. The output terminals themselves are sometimes called *outputs*.

The output signal in a given situation must conform to certain requirements. The amplitude might have to be within specified limits. A load of a certain impedance might have to be connected to the output terminals of a circuit for proper operation. *See also* OUTPUT IMPEDANCE.

OUTPUT DEVICE

In a computer system, an *output device* consists of any peripheral equipment that transfers data out of the computer. Examples of output devices include a video display terminal, a printer, card-punching devices, and the modulator sections of modems. *See also* MODEM, PRINTER, TERMINAL, and VIDEO DISPLAY TERMINAL.

In any electronic system, a circuit from which an output signal is obtained can be called an *output device*. Examples of such output devices are a power transformer, a signal transformer, and a coupling network. *See also* OUTPUT.

OUTPUT IMPEDANCE

The *output impedance* of a circuit is the load impedance which, when connected to the output terminals, results in optimum transfer of power. The output impedance of a circuit is normally a pure resistance. If a load contains reactance that reactance must, for best operation, be tuned out by means of an equal and opposite reactance connected in series with the load.

The resistance of a load should always be matched to the output resistance of the driving circuit. Removing the reactance might, by itself, not be sufficient. A transformer is required for resistance matching. For example, if an audio amplifier is designed to work into an 8-ohm load, a step-up transformer must be used for best results with 16-ohm speakers, and a step-down transformer should be used with 4-ohm speakers.

When the output impedance of a device differs from the load impedance, some of the electromagnetic field is reflected back from the load toward the source. This might or might not cause significant deterioration in the performance of the driving circuit. In the above audio-amplifier example, the transformer can, in some cases, be removed with only a slight reduction in system efficiency. However, in many instances, the impedance match must be nearly perfect. *See also* IMPEDANCE, IMPEDANCE MATCHING, INPUT IMPEDANCE, and REACTANCE.

OVERCURRENT PROTECTION

Overcurrent protection is any means of preventing excessive current from flowing through a circuit. The most common forms of overcurrent-protection devices are the circuit breaker and the fuse (*see* CIRCUIT BREAKER, and FUSE).

Some power supplies have built-in overcurrent protection. If the load resistance drops to a value that would cause excessive current to flow, the power supply automatically inserts an effective resistance in series with the load, preventing the current from rising above the maximum rated value. When the load resistance returns to normal, the power supply continues to deliver the rated current without the need for resetting a circuit breaker or replacing a blown fuse. *See also* CURRENT LIMITING, and CURRENT REGULATION.

OVERDAMPING

In an analog meter, the damping is the rapidity with which the needle reaches the actual current reading. Critical damping is the smallest amount of damping that does not result in needle overshoot (*see* CRITICAL DAMPING, and DAMPING). If the damping is greater than the critical value, a meter is said to be *overdamped.*

When a meter is overdamped, it responds slowly to changes in the current. This is desirable in some applications. For example, the value of a parameter can fluctuate rapidly to some extent, while the average value remains constant. An overdamped meter is preferable in such a case.

OVERFLOW

The capacity of any memory device is limited. When the amount of data is too great for storage in a memory, an *overflow condition* occurs. This can be observed in handheld calculators, minicomputers, and large computers.

Overflow can result in various responses from a calculator or computer. Generally, the excess data is ignored or compensated for in some way; or, an "error" signal can be generated. For example, a handheld calculator can have room for 10 digits. Keying in 3.6666666666 (an 11-digit number) would then result in a rounded-off display of 3.666666667, or perhaps a truncated display of 3.666666666. An attempt to divide by zero results, with most calculators, in an "error" signal.

Sophisticated computers will inform an operator if a memory overflow occurs. *See also* MEMORY.

OVERLOADING

Overloading is the result of an attempt to draw too much current from a source of power, or to apply excessive voltage to a load. A familiar example of overloading occurs when the frying pan, the oven, the coffee pot, and an air conditioner are all turned on at once—and the fuse or circuit breaker opens the circuit! Another, hopefully less familiar, example is the inadvertent connection of a 117-volt appliance to a 234-volt power source.

Still another example of overloading is the improper adjustment of the output matching circuit in a radio transmitter, resulting in excessive collector or plate current in the final-amplifier stage, and reduced radio-frequency output power and efficiency.

In a radio receiver, an incoming signal can be so strong that the front end is driven into nonlinear operation or desensitization. This condition is sometimes called *overloading.* It results in false signals, intermodulation, or a general reduction in sensitivity. *See also* DESENSITIZATION, and FRONT END.

OVERSHOOT

In an analog metering device, insufficient damping results in a condition called *overshoot* (*see* CRITICAL DAMPING, and DAMPING). When the current changes dramatically, the meter needle goes past the actual new reading, and then returns to the correct position. In severe cases of overshoot, the meter needle can oscillate back and forth several times before coming to rest at the proper position.

In a pulse waveform, a condition called *overshoot* often occurs in the rise or decay. This might or might not affect the operation of a circuit. The overshoot can usually be prevented by lengthening the rise and/or decay time. *See also* DROOP.

OVERTONE OSCILLATOR

A crystal oscillator can be made to operate at one of the harmonic frequencies of the crystal, rather than at the fundamental frequency of the crystal. For example, a crystal that is cut for 3.500 MHz will also produce outputs at 7.000 MHz, 10.500 MHz, 14.000 MHz and so on.

Normally, harmonic energy is not desired. But by providing a tuned circuit in the output of a crystal oscillator, harmonics can be deliberately used to get a higher-frequency output than is obtainable with fundamental-frequency crystal oscillators.

The most common way to get output on a harmonic frequency is to tune the output to the desired harmonic. This makes an *overtone oscillator.* The main advantage of this scheme is that it allows the generation of signals at higher frequencies than is possible with fundamental-frequency crystals. Some crystals are made especially for use with overtone oscillators. *See also* CRYSTAL OSCILLATOR, and HARMONIC.

OVERVOLTAGE PROTECTION

In a regulated power supply, the regulation system can fail. If this happens, and the power supply does not have an overvoltage-protection circuit, the voltage will rise dramatically. In some cases, the voltage can increase to more than twice the normal value. This can cause damage to equipment connected to the supply.

Most overvoltage-protection circuits consist of a Zener diode for sensing the output voltage, and a silicon-controlled rectifier for short-circuiting the output terminals when overvoltage occurs. This causes the regulator circuit to switch off the supply, or the fuse can blow. A schematic diagram of this type of overvoltage-protection circuit is shown.

There are other methods of obtaining overvoltage protection besides that shown in the illustration. One common alternative method is called *remote sensing*. The remote-sensing circuit consists of a device that has a certain threshold voltage, shutting down the supply when the threshold is exceeded. A bipolar transistor can be used for this purpose. The threshold voltage depends on the base bias. Under normal conditions the transistor, which is connected in series with the supply output, conducts. If the voltage becomes excessive, the transistor opens the circuit. *See also* REGULATED POWER SUPPLY.

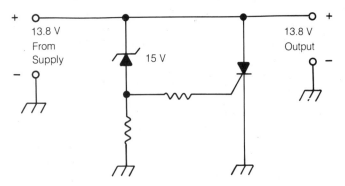

OVERVOLTAGE PROTECTION: The Zener diode conducts at 15 V and causes the SCR to shut down the power supply.

OWEN BRIDGE

The *Owen bridge* is a device for measuring unknown values of inductance. The Owen bridge is noted for its extremely wide range—it can determine any inductance from less than 0.1 nH to more than 100 H. This covers essentially all possible inductance values.

OXIDATION

Oxidation is a form of corrosion. A substance becomes oxidized when it combines with oxygen. Many different materials undergo oxidation when exposed to the air. The oxidation process is similar to combustion, but oxidation occurs more slowly.

Most metals will oxidize if exposed to the atmosphere for long periods. Some metals oxidize more rapidly than others. Iron, for example, rusts easily, and deteriorates much more quickly than chromium. For all materials, the oxidation process is hastened by high temperature and high humidity. This is why metals corrode so rapidly in tropical or subtropical regions. Electrical connections should be soldered and, if possible, insulated, if they are to be located outdoors.

Oxygen is not the only chemical that can cause corrosion. Chlorine is highly corrosive. Salt (sodium chloride) greatly exacerbates corrosion, because most metals react readily with chlorine. Marine environments are notorious for causing rapid deterioration of electrical and mechanical equipment because of the "salt air" (see CORROSION).

Oxidation is used to advantage in the manufacture of certain kinds of electronic components. An example is the copper-oxide rectifier; another is the electrolytic capacitor. An oxide coating on some materials provides protection against further corrosion. *See also* COPPER-OXIDE DIODE, and ELECTROLYTIC CAPACITOR.

PA

See POWER AMPLIFIER.

PACKET ASSEMBLER/ DISASSEMBLER

In packet radio, a *packet assembler/disassembler (PAD)* converts computer data into packets for transmission, and also changes received packets back into computer data. It is analogous to a radioteletype (RTTY) terminal unit.

A typical ham packet radio station has a computer terminal or a personal computer (PC) used directly by the operator, a PAD to process the data to and from the computer, a *modem* (modulator/demodulator), and a transceiver or transmitter/receiver combination. *See also* MODEM, and PACKET RADIO.

PACKET RADIO

In recent years, personal computers (PCs) have become common household appliances. Many hams own PCs. Computers and ham radio go naturally together in a rapidly advancing technology known as *packet radio*.

What is Packet Radio?   A *packet* is a message sent from a PC at a source station to another PC at a destination station. Any ham who has a PC can interface it to a transceiver so that his or her PC can communicate with other hams' PCs. It is not necessary for the operator at the destination station to be present for a message to be received; the computer can store it. For this reason, packet radio is a form of time-shifting communications.

Packet radio provides rapid, error-free transmission and reception of messages. Complex and massive antenna systems, high-powered amplifiers or sophisticated radios are not required. A modest setup can serve as a packet station. All you need are a PC, a transceiver and a terminal node controller.

Packet networks are getting larger, easier to access, and more sophisticated all the time. If you tap in at one point, you can take advantage of a network.

What is it Good For?   Packet radio is like a fast mail service. It's similar to a facsimile network of the type you can access via the telephone. But ham radio is toll free. Packet radio will ultimately make it possible for hams to share PCs, and even to interconnect many PCs to form *supercomputers*.

Packet radio is fast. The data rate is sufficiently high so that long messages can be sent via meteor-scatter propagation (*see* METEOR SCATTER). Packet radio is self-correcting. There is minimal chance for error.

Reliability and ease of use are important features of packet-radio networks. The day is rapidly approaching when you will be able to sit down at your PC and address a message to any packet station, anywhere in the world, and be sure that the network will automatically get it there, no matter what the propagation conditions at the time.

Standardization of Computer Communications   In order to make a packet network function, all the computers must use protocols that adhere to a standard model. The universal standard is called the *open systems interconnection reference model (OSI-RM)*.

OSI-RM is arranged into seven layers, ranging from the physical level to the application level. Amateur packet-radio networks have not yet come anywhere near making use of the full capabilities of OSI-RM. But hams have a long-standing reputation for rapid technological advancement. *See* OPEN SYSTEMS INTERCONNECTION REFERENCE MODEL.

Contemporary Network Protocols   Packet-radio technology is evolving toward the ultimate goal of a worldwide system. One of the most recent developments is *firmware* developed by Mike Busch, W6IXU and Ron Raikes, WA8DED. Firmware is computer programming that is installed permanently in a system. It's like software, except that it is not as easily changed (*see* FIRMWARE, and SOFTWARE). The firmware developed by W6IXU and WA8DED has become known as *NET/ROM*. It is protected by copyright, as is all commercial software and firmware.

A NET/ROM network is convenient to use. The computers do all of the routing work, once the source station operator knows the local *node* of the desired destination station. The operator only needs to key in the destination-station information in order to establish the route. NET/ROM also keeps the connection intact by means of rerouting, in case there is some disruption along a given route.

There are other protocols in existence, in addition to NET/ROM. These include KA-Node, ROSE, TCP/IP and TexNet. The ROSE protocol is described under the article RATS OPEN SYSTEMS ENVIRONMENT. For comprehensive information about various existing protocols, and new ones that are always arising, ham magazines and packet-radio clubs are recommended as information sources.

The Future of Packet Radio   It is difficult to predict all the ways in which packet radio might be used in the future. But one especially interesting possibility involves the linking-up of many hams' PCs into a supercomputer.

PC users and hams might find a common interest in the form of "packet-radio supercomputer clubs" once a geostationary ham satellite is in orbit. This will encourage PC owners to obtain amateur radio licenses, increasing the ham population and the significance of ham radio in society.

For Further Information   There are several good books on packet radio and networking. These are advertised in the ham magazines. There's nothing quite like seeing packet radio in operation. If you know anyone in your area who is involved with packet radio, you should make an appointment to get together if he or she is willing, and see packet working in real time.

For information on specific topics relating to packet radio, please refer to the following articles in this encyclopedia: AX.25, BACKBONE, BELL 103 AND 202, BIT STUFFING, BULLETIN-BOARD SYSTEM, CONNECTION PROTOCOL, CONNECTIONLESS PROTOCOL, DIGIPEATER, EIA-232-D, END-TO-END ACKNOWLEDGMENT, FILE TRANSFER PROTOCOL, FRAME, MAIL FORWARDING, MAILBOX, MESSAGE HEADER, NETWORK, NETWORKING, NETWORK NODE CONTROLLER, NODE-TO-NODE ACKNOWLEDGMENT, OCTET, OPEN SYSTEMS INTERCONNECTION REFERENCE MODEL, PACKET ASSEMBLER/DISASSEMBLER, PACKET-RADIO BULLETIN-BOARD SYSTEM, POLLING PROTOCOL, PROTOCOL, RATS OPEN SYSTEMS ENVIRONMENT (ROSE), RECEIVE STATE VARIABLE, ROSE X.25 PACKET SWITCH, SECONDARY STATION IDENTIFIER, SERIAL LINE INTERFACE PROTOCOL, SIMPLE MAIL TRANSFER PROTOCOL, SLIDING WINDOW PROTOCOL, STORE-AND-FORWARD, SYNCHRONOUS DATA LINK CONTROL, TERMINAL EMULATION SOFTWARE, TERMINAL NODE CONTROLLER, TRANSMISSION CONTROL PROTOCOL, TRANSPARENT MODE, TURNAROUND TIME, UNCONNECTED PACKET, and VIRTUAL CIRCUIT.

PACKET-RADIO BULLETIN-BOARD SYSTEM

A *bulletin-board system* (BBS) is a set of stored messages, accessible by means of a personal computer (PC). Many PC users are familiar with BBSs. They use modems to connect their PCs to telephone lines, via which the BBS is accessed. There is usually a telephone toll charge associated with the use of a BBS. *See* BULLETIN-BOARD SYSTEM.

Ham radio makes it possible to communicate over long distances without telephone lines, and therefore without tolls. A logical outgrowth of packet radio and ham radio has been the *packet-radio bulletin-board system* (PBBS).

With a terminal node controller (TNC) to connect a personal computer to a radio transceiver, you have a packet radio station, and can access PBBSs. Messages can be left for other computer users, or retrieved, in a very short time. It's like a fast mail service.

The PBBS is a form of *time-shifting communications*. It is a means of leaving messages for other hams when they are not necessarily at their stations. Because of this, there is often some delay in transferring the message, simply because the operator of the destination station must check the PBBS before he or she can get the message (hence the term "time shifting," which sounds better than "time delayed"). The PBBS is a "public" storage system; messages for a large number of hams are left on the bulletin board.

As packet radio gets more sophisticated, PBBSs can be expected to become more versatile. It is already possible for different PBBSs to exchange messages. Eventually, there might be a single huge world PBBS (WPBBS?), with individual sub-PBBSs for countries, states, provinces, counties, prefectures, etc.

Sometimes, you might want to leave a message for a ham who has a packet radio station, but not do it on the general PBBS. A private way to leave messages for an individual ham is the *mailbox*. *See also* MAILBOX, PACKET RADIO, TERMINAL NODE CONTROLLER, and TIME-SHIFTING COMMUNICATIONS.

PAD

A *pad* is an attenuator network that displays no reactance, and exhibits constant input and output impedances. Pads are often used between radio transmitters and external linear amplifiers to regulate the amplifier drive. Pads are also used in some receiver front ends to reduce or prevent overloading in the presence of extremely strong signals. Such pads can be switched in or out of the front end as desired. Pads are sometimes used for impedance matching.

A typical attenuator pad configuration is shown in the illustration. The resistors are noninductive, and must be capable of dissipating the necessary amount of power. In a receiver front end, small resistors (¼-watt or ⅛-watt) are satisfactory. For external power-amplifier applications, the resistors might have to dissipate as much as 100 watts or more. *See also* ATTENUATOR.

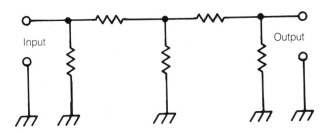

PAD: An attenuator pad for unbalanced lines.

PADDER CAPACITOR

A *padder capacitor* is a small, variable capacitor used to adjust the frequency of an oscillator circuit. The padder capacitor is placed in parallel with the tuning capacitor in the tank circuit of the oscillator. The value of the padder capacitor is much smaller than that of the main tuning capacitor. Most padders have a maximum capacitance of only a few picofarads. The padder allows precise calibration of the tuning dial or frequency readout of a receiver or transmitter.

When a ganged variable capacitor is used to tune more than one resonant circuit at a time, padder capacitors can be placed across each section of the ganged capacitor. The padders are adjusted until proper tracking is obtained (*see* GANG CAPACITOR). Some gang capacitors have built-in padders.

The adjustment is made with a plastic device resembling a miniature screwdriver. The adjusting tool must have an insulating shaft to prevent detuning from hand-capacitance effects.

PADDING

Padding is the process of precisely adjusting a circuit by means of small, series or parallel components. An example is the use of a padder capacitor for adjustment of oscillator frequency (*see* PADDER CAPACITOR).

In computer practice, padding is a process in which "blanks" or other meaningless data characters are added to a file to make the file a certain standard size. For example, the meaningful data in a computer file can comprise 12 kilobytes, but the standard file size might be 16 kilobytes. The file is padded by adding 4 kilobytes of "blanks." The "blanks" can be added anywhere in the file; the most common location is at the beginning or end.

When an attenuator pad is inserted into a circuit, the process is sometimes called *padding* (*see* PAD).

PANORAMIC RECEIVER

A *panoramic receiver* is a device that allows continuous visual monitoring of a specified band of frequencies. Most receivers can be adapted for panoramic reception by connecting a specialized form of spectrum analyzer into the first intermediate-frequency chain (see A in illustration).

A monitor screen displays the signals as vertical pips along a horizontal axis. The signal amplitude is indicated by the height of the pip. The position of the pip along the horizontal axis indicates its frequency. The frequency at which the receiver is tuned appears at the center of the horizontal scale. The number of kilohertz per horizontal division can be set for spectral analysis of a single signal or a narrow frequency band (B) or the scale can be set for observation of a wide range of frequencies (C). The maximum possible range is limited by the selectivity characteristics of the radio-frequency receiver stages. *See also* SPECTRUM ANALYZER.

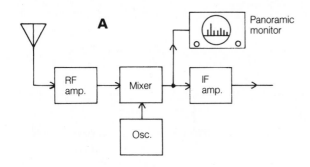

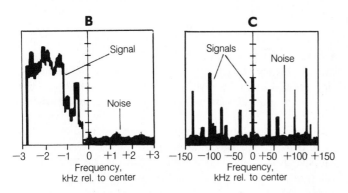

PANORAMIC RECEIVER: At A, installation in superheterodyne receiver; at B, a narrowband display; at C, a wideband display.

PAPER CAPACITOR

A *paper capacitor* is a capacitor that uses a thin film of paper as the dielectric material. The paper is soaked with wax or some other water-resistant material. Paper capacitors are usually cylindrical in shape. The paper is placed between strips of foil, and the entire assembly is rolled up to form the capacitor. Values range from about 1000 pF to 0.5 μF. Paper capacitors are nonpolarized.

Paper capacitors were once common from the very-low to the high frequencies. Paper capacitors are not generally suitable for use at very high frequencies and above because paper dielectrics become lossy at these frequencies. Paper capacitors were used at low to moderate voltage levels. They are rarely used today. *See also* CAPACITOR.

PAPER TAPE

Paper tape is a permanent data-storage medium. Paper tape has been largely replaced by electronic memory devices in modern equipment, but occasionally it can still be found.

Each character is punched into a paper tape as a set of small holes in a row perpendicular to the edges of the tape. This is done by a device called a *reperforator*.

Paper tape can be read mechanically, at a rate of approximately 60 characters per second. Electronic devices, using optical coupling, can read the tape much faster. Paper tape has the advantage of being a permanent form of memory. It cannot be erased by accidental removal of power or exposure to magnetic fields. However, paper tape tends to be bulky for storage of large amounts of data. The disk or magnetic-tape storage media are generally used for large files. *See also* DISKETTE, FLOPPY DISK, and MAGNETIC TAPE.

PARABOLOID ANTENNA

A *paraboloid antenna* is a form of dish antenna in which the reflecting surface is a geometric paraboloid in three dimensions. A paraboloid is the surface that results from the rotation of a parabola about its axis.

The paraboloid antenna has a focal point, at which rays arriving parallel to the antenna axis converge. This point is, in theory, geometrically perfect. This distinguishes the paraboloid antenna from the spherical antenna, in which the rays converge almost, but not quite, perfectly. The distance of the focal point from the center of the surface depends on the degree of curvature of the paraboloid.

Paraboloid reflectors can be constructed from wire mesh, or they can be made of sheet metal. Paraboloid antennas are generally used at ultra-high frequencies and above. Wire mesh is satisfactory at the longer wavelengths, but at very short wavelengths, sheet metal is preferable. The driven element is located at the focal point of the paraboloid. *See also* DISH ANTENNA.

PARALLEL DATA TRANSFER

Information can be transmitted bit by bit, along a single line, or it can be transmitted simultaneously along two or more lines. The first method is known as *serial data transfer* (see SERIAL DATA TRANSFER). The latter method is called *parallel data transfer*. This term arises from the fact that the transmission lines each carry portions of the same data.

In computer practice, parallel data transfer refers to the transmission of all bits in a word at the same time, over individual parallel lines; however, different words are generally sent one after the other.

Parallel data transfer has the advantage of being more rapid than serial transfer. However, more lines are required, in proportion to the factor by which the speed is increased. It takes ten lines, for example, to cut the data transmission time from a serial value of 60 seconds to a parallel value of 6 seconds.

PARALLEL LEADS

When two or more wire leads are used together in a circuit, with their ends joined, the leads are said to be *parallel*. Parallel leads are sometimes used in radio-frequency circuits to reduce the inductance per unit length. Parallel leads also improve the conductivity at radio frequencies over that of a single conductor, by reducing losses caused by skin effect. *See also* LITZ WIRE, and SKIN EFFECT.

PARALLEL RESONANCE

When an inductor and capacitor are connected in parallel, the combination will exhibit resonance at a certain frequency (*see* RESONANCE, and RESONANT FREQUENCY). This condition occurs when the inductive and capacitive reactances are equal and opposite, thereby canceling each other, and leaving a pure resistance.

The resistive impedance that appears across a parallel-resonant circuit is extremely high. In theory, assuming zero loss in the inductor and capacitor, the impedance at resonance is infinite. In practice, because there is always some loss in the components, the impedance is finite, but it can be on the order of hundreds of kilohms. The exact value depends on the Q factor, which in turn depends on the resistive losses in the components and interconnecting wiring (*see* IMPEDANCE, and Q FACTOR). Below the resonant frequency, the circuit appears capacitively reactive; above the resonant frequency it is inductively reactive. Resistance is also present under nonresonant conditions, making the impedance complex.

Parallel-resonant circuits are commonly used in the output configurations of radio-frequency oscillators and amplifiers. They can also be seen in some input circuits and coupling-transformer arrangements.

A section of transmission line can be used as a parallel-resonant circuit. A quarter-wave section, short-circuited at the far end, behaves like a parallel, and resonant, inductance and capacitance. A half-wave section, open at the far end, also exhibits this property (*see* LINE TRAP). Such resonant circuits can be used in place of inductance-capacitance (LC) circuits. This is often done at very-high and ultra-high frequencies, where the lengths of quarter-wave or half-wave sections are reasonable.

PARALLEL TRANSISTORS

Two or more bipolar or field-effect transistors can be connected in parallel to increase the power-handling capability. All of the emitters or sources, bases or gates, and collectors or drains, are connected together. All of the transistors are of identical manufacture.

Ideally, the power-handling capability of a parallel combination of *n* transistors is *n* times that of a single transistor of the same kind. When two or three transistors are connected in parallel, this is essentially the case. However, in practice, certain limitations exist; it is not feasible to tie many transistors together, expecting an effective high-power amplifier to be the result. It is always better, if possible, to use one high-power transistor rather than several small ones.

The main reason that massive parallel combinations of bipolar transistors do not work well is that the input impedance is reduced in such a configuration, and most bipolar power transistors exhibit low input impedances to begin with. If ten transistors are connected in parallel, for example, the input impedance becomes just one-tenth of the value for a single transistor. This makes it necessary to use a cumbersome and lossy input transformer. The input capacitance also increases in a parallel combination, as compared with a single unit.

In any parallel combination of transistors, it is advantageous to place small resistors in the emitter or source leads of each of the units. This prevents the phenomenon called current

hogging, in which one transistor does all the work while the other one does practically no work (*see* CURRENT HOGGING).

Parallel combinations of two or three transistors can be used with excellent results in a power amplifier, if care is exercised in the design of the circuit. Many well-designed, commercially manufactured audio-frequency and radio-frequency power amplifiers use parallel-transistor circuits. *See also* FIELD-EFFECT TRANSISTOR, POWER AMPLIFIER, and TRANSISTOR.

PARAMETRIC AMPLIFIER

A *parametric amplifier* is a form of radio-frequency amplifier that operates from a high-frequency alternating-current source, rather than the usual direct-current source. Some characteristic of the circuit, such as reactance or impedance, is made to vary with time at the power-supply frequency. Parametric amplification is commonly used with electron-beam devices at microwave frequencies. The traveling-wave tube is an example of such a device (*see* TRAVELING-WAVE TUBE).

Parametric amplifiers are characterized by their low noise figures, which may, in some instances, be less than 3 dB. The gain is about the same as that of other types of amplifiers; typically, it ranges from 15 to 20 dB.

PARAPHASE INVERTER

A *paraphase inverter* is a form of circuit that provides two outputs, differing in phase by 180 degrees. A transistor, field-effect transistor, or tube can be used. The paraphase inverter offers a simple and inexpensive way to obtain two signals in opposing phase.

PARASITIC ARRAY

An antenna with one or more parasitic elements is called a *parasitic array*. Parasitic arrays are commonly used at the high, very-high, and ultra-high frequencies for obtaining antenna power gain.

Common examples of parasitic arrays are the quad and the Yagi. Other configurations also exist. *See also* PARASITIC ELEMENT, QUAD ANTENNA, and YAGI ANTENNA.

PARASITIC ELEMENT

In an antenna, a *parasitic element* is an element that is not directly connected to the feed line. Parasitic elements are used for the purpose of obtaining directional power gain. Generally, parasitic elements can be classified as either directors or reflectors. Directors and reflectors work in opposite ways.

Parasitic elements operate by electromagnetic coupling to the driven element. The principle of parasitic-element operation was first discovered by a Japanese engineer named Yagi; the Yagi antenna is named after him. Yagi found that elements parallel to a radiating element, at a specific distance from it, and of a certain length, caused the radiation pattern to show gain in one direction and loss in the opposite direction. Nowadays, Yagi's principle is used at high frequencies by amateur and commercial radio operators.

At high frequencies, parasitic elements are often used in directional antennas. The most common of these are the quad and the Yagi. Such antennas are known as *parasitic arrays*. They exhibit a unidirectional pattern. *See also* DIRECTOR, QUAD ANTENNA, REFLECTOR, and YAGI ANTENNA.

PARASITIC OSCILLATION

In a radio-frequency power amplifier, oscillation sometimes occurs. This oscillation is undesirable, and it can occur at frequencies much different from the actual operating frequency. Such oscillations are called *parasitic oscillations*, or simply *parasitics.*

Parasitics can be eliminated by providing a certain amount of negative feedback in a power amplifier. This is accomplished by a neutralization circuit. A parasitic suppressor can also be used to choke off parasitics in some instances. *See also* NEUTRALIZATION, and PARASITIC SUPPRESSOR.

PARASITIC SUPPRESSOR

A radio-frequency amplifier sometimes oscillates at a frequency far removed from the operating frequency. Such oscillation is called *parasitic oscillation* (see PARASITIC OSCILLATION).

When parasitic oscillations are at a frequency much higher or lower than the operating frequency, the unwanted oscillation can sometimes be choked off by means of parasitic suppressors. This method is especially common for eliminating parasitics in medium-frequency and high-frequency power amplifiers. The parasitic choke consists of a resistor and a small coil connected in parallel, and placed in series with the collector, drain, or plate lead of the amplifier (see illustration). The resistor is typically 50 to 150 ohms; it is a noninductive carbon type. The coil consists of three to five turns of wire, wound on the resistor. In an amplifier having two or more transistors or tubes in parallel, suppressors are installed at each device, individually.

Low-frequency parasitics are not affected by the type of suppressor shown. To eliminate low-frequency parastics in a radio-frequency power amplifier, neutralization is usually necessary. *See also* NEUTRALIZATION.

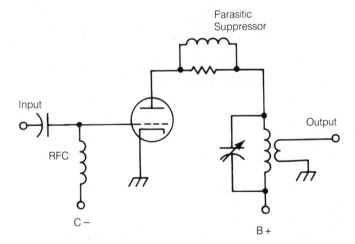

PARASITIC SUPPRESSOR: Chokes off VHF oscillations in power amplifier.

PARITY

Parity is an expression that indicates whether the sum of the binary digits in a code word is even or odd. Accordingly, if the sum is even, the parity is called *even*; if the sum is odd, the parity is called *odd*. In certain situations, it is necessary to have the digits in all code words add up to an even number; in other cases, the sum must always be odd.

Suppose, for example, that a code consists of five-digit words. Some examples might be 01001, 11000, and 11101, if the parity is even; in the case of odd parity, some code words might be 10000, 10101, and 00111.

If the sum of the digits in a word is incorrect (odd when it should be even, or vice-versa), an extra bit can be added to the word to correct the discrepancy. Some codes have an extra bit that can be 0 (if the parity of the word is correct) or 1 (if not).

In the previous example, you might add a sixth digit at the end of each word as a parity bit. Suppose that the parity is even. Then, the first three words, as given, would have a 0 added at the end, and they would become, respectively, 010010, 110000, and 111010. The second three words, however, would require the addition of a 1, becoming 100001, 101011, and 001111. With the addition of the parity bit, the sum of the digits in all of the words is made to be even.

In a data-processing or transmission system, a parity check is sometimes made as a test for accuracy. If the parity is wrong, the state of the parity bit can be reversed, or the receiver can "ask" the transmitter to repeat the word.

PARITY BIT

In a binary code word, it is sometimes necessary to have the parity always be even, or always be odd. A *parity bit* is an extra bit added to the binary word, in order to ensure that the parity is correct. The parity bit might be 0, or it might be 1, depending on which is needed in a given word in order to get the necessity parity. Sometimes, the addition of parity bits is used as a means of checking the accuracy of a received digital signal. *See* PARITY.

PASSBAND

In a selective circuit, the *passband* is the band of frequencies at which the attenuation is less than a certain value, with respect to the minimum attenuation. This value is usually specified as 3 dB, representing half the power at the maximum-gain frequency.

In a superheterodyne receiver that has a narrow bandpass filter, the passband can be as small as a few hundred hertz for continuous-wave signals, or as wide as 5 to 10 kHz for amplitude-modulated signals. In a receiver designed for reception of frequency-modulated signals, the passband can be as wide as 100 kHz or more. In a television receiver, the passband is about 6 MHz in width.

If the passband is too wide for a given emission mode, excessive interference is received, reducing the sensitivity of the receiver. If the passband is too narrow, the full signal cannot be received, and again, the sensitivity is degraded. There exists an optimum passband for every situation.

For continuous-wave signals of moderate speed, the passband can be as narrow as 50 to 100 Hz. For single-sideband and slow-scan television, 2 to 3 kHz is best. For amplitude-modulation, 6 kHz is standard for voice signals, about 10 kHz for music signals, and 6 MHz for fast-scan television. For frequency modulation, phase modulation, and pulse modulation, the optimum passband depends on the deviation, phase change, or pulse rate. *See also* BANDPASS FILTER, and BANDPASS RESPONSE.

PASSIVE COMPONENT

In any electronic circuit, some components require no external source of power for their operation. Such components are called *passive*. Passive components include semiconductor diodes (in most cases), resistors, capacitors, inductors, and the like. Components that require an external source of power, and/or produce gain in a circuit, are called *active* (*see* ACTIVE COMPONENT).

Some components can act as either active or passive devices, depending on their application. The varactor diode is an example of such a device. *See also* VARACTOR DIODE.

PASSIVE FILTER

A *passive filter* is a form of selective filter that does not require an external source of power for its operation. Passive filters can be of the bandpass, band-rejection, highpass, or lowpass variety (*see* BANDPASS FILTER, BAND-REJECTION FILTER, HIGHPASS FILTER, and LOWPASS FILTER). Passive filters are commonly made using inductors, resistors, and/or capacitors. Crystal and mechanical filters are also passive (*see* CRYSTAL LATTICE FILTER, and MECHANICAL FILTER).

Passive filters are commonly used in radio-frequency circuits. At audio frequencies, active filters, utilizing integrated-circuit operational amplifiers, are often seen. *See also* ACTIVE FILTER.

PASSIVE SATELLITE

A satellite intended for over-the-horizon radio communication can operate in two different ways. The most common type of satellite is the active satellite, which picks up the signals, converts them to another band of frequencies, and retransmits them (*see* ACTIVE COMMUNICATIONS SATELLITE, and MOONBOUNCE). However, passive satellites can also be used. Passive satellites simply reflect incident electromagnetic signals, without the need for any electronic apparatus.

In the 1960s, the Echo satellites were put into orbit around the earth. These were passive satellites. These devices were large, spherical, and metallic. They reflected radio signals in much the same way as the ground surface does. Another, more familiar but less often-used, passive satellite is our own moon. The main advantage of passive satellites is that they carry no equipment that might malfunction. The disadvantage of the passive satellite is that the reflected signals tend to be very weak.

PATCH CORD

A *patch cord* is a length of conductor or cable, usually equipped with plugs on both ends, and used for the purpose of making temporary connections between circuits. Patch cords are useful in the test lab. Patch cords were once used by telephone operators. Today, computers have largely replaced operators with patch cords in the telephone industry.

Patch cords can be configured in an immense variety of different ways, because of the proliferation of connector types used in the electronics industry. Patch cords with different connectors on either end can serve as adaptors (*see* ADAPTOR, and CONNECTOR). The general experience of electronic technicians seems to be that there are never enough patch cords, or the existing cords are too short or lack the proper connectors. A good laboratory should be thoroughly stocked with a complete repertoire of these cords. The cords should also be periodically checked to ensure that the electrical connections are intact.

PEAK

In a waveform or other changing parameter, a *peak* is an instantaneous or local maximum or minimum. The value of the parameter falls off on either side of a maximum peak, and rises on either side of a minimum peak.

The term *peak* is used in many different situations. A local maximum or minimum in a mathematical function can be called a *peak*. In the response curve of a bandpass filter, the frequency of least attenuation is often called the *peak frequency* or *peak point*. In the adjustment of device, the peak setting is the control position that results in the maximum value for a certain parameter. An example of this is the adjustment of the plate-tuning capacitor for maximum output power in a radio-frequency transmitter.

A waveform peak usually has infinitesimal, or nearly zero, duration. In some cases a peak can be sustained for a certain length of time; the square wave, for example, has peaks that last for one-half cycle. A peak can be either positive or negative in terms of polarity or direction of current flow. *See also* PEAK CURRENT, PEAK ENVELOPE POWER, PEAK POWER, PEAK-READING METER, PEAK VALUE, and PEAK VOLTAGE.

PEAK CURRENT

Peak current is the maximum instantaneous value reached by the current in an alternating or pulsating waveform. In a symmetrical alternating-current sine wave, the peak current is the same in either direction, and is reached once per cycle in either direction. In nonsymmetrical waveforms, the peak current can be greater in one direction than in the other. The peak current can be sustained for a definite length of time during each cycle, but more often the peak current is attained for only an infinitesimal time.

The peak current in a sinusoidal waveform is 1.414 times the root-mean-square (RMS) current, assuming there is no direct-current component associated with the signal. For other waveforms, the peak-to-RMS current ratio is different. *See also* ROOT MEAN SQUARE.

PEAK ENVELOPE POWER

In an amplitude-modulated or single-sideband signal, the radio-frequency carrier output power of the transmitter varies with time. The input power also fluctuates. The maximum instantaneous value reached by the carrier power is called the *peak envelope power*. For any form of amplitude modulation, the peak envelope power is always greater than the average power. The peak envelope power differs from the actual peak waveform power. The peak waveform power is determined according to the product of the instantaneous waveform voltage and current; the peak envelope power is the maximum root-mean-square waveform power (*see* PEAK POWER). Nevertheless, the terms *peak envelope power* and *peak power* are often used interchangeably.

In the case of frequency modulation or phase modulation, the carrier amplitude does not change, and the peak envelope power is the same as the average power. In a continuous-wave (CW) morse-code signal, the peak envelope power is the key-down average power, which is approximately twice the long-term average power.

The peak envelope power is not shown by ordinary metering devices when signals are voice-modulated. This is because the actual transmitter input and output power fluctuate too rapidly for any meter needle to follow. Special meters exist for measurement of peak envelope power in such cases (*see* PEAK-READING METER).

An oscilloscope can be used to show the modulation envelope of a transmitter; the peak envelope power can then be determined by comparison with a known, constant power level. Such a constant transmitter output can be obtained in a conventional amplitude-modulated transmitter by cutting off the modulating audio. In a single-sideband transmitter, a sine-wave tone can be supplied to the audio input, at a level sufficient to maximize the transmitter output without envelope distortion.

To determine the peak-envelope output power, a watt-meter is first used to read the continuous output as obtained in the above manner. The constant-amplitude signal is then displayed on an oscilloscope, and the sensitivity of the scope adjusted until the display indicates plus or minus 1 division. Then, normal modulation is applied. The peaks of the envelope can be seen, reaching a level of plus or minus x divisions, where x is greater than 1.

The oscilloscope indicates voltage, not power; thus, the ratio of the peak envelope power P to the constant-carrier power Q is:

$$P/Q = x^2$$

and the peak envelope power is given by:

$$P = x^2Q$$

Peak envelope power is often specified for single-sideband signals. The ratio of the peak envelope power to the average power in a single-sideband emission depends on the modulating waveform. In a single-sideband signal, the peak-envelope power for a human voice signal is approximately two to three times the average power. *See also* ENVELOPE, and SINGLE SIDEBAND.

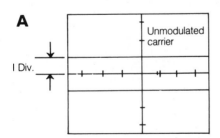

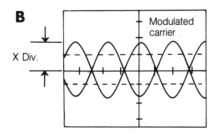

PEAK ENVELOPE POWER: At A, unmodulated carrier. At B, fully modulated carrier showing peaks.

PEAK INVERSE VOLTAGE

The *peak inverse voltage* in a circuit is the maximum instantaneous voltage that occurs with a polarity opposite from the polarity of normal conduction. The term is often used with respect to rectifier diodes. The peak inverse voltage across a diode is the maximum instantaneous negative anode voltage. This voltage is sometimes called the *peak reverse voltage.*

Rectifier diodes are rated according to their maximum peak-inverse-voltage or peak-reverse-voltage (PIV or PRV) tolerance. If the inverse voltage significantly exceeds the PIV rating for a diode, avalanche breakdown occurs and the diode conducts in the reverse direction. Certain diodes, called *Zener diodes*, are deliberately designed to be operated near, or at, their avalanche points. *See also* AVALANCHE BREAKDOWN, RECTIFIER DIODE, and ZENER DIODE.

PEAK LIMITING

In a fluctuating or alternating waveform, it is sometimes desirable to prohibit the peak value from exceeding a certain level (*see* PEAK VALUE). This is called *peak limiting.*

Peak limiting can be accomplished in a variety of different ways. A pair of back-to-back semiconductor diodes results in peak limiting of low-voltage alternating-current signals; this technique is often used in the audio stages of communications receivers to prevent "blasting" (*see* AUDIO LIMITER).

In a modulated transmitter, peak limiting can be used to increase the ratio of average power to peak-envelope power. This technique is sometimes also called *radio-frequency clipping.* Peak limiting of an amplitude-modulated signal can be done indirectly by modifying the audio waveform; this is called *speech clipping. See also* RF CLIPPING, and SPEECH CLIPPING.

PEAK POWER

The *peak power* of an alternating-current signal is the maximum instantaneous power. The peak power is equivalent to the peak voltage multiplied by the peak current when the phase angle is zero (no reactance).

In a sinusoidal alternating-current signal, the peak power, P_p, can be determined from the root-mean-square (RMS) voltage, E_{RMS}, and the RMS current, I_{RMS}, according to the formula:

$$P_p = 2E_{RMS} I_{RMS}$$

where P_p is given in watts, E_{RMS} is given in volts, and I_{RMS} is given in amperes. The peak power, in terms of the peak voltage, E_p, and the peak current, I_p, is:

$$P_p = E_pI_p$$

In a modulated radio-frequency signal, the peak envelope power is sometimes called the *peak power. See also* PEAK ENVELOPE POWER.

PEAK-READING METER

A typical meter averages voltage, current or power. It cannot respond fast enough to indicate the peak values. When you talk into a single-sideband (SSB) transceiver and watch the output wattmeter, you can see the needle kick up. Although the meter appears to reach peaks when you utter certain syllables, these are not the true maxima of the peak-envelope power (PEP). To obtain PEP readings, a *peak-reading meter* is needed.

A peak-reading meter senses the instantaneous value, and registers the highest level within a certain span of time. A peak-reading radio-frequency (RF) power meter is a useful instrument in a ham shack if the equipment is operating at, or near, the legal maximum power.

An oscilloscope can be used as a peak-reading meter. Almost any oscilloscope can follow the instantaneous amplitude changes of voice audio. But an oscilloscope measures *voltage*. Instantaneous power is proportional to the square of the instantaneous voltage. This must be taken into account when an oscilloscope is used to measure peak power.

PEAK-TO-PEAK VALUE

In a fluctuating or alternating waveform, the *peak-to-peak value* is the difference between the maximum positive instantaneous value and the maximum negative instantaneous value. Peak-to-peak is sometimes abbreviated as *p-p* or *pk-pk*.

The drawing illustrates peak-to-peak values for an alternating-current sine wave (A), a rectified sine wave (B), and a sawtooth wave (C). The peak-to-peak value of a waveform is a measure of signal intensity at a given frequency. Signal voltage is often given in terms of the peak-to-peak value. Note that changes in the direct-current component of a signal have no effect on the peak-to-peak value.

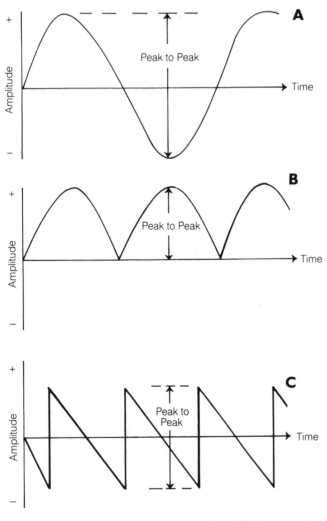

PEAK-TO-PEAK VALUE: For sine wave (A), rectified sine wave (B), and sawtooth wave (C).

Peak-to-peak values for any waveform are readily observed on an oscilloscope display. This is possible even if the frequency is too high to allow viewing of the waveform, or if waveform irregularity makes it difficult to determine the average value.

For an alternating-current sinusoidal waveform, the peak-to-peak value is equal to twice the peak value, and 2.828 times the root-mean-square value. *See also* PEAK VALUE, and ROOT MEAN SQUARE.

PEAK VALUE

The *peak value* of a fluctuating or alternating waveform is the maximum instantaneous value. Some waveforms reach peaks in two directions of polarity or current flow (*see* PEAK CURRENT, and PEAK VOLTAGE). These peaks can be equal in absolute magnitude, but this is not necessarily so. The peak value can last for a definite length of time; this is the case for a square wave. The peak value of a sine wave lasts for a theoretically infinitesimal (zero) time. The illustration at A illustrates peak values for a sine-wave alternating-current waveform. The peak value for the same waveform with the addition of a direct-current component is shown at B.

Peak values of power can be specified in either of two ways. The maximum instantaneous waveform power is the product of the peak voltage and the peak current (*see* PEAK POWER). The peak envelope power is the maximum carrier power in an amplitude-modulated or single-sideband signal (*see* PEAK ENVELOPE POWER).

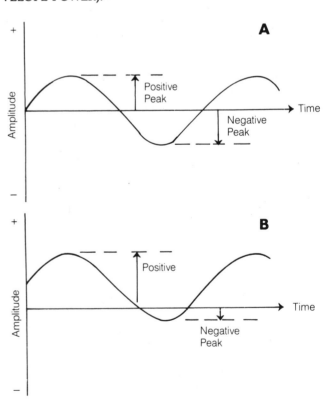

PEAK VALUE: Positive and negative peak values might be equal (A) or different (B).

PEAK VOLTAGE

The *peak voltage* of a fluctuating or alternating waveform is the instantaneous maximum positive or negative voltage. The peak positive voltage is often equal in magnitude to the peak negative voltage, but not always. The peak voltage can be sustained for a certain length of time, but in most cases it is attained only for an infinitesimally short time.

For a sinusoidal alternating-current wave, having no direct-current component, the peak voltage is equal to 1.414 times the root-mean-square voltage. For other types of waveforms, the relationship is different. *See also* ROOT MEAN SQUARE.

PEN RECORDER

A *pen recorder* is a device for graphically plotting the magnitude of a parameter as a function of time. The device is a form of analog meter. Pen recorders are used for long-term monitoring of a variable quantity, or whenever a permanent record is needed. Pen recorders are used for plotting such functions as heartbeat, temperature, barometric pressure, and wind speed versus time.

The pen recorder works very much like an ordinary analog meter. The needle, to which a pen is attached, moves back and forth according to the magnitude of the measured parameter. A paper is guided under the pen at a certain speed, corresponding to the required number of seconds, minutes, or hours per division. The paper is usually attached to a rotating drum.

A pen recorder cannot be used to plot functions that vary with extreme rapidity because the metering device cannot respond fast enough. For detailed permanent records of rapidly fluctuating or complicated waveforms, an oscilloscope, equipped with a camera, is preferable. *See also* OSCILLOSCOPE, and OSCILLOSCOPE CAMERA.

PENTODE TUBE

A *pentode tube* is a vacuum tube that has five elements. In addition to the cathode and the plate or anode, there are three grids. The three grids serve different purposes.

The *control grid* operates as the input-signal electrode in most pentode-tube circuits. The *screen grid* serves the same function as in the tetrode tube (*see* TETRODE TUBE). The third grid, the *suppresser*, increases the gain of the pentode over that usually obtainable with the tetrode. It does this by repelling secondary electrons from the plate, directing them back to the plate so that they can contribute to the signal output (*see* SECONDARY EMISSION).

The control grid in the pentode tube is usually biased negatively with respect to the cathode. The screen grid is biased at a voltage more positive than the cathode, but less positive than the plate. The suppressor grid is usually biased at the same voltage as the cathode. In fact, many pentode tubes have an internal short circuit between the cathode and the suppressor grid.

The three grids are, in the physical construction of the pentode, roughly concentric. The control grid is closest to the cathode. The screen grid is immediately outside the control grid. The suppressor is closest to the plate. *See also* TUBE.

PEP

See PEAK ENVELOPE POWER.

PERIGEE

Any earth-orbiting satellite follows either of two types of path through space. A circular orbit has the center of the earth at the center of the orbit. When a satellite follows an elliptical orbit, which is far more common, the center of the earth is at one focus of the ellipse. The extent to which the orbit differs from a circle is called the *eccentricity*.

Whenever a satellite is placed into orbit around the earth, or around any other large celestial body, there is almost always some deviation from a perfectly circular path. This is because the circular orbit represents a very special case, and attaining such an orbit requires precise speeds and launch trajectories. Perfection is difficult to attain.

Whenever a satellite has an elliptical orbit, its altitude varies. The minimum altitude is called the *perigee* of the satellite. It occurs once for every complete orbit. At perigee, the satellite travels faster than at any other point in the orbit.

Lunar Perigee   The moon's orbit around the earth is an elliptical orbit. The distance between the earth and the moon varies between about 225,000 and 253,000 miles. The first of these two numbers represents the moon's perigee.

The eccentricity of the moon's orbit is not very great. But it is enough to affect moonbounce communications, also called earth-moon-earth or EME (*see* MOONBOUNCE). The moon actually looks a little bit bigger at perigee than at apogee, or the point in the orbit where the distance between the earth and moon is the greatest (*see* APOGEE). Moonbounce is easiest when the moon is at perigee, and is most difficult when it is at apogee.

Perigee of a Communications Satellite   Most ham satellites have orbits that are at least somewhat eccentric. A highly eccentric orbit offers some advantages and some disadvantages. Interestingly, satellite communications is most difficult when the satellite is at or near perigee, if it has a very elongated orbit. This is because the satellite moves very fast then, and its position in the sky changes rapidly. The antenna at the ground station must be constantly turned to follow the satellite during this time.

Near apogee, the signals from the satellite are weaker, and more power is needed to reach it. But it moves more slowly across the sky, and the antennas need not be moved as often to track the satellite. *See also* ACTIVE COMMUNICATIONS SATELLITE, and OSCAR.

PERMANENT MAGNET

Certain metals, notably iron and nickel, are affected by magnetic fields in a unique way. The magnetic dipoles within such substances can, in the presence of a sufficiently intense magnetizing force, become permanently aligned. This results in a so-called *permanent magnet*—a piece of material that is constantly surrounded by a magnetic field.

A piece of iron or steel can be permanently magnetized by stroking it with another permanent magnet. This action causes the magnetic dipoles, which are normally aligned at random, to be oriented more or less in a common direction. Their effects then average out to create a magnetic field around the object. Normally, the effects of the dipoles average out to a zero magnetic field.

Permanent magnets are used in speakers, microphones, meters, and certain types of transducers. In the laboratory or the factory, tools (such as screwdrivers) are sometimes weakly magnetized for convenience in dismantling or assembling electronic equipment. Tiny permanent magnets are the medium by which a magnetic tape stores its information. *See also*

D'ARSONVAL, DYNAMIC LOUDSPEAKER, DYNAMIC MI-
CROPHONE, MAGNETIC FIELD, and MAGNETIC RE-
CORDING.

PERMEABILITY

Certain materials affect the concentration of the lines of force in
a magnetic field (see MAGNETIC FIELD, and MAGNETIC
FLUX). Some substances cause the lines of flux to move farther
apart, resulting in a decrease in the intensity of the field, compared with its intensity in a vacuum. Other substances cause
the density of the magnetic flux to increase.

The permeability of a substance is an expression of the extent to which that material affects the density of the magnetic
flux. Free space (a vacuum) is assigned the permeability value 1.
Materials that reduce the flux density, known as *diamagnetic
materials*, have permeability less than 1. Substances that cause
an increase in the flux density are known as *ferromagnetic* and
paramagnetic materials. They have permeability factors greater
than 1. The paramagnetic materials have values slightly larger
than 1; the ferromagnetics have values much greater.

The table lists several common types of materials, along
with their approximate permeability values at room temperature. The permeability of a given substance can change substantially with fluctuations in the temperature. The permeability factors of some materials, notably the ferromagnetics, can
vary, depending on the intensity of the magnetic field. For these
reasons, the values in the table should not be taken as precise.
The table reflects the fact that no substances have permeability
values much smaller than 1, although some have values far in
excess of 1. Certain nickel-iron alloys exhibit permeability factors of more than 1,000,000.

Substances with high magnetic-permeability values are
used primarily for the purpose of increasing the inductances of
coils. When a high-permeability core is inserted into a coil, the
inductance is multiplied by the permeability factor. This effect
is useful in the design of transformers and chokes at all frequencies. *See also* CHOKE, FERRITE CORE, LAMINATED
CORE, POWDERED-IRON CORE, and TRANSFORMER.

PERMEABILITY: PERMEABILITY
FACTORS OF SOME COMMON SUBSTANCES.

Substance	Permeability (Approx.)
Aluminum	Slightly more than 1
Bismuth	Slightly less than 1
Cobalt	60–70
Ferrite	100–3000
Free Space	1
Iron	60–100
Iron, refined	3000–8000
Nickel	50–60
Permalloy	3000–30,000
Silver	Slightly less than 1
Steel	300–600
Super-permalloys	100,000–1,000,000
Wax	Slightly less than 1
Wood, dry	Slightly less than 1

PERMEABILITY TUNING

In a resonant radio-frequency circuit, tuning can be accomplished by varying the value of either the inductor or the capacitor. The value of the inductor can be adjusted by changing the
number of turns, or by changing the magnetic permeability of

the core. The latter method of inductor tuning is known as *permeability tuning*.

Permeability tuning is accomplished by moving a powdered-iron core in and out of the coil (see illustration). A
threaded shaft is attached to the core, and this shaft is rotated to
allow precise positioning of the core. The farther into the coil
the core is set, the larger the inductance, and the lower the resonant frequency becomes. Conversely, as the core gets farther
outside the coil, the inductance gets smaller, and the resonant
frequency becomes higher.

The main advantage of permeability tuning, as compared
with capacitor tuning, is that the control adjustment is more linear. This allows the use of a dial calibrated in linear increments.
Permeability tuning also has the advantage that the capacitor in
the tuned circuit can be fixed, rather than variable. This reduces
the effects of external capacitance, and allows the use of a capacitor having optimum temperature coefficient. *See also* CA-
PACITOR TUNING, and INDUCTOR TUNING.

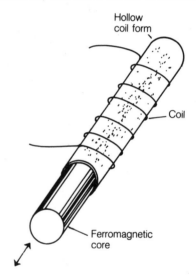

PERMEABILITY TUNING: The core moves in and out of coil.

PERMEANCE

In a magnetic circuit, *permeance* is an expression of the ease with
which a magnetic field is conducted. Permeance is the reciprocal of reluctance. Permeance in a magnetic circuit is analogous
to conductance in an electric circuit. Permeance is generally
measured in webers per ampere. *See also* RELUCTANCE.

PERMITTIVITY

Permittivity is an expression of the absolute dielectric properties
of a material or medium. The dielectric constant is determined
from the permittivity.

Permittivity is represented by the lowercase Greek letter
epsilon (ϵ). The permittivity of free space is represented by the
symbol ϵ_o. The quantity is usually expressed in farads per
meter; the permittivity of free space, in these units, is:

$$\epsilon_o = 8.85 \times 10^{-12} \text{ farad per meter}$$

Given a substance with permittivity ϵ farad per meter, the dielectric constant, k, is determined according to the formula:

$$k = \epsilon/\epsilon_o$$
$$= \epsilon/(8.85 \times 10^{-12})$$

See also DIELECTRIC, and DIELECTRIC CONSTANT.

PERSISTENCE TRACE

Some oscilloscopes are equipped with a device that "freezes" the trace on the screen for prolonged viewing. This feature is called a *persistence trace*. A single sweep of the electron beam is stored in a memory circuit, and the waveform is shown on the screen for as long as the operator wants.

Persistence trace oscilloscopes is especially useful when analyzing waveforms of extremely low frequency. Without such a feature, the display appears as nothing but a spot that jumps as it moves, slowly, from left to right across the screen. The persistence trace reveals the details. The persistence trace can also be used with an oscilloscope camera when the sweep frequency is relatively low. *See also* OSCILLOSCOPE, and OSCILLOSCOPE CAMERA.

PHASE

Phase is a relative quantity, describing the time relationship between or among waves having identical frequency. The complete wave cycle is divided into 360 equal parts, called *degrees of phase*. The time difference between two waves can then be expressed in terms of these degrees. Phase is sometimes also expressed in radians. One radian corresponds to about 57.3 degrees of phase.

One wave can occur sooner than the other by as much as 180 degrees. The earlier wave is called the *leading wave*. A disturbance can occur as much as 180 degrees later than its counterpart; the later wave is called the *lagging wave* (*see* LAGGING PHASE, and LEADING PHASE). When waves are exactly 180 degrees different in phase, they are said to be perfectly out of phase, or in phase opposition. For specific information about various aspects of phase, see the articles immediately following.

PHASE ANGLE

Phase angle is an expression, given in degrees, for the relative difference in phase between two signals (*see* PHASE). Phase angle is often given to indicate the difference in phase between the current and the voltage in a circuit containing reactance (*see* LAGGING PHASE, and LEADING PHASE).

In a dielectric material, the phase angle is the extent to which the current leads the voltage. The phase angle, indicated by ϕ, is equal to the complement of the loss angle θ. That is:

$$\phi = 90 - \theta$$

The phase angle can vary from 0 to 90 degrees. The larger the phase angle, the lower the dielectric loss (*see* DISSIPATION FACTOR).

Phase angle also indicates the lossiness of an inductor. In a perfect inductive reactance, the current lags the voltage by 90 degrees. The phase angle ϕ can, again, vary from 0 to 90 degrees; the larger the phase angle, the lower the loss in the inductor.

A phase angle of 0 degrees indicates that the current and voltage are in phase. This occurs only when there is no reactance in the circuit. *See also* CAPACITIVE REACTANCE, IMPEDANCE, and INDUCTIVE REACTANCE.

PHASE BALANCE

Phase balance is an expression of the relative symmetry of a square wave. The phase-balance angle is defined as the difference between 180 degrees and the measured phase angle between the centers of the positive and negative pulses.

The period, T, of the waveform, is the length of one complete cycle, from the midpoint of one positive pulse to the midpoint of the following positive pulse. This interval is divided into 360 degrees. The midpoint of the negative pulse is then located; ideally, it should be at the 180-degree point. If the midpoint of the negative pulse occurs at x degrees, then the phase-balance angle is $180 - x$. In a square wave with zero transition time, the phase-balance angle is 0 degrees because $x = 180$.

In a phased array, the condition of signal cancellation, resulting in a directional null, is sometimes called *phase balance*. As seen from a distant point along such a null, the currents in the various phased elements oppose each other in phase, resulting in a field strength of zero. *See* PHASED ARRAY.

Phase opposition can occasionally be spoken of as phase balance. This is the condition wherein two signals of identical frequency differ in phase by 180 degrees. *See also* PHASE, and PHASE OPPOSITION.

PHASE COMPARATOR

A *phase comparator* is a part of a phase-locked-loop circuit (*see* PHASE-LOCKED LOOP). The phase comparator does exactly what its name implies: it compares two signals in terms of their phase. The output voltage of the phase comparator depends on whether the signals are in phase or not. If the signals are not in phase, the voltage from the comparator causes the phase of an oscillator signal to be shifted back and forth until its signal is exactly in phase with the signal from a reference oscillator.

In some types of radiolocation and radionavigation systems, the relative phase of two signals is used to determine the position of a ship or airplane. The relative phase is determined by the phase comparator. The output of the phase comparator is fed to a computer, which determines the position. *See also* RADIOLOCATION, and RADIONAVIGATION.

PHASED ARRAY

A *phased array* is an antenna having two or more driven elements. The elements are fed with a certain relative phase, and they are spaced at a certain distance, resulting in a directivity pattern that exhibits gain in some directions and little or no radiation in other directions.

Phased arrays can be very simple, consisting of only two elements. Two examples of simple pairs of phased dipoles are shown in the illustration on pg. 356. At A, the two dipoles are spaced one-quarter wavelength apart in free space, and they are fed 90 degrees out of phase. The result is that the signals from the two antennas add in phase in one direction, and cancel in the opposite direction, as shown by the arrows. In this particular case, the radiation pattern is unidirectional. However, phased arrays can have directivity patterns with two, three, or even four different optimum directions. A bidirectional pattern can be obtained, for example, by spacing the dipoles at one wavelength, and feeding them in phase, as shown at B.

More complicated phased arrays are sometimes used by radio transmitting stations. Several vertical radiators, arranged in a specified pattern and fed with signals of specified phase, produce a designated directional pattern. This is done to avoid interference with other broadcast stations on the same channel.

Phased arrays can have fixed directional patterns, or they can have rotatable or steerable patterns. The pair of phased dipoles (A) can, if the wavelength is short enough to allow construction from metal tubing, be mounted on a rotator for 360-degree directional adjustability. With phased vertical antennas, the relative signal phase can be varied, and the directional pattern thereby adjusted to a certain extent. *See also* PHASE.

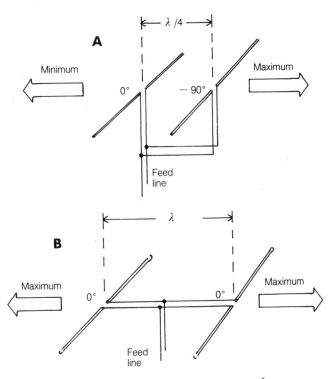

PHASED ARRAY: At A, unidirectional; at B, bidirectional.

PHASE DISTORTION

When the output of an amplifier fluctuates in phase, even though the input does not, the circuit is said to introduce *phase distortion* into the signal.

Phase distortion does not always cause problems in radio equipment. When phase distortion occurs, the effect is similar to phase modulation or frequency modulation of the signal. The signal can have a buzzing or fluttering sound, as heard in a frequency-modulation receiver. In an amplitude-modulation or code receiver, phase distortion is usually not noticeable.

Phase distortion sometimes occurs when a phase-locked-loop circuit malfunctions. If the circuit is not able to lock properly, the frequency and/or phase of the signal will continually shift as the device "attempts" to stabilize. In a phase-modulation or frequency-modulation transmitter or receiver, phase distortion is extremely undesirable because it interferes with the transmitted information. *See also* PHASE, and PHASE-LOCKED LOOP.

PHASE/FREQUENCY DETECTOR

In a *phase-locked loop (PLL)*, there is a circuit that detects changes in the phase and/or the frequency of the signal. This circuit is a *phase/frequency detector*.

The output of the phase/frequency detector has a polarity that depends on changes in the phase or frequency of the incoming signal. For instance, an increase in the frequency might cause a negative output, but a decrease in frequency causes a positive output. The output is used to keep a *voltage-controlled oscillator* (VCO) locked on the correct frequency.

A phase/frequency detector can be used to demodulate frequency-modulated (FM) or phase-modulated signals. This is because the positive/negative fluctuations in the circuit output follow the frequency or phase changes in the signal. The output of the phase/frequency detector is therefore a waveform that is the same as, or is similar to, the waveform of the audio at the transmitter. *See also* FREQUENCY MODULATION, PHASE-LOCKED LOOP, PHASE MODULATION, and VOLTAGE-CONTROLLED OSCILLATOR.

PHASE I SATELLITE

The most primitive, early ham satellites, called *OSCAR*, used only batteries, and did not have solar panels to keep them in operation beyond the battery life. Today, these are known as the *Phase I satellites*. *See also* ACTIVE COMMUNICATIONS SATELLITE, OSCAR, PHASE II SATELLITE, and PHASE III SATELLITE.

PHASE II SATELLITE

Some ham satellites, called *OSCAR*, orbit the earth in nearly circular, low paths. The orbital period for such a satellite is generally a few hours, and might be as short as about 90 minutes. The *inclination*, or angle of orbit relative to the equator, is usually quite high so that the satellite covers most of the earth over time. A satellite of this type is called a *Phase II ham satellite*.

Phase II satellites use solar panels to keep the batteries operating for years. This contrasts them with the earlier Phase I satellites, whose batteries, lacking solar panels, dictated the operational life of the whole machine. *See also* ACTIVE COMMUNICATIONS SATELLITE, OSCAR, PHASE I SATELLITE, and PHASE III SATELLITE.

PHASE III SATELLITE

The term *Phase III satellite* refers to any OSCAR with a very elongated, elliptical orbit. The high orbital eccentricity produces a period of several hours near apogee, during which the satellite moves slowly in the sky. This allows communications without the need for constantly adjusting the antenna direction.

Ham satellites in geostationary orbit fall into the category of Phase III. *See also* ACTIVE COMMUNICATIONS SATELLITE, APOGEE, ELLIPTICAL ORBIT, OSCAR, PHASE I SATELLITE, and PHASE II SATELLITE.

PHASED VERTICALS

See PHASED ARRAY.

PHASE INVERTER

A *phase inverter* is a circuit that literally flips the waveform upside down. Phase inversion can be accomplished by most ordinary single-ended amplifier circuits. Transformers can also be used.

Phase inversion differs from a phase delay of 180 degrees. A certain delay time will result in 180 degrees of phase delay at some frequency (f) and the equivalent of 180-degree delay at all odd harmonics $3f$, $5f$, $7f$, and so on. But a phase inverter will operate effectively at all frequencies.

The paraphase inverter circuit, having a single-ended input and a push-pull output, is sometimes called a *phase inverter*.

This circuit produces two output signals, in phase opposition, with respect to each other. Such a circuit can be used to drive a push-pull or push-push amplifier, without the need for an input transformer. *See also* PARAPHASE INVERTER, and PHASE.

PHASE-LOCKED LOOP

A *phase-locked loop* is a circuit that produces a signal with variable frequency. The phase-locked loop is extensively used in radio equipment, both for transmitting and receiving.

The phase-locked loop (PLL), has a voltage-controlled oscillator, the frequency of which can be varied by means of a varactor diode. The voltage-controlled oscillator is set for a frequency close to the intermediate frequency. A phase comparator causes the voltage-controlled oscillator to stabilize at exactly the intermediate frequency. If the voltage-controlled-oscillator frequency departs from the intermediate frequency, the phase comparator produces a voltage that brings the oscillator back to the correct frequency (*see* PHASE COMPARATOR). If the signal frequency changes, the voltage-controlled oscillator follows along, provided that the change is not too rapid. The drawing shows a block diagram of a phase-locked-loop circuit.

A PLL requires a certain amount of time to compensate for changes in the frequency of the incoming signal. If the incoming-signal frequency is modulated at more than a few hertz, the PLL will not lock continuously, but it will produce a fluctuating error voltage. The PLL can therefore be used for the purpose of detecting frequency-modulated or phase-modulated signals.

Modern PLL circuits are housed in single integrated-circuit packages. Phase-locked-loop circuits are used in most frequency-modulation communications equipment at the high, very-high, and ultra-high frequencies today. *See also* FREQUENCY MODULATION, PHASE MODULATION, and VOLTAGE-CONTROLLED OSCILLATOR.

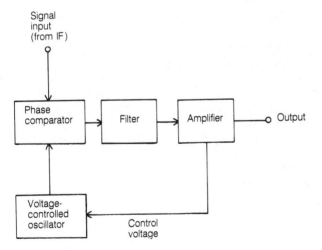

PHASE-LOCKED LOOP: Feedback keeps the signal on frequency.

PHASE MODULATION

Phase modulation is a method of conveying information via radio-frequency carrier waves. The instantaneous phase of the carrier is shifted in accordance with the modulating waveform.

Phase modulation is very similar, in practice, to frequency modulation. This is because any change in the instantaneous phase of a carrier also results in an instantaneous fluctuation in the frequency, and vice-versa.

In phase modulation, the extent of the phase shift is directly proportional to the amplitude of the modulating signal. The rapidity of the phase shift is directly proportional to both the amplitude and the frequency of the modulating signal. This differentiates phase modulation from frequency modulation; the result is a difference in the frequency-response characteristics.

Many frequency-modulated transmitters actually use phase modulation of one of the amplifier stages. When this is done, the high frequencies appear exaggerated at the receiver unless an audio-frequency lowpass filter is used at the transmitter. The output of such a filter must decrease in direct proportion to the modulating frequency. When this modification has been made at the transmitter, it is impossible to distinguish between phase modulation and true frequency modulation at the receiver. *See also* FREQUENCY MODULATION, and PHASE.

PHASE OPPOSITION

When two signals are exactly 180 degrees different in phase, they are said to be in *phase opposition*. Phase opposition results in a net amplitude that is the absolute value of the difference in the amplitudes of two signals having identical frequency. The phase of the resultant signal is the same as the phase of the stronger of the two constituent signals. If the two signals have the same amplitude, they combine in phase opposition to produce no signal. This effect is called *phase cancellation*.

Two signals can be in phase opposition for either of two reasons: one signal can be delayed, with respect to the other by one-half cycle, or one signal can be inverted (upside down), with respect to the other signal. A phase delay of 180 degrees can be obtained with a half-wavelength delay line. Phase inversion can be accomplished with an amplifier circuit or a transformer. *See also* PHASE INVERTER.

PHASE REINFORCEMENT

Two signals of identical frequency are said to be in *phase reinforcement* when they have the same phase. Waves in phase reinforcement add in amplitude arithmetically. That is, the amplitude of the resultant signal is equal to the arithmetic sum of the amplitudes of the constituent signals. *See also* PHASE.

PHASE SHIFT

A *phase shift* is a change in the phase of a signal. Phase shift can occur over a small fraction of one cycle, or it can occur over a span of many cycles. Phase shift is normally measured in degrees. One degree of phase shift corresponds to 1/360 (0.00278) cycle.

Certain electronic circuits shift the phase of an input signal. Phase shift is normally expressed as either positive or negative in such cases. A negative phase shift refers to a delay of up to one-half cycle (see A in the illustration). A positive phase shift is a delay of more than one-half cycle, and up to one full cycle, as at B. Generally, a phase delay of one full cycle is equivalent to zero phase shift.

A phase shift of more than 360 degrees can generally be expressed as equivalent to some value of shift smaller than 360 degrees. For example, if a signal is delayed by 500 degrees, the result is effectively equivalent to a delay of $500 - 360 = 140$ degrees.

Phase shift can be used to convey information via radio-frequency carrier waves. In such a system, the instantaneous phase of a carrier wave is made to follow the waveform of some input signal, such as a human voice. This is called *phase modulation*. See also PHASE, PHASE ANGLE, and PHASE MODULATION.

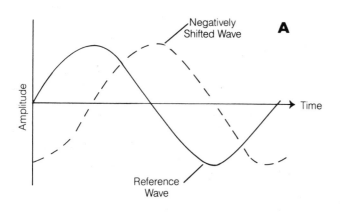

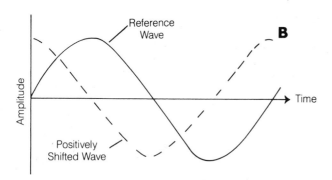

PHASE SHIFT: At A, negative shift; at B, positive shift.

PHASE-SHIFT KEYING

Phase-shift keying is a method of transmitting digital information. It is similar to frequency-shift keying (see FREQUENCY-SHIFT KEYING), except that the phase, not the frequency, is shifted.

In phase-shift keying, the carrier is transmitted at a constant amplitude. During the space, or key-up, condition, the wave is at one phase; during the mark, or keydown, condition, the phase is altered by a predetermined amount.

Phase-shift keying can be used in place of frequency-shift keying in certain applications. The primary advantage of phase-shift keying is that it can be accomplished in an amplifier stage, whereas frequency-shift keying cannot. See also PHASE.

PHASING

Phasing is the technique by which a phased array is made to have a certain directional characteristic (see PHASED ARRAY). A device that accomplishes this phasing is sometimes called a *phasing harness.*

Depending on the relative phase of the elements in a phased antenna, and on their spacing, the radiation pattern can have one, two, three, or more major lobes. Phasing harnesses usually consist of simple delay lines. A transmission line, measuring an electrical quarter wavelength, produces a delay of 90 degrees. A half-wave section of line causes the signal to be shifted in phase by 180 degrees. See also PHASE.

PHONETIC ALPHABET

The *phonetic alphabet* is a set of 26 words, one for each letter of the alphabet, used by radiotelephone operators for the purpose of clarifying messages under marginal conditions. The words are deliberately chosen so that they are not easily confused with other words in the list. The table shows the most commonly used English phonetic alphabet.

When it is necessary to spell out a word in a radiotelephone communication, the operator will say, for example, "City of Miami. I spell: Mike, India, Alpha, Mike, India."

Phonetics should be used only when necessary. Otherwise, the receiving operator can be confused by them. Phonetics should not be used to spell out common words (such as prepositions, verbs, and the like).

PHONETIC ALPHABET: PHONETIC ALPHABET RECOMMENDED BY THE INTERNATIONAL TELECOMMUNICATION UNION (ITU).

Letter	Word	Letter	Word
A	Alpha	N	November
B	Bravo	O	Oscar
C	Charlie	P	Papa
D	Delta	Q	Quebec
E	Echo	R	Romeo
F	Foxtrot	S	Sierra
G	Golf	T	Tango
H	Hotel	U	Uniform
I	India	V	Victor
J	Juliet	W	Whiskey
K	Kilo	X	X ray
L	Lima	Y	Yankee
M	Mike	Z	Zulu

PHOSPHOR

A *phosphor material* is a substance that glows when bombarded by a beam of electrons or other high-speed subatomic particles. Phosphor materials will also usually glow when exposed to ultraviolet rays, X rays, or gamma rays.

Phosphor materials are available in a wide variety of different glow colors. A common color is yellow-green, which is used in oscilloscopes, radar, and video-display terminals. White phosphors are used in black-and-white television receivers. Color television receivers utilize tricolor phosphors — red, blue, and green. Phosphors can be found in almost any color.

Various different phosphors have different persistence characteristics. The persistence is a measure of how long the phosphor will continue to glow after the radiation has been removed. The optimum degree of persistence depends on the application. For example, in television reception, a short persistence is needed; in radar, a longer persistence is desirable.

Common phosphor materials include zinc, silicon, and potassium compounds. See also CATHODE-RAY TUBE.

PHOTOCELL

The term *photocell* is used to describe any of a variety of devices that convert light energy into electrical energy. Photocells are often called *photoelectric cells.*

There are two kinds of photocells: those that generate a current all by themselves in the presence of light, and those that simply cause a change in effective resistance when the intensity of the light is varied. The first type of photocell is called a *photovoltaic cell* or *solar cell*. The latter type of photocell can be found in a variety of different forms. *See also* PHOTOCONDUCTIVITY, PHOTODIODE, PHOTOFET, PHOTORESISTOR, PHOTOTRANSISTOR, PHOTOVOLTAIC CELL, and SOLAR CELL.

PHOTOCONDUCTIVITY

Certain materials exhibit a resistance, or conductivity, that varies in the presence of visible light, infrared, or ultraviolet. Such substances are called *photoconductive materials.* The property of changing resistance in accordance with impinging light intensity is called *photoconductivity.*

In general, a photoconductive substance has a certain finite resistance when there is no visible light falling on it. As the intensity of the visible light increases, the resistance decreases. There is a limit, however, to the extent that the resistance will continue to decrease as the light gets brighter and brighter.

Photoconductivity occurs in almost all materials to a certain extent, but it is much more pronounced in semiconductors. When light energy strikes a photoconductive material, the charge-carrier mobility increases. Thus, current can be more easily made to flow when a voltage is applied. The more photons are absorbed by the material for a given electromagnetic wavelength, the more easily the material conducts an electric current.

Photoconductive materials are used in the manufacture of photoelectric cells. Some examples of photoconductive substances are germanium, silicon, and the sulfides of various other elements. *See also* PHOTODIODE, PHOTOFET, PHOTORESISTOR, and PHOTOTRANSISTOR.

PHOTODIODE

A *photodiode* is, as its name suggests, a diode that exhibits a variable effective resistance depending on the intensity of visible light that lands on its P-N junction. Photodiodes are quite common. In fact, most ordinary diodes having P-N junctions are photodiodes (*see* DIODE, and P-N JUNCTION). The reason that not all diodes can be used as photodiodes is simply that the housing is opaque, or that the P-N junction is situated between two opaque pieces of semiconductor material. Thus, no light can reach the junction.

A photodiode is specially designed so that its P-N junction is readily exposed to visible light. This necessitates the use of a large surface area for the junction. A transparent enclosure is also needed. The conductivity of a photodiode generally increases in the reverse direction as the impinging light gets brighter, but there is a limit to the drop in forward resistance. This limiting value is called the *saturation value.*

A photodiode is connected in series with the circuit to be controlled, and is reverse-biased. Care must be exercised to ensure that the photodiode does not carry too much current. *See also* PHOTOTRANSISTOR.

PHOTOELECTRIC CELL

See PHOTOCELL.

PHOTOELECTRIC EFFECT

Most materials have electrical resistance that changes if visible light shines on them. Ultraviolet and X rays have the same effect, called *photoelectric effect.*

Usually, in a conductor or semiconductor, the resistance decreases when radiant energy strikes the substance. This is because photons "knock" electrons out of their orbits, allowing the electrons to move more easily from atom to atom. In practice, photoelectric effect is most noticeable and useful in semiconductor junctions.

In some semiconductor diodes, electromagnetic energy will produce a potential difference between the P-type and N-type materials. This is sometimes called *photoelectric effect,* although it is more accurately termed *photovoltaic effect. See also* PHOTOCELL, PHOTOCONDUCTIVITY, PHOTODIODE, PHOTOFET, PHOTOMOSAIC, PHOTOMULTIPLIER, PHOTON, PHOTORESISTOR, PHOTOTRANSISTOR, and PHOTOVOLTAIC CELL.

PHOTOFET

A *photoFET* is a field-effect transistor that exhibits photoconductive properties (*see* FIELD-EFFECT TRANSISTOR, and PHOTOCONDUCTIVITY). Most ordinary field-effect transistors would be suitable for use as photoFETs, except that their housings and construction are opaque, and it is therefore impossible for any light to reach the P-N junction inside the device.

The photoFET must be constructed in such a way that its P-N junction, forming the boundary between the gate and the channel, has the largest possible area. The P-N junction must be situated so that light can fall on it. The enclosure for the device must be transparent. PhotoFETs differ from other kinds of photocells primarily because of their higher impedance.

PHOTOMOSAIC

In a camera tube, the *photocathode* is divided into a grid of small, photosensitive dots. This grid of dots is called the *photomosaic.*

Each small spot on the photomosaic receives a certain amount of visible-light energy, depending on the image received. A lens focuses the image onto the flat photomosaic. An electron beam then scans the photomosaic, and the beam is modulated according to the amount of light falling on each spot. *See also* COMPOSITE VIDEO SIGNAL, IMAGE ORTHICON, and VIDICON.

PHOTOMULTIPLIER

A *photomultiplier* is a vacuum-tube device that generates a variable current, depending on the intensity of the light that strikes it. The photomultiplier gets its name from the fact that it literally multiplies its own output, thereby obtaining extremely high sensitivity. Photomultipliers are used to measure light intensity at low levels.

The photomultiplier consists of a photocathode, which emits electrons in proportion to the intensity of the light impinging on it. These electrons are focused into a beam, and this beam strikes an electrode called a *dynode* (*see* DYNODE). The dynode emits several secondary electrons for each electron that strikes it. Generally, the dynode produces four or five secondary electrons for each primary electron (*see* SECONDARY EMISSION). The resulting, intense electron beam is collected by the anode.

A photomultiplier tube can have several dynodes, resulting in a very large amount of gain. The amount by which the sensitivity can be improved, by cascading one dynode after another, is limited by the amount of background electron emission from the photocathode. This background emission is called the *dark noise* or *photocathode dark noise*.

PHOTON

A *photon* is a particle of electromagnetic radiation. We normally think of a photon as a packet of visible-light energy, but actually all electromagnetic radiation is made up of particles.

Isaac Newton was the first scientist to formulate, in detail, a particle theory of light. Modern scientists can observe both particle-like and wave-like properties of electromagnetic radiation. The energy contained in a single photon depends on the wavelength.

If e represents the energy of one photon, in ergs, and the frequency in hertz is f, then:

$$e = hf$$

where h is Planck's constant, 6.62×10^{-27} erg-seconds. If the wavelength is given by λ, in meters, then:

$$e = (3 \times 10^8)h/\lambda$$

Photons can be observed to exert pressure on objects, as a barrage of particles would.

PHOTORESISTOR

A *photoresistor* is a device that exhibits a variable resistance, depending on the amount of light that strikes it. Normally, the resistance is a certain finite value in total darkness; as the light intensity increases, the resistance decreases. But this only occurs up to a certain point. If the intensity of the light increases further than the maximum limit of the device, the resistance does not drop any further. *See also* PHOTOCONDUCTIVITY.

PHOTOTRANSISTOR

A *phototransistor* is a bipolar transistor that varies in effective resistance as the intensity of radiant light changes (*see* PHOTOCONDUCTIVITY). Transistors are normally enclosed in opaque packages, for the purpose of eliminating noise caused by the action of ambient light in the collector-base junction. Phototransistors have transparent or translucent packages, so that light can reach this junction. Phototransistors are constructed in such a way that the base-collector P-N junction can receive the maximum possible amount of light.

The most common phototransistors are of the npn type, and are manufactured from silicon. Most phototransistors do not have terminals at the base electrode, but only at the emitter and the collector. Phototransistors are sometimes biased by means of a light-emitting diode near the device; but base bias is not necessary and, in fact, can even be detrimental to the performance of a phototransistor.

Phototransistors are used in much the same way as photodiodes and photoresistors. The effective conductivity increases as the light intensity increases. This occurs, however, only up to a certain point, known as the *saturation point*. *See also* PHOTODIODE, and PHOTORESISTOR.

PHOTOVOLTAIC CELL

A *photovoltaic cell* is a semiconductor device that generates a direct current when it is exposed to visible light. Photovoltaic cells generally consist of a P-N junction that has a large surface area, and a transparent housing to allow light to enter easily (see A in illustration).

Most photovoltaic cells are made from silicon. Some are made from selenium. Photovoltaic cells require no external bias; they generate their electricity by themselves. When light (and, in some cases, infrared or ultraviolet energy) impinges on the P-N junction, electron-hole pairs are produced. The intensity of the current from the photovoltaic cell, under constant load conditions, varies in linear proportion to the brightness of the light striking the device, up to a certain point. Beyond that point, the increase is more gradual, and it finally levels off at a maximum current, called the *saturation current*, as at B.

The ratio of the available output power to the light power striking a photovoltaic cell is called the *efficiency* or the *conversion efficiency* of the cell. Most photovoltaic cells have relatively low conversion efficiency, about 10 to 15 percent. Advances in the state of the art are expected to improve this figure in the future.

Photovoltaic cells are used in many different electronic devices, including modulated-light systems and solar-power generators (*see* MODULATED LIGHT, and SOLAR POWER).

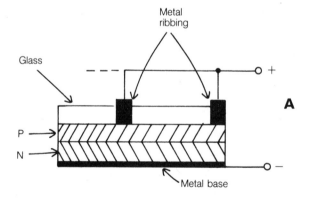

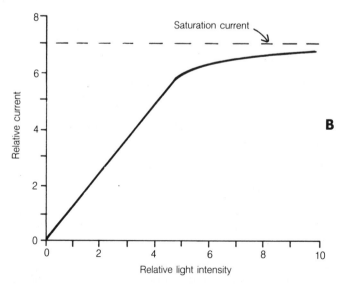

PHOTOVOLTAIC CELL: At A, cross-sectional view. At B, relative output current as a function of light intensity.

One important application of the photovoltaic cell is its power-generating capability. Photovoltaic cells can, in conjunction with storage batteries, provide continuous electric power from sunlight. This is especially true in the lower latitudes, and at the higher elevations, where sunshine is relatively abundant. At present, solar power, obtained from photovoltaic cells, is still relatively expensive, but its cost is decreasing.

PHYSICAL LAYER

See OPEN SYSTEMS INTERCONNECTION REFERENCE MODEL.

PICO

Pico is a prefix multiplier meaning one trillionth (10^{-12}). The abbreviation for pico is the small letter p.

Some electrical units, such as the farad, are extremely large in practice. At the high, very-high, and ultra-high radio frequencies, capacitances are often specified in picofarads (pF). A capacitance of 1 pF is equivalent to a millionth of a microfarad ($1\ pF = 10^{-6}\ \mu F$). *See also* PREFIX MULTIPLIERS.

PICTURE SIGNAL

A *picture signal*, sometimes called a *video signal*, refers to any of a variety of different methods of conveying visual images by modulating electromagnetic waves. The picture signal most familiar to us is the television signal.

A television signal consists of a composite video signal that modulates a radio-frequency carrier (*see* COMPOSITE VIDEO SIGNAL, and TELEVISION). Some television signals convey color as well as brightness information; others convey only brightness.

The standard fast-scan television signal in the United States is defined by the Federal Communications Commission. The channel bandwidth is nominally 6 MHz. The picture and sound carriers are separated by 4.5 MHz within this passband. The picture carrier is near the bottom of the passband, and the sound carrier is almost at the very top. The picture carrier is amplitude-modulated, and the sound carrier is frequency-modulated. The illustration is a spectral representation of the typical fast-scan television signal.

The number of scanning lines in a fast-scan picture signal varies from country to country. In the United States, 525 lines are used, with a horizontal scanning frequency of 15.734264 MHz for color signals and 15.75 MHz for black-and-white signals. The picture scanning occurs from left to right and from top to bottom, exactly as you would read a single-column book. The frame-scanning rate is 59.94 Hz for color and 60 Hz for black-and-white.

In recent years, technology has improved with respect to television picture transmission, and some thought has been given to increasing the number of scanned lines per frame. This would provide a picture with greater resolution than is now possible. Since the advent of cable television transmission, which has almost eliminated reception problems that are associated with airwave propagation, high-resolution television has become entirely practical.

Fast-scan television is not the only means of transmitting a picture. Radio amateurs commonly use a method of transmission called *slow scan* (*see* SLOW-SCAN TELEVISION). The resolution of slow scan is not as good as that of fast scan, and motion information is not conveyed, but the bandwidth is much narrower. A method of picture transmission that gives excellent resolution, but requires much more time per frame, is called *facsimile (fax)*. This mode is extensively used by weather satellites (*see* FACSIMILE).

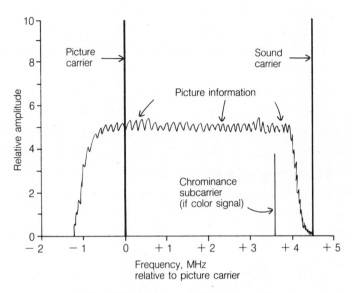

PICTURE SIGNAL: Typical AM fast-scan signal.

PICTURE TUBE

A *picture tube* is a special form of cathode-ray tube, used for the purpose of receiving fast-scan television signals (*see* CATHODE-RAY TUBE). Television picture tubes can be found in an enormous variety of sizes; some are smaller than a low-wattage incandescent light bulb, and others are larger than a person and weigh hundreds of pounds. The two main types of picture tubes are the black-and-white tube and the color tube.

The picture tube, also known as the *kinescope*, operates as shown in the illustration on pg. 362. An electron gun or guns, located at the back of the tube, produces a stream of electrons. The gun consists of a cathode and a set of focusing anodes (*see* ELECTRON-BEAM GENERATOR). Deflecting electrodes cause the beam to scan the screen in the characteristic 525-line format, with a frame-repetition rate of 59.94 Hz for color and 60 Hz for black-and-white.

The signal modulation is applied to the control grid or grids, resulting in fluctuations in the intensity of the electron beam as it scans the lines. As the beam strikes the phosphor screen, this modulation produces lighter and darker regions. From a distance, you see a picture.

In black-and-white television, there is only one electron gun, one set of deflecting plates, and one set of modulating grids. This is because only the brightness of the signal varies. The phosphor material glows white when the electron beam strikes it.

In color television, the phosphor screen consists of an array of dots or vertical phosphors that glow red, blue, and green in various brightness combinations. The color picture tube contains three electron guns, three sets of deflecting coils, and three sets of modulating grids: one each for the red, blue, and green components. *See also* COLOR TELEVISION, and TELEVISION.

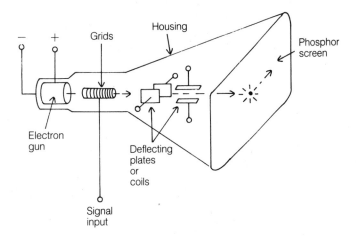

PICTURE TUBE: Simplified rendition of black-and-white tube.

PIERCE OSCILLATOR

The *Pierce oscillator* is a form of crystal oscillator. The Pierce configuration can be recognized by the connection of the crystal between grid and plate of a tube circuit (A) the base and collector of a bipolar-transistor circuit (B) or between the gate and drain of a field-effect-transistor circuit (C).

The Pierce circuit is used extensively in radio-frequency applications. The main advantage of the Pierce oscillator is that the crystal acts as its own tuned circuit. This eliminates the need for an adjustable inductance-capacitance tank circuit in the output. *See also* CRYSTAL OSCILLATOR.

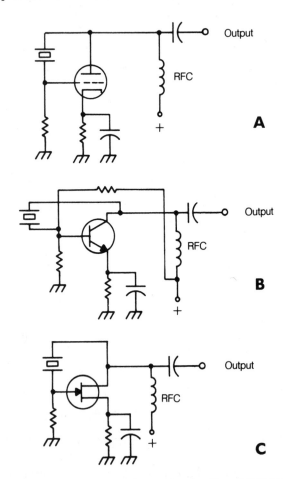

PIERCE OSCILLATOR: Tube (A), bipolar (B) and FET (C).

PIEZOELECTRIC EFFECT

Certain crystalline or ceramic substances can act as transducers at audio and radio frequencies. When subjected to mechanical stress, these materials produce electric currents; when subjected to an electric voltage, the substances will vibrate. This effect is known as the *piezoelectric effect*. Piezoelectric substances include such materials as quartz, Rochelle salts, and various artificial solids.

Piezoelectric devices are used at audio frequencies as pickups, microphones, earphones, and beepers or buzzers. At radio frequencies, piezoelectric effect makes it possible to use crystals and ceramics as oscillators and tuned circuits. *See also* CERAMIC, CERAMIC FILTER, CERAMIC MICROPHONE, CRYSTAL, CRYSTAL CONTROL, CRYSTAL-LATTICE FILTER, CRYSTAL OSCILLATOR, and CRYSTAL MICROPHONE.

PINCHOFF

In a junction field-effect transistor, *pinchoff* is the condition in which the channel is completely severed by the depletion region (*see* FIELD-EFFECT TRANSISTOR, and METAL-OXIDE-SEMICONDUCTOR FIELD-EFFECT TRANSISTOR). Pinchoff results in minimum conductivity from the source to the drain.

Pinchoff is the result of a large negative gate-to-source voltage in an N-channel junction field-effect transistor. In the P-channel device, pinchoff results from a large positive gate-to-source voltage. The minimum gate-to-source potential that results in pinchoff depends on the particular type of field-effect transistor. It also depends on the voltage between the source and the drain.

In the N-channel device, as the gate voltage becomes more negative, the drain voltage required to cause pinchoff becomes less and less positive. This is because the potential difference between the gate electrode and the channel is greater near the drain than near the source, and this effect is exaggerated as the drain-to-source voltage is increased. If the drain-to-source voltage is made sufficiently large, in fact, pinchoff will occur with zero gate-to-source bias. The graph shows the gate-to-source pinchoff voltages for a hypothetical N-channel field-effect transistor. The different curves represent the pinchoff voltages for various values of the drain-to-source voltage.

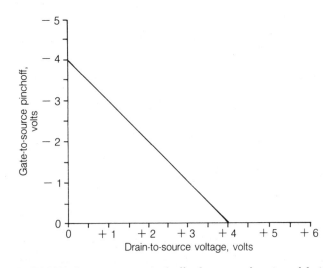

PINCHOFF: Gate-to-source pinchoff voltage, as a function of drain-to-source voltage, for a hypothetical FET.

In class-A and class-AB amplifiers, a field-effect transistor is normally biased at a lower value than that required to produce pinchoff. For operation in class B, the device is biased at, or slightly beyond, the pinchoff voltage. In class-C operation, a field-effect transistor is biased considerably beyond pinchoff (*see* CLASS-A AMPLIFIER, CLASS-AB AMPLIFIER, CLASS-B AMPLIFIER, and CLASS-C AMPLIFIER).

The field-effect transistor depicted is normally conductive through the channel; application of larger and larger bias voltages will eventually result in pinchoff. This is known as *depletion-mode operation*. Some types of field-effect transistors are normally nonconductive under conditions of zero bias. For conduction to occur in this kind of device, a certain gate-to-source bias must be applied. These field-effect transistors are called *enhancement-mode devices. See also* DEPLETION MODE, and ENHANCEMENT MODE.

PINCHOFF VOLTAGE

In a depletion-mode field-effect transistor, the *pinchoff voltage* is the smallest gate-to-source bias voltage that results in complete blocking of the channel by the depletion region. The pinchoff voltage depends on the kind of field-effect device, and also on the voltage between the drain and the source. *See* PINCHOFF.

PI NETWORK

A *pi network* is an unbalanced circuit, used especially in the construction of attenuators, impedance-matching circuits, and various forms of filters. Two parallel components are connected on either side of a single series component. The drawing at A illustrates a pi-network circuit consisting of noninductive resistors. An inductance-capacitance impedance-matching circuit is shown at B. The pi network gets its name from its resemblance, at least in diagram form, to the Greek letter pi (π).

In a pi-network impedance-matching circuit, the capacitor nearer the generator or receiver is called the *tuning capacitor*, and the capacitor nearer the load or antenna is called the *loading capacitor*. By properly adjusting the values of the two capacitors and the inductor, matching between nonreactive impedances can be obtained over a fairly wide range (in practice).

Pi-network impedance-matching circuits are extensively used in the output circuits of tube-type and transistor-type radio-frequency power amplifiers. In such circuits, the plate or collector impedance is often several times the actual load impedance. The loading-capacitor value is set so that, when the tuning capacitor is adjusted, the "dip" in the plate or collector current results in the optimum operation of the final-amplifier tube or transistor.

In certain applications, a series inductor can be added to a pi-network impedance-matching circuit. This allows the cancellation of capacitive reactances in the load. Such a circuit (C) is called a *pi-L network*. Similarly, if inductive reactance exists in the load, a capacitor can be included in series with the network. This is known as a *pi-C circuit* (D). *See also* IMPEDANCE MATCHING.

PLANAR ARRAY

A *planar array* is a form of phased antenna system, in which three or more coplanar elements are fed with a certain phase relationship to obtain directional gain. A planar array can consist

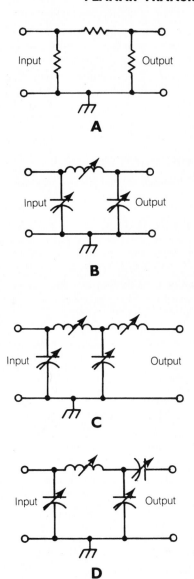

PI NETWORK: At A, an attenuator; at B, a basic pi network for impedance matching; at C, a pi-L network; at D, a pi-C network.

of linear elements or looped elements; it might have a reflecting device or parasitic elements. The billboard antenna, and various forms of broadside arrays, are forms of planar arrays (*see* BILLBOARD ANTENNA, BROADSIDE ARRAY, and PHASED ARRAY).

Planar arrays can exhibit the major lobe in a direction or directions perpendicular to the plane containing the elements, but this need not necessarily be the case. Some planar arrays, such as the Mills Cross or Christiansen antenna, have steerable main lobes. *See also* MILLS CROSS.

PLANAR TRANSISTOR

A *planar transistor* is a special form of bipolar transistor. It can be of either the npn or the pnp type. The planar transistor gets its name from the fact that all three elements — the emitter, the base, and the collector — are fabricated from a single layer of material.

The emitter is at the center, the base surrounds the emitter, and the collector is at the outside and bottom. The entire structure is covered with a flat film of silicon dioxide or other insulating material. This protects the transistor.

The collector has by far the greatest volume. The base-collector junction has much larger surface area, because of the concentric construction, than the emitter-base junction. This gives the planar transistor the ability to dissipate relatively large amounts of heat at the base-collector junction. Planar design is well suited to the fabrication of power transistors. *See also* TRANSISTOR.

PLANE REFLECTOR

A *plane reflector* is a passive antenna reflector that is commonly used for transmitting and receiving at ultra-high and microwave frequencies. The reflector gets its name from the fact that it is perfectly flat and electrically continuous. A plane reflector is positioned ¼ wavelength behind a set of dipole antennas, producing approximately 3 dB of power gain.

The direct, or forward, radiation from the antenna is not affected. The radiation in the opposite direction encounters the metal plane, where the wave is reversed in phase by 180 degrees and reflected back in the forward direction. Because the plane is ¼ wavelength away from the radiating element, the total additional path distance for the reflected wave is ½ wavelength (180 degrees), with respect to the direct wave. The extra distance and the phase reversal, together, result in phase reinforcement between the direct wave and the reflected wave.

Plane reflectors can be made of wire mesh, screen, a set of parallel bars, or a solid sheet. If wire screen or mesh is used, the spacing between the wires or bars should be less than about 0.05 wavelength for best results.

The plane reflector can be used with any number of radiating elements. A large array of dipoles with a flat reflector is called a *billboard antenna*. Plane reflectors can be used with helical or loop antennas, as well as with dipole elements. In all cases, the extra power gain is 3 dB, corresponding to a doubling of the effective radiated power compared to an identical antenna without the reflector. All antennas that incorporate plane reflectors are a form of planar array. *See also* BILLBOARD ANTENNA, and PLANAR ARRAY.

PLATE

The anode of a vacuum tube is often called the *plate*. The plate is supplied with a positive voltage so that it attracts electrons from the cathode. The plate of a tube is analogous to the collector of a bipolar transistor, and to the drain of a field-effect transistor.

The plate of a tube can be shaped like a cylinder, a rectangular prism, or some other configuration. In some vacuum tubes, the plate is ribbed or finned to increase its surface area. This, in turn, increases its ability to dissipate power.

Tube plates are subject to heating to a greater or lesser extent, depending on the type of tube and the circuit in which it is used. Some tubes have plates made from carbon or a metal capable of withstanding extreme temperatures. *See* PLATE DISSIPATION, PLATE POWER INPUT, PLATE RESISTANCE, and TUBE.

The term *plate* is sometimes used to describe the electrodes in a capacitor, or the anode of an electrochemical cell. *See also* ANODE, CAPACITOR, CELL, and ELECTRODE.

PLATE DISSIPATION

In a vacuum tube amplifier or oscillator circuit, there is a difference between the plate power input (*see* PLATE POWER INPUT) and the circuit power output. This power difference is largely spent as heat in the plate of the tube. The power that is used up in heating the plate of the tube is the plate dissipation.

In general, for a given amount of plate power input, the plate dissipation increases as the efficiency decreases. The class-A amplifier results in a plate dissipation of more than 50 percent of the total plate power input; the class-C amplifier can result in a plate dissipation of as little as 20 percent of the total plate power input (*see* CLASS-A AMPLIFIER, CLASS-AB AMPLIFIER, CLASS-B AMPLIFIER, and CLASS-C AMPLIFIER). These figures are based on the assumption that the amplifier is properly adjusted. Improper adjustment will substantially increase the plate dissipation. This can occur, for example, when a radio-frequency power amplifier is not tuned to resonance in the output circuit (*see* POWER AMPLIFIER).

Vacuum tubes are rated according to the maximum amount of power they can safely dissipate without the likelihood of damage. The continuous rating is smaller than the intermittent-duty rating (*see* DUTY CYCLE). Cooling systems can, in some cases, dramatically increase the plate-dissipation rating of a tube (*see* COOLING).

PLATE POWER INPUT

The *plate power input* to a tube circuit is the product of the direct-current plate voltage and the plate current. Plate power input is the sum of the output power and the power dissipated as heat in the tank circuit, conductors, and the plate of the tube itself.

The plate power input to an amplifier is always greater than the output. The ratio of the output to the plate power input is called the *efficiency*. In class-A amplifiers, the plate power input can be more than twice the output. In class-C amplifiers, the output is almost as large as the input (*see* CLASS-A AMPLIFIER, CLASS-AB AMPLIFIER, CLASS-B AMPLIFIER, and CLASS-C AMPLIFIER).

In the determination of plate power input, the plate voltage is usually measured with a direct-current voltmeter, and the plate current is measured with a direct-current ammeter.

PLATE RESISTANCE

The *plate resistance* of a vacuum tube is an expression of the internal resistance of the device. When there is no reactance in the output circuit—that is, it is resonant—the output impedance of the tube is equal to the plate resistance. Plate resistance is abbreviated R_P.

Most vacuum tubes have plate-resistance ratings of hundreds, or even thousands, of ohms. This usually necessitates the use of an impedance transformer between the plate circuit of the tube and the load. For example, an antenna system can show a resistive impedance of 50 ohms, and the final-amplifier tube of a transmitter has a plate resistance of 2,500 ohms. This would require a transformer with an impedance-transfer ratio of 25:1; the turns ratio in such a case would be 5:1.

The plate resistance, R_P, of a tube under nonreactive conditions is the quotient of the plate voltage, E_P, and the plate current, I_P:

$$R_P = E_P / I_P$$

This value is sometimes called the *static plate resistance*. When a signal is applied, the instantaneous plate resistance can differ from the static value. Under such conditions, the instantaneous, or dynamic, plate resistance is equal to the rate of change in the plate voltage divided by the rate of change in the plate current:

$$R_p = dE_p/dI_p$$

PLL

See PHASE-LOCKED LOOP.

PLUG

Any male connector can be called a *plug*. Examples of plugs include the standard ¼-inch phone plug, lamp-cord plugs, phono plugs, Jones plugs, and many others. A plug mates with a female jack of the same size and number of conductors. Sometimes the female socket or jack, together with the male connector, is called a *plug*.

Plugs always have their conductors exposed when they are disconnected. Therefore, plugs should not be installed in the part of the circuit normally carrying the voltage from the power supply. This precaution will minimize the shock hazard, and prevent possible inadvertent short circuits.

PLUG-IN COMPONENT

A *plug-in component* is any component that can be removed for easy replacement. Some transistors are of the plug-in type. Many integrated circuits are installed in sockets, and thus they are plug-in components.

In certain types of equipment, especially test apparatus, plug-in components are used to allow the adjustment of certain parameters. For example, a common type of radio-frequency wattmeter has a socket in which various measuring elements can be inserted. The elements are plug-in components. One element can be used for a full-scale indication of 25 watts in the frequency range 25 to 200 MHz, for example; another element will provide a full-scale indication of 500 watts at 200 to 500 MHz. The element is chosen for maximum meter readability in the appropriate frequency range.

In older radio receivers, bandswitching was done by means of plug-in inductors. For the low frequencies, an inductor having many turns was used. This plug-in inductor could be removed, and a smaller one put in its place, for reception of medium and high frequencies. This technique of bandchanging is not used nowadays. Rotary switches have proven considerably more convenient for the purpose.

PL-259

See UHF CONNECTOR.

PMOS

See METAL-OXIDE-SEMICONDUCTOR LOGIC FAMILIES.

P-N JUNCTION

A *P-N junction* is the boundary between layers of P-type and N-type semiconductor materials. A P-N junction tends to conduct when the N-type material is negative, with respect to the P-type material, but it exhibits high resistance when the polarity is opposite.

A P-N junction is said to be forward-biased when the N-type material is negative with respect to the P-type material (see A in illustration). The junction conducts well in this mode. The P-type material, having a deficiency of electrons, receives electrons from the N-type material (*see* N-TYPE SEMICONDUCTOR, and P-TYPE SEMICONDUCTOR).

A P-N junction is said to be reverse-biased when the N-type material is positive with respect to the P-type material as at B. In this condition, electrons are pulled from the P-type material, which already has a deficiency of electrons; and electrons accumulate in the N-type material, which already has an excess of electrons. The result is a zone that has extremely high resistance. This region occurs immediately on either side of the boundary between the two different semiconductor layers. Therefore, the P-N junction does not conduct well in the reverse direction; the resistance can be millions of megohms or more. If the reverse bias is increased to larger and larger values, however, the P-N junction suddenly begins to conduct at a certain bias level, known as the avalanche voltage (*see* AVALANCHE BREAKDOWN).

Because a P-N junction conducts well in one direction, but not in the other, devices consisting of P-N junctions are useful as detectors and rectifiers (*see* DETECTION, DIODE ACTION, and RECTIFICATION). The P-N junction of a semiconductor device can be capable of accumulating and depleting the insulating layer very rapidly, perhaps on the order of millions or billions of times per second. In such instances, the resulting diode can be used for detection at megahertz and even gigahertz frequencies. Other P-N junction devices can follow alternating-current frequencies of only a few tens or hundreds of kilohertz. The ability of a P-N-junction diode to handle high frequencies depends largely on the surface area of the junction. It also depends on the thickness of the depletion layer that results from a given amount of reverse voltage. *See also* DIODE CAPACITANCE.

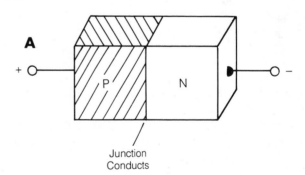

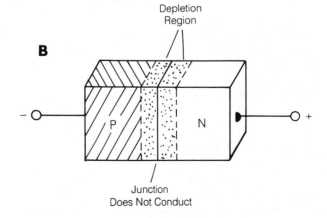

P-N JUNCTION: Forward bias (A) and reverse bias (B).

PNP TRANSISTOR

A pnp transistor is a three-layer, bipolar semiconductor device. A layer of N-type semiconductor material is sandwiched in between two layers of P-type material (see A in illustration). One of the P-type layers is thicker than the other. The thicker P-type layer is designated the collector of the transistor, and the thinner layer is the emitter. The N-type layer in the center is the base of the transistor.

The pnp transistor is normally biased with the base negative, with respect to the emitter. The collector is biased negatively, with respect to the base. This biasing configuration is illustrated schematically at B.

For class-B or class-C amplifier operation, the base can be biased at the same level as the emitter, or perhaps even positively with respect to the emitter. This places the transistor at, or beyond, cutoff (see CLASS-B AMPLIFIER, and CLASS-C AMPLIFIER).

The pnp transistor uses a biasing polarity that is exactly opposite to that of the npn transistor. See also NPN TRANSISTOR.

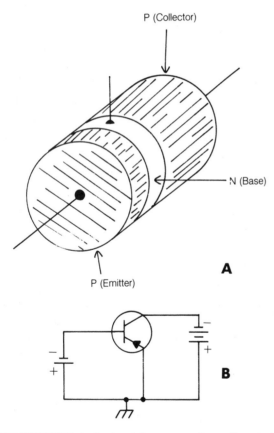

PNP TRANSISTOR: At A, pictorial representation; at B, typical biasing.

PNPN TRANSISTOR

See FOUR-LAYER TRANSISTOR.

POINT-CONTACT JUNCTION

A *point-contact junction* is a form of semiconductor junction. A fine wire, called a *cat's whisker*, is placed in contact with a piece of semiconductor material, such as germanium or silicon. The semiconductor material forms the anode of the junction, and the cat's whisker forms the cathode. The point-contact junction conducts well when the anode is positive with respect to the cathode, but the device conducts poorly when the anode is negative, with respect to the cathode. This one-way conductivity results in diode action.

The point-contact method is used in the fabrication of various types of diodes and transistors. A point-contact junction generally has much lower capacitance than a P-N junction, because the surface area is much smaller. This makes the point-contact junction useful at higher frequencies. However, the point-contact junction cannot carry much current. Therefore, point-contact devices are limited to low-power applications. See also P-N JUNCTION, and POINT-CONTACT TRANSISTOR.

POINT-CONTACT TRANSISTOR

Two *point-contact junctions* can be combined to form a transistor (see POINT-CONTACT JUNCTION). A piece of germanium or silicon serves as the base, and two fine wires (cat's whiskers) are placed in contact with the wafer to form the emitter and collector. The result is a device that behaves like an npn bipolar transistor.

Point-contact transistors were the earliest form of semiconductor transistor. Such devices are still occasionally used, although junction-type transistors are now more common. The point-contact transistor works well at very-high and even ultra-high frequencies because the point contact junctions exhibit low capacitance. However, point-contact transistors cannot carry much current, and are therefore used only in receivers or local-oscillator circuits. See also TRANSISTOR.

POINT-TO-POINT WIRING

Point-to-point wiring is a form of circuit wiring used mostly in vacuum-tube circuits. Point-to-point wiring is not often seen in modern solid-state circuits; printed-circuit boards are much more common nowadays (see PRINTED CIRCUIT). Point-to-point wiring consists of individual hookup-wire links, soldered among various points in a circuit. The beginning and ending points of each wire usually consist of tie-strip terminals. Tie strips are also known as *terminal strips* (see TIE STRIP).

In some applications, especially at frequencies below about 30 MHz when high radio-frequency power levels are involved, point-to-point wiring is still used. This is because individual wire conductors can withstand higher current levels than the foil runs of printed circuits. The main advantage of point-to-point wiring is its ability to handle large amounts of current. The main disadvantage of point-to-point wiring is that it requires more physical space than circuit-board wiring.

A modern form of point-to-point wiring, having greater compactness and convenience than the old-fashioned form, is called *wire wrapping*. See also WIRE WRAPPING.

POLARITY

Polarity refers to the relative voltage between two points in a circuit. One pole, or voltage point, is called *positive*, and the other pole is called *negative*. The polarity affects the direction in which the current flows in the circuit.

Physicists consider current to flow from the positive pole to the negative pole. The actual movement of electrons is from the negative pole to the positive pole, since electrons carry a negative charge.

In magnetism, polarity refers to the orientation of the north and south magnetic poles.

POLARIZATION

Polarization is an expression of the orientation of the lines of flux in an electric, electromagnetic, or magnetic field. Polarization is of primary interest in electromagnetic effects. The polarization of an electromagnetic field is considered to be the orientation of the electric flux lines.

An electromagnetic field can be horizontally polarized, vertically polarized, or diagonally polarized (*see* HORIZONTAL POLARIZATION, and VERTICAL POLARIZATION). The polarization is generally parallel with the active element of an antenna; thus, a vertical antenna radiates and receives fields with vertical polarization, and a horizontal antenna radiates and receives fields having horizontal polarization.

Electromagnetic fields can have polarization that continually changes. This kind of polarization can be produced in a variety of ways (*see* CIRCULAR POLARIZATION, and ELLIPTICAL POLARIZATION).

Polarization effects occur at all wavelengths, from the very low frequencies to the gamma-ray spectrum. Polarization effects are quite noticeable in the visible-light range. Light that has horizontal polarization, for example, will reflect well from a horizontal surface (such as a pool of water), and vertically polarized light reflects poorly off the same surface. Simple experiments, conducted with polarized sunglasses, illustrate the effects of visible-light polarization. *See also* ELECTRIC FIELD, ELECTRIC FLUX, and ELECTROMAGNETIC FIELD.

In an electrochemical cell or battery, the electrolysis process sometimes results in the formation of an insulating layer of gas on one of the plates or sets of plates. This effect, called *polarization*, causes an increase in the internal resistance of the cell or battery. This, in turn, limits the amount of current that the device can deliver. If the polarization is severe, the cell or battery can become totally useless. *See also* CELL.

POLARIZATION MODULATION

Information can be impressed on an electromagnetic wave by causing rapid changes in the polarization of the wave (*see* POLARIZATION). This method of modulation is known as *polarization modulation*. Polarization modulation can theoretically be achieved at any wavelength from the very low frequencies through the gamma-ray spectrum. However, polarization modulation is most often used at microwave and visible-light wavelengths.

Two antennas are oriented at right angles, with respect to each other. Under zero-modulation conditions, one antenna (for example, the vertical one), receives all of the signal, while the other antenna gets none. When modulation is applied, some of the signal is applied to the horizontal antenna at the expense of the signal applied to the vertical antenna. The total signal power reaching both antennas remains constant. Under maximum instantaneous modulation conditions, the horizontal antenna gets all of the power and the vertical antenna gets none. This causes the polarization of the transmitted signal to shift by values up to 90 degrees, in accordance with the amplitude of the modulating waveform.

When the signal from the antenna encounters a receiving antenna having linear polarization, the instantaneous signal strength fluctuates in sync with the modulating waveform. This will occur regardless of the orientation of the receiving antenna. For best results, the receiving antenna should be parallel with one of the transmitting-antenna elements.

The polarization of visible light can be modulated in a variety of different ways. A pair of light sources, equipped with mutually perpendicular polarizing filters, can be amplitude-modulated in a manner similar to that described above. The receiving apparatus must then incorporate a polarizing filter for retrieval of the information.

POLARIZED COMPONENT

A *polarized component* is an electronic component that must be installed with its leads in a certain orientation. If a polarized component is installed incorrectly, the circuit will not work properly.

Examples of polarized components include diodes, certain types of capacitors, and, of course, all direct-current power sources. *See also* NONPOLARIZED COMPONENT.

POLE

A *pole* is a point toward which the flux lines of an electric or magnetic field converge, or from which the flux lines diverge. Electric poles can be either positive, representing a deficiency of electrons, or negative, representing an excess of electrons. Magnetic poles can be either north or south.

Electric poles can exist alone—that is, a single positive or negative pole can be isolated—but magnetic poles are always paired.

In a power supply, the positive terminal is sometimes called the *positive pole*, and the negative terminal is the *negative pole*. The manner in which the supply is connected to the load, or circuit, is called the *polarity. See also* POLARITY.

POLLING PROTOCOL

In a packet radio network, it is useless to have a station work through a node if that station has nothing to transmit. In fact, it is worse than useless because it slows down the operation of the whole network.

Early packet-radio experimenters developed a *polling protocol*. In this operating environment, a *controller* checks stations at intervals, according to a predefined list, to see if each respective station has anything to send. If a station has data, it sends the data, and then the controller moves to the next station on the list. If a station is polled and has nothing to send, the controller moves immediately to the next station on the list.

A polling protocol optimizes network efficiency by minimizing the proportion of time during which no data is being transferred. *See also* PACKET RADIO, and PROTOCOL.

POLYETHYLENE

Polyethylene is a form of plastic. The full name of the substance is *polymerized ethylene*. This material is noted for its resistance to moisture, its low dielectric loss, and its high dielectric strength. Polytheylene is a tough, fairly flexible, semi-transparent plastic.

Polyethylene is widely used in electronics as an insulator, especially in coaxial cables and twin-lead transmission lines. The polyethylene can be used either in solid form or in foamed form. Solid polyethylene lasts longer than the foamed variety, and also is less susceptible to contamination. Foamed polyethylene has somewhat lower dielectric loss, and a lower dielectric constant.

POLYPHASE RECTIFIER

A *polyphase rectifier* is a device used to convert three-phase alternating current to direct current (*see* THREE-PHASE ALTERNATING CURRENT). Polyphase rectifiers operate according to essentially the same principles as single-phase rectifiers. All of the polyphase circuits make use of diodes; either solid-state or tube-type diodes can be used.

The main advantage of polyphase rectifiers over single-phase rectifiers is the fact that the ripple frequency is higher, resulting in more pure direct current after filtering.

POLYSTYRENE CAPACITOR

Polystyrene is well suited as a dielectric material for the manufacture of capacitors. The electrical characteristics are excellent at radio frequencies. Polystyrene capacitors are relatively inexpensive.

Polystyrene capacitors are noted for their stability under variable-temperature conditions, and for this reason, they are preferred for use in variable-frequency oscillators at the low, medium, and high frequencies. Polystyrene capacitors can be made to have very low tolerance values (high precision). Polystyrene capacitors are physically rugged and moisture-resistant. However, they are not tolerant of oils and petroleum-based solvents.

While polystyrene itself has low dielectric loss up through the ultra-high-frequency range, the electrode resistivity limits the maximum frequency at which polystyrene capacitors can be used. Polystyrene capacitors are used in low-voltage applications.

PORCELAIN

Porcelain is a ceramic material having excellent dielectric properties. Porcelain appears as a hard, white, dull substance when not glazed; it is shiny when glazed. Porcelain has high tensile strength, but it is brittle.

Porcelain is widely used in the manufacture of insulators for direct current, as well as radio-frequency alternating current; it can withstand very high temperatures and voltages. The dielectric constant of porcelain decreases somewhat as the frequency goes up; although it is about 5.4 at 1 kHz, it decreases to 5.1 at 1 MHz and 5.0 at 100 MHz.

The dielectric loss of dry porcelain is quite low. Some trimmer capacitors use porcelain as the dielectric. Many air-variable capacitors, especially those designed for transmitting, incorporate porcelain for insulating purposes. Porcelain is also occasionally used as a coil-winding form, especially in power amplifiers. *See also* CERAMIC, and CERAMIC CAPACITOR.

PORT

In digital electronics, a *port* is any point where data goes into and/or comes out of a device. For example, the plug-ins on the back of a personal computer (PC) are ports. One is for the monitor, another is for the printer, another is for the modem, and so on.

A port that is used only for data input is called an *input port*.

A port that is used only for data output is called an *output port*. A port that is used both for data input and data output is called an *input/output (I/O) port*.

Ports are sometimes categorized as *parallel* or *serial*. These terms refer to the nature of the data transfer. *See* PARALLEL DATA TRANSFER, and SERIAL DATA TRANSFER.

PORTABLE EQUIPMENT

Portable communications equipment is any apparatus that is designed for operation in remote locations. Portable equipment is battery-operated, and can be set up and dismantled in a minimum time. Some portable equipment can be operated while being carried; an example is the handy-talkie or walkie-talkie.

With the improvement in miniaturization and semiconductor technology, portable equipment can have complex and sophisticated features. A few decades ago, the mention of a microcomputer-controlled walkie-talkie would have elicited ridicule. Today, such devices are commonplace.

Portable equipment must be compact and light in weight. For this reason, such equipment always has modest power requirements because a high-power battery is invariably bulky and heavy.

In addition to being small and light, portable equipment must be designed to withstand physical abuse (such as vibration, temperature and humidity extremes) and prolonged use without complicated servicing.

Portable equipment is sometimes used in mobile applications. *See also* MOBILE EQUIPMENT.

POSITIVE CHARGE

Positive charge, or *positive electrification*, is the result of a deficiency of electrons in the atoms of which an object is composed. Friction between two objects can cause an electron imbalance, resulting in positive charge on one of the bodies. When an atom has fewer electrons than protons that atom is said to be *positively charged*.

The smallest unit of positive charge is carried by a single proton. The smallest unit of negative charge is carried by the electron. The proton and electron have opposite charge polarity, but equal charge quantity. The terms positive and negative are chosen arbitrarily.

Electric lines of flux are considered to originate at positive-charge poles and terminate at negative-charge poles. An electric current, likewise, is theoretically considered to flow from the positive pole of a circuit to the negative pole, even though current often consists of electrons that move away from the negative-charge terminal. *See also* CHARGE, and NEGATIVE CHARGE.

POSITIVE LOGIC

The way in which logical signals are defined is a matter of convention. Normally, the logic 1 is the more positive of the voltage levels in a binary circuit, and the logic 0 is the more negative. Thus, the logic 1 is high and the logic 0 is low. This is called *positive logic*.

In some digital circuits, the logic 1 is the more negative (low) and the logic 0 is the more positive (high) of the voltage levels. This is known as *negative logic* (*see* NEGATIVE LOGIC).

From a practical standpoint, it makes no difference whether a circuit uses positive logic or negative logic, as long as the same form is consistently used throughout the circuit. *See also* LOGIC.

POT CORE

A special form of ferromagnetic coil-winding form, designed to provide large inductances for relatively few turns of wire, is known as a *pot core*. The pot core gets its name from the fact that it is shaped something like a pot. The core consists of two identical sections that are fastened together by means of a nut and bolt through the center. The coil-winding space is toroidal in shape, and the coil is wound in circular turns.

Pot cores can be found with various different permeability factors, and they are made in a wide variety of physical sizes. The magnetic field of the coil is contained entirely within the core, as is the case with toroidal coils (*see* TOROID). This allows the pot-core coil to be placed adjacent to other components with practically no interaction. The inductance of a pot-core coil can be as much as 1 henry or more, with a very reasonable number of coil turns. The pot-core can handle quite large amounts of current, in proportion to its physical size. Pot-core coils exhibit low loss and high *Q* factor. They are used mostly at audio frequencies, in passive filters and as transformers.

POTENTIAL DIFFERENCE

Two points in a circuit are said to have *a difference of potential* when the electric charge at one point is not the same as the electric charge at the other point. Potential difference is measured in volts (*see* VOLTAGE).

When two points having a potential difference are connected by a conducting or semiconducting medium, current flows. Physicists consider the current to flow from the more positive point to the more negative point. If the charge carriers are electrons, they actually move from the more negative point to the more positive point. If the charge carriers are holes, they move from positive to negative. *See also* CHARGE, ELECTRON, and HOLE.

POTENTIAL ENERGY

Potential energy is an expression of the capability of a body to produce useful energy. For example, if a weight is lifted up several feet, it gains potential energy; when the weight is dropped, the energy output is realized.

Potential energy can be harnessed for specific purposes. For example, in a hydroelectric power plant, the potential energy of the water at the top of the dam is converted into electricity. Potential energy cannot always be harnessed conveniently, and the conversion from potential energy into useful energy is never 100-percent efficient. *See also* ENERGY.

POTENTIAL TRANSFORMER

A small transformer, having multiple taps, is sometimes used to extend the range of an alternating-current voltmeter. Such a transformer is called a *potential transformer*.

The potential transformer is, of course, not the only means of adjusting the full-scale range of an alternating-current voltmeter. Shunt or series resistors can be used to increase the full-scale range. The range can be decreased by reducing the value of the series resistance. Alternatively, a direct-current amplifier, following the rectifier and filter, can be used to decrease the full-scale range. *See also* VOLTMETER.

POWDERED-IRON CORE

Powdered-iron cores are used in coils and transformers for the purpose of increasing the value of inductance for a given number of turns. This makes such components much more compact than would otherwise be possible. Powdered-iron cores are used in audio-frequency and radio-frequency applications.

Powdered-iron cores are found in three basic configurations: the solenoid, the toroid, and the pot core. The solenoidal core allows some of the magnetic field to exist outside the windings of the coil. The toroid and pot cores confine the field to the inside of the component.

Powdered-iron cores increase the inductance of a coil because they have high permeability (*see* PERMEABILITY). This causes the magnetic-field intensity within the material to increase, just as it would if the number of coil turns were increased. Powdered-iron cores can be found with various values of permeability for different applications. A material similar to powdered iron with extremely high permeability is known as *ferrite* (*see* FERRITE CORE).

POWER

Power is the rate at which energy is expended or dissipated. Power is expressed in joules per second, more often called *watts*. In a dc circuit, the power is the product of the voltage and the current. A source of E volts, delivering I amperes to a circuit, produces P watts, as follows:

$$P = EI$$

From Ohm's law, you can find the power in terms of the current and the resistance, R, in ohms:

$$P = I^2R$$

Similarly, in terms of the voltage and the resistance:

$$P = E^2/R$$

In an ac circuit, the determination of power is more complicated than in the case of direct current. Assuming there is no reactance in the circuit, the power is determined in this manner, using root-mean-square values for the current and the voltage (*see* ROOT MEAN SQUARE).

If reactance is present in an alternating-current circuit, some of the power appears across the reactance, and some appears across the resistance. The power that appears across the reactance is not actually dissipated in real form. For this reason, it has been called *imaginary* or *reactive power*. The power appearing across the resistance is actually dissipated in real form, and it is thus called *true power*. The combination of real and reactive power is known as the *apparent power* (*see* APPARENT POWER, and REACTIVE POWER).

The proportion of the real power to the apparent power in an alternating-current circuit depends on the proportion of the resistance to the total or net impedance. The smaller this ratio, the less of the power is real power. In the case of a pure reactance, none of the power is real. *See also* POWER FACTOR.

POWER AMPLIFIER

In amateur radio, the term *power amplifier* is used mainly in reference to radio-frequency (RF) transmitting amplifiers. All transmitters have a final amplifier stage, and this is always a power amplifier. The output power might be 1500 W, or a fraction of 1 W, or anything in between.

Power amplifiers with outputs of about 200 W or less usually work with bipolar transistors. Some outboard amplifiers can produce 1500 W with transistors, although vacuum tubes are more commonly seen at these power levels.

The power amplifier gets its name, not from the fact that it produces high power (some don't), but from the fact that it has more power output than power input, and also because the circuit is specifically intended, and designed, to increase signal power. Some amplifiers amplify signal current or voltage, but not power. There are various types of power amplifiers for the different modes and frequencies of ham operation.

Linearity Most external power amplifiers, meant to boost the output of a transceiver to the full legal limit (1500 W) or something near it, are designed so that the instantaneous output is a direct function of the instantaneous input. These amplifiers are called *linear amplifiers*. The quality of amplification, that ensures there will be no distortion on the envelope of a single-sideband signal, is called *linearity*.

Linearity is not important for digital modes such as continuous-wave (CW) or frequency-shift keying (FSK), or for frequency-modulated (FM) signals. The only precaution involves making sure that key clicks are not introduced onto a CW signal if class-C amplification is used. Many FM transceivers, designed for use at 144 MHz and above, use nonlinear, class-C final amplifiers. *See* CLASS-C AMPLIFIER, KEY CLICK, and LINEAR AMPLIFIER.

Considerations at HF Power transistors are used at moderate output levels nowadays at high frequencies (HF): the bands up through 29.7 MHz. Tubes are rarely seen in new equipment in this frequency range, except at the kilowatt-plus level. Most circuits operate in class-AB or class-B so that they have linear characteristics. In some CW-only transmitters, especially low-power (QRP) transistorized ones, class-C final amplifiers are used. Sometimes, class-B push-pull circuits are employed. *See* CLASS-AB AMPLIFIER, CLASS-B AMPLIFIER, and PUSH-PULL AMPLIFIER.

Broadband versus Tuned Amplifiers In recent years, power amplifiers have been designed that require no tuning on the part of the operator. In the past, tuned-output amplifiers were the norm. Broadband amplifiers have certain advantages over tuned types. In the tuned amplifier, unless the collector or plate current is "dipped" by tuning the output capacitor, the efficiency is very low, and most of the power is dissipated by the collector or plate of the output device. This can destroy the transistor(s) or tube(s). The broadband amplifier is easier to use. Changing frequencies or bands requires that the operator simply adjust the tuning dial, and nothing else. Broadband amplifiers are less likely to oscillate than tuned amplifiers, although neutralization, and the use of parasitic suppressors, are advisable for any power amplifier at any frequency. *See* NEUTRALIZATION, PARASITIC OSCILLATION, and PARASITIC SUPPRESSOR.

There are some advantages to tuned amplifiers. Broadband power amplifiers offer no attenuation of spurious signals or harmonics within the entire design range of the transmitter. Tuned amplifiers are generally somewhat more efficient than broadband types.

Considerations at VHF, UHF and Microwave Frequencies Obtaining high RF power becomes progressively more difficult as the frequency increases, particularly above about 200 MHz. Vacuum tubes are far easier to work with than transistors for full-legal-limit amplifiers at these frequencies. Spurious oscillation can also be a problem; a VHF, UHF, or microwave power amplifier might generate signals at HF, unless precautions are taken to prevent it. As at HF, the use of neutralization and parasitic suppressors is good engineering practice.

As digital modes become increasingly common, class-C amplifiers will be used more often at VHF and above. These circuits are easier to design, especially at VHF and above, than linear amplifiers.

The tuned circuits at frequencies well into UHF, and at microwavelengths, take the form of strips of transmission line, rather than coils and capacitors. These strips can be etched right onto a printed circuit. Alternatively, cavity resonators can be used. *See* CAVITY RESONATOR, and STRIPLINE.

POWER BUDGET

In a communications satellite, such as OSCAR, the power generated by the solar panels varies. It sometimes exceeds that available from the rechargeable battery, and it sometimes is less than that available from the battery. This balance, known as the *power budget*, might be either positive or negative.

A *positive power budget* means that the solar panels are generating more power than the battery, so the battery can "rest." This occurs when the panels are getting maximum sunlight. A *negative power budget* means that the solar panels are generating less power than the battery, so current is drawn from the battery. This occurs when the solar panels are at a bad angle, with respect to the sun, or when the satellite is on the dark side of the earth.

When the power budget of a satellite is negative for a long period of time, such as when apogee occurs on the dark side of the earth, shutdown is sometimes required so that the battery does not completely discharge. *See also* ACTIVE COMMUNICATIONS SATELLITE, and OSCAR.

POWER DENSITY

Power density is one means of expressing the strength of an electromagnetic field. Power density is expressed in watts per square meter or watts per square centimeter, as determined in a plane oriented parallel to both the electric and magnetic lines of flux.

In the vicinity of a source of electromagnetic radiation, the power density decreases according to the law of inverse squares. *See also* FIELD STRENGTH, and INVERSE-SQUARE LAW.

POWER FACTOR

In an alternating-current circuit containing reactance and resistance, not all of the apparent power is true power. Some of the power occurs in reactive form (*see* APPARENT POWER, POWER, REACTIVE POWER, and TRUE POWER). The ratio of the true power to the apparent power is known as the *power factor*. Its value can range from 0 to 1.

Power factor can be determined according to the phase angle in a circuit. The phase angle is the difference, in degrees, between the current and the voltage in an alternating-current circuit. If the phase angle is represented by ϕ, then the power factor, PF, is:

$$PF = \cos \phi$$

The power factor, as a percentage, is:

$$PF = 100 \cos \phi$$

The power factor in an ac circuit can also be determined by plotting the impedance on the complex plane. If the resistance is R ohms and the reactance is X ohms, the phase angle is:

$$\phi = \arctan (X/R)$$

and thus the power factor is:

$$PF = \cos (\arctan (X/R))$$

As a percentage:

$$PF = 100 \cos (\arctan (X/R))$$

See also IMPEDANCE, and PHASE ANGLE.

POWER GAIN

Power gain is an increase in signal power between one point and another. Power gain is used as a specification for power amplifiers. If the input to a power amplifier, in watts, is given by P_{IN} and the output, in watts, is given by P_{OUT}, then the power gain in decibels is:

$$G = 10 \log_{10} (P_{OUT}/P_{IN})$$

See also DECIBEL, and GAIN.

POWER LINE

A utility line, carrying electricity for public use, is called a *power line* or a *power-transmission line*. The small line, usually carrying 234 volts, that runs from a house to a nearby utility pole, is a power line. The high-tension wires that carry several hundred thousand volts are also power lines.

The term *power line* is technically imprecise because power does not actually travel from place to place. It is the electrons in the conductors, and the electromagnetic field, that travel. Power is dissipated, or used, at the terminating points of the power line (see POWER). Some power is dissipated in the line because of losses in the conductors and insulating materials. This dissipation is, of course, undesirable because power lost in the line represents power that cannot be used at the terminating points.

POWER-LINE NOISE

Power lines, in addition to carrying the 60-Hz alternating current that they are intended to transmit, also carry other currents. These currents have a broadband nature. They result in an effect called *power-line noise.*

The currents usually occur because of electric arcing at some point in the circuit. The arcing can originate in appliances connected to the terminating points; it can occur in faulty or obstructed transformers; it can even occur in high-tension lines as a corona discharge into humid air. The broadband currents cause an electromagnetic field to be radiated from the power line because they flow in the same direction along all of the line conductors.

Power-line noise sounds like a buzz or hiss when it is picked up by a radio receiver. Some types of power-line noise can be greatly attenuated by means of a noise blanker. For other kinds of noise, a limiter can be used to improve the reception. Phasing can sometimes be used in an antenna system to reduce the level of received power-line noise. *See also* NOISE LIMITER, and NOISE-REDUCING ANTENNA.

POWER LOSS

Power loss occurs in the form of dissipation, usually resulting in the generation of heat. Power loss occurs to a certain extent in all electrical conductors and dielectric materials. Power loss can also occur in the cores of transformers and inductors.

The power loss in a given circuit is usually specified in decibels. If the input power is represented by P_{IN} and the output power by P_{OUT}, then the loss L, in decibels, is:

$$L = 10 \log_{10} (P_{IN}/P_{OUT})$$

A power loss is equivalent to a negative power gain. That is, if the loss (according to the previous formula) is x dB, then the gain is $-x$ dB. *See* DECIBEL, and LOSS.

In a power-transmission line, loss occurs because of the finite resistance of the conductors, and also to some extent because of losses in the insulating and/or dielectric materials between the conductors. Power loss in a transmission line is sometimes given in watts, as the difference between the available power at the line input and the actual power dissipated in the load.

POWER MEASUREMENT

The power used by a device, or radiated by an antenna, can be measured in various different ways. The most direct method is to use a wattmeter. A watt-hour meter can also be used; the device is allowed to run for a certain time, and the meter reading is then divided by that period of time (see WATTMETER, and WATT-HOUR METER).

Direct-current power can be measured by means of a voltmeter and ammeter. The voltmeter is connected in parallel with the power-dissipating device, and the ammeter is connected in series. The power in watts is the product of the voltage across the device, in volts, and the current through the device, in amperes.

In certain situations, power can be measured by determining the amount of heat that is liberated. This method works only for devices that dissipate all of their power as heat, such as resistors. The total heat energy can be determined, over a period of time, and the dissipated power then obtained by dividing the total heat energy by the time period.

POWER SENSITIVITY

Certain amplifier circuits, notably class-A circuits using vacuum tubes or field-effect transistors, consume essentially no power while producing measurable output power (see CLASS-A AMPLIFIER). Such amplifiers theoretically draw no current, and need only an alternating voltage at the grid or gate in order to function. The performance of such an amplifier is sometimes expressed by a parameter called *power sensitivity.*

If the root-mean-square (RMS) input-signal voltage is E, and the RMS output power is P, then the power sensitivity S is given by the ratio:

$$S = P/E$$

Power sensitivity is expressed in watts per volt.

The power sensitivity of a class-A amplifier does not vary with the driving voltage, as long as the amplifier is correctly biased and is not overdriven. If the amplifier is driven into the nonlinear portion of the characteristic curve, the power sensitivity decreases.

POWER SUPPLY

A *power supply* is a circuit that provides a transmitter, receiver, computer or other device with the proper voltages and currents. Practically all active electronic devices require direct current (dc). Solid-state equipment commonly operates from 6 V dc to 12 V dc; tubes use anywhere from about 100 V dc to 4 kV dc in amateur equipment.

Batteries and Generators Sometimes the power supply for part or all of a ham station is a battery. This is often the case for portable and mobile equipment. Sometimes solar cells supplement a power supply. *See* BATTERY, SOLAR CELL, and SOLAR POWER. A gasoline-powered generator can be used as the main power supply for an entire station; secondary supplies convert the alternating current (ac) from the generator to dc for each particular device. *See* GENERATOR.

The Power Transformer The *power transformer* converts the utility ac to the proper voltage for rectification and filtering. This transformer might be either step-up (for high voltages) or step-down (for low-voltage supplies). *See* POWER TRANSFORMER.

The Rectifier In modern ham equipment, rectification is always done by means of semiconductor diodes. These are available in various voltage, peak-inverse-voltage, and current ratings. Older equipment, using rectifier tubes, is still occasionally used; such are regarded as antiques. *See* PEAK INVERSE VOLTAGE, and RECTIFIER DIODE.

The diodes convert ac to pulsating dc by letting current pass in only one direction. There are several different configurations for connecting the diodes to get pulsating dc output. *See* BRIDGE RECTIFIER, FULL-WAVE RECTIFIER, FULL-WAVE VOLTAGE DOUBLER, HALF-WAVE RECTIFIER, and VOLTAGE MULTIPLIER.

The Filter The pulsations, or ripple, in the dc output of a rectifier must usually be smoothed out for proper operation of electronic equipment. Two methods are used to accomplish this: large capacitances in parallel, and large inductances in series, along with capacitances in parallel. This part of a power supply is called the *filter*.

In high-voltage supplies, the filter capacitor(s) can hold the charge for some time after the power has been switched off. *Bleeder resistors* across the capacitors speed up the discharge process, minimizing the danger to personnel who service the equipment. *See also* BLEEDER RESISTOR, and FILTER CAPACITOR.

Filters with both inductors (chokes) and capacitors can have the series choke ahead of the capacitor; this is called a *choke-input filter*. If the capacitor comes first, it is called a *capacitor-input filter*. Low-voltage supplies often use one or two large-value electrolytic capacitors as the filter, with no chokes.

Regulation *Regulation* in low-voltage supplies is often done by *Zener diodes*. These limit the voltage that can exist across them (*see* ZENER DIODE). Regulation in a low-voltage supply

can also be done using a series pass transistor that "bucks" the supply if its output gets too large. Integrated-circuit regulators are also available.

In high-voltage supplies, Zener diodes can be used in series. Sometimes gas-filled regulator tubes are used instead. Special transformers are also available. Many vacuum tubes will work all right with unregulated power supplies. *See* REGULATED POWER SUPPLY, REGULATING TRANSFORMER, and VOLTAGE REGULATION.

Current Limiting If the load draws too much current from a power supply, the regulator can be ruined, causing excessive output voltage and damaging equipment connected to the supply. This is prevented by *current limiting*. A resistor-transistor circuit is connected so that, if the load tries to draw too much current, the ability of the supply to provide that current is reduced. A transistor in series is partly cut off by the excessive current, introducing resistance in series with the supply output.

Some current limiters keep the current from exceeding a certain maximum. Others, in the event of a load that tries to draw excessive current, will reduce the available output current of the supply. This is called *foldback*.

Some power supplies shut down if a load tries to draw excessive current. Fuses and circuit breakers can also be used as an extra safety measure. *See* CURRENT LIMITING.

Overvoltage Protection If the regulating circuitry in a power supply fails, the output voltage can more than double, and cause damage to equipment that is connected to the supply. This is prevented by a circuit that shuts down the supply in the event of regulator failure.

One way to do accomplish *overvoltage protection* is to use a sensing device to trip a breaker if excessive voltage appears at the supply output. The supply will then shut down, and will not function until it is repaired.

Another way is by means of a Zener diode, or set of Zener diodes, rated at the normal output voltage of the supply. This acts as a backup regulator. However, if this method is used, there must be a way for the operator to know that the main regulator has failed so that it can be repaired.

Circuit Breakers and Fuses If a short circuit occurs at the output of a power supply, or in the event of malfunction in a major component within the supply, the circuit should shut down completely. This is ensured by means of a *circuit breaker* or *fuse* in series with the ac line to the supply. *See* CIRCUIT BREAKER, and FUSE.

Personal Safety Low-voltage power supplies for use with solid-state equipment (usually 6 V or 12 V) present a danger only before the power transformer, where utility ac exists. But some solid-state devices require moderately high voltages. Anything over about 25 V should be considered potentially hazardous.

High-voltage supplies should *never* be serviced when the power is on. Before opening the supply cabinet, switch it off, unplug it from the utility mains, and wait several minutes. Once the supply cabinet is opened, precautions *must* be taken to ensure that the filter capacitors have fully discharged before any work is done. Never rely on the bleeder resistors for this. Always short-circuit all filter capacitor terminals to the chassis, using a tool with a massive, insulated handle (such as a large screwdriver), before touching any of the components in a power supply whose output is more than 25 V. It is wise to wear heavy, insulted gloves whenever possible, as work is done on a high-voltage power supply.

If you have the slightest hesitancy about the safety of high-voltage equipment, it should not be used until a qualified engineer or technician has examined it. While hams should be encouraged to acquire some technical skill, there is no good justification for undertaking potentially dangerous work. Power supplies need not be a hazard if they are treated with respect.

POWER TRANSFORMER

A *power transformer* is a device that is designed especially for changing the voltage of alternating-current electricity. Power transformers can be found in all power supplies designed to be used with household current. Such transformers can be very small, supplying a relatively low voltage, or they can be large, supplying hundreds or even thousands of volts. Massive power transformers are used in the distribution of alternating-current power for utility purposes. Such transformers often must deal with many thousands of volts, and large amounts of current (*see* TRANSFORMER).

The basic construction of a typical power transformer is shown in the illustration. Two coils of wire are wound around a ferromagnetic core. The coil receiving the input current is called the *primary*, and the coil from which the output is taken is called the *secondary*. The core of a power transformer is usually made from laminated iron (*see* LAMINATED CORE, TRANSFORMER PRIMARY, and TRANSFORMER SECONDARY).

The alternating current in the primary winding generates a fluctuating magnetic field in the core. The magnetic field causes alternating currents to flow in the secondary winding. The voltage across the secondary winding can be larger or smaller than the voltage applied to the primary, depending on the primary-to-secondary turns ratio.

Assume that a given power transformer is 100-percent efficient and has P turns in the primary and S turns in the secondary. Also, suppose that x volts are applied to the primary winding. Then, voltage y, appearing across the secondary winding, is:

$$y = (S/P)x$$

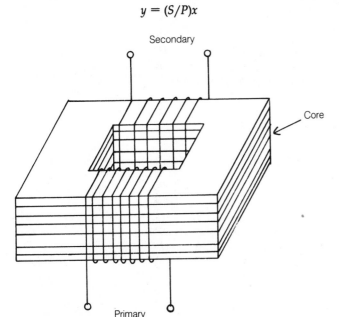

POWER TRANSFORMER: A simplified pictorial diagram.

If the same transformer is connected to a circuit that draws x amperes from the secondary winding, the current y flowing in the primary can be determined according to the previous formula.

Of course, no transformer is completely efficient. This results in slightly lower secondary-winding voltages than the above formula would indicate; it also necessitates that the primary-winding current be greater than the above formula dictates. *See also* TRANSFORMER EFFICIENCY.

POWER TRANSISTOR

In ham radio work, transmitting amplifiers are increasingly making use of transistors. It is common nowadays to see a 200-W transmitter that uses all transistors, even in the final amplifier circuit. A *power transistor* is, technically, any transistor that can deliver at least 1 W continuous output. Practically all ham transmitters produce this much radio-frequency (RF) power output, or more. Power transistors are available in two main forms: bipolar and field-effect (FET).

Bipolar Types   *Bipolar power transistors* work in the same way as their smaller bipolar cousins. The main difference is that the power transistor is physically larger, and must be used with a heatsink (*see* HEATSINK).

Bipolar power transistors typically operate from 12 V or so. But the power units draw more current than ordinary bipolar transistors. If an amplifier has 100 W output at 50-percent efficiency, its collector will draw 16.7 A from a 12-V supply. Efficiency of 50 percent is typical for a linear power amplifier using a bipolar transistor in class AB or class B. Power transistors also work well in class B push pull.

Transmitters that use only CW, FM, or FSK, and whose final amplifiers therefore need not be linear, can use bipolar power transistors in class C at an efficiency of about 65 percent. *See* CLASS-AB AMPLIFIER, CLASS-B AMPLIFIER, CLASS-C AMPLIFIER, and PUSH-PULL AMPLIFIER.

Amplifiers can be built that deliver the full legal power limit, using only bipolar power transistors. But design can be tricky, and the large current requires a heavy-duty, expensive power supply. As technology improves, bipolar power amplifiers at 1.5 kW will become more cost-effective, and therefore more common.

Bipolar power transistors, especially combinations of several in parallel, have very low input and output impedances. This makes it necessary to use transformers to provide a match from the driver, and to a 50-ohm antenna system.

FET Types   The other major type of power transistor is a *metal-oxide-semiconductor field-effect transistor (MOSFET)*. The most common power MOSFET is called a *double-diffused (DMOSFET)*. It gets this name from the way it is manufactured.

Power FETs have higher input impedance than bipolar power transistors. This is an advantage in some applications. Power MOSFETs can also switch faster. Efficiency of FET amplifiers is often superior at VHF this reason. The circuit design differs from that of a bipolar circuit, in the same ways as low-power FET circuit design differs from low-power bipolar circuit design. These differences can be seen in the schematics that accompany the articles CLASS-A AMPLIFIER, CLASS-AB AMPLIFIER, CLASS-B AMPLIFIER, and CLASS-C AMPLIFIER.

Technology in this field is advancing rapidly, and improvements can be expected in power MOSFET devices over the next few years. Presently, power amplifiers using MOSFET devices are rather costly, especially at higher power levels. *See also* METAL-OXIDE-SEMICONDUCTOR FIELD-EFFECT TRANSISTOR.

POWER TUBE

A *power tube* is a vacuum tube designed for use in power amplifiers. A power tube can be a triode device, a tetrode, or a pentode. Power tubes have large plates because the plate is where most of the dissipation occurs. The plate of a power tube can be finned, thus maximizing the heat-radiating surface area. A power tube having a plate-dissipation rating of more than a few watts is usually cooled by means of a fan or heatsink (*see* AIR COOLING, COOLING, HEATSINK, and PLATE DISSIPATION).

Power tubes can be used at frequencies from direct current well into the microwave range. At the low, medium, and high frequencies, single tubes can deliver several thousand watts of useful radio-frequency output power. The biggest power tubes are as large as a man, are housed in room-sized enclosures, and are cooled by elaborate air or fluid circulation devices.

Power tubes typically require rather high plate voltages, on the order of several kilovolts in some cases. For this reason, extreme precautions must be taken when servicing equipment that contains power tubes. Lethal voltages are present. The power supply should be switched off before any work is done; it should be made certain that the filter capacitors have completely discharged.

Power tubes are usually operated in class AB, class B, or class C. The push-pull configuration is sometimes used. *See also* CLASS-AB AMPLIFIER, CLASS-B AMPLIFIER, CLASS-C AMPLIFIER, and PUSH-PULL AMPLIFIER.

POYNTING VECTOR

In an electromagnetic field, the electric lines of flux are perpendicular to the magentic lines of flux at every point in space (*see* ELECTROMAGNETIC FIELD). The electric field and the magnetic field have magnitude, as well as direction, and thus they can be represented as vector quantities. Generally, the electric-field vector at a given point in space is called E, and the magnetic-field vector is called H.

The cross product of the E and H vectors results in a vector perpendicular to both, and having length equal to the product of the lengths of E and H.

The vector $E \times H$ is called the *Poynting vector* for an electromagnetic field. The Poynting vector is significant because its direction indicates the direction of propagation of the field, and its length indicates the intensity, or field strength, in terms of power per unit area in the surface containing the electric and magnetic vectors. *See also* ELECTRIC FIELD, ELECTRIC FLUX, FIELD STRENGTH, MAGNETIC FIELD, and MAGNETIC FLUX.

PREAMPLIFIER

A *preamplifier* is a high-gain, low-noise amplifier, that is intended to boost the amplitude of a weak signal. In radio-frequency receivers, preamplifiers are sometimes used to improve the sensitivity. In audio applications, the amplifier immediately following the pickup or microphone is called the *preamplifier*.

Some modern radio-frequency preamplifiers use field-effect transistors. This type of device draws almost no power, and has a high input impedance that is ideally suited to weak-signal work. The gallium-arsenide field-effect transistor, or GaAsFET, is often used (*see* FIELD-EFFECT TRANSISTOR, and GALLIUM-ARSENIDE FIELD-EFFECT TRANSISTOR). For audio-frequency work, either bipolar or field-effect transistors can provide good results. Field-effect transistors are preferable if noise reduction is a major consideration.

The illustration at A shows a typical circuit suitable for use at radio frequencies. The amplifier operates in class A. Input tuning is used to reduce the noise pickup and provide some selectivity against signals on unwanted channels. Impedance-matching networks are employed both at the input and at the output, to optimize the transfer of signal through the circuit and to the receiver. This preamplifier will produce about 10 dB to 15 dB gain, depending on the frequency.

A preamplifier suitable for use at audio frequencies is shown at B. This circuit has a broadband response because it is intended for high-fidelity work. The gain is essentially constant, at about 15 dB to 20 dB, up to frequencies well beyond the range of human hearing.

In the design of any preamplifier, regardless of the intended frequencies of operation, it is important that linearity be as good as possible. Nonlinearity in a preamplifier results in intermodulation distortion at radio frequencies and harmonic distortion at audio frequencies. *See also* CLASS-A AMPLIFIER, INTERMODULATION, LINEARITY, and NONLINEARITY.

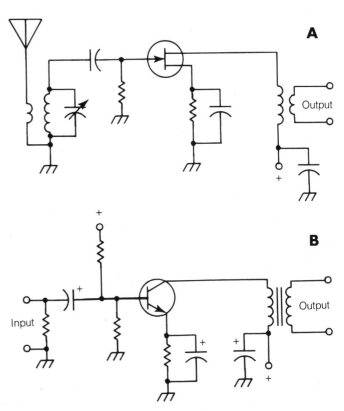

PREAMPLIFIER: At A, an RF preamp using a JFET. At B, an audio preamp using a bipolar transistor.

PRECESSION

The earth's axis is tilted by 23.5 degrees, with respect to the plane of its orbit around the sun. Right now, the axis is aligned so that the star Polaris is almost exactly over the geographic North Pole. But over a period of thousands of years, the orientation of the earth's axis changes, with respect to the heavens, although its tilt is always 23.5 degrees, relative to the plane of its orbit around the sun. This slow "wobbling," similar to the wobbling of a spinning top, is called *precession*.

Precession occurs not only with the rotation of the earth (as well as other planets), but with orbits too. The apogee of any earth-orbiting satellite gradually revolves around the earth. This affects the usefulness of a satellite over time, if the satellite is of the Phase III type, with an elongated orbit.

Phase III satellites are launched into orbits designed so that their apogees occur over those parts of the earth most desiring to use the satellite. The satellites work best when they are at apogee because they move slowly across the sky then, and it's easy to keep the ground-station antennas aimed.

Over a period of years, precession causes the apogee of any Phase III satellite to change position. This means that, eventually, hams once able to use the satellite easily will find it less and less convenient. *See also* ACTIVE COMMUNICATIONS SATELLITE, APOGEE, OSCAR, and PHASE III SATELLITE.

PRECIPITATION STATIC

Precipitation static is a form of radio interference that is caused by electrically charged water droplets or ice crystals as they strike objects. The resulting discharge produces wideband noise that sounds similar to the noise generated by electric motors, fluorescent lights, or other appliances.

Precipitation static is often observed in aircraft flying through clouds containing rain, snow, or sleet. But occasionally, precipitation static occurs in base or mobile radio installations. This is especially likely to happen when it is snowing; then the noise is called *snow static*. Dust storms can also cause precipitation static. Precipitation static can be quite severe at times, making radio reception difficult, especially at the very low, low, and medium frequencies.

A noise blanker or limiter is effective in reducing interference caused by precipitation static. A means of facilitating discharge, such as an inductor between the antenna and ground, can be helpful. Improvement can also be obtained by blunting any sharp points in antenna elements. *See also* NOISE LIMITER.

PRECISION POTENTIOMETER

A *precision potentiometer* is a variable resistor that has very accurate calibration and ease of resettability. Precision potentiometers are usually constructed using wirewound elements, in a manner similar to a rheostat (*see* RHEOSTAT, and WIREWOUND RESISTOR). Because of this, such devices are not suitable for use in situations where they must carry radio-frequency currents. The windings exhibit significant inductive reactance, as well as resistance, at radio frequencies.

Some precision potentiometers have multiturn shafts. This makes it possible to set them almost exactly to any desired value within the resistance range.

PREEMPHASIS

Preemphasis is the deliberate accentuation of the higher audio frequencies in a frequency-modulated transmitter. Preemphasis is obtained in a transmitter by inserting a highpass filter in the audio stages.

In a phase-modulated transmitter, the preemphasis occurs automatically, because of the nature of phase modulation as compared with frequency modulation (*see* FREQUENCY MODULATION, and PHASE MODULATION).

Preemphasis in a frequency-modulated transmitter, in conjunction with deemphasis at the receiver, improves the signal-to-noise ratio at the upper end of the audio range. *See also* DEEMPHASIS.

PREFIX MULTIPLIERS

Physical units are often subdivided or multiplied for purposes of mathematical convenience. Special prefixes, known as prefix multipliers, indicate the constant factor by which the unit is changed.

Generally, prefix multipliers are given in increments of three, such as 10^3, 10^6, and so on for large units, and 10^{-3}, 10^{-6}, and so on for small quantities. However, intermediate values are sometimes given.

As exponents become more and more negative, the unit size diminishes. As the exponents become more and more positive, the unit size increases. Each exponential increment represents one order of magnitude (a factor of 10).

THE MOST COMMONLY USED PREFIX MULTIPLIERS.

Prefix	Abbreviation	Multiple
atto-	a	10^{-18}
femto-	f	10^{-15}
pico-	p	10^{-12}
nano-	n	10^{-9}
micro-	μ or u	10^{-6}
milli-	m	10^{-3}
centi-	c	10^{-2}
deci-	d	10^{-1}
deca-	D or da	10
hecto-	h	10^{2}
kilo-	k	10^{3}
mega-	M	10^{6}
giga-	G	10^{9}
tera-	T	10^{12}
peta-	P	10^{15}
exa-	E	10^{18}

PRESCALER

In order for a *frequency counter* to measure very high, ultrahigh, or microwave frequencies, the input frequency is usually divided down by some large factor (such as 1000, 10,000, 100,000, or more). This makes it possible for the counter circuit to measure the frequency. The counter display takes the divider factor into account. The circuit arrangement that does this is called a *prescaler*. *See also* FREQUENCY COUNTER, and FREQUENCY DIVIDER.

PRESELECTOR

A tuned circuit in the front end of a radio receiver is called a *preselector*. Most receivers incorporate preselectors. The preselector provides a bandpass response that improves the signal-to-noise ratio, and also reduces the likelihood of receiver overloading by a strong signal far removed from the operating frequency. The preselector also provides a high degree of image rejection in a superheterodyne circuit. Most preselectors have a 3-dB bandwidth that is a few percent of the received frequency.

A preselector can be tuned by means of tracking with the tuning dial. This eliminates the need for continual readjustment of the preselector control, thereby making the receiver more convenient to operate. The tracking type of preselector, however, requires careful design and alignment.

Some receivers incorporate preselectors that must be adjusted independently. Although this is less convenient from an operating standpoint, the control need be reset only when a large frequency change is made. The independent preselector simplifies receiver design. *See also* FRONT END.

PRESENTATION LAYER

See OPEN SYSTEMS INTERCONNECTION REFERENCE MODEL.

PRIMARY EMISSION

When electrons emanate directly from an object, such as an electron gun or the cathode of a vacuum tube, the emission is said to be *primary*. *Primary emission* results from the application of a negative charge to an electrode. The greater the applied charge, assuming all other parameters remain constant, the greater is the primary emission.

PRINTED CIRCUIT

A *printed circuit* is a wiring arrangement that is fabricated by means of foil runs on a circuit board. Printed circuits can be mass-produced inexpensively and efficiently. Printed circuits allow extreme miniaturization and high reliability. Most electronic devices today are built using printed-circuit technology, although high-power circuits still use point-to-point wiring methods. The photograph shows a typical printed circuit.

Printed circuits are fabricated by first drawing an etching pattern. This pattern is then photographed and reproduced on clear plastic. The plastic is placed over a copper-coated glass-epoxy or phenolic board, and the assembly undergoes a photochemical process (*see* ETCHING). This can be done on one side of the board only, resulting in what is called a *single-sided printed circuit*. If the circuit is too complicated to be drawn on a single plane surface without crossing foil runs, the etching is done on both sides of the board. Then, the board is called a *double-sided printed circuit*.

The use of printed circuits has vastly enhanced the ease with which electronic equipment can be serviced. Printed circuits allow modular construction, so that an entire board can be replaced in the field and repaired in a fully equipped laboratory. *See also* CARD, and MODULAR CONSTRUCTION.

PRINTED CIRCUIT: A modern printed-circuit board.

PRINTER

A *printer* is any device that produces a permanent copy of alphanumeric data. Printers are widely used in conjunction with computers, terminal units, word processors, and many other electronic systems.

Most modern printers are of either the daisy-wheel type or the dot-matrix type (*see* DAISY-WHEEL PRINTER, DOT MATRIX PRINTER, and LASER PRINTER). The dot-matrix printer can produce hard copy at an extremely rapid rate; typical speeds are 60 characters per second or more. The daisy-wheel printer is somewhat slower, but produces letter-quality copy.

Older-style printers include the ball type, such as is used in many electric typewriters, and the key-strike type, used in most mechanical typewriters. These types of printers are slower than the daisy-wheel and dot-matrix printers.

PRIVATE LINE ®

Private Line ® is a trademark of Motorola, Inc. The term is commonly used to describe a radiotelephone tone-squelch system. Private-Line (PL) systems are sometimes used in repeaters to limit the access to certain persons. A PL system can also be used to keep a receiver from responding to unnecessary or unwanted signals. For the receiver squelch to open in the presence of a signal, that signal must be modulated with a tone that has a certain frequency. Tone-squelch systems are most often used with frequency modulation. Private-line communications is used by radio amateurs in certain parts of America, as well as other countries.

PRIVATE LINE®: SUBAUDIBLE-TONE FREQUENCIES.

Designation	Freq., Hz.	Designation	Freq., Hz.
XZ	67.0	2B	118.8
XA	71.9	3Z	123.0
WA	74.4	3A	127.3
XB	77.0	3B	131.8
SP	79.7	4Z	136.5
YZ	82.5	4A	141.3
YA	85.4	4B	146.2
YB	88.5	5Z	151.4
ZZ	91.5	5A	156.7
ZA	94.8	5B	162.2
ZB	97.4	6Z	167.9
1Z	100.0	6A	173.8
1A	103.5	6B	179.9
1B	107.2	7Z	186.2
2Z	110.9	7A	192.8
2A	114.8	MI	203.5

Table 1

PRIVATE LINE: TONE-SQUELCH BURST FREQUENCIES.

Freq., Hz.	Freq., Hz.
1600	2150
1650	2200
1700	2250
1750	2300
1800	2350
1850	2400
1900	2450
1950	2500
2000	2550
2100	

Table 2

In the United States, most PL systems utilize one of 32 standard subaudible-tone frequencies (see Table 1). In some systems, nonstandard tones are used that do not appear in the table. Each tone is designated by a pair of letters.

In a frequency-modulation PL system, the carrier is usually continuously modulated with the deviation of approximately about 0.5 to 1.5 kHz by a sine-wave tone. The tone cannot be heard on the signal as it is received because the frequency is below the response cutoff of the voice-amplifier chain.

In European countries, and also in some U.S. communications systems, tone bursts of higher frequency are often used for tone-squelch purposes. The most common burst-tone frequencies appear in Table 2. The burst can consist of just one tone, or it can consist of several tones in sequence, each having a different frequency.

Burst tones must have a certain frequency, duration, and deviation. Tone-burst systems are used by some police departments and business communications systems in the United States; in recent years, tone-burst squelch has gained popularity among radio amateurs as well.

PROBE

A *probe* is a device intended for picking off a signal from a specific point in a circuit for test purposes. The probe normally does not disturb the operation of the circuit.

Probes can be used to measure direct-current voltages. Probes are used with ohmmeters to determine the resistance between two points in a circuit. Probes can also be used with oscilloscopes to determine the voltage, waveform characteristics, frequency, and phase of an alternating-current signal. Oscilloscope probes are also extensively used in the testing and troubleshooting of digital equipment.

Most probes have extremely high impedance. An insulated handle prevents shock to personnel and reduces the effects of hand capacitance. Probe cables should be as short as possible, and should be shielded to prevent the pickup of hum or radio-frequency interference.

PRODUCT DETECTOR

Single-sideband (SSB), continuous-wave (code or CW), and frequency-shift-keyed (FSK) signals require a special form of detector for reception. Such a detector must supply an unmodulated local signal, and must provide for some means of mixing this signal with the incoming signal. This type of detector is called a *product detector*. The product detector works according to the same principles as the heterodyne detector. Generally, the term *heterodyne detector* is used for direct-conversion reception, and the term *product detector* is used with superheterodyne reception, although the terms can sometimes be used interchangeably (*see* DIRECT-CONVERSION RECEIVER, HETERODYNE DETECTOR, and SUPERHETERODYNE RECEIVER).

Product detectors can be either passive or active. The illustration shows schematic diagrams of a passive product detector (A) and an active product detector (B). These circuits are only examples of product detectors; many different configurations are possible.

In single-sideband reception, the local oscillator must be set to precisely the same frequency as the suppressed carrier of the incoming intermediate-frequency signal. This requires that the local oscillator be very stable, for otherwise, severe distortion

will result. In code and FSK reception, the local oscillator is set to a slightly different frequency than the incoming intermediate-frequency signal. The difference is usually between 300 Hz and 3 kHz, creating an audible beat note or notes.

Product detectors can be used to receive ordinary amplitude-modulated signals, but an annoying beat note or flutter usually occurs as the local-oscillator signal competes with the unsuppressed carrier. For amplitude modulation, the envelope detector gives much better results than the product detector (*see* ENVELOPE DETECTOR).

A product detector can be used to receive narrow-band frequency-modulated signals, but the same problems are encountered in this case, as with amplitude modulation. For frequency modulation, the discriminator, phase-locked loop, or ratio detector are the most preferable circuits (*see* DISCRIMINATOR, PHASE-LOCKED LOOP, and RATIO DETECTOR).

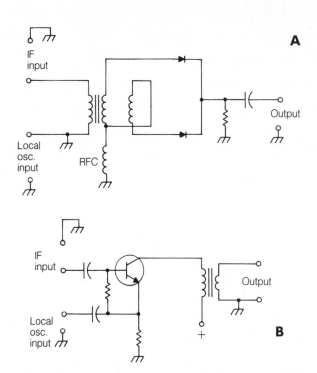

PRODUCT DETECTOR: At A, passive; at B, active.

PROGRAM

A *program* is a set of data bits, comprising a unit of information and/or instructions to perform a specific task. In computer practice, a program tells the device what to do. One program, called the *assembler*, is used to derive the machine language of the computer from the assembly language (*see* HIGHER-ORDER LANGUAGE, and MACHINE LANGUAGE). Another program, the compiler, converts higher-order language to machine language.

For a given higher-order language, the assembler and compiler do not change until a different higher-order language is used. The higher-order program can be changed at will by the operator (*see* COMPUTER PROGRAMMING).

A radio or television broadcast, intended for reception by the public and conveying a message, is called a *program*. Examples of such a program include the news/weather/sports broadcast, a movie broadcast, and a single episode in a continuing series.

PROGRAMMABLE READ-ONLY MEMORY

A *programmable read-only memory (PROM)* is a form of read-only memory (ROM) used in computers. The information in the PROM is stored permanently. Data is not lost if power is removed from the unit. In its original form, the PROM integrated circuit is filled with logic blanks. These can be either zeroes or ones; it makes no difference. A device called a *programmer* allows any desired information to be written into the PROM, provided the data does not exceed the storage capacity of the integrated circuit.

Some PROM devices can be erased and reprogrammed with a different set of data. Such a device is called an EPROM, for erasable PROM. *See also* READ-ONLY MEMORY.

PROGRAMMABLE UNIJUNCTION TRANSISTOR

A *programmable unijunction transistor (PUT)* is a form of unijunction transistor with a variable triggering voltage. The triggering voltage is set by means of an external voltage divider. Such a divider can consist of resistors, or it can consist of independent sources.

The PUT is similar to the silicon-controlled rectifier. The PUT, silicon-controlled rectifier, and conventional unijunction transistor are all used in similar applications. *See also* SILICON-CONTROLLED RECTIFIER, and UNIJUNCTION TRANSISTOR.

PROGRAMMING

See COMPUTER PROGRAMMING, and PROGRAM.

PROM

See PROGRAMMABLE READ-ONLY MEMORY.

PROPAGATION

Propagation is the transfer of energy through a medium or through space. Certain disturbances, such as sound waves or electric currents, can propagate only through a material substance. Electromagnetic fields can propagate through empty space (*see* ELECTRIC FIELD, ELECTROMAGNETIC FIELD, and MAGNETIC FIELD).

Electromagnetic-wave propagation occurs in straight lines through a perfect vacuum in the absence of intervening forces or effects. In and around the atmosphere of the earth, however, electromagnetic waves are often propagated in bent paths. The science of radio-wave propagation is sophisticated. The following several articles describe the most important aspects of wave propagation at radio frequencies.

PROPAGATION CHARACTERISTICS

Radio waves are affected in a variety of ways as they travel through the atmosphere of the earth. The effects vary with the wavelength. The troposphere, or lowest part of the atmosphere, causes electromagnetic waves to be bent or scattered at some frequencies. The ionized layers (*see* IONOSPHERE) affect radio waves at some frequencies.

A general description of propagation characteristics, for various electromagnetic frequencies, follows.

Very Low Frequencies (Below 30 kHz) At very low frequencies, propagation takes place mainly via waveguide effect between the earth and the ionosphere. Surface-wave propagation also occurs for considerable distances. With high-power transmitters and large antennas, communication can be realized over distances of several thousand miles at frequencies between 10 kHz and 30 kHz. The earth-ionosphere waveguide has a lower cutoff frequency slightly below 10 kHz. For this reason, signals much below 10 kHz suffer severe attenuation and do not propagate well.

Antennas for very-low-frequency transmitting must be vertically polarized. Otherwise, surface-wave propagation will not take place, and the proximity of the ground tends to short-circuit the electromagnetic field.

Propagation at very low frequencies is remarkably stable; there is very little fading. Solar flares occasionally disrupt communication in this frequency range, by making the ionosphere absorptive and by raising the cutoff frequency of the earth-ionosphere waveguide.

Because the surface-wave mode is very efficient at frequencies below 30 kHz, it is possible that the very-low-frequency band might be used for over-the-horizon communication on planets that lack ionized layers concentrated enough to return signals at higher frequencies. The moon and Mars are examples of such planets. *See* SOLAR FLARE, SURFACE WAVE, and WAVEGUIDE PROPAGATION.

Low Frequencies (30 kHz to 300 kHz) In the low-frequency range, propagation occurs in the surface-wave mode, and also as a result of ionospheric effects. Toward the lower end of the band, wave propagation is similar to that in the very-low-frequency range. As the frequency is raised, surface-wave attenuation increases. Although a surface-wave range of more than 3,000 miles is common at 30 kHz, it is unusual for the range to be greater than a few hundred miles at 300 kHz. Surface-wave propagation requires that the electromagnetic field be vertically polarized.

Ionospheric propagation at low frequencies usually occurs via the E layer. This increases the useful range during the nighttime hours, especially toward the upper end of the band. Intercontinental communication is possible with high-power transmitters.

Solar flares can disrupt communication at low frequencies. Following such a flare, the D layer becomes highly absorptive, preventing ionospheric communication. *See* D LAYER, E LAYER, SOLAR FLARE, and SURFACE WAVE.

Medium Frequencies (300 kHz to 3 MHz) Propagation at medium frequencies occurs by means of the surface wave, and by E-layer and F-layer ionospheric modes. Near the lower end of the band, surface-wave communications paths can be up to several hundred miles. As the frequency is raised, the surface-wave attenuation increases. At 3 MHz, the range of the surface wave is limited to about 150 or 200 miles.

Ionospheric propagation at medium frequencies is almost never observed during the daylight hours because the D layer prevents electromagnetic waves from reaching the higher E and F layers. During the night, ionospheric propagation occurs mostly via the E layer in the lower portion of the band, and primarily via the F layer in the upper part of the band. The communications range increases as the frequency is raised. At 3 MHz, worldwide communication is possible over darkness paths.

As at other frequencies, medium-frequency propagation is severely affected by solar flares. The 11-year sunspot cycle and the season of the year also affect propagation at medium frequencies. Propagation is usually better in the winter than in the summer. *See* D LAYER, E LAYER, F LAYER, SOLAR FLARE, SUNSPOT CYCLE, and SURFACE WAVE.

High Frequencies (3 MHz to 30 MHz)

Propagation at the high frequencies exhibits widely variable characteristics. Effects are much different in the lower part of this band than in the upper part. Consider the lower portion as the range 3 MHz to 10 MHz, and the upper portion as the range 10 MHz to 30 MHz.

Some surface-wave propagation occurs in the lower part of the high-frequency band. At 3 MHz, the maximum range is 150 to 200 miles; at 10 MHz, it decreases to about the radio-horizon distance, perhaps 15 miles. Above 10 MHz, surface-wave propagation is essentially nonexistent.

Ionospheric communication occurs mainly via the F layer. In the lower part of the band, there is very little daytime ionospheric propagation because of D-layer absorption; at night, worldwide communication can be had. In the upper part of the band, this situation is reversed; ionospheric communication can occur on a worldwide scale during the daylight hours, but the maximum usable frequency often drops below 10 MHz at night. Of course, the transition is gradual, and the range in which it occurs is variable.

Communications in the lower part of the high-frequency band are generally better during the winter months than during the summer months. In the upper part of the band, this situation is reversed.

Some E-layer propagation is occasionally observed in the upper part of the high-frequency band. This is usually of the sporadic-E type. This can occur even when the F-layer maximum usable frequency is below the communication frequency.

Near 30 MHz, there is often no ionospheric communication. This is especially true when the sunspot cycle is at or near a minimum. The extreme upper portion of the high-frequency band then behaves similarly to the very high frequencies.

Solar flares cause dramatic changes in conditions in the high-frequency band. Sometimes, ionospheric communications are almost totally wiped out within a matter of minutes by the effects of a solar flare. *See* D LAYER, E LAYER, F LAYER, MAXIMUM USABLE FREQUENCY, SOLAR FLARE, SPORADIC-E PROPAGATION, SUNSPOT CYCLE, and SURFACE WAVE.

Very High Frequencies (30 MHz to 300 MHz)

Propagation in the very-high-frequency band occurs along the line-of-sight path, or via tropospheric modes. Ionospheric F-layer propagation is rarely observed, although it can occur at times of sunspot maxima at frequencies as high as about 70 MHz. Sporadic-E propagation is fairly common, and can occur up to approximately 200 MHz.

Tropospheric propagation can occur in three different ways: bending, ducting, and scattering. Communications range varies, but can often be had at distances of several hundred miles.

Meteor-scatter and auroral propagation can sometimes be observed in the very-high-frequency band. The range for meteor scatter is typically a few hundred miles; auroral propagation can sometimes produce communications over distances of up to about 1,500 miles. Code transmission must usually be employed for auroral communications; the moving auroral curtains introduce phase modulation that makes voices practically unintelligible.

Repeaters are used extensively at very high frequencies to extend the range of mobile communications equipment. A repeater, located in a high place, can provide coverage over an area of thousands of square miles.

At the very high frequencies, satellites are often used to provide worldwide communication on a reliable basis. The satellites are actually orbiting repeaters. The moon, a passive satellite, is sometimes used, mostly by amateur radio operators. *See* AURORAL PROPAGATION, E LAYER, F LAYER, METEOR SCATTER, MOONBOUNCE, REPEATER, SATELLITE COMMUNICATIONS, TROPOSPHERIC PROPAGATION, and TROPOSPHERIC-SCATTER PROPAGATION.

Ultra High Frequencies (300 MHz to 3 GHz)

Propagation in the ultra-high-frequency band occurs almost exclusively via the line-of-sight mode and via satellites and repeaters. No ionospheric effects are ever observed. Auroral, meteor-scatter, and tropospheric propagation are occasionally observed in the lower portion of this band. Ducting can result in propagation over distances of several hundred miles.

The main feature of the ultra-high-frequency band is its sheer immensity—2,700 MHz of spectrum space. Another advantage of this band is its relative immunity to the effects of solar flares. Because ultra-high-frequency energy has nothing to do with the ionosphere, disruptions in the ionosphere have no effect. However, an intense solar flare can interfere to some extent with ultra-high-frequency circuits, because of electromagnetic effects resulting from fluctuations in the magnetic field of the earth. *See* AURORAL PROPAGATION, LINE-OF-SIGHT COMMUNICATION, METEOR SCATTER, MOONBOUNCE, REPEATER, TROPOSPHERIC PROPAGATION, and TROPOSPHERIC-SCATTER PROPAGATION.

Microwaves (Above 3 GHz)

Microwave energy travels essentially in straight lines through the atmosphere. Microwaves are affected very little by such phenomena as temperature inversions and scattering. Microwaves are unaffected by the ionosphere.

The primary mode of propagation in the microwave range is the line-of-sight mode. This facilitates repeater and satellite communications. The microwave band is much more vast than even the ultra-high-frequency band. The upper limit of the microwave range, in theory, is the infrared spectrum. Current technology has produced radio transmitters capable of operation at frequencies of more than 300 GHz.

At some microwave frequencies, atmospheric attenuation becomes a consideration. Rain, fog, and other weather effects cause changes in the path attenuation as the wavelength becomes comparable with the diameter of water droplets. *See* ACTIVE COMMUNICATIONS SATELLITE, LINE-OF-SIGHT COMMUNICATION, and REPEATER.

Infrared, Visible Light, Ultraviolet, X Rays, and Gamma Rays

Propagation at infrared and shorter wavelengths occurs in straight lines. The only factor that varies is the atmospheric attenuation. Water in the atmosphere causes severe attenuation of infrared radiation between the wavelengths of approximately 4,500 and 8,000 nanometers (nm). Carbon dioxide interferes with the transmission of infrared radiation at wavelengths ranging from about 14,000 nm to 16,000 nm. Rain, snow, fog, and dust interfere with the transmission of infrared radiation.

Light is transmitted fairly well through the atmosphere at all visible wavelengths. But rain, snow, fog, and dust interfere considerably with the transmission of visible light through the air.

Ultraviolet light at the longer wavelengths can penetrate the air with comparative ease. At shorter wavelengths, attenuation increases. Rain, snow, fog, and dust interfere.

X rays and gamma rays do not propagate well for long distances through the air. This is primarily because the mass of the air, over propagation paths of great distance, is sufficient to block these types of radiation.

PROPAGATION DELAY

See PROPAGATION TIME.

PROPAGATION FORECASTING

Ionospheric propagation conditions vary considerably from day to day, month to month, and year to year at frequencies below 60 MHz. To some extent, these changes are predictable.

The sun is constantly monitored during the daylight hours for fluctuations in the intensity of radiation at various electromagnetic frequencies, especially 2800 MHz. The 2800-MHz solar flux, as it is called, provides a good indicator of propagation conditions for the ensuing several hours. A sudden increase in the solar flux means that a flare is occurring. Within a short time, propagation conditions usually become disturbed following a flare. Flares sometimes wipe out F-layer communications completely, but the E layer can become sufficiently ionized to allow propagation over moderate distances. At higher latitudes, auroral propagation can usually be observed after a solar flare (*see* AURORAL PROPAGATION, SOLAR FLARE, SOLAR FLUX, and SPORADIC-E PROPAGATION).

Regular variations in propagation conditions occur. At frequencies below about 10 MHz, propagation is generally better at night than during the day. Above about 10 MHz, this situation is reversed. The winter months are generally better for propagation conditions below 10 MHz, while the summer months bring better conditions above 10 MHz.

The 11-year sunspot cycle produces regular changes in the maximum usable frequency (MUF). The MUF is highest at the time of the sunspot maximum. Occasionally the MUF can reach 60 MHz or perhaps even 70 MHz. When the sunspot level is at or near a minimum, the MUF can fall below 10 MHz (*see* MAXIMUM USABLE FREQUENCY, and SUNSPOT CYCLE).

Tropospheric-propagation conditions can be forecast to some extent simply by observing weather maps. At very high frequencies, long-distance tropospheric bending or ducting often occurs along a frontal line (*see* TROPOSPHERIC PROPAGATION, and TROPOSPHERIC-SCATTER PROPAGATION).

The likelihood of meteor-scatter propagation is, logically, increased when a meteor shower occurs. Astronomical publications often provide information about meteor showers. Several different meteor showers occur at the same time every year (*see* METEOR SCATTER).

Propagation-forecast bulletins are regularly transmitted by the National Bureau of Standards time-and-frequency stations, WWV and WWVH. Long-term propagation forecasts are also issued each month in amateur-radio magazines and other publications for the electronics hobbyist and professional. *See also* PROPAGATION CHARACTERISTICS, and WWV/WWVH/WWVB.

PROPAGATION MODES

Electromagnetic communications occur as a result of several different forms of wave propagation. In free space, propagation occurs in straight lines, in directions perpendicular to the electric and magnetic flux vectors (*see* ELECTROMAGNETIC FIELD). On and near the surface of the earth, however, electromagnetic radiation sometimes follows curved or irregular paths. The effects vary, depending on the wavelength.

The most common propagation modes, and the frequency bands at which they are most likely to be observed, are listed in the table. For more detail concerning these propagation modes, consult the following articles: AURORAL PROPAGATION, E LAYER, F LAYER, IONOSPHERE, LINE-OF-SIGHT COMMUNICATION, METEOR SCATTER, MOONBOUNCE, OPTICAL COMMUNICATION, PROPAGATION CHARACTERISTICS, REPEATER, SPORADIC-E PROPAGATION, SURFACE WAVE, TROPOSPHERIC PROPAGATION, TROPOSPHERIC-SCATTER PROPAGATION, and WAVEGUIDE PROPAGATION.

PROPAGATION MODES: THE MOST COMMON PROPAGATION MODES, AND THE BANDS OF FREQUENCIES AT WHICH THEY ARE MOST LIKELY TO BE OBSERVED. ABBREVIATIONS ARE AS FOLLOWS: VLF = BELOW 30 kHz; LF = 30 kHz TO 300 kHz; MF = 300 kHz TO 3 MHz; HF = 3 MHz TO 30 MHz; VHF = 30 MHz TO 300 MHz; UHF = 300 MHz TO 3 GHz; MICROWAVE = ABOVE 3 GHz.

Mode	Band(s)
Auroral	HF (upper)
	VHF (all)
E-layer	VLF, LF, HF, VHF
F-layer	LF, MF, HF
Line-of-sight	All bands
Meteor Scatter	VHF
Moonbounce	VHF, UHF
Repeater	VHF, UHF
Satellite	VHF, UHF,
	Microwave
Sporadic-E	VHF
Surface-wave	VLF, LF, MF, HF
Tropospheric	VHF, UHF
Waveguide	VLF

PROPAGATION SPEED

The speed with which a disturbance propagates depends on many factors. Electromagnetic waves in free space travel at 186,282 miles per second (299,792 kilometers per second). In media other than free space, this speed can be reduced considerably.

In air, electromagnetic waves travel at essentially the same speed as in free space. However, in a transmission line, the speed with which an electromagnetic field propagates is much slower. *See also* VELOCITY FACTOR.

PROPAGATION TIME

The *propagation time*, also known as the *propagation delay*, is the time required for a wave disturbance to travel from one point to another. The propagation time depends on the speed with which the disturbance travels, and on the distance between the two points of interest.

In general, if the speed of propagation is given by c and the distance by d, then the propagation time is:

$$t = d/c$$

For electromagnetic waves traveling via ionospheric modes, the propagation time is somewhat longer than you would calculate by setting $c = 186,282$ miles per second or $299,792$ kilometers per second (the speed of light in free space). The reason for this is that the waves require some time to be bent in the ionospheric layers. Similar effects occur in tropospheric propagation. *See also* PROPAGATION SPEED.

PROTECTIVE GROUNDING

In an antenna system, a substantial voltage can develop between the active elements and the ground. This often happens in the proximity of a thundershower. It can also result from electromagnetic interaction with nearby power lines. This voltage is often large enough to present a shock hazard.

To keep an antenna system at dc or 60-Hz ground potential, *protective grounding* is used. The most common method of protective grounding is the installation of radio-frequency chokes between the antenna feed line and ground.

In an unbalanced system, one choke is connected as shown at A in the illustration. In a balanced system, two chokes are needed (B). The chokes should have about 10 times the impedance of the antenna system at the lowest operating frequency.

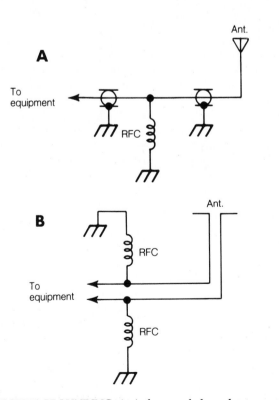

PROTECTIVE GROUNDING: At A, for an unbalanced antenna system; at B, for a balanced system.

For example, in a 50-ohm unbalanced system, the choke should present 500 ohms of inductive reactance at the lowest frequency on which operation is contemplated. The inductor windings must be made of wire that is large enough to handle several amperes of current.

When an antenna is not in use, it is wise to disconnect it entirely from radio equipment, while the protective-grounding circuit is left intact. This will prevent damage to radio equipment in the event of a direct lightning strike. Alternatively, an antenna grounding switch can be used (*see* LIGHTNING PROTECTION).

When several pieces of equipment are used together, dangerous potential differences can build up between or among the chassis of the individual units unless adequate grounding precautions are taken. In any electronic installation, all equipment chassis should be grounded by means of a common bus. *See also* GROUND BUS.

PROTOCOL

In communications of any kind, *protocol* refers to an established set of procedures to be used. In traffic nets, for example, protocol must be followed by each operator in order to have an efficient flow of messages. The alternative would be confusion and chaos.

In packet radio, protocol is carried out by computers, rather than by the operators. The standard general packet communications model is called the *Open Systems Interconnection Reference Model (OSI-RM)*. There are other, more specific protocols for various different communications functions. *See* CONNECTION PROTOCOL, CONNECTIONLESS PROTOCOL, FILE TRANSFER PROTOCOL, OPEN SYSTEMS INTERCONNECTION REFERENCE MODEL, PACKET RADIO, POLLING PROTOCOL, RATS OPEN SYSTEMS ENVIRONMENT, SERIAL LINE INTERFACE PROTOCOL, SIMPLE MAIL TRANSFER PROTOCOL, SLIDING WINDOW PROTOCOL, TRANSMISSION CONTROL PROTOCOL, and VIRTUAL CIRCUIT.

PROTOTYPE

In the design and manufacture of electronic apparatus, a test unit is generally built before production begins. The test unit is perfected by experimentation, and also to some extent by trial and error. Two or more test units, known as *prototypes*, can be built, with each succeeding prototype designed to more closely resemble the actual production unit. The construction of prototypes allows equipment to be debugged before mass manufacture.

PROXIMITY EFFECT

A conductor or component carrying a high-frequency alternating current can induce a current of the same frequency in a nearby conductor or component. The intensity of the current in a component can be affected by nearby objects. These effects are called *proximity effects*.

Proximity effect can occur as a result of electric coupling between or among objects. Proximity effect can also result from magnetic coupling or from electromagnetic-field propagation. Body capacitance is an example of an electric proximity effect (*see* BODY CAPACITANCE). All transformers operate because of magnetic proximity effects (*see* MUTUAL INDUCTANCE, and TRANSFORMER).

Proximity effect can be manifested in other ways, and used for specific purposes. For example, certain intrusion detectors make use of ultrasonic-wave disturbances caused by nearby moving objects. Another form of intrusion detector senses the infrared radiation given off by a person's body.

PSEUDONOISE

In *spread spectrum* communications, the frequency of the carrier wave is varied rapidly, according to a predetermined sequence. This sequence is sufficiently complicated so that it appears to be random. Of course, it is not really random because it follows a specific pattern.

A sequence of this type is known as *pseudonoise*, abbreviated *PN*. The prefix *pseudo* means "resembling." In a receiver not designed to receive a spread-spectrum signal, if anything is noticed at all, it will be what appears as a slight increase in the noise level. *See* SPREAD SPECTRUM.

P-TYPE SEMICONDUCTOR

A *P-type semiconductor* is a substance that conducts mostly by the transfer of electron deficiencies among atoms. The electron deficiencies are called *holes* and carry a positive charge. The P-type semiconductor gets its name from the fact that it conducts mainly via positive carriers.

P-type material is manufactured by adding certain impurities to the semiconductor base, which can consist originally of pure germanium, silicon, or other semiconductor element. The impurities are called *acceptors*, because they are electron-deficient. The carrier mobility in a P-type material is generally less than that of N-type material.

P-type material is used in conjunction with N-type material in most semiconductor devices, including diodes, transistors, field-effect transistors, and integrated circuits. *See also* DOPING, ELECTRON, HOLE, N-TYPE SEMICONDUCTOR, and P-N JUNCTION.

PULLING

When a capacitor or inductor is placed in parallel with a piezoelectric crystal, the natural frequency of the crystal is lowered. This effect is called *pulling*.

Pulling is routinely used in crystal oscillators and filters for the purpose of adjusting the crystal frequency to exactly the desired value. The frequency of the average crystal can be pulled down by less than 0.1 percent of its natural frequency (approximately 1 kHz/MHz).

When two variable-frequency oscillators are connected to a common circuit, such as a mixer, one of the oscillators can drift into tune with the other. This effect is sometimes called *pulling*. It is most likely to occur when tuned circuits are too tightly coupled. *See also* CRYSTAL, CRYSTAL-LATTICE FILTER, CRYSTAL OSCILLATOR, VARIABLE CRYSTAL OSCILLATOR, and VARIABLE-FREQUENCY OSCILLATOR.

PULSATING DIRECT CURRENT

In a direct-current power supply that has the filter disconnected, the output voltage pulsates. If the alternating-current frequency is 60 Hz at the supply input, then the pulsating direct current has a frequency of either 60 Hz or 120 Hz. The half-wave rectifier circuit produces pulsating direct current with a frequency of 60 Hz; the bridge and full-wave circuits produce 120-Hz pulsating direct current.

In general, 120-Hz pulsating direct current is easier to filter than 60-Hz pulsating direct current. This is simply because the filter capacitors and inductors must hold their charge for only half as long when the frequency is 120 Hz, as compared to when the ripple frequency is 60 Hz. *See also* BRIDGE RECTIFIER, FULL-WAVE RECTIFIER, and HALF-WAVE RECTIFIER.

PULSE

A *pulse* is a burst of current, voltage, power, or electromagnetic energy. A pulse can have an extremely short duration (as little as a fraction of a nanosecond), or it can have a long duration (thousands or millions of centuries).

The amplitude of a pulse is expressed either in terms of the maximum instantaneous (peak) value, or in terms of the average value (*see* AVERAGE VALUE, and PEAK VALUE). The average and peak amplitudes are sometimes, but not always, the same.

The duration of a pulse can be expressed either in actual terms or in effective terms. The actual duration of a pulse is the length of time between its beginning and ending. The effective duration can be shorter than the actual duration (*see* AREA REDISTRIBUTION). Pulse duration is also called *pulse width*.

The interval, or length of time between pulses, has meaning if the pulses are recurrent. The pulse interval is the length of time from the end of one pulse to the beginning of another. The pulse interval can be zero in some cases. An example is the output of a full-wave rectifier.

In electronics, a pulse generally has a well-defined waveshape, such as rectangular, sawtooth, sinusoidal, or square. These pulse shapes are the most common. However, a pulse can have a highly irregular shape, and the number of possible configurations is infinite.

A pulse of current or voltage normally maintains the same polarity from beginning to end, unless droop or overshoot occur. *See also* DROOP, and OVERSHOOT.

PULSE AMPLIFIER

A specialized form of wideband amplifier, characterized by the ability to respond to extremely rapid changes in amplitude, is called a *pulse amplifier*. The pulse amplifier must be capable of handling pulses of short duration with a minimum of distortion. The output pulse should have the same waveshape and the same duration as the input pulse.

Pulse amplifiers are used in such applications as digital circuits, pulse-modulated transmitters, and radar transmitters. *See also* PULSE.

PULSE GENERATOR

A *pulse generator* is a circuit specifically designed for producing electronic pulses. Commercially manufactured pulse generators are available for a variety of laboratory test purposes.

A typical pulse generator can produce rectangular, sawtooth, sinusoidal, and square pulses. The pulse amplitude, duration, and frequency are independently adjustable. Some sophisticated devices can reproduce pulses of any desired waveshape. This is done via computer. The operator draws the desired pulse function, using a light pen, on a video-monitor screen. The computer then produces a pulse or pulse train having that waveshape. *See also* PULSE.

PULSE MODULATION

Pulse modulation involves the transmission of intelligence by means of varying the characteristics of a series of electromagnetic pulses. Pulse modulation can be accomplished by varying the amplitude, the duration, the frequency, or the position of the pulses. Pulse modulation can also be obtained by means of coding. A brief description of each of these pulse-modulation methods follows:

Pulse-Amplitude Modulation A complex waveform can be transmitted by varying the amplitude of a series of pulses. Generally, the greater the modulating-signal amplitude, the greater the pulse amplitude at the output of the transmitter. However, this can be reversed. The drawing illustrates pulse-amplitude modulation at A.

Pulse-Duration Modulation The effective energy contained in a pulse depends not only on its amplitude, but also on its duration. In pulse-duration modulation, the width of a given pulse depends on the instantaneous amplitude of the modulating waveform. Usually, the greater the amplitude of the modulating waveform, the longer the transmitted pulse, but this can be reversed. Pulse-duration modulation can also be called *pulse-length* or *pulse-width modulation*. The principle is illustrated at B.

Pulse-Frequency Modulation The number of pulses per second can be varied in accordance with the modulating-waveform amplitude. The duration and intensity of the individual pulses remains constant in *pulse-frequency modulation*. However, the effective signal power is increased when the pulse frequency is increased.

Usually, the pulse frequency increases as the modulating-signal amplitude increases; but this can be reversed. Pulse-frequency modulation is illustrated at C.

Pulse-Position Modulation Pulse modulation can be obtained even without varying the frequency, amplitude, or duration of the pulses. The actual timing of the pulses can be varied, as shown at D. This is known as *pulse-position modulation*. It can also be called *pulse-interval* or *pulse-time modulation*.

Pulse-Code Modulation All of the foregoing methods of pulse modulation are analog methods. That is, the amplitude, duration, frequency, or position of the pulses can be varied in a continuous manner. A digital method of pulse modulation is called *pulse-code modulation*. In pulse-code modulation, some parameter (usually amplitude) of the pulses can achieve only certain values. The drawing at E illustrates the concept of pulse-code modulation, in which the pulse amplitude can attain any of eight discrete levels.

A special form of pulse-code modulation makes use of the derivative of the modulating-signal waveform. This is known as *delta modulation*. The pulse amplitude (or other parameter) can attain only certain values, according to the derivative of the modulating-waveform amplitude function, as at F.

Pulse modulation is used for a variety of communications purposes. It is especially well-suited to use with communications systems incorporating time-devision multiplexing. *See also* TIME-DIVISION MULTIPLEX.

PULSE TRANSFORMER

A *pulse transformer* is a special kind of transformer designed to accommodate rapid rise and decay times with a minimum of distortion. A normal transformer can introduce distortion into a square pulse by slowing down the rise and decay times.

Pulse transformers generally have windings with less inductance than the windings of a transformer for sine waves. The core material of a pulse transformer usually has lower permeability than the core material in a typical transformer. *See also* TRANSFORMER.

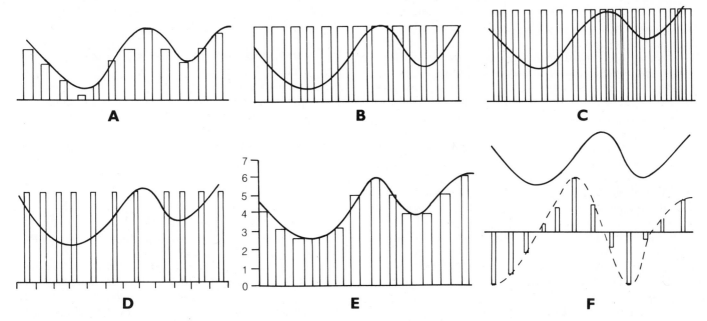

PULSE MODULATION: Modulation of pulse amplitude (A), duration (B), frequency (C), and position (D). At E, pulse-code modulation. At F, delta modulation.

PUSHDOWN STACK

A first-in/last-out memory is called a *pushdown stack*. It is a read-write memory; information can be both stored and retrieved. The pushdown stack differs from the first-in/first-out memory (*see* FIRST-IN/FIRST-OUT). In the pushdown stack, the data bits that are inserted first must be retrieved last; the bits inserted last are recalled first.

The illustration shows the principle of the pushdown stack. The term describes the operation quite well; data bits are stored and retrieved as if they are "stacked" in a confined column. As the data is read from the pushdown stack, the remaining bits all move one space closer to the input.

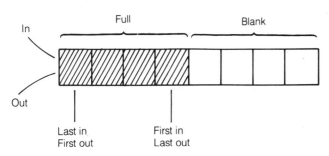

PUSHDOWN STACK: Also known as *first-in/last-out*.

PUSH-PULL AMPLIFIER

A *push-pull amplifier* is a specialized, low-distortion amplifier that is often used at audio frequencies and is sometimes used at radio frequencies. The drawing illustrates push-pull amplifiers incorporating tubes (A), bipolar transistors (B) and field-effect transistors (C). These are audio-frequency circuits. The radio-frequency circuits differ only in that they can incorporate a tuned input and output.

Any push-pull amplifier requires a pair of active devices. The emitters or sources are grounded through small resistors. The resistors help to prevent a phenomenon called *current hogging*, in which one device tends to do most of the work in the circuit (*see* CURRENT HOGGING). The bases or gates receive the input signal in phase opposition; they are connected to opposite ends of a transformer secondary. The center tap at the transformer helps to balance the system, and also provides a point at which bias can be applied. The collectors or drains are connected to opposite ends of the primary winding of the output transformer. As with the input, the center tap helps to balance the system, and also provides the point at which voltage is applied.

The input signal can be applied at the emitters or sources of the active devices, leaving the bases or gates at ground potential. This configuration also provides push-pull operation (*see* PUSH-PULL, GROUNDED-GRID/BASE/GATE AMPLIFIER).

Push-pull amplifiers are operated either in Class AB or in Class B, and are always used as power amplifiers (*see* CLASS-AB AMPLIFIER, CLASS-B AMPLIFIER, and POWER AMPLIFIER). The main advantage of a push-pull amplifier is the fact that it effectively cancels all of the even harmonics in the output circuit. This reduces the distortion in audio-frequency applications, and enhances the even-harmonic attenuation in a radio-frequency circuit.

A push-pull radio-frequency amplifier is sometimes used to multiply the frequency of a signal by a factor of 3. Such a circuit is called a *tripler*. The odd harmonics are not attenuated in the push-pull configuration; in fact, if the output circuit is tuned to an odd-harmonic frequency, the push-pull circuit will favor the harmonic over the fundamental frequency (*see* TRIPLER).

Push-pull circuits can be used in oscillators, modulators, and passive devices, as well as in power amplifiers. *See also* PUSH-PULL CONFIGURATION.

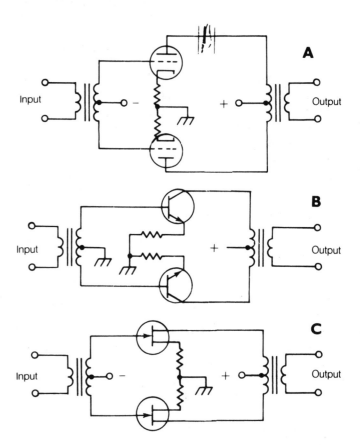

PUSH-PULL AMPLIFIER AND PUSH-PULL CONFIGURATION: Circuits with triodes (A), bipolar transistors (B) and FETs (C).

PUSH-PULL CONFIGURATION

Any circuit in which both the input and the output are divided between two identical devices, each operating in phase opposition with respect to the other, is called a *push-pull circuit*. The push-pull configuration requires two transformers, one at the input and one at the output. The secondary of the input transformer, and the primary of the output transformer, are usually center-tapped. The previous illustration shows three examples of push-pull configuration; in this case the circuits are audio-frequency power amplifiers.

The term *push-pull* is well-chosen because it graphically illustrates the operation of the circuit. While one side of the circuit carries current in one direction, the other side carries current in the opposite direction.

In the push-pull configuration, the even harmonics are effectively canceled, but the odd harmonics are reinforced. This makes the push-pull circuit useful as a radio-frequency tripler. The push-pull configuration is widely used in audio-frequency power amplifiers, and to some extent in radio-frequency amplifiers.

In push-pull, it is important that balance be maintained between the two halves of the circuit. Otherwise, distortion will occur in the output, and current hogging can occur. *See also* CURRENT HOGGING, PUSH-PULL AMPLIFIER, PUSH-PULL, GROUNDED-GRID/BASE/GATE AMPLIFIER, and TRIPLER.

PUSH-PULL, GROUNDED-GRID/BASE/GATE AMPLIFIER

In a push-pull power amplifier, the input is usually applied at the grids, bases, or gates of the active devices, as shown in the previous illustration (*see* PUSH-PULL AMPLIFIER). However, the input can be applied at the cathodes, emitters, or sources. This kind of push-pull circuit is shown here using tubes (A), bipolar transistors (B), and field-effect transistors (C). These are radio-frequency power amplifiers that have tuned output circuits.

The amplifiers shown are somewhat more stable than the normal push-pull type. The circuits shown here do not normally require neutralization. The main disadvantage of the grounded-grid, grounded-base, or grounded-gate configuration is that they require more driving power than the grounded-cathode, grounded-emitter, or grounded-source amplifiers. *See also* POWER AMPLIFIER.

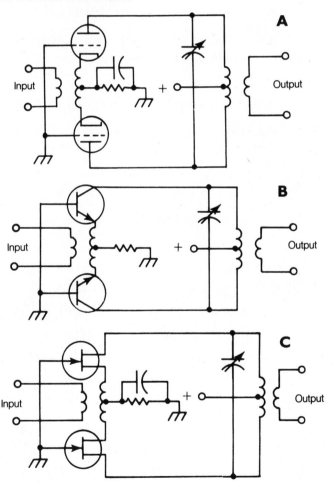

PUSH-PULL, GROUNDED GRID/BASE/GATE AMPLIFIER: Circuits with triodes (A), bipolar transistors (B), and FETs (C).

PUSH-PUSH CONFIGURATION

A *push-push circuit* is a form of balanced circuit, incorporating two identical passive or active halves. The push-push configu-ration is similar to push-pull (*see* PUSH-PULL CONFIGURA-TION), but there is one important distinction. In a push-push circuit, the inputs of the devices are connected in phase opposition, just as they are in push-pull, but the outputs are connected in parallel. The illustration shows examples of push-push radio-frequency amplifiers using tubes (A), bipolar transistors (B) and field-effect transistors (C).

The push-push circuit tends to cancel all of the odd harmonics (including the fundamental frequency), while reinforcing the even harmonics. This is just the opposite of the push-pull configuration. The push-push circuit is not often used at audio frequencies, but it can be used as a radio-frequency doubler or quadrupler.

Perhaps the simplest form of push-push circuit is the full-wave rectifier. This is, in effect, a frequency doubler. With 60-Hz alternating-current input, the full-wave rectifier produces a 120-Hz pulsating direct current output. An identical circuit can be used as a passive radio-frequency doubler. *See also* DOUBLER, FULL-WAVE RECTIFIER, and QUADRUPLER.

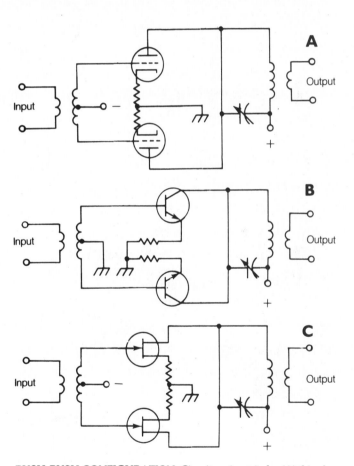

PUSH-PUSH CONFIGURATION: Circuits using triodes (A), bipolar transistors (B), and FETs (C).

PYRAMID HORN ANTENNA

A *pyramid horn* is an antenna that is used for transmitting and receiving microwaves. The pyramid horn gets its name from the fact that it is shaped like a pyramid; the faces are flat, and they converge to a single point, called the *apex*. The flare angle can vary, but it is usually about 30 to 45 degrees, with respect to the center axis. The pyramid horn is fed directly at the apex by a waveguide. *See also* HORN ANTENNA.

Q FACTOR

For a capacitor, inductor, or tuned circuit, the *Q factor* (also known simply as the *Q*) is a figure of merit. The higher the *Q* factor, the lower the loss and the more efficient the component or tuned circuit. In tuned circuits, the *Q* factor is directly related to the selectivity.

In the case of a capacitor or inductor, the *Q* factor is given in terms of the reactance and the resistance. If *X* represents the reactance and *R* represents the resistance, in a series circuit, both given in ohms, then:

$$Q = X/R$$

for an inductor, and:

$$Q = -X/R$$

for a capacitor (capacitive reactance is considered negative).

Because both inductive and capacitive reactance vary with the frequency, the *Q* factor of an inductor or capacitor depends on the frequency (*see* CAPACITIVE REACTANCE, and INDUCTIVE REACTANCE).

In a series-resonant inductance-capacitance (LC) circuit, the *Q* factor is given by the same formula as above, where *X* represents the absolute value of the reactance of either the coil or the capacitor (the absolute values are equal at resonance) and *R* represents the resistive impedance of the circuit at resonance.

In a parallel-resonant LC circuit, the *Q* factor is given by:

$$Q = R/X$$

where *R* represents the resistive impedance of the circuit at resonance and *X* represents the absolute value of the reactance of either the inductor or the capacitor (again, the absolute values are identical at resonance).

The *Q* factor of a tuned circuit can be determined in another way. Suppose a series or parallel LC circuit has a resonant frequency *f* Hz. Further suppose that the 3-dB attenuation points are separated by *g* Hz (that is, the 3-dB bandwidth is *g* Hz). Then the *Q* factor of the tuned circuit is given by:

$$Q = f/g$$

This method of determining the *Q* is convenient from an experimental standpoint. The narrower the 3-dB bandwidth, the higher the factor at a given resonant frequency. For a constant bandwidth, the *Q* factor increases as the resonant frequency is increased.

The *Q* factor of a trap circuit can also be determined according to the 3-dB bandwidth. The only difference is that the 3-dB points are taken with respect to the frequency of maximum attenuation, instead of with respect to the frequency of least attenuation. *See also* BANDWIDTH, RESONANCE, and TRAP.

Q MULTIPLIER

In early superheterodyne receivers, a circuit called a *Q multiplier* was occasionally used to enhance the selectivity (that is, to reduce the bandwidth). Such circuits can still be found in some superheterodyne receivers, although crystal-lattice filters, ceramic filters, and mechanical filters are more common today.

The multiplier consists of a tuned intermediate-frequency amplifier in which some of the output signal is fed back to the input through the tuned circuits. This forces the signal to effectively pass through the resonant networks several times.

A potentiometer, connected in the feedback path, is used to control the regeneration. There is a limit to how large the effective *Q* factor can become before oscillation occurs in the circuit. In a well-designed *Q* multiplier, the selectivity can be made very sharp before oscillation begins. *See also* Q FACTOR, and SELECTIVITY.

QRP

The abbreviation *QRP* is one of the ham radio *Q signals*, used mainly in code (CW) communications. (*See* Q SIGNAL.) It means "Reduce power."

The abbreviation QRP has, in recent years, come to mean "low power" and/or "low-power operation." This is a relative thing, of course; to some hams, 100 watts is QRP, while to others, only levels less than a few milliwatts are true QRP. The most accepted figure is probably 5 watts output or less on the bands below 30 MHz.

Operating QRP is a challenge that appeals to many hams. To the QRP enthusiast, there's nothing quite like working some rare DX (foreign) station using less than 5 watts. This is especially true if the contact is made when numerous other, high-powered stations are also calling.

One of the biggest advantages of QRP is that it allows the use of battery power. This makes it possible to go camping in the wilderness, or boating, or practically anyplace, and carry a complete ham radio station in a small suitcase. A QRP station can also be powered by solar panels or other alternative energy sources.

Another advantage of QRP is that it reduces the tendency for problems to occur with electromagnetic interference (EMI). *See* ELECTROMAGNETIC INTERFERENCE.

The easiest bands on which to make DX contacts using QRP are 6, 10, 12 and 15 meters. This is because of the propagation characteristics on these bands. I have had the experience of working a New Zealand station on 28 MHz CW, using just 3 watts output to a three-element Yagi antenna on the roof of a three-story apartment house. The signal report was 579 (*See* RST SYSTEM)! With this same station, contacts in Europe and South America were made almost at will, in large numbers, during the peak times of the day.

QRP can yield exciting and fascinating contacts on any band. Some hams have worked hundreds or even thousands of miles using less than 5 watts on 160 meters.

QRQ

The signal *QRQ* means "Increase code sending speed." It also refers to high-speed code (CW) in general. Nowadays, keyboard keyers allow an operator to send CW at a speed limited only by typing ability. Buffers eliminate any need to worry about spacing. A good typist can therefore send CW at 40 words per minute (wpm) or more, with minimal effort.

Machines can "copy" CW and display the text on a screen. This, too, can be carried out at speeds considerably greater than most people can read in their heads or take down on paper. However, for high-speed printed communications, radioteletype (RTTY), using Baudot or ASCII codes and frequency-shift keying, are preferred because these methods reduce the number of teleprinter errors. The AMTOR modes provide greater accuracy still. *See* AMTOR, ASCII, BAUDOT, FREQUENCY-SHIFT KEYING, and RADIOTELETYPE.

There are numerous hams who enjoy communicating via "QRQ" morse code, sending with keyboards, but reading the code in their heads. Some of these operators can send and read at speeds of 60 wpm or more. The best can send and read up to about 100 wpm. You can find these hams in the lower parts of the General/Advanced CW band segments. They send CQ at 40-45 wpm, increasing speed once a QSO gets underway.

Q SIGNAL

In radiotelegraphy, certain statements, phrases, or words are made very often. It becomes tedious to send complete sentences, phrases, and words in Morse code, especially if they are repeated. To streamline code operation, therefore, a set of abbreviations called *Q signals* has been devised.

Each Q signal consists of the letter Q, followed by two more letters. A Q signal followed by a question mark indicates a query; if no question mark follows the signal, or if data follows, it indicates a statement. For example, "QRM?" means "Are you experiencing interference?" and "QRM" means "I am experiencing interference." As another example, "QTH?" means "What is your location?" and "QTH ROCHESTER, MN" means "My location is Rochester, Minnesota."

The table is a comprehensive list of Q signals. Although Q signals were originally designed for radiotelegraph use, many radioteletype and radiotelephone operators also use them. The Q signals can streamline radioteletype operation in the same way that they make code operation more convenient. There is some question as to how beneficial (if not detrimental) the Q signals are in voice communication.

Q SIGNAL: QUERY AND STATEMENT INFORMATION.

Signal	Meaning
QRA	What is the name of your station? The name of my station is —.
QRB	Approximately how far from my station are you? I am about — miles or kilometers from you.
QRD	From where to where are you going? I am going from — to —.
QRG	What is my exact frequency, or that of —? Your frequency, or that of —, is —.
QRH	Does my frequency vary? Your frequency varies.
QRI	How is the tone of my signal? The tone of your signal is: 1 (good), 2 (fair or variable), 3 (poor)
QRK	How readable are my signals? Your signals are 1 (unreadable), 2 (somewhat readable), 3 (readable with difficulty), 4 (readable with almost no difficulty), 5 (perfectly readable).
QRL	Are you busy? I am busy.
QRM	Are you experiencing interference? I am experiencing interference.
QRN	Are you bothered by static? I am bothered by static.
QRO	Shall I increase transmitter power? Increase transmitter power.
QRP	Shall I decrease transmitter power? Decrease transmitter power.
QRQ	Shall I send faster? Send faster, or speed up to — words per minute.
QRS	Shall I send more slowly? Send more slowly, or at — words per minute.
QSD	Is my keying bad? Your keying is bad.
QSG	Shall I send more than one message? Send more than one (or —), messages.
QSK	Can you hear me between your signals? Or, do you have full break-in capability? I can hear you between my signals. Or, I have full break-in capability.
QSL	Can you acknowledge receipt of my message? I acknowledge receipt of your message.
QSM	Shall I repeat my message? Repeat your message.
QSN	Did you hear me on — MHz? I heard you on — MHz.
QSO	Can you communicate with —? I can communicate with —.
QSP	Will you send a message to —? I will send a message to —.
QSQ	Is there a doctor there? Or, is — there? A doctor is here. Or, — is here.
QSU	On what frequency shall I reply? Reply on — MHz.
QSV	Shall I send a series of Vs? Send a series of Vs.
QSW	On which frequency will you transmit? I will transmit on — MHz.
QSX	Will you listen for —? I will listen for —.
QSY	Shall I change the frequency of my transmitter and/or receiver? Change the frequency of your transmitter and/or receiver.

Q SIGNAL: QUERY AND STATEMENT INFORMATION Continued.

Signal	Meaning
QRT	Shall I stop transmitting? Stop transmitting.
QRU	Have you any information for me? I have no information for you.
QRV	Are you ready? I am ready.
QRW	Shall I tell — that you are calling him/her? Tell — that I am calling him/her.
QRX	When will you call me again? Call me again at —.
QRY	What is my turn? Your turn is number — in order.
QRZ	Who is calling me? You are being called by —.
QSA	How strong are my signals? Your signals are: 1 (almost imperceptible), 2 (weak), 3 (fairly strong), 4 (strong), 5 (very strong).
QSB	Are my signals varying in strength? Your signals are varying in strength.
QTN	When did you leave —? I left — at —.
QTO	Are you airborne? I am airborne.
QTP	Do you intend to land? I intend to land.
QTR	What is the correct time? The correct time is — UTC.
QTX	Will you stand by for me? I will stand by for you until —.

Signal	Meaning
QSZ	Shall I send each word or group of words more than once? Send each word or group of words — times.
QTA	Shall I cancel message number —? Cancel message number —.
QTB	Does your word count agree with mine? My word count does not agree with yours.
QTC	How many messages do you have to send? I have — messages to send.
QTE	What is my bearing relative to you? Your bearing relative to me is — degrees.
QTH	What is your location? My location is —.
QTJ	What is your speed? My speed is — miles or kilometers per hour.
QTL	In which direction are you headed? I am headed toward — or at bearing — degrees.
QUA	Do you have information concerning —? I have information concerning —.
QUD	Have you received my urgency signal, or that of —? I have received your urgency signal, or that of —.
QUF	Have you received my distress signal, or that of —? I have received your distress signal, or that of —.

QSL BUREAU

For hams who work DX, exchanging QSL cards in large numbers can become a nuisance. A *QSL bureau* is a means of saving some of this work. To use a QSL bureau for receiving QSL cards, you inform stations you contact to "QSL via bureau," or "QSL via (callsign)." All cards will be sent to the bureau. You collect them periodically. This is called an *incoming QSL bureau*.

An *outgoing QSL bureau* works the other way around. You send cards to the bureau; the bureau mails them and bills you for postage plus a nominal fee for their work.

One of the largest QSL bureaus in the United States is at American Radio Relay League (ARRL) Headquarters, 225 Main Street, Newington, CT 06111. They will provide details of their services, if you send them a request for the *QSL Bureau Reprint* and a self-addressed, stamped, business-sized envelope. Most foreign countries have similar bureaus. *See also* QSL CARD.

QSL CARD

In order to qualify for various operating awards, and also for personal satisfaction, many hams collect and send *QSL cards*. The signal *QSL* means "contact confirmed." A typical QSL card has the station callsign in prominent, large print, along with the operator's name, handle (nickname) if any, and full address. Some hams like to try to "work all counties in the U.S.," so a good QSL card also indicates the county in which the station is located.

Information on a QSL card includes the callsign of the station for which contact is being confirmed, date and time of contact, the frequency, and the mode(s) of operation. Signal reports, station equipment, and other data might also be included.

There are several small companies that print QSL cards in large batches at a nominal cost. These companies advertise in the ham magazines. Postcards can serve as QSL cards if you don't send them often. Print your callsign neatly on the front using an indelible marker, and include all the appropriate information on the back side.

QUAD ANTENNA

A *quad antenna* is a form of parasitic array. It operates according to the same principles as the Yagi antenna, except that full-wavelength loops are used instead of half-wavelength straight elements.

A full-wavelength loop has approximately 2 dB gain, in terms of effective radiated power, compared to a half-wavelength dipole. This implies that a quad antenna should have 2 dB gain over a Yagi with the same number of elements. Experiments have shown that this is true.

A two-element quad antenna might consist of a driven element and a director, or a driven element and a reflector. A three-element quad has one driven element, one director, and one reflector. The director has a circumference of approximately 0.97 electrical wavelength; the driven element measures exactly 1 wavelength around; the reflector measures about 1.03 wavelength in circumference (*see* DIRECTOR, DRIVEN ELEMENT, and REFLECTOR). The lengths of the director and reflector depend, to some extent, on the element spacing; therefore, these values should be considered approximate.

Additional director elements can be added to form quad antennas having any desired numbers of elements. Provided that optimum spacing is used, the gain increases as the number of elements increases. Each succeeding director should be slightly shorter than its predecessor. Long quad antennas are practical at very-high and ultra-high frequencies, but they tend to be mechanically unwieldy at high frequencies.

A complete discussion of all the design parameters of quad-antenna construction is beyond the scope of this book. However, a simple two-element quad antenna can be constructed according to the dimensions shown in the illustration. Such an antenna will provide approximately 7 dBd forward gain. The elements are square. The length of each side of the driven element, in feet, is given by:

$$L_d = 251/f$$

where f is the frequency in megahertz. The length of each side of the reflector is given by:

$$L_r = 258/f$$

The element spacing, in feet, is given by:

$$s = 200/f$$

QUAD ANTENNA: Dimensions are covered in the text.

Geometrically, the quad antenna shown has an almost perfect cube shape. For this reason, two-element quad antennas are often called cubical quads. *See also* PARASITIC ARRAY, and YAGI ANTENNA.

QUADRIFILAR WINDING

A set of four coil windings, usually oriented in the same sense and having the same number of turns on a common core, is called a *quadrifilar winding*. Quadrifilar windings are sometimes used in the manufacture of broadband balun transformers. Quadrifilar windings on a toroidal core can be connected to obtain many different impedance-transfer ratios. *See also* BALUN.

QUADRUPLER

A *quadrupler* is a circuit that multiplies a radio-frequency signal by a factor of 4. The quadrupler generally consists of a push-push amplifier with tuned input and output circuits. The input circuit is tuned to the frequency of the incoming signal, while the output circuit is tuned to the fourth harmonic of the incoming-signal frequency (*see* PUSH-PUSH CONFIGURATION).

A power-supply circuit that multiplies the incoming voltage by 4 is sometimes called a *quadrupler*. Such a power supply is used when the secondary of the power transformer will not provide adequate voltage, and when the regulation need not be especially precise. *See also* VOLTAGE MULTIPLIER.

QUAGI ANTENNA

An optimally designed, two-element quad antenna, consisting of a driven element and a reflector, offers about 2 dB gain over a similar yagi antenna. (*See* QUAD ANTENNA, and YAGI ANTENNA).

A few years ago, someone came up with the idea of placing Yagi-type director elements in front of a two-element quad, creating a long Yagi/quad hybrid that might have 2-dB gain over a Yagi of similar boom length. Working models were built and tested at very high frequencies (VHF), especially 144 MHz and above. The so-called *quagi antenna* was indeed found to produce slightly more gain, for a given boom length, than a long Yagi.

The illustration on pg. 390 shows the basic concept of the quagi. The driven-element circumference, in feet, is given by $C_d = 1020/f$, where f is in megahertz. This is an approximate formula for elements constructed of metal tubing at VHF. Precise resonance must be found by experimentation.

The reflector circumference should be about $C_r = 1050/f$. The reflector is about 0.2 wavelength behind the driven element. Optimum circumference for the reflector, and the best spacing between the driven element and the reflector, must be found by experimentation.

The directors in a quagi are linear elements, parallel to the plane of polarization from the driven-element loop. It is crucial that these elements be aligned with the electric lines of flux from the driven element; otherwise the directors will have no effect. Director lengths and spacings are similar to those for long Yagi antennas. Quagi antennas have become quite popular among hams at VHF, especially for satellite and tropospheric modes of communication.

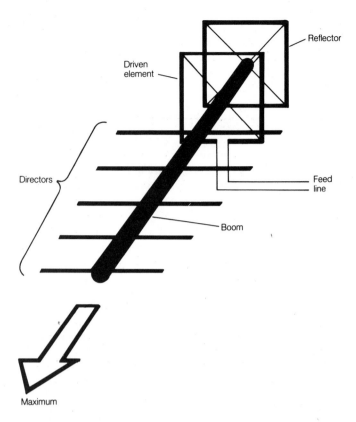

QUAGI ANTENNA: The configuration of a quagi with several directors.

QUALITATIVE TESTING

A *qualitative test* or *expression* involves the determination of equipment performance in general, without considering exact numerical values. An example of a qualitative test of a radio receiver might consist of such exercises as: (1) Switch it on. Does it receive signals? (2) Does it appear to be as sensitive as it should be, or as sensitive as a second receiver used for comparison purposes? (3) Does it seem to be calibrated correctly? (4) Does the selectivity appear adequate?

The above are just a few examples of qualitative tests. More precise testing, in which numerical measurements are made, is called *quantitative testing. See also* QUANTITATIVE TESTING.

QUALITY CONTROL

Quality control is a part of the manufacturing process of any goods. It is also known as *quality assurance*. Quality control is necessary in order to ensure that electronic (or other) equipment performs according to the claimed specifications.

The most common methods of quality control are: the complete testing of each unit at various stages during, and after, manufacture; the complete testing of each unit only after manufacture is completed; the complete testing of a certain proportion of units at various stages during, and after, manufacture; and the complete testing of a certain proportion of units only after manufacture is completed.

The first or second method is preferable, but when it is not possible to test every unit, a less rigorous quality-control procedure can be used. The most comprehensive quality-control procedures are used with complicated devices, equipment that must be calibrated to a high degree of accuracy, or equipment in which human lives or safety is involved.

A person who devises and performs quality-control tests is called a quality-control engineer. All manufacturing companies have one or more such engineers. In some companies, the lead quality-control engineer actually has the power to shut down a production line if manufactured units do not meet the specifications.

QUANTITATIVE TESTING

Quantitative testing involves the evaluation of equipment performance in concrete, numerical terms. This is done with standard test equipment (such as voltmeters, oscilloscopes, wattmeters, and spectrum analyzers).

Examples of quantitative tests for a radio receiver are: (1) How sensitive is it, in terms of the number of microvolts required to produce a specific signal-to-noise ratio at the speaker terminals? (2) What is the actual frequency coverage? (3) How accurate, in terms of percentage discrepancy, is the frequency calibration? (4) How much current does the unit require, in amperes?

Quantitative testing is invariably subject to some error. The accuracy of the measuring instruments is not perfect; the engineer or technician using the apparatus may not interpolate analog readings just right. The maximum measurement error, as a percentage of the total quantity, must be known if quantitative data are to have any meaning. *See also* ACCURACY, ANALOG METERING, ERROR, INTERPOLATION, and QUALITATIVE TESTING.

QUARTER WAVELENGTH

A *quarter wavelength* is the distance that corresponds to 90 degrees of phase as an electromagnetic disturbance is propagated. In free space, it is related to the frequency by a simple equation:

$$\lambda/4 = 246/f$$

where $\lambda/4$ represents a quarter wavelength in feet, and f represents the frequency in megahertz. If $\lambda/4$ is expressed in meters, then the formula is:

$$\lambda/4 = 75/f$$

for the frequency (f) given in megahertz.

In media other than free space, electromagnetic waves propagate at speeds less than their speed in free space. The wavelength of a disturbance having a given frequency depends on the speed of propagation. In general, if v is the velocity factor in a given medium (*see* VELOCITY FACTOR), then:

$$\lambda/4 = 246v/f$$

in feet, and:

$$\lambda/4 = 75v/f$$

in meters.

The quarter wavelength is important from an electrical standpoint. Conductors and transmission lines have special properties when they have length $\lambda/4$. *See also* CYCLE, MONOPOLE ANTENNA, QUARTER-WAVE TRANSMISSION LINE, and WAVELENGTH.

QUARTER-WAVE MATCHING SECTION

See QUARTER-WAVE TRANSMISSION LINE.

QUARTER-WAVE TRANSMISSION LINE

A *quarter-wave transmission line* is any section of transmission line that measures an electrical quarter wavelength. If the velocity factor of a certain type of line is given by v (*see* VELOCITY FACTOR), then a quarter-wave transmission line measures:

$$L = 246v/f$$

in feet, where f is the frequency in megahertz. The length in meters is:

$$L = 75v/f$$

A quarter-wave section of transmission line can be used for the purpose of impedance transformation. Suppose a quarter-wave line has a characteristic impedance of Z_0 ohms. Also imagine that a purely resistive impedance of Z_1 ohms is connected to one end of this section. Then, at the other end, the impedance Z_2 will also be a pure resistance, with a value of:

$$Z_2 = Z_0^2/Z_1$$

Quarter-wave sections are used in a variety of antenna systems, especially phased arrays, for the purpose of impedance matching. Quarter-wave sections can also act as series-resonant or parallel-resonant tuned circuits. *See also* CHARACTERISTIC IMPEDANCE, COAXIAL TANK CIRCUIT, and QUARTER WAVELENGTH.

QUARTZ

Quartz is a naturally occurring, clear mineral, with characteristic hexagonal cleavages. Quartz has a higher index of refraction and a greater hardness than glass. Quartz has a characteristic sparkle, not unlike that of diamond. Quartz can be grown artificially. Chemically, it is composed of silicon dioxide.

Quartz is used in electronics because of its piezoelectric properties. When a piece of quartz is subjected to electric currents, it will produce vibrations. The vibration frequency is determined by the size and shape of the crystal. The frequency can be from audio up to several megahertz.

Probably the most important use of quartz in electronics is in the manufacture of oscillator crystals. Quartz crystals can be made to oscillate at frequencies of several megahertz, and the frequency is extremely stable. *See* CRYSTAL.

QUARTZ CRYSTAL

See CRYSTAL.

QUICK-BREAK FUSE

A *quick-break fuse*, also known as a *fast-blow fuse*, is designed to break a circuit almost instantaneously if excessive current is drawn. Quick-break fuses are used in circuits containing components that are easily destroyed, in a short time, by excessive voltages or currents. Such components include many solid-state devices, such as bipolar transistors, diodes, and field-effect transistors.

A typical quick-break fuse consists of a piece of wire enclosed in a cartridge. The wire is of such a gauge that it will immediately melt with current greater than a precise magnitude. Quick-break fuses are available for many different applications; some blow with current flow less than 1 mA, and others will tolerate 10 to 20 A or more. In some circuits, it is desirable to have a fuse that will break the circuit rather slowly. *See also* FUSE, and SLOW-BLOW FUSE.

QUICK CHARGE

A rechargeable battery, such as a lead-acid type or a nickel-cadmium type, may be recharged at various current levels. In general, within reason, a rechargeable battery having x ampere hours capacity can be charged at I amperes for t hours, as long as $x = tI$. When the time (t) is relatively short, the charging process is called *quick charging*.

In most cases, the shorter the charging period, the less readily a battery will hold the charge. Therefore, it is desirable, whenever possible, to use a slow-charging process (trickle charge). But this is not always convenient. A quick charge can provide short-term operation of a battery in an emergency situation. *See also* TRICKLE CHARGE.

RACK AND PINION

A special form of gear drive, intended to convert rotary motion to linear motion or vice-versa, is called a *rack and pinion*. Rack-and-pinion drives are used in some tuning-dial mechanisms.

The control shaft is attached to the circular gear. The indicator needle is attached to the straight gear. As the shaft is rotated, the needle thus moves back and forth. Several rotations of the control shaft are necessary to move the indicator needle from one end of the scale to the other.

Rack-and-pinion dial mechanisms provide high resolution and accurate resettability. There are several other commonly used methods of constructing dial-readout systems. *See also* DIAL SYSTEM.

RADAR

Almost since the discovery of radio, it has been known that electromagnetic waves are reflected from many kinds of objects, especially metallic objects. Just before and during World War II, this property of radio waves was put to use for the purpose of locating aircraft. The term *radar* is a contraction of the full technical description, *r*adio *d*etection *a*nd *r*anging. Since the war, it has been found that radar is useful in a great variety of applications, such as measurement of automobile speed (by the police), weather forecasting (rain reflects radar signals), and even the mapping of the moon and the planet Venus. Radar is extensively used in aviation, both commercial and military.

A radar system consists of a transmitter, a highly directional antenna, a receiver, and an indicator or display. The transmitter produces intense pulses of microwave electromagnetic energy at short intervals. The pulses are propagated outward in a narrow beam from the antenna, where they strike objects at various distances. The reflected signals, or echoes, are picked up by the antenna shortly after the pulse is transmitted. The farther away the reflecting object, or target, the longer the time before the echo is received. The transmitting antenna is rotated so that all azimuth bearings can be observed.

As the radar antenna rotates, and echoes are received from various directions, the electron beam in the cathode-ray tube is swept around and around. The position of the sweep coincides exactly with the position of the antenna. The period of rotation can vary, but it is typically from 1 to 10 seconds. The transmitted pulse frequency must be high enough so that targets will not be missed as the antenna rotates.

When an echo is received, it appears as a spot on the screen, depending on the shape and size of the target. The cathode-ray-tube phopshor has persistence, so a blip will remain visible for about one antenna rotation. This makes it easy to see the echoes on the screen, and it also facilitates comparison of the relative locations of targets in a group (*see* PERSISTENCE TRACE).

The maximum range of a radar system depends on several factors: the height of the antenna above the ground, the nature of the terrain in the area, the transmitter output power and an-tenna gain, the receiver sensitivity, and the weather conditions in the vicinity. Airborne long-range radar can detect echoes from several hundred miles away under ideal conditions. A low-power radar system, with the antenna at a low height above the ground, can receive echoes from only about 50 miles.

The fact that rain reflects radar echoes is a nuisance to aviation personnel, but it is invaluable for weather forecasting and observation. Radar has made it possible to detect and track severe thunderstorms and hurricanes. The double-vortex type thunderstorm, which is likely to produce tornadoes, causes a hook-shaped echo on radar. The eye of the hurricane is easy to distinguish on radar.

A special form of radar, called *Doppler radar*, is used to measure the speed of an approaching or retreating target. As its name implies, Doppler radar operates by means of Doppler effect. *See also* DOPPLER EFFECT.

RADIAL

A *radial* is a conductor that is used to enhance the ground system of an unbalanced, vertical monopole antenna. Radials can be constructed from wire or metal tubing. They generally measure ¼ wavelength or more.

When a vertical monopole antenna is mounted at ground level, the earth itself can be quite lossy, and will not serve as a very good radio-frequency ground. The ground conductivity is improved by the installation of radials, as shown in the illustration. The radials are run outward from the base of the antenna, and are connected to the shield part of the coaxial feed line. The improved ground conductivity results in better efficiency for any type of ground-mounted monopole, including the conical and Marconi types (*see* CONICAL MONOPOLE ANTENNA, MARCONI ANTENNA, and MONOPOLE ANTENNA).

The radials can be installed just above the ground, or buried just under the surface. The greater the number of radials for a given length, the better the antenna will work. Also, the longer

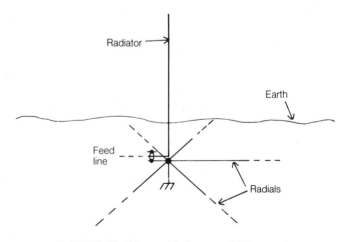

RADIAL: Radials provide improved RF ground.

the radials for a given number, the better. Some broadcast stations have 360 radials, one for every azimuth degree, each measuring 1/2 wavelength at the broadcast frequency.

If a vertical monopole is placed well above the ground level, forming a ground-plane antenna, there need only be three or four radials of exactly 1/4 wavelength (*see* GROUND-PLANE ANTENNA). The radials in the groundplane antenna sometimes run down at a slant, rather than horizontally outward; such radials are called *drooping radials* (*see* DROOPING RADIAL).

Radials can be used to obtain a good radio-frequency ground for a set of communications equipment in any situation. A large number of radials measuring at least 1/4 wavelength, attached to a station ground rod, can improve transmitting efficiency and reduce received noise.

RADIAN

A *radian* is the angle measure subtended around the perimeter of a circle, over a length equal to the radius of the circle. Because the circumference of a circle is 2π times the radius, there are 2π radians in a full circle of 360 degrees.

The radian is used as an angle of measure by mathematicians and physicists. This is because, in many situations, the radian arises naturally, while the degree was fabricated by man and is not based on pure mathematics. To 10 significant digits, a radian is equivalent to 57.29577951 degrees.

For most practical purposes, degrees can be obtained from radians by multiplying by 57.3; an angle measure in radians can be obtained from degrees by multiplying by 0.0175.

RADIATION LOSS

In a radio-frequency feed line, there is always some loss; no feed line is 100-percent efficient. The conductor resistance and the dielectric loss contribute to the dissipation of power in the form of heat. Some energy can also be radiated from the line, and this energy, because it cannot reach the antenna, is considered loss.

In an unbalanced feed line, such as a coaxial cable, radiation loss can occur because of inadequate shield continuity. Radiation loss can also occur because of currents induced on the outer shield by the field surrounding the antenna. Radiation loss in a coaxial feed line can be minimized by using cable with the highest possible degree of shielding continuity, and also by ensuring that the ground system is adequate at radio frequencies. A balun, placed at the feed point, where a coaxial cable joins a balanced antenna system, can also be helpful.

In a balanced feed line, such as open wire or twin-lead, radiation loss results from unbalance between the currents in the line conductors. Ideally, the currents should be equal in magnitude and opposite in phase everywhere along a balanced line. If this is not the case, radiation loss will occur. Imbalance in a parallel-conductor line can result from a physically or electrically asymmetrical antenna, lack of symmetry in the antenna/feed configuration, or poor feed-line installation. *See also* BALANCED LINE.

RADIATION RESISTANCE

When radio-frequency energy is fed into an electrical conductor, some of the power is radiated. If a nonreactive resistor were substituted for the antenna, in combination with a capacitive or inductive reactance equivalent to the reactance of the antenna, the transmitter would behave in precisely the same manner as it would when connected to the actual antenna. The resistor would dissipate the same amount of power as the antenna would radiate. For any antenna, there exists a resistance, in ohms, for which this can be done. The value of such a theoretical resistor is called the *radiation resistance* of the antenna.

Radiation resistance depends on several factors. The main consideration is the length of the antenna, as measured in free-space wavelengths. The presence of objects near the antenna (such as trees, buildings, and utility wires) can also affect the radiation resistance.

Suppose that you place an infinitely thin, perfectly straight, lossless vertical monopole antenna over perfectly conducting ground. Further suppose that there are no objects in the vicinity to affect the radiation resistance. Then, the radiation resistance, as a function of the vertical-antenna height in wavelengths, is as shown at A in the illustration.

For a quarter-wavelength vertical, the radiation resistance is approximately 37 ohms. As the conductor length decreases, the radiation resistance also decreases, becoming zero when the

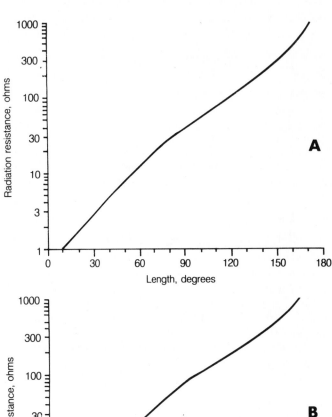

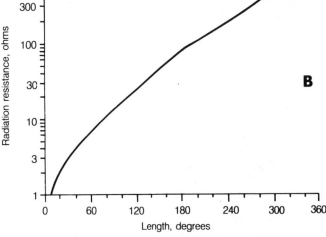

RADIATION RESISTANCE: At A, for a monopole over perfect ground; at B, for a center-fed radiator in free space.

conductor vanishes. As the conductor becomes longer than a quarter wavelength, the radiation resistance increases, and becomes larger without limit as the height approaches a half wavelength. These are theoretical values. In practice, the radiation resistance is somewhat lower than the figures given in A. However, the graph represents a good approximation for most practical purposes.

Suppose that you place a conductor in free space and feed it at the exact center. Again, it is assumed that the conductor is perfectly straight, infinitely thin, lossless, and far from objects that might affect it. Then, the value of the radiation resistance, as a function of the antenna span in wavelengths, is given by the graph at B.

When the conductor measures a half wavelength, the radiation resistance is approximately 73 ohms. As the conductor becomes shorter, the radiation resistance gets smaller, approaching zero. As the conductor is lengthened, the radiation resistance increases without bound as the full-wavelength value is approached. As with the vertical system, the values shown are theoretical, but they represent a fair approximation for most center-fed vertical or horizontal antenna systems.

In practice, it is desirable to have a radiation resistance that is very large. This is because the efficiency of an antenna depends on the ratio of the radiation resistance to the total system resistance. If the radiation resistance is R and the loss resistance is S, the total system resistance, Z, is given by:

$$Z = R + S$$

and the efficiency, in percent, is:

$$Eff \text{ (\%)} = 100 \ (R/Z)$$

The loss resistance in an average antenna system is a few ohms, but can be as high as 30 to 50 ohms or more. If the radiation resistance is very low, therefore, an antenna tends to be rather inefficient. *See also* ANTENNA EFFICIENCY.

RADIO AMATEUR SATELLITE CORPORATION

The *Radio Amateur Satellite Corporation (AMSAT)* is a nonprofit institution established for the purpose of placing ham radio transponders on orbiting satellites, and promoting their use. AMSAT publishes literature, including a periodical, *AMSAT Satellite Journal*. The organization can be reached by writing to AMSAT, P.O. Box 27, Washington, DC 20044.

The U.S. ham satellites are called *OSCAR* (*o*rbiting *s*atellite *c*arrying *a*mateur *r*adio). *See* ACTIVE COMMUNICATIONS SATELLITE, and OSCAR.

RADIO ASTRONOMY

In the 1930s Karl Jansky, an employee of the Bell Laboratories, discovered that the center of the Milky Way produces radio-frequency energy at a wavelength of 15 meters. Jansky was the first scientist to discover that radio waves come from the cosmos. Many others have followed him; a few are named here, but it is regrettably impossible to name all of the radio astronomers who have added so much to our understanding of the universe.

A few years after Jansky's discovery, a radio amateur, Grote Reber, built a 31-foot dish antenna for the purpose of receiving radio noise from space. Reber also found radio waves coming from the Milky Way, at various wavelengths down to about 2 meters. With his modest apparatus and a lot of hard work, Reber was able to make a rough radio map of our galaxy. By 1940, astronomers began to get interested, and the science of radio astronomy was born.

Modern radio astronomers use large dish antennas; some are several hundred feet across. A phased antenna system, called the *interferometer*, allows radio astronomers to view the sky with a high degree of resolution. Systems (such as the Mills Cross in Australia) are also used (*see* INTERFEROMETER, and MILLS CROSS). Most of the observations are carried out at ultra-high and microwave frequencies, where the wavelengths are short enough to allow good resolution and the electromagnetic fields are not greatly affected by the atmosphere of the earth.

Radio astronomy has provided a wealth of knowledge that could not have been discovered by optical means. Quasars, pulsars, and radio galaxies, for example, would not have been found by optical telescopes.

An offshoot of radio astronomy has developed in recent years. High-powered transmitters, in conjunction with high-gain antennas, have made it possible to receive echoes from the nearer planets. The distances of these objects can be precisely measured, and their surfaces mapped, by a technique called radar astronomy.

Perhaps the most significant contribution that radio astronomy has made, however, was inadvertent. In 1965, two Bell Laboratories employees, Arno Penzias and Robert Wilson, found that a weak radio-noise signal seemed to be coming from all directions in space. They had discovered the remnants of radiation from the very beginning of the universe 20,000,000,000 years ago. At that time, it is believed, all the matter in the universe exploded in an incredible cataclysm called the *big bang*. *See also* RADIO TELESCOPE.

RADIO FREQUENCY

An electromagnetic disturbance is called a *radio-frequency (RF) wave* if its wavelength falls within the range of 30 km to 1 mm. This is a frequency range of 10 kHz to 3000 GHz.

The radio-frequency spectrum is split into eight bands, each representing one order of magnitude in terms of frequency and wavelength. These bands are the very-low, low, medium, high, very-high, ultra-high, super-high, and extremely-high frequencies. They are abbreviated, respectively, as VLF, LF, MF, HF, VHF, UHF, SHF, and EHF. Super-high-frequency and extremely-high-frequency RF waves are sometimes called *microwaves* (*see* BAND).

Radio-frequency waves propagate in different ways, depending on the wavelength. Some waves are affected by the ionosphere, troposphere, or other environmental factors (*see* PROPAGATION CHARACTERISTICS).

Radio frequencies represent a sizable part of the electromagnetic spectrum. As the wavelength becomes shorter than 1 mm, the infrared, then visible light, ultraviolet, X rays, and gamma rays are encountered. *See also* ELECTROMAGNETIC SPECTRUM.

RADIO-FREQUENCY INTERFERENCE

See ELECTROMAGNETIC INTERFERENCE.

RADIO-FREQUENCY PROBE

A *radio-frequency probe* is a special pickoff device, intended for sampling radio-frequency signals. A radio-frequency probe resembles an ordinary alternating-current or direct-current probe (*see* PROBE), except that precautions are taken to prevent signal leakage.

A radio-frequency probe must have a shielded cable, and the cable length must be as short as possible. The probe impedance is high, generally about 1 megohm. If the probe is designed to be held with the hand, the handle must be adequately insulated and shielded to prevent hand capacitance from affecting the impedance.

Radio-frequency probes are sometimes inserted into the tuned coil of a transmitter final amplifier, for such purposes as signal monitoring or the operation of radio-frequency-actuated devices. This type of probe consists of a short, stiff wire or a small coil, which allows some energy to be picked up without affecting the tank-circuit impedance.

RADIO-FREQUENCY TRANSFORMER

At radio frequencies, transformers are often used for the purpose of impedance matching. Radio-frequency transformers are also used for coupling between or among amplifiers, mixers, and oscillators. A special form of radio-frequency transformer, known as a *balun*, is sometimes used in antenna systems to match impedances and optimize the balance (*see* BALUN).

A radio-frequency transformer can consist of solenoidal windings with an air core, solenoidal windings with a powdered-iron or ferrite core, or toroidal windings with a powdered-iron or ferrite core.

The air-core transformer presents the least loss, and is therefore the most efficient; however, air-core transformers are rather bulky and are easily affected by nearby metallic objects. Solenoidal transformers with powdered-iron or ferrite cores are less bulky and almost as efficient as the air-core type. The solenoidal core can be moved in and out of the coil to vary the inductances of the windings (*see* PERMEABILITY TUNING). Toroidal transformers offer the advantage of being immune to the presence of nearby metallic objects because all of the magnetic flux is contained within the powdered-iron or ferrite core. This facilitates miniaturization (*see* TOROID).

When no reactance is present, the impedance-transfer ratio of a radio-frequency transformer is equal to the square of the turns ratio. Thus, a 2:1 turns ratio results in an impedance-transfer ratio of 4:1; a turns ratio of 3:1 produces a 9:1 impedance-transfer ratio. *See also* IMPEDANCE TRANSFORMER.

RADIOLOCATION

Radiolocation is a process by which the position of a vehicle, aircraft, or ocean-going vessel is determined. A radiolocation system can operate in different ways. The simplest method is the directional method. Two or three fixed receiving stations, some distance from each other, are used. Direction-finding equipment is used at each of these stations, in conjunction with an omnidirectional transmitter aboard the vessel, to establish the bearings of the vessel, with respect to each station. The vessel location corresponds to the intersection point of great circles drawn outward from the receiver points in the appropriate directions (*see* DIRECTION FINDER, and GREAT CIRCLE).

This method of radiolocation requires that the vessel be equipped with a transmitter, and that the captain of the vessel wants to be located. If the captain does not want to be located, or if the radio equipment aboard the vessel is not working, other means of radiolocation are necessary. Radar can be used to locate such vessels (*see* RADAR). Sometimes enemy craft can be located by visual or infrared apparatus. In recent years, satellites have been developed that can locate enemy ships and missiles.

A special form of over-the-horizon radiolocation device, operating in the high-frequency radio band, has been recently developed to detect and locate launchings of intercontinental ballistic missiles. Because of its sound in radio receivers, this device has been called a *woodpecker*. *See also* RADIONAVIGATION, and WOODPECKER.

RADIONAVIGATION

Radionavigation refers to the use of radio apparatus, by personnel aboard moving vessels, for the purpose of determining position and plotting courses.

There are many means of radionavigation. The intersecting-line method is probably the simplest. Two or three land-based transmitters are needed. Their locations must be accurately known. A direction-finding device on the vessel is used to determine the bearings of each of the transmitters (*see* DIRECTION FINDER). After the bearings have been determined, great circles are drawn on a map outward from each transmitter so that the lines all intersect at a common point. That point represents the position of the vessel.

An alternative intersecting-line method requires a transmitter aboard the vessel. The position is determined by land-based personnel with the aid of direction finders. The land stations, in communication with each other and with the captain of the vessel, can inform the captain of his position (*see* RADIOLOCATION).

Modern radionavigation systems use low-frequency, land-based, transmitters in conjunction with computers aboard the vessel. The signals or pulses from the land-based transmitters arrive at the vessel in varying phase as the vessel follows its course. If the ship or plane strays from its course, the computer indicates the error. These systems are known as *hyperbolic radionavigation systems* (*see* LORAN).

Aircraft radionavigation can be performed by radar, or with direction-finding apparatus at the very-high or ultra-high frequencies. Air-traffic controllers constantly monitor the skies via radar. An airline pilot is automatically told if the plane is off course. *See also* RADAR.

RADIOTELEPHONE

Radiotelephone refers to any form of two-way communication by voice. The earliest form of radiotelephone was amplitude modulation *See* AMPLITUDE MODULATION, EMISSION CLASS, FREQUENCY MODULATION, PULSE MODULATION, and SINGLE SIDEBAND.

RADIO TELESCOPE

A *radio telescope* is a sensitive, highly directional radio receiver, intended for intercepting and analyzing radio-frequency noise from the cosmos. The first radio telescopes were built by Karl

Jansky and Grote Reber in the 1930s. Modern radio telescopes use the most advanced antennas, receiver preamplifiers, and signal processing techniques. Radio astronomy is generally carried out at wavelengths shorter than about 10 m, but especially at ultra-high and microwave frequencies.

A block diagram of a radio telescope is illustrated. The installation consists of an antenna, a feed line, a preamplifier, the main receiver, and a signal processor and recorder.

A radio-telescope antenna can consist of a large dish or an extensive phased array. The dish antenna offers the advantage of being fully steerable (usually), so that it can be pointed in any direction without changing the resolution or sensitivity. The phased array, which can consist of a set of yagis, dipoles, or dish antennas, provides better resolution because it can be made very large. However, the steerability is not as flexible; phased arrays can usually be steered only along the meridian (north and south). The rotation of the earth must then be used to obtain full coverage of the sky.

Two or more large antennas, spaced a great distance apart, can be used to obtain extreme resolution with a radio telescope. This technique is called *interferometry* (*see* INTERFEROMETER). The interferometer allows the radio astronomer to probe into the details of celestial radio sources. In some cases, what was originally thought to be a single source has been found to contain multiple components.

The antenna feed line is usually a coaxial cable at very-high and ultra-high frequencies. A waveguide is used for microwave reception. The lowest possible loss, and the highest possible shielding continuity, must be maintained. This minimizes interference from earth-based sources, and ensures that the antenna directivity and sensitivity are optimum.

The radio-telescope receiver differs somewhat from an ordinary communications receiver. Sensitivity is of the utmost importance. A special preamplifier is needed (*see* PREAMPLIFIER). The components can be cooled down to just a few degrees Kelvin to reduce the noise in the circuit (*see* SUPERCONDUCTIVITY).

The main receiver of a radio telescope must have a relatively large response bandwidth, in contrast to a communications receiver, which usually exhibits a very narrow bandwidth. The wavelength must be adjustable over a wide and continuous range.

In a radio telescope, a speaker can be used for effect, but is of little practical value in most cases because cosmic radio emissions sound like little more than hiss or static. A pen recorder and/or oscilloscope, and perhaps a spectrum analyzer, are used to evaluate the characteristics of the cosmic noise (*see* OSCILLOSCOPE, PANORAMIC RECEIVER, and SPECTRUM ANALYZER).

Most radio telescopes are located well away from sources of manmade noise. A city is a poor choice as a location to build a radio telescope. Some radio-telescope observatories, such as the one at Green Bank, West Virginia, are protected against manmade noise by law. The Federal Communications Commission has established a radio quiet zone near this installation. This allows the radio astronomers at Green Bank to carry on their research under the best manageable conditions.

Radio telescopes can ''see'' things that cannot be detected with optical telescopes. For example, the hydrogen line at the 21-cm wavelength can be easily observed with a radio telescope, although it is invisible with optical apparatus. It is

thought that the 21-cm line might serve as a marker for possible interstellar communication—messages from other worlds.

The radio telescope, with its high-gain, high-resolution antennas, can be used to transmit signals as well as receive them. This has allowed astronomers to precisely ascertain the distances to, the motions of, and the surface characteristics of the moon, Venus, and Mercury. When a radio telescope is used in this way, it is called a *radar telescope*.

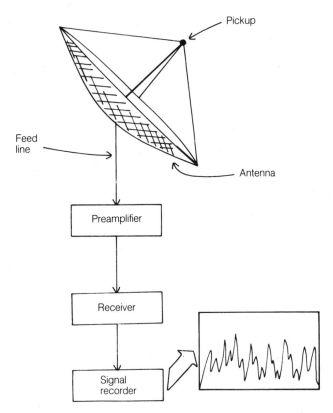

RADIO TELESCOPE: A simplified block diagram.

In 1974, Dr. Frank Drake used the massive radio telescope at Arecibo, Puerto Rico, to transmit the first message to the cosmos. This transmission has already passed some of the nearer stars. A binary code was used, telling of our world and our society. Perhaps it has been received by now. Our radio telescopes might someday receive a reply. *See also* RADIO ASTRONOMY.

RADIOTELETYPE

Radioteletype is a mode of communications in which printed messages are sent and received via radio waves. For this to be possible, a special device is needed to properly modulate a transmitter so that the printed characters are conveyed. A demodulator is needed to convert the audio output of the receiver into the electrical impulses that drive the teleprinter or cathode-ray-tube display. Both the modulator and demodulator are usually combined into a single circuit called a *terminal unit*. The terminal unit is a form of modem (*see* MODEM).

The most common method of transmitting radioteletype signals is frequency-shift keying. Two different carrier frequencies are used, called the *mark (keydown)* and the *space (key-up) conditions*. The ASCII or Baudot codes are used with frequency-shift keying (*see* ASCII, BAUDOT, and FRE-QUENCY-SHIFT KEYING). Morse code is sometimes used to transmit and receive radioteletype signals.

The illustration is a simple block diagram of a radioteletype station using a single-sideband transceiver. The received signals are fed into the terminal-unit demodulator from the speaker terminals of the transceiver. The modulator generates a pair of audio tones, which are fed into the microphone terminals of the transceiver. This arrangement can provide communication via ASCII or Baudot at speeds up to approximately 300 baud.

Commercially manufactured radioteletype equipment, including terminal units, cathode-ray-tube displays, and printers, is available. A radioteletype station is, however, not hard to build from scratch, or "junk-box," components. Many amateur radioteletype enthusiasts enjoy building their own terminal units and using discarded machines.

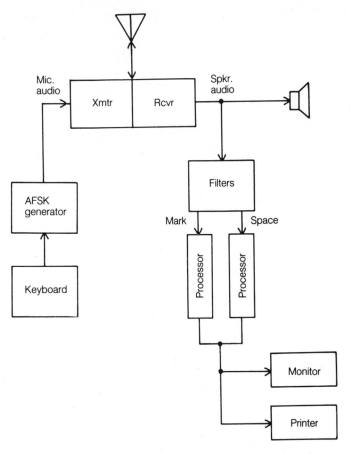

RADIOTELETYPE: Block diagram of a radioteletype station.

RAG CHEWING

Many hams like to carry on long conversations during a QSO, or contact. There are as many subjects to talk about, as there are in any ordinary conversation. Sometimes, a contact can go on for hours. This pastime is called *rag chewing*.

When three or more hams get involved in a rag chew, each one makes a contribution in a rotating sequence. This is done often in informal nets. It is called a *round table*.

Rag chewing is an integral, and important, part of ham radio to those who do it. Lifetime friends can be made that way.

It can be fun. Confirmed rag chewers like to take a little time, getting to know the person on the other radio, rather than being concerned with equipment, propagation conditions, and contact volume and efficiency. Conversely, contesters and DXers enjoy the challenges that their pastimes offer.

There are frequencies on which rag chewing is all right, and those on which it is not. In some cases, rag chewing is frowned upon; for example, it should not be done on a repeater. Common rules of courtesy and etiquette apply to rag chewing, as to all ham radio communications. If you might get in the way of someone else's need for a frequency or repeater, your rag chew should go simplex on some unoccupied frequency.

Rag chewing with rare DX will earn you many enemies in a very short time — if the DX station will even engage you. There are some modes on which rag chewing is difficult or impossible for physical reasons. These include meteor scatter, low-orbit OSCAR communications, and moonbounce.

RAM

See RANDOM-ACCESS MEMORY.

RAMP WAVE

A *ramp wave*, or *ramp function*, is a form of sawtooth wave in which the amplitude rises steadily and drops abruptly (*see* SAWTOOTH WAVE). The rise in amplitude is linear, with respect to time, so that the increasing part of the wave looks straight on an oscilloscope display. The ramp wave is usually of only one polarity, with the minimum amplitude being zero at the beginning of the linear rise.

The ramp wave is ideally suited for applications that require scanning. For example, a ramp-wave generator is used in an oscilloscope, spectrum analyzer, television receiver, or camera tube to make the electron beam move across the screen at a constant speed. A ramp wave is also used to operate a device called a *sweep generator* (*see* SWEEP GENERATOR).

RANDOM-ACCESS MEMORY

A form of memory that allows inputting and removal of data in any order, regardless of the particular data addresses, is called *random-access memory (RAM)*. An integrated circuit that contains this type of memory circuitry is sometimes called a *RAM*.

Random-access memory is extensively used in computers and other data-processing systems. The RAM offers much greater ease and convenience of retrieval than other forms of memory.

RANDOM NOISE

Certain forms of electromagnetic noise, especially manmade noise, exhibit a definite pattern of amplitude versus time. Other types of noise show no apparent relationship between amplitude and time; such noise is called *random noise*. An example of non-random noise is the impulse type. Example of random noise are sferics, thermal noise, and shot noise.

Random noise is more difficult to suppress than noise having an identifiable waveform. If noise shows discernible amplitude-versus-time patterns, you can design a circuit that "knows" the noise behavior, and that can act accordingly to get rid of it. No circuit exists that "knows" the patterns of random noise: there are no patterns!

A limiting circuit, which at least allows the desired signal to compete with the noise, is a good defense against strong random noise. The use of narrowband emission, in conjunction with a narrow receiver bandpass, is also helpful in dealing with strong random noise. The use of frequency modulation can give better results than amplitude modulation if the receiver is equipped with an effective limiter or ratio detector. Directional or noise-cancelling antenna systems can sometimes improve communications in the presence of random noise. *See also* NOISE, and NOISE LIMITER.

In a radio broadcasting or communications system, the distance over which messages can be sent and received is sometimes called the *range*. The communications range depends on the transmitter power, the type and location of the antennas, the receiver sensitivity, and various propagation factors. *See also* PROPAGATION CHARACTERISTICS.

In radar, the distance from the station to a target is called the *range*. Most radar systems display the range radially from the center of a circular screen. *See also* RADAR.

The term *range* is used in a variety of other situations in electronics. We can speak, for example, of the useful range of a meter, the range over which the value of a component varies or fluctuates, or the range of power-supply voltage that a circuit will tolerate without malfunctioning. A site for measuring the characteristics of an antenna, such as the gain, front-to-back ratio, or front-to-side ratio, is called an *antenna test range*. *See also* TEST RANGE, and TOLERANCE.

RASTER

In a television receiver tuned to a channel on which there is no signal, the rectangle of light, composed of horizontally scanned lines, is called the *raster*.

In the United States, the television-broadcast raster is standardized at 525 lines per frame. The aspect ratio, or ratio of width to height, is 4 to 3. In some other countries, the number of lines per frame is 625, rather than 525; however, the aspect ratio is almost universally 4 to 3.

In recent years, because of improved television broadcasting and receiving techniques, some thought has been given to increasing the number of lines per frame to get better resolution. *See also* TELEVISION.

RATIO-ARM BRIDGE

Ratio-arm bridge is a device for measuring resistance. The bridge gets its name from the fact that it compares the value of a standard resistance with that of the unknown resistance by electrically determining their ratio.

RATIO DETECTOR

The *ratio detector* is a special form of detector for frequency-modulated or phase-modulated signals. The ratio detector, unlike other types of detectors for these two modes, is insensitive to changes in amplitude. This eliminates the need for a preceding limiter stage. The ratio detector was developed by RCA. The ratio detector is used in most modern high-fidelity and television receivers. The ratio detector is not as commonly used in communications equipment.

A typical passive ratio detector is shown schematically in the illustration. A transformer, with tuned primary and secondary windings, splits the signal into two components. A change

in amplitude will result in equal and opposite fluctuations in both halves of the circuit; thus they cancel out. But if the frequency changes, an instantaneous phase shift is introduced, unbalancing the circuit and producing output.

The ratio detector is not as sensitive as the discriminator, but additional amplification can be incorporated to make up for this slight deficiency, which amounts to approximately 3 dB. *See also* DISCRIMINATOR, FREQUENCY MODULATION, and PHASE-LOCKED LOOP.

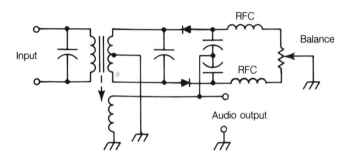

RATIO DETECTOR: For frequency or phase modulation.

RATS OPEN SYSTEM ENVIRONMENT (ROSE)

The acronym *RATS* stands for *Radio Amateur Telecommunications Society*, a ham radio club located in New Jersey. The *RATS Open Systems Environment* is a protocol that is popular with users of packet radio. When hams talk about the RATS Open Systems Environment, it is usually abbreviated to another acronym-within-an-acronym, *ROSE*.

In ROSE, operation is streamlined, so that it is not necessary for each packet to have a header to guide the message through the network. It is somewhat like the difference between being able to dial a long-distance phone number directly, rather than having to go through an operator.

In ROSE, *node-to-node acknowledgment* is used. But during operation, communications proceeds as if you were directly linked to the station at the other end of the circuit. A single packet at the beginning of the communications session establishes the circuit path for the duration of the session, placing this path in the memories of all the computers along the way. The path information is erased by a single command when communications ends. This path is known as a *virtual circuit* because, although it might go through several nodes between the two communicating stations, it seems to be a direct link.

The ROSE system contains switches whose locations are encoded via telephone area codes and dialing prefixes. Thus, these codes have six numerals. For example, suppose you want to access a ROSE switch serving South Miami Beach, Florida. The area code for South Florida is 305. There are several different dialing prefixes for South Miami Beach; one is 672 and another is 534. You would use 305672 and 305534, respectively, for these ROSE switches. Obviously, this numbering scheme allows for thousands of ROSE switches throughout the United States. *See also* CONNECTION PROTOCOL, NODE, NODE-TO-NODE ACKNOWLEDGMENT, PACKET RADIO, PROTOCOL, and VIRTUAL CIRCUIT.

RC CIRCUIT

See RESISTANCE-CAPACITANCE CIRCUIT.

RC CONSTANT

See RESISTANCE-CAPACITANCE TIME CONSTANT.

RC COUPLING

See RESISTANCE-CAPACITANCE COUPLING.

REACTANCE

Reactance is the opposition, independent of resistance, that a circuit or component offers to alternating currents. Reactance is exhibited by inductors, capacitors, antennas, and transmission lines. Waveguides and cavities also contain reactance. The unit of reactance is the ohm, and the symbol for reactance is X.

Reactances do not actually dissipate power. Instead, reactances affect the phase relationship between the current and the voltage in an alternating-current circuit. In a circuit having inductive reactance, the current lags behind the voltage by an amount ranging from 0 to 90 degrees. In a capacitive reactance, the current leads the voltage by an amount ranging from 0 to 90 degrees. The exact phase angle depends on the amount of resistance, as well as reactance, in a circuit or component (*see* PHASE ANGLE).

Mathematically, reactance can be either positive or negative. An inductive reactance is considered positive, and a capacitive reactance is considered negative. This is a matter of convention (*see* CAPACITIVE REACTANCE, and INDUCTIVE REACTANCE). In engineering mathematics, reactance takes imaginary values. The ohmic quantities are multiplied by the *j* factor (*see* J OPERATOR). This results in a model that accurately reflects the way in which reactances behave.

When resistance and reactance occur together, the result is a complex impedance. The resistance forms the real part of a complex impedance, and the reactance forms the imaginary part. A complex impedance can be represented in the form:

$$Z = R + jX \text{ or } Z = R - jX$$

where R and X represent resistance and reactance and $j = \sqrt{-1}$. *See also* COMPLEX NUMBER, and IMPEDANCE.

REACTANCE GROUNDING

When a circuit point is grounded through a capacitor or inductor, that point is said to be *reactively grounded*. Reactive grounding is used for various purposes in radio-frequency circuits.

A capacitor, connected between a circuit point and ground, allows some signals to pass while others are shorted to ground. Such a capacitor is called a *bypass* (*see* BYPASS CAPACITOR).

An inductor can be connected between a circuit point and chassis ground, when it is desired to keep the point at a stable dc voltage while allowing radio-frequency signals to exist there. Transistors and tubes, especially in power amplifiers, are often biased in this manner. An antenna can be kept at dc ground, while still accepting radio-frequency energy, by means of reactance grounding (*see* PROTECTIVE GROUNDING).

REACTANCE MODULATOR

A *reactance modulator* is a circuit that produces a variable reactance from an audio-frequency input signal. The variable reactance is coupled to the crystal or tuned circuit of an oscillator, resulting in frequency modulation or phase modulation.

A reactance modulator can be built using a bipolar transistor, a field-effect transistor or a vacuum tube. However, the most common reactance-modulator circuit today uses a device called a varactor diode. The varactor exhibits a variable capacitance, depending on the amount of reverse bias applied to it (*see* VARACTOR DIODE).

The illustration is a schematic diagram of a crystal oscillator incorporating a varactor diode to obtain frequency modulation. The capacitance of the diode pulls the crystal frequency by an amount that varies with the instantaneous audio-frequency amplitude. The deviation produced by this method is fairly small, but frequency multipliers can be used to increase it. *See also* FREQUENCY MODULATION, and PHASE MODULATION.

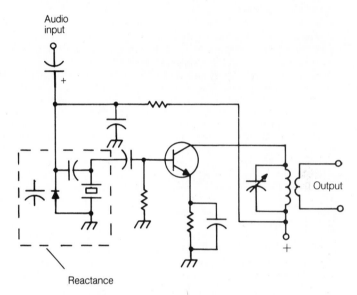

REACTANCE MODULATOR: Generates frequency-modulated signals.

REACTIVE CIRCUIT

An alternating-current circuit is called *reactive* if it contains inductance and/or capacitance, in addition to dc resistance. In a *reactive circuit*, the resistance dissipates power, but the reactance does not. Examples of reactive circuits include antennas, transmission lines, waveguides, and tuned circuits. At radio frequencies, most circuits contain some reactance as well as resistance, and thus they must be considered reactive circuits.

A reactive circuit can become nonreactive at one or more frequencies. This occurs when there are equal amounts of inductive and capacitive reactance. Such a condition is known as *resonance*. *See also* CAPACITIVE REACTANCE, IMPEDANCE, INDUCTIVE REACTANCE, REACTIVE POWER, and RESONANCE.

REACTIVE POWER

In an ac circuit, power can be dissipated in a resistance, but not in a reactance. Nevertheless, voltages and currents do occur in reactances. The product of the voltage and the current in a reactance is called *reactive power*. Reactive power is the difference between the apparent power and true power in a complex impedance (*see* APPARENT POWER, and TRUE POWER). Reactive power is sometimes spoken of as reactive volt-amperes, or imaginary power.

The proportion of apparent power that is reactive depends on the phase angle in the circuit. The phase angle, in turn, depends on the ratio of reactance to resistance. The larger the phase angle, the more of the apparent power is reactive. In a pure resistance, all of the apparent power is true power. In a pure capacitance or inductance, none of the apparent power is true power (*see* POWER FACTOR).

Reactive power, if not recognized, can cause confusion. A radio-frequency wattmeter, installed in an antenna system containing reactance, will exhibit an exaggerated reading. On a directional wattmeter (reflectometer), the reactive power appears as reflected power and the apparent power appears as forward power. The true power (that which can actually be radiated by the antenna) is the difference between the forward and reflected readings. *See also* REFLECTOMETER.

READ-MAINLY MEMORY

A form of memory, similar in practice to the read-only memory, is known as a *read-mainly memory (RMM)*. The contents of RMM are nonvolatile unless deliberately changed by a special process. The process is much too slow, however, to allow the RMM to be used as a read-write memory.

The read-mainly memory is essentially the same as the erasable programmable read-only memory (EPROM). *See also* PROGRAMMABLE READ-ONLY MEMORY, READ-ONLY MEMORY, and READ-WRITE MEMORY.

READ-ONLY MEMORY

A *read-only memory (ROM)* is a common form of digital memory used in computers, calculators, and data-processing devices. An integrated circuit containing such memory is itself called a *ROM*.

The ROM is nonvolatile, retaining its data even when power is removed. Some types of ROM can be programmed as desired, but never changed after that; this is called *programmable read-only memory (PROM)*. Some types of ROM can be erased and reprogrammed at will; this is known as *erasable PROM (EPROM)*. *See also* PROGRAMMABLE READ-ONLY MEMORY.

READ-WRITE MEMORY

Any form of memory in which it is possible to both store and retrieve information is called a *read-write memory*. The contents of a read-write memory can be changed quickly. This makes read-write memory different from the read-mainly or read-only memory (*see* READ-MAINLY MEMORY, and READ-ONLY MEMORY).

Read-write memory can be either nonvolatile or volatile; that is, it might or might not be retained if power is removed. In a calculator or computer, a read-write memory used for temporary storage of numbers in the intermediate stages of a computation is called *scratch-pad memory* (*see* SCRATCH-PAD MEMORY).

There are various forms of read-write memory, used for different purposes. The first-in/first-out (FIFO) memory allows removal of data in the same order it is written. The pushdown stack, or last-in/first-out memory, facilitates removal of data in opposite order from the way it is written. A random-access memory allows retrieval of information in any order. *See also* FIRST-IN/FIRST-OUT, PUSHDOWN STACK, and RANDOM-ACCESS MEMORY.

REAL POWER

See TRUE POWER.

REAL-TIME OPERATION

In any broadcast, communications, or data-processing system, operation done "live" is called *real-time operation*. The term *real-time* applies especially to computers. Real-time data exchange allows a computer and the operator to converse. This is not true of prerecorded modes.

Real-time operation is a great convenience when it is necessary to store and verify data within a short time. This is the case, for example, when making airline reservations, checking a credit card, or making a bank transaction. But real-time operation is not always necessary. It is a waste of expensive computer time to write a long program at an active terminal. Long programs are best written off-line, tested in real time, and then debugged off-line.

In a large computer system, real-time operation can be obtained from many terminals simultaneously. The most common method of achieving this is called *time sharing*. The computer pays attention to each terminal for a small increment of time, constantly rotating among the terminals at a high rate of speed. In a powerful system, this provides great convenience for a large number of operators. *See also* TIME SHARING.

RECEIVER

A *receiver* is a device that converts electromagnetic waves, having come from some transmitter, into the original messages that were sent by that transmitter. In the broadest sense, a receiver is a kind of *transducer*.

Receivers operate over a specific range of electromagnetic frequencies. For example, a high-frequency (HF) ham receiver works from about 1.8 through 29.7 MHz; a 2-meter receiver operates from approximately 144 to 148 MHz. Some receivers work from the very-low-frequency (VLF) range well into the ultra-high-frequency (UHF) spectrum.

Many receivers can pick up several of the common ham emissions. *See* EMISSION CLASS. Others, such as those in FM handheld and mobile units, can operate in only one mode.

The illustration shows a block diagram of a typical receiver. Each stage shown is covered briefly in this article. Receiver design can vary from this general form; it is intended only as a basic overview.

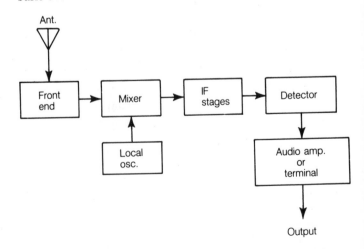

RECEIVER: A general block diagram for a communications receiver.

Front End The *front end* of a receiver consists of the first radio-frequency (RF) amplifier, and often includes bandpass filters between that amplifier and the antenna. The dynamic range and sensitivity of a receiver are determined by the performance of the front end. These two characteristics are among the most important for any receiver. Low-noise, high-gain amplifiers are the rule here. Field-effect transistors are most often used. *See* DYNAMIC RANGE, FRONT END, and SENSITIVITY.

Mixer The most common type of receiver, the *superheterodyne*, uses one or two stages that convert the variable signal frequency to a constant *intermediate frequency*. Such a conversion stage is known as a *mixer*. *See* INTERMEDIATE FREQUENCY, MIXER, and SUPERHETERODYNE RECEIVER. The mixer is a nonlinear circuit that combines the signal with a carrier from a local oscillator (LO). The output is either the sum or the difference of the signal frequency and the LO frequency. *See* LOCAL OSCILLATOR.

Mixing always produces false signals called *images*. In a well-designed receiver, the image signals are very weak, and are not on frequencies that cause interference to desired signals. *See* IMAGE FREQUENCY, and IMAGE REJECTION. Some receivers convert the signal frequency directly into audio information. *See* DIRECT-CONVERSION RECEIVER.

IF stages The *intermediate-frequency (IF) stages* of a receiver are where much, if not most, of the gain occurs. These stages are also where the best possible selectivity is obtained. The IF is a constant frequency. This simplifies the design of the amplifiers to produce optimum gain and selectivity. Crystal-lattice filters or mechanical filters are commonly used in these stages to obtain the desired bandwidth and skirt response. *See* CRYSTAL-LATTICE FILTER, INTERMEDIATE-FREQUENCY AMPLIFIER, MECHANICAL FILTER, SELECTIVITY, and SKIRT SELECTIVITY.

Detector The *detector* extracts the information from the signal. The circuit details depend on the type of emission to be received. The most common voice detectors for ham operation are the *product detector* for CW, FSK, and SSB, and the *discriminator, phase-locked loop (PLL)* or *ratio detector* for FM. *See* DETECTION, DETECTION METHODS, DISCRIMINATOR, PHASE-LOCKED LOOP, PRODUCT DETECTOR, and RATIO DETECTOR.

Post-detector Circuits Following the detector, one or two stages of audio amplification might be used, to boost the audio to a level suitable for listening with a speaker or headset. Alternatively, the signal can be fed to a printer, FAX machine, slow-scan television picture tube, or computer via a terminal unit. This device is a decoding circuit that converts the detector output to impulses suitable for these machines.

In amateur fast-scan television, the detector output is a video signal. This is fed to the cathode-ray tube for display. The amateur fast-scan television receiver is often a consumer-type television set with a frequency converter between the antenna and the front end.

RECEIVER INCREMENTAL TUNING

In a radio transceiver, it is often desirable to set the receiver to a slightly different frequency from that of the transmitter. This is especially true in code (CW) operation, but it can also be necessary at times in other modes. A *receiver-incremental-tuning (RIT) control* accomplishes this. The RIT moves the receiver frequency without changing the transmitter frequency. Receiver incremental tuning is also known as *receiver offset tuning*.

The most common method of obtaining RIT is to change the frequency of the variable-frequency oscillator (VFO). Because most transceivers use a common VFO for both the transmitter and the receiver, it is necessary to disconnect the RIT while transmitting. This is done either by relays or by electronic switching methods.

Generally, an RIT control allows a receiver to be tuned approximately 5 kHz above or below the transmitter frequency. If a larger split is needed, separate VFO units can be used, one for the receiver and one for the transmitter. *See also* TRANSCEIVER.

RECEIVE STATE VARIABLE

In packet radio, some means is needed to ensure that messages are sent in the right sequence. This is done by assigning numbers to the packets. One set of numbers is assigned to packets as they are sent. Numbers are also assigned to the packets at the receiving end. A number, so assigned, is called the *receive state variable*.

When the receive state variable is compared with the send sequence number, the computers can ascertain that the packets have been sent in the right order. *See also* PACKET RADIO.

RECEIVING ANTENNA

Any transmitting antenna will function well for receiving purposes within the same range of frequencies. However, receiving antennas need not be as sophisticated or large as transmitting antennas in order to function well, especially at very low, low, medium, and high frequencies. Because of the nature of electromagnetic waves, tiny antennas can function very well for reception at these frequencies, even though they are virtually useless for transmission. At very-high, ultra-high, and microwave frequencies, large dish antennas are necessary.

The simplest type of receiving antenna is a random wire, connected to the receiver input terminals either directly or through a tuning network. The antenna might consist of a simple end-fed wire, or a horizontal conductor can be fed by a lead-in wire (see illustration). Such an antenna is usually made as long as possible. It will work well at all radio frequencies into the very-high or even ultra-high range. Such an antenna can provide marginal performance for transmitting.

Tuned receiving antennas provide a high degree of selectivity and some noise suppression. The most common tuned receiving antennas are the ferrite rod and the loop (*see* FERRITE-ROD ANTENNA, and LOOP ANTENNA). These antennas can be used at frequencies from about 30 MHz down to 10 kHz with excellent results. Despite their physically tiny size at the longer wavelengths, the ferrite rod and the loop perform very well for receiving at the very low, low, and medium frequencies. Marginal performance can be had up to about 30 MHz. Above that frequency, ferrite becomes rather lossy, and open-loop antennas are not significantly smaller than a half-wave dipole or full-wave loop.

At very high frequencies and above, the same antenna is generally used for receiving as for transmitting because physical size is not a major problem. When the same antenna is used for receiving and transmitting, the directional characteristics work to advantage in both modes.

Receiving antennas will often work fairly well if located indoors. However, it is always best, if possible, to put them outdoors, especially if the receiver is inside a concrete-and-steel structure. *See also* INDOOR ANTENNA, and OUTDOOR ANTENNA.

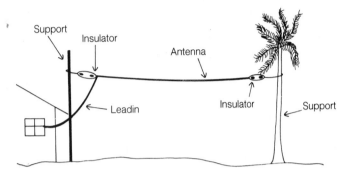

RECEIVING ANTENNA: A simple random-length wire.

RECHARGEABLE CELL

Some electrochemical cells can accept a charge and deliver the current at a later time. Such cells are called *rechargeable* because they can be used over and over. Rechargeable cells offer an obvious advantage over one-time cells. Although rechargeable cells are more expensive than the one-time variety, and a charging unit is needed, the additional expense is more than offset by the reusability.

There are many different kinds of rechargeable cells, but the most common are the *lead-acid type* and the *nickel-cadmium type*. Lead-acid cells are used in automobile batteries. Nickel-cadmium cells are used in a wide variety of electronic devices, such as radio receivers and low-power transmitters, calculators, and memory-backup systems. The lead-acid cell is a high-current cell; the nickel-cadmium cell is designed for low-current use. *See also* CHARGING, LEAD-ACID BATTERY, NICKEL-CADMIUM BATTERY, QUICK CHARGE, and TRICKLE CHARGE.

RECIPROCITY THEOREM

The *Reciprocity Theorem* is a rule that describes the behavior of currents and voltages in certain multibranch circuits. Suppose you have a passive, linear, bilateral circuit with two or more separate, current-carrying branches. Suppose a power supply, delivering E_x volts, is placed across branch X of this circuit, resulting in a current I_x in branch X, and a current I_y in branch Y. We assume that the power supply has zero internal resistance. Then, the voltage E_y across branch Y is found from the current I_y and the resistance R_y according to Ohm's Law:

$$E_y = I_y R_y$$

Now imagine that you disconnect the power supply having voltage E_x from branch X, and connect a different supply, with voltage E_y and zero internal resistance, across branch Y. Suppose the same current as before, I_y, flows through resistance R_y.

According to the Reciprocity Theorem, under these circumstances, the current through the resistance R_x will be the same as before, or I_x. Therefore, the voltage across branch X will be:

$$E_x = I_x R_x$$

which is precisely the same voltage E_x of the original supply.

The Reciprocity Theorem is a direct result of Kirchhoff's laws and Ohm's law. This theorem can be used to advantage in determining currents and voltages in complicated networks. *See also* KIRCHHOFF'S LAWS, and OHM'S LAW.

RECOMBINATION

When a charge is applied to a semiconductor material, excess holes and/or electrons are present. When the charge is removed, equilibrium is restored, assuming there is no other disturbance present that might produce excess holes and/or electrons. The process of restoring charge-carrier equilibrium is called *recombination*. The excess holes are filled in by electrons.

Recombination takes a certain amount of time. It occurs exponentially, as do most decay processes. Recombination time is also called the *lifetime*. For high-speed switching and in ultra-high-frequency applications, it is necessary to have a very short recombination time in a semiconductor device. In other applications, this can not be especially important.

Recombination time varies for different semiconductor materials; it can range from a few microseconds to less than 1 nanosecond, depending on the impurity substances that have been added to the semiconductor. *See also* CARRIER MOBILITY, DOPING, ELECTRON, HOLE, N-TYPE SEMICONDUCTOR, P-N JUNCTION, and P-TYPE SEMICONDUCTOR.

RECORDING HEAD

A *recording head* is an electromagnetic transducer that converts alternating electric currents to fluctuating magnetic fields and vice-versa. A recording head is used for the purpose of writing data onto, or retrieving information from, magnetic tapes or disks. Every tape recorder or disk recorder contains one or more such transducers. Usually, there are two: one for recording and one for playback.

The recording head is placed adjacent to a tape or disk. The tape or disk is drawn past the head at a uniform, precise speed. If the head is to be used for recording, an electric current is supplied to it. Fluctuating magnetic fields, which occur in synchronization with the alternating currents, cause polarization of the fine magnetic particles in the tape or disk. If the head is to be used for playback or information retrieval, the polarized particles cause electric currents to flow in the recording head. *See also* MAGNETIC RECORDING.

RECORDING TAPE

Recording tape is a linear medium on which information can be stored. The most common and familiar form of recording tape is the magnetic variety. Magnetic tape is used by almost everyone for some purpose (*see* MAGNETIC TAPE).

There are other, less familiar forms of recording tape. In very old computer terminals and teletype systems, paper tape is used to record printed data. A machine punches holes in the tape in rows perpendicular to the edges of the tape. Each row of holes represents one character (*see* PAPER TAPE).

Some pen recorders use a long, thin strip of paper, marked in rectangular coordinates, for storing the information. This is a form of recording tape. In some older morse-code machines, optical paper tape is used. This tape is marked with a long, black, crooked line. During silent periods, the black line is on one side of the strip, and it blocks the light beam in an optical coupler. During key-down periods, it is on the other side, allowing the beam to pass through the paper or reflect from it, actuating a tone generator.

In recent years, disks have largely replaced tapes for recording purposes. This is because disks allow faster access to information. On a long roll of recording tape, two files of data can be separated by thousands of feet. To access two such data files in succession, it is necessary to move the entire tape through the recorder. But on a disk, the maximum possible separation is the diameter of the disk—never more than a few inches.

RECTANGULAR RESPONSE

A bandpass or band-rejection filter is usually designed to have what is called a *rectangular response*. In theory, the rectangular bandpass response is characterized by zero attenuation between the cutoff frequencies, and infinite attenuation outside the bandpass; a rectangular band-rejection response is characterized by infinite attenuation between the cutoff points and zero attenuation at all other frequencies.

In a radio receiver, a rectangular bandpass response provides the best possible signal-to-noise ratio. Modulated signals have well-defined bandwidth. The optimum receiver response is therefore rectangular, having a bandwidth equal to that of the incoming signal.

The ideal rectangular response cannot be realized in practice, but engineers have devised filters that have sharp cutoff, high attenuation in the rejection range, and uniform, low attenuation within the pass range. The response of such a filter is said to be rectangular, even though there is some deviation from the ideal.

A single resonant circuit provides a peaked response, while a group of tuned circuits or resonant devices, all having a slightly different natural frequency, can be connected together to produce a rectangular response. Piezoelectric ceramics, and crystals are widely used at radio frequencies to obtain a rectangular response. Mechanical filters are also quite common. *See also* BANDPASS FILTER, BANDPASS RESPONSE, BAND-REJECTION FILTER, BAND-REJECTION RESPONSE, CERAMIC FILTER, CRYSTAL-LATTICE FILTER, and MECHANICAL FILTER.

RECTIFICATION

The conversion of alternating current to direct current is called *rectification*. Rectification can be accomplished in a variety of ways, but the most common methods are the use of diodes or commutators. Diodes are generally used for electronic-circuit rectification; commutators are used in motors and generators when direct current is needed (*see* COMMUTATOR, DIODE, and DIODE ACTION).

The diode conducts current in only one direction. Therefore, when a diode is placed in a circuit where alternating current flows, the result is pulsating direct current. Rectification is used in all direct-current power supplies designed for operation with ordinary household current. *See also* RECTIFIER, RECTIFIER CIRCUITS, and RECTIFIER DIODE.

At radio frequencies, a process similar to rectification is used to demodulate certain types of signals. This process is called *envelope detection. See* ENVELOPE DETECTOR.

RECTIFIER

A *rectifier* is a device that allows current to flow in only one direction. Rectifiers can be mechanical or electronic. In some motors and generators, a mechanical device called a *commutator* is used as a rectifier (*see* COMMUTATOR). In electronic circuits, semiconductor devices or vacuum tubes are used (*see* RECTIFIER DIODE).

A complete circuit that converts alternating current to pulsating direct current is called a *rectifier circuit*. There are several different types of rectifier circuits. Electronic rectifier circuits generally consist of a transformer and one or more diodes. *See also* RECTIFIER CIRCUITS.

RECTIFIER CIRCUITS

Rectifier circuits are used for the purpose of converting alternating current to pulsating direct current. Rectifier circuits are found in all direct-current power supplies designed to operate from the utility mains. Rectifier circuits are also used in some radio-frequency-actuated devices.

The simplest type of rectifier circuit is the half-wave circuit. A single diode is placed in series with the alternating-current line. The diode allows the current to flow in only one direction (*see* HALF-WAVE RECTIFIER). A half-wave circuit results in a 60-Hz pulsation frequency with 60-Hz alternating-current input. Half-wave circuits are used in applications where some ripple can be tolerated, and in which voltage regulation need not be especially good.

The bridge and full-wave rectifier circuits are probably the most popular (*see* BRIDGE RECTIFIER, and FULL-WAVE RECTIFIER). Both circuits require transformers in order to function properly. With 60-Hz alternating current, the output of a full-wave rectifier has a pulsation frequency of 120 Hz. The bridge and full-wave circuits are the most common in modern electronic equipment. The output is fairly easy to filter, and good regulation is easily obtained.

In three-phase systems, various different configurations are used to accomplish rectification. Special transformers are required for three-phase rectifier circuits. The three-phase system offers the advantage of minimal ripple and excellent regulation (*see* POLYPHASE RECTIFIER).

The output of a rectifier circuit must normally be filtered before it is suitable for use by electronic devices. Filtering can be accomplished by means of chokes in series and capacitors in parallel with the rectifier output. *See also* FILTER, FILTER CAPACITOR, POWER SUPPLY, and RECTIFIER DIODE.

RECTIFIER DIODE

A *rectifier diode* is a semiconductor diode designed for use in rectifier circuits. Rectifier diodes are generally fabricated from silicon. They are characterized by a fairly large P-N-junction surface area, resulting in high capacitance under reverse-bias conditions. Rectifier diodes have replaced tubes in most modern power-supply circuits.

Silicon rectifier diodes are available with various peak-inverse-voltage (PIV) ratings, ranging from a few volts to several thousand volts. In high-voltage supplies, two or more rectifier diodes can be connected in series to increase the PIV rating of the combination. A large-value equalizing resistor is connected in shunt with each diode to distribute the voltage among them. An 0.01-μF capacitor should also be connected in parallel with each diode to protect against transients.

Some rectifier diodes can carry several amperes of current in the forward direction, and a few are available for use at 300 to 500 amperes. High-current rectifier diodes require heatsinks. Two or more diodes can be connected in parallel to increase the forward-current-handling capacity (a small-value resistor is connected in series with each diode to prevent current hogging).

The primary advantage of rectifier diodes over rectifier tubes is compactness. Rectifier diodes also require no warm-up and no filament power. *See also* PEAK INVERSE VOLTAGE, P-N JUNCTION, POWER SUPPLY, and RECTIFIER CIRCUIT.

REDUNDANCY

Redundancy refers to the use of extra devices or methods for the purpose of guarding against possible malfunction of a system. An example of component redundancy is the use of two diodes in parallel in a power supply when only one is actually needed for circuit operation. If one diode burns out, the other diode can take over. Component redundancy is used in computer systems so that operation can continue even if part of the system fails (*see* GRACEFUL DEGRADATION).

In data storage, several copies of the information are usually made to guard against damage or loss. For example, two or three backup copies can be made for each disk in a word-processor file. This is called *redundancy*.

In mathematics, the term has a different meaning. *Redundancy* in an equation refers to the use of unnecessary parameters or variables. In a computer program, redundancy results in less-than-optimum efficiency, and therefore unnecessary consumption of computer time.

In data transmission, redundancy can be used as a means of checking the accuracy of the information at the receiving end. This is also known as *parity*. *See also* PARITY.

REED RELAY

A *reed relay* is a high-speed relay used for such purposes as code-transmitter keying and rapid switching. Reed relays are sometimes used in logic circuits and low-frequency oscillators.

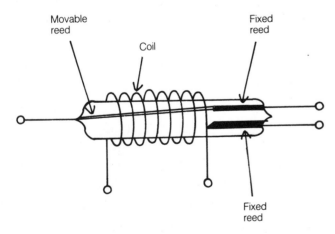

REED RELAY: A fast-switching relay.

A typical single-pole, double-throw reed relay is shown in the illustration. The relay assembly is enclosed in a glass envelope, which can be evacuated or filled with an inert gas to prevent corrosion of the contacts. The glass envelope is surrounded by a coil of wire, to which a voltage is applied to actuate the relay. If multiple-pole operation is desired, several reed envelopes can be surrounded by a single coil.

The metal reed is pulled from one contact to the other by the magnetic field created by the current in the coil. The small gap and the flexibility of the reed result in the short switching time. Some reed relays can be actuated at a rate of more than 1 kHz (1000 openings and closings per second). Reed relays have excellent longevity and reliability.

Reed relays can be either dry or mercury-wetted. Mercury wetting eliminates contact bounce that sometimes occurs in dry-reed relays. *See also* DRY-REED SWITCH, and RELAY.

REED SWITCH

See DRY-REED SWITCH.

REFERENCE ANTENNA

For antenna power gain to have meaning, a *reference antenna* is needed. Antenna power gain is generally measured in decibels relative to a reference dipole. Sometimes the reference antenna is an isotropic radiator (*see* ANTENNA POWER GAIN, dBd, dBi, DIPOLE ANTENNA, and ISOTROPIC ANTENNA).

For the measurement of transmitted-signal gain, the reference antenna and the test antenna are set up side by side. An accurately calibrated field-strength meter is positioned at a distance of at least several wavelengths from both antennas. The reference and test antennas are oriented so that the field-strength meter is located in the favored direction (major lobe) of both antennas. A signal of P watts is applied to the reference antenna, and the field-strength meter is set to show a relative indication of 0 dB. Then, the same signal of P watts is applied to the test antenna, and the gain is read from the field-strength meter. *See* TEST RANGE.

To measure received-signal gain, a similar arrangement is used. The reference and test antennas are placed side by side with their major lobes oriented toward a signal-generator/antenna several wavelengths distant. A field-strength meter or receiver, equipped with an accurate S meter, is connected to the reference antenna, and the receiver gain is adjusted until a relative indication of 0 dB is obtained. Then, the receiver is connected to the test antenna. A perfect impedance match must be maintained for both antennas. The gain of the test antenna can then be read from the receiver S meter. *See* S METER.

REFERENCE FREQUENCY

In the operation and calibration of radio receivers, a reference frequency is sometimes used. A standard signal generator, such as a crystal oscillator or broadcast station whose frequency is precisely known, can be used for this purpose.

Consider an example of the use of a reference frequency in the operation of a shortwave receiver. Suppose you are listening to signals in the band 7.0 to 7.5 MHz. The accuracy of the dial calibration might not be known. However, we do know that the time-and-frequency station, CHU, is operating at 7.335 MHz. Tune the receiver back and forth in the vicinity of this frequency, as shown on the dial, and finally you hear the intermittent tones of CHU. The receiver dial shows 7.328 MHz; the indication is 7 kHz (0.007 MHz) too low. Adjust the dial-calibration control until the dial shows 7.335 MHz. You can then be confident that the calibration is reasonably accurate over the range 7.0 to 7.5 MHz.

A reference-frequency source is considered primary if it is received directly from a standard-frequency broadcast station, such as CHU or WWV/WWVH. If some other source, such as a commercial broadcast station or crystal calibrator is used, the reference frequency is considered secondary. *See also* CHU, FREQUENCY CALIBRATOR, FREQUENCY METER, and WWV/WWVH/WWVB.

REFERENCE LEVEL

When signal strength is measured in relative terms, a *reference level* must be specified. A reference level can be established for current, field strength, flux density, power, voltage, or any other quantity that varies. The reference level is assigned a gain factor of 1, corresponding to 0 dB.

Certain standard reference levels are used. A common reference level for current is the milliampere; for power, the milliwatt; for voltage, the millivolt. Other standard values, however, can be specified. If the parameter exceeds the reference level, the gain factor is greater than 1 and the gain figure in decibels is positive. If the parameter is smaller than the reference level, the gain factor is less than 1 and the gain figure in decibels is negative. *See also* DECIBEL, and GAIN.

REFERENCE ORBIT

In satellite communications, it is necessary to know where in the sky the satellite will be if you plan to use it. This depends on the groundtrack of the satellite, and also on its altitude above the surface of the earth.

It can become rather complicated to calculate satellite positions without the aid of specialized nomographs or, preferably, computers, and appropriate software.

Programs are available that allow a home personal computer (PC) to accurately predict antenna coordinates versus time, for any of the existing OSCAR satellites. But for these programs to work, a *reference orbit* is needed.

Reference orbits are specified for each day coordinated universal time (UTC). The first ascending node after 0000 UTC marks the beginning of the reference orbit for that day. Reference orbit specifications are published in the ham magazines, and are also broadcast by station W1AW. *See also* ACTIVE COMMUNICATIONS SATELLITE, ASCENDING NODE, COORDINATED UNIVERSAL TIME, and W1AW.

REFLECTANCE

When an electromagnetic field strikes a surface or boundary, some of it can be reflected. The ratio of the reflected energy or field intensity to the incident intensity is known as *reflectance*. If the incident field intensity (known as *incident* or *forward power*) is P_1 and the reflected field intensity (called the *reflected power*) is P_2, then the reflectance factor R is:

$$R = P_2/P_1$$

In percent, reflectance is given by:

$$R \text{ (percent)} = 100P_2/P_1$$

As an example, consider a mirror. When visible light strikes a mirror, almost all of the energy is reflected. Therefore, the reflectance factor is almost 1 (100 percent). A pane of glass reflects some light, but most of it is transmitted. Thus, the reflectance of glass is just a few percent. The reflectance of a surface can vary, depending on the angle at which the light strikes.

In a power-transmission line and load system, some of the electromagnetic field, traveling from the generator to the load, can be reflected upon reaching the load. This results in less than optimum current in the load. The ratio of actual current to optimum current is called *reflectance* or, more commonly, the *reflection factor*. *See also* REFLECTED POWER, REFLECTION COEFFICIENT, and REFLECTION FACTOR.

REFLECTED IMPEDANCE

In an *impedance transformer* having a certain load connected to the secondary winding, the impedance across the primary is called reflected impedance (*see* IMPEDANCE TRANSFORMER). If the secondary, or load, impedance does not contain reactance, then the reflected impedance will contain no reactance. This is generally the situation with impedance transformers. However, if the load does contain reactance, then the reflected impedance will also contain reactance (*see* CAPACITIVE REACTANCE, IMPEDANCE, and INDUCTIVE REACTANCE).

In a transmission line operating at radio frequencies, the reflected impedance is the impedance at the transmitter end of the line. In the ideal case, where the antenna exhibits a pure resistance equal to the characteristic impedance of the line, the reflected impedance is the same as the load impedance for any line length. If the load contains reactance, or if the resistance of the load differs from the characteristic impedance of the line, the reflected impedance will vary, depending on the line length.

Certain transmission lines have special properties. Neglecting line loss, a transmission line measuring a multiple of a half wavelength will always result in a reflected impedance identical to the load impedance; a line measuring an odd multiple of a quarter wavelength will invert the resistance and reactance. *See also* CHARACTERISTIC IMPEDANCE, HALF-WAVE TRANSMISSION LINE, MATCHED LINE, MISMATCHED LINE, and QUARTER-WAVE TRANSMISSION LINE.

REFLECTED POWER

Ideally, the impedance of the load in a power-transmission system should be a pure resistance, and should have the same ohmic value as the characteristic impedance of the line. This is not always the case. When the load impedance differs from the characteristic impedance of the line, the line is said to be *mismatched*.

The electromagnetic field, traveling along a mismatched line from the generator to the load, is not all absorbed by the load. Some of the electromagnetic field is reflected at the feed point, and travels back toward the generator. Because the electromagnetic field causes power to be dissipated in a resistance, the forward-moving field is sometimes called forward power and the backward-moving field is called *reflected power*.

The amount, or proportion, of incident power that is reflected depends on the severity of the mismatch between the line and the load. The greater the mismatch, the more of the incident power is reflected at the feed point. In the extreme, when the load is a pure reactance, or if it is a short circuit or open circuit, all of the incident power is reflected. The degree of mismatch is expressed by a figure known as the *standing-wave ratio* (*see* STANDING-WAVE RATIO).

If the standing-wave ratio is represented by S, the forward power by P_1, and the reflected power by P_2, then:

$$P_2/P_1 = [(S - 1)/(S + 1)]^2$$
$$= (S^2 - 2S + 1)/(S^2 + 2S + 1)$$

The proportion of reflected power to forward power is equal to the square of the reflection coefficient (*see* REFLECTION COEFFICIENT).

Reflected power can be directly read in a radio-frequency transmission line with a calibrated reflectometer (*see* REFLECTOMETER). If the reflectometer is not calibrated in forward and reflected watts, the reflected power can be determined from the standing-wave ratio observed on the meter.

When reflected power reaches the transmitter end of a radio-frequency transmission line, all of the electromagnetic field is reflected again back toward the load, assuming the transmitter is adjusted for optimum output. If the transmitter output circuit is not adjusted to compensate for the reactance at the input end of the line, some of the reflected power is absorbed by the final amplifier. This is observed as a reduction in power output, although the final-amplifier dc input is just as great, or greater, than it would be if the output circuit were adjusted properly. *See also* MATCHED LINE, MISMATCHED LINE, and REFLECTION FACTOR.

REFLECTED WAVE

In line-of-sight communication, the signal at the receiving antenna consists of two components. One component, the direct wave, follows the path through the air between the two antennas (*see* DIRECT WAVE). The other component consists of one or more waves that are reflected from the ground and/or other objects, such as buildings. They are called the *reflected wave* or *waves*.

The direct wave and the reflected wave(s) add together in varying phase. The reflected-wave signal is usually weaker than the direct wave, but not always. In certain situations, the direct-wave and reflected-wave signals are equal in amplitude and opposite in phase at the receiving antenna. This renders communication difficult or impossible. The remedy is to move either the transmitting antenna or the receiving antenna a fraction of a wavelength, so that the coincidence is eliminated. This kind of phase problem is observed primarily at frequencies above about 10 MHz and is especially acute at microwave frequencies.

The effects of reflected waves can be noticed in a mobile frequency-modulation receiver such as a car stereo. As the received station becomes weak, it begins to flutter. This is the result of alternate phase reinforcement and phase cancellation between the direct-wave and reflected-wave components. The faster the car moves, the more rapid the flutter. If the car comes to a stop in a "dead zone" where total phase cancellation occurs, the signal will disappear until the car is moved again. *See also* LINE-OF-SIGHT COMMUNICATION.

REFLECTION COEFFICIENT

In a power transmission line, not all of the incident power is absorbed unless a perfect match exists (*see* MATCHED LINE, and

MISMATCHED LINE). When there is a mismatch, not all of the available or "forward" current is absorbed by the load; some current is "reflected." The same thing happens with the voltage. The ratio of reflected current to forward current, or reflected voltage to forward voltage, is called the *reflection coefficient*. The reflection coefficient is represented by the lowercase Greek rho (ρ).

Let Z_0 be the characteristic impedance of a transmission line and let R be the ohmic value of a purely resistive load. Then, the reflection coefficient is:

$$\rho = (R - Z_0)/(R + Z_0)$$

if R is larger than Z_0, and:

$$\rho = (Z_0 - R)/(R + Z_0)$$

if R is smaller than Z_0. The above formulas are accurate only for loads that contain no reactance.

The reflection coefficient can also be derived from the standing-wave ratio, S:

$$\rho = (S - 1)/(S + 1)$$

This second formula is accurate whether or not the load contains reactance. *See also* REFLECTED POWER, and STANDING-WAVE RATIO.

REFLECTION FACTOR

In a matched transmission line, all of the incident current is absorbed by the load. This is not the case in a mismatched line (*see* MATCHED LINE, and MISMATCHED LINE). The reflection factor is an expression of the ratio of the current actually absorbed by a load, compared with the current that would be absorbed under perfectly matched conditions. The reflection factor can also be called *reflectance*.

Let R be the ohmic value of a purely resistive load, and let Z_0 be the characteristic impedance of the transmission line. Then the reflection factor, F, is given by:

$$F = \sqrt{4RZ_0}/(R + Z_0)$$

This formula holds only if the load impedance is a pure resistance. *See also* REFLECTED POWER, and REFLECTION COEFFICIENT.

REFLECTION LAW

If a particle or wave disturbance encounters a flat (plane) reflective surface, the angle of incidence is equal to the angle of reflection, measured with respect to the normal to the surface. If the surface is not flat, then the angles of incidence and reflection are identical as measured, with respect to the normal to a plane tangent to the surface at the point of reflection. This fact is sometimes called the *reflection law* or the *law of reflection*.

The reflection law holds for radio waves impinging against the ground, light striking a mirror, sound waves bouncing off a wall or baffle, and all other situations in which reflection occurs.

REFLECTOMETER

A *reflectometer* is a device that is installed in a transmission line for the purpose of measuring the standing-wave ratio (*see* STANDING-WAVE RATIO). Some reflectometers are calibrated in forward watts and reflected watts. A reflectometer can also be called a *directional coupler, directional power meter,* or a *monimatch.*

Many different kinds of reflectometers are available from commercial manufacturers for use in coaxial transmission lines. A simple uncalibrated reflectometer, for use at frequencies below 300 MHz, can be constructed from parts that are easy to find. A schematic diagram of a simple reflectometer is shown at B.

An uncalibrated reflectometer is first adjusted so that the meter reads full scale with the switch in the FORWARD position. Then, the switch is moved to the REVERSE or RE-FLECTED position to read the standing-wave ratio. A cali-brated reflectometer shows the forward power and the reflected power directly in watts; the standing-wave ratio, S, can be calculated from these readings according to the formula:

$$S = (1 + \sqrt{P_2/P_1})/(1 - \sqrt{P_2/P_1})$$

where P_1 is the forward power and P_2 is the reflected power, both specified in the same units.

When the load is a pure resistance with an ohmic value equal to the characteristic impedance of the line, the reflec-tometer will indicate zero with the switch in the REFLECTED position. If there is a mismatch, the reflected-power reading will not be zero. If the mismatch is very severe, the reflected-power reading will be almost as great as the forward-power reading. When a calibrated reflectometer is used, the true power is determined by subtracting the reflected-power read-ing from the forward-power reading. *See also* APPARENT POWER, MATCHED LINE, MISMATCHED LINE, RE-FLECTED POWER, and TRUE POWER.

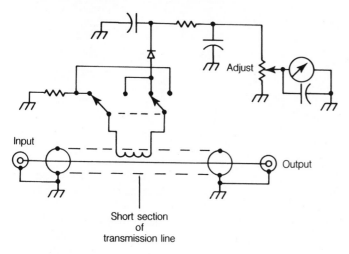

REFLECTOMETER: A schematic of an uncalibrated reflectometer.

REFLECTOR

A *reflector* is a form of parasitic element used in a directional an-tenna array. The reflector results in signal gain in some direc-tions at the expense of gain in other directions. Parasitic arrays increase the efficiency of communications by maximizing the effective radiated power, and also by reducing interference from unwanted directions. *See* ANTENNA POWER GAIN, and PARASITIC ARRAY.

An example of the operation of a reflector is found in most Yagi antennas. A half-wave dipole has a gain of 0 dBd (*see* dBd) in free space. A reflector, slightly longer than the dipole and po-sitioned about a quarter wavelength from it, produces gain.

The behavior of parasitic arrays in general, and reflectors in particular, is very complex. The gain depends on the spacing between the driven element and the reflector, and also on the length of the reflector. The forward gain and the front-to-back ratio are optimized under slightly different conditions. Many excellent references are available that cover the design of para-sitic antennas in detail. *See also* DIRECTOR, and DRIVEN ELE-MENT.

REFRACTED WAVE

When a radio wave enters the ionosphere, it is *refracted* (bent), in much the same way that light is refracted in a lens. Often, this refraction is enough to return the wave to earth at some point far away from the transmitter. This is sometimes mistakenly thought of as "reflection" of radio waves from the ionized layers. The phenomenon is most common at frequencies below about 30 MHz, although it has been noticed at ham wave-lengths as short as 2 meters (the 144-MHz band). *See* E LAYER, F LAYER, and IONOSPHERE.

Radio waves at very-high and ultra-high frequencies (VHF and UHF), or 30 MHz and above, are sometimes bent by the lower atmosphere. These are also refracted waves. The most common cause of wave refraction at VHF and UHF is *tropo-spheric bending. See* TROPOSPHERIC PROPAGATION. For details on radio-wave propagation at various frequencies, *see* PROPAGATION CHARACTERISTICS.

REGENERATION

Regeneration is another name for positive feedback. Regenera-tion is deliberately used in oscillators, certain amplifiers, and some detectors. *See also* FEEDBACK, OSCILLATOR, and RE-GENERATIVE DETECTOR.

REGENERATIVE DETECTOR

A *regenerative detector* is a form of demodulator that can be used for receiving code signals, frequency-shift keying, and single sideband. The regenerative detector can also increase the gain for the reception of amplitude-modulated signals, provided an envelope detector is placed after the circuit. Regenerative de-tectors are not often used today, although they can still be found in simple direct-conversion receivers.

The regenerative detector consists of an amplifier circuit, with some of the output applied in phase to the input. This in-creases the gain of the amplifier. If the positive feedback is made large enough, the circuit will oscillate. Incoming signals then mix with the resulting carrier, producing beat notes that can be heard in the speaker or headset. This makes code, fre-quency-shift-keyed, or single-sideband signals discernible.

Regenerative detectors were extensively used in early radio receivers. In many circuits, the regenerative detector was incor-porated into the front end of the receiver. This resulted in the transmission of a carrier from the receiver antenna; this carrier could travel several miles and cause interference to other re-ceivers on or near the same frequency. Contemporary circuits never use this arrangement; the regenerative detector is always preceded by at least one tuned circuit to avoid transmission of the carrier. *See also* DIRECT-CONVERSION RECEIVER.

REGENERATIVE RECEIVER

See REGENERATIVE DETECTOR, and SUPERREGENERA-TIVE RECEIVER.

REGENERATIVE REPEATER

A *regenerative repeater* is a circuit that reshapes pulses. In teletype or code transmission, the pulses often become distorted in transit. This is particularly true for signals that have been propagated via the ionosphere. The regenerative repeater restores the pulses to their original shape and duration.

A regenerative repeater "knows" the appropriate length, spacing, and shape for a pulse. For example, the duration of a morse-code dot at a given speed might be 0.1 second. At this same speed, the space between code elements in a character is also 0.1 second; the length of a dash is 0.3 second; the space between characters is 0.3 second; the space between words is 0.7 second. If a code signal having some distortion, but the appropriate speed, is applied to the input of the regenerative repeater, perfect code will appear at the output.

Regenerative repeaters can reduce the error rate in digital communications, especially when a machine is used to receive the signals. The higher the data speed, the greater is the chance for distortion in the signal. Therefore, regenerative repeaters become more useful as the data speed is increased.

REGISTER

A short-term memory-storage circuit is sometimes called a *register*. A register can store any kind of information, and usually has a small capacity. In a computer, a number of registers store data for the central processing unit. A register configuration can be first-in/first-out (FIFO), pushdown stack, or random access (*see* FIRST-IN/FIRST-OUT, PUSHDOWN STACK, and RANDOM-ACCESS MEMORY).

The reading of an instrument, in specified units, is sometimes called the *register*. A meter reading might have to be multiplied by a certain value in order to obtain the correct register (*see* REGISTER CONSTANT).

REGISTER CONSTANT

In some meters and controls, the indication must be multiplied by, or added to, a constant in order to obtain the actual value. This constant is called the *register constant.*

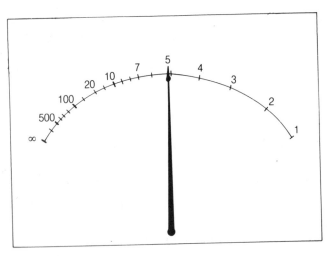

REGISTER CONSTANT: Multiplier is indicated by switch position.

The illustration shows the scale of a typical ohmmeter. Note that half scale is near the point marked 5. If the register constant is 1, then a half-scale reading indicates approximately 5 ohms. As the value of the resistance increases, it gets more and more difficult to read the meter with a register constant of 1. Therefore, a switch is incorporated into the ohmmeter, allowing you to select various register constants. This particular unit has register constants of 1, 10, 100, 1,000, 10,000, 100,000, and 1,000,000. When the meter needle is at the point marked 5 on the scale, the actual resistance is 5 ohms with the switch in the $R \times 1$ position, 50 ohms in the $R \times 10$ position, 500 ohms in the $R \times 100$ position, and so on.

In certain meters, indicators, and dial systems, it is sometimes necessary to add a constant to the reading in order to obtain the actual value. For example, a radio transceiver, operating in the range 140.000 MHz to 149.999 MHz, can have a frequency readout that shows values from 0.000 to 9.999. The actual frequency is obtained by adding 140. The value 140 is, therefore, a register constant.

REGULATED POWER SUPPLY

A direct-current power supply is said to be regulated if it has a circuit designed for maintaining a precise output voltage under varying load conditions. Many types of electronic equipment will function properly only if the supply voltage is within a narrow range. An especially common operating range is 13.8 V with a tolerance of plus or minus 10 percent.

There are several different methods for obtaining voltage regulation. Zener diodes are sometimes used in low-voltage supplies. Special tubes are more often used in high-voltage power supplies.

In some regulated supplies, failure of the regulating circuit will result in a dramatic rise in the output voltage. This can damage equipment connected to the supply. A well-designed power supply should have some means of overvoltage protection. *See also* OVERVOLTAGE PROTECTION, POWER SUPPLY, and VOLTAGE REGULATION.

REGULATING TRANSFORMER

There are two types of regulating transformers, both designed to maintain a constant output voltage under varying load conditions. Some transformers use resonance and core saturation to maintain a constant output voltage. This type of transformer actually varies in efficiency as the load changes. As the load resistance decreases, the transformer operates more efficiently, compensating for the drop in voltage that would otherwise occur.

A small transformer can be connected in series with the secondary winding of the main transformer in a power supply, for the purpose of adjusting the voltage. This type of arrangement is shown in the schematic. The small transformer is called a regulating transformer. When the current from the regulating transformer is in phase with the current from the main transformer, the output voltage is increased. When the phase of the current from the regulating transformer is opposite to that of the main transformer, the output voltage is decreased. By applying a small voltage of the appropriate phase to the primary of the regulating transformer, the output voltage can be held constant under fluctuating load conditions. *See also* VOLTAGE REGULATION.

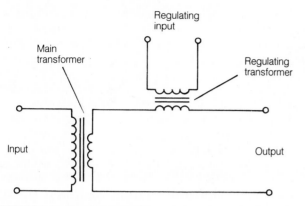

REGULATING TRANSFORMER: The winding is in series with the secondary of the power transformer.

REGULATION

Regulation refers to the maintenance of a constant condition in a circuit. Any parameter, such as current, voltage, resistance, or power, can be regulated. The extent to which a parameter varies can be expressed mathematically as a percentage. This figure is called the *regulation*. Let the optimum, or rated, value of a parameter be represented by P, and suppose that the actual value of the parameter fluctuates between the minimum value P_1 and the maximum value P_2, such that $P_1 < P < P_2$. Then, the regulation (R), in percent, is given by:

$$R = 100(P_2 - P_1)/P$$

Regulation is sometimes expressed as a plus-or-minus figure. In the above example, we might specify the regulation in this manner if the fluctuation is the same above and below the ideal; that is, if:

$$P_2 - P = P - P_1$$

The plus-or-minus regulation in percent, is:

$$R\pm = \pm 100(P_2 - P)/P$$
$$= \pm 100(P - P_1)/P$$

Notice that the plus-or-minus regulation, as determined by the second method, is exactly half the regulation as determined by the first method. *See also* CURRENT REGULATION, and VOLTAGE REGULATION.

REGULATOR

A *regulator* is a circuit or device that maintains a parameter at a constant value. Many different devices can serve as regulators; a few examples are diodes, transformers, transistors, tubes, and varactors.

Some regulators operate on a "brute-force" principle, by clipping or limiting the output of a power supply. An example of a "brute-force" regulator is the Zener diode (*see* ZENER DIODE). Other regulators consist of a sensing circuit that determines whether the value of the parameter is too high or too low, and a control circuit that compensates for any error. An example of this kind of regulator is the phase-locked-loop circuit. *See also* CURRENT REGULATION, and VOLTAGE REGULATION.

REINARTZ CRYSTAL OSCILLATOR

A *Reinartz crystal oscillator* is a special form of oscillator, characterized by high efficiency and little or no output at frequencies

other than the fundamental. The Reinartz configuration can be used in oscillators with field-effect transistors, bipolar transistors, or vacuum tubes.

The illustration is a schematic diagram of a Reinartz oscillator that employs a dual-gate MOSFET. A tank circuit is inserted in the source line. This resonant circuit is tuned to approximately half the crystal frequency. The result is enhanced positive feedback, which allows the circuit to oscillate at a lower level of crystal current than would otherwise be possible. *See also* CRYSTAL OSCILLATOR.

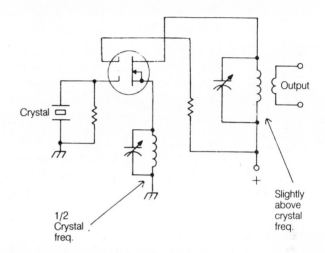

REINARTZ CRYSTAL OSCILLATOR: Generates a pure, single-frequency signal.

REINITIALIZATION

Under certain circumstances, a microcomputer can operate improperly because of misprogramming or stray voltages. When this happens, the microcomputer becomes useless until it is *reinitialized*.

Reinitialization consists of setting all of the microcomputer lines to the low or zero condition. Most microcomputers are automatically reinitialized every time power is removed and reapplied. However, not all microcomputers have this feature; a specific procedure must be followed to reinitialize such devices.

The deliberate removal and reapplication of power, for the purpose of reinitialization, is called a *cold boot*. The reinitialization of a microcomputer without total removal of power is called a *warm boot*. *See also* INITIALIZATION, and MICROCOMPUTER.

RELATIVE MEASUREMENT

The magnitude of a quantity can be measured by comparing it with a known standard. This is called a *relative measurement*. Relative measurements are often made for current, power, radiant energy, sound, and voltage.

A relative measurement can be expressed as a numerical ratio between the unknown and the standard. But the most common method of relative measurement is the decibel system. *See* dBa, dBd, dBi, and DECIBEL.

RELATIVE-POWER METER

An uncalibrated voltmeter, installed at the output of a transmitter or in a transmission line, can be used for relative measurement of power. Such a device is called a *relative-power meter*.

The sensitivity of a relative-power meter usually varies with frequency. The presence of standing waves on a transmission line will also affect the sensitivity of a meter. The power at a given frequency is proportional to the square of the meter reading. Thus, if the power is doubled, the meter reading increases by a factor of 1.414; if the power is quadrupled, the meter reading increases by a factor of 2, and so on.

Relative-power meters are extensively used in low-power and medium-power transmitters, especially those intended for Amateur Radio or CB operation. A relative-power meter makes transmitter adjustment easy, and also serves to indicate whether or not the unit is functioning. Many transceivers use the receiver S meter as a relative-power indicator in the transmit mode.

RELAXATION OSCILLATION

A *relaxation oscillation* is a form of oscillation that results from the gradual buildup of charge in a capacitor, followed by a sudden discharge through a resistance. Various simple circuits can be designed for relaxation oscillation. Such oscillators are often used for audio-frequency applications in which the wave shape is not important. The neon-lamp oscillator is the most common relaxation circuit (*see* NEON-LAMP OSCILLATOR).

At radio frequencies, and in audio applications that require a pure sine wave, relaxation oscillators are not used. Crystal-controlled or tuned oscillators are necessary if precision or high-frequency signals are needed. *See also* OSCILLATION, and OSCILLATOR.

RELAY

A *relay* is a remote-control switch that is actuated by a direct or alternating current. Relays allow control of a signal from a distant point, without having to route the signal path to the control point. Relays are extensively used in the switching of radio-frequency signals. Relays are also useful when it is necessary to switch a very large current or voltage; a smaller (and safer) control voltage can be used to actuate the relay. Relays can have one or two throw positions, and one or more poles.

A relay consists of an iron-core solenoidal coil to which the control voltage is applied, an armature that is attracted to the solenoid by the magnetic field set up when current flows through the coil, and a set of contacts that carry the signal current. A small relay, used for switching radio-frequency signals, is shown in the photograph.

Relays vary greatly in specifications. Some relays are designed for high-speed or frequent switching; the reed relay is made especially for such applications (*see* REED RELAY). Other relays are intended for occasional switching of large amounts of current. Relays are rated according to the number of actuations that can be accomplished per minute or per second. Some relays operate with a small control current, in some cases less than 50 mA; others require several amperes of control current. Certain relays will operate from either alternating or direct control currents, while others must have direct current. Relay contacts are rated according to maximum allowable signal current and voltage.

In modern devices, electronic switching has largely replaced the relay for control purposes. Diodes, silicon-controlled rectifiers, and transistors can be used in place of relays in many applications. Tubes are also sometimes used for switching. Because semiconductor devices and vacuum tubes have no moving parts, they offer better reliability and higher switching speed than relays. *See also* SILICON-CONTROLLED RECTIFIER, SWITCHING DIODE, and SWITCHING-TRANSISTOR.

In a communications system, an intermediate link between stations is sometimes called a *relay*. There can be several relays in a single system. In the early days of radio, communications range was limited to only a few miles. In order to send a message over a great distance, relays were used. The originating station sent the message to the first intermediate station; the first intermediate station sent the message to the second intermediate station; this process was continued until the message reached the receiving station. Today, relays are still sometimes used at the very-high, ultra-high, and microwave frequencies. The relaying is usually done automatically by means of a repeater or satellite. *See also* ACTIVE COMMUNICATIONS SATELLITE, and REPEATER.

RELAY: A typical miniature relay.

RELIABILITY

Reliability is a measure of the probability that a circuit or system will operate according to specifications for a certain period of time. Reliability is obviously related, inversely, to the failure rate and the hazard rate of the components in a circuit (*see* FAILURE RATE, and HAZARD RATE).

Reliability is sometimes expressed as a percentage, based on the proportion of units that remain in proper operation after a given length of time. Suppose, for example, that 1,000,000 units are placed in operation on January 1, 1990. If 900,000 units are operating properly on January 1, 1991, the reliability is 90 percent per year. It is assumed that the units are operated under electrical and environmental conditions that are within the allowable specifications.

Reliability is largely a function of the design of electronic apparatus. Even if the components are of the best possible quality, equipment failure is likely if the design is poor. Reliability can be optimized by a good quality-assurance procedure (*see* QUALITY CONTROL).

The reliability of electronic equipment can be tested in the laboratory, provided a large number of units is used. Reliability testing is part of the overall quality-control procedure in a good engineering scheme.

RELUCTANCE

Magnetic resistance is known as *reluctance*. Reluctance is generally measured in ampere turns per weber. The greater the number of ampere turns required to produce a magnetic field of a given intensity, the greater the reluctance (*see* AMPERE TURN, and WEBER).

Magnetic reluctance is a property of materials that oppose the generation of a magnetic field. The unit of reluctance is the rel. Reluctance is usually abbreviated by the capital letter R.

When two or more reluctances are combined in series, the total reluctance is found by adding the individual reluctances. If $R_1, R_2, R_3, \ldots, R_n$ represent the individual reluctances, and R is the overall total reluctance, then:

$$R = R_1 + R_2 + R_3 + \ldots + R_n$$

in series. In parallel, the same reluctances total:

$$R = 1/(1/R_1 + 1/R_2 + 1/R_3 + \ldots + 1/R_n)$$

Reluctances add in the same way as resistances in series and in parallel.

Suppose that F is the magnetomotive force in a magnetic circuit, ϕ is the magnetic flux, and R is the reluctance. Then, the three parameters are related according to the equation:

$$F = \phi R$$

This formula is known as Ohm's law for magnetic circuits. *See also* MAGNETIC FLUX, MAGNETOMOTIVE FORCE, and OHM'S LAW.

RELUCTIVITY

Reluctivity is a measure of the reluctance of a material per unit volume. Reluctivity is usually expressed in rels per cubic centimeter. Reluctivity is the reciprocal of magnetic permeability. *See also* PERMEABILITY, and RELUCTANCE.

REMANENCE

Certain materials remain magnetized after the removal of an applied magnetic field which has resulted in saturation, or maximum magnetization. All ferromagnetic substances have *remanence*. Diamagnetic and paramagnetic substances do not have remanence.

The remanence of a ferromagnetic material is related to the degree of magnetic hysteresis. When a substance remains magnetized, it is more difficult for the magnetic field to be reversed than it would be if the substance did not remain magnetized. The greater the remanence of the material in an inductor core, for example the lower is the maximum frequency at which the coil can be efficiently operated. *See also* FERROMAGNETIC MATERIAL.

The residual magnetism in a material is sometimes called the *remanence* of that material. Residual magnetism is expressed as a percentage. If the flux density is given by B_1 at saturation and by B_2 after the removal of the magnetic field, then the residual magnetism B_r, in percent, is:

$$B_r = 100(B_2/B_1)$$

Residual magnetism is dependent on the type of material, and on the shape and size of the sample.

REMOTE CONTROL

It is not necessary to be physically present at the location of a ham station, in order to operate that station. A repeater is an example of a ham station that does not always have an operator actually at the controls (*see* REPEATER). A ham shack can be located hundreds, or even thousands, of miles away from its operator, provided a suitable *remote control system* is used.

Advantages and Disadvantages   Remote control has advantages for the city dweller. A ham station can be located at a rural place, where large antennas can be erected, and where radio-frequency interference (RFI) will not be as likely. Then, the operator can work the station from almost anywhere: a downtown condo, a car, or even from a handheld unit on a bus or train. No matter where the remote location might be, the fixed station is always in the ideal location. Imagine working DX on 80-meter CW from a bus, using a laptop computer and a handheld 2-meter transceiver! It can be done with remote control.

The main disadvantage of remote control is that, if something goes wrong at the remote station, it cannot be conveniently repaired. Another problem is that, if the remote control is via long-distance telephone, the phone bill can get very high. And finally, no matter how sophisticated remote control linkups become, there will always be some sacrifices that result from the fact that the operator is not physically present at the remote station.

Equipment at Local Site   At the operator's location, also known as the *local site*, a computer allows use of peripheral apparatus with the telephone or radio linkup unit. A microphone, keyer, speaker and modem (if telephone control is used) or terminal (if radio control is used) are standard. In addition, a telephone set or transceiver is needed. An antenna is needed for the link if radio control is used.

Equipment at Remote Site   At the *remote site*, where the large fixed station is located, there is a telephone or transceiver to handle the commands from, and send telemetry to the local site. There is a modem or terminal. With a telephone link, an automatic answerer is needed. With a radio link, a link antenna is necessary. There is a computer, similar to or identical with the one at the local site. There is a station (main) transceiver, connected to the main antenna. There might also be rotators, antenna switches, and transmatches.

For Further Information   Space does not allow construction details to be given here. The ham magazines frequently publish articles dealing with remote-control techniques and construction. Publishers of these magazines are usually willing to assist hams looking for information on specific topics.

An interesting way to work ham radio from practically any location is via packet radio. *See* PACKET RADIO.

REPEATER

A *repeater* is a receiver/transmitter installation that picks up a signal and retransmits it. In amateur work, repeaters are found mostly at VHF and UHF. Since about 1970, repeaters have completely revolutionized the nature of most communications on the 144-MHz and 440-MHz bands.

A repeater receives a signal on one frequency, and retransmits it at the same time, usually in the same ham band, but on a somewhat higher or lower frequency. Repeaters are common for two-way FM voice communications in government and commercial, as well as in amateur, radio. Increasingly, repeaters are being used with *packet radio*. A packet repeater is called a *digipeater* because it handles digital forms of communication. *See* DIGIPEATER, and PACKET RADIO.

A repeater on a satellite, called a *transponder*, works in much the same way as a land-based repeater, but there are some important differences. *See* ACTIVE COMMUNICATIONS SATELLITE, and TRANSPONDER.

Purpose and Uses   Repeaters increase the range between mobile and portable stations. This is probably their most important advantage. Any two transceivers that are within the repeater range can communicate with each other, even if they could not do so on *simplex*, that is, directly. *See* SIMPLEX.

Illustration A shows the principle of repeater communication. Ideally, a repeater is in a high place, such as on a mountain or tall building. The repeater transmitter can use high power. An ominidirectional gain antenna enhances the receiving and transmitting range of the repeater. The radius of reliable repeater coverage can be in excess of 100 miles when all these factors are optimized. Therefore, two hams using mobile equipment might have a QSO (conversation) over a distance of 200 miles, with almost 100-percent reliability. Without the repeater, the range would nominally be about 30 to 50 miles over flat terrain, and less over irregular terrain.

There are some problems with repeaters. Most significant is the fact that only one QSO can occur on a given repeater at a time. For this reason, repeater QSOs must be kept short so that repeater time is not "hogged." Another problem is that everything said over a repeater is usually heard by several (or many) people. Private conversations have no place on a repeater.

In some heavily populated areas, the 2-meter band is "saturated" with repeaters. This is even becoming troublesome at ¾ meters in dense urban regions. Sometimes two repeaters come into conflict with each other. This is especially likely in the event of unusual propagation conditions, such as "tropo" openings. In other instances, it can be difficult to find a good simplex frequency, just because almost every channel has a repeater on it. This last problem can usually be solved by choosing an "oddball" frequency that is not a standard channel.

How a Repeater Works   A repeater has a receiver, a transmitter, an antenna and an isolating network called a *duplexer*. The incoming signal is picked up by the antenna and fed to the receiver. The demodulated output of the receiver goes directly into the audio input of the transmitter. The transmitter retransmits the signal at the same time, using the same antenna as the receiver. The transmitter signal differs in frequency from the received signal by a certain amount, called the *offset* or *split*, that varies depending on the band. The split is necessary because it is impossible to have a transmitter and receiver working into the same antenna, on the same frequency, at the same time. Illustration B is a block diagram of a repeater.

The split is around 0.5 to 1.0 percent of the transmitter frequency. This is sufficient so that the duplexer can keep the repeater transmitter from overloading the receiver. On 2 meters, 600 kHz is the most common split. At 440 MHz, the split is usually 5.00 MHz. Some repeaters use other *nonstandard splits*. Most modern transceivers have a provision for working nonstandard splits, as well as for standard splits.

Operation   Repeater operation is governed by special Federal Communications Commission (FCC) regulations, but common courtesy is also important. Conversations should be short and to the point. If a long "ragchew" is desired, a simplex frequency should be found, and the contact carried out on that frequency so that other hams can use the repeater.

Emergency communications always get priority on any repeater. Some repeaters are equipped with telephone hookups. This is known as *autopatch* (*see* AUTOPATCH). Some repeaters require a subaudible tone of a certain frequency, sent along with the voice, in order to be accessed, or "brought up." This is called *tone squelch* or *private line* (*see* PRIVATE LINE).

Repeater operation has proliferated in recent years to the point that, especially on 2 meters, a ham with a mobile rig is almost always within range of at least one repeater, from anywhere in the continental United States.

Various publications, advertised in the ham magazines, are available that list all repeater locations and frequencies in the United States.

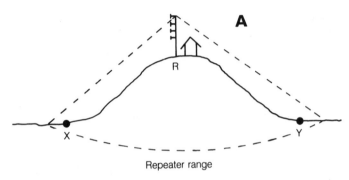

Repeater range

REPEATER: A. Two stations, X and Y, are both within the range of the repeater R, which allows communications that would not be possible on simplex (the vertical scale is exaggerated).

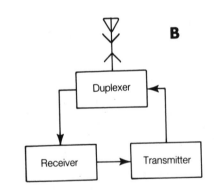

REPEATER: B. A block diagram of a repeater.

RESET

When a device is initialized, or placed in some prescribed state, the action is called *reset* or *reset action*. In a microcomputer, reset action is called *reinitialization*. In a calculator, the CLEAR command constitutes resetting. A count-down timer is reset by adjusting it to show the appropriate number of seconds, minutes, or hours. A count-up timer is reset by placing it at 0. *See also* INITIALIZATION, and REINITIALIZATION.

RESIDUAL MAGNETISM

See REMANENCE.

RESISTANCE

No electrical conductor is perfect. The transfer of charge carriers is always finite per unit time. The opposition that a substance offers to the flow of electric current can be precisely defined as the number of volts required for each ampere of current (*see* OHM'S LAW). This parameter is known as *electrical resistance.*

Resistance can be specified either for direct currents or alternating currents. In ac circuits, the average resistance can differ from the instantaneous or dynamic resistance (*see* DYNAMIC RESISTANCE).

The standard unit of resistance is the ohm, often abbreviated by the uppercase Greek letter omega (Ω). In equations, the symbol for resistance is R. The range of possible resistance values, in ohms, is represented by the set of positive real numbers. In theory, it is possible to have resistances that are zero or infinite as well. Resistance is the mathematical reciprocal of conductance. The higher the conductance, the lower the resistance, and vice-versa (*see* CONDUCTANCE).

The resistance of wire is usually measured in ohms per unit length. Silver has the lowest resistance of any wire. Copper and aluminum also have very low resistance per unit length. The dc resistance of a wire depends on the diameter.

Resistance is often introduced into a circuit deliberately, to limit the current and/or to provide various levels of voltage. This is done with components called *resistors* (*see* RESISTOR).

An inductor has extremely low resistance for dc, but it can exhibit a considerable amount of resistance for ac. This effective resistance is known as inductive reactance; it increases as the alternating-current frequency becomes higher. A capacitor has practically infinite resistance for dc, but it will exhibit a finite effective resistance for ac of a high enough frequency. This effective resistance is called capacitive reactance, and it decreases as the ac frequency gets higher. An ac circuit can contain both resistance and reactance. The combination of resistance and reactance is represented by a complex number, and is known as *impedance* (*see* CAPACITIVE REACTANCE, IMPEDANCE, INDUCTIVE REACTANCE, and REACTANCE).

The resistance of a material is, to a certain extent, affected by temperature. For most substances, the resistance increases with increasing temperature. A few materials exhibit very little change in resistance with temperature fluctuations; some substances become better conductors as the temperature increases. The behavior of a resistance under varying temperature conditions is known as the temperature coefficient (*see* TEMPERATURE COEFFICIENT). Certain materials notably the highly conductive metals, attain extremely low resistance values at temperatures near absolute zero (*see* SUPERCONDUCTIVITY).

A radio transmitting antenna offers a certain inherent opposition to the radiation of electromagnetic energy. This property is called *radiation resistance*, and is a function of the physical size of the antenna in wavelengths (*see* RADIATION RESISTANCE).

Resistance can be defined for phenomena other than electric currents. For example, a mechanical device can show a certain amount of resistance to changes in position; a pipe resists the flow of water to a greater or lesser degree. The opposition that a substance offers to the flow of heat is called *thermal resistance*; its reciprocal is thermal conductivity.

RESISTANCE-CAPACITANCE CIRCUIT

A circuit that contains only resistors and capacitors, or only resistance and capacitive reactance, is called a *resistance-capacitance (RC) circuit*. Such a circuit can be extremely simple, consisting of just one resistor and one capacitor, or it might be a complicated network of components.

An RC circuit can be used for the purpose of highpass or lowpass filtering. An example of an RC highpass filter is illustrated at A in the illustration. A lowpass network is shown at B. The RC filters have a less well-defined cutoff characteristic than inductance-capacitance (LC) circuits (*see* HIGHPASS FILTER, and LOWPASS FILTER). The RC-type highpass or lowpass filter is generally used at audio frequencies, and the LC type is more often used at radio frequencies.

An RC filter is commonly used in a code transmitter to obtain shaping of the keying envelope (*see* KEY CLICK, and SHAPING). Simple RC circuits provide filtering in low-current power supplies. More complicated RC circuits can be used to measure unknown capacitances.

Every RC circuit displays a certain charging and discharging characteristic. This property depends on the circuit configuration, and also on the values of the resistor(s) and capacitor(s). *See also* RESISTANCE-CAPACITANCE TIME CONSTANT.

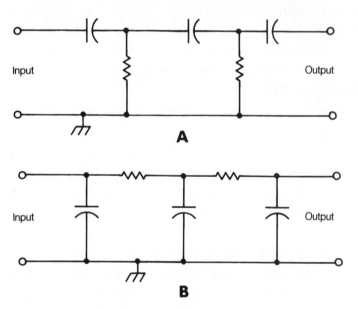

RESISTANCE-CAPACITANCE CIRCUIT: At A, an RC highpass filter; at B, an RC lowpass filter.

RESISTANCE-CAPACITANCE COUPLING

Resistance-capacitance coupling is a form of capacitive coupling that is sometimes used in audio-frequency and radio-frequency circuits. Biasing resistors are used to provide the proper voltages and currents at the output of the first stage and at the input of the second stage. The signal is transferred via a series blocking capacitor.

Resistance-capacitance coupling is simple and inexpensive. However, if a high degree of interstage isolation is required, or if selectivity is needed, transformer coupling is preferable. *See also* CAPACITIVE COUPLING, and TRANSFORMER COUPLING.

RESISTANCE-CAPACITANCE TIME CONSTANT

The length of time required for a resistance-capacitance (RC) circuit to charge and discharge depends on the values of the resistance and capacitance. The larger the resistance or capacitance, the longer it takes for a circuit to charge and discharge.

Charging and discharging are actually exponential processes. The full-charge condition is theoretically never reached from a zero-voltage start; the zero-charge condition is theoretically never reached from a partial-charge start. For practical purposes, however, we can speak of zero charge and full charge.

The period of time required for an RC circuit to reach 63 percent of the full-charge condition, starting at zero charge is called the *RC time constant*. The RC time constant can also be defined as the length of time required for an RC circuit to discharge to 37 percent of the full-charge condition, starting at full charge.

The time constant *t*, in seconds, of an RC circuit having an effective resistance of *R* ohms and an effective capacitance of *C* farads is:

$$t = RC$$

The RC time constant is of importance in the design of some power-supply filtering circuits, code-transmitter shaping circuits, and other RC-filtering devices. *See also* RESISTANCE-CAPACITANCE CIRCUIT.

RESISTANCE-INDUCTANCE CIRCUIT

A circuit that contains only resistors and inductors, or only resistance and inductive reactance, is called a *resistance-inductance (RL) circuit*. Such a circuit can be extremely simple, consisting of just one resistor and one inductor, or it might be a complicated network of components.

An RL circuit can be used for the purpose of highpass or lowpass filtering. An example of an RL highpass filter is illustrated at A. A lowpass network is shown at B. The RL filters

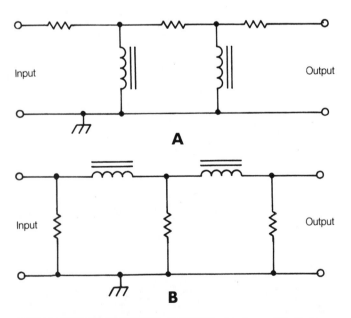

RESISTANCE-INDUCTANCE CIRCUIT: At A, an RL highpass filter; at B, an RL lowpass filter.

have a less well-defined cutoff characteristic than inductance-capacitance (LC) circuits (*see* HIGHPASS FILTER, and LOWPASS FILTER). The RL-type highpass or lowpass filter is generally used at audio frequencies, and the LC type is more often used at radio frequencies. Resistance-inductance circuits can be used to measure unknown values of inductance; there are several different types of RL circuit designed for this purpose.

Every RL circuit displays a certain charging and discharging characteristic. This depends on the circuit configuration, and also on the values of the inductor(s). *See also* RESISTANCE-INDUCTANCE TIME CONSTANT.

RESISTANCE-INDUCTANCE TIME CONSTANT

The length of time required for a *resistance-inductance (RL) circuit* to charge and discharge depends on the values of the resistances and inductances. The larger the resistance and inductance, the longer it takes for a circuit to charge and discharge.

Charging and discharging are actually exponential processes. The full-charge condition is theoretically never reached for a zero-voltage start; the zero-charge condition is theoretically never reached from a partial-charge start. For practical purposes, however, we can speak of zero charge and full charge.

The period of time required for an RL circuit to reach 63 percent of the full-charge condition, starting at zero charge, is called the *RL time constant*. The *RL time constant* can also be defined as the length of time required for an RL circuit to discharge to 37 percent of the full-charge condition, starting at full charge.

Any resistance-inductance circuit can, ideally, be reduced to a simple equivalent combination of one resistor and one inductor. The simpler equivalent circuit has the same charging and discharging times as the more complicated circuit. The time constant (*t*), in seconds, of an RL circuit having an effective resistance of *R* ohms and effective inductance of *L* henrys, is:

$$t = RL$$

See also RESISTANCE-INDUCTANCE CIRCUIT.

RESISTANCE LOSS

In any electrical or electronic circuit, some power is lost in the wiring because of resistance in the conductors. In an antenna system, some power is lost in the earth because of ground resistance. The amount of power lost because of resistance is called the *resistance loss*. It can also be called *I²R loss* or *ohmic loss*.

Resistance loss affects the efficiency of a circuit. The efficiency is 100 percent only if the resistance loss is zero. As the resistance loss increases in proportion to the load resistance, the efficiency goes down. If the loss resistance is *R* and the load resistance is *S*, the efficiency *E* in percent is:

$$E = 100S/(R + S)$$

The resistance loss in any circuit is directly proportional to the resistance external to the load. The loss is also proportional to the square of the current that flows. Mathematically, if *R* is the resistance in ohms and *I* is the current in amperes, the resistance power loss (*P*) in watts, is:

$$P = I^2R$$

Resistance loss, unless deliberately introduced for a specific purpose, should be kept to a minimum. This is especially true in ac power transmission, and in radio-frequency transmitting antennas.

In a power-transmission system, resistance loss is minimized by using large-diameter conductors, and by ensuring that splices have excellent electrical conductivity. The loss is also minimized by using the highest possible voltage, thereby reducing the current for delivery of a given amount of power.

In a transmitting antenna system, the resistance loss is reduced by using large-diameter wire or metal tubing for the radiating conductors, by using a heavy-duty tuning network and feed system, by providing excellent electrical connections where splices are necessary, and by ensuring that the ground conductivity is as high as possible. The resistance loss in a transmitting antenna can also be reduced by deliberately maximizing the radiation (load) resistance. *See also* ANTENNA EFFICIENCY, RADIAL, and RADIATION RESISTANCE.

RESISTANCE MEASUREMENT

The value of a resistance can be measured in various ways. All methods of *resistance measurement* involve the application of voltage across the resistive element, and the direct or indirect measurement of the current that flows as a result of the applied voltage.

The most common method of resistance measurement is the use of a device called an *ohmmeter*. The voltage source is usually a cell or battery, which provides between 1.5 and 12 volts in most cases. The ohmmeter is used to measure moderate values of resistance. To determine extremely high values of resistance, a special form of ohmmeter, known as a *megger*, is used. The megger has a high-voltage generator instead of the low-voltage cell or battery (*see* MEGGER, and OHMMETER).

An alternative means of resistance measurement is the *comparison* or *balance method*. The balance method gives more accurate results than the ohmmeter when the resistance is extremely low or high. The resistance-balance circuit is known as a *ratio-arm bridge*. The unknown resistance is compared with a known, variable, resistance, resulting in a balanced condition when a specific ratio is achieved.

RESISTIVE LOAD

A load is said to be resistive, or purely resistive, when it contains no reactance for alternating current at a specific frequency. A load can be resistive for either of two quite different reasons: it can contain no reactive components, or it can consist of equal, but opposite reactances.

A simple, noninductive resistor presents a load that has essentially zero reactance. This is true at all frequencies up to the point at which the lead lengths become a substantial fraction of a wavelength. Noninductive resistors are used for the purpose of testing radio transmitters off the air; such a load is called a *dummy load* or *dummy antenna* (*see* DUMMY ANTENNA).

A tuned circuit, consisting of a capacitor and an inductor, is resistive at one specific frequency, known as the *resonant frequency*. The same is true of an antenna system. This type of resistive load is clearly different from the noninductive resistor. When the capacitive and inductive reactances cancel each other, a tuned circuit or antenna is said to be *resonant*. The remaining resistance is the result of losses in a tank circuit, or the

combination of radiation resistance and resistance loss in an antenna.

A resistive load is desirable in any power-transfer circuit. If the load contains reactance, some of the incident power is reflected upon reaching the load. Ideally, the load resistance should be identical to the characteristic impedance of the feed line. *See also* CAPACITIVE REACTANCE, IMPEDANCE, INDUCTIVE REACTANCE, RADIATION RESISTANCE, REFLECTED POWER, RESISTANCE LOSS, and RESONANCE.

RESISTIVITY

The resistance of a flat surface can be expressed per unit area, and the resistance of a solid can be expressed per unit volume. These parameters are called *resistivity*. The resistivity of a surface is specified in ohms; the resistivity of a solid is usually given in ohm-centimeters or ohm-inches.

Surface resistivity and volume resistivity depend on the type of material, and also on the temperature. Metals such as aluminum, copper, and silver have very low resistivity. Various other elements, compounds, and mixtures are designed to have certain amounts of resistivity for different applications. The dielectric substances have high resistivity.

The term *resistivity* is occasionally used loosely as a synonym for resistance, although this is technically imprecise. The resistance of a conductor per unit length can be called *resistivity*. *See also* RESISTANCE.

RESISTOR

A *resistor* is an electronic component that is deliberately designed to have a specific amount of resistance (*see* RESISTANCE). Resistors are available in many different forms. An often-used type of resistor in electronic devices is the carbon variety. Values range from less than 1 ohm to millions of ohms; power-dissipation ratings are usually ¼ watt or ½ watt. The photograph shows a ¼-watt resistor and a ½-watt resistor, next to a dime for size comparison. Some carbon resistors have power-dissipation ratings of more than ½ watt or less than ¼ watt.

When a fairly large amount of current must be handled, resulting in a power dissipation of several watts, wirewound resistors are used. Such resistors consist of a coil of nichrome or other high-resistance wire wound on a heat-resistant core. Wirewound resistors have inductance as well as resistance.

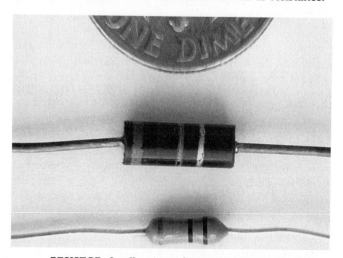

RESISTOR: Small resistors for use on PC boards.

Variable resistors are extensively used in electronic apparatus. The most common type of variable resistor is the potentiometer. For high-current applications, a wirewound type of potentiometer, known as a *rheostat*, is used.

RESISTOR-CAPACITOR-TRANSISTOR LOGIC

Resistor-capacitor-transistor logic (RCTL) is a variation of resistor-transistor logic. In RCTL, bypass capacitors are connected in parallel with the base resistors. This results in higher operating speed than is possible with conventional resistor-transistor logic.

The RCTL scheme was one of the earliest forms of high-speed logic. The power consumption is low to moderate. This type of logic circuit exhibits sensitivity to noise, as does conventional resistor-transistor logic. *See also* DIODE-TRANSISTOR LOGIC, DIRECT-COUPLED TRANSISTOR LOGIC, EMITTER-COUPLED LOGIC, HIGH-THRESHOLD LOGIC, INTEGRATED INJECTION LOGIC, METAL-OXIDE-SEMICONDUCTOR LOGIC FAMILIES, RESISTOR-TRANSISTOR LOGIC, TRANSISTOR-TRANSISTOR LOGIC, and TRIPLE-DIFFUSED EMITTER-FOLLOWER LOGIC.

RESISTOR COLOR CODE

On most carbon-composition resistors, it is not possible to print the value in ohms, because the component is so small and because heat can cause printed characters to fade. A standard color code has therefore been adopted for use on such resistors. The color coding gives the value of the resistor in ohms, and the tolerance above and below the indicated value. On some resistors, the expected failure rate is also shown.

Most resistors have three or four color bands. An example is shown in the illustration. The bands are read by placing the resistor so that the bands are to the left of center, as illustrated in the figure; the bands are read from left to right.

The first (left-most) band specifies a digital value ranging from 0 through 9. The second band indicates a digital value ranging from 0 through 9. The color of the band indicates the digit, according to Table A. The third band designates the power of 10 by which the two-digit number (according to the first two bands) is to be multiplied. These powers of 10 are given in Table B.

There might or might not be a fourth band. If there is no fourth band, then the resistor value can be as much as 20 percent greater than or less than the indicated value. A silver band indicates that the tolerance is plus or minus 10 percent. A gold band indicates that the tolerance is plus or minus 5 percent.

A fifth band might exist; this indicates the failure rate. The failure rate is expressed as a percentage of units expected to malfunction within 1,000 hours of operation at the maximum rated power dissipation. Table C indicates the colors for failure rate.

As an example, consider that a resistor has the following sequence of color bands, from left to right: yellow, violet, red, silver, red. The first two bands indicate the digits 47. The third band designates that the number 47 is to be multiplied by 100; thus, the rated resistance is 4700 ohms. The fourth band designates a tolerance of plus or minus 10 percent (470 ohms above or below the rated value). The fifth band indicates an expected failure rate of 0.1 percent per 1,000 hours; in other words, one

out of 1,000 resistors can be expected to fail within this length of time.

Some precision resistors have 5 color bands in which the first three bands are significant figures and the fourth is the multiplier. Then the fifth is the tolerance. Example: a resistor colored green yellow red red gold is 54,200 ohms ±5 percent. *See also* RESISTOR.

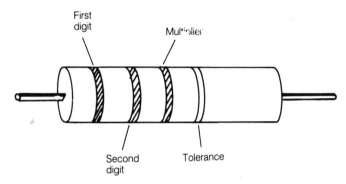

RESISTOR COLOR CODE: Placement of bands.

RESISTOR COLOR CODE:
AT A, DIGITAL COLOR BANDS. AT B, MULTIPLIER BANDS.
AT C, FAILURE-RATE BANDS.

Color	(A) Digit	Color	(B) Multiplier
Black	0	Black	1
Brown	1	Brown	10
Red	2	Red	10^2
Orange	3	Orange	10^3
Yellow	4	Yellow	10^4
Green	5	Green	10^5
Blue	6	Blue	10^6
Violet	7		
Gray	8		
White	9		

(C) Color	Failure rate %/1000 hours
Brown	1
Red	0.1
Orange	0.01
Yellow	0.001

RESISTOR-TRANSISTOR LOGIC

Resistor-transistor logic (RTL) is a form of bipolar digital logic circuit that uses, as its name implies, resistors and transistors. Resistor-transistor logic was among the first bipolar forms of logic to be extensively used in the manufacture of digital integrated circuits. The operating speed is moderate, but it can be increased by placing capacitors in shunt across the resistors (*see* RESISTOR-CAPACITOR-TRANSISTOR LOGIC).

The power-dissipation rate of RTL is fairly high. The design is quite simple and economical. However, RTL is susceptible to noise impulses. *See also* DIODE-TRANSISTOR LOGIC, DIRECT-COUPLED TRANSISTOR LOGIC, EMITTER-COUPLED LOGIC, HIGH-THRESHOLD LOGIC, INTEGRATED INJECTION LOGIC, METAL-OXIDE-SEMICONDUCTOR LOGIC FAMILIES, TRANSISTOR-TRANSISTOR LOGIC, and TRIPLE-DIFFUSED EMITTER-FOLLOWER LOGIC.

RESONANCE

Resonance is a condition in which the frequency of an applied signal coincides with a natural response frequency of a circuit or object. In radio-frequency applications, resonance is a circuit condition in which there are equal amounts of capacitive reactance and inductive reactance. There must be nonzero reactances in a circuit for resonance to be possible; a purely resistive circuit cannot have resonance. Resonance occurs at a specific, discrete frequency or frequencies.

In a parallel-tuned or series-tuned inductance-capacitance (LC) circuit, the reactances balance at just one frequency. A parallel-tuned circuit exhibits maximum impedance at the resonant frequency; neglecting conductor losses, the resonant impedance of such a circuit is infinite, and decreases as the frequency departs from resonance. In a series-tuned circuit, neglecting conductor losses, the impedance at resonance is theoretically zero (*see* CAPACITIVE REACTANCE, IMPEDANCE, and INDUCTIVE REACTANCE). The impedance of a series-tuned circuit rises as the frequency departs from resonance (*see* RESONANCE CURVE).

In an antenna radiator, resonance occurs at an infinite number of frequencies; the lowest of these is the fundamental frequency, and the integral-multiple frequencies are the harmonics (*see* ANTENNA RESONANT FREQUENCY, FUNDAMENTAL FREQUENCY, and HARMONIC). The impedance of an antenna at resonance consists of radiation resistance and loss resistance. This is a finite value that depends on many factors (*see* RADIATION RESISTANCE).

Radio-frequency resonance can occur within a metal enclosure known as a *cavity*. Cavities are used as tuned circuits at ultra-high and microwave frequencies. A cavity exhibits resonance at an infinite number of frequencies, just as an antenna does (*see* CAVITY RESONATOR). A length of transmission line exhibits resonance on the frequency at which it measures ¼ electrical wavelength, and also on all integral multiples of this frequency (*see* HALF-WAVE TRANSMISSION LINE, and QUARTER-WAVE TRANSMISSION LINE).

In radio-frequency applications, resonance is always the result of equal and opposite reactances. If the inductance and capacitance are known, the fundamental resonant frequency of any circuit can be calculated. *See also* PARALLEL RESONANCE, RESONANT CIRCUIT, RESONANT FREQUENCY, and SERIES RESONANCE.

RESONANCE CURVE

A *resonance curve* is any graphic representation of the resonant properties of a circuit or object. Resonant curves are almost always plotted on the Cartesian plane, with the frequency as the independent variable. The dependent variable can be any characteristic that displays a peak or dip at resonance. In radio-frequency circuits, such parameters include the attenuation, current, absolute-value impedance, and voltage. Two examples of resonant curves are shown in the illustration.

At A, the absolute-value impedance of a lossless parallel-resonant inductance-capacitance (LC) circuit is plotted against the frequency. The absolute-value impedance of a lossless parallel-resonant circuit is given by:

$$Z = X_L X_C / (X_L + X_C)$$

where X_L and X_C represent the inductive and capacitive reactances, respectively. The curve at A is based on an inductance of 1 μH and a capacitance of 1 μF, resulting in a resonant frequency of about 159 kHz. The absolute-value impedance is theoretically infinite at resonance. In a practical circuit, losses in

the components and wiring result in a finite, but very large, absolute-value impedance at resonance.

At B, the absolute-value impedance is plotted as a function of frequency for a lossless series-resonant circuit, having an inductance of 1 μH and a capacitance of 1 μF. This yields the same resonant frequency as the parallel configuration, but the absolute-value impedance shows a minimum, rather than a maximum. In a lossless series-resonant circuit, the absolute-value impedance is:

$$Z = X_L + X_C$$

At the resonant frequency, the absolute-value is theoretically zero. In a practical circuit, losses in the components and wiring result in a finite, but very small, absolute-value impedance. *See also* CAPACITIVE REACTANCE, IMPEDANCE, INDUCTIVE REACTANCE, PARALLEL RESONANCE, RESONANCE, RESONANT CIRCUIT, RESONANT FREQUENCY, and SERIES RESONANCE.

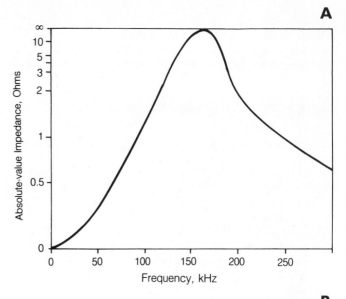

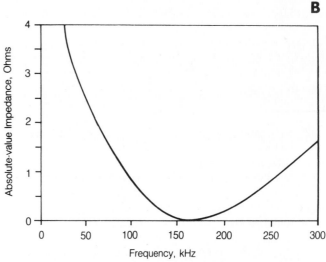

RESONANCE CURVE: At A, a hypothetical parallel-resonant circuit. At B, a hypothetical series-resonant circuit.

RESONANT CIRCUIT

A circuit is considered resonant if it contains finite nonzero reactances that cancel each other. A parallel or series inductance-capacitance (LC) circuit, is an example of a *resonant circuit* at a specific frequency. Antennas, lengths of transmission line, and cavities are resonant circuits at a multiplicity of frequencies. Piezoelectric crystals have resonant properties similar to those of parallel LC circuits. Some ceramics and metal objects act as resonant circuits under certain conditions.

All resonant circuits exhibit variable attenuation, depending on frequency. Resonant circuits are extensively used in audio-frequency and radio-frequency design, for such purposes as obtaining selectivity, impedance matching, and notching. Related articles include ANTENNA RESONANT FREQUENCY, CAPACITIVE REACTANCE, CAVITY RESONATOR, HALF-WAVE TRANSMISSION LINE, IMPEDANCE, INDUCTIVE REACTANCE, NOTCH FILTER, PARALLEL RESONANCE, Q FACTOR, QUARTER-WAVE TRANSMISSION LINE, RESONANCE, RESONANCE CURVE, RESONANT FREQUENCY, SELECTIVITY, SERIES RESONANCE, TANK CIRCUIT, TRAP, and TUNED CIRCUIT.

RESONANT FREQUENCY

The *resonant frequency* of a tuned inductance-capacitance circuit depends on the values of the components. The larger the product of the inductance and capacitance, the lower the resonant frequency will be. If a series-tuned or parallel-tuned circuit has an inductance of L henrys and a capacitance of C farads, then the resonant frequency f, in hertz, is given by the formula:

$$f = 1/(2\pi\sqrt{LC})$$

This formula also holds for values of L in microhenrys, C in microfarads, and f in megahertz.

The resonant frequency of a piezoelectric crystal depends on the thickness of the crystal, the manner in which it is cut, and, in some cases, the temperature. The presence of a coil or capacitor in series or parallel with a crystal also affects the resonant frequency to some extent (*see* CRYSTAL).

The resonant frequencies of an antenna radiator, assuming that no loading reactances are present, can be determined from the length of the radiator. The fundamental frequency f, in megahertz, for a free-space radiator can be determined from the length s in feet according to the formula:

$$f = 468/s$$

For s in meters:

$$f = 143/s$$

The radiator is resonant at all positive integral multiples of the fundamental frequency (f). For a radiator operating against a ground plane, the fundamental resonant frequency (f), in megahertz, is given according to the length (s), in feet, by:

$$f = 234/s$$

For s in meters:

$$f = 71/s$$

The radiator is resonant at all positive integral multiples of the fundamental frequency (f).

The resonant frequencies of any antenna radiator are affected by the presence of reactive elements. An inductor in series with an antenna radiator causes the resonant frequency to be lower than the above formulas would indicate. A capacitor in series will result in a higher resonant frequency than the

above formulas would yield (*see* CAPACITIVE LOADING, and INDUCTIVE LOADING).

A cavity exhibits resonance at frequencies that depend on its length. The same is true for half-wave and quarter-wave section of transmission line, when the velocity factor is taken into account. *See also* CAVITY RESONATOR, HALF-WAVE TRANSMISSION LINE, QUARTER-WAVE TRANSMISSION LINE, RESONANCE, and RESONANT CIRCUIT.

RESPONSE TIME

The length of time between the occurrence of an event, and the response of an instrument or circuit to that event, is called response time. *Response time* is of importance in switching circuits and measuring devices, such as meters.

If the response time of a switching circuit is not fast enough, the switch will fail to actuate properly in accordance with an applied signal. Similarly, if the response time of a measuring instrument is too slow, the instrument will not give accurate readings.

If the response time of a switching device is much more rapid than necessary, undesirable effects can sometimes be observed. An example of this is an excessively fast time constant in the shaping circuit of a code transmitter; this can result in key clicks (*see* KEY CLICK, and SHAPING). In a measuring instrument, a fast response time can be undesirable if average values must be determined. *See also* DAMPING, and OVERDAMPING.

RESUSCITATION

See CARDIOPULMONARY RESUSCITATION, MOUTH-TO-MOUTH RESUSCITATION, and MOUTH-TO-NOSE RESUSCITATION.

RETRACE

In a cathode-ray tube, the electron beam normally scans from left to right at a predetermined, and sometimes adjustable, rate of speed. At the end of the line, when the beam reaches the right-hand side of the screen, the beam moves rapidly back to the left-hand side to begin the next line or trace. This rapid right-to-left movement is called the *retrace* or *return trace*.

The forward trace creates the image that is displayed on the screen. The retrace is usually blanked out, so that it will not interfere with the viewing of the image. This is called *retrace blanking*. The ratio of the retrace time to the trace time is called the *retrace ratio*. *See also* CATHODE-RAY TUBE, OSCILLOSCOPE, PICTURE TUBE, and TELEVISION.

REVERSE BIAS

When a semiconductor P-N junction is biased in the nonconducting direction, the junction is said to be *reverse-biased*. Under conditions of reverse bias, the P-type semiconductor is negative with respect to the N-type semiconductor. This creates an ion-depletion region at the junction (*see* P-N JUNCTION).

If the reverse bias at a P-N junction is made large enough, the junction begins to conduct. This is called *avalanche breakdown* (*see* AVALANCHE BREAKDOWN). Some diodes are designed to take advantage of this effect; these devices are called *Zener diodes*. The minimum reverse voltage at which avalanche breakdown occurs is called the *avalanche voltage* or *peak inverse voltage*.

A vacuum tube is reverse-biased when the anode (plate) is negative, with respect to the cathode. The resistance of a tube under reverse-bias conditions is extremely high. However, if the voltage becomes too great, conduction will occur. The minimum voltage at which reverse-bias conduction occurs is called the *reverse-breakdown voltage* or *peak inverse voltage*.

Reverse bias can occur during half of an applied alternating-current cycle; this is the case in detection and rectification. Reverse bias can be applied deliberately by an external voltage source. This is done with varactor diodes and Zener diodes. *See also* DIODE, DIODE ACTION, FORWARD BIAS, PEAK INVERSE VOLTAGE, VARACTOR DIODE, and ZENER DIODE.

RF

See RADIO FREQUENCY.

RF AMMETER

An *RF ammeter* is a device that measures radio-frequency (RF) current. An RF ammeter is sometimes used in an antenna system for the indirect determination of the RF power output from a transmitter.

When a transmission line is perfectly matched (*see* MATCHED LINE), and the antenna impedance is a pure resistance of Z ohms, the RF power (*P*), in watts, can be determined from the current (*I*), according to the formula:

$$P = I^2Z$$

This formula holds only with a perfect match. *See also* RF CURRENT, and RF POWER.

RF AMPLIFIER

There are various different types of amplifiers for radio-frequency energy. They can be broadly classified as signal amplifiers and power amplifiers.

Signal amplifiers are intended for receiving. A signal amplifier draws essentially no power from the source. The input impedance is extremely high. Signal amplifiers are used in the front ends of all radio-frequency receivers (*see* FRONT END, and PREAMPLIFIER).

Radio-frequency power amplifiers are used in the driver and final stages of transmitters. Power amplifiers are usually of the class-AB, class-B, or class-C type (*see* CLASS-AB AMPLIFIER, CLASS-B AMPLIFIER, CLASS-C AMPLIFIER, and POWER AMPLIFIER).

Radio-frequency amplifiers can have narrow bandwidth or they can be broadbanded. Narrow-band RF amplifiers offer superior unwanted-signal rejection in receivers, and provide attenuation of harmonics and spurious signals in transmitters. Broadbanded amplifiers are simpler to operate because no tuning is required, but undesired signals can be amplified along with the primary signal. Highpass and/or lowpass filters are often used in conjunction with broadbanded RF amplifiers to ameliorate this problem. *See also* HIGHPASS FILTER, and LOWPASS FILTER.

RF CHOKE

A radio-frequency (RF) choke is an inductor used for the purpose of blocking RF signals while allowing lower-frequency and dc signals to pass. Such chokes are often used in electronic circuits when it is necessary to apply an audio-frequency or direct-current bias to a component without allowing RF to enter or leave.

Radio-frequency chokes typically have inductance values in the range of about 100 μH. The exact value depends on the impedance of the circuit, and on the frequency of the signals to be choked. If the impedance of a circuit is Z ohms at the frequency to be choked, an RF choke is usually selected to have an impedance of approximately 10Z ohms.

Radio-frequency chokes are commercially manufactured in a variety of configurations. Most have powdered-iron or ferrite cores and solenoidal windings. Some have air cores or toroidal windings. *See also* CHOKE, and INDUCTIVE REACTANCE.

RF CLIPPING

The intelligibility of an amplitude-modulated or single-sideband speech signal can be increased by either of two basic methods: audio compression or radio-frequency-envelope compression. The latter of these two methods is usually called *RF clipping* or *RF speech processing*.

In RF speech clipping, the signal envelope is increased in average amplitude by a combination of an amplifier, clipper, and filter. The amplifier boosts the signal voltage so that the minima are quite strong; the clipper cuts off the maxima, resulting in a signal with much greater average amplitude than the original. The filter, having a bandpass response with a bandwidth corresponding to the minimum needed for transfer of the modulation information, eliminates the splatter that would otherwise be caused by the clipping. The illustration is a block diagram of a simple RF clipping circuit that can be installed in the intermediate-frequency chain of a single-sideband transmitter.

When RF clipping is used, the readability of a signal can be improved when conditions are marginal, as compared with the readability when no clipping is used. However, if the clipping is excessive, or if the filtering is inadequate, the signal can be distorted to such an extent that the intelligibility is made worse, not better. *See also* SPEECH CLIPPING, and SPEECH COMPRESSION.

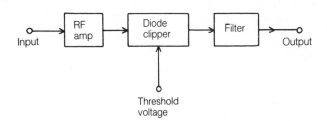

RF CLIPPING: A simple scheme for RF clipping.

RF CURRENT

An alternating current having a frequency of at least 10 kHz is known as a *radio-frequency (RF) current*. The alternating component of a direct current that pulsates in magnitude at a rate of at least 10 kHz can also be called an *RF current*.

The intensity of an RF current is measured in root-mean-square amperes (*see* ROOT MEAN SQUARE). In a circuit having a nonreactive impedance of Z ohms and an RF power level of *P* watts, the RF current *I* is given by:

$$I = \sqrt{P/Z}$$

Radio-frequency current is measured by means of an RF ammeter. *See also* RF AMMETER.

RFI

See ELECTROMAGNETIC INTERFERENCE.

RF POWER

At radio frequencies, power can be manifested in three ways: It can be radiated, it can be dissipated as heat and/or light, and it can appear in reactive form.

Power can be radiated or dissipated only across a resistance; this is called true power (*see* TRUE POWER). A current and voltage can exist across a reactance, but no radiation or dissipation can occur in a reactance. This form of power is called *reactive power* (*see* REACTIVE POWER). The sum of the true power and the reactive power is known as the *apparent power* (*see* APPARENT POWER).

Power at radio frequencies is usually measured by means of a wattmeter in a transmission line. The wattmeter indicates the apparent power. If reactive power is present in a transmission line and antenna system, it causes an exaggerated reading on the wattmeter. *See also* REFLECTED POWER, REFLECTOMETER, and WATTMETER.

RF VOLTAGE

An *RF voltage* is an alternating potential difference between two points, having a frequency of 10 kHz or more. If a voltage fluctuates at a rate of 10 kHz or more, but maintains the same polarity at all times, the alternating component constitutes an RF voltage.

The intensity of RF voltage is measured in root-mean-square volts (*see* ROOT MEAN SQUARE). In a circuit having a nonreactive impedance of Z ohms and an RF power level of P watts, the RF voltage E is given by:

$$E = \sqrt{PZ}$$

A radio-frequency voltage is measured by means of an RF voltmeter. An oscilloscope can be used to measure the peak or peak-to-peak RF voltage; the root-mean-square value is obtained by multiplying the peak value by 0.707 or the peak-to-peak value by 0.354. *See also* OSCILLOSCOPE, and RF VOLTMETER.

RF VOLTMETER

At radio frequencies, the measurement of voltage is somewhat more difficult than at the 60-Hz ac line frequency. Radio-frequency voltage can be measured directly by means of an oscilloscope, but a meter can only work indirectly.

Most radio-frequency voltmeters operate by rectifying and filtering the signal, obtaining a direct-current potential that is proportional to the radio-frequency voltage. An uncalibrated reflectometer uses an RF voltmeter of this type (*see* REFLECTOMETER).

An RF voltmeter is usually frequency-sensitive. This is because the rectifier diode exhibits a capacitance that depends on the frequency of the applied voltage. When using an RF voltmeter, therefore, it is necessary to be sure that the frequency is within prescribed limits.

An RF voltmeter can be used indirectly to determine RF power. If the load contains no reactance and has an impedance of Z ohms, then the RF power P, in watts, that is dissipated and/or radiated by the load is given by:

$$P = E^2/Z$$

where E is the voltmeter reading in root-mean-square volts. *See also* RF VOLTAGE.

RF WATTMETER

See REFLECTOMETER, and WATTMETER.

RHEOSTAT

A *rheostat* is a form of variable resistor. A rheostat consists of a solenoidal or toroidal winding of resistance wire, such as nichrome, two fixed end contacts, and a sliding or rotary contact. A solenoidal rheostat is shown in the illustration at A; a rotary or toroidal type is shown at B.

Rheostats have inductance as well as resistance, and they are therefore not suitable for radio-frequency use. The rheostat is not continuously adjustable; the resistance is determined by a whole number of wire turns, presenting a finite number of discrete values.

Rheostats can be made to dissipate large amounts of power, and this makes them useful for voltage dropping in high-current circuits.

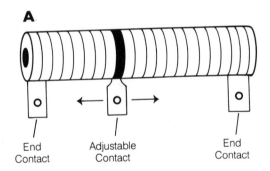

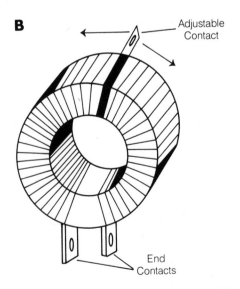

RHEOSTAT: At A, solenoidal; at B, toroidal.

RHOMBIC ANTENNA

A *rhombic antenna* is a form of longwire antenna that exhibits gain in one or two fixed directions. Rhombic antennas are used mostly at the high frequencies (3 to 30 MHz), and are constructed of wire. The rhombic antenna gets its name from the fact that it is shaped like a rhombus (see illustration).

The amount of gain of a rhombic antenna depends on the physical size in wavelengths, and also on the corner angles. The larger the rhombic antenna in terms of the wavelength, the more elongated the rhombus must be in order to realize optimum gain. A rhombic antenna having legs measuring 1/2 wavelength can produce approximately 2 dBd of power gain; this figure increases to 5 dBd for legs of 1 wavelength, 10 dBd for legs of 3 wavelengths, and 12 dBd for legs of 5 wavelengths, provided that the corner angles are optimized (a detailed discussion of rhombic-antenna design is beyond the scope of this book, but many references are available).

The rhombic antenna shown can be fed at either corner at which the angle between the wires is less than 90 degrees. The wires at the opposite corner are simply left free, resulting in the bidirectional pattern of radiation and reception (shown by the double arrow). The feed-point impedance varies with the frequency, but open-wire line can be used in conjunction with a transmatch for efficient operation at all frequencies at which the length of each leg of the rhombus is 1/2 wavelength or greater.

A 600-ohm, noninductive, high-power resistor can be connected at the far end to obtain a unidirectional pattern (shown by the single arrow). The addition of the terminating resistor results in a nearly constant feed-point impedance of 600 ohms, at all frequencies at which the leg length is 1/2 wavelength or more.

The main advantages of the rhombic antenna are its power gain, and the fact that it can produce this gain over a wide range of frequencies. The principal disadvantages are that it cannot be rotated, and that it requires a large amount of real estate. In recent years, the quad and yagi antennas have become much more common than the rhombic. *See also* QUAD ANTENNA, and YAGI ANTENNA.

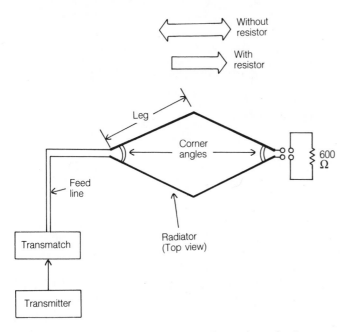

RHOMBIC ANTENNA: Provides directivity and gain.

RIBBON CABLE

Ribbon cable is a flat, flexible, multiconductor cable. Ribbon cables are used in situations where the interconnected components must be frequently moved. Ribbon cables are found in computers, electronic printers, and typewriters, and other devices with movable or often-replaced parts.

Ribbon cables are generally made using stranded, specially treated wire. This allows the cable to be flexed thousands of times without conductor breakage.

RIGHT ASCENSION

One of the two celestial coordinates used by astronomers, and also of importance in satellite communications, is *right ascension*. The other is *declination* (*see* DECLINATION).

Right ascension is an expression for celestial longitude that does not depend on the time of day. Ordinary celestial longitude lines sweep across the heavens because of the rotation of the earth. Right ascension lines are fixed in the heavens.

In the sky, the *celestial equator* is a huge, imaginary circle over the earth's equator. The *ecliptic* is another imaginary circle, formed by the plane of the earth's orbit around the sun. The sun is always someplace on the ecliptic.

From the earth's surface, the celestial equator and the ecliptic both appear as great circles around the celestial sphere, or the illusory sphere of the heavens. But they are 23.5 degrees askew relative to each other because of the tilt of the earth's axis. The ecliptic and the celestial equator intersect at two points in the sky. One of these points is the vernal equinox, the position of the sun on March 20 every year.

The *right ascension coordinates* are assigned by measuring in degrees eastward, along the celestial equator, from the vernal equinox. Right ascension can range from 0 to 360 degrees. In the sky, each right ascension line appears as a great circle running through the designated point on the celestial equator, northward through the north celestial pole, then southward across the celestial equator again, through the south celestial pole, and back to the starting point.

For a satellite orbit, the position of the *ascending node* is often specified in right-ascension coordinates. This number is called *RAAN*, for *r*ight *a*scension of *a*scending *n*ode. *See also* ACTIVE COMMUNICATIONS SATELLITE, and ASCENDING NODE.

RING COUNTER

A *ring counter* is a form of shift register in which data bits move in circular or loop fashion. If a data bit reaches the end of the ring counter, the next pulse will cause the data bit to revert to the beginning of the loop. This circular data movement can occur in either a left-to-right (clockwise) or right-to-left (counterclockwise) direction.

If a ring counter has *n* data positions, the information pattern is repeated every *n* pulses, as each data bit completes the circle exactly once. *See also* SHIFT REGISTER.

RINGING

When an alternating-current signal is applied to a tuned circuit at the resonant frequency, circulating currents are set up in the inductance and capacitance. These circulating currents continue for a short time after the signal is removed. This effect is called *ringing*.

The amount of time for which ringing will continue depends on the Q factor of the tuned circuit; the higher the Q, the longer it takes for the ringing to decay (*see* Q FACTOR). Ringing can be either desirable or undesirable.

In a class-B or class-C radio-frequency power amplifier, the tuned output circuit stores energy during the conduction period and releases it during the nonconduction period of the transistor or tube. This is a form of ringing called the *flywheel effect* (*see* CLASS-B AMPLIFIER, CLASS-C AMPLIFIER, and FLYWHEEL EFFECT). It results in a nearly pure sine-wave output signal from the amplifier, even though the amplifier does not conduct for the full cycle. The flywheel effect also enhances the operation of most types of tuned oscillators.

In an audio filter having extremely narrow bandwidth, ringing limits the maximum data speed that can be received. If the bandpass is too small, the maximum speed will be too slow, and reception will be impaired. Less frequently, the same problem can be encountered in intermediate-frequency bandpass filters. Ringing in a bandpass filter can be reduced by designing the circuit so that the response is rectangular. *See also* BANDPASS FILTER, BANDPASS RESPONSE, and RECTANGULAR RESPONSE.

RING MODULATOR

A *ring modulator* is a circuit that is used for mixing or modulating of radio-frequency signals. The ring modulator consists of four semiconductor diodes connected in a loop, allowing current to circulate around and around. The signal inputs and outputs are taken from the apex points of the loop.

The ring modulator is a passive circuit. The main advantage of the circuit is its simplicity. The ring modulator does not produce gain, but this is not a major problem in most applications because the ring modulator can be followed by a simple amplifier circuit.

The ring modulator can be used as a balanced modulator for generating single-sideband signals; it can be used as a mixer in superheterodyne circuits; it can be used as a product detector. *See also* BALANCED MODULATOR, DOUBLE BALANCED MIXER, MIXER, MODULATOR, and PRODUCT DETECTOR.

RIPPLE

Ripple is the presence of an alternating-current component in a direct-current signal. Usually, the term refers to the residual 60-Hz or 120-Hz ac component in the output of a dc power supply. The ripple in the output of a power supply can result in modulation of a radio-frequency transmitter; power-supply ripple can also cause apparent modulation of received signals. This modulation is, itself, sometimes called *ripple*.

The ripple in the output of a power supply can be expressed quantitatively in two ways. The actual ripple voltage is determined by eliminating the direct-current component, and then determining the root-mean-square value of the remaining alternating current (*see* ROOT MEAN SQUARE). We might call this voltage E_r, and the steady direct-current voltage E_d. We can then express the ripple as a percentage, R_p:

$$R_p = 100 \ E_r/E_d$$

These expressions of ripple are shown in the illustration.

Ripple can be reduced by the use of large-value capacitors

and chokes in the filter of a power supply. The amount of filtering depends on whether the ripple frequency is 60 Hz or 120 Hz; less filtering is needed at 120 Hz. The amount of filtering also depends on the load resistance. The lower the load resistance for a given power supply, the more filtering is required. *See also* FILTER, FILTER CAPACITOR, and POWER SUPPLY.

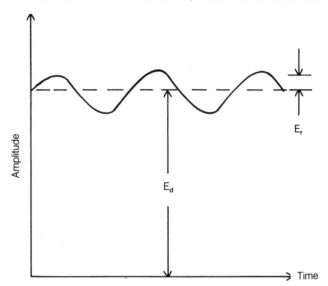

RIPPLE: Expressed as $100 \times E_r/E_d$.

RISE

The increase in amplitude of a pulse or waveform, from zero to full strength, is called the *rise*. In acoustics, and in automatic-gain/level-control systems, rise is called *attack* (*see* ATTACK, AUTOMATIC GAIN CONTROL, and AUTOMATIC LEVEL CONTROL). The rise of a pulse or waveform can sometimes appear instantaneous, such as in a square wave; but it is never actually so. A certain finite amount of time is required for the rise.

The rise in output of some devices, such as high-wattage incandescent bulbs, can be watched and noticed. The rise in amplitude from other devices, such as neon lamps or light-emitting diodes, is too rapid to be seen. But the rise, no matter how rapid, is never instantaneous. The opposite of rise (the drop in amplitude from full strength to zero) is called *decay*. *See also* DECAY, DECAY TIME, and RISE TIME.

RISE TIME

The *rise time* of a pulse or waveform is the time required for the amplitude to rise to a certain percentage of the final value. The time interval begins at the instant the amplitude begins to rise, and ends when the determined percentage has been attained.

The rise of a pulse or waveform often proceeds in a logarithmic manner, as does the decay. In theory, the amplitude never reaches the final value; it is always rising a little bit, until decay begins. In practice, a point is reached at which the amplitude can be considered maximum. This point can be chosen for the determination of the rise time interval. *See also* DECAY, DECAY TIME, RISE, and TIME CONSTANT.

RL CIRCUIT

See RESISTANCE-INDUCTANCE CIRCUIT.

RMM

See READ-MAINLY MEMORY.

RMS

See ROOT MEAN SQUARE.

ROLL

In a television picture, *roll* is the result of a lack of vertical synchronization. Roll is sometimes called *flip-flop* or *rolling*.

Rolling can generally be eliminated by changing the vertical-hold adjustment in a television receiver. Adjusting the antenna can help by improving reception in general. *See also* TELEVISION.

ROM

See READ-ONLY MEMORY.

ROOF MOUNT

A *roof mount* is a bracket that is used for attaching an antenna mast to a peaked roof. Roof mounts are available in most electronics and hardware stores. Roof mounts are generally used for mounting of television-antenna masts, and masts for smaller amateur and CB antennas.

A typical roof mount is shown in the illustration. This roof mount accepts a mast having a diameter of about 1 inch to 3 inches. Guy wires are required to keep the mast vertical. *See also* GUYING.

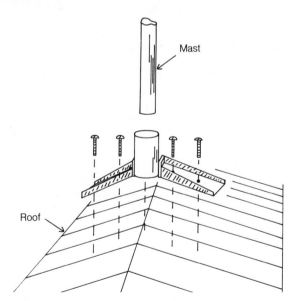

ROOF MOUNT: Secures the mast base to the roof.

ROOT MEAN SQUARE

The current, power, or voltage in an alternating-current signal can be determined in various ways. The most common method of expressing the effective value of an alternating-current waveform is the *root-mean-square (RMS) method*. The root-mean-square current, power, or voltage is an expression of the effective value of a signal.

The root-mean-square current, power, or voltage is determined via the following procedure. First, the amplitude is squared so that the negative and positive halves of a waveform are made identical. Then, the value is averaged over time. Finally, the square root of the average square value is determined. Mathematically, the RMS value is an expression of the direct-current effective magnitude of an alternating-current or pulsating-direct-current waveform.

For a sine wave, the RMS value is 0.707 times the peak value, or 0.354 times the peak-to-peak value. For a square wave, the RMS value is the same as the peak value, or half the peak-to-peak value. For other waveforms, the ratio varies. *See also* PEAK-TO-PEAK VALUE, and PEAK VALUE.

ROSE

See RATS OPEN SYSTEMS ENVIRONMENT.

ROSE X.25 PACKET SWITCH

In the *RATS Open Systems Environment (ROSE)* for a packet radio network, the *network node controller (NNC)* is sometimes called a *ROSE X.25 packet switch*. *See* NETWORK NODE CONTROLLER, PACKET RADIO, and RATS OPEN SYSTEMS ENVIRONMENT.

ROTARY SWITCH

A *rotary switch* is a switch that is thrown or actuated by turning. Rotary switches are generally used in multiple-throw applications. A rotary switch can have as many as 10 or even 20 throw positions; some rotary switches have several poles as well (*see* MULTIPLE-POLE, and MULTIPLE-THROW).

In circuits carrying high voltages, especially at radio frequencies, rotary switches provide wide separation between contacts. This reduces the chance of arcing. Rotary switches can have long shafts so that the control knob is several inches away from the contacts.

Most rotary switches are constructed on ceramic or phenolic wafers. These materials can withstand the high temperatures that are present when the current is high. Ceramic and phenolic also have excellent dielectric properties, and this minimizes the chance of arcing between contacts. Rotary switches are available in various sizes for different purposes. *See also* WAFER SWITCH.

ROTATABLE ANTENNA

A *rotatable antenna* is, in general, any directional antenna that is small and light enough to be placed on a tower and turned with a rotator. These antennas are commonly used on the ham bands at 14 MHz and above (20 meters and shorter wavelengths).

Rotatable antennas are occasionally seen at frequencies as low as 7 MHz (40 meters). These arrays are large and expensive. They are usually Yagi, quad or log-periodic antennas.

Rotatable Yagi and quad antenna systems have been built for use at 3.5 MHz, but the dimensions are monstrous and the cost is astronomical. The types of antennas most often used with rotators include the following, each represented by an article in this book: BILLBOARD ANTENNA, DIPOLE ANTENNA, DISH ANTENNA, HELICAL ANTENNA, HORN ANTENNA, LOG-PERIODIC ANTENNA, LOOP ANTENNA, QUAD ANTENNA, and YAGI ANTENNA.

ROTATOR

A *rotator* is a slow-operating motor that is capable of withstanding and providing a large amount of torque. Antenna rotators are commercially manufactured in many different sizes and configurations for various purposes. Most turn at the rate of one rotation per minute, or about 6 degrees per second. Small rotators are used with directional television receiving antennas; larger rotators are used in amateur and commercial radio installations in conjunction with log-periodic, quad, or yagi antennas.

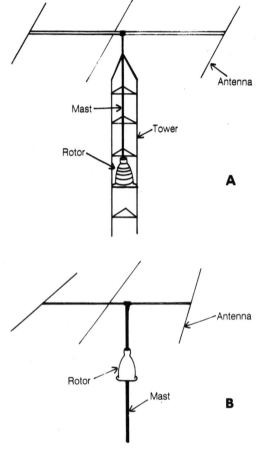

ROTATOR: On a tower (A) and on a mast (B).

A rotator is sometimes installed inside a tower (A in the illustration). This arrangement prevents excessive lateral strain on the mechanism; the only forces occur vertically downward (because of the weight of the antenna) and in a clockwise or counterclockwise direction (torque). Smaller antenna systems can allow the less rugged installation method (shown at B). There is some lateral strain on the mechanism when this installation scheme is used.

When choosing a rotator for a particular antenna system, it is important that the device be large enough and rugged enough. However, the use of an excessively heavy rotator is a waste of money. Small television-antenna rotators, available in most hardware stores, can be used with small amateur and CB antennas. For larger antennas at frequencies below the 27-MHz (11-meter) band, more rugged rotators can be found in amateur-radio stores or ordered via catalogs.

Most rotators are designed for the purpose of turning an antenna in the azimuth plane. However, some rotators are made especially for moving an antenna in the elevation plane. In a satellite-communications system, one of each type of rotator is used in an azimuth-elevation (az-el) configuration (*see* AZ-EL). This arrangement can be used with groups of yagi antennas, helical arrays, or dish antennas.

RS-232-C

See EIA-232-D.

RST SYSTEM

The *RST system* is a scheme used by radiotelegraph operators for telling each other about the quality of reception. The acronym *RST* stands for *Readability-Strength-Tone*. A variation of the system, expressing only the readability and the signal strength, is used by radiotelephone operators.

RST SYSTEM: THE RST
(READABILITY-STRENGTH-TONE) SYSTEM.

Readability

1 — Unreadable signals
2 — Barely readable signals
3 — Readable, but with difficulty
4 — Almost perfectly readable
5 — Readable with no difficulty

Strength	
1 — Faint signals	6 — Good signal strength
2 — Very weak signals	7 — Moderately strong signals
3 — Weak signals	8 — Strong signals
4 — Fair signals	9 — Very strong signals
5 — Fairly good signals	

Tone (Code only)	
1 — Ac Note, 60 Hz	6 — Definite trace of ripple
2 — Rough ac, 60 Hz or 120 Hz	7 — Some ripple, but not much
3 — Rectified but unfiltered note	
4 — Rough, some filtering	8 — Barely detectable ripple
5 — Markedly rippled note	9 — No detectable ripple

A radiotelegraph operator gives the RST as a three-digit number, according to the definitions shown in the table. Radiotelephone operators use only the readability and strength figures. In a radiotelegraph conversation, the operator at the other end might send to you, "RST 389," which means, "Your signals are readable with difficulty; they are strong; the tone is perfectly pure." The radiotelephone operator would tell you, "You are 3 by 8." The marginal readability, in spite of the signal strength, implies interference in this situation.

SAFETY FACTOR

Many electrical and electronic components are rated according to the maximum current, voltage, or power they can handle. For example, a resistor can be rated at 1/2 watt; a vacuum tube can be rated at 100 watts of continuous plate dissipation; a diode might be rated at 500 peak inverse volts. These ratings incorporate a factor called the safety factor.

The safety factor allows for the possibility that a component or device might be operated at, or near, the maximum limit of its rated capability. You might insist that a 1/2-watt resistor actually dissipate 1/2 watt; you might design an amplifier so that its tubes are run at their maximum level of rated plate dissipation; you might design a rectifier circuit using diodes that are operated at their maximum peak-inverse-voltage rating. The ratings are purposely under-quoted by the manufacturer for this reason; the actual limits can be from 10 to 20 percent higher than the rated limits. The above-mentioned resistor, for instance, might really be able to handle 0.6 watt before it burns out. The manufacturer includes a safety factor in the ratings. This improves the reliability of the component, because it provides a buffer zone that allows for slight variation in actual ratings from one component to another.

When you design a circuit, you should never subject components to the maximum rated current, voltage, or power. You should incorporate a safety factor of 10 to 20 percent, in addition to that of the manufacturer. This will ensure that the components will remain operational for the longest possible time, thus greatly minimizing service problems with the equipment.

Some ratings obviously do not incorporate safety factors. An example is a 100-watt incandescent bulb or an 0.1-μF capacitor. Component values of this type can be in error by a certain percentage; the maximum possible error is called the tolerance. If the component value in a particular application is critical, you might want to use a high-precision component. You would then install a component with a tolerance somewhat smaller (more precise) than you think necessary. The difference between the necessary component tolerance and the tolerance of the component you use would then constitute a tolerance safety factor. *See also* TOLERANCE.

SAFETY GROUNDING

When several different pieces of electrical or electronic apparatus are used together, a potential difference can develop between their respective chassis. This can present a shock hazard when moderate or high voltages are used. Experienced electricians and engineers will tell you: "never touch two grounds at the same time." You might already know what they mean!

Dangerous potential differences among various constituents of a system are eliminated by bonding all chassis together electrically, and connecting them to a common ground. This is called *safety grounding*. The ground-bus method is the most often-used means of accomplishing safety grounding. Safety grounding should always be used when dangerous voltages

might occur. Ground loops, which can cause problems in radio-frequency equipment, are avoided by the bus system. *See also* GROUND BUS.

SAG

When a wire is strung between two points, some *sag* is inevitable. In longwire antenna systems, sag is important, because it can, if severe, modify the radiation pattern from the system (*see* LONGWIRE ANTENNA, RHOMBIC ANTENNA, and V BEAM).

A sagging wire describes a curve called a *catenary*, which is very similar to the parabola. For two supports at the same height, wire sag can be calculated from the following formula:

$$s = 0.612 \sqrt{mn - m^2}$$

where s is the sag at the center of the span, m is the distance between the two supports, and n is the length of the cable (all measurements are in feet). The sag is defined as the distance at the center of a span between the actual wire position and the position if the wire were perfectly straight. The above formula is based on the assumption that the sagging wire assumes the geometric form of a parabola.

The problem of sag is increased by the use of elastic wire, and also by icing or severe winds. *See also* ICE LOADING, and WIND LOADING.

SAINT ELMO'S FIRE

During a thunderstorm, the ends of long metal objects (such as antenna elements, flag poles, and airplane wings) can acquire a strange glow. This glow results from the accumulation of electric charge on the object with respect to the surrounding environment. It is called *Saint Elmo's Fire*.

Saint Elmo's Fire was observed on the masts of sailing ships, even in ancient times. Because we have known about the nature of electricity for only the past two centuries, the phenomenon caused bewilderment and even fear in people who saw it. There is some good reason to be afraid of Saint Elmo's Fire: lightning is likely to strike an object when it is aglow with an excess of electric charge. *See also* LIGHTNING.

SAMPLE-AND-HOLD CIRCUIT

The term *sample-and-hold (S/H or SH)* refers to a simple circuit used in various digital applications. In its basic form, the S/H circuit consists of a capacitor and a switch, as shown in the illustration. The switch has two positions, called *sample mode* and *hold mode*.

The input voltage is "sampled" and stored in the capacitor as long as the switch is in the sample-mode position. When the circuit is switched to the hold mode, the charge on the capacitor is maintained at the input voltage level.

The most common use for the S/H circuit is in analog-to-digital or digital-to-analog conversion. *See* ANALOG-TO-DIGITAL CONVERTER, and DIGITAL-TO-ANALOG CONVERTER.

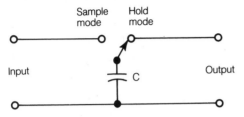

SAMPLE-AND-HOLD CIRCUIT: The capacitor C holds the input voltage when the switch is moved to "hold."

SAMPLING OSCILLOSCOPE

An oscilloscope can be used to obtain a prolonged view of a regular, repeating waveform simply by adjusting the sweep rate to correspond with the period of the wave. Most oscilloscopes can perform this function automatically, by the use of a triggering circuit (*see* TRIGGERING). If the waveform is very irregular, however, this cannot be done. A different method must be used for the analysis of irregular, repeating waveforms or waveforms of extreme frequency. A sampling oscilloscope provides one means for evaluating such waveforms.

The *sampling oscilloscope* measures the instantaneous voltage of a waveform at various moments. The samples are taken at the rate of one per cycle, until the entire waveform has been sampled at intervals. The signals are amplified and stored. Then, they are displayed simultaneously on the cathode-ray screen. The display appears segmented or digitized because a finite number of samples are used. The result is an approximation of the waveform. The accuracy of this approximation depends on the number of samples taken in a given amount of time, and also on the complexity of the waveform itself. *See also* OSCILLOSCOPE.

SAMPLING THEOREM

When a signal is evaluated by sampling, it is necessary to take at least a certain number of samples per cycle. Otherwise, ambiguity will result, and you cannot be certain that the waveform actually is what you think it is. For example, it is apparent that the sampling oscilloscope must obtain at least a certain number of samples per cycle in order to show accurately what the waveform looks like. The minimum sampling rate for accurate representation of a waveform is defined by a rule known as the *sampling theorem*. The sampling theorem was developed by the engineer Nyquist.

Suppose that the highest-frequency component of a signal has a frequency of f Hz. According to the sampling theorem, you need at least two samples per cycle for this component. That is, the sampling rate must be at least $2f$. The maximum allowable interval between samples is therefore $1/2f$ seconds.

In practice, it is a good procedure to obtain more than the minimum number of samples as defined by this theorem. The greater the number of samples, the more accurate the representation will be. It is rarely necessary, however, to obtain more than 10 to 15 samples for each cycle of the highest-frequency component. *See also* SAMPLING OSCILLOSCOPE.

SATELLITE COMMUNICATIONS

See ACTIVE COMMUNICATIONS SATELLITE, and OSCAR.

SATELLITE TRANSPONDER MODES

Ham satellites operate using *full-duplex transponders*. This means that transmission to, and reception from, a ham satellite is done simultaneously, by sending the uplink signal on one band and receiving the downlink signal on another band. In most cases, the frequencies are in the very-high (VHF) or ultra-high (UHF) range. *See* DOWNLINK, TRANSPONDER, and UPLINK.

There are several different frequency-band combinations for uplink and downlink. These are called *satellite transponder modes* or *simply modes*. The modes are designated by letters of the alphabet. The table indicates the satellite transponder modes in use as of this writing. *See also* ACTIVE COMMUNICATIONS SATELLITE, and OSCAR.

SATELLITE TRANSPONDER MODES

Mode	Transponder input (Uplink band)	Transponder output (Downlink band)
A	2 m	10 m
B	3/4 m	2 m
J	2 m	3/4 m
L	1/4 m	3/4 m
S	1/4 and 3/4 m	1/8 m
LJ	1/4 and 2 m	3/4 m
K	15 m	10 m

Bands: 1/8 m = 13 cm = 2401 MHz, 1/4 m = 24 cm = 1269 MHz, 3/4 m = 70 cm = 435–436 MHz, 2 m = 145 MHz, 10 m = 29 MHz, 15 m = 21 MHz.

SATURABLE REACTOR

A *saturable reactor* is an inductor with a ferromagnetic core having special properties. The core of the saturable reactor achieves magnetic saturation at a fairly low level of coil current. This makes it possible to change the effective permeability by passing dc through the windings.

The saturable reactor exhibits its maximum inductance, and therefore the maximum inductive reactance, when there is no direct current passing through the coil. As the current increases, the inductive reactance decreases, reaching a minimum when saturation occurs in the core material.

The saturable reactor is used in a circuit called a *magnetic amplifier*, which is a form of voltage amplifier. *See also* MAGNETIC AMPLIFIER.

SATURATED LOGIC

All digital logic circuits operate in two defined states, generally called *high* and *low*. A variety of different schemes has been developed to accomplish digital-logic switching. Most of the methods use groups of bipolar transistors or metal-oxide-semiconductor devices. Sometimes other components, such as resistors and capacitors, are included.

Although there are only two possible logic states, many semiconductor devices are not always "all the way on" or "all the way off." Some forms of logic do operate this way—the devices are either at cutoff or at saturation—and this form of circuit is known as *saturated logic* (*see* CUTOFF, and SATURATION).

Saturated logic offers definite advantages. It is relatively immune to the effects of noise and external analog signals. The error rate is therefore quite low. Saturated logic circuits are often simpler and less expensive than other forms. However, saturated forms of logic generally consume more power than their nonsaturated counterparts, and the switching speed is somewhat slower. *See also* DIODE-TRANSISTOR LOGIC, DIRECT-COUPLED TRANSISTOR LOGIC, HIGH-THRESHOLD LOGIC, RESISTOR-CAPACITOR-TRANSISTOR LOGIC, RESISTOR-TRANSISTOR LOGIC, and TRANSISTOR-TRANSISTOR LOGIC.

SATURATION

In a switching or amplifying device, the fully conducting state is called *saturation*. The term is especially used in bipolar-transistor and field-effect-transistor circuitry. Saturation can also be mentioned in reference to the operation of a vacuum tube.

An npn bipolar transistor becomes saturated when the base voltage is sufficiently positive relative to the emitter; a pnp transistor becomes saturated when the base voltage is sufficiently negative relative to the emitter. An N-channel junction field-effect transistor becomes saturated when the gate voltage is sufficiently positive relative to the source; a P-channel junction field-effect transistor becomes saturated at high negative gate voltages relative to the source. A vacuum tube conducts better and better as the grid voltage gets less and less negative; the actual saturation voltage can be either positive or negative, depending on the tube characteristics.

The voltage at which saturation occurs is called the *saturation point* or *saturation voltage*. The current that flows under these conditions is called the *saturation current* (*see* SATURATION CURRENT, and SATURATION VOLTAGE). In the characteristic curve of an amplifying or switching device, saturation is indicated by a leveling off of the collector, drain, or plate current as the base, gate, or grid voltage changes. The saturation voltage and current are affected by the voltage at the collector, drain, or plate of the amplifying or switching device. *See also* TRANSISTOR, NPN TRANSISTOR, PNP TRANSISTOR, and TUBE.

In a field-effect transistor, the condition of pinchoff is sometimes called *saturation* (*see* FIELD-EFFECT TRANSISTOR, and PINCHOFF). In a ferromagnetic substance, the flux density generally increases as the magnetizing force increases. However, there is a limit to the flux density that can be obtained in a ferromagnetic material. When this limit is reached, further increases in the magnetizing force do not produce an increase in the flux density. This condition is called *saturation*. *See also* FERROMAGNETIC MATERIAL.

SATURATION CURRENT

When an amplifying or switching device becomes saturated, a certain current flows through its collector, drain, or plate circuit. This current represents the maximum possible current that can flow in the circuit for a fixed collector, drain, or plate voltage.

Generally, the saturation current of a transistor or tube increases as the collector, drain, or plate voltage is made larger. This is simply because large voltage causes greater current flow than small voltages. *See also* SATURATION.

SATURATION CURVE

A *saturation curve* is a graphical representation of saturation in an amplifying or switching device, or in a ferromagnetic material.

The condition of saturation can be affected by parameters (such as the collector, drain, or plate voltage) in an amplifying or switching transistor or tube. In a ferromagnetic substance, the temperature can have an effect. Sometimes several different saturation curves are graphed on a single set of coordinates, one curve each for specific values of the parameter. This kind of graph is called a *family of saturation curves*. *See also* SATURATION.

SATURATION VOLTAGE

In an amplifying or switching transistor or tube, the current in the collector, drain, or plate circuit changes with changing voltage at the control electrode. In an npn transistor or N-channel field-effect transistor, the current rises with increasing positive voltage at the base or gate. In a pnp transistor or P-channel field-effect transistor, the current rises with increasing negative voltage at the base or gate. In a vacuum tube, the current rises as the voltage at the grid becomes less negative or more positive.

As the voltage is changed so that the collector, drain, or plate current rises, a point is eventually reached at which no further increase occurs. The voltage at the base, gate, or grid, measured with respect to the potential at the emitter, source, or cathode, is called the *saturation voltage*. The saturation voltage depends on the type of device, and also on the collector-emitter, drain-source, or plate-cathode voltage. The saturation voltage is sometimes called the *saturation bias* or *saturation point*. *See also* SATURATION, and SATURATION CURRENT.

SAWTOOTH WAVE

A waveform that gradually rises and abruptly falls, or viceversa, is called a *sawtooth wave*. When displayed on an oscilloscope, a sawtooth wave appears to have perfectly straight, or linear, rise and decay traces. The illustration on pg. 428 shows three examples of sawtooth waves. Sawtooth waves can reverse polarity, such as the example at A, or they can maintain one polarity, as shown at B and C.

The sawtooth wave is used for scanning in cathode-raytube devices. The gradual-rise, rapid-fall wave, called a *ramp wave*, facilitates controlled scanning from left to right and an almost instantaneous return trace from right to left (*see* RAMP WAVE, and RETRACE). Sawtooth waves are used for various test purposes. Modern audio signal generators can produce sawtooth waves as well as square waves and sine waves.

Some sawtooth waves have nonzero rise and fall times. An example of such a wave is shown at C. If the rise and fall times are identical, the wave is called a *triangular wave*.

The period of a sawtooth wave is the length of time from any point on the waveform to the same point on the next pulse in the train. This period is divided into 360 degrees of phase, with the zero point usually corresponding to the moment of instantaneous change, as at A and B, or the end of the more rapid change (C). Alternatively, the zero-degree point can be set at the time of zero polarity and increasing amplitude. The frequency of a sawtooth wave, in hertz, is the reciprocal of the period in seconds.

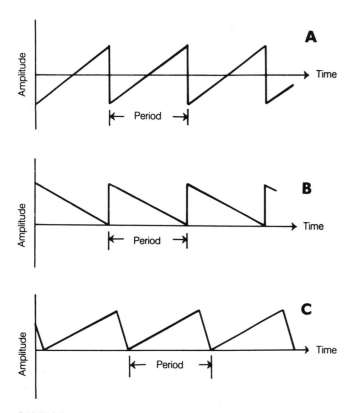

SAWTOOTH WAVE: Slow-rise, fast-decay (A); fast-rise, slow-decay (B); slow-rise, moderate decay (C).

SCAN CONVERTER

In *slow-scan television (SSTV)* communications, an integral part of the station is a device that converts the audio from a single-sideband (SSB) receiver to a sequence of video images, and the outgoing images into audio that modulates an SSB transmitter. This piece of equipment is the *scan converter*.

A scan converter consists of two data converters, a memory, a tone generator and a detector. They are interconnected, as shown in the figure for the article SLOW-SCAN TELEVISION.

When you listen to an SSTV signal with an SSB receiver, you hear a rapid sequence of audio tones. These contain the video information, but this audio can't be fed directly into a television set. The parameters for SSTV are considerably different than those for fast-scan (conventional) TV. The scan converter changes them over.

Similarly, the output of a camera cannot be directly used to modulate an SSB transmitter. The scan converter changes the video image to a rapid sequence of audio tones that are fed to the microphone input of the SSB transmitter.

Scan converters are commercially available, and are advertised in the ham magazines. If you're interested in buying one, it is best to look through the most recent ham-radio periodicals obtainable, because the technology is rapidly advancing and the quality is constantly improving. *See also* COLOR SLOW-SCAN TELEVISION, SLOW-SCAN TELEVISION, and TELEVISION.

SCAN FUNCTIONS

A *scan function* is a mode of scanning in a cathode-ray-tube system. Different scan functions are used for different purposes. The waveform used to actuate scanning is often referred to as the scan function. The scan waveform in an oscilloscope or television system, for example, is a ramp wave (*see* RAMP WAVE).

In radar, there are several different scan functions that can be used. The most common is the circular scan. In television, a left-to-right, top-to-bottom scan function is used. In most oscilloscopes, a left-to-right, single-line scan function is used. *See also* OSCILLOSCOPE, RADAR, RASTER, and TELEVISION.

In an automatic scanning receiver, the scanning apparatus can work in various different ways. These modes are called *scan functions* or *scan modes*. The scanner can search for an occupied channel among empty channels, stopping when a signal is encountered and remaining on that channel until the signal disappears. This is called the *busy scan mode*. The scanner can stop at an occupied channel and remain there only for a predetermined length of time, such as 5 seconds; this is called *free scanning*. The scanner can search for an empty channel among busy ones; this is called the *vacant scan mode*.

SCANNING

Scanning is a term that describes two quite different electronic processes. In a television or facsimile system, the picture is *scanned* both in the camera tube and in the receiver. The picture is generally scanned from left to right and top to bottom, in the same way you read the lines of this page. Scanning is controlled by an actuating circuit having a ramp waveform (*see* RAMP WAVE). In an oscilloscope, the electron beam in the cathode-ray tube scans from left to right along a single line. The scanning rate can be controlled, and is measured in terms of the time per graticule division (about 1 cm). Scanning in an oscilloscope is sometimes called *sweeping*. As in television and facsimile, a ramp wave is used to produce the scanning effect in an oscilloscope. *See also* CATHODE-RAY TUBE, FACSIMILE, OSCILLOSCOPE, RASTER, and TELEVISION.

In computers or microcomputer-controlled devices, *scanning* refers to the sampling of data in a continuous manner. If there are n data channels, having addresses 1, 2, 3, . . . , n, then scanning can proceed from channel 1 upward to channel n, or from channel n downward to channel 1. The process of scanning can occur just once across the range of channels, or it can be repeated over and over. A good example of this kind of scanning is illustrated by the operation of an automatic scanning receiver, which constantly checks communications channels for signals. The scanning rate is designated in channels per second.

SCANNING DENSITY

In *facsimile (FAX)* image communications, the number of lines per inch is called the *scanning density*. In general, the greater the scanning density (the more image lines per inch), the better the quality of the image. But high scanning density means that it takes more time to send a given image, and/or more spectrum space, compared with a lower scanning density.

Typical scanning density for a FAX signal is in the range of about 80-200 lines per inch (lines/in). A common standard is 96 lines/in; another is 166 lines/in.

For amateur purposes, FAX is often converted into slow-scan television (SSTV) video for display on a video monitor. The resolution is then limited by the quality of the monitor unit, unless a "zoom" feature is included in the converter software. *See* FACSIMILE, and SLOW-SCAN TELEVISION.

SCATTERING

When any type of disturbance, such as sound waves, electromagnetic waves, or particle radiation, passes through a material substance, *scattering* can occur. The effects of scattering can be unnoticeable, or they might be very evident. The atmosphere, for example, scatters blue light to some extent. This is why the sky looks blue to us. Water has the same effect; this is why a swimming pool has such a vivid blue appearance.

Radio waves at very-high and ultra-high frequencies are sometimes scattered by the molecules of the atmosphere. This is called *tropospheric-scatter propagation*. The ionosphere scatters radio waves at low, medium, high, and sometimes very-high frequencies. *See also* PROPAGATION CHARACTERISTICS, and TROPOSPHERIC-SCATTER PROPAGATION.

SCHEMATIC DIAGRAM

A *schematic diagram* is a technical illustration of the interconnection of components in a circuit. Most schematic diagrams include component values, and perhaps tolerances. Standard symbols are used. A schematic diagram does not indicate the physical arrangement of the components on the chassis or circuit board; it shows only how the components are interconnected.

Schematic diagrams are designed so that they are easy to read. Although a schematic diagram can look complicated to the beginner in electronics, this kind of representation is far simpler than a pictorial diagram. In the case of a complex device, such as a superheterodyne radio receiver or transceiver, a pictorial diagram would be utterly impractical, although a schematic diagram of such a device can usually be placed on one page.

SCHEMATIC SYMBOLS

In electronics, standard symbols are used to represent components in circuit diagrams. These symbols are used in amateur radio as well as commercial engineering applications. *See the* appendix SCHEMATIC SYMBOLS.

SCHERING BRIDGE

The *Schering bridge* is a circuit for the measurement of capacitance. The Schering bridge operates by comparing an unknown capacitance with a standard, known capacitance. Balance is indicated either by a meter or by a set of headphones. The Schering bridge is used for the determination of relatively large capacitances.

SCHMITT TRIGGER

A *Schmitt trigger* is a form of multivibrator circuit that is actuated by means of an input signal having a constant frequency. The Schmitt trigger produces rectangular waves, regardless of the input waveform. The schematic illustrates a typical Schmitt trigger circuit.

The circuit remains off until a specified rise threshold voltage is crossed; then it is actuated, and the output voltage abruptly rises. When the input voltage falls back below the fall triggering level, the output voltage drops to zero almost instantly. The rise and fall thresholds are generally different.

The Schmitt trigger is used in applications where square waves with a constant amplitude are needed. A Schmitt trigger

can also be used to convert sine waves to square waves. The Schmitt trigger is a form of bistable multivibrator, or flip-flop. *See also* FLIP-FLOP.

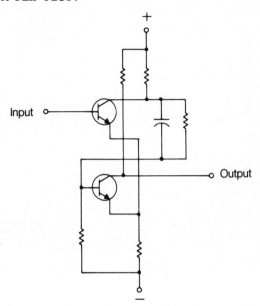

SCHMITT TRIGGER: A form of multivibrator circuit.

SCHOTTKY BARRIER DIODE

A *Schottky barrier diode*, also known as a *Schottky diode* or *hot-electron diode*, is a solid-state diode in which the junction is formed by metal-semiconductor contact. The Schottky barrier diode is usually fabricated from lightly doped N-type material and a metal, such as aluminum. The metal is evaporated or sputtered onto the semiconductor. The hot-carrier diode is a form of Schottky barrier diode (*see* HOT-CARRIER DIODE, and POINT-CONTACT JUNCTION).

Schottky barrier diodes are characterized by extremely rapid switching capability. The reverse-bias capacitance is very low. Schottky barrier diodes are useful at very-high and ultra-high frequencies because of their high-speed properties. They can be used as mixers, harmonic generators, detectors, and in other applications requiring diodes. The Schottky-diode design is sometimes used in the manufacture of transistors; this type of semiconductor device is known as a *Schottky transistor*.

SCHOTTKY LOGIC

Schottky logic is any form of logic that incorporates Schottky diodes or transistors. Schottky transistors are used especially in the manufacture of integrated-injection-logic devices. Schottky logic is characterized by high switching speed. *See also* INTEGRATED INJECTION LOGIC.

SCIENTIFIC NOTATION

Very large or small numbers are cumbersome to write in the conventional fashion. For example, if you write out the number 7 googol, you have to write a 7 followed by 100 zeroes. *Scientific notation* provides a shortcut for expressing extreme numerical values.

A number in scientific notation consists of a decimal number with a value of at least 1, but less than 10, followed by a power of 10. The decimal number tells us the first few digits of the actual value as you would write it in longhand, or conventional, form. The power of 10 tells us the factor by which the decimal number is to be multiplied. The exponent is always an integer; it can be positive or negative.

The decimal part of a scientific-notation expression can have any number of significant figures, depending on the accuracy you need. For example, you might write 2,345,678 as 2.3×10^6; this is an approximation accurate to two significant figures.

Many small calculators, and almost all computers, have the capability to work in scientific notation. This makes it possible for them to handle much larger numbers than would otherwise be possible.

When two numbers are multiplied or divided in scientific notation, the decimal numbers are first multiplied or divided by each other. Then, the powers of 10 are added (for multiplication) or subtracted (for division). Finally, the product or quotient should be reduced to standard form. That is, the decimal part of the expression should be at least 1, but less than 10. For example:

$$3 \times 10^2 \times 7 \times 10^3 = 21 \times 10^5 = 2.1 \times 10^6$$

When working with scientific notation, it is important to be aware of the number of significant figures in the expressions. Values are often expressed approximately, not precisely.

SCR
See SILICON-CONTROLLED RECTIFIER.

SCRAMBLER

A *scrambler* is a device that is sometimes used to encode signals, making it difficult for unauthorized persons to intercept them. Scramblers can be used for digital or analog signals of any kind, and at any speed.

A common type of scrambler, used with voice signals, inverts the frequency characteristics of the voice. The average human voice can be transmitted within a passband about 2.7 kHz wide, having frequency components between 300 Hz and 3 kHz. The scrambler turns the voice frequencies "upside down." A given impulse, having a frequency f, in hertz, is converted to a frequency g, in hertz, such that:

$$g = 3300 - f$$

When a human voice signal is applied to the input of such a circuit, the resulting output signal is utterly unreadable. (You are familiar with this effect if you have ever tried to listen to a lower-sideband signal with a receiver set for upper sideband, or vice-versa.) However, by passing the unreadable "monkey chatter" through a second, identical scrambler circuit, the voice is restored to its original form.

The foregoing is just one method of data scrambling. It is actually an oversimplification, and is described for illustrative purposes only. Many different scrambling methods are used by various commercial and government communications services. Most scramblers are more complex than the type described above, and can incorporate such devices as digital-to-analog and analog-to-digital converters, multiple-frequency operation, and other sophisticated schemes, making it essentially impossible for unauthorized persons to decode the signals.

SCRATCH-PAD MEMORY

A low-capacity, random-access memory, used in computers and some calculators for temporary data storage, is called a *scratch-pad memory*. Scratch-pad memory is used in complex arithmetic computations involving more than one operation.

Suppose, for example, that you want to calculate the value of the following expression:

$$x = (3.443 - 6.474)(4.001 + 8.211)$$

If you have a calculator with a scratch-pad memory, you would first compute:

$$y = 3.443 - 6.474 = -3.031$$

and store this number in the memory. You would then compute:

$$z = 4.001 + 8.211 = 12.212$$

Then, you would recall the number y from the memory and multiply z by y, obtaining x:

$$zy = (12.212)(-3.031) = -37.014572 = x$$

If you are working in scientific notation and are concerned about significant figures (*see* SCIENTIFIC NOTATION), then you would obtain:

$$x = -3.701 \times 10^1$$

If your calculator does not have a scratch-pad memory, you need a paper and pencil (a real scratch pad)!

In computers and programmable calculators, there can be several different scratch-pad memories which are used automatically, as necessary, in more complex computations. *See also* MEMORY, and RANDOM-ACCESS MEMORY.

SECOND

The *second* is the standard international unit of time. A second corresponds to $1/86,400$, or 1.1574×10^{-5}, mean solar day. Using modern atomic clocks, a more precise definition has been formulated: The second is represented by 1,420,405,752 oscillations of the hydrogen maser. Seconds are transmitted over time-and-frequency standard stations, such as CHU and WWV/WWVH (*see* CHU, and WWV/WWVH/WWVB).

In angular measure, the second is $1/3600$, or 2.778×10^{-4}, angular degree. This unit is also known as the *arc second* or the *second of arc*. It is abbreviated by a quotation mark or a double apostrophe ("). The arc second is used by astronomers as an indicator of the resolving power of an optical or radio telescope. The angular size of a celestial object can be expressed in seconds of arc.

SECONDARY
See TRANSFORMER SECONDARY.

SECONDARY EMISSION

When an electrode is bombarded by high-speed electrons, other electrons are knocked from the atoms of the electrode. Such electron emission is known as *secondary emission*.

The photomultiplier tube operates via secondary emission. A single electron can knock several electrons free from an electrode under certain conditions. This makes it possible to magnify the intensity of an electron beam by a large factor (*see* DYNODE, and PHOTOMULTIPLIER).

Secondary emission can be undesirable. In a tetrode vacuum tube, the electrons can be accelerated to such speed that secondary emission occurs from the plate of the tube. This reduces the efficiency of the tube, because the secondary electrons represent lost plate current. A negatively charged electrode, placed between the plate and the screen grid, reduces this effect. Such a grid is called a *suppressor* because it suppresses the secondary emission by driving the electrons back toward the plate. *See also* PENTODE TUBE, TETRODE TUBE, and TUBE.

SECONDARY FAILURE

When an electronic component malfunctions, it can cause other parts of a circuit or system to malfunction also. The failure of the other components is known as *secondary failure*.

Secondary failure often represents a much more serious problem than the original malfunction. Suppose that a regulated power supply, having no provision for over-voltage protection, malfunctions and produces 25 V at the output instead of the intended 13.8 V. The single component that causes this failure—the pass transistor—might cost $1.00 to replace. But the damage resulting to a radio transceiver, connected to the supply when the overvoltage condition occurs, can result in a far greater repair cost.

In well-designed electronic equipment, the possibility of secondary failure is addressed and minimized. In the above situation, for example, an overvoltage-protection circuit would prevent the secondary damage to the transceiver.

In the troubleshooting and repair of electronic apparatus, a technician must always be aware of the possibility of secondary failure. If a component has burned out because of secondary failure, merely replacing that component will not solve the problem, and it will probably recur. *See also* TROUBLE-SHOOTING.

SECONDARY RADIATION

Electromagnetic fields cause electrons to move back and forth in an electrical conductor. This is how a receiving antenna works; alternating current is generated by the passing waves. The alternating current is not, however, absorbed completely by the receiver. The current causes its own electromagnetic field to be transmitted from the receiving antenna; approximately half of the energy in a receiving antenna is reradiated into space. This radiation is called *secondary radiation*.

Secondary radiation takes place from all electrical conductors. Utility wires, fences, and even metal clothes lines all radiate electromagnetic fields as a result of excitation by external fields.

SECONDARY STANDARD

When you make a measurement of any parameter, you must have a standard. This standard must, in turn, be calibrated or set against a universal standard. Universal standards are rarely used directly. The standards you use are called *secondary* because they are not the original standards.

As an example, suppose you wish to find out what time it is. We look at our digital quartz watch, which tells us it is 5:00:00 P.M. (17 hours, 0 minutes, 0 seconds) local time. The watch, although perhaps a very accurate timepiece, probably does not exactly agree with the absolute time standard of WWV/WWVH. There is almost certainly an error of at least a few tenths of a second. The watch is a secondary time standard. The broadcasts of WWV/WWVH are a primary time standard.

As another example, consider the measurement of frequency. We might maintain a crystal oscillator under the most ideal possible conditions of temperature and humidity, and use it to calibrate our frequency-measuring apparatus. But the oscillator nevertheless represents a secondary standard. The primary standard is the transmission from WWV/WWVH (*see* REFERENCE FREQUENCY).

Primary or absolute standards are determined by agreement among the societies of the world. All parameters and standards are based on this agreement, which is known as the *Standard International (SI) system* of units. In the United States, absolute units are maintained by the National Bureau of Standards. *See also* NATIONAL BUREAU OF STANDARDS.

SECONDARY STATION IDENTIFIER

In packet radio, a single callsign is often used by more than one station. The callsign might be used to represent *digipeaters* as well as a ham's primary fixed station. When this is done, some means is needed to differentiate among the stations. A *secondary station identifier (SSID)* does this.

The SSID is a numeral from 0 to 15, allowing up to 16 stations to be used under a single callsign. If there is only one station for the callsign, the SSID is set to zero, or is dropped altogether. The home station generally has SSID zero. The SSID is sent as part of the *address field*.

If you see the callsign W1GV-4 while operating packet radio, you know that this probably refers to the fifth station working under the callsign W1GV. (There are most likely at least four others: W1GV-0, -1, -2, and -3.) *See also* DIGI-PEATER, and PACKET RADIO.

SECONDARY WINDING

See TRANSFORMER SECONDARY.

SECOND BREAKDOWN

Second breakdown is a phenomenon that occurs in some bipolar power transistors under certain adverse conditions. Second breakdown can occur as a result of flaws in manufacture. It can also take place if the device is subjected to excessive voltage or current.

Second breakdown causes a dramatic fall in the output impedance of a transistor. The emitter-base voltage, and any input signal, no longer affects the current through the device, and the output therefore drops to practically zero. The collector becomes effectively short-circuited to the emitter. A transistor can be permanently damaged by the occurrence of second breakdown. The probability of this happening can be reduced by ensuring that transistors are operated well within their ratings. *See also* POWER TRANSISTOR, and TRANSISTOR.

SELCAL

See SELECTIVE CALLING.

SELECTANCE

Selectance is an expression of the selectivity of a receiver or a resonant circuit (*see* SELECTIVITY). In a receiver, selectance is given as a sensitivity ratio between two channels, one desired and the other undesired. Suppose the receiver is tuned to a channel at f_1 MHz, and the sensitivity is found to be E_1 microvolts for 10-dB signal-to-noise ratio. If a signal is applied at some frequency f_2 MHz, different from f_1, the sensitivity will be less; the voltage E_2, needed to produce a 10-dB signal-to-noise ratio, will be larger than E_1. The selectance S can be expressed in different ways. The general form is:

$$S = (E_2 - E_1)/(f_2 - f_1)$$

if $f_2 > f_1$, and:

$$S = (E_2 - E_1)/(f_1 - f_2)$$

if $f_2 < f_1$. These expressions are given in microvolts per megahertz. Of course, other units of voltage and frequency might be used.

A specific expression of selectance is obtained by determining the frequency f_2 at which $E_2 = 2E_1$. Then:

$$S = (2E_1 - E_1)/(f_2 - f_1) = E_1/(f_2 - f_1)$$

if $f_2 > f_1$, and:

$$S = (2E_1 - E_1)/(f_1 - f_2) = E_1/(f_1 - f_2)$$

if $f_2 < f_1$.

In the determination of selectance, there will be two frequencies f_2; one larger than f_1 and the other smaller than f_1. The selectance for $f_2 > f_1$ can differ from that in the case $f_2 < f_1$.

Selectance can be expressed for resonant circuits. If the voltage is E_1 at the resonant frequency f_1, and E_2 at some nonresonant frequency f_2, then selectance is specified in the same ways as outlined above. There are, again, two selectance figures: one for $f_2 > f_1$ and one for $f_2 < f_1$. These two figures can differ for a fixed ratio E_2/E_1.

The selectance of a radio receiver determines the degree of adjacent-channel rejection. The greater the selectance, the less the possibility of adjacent-channel interference. The selectance of a resonant circuit depends on the Q factor. The higher the Q, the greater the selectance. *See also* ADJACENT-CHANNEL INTERFERENCE, and Q FACTOR.

SELECTIVE CALLING

Selective calling is an automatic method of sending messages to specific points, without causing interference to other stations.

The most common example of selective calling is the telephone. When you call a particular number, only the phone or phones at that number will ring (*see* TELEPHONE). Selective calling can be done in a similar fashion in wireless communications. In radio, selective calling is known as *SELCAL*. It is used in radioteletype systems and in radiotelephones.

Suppose you have 10 different messages intended for 10 different stations. Without selective calling, you would have to send all 10 messages to all ten stations, resulting in a great waste of time. With selective calling, however, this waste does not occur.

Selective calling for radioteletype involves the use of specific opening (start-up) and closing (shut-down) codes for each station. The opening and closing codes can be the, same, such as the call letters of the station. When the SELCAL unit "hears" the appropriate sequence of characters, the printer is actuated. When the closing sequence is received, the printer stops. If the opening and closing codes are identical, then the printer is alternately turned on and off each time the SELCAL unit "hears"

the code. It is generally best to use different opening and closing codes in a SELCAL system. Otherwise, the accidental transmission of one extra code sequence will throw off the system indefinitely, causing it to be up when it should be down, and down when it should be up.

Radioteletype terminal units, incorporating SELCAL as well as many other features, can be purchased, already assembled, for amateur or commercial use. SELCAL is also used in selective signalling on radiotelephone circuits, particularly between ground stations and aircraft. Most commercial air-carriers (airlines) have frequencies that are used for conducting business communications (having nothing to do with the safe operation of the aircraft). Rather than have all planes monitor all communications all the time (which could be distracting), selective signalling is used to alert only the particular aircraft involved. This type of signalling uses tone-coded squelch systems such that the proper sequence of tones must be received by the equipment on the aircraft in order for the receiver's audio circuits to become activated. The equipment can then sound a tone, flash a light, or both, to alert the crew that a message requires their attention. *See also* PACKET RADIO, PRIVATE LINE, and RADIOTELETYPE.

SELECTIVE FADING

When a signal is propagated via the ionosphere, fading is commonly experienced at the receiving station (*see* FADING, and PROPAGATION CHARACTERISTICS). This fading usually does not occur at the same time at all frequencies. For a given communications circuit, fading will not normally be observed at 5 MHz and 7 MHz at the same time, for example. This is because fading is a phase-related phenomenon; frequency and phase changes are interdependent.

Normally, the effects of frequency are not noticeable for channels separated by just a few hundred hertz. However, under certain conditions, fading can take place several seconds apart at two frequencies that are very close together. If this occurs, a signal can become distorted. For example, a single-sideband signal can get "bassy" and then "tinny" as the fade moves across the occupied channel.

The narrower the bandwidth of a signal the less likely it is to be affected by selective fading. Code (continuous-wave type-A1 emission) signals are essentially unaffected by selective fading because the bandwidth is very small. But radioteletype (F1 emission) can be affected; single-sideband transmissions are often distorted; amplitude-modulated signals can be severely affected; and wideband frequency-modulated signals can be mutilated beyond recognition. The effects of selective fading can be reduced by using the minimum bandwidth needed to carry on communications. Selective fading can also be alleviated, to some extent, by the use of diversity receiving systems. *See also* DIVERSITY RECEPTION.

SELECTIVITY

Selectivity is the ability of a radio receiver to distinguish between a signal at the desired frequency and signals at other frequencies. Selectivity is an extremely important criterion for any receiver.

In the front end of a receiver, selectivity is important because it reduces the chances of overloading or image reception (*see* FRONT END, and PRESELECTOR). In the intermediate-frequency stages of a superheterodyne receiver, selectivity is important because it reduces the probability of adjacent-channel interference (*see* ADJACENT-CHANNEL INTERFERENCE). Selectivity is obtained by the use of bandpass filters (*see* BANDPASS FILTER, and BANDPASS RESPONSE).

The selectivity of the intermediate-frequency chain in a superheterodyne receiver is often expressed mathematically. The bandwidths are compared for two voltage-attenuation values, usually 6 dB and 60 dB. For example, after you purchase a new receiver, you can read in the specifications:

Selectivity: $\pm$ 8 kHz or more at -6 dB

± 16 kHz or less at -60 dB

This gives you an idea of the shape of the bandpass response.

A single receiver can have several different bandpass filters, each with a different degree of selectivity. The ideal amount of selectivity depends on the mode of communication that is contemplated. The above figures are typical of a very-high-frequency receiver for frequency-modulation communications. For amplitude-modulated signals, the plus-or-minus bandwidth at the 6-dB points is usually 3 to 5 kHz. For single-sideband, it can be roughly 1 to 1.5 kHz, representing an overall 6-dB bandwidth of 2 to 3 kHz. For code reception, the overall bandwidth figure can be as small as 30 to 50 Hz at lower speeds, and perhaps 100 Hz at higher speeds. Examples of receiver selectivity curves for various emission modes are shown in the illustration. These examples are representative, and are not intended as absolute standards.

The ratio of the 60-dB selectivity to the 6-dB selectivity is called the *shape factor*. It is of interest, because it denotes the degree to which the response is rectangular. A rectangular response is the most desirable response in most receiving applications. The smaller the shape factor, the more rectangular the response (*see* RECTANGULAR RESPONSE, and SHAPE FACTOR). Filters (such as the ceramic, crystal-lattice, and mechanical types) can provide a good rectangular response in the intermediate-frequency chain of a receiver (*see* CERAMIC FILTER, CRYSTAL-LATTICE FILTER, and MECHANICAL FILTER).

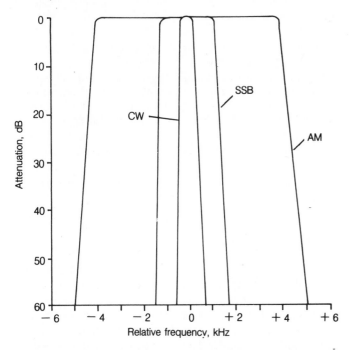

SELECTIVITY: Passbands for various reception modes.

Selectivity is usually provided in the radio-frequency and intermediate-frequency stages of a receiver, but additional selectivity is sometimes added in the audio-frequency amplifier chain. This is especially useful for reception of code and radioteletype signals. Many different commercially manufactured audio filters are available for enhancing the selectivity of a receiver.

SELENIUM

Selenium is an element with atomic number 34 and atomic weight 79. Selenium is a semiconductor material that exhibits variable resistance, depending on the amount of ambient light. Selenium is used in the manufacture of photocells, and also in certain rectifier diodes. *See also* PHOTOCELL, and SELENIUM RECTIFIER.

SELENIUM RECTIFIER

A *selenium rectifier* is a diode device made by joining selenium and some other metal, usually aluminum. Electrons flow easily from the metal to the selenium, but the resistance in the opposite direction is high (see illustration at A). Thus, the metal forms the cathode, and the selenium forms the anode of the rectifier. A selenium rectifier can be recognized by its heatsink, which makes it look similar to a stack of cards (B).

Selenium rectifiers are useful only at low frequencies, such as the 60-Hz line frequency. Although they are light in weight, they tend to be physically large because of the heatsink that is needed. The forward drop is about 1 V in normal operation. Selenium rectifiers are especially resistant to transient spikes on the power line. This is their main advantage over other types of rectifier diodes. *See also* RECTIFIER DIODE.

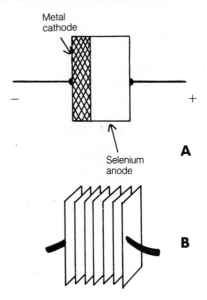

SELENIUM RECTIFIER: At A, construction; at B, appearance.

SELF BIAS

Self bias is a means of providing effective negative grid or gate bias in a vacuum tube or field-effect transistor. Although negative grid bias can be obtained from a special power supply, it can also be provided by elevating the cathode above direct-current ground potential. A noninductive resistor and a capacitor are used for this purpose.

There is a limit to the amount of effective negative gate bias that can be obtained by the method shown. The bias voltage is the product of the current (which is essentially constant) and the ohmic value of the resistor. The greater the bias voltage that is needed, the larger the resistor must be, and thus the more power it must dissipate. Most noninductive resistors have rather low dissipation ratings.

Self bias is often used in low-power tube amplifiers in class A, class AB, or class B. Self bias cannot normally be used to obtain sufficient bias for class-C operation. Self bias can be used in audio-frequency or radio-frequency circuits.

Biasing can be obtained in circuits using bipolar transistors, by means of the same technique described here. In solid-state circuits, the method is also called *automatic bias*. See also AUTOMATIC BIAS.

SELF OSCILLATION

An amplifier circuit will sometimes oscillate if positive feedback occurs. This type of oscillation degrades the performance of the amplifier. In audio-frequency circuits, *self oscillation* causes a fluttering or popping sound known as *motorboating*. In radio-frequency circuits, it can result in the emission of signals on undesired frequencies, and is called *parasitic oscillation*. See also FEEDBACK, and PARASITIC OSCILLATION.

SELF RESONANCE

Most electronic components exhibit a certain amount of capacitive reactance and inductive reactance. This can be, of course, because a component is deliberately designed to show reactance. But the physical dimensions of a component, especially the lead lengths, result in some reactance of both types, even if the component is not a capacitor or an inductor.

Self resonance usually occurs at ultra-high or microwave frequencies. In some larger capacitors and inductors, self resonance can exist at the high or very high frequencies. Self resonance can be fairly pronounced or almost indistinguishable. It is of primary concern in the design of circuits for operation above approximately 300 MHz.

Inductors, because of their inherent interwinding capacitance, are the most likely type of component to have significant self-resonant effects. Some inductors are deliberately manufactured to be self-resonant. The resonance can take the form of a bandpass, band-rejection, highpass, or lowpass response. *See also* BANDPASS RESPONSE, BAND-REJECTION RESPONSE, HIGHPASS RESPONSE, LOWPASS RESPONSE, and RESONANCE.

SELSYN

A *selsyn* is an indicating device used to show the direction in which an object is pointing. A selsyn is a synchronous motor system, with the transmitting unit at the movable device and the receiving unit where the operator can easily see it. A common application of the selsyn is as a direction indicator for a rotatable antenna.

One advantage of a selsyn over other directional indicators in an antenna system is that the selsyn will respond to unexpected changes in the antenna direction, such as might occur if the rotor slips. Assuming the antenna boom-to-mast joint does not slip, the selsyn will always show the antenna direction.

Another advantage of the selsyn is that the indicator needle normally rotates the same number of degrees as the antenna itself. Thus, a selsyn for azimuth bearings will turn 360 degrees; a selsyn for elevation bearings will turn 90 degrees. This gives the station operator a good mental picture of the antenna position. For azimuth indication, a great-circle map of the world, centered at the geographical location of the station, can be placed in the indicator (see illustration). The station operator can then observe, with reasonable accuracy, the part of the world toward which the antenna is beaming. *See also* AUTOSYN, and GREAT CIRCLE.

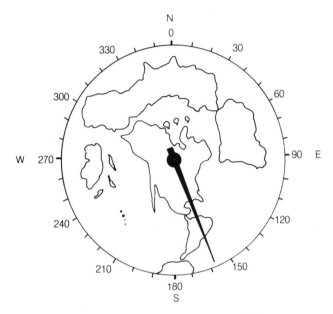

SELSYN: An antenna-rotator direction indicator.

SEMIAUTOMATIC KEY

A *semiautomatic key*, also popularly known as a *bug*, is a mechanical device used for sending morse code at high speeds. The semiautomatic key has been largely replaced, in recent years, by various forms of electronic keyers. Some radio operators, however, still use semiautomatic keys.

With a straight key, most operators find it difficult to send more than about 25 words per minute. This is because, at that speed, the dot frequency is 10 Hz, and it is not easy to pump a straight key up and down that fast!

The semiautomatic key uses a weight, lever, and spring combination to produce mechanical oscillation that forms the dots automatically. The speed is adjusted by moving the weight along the lever. The dashes must be made manually. The illustration shows a typical semiautomatic key.

To form a string of dots, the lever, or paddle, is pressed toward the right and held in position. To make dashes, the lever is intermittently pressed toward the left. This is how a "right-handed" bug works. For "left-handed" operation, these directions are reversed. Semiautomatic keys are usually made for "right-handed" operation, but a few are available for "left-handed" operation.

The dashes must be made manually, but since their frequency is just half that of the dots, this is not especially hard to do at speeds as high as 30 to 40 words per minute. Some operators can send as fast as 50 words per minute using a semiautomatic key. Code sent with a bug has a tempo that varies from one operator to another. The bug is a mechanical device, and different operators make the dashes in various, subtly different

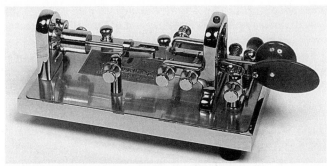

SEMIAUTOMATIC KEY: A semiautomatic key, the original "Presentation" model. THE VIBROPLEX COMPANY, INC.

ways. *See also* AMERICAN MORSE CODE, INTERNATIONAL MORSE CODE, and KEYER.

SEMI BREAK-IN OPERATION

Semi break-in operation is a form of switching scheme used in many communications installations. Semi break-in can be used for code, radioteletype, or voice operation. In voice systems, it is called *voice-operated transmission* or *VOX* (*see* VOX).

In semiautomatic code or radioteletype break-in, the transmitter is actuated the moment the operator presses the key. A relay or electronic switch performs the changeover from the receive mode to the transmit mode. When a pause occurs in the transmission, the transmitter stays on for a predetermined length of time. This delay is adjustable from about 0.1 second to 3 or 5 seconds. Most code operators do not like to have the receiver click on between characters or words of a code transmission, and they set the delay according to the speed at which they most often send the code. In radioteletype operation, the delay is not of much concern because the carrier is on all the time.

Semi break-in operation does not allow the code operator to hear between dots and dashes, as full break-in does. However, semi break-in is easier and less expensive to implement. Some operators prefer semi break-in over full break-in. *See also* BREAK-IN OPERATION.

SEMICONDUCTOR

A *semiconductor* is an element or compound that exhibits a moderate or large resistance. The resistance of a semiconductor is not as low as that of an electrical conductor, but it is much lower than that of an insulator or dielectric. The resistance of a semiconductor depends on the particular impurities, or dopants, added to it (*see* DOPING). Semiconductors conduct electricity via two forms of charge carrier: the electron and the hole (*see* ELECTRON, and HOLE).

Common semiconductor elements include germanium, selenium, and silicon. Other semiconductor elements are arsenic, antimony, boron, carbon, sulfur, and telurium. Many compounds are classified as semiconductors; the most often-used of these include gallium arsenide, indium antimonide, and various metal oxides.

Semiconductors have revolutionized electronics. A semiconductor device can perform the function of a vacuum tube having hundreds of times its volume. Integrated circuits, manufactured on wafers of semiconductor material, have increased the miniaturization even more. For more specific information on various semiconductor devices and applications, please refer to articles according to subject.

SENSE AMPLIFIER

Many types of memory circuits retain logic states at very low voltage levels. A logic circuit requires a difference of several volts—normally 4 or 5 V—between the low and high conditions. A sense amplifier is a circuit that magnifies the relatively small voltages of the memory so that they differ by the required amount for use by the logic circuit.

A sense amplifier is a straightforward, low-level, direct-current amplifier. It is advantageous for the sense amplifier to draw as little current as possible from the memory. A sense amplifier must therefore have a high input impedance. Sense amplifiers are usually low-noise circuits; this reduces the chances of interference from stray signal sources. *See also* DC AMPLIFIER.

SENSITIVITY

Sensitivity is a general term that applies to many different electronic devices. In general, sensitivity is an expression of the change in input that is necessary to cause a certain change in the output of a device. Usually, sensitivity is measured in terms of voltage.

In an amplifier or a radio receiver, sensitivity is a measure of the ability of the device to distinguish between a signal and noise at the output. Sensitivity can also be given for microphones and other forms of transducers, and for such electromechanical devices as relays and meters.

In an amplifier, sensitivity is expressed in terms of the input-signal voltage that is needed to produce a certain signal-to-noise ratio, in decibels, at the output. The sensitivity is indirectly related to the gain of the amplifier circuit or amplifier chain. But the noise figure is just as important (*see* GAIN, NOISE FIGURE, and SIGNAL-TO-NOISE RATIO). Sensitivity is limited absolutely by the level of the noise that is generated in the amplifier, especially the first amplifier in a chain.

In a microphone or other transducer, sensitivity is expressed as the acoustic pressure, light intensity, or other flux required to produce a certain amount of signal at the output. Alternatively, the sensitivity can be expressed as the minimum actuating flux that results in satisfactory circuit operation. The sensitivity of a relay is the smallest coil current that will reliably close the armature contacts.

In a radio receiver, sensitivity is expressed as the number of microvolts at the antenna terminals that is required to produce a certain signal-to-noise ratio or level of noise quieting at the speaker. In amplitude-modulation, continuous-wave, frequency-shift-keying, and single-sideband applications, the sensitivity is usually given as the number of microvolts needed to produce a 10-dB signal-to-noise ratio. In frequency-modulation equipment, the sensitivity is specified either as the number of microvolts needed to cause 20-dB noise quieting, or as the number of microvolts that results in a 12-dB signal-to-noise-and-distortion ratio (*see* NOISE QUIETING, SINAD).

The sensitivity of a piece of equipment is usually given in the table of specifications. Sensitivity is an important criterion for any amplifier or receiver, but it is not the only one. The audio distortion, dynamic range, image rejection, intermodulation distortion, and selectivity are also important figures of merit for receivers and amplifiers. *See also* DISTORTION, DYNAMIC RANGE, FRONT END, IMAGE REJECTION, INTERMODULATION, and SELECTIVITY.

SEPARATOR

A *separator* is a layer of material, in a battery or capacitor, intended to prevent the negative and positive poles of the cells, or the opposite plates of the capacitor, from coming into direct physical contact.

In a battery, the separator is soaked with an electrolyte substance, so current conduction occurs through it. But a potential difference, ranging from 1 to 2 V, is present across the separator material (*see* BATTERY, and STORAGE BATTERY). If the separator should fail and the poles come into contact, the action of one cell will be lost, and the battery voltage will drop by 1 to 2 V.

In a capacitor, the separator is usually called the *dielectric*. It is a nearly perfect insulator. If the separator fails, and the plates come into contact, the capacitor is short-circuited, and no longer will hold a charge (*see* CAPACITOR, and DIELECTRIC).

In an air-dielectric feed line, separators are used to keep the conductors apart. In this application, the separators are usually called *spacers*. *See also* SPACER.

SEQUENTIAL ACCESS MEMORY

A *sequential-access memory* is a form of memory in which data channels can be recalled only in a certain predetermined order. Any form of memory can have sequential access. The most common forms of sequential-access memory are the first-in/first-out (FIFO) memory and the pushdown stack (*see* FIRST-IN/FIRST-OUT, and PUSHDOWN STACK).

If the addresses in a memory are numbered, such as m_1, m_2, m_3, . . ., m_n, then the most common sequence for recall is upward from m_1 to m_n. A downward sequence, from m_n to m_1, is somewhat less common. Still less common are odd sequences in which the addresses do not ascend or descend in numerical order.

The data from a sequential-access memory can sometimes be recalled in two or more specific sequences. If the data can be recalled in any sequence possible, then the memory is called a *random-access memory*. *See also* RANDOM-ACCESS MEMORY.

SEQUENTIAL LOGIC

In digital electronics, the term *sequential logic* refers to any circuit or circuit element whose output depends both on present and past conditions.

A sequential-logic circuit has one or more inputs, and one output. The present state of the output depends on the present states of the inputs, and also on the immediately previous state of the output. An example of a sequential-logic circuit is a *flip-flop*, also called a *bistable multivibrator* or *latch*. *See also* DIGITAL, LOGIC, FLIP-FLOP, LOGIC CIRCUIT, and LOGIC GATE.

SERIAL DATA TRANSFER

Information can be transmitted simultaneously along two or more lines, or it can be sent bit by bit along a single line. The first method is known as *parallel data transfer* (*see* PARALLEL DATA TRANSFER). The latter method is called *serial data transfer*. The term *serial* is used because the data is sent in numerical order, according to some predetermined sequence.

In computer practice, serial data transfer requires that a word be split up into all of its constituent bits, then sent over the line, and reassembled at the other end of the line in the same order as originally.

Serial data transfer requires only one transmission line, while parallel data transfer requires several lines. However, serial data transfer is somewhat slower than parallel transfer.

SERIAL LINE INTERFACE PROTOCOL

In packet radio, *serial line interface protocol (SLIP)* transfers information between the network layer and the physical layer. These are also known as levels 1 and 3 of the *open systems interconnection reference model (OSI-RM)*. The SLIP itself is at level 2, or the link layer of OSI-RM.

The data elements in SLIP are sent in succession, one after the other. This is known as *serial data transfer*. *See also* OPEN SYSTEMS INTERCONNECTION REFERENCE MODEL, PACKET RADIO, PROTOCOL, and SERIAL DATA TRANSFER.

SERIES-PARALLEL CONNECTION

A *series-parallel connection* is a method of increasing the power-handling capacity of a component. Series-parallel connections are often used with resistors to obtain a higher wattage capability. Series-parallel connections can also be used with other components, such as inductors, capacitors, and diodes.

Series-parallel connections can be made in two basic configurations. One method is as follows: A certain number (n, for example) of components can be connected in series, and n of these combinations assembled; the resulting n combinations are then connected in parallel. This method is the most common, and is shown at A in the illustration for $n = 3$. Alternatively, n components can be connected in parallel, and n of these combinations made up; then these n combinations are connected in series. This scheme is shown at B for $n = 3$. In both of these arrangements, if the value of each resistor is R ohms, then the value of the composite is R ohms. If the power-dissipation rating of each resistor is P watts, then the composite will be capable of dissipating 9P watts. In general, the power-dissipation rating is increased by a factor of n^2.

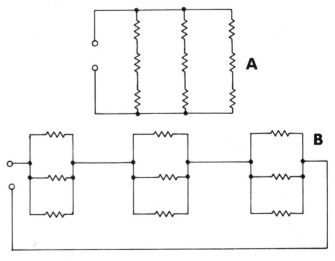

SERIES-PARALLEL CONNECTION: Series sets in parallel (A); parallel sets in series (B).

Series-parallel configurations can be assembled in other ways, where the number of parallel elements differs from the number of series elements. This causes a change in the composite value, depending on the particular arrangement used. *See also* PARALLEL CONNECTION, and SERIES CONNECTION.

SERIES REGULATOR

In a power supply, voltage regulation can be obtained by means of a component (usually a transistor) connected in series with the output. Such a component is called a *series regulator*; this method of regulation is known as *series regulation*.

The transistor is called a *pass transistor*. A Zener diode determines the output voltage; normally the input voltage is about 1.5 to 2 times the output voltage for best regulation. A standard 13.8-V power supply, for example, uses a 14-V Zener diode, and an input voltage of 20 to 25 V.

Integrated-circuit series regulators, containing the transistor, Zener diode, and associated components, are available for various regulated voltage levels.

With any regulation circuit, it is a good idea to have overvoltage protection. *See also* OVERVOLTAGE PROTECTION, and SHUNT REGULATOR.

SERIES RESONANCE

When an inductor and capacitor are connected in series, the combination will exhibit resonance at a certain frequency (*see* RESONANCE, and RESONANT FREQUENCY). This condition occurs when the inductive and capacitive reactances are equal and opposite, thereby cancelling each other and leaving a pure resistance.

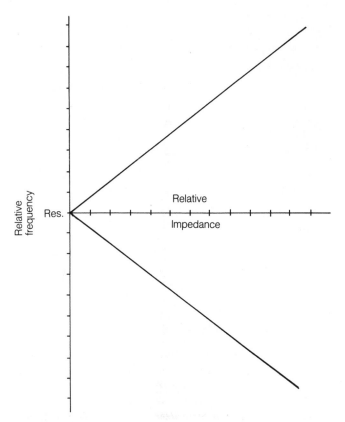

SERIES RESONANCE: Impedance is lowest at resonance.

The resistive impedance that appears across a series-resonant circuit is very low. In theory, assuming that the inductor and capacitor are lossless components, the impedance at resonance is zero (see illustration). But in practice, because there is always some loss in the components, the impedance is greater than zero, but can be less than 1 ohm. The exact value depends on the Q factor; the higher the Q, the lower the impedance at resonance. The Q factor is a function of the losses in the components (*see* IMPEDANCE, Q FACTOR).

Below the resonant frequency, a series-resonant circuit shows capacitive reactance. Above the resonant frequency, it shows inductive reactance. Resistance is also present under nonresonant conditions, because of the losses in the components; therefore, the impedance is complex at frequencies other than the resonant frequency.

Series-resonant circuits are sometimes used as traps (*see* TRAP). They are also used in some oscillators and amplifiers for tuning purposes.

A section of transmission line can be used as a series-resonant circuit. A quarter-wave section, open at the far end, behaves like a series-resonant circuit. A half-wave section, short-circuited at the far end, also exhibits this property. Such resonant circuits can be, and sometimes are, used in place of coil-capacitor (LC) combinations, especially at the very-high and ultra-high frequencies, where the physical lengths of such sections are reasonable.

SERVICE MONITOR/GENERATOR

A *service monitor/generator* is a self-contained unit that is used in the troubleshooting, alignment, and repair of electronic equipment, particularly radio-frequency devices, such as amplifiers, receivers, and transmitters.

This device can operate from the very low frequencies up to the microwave range. The frequency is set by means of the set of 10-position rotary switches, for the testing of transmitters. A meter indicates the amount by which the signal frequency departs from the channel center. An oscilloscope allows the technician to observe the modulation characteristics of the signal. A switch provides high sensitivity or low sensitivity, as desired. Another switch selects wideband or narrowband response.

For the testing of receivers, a built-in signal generator can be adjusted to any frequency by means of the rotary switches. The level of the output is set by a control knob with a dial calibrated in microvolts. The signal can be amplitude-modulated or frequency-modulated at various audio frequencies and levels. A service monitor is used in conjunction with other test equipment when necessary.

SERVOMECHANISM

A *servomechanism* is a form of feedback-control device. Servomechanisms are used in the remote control or automatic control of mechanical apparatus (such as motors, steering mechanisms, radio controls, and valves). A servomechanism is sometimes called a *servo*. A complete electromechanical system, incorporating one or more servomechanisms, is called a *servo system* (*see* SERVO SYSTEM).

An example of a servomechanism is an automatically tuned antenna. A rotary inductor is adjusted by means of an electric motor, which is in turn moved by the output signal from a sensing circuit. The sensing circuit and motor form the servomechanism; the complete antenna-tuning unit is a servo system.

When the transmitter is actuated and radio-frequency energy is applied to the antenna at a frequency f, the antenna is initially tuned to some higher or lower frequency g. Therefore, there is reactance as well as resistance present at the feed point. If $f > g$, the reactance is inductive, and the current lags the voltage. If $f < g$, the reactance is capacitive, and the current leads the voltage. The sensing circuit detects either condition, and causes the motor to turn the coil in the required direction to bring the current and the voltage into phase. Then the rotary inductor remains fixed in that position, until a change is made in the transmitter frequency. The response time depends on the difference between the initial frequencies f and g.

Servomechanisms can be operated in conjunction with computers. For example, a computer program can tell a servomechanism to move according to a certain function. Several servomechanisms, when interconnected and controlled by a computer, can do quite complicated things, such as cook a meal automatically. This type of servo system is called a *robot*. The technology of robot design and construction is known as *robotics*.

SERVO SYSTEM

A complete electromechanical set of one or more servomechanisms, including all of the associated circuits and hardware, and intended for a specific task or purpose, is called a *servo system*. Servo systems can be used for a variety of purposes, such as temperature regulation, frequency control, voltage regulation, or transmitter tuning.

In recent years, servo systems have become more and more complex and sophisticated. Small computers can be used to control a servo system consisting of several different servomechanisms. For example, an unmanned plane can be programmed to take off, fly a specific mission, return, and land (such an airplane is called a *drone*). Servo systems can be programmed to do assembly-line work and other mundane things. Such servo systems are called *robots*. *See also* SERVOMECHANISM.

SESSION LAYER

See OPEN SYSTEMS INTERCONNECTION REFERENCE MODEL.

SETTLING TIME

When a digital meter is used to measure current, frequency, power, resistance, or voltage, a certain amount of time must elapse before the meter comes to rest and shows the value. The length of time between the connection of the meter leads and the stabilization of the reading is called the *settling time*.

Settling time varies, depending on the type of instrument and on the number of significant figures that it shows. In general, the more accurate the meter, the longer the settling time. Typical digital test meters have a settling time of about 1 to 2 seconds.

In a servo system, equilibrium is reached after a certain length of time. When a parameter changes, the servomechanism does not compensate for it instantaneously (although it can do so very fast). The length of time required for the system to reestablish equilibrium is called *settling time*. *See also* DIGITAL METERING, SERVOMECHANISM, and SERVO SYSTEM.

SEVEN-SEGMENT DISPLAY

When seven independently controlled, bar-shaped light-emitting diodes, liquid crystals, or other two-state visual devices are arranged in the pattern shown in the illustration, the combination is known as a *seven-segment display*. The seven-segment display is universally used in digital devices of all kinds because most English letters, and all numerals, can be represented by some combination of states.

Although the seven-segment display is the most commonly used configuration in digital clocks, calculators, meters, and other indicators, there are other methods of obtaining an alphanumeric display. The dot-matrix method, for example, is often used, especially in languages having characters that cannot be represented on a conventional seven-segment display.

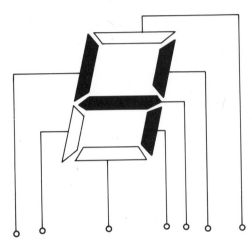

SEVEN-SEGMENT DISPLAY: Shows all numerals and some letters of the alphabet.

SEVEN-UNIT TELEPRINTER CODE

The *seven-unit code* is identical to the Baudot teleprinter code (*see* BAUDOT), except that each character is preceded by a start signal and followed by a stop signal. Thus, instead of five bits or units per character, there are seven. The start unit is a space, and the stop unit is a mark (*see* MARK/SPACE).

The advantage of the seven-unit teleprinter code is that it provides synchronization between the transmitter and receiver. If some data bits are missed, a single character can be lost, but the error will not continue for several characters.

SFERICS

Electromagnetic noise is generated in the atmosphere of our planet, mostly by lightning discharges in thundershowers. This noise is called *sferics*. In a radio receiver, sferics produce a faint background hiss or roar, punctuated by bursts of sound you call static.

A potential difference of about 300,000 volts exists between the surface of the earth and the ionosphere. The earth and ionosphere therefore act like a huge capacitor, with the troposphere and stratosphere serving as the dielectric (*see* ATMOSPHERE). Sometimes this dielectric develops "holes," or pockets of imperfection, where discharge takes place. Such "holes" are the thundershowers. There are normally about 700 to 800 such areas at any given time, concentrated mostly in the tropics.

Sand storms, dust storms, and volcanic eruptions also produce some lightning, contributing to the overall sferics level.

An individual lightning stroke produces a burst of electromagnetic energy from the very low frequencies through the microwave spectrum—and even in the infrared, visible, and ultraviolet ranges (*see* LIGHTNING).

The current that flows across the atmospheric capacitor averages 1500 amperes. Since the potential difference across this capacitor is 300,000 volts, the sferics generator on our planet is a 450-megawatt, wideband radio transmitter! At very low and low frequencies, the sferics propagate around the world, as the electromagnetic field is trapped between the "plates" of the capacitor. As the frequency increases, sferics become a more local phenomenon until, at very high frequencies and above, sferics travel only a few miles. This is why the general level of background noise decreases as the frequency gets higher, making low-noise circuit design of such importance in the very-high-frequency, ultra-high-frequency, and microwave spectra.

Sferics are not confined to the earth. Some noise is generated by the immensely powerful storms in the atmosphere of the giant planet Jupiter. Astronomers have heard this noise with radio telescopes. Sferics probably also occur on Saturn, and perhaps on Uranus, Neptune, Venus, and even Mars. In the cases of Venus and Mars, dust storms and volcanic eruptions would be the cause of sferics.

Not all of the background hiss that you hear in our radio receivers is the result of sferics; the sun generates a great deal of electromagnetic noise, as does the center of our galaxy, a powerful object in the constellation Cygnus, and many other celestial sources. Noise of extraterrestrial origin is not, technically, categorized as sferics.

You can hear the sferics from a distant thundershower on the standard AM broadcast band. If you have a general-coverage communications receiver, you can listen at progressively higher frequencies as the storm or storm system approaches. When you hear the static bursts at 30 MHz, the storm is probably less than 100 miles away.

All storm systems produce sferics, although the amount of noise varies. Using directional antenna systems and special receivers, some meteorologists have been able to locate and track large storm systems. A receiver designed especially for listening to atmospheric noise is called a *sferics receiver*. In normal radio communications, however, sferics are just a nuisance because they cause interference. *See also* INTERFERENCE, INTERFERENCE FILTER, INTERFERENCE REDUCTION, NOISE, and NOISE LIMITER.

SHADOW EFFECT

Electromagnetic fields sometimes pass around obstructions with very little or no attenuation; under other circumstances the obstruction causes a large amount of attenuation. The extent to which this shadow effect occurs depends on the size, or diameter, of the obstruction relative to the wavelength of the electromagnetic field.

Generally, the shadow effect is produced by objects having a diameter of at least several wavelengths. When the diameter of the obstruction is small compared with the wavelength of the electromagnetic field, diffraction occurs and the field is essentially unaffected (*see* DIFFRACTION). As the obstruction becomes larger, or the wavelength becomes shorter, less and less diffraction is observed. The shape of the object has some effect,

as well: a rounded or smooth object casts a deeper shadow, for a given ratio of wavelength to obstruction diameter, than does an object having sharp corners.

Shadow effect is most likely to affect radio communications at very high frequencies and above, where buildings, hills, and other obstructions measure many wavelengths across.

SHAPE FACTOR

The attenuation-versus-frequency response of a bandpass filter, especially in a communications receiver, is often evaluated according to an expression known as the *shape factor*. The shape factor determines the extent to which the response is rectangular.

The shape factor is generally given as the ratio of the bandwidth of a filter at the —60-dB points to the bandwidth at the —6-dB points. The signal levels are measured in terms of voltage attenuation in the intermediate-frequency chain.

Various shape factors are shown graphically in the illustration. In general, the smaller the shape factor, the more rectangular the response. (A shape factor of 1:1 indicates a perfectly rectangular response—a theoretical ideal, but not obtainable in practice.)

A rectangular response is desirable in a bandpass filter because a signal generally has a well-defined band over which the information is carried. It is essential that all of the signal components get through the filter, but it is undesirable for frequencies outside the signal band to be passed. *See also* BANDPASS RESPONSE, BANDWIDTH, RECTANGULAR RESPONSE, and SELECTIVITY.

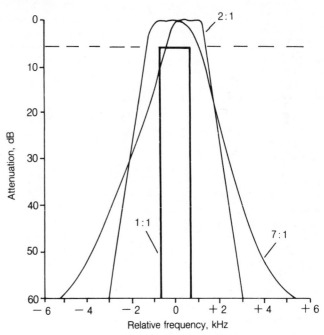

SHAPE FACTOR: Various bandpass curves.

SHAPING

In continuous-wave (code) transmission, the rise and decay times must be fast enough to convey the necessary information at the required speed. However, if the rise or decay times are too rapid, key clicks will occur (*see* KEY CLICK). The envelope of a code signal is called *the shaping of the signal*. The process of tailoring, or adjusting, the envelope is also called *shaping*.

Shaping is usually accomplished by means of a simple resistor-capacitor combination in the keying circuit of a code transmitter. The resistor is connected in series with the keying line, and the capacitor in parallel, as shown at A in the illustration. This type of circuit is suitable for use in blocked-gate or blocked-base circuits. In the case of source keyed or emitter-keyed circuits, a large-value choke can be substituted for the resistor, as at B.

In general, the larger the value of the resistor or choke, the slower the rise time for a given capacitance. The larger the value of the capacitor, the slower both the rise and decay times will be. The values are selected so that the shaping is optimum for the range of code speeds to be transmitted. Usually, the rise time is on the order of a few milliseconds, and the decay time is approximately twice the rise time.

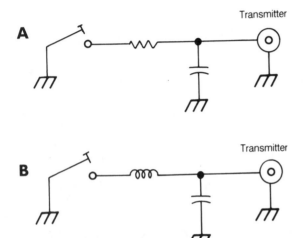

SHAPING: At A, the RC method; at B, the LC method.

SHELF LIFE

A battery will, of course, last longer if it is not used than if it is used in a circuit or device. Even if a battery is kept in storage, it will not retain its charge or charge-holding capacity forever. Eventually, a cell or battery will become unusable even if it is stored under the best possible conditions. The length of time during which a cell or battery is usable is called the *shelf life*. Ordinary dry cells, and most other types of cells and batteries, have shelf lives of 2 or 3 years.

The shelf life can be specified for any type of component. Some devices (such as resistors, diodes, and transistors) will last for decades (or even centuries!) in storage under ideal conditions. Other components (such as electrolytic capacitors, oil-filled capacitors, potentiometers, switches, and some tubes) have more meaningful shelf lives.

SHIELD CONTINUITY

See SHIELDING EFFECTIVENESS.

SHIELDING

The intentional blocking of an electric, electromagnetic, or magnetic field is known as *shielding*. The deliberate blocking of high-energy radiation, such as X rays, gamma rays, or subatomic particles, is also called *shielding*. Different types of fields or rays are shielded in different ways.

Electric shielding is also called *electrostatic shielding* or *Faraday shielding*. A grounded metal screen or plate will block the lines of flux of an electric field. Electromagnetic shielding requires a completely enclosed cage or box made of a conducting material (*see* ELECTROMAGNETIC SHIELDING, and ELECTROSTATIC SHIELDING). Magnetic shielding necessitates the use of a ferromagnetic metal enclosure, such as iron or steel, which will block the lines of flux of a nonfluctuating magnetic field. *See also* ELECTRIC FIELD, ELECTROMAGNETIC FIELD, and MAGNETIC FIELD.

SHIELDING EFFECTIVENESS

No shielding barrier stops 100 percent of the field or radiation that it is designed to block, although in many cases the figure is very close to 100 percent. The percentage of energy that is blocked is called the *shielding effectiveness*. In electronics, this term is used mostly with electromagnetic shielding, and is of particular importance in coaxial cables.

The effectiveness of an electromagnetic shield is a function of the continuity, or the physical completeness, of the barrier. The shield continuity is defined as the percentage of the total surface area of the shield enclosure that is actually covered by metal. If the total surface area is A, and the shield surface area is B, then the continuity C is given in percent by the equation

$$C = 100B/A$$

A solid metal enclosure, with absolutely no holes or gaps and with an excellent ground, provides 100-percent shielding continuity, and practically 100-percent shielding effectiveness if a good ground is used. A screen or wire cage offers less shielding continuity, but the effectiveness will be near 100 percent as long as the holes or gaps are very small, compared to the wavelength of the electromagnetic field.

In a coaxial cable, the shield continuity is of increasing importance as the frequency becomes higher. Some types of coaxial cable, designed for use at audio frequencies only, have very poor shield continuity. Some radio-frequency cables have marginal shield continuity; such cables are satisfactory for most high-frequency applications, but are not suitable for use at very high frequencies and above. The best shield continuity is afforded by hard line, which has a solid metal outer conductor. Double-shielded coaxial cable is also good for use at very-high and ultra-high frequencies.

Even if the shield continuity of a coaxial cable is 100 percent, the shielding effectiveness can be poor if an installation is faulty. Radio-frequency currents can be induced on the outer conductor of any coaxial cable. These so-called "antenna currents" can be minimized, and the shielding effectiveness thereby optimized, by taking these precautions: (1) ensure that the ground system is adequate; (2) use a balun at the feed point if the antenna is balanced; (3) avoid using resonant lengths of coaxial cable in the feed line; (4) run the cable away from the antenna at right angles to the radiating element for a distance of at least ¼ wavelength. *See also* ELECTROMAGNETIC SHIELDING.

SHIFT

Shift is the difference between the mark and space signals in a radioteletype transmission. When F1-type emission (frequency-shift keying) is used, the mark and space frequencies can both be on the order of hundreds of megahertz, but the difference is usually less than 900 Hz. The most common values are 170 Hz, 425 Hz, and 850 Hz (*see* FREQUENCY-SHIFT KEYING). When F2-type emission (audio frequency-shift keying) is used, the audio frequencies are normally 2125 Hz and 2295 Hz, with a shift of 170 Hz. Occasionally, tones of 1275 Hz and 1445 Hz are used. Sometimes, audio tone pairs of 2125 Hz and 2975 Hz, or 1275 and 2125 Hz, are used, giving 850-Hz shift. In two-way modem systems, standard frequencies are 1270 Hz (mark) and 1070 Hz (space) in one direction, and 2225 Hz (mark) and 2025 Hz (space) in the other. Other shift values are less common.

In a memory or other circuit containing data in specific order, a shift is the displacement of each data bit one place to the right or one place to the left. In the case of a shift to the right, the far right-hand bit moves to the extreme left; in the case of a shift to the left, the far left-hand bit moves to the extreme right. This movement of data is also known as a *cyclic shift*. See CYCLIC SHIFT, and SHIFT REGISTER.

SHIFT REGISTER

A *shift register* is a form of digital memory circuit that is extensively used in calculators, computers, and data-processing systems. Information is fed into the shift register at one "end," and emerges from the other "end." These "ends" are usually called the *left* and the *right*.

A clock provides timing pulses for the shift register. Each time a clock pulse is received, every data bit moves one place to the right or left. As data moves to the right, the right-hand bits leave the shift register one by one (see illustration). The opposite happens with the left-hand movement of data. Some shift registers can operate in only one direction, and are therefore called *unidirectional*. Other shift registers can operate in either direction, and are called *bidirectional*.

There are two basic categories of unidirectional or bidirectional shift register. These are known as the *dynamic type* and the *static type*. The dynamic shift register uses temporary storage methods. Data is lost if the clock operates at an insufficient rate of speed. In the static shift register, flip-flops are used to store the information (*see* FLIP-FLOP). The clock rate can be as slow as desired because the data remains as long as power is supplied to the flip-flops. Shift registers are available in integrated-circuit form.

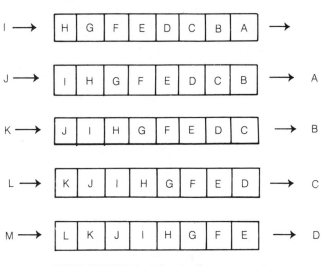

SHIFT REGISTER: A first-in/first-out circuit.

SHOCK HAZARD

A *shock hazard* exists whenever there is a sufficient potential difference between two exposed objects to cause harmful or lethal currents in a person who touches both objects at the same time. Even the 117-V potential in household appliances can, under some circumstances, be lethal. Sometimes a hazard can exist at far lower voltages.

A current of just a few milliamperes causes pain. If a current of approximately 100 to 300 mA flows through the heart, there is a high probability that fibrillation will occur (*see* HEART FIBRILLATION). Fibrillation is possible even with currents as low as 30 mA in some cases. This can cause death. Large amounts of current can result in severe burns.

Generally, a shock hazard is considered to exist if there is sufficient voltage to cause a current of 5 mA to flow through a resistance of 500 ohms. The resistance of the human body can be much less than 500 ohms, especially if water is present. Sweaty hands, for example, greatly lower the effective resistance of the body. *See also* ELECTRIC SHOCK.

SHOCKLEY DIODE

A *Shockley diode* is a form of switching diode. It has four semiconductor layers, instead of the usual two. *See* FOUR-LAYER SEMICONDUCTOR.

SHORE EFFECT

Radio waves at the very low, low, and medium frequencies travel more rapidly over salt water than over land, because the surface-wave conductivity of sea water is better than that of land (*see* SURFACE WAVE). This difference results in a change of direction in the path of a radio wave crossing a shoreline.

The extent of the bending depends on the angle with which the radio waves cross the shoreline. No refraction occurs when the waves cross at a right angle. As the crossing angle becomes smaller, the refraction increases. Radio waves traveling parallel to the shore tend to curve inland.

At the very-high and ultra-high frequencies, another type of shore effect occurs. During the daylight hours, the air over the land is heated and becomes less dense than the air over the water. As a result, radio waves are refracted upon crossing the shoreline. The effect is most pronounced at large angles of incidence. At night, the temperature gradient is reversed, and the sense of the refraction is therefore inverted. The shore effect is increased as the difference in temperature gets larger. *See also* TROPOSPHERIC PROPAGATION.

SHORT CIRCUIT

A *short circuit* occurs when current is partially or wholly diverted from its normal or proper path. It is called a *short circuit* because the path of the current is often shorter than it otherwise would be. A short circuit is sometimes called a *short out*, or simply a *short*.

In a 117-V household utility system, or in a power supply, a short circuit causes the tripping of a circuit breaker or the blowing of a fuse. This is known as short-circuit protection (*see* CIRCUIT BREAKER, FUSE, and SHORT-CIRCUIT PROTECTION). In an audio-frequency or radio-frequency system, a short circuit causes a reduction in the level of the signal at the output. A short circuit is sometimes deliberately inserted into a circuit for a specific purpose.

When a single component is short circuited either directly or by another component, the connection is called a *shunt. See also* SHUNT.

SHORT-CIRCUIT PROTECTION

When a short circuit occurs in the utility lines or in the output of a power supply, overheating will occur unless there is some means of protection for the components. This overheating can damage wiring, transformers, and series-connected resistors and chokes. If the current flow is large enough, fire can result.

The most common methods of short-circuit protection are the circuit breaker and the fuse. These devices open the circuit when the current exceeds a certain level (*see* CIRCUIT BREAKER, and FUSE).

Current limiting, also called *foldback*, is provided in some power supplies to protect the components against a short circuit. The current-limiting circuit inserts an effective resistance in series with the output of the supply (*see* CURRENT LIMITING).

SHORT-PATH PROPAGATION

At the very low, low, medium, and high frequencies, the ionosphere affects radio waves and often results in worldwide communications. This is sometimes also observed in the lower part of the very-high-frequency band. In long-distance ionospheric communication, the waves travel over, or near, great-circle paths (*see* GREAT CIRCLE).

There are two possible great-circle paths between any two points on the globe. Both paths lie along the same great circle, but one path is generally much shorter than the other. The shorter path is known, appropriately, as the *short path*. It is this path that usually results in the best signal propagation between the two points.

The term *short-path propagation* generally applies only to signals that are returned by the ionosphere. The direct wave, ground wave, or signals that travel via tropospheric modes, are short-path in an obvious sense, but they are not short-path ionospheric propagation. For example, if you communicate with a friend via CB, and your two stations are only 1 mile apart, the propagation occurs over the short path, obviously, but it is not usually referred to as short-path propagation.

Sometimes radio waves will travel the long way around the world to get from one station to another. This is called *long-path propagation*. It is most likely to be observed between stations that are separated by a great distance. Sometimes a signal is propagated via both the short path and the long path at the same time. This takes place most often at high frequencies, especially between 5 and 25 MHz, and is the result of interaction of the signals with the F layer of the ionosphere. *See also* F LAYER, IONOSPHERE, LONG-PATH PROPAGATION, and PROPAGATION CHARACTERISTICS.

SHORT WAVES

The high-frequency band, from 3 to 30 MHz (100 m to 10 m), is sometimes called the *shortwave band*. The wavelengths are not really very short; compared to microwaves, for example, the short waves are extremely long.

The term *shortwave* originated in the early days of radio, when practically all communication and broadcasting was done at frequencies below 1.5 MHz (200 m). It was thought that

the higher frequencies were useless. The range "200 meters and down"—that is, 1.5 MHz and above—was given to the radio amateurs. In fact, radio amateurs were restricted to the frequencies above 1.5 MHz. Within a few years, the radio amateurs discovered that these frequencies were anything but useless. The shortwave-radio era was thus born, and the short waves are still extensively used today.

A high-frequency, general-coverage, communications receiver is sometimes still called a *shortwave receiver*. Most shortwave receivers cover the range from 1.5 MHz through 30 MHz. Some also operate in the standard broadcast band at 535 kHz to 1.605 MHz. A few shortwave receivers can operate below 535 kHz, and into the so-called "longwave band."

SHORTWAVE LISTENING

Anyone can build or obtain a shortwave or general-coverage receiver, install a modest random-wire antenna, and listen to signals from all around the world. This interesting hobby is called *shortwave listening*. Millions of people in the United States enjoy this hobby.

There are many commercially manufactured shortwave receivers on the market today, some for very modest prices. A wire antenna costs practically nothing. Most electronics stores carry one or more models of shortwave receiver, along with antenna equipment, for a complete installation. Most electronics stores and book stores carry periodicals and books for the beginner, as well as for the seasoned shortwave listener.

A shortwave listener need not obtain a license to receive signals. At the time of writing, no formal licensing procedure exists for class-D Citizen's-Band (CB) radio transmissions, which are made on the shortwave bands in a narrow segment at about 27 MHz. In general, however, a license is required if shortwave transmission is contemplated. Shortwave listeners often get interested enough in communications to obtain amateur radio licenses.

SHOT EFFECT

In a transistor or tube, or in any current-carrying medium, the individual charge carriers cause noise impulses as they move from atom to atom. The individual impulses are extremely faint, but the large number of moving carriers results in a hiss that can be heard in any amplifier or radio receiver. This effect is called *shot effect*. The noise is known as *shot-effect noise* or *shot noise*.

Shot-effect noise limits the ultimate sensitivity that can be obtained in a radio receiver. This is because a certain amount of noise is always produced in the front end, or first amplifying stage (*see* FRONT END), and this noise is amplified by succeeding stages along with the desired signals. The amount of shot-effect noise that a device produces is roughly proportional to the current that it carries. In recent years, low-current solid-state devices such as the gallium-arsenide field-effect transistor (GaAsFET) have been developed for optimizing the sensitivity of a receiver front end.

Shot-effect noise is not the only form of noise that is generated in an electronic circuit. All atoms and subatomic particles are constantly in motion. The rate of motion is proportional to the temperature of the conducting media. The higher the temperature, the more the particles move, and the more electrical noise they make. This noise is called *thermal noise*. It, too, limits the sensitivity that can be obtained in the front end of a radio receiver. *See also* NOISE FIGURE, SENSITIVITY, SIGNAL-TO-NOISE RATIO, and THERMAL NOISE.

SHOT NOISE

Shot noise is a term that is used to describe three different forms of noise. Whenever current flows, the movement of the electrons or holes results in electrical noise. This is called *shot effect*, and the noise is known as *shot-effect noise* or *shot noise* (*see* SHOT EFFECT). All semiconductor devices produce this kind of noise; the amount of noise depends on many factors, primarily the current and the area of the P-N junction(s), through which the current flows.

In a vacuum tube, noise is produced when the electrons bombard the plate. A smaller amount of noise also results from electrons striking the screen grid in a pentode or tetrode tube. This noise is called *shot noise*.

Whenever an electric spark occurs, an electromagnetic field is produced. Many different types of appliances, as well as internal-combustion engines employing spark plugs, produce sparks. The noise is usually called *impulse noise*, but it can sometimes be referred to as *shot noise*. *See also* IMPULSE NOISE.

SHUNT

The term *shunt* refers to a parallel connection of one component across another component or group of components. When one component is placed across another for a specific purpose, the inserted component is sometimes called a *shunt*.

A shunt can be any type of component. Examples of shunt components include bypass capacitors, bleeder resistors, diode limiters, and various types of voltage regulators. Shunting resistors or coils are often used in ammeters to increase the indicated current range.

SHUNT FEED

Shunt feed is a method of supplying radio-frequency energy to a grounded mast or tower. The principle of shunt feed is the same as that of the gamma match (*see* GAMMA MATCH), except that the system is unbalanced rather than balanced.

The mast or tower should be well-grounded at the base; a set of radial wires will enhance performance (*see* RADIAL). The shunt element can be a wire, held away from the mast or tower by means of standoff insulators. Aluminum or copper tubing can also be used. The spacing distance should be approximately 1 to 3 feet. The tapped coil at the base can be a solenoidal air-core inductor or a toroidal inductor, but it must be capable of carrying the required current without suffering damage.

For purposes of shunt feeding, a mast or tower should measure at least ¼ electrical wavelength in height. Because a sizable antenna or set of antennas is often mounted at the top of a mast or tower, the physical height can be considerably less than ¼ wavelength. Guy wires result in greatly increased electrical height.

Adjustment of the shunt-feed system is done by trial and error. The coil is first disconnected entirely from the system, and the feed point is directly connected to the transmission line. The shunt point is moved up and down the mast or tower by means of a shorting wire, until the standing-wave ratio is minimum, indicating zero reactance. The coil is then inserted, and the tap is adjusted for a perfect match. As an alternative to the

use of the matching inductor, the spacing between the tower and the shunt wire or tubing can be varied until the standing-wave ratio is 1:1.

Shunt feed is especially useful in the lower part of the high-frequency band, and in the medium-frequency band. It is used extensively by radio amateurs in the 160-meter and 80-meter (1.8-MHz and 3.5-MHz) bands. Many amateur radio operators have obtained amazing results in DX (distance) work using shunt-fed towers on these bands.

SHUNT FEEDBACK

Shunt feedback is a circuit design technique for minimizing interstage impedance fluctuations. It is especially useful in radio-frequency transmitters. Two bipolar transistors are directly coupled, as shown in the schematic. The input to the first transistor is supplied through resistor R_{IN}. The input impedance, Z_{IN}, is constant, with $Z_{IN} = R_{IN}$ under a wide range of output load impedances.

The voltage gain of the shunt-feedback circuit is equal to the ratio of the values of the feedback resistor, R_F, and the input resistor, R_{IN}: $Gain = R_F/R_{IN}$. The phase of the output signal is opposite to the phase of the input signal. *See also* BUFFER STAGE, and EMITTER DEGENERATION.

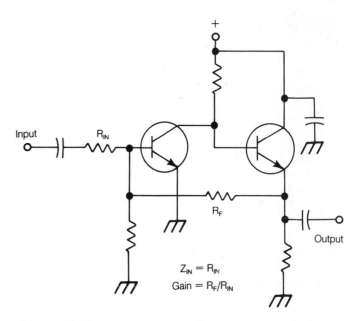

SHUNT FEEDBACK: The feedback resistor, R_F, and the input resistor, R_{IN}, set the gain.

SHUNT REGULATOR

A *shunt regulator* is a voltage-regulation device, connected in parallel with the output of a power supply. Various types of transistors and tubes can be used as shunt regulators. In low-current power supplies, zener diodes are sometimes used for shunt regulation. *See also* OVERVOLTAGE PROTECTION, SERIES REGULATOR, VOLTAGE REGULATION, and ZENER DIODE.

SHUNT RESISTOR

The current-indicating range of an ammeter can be increased by placing small-value resistors across the coil of the ammeter.

Such resistors can be noninductive, or they can consist of coils of resistance wire (such as nichrome). When shunt resistors are used, care must be exercised to ensure that they can handle the current without overheating. *See also* AMMETER.

SIDEBAND

When a carrier is modulated in any manner, for conveying intelligence of any sort and at any speed, *sidebands* are produced. Sideband signals occur immediately above and below the carrier frequency.

Sideband signals are the result of mixing between the carrier and the modulating signal. The greater the data speed, the higher the frequency components of the modulation signal, and the farther from the carrier the sidebands will appear (see illustration).

The data speed is not the only factor that affects the overall band of frequencies occupied by a carrier and its sidebands. Other influences on bandwidth include the type of modulation, the percentage or index of the modulation, and the efficiency of the data-transmission method. *See also* AMPLITUDE MODULATION, BANDWIDTH, EMISSION CLASS, FREQUENCY MODULATION, MODULATION, MODULATION METHODS, PHASE MODULATION, and SINGLE SIDEBAND.

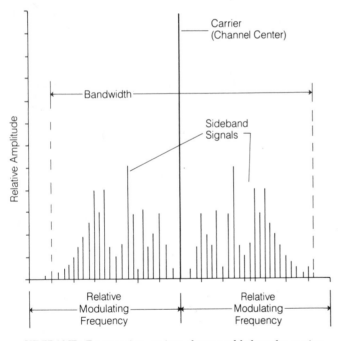

SIDEBAND: Frequencies are just above and below the carrier.

SIDESCATTER

Sidescatter is a form of E-layer and F-layer ionospheric propagation, similar to *backscatter*. In ionospheric wave propagation, the angle of incidence is about the same as the angle of return for most of the signal energy. But some energy is scattered in all directions. A small amount of the electromagnetic field falls within the skip zone. Sidescatter refers to that energy that goes off more or less "sideways," with respect to the signal entering the ionized layer, and comes back to earth within the skip zone.

Sidescatter signals have a characteristic weak, watery sound. This mode of propagation is sometimes confused with other weak-signal modes. *See also* BACKSCATTER, E LAYER,

F LAYER, PROPAGATION CHARACTERISTICS, and SKIP ZONE.

SIDESTACKING

Sidestacking is a method of combining two or more parasitic arrays or phased arrays to obtain greater power gain than would be possible with a single antenna.

The sidestacking technique involves the connection of antennas in the horizontal plane, with the elements (and thus the polarization) oriented vertically. The antennas are fed in phase by means of a harness. The technique is used mostly at the very-high and ultra-high frequencies.

The gain that can be obtained with sidestacking is the same as the gain that results from vertical stacking. Assuming identical antennas are used, the power gain increases by 3 dB every time the number of antennas is doubled. *See also* STACKING.

SIDETONE

In a code transmitter or transceiver in which receiver muting is used, the actual transmitted signal cannot be heard. Even if receiver muting is not used, the transmitter and receiver are often operated on frequencies separated by an amount that results in an inaudible tone, or a tone with an objectionably high or low pitch. In such situations, a *sidetone* is used.

A sidetone is generated by a simple audio oscillator, such as a multivibrator or relaxation circuit. Most keyers have built-in sidetone oscillators, some with adjustable pitch. Many code transmitters and transceivers also have built-in sidetone oscillators, which are keyed along with the transmitter. *See also* SIDETONE OSCILLATOR.

SIDETONE OSCILLATOR

A *sidetone oscillator* is a simple audio-oscillator circuit used with keyers and code transmitters for the purpose of generating a monitor tone or sidetone (*see* SIDETONE). Many different types of circuit will function well as a sidetone oscillator.

A sidetone oscillator should have certain characteristics. First, it is imperative that the oscillator function every time the key is closed; that is, the oscillator must start instantly 100 percent of the time. Second, the frequency (pitch) of the tone must not change as the circuit is keyed. Third, the tone should have a pleasing sound that is not tiring to the ear. A sine-wave tone is undesirable because it causes fatigue; an excessively raw or "buzzing" sidetone is equally objectionable. If a transmitter/ keyer combination has no sidetone oscillator, a radio-frequency-actuated device can be built.

SIEMENS

The *siemens* is the unit of electrical conductance. It is also sometimes called the *mho*. Given a resistance of R ohms, the conductance S in siemens is simply $1/R$. *See also* MHO.

SIGNAL DIODE

A semiconductor diode, used for low-power applications at audio or radio frequencies, is sometimes called a *signal diode*. Signal diodes are used for envelope detection, peak clipping or limiting, mixing, and many other purposes.

Signal diodes are characterized by relatively small P-Njunction surface area. Signal diodes have small to moderate junction capacitance. The current-handling capacity and peak-inverse-voltage (PIV) rating are usually low. *See also* DIODE, POINT-CONTACT JUNCTION, and P-N JUNCTION.

SIGNAL GENERATOR

A *signal generator* is an oscillator, often equipped with a modulator, that is used for the purpose of testing audio-frequency or radio-frequency equipment. Most signal generators are intended for either audio-frequency or radio-frequency applications, but not both.

In its simplest form, a signal generator consists of an oscillator that produces a sine wave of a certain amplitude in microvolts or millivolts, and a certain frequency in hertz, kilohertz, or megahertz. Some audio-frequency signal generators can produce several different types of waveforms. The more sophisticated signal generators for radio-frequency testing have amplitude modulators and/or frequency modulators.

For the testing, adjustment, and servicing of radio-frequency transmitters and receivers, a combined signal generator and monitor is often used. *See also* FUNCTION GENERATOR, and SERVICE MONITOR/GENERATOR.

SIGNAL MONITOR

A *signal monitor* is a test instrument that is used for analyzing radio-frequency signals. An ordinary radio receiver can be used as a signal monitor in many situations. The technician simply listens to the characteristics of the incoming signal. For a more detailed view of the signal than can be obtained by ear, a panoramic receiver can be used.

An oscilloscope is often used for the purpose of signal monitoring. In the test laboratory, a service monitor/generator is used. A spectrum analyzer is a form of signal monitor. *See* OSCILLOSCOPE, PANORAMIC RECEIVER, SERVICE MONITOR/GENERATOR, and SPECTRUM ANALYZER.

SIGNAL-PLUS-NOISE-TO-NOISE RATIO

In a receiver, the sensitivity is sometimes specified in terms of the ratio of the audio signal-plus-noise strength to the noise strength for a given input. This is called the *signal-plus-noise-to-noise ratio*. It is abbreviated $(S + N)/N$ or $(S + N):N$. (The parentheses are mathematically important, but they are often omitted, giving the imprecise expressions $S + N/N$ or $S + N:N$.)

The $(S + N)/N$ ratio is always specified in decibels. A certain signal level is required to cause, say, a $(S + N)/N$ ratio of 10 dB. If the root-mean-square (RMS) signal level in microvolts is E_s, and the RMS noise level in microvolts is E_n, then:

$$(S + N)/N = 20 \log_{10} ((E_s + E_n)/E_n)$$

The $(S + N)/N$ ratio is always greater than the signal-to-noise ratio. However, the latter is more frequently used. *See* SIGNAL-TO-NOISE RATIO.

SIGNAL-TO-IMAGE RATIO

In a superheterodyne receiver, images are always present. The image signals should, ideally, be rejected to the extent that they do not interfere with desired signals. However, this is not always the case.

The *signal-to-image (S/I) ratio* is usually given in decibels at the output of a receiver. *See also* IMAGE REJECTION, and SUPERHETERODYNE RECEIVER.

SIGNAL-TO-INTERFERENCE RATIO

In radio reception, the signal-to-interference ratio is sometimes mentioned. This ratio is measured in decibels, and is given as S, where:

$$S = 20 \log_{10} (a/b)$$

where a is the strength of the desired signal in microvolts at the antenna terminals, and b is the sum of the strengths of all of the interfering signals in microvolts at the antenna terminals.

The signal-to-interference ratio is important as a specification of sensitivity and selectivity in a radio receiver. *See also* SELECTIVITY, SENSITIVITY, SIGNAL-PLUS-NOISE-TO-NOISE RATIO, and SIGNAL-TO-NOISE RATIO.

SIGNAL-TO-NOISE RATIO

The sensitivity of a communications receiver is often specified in terms of the audio signal-to-noise ratio that results from an input signal of a certain number of microvolts. This ratio is abbreviated S/N or $S:N$.

If the root-mean-square (RMS) signal strength at the antenna terminals of a receiver is E_s, given in microvolts, and the RMS noise level is E_n, also in microvolts, then the ratio S/N, in decibels, is:

$$S/N = 20 \log_{10} (E_s/E_n)$$

Usually, the sensitivity is specified as the signal strength in microvolts that is necessary to cause a S/N ratio of 10 dB, or 3.16:1.

Modern communications receivers generally require about $0.5\ \mu V$, or less, to produce a S/N ratio of 10 dB at the high frequencies in the continuous-wave or single-sideband modes. For amplitude modulation, the rating is usually $1\ \mu V$ or less. The S/N sensitivity of a communications receiver is usually mentioned in the table of specifications. In the case of frequency-modulation (FM) receiver, the noise-quieting figure or the SINAD figure are standard. *See also* NOISE QUIETING, and SINAD.

SILICON

Silicon is an element with atomic number 14 and atomic weight 28. In its pure state, silicon appears as a light-weight metal similar to aluminum. Silicon is a semiconductor substance; it conducts electric currents better than a dielectric, but not as well as the excellent conductors (such as silver and copper).

Silicon is found in great quantities in the crust of the earth. In its natural state, silicon is almost always combined with oxygen or other elements. Pure silicon is extracted from the various compounds for use in the manufacture of semiconductor diodes, transistors, integrated circuits, and other devices.

Silicon has replaced germanium in the manufacture of many types of semiconductor components because silicon can withstand higher temperatures than germanium. However, germanium is still used in some devices. *See also* GERMANIUM, and SEMICONDUCTOR.

SILICON-CONTROLLED RECTIFIER

The silicon-controlled rectifier (abbreviated SCR) is a four-layer semiconductor device that is used primarily for power control. The SCR is a commonly used form of thyristor. The terms SCR and thyristor are sometimes used interchangeably (*see* THYRISTOR).

The SCR has three electrodes, called the *cathode, anode,* and *gate.* When a forward bias is applied between the cathode and the anode, no current flows until a pulse is applied to the gate. Then, the SCR continues to conduct until the bias between the cathode and anode is reversed or reduced below a certain threshold value.

Silicon-controlled rectifiers are available with many different voltage and current ratings for various applications. The devices are widely used in power-control circuits for ac or dc.

SILICONE

Silicone is a polymerized material consisting of silicon and oxygen atoms. It is an excellent electrical insulating material, but is a good conductor of heat. Silicone can withstand very high temperatures.

Silicone is commercially available in various forms. One common product comes in a small tube (like toothpaste), is white in color, and has the consistency of petroleum jelly.

Silicone is commonly used as a heat-transfer agent for semiconductor power transistors and diodes. The silicone is applied between the heat-conducting metal base of the device and a metal heatsink. This ensures efficient transfer of heat away from the semiconductor device. Small particles of dust or dirt would otherwise interfere with the heat bond.

SILICON UNILATERAL SWITCH

A *silicon unilateral switch* (SUS) is a four-layer semiconductor device, identical to the silicon-controlled rectifier (SCR), except that a zener diode is placed in the gate lead. The zener diode reduces the sensitivity of the switch, by blocking trigger pulses having voltages smaller than a certain predetermined level.

The SUS is less likely than an ordinary SCR to be triggered accidentally by stray noise pulses. *See also* SILICON-CONTROLLED RECTIFIER, and ZENER DIODE.

SILVER-MICA CAPACITOR

A *silver-mica capacitor* is a form of mica capacitor (*see* MICA CAPACITOR) in which silver foil or plating is used on the mica sheet. Silver-mica capacitors have basically the same characteristics as other mica capacitors. The loss is somewhat lower when silver is used, as compared with the loss when other metals are employed; this difference arises from the fact that silver is an excellent electric conductor.

SILVER SOLDER

Silver solder is a hard solder that is sometimes used in the manufacture of electronic devices. Silver solder consists of copper, zinc, and silver.

Silver solder melts at a higher temperature than ordinary solder, which is composed of tin and lead. This makes silver solder an advantage in circuits where large amounts of current must be handled. Silver solder maintains a better electrical bond than tin-lead solder, primarily because the metals used are better conductors. *See also* SOLDER.

SIMPLE MAIL TRANSFER PROTOCOL

In packet radio, *simple mail transfer protocol* (SMTP) refers to a procedure for automatically sending messages. To use SMTP, a certain format is required for the message. The formatting can be done by the operator, or by means of software known as *Bdale's mailer.* The operator writes the message on a personal computer (PC), and saves the message on a disk. The SMTP system then works constantly to establish a communications route to the destination station. Once the route has been set up, the message is sent automatically. *See also* PACKET RADIO.

SIMPLEX

In a two-way communications system, both transmitters and receivers are often operated on a single frequency. This is known as *simplex* or *simplex operation.* The two stations communicate directly with each other; no repeater or other intermediary is used.

At very-high and ultra-high frequencies, simplex operation might not provide enough communications range. This is especially true if one or both stations are mobile. To increase the effective range, repeaters are used. Some repeaters are placed on satellites, greatly increasing the communications range (*see* ACTIVE COMMUNICATIONS SATELLITE, and REPEATER).

In simplex operation, it is possible for only one station to transmit at a time. This is because neither station can receive signals at the same time, and on the same frequency, as they are transmitting. If it is necessary to send and receive data simultaneously, two different frequencies must be used. This is called *duplex operation* (*see* DUPLEX OPERATION).

It is not normally possible for one station to interrupt the other station in simplex operation. However, it can be accomplished if continuous-wave or single-sideband, suppressed-carrier emissions are used. This is called *break-in operation.* It requires the use of a special switch and muting system. The receiver is activated during brief pauses in a transmission (*see* BREAK-IN OPERATION).

SINAD

The term *SINAD* is an abbreviation of the words *signal* to *noise* and *distortion.* This expression is frequently used to define the sensitivity of a frequency-modulation receiver at the very-high and ultra-high frequencies. The SINAD figure takes into account not only the quieting sensitivity of a receiver, but its ability to reproduce a weak signal with a minimum of distortion.

Usually, the SINAD sensitivity of a receiver is given as the signal strength in microvolts at the antenna terminals that results in a signal-to-noise-and-distortion ratio of 12 dB at the speaker terminals. The signal is modulated with a 1-kHz sine-wave tone, at a deviation of plus-or-minus 3 kHz. This is the standard test modulation for a frequency-modulated system (*see* STANDARD TEST MODULATION).

The SINAD sensitivity of a receiver is measured by using a calibrated signal generator and a distortion analyzer or SINAD meter. *See also* NOISE QUIETING, and SINAD METER.

SINAD METER

A distortion analyzer, designed especially for the purpose of measuring the signal-to-noise-and-distortion (SINAD) sensitivity of a frequency-modulation receiver, is called a *SINAD meter* or *SINAD distortion analyzer*.

The SINAD meter contains a notch filter centered at an audio frequency of 1 kHz. For determination of the SINAD sensitivity of a receiver, the procedure is as follows:

1. The meter is connected to the speaker terminals of the receiver.
2. The receiver squelch is opened, and the volume set near the middle of the control range.
3. The meter-level control is set for a meter reading of 0 dB. Many SINAD meters have an automatic level control that sets the meter at 0 dB regardless of the volume-control setting of the receiver.
4. A signal generator is connected to the receiver antenna terminals, and the generator is set to the same frequency or channel as the receiver.
5. The signal is modulated at an audio frequency of 1 kHz, with a deviation of plus-or-minus 3 kHz.
6. The signal level is adjusted until the SINAD meter indicates 12 dB. The SINAD sensitivity, in microvolts, is then read from the calibrated scale of the signal-generator amplitude control.

Most SINAD meters can be used for measuring the quieting sensitivity, as well as the SINAD sensitivity, of a frequency-modulation receiver. *See also* DISTORTION ANALYZER, NOISE QUIETING, and SINAD.

SINE

The *sine function* is a trigonometric function. In a right triangle, the sine is equal to the length of the far or opposite side, divided by the length of the hypotenuse. In the unit circle $x^2 + y^2 = 1$, plotted on the Cartesian (x,y) plane, the sine of the angle θ, measured counterclockwise from the x axis, is equal to y. The sine function is periodic, and begins with a value of 0 at $\theta = 0$. The shape of the sine function is identical to that of the cosine function (*see* COSINE), except that the sine function is displaced to the right by 90 degrees.

The sine function represents the waveform of a pure, harmonic-free, alternating-current disturbance. In mathematical calculations, the sine function is abbreviated sin. *See also* SINE WAVE, and TRIGONOMETRIC FUNCTION.

SINE WAVE

A *sine wave*, also called a *sinusoidal waveform*, is an alternating-current disturbance that has only one frequency. The harmonic content, and the bandwidth, are theoretically zero (*see* BANDWIDTH, and HARMONIC). The sine wave gets its name from the fact that the amplitude-versus-time function is identical to the trigonometric sine function. The cosine function also is a perfect representation of the shape of a sine wave (*see* COSINE).

When displayed on an oscilloscope, the sine wave has a characteristic shape as shown in the illustration. Each cycle is represented by one complete alternation as shown. The cycle of a sine wave is divided into 360 electrical degrees.

The waveform produced by an oscillator is never a perfect sine wave. There is always some harmonic content. However, the amount of harmonic energy is often so small that, when the waveform is displayed on an oscilloscope, it appears as a perfect sine wave. *See also* SINE.

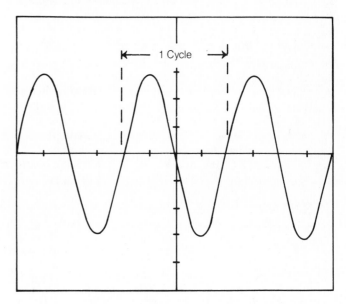

SINE WAVE: Characteristic shape as seen on oscilloscope.

SINGLE BALANCED MIXER

A *single balanced mixer* is a mixer circuit that is easily built for a minimum amount of expense. The circuit operates in a manner similar to a balanced modulator (*see* BALANCED MODULATOR, MIXER, and MODULATOR). The input and output ports are not completely isolated in the single balanced mixer; some of either input signal leaks through to the output. If isolation is needed, the double balanced mixer is preferable (*see* DOUBLE BALANCED MIXER).

A typical single-balanced-mixer circuit is shown in the illustration. The diodes are generally of the hot-carrier type. This circuit will work at frequencies up to several gigahertz. The circuit shown is a passive circuit, and therefore some loss will occur. However, the loss can be overcome by means of an amplifier following the mixer. *See also* HOT-CARRIER DIODE.

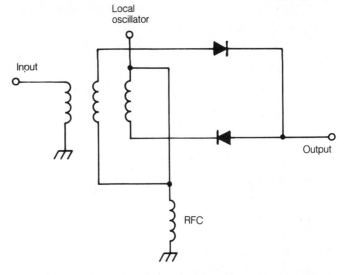

SINGLE BALANCED MIXER: An example of a passive circuit.

SINGLE-CONVERSION RECEIVER

A *single-conversion receiver* is a form of superheterodyne receiver that has one intermediate frequency. The incoming signal is heterodyned to a fixed frequency. Selective circuits are employed to provide discrimination against unwanted signals. The output from the mixer is amplified and fed directly to the detector. A block diagram of a single-conversion receiver is illustrated.

The main advantage of the single-conversion receiver is its simplicity. However, the intermediate frequency must be rather high — on the order of several megahertz for a typical high-frequency communications receiver — and this limits the degree of selectivity that can be obtained. In recent years, excellent ceramic and crystal-lattice filters have been designed, improving the performance of single-conversion receivers. *See also* DOUBLE-CONVERSION RECEIVER, INTERMEDIATE FREQUENCY, MIXER, and SUPERHETERODYNE RECEIVER.

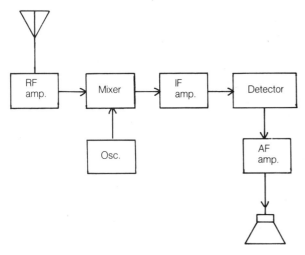

SINGLE-CONVERSION RECEIVER: Has one mixer and one IF.

SINGLE IN-LINE PACKAGE

The *single in-line package (SIP)* is a form of housing for integrated circuits. A flat, rectangular box containing the chip is fitted with lugs along one side, as shown in the illustration. There can be just a few pins, or as many as 12, or even 15 pins.

The single in-line package is very easy to install in, and to remove from, a circuit board. *See also* DUAL IN-LINE PACKAGE, and INTEGRATED CIRCUIT.

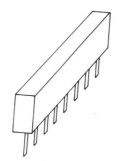

SINGLE IN-LINE PACKAGE: All pins are along one edge.

SINGLE SIDEBAND

Single sideband is a form of amplitude modulation. An ordinary amplitude-modulated signal consists of a carrier and two sidebands, one above the carrier frequency and one below the carrier frequency (*see* AMPLITUDE MODULATION). A single-sideband signal results from the removal of the carrier and one of the sidebands.

Single-sideband, suppressed-carrier emission, also called *J3E emission*, provides greater communications efficiency than ordinary amplitude modulation, or type A3 emission. This is because 2/3 (67 percent) of the power in an amplitude-modulated signal is taken up by the carrier wave, which conveys no intelligence. In a single-sideband signal, all of the power is concentrated into one sideband, and this yields an improvement of about 8 dB over A3 emission. A single-sideband signal has a bandwidth of approximately half that required for amplitude modulation. Therefore, it is possible to get twice as many J3E signals as A3 signals into a given amount of spectrum space.

In order to obtain J3E emission, a balanced modulator must be used, followed by a filter or phasing circuit. The balanced modulator produces a double-sideband signal with a suppressed carrier (*see* BALANCED MODULATOR, and DOUBLE SIDEBAND). The filter or phasing network then removes either the lower sideband or the upper sideband. If the lower sideband is removed, the resulting J3E signal is called an *upper-sideband (USB) signal*. If the upper sideband is removed, a lower-sideband (LSB) signal results (*see* LOWER SIDEBAND, and UPPER SIDEBAND).

Most voice communication at the low, medium, and high frequencies is carried out via single sideband. Generally, the lower sideband is used at frequencies below about 10 MHz; the upper sideband is preferred at frequencies higher than 10 MHz. This is simply a matter of convention; either sideband will provide equally good communication at a particular frequency.

Reception of single-sideband signals requires a product detector. The product detector contains a local oscillator and a mixer. The local oscillator supplies the "missing" carrier, so that the signals are intelligible (*see* PRODUCT DETECTOR).

The tuning of a single-sideband receiver is rather critical. Stability is therefore extremely important, both for the transmitter and the receiver. The receiver local-oscillator frequency must be within a few hertz of the suppressed-carrier frequency of the intermediate-frequency signal. If the receiver is too far off frequency, the result is an unnatural, sometimes humorous-sounding, and perhaps even unintelligible, garbled noise. If the receiver is set for the wrong sideband, a completely unreadable signal will be heard (*see* MONKEY CHATTER).

SINGLE-SIGNAL RECEPTION

In a receiver having a product detector, it is an advantage to have single-signal reception. Most modern superheterodyne receivers have this feature. Direct-conversion receivers and regenerative receivers generally do not.

In a receiver with single-signal reception, the signal can be heard only on one side of zero beat. In the case of A1 or F1 type emission, a beat note or pair of notes is audible on either the upper or lower side of zero beat, but not on both sides. As the receiver is tuned through zero beat, the pitch of the sound drops until the note or notes disappear. The notes do not recur as tuning is continued. If the receiver does not have single-signal reception, the beat notes will reappear on the other side of zero beat, gradually rising in pitch. A similar effect is observed with single sideband type emission.

The advantage of single-signal reception over so-called *double-signal reception* is that it cuts the interference among signals in half. With single-signal reception, the selectivity can be much sharper, while still maintaining intelligibility, as compared with double-signal reception. *See also* PRODUCT DETECTOR, and SUPERHETERODYNE RECEIVER.

SINGLE THROW

A relay or switch is called *single-throw* if it simply connects and disconnects a circuit or group of circuits. The single-throw device cannot be used for switching a common line between two or more other lines.

Single-throw switches are commonly used for power switching. A single-pole switch or relay can be used as a circuit interrupter in any system. A single-throw device can have any number of poles, facilitating the opening and closing of several circuits at once. A single-pole, single-throw switch is often called an *SPST switch*; a double-pole, single-throw switch is called a *DPST switch*. *See also* MULTIPLE THROW.

SINGLE TUNED CIRCUIT

The most common type of preselector used in modern high-frequency (HF) receivers is the *single tuned circuit*. This is what its name implies: an inductance-capacitance (LC) circuit, usually parallel-resonant. It is generally placed at the front end of the receiver.

This type of preselector offers the advantage of simplicity. It's also easy to tune over a wide range of frequencies. But a single tuned circuit is not very good at rejecting images or strong out-of-band signals. A multiple tuned circuit, such as a *helical filter*, is far better for this purpose. *See also* FRONT END, HELICAL FILTER, and PRESELECTOR.

SINGLE-WIRE LINE

A radio antenna can be fed with a single wire for either receiving or transmitting purposes. The single wire forms an unbalanced feed line. Single-wire lines are sometimes used at medium and high frequencies, especially for reception.

A single-wire feed line exhibits a free-space characteristic impedance of approximately 600 ohms to 800 ohms. The return circuit is provided by the earth; therefore, an excellent ground system is needed if single-wire line is used.

A single-wire feed line inevitably picks up or radiates a considerable amount of electromagnetic energy. This makes the single-wire line unsuitable for use with directional antennas. If the feed line must be kept from intercepting or radiating signals, a balanced or shielded line must be used. The most common such lines are open wire, twin-lead, four-wire line, and coaxial cable (*see* FEEDLINE).

Single-wire line can be used to feed a horizontal or vertical radiator at either end, in the center, or off center. The true Windom antenna uses a single-wire line in an off-center configuration. *See also* WINDOM ANTENNA.

SKEW

In video transmission, *skew* is distortion that results from failure of the receiver to be exactly synchronized with the transmitter.

Skew results from a constant discrepancy in the horizontal scanning rates of the receiver and transmitter. Skew can also occur in an improperly operating video recording system.

Skew can be detected in a facsimile or television system by transmitting a set of vertical, straight lines. If the receiver and transmitter are in precise synchronization, then the received picture will be identical to this pattern. However, if skew is present, the lines will appear slanted. If the receiver scans faster than the transmitter, the lines are tilted clockwise. If the receiver scan rate is slower than that of the transmitter, the lines appear tilted counterclockwise. The severity of the skew is given in degrees. The perfectly vertical lines represent a skew of 0 degrees.

If the synchronization discrepancy becomes nonconstant, or if it is especially severe, a picture becomes completely unrecognizable. This effect is familiar to all television viewers who have inadvertently misadjusted the horizontal-hold control. *See also* HORIZONTAL SYNCHRONIZATION.

SKIN EFFECT

In a solid wire conductor, direct current flows uniformly along the length of the wire. The number of electrons passing through a given cross section of the wire does not depend on whether the cross section is near the surface of the wire or near the center. The same holds for alternating currents at relatively low frequencies. The conductivity of the wire is proportional to the cross-sectional area, which is in turn proportional to the square of the diameter.

At radio frequencies, the conduction in a solid wire becomes nonuniform. Most of the current tends to flow near the outer surface of the wire. This is called *skin effect*. It increases the effective ohmic resistance of a wire at radio frequencies. The higher the frequency becomes, the more pronounced the skin effect will be. At high and very high frequencies, the conductivity of a wire is more nearly proportional to the diameter than to the cross-sectional area.

In the design of radio-frequency transmitting antennas, tubing is generally used rather than wire, if such construction is mechanically feasible. A large-diameter tubing conducts at the high and very-high frequencies almost as well as a solid metal rod of the same diameter. The use of tubing thus provides very low ohmic loss in an antenna system, at reasonable cost.

SKIP

Long-distance ionospheric propagation is sometimes called *skip*, although this is a technically inaccurate use of the term. Skip is actually the tendency for signals to pass over a certain geographical region. At high frequencies, skip is sometimes observed. This effect is shown in the illustration. A transmitting station X is heard by station Z, located thousands of miles distant, but not by station Y. The ionization of the F layer is insufficient to bend the signals back to earth at the sharper angle necessary to allow reception by station Y. *See also* PROPAGATION CHARACTERISTICS, and SKIP ZONE.

SKIP ZONE

When skip occurs in an ionospheric F-layer communication circuit, the signals from a transmitting station cannot be received by other stations located within a certain geographical area. This "dead" area is called the *skip zone*.

Under most conditions, the skip zone begins at a distance of about 10 to 15 miles from the transmitting station, or the limit of the range provided by the direct wave, the reflected waves, and the surface wave (*see* DIRECT WAVE, REFLECTED WAVE, and SURFACE WAVE). The outer limit of the skip zone varies considerably, depending on the operating frequency, the time of day, the season of the year, the level of sunspot activity, and the direction in which transmission is attempted. A pictorial example is shown in the illustration.

At the very low, low, and medium frequencies, a skip zone is never observed. In the high-frequency spectrum, however, a skip zone is often present. In the upper part of the high-frequency band, a skip zone is almost always observed.

At times, certain conditions arise that allow signals to be heard from points within the skip zone. Densely ionized areas can form in the E layer, causing propagation over shorter paths than would normally be expected (*see* E LAYER, and SPORADIC-E PROPAGATION). Auroral propagation and backscatter sometimes allow communication between stations that would otherwise be isolated by the skip zone (*see* AURORAL PROPAGATION, and BACKSCATTER). Above about 20 MHz, tropospheric effects can partially "fill in" the skip zone at times (*see* TROPOSPHERIC PROPAGATION).

If the frequency of operation is increased, the skip zone widens. The outer limit of the skip zone can be several thousand miles away. The same widening effect takes place at a constant frequency as darkness falls. At frequencies above a certain maximum, the outer limit of the skip zone disappears entirely, and no F-layer propagation is observed. *See also* F LAYER, IONOSPHERE, PROPAGATION CHARACTERISTICS, and SKIP.

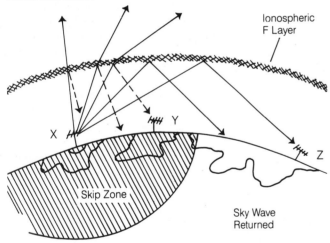

SKIP ZONE: Z can hear X, but Y cannot hear X.

SKIRT SELECTIVITY

Skirt selectivity is an expression of the cutoff sharpness of a receiver bandpass filter, and also of the maximum or ultimate attenuation. Good skirt selectivity is an important feature of any communications receiver.

Skirt selectivity can be expressed qualitatively in terms of relative "steepness." Skirt selectivity can be specified quantitatively according to the shape factor and ultimate attenuation of the response (*see* SHAPE FACTOR).

In general, steep skirts represent a nearly rectangular bandpass response. This is desirable in most receiving situations. Steep skirts provide excellent adjacent-channel rejection, and

optimize the signal-to-noise ratio. *See also* ADJACENT-CHANNEL INTERFERENCE, RECTANGULAR RESPONSE, and SELECTIVITY.

SKY WAVE

An electromagnetic signal is called a *sky wave* if it has been returned to the earth by the ionosphere. Sky-wave propagation can be caused by either the E layer or the F layer of the ionosphere. Sky-wave propagation can also result from reflection off the auroral curtains, or from scattering by the ionized trails left by meteors (*see* AURORAL PROPAGATION, E LAYER, F LAYER, METEOR SCATTER, and SPORADIC-E PROPAGATION).

Sky waves are observed at various times of day at different frequencies. It is this form of propagation that is responsible for most of the long-distance communication that takes place on the so-called "shortwave bands." Skywave propagation occurs almost every night on the standard amplitude-modulation broadcast band. Sky waves are affected by the season of the year and the level of sunspot activity, as well as the time of day and the operating frequency. *See also* PROPAGATION CHARACTERISTICS.

SLEW RATE

In a closed-loop operational-amplifier circuit, there is a limit to how rapidly the output voltage can change as the input voltage changes. This maximum limit, under linear operating conditions, is called the *slew rate*. It is measured in volts per second, volts per millisecond, or other units of voltage per unit time.

The slew rate determines the maximum frequency at which an operational amplifier will function in a linear manner. If the frequency is increased so that the instantaneous rate of change of the input signal exceeds the slew rate, distortion will occur, and the operational amplifier will not perform according to the theoretical specifications.

SLIDE POTENTIOMETER

A *slide potentiometer* is a form of variable resistor, designed for low-current applications in direct-current and audio-frequency circuits. Most potentiometers are of the rotary type, and are adjusted by turning a knob, shaft, or metal plate. The slide potentiometer is set by means of a lever that moves back and forth, or up and down.

Slide potentiometers give visual reinforcement of a control position in certain situations. A common example is the graphic equalizer.

SLIDE SWITCH

A *slide switch* is a small, one-pole or two-pole switch that is often used in communications equipment. Slide switches are employed mostly in situations not requiring frequent changing of the switch position.

The slide switch has two positions, and can be either a single-throw or a double-throw device. The switch is thrown by pressing a lever up, down, or sideways. Some slide switches are equipped with a plate that can be screwed down over the lever to prevent accidental changing of the switch position.

SLIDING WINDOW PROTOCOL

In packet radio, *sliding window protocol* is a means of distributing the load of communications traffic. It prevents any given route from becoming overloaded, by automatically routing packets through various nodes in the network. Sliding window protocol works at the transport layer of the open systems interconnection reference model (OSI-RM).

Sometimes, a route changes while messages are being sent from one station to another. This might occur because of interference, equipment failure, or changes in propagation. It can result in packets being partially or completely lost. The sliding window protocol keeps track of such events, and ensures that lost packets are retransmitted by the appropriate node(s). *See also* NODE, OPEN SYSTEMS INTERCONNECTION REFERENCE MODEL, PACKET RADIO, and PROTOCOL.

SLOT COUPLING

Slot coupling is a method of coupling between a waveguide and a coaxial transmission line. Slot coupling is used at radio frequencies in the ultra-high and microwave range.

Slot coupling is accomplished by cutting identical rectangular openings in the waveguide and the coaxial-cable outer conductor. The rectangular openings are placed in precise alignment. The waveguide and the coaxial cable can be either parallel or perpendicular, with respect to each other.

Slot coupling can be used to transfer radio-frequency energy from a coaxial cable into a waveguide, or to transfer energy from a waveguide into a coaxial cable. Thus, slot coupling is useful for transmitting and receiving with a single physical installation. Slot coupling can also be used to provide signal transfer between two sections of waveguide. *See also* WAVEGUIDE.

SLOW-BLOW FUSE

It is often desirable to have a fuse that does not blow instantly, but instead allows some time before breaking a circuit. This is the case, for example, in equipment that draws a heavy initial current surge when the power is first applied. A *slow-blow fuse* is designed for this type of application.

The slow-blow fuse has a wire filament that softens if too much current passes through it. The wire is held taut by a spring. If the current exceeds the fuse rating for several seconds, the wire gets soft enough for the spring to pull it apart, breaking the circuit. The photograph shows a typical cartridge-type slow-blow fuse.

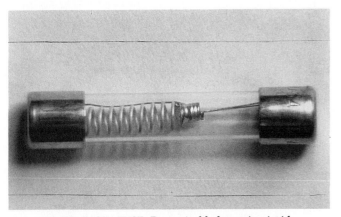

SLOW-BLOW FUSE: Recognizable by spring inside.

In low-current solid-state devices, slow-blow fuses might not react fast enough to protect the components in case of a malfunction. For this sort of situation, fast-blowing or quick-break fuses are used. See also QUICK-BREAK FUSE.

SLOW-SCAN TELEVISION

A *fast-scan television signal*, of the kind used in television broadcasting, takes 6 MHz of spectrum space to send. This is because of the high resolution of the image, along with the rapid changes that must be conveyed to give realistic video. It's almost the amount of spectrum between the amateur 40-meter and 20-meter bands, or between 75 meters and 30 meters. For this reason, fast-scan television is practical only at very-high frequencies (VHF) and above, where 6 MHz is a reasonable percentage of the carrier frequency.

But if some quality or qualities of the signal are compromised, it becomes possible to send a video image in a much narrower band. *Slow-scan television (SSTV)* accomplishes this, mainly by slowing down the rate at which pictures are sent. An SSTV signal is sent within just 3 kHz of spectrum, the same as needed by a voice signal. This makes SSTV practical on all of the amateur bands. With modest transmitters and antennas, SSTV DX contacts are common on the high-frequency (HF) ham bands.

Technical Summary A fast-scan signal has 30 complete picture frames per second. A slow-scan signal sends one frame every 8 seconds. This is 240 times slower, and allows a dramatic reduction in bandwidth.

There are 120 lines per frame in a slow-scan signal, compared with 525 in a fast-scan signal. This further reduces the bandwidth of SSTV. The modulation is obtained by inputting audio signals into a single-sideband (SSB) transmitter. A tone of 1.500 kHz corresponds to black; 2.300 kHz gives white. Intermediate audio frequencies produce shades of gray.

Synchronization signals are sent at 1.200 kHz. These are short bursts, lasting 0.030 seconds (30 ms) for vertical sync and 0.005 seconds (5 ms) for horizontal sync.

Sometimes, SSTV is sent along with a voice. The picture is sent on one sideband, say the upper (USB), while the voice is sent on the other sideband, say the lower (LSB). The resulting signal is a continuous voice narration, with still pictures every 8 seconds, something like a slide show.

Modern SSTV equipment makes it possible to use an ordinary television set to display the video images. Although there are fewer scan lines in SSTV than in the TV receiver, a converter adjusts for this, yielding a picture that is just about as detailed (except for the lack of motion) as a TV broadcast signal under good band conditions.

Color SSTV is possible and has been done by some amateurs. The differences between color SSTV and black-and-white SSTV are similar to the differences between color fast-scan and black-and-white fast-scan TV. *See also* COLOR TELEVISION, and COMPOSITE VIDEO SIGNAL.

Equipment Needed An SSTV station needs a transceiver with SSB capability, a *scan converter*, a TV set and a camera. The scan converter consists of several components, interconnected as shown in the block diagram. Scan converters are commercially available from many ham manufacturers and dealers, as are cameras.

The color SSTV scan converter is more sophisticated than the black-and-white converter shown in the diagram. Basically, there are three memories, instead of just one: a red memory, a green memory and a blue memory. The TV set and the camera must be able to display and pick up color signals. Color scan converters are available commercially. As you might expect, they cost a little more than black-and-white scan converters.

SMALL SIGNAL

A *small signal* is an alternating-current signal which, when applied to the input of an amplifier or other device, will not cause nonlinearity. A circuit designed to operate with small signals is called a *small-signal* or *low-level circuit*.

All radio receivers contain small-signal circuits at the front end, intermediate-frequency chain (in superheterodyne receivers), and audio stages. The detector and mixer stages are nonlinear, and are therefore not small-signal circuits. Radio transmitters usually have small-signal circuits at all stages except for the mixers, modulator, driver, and final amplifier.

Small-signal circuits do not require significant input-signal power for their operation. Small-signal amplifiers must be operated in class A, and generally have a high input impedance. Small-signal probes, likewise, must have a high input impedance to minimize the current drawn. *See also* CLASS-A AMPLIFIER.

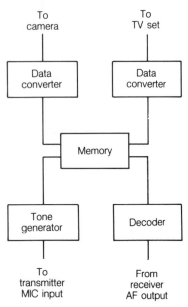

SLOW-SCAN TELEVISION: A scan converter for an SSTV station.

SMITH CHART

The *Smith chart* is a special form of coordinate system that is used for plotting complex impedances (*see* IMPEDANCE). Smith charts are especially useful for determining the resistance and reactance at the input end of a transmission line when the resistance and reactance at the antenna feed point are known. The Smith chart is named after the engineer P.H. Smith, who first devised and used it.

The resistance coordinates on the Smith chart are eccentric circles, mutually tangent at the bottom of the circular graph. The reactance coordinates are partial circles having variable diameter and centering. Resistance and reactance values can be assigned to the circles in any desired magnitude, depending on the characteristic impedance of the transmission line. The illus-

tration is an example of a Smith chart intended for analysis of feed systems in which the characteristic impedance of the line is 50 ohms ($Z_o = 50$).

Complex impedances appear as points on the Smith chart. Pure resistances (impedances having the form $R + j0$, where R is a nonnegative real number) lie along the resistance line; the top of the line represents a short circuit and the bottom represents an open circuit. Pure reactances (having the form $0 + jX$, where X is any real number except 0) lie on the perimeter of the circle, with inductance on the right and capacitance on the left. Impedances of the form $R + jX$, containing finite, nonzero resistances and reactances, correspond to points within the circle. Several different complex impedance points are illustrated.

The Smith chart can be used to determine the standing-wave ratio (SWR) on a transmission line, if the characteristic impedance of the line and the complex antenna impedance are known (*see* STANDING-WAVE RATIO). For determination of SWR, a set of concentric circles is added to the Smith chart. These circles are called *SWR circles*.

The center point of the chart corresponds to an SWR of 1:1. Higher values of SWR are represented by progressively larger circles. The radii of the SWR circles on a given Smith chart can be determined according to points on the resistance line. In the example, the 2:1 SWR circle passes through the 100-ohm point on the resistance line; the 4:1 SWR circle passes through the 200-ohm point on the resistance line, and soon. In general, for any SWR value $x:1$, the $x:1$ SWR circle passes through the point on the resistance circle corresponding to $50x$ ohms.

The Smith chart illustrates why an SWR of 1:1 can be obtained only if the impedance of the load is a pure resistance equal to the characteristic impedance of the feed line. The SWR cannot be 1:1 if reactance exists in the load. A given SWR greater than 1:1 can occur in infinitely many ways, corresponding to the infinite number of points on a circle. An SWR of "infinity" exists when the load is a short circuit, an open circuit, a pure capacitive reactance, or a pure inductive reactance.

For a thorough discussion of the use of the Smith chart in feed-line applications, a book on antenna theory or communications engineering should be consulted. *See also* CAPACITIVE REACTANCE, and INDUCTIVE REACTANCE.

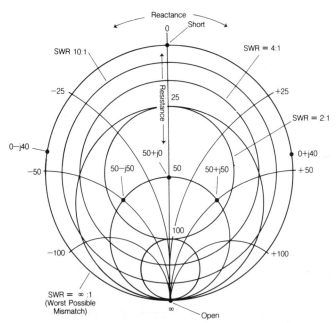

SMITH CHART: Denotes complex impedances.

S METER

An *S meter* is a device in a radio receiver that indicates the relative or absolute amplitude of an incoming signal. Many types of receivers are equipped with such meters. There are many different styles of S meter, but they can be categorized as either uncalibrated or calibrated.

Uncalibrated S meters can consist either of an analog meter or a digital meter (*see* ANALOG METERING, and DIGITAL METERING). The meter indicates relative signal strength. The signal for driving the meter can be obtained from the IF stages of the receiver. The most common method of obtaining this signal is by monitoring the automatic-gain-control (AGC) voltage. The stronger the signal, the greater the AGC voltage, and the higher the meter reading.

Uncalibrated S meters are found in many FM stereo tuners, and also in most FM communications equipment. Some general-coverage communications receivers use uncalibrated S meters.

Calibrated S meters are generally of the analog type. This facilitates marking the scale in definite increments. The standard unit of signal strength is the *S unit* (*see* S UNIT), which is a number ranging from 0 to 9 or from 1 to 9. A meter indication of S9 is defined as resulting from a certain number of microvolts at the antenna terminals. Most manufacturers agree on the figure of 50 μV for a meter reading of S9, as determined at a frequency in the center of the coverage range of the receiver. Some manufacturers use a different standard level to represent S9.

Because many receivers exhibit greater gain at the lower frequencies than at the higher frequencies, a reading of S9 often results from a weaker signal at the low frequencies (perhaps 40 μV) and a stronger signal at the higher frequencies (perhaps 60 μV). Each S unit below S9 represents a signal-strength change of 3 dB or 6 dB, depending on the manufacturer. This value is independent of the frequency to which the receiver is tuned.

Calibrated S meters are usually marked off in decibels above the S9 level. The meter readings above S9 are thus designated S9 + 10 dB, S9 + 20 dB, and so on; most meters can register up to S9 + 30 dB. These scales are accurate for typical communications. If signal-strength levels must be known with great accuracy, a laboratory test instrument should be used for measurement.

A receiver S meter provides a mental picture of how strong an incoming signal is. All signal-strength values are subjective, however, because band conditions can vary. A 50-μV signal might be masked by sferics (static) at 1.8 MHz, while the same signal stands out as that of a local station at 28 MHz. Although many radio operators define signal strength solely on the basis of the S-meter indication, others use their own judgment, taking the meter reading into account as a subjective quantity. Relative signal-strength information is exchanged among radio operators by means of an S (strength) number ranging from 1 to 9. *See also* RST SYSTEM.

SMOOTHING

Smoothing is the elimination of rapid fluctuations in the strength of a current or voltage. A good example of smoothing is the removal of the ripple in the output of a power supply. Smoothing can also be called *filtering*. Smoothing is usually accomplished by means of a capacitor (*see* FILTER CAPACITOR, and RIPPLE).

In an envelope detector, a small capacitor is used to eliminate the radio-frequency fluctuations of the carrier wave, leaving only the audio-frequency signals. This process is called *smoothing* (*see* DIODE DETECTOR).

Smoothing is used in automatic-gain-control (AGC) systems of all kinds. A capacitor and resistor are used to smooth out the audio-frequency components of a received or transmitted signal, while still providing a fast enough time constant to allow effective compensation for changes in the signal intensity (*see* AUTOMATIC GAIN CONTROL, and RESISTANCE-CAPACITANCE TIME CONSTANT).

SNAP DIODE

See STEP-RECOVERY DIODE.

SNOW STATIC

See PRECIPITATION STATIC.

SOCKET

A *socket* is a form of jack into which a device with many prongs is designed to fit. Sockets are used in some electronic circuits for easy installation and replacement of integrated circuits. *See also* JACK.

SOFTWARE

In a computer system, the programs are called *software*. Software can exist in written form, as magnetic impulses on tapes or disks, or as electrical or magnetic bits in a computer memory. Software also includes the instructions that tell personnel how to operate the computer.

There are several types of computer-programming languages, each with its own special purpose. The most primitive form of software language is called *machine language* because it consists of the actual binary information used by the electronic components of the computer.

Software can be programmed temporarily into a memory, or it can be programmed permanently by various means. When software is not alterable (that is, it is programmed permanently), it is called *firmware*. *See also* COMPUTER, COMPUTER PROGRAMMING, FIRMWARE, HARDWARE, HIGHER-ORDER LANGUAGE, MACHINE LANGUAGE, MICROCOMPUTER, and MINICOMPUTER.

SOLAR CELL

A *solar cell* is a photovoltaic cell (*see* PHOTOVOLTAIC CELL) that is designed for supplying electricity directly from the light of the sun. Solar cells are generally made of silicon, and resemble semiconductor diodes. The P-N junction, which has a very large surface area, produces direct current when sunlight strikes it.

In direct sunlight at noon on a midsummer day, a typical solar cell will produce about 0.1 watt for each square inch of surface area. The power output is decreased if clouds obscure the sun, or if the sun is very close to the horizon. However, solar cells will produce some power even on a cloudy day, or just after sunrise or before sunset.

Most solar-power systems have some means of storing the energy obtained during the hours of maximum sunlight, for use at night or when the sunlight is very weak. Energy storage can be done by means of storage batteries; many active communications satellites and other space vehicles use this method of obtaining power from the sun. A solar-energy system can also be interconnected with commercial power sources. Power is sold to the utility company in times of excess, and bought back during times of shortage.

Solar-cells can be used in conjunction with ammeters for light-intensity determination. Many light meters operate on this principle. Solar cells can also be used to receive modulated-light signals (*see* MODULATED LIGHT, and OPTICAL COMMUNICATION).

Most solar cells are rather inefficient. They convert less than 20 percent of the incident solar power into electrical power. New types of solar cells are being developed, however, that might provide much better efficiency and lower production cost. Solar cells are available in large batteries called *solar panels*, which can be connected in series-parallel to obtain large amounts of electrical power. *See also* SOLAR POWER.

SOLAR FLARE

A *solar flare* is a violent storm on the surface of the sun. Solar flares can be seen with astronomical telescopes equipped with projecting devices to protect the eyes of observers. A solar flare appears as a bright spot on the solar disk, thousands of miles across and thousands of miles high. Solar flares also cause an increase in the level of radio noise that comes from the sun (*see* SOLAR FLUX).

Solar flares emit large quantities of high-speed atomic particles. These particles travel through space and arrive at the earth a few hours after the occurrence of the flare. Because the particles are charged, they are attracted toward the geomagnetic poles. Sometimes a geomagnetic disturbance results (*see* GEOMAGNETIC, FIELD, and GEOMAGNETIC STORM). Then, you see the aurora at night, and experience a sudden, dramatic deterioration of ionospheric radio-propagation conditions. At some frequencies, communications can be completely cut off within a matter of a few seconds. Even wire communications circuits are sometimes affected.

Solar flares can occur at any time, but they seem to take place most often near the peak of the 11-year sunspot cycle. Scientists do not know exactly what causes solar flares, but they are evidently associated with sunspots, which are another type of solar storm. *See also* PROPAGATION CHARACTERISTICS, SUNSPOT, and SUNSPOT CYCLE.

SOLAR FLUX

The amount of radio noise emitted by the sun is called the *solar radio-noise flux*, or simply the *solar flux*. The solar flux varies with frequency. However, at any frequency, the level of solar flux increases abruptly when a solar flare occurs (*see* SOLAR FLARE). This makes the solar flux useful for propagation forecasting: a sudden increase in the solar flux indicates that ionospheric propagation conditions will deteriorate within a few hours.

The solar flux is most often monitored at a wavelength of 10.7 cm, or a frequency of 2800 MHz. At this frequency, the troposphere and ionosphere have no effect on radio waves, making observation easy.

The 2800-MHz solar flux is correlated with the 11-year sunspot cycle. On the average, the solar flux is higher near the peak of the sunspot cycle, and lower near a sunspot minimum. *See also* SUNSPOT, and SUNSPOT CYCLE.

SOLAR POWER

Electrical power can be obtained directly from sunlight by means of solar cells. Most modern solar cells can produce about 0.1 watt for each square inch of surface area exposed to bright sunlight. This is about 15 watts per square foot, or 150 watts per square meter, of photovoltaic-cell surface area. Large solar panels can be assembled, having hundreds of square feet of

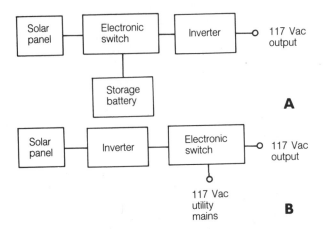

SOLAR POWER: Block diagrams of a stand-alone system (A) and an interactive system (B).

A

B

SOLAR POWER: Photos of a solar-powered "three-wheeler" (A) and an electric car (B).

surface area, and producing thousands of watts of power in direct sunlight. This makes the sun a potential source of power for residences and businesses. Even motor vehicles can be powered by sunlight (see photographs).

There are basically two types of solar-power system: the stand-alone system and the interconnected system. A stand-alone system uses batteries to store solar energy during the hours of daylight; the energy is released by the batteries at night. An interconnected system operates in conjunction with the utility company; energy is sold to the utility during times of daylight and minimum usage, and is bought back at night or during times of heavy usage.

Solar cells produce direct current, and batteries store direct current. But the typical household appliance will not operate from direct current; 117-V alternating current, having a frequency close to 60 Hz, is required. All solar-power systems, therefore, must incorporate power inverters (see INVERTER). The inverter produces 117-V alternating current from the direct current of the solar cells and/or batteries. A typical stand-alone solar-power system is shown at A in the illustration; an interconnected system is shown at B.

Large solar panels are expensive; a typical stand-alone or interconnected solar-power system costs several thousand dollars. However, if conventional fossil fuels continue to increase in price and decline in supply, solar power will become an attractive alternative for more and more people. *See also* PHOTOVOLTAIC CELL, and SOLAR CELL.

SOLDER

Solder is a metal alloy that is used for making electrical connections between conductors. There are several types of solder, intended for use with various metals and in various applications.

The most common variety of solder consists of tin and lead, with a rosin core. Some types of solder have an acid core. In electronic devices, rosin-core solder should be used; acid-core solder will result in rapid corrosion. Acid-core solder is more commonly used for bonding sheet metal.

The ratio of tin to lead in a rosin-core solder determines the temperature at which the solder will melt. In general, the higher the ratio of tin to lead, the lower the melting temperature. For general soldering purposes, the 50:50 solder can be used. For

SOLDER: COMMON TYPES OF SOLDER

Solder type	Melting point °F/°C	Principal uses
Tin-lead 50:50, Rosin-core	430/220	Electronic circuits
Tin-lead 60:40, Rosin-core	370/190	Electronic circuits, low-heat
Tin-lead 63:37, Rosin-core	360/180	Electronic circuits, low-heat
Tin-lead 50:50, Acid-core	430/220	Nonelectronic metal bonding
Silver	600/320	High-current, High-heat

heat-sensitive components, 60:40 solder is better because it melts at lower temperature. An ordinary soldering gun or iron can be used for applying and removing all types of tin-lead solder. *See* DESOLDERING TECHNIQUE, SOLDERING GUN, SOLDERING IRON, and SOLDERING TECHNIQUE.

Tin-lead solders are suitable for use with most metals except aluminum. For soldering to aluminum, a special form of solder, called *aluminum solder*, is available. It melts at a much higher temperature than tin-lead solder, and requires the use of a blowtorch for application.

In high-current applications, a special form of solder, known as *silver solder*, is used because it can withstand higher temperatures than tin-lead solder. A blowtorch is usually necessary for applying silver solder. Silver solder must be applied in a well-ventilated area because it produces hazardous fumes. The table lists the most common types of solder, and their principal characteristics and applications.

SOLDERING GUN

A *soldering gun* is a quick-heating soldering instrument. It is called a *gun* because of its shape. A trigger-operated switch is pressed with the finger, allowing the element to heat up within a few seconds.

Soldering guns are convenient in the assembly and repair of some kinds of electronic equipment. Soldering guns are available in various wattage ratings for different electronic applications.

Some electronic engineers and technicians prefer soldering irons to soldering guns. Soldering irons are generally easier to use with extremely miniaturized equipment, such as handheld and compact radio transceivers. Soldering guns are most often used in the assembly and repair of large, point-to-point wired apparatus, such as power amplifiers. *See also* SOLDERING IRON, and SOLDERING TECHNIQUE.

SOLDERING IRON

A *soldering iron* is a device that is used by engineers and technicians in the assembly and maintenance of electronic apparatus. A typical soldering iron consists of a heating element and a handle. The soldering iron requires from about 1 minute to 5 minutes to warm up after it is turned on; the larger the iron, the longer the warm-up time. Soldering irons are available in a wide range of different wattages. The smallest irons are rated at only a few watts, and are used in highly miniaturized equipment; the largest irons draw hundreds of watts, and are used for such purposes as wire splicing and sheet-metal bonding.

Some electronic engineers and technicians prefer soldering guns to soldering irons. Soldering guns are often more convenient in situations where moderately high heat is needed, and where point-to-point wiring is used. The soldering gun heats up and cools down very quickly. *See also* SOLDERING GUN, and SOLDERING TECHNIQUE.

SOLDERING TECHNIQUE

Most electrical and electronic circuits have soldered connections. In the assembly of such equipment, or in component replacement, the proper soldering technique assures optimum performance with a minimal chance for connection failure.

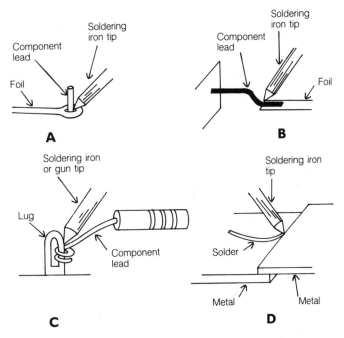

SOLDERING TECHNIQUE: For PC boards (A, B), point-to-point (C), and sheet metal (D).

Choosing Solder In electronic-circuit work, a tin-lead solder of 50:50 or 60:40 ratio, having a rosin core, should be used. The 50:50 type is suitable for most wiring, but if heat-sensitive components are to be wired on a printed-circuit board, the 60:40 type will reduce the chance of heat damage (*see* SOLDER) because it melts at a lower temperature. Acid-core solder is not suitable for wiring electronic components, but can be used to bond sheet metal that will not readily adhere to rosin-core solder.

The Soldering Instrument The proper soldering instrument for a given soldering application is, to some extent, a matter of choice. Soldering irons of low wattage (10 to 40 W) are generally best for printed-circuit wiring or highly miniaturized apparatus. For point-to-point electronic wiring, or for splicing of small wires, a soldering gun or iron can be used; the power rating should be between about 50 and 150 W. For heavy-gauge wire splicing or sheet-metal bonding, a high-wattage iron (200 W or more) is best.

In all cases, it is essential that the soldering instrument provide sufficient heat to prevent "cold" solder joints. However, too much heat can result in component damage (*see* SOLDERING GUN, and SOLDERING IRON).

Soldering on Printed-Circuit Boards Most printed-circuit soldering is done from the noncomponent (foil) side of the printed-circuit board. The component lead is inserted through the appropriate hole, and the soldering iron is placed so that it heats both the foil and the component lead (A in illustration). If the component is heat-sensitive, a needle-nosed pliers should be used to grip the lead on the component (nonfoil) side of the board while heat is applied. The solder is allowed to flow onto the foil and the component lead after the joint becomes hot enough to melt the solder; heating normally takes only 1 to 3 seconds. The solder should completely cover the foil dot or square in which the component lead is centered. Excessive solder should not be used. After the joint has cooled, the com-

ponent lead should be snipped off, flush with the solder, using a diagonal cutter.

If the circuit board is double-sided (foil on both sides, rather than on just one side), it will usually have plated-through holes so that the soldering procedure described above will be adequate. However, if the holes are not plated-through, solder must be applied, as described above, to the foil and component lead on either side of the circuit board.

Some printed-circuit components are mounted on the foil side of the board. This is the case, for example, with flat-pack integrated circuits. In such cases, the component lead and the circuit board foil are first coated, or "tinned," with a thin layer of solder. The component lead is then placed flat against the foil, and the iron is placed in contact with the component lead as at B. The heat melts the solder by conduction.

Soldering a Point-to-Point Connection Most point-to-point wiring is accomplished by means of tie strips, where one or more wires terminate (*see* TIE STRIP). In tie-strip wiring, the lug should first be coated with a thin layer of solder. The actual soldering of the connection should not be carried out until all of the wires have been attached to the lug. Wires are wrapped two or three times around the lug using a needle-nosed pliers. Excess wire is cut off using a diagonal cutter. When all wires have been attached to the lug, the soldering instrument is held against each wire coil, one at a time, and the connection is allowed to heat up until the solder flows freely in between the wire turns each time, adhering to both the wire and the lug as at C. Sufficient solder should be used so that the connection is completely coated. However, solder should not be allowed to ball up or drip from the connection.

If a heat-sensitive component is wired to a tie strip, a needle-nosed pliers should be used to conduct heat away from the component lead as long as heat is applied to the connection.

Splicing Wires In temporary or permanent wire splicing, the wires should be twisted together and soldered according to the procedure described under WIRE SPLICING.

Bonding Sheet Metal When soldering sheet metal, rosin-core solder should be used if possible, but sometimes the rosin does not allow a good enough mechanical bond. Then, acid-core solder can be used. Special solder is available for use with aluminum, which does not readily adhere to most other types of solder.

The edges to be bonded should be sanded with a fine emery paper, and cleaned with a noncorrosive, grease-free solvent such as rubbing alcohol. A high-wattage iron or blow-torch should be used to heat the metal while both sheets are "tinned" with a thin layer of solder. Then, the sheets should be secured in place. Sufficient heat should be applied so that the solder will flow freely. Some additional solder should be applied on each side of the bond, working gradually along the length of the bond from one side to the other (*see* D in illustration). The bond will require some time to cool, and it should be kept free from stress until it has cooled completely. Water or other fluids should not be used in an attempt to hasten the cooling process.

"Cold" Solder Joints If sufficient heat has not been applied to a solder connection, a "cold" joint can result. A properly soldered connection will have a shiny, clean appearance. A "cold" joint looks dull or rough. Many equipment failures occur simply because of "cold" solder joints, which exhibit high resistance.

If a "cold" joint is found, the solder should be removed as much as possible, using a wire braid (solder wick). Then, the connection should be resoldered.

Removing Solder   *See* DESOLDERING TECHNIQUE, and SOLDER WICK.

SOLDER WICK

Solder wick is a specially treated wire braid that is used to draw solder away from a connection for desoldering purposes. Solder wick appears somewhat similar to braided grounding wire, except that solder wick is smaller in size.

Solder wick is commercially available in easy-to-use rolls and in various different gauges for different purposes. *See also* DESOLDERING TECHNIQUE.

SOLID-STATE ELECTRONICS

During the 1960s, the transistor came into widespread use as an active device in amplifiers, oscillators, and mixers. The semiconductor diode replaced the diode tube for rectification and envelope detection. Since that time, many different kinds of semiconductor devices have been developed. All of these devices are called *solid-state components*. The science involved with the development and application of these components, and the design of circuits using them, is known as *solid-state electronics*.

Solid-state components get their name from the fact that the flow of charge carriers is entirely confined to solid substances. A vacuum tube is not a solid state device because electrons flow through free space. A transistor, though, operates via charge carriers that flow inside semiconductor material. There are two types of charge carriers in a solid-state device: the negatively charged electron and the positively charged hole (*see* ELECTRON, and HOLE).

The Diode   The *solid-state diode* consists of a junction between an N-type semiconductor and a P-type semiconductor, or a semiconductor and a wire. In either type of junction, current flows readily in one direction, provided the voltage is large enough (about 0.3 to 0.6 V). In the other direction, almost no current flows (*see* N-TYPE SEMICONDUCTOR, P-N JUNCTION, and P-TYPE SEMICONDUCTOR).

Many types of solid-state diodes are available for various different purposes. *See also* DETECTION, DIODE, DIODE ACTION, DIODE CAPACITANCE, DIODE CLIPPING, DIODE DETECTOR, DIODE FEEDBACK RECTIFIER, DIODE FIELD-STRENGTH METER, DIODE MATRIX, DIODE MIXER, DIODE OSCILLATOR, DIODE-TRANSISTOR LOGIC, DIODE TYPES, DOUBLE BALANCED MIXER, GUNN DIODE, LIGHT-EMITTING DIODE, MIXER, PHOTOVOLTAIC CELL, POINT-CONTACT JUNCTION, RECTIFICATION, RECTIFIER DIODE, SINGLE-BALANCED-MIXER, VARACTOR-DIODE, and ZENER DIODE.

Switching and Regulating Devices   Bipolar transistors, zener diodes, and certain integrated circuits can be used for current and voltage regulation (*see* SERIES REGULATOR, and SHUNT REGULATOR). Light dimmers and motor-speed controls use four-layer semiconductor devices such as silicon-controlled rectifiers and thyristors (*see* FOUR-LAYER SEMICONDUCTOR, SILICON-CONTROLLED RECTIFIER, and THYRISTOR).

Bipolar Transistors   *Bipolar transistors* are found in two basic types: the npn and pnp. Bipolar transistors can be used as oscillators, amplifiers, detectors, mixers, modulators, and switching and logic devices. They behave in a manner similar to vacuum tubes, except that they require much less power for operation, are physically smaller, and amplify current rather than voltage. *See also* BIPOLAR TRANSISTOR, NPN TRANSISTOR, and PNP TRANSISTOR.

Field-Effect Transistors   *Field-effect transistors* are made of either N-type and P-type material, or of certain metal oxides that behave in a similar manner. New, improved types of field-effect transistors are constantly being developed. They are generally used in oscillators and low-noise amplifiers. *See also* FIELD-EFFECT TRANSISTOR, GALLIUM ARSENIDE FIELD-EFFECT TRANSISTOR, METAL-OXIDE-SEMICONDUCTOR FIELD-EFFECT TRANSISTOR, and VERTICAL METAL-OXIDE-SEMICONDUCTOR FIELD-EFFECT TRANSISTOR.

Integrated Circuits   The variety of currently available integrated circuits is tremendous. An integrated-circuit catalog should be consulted to obtain details concerning specific integrated circuits. Some of the functions of integrated circuits are amplification, counting, logic operations, oscillation, and switching. *See* INTEGRATED CIRCUIT.

Other Solid-State Devices   Semiconductor devices are not the only types of solid-state components. Resistors, capacitors, inductors, and other common components are also solid-state. An entire circuit is considered solid-state only if all of its components are solid-state components (an exception is sometimes made for devices that contain a picture tube or camera tube).

Advantages of Solid-State Technology   Solid-state semiconductor components generally require far less power for their operation than did the older vacuum tubes. Solid-state components have no heaters or filaments, and are also physically smaller than vacuum tubes. A single integrated circuit, smaller in size than a dime, can house the equivalent of a whole building full of vacuum tube circuitry. Solid-state equipment is vastly more reliable than vacuum-tube equipment. Solid-state electronics has given us things that we now take for granted (such as battery-powered television sets, handheld calculators, and home computers) that were unheard of just a few decades ago.

Further Information   Solid-state devices are used in so many different applications, and are found in so many different forms, that the science of solid-state technology can only be touched upon here. For detailed information, please consult the appropriate article according to the device, circuit, or subject. A textbook on solid-state technology is recommended for a complete and thorough discussion of this science.

SOS

See DISTRESS SIGNAL.

SOUND BAR

When an amplitude-modulation or single-sideband signal interferes with a television signal, *sound bars* can occur. Sound bars appear as horizontal bands of light and dark across the picture when a television signal is received. *See also* ELECTROMAGNETIC INTERFERENCE.

SOURCE

The term *source* is used to refer to the emitting electrode of a field-effect transistor. The output of a field-effect-transistor amplifier is sometimes taken from the source. The input signal can be applied to the source. The source of a field-effect transistor corresponds to the cathode of a tube or the emitter of a transistor (*see* FIELD-EFFECT TRANSISTOR).

The originating generator of a signal is called the *source* of the signal. You might, for example, speak of a radio-frequency source, a sound source, or a light source.

SOURCE COUPLING

When the output of a field-effect-transistor amplifier or oscillator is taken from the source circuit, or when the input to a field-effect-transistor amplifier is applied in series with the source, the amplifier or oscillator is said to use *source coupling*. Source coupling can be capacitive, or it can make use of transformers.

Source coupling generally results in a low input or output impedance, depending on whether the coupling is in the input or output of the amplifier stage. Source coupling is often used with grounded-gate amplifiers. If the following stage needs a low driving impedance, source coupling can be used; this is called a *source follower*. *See also* SOURCE FOLLOWER.

SOURCE FOLLOWER

A *source follower* is an amplifier circuit in which the output is taken from the source circuit of a field-effect transistor. The output impedance of the source follower is low. The voltage gain is always less than 1; in other words, the output signal voltage is smaller than the input signal voltage. The source-follower circuit is generally used for impedance matching. The amplifier components are often less expensive, and offer greater bandwidth, than transformers. Source-follower circuits are useful because they offer wideband impedance matching at low cost. *See also* SOURCE COUPLING.

SOURCE KEYING

In a field-effect-transistor oscillator or amplifier, *source keying* is accomplished by placing the key in series with the source. This is done for the purpose of obtaining code transmission. Source keying is analogous to emitter keying in a bipolar-transistor circuit (*see* EMITTER KEYING).

In a source-keyed oscillator, the key-up condition cuts off the power and stops the oscillation. In a source-keyed amplifier, only a negligible amount of signal leaks through when the key is up.

Source keying is the most common method of keying a field-effect-transistor oscillator or amplifier. Source keying is always done either in the low-level amplifier stages of a transmitter, or in the local-oscillator circuit. Source keying is sometimes done at more than one stage simultaneously.

A shaping circuit slows down the source voltage drop when the key is closed. This allows a controlled signal rise time. When the key is released, the capacitor discharges through the resistor, slowing down the decay time of the signal. The resistor and capacitor values are chosen for the desired rise and decay times in the range of keying speeds to be used. *See also* KEY CLICK, and SHAPING.

SOURCE MODULATION

Source modulation is a method of obtaining amplitude modulation in a radio-frequency amplifier. The carrier is applied to the gate of the device in most cases. Sometimes the carrier is applied to the source. The audio signal is applied in series with the source; this can be done across a resistor, or by means of an audio-frequency transformer. Source modulation is the counterpart of emitter modulation in a bipolar-transistor amplifier (*see* EMITTER MODULATION).

As the audio-frequency signal at the source swings negative, the instantaneous radio-frequency output voltage rises if an N-channel field-effect transistor is used. If a P-channel device is used, the instantaneous signal output voltage increases when the audio signal swings positive. Under conditions of 100-percent modulation, the instantaneous amplitude just drops to zero at the negative radio-frequency signal peaks.

Source modulation requires very little audio power for 100-percent amplitude modulation, provided that the modulation is done at a low-level stage and not in the final amplifier circuit of the transmitter. *See also* AMPLITUDE MODULATION.

SOURCE RESISTANCE

The *source resistance* of a field-effect transistor is the effective resistance of the source in a given circuit. The source resistance depends on the bias voltages at the gate and the drain. It also depends on the input-signal level and on the characteristics of the particular field-effect transistor used. The source resistance can be controlled, to a certain extent, by inserting a resistor in series with the source lead.

The external resistor in the source circuit of a field-effect-transistor oscillator or amplifier is usually called the *source resistor*, although its value is sometimes called the *source resistance*. A source resistor can be used for impedance-matching purposes, for biasing, for current limiting, or for stabilization. The source resistor often has a capacitor connected across it to bypass radio-frequency energy to ground. *See also* FIELD-EFFECT TRANSISTOR, SOURCE FOLLOWER, SOURCE STABILIZATION, and SOURCE VOLTAGE.

SOURCE STABILIZATION

Source stabilization is a method of providing bias control in a power-FET circuit using two or more FETs in parallel. A resistor, having a value that depends on the circuit application and input impedance, is connected in series with the source of each FET. This reduces the effects of minor differences in the characteristics of the FETs. Bypass capacitors can be connected across each resistor if it is necessary to keep the sources at signal ground.

If the channel current increases because of a temperature rise in an FET, and there is no resistor in series with the source lead, the change in current will reduce the gain of the amplifier, and thereby reduce its efficiency. If a source-stabilization resistor is connected in series with the current path, however, any increase in the current will cause an increase in the voltage drop across the resistor. This will change the bias in such a way as to stabilize the current through the channel.

Stabilization is more important in bipolar-transistor circuits than in FET circuits. In fact, stabilization is used almost universally in bipolar-transistor amplifiers and oscillators, while in the case of the FET, stabilization is usually necessary only in power amplifiers. *See also* COMMON CATHODE/EMITTER/SOURCE, and EMITTER STABILIZATION.

SOURCE VOLTAGE

In a field-effect-transistor circuit, the *source voltage* is the direct-current potential difference between the source and ground. If the source is connected directly to chassis ground, or is grounded through an inductor, then the source voltage is zero. But if a series resistor is used for stabilization, as is the case in some common-source power amplifiers, the source voltage is positive in an N-channel device and negative in a P-channel device.

If *I* is the channel current under no-signal conditions and *R* is the value of the source resistor (assuming there is a source resistor), then the source voltage *V* is obtained by Ohm's law as:

$$V = IR$$

A capacitor, placed across the source resistor, keeps the source voltage constant under conditions of variable input signal. If no such capacitor is used, the instantaneous source voltage will vary along with the input signal because of changes in the instantaneous current through the channel. *See* SOURCE STABILIZATION.

In common-gate and common-drain circuits, it is usually necessary to provide a voltage at the source. In a common-gate amplifier, the source voltage is normally positive in the N-channel case and negative in the P-channel case. In a common-drain configuration, the source voltage must be negative if an N-channel device is used, and positive if a P-channel device is used. In these situations, the source voltage is provided directly by the power supply. *See also* COMMON CATHODE/EMITTER/SOURCE, COMMON GRID/BASE/GATE, COMMON PLATE/COLLECTOR/DRAIN, and FIELD-EFFECT TRANSISTOR.

SO-239

See UHF CONNECTOR.

SPACE

See MARK/SPACE.

SPACE COMMUNICATIONS

Years ago, all long-distance radio communications took place via ionospheric or tropospheric propagation. If conditions were unfavorable (and they often were), a communications circuit became unreliable, and perhaps even unusable (*see* PROPAGATION CHARACTERISTICS). We no longer have this problem because of the advent of space communications.

Space communications is, in general, the transmission and reception of signals over paths that lie partially or entirely above the atmosphere of our planet. Space communications

can be categorized in three ways: earth-to-earth, earth-to-space, and space-to-space.

Earth-to-Earth Communications   Earth-to-space communications has actually been possible since the early days of radio. Radio amateurs bounced signals off the moon when it was discovered that radio waves at very high frequencies would travel through the ionosphere unaffected. Some radio amateurs still converse this way today (*see* MOONBOUNCE).

Since the first satellite was launched into space in the late 1950s, satellite communications has become increasingly widespread. The earliest artificial communications satellites were passive devices, launched in the 1960s and known as the Echo satellites. Since then, active repeaters have been placed aboard satellites. Today, satellites provide constant communications capability among almost all points on the globe (*see* ACTIVE COMMUNICATIONS SATELLITE).

Earth-to-Space and Space-to-Space Communications   As we make our first tentative ventures to the other planets—and perhaps someday even to other stars or galaxies—earth-to-space and space-to-space communications will become more and more important. There will be certain difficulties to overcome.

The first, and most obvious, difficulty arises from the sheer distances involved. The planet Pluto is about 4,000,000,000 miles (or 6,000,000,000 kilometers) from the earth. The nearest star is over 6,000 times further away than Pluto! High-power transmitters and extremely directional antenna systems will be necessary to carry out communications over such great distances.

When the Apollo astronauts visited the moon, their conversations with earth-based personnel were complicated by the fact that the earth and the moon are separated by about 1.34 light seconds. This caused a delay after every transmission before a reply was received. This 2.68-second round-trip propagation delay might have been bothersome, but it did not seriously affect communications. But when astronauts go farther into space, two-way conversations will become difficult, and ultimately impossible because the speed of radio signals is only 186,282 miles (299,792 kilometers) per second. This problem has already been experienced, in a sense, by technicians controlling the distant space probes. Pluto is about 6 light hours away; the nearest star is over 4 light years from us.

If space vessels achieve very high speeds, Doppler shifts will cause changes not only in the frequency of transmission, but also in the rate at which the modulation is received. At speeds approaching the speed of light, relativistic effects will also become significant. The frequency and data-transmission rate can be altered so greatly that special equipment is needed to receive and demodulate the signals.

The Future of Space Communications   Some scientists believe that civilizations might exist on planets in orbit around other stars, and that some of these civilizations might attempt to transmit radio signals into space. Attempts have already been made to receive such signals. Dr. Frank Drake has even made a transmission of his own (*see* RADIO ASTRONOMY, and RADIO TELESCOPE.)

Satellite-communications technology can be expected to improve continually in the coming years. Someday it will be possible for two people on opposite sides of the world to converse via satellite, using handheld radios, at any time of the day or night.

There might be as-yet undiscovered modes of communication that will prove useful for space travelers. Some of these modes might eliminate the time-lag, Doppler, and relativistic problems associated with long-distance communications.

SPACE OPERATION

In space communications, *space operation* refers to the operation of equipment not on the surface of the earth. Satellite transponders are the most common example of equipment in space operation.

Hams have operated from on board the Space Shuttle. This is a form of space operation in which an operator is actually in attendance. *See also* ACTIVE COMMUNICATIONS SATELLITE.

SPACER

A *spacer* is an insulator that is used for keeping the conductors of a transmission line separated. Spacers are used in air-dielectric coaxial and open-wire lines to maintain the proper conductor separation. In a coaxial line, spacers are shaped like disks or flattened spheres. In an open-wire line, the spacers are rod-shaped.

Spacers are positioned at regular intervals along a feed line. The spacers have a small effect on the dielectric constant of the medium between the conductors; this tends to lower the characteristic impedance of the line by a few percent. The more spacers that are used, the greater this effect becomes.

Spacers should be made of a low-loss, waterproof dielectric, such as plastic or glass. When making an open-wire transmission line, the smallest possible number of spacers should be used.

SPADE LUG

A *spade lug* is a simple connector that is sometimes used in conjunction with binding posts. The spade lug eliminates the problem of wire slippage when binding posts are used. The spade lug also improves the quality of the connection, because the metal-to-metal surface area is increased.

Spade lugs can be purchased at most electronic stores in various sizes. The spade lug is simply soldered or crimped onto the end of the wire. *See also* BINDING POST.

SPAGHETTI

In the construction of electronic equipment, it is sometimes necessary to insulate component leads to prevent a possible short circuit with other wiring. This is often the case in point-to-point wiring when a component must be mounted between two tie strips located a considerable distance apart. Plastic tubing, called *spaghetti*, is used to insulate component leads.

Spaghetti is available in many different sizes for different gauges of wire. In circuit wiring, the smallest size of spaghetti should be used that will fit easily over the component lead. The spaghetti is simply cut to the necessary length with scissors.

SPARK

When the voltage gets very large between two points that are separated by an air dielectric, a momentary discharge, or arc, will occur through the air. This discharge is called a *spark*. If a spark continues for more than a brief instant, or through some medium other than the air, it is usually called an *arc* (see ARC). The exact voltage that will cause a spark depends on the shapes of the charged objects, the distance between them, and amount of impurities (such as water vapor, dust, and pollutants in the air).

If a spark carries a great amount of electric charge, flammable materials in the path of the spark can be ignited. Some materials, such as gunpowder or gasoline, will explode if a spark passes through them. Internal combustion engines use gasoline explosions, ignited by sparks, to generate mechanical energy. Lightning causes many forest fires each year, in addition to numerous fires in homes and businesses (see LIGHTNING). If lightning strikes a tank containing a flammable liquid, the result can be disastrous.

SPEAKER

A *speaker*, also known as a *loudspeaker*, is a form of electroacoustic transducer. The speaker converts alternating electric currents into sound waves having the same frequency characteristics and the same waveforms, within the range of human hearing (approximately 16 Hz to 20 kHz).

A coil of wire, called the *voice coil*, is held in the field of a permanent magnet. When alternating currents flow in the coil, a fluctuating magnetic field is produced around the coil. This field interacts with the field from the permanent magnet to produce back-and-forth forces on the coil. The coil is mounted so that it can move as these forces occur. The coil is physically attached to a cone-shaped diaphragm, which vibrates along with the coil. The moving diaphragm produces sound waves in the air.

Electrostatic speakers operate via electric fields, rather than magnetic fields. Some speakers use electromagnets, rather than permanent magnets, to produce the stationary magnetic field. In public-address systems and some high-fidelity midrange and high-frequency applications, horn-shaped speakers are used. Various other forms of transducers are used for SONAR and other acoustic devices. *See also* TRANSDUCER.

SPECIFICATIONS

The *specifications* of an electronic device are the operating characteristics, expressed in tabular form. Specifications give information of importance to operators of the equipment.

In a radio receiver, important specifications include the sensitivity, selectivity, tuning range, emission modes that can be received, and frequency stability. In a transmitter, the specifications might include such factors as the frequency range, emission types, frequency stability, and radio-frequency power output. The table is a list of specifications for a commercially manufactured FM transceiver. For test instruments, the degree of accuracy is the most important specification. For power supplies, the voltage output, regulation, and current-delivering capability are important specifications. These examples are, however, only representative of the many different specifications that can be expressed for a particular piece of electronic equipment.

SPECIFIC RESISTANCE

The resistance of an electrical conductor is sometimes expressed in terms of a factor called the *specific resistance*. Specific resistance is given in ohms per foot per circular mil.

SPECIFICATIONS: SPECIFICATIONS FOR A TYPICAL
(BUT HYPOTHETICAL) VHF FM TRANSCEIVER.

General

Frequency range	144-148 MHz
Display type	LED
Frequency control	Microcomputer PLL VCO
Emission type	F3
Memory channels	8
Temperature tolerance range	−20 to +60 degrees C
Power-supply requirements	12 to 15 V dc, 5 A
Semiconductors	17 IC, 20 FET, 29 Tr, 59 Di
Dimensions	HWD 2.5 × 6 × 9 in
	(64 × 152 × 229 mm)
Weight	3 lbs (1.4 kg)

Receiver section

Intermediate frequencies	17.0 MHz, 455 kHz
Sensitivity	Better than 0.35 μV for 20−dB
	noise quieting
Selectivity	Plus/minus 5 kHz at −6 dB
	Plus/minus 15 kHz at −60 dB
Audio output	2 watts or more
Speaker requirements	Impedance 4-8 ohms

Transmitter section

RF output	15 watts
Frequency deviation	Plus/minus 5 kHz
Spurious radiation	Less than −60 dB
Antenna requirements	50 ohms resistive, SWR < 2:1
Microphone requirements	Impedance 300−600 ohms

The specific resistance of a conductor depends on the material and the temperature. The specific resistance of a material is generally expressed for direct current. *See also* RESISTANCE.

SPECTRAL DENSITY

A radio-frequency signal or other electromagnetic emission often contains more than one wavelength component. For example, an amplitude-modulated signal contains not only the carrier frequency, but sideband frequencies as well. *Spectral density* is an expression of the concentration of energy in terms of frequency.

In general, for a given signal power, the spectral density increases as the bandwidth of the signal gets smaller. The maximum possible spectral density exists when a signal is a steady carrier (emission type A0 or F0). The faster the rate data transmission, the lower the spectral density. Also, the spectral density decreases as the efficiency of data transmission decreases.

Spectral density is given in terms of the average number of watts per kilohertz of spectrum space at the fundamental frequency of the signal. Spectral density can be expressed in relative terms, as the percentage of the total signal power per kilohertz of spectrum space at the fundamental frequency.

SPECTRAL PURITY

Any transmitted signal should contain energy at frequencies where it is needed, but not at other frequencies. An ideal single-sideband (SSB) signal has all of its energy within a band of frequencies 3 kHz wide, and no energy at any other frequency. An ideal code (CW) signal has its energy concentrated within a few hertz of spectrum space. In practice, ideals are never fully realized. But it is good engineering practice to get a transmitter to work as close to the ideal as possible, within limits imposed by technology.

The extent to which a signal conforms to the ideal distribution of frequencies is called *spectral purity*. This factor can be scrutinized by looking at narrowband and wideband *spectrum analyzer* displays (*see* SPECTRUM ANALYZER).

The spectral purity of a transmitter output can be degraded by any of the following causes, represented by articles in this book: FLAT TOPPING, HARMONIC, KEY CLICK, NONLINEARITY, PARASITIC OSCILLATION, SPLATTER, and SPURIOUS EMISSIONS.

SPECTRUM ANALYZER

A *spectrum analyzer* is an electronic test instrument that graphically displays the spectral energy distribution of radio-frequency signal generator. Engineers use spectrum analyzers extensively in the design, alignment, and troubleshooting of radio transmitters and receivers.

A typical spectrum analyzer consists of an oscilloscope and a circuit that provides the spectral display for the cathode-ray tube. The photograph is of a typical spectrum analyzer, suitable for work with equipment at frequencies up to 1 GHz. The frequency is displayed horizontally from left to right. Signal amplitude is displayed vertically. In the spectral display shown, there are four signals on the screen, in addition to the local-oscillator signal at the extreme left (the second signal is barely visible above the noise floor).

Some spectrum analyzers are designed for the audio frequency range. These include narrowband analyzers of fixed bandwidth, such as 5 Hz, and devices having a bandwidth that is a constant percentage of the frequency of the signal being evaluated. Examples of the latter are octave, half-octave, and 1/3-octave analyzers. Real-time analyzers (RTAs) commonly have visual displays of response in 1/3-octave bands covering the audio range.

To operate the spectrum analyzer, a technician first chooses, by means of a selector switch, the band pass of frequencies to be displayed. This band can cover the whole electromagnetic spectrum from direct current to 1 GHz or more, or it can cover any desired part of this range. The gain (vertical-scale sensitivity) of the circuit is adjusted as desired. The resolution and sweep rate must also be adjusted for the application intended.

The spectrum analyzer is useful for determining the spurious-signal and harmonic content of the output of a radio transmitter. In the United States, radio equipment must meet certain government-imposed standards regarding spectral purity. The spectrum analyzer gives an immediate indication of whether or not a transmitter is functioning properly. The spectrum analyzer can also be used to observe the bandwidth of a modulated signal. Such improper operating conditions as splatter or overmodulation can be easily seen as excessive bandwidth.

Spectrum analyzers are often used in conjunction with sweep generators, for evaluating the characteristics of bandpass, band-rejection, highpass, or lowpass filters. You might, for example, need to adjust a helical filter to obtain the desired bandpass at the front end of a radio receiver. You might wish to determine whether a lowpass filter is providing the right cutoff frequency and attenuation characteristics. The sweep generator produces a radio-frequency signal that varies in frequency, exactly in synchronization with the display of the spectrum analyzer (*see* SWEEP-FREQUENCY FILTER OSCILLATOR, and SWEEP GENERATOR).

Some radio receivers are equipped with narrow-band spectrum analyzers for monitoring an entire communications band at once. Such a receiver is called a *panoramic receiver. See also* PANORAMIC RECEIVER.

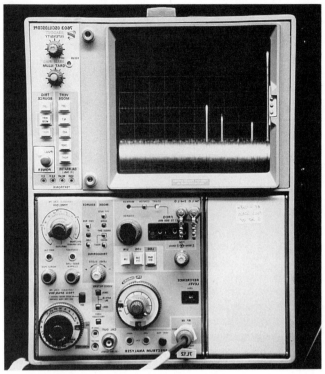

SPECTRUM ANALYZER: Displays amplitude versus frequency.

SPECTRUM MONITOR

A *spectrum monitor* is a narrow-band spectrum analyzer that can be connected in the intermediate-frequency chain of a receiver to observe signals at or near the operating frequency. The spectrum monitor converts any ordinary superheterodyne receiver into a panoramic receiver.

The spectrum-monitor display is contered at the intermediate frequency of the receiver. Signals appear as pips to the left or right of the center of a cathode-ray-tube screen. The bandwidth is adjustable. This makes it possible to observe an entire communications band, such as 7.000 to 7.300 MHz, at once; or the operator can choose to "zero in" on one signal and observe its modulation characteristics. *See also* PANORAMIC RECEIVER, and SPECTRUM ANALYZER.

SPEECH CLIPPING

Speech clipping is a method of increasing the average power of a voice signal without increasing the peak power. This can be done either in the audio-frequency microphone-amplifier stages of a transmitter, or in the intermediate-frequency or radio-frequency (RF) stages. The latter method is usually called *RF clipping* (see RF CLIPPING).

To accomplish audio-frequency speech clipping, a voice signal is first amplified. Then, the signal is passed through a limiting device, which has a cutoff voltage substantially below the peak voltage of the amplified voice signal. The output of the limiter is then amplified so that the clipped peak voltage is the same as the unclipped peak voltage ahead of the limiter.

Speech clipping, if done properly, can improve the intelligibility of an amplitude-modulated or single-sideband signal. If done improperly, however, speech clipping can cause severe distortion, and actually reduce the intelligibility as well as cause objectionable splatter (*see* SPLATTER). Audio-frequency speech clipping can be accomplished only to a certain extent before distortion occurs. If the clipping threshold is made too low, or the gain of the first amplifier is made too high, the resulting distortion will partially or totally offset the effects of the average-power increase.

Whenever speech clipping is used in a transmitter, a bandpass filter, having steep skirts and high ultimate attenuation, is an absolute necessity. A spectrum analyzer or panoramic receiver should always be used to check the output of a transmitter in which a speech clipper is used.

Because speech clipping increases the average power in a transmitter, the final amplifier is forced to run at a higher duty cycle than would be the case without speech clipping. This can place excessive strain on the transistors or tubes in that circuit. Before any type of speech clipper is used in a transmitter, therefore, it is a good idea to check the ratings of the final amplifier to avoid possible damage.

Speech clipping is not the only method that can be used to increase the ratio of average power to peak power in a voice transmitter; speech compression is also used for this purpose. *See also* SPEECH COMPRESSION.

SPEECH COMPRESSION

Speech compression is a method of increasing the average power in a voice signal, without increasing the peak power. Speech compression is used in many amplitude-modulated and single-sideband communications transmitters.

The speech compression circuit operates in the same way as an automatic-level-control (ALC) circuit (*see* AUTOMATIC LEVEL CONTROL). But speech compression carries the process farther than ordinary ALC. Although ALC is typically used only to prevent overmodulation, speech compression uses additional amplification of the low-level components of a voice. The result is greatly reduced amplitude range, but the intelligibility of the signal is often considerably improved because more effective use is made of the modulating-voice signal. A speech-compression circuit is sometimes called an *amplified-ALC (AALC) circuit.* The illustration is a block diagram of an audio speech-compression circuit.

Speech compression is usually done in the audio (microphone-amplifier) circuits of a transmitter. But it can be done in the intermediate-frequency or radio-frequency (RF) stages. Accordingly, you can speak of either audio speech compression or RF speech compression. Radio-frequency speech compression is often called *envelope compression.*

Speech compression, like speech clipping, can cause problems if it is not done correctly. There is a certain maximum increase in the ratio of average power to peak power that can be realized without objectionable envelope distortion. Too much speech compression will actually degrade the intelligibility of a voice signal.

When speech compression is used in a transmitter, the transmitter output should be monitored with a spectrum analyzer or panoramic receiver to ascertain that splatter is not taking place (*see* SPLATTER). Also, precautions should be taken to ensure that the average-power increase will not put too much strain on the final amplifier. *See also* SPEECH CLIPPING.

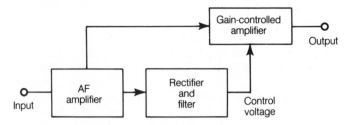

SPEECH COMPRESSION: An audio compression circuit.

SPEECH PROCESSING
See SPEECH CLIPPING, and SPEECH COMPRESSION.

SPEED OF TRANSMISSION

The speed at which digital data is transmitted can be expressed in a variety of different ways. The most common methods are the *baud rate* and *the word-per-minute (WPM) rate*.

A baud consists of one pulse or element of a digital signal. The number of bauds sent in 1 second is called the *baud rate*. A data word is usually considered to consist of six characters, including spaces if present. Thus, the speed in words per minute is equal to the number of characters and spaces in $\frac{1}{6}$ minute (10 seconds). Both the baud rate and the WPM rate are determined by averaging over a period of time.

The ratio of the baud rate to the WPM rate is not exactly the same for all types of signals. Some codes are a little more efficient, and require fewer bauds per word, than other codes. In ASCII code, the baud rate is the same as the WPM rate. In BAUDOT code, the baud rate is approximately 75 percent of the WPM rate. In the International Morse code, the baud rate is approximately 83 percent of the WPM rate. *See also* ASCII, BAUDOT, BAUD RATE, INTERNATIONAL MORSE CODE, and WORDS PER MINUTE.

SPIDERWEB ANTENNA

A *spiderweb antenna* is a broadband form of antenna, consisting of several dipoles connected in parallel at a common feed point. The dipoles have varying lengths, and all lie in a single horizontal plane. The dipoles run radially outward from the feed point.

The bandwidth of a spiderweb antenna depends on the lengths of the shortest and longest dipoles. If the longest dipole measures x feet in length, and the shortest dipole measures y feet, then the lower and upper limit frequencies f and g, in megahertz, are approximately:

$$f = 468/x$$
$$g = 468/y$$

If x and y are given in meters, then:

$$f = 143/x$$
$$g = 143/y$$

Spiderweb antennas can be built to operate at any frequency. In the shortwave bands, such antennas are sometimes constructed from wire, using multiple supports in a radial pattern around a central support. The ends of the wires can also be run downward at a slant, terminating at ground stakes, and using insulators to provide the proper resonant lengths. At very high frequencies, spiderweb antennas are made from aluminum tubing for broadband television reception. *See also* DIPOLE ANTENNA.

SPIN MODULATION

When communicating through a Phase III amateur satellite, the downlink signal fades slightly three times per second. This is more noticeable on a linearly polarized antenna than on a circularly polarized one.

In order to stabilize the satellite in orbit, it is made to tumble at one revolution per second. There are three downlink antennas on the satellite, and their beams sweep around as the satellite tumbles. Therefore, three beams sweep past any given earth location every second. The resulting 3-Hz fading is called *spin modulation*.

Normally, spin modulation is not severe enough to have a degrading effect on signals via satellite. It is best to use a circularly polarized antenna system when working through a satellite, because it minimizes the effects of spin modulation, as well as polarization changes arising from other causes. *See also* ACTIVE COMMUNICATIONS SATELLITE, OSCAR, and PHASE III SATELLITE.

SPLATTER

Splatter is a term that is used to describe the effects of a voice signal having too much bandwidth. Splatter can occur with amplitude-modulation, frequency-modulation, phase-modulation, or single-sideband emission types.

When the amplitude, frequency, or phase of a signal is varied, sidebands are produced (*see* SIDEBAND). The more rapid the instantaneous change in amplitude, frequency, or phase becomes, the farther the sidebands occur from the carrier frequency. If modulation is excessive, peak clipping will occur in amplitude modulation or single sideband, and over-deviation will occur in frequency modulation and phase modulation. This causes sidebands to be generated at frequencies too far removed from the carrier.

In a receiver, splatter sounds like a crackling noise at frequencies up to several hundred kilohertz from the carrier frequency of the transmitter. Splatter can cause serious interference to numerous communications circuits.

In an amplitude-modulated or single-sideband transmitter, splatter can be eliminated by:

1. Avoiding overmodulation
2. Ensuring that the audio-amplifier circuits are operating with a minimum of distortion
3. Ensuring proper bias and drive for transmitter amplifiers, especially the final amplifier
4. Ensuring that any external linear amplifiers are operating in a linear manner
5. Proper operation of speech-processing circuits and automatic level control

In a frequency-modulated or phase-modulated transmitter, splatter can be avoided by:

1. Keeping the deviation within rated system limits
2. Ensuring that the audio-amplifier stages are operating with minimum distortion

A form of splatter is sometimes observed in continuous-wave (code) transmissions. If the rise and/or decay times are too rapid, clicks can be heard at frequencies far removed from the carrier frequency. These sidebands are called *key clicks*. *See also* KEY CLICK, and SHAPING.

SPONGE LEAD

Sponge lead is a black or gray, lead-saturated sponge material that is used for storing metal-oxide-semiconductor (MOS) components. Sponge lead is conductive, and this reduces the chances for static charges to build up in a MOS component during storage.

Many integrated circuits and transistors use MOS technology. These components are extremely susceptible to destruction by small static charges. The leads of MOS components should always be inserted into sponge lead when such devices must be stored for any period of time. Sponge lead is available at most electronics stores. *See also* METAL-OXIDE SEMICONDUCTOR.

SPORADIC-E PROPAGATION

At certain radio frequencies, the ionospheric E layer occasionally returns signals to earth. This kind of propagation tends to be intermittent, and conditions can change rapidly. For this reason, it is known as *sporadic-E propagation.*

Sporadic-E propagation is most likely to occur at frequencies between approximately 20 and 150 MHz. Rarely, it is observed at frequencies as high as 200 MHz. The propagation range is on the order of several hundred miles, but occasionally communication is observed over distances of 1,000 to 1,500 miles.

The standard frequency-modulation broadcast band is sometimes affected by sporadic-E propagation. The same is true of the lower television channels, especially channel 2. Sporadic-E propagation often occurs on the amateur bands at 21 MHz through 148 MHz.

Sporadic-E propagation is sometimes thought to exist when, in fact, tropospheric propagation is taking place. The confusion can also occur in the opposite sense. *See also* E LAYER, IONOSPHERE, PROPAGATION CHARACTERISTICS, and TROPOSPHERIC PROPAGATION.

SPREAD SPECTRUM

Until recently, efforts have been made to minimize the bandwidth occupied by a transmitted signal. This is because the *signal-to-noise ratio* improves, all other things being equal, as the bandwidth of a signal is reduced.

Part of the reason single sideband (SSB) works better than plain amplitude modulation (AM) is that the SSB signal takes up half as much spectrum as the AM signal, while conveying the same information. The receiver can use a filter with half the bandwidth, letting in only half the noise, while still admitting all the signal intelligence at the same speed. *See* AMPLITUDE MODULATION, SINGLE SIDEBAND.

The signal frequency need not be constant. It can be rapidly varied at the transmitter, and the receiver can follow along. This dilutes the energy over a wide band of frequencies, but the receiver still "thinks" it is picking up a narrowband signal. This is known as *spread spectrum* because the signal energy is spread out over a portion of electromagnetic spectrum.

Advantages of Spread Spectrum Spread spectrum probably arose because of a desire to encode signals, so only authorized personnel could receive them. If a signal frequency is varied according to a complicated function, it can be received if, but only if, the receiver "knows" that function and synchro-

nizes its frequency with that of the transmitter. This is one advantage of spread spectrum.

Another advantage is that catastrophic QRM (total inferference), such as occurs when someone tunes up on top of a conversation (QSO) in progress, is almost impossible. The QSO frequency is changing all the time. It might fluctuate all over a band, a range of several megahertz or tens of megahertz.

As a band becomes occupied with more and more spread-spectrum signals, the only degradation is a gradual increase in the noise level. Ultimately, there is a limit to the number of spread-spectrum QSOs that a band can handle. This limit is about the same as it would be if all the signals were constant in frequency. But the probability of one QSO being wiped out, even momentarily, by another is practically zero.

Limitations of Spread Spectrum Spread spectrum requires specialized equipment that can maintain exact alignment. This is expensive, or else requires that the ham be willing to do some technical work.

Random QSOing on a band, where someone calls CQ and waits for an answer, requires that someone else be on the same frequency at the same time. With spread spectrum, this would not be likely. The time might come when certain frequency-vs-time functions are standard for spread spectrum, so that one might listen on a band while switching a terminal unit among any of, say, 200 different function modes. But that would not be any better than a band divided up into 200 fixed channels. Spread spectrum is intended mainly for those who wish to maintain QSO privacy.

At the time of this writing, spread spectrum is legal only above 420 MHz, and with no more than 100 W PEP transmitter output. Users of this mode must make identification so an ordinary receiver can pick it up in the band being used. Rules require that logs be kept for all spread-spectrum transmissions.

Obtaining Spread Spectrum The fundamental signal for spread spectrum—that is, the mode of the signal whose frequency is made to fluctuate—should concentrate the modulation energy into the narrowest possible band. That is, SSB is preferable to AM or FM for this signal, and digital modes are better still.

A common way of getting the frequency to change, thereby spreading the spectrum, is known as *frequency hopping.* In this scheme, the transmitter has a list of channels that it follows in a certain order. The receiver must be programmed with this same list, in the same order, and must be synchronized with the transmitter. The list is a pseudorandom sequence that repeats over and over. An example might be 420.4, 420.2, 420.7, 420.1, 420.8, 420.5, 420.9, 420.6, 421.0 and 420.3 MHz, with each frequency being sent for 0.005 second, and the whole sequence repeated again and again by the transmitter, synchronized with a clock that has been initialized against some primary standard (such as WWV/WWVH).

The *dwell time* is the interval at which the frequency changes occur. It should be short enough so that a signal will not be noticed, and not cause interference, on any single frequency. In the foregoing example it is 0.005 second, or 5 milliseconds.

In the above example, there are 10 frequencies. Therefore, the energy is diluted on any one frequency by a factor of 10, as compared with the way it would be if its frequency were constant. This is only 10 dB. In practice, there should be many more frequencies, so that the energy is diluted much more, to the point that, if someone tunes to any frequency in the sequence, the signal will not be noticeable. This requires hundreds, or even thousands, of frequencies in sequence.

Another method of getting spread spectrum is to modulate the fundamental signal at a slow rate (20 Hz, for example) with a sine wave that guides it up and down over the range of the band. The receiver will be able to pick the signal up only if it follows at the same frequency, and in the same phase. The frequency of this sine wave might itself be gradually increased and decreased. The waveform might be some distorted shape, such as a ramp, or the output of a full-wave rectifier, or something completely "off-the-wall." This can lend various dimensions to the frequency-vs-time function that is used. Almost anything will work, as long as the receiver is programmed to follow right along with the transmitter.

Details of all the schemes for getting spread spectrum is beyond the scope of this article. Books are available on the subject. Technical and professional journals are a good source of up-to-date information on this rapidly advancing technique. A good university library is the best source of information for those interested in designing a station to work spread spectrum.

SPURIOUS EMISSIONS

Spurious emissions, also known as *spurious radiation* or *spurious signals*, are emissions from a radio transmitter that occur at frequencies other than the desired frequency. Harmonic emissions are a particular example of spurious radiation. Harmonics are signals that occur at integral multiples of the fundamental frequency of a radio transmitter (*see* HARMONIC, and HARMONIC SUPPRESSION).

Spurious emissions sometimes occur as a result of parasitic oscillation (*see* PARASITIC OSCILLATION, and PARASITIC SUPPRESSOR). Spurious emissions can also result from inadequate selectivity in the output stages of a transmitter.

In the United States, the maximum allowable level of spurious emissions is dictated by the Federal Communications Commission. This protects the various radio services from interference that could be caused by improperly operating radio transmitters. A transmitter can be checked for spurious emissions by means of a device called a *spectrum analyzer*. *See also* SPECTRUM ANALYZER.

SPURIOUS RESPONSES

A radio receiver sometimes picks up signals that exist at frequencies other than the frequency to which the receiver is tuned. Such false signals are known as *spurious responses.*

Spurious responses in a receiver can occur as a result of mixing between or among two or more external signals. Such mixing can occur in the front end of a receiver, especially if one signal is strong enough to cause nonlinear operation of this stage. Spurious signals can sometimes result from mixing in nonlinear junctions external to the receiver (*see* FRONT END, and INTERMODULATION).

Image signals in a superheterodyne receiver are an example of spurious responses. These signals can cause severe interference in an improperly designed receiver (*see* IMAGE FREQUENCY, and IMAGE REJECTION). A receiver can be checked for image responses by using a signal generator and for intermodulation distortion by using two signal generators.

SQUARE-LAW DETECTOR

A *square-law detector* is a form of envelope detector (*see* ENVELOPE DETECTOR). The square-law detector gets its name from the fact that the root-mean-square output-signal current is proportional to the square of the root-mean-square input-signal voltage.

The square-law detector operates not by rectification, but because of the nonlinear response of the circuit. This type of detector is sometimes called a *weak-signal detector* because there is no threshold level above which the signal strength must be in order to accomplish demodulation.

SQUARE-LAW METER

Some analog meters respond to an applied quantity according to the square of that quantity. This is the case, for example, in wattmeters that actually measure the voltage across a particular resistance. Such a meter is called a *square-law meter.*

The square-law meter has a characteristic nonlinear scale. The divisions become progressively closer together toward the right-hand end of the scale, and are widely spaced near the left-hand end. *See also* ANALOG METERING.

SQUARE-LAW RESPONSE

A circuit or device is said to have a *square-law response* when the output is proportional to the square of the input, or when the deflection of a meter is proportional to the square root of the input signal magnitude. A square-law curve is parabolic.

In general, if x represents the input quantity for a square-law circuit and y represents the output magnitude, then:

$$y = kx^2 + c$$

where k and c are constants. In the case of a square-law meter, if x represents the input-signal magnitude and y represents the meter-needle deflection, then:

$$x = ky^2 + c$$

where k and c, again, are constants. In both cases, the value of k depends on the units specified for input and output or scale deflection, and the value of c depends on the bias of the circuit or the starting point of the meter scale. *See also* SQUARE-LAW DETECTOR, and SQUARE-LAW METER.

SQUARE WAVE

A *square wave* is a special form of alternating-current or pulsating direct-current waveform. The amplitude transitions, both rise and decay, occur instantaneously. When displayed on an oscilloscope, a square wave appears as two parallel, dotted lines, sometimes with faint vertical traces connecting the ends of the line segments (see illustration at A on pg. 466).

Square waves can have equal positive and negative peaks, or the peaks might be unequal. Unequal peaks result from the combination of an alternating square-wave current and a direct current. If the magnitude of the direct-current component is exactly equal to the peak magnitude of the square-wave current, a train of pulses results as shown at B. If the direct-current amplitude exceeds the peak amplitude of the square wave, both peaks of the waveform have the same polarity, as at C.

The period of a square wave is the length of time from any point on the waveform to the same point on the next pulse in the train. This period is usually considered to begin at the instant during which the amplitude is increasing positively (B); sometimes some other point is considered to mark the beginning of the period (C). The period is divided into 360 degrees of phase. The frequency of a square wave, in hertz, is equal to the reciprocal of the period in seconds.

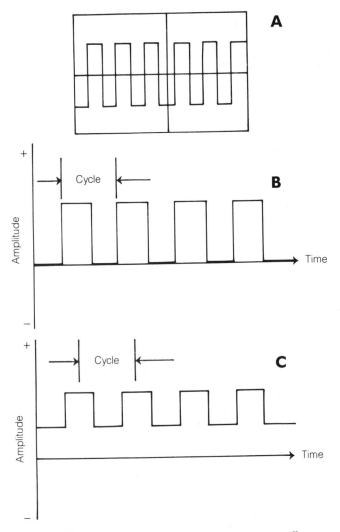

SQUARE WAVE: At A, a square wave as seen on an oscilloscope screen. At B and C, with dc components.

SQUELCH

When it is necessary to listen to a radio receiver for long periods, and signals are present infrequently, the constant hiss or roar becomes fatiguing to the ears. A *squelch circuit* is used to silence a receiver when no signal is present, while allowing reception of signals when they do appear.

Squelch circuits are most often used in channelized units, such as Citizen's-Band and very-high-frequency communications transceivers. Squelch circuits are less common in continuous-tuning radio receivers. Most frequency-modulation receivers use squelching systems.

A typical squelching circuit is shown in the diagram. This squelch circuit operates in the audio-frequency stages of a frequency-modulation receiver. The squelch is actuated by the incoming audio signal. When no signal is present, the rectified hiss produces a negative direct-current voltage, cutting off the field-effect transistor and keeping the noise from reaching the output. When a signal appears, the hiss level is greatly reduced, and the field-effect transistor conducts, allowing the audio to reach the output. The cut-off squelch circuit is said to be *closed*; the conducting squelch is said to be *open*.

In most receivers, the squelch is normally closed when no signal is present. A potentiometer facilitates adjustment of the squelch, so that it can be opened if desired, allowing the receiver hiss to reach the speaker. This control also provides for adjustment of the squelch sensitivity. Most squelch controls are open when the knob is turned fully counterclockwise. As the knob is rotated clockwise, the receiver hiss abruptly disappears; this point is called the *squelch threshold*. At this point, even the weakest signals will open the squelch. As the squelch knob is turned farther clockwise, stronger and stronger signals are required to open the squelch.

In some receivers, the squelch will not open unless the signal has certain characteristics. This is called *selective squelching*. It is used in some repeaters and receivers to prevent undesired signals from being heard. The most common methods of selective squelching use subaudible-tone generators or tone-burst generators. Selective squelching is also known as *private-line (PL) operation. See also* PRIVATE LINE, and SQUELCH SENSITIVITY.

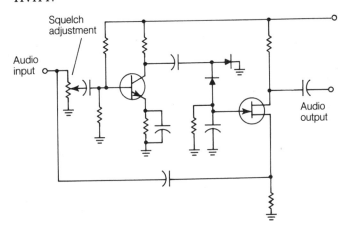

SQUELCH: An audio squelch circuit.

SQUELCH SENSITIVITY

The *squelch sensitivity* of a squelched receiver is the signal level, in microvolts at the antenna terminals, that is required to keep the squelch open (*see* SQUELCH). The squelch sensitivity of a receiver depends on several factors.

Virtually all squelch controls allow for some adjustment of the squelch sensitivity by means of a control knob. The squelch sensitivity is greatest (the least signal is required to open it) at the threshold setting of the control.

An unmodulated carrier will usually open a squelch at a lower level than will a modulated carrier. If a frequency-modulated signal has excessive deviation, the squelch can cut off on modulation peaks. This effect is known as *squelch blocking*.

The squelch sensitivity of a frequency-modulation receiver is related to its noise-quieting sensitivity. The better the quieting sensitivity, the less signal is required to actuate the squelch at the threshold.

Squelch sensitivity is measured using a calibrated signal generator. The signal is modulated at a frequency of 1 kHz, with a deviation of plus-or-minus 3 kHz. The generator output is increased until the squelch remains continuously open at the threshold setting. *See also* NOISE QUIETING.

STABILITY

Stability is an expression of how well a component, circuit, or system maintains constant operating conditions over a period of time. You can speak, for example, of the frequency stability of a radio receiver, the stability of the output power of a radio transmitter, or the voltage-output stability of a power supply. The stability of an electronic component is usually called *tolerance*, and the stability of a power supply is known as the *regulation* (see REGULATION, and TOLERANCE).

Stability can be expressed mathematically in absolute terms or as a percentage. Suppose the intended, or desired, value of a parameter is k units. Suppose that the actual value fluctuates between values of x units and y units, where $x < k < y$. Then, the absolute stability is represented by the larger of two values S_1 and S_2, such that:

$$S_1 = k - x$$
$$S_2 = y - k$$

Under the same circumstances, the percentage stability is represented by the larger of two values:

$$SP_1 = 100 \ (k - x)/k$$
$$SP_2 = 100 \ (y - k)/k$$

Alternatively, the stability percentage can be given by:

$$SP = 100 \ (y - x)/k$$

This denotes the maximum fluctuation in the parameter.

Stability is obviously a desirable trait in any component, circuit, or system. Receiver frequency stability should be on the order of a few hundred hertz over a period of hours following initial warmup. The same holds for radio transmitters. The output voltage of a power supply should not change significantly over a period of time, as long as the supply is operated within rated limits.

In some circuits, such as radio-frequency power amplifiers, stability is used to describe normal operating conditions. If the gain constantly changes, or if an amplifier breaks into oscillation, the circuit is said to be unstable. Good circuit design and proper operation are obviously important for maintaining normal function.

The stability of a device is affected by the environment in which it is operated. Temperature changes often cause a change in the characteristics of a component, circuit, or system. Mechanical vibration, the presence of external electromagnetic fields, and other factors can also affect stability. Circuit design, and the quality of the components used, are important. In recent years, electronic circuit stability has greatly improved because of advances in the state of the art.

STABISTOR

A *stabistor* is a form of semiconductor diode that has a specific forward-conduction (breakover) voltage. All ordinary diodes have this property; a germanium diode breaks over at about 0.3 V and a silicon diode at 0.6 V. The stabistor characteristically has a much higher breakover voltage.

If a forward voltage is applied to a stabistor, the device will not conduct until the breakover voltage is reached. At forward voltages greater than the breakover voltage, the stabistor exhibits a very low resistance. A stabistor behaves, in the forward-biased mode, very much like a Zener diode behaves in the reverse-biased mode. The stabistor is thus used in many of the same applications as the Zener diode. *See also* ZENER DIODE.

STACK

A *stack* is a computer memory that is used for the purpose of short-term data storage. In a stack-type memory, information must be entered and retrieved in a specific sequence, such as first-in/first-out or first-in/last-out. Random access is not possible.

A magnetic-core memory is often called a *stack. See also* CORE MEMORY, FIRST-IN/FIRST-OUT, MEMORY, PUSHDOWN STACK, SEQUENTIAL ACCESS MEMORY, and SHIFT REGISTER.

STACKING

Stacking is a method of combining two or more identical antennas for the purpose of obtaining directional characteristics, especially enhanced effective-power gain. Stacking can be done with any type of antenna, but it is most often done with dipoles, halos, or Yagis (see DIPOLE ANTENNA, HALO ANTENNA, and YAGI ANTENNA).

The usual stacking configuration consists of two, three, or four horizontally polarized antennas placed vertically above each other and spaced apart by at least ½ wavelength. The optimum spacing is between ¾ and 2 wavelengths. The individual antennas, if they are directional, each point in the same direction. The antennas are fed in phase. If two antennas are used, the power gain is about 3 dB over that of a single antenna. If three antennas are used, 4.8 dB gain can be realized; if four antennas are used, 6 dB gain is possible compared with the gain of a single antenna.

The above gain figures are theoretical. In practice, the gain will probably be slightly less. To optimize the gain, it is essential that the phasing harness be properly designed to provide the antennas with signals that are exactly in phase. The impedances must also be matched by means of quarter-wave feeder sections.

Vertical dipoles are sometimes stacked in collinear fashion to obtain gain in the horizontal plane. Other types of antennas can be stacked either vertically or horizontally in various ways.

STAGGER TUNING

Stagger tuning is a method of aligning a bandpass filter or the intermediate-frequency chain of a receiver or transmitter. In a stagger-tuned filter, there are several tuned circuits, each set for a slightly different frequency. In the stagger-tuned amplifier chain, each stage is tuned to a higher or lower frequency than the stage immediately before or after it.

In a bandpass filter, stagger tuning results in a more nearly rectangular response than would be obtained if all of the resonant circuits or devices were tuned to the same frequency. Stagger tuning broadens the bandpass response and provides steep skirts (see BANDPASS RESPONSE, RECTANGULAR RESPONSE, and SKIRT SELECTIVITY).

In a multistage, radio-frequency amplifier, stagger tuning reduces the possibility of interstage oscillation. If all of the tuned circuits were set for exactly the same resonant frequency, the positive feedback among the various amplifiers might be enough to cause instability (see FEEDBACK).

STANDARD ANTENNA

A *standard antenna* is an antenna used for comparison purposes in the determination of effective power gain. Such an antenna can be either a dipole or an isotropic radiator. *See* ANTENNA POWER GAIN, DIPOLE ANTENNA, and ISOTROPIC ANTENNA. An unbalanced, end-fed radiator having an overall length of exactly 13 feet (4 meters) is sometimes called a *standard antenna.*

STANDARD CAPACITOR

A *standard capacitor* is a capacitor that is manufactured to have a known, stable value. Standard capacitors are used in precision test equipment, particularly frequency meters and impedance bridges, to ensure the highest possible degree of accuracy. The tolerance of a standard capacitor is very low. The temperature coefficient is practically zero. *See also* ACCURACY, TEMPERATURE COEFFICIENT, and TOLERANCE.

STANDARD INDUCTOR

A *standard inductor* is an inductor that is manufactured to have a precise, known, and essentially unchanging value. Standard inductors are used in precision test equipment, especially frequency-measuring apparatus or impedance bridges. The tolerance is very low, and the temperature coefficient is essentially zero. *See also* ACCURACY, TEMPERATURE COEFFICIENT, and TOLERANCE.

STANDARD RESISTOR

A *standard resistor* is a noninductive, fixed-value resistor having a precise value that does not change significantly with time. Standard resistors are commonly used in precision electronic circuits, especially resistance-measuring bridges. The standard resistor has a low tolerance and a temperature coefficient of practically zero. *See also* ACCURACY, TEMPERATURE COEFFICIENT, and TOLERANCE.

STANDARD TEST MODULATION

In the evaluation of the distortion characteristics of radio receivers, a modulated signal must be used. Engineers and technicians use a signal generator that produces modulation at a certain frequency and extent. This is called *standard test modulation.*

For amplitude-modulation receivers, a test signal is 100-percent modulated by a pure sine-wave audio tone having a frequency of 1 kHz. For frequency-modulation and phase-modulation communications receivers, the carrier is modulated at a deviation of plus or minus 3 kHz by a pure sine-wave tone of 1 kHz. For single-sideband equipment, two sine-wave tones of equal amplitude, both within the passband (between 300 Hz and 3 kHz), and spaced 1 kHz apart, are used (a common combination is 1 kHz and 2 kHz). The audio output of the receiver is then analyzed with a distortion analyzer and/or oscilloscope. *See also* DISTORTION ANALYZER.

STANDARD TIME THROUGHOUT THE WORLD

The common worldwide time standard is based on the solar time at the Greenwich Meridian. This meridian is assigned 0 degrees of longitude and passes through the city of Greenwich, England. This common time is known as *Coordinated Universal Time,* and is abbreviated *UTC* (*see* COORDINATED UNIVERSAL TIME). The various countries of the world have their own local standard times, which differ from UTC.

The continental United States is divided into four time zones, known as the Eastern, Central, Mountain, and Pacific zones. During the part of the year in which standard time is used, these zones are behind UTC by 5, 6, 7, and 8 hours, respectively. During the part of the year in which daylight-savings time is used, these zones run behind UTC by 4, 5, 6, and 7 hours, respectively. Some states and counties do not use daylight time, and thus their time lags behind UTC by the same amount all year.

In general, locations in the western hemisphere of the world run from 0 to 12 hours behind UTC, while places in the eastern hemisphere run from 0 to 12 hours ahead of UTC. An approximate idea of the local standard time at any given location can be determined from its longitude. Consider west longitude values to be negative, ranging from 0 to −180 degrees. Consider east longitude values to be positive, from 0 to +180 degrees.

For the western hemisphere, divide the west longitude in degrees by 15; you will get a number between 0 and −12. Round this number off to the nearest whole integer, and add the result to the time in UTC. Use 24-hour time, in the manner of the military: 0000 for 12:00 midnight, 0100 for 1:00 A.M., and so on up to 2400 for 12:00 midnight the following night. If the final value is less than 0000, add 2400 and subtract one day.

For the eastern hemisphere, divide the east longitude in degrees by 15; you will get a number between 0 and +12. Round this number off to the nearest whole integer, and add the result to the time in UTC. If the final value is greater than 2400, subtract 2400 and add one day. This will give the approximate local time in the part of the world of interest.

The above scheme is not completely reliable, because time zones are not strictly based on longitude. Some peoples have chosen to use the same standard time everywhere in their country, even though the land spans much more than 15 degrees of longitude. Other peoples run a fractional number of hours ahead of UTC, or behind UTC (Tonga, for example, runs 12 hours and 20 minutes ahead of UTC). However, in astronomical terms, the above method is accurate.

STANDING WAVE

A *standing wave* is an electromagnetic wave that occurs on a transmission line or antenna radiator when the terminating impedance differs from the characteristic impedance.

Standing waves are generally present in resonant antenna radiators. This is because the terminating impedance at the end of a typical radiator, which has a finite physical length, is a practically infinite, pure resistance. This means that essentially no current flows there. This sets up standing-wave patterns, as shown in the illustration.

The pattern of standing waves on a half-wave resonant radiator, fed at the center, is shown at A. The pattern of standing waves on a full-wave resonant radiator, fed at the center, is shown at B. The pattern of standing waves on a full-wave resonant radiator, fed at one end, is shown at C. Notice the difference between the pattern for current and the pattern for voltage.

Some antenna radiators do not exhibit standing waves; instead, the current and voltage are uniform all along the length of the radiating element. This is the case in antennas that have a terminating resistor with a value that is equal to the characteristic impedance of the radiator (about 600 ohms). Some longwire and rhombic antennas use such resistors to obtain a unidirectional pattern (*see* LONGWIRE ANTENNA, and RHOMBIC ANTENNA).

In a transmission line that is terminated in a pure resistance having a value equal to the characteristic impedance of the line, no standing waves occur. This is a desirable condition because standing waves on a transmission line contribute to power loss. *See also* CHARACTERISTIC IMPEDANCE, and STANDING-WAVE RATIO.

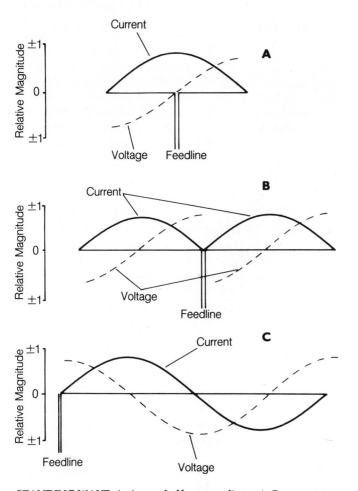

STANDING WAVE: At A, on a half-wave radiator. At B, on a center-fed, full-wave radiator. At C, on an end-fed, full-wave radiator.

STANDING-WAVE RATIO

On a transmission line terminated in an impedance that differs from the characteristic impedance (*see* CHARACTERISTIC IMPEDANCE), a nonuniform distribution of current and voltage occurs along the length of the line. The greater the mismatch, the more nonuniform this distribution becomes.

The ratio of the maximum voltage to the minimum voltage, or the maximum current to the minimum current, is called the *standing-wave ratio* on the transmission line. The standing-wave ratio is 1 : 1 if the current and voltage are in the same proportions everywhere along the line. Standing-wave ratio is usually abbreviated *SWR*.

The SWR can be 1 : 1 only when a transmission line is terminated in a pure resistance having the same ohmic value as the

characteristic impedance of the line. If the load contains reactance, or if the resistive component of the load impedance is not the same as the characteristic impedance of the line, the SWR cannot be 1 : 1. Then, standing waves will exist along the length of the line (*see* STANDING WAVE). In theory, there is no limit to how large the SWR can be. In the extreme cases of a short circuit, open circuit, or pure reactance at the load end of the line, the SWR is theoretically infinite. In practice, line losses and loading effects prevent the SWR from becoming infinite, but it can be as great as 100 : 1 or more.

The SWR is important as an indicator of the performance of an antenna system, because a high SWR indicates a severe mismatch between the antenna and the transmission line. This can have an adverse effect on the performance of a transmitter or receiver connected to the system. An extremely large SWR can result in significant signal loss in the transmission line. If a high-power transmitter is used, the large currents and voltages caused by a high SWR can actually damage the line. *See also* IMPEDANCE, REFLECTED POWER, REFLECTOMETER, SMITH CHART, and STANDING-WAVE-RATIO LOSS.

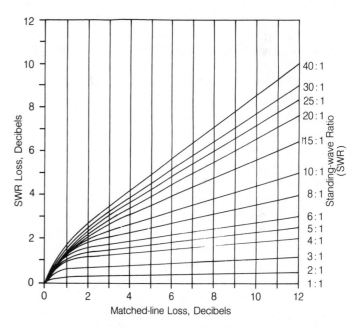

STANDING-WAVE-RATIO LOSS: Extra loss produced by standing waves on transmission line.

STANDING-WAVE-RATIO LOSS

No transmission line is lossless. Even the most carefully designed line has some loss because of the ohmic resistance of the conductors and the imperfections of the dielectric material. In any transmission line, the loss is smallest when the line is terminated in a pure resistance equal to the characteristic impedance of the line: that is, when the standing-wave ratio (SWR) is 1 : 1.

If the SWR is not 1 : 1, the line loss increases. This additional loss is called *standing-wave-ratio (SWR) loss*. The illustration shows the SWR loss that occurs for various values of matched-line loss and SWR. The SWR loss is generally not serious until the SWR becomes greater than 2 : 1. *See also* STANDING-WAVE RATIO.

STANDING-WAVE-RATIO METER
See REFLECTOMETER.

STANDOFF INSULATOR

A *standoff insulator* is an insulator that is used to hold an electrical conductor or feed line away from a surface. Standoff insulators are used with open-wire or twin-lead feed lines.

A typical standoff insulator, suitable for use with television twin-lead, is shown at A in the illustration. The feed line is run through a slot in the insulating material. This type of standoff insulator is commonly available, in various lengths, at most hardware and electronics shops.

A standoff insulator for use with open-wire line must usually be homemade. One method of construction is illustrated at B. A board is painted or coated with wax to protect it against rotting. Two cone-shaped insulators are attached to one end of the board. The entire board is clamped to the tower with U bolts. The feed-line wires are attached to the screws at the ends of the cone-shaped insulators.

Standoff insulators keep the electromagnetic field in an open-wire or twin-lead line from being affected by objects near which the line passes. This keeps the line loss at a minimum, and keeps the characteristic impedance constant all along the length of the line. Standoff insulators are not generally used with coaxial cables. The electromagnetic field is entirely contained within the shield of such a cable, and standoff insulators are therefore not necessary.

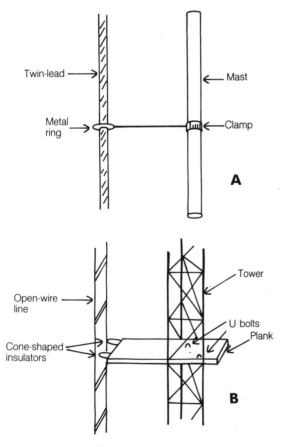

STANDOFF INSULATOR: At A, on a mast; at B, on a tower.

STATIC
See SFERICS, STATIC CHARACTERISTIC, and STATIC ELECTRICITY.

STATIC CHARACTERISTIC

The steady-state behavior of a component, circuit, or system is known as the *static characteristic*. The static characteristic of a component is determined using direct currents that do not fluctuate rapidly with time. For example, you can determine the characteristic curve of a transistor or tube by making measurements as you apply various direct currents or voltages. The static characteristic of a circuit or system is an expression of its behavior with no signal input or with a direct-current signal input.

When a component carries an alternating current or a fluctuating direct current, the characteristics can differ from the static condition. In fact, this is usually the case. A diode does not function the same way at 1 GHz as it does with the application of direct current. Even such simple components as resistors, capacitors, or wire conductors behave differently for alternating or fluctuating currents than for direct currents. The behavior of a component under rapidly changing conditions is called the *dynamic characteristic. See also* DYNAMIC CHARACTERISTIC.

STATIC ELECTRICITY

The existence of a potential difference between two objects without the flow of current, or the existence of an excess or deficiency of electrons on an object, is called *static electricity*. This expression arises from the fact that no charge transfer takes place.

A static electric charge generates an electric field, the intensity of which depends on the potential difference and distance between two objects, or the quantity of charge and its concentration on a single object (*see* CHARGE, and ELECTRIC FIELD).

If a static electric potential between two objects becomes large enough, a discharge will occur. The greater the separation distance between the objects, the larger the potential difference must be in order to cause the discharge. *See also* LIGHTNING, and SPARK.

STATIC INPUT

In a digital-logic circuit, a *static input*, also known as a *level-triggered input*, produces an output state that depends on whether the input is logic low (0) or logic high (1). It is triggered by the actual signal level, rather than by a change of signal level. As long as the input remains at the same level (0 or 1), the output stays constant. *See also* DYNAMIC INPUT.

STATIC MEMORY

A *static memory* is an electronic memory that retains its information as long as power is supplied. The data does not change position unless a command is given to that effect. Many integrated-circuit memory devices are of the static type. Some static memory devices draw almost no current, and thus memory can be stored for months or even years by means of the power from a small battery. *See also* MEMORY.

STATION ACCESSORIES

The heart of any ham station is the *transceiver* or combination of *transmitter* and *receiver. See* RECEIVER, TRANSCEIVER, and TRANSMITTER. A station might have several rigs, such as a receiver and transmitter for 160 through 10 meters, and individual transceivers for 144 MHz, 222 MHz, and 432 MHz.

Of course, any ham station needs antennas and/or power supplies. These are standard equipment. *See* ANTENNA, and POWER SUPPLY. Besides standard equipment, most stations have *accessories* or *peripherals. See* ACTIVE COMMUNICATIONS SATELLITE and related articles, AMATEUR TELEVISION, AUDIO FILTERING, COMPUTER, DIGITAL SIGNAL PROCESSING, DUMMY ANTENNA, FACSIMILE, FREQUENCY COUNTER, KEYER, MODEM, MONITOR, OSCILLOSCOPE, PACKET RADIO and related articles, PREAMPLIFIER, PRINTER, RADIOTELETYPE, REFLECTOMETER, REMOTE CONTROL, SLOW-SCAN TELEVISION, SPECTRUM MONITOR, TERMINAL, TRANSMATCH, and WATTMETER.

STATION ASSEMBLY

Every ham takes pride in his or her station. The way your station is assembled is a major factor in determining how much enjoyment you get from the hobby. A neat, efficient station is fun to operate. A messy, poorly wired, or cramped arrangement is no fun, and can even be unsafe.

Safety First   The most important consideration is operator safety. There should be minimal risk of shock from high-voltage wires. No conductor that carries more than 24 V should ever be exposed.

Station wiring must always conform to good electrical standards. If in doubt, an electrician can be consulted. Any ham station should be able to pass a fire inspection.

Antenna feed lines should be disconnected from the radio equipment, and connected to a ground rod, ideally outside of the house or apartment building, whenever the station is not in use. This minimizes the danger of equipment damage or personal injury from lightning. The importance of lightning protection *cannot be overemphasized* (*see* LIGHTNING, LIGHTNING PROTECTION, and LIGHTNING ROD). "Lightning arrestors" should *not* be relied on to give effective protection (*see* LIGHTNING ARRESTOR).

The Ground System   For good station performance, you should always try to get the best possible ground system. This is important not only for safety reasons, but to minimize electromagnetic interference (*see* ELECTROMAGNETIC INTERFERENCE). In recent years, the proliferation of personal computers (PCs) and other sophisticated, solid-state devices and accessories has made it much more likely that different pieces of equipment in a ham station will interfere with one another.

All equipment should be grounded to a common bus (*see* GROUND BUS), which is in turn connected to an earth ground. The best electrical ground is a long rod driven into highly conductive soil. This might not, however, be a good radio-frequency ground. A ground plane, or a system of radials, might be needed in addition to the electrical ground (*see* GROUND CONNECTION, GROUND PLANE, and RADIAL).

Coaxial cable, or "coax," is best for equipment interconnection. The cables should have good shield continuity. A high-quality transmission-line type of coax, such as RG-58/U, is sufficient for most interconnection purposes. When multiple conductors are used, the manufacturer usually supplies the appropriate type of cable.

The Heart of the Station   The operating table or desk should be sturdy. You don't want it to collapse under the weight of the station equipment. The combined mass of the radios and accessories can easily exceed 100 pounds, and in large stations, might be 200 pounds or more.

The desk should be at least 28 inches deep, and preferably 30 to 36 inches. This allows plenty of room for your arms, and for papers (such as logs and data sheets). Some operators place the equipment on a shelf elevated 6 to 10 inches above the main desk surface.

Of course, a good lighting system should be used. Some hams like indirect lighting, such as is provided by a halogen torchiere lamp; others prefer direct, close-in illumination, of the type given off by a high-intensity lamp.

The main part of a ham station is the "rig" or radio. A station might have several rigs. These are usually *transceivers*, although separate transmitters and receivers are sometimes seen (*see* RECEIVER, TRANSCEIVER, and TRANSMITTER). This unit, or units, should be placed at the most accessible position in the station, with accessories around it. All equipment should have enough space around it to allow for proper ventilation.

Linear Amplifiers   If you have one or more *linear amplifiers* (*see* LINEAR AMPLIFIER), it/they should be placed where the controls can be reached easily. Many "linears" have separate power supplies. These can be placed out of sight and out of the way of the operating position, such as on the floor under the operating table.

Linear amplifiers have large current requirements. The electrical circuit must be adequate to provide this current. Most "bedroom" circuits supply 15 amperes at 117 volts. This is not sufficient to power a full-legal-limit amplifier. Ideally, 234-volt circuits, such as those used for laundry machines or kitchen ranges, should be used. It might be necessary to have a special circuit installed, by a professional electrician, to provide adequate power for your linear amplifier.

High power increases the chances for electromagnetic interference, both to station accessories and to home entertainment equipment, such as hi-fi sets, television receivers and even telephones. *See* ELECTROMAGNETIC INTERFERENCE.

Antennas   A station can only work as well as its antenna system will allow. There are many different types of antennas, both commercially manufactured and "homebrew," that you can choose from. *See* ANTENNA.

Accessories   *Station accessories* include various devices and special station arrangements. Please refer to the following articles for information: ACTIVE COMMUNICATIONS SATELLITE and related articles, AMATEUR TELEVISION, AUDIO FILTERING, COMPUTER, DIGITAL SIGNAL PROCESSING, DUMMY ANTENNA, FACSIMILE, FREQUENCY COUNTER, KEYER, MODEM, MONITOR, OSCILLOSCOPE, PACKET RADIO and related articles, PREAMPLIFIER, PRINTER, RADIOTELETYPE, REFLECTOMETER, REMOTE CONTROL, SLOW-SCAN TELEVISION, SPECTRUM MONITOR, TERMINAL, TRANSMATCH, and WATTMETER.

Mobile and Portable Stations A *mobile* station is any ham radio station operated from a moving vehicle. This might be a car, truck, boat, airplane, train, or even a balloon, bicycle, or horse and buggy! *See* MOBILE ANTENNA, and MOBILE EQUIPMENT.

A *portable station* is a ham radio station operated from a temporary location, and/or a station that can be easily packed up and moved from place to place. *See* PORTABLE EQUIPMENT.

STATOR

In a variable capacitor, the plate or set of plates that remains fixed is called the *stator*. The stator is usually connected to a terminal that is insulated from the frame of the capacitor. In a motor, the stationary coil is called the *stator*. In a rotary switch, the nonmoving set of contacts is sometimes called the *stator*. The moving set of plates, coils, or contacts is called the *rotor*. *See also* AIR-VARIABLE CAPACITOR, MOTOR, and ROTARY SWITCH.

STEATITE

Steatite is a ceramic material made from magnesium silicate. Steatite has excellent dielectric properties, and is extensively used in insulators for radio-frequency equipment. The dielectric constant is approximately 5.8 throughout the radio-frequency spectrum. Steatite appears somewhat like porcelain. *See also* DIELECTRIC.

STEEL WIRE

Steel is extensively used in the manufacture of wire requiring high tensile (breaking) strength. In antenna systems, copper-clad steel wire is often used (*see* COPPER-CLAD WIRE). Steel is also commonly used in the manufacture of guy wire (*see* GUYING).

Steel wire is generally coated with zinc (galvanized) to prevent it from rusting and becoming weak. Steel wire is available in solid or stranded form, in the common American Wire Gauge (AWG) sizes. Steel wire has a maximum breaking strength of approximately twice that of annealed copper wire, or 1.5 times that of hard-drawn copper wire. *See also* AMERICAN WIRE GAUGE.

STENODE CIRCUIT

In a superheterodyne receiver, a resonant crystal circuit can be used to obtain a high degree of selectivity. This can be done in various ways. The stenode circuit is a special form of piezoelectric resonant circuit.

The stenode actually operates by cancelling out all signals at frequencies other than the desired frequency. The crystal is connected between the output of one intermediate-frequency amplifier and the input of the same stage or a previous stage, in such a way as to produce negative feedback. The negative feedback causes the gain of the stages to be greatly reduced, except at the crystal frequency, where the crystal acts as a trap to block the feedback. The advantage of a stenode circuit is that it is very stable. Oscillation is unlikely. The stenode circuit is simple to construct and align.

STEP-DOWN TRANSFORMER

A *step-down transformer* is a transformer in which the output voltage is smaller than the input voltage. The primary-to-secondary turns ratio is the same as the input-to-output voltage ratio. The input-to-output impedance ratio is equal to the square of the input-to-output voltage ratio. *See also* IMPEDANCE TRANSFORMER, TRANSFORMER, and VOLTAGE TRANSFORMER.

STEP-RECOVERY DIODE

A step-recovery diode, also called a *charge-storage* or *snap diode*, is a special form of semiconductor device that is used chiefly as a harmonic generator. When a signal is passed through a step-recovery diode, literally dozens of harmonics occur at the output. A step-recovery diode, connected in the output circuit of a very-high-frequency transmitter, can be used in conjunction with a resonator to obtain signals at frequencies up to several gigahertz. Step-recovery diodes are also used to square off the rise and decay characteristics of digital pulses.

A step-recovery diode is manufactured in a manner similar to the PIN diode. Charge storage is accomplished by means of a very thin layer of intrinsic semiconductor material between the P-type and N-type wafers. The charge carriers accumulate very close to the junction.

When the step-recovery diode is forward-biased, current flows in the same manner as it would in any forward-biased P-N junction. However, when reverse bias is applied, the conduction does not immediately stop. The step-recovery diode stores a large number of charge carriers while it is forward-biased, and it takes a short time for these charge carriers to be drained off. Current flows in the reverse direction until the charge carriers have been removed from the P-N junction, then the current falls off with extreme rapidity. The transition time between maximum current and zero current can be less than 10^{-10} second. It is this rapid change in the current that produces the harmonic energy. *See also* P-N JUNCTION.

STEP-UP TRANSFORMER

A *step-up transformer* is a transformer in which the output voltage is larger than the input voltage. The input-to-output voltage ratio is equal to the primary-to-secondary turns ratio. The input-to-output impedance ratio is equal to the square of the input-to-output voltage ratio. *See also* IMPEDANCE TRANSFORMER, TRANSFORMER, and VOLTAGE TRANSFORMER.

STEP-VOLTAGE REGULATOR

A *step-voltage regulator* is a special form of alternating-current regulating transformer that allows manual adjustment of the output voltage to compensate for changes in the load (*see* REGULATING TRANSFORMER).

The step-voltage regulator is connected in series with the secondary winding of the main transformer, just as is the case with an ordinary series regulating transformer. The phase can either add to, or buck against, the output of the main transformer. Regulation is accomplished without interrupting the power to the load, and without creating transient surges.

STERBA ANTENNA

A *Sterba antenna*, also called the *Sterba-curtain array*, is a directional array that exhibits gain with respect to a dipole antenna. The Sterba-curtain antenna consists of several elements fed by transposed sections of transmission line. The feed point is normally at the end of the array.

The two end elements of the Sterba curtain array are ¼ wavelength long, providing an impedance transformation for the input and a termination for the last section. When fed, the input is at a current maximum.

The Sterba antenna has a bidirectional response pattern, with the maxima perpendicular to the plane of the elements; therefore, it is a broadside array (*see* BROADSIDE ARRAY). A screen could be placed ¼ wavelength behind the array to provide a unidirectional response.

Sterba-curtain arrays can be used at any frequency up through VHF. They were once popular for point-to-point communications systems on the HF bands, where their directional characteristics provided an advantage. The array can be mounted with the elements vertical or horizontal.

STOPBAND

A selective filter causes attenuation of energy at some frequencies. The band of frequencies at which a filter has high attenuation is called the *stopband*. Generally, the stopband is that band for which the filter blocks at least half of the applied energy. In other words, the stopband is that part of the spectrum for which the power attenuation is 3 dB or more. Sometimes a 6-dB attenuation figure is specified instead.

We can speak of the stopband of any kind of filter. But this term is most often used with band-rejection filters. *See also* BANDPASS FILTER, BANDPASS RESPONSE, BAND-REJECTION FILTER, BAND-REJECTION RESPONSE, HIGHPASS FILTER, HIGHPASS RESPONSE, LOWPASS FILTER, and LOWPASS RESPONSE.

STORAGE

Storage refers to the retention of data or information, either analog or digital, and by any means. The electrical or magnetic storage of digital information is usually called *memory*.

The deliberate accumulation of an electric charge is called *storage*. Electric charge is stored in such things as batteries and capacitors. *See also* MEMORY, STORAGE BATTERY, STORAGE CELL, and STORAGE OSCILLOSCOPE.

STORAGE BATTERY

A *storage battery*, also known as a *secondary battery*, is a battery capable of accepting and releasing a large quantity of electric charge. The battery contains two or more cells that each have a positive plate and a negative plate immersed or embedded in an electrolyte material. A common automobile battery is an example of a storage battery.

The charge in a storage battery is usually expressed in terms of ampere hours. An ampere hour is equivalent to 3,600 coulombs. A typical automotive storage battery can hold approximately 50 to 100 ampere hours of electric charge.

Storage batteries can be charged in a short time at a high level of current, or over a longer period of time using correspondingly less current. *See also* QUICK CHARGE, and TRICKLE CHARGE.

STORAGE CELL

A *storage cell* is a single electrochemical cell that is capable of accepting and releasing a quantity of electric charge. Storage cells supply approximately 1 V to 1.5 V. The amount of charge that can be stored depends on the physical size of the cell, and on the materials that make up the cell.

Two or more storage cells, connected in series, form a storage battery. *See also* CELL, and STORAGE BATTERY.

STORAGE OSCILLOSCOPE

In the design, alignment, and servicing of electronic equipment, it is sometimes necessary to observe a waveform for a long period of time, or to compare two waveforms that occur at different times. A storage oscilloscope is designed with this in mind. The storage oscilloscope can be operated in the same way as a conventional oscilloscope if desired.

The heart of the storage oscilloscope is a special cathode-ray tube that can hold an electric charge in a pattern corresponding to the waveform displayed. There are two electron guns, called the *writing* or *main gun* and the *flooding gun*.

A fine grid of electrodes, just inside the phosphor screen, is saturated with negative charge prior to storing a waveform. Then, the oscilloscope is adjusted until the waveform appears as desired. The saturating charge is removed from the grid, and it acquires localized positive charges wherever it is struck by the electron beam from the main gun. This positive charge decays very slowly unless it is deliberately removed.

When a stored waveform is to be observed, the grid of electrodes is irradiated by electrons from the flooding gun. The grid prevents these electrons from reaching the phosphor, except in the areas of localized positive charge. Thus, the original waveform is reproduced on the phosphor screen. The waveform can be viewed for as long as the operator wants, up to several minutes. The stored waveform is erased by saturating the grid with negative charge. *See also* OSCILLOSCOPE.

STORAGE TIME

When the voltage at the base of a bipolar transistor is beyond the saturation level, an excess current flows in the emitter-base circuit. This current is greater than the amount required to keep current flowing in the collector circuit. Because of this excess current, an electrical charge accumulates in the base of the transistor.

If the bias is suddenly removed from the base of the transistor, the accumulated charge must dissipate before the transistor will come out of saturation. This requires a small amount of time, known as the *storage time*, abbreviated t_s.

Storage time depends on several factors, including the size of the base electrode, the type and concentration of impurity in the semiconductor material, and the extent to which the transistor is driven beyond saturation. The storage time limits the speed at which switching can be accomplished by means of saturated logic. *See also* SATURATION, SATURATED LOGIC, and SATURATION VOLTAGE.

STORE

A memory file is sometimes called a *store*. A store can contain any amount of information, usually specified in bytes, kilobytes, or megabytes. The act of putting information into memory is called a *STORE operation. See also* BYTE, FILE, MEMORY, and STORING.

STORE-AND-FORWARD

Store-and-forward refers to communications in which messages are held, or stored, for transmission at a later time. The most common store-and-forward scheme is *packet radio*. "Personal" ham communications is done in real time. This means that two or more operators converse with each other as if they were on the telephone. Contacts of this type are called *QSOs*. There are times when direct QSOs aren't convenient. Store-and-forward communications provides an alternative in these situations. *See* PACKET RADIO.

STORING

Storing is the process of putting information into a memory circuit (*see* MEMORY). Storing is usually accomplished by a simple sequence of operations, or even a single operation, such as pushing a button.

For example, in a small electronic calculator, you might wish to store a number in the memory. Simply press a button, perhaps labeled M IN (memory input), and the displayed number is stored in the memory. If there is more than one memory, you must press two buttons, such as M IN and the memory address (*see* ADDRESS).

In computers, the storing process usually involves several operations. You must tell the computer which file you want to store, and where you want to store it.

STRAIN

When an object is subjected to physical forces, that object can change in shape. *Strain* is an expression of the extent to which an object changes shape under stress (*see* STRESS).

Strain is an important consideration for wires run in long spans. Some metals change shape quite easily because they are highly elastic. An example of such a substance is annealed, soft-drawn, copper wire. A long antenna of solid, soft-drawn copper wire will stretch. The stretching is clearly undesirable if an antenna is to remain resonant at the same wavelength.

Some materials are relatively strain-resistant. Hard-drawn or stranded copper wire stretches much less than solid, soft-drawn copper wire. Steel wire stretches less than copper. This is part of the reason why copper-clad steel wire is preferred in the construction of wire antennas (*see* COPPER-CLAD WIRE, and WIRE ANTENNA).

Strain is of importance in other applications besides the construction of wire antennas. A piezoelectric device operates by producing electrical impulses as a result of mechanical strain, but the crystal will break if it is put under too much strain. Electrostriction and magnetostriction are examples of strain that occurs in various substances in the presence of strong electric or magnetic fields (*see* ELECTROSTRICTION, and MAGNETOSTRICTION).

STRAIN INSULATOR

Insulators are often used in guying systems to eliminate resonances that can affect the radiation pattern of an antenna. Such insulators are subjected to tremendous stress in the guy wires of a large tower. This stress can pull a conventional insulator apart. This can cause a catastrophe!

A *strain insulator* is designed so that the stress compresses the material instead of pulling at it. Most substances can endure far more stress in the form of compression rather than tension. Strain insulators are thus preferred in guying systems.

A typical strain insulator is shown. The two wire loops are oriented at right angles with respect to each other. Stress causes the insulator to be squeezed. If the stress becomes too great, and the insulator breaks, the wire span will not fall because the wire loops interlock.

Strain insulators have a fairly large amount of leakage capacitance, and are not used in active wire-antenna elements unless conventional insulators are unable to withstand the stress. *See also* GUYING, and INSULATOR.

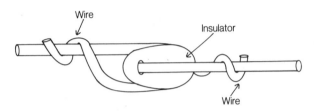

STRAIN INSULATOR: Also called an *egg insulator*.

STRANDED WIRE

Several thin wires can be wound, or spun, together to form a single, larger conductor. Wire is often made in this way. Such wire is called *stranded*. Most stranded wires have seven individual conductors; some have 19 conductors.

Stranded wire has some advantages over solid wire in electrical and electronic work. Stranded wire is less likely to break when subjected to flexing. Stranded wire adheres more readily to solder. When used in antennas, stranded wire stretches less than solid wire. Stranded wire is less susceptible to radio-frequency skin effect than solid wire because the total surface area is larger than that of a solid conductor of the same diameter (*see* SKIN EFFECT).

Stranded wire also suffers from certain shortcomings compared with solid wire. Uninsulated stranded wire corrodes more quickly than uninsulated solid wire. Stranded wire generally cannot be used in the winding of coils; finely stranded copper wire does not hold its position when bent. The cross-sectional area of stranded wire is less than that of solid wire having the same outside diameter; therefore, stranded wire has slightly higher direct-current resistance than solid wire.

The size of stranded wire can be specified in either of two ways. The gauge of the entire conductor can be specified, such as AWG #12. Alternatively, the number of strands and the gauge of each strand can be indicated. A wire made from seven strands of AWG #20 wire, for example, might be called *7x AWG #20*. Wire having very fine strands is sometimes called *Litzendracht (Litz) wire*. *See also* AMERICAN WIRE GAUGE, BIRMINGHAM WIRE GAUGE, BRITISH STANDARD WIRE GAUGE, and LITZ WIRE.

STRAPPING

Some printed circuits can be used for more than one purpose, depending on the way in which external wires are connected. For example, an amplifier and filter can be wired to operate as an oscillator, or an oscillator can be wired to operate at any of several fixed frequencies. The manner in which the wires are connected among designated points, for obtaining the desired operation, is called *strapping*.

In a magnetron tube, oscillation sometimes occurs at un-wanted frequencies. To prevent this, metal strips are connected among the resonator segments. This is known as *strapping. See also* MAGNETRON.

STRESS

Stress is any static force exerted against a body. Stress is normally measured in pounds or kilograms per unit of cross-sectional area. Stress can occur in three different ways: constriction (pushing in), tension (pulling apart), or laterally (sideways).

Stress causes distortion in the shape of an object. A dielectric material changes shape because of electrostriction. Wire stretches when it is subjected to tension. A metal-tubing antenna bends because of the lateral stress caused by wind. The change in the shape of a body, induced by stress, is called *strain. See also* STRAIN.

STRIPLINE

Stripline, also (somewhat imprecisely) called *microstrip*, is a method of constructing radio-frequency tuned circuits at short wavelengths. A section of transmission line, measuring a quarter wavelength or a half wavelength, exhibits resonant properties. If the wavelength is very short, a quarter or half wavelength can be smaller than the length or width of a typical printed-circuit board. Then, it is practical to construct tuned circuits from parallel foil runs.

An example of stripline construction is shown in the photograph. This circuit is a power amplifier that delivers about 40 watts of radio-frequency output at 440 MHz. At this frequency, the wavelength is about 75 centimeters in free space. The electrical quarter wavelength, considering the velocity factor of the circuit-board dielectric, is only 15 centimeters, or a little less than 6 inches.

Stripline construction results in a high Q factor and excellent efficiency. *See also* HALF-WAVE TRANSMISSION LINE, Q FACTOR, and QUARTER-WAVE TRANSMISSION LINE.

STRIPLINE: Foil strips on a PC board.

STRIP TRANSMISSION LINE

A specialized form of parallel-conductor radio-frequency transmission line, made of metal strips rather than wires, is called a *strip transmission line*. The strip transmission line is a balanced line (*see* BALANCED LINE).

The illustration shows the strip-transmission-line configuration. The spacing between the strips, as measured between the two inside surfaces, is d, and the width of each strip is w (it is assumed that the two strips are identical in thickness and width). The characteristic impedance of an air-dielectric strip transmission line is given by:

$$Z_o = 377d/w$$

Strip transmission lines are generally used at very-high and ultra-high frequencies.

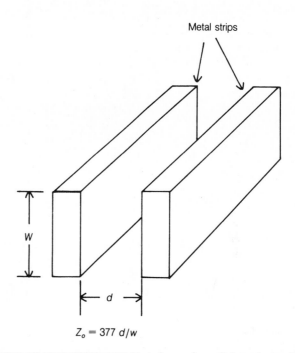

$$Z_o = 377\ d/w$$

STRIP TRANSMISSION LINE: A form of parallel-conductor line.

STUB

Obtaining a perfect match, or 1:1 standing-wave ratio (SWR), between an antenna element and a transmission line requires that the antenna reactance be tuned out, and that the remaining resistance match the characteristic impedance of the line. This can be done in several ways. *See* CHARACTERISTIC IMPEDANCE, and STANDING-WAVE RATIO.

One method of antenna matching, especially workable at short wavelengths (about 10 meters or less), makes use of a length of transmission line, shorted out at one end. This is called a *stub*. Stubs can also be used as tuned circuits for rejecting interference, or for providing a sharp bandpass.

The Universal Match When a stub is shorted at one end, it can serve as a universal matching network, provided that its length can be adjusted from zero to ½ wavelength. At intermediate points along the stub, intermediate values of impedance appear. If one end of the stub is shorted out and the other end is connected to an antenna element at the center (see illustration A), there will be some point at which a purely resistive impedance appears.

By adjusting the stub length, and by moving the tap point where the transmission line feeds the antenna/stub combination, a pure resistance of any value can be obtained. This allows for a 1:1 SWR with any feed line, whether its characteristic impedance is 50 ohms or 600 ohms, or anything in between. Coaxial as well as parallel-wire lines can be used.

Finding the proper stub length, and the right tap point, is a matter of trial and error, just as tuning a transmatch must be done by trial and error to get the minimum SWR.

It is rather inconvenient to use a universal stub on more than one band, or at frequencies very far apart within a given band. This is because adjustment must be made where the line joins the antenna. But a universal match will work over a reasonable range either side of its design frequency.

A Stub Band-rejection Filter   A 1/4-wave stub, open at one end, acts as a series-tuned inductance-capacitance (LC) circuit at the opposite end. When one end is shorted out, it acts as a parallel-tuned LC circuit at the opposite end. *See* QUARTER-WAVE TRANSMISSION LINE.

A band-rejection filter (*see* BAND-REJECTION FILTER, and BAND-REJECTION RESPONSE) can be designed using a quarter-wave stub, one end of which is open and the other end of which is connected at the input terminals of a receiver or at the output of a transmitter. This filter can then serve to "notch out" signals at some undesired frequency. The stub is cut to measure 1/4 wavelength at the frequency to be notched out, taking the velocity factor of the line into account (*see* VELOCITY FACTOR). See illustration B.

If the antenna terminals of the receiver or transmitter are balanced, a parallel-wire line should be used to make the stub. If the antenna terminals are unbalanced, a coaxial line should be used. The characteristic impedance of the stub is not particularly important; it need not be the same as the characteristic impedance of the antenna transmission line. The stub will work best if it is made from a line with the lowest possible loss per unit length. For a balanced system, open wire is ideal. For an unbalanced system, large-size foam-dielectric coaxial cable, or hard line, is preferable to the smaller sizes or the types with solid dielectric.

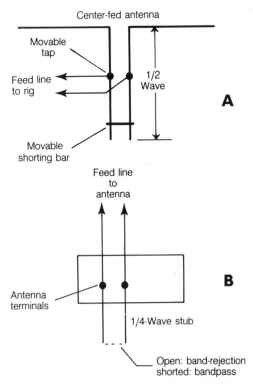

STUB: At A, a universal stub for antenna matching. At B, a quarter-wave stub for use as band-rejection or bandpass filter.

A Stub Bandpass Filter   If the far end of the stub in illustration B is shorted out, the band-rejection response becomes a bandpass response instead. *See* BANDPASS FILTER, and BANDPASS RESPONSE. The resonant frequency will remain the same. This arrangement can be used to provide extra attenuation of spurious signals at the output of a transmitter. It can also help to suppress the even harmonics (it will pass odd harmonics along with the fundamental). Such a filter is also useful in reduction of interference to a receiver, when that interference is caused by out-of-band signals. In regards to the choice of line type to use for the bandpass stub, the same considerations apply as for the band-rejection stub.

SUBAUDIBLE TONE

A *subaudible tone* is a tone that is used to modulate a communications transmitter for the purpose of obtaining private-line or tone-squelch operation. Subaudible tones all have frequencies below 300 Hz, which is the approximate lower audio cutoff for communications. *See* PRIVATE LINE, and TONE SQUELCH.

SUBBAND

A given frequency allocation can be subdivided for various types of radio transmission. A *subband* is literally a band within a band. Consider, for example, the amateur band at 144 to 148 MHz. The lowest 100 kHz of this band—that is, 144.000 to 144.100 MHz—is allocated for nonvoice emission types only. This 100-kHz subband is an example of a legislated subband.

In the amateur band at 1.8 to 2.0 MHz, there were, at one time, eight different subbands in the United States. Each subband was 25 kHz wide. Various parts of the country had their own set of power-output limitations for each subband.

Sometimes a subband is agreed upon among the operators who use the band, but it does not carry the force of law. In the non-voice subbands of the high-frequency amateur bands, for example, the users of A1 and F1 emissions stay out of each other's way by operating within different frequency ranges. *See also* BAND.

SUBHARMONIC

A *subharmonic* is a signal with a frequency that is an integral fraction of the frequency of a specified signal. Any signal has infinitely many subharmonics. If the main-signal frequency is f, then the subharmonic frequencies are $f/2$, $f/3$, $f/4$, and so on. Subharmonic signals occur at wavelengths that are integral multiples of the main-signal wavelength.

Although most alternating-current signals inherently contain some energy at harmonic frequencies (*see* HARMONIC), signals do not naturally contain energy at subharmonic frequencies. Subharmonics must be generated by means of a frequency divider (*see* FREQUENCY DIVIDER).

The term *subharmonic* is sometimes used, imprecisely, in reference to certain signals in the output of a frequency multiplier. Some radio transmitters, for example, use doublers, triplers, or quadruplers (*see* FREQUENCY MULTIPLIER). Suppose that a tripler is used so that the input frequency is f and the output frequency is $3f$. A small amount of energy will appear in the tripler output at frequencies of f and $2f$. These signals are occasionally referred to as subharmonics. The signal at frequency $2f$ is not, however, a true subharmonic of the signal at frequency $3f$.

SUBROUTINE

A *subroutine* is a computer program within a program, and it is intended to perform a specific function. A subroutine is often used when it is necessary to perform a certain calculation numerous times.

A computer program might have no subroutines, or just a few, or perhaps hundreds of them. Some subroutines might even have subroutines within them. When subroutines are contained within other subroutines, the programs are said to be *nested.* Generally, a subroutine can be called from any point within a computer program, as needed.

SUDDEN IONOSPHERIC DISTURBANCE

When a solar flare occurs, ionospheric propagation is almost always affected (*see* SOLAR FLARE). A short time after an intense solar flare, medium-frequency and high-frequency propagation can be almost totally wiped out. This is called a *sudden ionospheric disturbance.*

A solar flare causes a dramatic, abrupt increase in the number of charged subatomic particles emitted by the sun. These particles travel much slower than light, so they arrive at the earth a few hours after we see the solar flare. Astronomers can predict sudden ionospheric disturbances by watching the sun.

When the charged particles reach the earth, they cause an increase in the ionization density of the D layer (*see* D LAYER). This results in absorption of radio-frequency energy at medium and high frequencies. Long-distance propagation is seriously degraded by dense ionization in the D layer. At night, the ionization can be so dense that it is visible as an auroral display (*see* AURORA).

The magnetic field of the earth is also affected by the arrival of many charged particles from the sun. The magnetic field deflects large numbers of the particles toward the poles. Because the particles are accelerated, they produce a fluctuating magnetic field of their own. This magnetic field affects the overall geomagnetic field, producing a so-called "geomagnetic storm" (*see* GEOMAGNETIC FIELD, and GEOMAGNETIC STORM).

Many of our communications circuits today are via satellite. This has helped to ameliorate the problem of sudden ionospheric disturbances because the ultra-high-frequency and microwave spectra are not affected by the ionosphere. *See also* IONOSPHERE, and PROPAGATION CHARACTERISTICS.

S UNIT

The *S unit* is a unit of signal strength. Generally, 1 S unit represents a change in signal strength amounting to 6 dB of voltage, or a ratio of 2:1. A root-mean-square signal level of 50 μV is considered to represent 9 S units. A few engineers use other values to represent S9, or other voltage ratios to represent 1 S unit.

The S-unit scale ranges from 1 to 9. A signal strength of S8 means that the signal voltage at the antenna terminals is 50/2, or 25, μV; a signal strength of S7 means that the voltage is 25/2, or 12.5, μV, and so on. The signal strength is sometimes expressed in fractions of an S unit, such as S5.5. The illustration is a graph showing the actual RMS signal voltages for S units ranging from 1 to 9, according to the most common definition of the scale.

Many radio operators prefer to use their own intuition regarding signal strength. The strength (S) expression in the RST system is a subjective, rather than quantitative, variable. *See also* RST SYSTEM, and S METER.

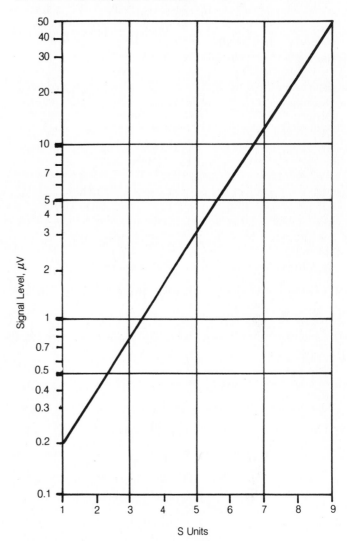

S UNIT: A graph of S units vs signal strength in a receiver.

SUNSPOT

A *sunspot* is a massive magnetic storm that occurs on the sun. Sunspots are visible through a telescope equipped with a special projection system. Sometimes sunspots are large enough to be seen with a simple pinhole projection device. The sun should never be viewed directly, even through dark film negatives because the ultraviolet light can damage the eyes. Sunspots appear as dark regions on the sun's surface.

A typical sunspot is several times the diameter of the earth. The darkest part of the spot is at, or near, the center; this is called the *umbra.* The periphery of a sunspot is less dark, and is known as the penumbra.

Sunspots are more numerous in certain years than in other years. The number of sunspots affects the density of the earth's ionosphere, and therefore we observe changes in propagation conditions over the years. The maximum usable frequency tends to be much higher when the sunspots are numerous, as compared to when they are sparse. *See also* IONOSPHERE, MAXIMUM USABLE FREQUENCY, PROPAGATION CHARACTERISTICS, and SUNSPOT CYCLE.

SUNSPOT CYCLE

In the past century, scientists have noticed that the number of sunspots is not constant, but changes from year to year. The fluctuation is fairly regular, and very dramatic. This fluctuation of sunspot numbers is called the *sunspot cycle*. It has a period of approximately 11 years (see illustration). The rise in the number of sunspots is generally more rapid than the decline, and the maxima and minima vary from cycle to cycle.

The sunspot cycle affects propagation conditions, especially at the high frequencies and in the very high frequency spectrum up to about 70 MHz for F-layer propagation and 150-200 MHz for sporadic-E propagation (*see* E LAYER, F LAYER, and SPORADIC-E PROPAGATION). When there are few sunspots, the maximum usable frequency is low, because the ionization of the upper atmosphere is not dense. At or near a sunspot maximum, the maximum usable frequency is high because the upper atmosphere is densely ionized. The changes in ionization result from the fact that the number of sunspots is correlated with the solar flux, and the solar flux directly affects the degree to which the upper atmosphere is ionized. *See also* MAXIMUM USABLE FREQUENCY, PROPAGATION CHARACTERISTICS, SOLAR FLUX, and SUNSPOT.

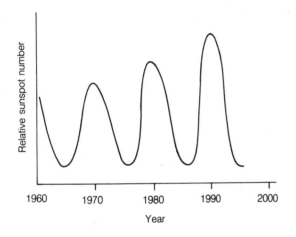

SUNSPOT CYCLE: Sunspot numbers versus year.

SUPERCONDUCTIVITY

The resistance of an electrical conductor depends on the temperature. Most materials exhibit increasing resistance with increasing temperature. At extremely low temperatures, near absolute zero, some conducting materials have extremely low resistance. This is known as *superconductivity*. When a closed loop of wire is cooled with liquid helium to a temperature of a few degrees Kelvin, an electric current will flow around and around the loop almost indefinitely. This is unheard of at ordinary temperatures. Magnetic fields will, however, interfere with this effect.

Superconductivity results in low levels of thermal and shot-effect noise in an electronic circuit. The ohmic losses are greatly reduced, and the Q factor is thereby increased. This makes it possible to build receiver preamplifiers that are very sensitive. Superconductivity might someday be used to make tiny antennas efficient for transmitting. Theoretically, it should be possible to make a paper clip an effective transmitting radiator for high-frequency use! *See also* Q FACTOR, SHOT EFFECT, and THERMAL NOISE.

SUPERHETERODYNE RECEIVER

A *superheterodyne receiver* is a receiver that uses one or more local oscillators and mixers to obtain a constant-frequency signal. A fixed-frequency signal is more easily processed than a signal that changes in frequency.

In the superheterodyne circuit, the incoming signal is first passed through a tunable, sensitive amplifier called the *front end* (see FRONT END). The output of the front end is mixed with the signal from a tunable, unmodulated local oscillator (*see* LOCAL OSCILLATOR, and MIXER). Either the sum or the difference between the frequencies of the two signals is then amplified. The sum or difference signal is called the *first intermediate-frequency (IF) signal* (*see* INTERMEDIATE FREQUENCY, and INTERMEDIATE-FREQUENCY AMPLIFIER). The signal is filtered to obtain a high degree of selectivity.

If the first IF signal is detected, we have a single-conversion receiver (*see* SINGLE-CONVERSION RECEIVER). Some receivers use a second mixer and second local oscillator, converting the first IF signal to a much lower frequency, called the *second IF*. This type of receiver is known as a *double-conversion circuit* (*see* DOUBLE-CONVERSION RECEIVER).

The superheterodyne receiver offers several advantages over direct-conversion receivers. Probably the most important advantage is the excellent selectivity that can be obtained, along with single-signal reception (*see* SELECTIVITY, and SINGLE-SIGNAL RECEPTION). Sensitivity is enhanced because the fixed IF amplifiers are easy to keep in optimum tune. A superheterodyne circuit can be used to receive signals having any emission type.

The main disadvantage of the superheterodyne circuit is that it sometimes receives signals that are not on the desired frequency. The signals can be external to the receiver, in which case they are called *images*, or they can come from the local oscillator or oscillators. If the local-oscillator frequencies are judiciously chosen, false signals are generally not a problem. *See also* DIRECT-CONVERSION RECEIVER, and IMAGE FREQUENCY.

SUPERHIGH FREQUENCY

The *superhigh-frequency (SHF) range* of the radio spectrum is the range from 3 GHz to 30 GHz. The wavelengths corresponding to these limit frequencies are 10 cm and 1 cm, and thus the superhigh-frequency waves are sometimes called *centimetric waves*. Most of the superhigh frequencies are allocated by the International Telecommunication Union, headquartered in Geneva, Switzerland.

Superhigh-frequency waves are not bent by the ionosphere of the earth. Centimetric waves behave very much like waves at higher frequencies. Superhigh-frequency waves can be focused by medium-sized parabolic or spherical reflectors. Because of their tendency to propagate in straight lines unaffected by the atmosphere, centimetric waves are extensively used in satellite communications. Centimetric waves are also used for such purposes as studio-to-transmitter links. Because the frequency of a centimetric signal is so high, wideband modulation is practical. *See also* ELECTROMAGNETIC SPECTRUM, and FREQUENCY ALLOCATIONS.

SUPERPOSITION THEOREM

The *Superposition Theorem* is a rule that describes the behavior of currents and voltages in linear multi-branch circuits. Suppose you have a linear circuit with three or more separate, current-carrying branches. Suppose that a power supply, delivering E_x volts, is placed across branch X of this circuit, while the terminals at E_y are shorted. This results in a certain current, call it I_{zx}, in branch Z. Now suppose we remove the voltage source from terminals E_x and short them, and connect a voltage E_y to terminals E_y. This will also cause a current, I_{zy}, to flow in branch Z. Currents I_{zx} and I_{zy} will not necessarily be the same; in fact, they probably will not be.

The superposition theorem tells us what current will flow in branch Z of the circuit when the supplies E_x and E_y are connected at the same time. It is simply the sum of currents I_{zx} and I_{zy}. That is, if the current in branch Z is I_z with both E_x and E_y connected, then:

$$I_z = I_{zx} + I_{zy}$$

This theorem can be used to advantage in determining currents and voltages in complicated circuits. *See also* KIRCHHOFF'S LAWS.

SUPERREGENERATIVE RECEIVER

A special form of regenerative detector is used in a circuit called a *superregenerative receiver*. The superregenerative circuit provides greater gain, and better sensitivity, than does an ordinary regenerative circuit. A superregenerative receiver is a form of direct-conversion circuit (*see* DIRECT-CONVERSION RECEIVER).

In the superregenerative detector, the positive feedback is periodically increased and decreased. This makes it possible to increase the feedback without resulting in unwanted oscillation of the detector. The changes in feedback cause fluctuations in the circuit gain, and this modulates incoming signals. However, if the feedback is varied at a rate of more than 20 kHz, the modulation cannot be heard, and it does not affect the sound of an incoming signal.

The amount of feedback is controlled by a varactor diode. The diode capacitance fluctuates with the applied ultrasonic signal. The changes in capacitance cause the amount of feedback to vary. The detector is adjusted by manipulating the series variable capacitor and the amplitude of the applied ultrasonic signal. *See also* REGENERATIVE DETECTOR.

SUPPRESSED CARRIER

In single-sideband transmission, the carrier wave is usually reduced greatly in amplitude. This is also sometimes done in double-sideband transmission. A signal is said to have a *suppressed carrier* in such instances.

The advantage of suppressed-carrier transmission is that all of the signal power goes into the sidebands, resulting in optimal efficiency. Suppressed-carrier signals are generated by means of a balanced modulator. *See also* BALANCED MODULATOR, and SINGLE SIDEBAND.

SURFACE WAVE

At some radio frequencies, the electromagnetic field is propagated over the horizon because of the effects of the ground. This is observed at very low, low, and medium frequencies, and to a limited extent at high frequencies. Surface-wave propagation does not occur at very high frequencies or above. This is because the ground becomes progressively more lossy as the frequency increases.

Surface-wave propagation occurs when the electromagnetic field is vertically polarized: that is, when the electric lines of flux are perpendicular to the surface of the earth. This requires a vertical radiator (*see* VERTICAL POLARIZATION). If the electromagnetic field is horizontally polarized, the electric lines of flux are short circuited in the ground, and surface-wave propagation does not occur.

When you listen to a broadcast station that is more than a few miles away during the daytime, the signal is reaching you because of surface-wave propagation. The same is true of daytime signals at low frequencies. At very low frequencies, surface-wave propagation occurs in conjunction with waveguide propagation (*see* WAVEGUIDE PROPAGATION).

The moon has no ionosphere. The same is true of many other planets. Over-the-horizon communication, without satellites, will nevertheless be possible on such planets if the ground conductivity is fairly good: surface-wave propagation might be used at low and very low frequencies for planet-wide communication. *See also* PROPAGATION CHARACTERISTICS.

SURFACE-WAVE TRANSMISSION LINE

Surface-wave propagation can be deliberately made to occur along wire conductors at very-high, ultra-high, and microwave frequencies. We never see surface-wave propagation along the ground at these wavelengths because the ground is not an excellent conductor (*see* SURFACE-WAVE PROPAGATION), but the situation is different when a good conductor is used. A single-wire line, designed to transfer surface-wave signals along its length, is called a *surface-wave transmission line*.

A surface-wave line consists of a wire conductor surrounded by a thick layer of low-loss dielectric material (see illustration). The signals are coupled into and out of the line by horns called *launchers*. The surface-wave line does not radiate significantly because the electromagnetic field travels in contact with the conductor. The line is unbalanced since there is only one conductor. This type of line is also called *G-line*, or *Goubau line*, after its inventor, Dr. George Goubau.

The surface-wave line is more efficient than coaxial cable or a parallel-wire line with two or four conductors. However, the wavelength must be short or the electromagnetic field will tend to be radiated rather than propagated along the line. Surface-wave transmission lines are not effective at the very low, low, medium, and high frequencies. Shielded or balanced feed lines are used at those wavelengths.

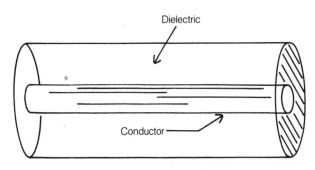

SURFACE-WAVE TRANSMISSION LINE: A single wire surrounded by dielectric.

SURGE

When power is first applied to a circuit or device that draws considerable current, a *surge* often occurs. For a moment, excessive current flows. This surge can result in damage to the circuit or appliance.

You have probably experienced the effects of a surge following a general power failure. People tend to leave appliances switched on during a power failure. When power is restored, thousands or millions of devices are connected to the utility mains, and a massive start-up current is demanded. The result is a momentary initial surge. If you are lucky, nothing happens, or perhaps fuses or circuit breakers blow or trip off.

In the event of a power failure, it is a good idea to switch off or unplug circuits and appliances that might be damaged by a surge—and that includes just about everything. If you have a breaker box, you should switch off all of the breakers, including the main breakers. Then, even if nobody else unplugs their appliances, yours will not be damaged. After power returns, switch on the main breakers first, and then the auxiliary breakers, one at a time.

Surges occur on a small scale in electronic circuits, especially those that draw significant current and operate from utility mains. *See also* SURGE SUPPRESSOR.

SURGE IMPEDANCE

See CHARACTERISTIC IMPEDANCE.

SURGE PROTECTION

All electronic equipment should have some means of protection against damage from surges. Surge protection can be accomplished by means of fuses, circuit breakers, or suppressors. *See also* CIRCUIT BREAKER, FUSE, SURGE, and SURGE SUPPRESSOR.

SURGE SUPPRESSOR

A *surge suppressor* is a device that is inserted in series or in parallel with the power line to a circuit, protecting the circuit from excessive voltages or currents (*see* SURGE).

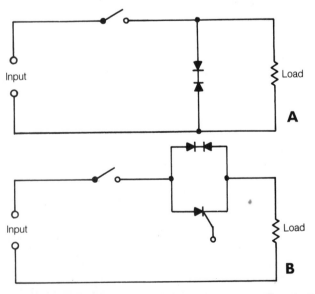

SURGE SUPPRESSOR: At A, using two selenium rectifiers. At B, using two selenium rectifiers and an SCR.

There are various kinds of surge suppressors. Perhaps the simplest is a Zener diode, with a voltage rating equal to the voltage for which the equipment is designed. Current-regulating and voltage-regulating circuits can be used for surge protection.

A common method of surge suppression is the use of selenium rectifiers connected back-to-back. Selenium rectifiers are sometimes used in conjunction with silicon-controlled rectifiers to obtain surge suppression. These two configurations are shown in the illustration. *See* CURRENT REGULATION, SELENIUM RECTIFIER, SILICON-CONTROLLED RECTIFIER, VOLTAGE REGULATION, and ZENER DIODE.

SUSCEPTANCE

Susceptance is the alternating-current equivalent of conductance, and the reciprocal of reactance. Susceptance is a mathematically imaginary quantity. Given an inductive reactance $X_L = jX$ the susceptance Y_L is given by:

$$Y_L = 1/X_L = 1/(jX) = -j\,(1/X)$$

where X is a positive real number.

Given a capacitive reactance $X_C = -jX$, the susceptance Y_C is given by:

$$Y_C = 1/X_C = 1/(-jX) = j(1/X)$$

where X is, again, a positive real number. *See also* REACTANCE.

SUSPENSION GALVANOMETER

A *suspension galvanometer* is a device that is sometimes used to measure weak currents. The device operates on the principle of interaction between magnetic fields.

A coil of wire is connected to two terminals. The coil is hung in the field of a permanent magnet. When a voltage is applied to the terminals, causing a current to flow in the coil, the coil turns as a result of interaction between the magnetic fields of the coil and the permanent magnet. The deflection of the coil is proportional to the current.

The sensitivity of the suspension galvanometer can be increased by attaching a mirror to the coil, and reflecting a beam of light off the mirror. The beam of light forms an optical lever.

SWEEP

Sweep is a continuous, usually linear, change of the value of a variable from one defined limit to another. The movement of an electron beam across the face of a television picture tube is an example of sweep. An oscillator or radio transmitter can be made to change in frequency; this is another example of sweep.

If the limits of parameter values are defined as A and B, sweep is usually characterized by a controlled change from limit A to limit B, and an almost instantaneous return from limit B to limit A. Sweep is sometimes called *scanning*. *See also* SCANNING.

SWEEP-FREQUENCY FILTER OSCILLATOR

A *sweep-frequency filter oscillator* is a specialized form of sweep generator (*see* SWEEP GENERATOR). The sweep-frequency filter oscillator produces a signal that varies rapidly in frequency between two defined and adjustable limits. The oscillator is used in conjunction with a spectrum analyzer for testing and alignment of selective filters (*see* SPECTRUM ANALYZER).

The oscillator frequency sweeps between two limits, f_1 and f_2, where $f_1 < f_2$. The frequency f_1 is well below the cutoff or passband; the frequency f_2 is well above the cutoff or passband. The sweep rate is very rapid so that the spectrum-analyzer display appears continuous.

The sweep-frequency filter oscillator can be used with all types of filters for checking the response. *See also* BAND-PASS RESPONSE, BAND-REJECTION RESPONSE, HIGHPASS RESPONSE, LOWPASS RESPONSE, and RECTANGULAR RESPONSE.

SWEEP GENERATOR

A *sweep generator* is a device that generates the fluctuating voltages or currents that operate the deflection apparatus in a cathode-ray tube. The changing voltages or currents cause the electron beam to move from left to right, causing a trace to appear on the screen. The voltage or current generally varies with time. *See also* CATHODE-RAY TUBE.

A wideband oscillator, with a frequency output that fluctuates continuously between two limits in a sweeping manner, is sometimes called a *sweep generator*. Sweep generators are used in conjunction with spectrum analyzers to evaluate the frequency-response characteristics of receivers, tuned circuits, and selective filters of various types. *See also* SWEEP, SPECTRUM ANALYZER, and SWEEP-FREQUENCY FILTER OSCILLATOR.

SWINGING CHOKE

A *swinging choke* is a special form of power-supply filter choke, designed to exhibit an inductance that varies inversely with the flow of current. When the current level is low, the swinging choke has maximum inductance, but the core saturates easily as the level of current rises.

In the average swinging choke, the inductance is halved when the current is tripled; the inductance is cut to $1/4$ its original value when the current increases by a factor of 9. Some swinging chokes have more or less variability. The low-current inductance of a swinging choke depends on the application, but is generally in the range of 1 to 15 H.

SWITCHING

Switching is a term broadly used in reference to the selection of components, circuits, bands, modes, or some other variable in an electronic environment. Switching can be categorized as either *manual* or *electronic*.

Manual Switching *Manual switches* include toggle devices, rotary devices, and in general, anything that the operator must manipulate consciously. In most equipment, the power switch is an example of a manual switch. So are the bandswitches on some receivers, transmitters and transceivers. A station might have several different antennas with individual feed lines, selectable by a manual coaxial switch at the operating position.

Manual switching works well in direct-current (dc) circuits,

and in some audio applications. But when a switch must handle radio frequencies (RF), especially higher frequencies, the wiring of the switch, and the switch contacts themselves, can introduce reactance into the system. This often causes trouble. In these cases, it is better to have the switching done electronically, or by hybrid means.

Electronic Switching Transistors, diodes and integrated circuits are capable of doing many switching operations per second. Examples include automatic transmit/receive switches that allow full break-in with a transmitter and receiver combination hooked up to a single antenna. Binary digital circuits, no matter how complex, are nothing more than arrays of electronic switches that can each have either of two states (high or low).

Electronic switches work by means of a change in the bias voltage across a diode, or across the P-N junction in a transistor, or in a metal-oxide-semiconductor (MOS) device. Some substances, and some arrangements, work faster than others. Some electronic switches can perform millons of operations per second; that is, they work up to many megahertz. Some even work in the gigahertz range—billions of operations per second.

Purely electronic switches, and switching networks, do their jobs without the conscious control of an operator. The transmit/receive switch, for example, works directly from the transmitter RF output. The operator doesn't have to think about the switching process.

Hybrid Switching Sometimes a manual switch controls an electronic switch, with the manual device handling a dc control voltage. This allows the control to be some distance away from where the actual switching takes place, without introducing inductance or capacitance into an RF circuit. It is called *hybrid switching*.

An example of hybrid switching is a remote-control coaxial switch, set up so that one feed line can be used with several antennas. There is a manual switch at the operating position, and this controls a coaxial switch located outdoors where the feed line branches off to the different antennas.

Related articles in this encyclopedia include COAXIAL SWITCH, CROSSBAR SWITCH, ROTARY SWITCH, SWITCHING DIODE, SWITCHING TRANSISTOR, TOGGLE SWITCH, and TRANSMIT/RECEIVE SWITCH.

SWITCHING DIODE

Semiconductor diodes are often used for switching purposes, especially at high speeds or at radio frequencies. Diodes conduct when they are supplied with a forward bias of a few volts; they do not conduct when no bias, or a reverse bias, is applied (*see* DIODE, and P-N JUNCTION).

For direct-current or low-frequency ac switching, most germanium or silicon diodes are satisfactory. For switching of radio-frequency signals, diodes must exhibit the smallest possible amount of capacitance. The PIN diode is often used for radio-frequency switching. An example of diode radio-frequency switching is shown on pg. 482. The radio-frequency signal can be directed along any of four paths by applying a direct-current forward bias to the appropriate diode. The bias is applied by remote control.

Diodes are used in some digital-logic devices for switching purposes. An example of this is found in diode-transistor logic, or DTL (*see* DIODE-TRANSISTOR LOGIC).

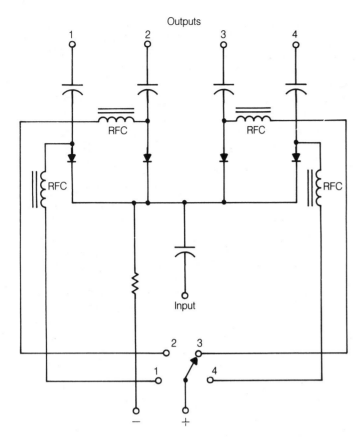

SWITCHING DIODE: Switching circuit for RF.

SWITCHING POWER SUPPLY

In a power supply, voltage or current regulation can be achieved in various ways. An unusual, but effective, method of regulation is the switching scheme. In the switching power supply, the output is not continuous, but instead it is switched on and off at a high rate of speed. The output duty cycle varies, depending on the load resistance. The supply is either fully on or fully off at any given instant. The output is smoothed by means of a conventional power-supply filter.

The load resistance is determined by means of a sensing device. If the load resistance decreases, the sensing device tells the switch to increase the duty cycle. If the load resistance increases, the duty cycle decreases. *See also* DUTY CYCLE, POWER SUPPLY, and REGULATION.

SWITCHING TRANSISTOR

Transistors are often used for switching purposes. A transistor conducts when the emitter base junction is provided with sufficient forward bias to cause saturation. If no bias, or a reverse bias, exists at the junction, the transistor will not conduct.

Transistors can be used for switching direct-current or alternating-current signals. Most transistors will operate well as switches for direct current and low-frequency alternating current. For switching radio-frequency signals, transistors should have the smallest possible capacitance.

The schematic illustrates direct-current and radio-frequency transistor switching schemes. In these examples, npn bipolar transistors are used. If pnp transistors were used, the polarities would be reversed. A positive voltage, applied at the base of a transistor, results in conduction.

Switching transistors are extensively used in digital integrated circuits. Hundreds, or even thousands, of individual switching transistors can be fabricated on a single chip. *See* DIODE-TRANSISTOR LOGIC, DIRECT-COUPLED TRANSISTOR LOGIC, EMITTER-COUPLED LOGIC, HIGH-THRESHOLD LOGIC, INTEGRATED INJECTION LOGIC, METAL-OXIDE-SEMICONDUCTOR LOGIC FAMILIES, RESISTOR-CAPACITOR-TRANSISTOR LOGIC, RESISTOR-TRANSISTOR LOGIC, TRANSISTOR-TRANSISTOR LOGIC, and TRIPLE-DIFFUSED EMITTER-FOLLOWER LOGIC.

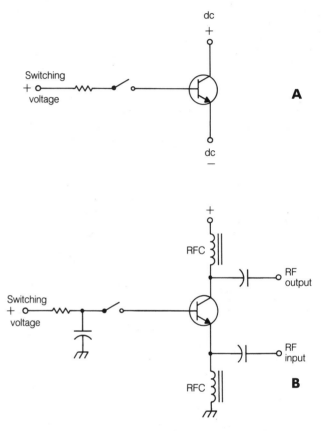

SWITCHING TRANSISTOR: At A, a switching circuit for dc; at B, a circuit for RF.

SWL

See SHORTWAVE LISTENING.

SWR

See STANDING-WAVE RATIO.

SWR LOSS

See STANDING-WAVE-RATIO LOSS.

SWR METER

See REFLECTOMETER.

SYMMETRICAL RESPONSE

A bandpass filter has a *symmetrical response* if the attenuation-versus frequency curve is balanced. Mathematically, a symmetrical response can be defined in terms of the center frequency.

Suppose that f is the center frequency of a bandpass filter. Let f_1 be some frequency below the center frequency, and let f_2 be some frequency above the center of the passband, such that $f_2 - f = f - f_1$. If the response is symmetrical, then the filter attenuation at frequency f_1 will be the same as the filter attenuation at f_2, no matter what values we choose for f_1 and f_2 within the above constraints. *See also* BANDPASS FILTER, and BANDPASS RESPONSE.

SYNCHRO

A *synchro* is a special type of motor, used for remote control of mechanical devices. A synchro consists of a generator and a receiver motor. As the shaft of the generator is turned, the shaft of the receiver motor follows along exactly.

In antenna rotors, a form of synchro can be used to indicate the rotor position. The generator is connected mechanically to the antenna mast, and the receiver is located in the station and connected to a directional pointer. This type of synchro is called a *selsyn* (*see* SELSYN).

Some synchro devices provide a digital indication of the angular position of the generator shaft. Some antenna rotor systems use this arrangement. Rather than a pointer, the display box has a digital readout of the azimuth bearing. Some synchro devices are programmable. The operator inputs a number into the synchro generator, and the receiver changes position accordingly.

SYNCHRODYNE

See DIRECT-CONVERSION RECEIVER.

SYNCHRONIZATION

When two signals or processes are exactly aligned, they are said to be in *synchronization*. Two identical waveforms, for example, are synchronized if they are in phase. Two clocks are synchronized if they agree more or less precisely.

In a television transmitting and receiving system, the electron beam in the picture tube must move in synchronization with the beam in the camera tube. Otherwise, the picture will appear to be split, rolling, or tearing (*see* HORIZONTAL SYNCHRONIZATION, and VERTICAL SYNCHRONIZATION). Pulses are sent by the transmitting station to keep the receiver in synchronization so that a clear picture is presented.

In an oscilloscope, the sweep rate can be synchronized with the frequency of the applied signal to obtain a motion-free display. This is called *triggering* (*see* TRIGGERING).

In some communications systems, the transmitter and receiver are synchronized against an external, common time standard. This is sometimes called *synchronous* or *coherent communications* (*see* SYNCHRONIZED COMMUNICATIONS).

SYNCHRONIZED COMMUNICATIONS

One of the most important problems in communications is the maximization of the number of signals that can be accommodated within a given band of frequencies. This has traditionally been done by attempting to minimize the bandwidth of a signal. There is a limit, however, to how small the bandwidth can be if the information is to be effectively received (*see* BANDWIDTH).

Digital signals, such as morse code, occupy less bandwidth than analog signals, such as voice. Perfectly timed morse code consists of regularly spaced bits, each bit having the duration of one dot. The length of a dash is three bits; the space between dots and/or dashes in a single character is one bit; the space between characters in a word is three bits; the space between words and sentences is seven bits. This makes it possible to identify every single bit by number, even in a long message (see A in illustration). Therefore, the receiver and transmitter can be synchronized so that the receiver "knows" which bit of the message is being sent at a given moment.

In synchronized or coherent morse code, the receiver and transmitter are synchronized so that the receiver hears and evaluates each bit individually. This makes it possible to use a receiving filter having extremely narrow bandwidth. The synchronization requires the use of an external, common frequency or time standard. The broadcasts of WWV/WWVH/WWVB are used for this purpose (*see* WWV/WWVH/WWVB). Frequency dividers are used to obtain the necessary synchronizing frequencies. A tone is generated in the receiver output for a particular bit, if, and only if, the average signal voltage exceeds a certain value over the duration of that bit, as at B. False signals, such as might be caused by filter ringing, sferics, or other noise, are generally ignored since they do not result in sufficient average voltage.

Synchronization can be used with other codes besides morse. For example, a system might be built for use with Baudot or ASCII teletype transmitters and receivers. Experiments with synchronized communications have shown that the improvement in signal-to-noise ratio, compared with nonsynchronized systems, is nearly 20 dB at low speeds (on the order of 15–20 words per minute). The reduced bandwidth of coherent communications allows proportionately more signals to be placed in any given band of frequencies.

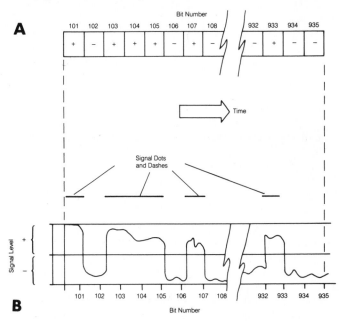

SYNCHRONIZED COMMUNICATIONS: High/low signals at A result in a signal at B.

SYNCHRONOUS DATA

Synchronous data is any data, usually digital, that is transmitted according to a precise time function. Baudot and ASCII are examples of codes that can be sent as synchronous data. Each bit has the same predetermined duration. Mechanically sent morse code can also be synchronous.

In some synchronous-data communications circuits, the receiver is kept in step with the transmitter by means of pulses sent at the beginning and/or end of each character, or at regular intervals. In radioteletype, some codes use such pulses to indicate the start and finish of each character. Television transmitters send out synchronizing pulses to keep the receiver scanning properly.

A special form of synchronous-data communication uses an independent reference standard to lock the receiver and transmitter precisely. This results in great improvement in the signal-to-noise ratio. *See also* SYNCHRONIZATION, and SYNCHRONIZED COMMUNICATIONS.

SYNCHRONOUS DATA LINK CONTROL

In the early days of packet radio, the protocol most often used was based on *synchronous data link control (SDLC)*. This was the protocol used by International Business Machines (IBM) at the link layer of *open systems interconnection reference model (OSI-RM)*. Today, the standard link-layer protocol is known as *AX.25*. *See also* AX.25, OPEN SYSTEMS INTERCONNECTION REFERENCE MODEL, and PACKET RADIO.

SYNCHRONOUS DETECTOR

A *synchronous detector* is a circuit that operates in a manner similar to a product detector (*see* PRODUCT DETECTOR). A local-oscillator signal is mixed with the incoming signal to recover the sidebands.

The synchronous detector uses a phase-locking system that keeps the local-oscillator output in exact synchronization with the carrier wave of the incoming signal. This ensures that the output of the detector will be identical with the original modulating signals.

Synchronous detectors are used in color television receivers to recover the chrominance sidebands with optimum precision. If the local-oscillator signal were not locked with the carrier or subcarrier, distortion would result from even the smallest frequency difference.

SYNCHRONOUS MULTIPLEX

Synchronous multiplex is a form of time-division multiplex, in which two or more signals are transmitted over a single circuit at the same time. The signals are split into discrete intervals having equal duration, and interwoven at a higher speed. Synchronous multiplexing is a form of serial data transfer (*see* SERIAL DATA TRANSFER, and TIME-DIVISION MULTIPLEX).

Suppose there are n different signals, designated $s_1, s_2, s_3, \ldots, s_n$, that are to be combined in series by means of synchronous multiplexing. Let each signal be divided into intervals having duration t. When the signals are recombined, they are sent one interval at a time, in sequence $s_1, s_2, s_3, \ldots, s_n, s_1, s_2, \ldots$. In order to maintain synchronization of the composite signal with each of the original signals, the information in each interval must be speeded up by a factor of n, resulting in interval durations of t/n.

TANGENT

The *tangent function* is a trigonometric function. In a right triangle, the tangent is equal to the length of the opposite side divided by the length of the adjacent side. In the unit circle $x^2 + y^2 = 1$, plotted on the Cartesian (x,y) plane, the tangent of the angle θ, measured counterclockwise from the positive x axis, is equal to y/x. For values of θ that are an odd multiple of 90 degrees, the tangent is not defined because for those angles, $x = 0$.

The tangent function is periodic and discontinuous. The discontinuities appear at odd multiples of 90 degrees. The tangent function ranges through the entire set of real numbers.

In mathematical calculations, the tangent function is abbreviated tan. Mathematically, the tangent function is always equal to the value of the sine divided by the value of the cosine:

$$\tan \theta = (\sin \theta)/(\cos \theta)$$

See also TRIGONOMETRIC FUNCTION.

TANGENTIAL SENSITIVITY

The sensitivity of a receiver is usually expressed in terms of the signal voltage that results in a certain signal-to-noise ratio or amount of noise quieting in the audio output (*see* NOISE QUIETING, SENSITIVITY, and SIGNAL-TO-NOISE RATIO). The most frequently used figures are a 10-dB signal-to-noise ratio or 20 dB of noise quieting. These specifications are not always a precise indication of how well a receiver will respond to weak signals. A more accurate indication of this is given by the tangential-sensitivity figure. Tangential sensitivity is sometimes called *threshold sensitivity* or *weak-signal sensitivity*.

Tangential sensitivity is defined as the signal level, in microvolts at the antenna terminals, that results in a barely discernible signal at the output under conditions of minimum external noise. Occasionally, the tangential sensitivity is defined as the signal level that produces a 3-dB signal-to-noise ratio, or 3 dB of noise quieting, at the output. These voltages are considerably smaller than the voltages required to produce a 10-dB signal-to-noise ratio or 20 dB of noise quieting. Two different receivers, both having the same sensitivity in terms of the 10-dB signal-to-noise or 20-dB noise-quieting figures, might differ in tangential sensitivity.

TANK CIRCUIT

A *tank circuit* is an electrical circuit that stores energy by passing it alternately between two reactances. A tank circuit operates at one frequency, and perhaps at integral multiples of that frequency. All parallel-resonant circuits are tank circuits. A resonant antenna, cavity, or length of transmission line can also act as a tank circuit if it is fed at a voltage loop. At resonance, the impedance of a tank circuit is theoretically infinite. In practice, there are some component losses, but the impedance is still extremely high.

Many radio-frequency power amplifiers employ parallel-resonant tuned circuits in the output. This circuit is often called the *tank circuit* or simply the *tank*. Either the inductor or capacitor, or both, are adjustable so that the tank circuit can be made resonant at the desired frequency. In the tuned power amplifier, tank-circuit resonance is indicated by a dip in the collector, drain, or plate current. *See also* PARALLEL RESONANCE, RESONANCE, and TUNED CIRCUIT.

TANTALUM CAPACITOR

A *tantalum capacitor* is a polarized device similar to an electrolytic capacitor. Tantalum capacitors characteristically have large values and small size, and this is their main advantage. Tantalum capacitors are extensively used in miniaturized equipment.

A tantalum electrode is placed in an electrolyte solution. This causes a thin layer of tantalum oxide to form on the surface of the electrode. Tantalum oxide is an insulating, or dielectric material, so a capacitor is formed by the electrolyte and the tantalum. Tantalum oxide has a very high dielectric constant— about three times that of the aluminum oxide found in conventional electrolytic capacitors. This makes it possible to obtain more capacitance per unit volume.

Tantalum capacitors are used at low voltages, and primarily at audio frequencies. When tantalum capacitors are used, it is necessary to observe the proper polarity, just as with conventional electrolytics. *See also* CAPACITOR, and ELECTROLYTIC CAPACITOR.

T ANTENNA

A *T antenna* is a wire antenna with a vertical radiator and a capacitance hat consisting of a horizontal wire. The vertical radiator is connected to the center of the horizontal element, as shown in the illustration on pg. 486. The T antenna radiates, and responds to, vertically polarized electromagnetic fields.

The main advantage of the T antenna is that it is shorter, for a given frequency of operation, than a straight quarter-wave vertical antenna. The resonant frequency of the T antenna depends on the height h of the vertical radiator and the radius r of the horizontal radiator. An approximate formula for determining the resonant frequency f, in megahertz, is:

$$f = 234/(h + 1.5r)$$

for h and r in feet, and:

$$f = 71/(h + 1.5r)$$

for h and r in meters.

For optimum performance, a system of radials should be used with the T antenna. The feed-point impedance varies from about 20 ohms to 35 ohms, depending on ratio h/r. The larger the ratio h/r, the higher the feed-point impedance will be. *See also* VERTICAL ANTENNA, and VERTICAL POLARIZATION.

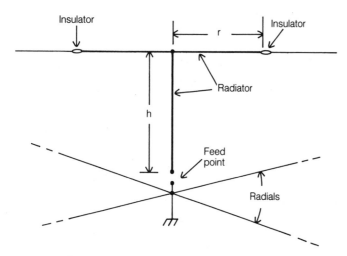

T ANTENNA: The horizontal wire forms a capacitance hat.

TAP

A potentiometer, rheostat, inductor, transformer primary, or transformer secondary can have one or more leads connected to an intermediate point in the element or winding, as well as at either end. Such a connection is called a *tap*. An element or winding can have several taps.

In a potentiometer, a tap is used to provide adjustment of the resistance of the device. The same is true of a rheostat. The tap position is adjustable.

In an inductor or transformer, a tap can be used to obtain various values of inductance or impedance. Some inductors have variable taps, but most coil taps are fixed. A center tap can be used to obtain two signals of opposite phase and equal amplitude. Center-tapped transformer secondaries are used for balanced-signal output at audio and radio frequencies. In a power transformer, a center-tapped secondary facilitates full-wave rectification. *See also* BALANCED OUTPUT, and FULL-WAVE RECTIFIER.

TAPERED FEED LINE

In a radio-frequency transmission line, the characteristic impedance can be made to vary with length. This is accomplished by means of a gradual change in the spacing between the conductors. Such a transmission line is called a *tapered line*.

A tapered line can be used for impedance-matching purposes. The principle is similar to that of the delta match (*see* DELTA MATCH). Suppose that an impedance Z_1 is to be matched to an impedance Z_2. A tapered line is connected between the two points. The characteristic impedance of the tapered line is equal to Z_1 at one end and Z_2 at the other end.

The main advantage of a tapered matching section over a stub type matching section is that no impedance "bumps" are created. Therefore, there are no standing waves on the line. This results in lower loss, and consequently the efficiency is better. Tapered parallel-wire sections can be tailor made for a particular antenna system.

TAUT-BAND SUSPENSION

The *taut-band suspension* is a fairly common type of analog meter movement. The principle is similar to the D'Arsonval movement (*see* D'ARSONVAL), except that a different method is used to obtain suspension and turning tension.

In a taut-band meter, a coil is rigidly attached to an indicating needle. The coil is placed within the field of a permanent magnet. The coil end of the indicating needle is suspended by two metal strips.

When a current passes through the coil, a magnetic field appears around the coil. This causes torque between the coil and the magnet. The needle therefore moves upscale until the torque is balanced by the resistance of the metal strips.

The main advantage of the taut-band suspension is that it is simpler and generally less expensive than the D'Arsonval-type suspension. Also, there are no pivot bearings to become loose or stick.

TCHEBYSHEV FILTER

See CHEBYSHEV FILTER.

TEE CONNECTION

When a conductor or cable feeds two other conductors or cables, the junction is sometimes called a *tee connection*. A tee is used for such purposes as splitting (when impedance matching is not critical), phasing, and the connection of stubs to a transmission line. The connection gets its name from the fact that is looks like a capital letter T in schematic diagrams.

Electrically, a tee is identical to a *wye* (Y). The wye expression is more often used in polyphase alternating-current systems. *See also* WYE CONNECTION.

TELECOMMAND

In amateur satellite operation, it is often necessary to adjust the operation of the satellite in some way. The most common example is a change in the operating mode of the transponder. The transponder mode is the specific pair of uplink and downlink frequency bands.

Satellite operation is adjusted via signals, on certain frequencies and encoded in specific ways, from a ground station. Such a signal is called a *telecommand*. *See also* ACTIVE COMMUNICATIONS SATELLITE, DOWNLINK, SATELLITE TRANSPONDER MODES, TRANSPONDER, and UPLINK.

TELEGRAPHY

Morse-code transmission is often called *telegraphy*. In telecommunication, telegraphy is regarded as the simplest mode of transmission because a code transmitter is easy to construct. Many radio operators regard telegraphy as the most efficient and accurate means of communication in high-frequency, ionospheric circuits. The human ear and brain make an excellent information processor! Telegraphy signals can still be heard in the high-frequency band today.

Proficiency in telegraphy is required for anyone who intends to operate amateur radio stations below 30 MHz in the United States. Telegraphy is also taught in the military. Telegraphy can be sent with a simple straight key, or with any of a variety of other keying devices. *See also* INTERNATIONAL MORSE CODE, KEYER, and SEMIAUTOMATIC KEY.

TELEMETRY

Telemetry is the transmission of quantitative information from one point to another by electromagnetic (radio) means. Telemetry is extensively used by weather balloons, satellites, and other environmental monitoring apparatus. Telemetry is used in space flights, both manned and unmanned, to keep track of all aspects of the equipment and the physical condition of astronauts.

A telemetry transmitter consists of a measuring instrument, an encoder that translates the instrument readings into electrical impulses, and a modulated radio transmitter with an antenna (see illustration at A). A telemetry receiver consists of a radio receiver with an antenna, a demodulator, and a recorder (B). A computer can also be used to process the data received.

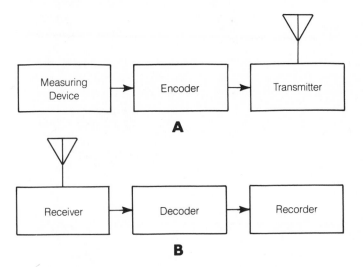

TELEMETRY: At A, a transmitter; at B, a receiver.

TELEPHONE

Any voice communications system constitutes a *telephone*, but the term is usually applied to a wire-and-radio system having many users, all of whom can contact any other. The earliest telephone consisted of a power supply, a variable-resistance transducer (microphone) for converting sound into electrical impulses, an earphone, and a long length of wire. The direct-current power supply was modulated, generating audio-frequency pulsations. The range was limited by the loss in the wire. Switching was done manually by operators at a central office.

Modern telephones use single-sideband-modulated carriers, digital multiplexing, and other sophisticated techniques for transmission. A large system is usually operated by a computer located at a master control center. The computer performs all switching functions. It also has a recorder-announcer that informs subscribers of disconnected or changed numbers, system overload, and other problems. Operators are still used to handle special problems.

Operating Features of Telephone Systems One of the most important considerations in designing and building a telephone system is the problem of switching. The complexity of the switching network depends on the number of users or subscribers. A telephone system normally has the following two characteristics:

1. Accessibility: any user can call any other user most of the time.
2. No redundancy: whenever a subscriber X calls a subscriber Y, only Y will receive the call.

Some telephone systems have various other features, such as:

- Conference calling: three or more subscribers can communicate on a single line.
- Call forwarding: a subscriber X can arrange to have his calls directed to any other subscriber Y.
- Call waiting: if subscriber X calls subscriber Y and the line is busy, subscriber Y is informed that someone is trying to call. Subscriber Y can then communicate with X at any time.
- Radio links: Subscribers can have mobile or portable telephone sets, linked by wireless radio into the system. An example of this is the cellular radio network.
- Local switching networks: A subscriber can have a single line split up into local lines (extensions). This arrangement is often used by businesses.

Interconnection A telephone system consists of one or more main lines, called *trunks*, one or more central-office switches, numerous branch lines, and the telephone sets themselves. There can also be operators, radio links, satellite links, and local switching systems. There can also be a provision for connecting with other systems.

When a subscriber X wishes to contact a subscriber Y, X dials a number consisting of several digits. The number of digits depends on the complexity of the switch. In the United States, telephone numbers have 10 digits, consisting of a three-digit area code and a seven-digit subscriber number. If the subscriber Y has a switching system, an operator answers the call and directs it to the appropriate extension.

Most telephone systems cannot operate properly if every subscriber attempts to use it at the same time. The system can handle a limited number of calls, based on the assumption that only a certain percentage of users will try to make calls at a given time. In areas of rapid population growth, or during holidays (such as Christmas), a master control center can become overloaded. Callers are then informed that all lines are busy.

Long-distance telephone circuits require the use of low-noise amplifiers and signal processors. The farther a signal travels along a wire, the worse the signal-to-noise ratio becomes. Thermal noise, shot-effect noise, sferics, and geomagnetic-field fluctuations all contribute to the signal-to-noise problem. Modern technology has largely overcome this problem, but long-distance circuits occasionally suffer when conditions are adverse.

For Further Information A detailed discussion of telephony is beyond the scope of this book. For more information, consult a textbook on telephony.

TELETYPE®

Teletype is a tradename of the Teletype Corporation. A teletype system can use wire transmission only, a combination of wire and radio circuits, or radio circuits exclusively. The transmission of printed material via radio-only circuits is called *radioteletype* and sometimes abbreviated *RTTY*. See RADIO-TELETYPE.

Teletype signals are digital, consisting of two levels or tones corresponding to on (high) and off (low) conditions. In wire systems, these conditions are represented by direct currents. In radio systems, the high and low conditions are represented by different carrier frequencies. The on state is called *mark*, and the off state is called *space* (*see* FREQUENCY-SHIFT KEYING, and MARK/SPACE).

Teletype systems normally use either the Baudot code or the ASCII code. The speed can vary (*see* ASCII, BAUDOT, BAUD RATE, and SPEED OF TRANSMISSION).

The heart of a teletype installation is the teleprinter, which resembles a typewriter. Messages are sent by typing on the keyboard. The message is printed out simultaneously in the transmitting station and at all receiving stations. For reception, the attending operator need do nothing at all, except make sure that the paper piles up neatly if the message is long!

Many teleprinters have a means of preloading and storing a message prior to transmitting it. In modern systems, this is done with electronic memory. In older teleprinters, messages are encoded on paper tape by a device called a *reperforator*. The reperforator punches holes in the tape according to a five-level or eight-level code (*see* PAPER TAPE). When the paper tape is run through the sending apparatus, the message goes out at the normal system speed.

Home computers can be used as teleprinters. There are various software packages available for this purpose. A teleprinter, conversely, can be interfaced with the telephone lines via a modem, and thereby interconnected with a computer (*see* MODEM).

Some teletype systems operate over telephone lines, making possible the quick sending of messages at moderate cost. Two-way communications can also be accomplished. This is called a *telex system*. It is extensively used by businesses throughout the world.

TELEVISION

Television is the transfer of moving visual images from one place to another. Television has existed for only a few decades, but it has greatly changed our lives since broadcasting stations first began to use it around 1950. Television is used not only for broadcasting, but for two-way communications. The video signals of television are normally sent and received along with audio signals.

Television systems can be categorized as either fast-scan or slow-scan. In broadcasting, fast-scan television is always used. Slow-scan television is used for communications when the available band space is limited (*see* SLOW-SCAN TELEVISION).

The Television Picture and Signal In order to get a realistic impression of motion, it is necessary to transmit at least 20 still pictures per second, and the detail must be adequate. A fast-scan television system provides 30 images, or frames, each second. There are 525 or 625 lines in each frame, running horizontally across the picture, which is 1.33 times as wide as it is high. Each line contains shades of brightness in a black-and-white system, and shades of brightness and color in a color system. The image is sent as an amplitude-modulated signal, and the sound is sent as a frequency-modulated signal.

Because of the large amount of information sent, the fast-scan television channel is wide. A standard television-broadcast channel in the North American system takes up 6 MHz of spectrum space. All television broadcasting is done at very high and ultra-high frequencies for this reason.

The Transmitter and Receiver A television transmitter consists of a camera tube, an oscillator, an amplitude modulator, and a series of amplifiers for the video signal. The audio system consists of an input device (such as a microphone), an oscillator, a frequency modulator, and a feed system that couples the radio-frequency output into the video amplifier chain. There is also, of course, an antenna or cable output. A simplified block diagram of a television transmitter is illustrated at A.

A television receiver is shown in simplified block form at B. The receiver contains an antenna, or an input having an impedance of either 75 ohms or 300 ohms, a tunable front end, an oscillator and mixer, a set of intermediate-frequency amplifiers, a video demodulator, an audio demodulator and amplifier chain, a picture tube with associated peripheral circuitry, and a speaker.

In order for a television picture to appear normal, the transmitter and the receiver must be exactly synchronized. The studio equipment generates pulses at the end of each line and at the end of each complete frame. These pulses are sent along with the video signal. In the receiver, the demodulator recovers the synchronizing pulses and sends them to the picture tube. The electron beam in the picture tube thus moves in exact synchronization with the scanning beam in the camera tube. If the synchronization is upset, the picture appears to "roll" or "tear."

New Trends in Television In recent years, more and more television transmission has been done via cable. In most major metropolitan areas, cable television is available to anyone who wants it. An increasing number of television stations also broadcast through geostationary satellites. People can select from dozens, or even hundreds, of different channels.

Fast-scan television is used by radio amateurs for communications at ultra-high and microwave frequencies. Because of the use of cable and satellite systems, television picture quality is much better than ever before. Some thought has been given to increasing the number of lines per frame in the television picture to get a sharper image.

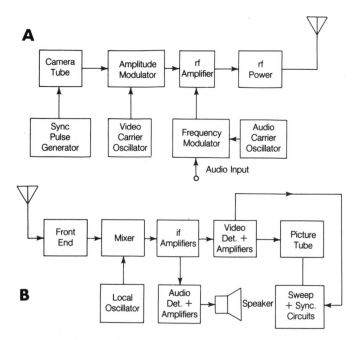

TELEVISION: At A, a TV transmitter; at B, a TV receiver.

Television can be used along with a computer. Experiments are being conducted in some areas, allowing viewers to interact with a program on television — for such things as shopping and opinion polls.

For further information about fast-scan television, consult the following articles: ASPECT RATIO, CAMERA TUBE, COLOR PICTURE SIGNAL, COLOR TELEVISION, COMPOSITE VIDEO SIGNAL, HORIZONTAL SYNCHRONIZATION, ICONOSCOPE, IMAGE ORTHICON, PICTURE SIGNAL, PICTURE TUBE, RASTER, VERTICAL SYNCHRONIZATION, and VIDICON.

TELEVISION BROADCAST BAND

In the United States, fast-scan television broadcasts are made on 68 different channels in the VHF and UHF ranges. Each channel is 6 MHz wide, including both the video and audio information.

There is no channel 1. Channels 2 through 13 are sometimes called the *very-high-frequency (VHF) television channels.* Channels 14 through 69 are called the *ultra-high-frequency (UHF) channels.*

Ionospheric propagation occasionally affects channels 2 through 6. Long-distance propagation can take place because of dense ionization in the E layer (*see* SPORADIC-E PROPAGATION) on these channels. Tropospheric propagation is occasionally observed on all of the VHF channels (*see* TROPOSPHERIC PROPAGATION). The UHF channels are unaffected by the ionosphere, although tropospheric effects can occur to a small extent.

Cable television signals are transmitted on a variety of wavelengths. Because the cable prevents signals from getting in or out, the entire electromagnetic spectrum can be used. Satellite television systems use frequencies other than those in standard television broadcasts. *See also* TELEVISION.

TEMPERATURE COEFFICIENT

Many electronic components are affected by fluctuations in temperature. Resistors and capacitors, especially, tend to change value when the temperature varies over a wide range. The tendency of a component to change in value with temperature variations is known as *temperature coefficient.*

If the value of a component decreases as the temperature rises, that component is said to have a negative temperature coefficient. If the value increases as the temperature rises, a component has a positive temperature coefficient. A few components exhibit relatively constant value regardless of the temperature; these devices are said to have a *zero temperature coefficient.* The temperature coefficient is usually expressed in percent per degree Celsius. For piezoelectric crystals, the temperature coefficient is often expressed in hertz or kilohertz per degree Celsius.

As an example, consider a capacitor that has a value of 100 pF at room temperature (20° C). If the component has a positive temperature coefficient of 0.1 percent per degree Celsius (0.1%/° C), then the value will rise by 0.1 pF as the temperature rises by 1° C. At 21° C, the value will be 100.1 pF; at 30° C, it will be 101 pF; at 120° C, it will be 110 pF.

Temperature coefficients are an important consideration in radio-frequency circuit design. This is especially true of oscillators. Sometimes, a component having a certain temperature co-efficient is deliberately placed in a circuit to offset the effects of the temperature coefficients of other components. This is known as *temperature compensation. See also* TEMPERATURE COMPENSATION.

TEMPERATURE COMPENSATION

Changes in temperature can cause instability in electronic circuits, especially oscillators. Although crystal oscillators are generally more stable than variable-frequency oscillators, either type of circuit will exhibit some drift as the ambient temperature rises or falls. To compensate for such changes, special components can be added to an oscillator circuit.

Suppose that a crystal has a negative temperature coefficient of 10 Hz/° C. This means that, for a rise in temperature of 1° C, the frequency will fall by 10 Hz; for a rise of 10° C, the frequency will drop by 100 Hz. Conversely, as the temperature falls, the frequency increases.

To compensate for this, a small capacitor, having a known negative temperature coefficient, can be installed in parallel with the crystal. As the temperature rises, the value of the capacitor decreases, effectively raising the frequency of the crystal-capacitor combination. As the temperature falls, the value of the capacitor gets larger, pulling the frequency down. If the capacitor is carefully selected, the temperature-coefficient effects of the capacitor will just offset those of the crystal, and the frequency will not change as the temperature fluctuates.

In any oscillator circuit, it is desirable to use components with temperature coefficients as close to zero as possible. This makes temperature compensation simpler than when components having large positive or negative temperature coefficients are used. It also makes sense to keep the temperature as constant as possible. *See also* TEMPERATURE COEFFICIENT.

TEMPERATURE DERATING

Some electronic equipment, especially power amplifiers, generate significant heat. If this heat becomes excessive, components can be damaged. The temperature is more likely to exceed the maximum limits if the weather is hot, as compared to when it is cold. A component can dissipate more power when the ambient temperature is low, and less power when it is hot. For this reason, some electronic devices must be operated at a reduced power level when the temperature is high. The hotter it gets, the more the power must be reduced. This deliberate reduction of power is called *temperature derating.*

Temperature derating is specified according to a graph of power level versus temperature. A radio transmitter, for example, might normally be run at 100 watts output. However, if the temperature exceeds a certain limit, the power input is reduced according to a derating curve. At some point, the curve drops to zero, indicating that the equipment should not be operated at all when the temperature gets that high.

Extreme temperatures do not usually occur in nature. However, inside a cabinet containing many heat-generating components, or inside an enclosed automobile on a hot, sunny day, the temperature can reach nearly 200 degrees Fahrenheit. This can have a significant effect on electronic equipment.

TEMPERATURE REGULATION

Many types of electronic equipment, especially complicated apparatus, is temperature-sensitive. Such equipment requires some form of temperature regulation so that it can function in an unchanging environment.

The most common form of temperature regulation consists of heating and air-conditioning systems. In buildings where computers are located, the temperature should be kept as constant as possible. Atmospheric humidity can also be regulated. These measures serve to minimize down time and service costs, as well as to optimize accuracy and reliability.

Temperature regulation is used to keep reference oscillators operating at constant frequency. Such oscillators are crystal-controlled, and the crystals are housed in temperature-regulated chambers called *ovens* (See CRYSTAL OVEN). All temperature-regulation devices use a heating or cooling system, or both, as well as a thermostat.

TENSILE STRENGTH

The amount of tension that a wire can withstand is known as the *tensile strength* of the wire. Tensile strength is sometimes called the *breaking load*. Tensile strength is usually indicated in pounds per square inch, or in kilograms per square centimeter, of cross-sectional area. Tensile strength can also be expressed in terms of the breaking load in pounds or kilograms. The larger the size of the wire for a given material, the greater the tensile strength.

Steel has the greatest tensile strength of common wire materials. The breaking load varies from about 75,000 to 150,000 pounds per square inch of cross-sectional area. For this reason, steel wire is preferred for use in tower-guying systems. Copper and aluminum wire have somewhat less tensile strength; typically, it ranges from 30,000 to 60,000 pounds per square inch for copper and 20,000 to 40,000 pounds per square inch for aluminum.

The average breaking load for various gauges of steel, copper, and aluminum wire is indicated in the accompanying table. *See also* WIRE.

TENSILE STRENGTH: TENSILE STRENGTH FOR VARIOUS AMERICAN WIRE GAUGE (AWG) SIZES OF STEEL, COPPER, AND ALUMINUM WIRE. ALL FIGURES ARE GIVEN IN POUNDS, AND ARE APPROXIMATE.

AWG No.	Steel wire	Copper wire	Aluminum wire
1	5,000–10,000	2,000–4,000	1,300–2,600
2	3,900–7,800	1,600–3,200	1,000–2,000
3	3,100–6,200	1,200–2,400	830–1,700
4	2,500–4,900	980–2,000	650–1,300
5	2,000–3,900	780–1,600	520–1,000
6	1,600–3,200	630–1,300	420–840
7	1,200–2,500	490–980	330–650
8	980–2,000	390–780	260–520
9	770–1,500	310–620	210–410
10	610–1,200	240–490	160–330
11	490–970	190–390	130–260
12	380–770	150–310	100–200
13	305–610	120–240	80–160
14	240–480	97–190	65–130
15	190–380	77–150	51–100
16	150–300	61–120	41–81
17	120–240	48–97	32–64
18	96–190	38–77	26–51
19	76–150	30–61	20–41
20	60–120	24–48	16–32

TERA

Tera is a prefix multiplier that means 1,000,000,000,000 (10^{12}). For example, 1 terahertz is 10^{12} Hz = 1,000,000 MHz (the wavelength of a signal at this frequency is just 0.3 mm). The abbreviation for tera is the capital letter T. *See also* PREFIX MULTIPLIERS.

TERMINAL

A *terminal* is a point at which two or more wires are connected together, or where voltage or power is applied or taken from a circuit. You can speak, for example, of the antenna terminals of a receiver or transmitter, or the speaker terminals of a stereo high-fidelity amplifier. Input and output terminals are generally equipped with binding posts or connectors.

In a large computer, the operating console or consoles are located separately from the actual computer circuitry. Such a console, which consists of a keyboard and a teleprinter or cathode-ray-tube display, is called a *terminal*. Some computer terminals can be connected to a telephone and used with a distant computer.

In a Teletype or radioteletype station, the keyboard, modem, and printer or cathode-ray-tube display are called the *terminal*. The modem alone is sometimes called a *terminal unit*. *See also* MODEM, and VIDEO DISPLAY TERMINAL.

TERMINAL EMULATION SOFTWARE

In digital communications, personal computers (PCs) are often used as terminals. A PC can be employed to send and receive AMTOR, morse code, packet radio and radioteletype (ASCII or BAUDOT).

For a PC to function as a communications terminal, special software is needed. The exact nature of the software depends on the parameters of the communications mode to be used. This software is commercially available on floppy disks, just like any PC software. It is called *terminal emulation software*. *See also* AMTOR, ASCII, BAUDOT, MORSE CODE, PACKET RADIO and related articles, RADIOTELETYPE, and SOFTWARE.

TERMINAL NODE CONTROLLER

In a packet radio station, a *terminal node controller (TNC)* interfaces between a communications terminal and a radio transceiver. A packet-radio terminal is usually a personal computer (PC) that uses special programming called *terminal emulation software*.

The TNC assembles packets composed on, and stored in, the terminal, and gets them in the proper form to be transmitted by the radio. The TNC also disassembles packets from the radio receiver, and puts them in the right form to be displayed and/or stored by the terminal.

A TNC contains memory circuits. This allows messages to be held for later transmission. A TNC can store an incoming message and send it out later; in this way, any packet station can operate as a *digipeater* or *node* in the network.

Using firmware known as *NET/ROM*, a TNC can be converted into a *network node controller*. *See* DIGIPEATER, NETWORK NODE CONTROLLER, NODE, PACKET RADIO, and TERMINAL EMULATION SOFTWARE.

TERMINAL RESISTANCE HEATING

When current flows through a terminal, heat is sometimes generated because of terminal resistance. This heating is normally very minor, but if excessive current flows, or if the terminal contacts are dirty, a large amount of power can be dissipated.

Consider a terminal that carries 10 A of current. If the resistance of the junction is just 0.1 ohm, the power dissipated will be 10 watts. This much power can raise a small terminal to a dangerously high temperature. Terminal resistance heating can cause damage to electrical and electronic equipment, and can also present a fire hazard. The preceding example is typical of high-current household appliances. You have probably noticed terminal resistance heating when pulling out the plug of some heavy appliance.

To minimize the chances and hazards of terminal resistance heating, all plugs, connectors, and terminals should be kept clean. Terminals should not be forced to carry more current than they can handle. Flammable materials should be kept well away from terminals that carry large currents.

TERMINAL STRIP

See TIE STRIP.

TERMINATION

The point at which a transmission line is connected to a load is called the *termination*. The load itself can also be called the *termination*. A termination can consist of any type of load, such as an antenna, a dummy antenna, a telephone set, or an electrical appliance. Sometimes a transmission line is terminated in a short or open circuit. In a radio-transmitting antenna system, the termination is usually called the *feed point*.

Ideally, all of the available power at a termination is radiated, absorbed, or dissipated by the load. In a radio-transmitting antenna, a cable-television system, a telephone line, an audio-frequency amplifier/speaker system, and many other types of circuits, this requires that the load impedance be purely resistive, with an ohmic value equal to the characteristic impedance of the transmission line. *See also* CHARACTERISTIC IMPEDANCE, IMPEDANCE, and IMPEDANCE MATCHING.

TERTIARY COIL

Some audio-frequency or radio-frequency transformers have a small coil, coupled to the main windings, that can be used for feedback purposes. Such a coil is called a *tertiary coil*. In a radio-frequency transformer, a tertiary coil is often called a *tickler coil*.

A tertiary winding can be used to facilitate either positive feedback or negative feedback. The Armstrong oscillator uses a tertiary coil to obtain the positive feedback needed to maintain oscillation. Some regenerative detectors use tertiary windings to obtain positive feedback. A negative-feedback tertiary winding is sometimes used to keep audio-frequency or radio-frequency power amplifiers from oscillating. *See also* FEEDBACK.

TESLA

The *tesla* is the standard international (SI) unit of magnetic flux, equivalent to 1 weber per square meter. Other units are also used quite often to express magnetic flux density. The most common of these is the gauss. A flux density of 1 tesla is equivalent to 10,000 gauss. *See also* GAUSS, MAGNETIC FLUX, and WEBER.

TEST MESSAGE

A *test message* or *test signal* is sent by a radio or television transmitter operator to give receiving operators time to adjust their equipment for optimum reception. There are various different test messages in common use.

In Morse-code communications, the test message "VVV" is used. The sending operator keeps repeating the message until the receiving operator tells him to stop. Sometimes, a continuous, unmodulated carrier is sent for a period of a few seconds.

In radioteletype, the test signal RYRYRY . . . is used in Baudot mode, and U*U*U* . . . is used in ASCII mode. These signals are used because they cause rapid transitions between the mark and space states. Another test message that is used in radioteletype is THE QUICK BROWN FOX JUMPS OVER THE LAZY DOG'S BACK 0123456789. This sentence contains all of the letters of the alphabet, and also a figures and letters shift.

In voice communications, the transmitting operator usually says something like "TESTING 1, 2, 3, 4." In single-sideband transmission, some operators whistle or blow into the microphone for test purposes.

In television communication, the audio and video must be tested separately. The audio signal is checked in the same manner as any voice transmitter. The video signal is checked by sending a specific pattern called a *test pattern*. There are several different kinds of television test patterns used for checking various aspects of a picture (*see* TEST PATTERN).

In any mode of telecommunications, test signals should not be sent unnecessarily because nonessential transmissions increase the chances of interference among stations.

TEST PATTERN

In television, special patterns are used to check the alignment of the transmitter and/or receiver. Various patterns facilitate the quantitative evaluation of such operating characteristics as resolution, horizontal and vertical linearity, brightness, contrast, and color reproduction. *Test patterns* are also used to help the receiving operator tune in the signal in a two-way television communications system.

If you switch on your television set very early in the morning, you can see stations transmitting video test patterns. Each station has its own distinctive pattern. The audio channel can have a steady tone, music, or other test signals. The test pattern is used by the station technicians to facilitate regular checking of the transmitter alignment. Test patterns can be picked up by a television camera, or they can be electronically generated.

Test patterns are used by technicians who repair television receivers. The most common of these patterns is the bar pattern. *See also* BAR GENERATOR.

TEST RANGE

A *test range* is a large area, usually located outdoors, in which transmitting and receiving antennas are tested for such properties as power gain and front-to-back ratio.

Testing of Transmitting Antennas A field-strength meter is placed at a distance of several wavelengths from the antenna under test. A reference antenna, such as a dipole or isotropic radiator, is placed near the antenna under test. The distance between the test antenna and the field-strength meter is identical to the distance between the reference antenna and the field-strength meter. The test and reference antennas are resonant at the same frequency.

For testing of antenna power gain, the reference antenna is supplied with a certain amount of radio-frequency power at its resonant frequency. The field-strength-meter sensitivity is adjusted until the scale reads 0 dB. Then the same amount of power, at the same frequency, is supplied to the test antenna. The reading on the scale of the field-strength meter, in decibels, indicates the power gain of the test antenna in dBd (if a dipole is used as the reference) or dBi (if an isotropic antenna is used as the reference).

For the testing of front-to-back ratio, if applicable, the test antenna is turned so that it points straight away from the field-strength meter. The scale of the field-strength meter is set to read 0 dB when a given amount of power is applied to the test antenna. Then, the test antenna is rotated so that it points directly at the field-strength meter. The reading on the meter scale, in decibels, indicates the front-to-back ratio. The antenna can be continuously turned, and meter readings taken at intervals of a few degrees, to get a complete directional pattern.

Testing of Receiving Antennas For the test-range layout for the evaluation of receiving-antenna performance, a low-power transmitter is located several wavelengths from the test and reference antennas. The test and reference antennas are both resonant at the same frequency, and are both located at the same distance from the transmitter.

For the testing of effective receiving gain, the test antenna is pointed directly at the transmitter, and the transmitter delivers a constant amount of power to its antenna. The receiver is switched between the reference antenna and the test antenna. The receiver has a calibrated S meter that allows precise determination of relative signal levels in decibels. The effective receiving gain of the test antenna is the increase, in decibels, that occurs in the signal when the receiver is switched from the reference antenna to the test antenna (if the signal is weaker with the test antenna, the gain is negative).

For measurement of front-to-back ratio, the receiver is connected to the test antenna, and signal levels are compared with the antenna pointed directly toward and directly away from the transmitter. A plot of the directional response can be obtained by checking the received signal level at numerous compass points.

For Further Information A complete discussion of all aspects of antenna testing cannot be given here. For detailed information, a textbook on antenna theory or communications theory should be consulted. Related articles in this book include: ANTENNA EFFICIENCY, ANTENNA PATTERN, ANTENNA POWER GAIN, dBd, dBi, DIPOLE ANTENNA, FIELD-STRENGTH METER, FRONT-TO-BACK RATIO, ISOTROPIC ANTENNA, REFERENCE ANTENNA, and S METER.

TETRODE TRANSISTOR

Some bipolar transistors have two leads connected to the base, for the purpose of reducing the lead inductance at the base (see illustration). This scheme can be used for either the npn or pnp

type of bipolar transistor. Some field-effect transistors have two gates, each with its own lead connected, as at B. Transistors with two base or gate leads, because they have four leads in all, are called *tetrode transistors.*

Tetrode bipolar transistors are used at very high frequencies where lead inductance and loss, especially at the input, must be kept to a minimum. Tetrode field-effect transistors, also known as *dual-gate field-effect transistors,* are often used as mixers and modulators. *See also* FIELD-EFFECT TRANSISTOR, and TRANSISTOR.

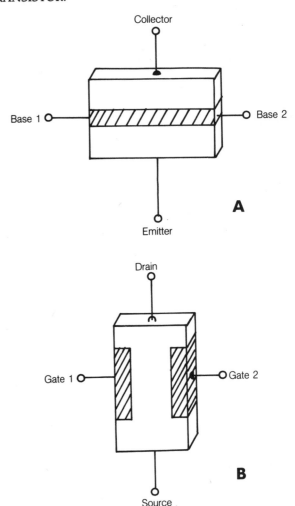

TETRODE TRANSISTOR: Bipolar (A) and FET (B) forms.

TETRODE TUBE

A *tetrode tube* is a vacuum tube with four elements. In addition to the cathode and the plate or anode, there are two grids. The two grids serve different purposes.

The control grid operates as the input-signal electrode in most tetrode-tube circuits. The second grid, known as the *screen grid,* can serve any of three functions. In an amplifier, the screen causes electrons, en route from the cathode to the plate, to be accelerated, resulting in enhanced gain. The screen can be used as an auxiliary plate. The screen grid can also be used as a second input grid in mixers and modulators.

The control grid in the tetrode tube is usually biased negatively with respect to the cathode. The screen grid is biased at a voltage more positive than the cathode, but less positive than the plate, in a tetrode amplifier circuit.

THD
See TOTAL HARMONIC DISTORTION.

THERMAL BREAKDOWN

In an electric conductor or semiconductor, the flow of current always generates some heat. This is because no conductor is completely lossless, and the resistance causes dissipation of power in the form of heat. If the flow of current is excessive, the dissipated power can cause the temperature to rise so high that the substance deteriorates. This is called *thermal breakdown*. It can occur in any device that carries current. *See also* THERMAL RUNAWAY.

An electric field in a dielectric substance, or a magnetic field in a ferromagnetic substance, causes a rise in the temperature of that substance. This is because no dielectric or ferromagnetic material is completely without loss. If the field strength becomes great enough, the dielectric or ferromagnetic material can get so hot that it is physically damaged. This is another form of thermal breakdown. *See also* DIELECTRIC, DIELECTRIC LOSS, EDDY-CURRENT LOSS, FERROMAGNETIC MATERIAL, and HYSTERESIS LOSS.

THERMAL CONDUCTIVITY

All substances conduct heat to some extent, but some conduct heat much better than others. The extent to which a substance can transfer heat efficiently from one place to another can be expressed quantitatively. This expression is known as the *thermal conductivity of the material.*

In general, substances are good thermal conductors if they are good electrical conductors, and vice-versa. Electrical insulators tend to be poor thermal conductors. But there are exceptions to this rule. For example, silicon, while a poorer electrical conductor than platinum, is a better thermal conductor. The best thermal insulator is a perfect vacuum; since there are no molecules, no conduction can take place. The best thermal conductor of the known elements is silver.

Thermal conductivity is expressed in watts, milliwatts, or microwatts per degree Celsius.

THERMAL NOISE

In all materials, the electrons are in constant motion. Because the electrons move in curved paths, they are always accelerating, even if they stay in the same orbital shell of a single atom (*see* ELECTRON, and ELECTRON ORBIT). This acceleration of charged particles produces electromagnetic fields over a wide spectrum of wavelengths. To some extent, the random movement of the positively charged atomic nuclei has the same effect. In electronic circuits, this charged-particle acceleration causes a form of noise known as *thermal noise.*

The level of thermal noise in any material is proportional to the absolute temperature. The higher the temperature gets, the more rapidly the charged particles are accelerated. The lower the temperature becomes, the slower the particles move. As the temperature approaches absolute zero—the coldest possible condition—the particle speed and acceleration approach zero, and so does the level of thermal noise.

Thermal noise imposes a limit on the sensitivity that can be obtained with radio-frequency receivers. This noise can be minimized by placing a preamplifier circuit in a bath of liquid gas, such as helium or nitrogen. Helium has the lowest boiling point of any element: just a few degrees Kelvin. Such cold temperatures cause the atoms and electrons to move very slowly, thus greatly reducing the thermal noise compared with the level at room temperature. This technique is used in the front ends of radio-telescope receivers.

Thermal noise is not the only form of noise that limits the sensitivity of receiving circuits. Some noise is caused by shot effect, and some noise comes from the environment outside the circuitry itself. *See also* CRYOGENICS, NOISE, SFERICS, SHOT EFFECT, and SUPERCONDUCTIVITY.

THERMAL PROTECTION

Thermal protection is a method of preventing damage to circuit components because of overheating that can result from an excessive long-term demand for current or power, or from the failure of an equipment-cooling system.

A thermal-protection device operates in a manner similar to a circuit breaker. However, rather than being actuated by the excessive current itself, a thermal-protection device is triggered by high temperature.

The illustration is a simplified diagram of a thermal-protection circuit that might be used in a power amplifier. A small thermostat is attached to, or placed in the vicinity of, the components to be protected. The thermostat is connected to a relay or switching transistor in the power-supply line of the protected components. If the temperature reaches the critical level, the thermostat changes state, removing power from the circuit. *See also* THERMAL BREAKDOWN, and THERMAL RUNAWAY.

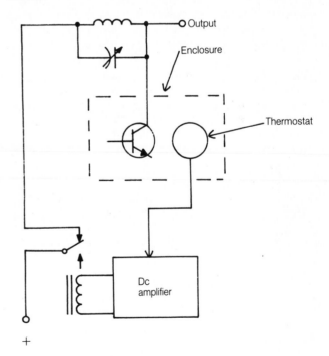

THERMAL PROTECTION: Protects against overheating.

THERMAL RUNAWAY

Some semiconductor devices, especially power transistors and high-current diodes, can be destroyed by a phenomenon called *thermal runaway*. Thermal runaway can occur only in a component that exhibits increased resistance with a rise in temperature. The problem begins as the current heats up the device, raising the resistance slightly. This increase in resistance might or might not produce a significant decrease in the flow of current, depending on the external circuitry. If the current does not decrease, the power dissipated in the component will increase in direct proportion to the resistance. This increased dissipation heats the device even faster, and a vicious circle ensues. If allowed to go on unchecked, the end result will be thermal breakdown.

Thermal runaway can be prevented by placing a small-value, high-power resistor in series with the diode or transistor. This reduces the current that flows through the component as the temperature rises. A thermal-protection circuit can also be used; it will shut off the power if thermal runaway starts to occur. *See also* THERMAL BREAKDOWN, and THERMAL PROTECTION.

THERMAL SHOCK

Thermal shock is a rapid, marked increase or decrease in the temperature of the environment in which an electronic circuit or component is operated. Thermal shock often causes dramatic changes in the characteristics of a device. This is usually the result of temperature-coefficient effects, although some electronic components completely fail if the temperature gets too cold or too hot.

Thermal shock is sometimes inflicted deliberately on a piece of equipment as part of a testing or troubleshooting procedure. Suppose that a transceiver is received at a service shop for repair, with the customer complaining that the radio will not operate at temperatures below freezing. The unit is subjected to thermal shock by being placed in a freezer for a few hours and then tested. If failure occurs, the radio is allowed to warm up again, or is deliberately warmed using a heater. The defective component can then be found using a cooling spray.

Some electronic equipment can withstand very little thermal shock without failure. A large computer, for example, requires fairly constant temperature conditions. Other circuits, such as mobile or portable devices, are designed to withstand severe thermal shock because they must function both indoors and outdoors. *See also* THERMAL STABILITY.

THERMAL STABILITY

Thermal stability is the capability of a circuit to function properly under conditions of variable temperature. The thermal stability of component values is expressed in terms of the temperature coefficient (*see* TEMPERATURE COEFFICIENT). The thermal stability of a circuit depends, to a large extent, on the temperature coefficients of the individual components.

Thermal stability can be quantitatively described in two ways. The simpler way is to state the temperature range in which the equipment will always operate normally. The specifications for a small computer might state an ambient-temperature range of 10 to 50 degrees Celsius, for example. A more comprehensive way of expressing thermal stability is to indicate how the various operating characteristics (such as frequency, power output, or sensitivity) change as a function of temperature.

Thermal stability is especially important in equipment that must be operated in diverse environments, where thermal shock is likely to be encountered. Thermal stability is not as important for equipment that is always used in a controlled environment. *See also* TEMPERATURE COMPENSATION, and THERMAL SHOCK.

THERMAL TIME CONSTANT

When the ambient temperature around an object changes suddenly, the object gradually gets hotter or cooler until the object and the environment are at the same temperature. The rapidity with which this occurs depends on the substance, and also on the mass of the object. The heating or cooling process is exponential, but the relative speed of change can be expressed in terms of the thermal time constant.

Suppose the ambient temperature around an object suddenly changes from T_1 to T_2 (in the Kelvin scale). The thermal time constant is the time t required for the object to reach a temperature of:

$$T_c = T_1 + 0.63 \, (T_2 - T_1)$$

if $T_2 > T_1$, that is, heating; or:

$$T_c = T_1 - 0.63 \, (T_1 - T_2)$$

if $T_2 < T_1$, that is, cooling.

The thermal time constant is sometimes expressed for electronic components (such as resistors and transistors). If the current suddenly changes, the temperature of the component will change exponentially, starting at T_1 and approaching a final value of T_2. The thermal time constant for an electronic component is determined as above.

THERMIONIC EMISSION

When an electrical conductor is heated to a high temperature in a vacuum, the electrons move with such speed that they are easily stripped from the material. This is called *thermionic emission*. It is the principle of operation of a vacuum-tube cathode.

Thermionic emission from a surface is expressed in watts per square meter or watts per square centimeter. The thermionic emission from a cathode depends on the temperature, the voltage supplied to the cathode, and the type of material from which the cathode is made. Oxide-coated cathodes generally have the greatest emission at a given absolute temperature and voltage. Specially treated nickel and tungsten also have high thermionic emissivity.

THERMISTOR

Most materials exhibit variable electrical conductivity with changes in temperature. A resistor, designed especially to change value with temperature, is called a *thermistor*. The term *thermistor* is a contraction of *therm*ally sensitive res*istor*.

Thermistors are made from semiconductor materials. The most common substances used are oxides of metals. The resistance can increase as the temperature rises, or it can decrease as the temperature rises. The resistance is a precise function of the temperature. Usually, the resistive temperature coefficient is large and negative (*see* TEMPERATURE COEFFICIENT). The construction of a typical thermistor is shown in the illustration.

Thermistors are used in various circuits for detecting temperature. The resistance-versus-temperature characteristic makes the thermistor ideal for use in thermostats and thermal protection circuits. Thermistors are operated at low current levels so that the resistance is affected only by the ambient temperature, and not by heating caused by the applied current itself. *See also* THERMAL PROTECTION.

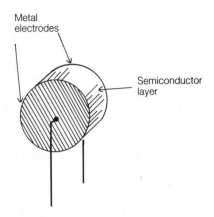

THERMISTOR: The semiconductor is between two metal plates.

THEVENIN'S THEOREM

In any linear circuit having more than one source and/or more than one load, simplification is possible. This simplification can be carried out theoretically, according to a rule known as *Thevenin's theorem*.

Suppose there is a circuit, no matter how complex, consisting of power sources and linear loads. Then, the circuit will operate in the same manner as a hypothetical circuit consisting of one source and one load of the appropriate values. This hypothetical circuit is called the *Thevenin equivalent circuit*.

Thevenin's theorem can be stated in another way. Suppose an impedance Z_x is connected into a linear system in which the open-circuit voltage is E and the existing impedance is Z. Then, the current I_x in the circuit will be:

$$I_x = E/(Z + Z_x)$$

Thevenin's theorem is used in network analysis for determination of current and voltage. *See also* KIRCHHOFF'S LAWS.

THICK-FILM AND THIN-FILM INTEGRATED CIRCUITS

An integrated circuit can be fabricated in various ways. Film processes are commonly used to obtain the passive components in an integrated circuit. There are two types of film integrated circuits, known as *thick-film (TF)* and *thin-film (tf)*.

Thick Films   A *thick-film circuit* is manufactured by depositing specially treated inks and pastes onto a ceramic wafer (substrate) by means of a photographic technique called *silk screening*. Several layers are generally applied. The layers are thicker than 0.001 mm. Each layer contains an intricate pattern of inks. The inks have different properties; some are excellent conductors, others are more or less resistive, and still others are dielectrics. The deposited materials form the passive circuit elements (wiring, capacitors, and resistors) in the thick-film circuit.

Active devices, such as transistors, are found in many thick-film integrated circuits. The active components are fabricated from microminiature semiconductor chips, and are welded or soldered to the passive components. Large-valued capacitors are usually added in discrete form, rather than thick-film form. A thick-film circuit containing active or discrete devices is called a *hybrid circuit*.

Thin Films   *Thin-film (tf) integrated circuits* are manufactured by means of depositing layers of conducting, semicon-

ducting, and insulating materials onto a ceramic wafer (substrate). The layers are less than 0.001 mm thick, and are formed by an evaporation process.

The materials used in the thin-film process are metals and metal oxides, rather than inks or pastes. Different combinations of resistive or semiconducting metals are used to obtain the wiring, resistors, and capacitors. As in the thick-film circuit, active components can be attached by means of soldering or welding. Large-valued capacitors are added as discrete microminiature components. A thin-film integrated circuit containing active or discrete components is called a *hybrid integrated circuit*.

Additive and Subtractive Processes   The layers in a thick-film or thin-film integrated circuit are usually applied one after the other to the insulating substrate. This method is known as the *additive process* because materials are deposited in a defined sequence, and only in the places where components are to be located. The interconnections, and the lower electrodes of capacitors, are deposited first. Then, the capacitor dielectric is applied, followed by the resistors, the top electrodes of the capacitors, and finally the discrete and active components.

In the subtractive process, layers of electronic material are first applied, and the components are then etched from the layers. The top layer is etched first, and then progressively lower layers. The etching is accomplished by photographic means, similar to the process used to make a printed-circuit board (*see* ETCHING). Finally, the discrete and active components are attached.

For Further Information   A textbook on semiconductor technology is recommended for a detailed discussion of thick-film and thin-film processes. *See also* INTEGRATED CIRCUIT.

THREE-PHASE ALTERNATING CURRENT

The alternating current that appears at 117-V utility outlets is single-phase current. We normally think of alternating current as a simple sine-wave current. However, the 234-V utility mains generally carry three such waves simultaneously. The waves are identical except that they are separated by 120 degrees of phase (1/3 cycle). The current is supplied through three wires, with each wire carrying a single sine wave differing in phase by 1/3 cycle, with respect to the currents in the other two wires (see illustration on pg. 496). Some three-phase circuits incorporate a fourth wire, which is kept at neutral (ground) potential.

Three-phase alternating current has certain advantages over single-phase current. When a three-phase current is rectified, the filtering process is much easier than when single-phase current is used, although the rectifier circuit is somewhat more complicated (*see* POLYPHASE RECTIFIER). Three-phase systems provide superior efficiency for the operation of heavy appliances, especially electric motors.

THREE-PHASE RECTIFIER

A three-phase rectifier is a special form of polyphase rectifier, used with three-phase alternating current. *See* POLYPHASE RECTIFIER.

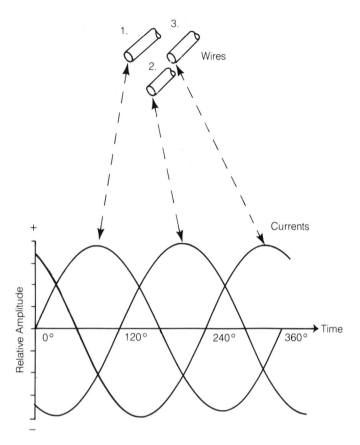

THREE-PHASE ALTERNATING CURRENT: The waves are 120 degrees apart.

THREE-QUARTER BRIDGE RECTIFIER

A *three-quarter bridge rectifier* is a form of alternating-current rectifier circuit identical to a bridge rectifier, except that one of the diodes is replaced by a resistor. The three-quarter bridge has three diodes and a conventional bridge rectifier has four. The three-quarter configuration performs in essentially the same way as a normal bridge rectifier in power supplies with unbalanced outputs. *See also* BRIDGE RECTIFIER.

THREE-STATE DEVICE

A *three-state device* is a logic circuit with several outputs, all but one of which are disabled at any given time. This is done by setting the outputs at high impedances. The high impedance differs from either of the standard logic states 0 (low) or 1 (high). It represents a sort of "third logic state;" hence the name for the device.

In a three-state device, any one, but only one at a time, of several different outputs is selected to drive a line. This output operates normally, and is always at either logic 0 or logic 1. The other outputs are held in the third state of high impedance, and have no effect on the line. They are therefore isolated from the circuit.

The most common application for three-state logic devices is in computers. They allow several different circuits to work on one set of lines. A particular circuit is connected to the line by setting all the outputs to the third state, except the one corresponding to the desired circuit.

THREE-WIRE SYSTEM

A *three-wire system* is a scheme for transmitting electric power, especially in alternating-current form. Three-wire systems are sometimes used in utility wiring. In recent years, the three-wire system has become increasingly popular because it affords better shock-hazard protection than the older two-wire system.

A three-wire system consists of one grounded (neutral) wire and two live wires. The live wires sometimes carry equal and opposite currents; in most three-wire systems, one of the live wires is kept at neutral potential, the same as the grounded wire, and the other carries 117 V of ac potential.

Three-wire electrical systems facilitate grounding of the chassis of radio equipment and appliances. This keeps lethal voltages from appearing at places where personnel might be electrocuted.

THRESHOLD

As the amplitude of a signal increases from zero, it is not detectable until it reaches a certain level. The minimum level at which a signal of any kind can be detected, either by the human senses or using electronic instrumentation, is known as the *threshold level.* You can speak of the threshold of hearing, the threshold of visibility, the threshold sensitivity of a meter or receiver, or the threshold sensitivity of some other device or component (*see* TANGENTIAL SENSITIVITY).

For certain adjustable controls, the point at which a defined transition takes place is called the *threshold setting.* A squelch control, for example, will cut off the internally generated receiver noise when it is set past a certain point. A regenerative detector will begin to oscillate if the feedback is increased past a certain threshold, called the *threshold of oscillation.*

Diodes, transistors, and tubes require a certain minimum forward voltage or current in order to conduct. This voltage or current is called the *breakover* or *threshold voltage* or *current.* For a germanium semiconductor diode, the threshold signal level is near 0.3 V. For a silicon diode, it is near 0.6 V. In the case of a vacuum tube, it varies from less than 1 V to several volts. The threshold property of a diode, transistor, or tube can be used for peak clipping (limiting) or for elimination of signals weaker than a certain amplitude. *See also* SPEECH CLIPPING, and THRESHOLD DETECTOR.

THRESHOLD DETECTOR

A *threshold detector* is a circuit that allows strong signals to pass, but blocks weaker signals. The transition, or threshold, is sharply defined. A squelch circuit is a form of threshold detector (*see* SQUELCH).

The simplest form of threshold detector consists of two diodes, connected in reverse parallel, with respect to each other, and in series with the signal path. Signals will not be passed until the peak amplitude exceeds the forward breakover voltage of either diode. If germanium diodes are used, the peak voltage must be 0.3 V or greater; for silicon diodes, a peak signal of 0.6 V is necessary.

A threshold detector can be used to obtain enhanced selectivity in an audio filter, as shown in the illustration. The first tuned circuit provides a sharp resonant audio response. The threshold detector cuts off the skirts at levels below 0.3 V or 0.6 V (depending on whether germanium or silicon diodes are used). The second tuned circuit, having values *L* and *C* identical to those of the first, reduces the waveform distortion caused by

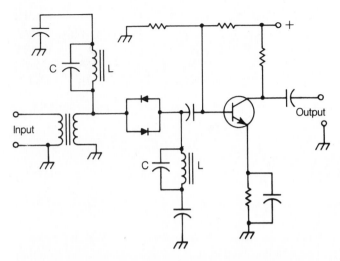

THRESHOLD DETECTOR: Steepens skirts in audio filter.

the threshold detector. The transistor circuit provides amplification, if needed.

THRESHOLD SENSITIVITY
See TANGENTIAL SENSITIVITY.

THROAT MICROPHONE
In high-noise environments, an ordinary microphone picks up too much background sound to be useful in voice communications. In some instances, a highly directional microphone can be used with adequate results. However, in extreme situations, a *throat microphone* is needed.

A throat microphone operates by conduction. It is worn around the neck of the user. The diaphragm is placed next to the operator's voice box. Vibration of the throat provides the audio input to the microphone. Because sound propagates much more efficiently by conduction than it does through the atmosphere, the background noise is greatly attenuated relative to the voice. *See also* MICROPHONE.

THROUGHPUT
The speed at which a computer can process a given amount of data is called the *throughput*. Throughput is generally measured in problems per minute. Suppose a computer is given n problems to solve, and it takes t minutes to process them and provide the answers. The throughput T is then:

$$T = n/t$$

in problems per minute.

The throughput varies, depending on the type of computer and the complexity of the problems. Therefore, throughput is a subjective expression of data-processing speed.

THYRECTOR
A *thyrector* is a semiconductor device, consisting essentially of two diodes connected in reverse series. The thyrector is used for protection of equipment against transients (*see* TRANSIENT).

A thyrector is connected across the alternating-current power terminals of a piece of electronic apparatus (see illustration). The thyrector does not conduct until the voltage reaches a

certain value. If the voltage across the thyrector exceeds the critical value, the device conducts. The conducting voltage does not depend on the polarity.

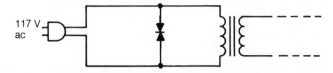

THYRECTOR: Useful for transient suppression.

THYRISTOR
A *thyristor* is a four-layer semiconductor device, similar to an ordinary rectifier except that it has a control electrode, called the *gate*, in addition to the anode and cathode.

The thyristor initially does not conduct. When a pulse is supplied to the gate electrode, the device conducts in one direction but not the other, in a manner that is identical to a diode. Conduction continues until the voltage between the anode and cathode is removed.

The thyristor behaves very much like a thyratron tube. The thyristor, however, operates much more rapidly than the thyratron. The forward-voltage drop of a typical thyristor is approximately 0.9 V.

Thyristors are used primarily for switching purposes. The most common type of thyristor is the silicon-controlled rectifier. *See also* SILICON-CONTROLLED RECTIFIER.

TICKLER COIL
See TERTIARY COIL.

TIE STRIP
A *tie strip*, also called a *terminal strip*, is a set of contacts used for connecting wires in a point-to-point-wired circuit (*see* POINT-TO-POINT WIRING). A tie strip consists of two or more lugs, attached to a strip of phenolic material. Some of the lugs can be connected to a mounting terminal, and thus grounded. The lugs are designed to adhere easily to solder. Tie strips are available in a wide variety of sizes for different applications.

TIME CONSTANT
Certain circuits change state exponentially as a function of time. Combinations of inductances, capacitances, and/or resistance exhibit this property. Materials heat up and cool down exponentially. The rate at which a change of state occurs, in a given situation, is known as the *time constant* (see illustration on pg. 498).

The time constant is generally defined as the time, in seconds, required for a change of state to reach 63 percent of completion. *See also* RESISTANCE-CAPACITANCE TIME CONSTANT, RESISTANCE-INDUCTANCE TIME CONSTANT, and THERMAL TIME CONSTANT.

TIME-DIVISION MULTIPLEX
Time-division multiplex is a method of combining analog signals for serial transfer (*see* SERIAL DATA TRANSFER). The signals are sampled at intervals, and interwoven for transmission. At the receiving end, the process is reversed; the signals are separated again (see illustration).

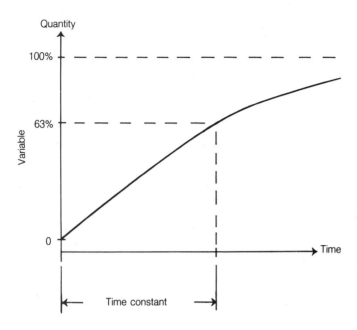

TIME CONSTANT: An expression for the speed of a transition.

Time-division multiplex is used for transmitting two or more channels of information over a single carrier. The speed of the multiplexed signal is increased, with respect to the individual channel speed, by a factor equal to the number of signals combined. For example, if 10 signals are to be multiplexed by time-division process, the data speed of each signal must be multiplied by 10 to maintain synchronization.

Time-division multiplexing results in an increase in the bandwidth of a signal. The increase takes place by a factor that is equal to the number of signals combined. This is because of the increased data speed.

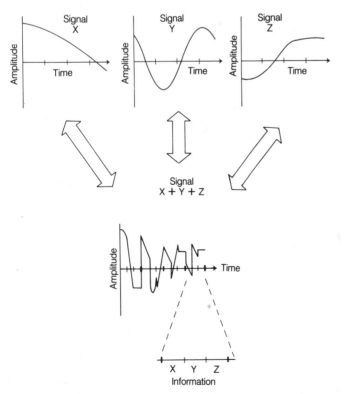

TIME-DIVISION MULTIPLEX: Successive bursts of X, Y, and Z are sent in rotation.

Time-division multiplex can be used with digital signals as well as analog signals. Digital time-division multiplex is usually called *synchronous multiplex* because the data from each channel is split bit by bit, in accordance with the speed of transmission. *See also* MULTIPLEX, and SYNCHRONOUS MULTIPLEX.

TIME DOMAIN

When time is displayed on the horizontal axis of an oscilloscope, the instrument is said to be operating in the *time domain*. Time is the independent variable in this mode. Time-domain operation is commonly used to display waveforms and modulation envelopes.

In time-domain operation, the sweep rate, or the speed at which the trace moves across the screen from left to right, is specified in seconds, milliseconds or microseconds per horizontal division. One division is equal to about a centimeter on a conventional scope. The vertical axis usually displays amplitude, on a linear scale or a logarithmic scale. *See also* FREQUENCY DOMAIN, and OSCILLOSCOPE.

TIME SHARING

Large computers can be used by many different operators at once, each at an independent terminal, by means of a technique called *time sharing*. The principle of time sharing is similar to that of time-division multiplex (*see* TIME-DIVISION MULTIPLEX).

In time sharing, the computer communicates with each terminal for a moment, then moves to the next terminal, and so on in a repeating sequence. If there is just one terminal, the computer spends all of its time communicating with that terminal; if there are *n* terminals, the computer works with each terminal for $1/n$th of the time.

The sequencing process is very rapid, so it seems to each operator that the computer is communicating only with that terminal. The computer appears to operate more slowly, however, when many terminals are connected, as compared to when only a few terminals are connected.

Time-sharing systems are often operated via telephone. A subscriber can work with a computer hundreds of miles away, simply by calling a certain number and using a telephone modem in conjunction with a terminal. *See also* MODEM, and TERMINAL.

TIME-SHIFTING COMMUNICATIONS

In a ham-radio *QSO*, or real-time contact, both operators are at their stations, and they conduct a conversation. The only difference between a voice QSO and a telephone conversation is that the ham contact is not usually full duplex; that is, one operator can't normally interrupt the other.

In radioteletype (RTTY), a message can be left at a station even when the operator is not there. If the equipment is on, and is tuned to the right frequency, the teleprinter or video display terminal will display and/or store messages. This is one form of *time-shifting communications*. It has been done in RTTY for decades.

In recent years, packet radio has become popular. One of the outstanding features of "packet" is that it can be a versatile mode of time-shifting communications. *See also* MAILBOX,

PACKET RADIO, PACKET-RADIO BULLETIN-BOARD SYSTEM, and RADIOTELETYPE.

TIME STANDARD

In order for clocks to be synchronized, they must be set according to a universal standard. For everyday purposes, most of us simply set our clocks according to the information broadcast by local radio stations (the news is often preceded by a tone that marks the hour to within a few seconds). For greater accuracy, the time signals from stations (such as CHU or WWV/WWVH) are used.

TIME ZONE

People prefer to have the sun rise and set at similar hours no matter what the longitude. For this reason, time zones have been devised. The earth rotates at an angular rate of 15 degrees per hour. The time zones of the world are thus about 15 degrees wide, so there is one zone for each hour of the day. The situation, in theory, is quite simple. However, in practice, it is complicated by the preferences of individual counties, states, and countries.

In the continental United States, there are four times zones, called the *Eastern, Central, Mountain,* and *Pacific.* Eastern Standard Time (EST) is 5 hours behind Coordinated Universal Time (UTC). Eastern Daylight Time (EDT) is 4 hours behind UTC. Central, Mountain, and Pacific time are 1, 2, and 3 hours behind Eastern time, respectively. *See also* COORDINATED UNIVERSAL TIME, and STANDARD TIME THROUGHOUT THE WORLD.

TIN

Tin is an element with atomic number 50 and atomic weight 119. In its pure form, tin is a soft, malleable, silver-colored metal. Tin is used in the manufacture of solder. The solder used in electronic circuits generally consists of 50 or 60 percent tin, and 40 to 50 percent lead. *See also* SOLDER.

TIN-LEAD SOLDER

See SOLDER.

TINNING

The leads of discrete electronic components are often covered with a thin layer of tin or solder. This practice, known as *tinning,* is done to prevent corrosion and to make the leads adhere easily to solder.

The tip of a soldering gun or iron should always be kept coated with a thin layer of solder. This is called *tinning.* Wires and sheet metal are generally tinned prior to splicing. *See also* SOLDERING TECHNIQUE, and WIRE SPLICING.

T MATCH

A *T match* is a device for coupling a balanced transmission line to a balanced antenna element, such as a half-wave, center-fed radiator. The radiating element consists of a single conductor. A rod or wire is run parallel to the radiator, and the feed line is connected to the center of the rod or wire. The ends of the rod or wire are connected to the radiator at equal distances from the ends of the radiator. The drawing illustrates a T match.

The impedance-matching ratio of the T match depends on physical dimensions, such as the spacing between the radiator and the rod or wire, the length of the matching element, and the relative diameters of the radiator and the matching element. The T match allows the low feed-point impedance of a yagi antenna driven element to be matched to a 75-ohm balanced line. If a balun is used at the feed point, the antenna can be fed with coaxial line. Sometimes a half-T, or gamma match, is used with a coaxial line (*see* GAMMA MATCH).

There are several variations of the T match that can be used when large impedance-step-up ratios are needed. If the rod or wire of a T match has the same length and diameter as the radiating element, the result is a folded dipole antenna. This configuration produces a step-up impedance ratio of 4:1, relative to a single element alone. Three conductors can be connected in parallel to obtain a step-up ratio of 9:1, or four conductors can be used to obtain a factor of 16:1. *See also* FOLDED-DIPOLE ANTENNA, IMPEDANCE MATCHING, and YAGI ANTENNA.

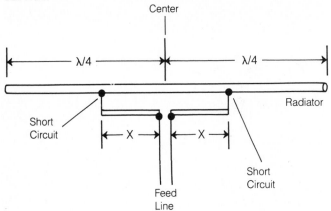

T MATCH: Provides a match between the radiator and balanced line.

T NETWORK

A *T network* is an unbalanced circuit, used in the construction of filters, attenuators, and impedance-matching devices. Two series components are connected on either side of a single parallel component.

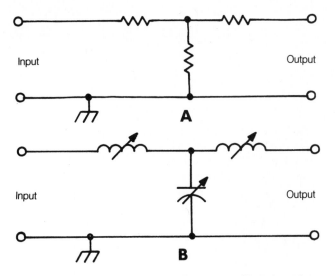

T NETWORK: At A, a resistive attenuator; at B, an impedance-matching circuit.

The diagram at A illustrates a T-network circuit consisting of non-inductive resistors. This circuit can be used as an attenuator. B shows a T-network circuit that can be used for impedance matching. The T network gets its name from its shape in the schematic diagrams.

In a T-network impedance-matching circuit, the capacitor and/or inductors can be adjustable to vary the impedance-transfer ratio. A fairly wide range of load impedances can be matched to a fixed source impedance. The T-network circuit at B might also be used as a lowpass filter. *See also* IMPEDANCE MATCHING.

TOGGLE SWITCH

A *toggle switch* is a lever-type panel switch that is often found in electronic equipment. Toggle switches usually have one or two poles, and one or two throw positions. Toggle switches are available in many sizes, ranging from subminiature to very large. The voltage and current ratings are low to moderate. Toggle switches are used for switching something on and off, or for selecting between two circuits or states.

TOLERANCE

The actual value of a fixed component is never exactly the same as the rated value; there is always some error. When components are mass produced, some end up having values less than the rated value, and others end up with larger values. The manufacturer will guarantee, however, that a component from a given lot will have a value that is within a certain range. This range is expressed as a figure called *tolerance*.

For a component with a rated value r, the tolerance T is expressed as a plus-or-minus percentage that is the larger of the following two figures:

$$T_1 = \pm 100(r - s)/r$$
$$T_2 = \pm 100(t - r)/r$$

where s represents the lowest expected value and t represents the highest expected value. Although it is true that $s < r < t$, it might not be true that $r - s = t - r$. The tolerance T represents the greatest possible departure of the actual value from the rated value.

If you have a component with a rated value r and a tolerance of plus-or-minus T percent, you can expect that the component will have a value larger than:

$$s = r - rT/100$$

and smaller than:

$$t = r + rT/100$$

Tolerance is an important consideration in the design and manufacture of any electronic device. Some equipment requires components with low tolerance, that is, high precision. But the use of low-tolerance components where they are not needed is a waste of money, because in general, the price of a component goes up as the tolerance goes down. In electronic measurement, the counterpart of tolerance is known as *accuracy*. *See also* ACCURACY.

TONE BURST

See PRIVATE LINE®, and TONE SQUELCH.

TONE SQUELCH

Some receivers are equipped with a special squelching circuit that stays closed unless an incoming signal is modulated with a tone of a certain frequency and/or duration. This type of squelch is called a *tone squelch*.

Tone squelching is used mostly in frequency-modulated communications systems operating at very high and ultra-high frequencies. The tone can be continuous and low-pitched, so it cannot be heard by the receiving operator. This is known as *subaudible-tone squelching*. Alternatively, a brief tone, having a midrange audio frequency, can be used at the beginning of the transmission. This is called *tone-burst squelching*.

Tone squelching is sometimes used in repeater-input receivers. This prevents unauthorized use of a repeater because only the properly modulated signals can actuate it. This use of tone squelching is called *Private-Line operation*. *See also* PRIVATE LINE®.

TOP-FED VERTICAL ANTENNA

Usually, a vertical antenna is fed at the base, and worked against an earth ground or a ground plane (*see* GROUND-PLANE ANTENNA, and VERTICAL ANTENNA). But there are some instances in which a vertical antenna can be fed at the top instead.

The illustration shows a *top-fed vertical antenna*. In this arrangement, a technique called *zepp feed* is used. The antenna length is ½ wavelength. The parallel-wire feed line should be of the open-wire type, not TV ribbon (*see* ZEPPELIN ANTENNA). The line can be any length if a *transmatch* having a balanced output is used.

Top feed is convenient for hams who live in high-rise buildings, if the landlord or condominium association will allow outdoor antennas. If there is any doubt about that, one should check the lease, or consult the condo association, before erecting any antenna. Also, standard safety precautions must be observed to minimize hazards from utility wiring and lightning.

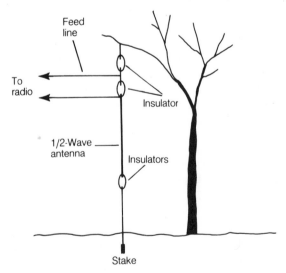

TOP-FED VERTICAL ANTENNA: A simple Zepp-fed vertical wire can be suspended from a tree branch.

TOP FEED

Top feed is a method of feeding a vertical antenna at the top, rather than the bottom, of the radiating element. This scheme is used when a transmitter or receiver must be located well above the ground and nearby support structures. Top feed can also be used as an alternative, in any situation, to other feed methods. Top feeding can be accomplished in various ways.

A dweller in a high-rise building can install a top-fed, more-or-less vertical wire antenna out the window. The grounding system can consist of the cold-water plumbing in the building. The antenna is tuned to resonance at the desired frequency by means of a transmatch.

Another method of top feed, in this case for a halfwave vertical element, is shown in the drawing. The antenna is suspended from a support such as a tree branch. This feed system is similar to that used in the zeppelin antenna (*see* ZEPPELIN ANTENNA).

TOP LOADING

In a vertical antenna system, *top loading* refers to a method of physically shortening the radiator. The technique works on the same principle as *base loading* or *center loading*. An inductor is placed at or near the top of the vertical part of the antenna. A *capacitance hat* is used above the coil to get resonance at the desired frequency (*see* CAPACITIVE LOADING). The radius of the capacitance hat can be up to 0.1 wavelength.

Top loading, in conjunction with a low-loss coil and a good ground system, can allow a vertical radiator to work well even when its height is considerably less than $\frac{1}{4}$ wavelength.

Top loading can be used to make a vertical antenna work on two bands far apart in frequency. For example, a 16-foot vertical, with a coil and capacitance hat at the top, can work at 14 MHz, where 16 feet is $\frac{1}{4}$ wavelength, and at 1.8 MHz via top loading. At 14 MHz, the coil is a choke; the capacitance hat is isolated from the antenna. At 1.8 MHz, the coil and capacitance hat bring the system to resonance.

TOROID

A *toroid* is a coil wound on a doughnut-shaped ferromagnetic core (see illustration). Toroid coils and cores have become increasingly popular in recent years, replacing the traditional solenoidal coils and cores.

The main advantage of the toroidal coil is that all of the magnetic flux is contained within the core material. Therefore, a toroid is not affected by surrounding components or objects. Also, nearby components are unaffected by the coil field. Another advantage of the toroid is that it has a smaller physical size, for a given amount of inductance, than a solenoidal coil. Still another advantage is that fewer coil turns are required to obtain a given inductance, compared with a solenoidal coil. This enhances the *Q* factor.

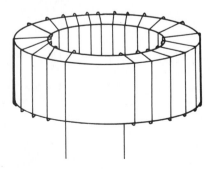

TOROID: The form is donut-shaped.

Toroidal coils can be used for winding simple inductors, chokes, and transformers. They are used at audio frequencies, and at radio frequencies up to several hundred megahertz. Ferromagnetic toroidal cores are available in a tremendous variety of sizes and permeability ratings, for use at various frequencies and power levels. *See also* INDUCTOR.

TOTAL HARMONIC DISTORTION

The *total root-mean-square (rms) harmonic voltage* in a signal, as a percentage of the voltage at the fundamental frequency, is called *total harmonic distortion (THD)*. Total harmonic distortion can vary from zero percent (no harmonic energy) to theoretically infinite (no fundamental-frequency energy).

The THD figure is often used as an expression of the performance of an audio-frequency amplifier. The amplifier is provided with a pure sine-wave input, and the harmonic content is measured using a distortion analyzer or SINAD voltmeter (*see* DISTORTION ANALYZER).

Ideally, an audio-amplifier circuit should have as little THD as possible at the rated audio-output level. Most amplifiers are rated according to the amount of root-mean-square power they can deliver to a speaker of a certain impedance (usually 8 ohms) with less than a specified THD (usually 10 percent).

TOTAL REGULATION

The output current or voltage of a power supply is affected by several factors. The load resistance and the input voltage are the most significant, but other variables, such as temperature or mechanical vibration, can also have an effect. Sometimes these variables partially cancel each other, but in some cases their effects can be additive. *Total regulation* is an expression of the worst case, in which all of the contributing factors work together to increase or decrease the output of the power supply.

Total regulation is determined by first determining how each variable affects the output of the supply. Then, the variables are adjusted so that they all tend to reduce the supply current or voltage. Finally, the variables are all adjusted so that they tend to maximize the output current or voltage. The total regulation is calculated in terms of the nominal, minimum, and maximum currents or voltages. *See also* CURRENT REGULATION, REGULATION, and VOLTAGE REGULATION.

TOUCHTONE®

In recent years, an increasing number of telephone sets have been using pushbutton dialing systems instead of the rotary dialer. The buttons actuate tone pairs that cause automatic dialing of the numbers. This is called *Touchtone® dialing* (the term is a trademark of American Telephone and Telegraph Company).

The original Touchtone dial, still found on most telephone sets, has 12 buttons corresponding to digits 0 through 9, the star symbol (*) and the pound symbol (#). Some dialers have four additional keys, designated A, B, C, and D. The tone-pair frequencies for the 16 designators are listed in the table.

Although the original push-button keypads were called *Touchtone pads*, other manufacturers have produced similar arrangements known by different names. All use the common combinations of frequencies listed in the table in order to access the telephone systems. These systems offer what is called *DTMF (Dual-Tone, Multiple Frequency) access*. Proper operation requires that both tones be present. The DTMF system can be used to remotely control distant objects or electronic equipment.

Tone-dialing operation has several advantages over the older rotary dial. Tone dialing is much faster, and it can be done automatically with extreme speed. The tones can also be transmitted into a telephone set from an external source, which is not possible with a rotary dial. Many radio repeaters, for example, are interconnected with the telephone lines so that mobile radio operators, equipped with Tone keypads in their transceivers, can gain access to the lines. *See also* AUTOPATCH.

TOUCHTONE®: AUDIO FREQUENCIES

Low group, HZ	High group, HZ 1209	1336	1477	1633
697	1	2	3	A
770	4	5	6	B
852	7	8	9	C
941	*	0	#	D

TOWER

A reinforced structure, intended for support of power lines or antenna systems, is called a *tower.* Towers vary in height and size from the small assemblies used for home television reception to massive structures that support broadcasting antennas or high-tension power lines.

Types of Towers Towers can be categorized as either self-supporting or guyed. The self-supporting towers require more rigid construction than the guyed types. One common method of tower construction uses three vertical members having a triangular cross-section. Reinforcing rods are welded to the three vertical members at intervals. A less common scheme is the use of a single, heavy vertical mast, similar to a flagpole. A guyed tower can have one or more sets of supporting wires (*see* GUYING).

Self-supporting towers are sometimes equipped with tilt-over devices. This allows an antenna to be installed with comparative ease at the top of the tower. Other self-supporting towers can have telescoping sections for height adjustment; this is known as the *crank-up configuration.*

Passive and Active Towers A tower can be used either as a support for one or more antennas, or as an antenna itself. In the former case, the tower is called *passive;* in the latter case, it is called *active.* At longer wavelengths, active towers are common, but at the shorter wavelengths, towers are used only to put the antennas at the height necessary for reliable communications.

Sometimes a passive tower, supporting antennas of a short wavelength, is fed with radio-frequency energy at a longer wavelength. This is a popular practice among radio amateurs. A tower, supporting antennas for 14, 21, 28, 144, and 220 MHz, for example, can be used as an active radiator at 1.8, 3.5, and 7 MHz. Active towers produce vertically polarized fields. Shunt feed is a common method of feeding a tower directly (*see* SHUNT FEED).

Safety Any massive antenna structure is potentially dangerous. When installing or maintaining a tower, personal safety is the most important consideration. The choice of a tower depends on the environment to be expected, the size of the antennas to be supported, and the height desired. Any tower should be periodically checked for signs of deterioration that might weaken the structure and eventually result in collapse.

Tall structures are more frequently hit by lightning than shorter ones, and for this reason, lightning protection is imperative with any tower system (*see* LIGHTNING, and LIGHTNING PROTECTION). An excellent electrical ground should be provided at the base of a tower to prevent dangerous induced voltages.

Tower climbing should be done by professionals only. Backyard towers can present a special hazard to children, who might attempt to climb the structure. Some means should be devised to prevent this from happening.

A professional structural engineer should be consulted before installing any tower. All applicable safety codes, as well as zoning laws, must be strictly followed.

TRAFFIC HANDLING

An important justification for the existence of amateur radio is the public-service communications that hams provide. As people gain experience in the hobby, they tend to focus most of their attention on one or two specialties within the hobby. Messages sent via ham radio are called *traffic. Traffic handling* consists of originating, passing and delivering messages. Some hams do almost all of their operating in this vein of public service.

Traffic Nets Most traffic handling is carried on in an organized manner by groups of stations that meet regularly on specific frequencies. These groups are known as *traffic nets.* Most of the messages on these nets are of minor importance. The activity is done largely to sharpen individual operating skills for use in emergencies.

Some traffic nets operate using morse code (CW). Others use voice communications, primarily single sideband (SSB) on the bands 1.8 through 29.7 MHz, and frequency modulation (FM) on the bands at 50 MHz and above. Much traffic-net operation is found on the 80-meter (3.5-MHz) CW band, the 75-meter (3.8-MHz) SSB band, and the 2-meter (144-MHz) FM band.

In recent years, an increasing amount of traffic handling has been done in digital modes at all frequencies. Packet radio lends itself perfectly to high-volume, high-speed, accurate and efficient traffic handling. *See* PACKET RADIO and related articles.

The Message Form A *radio message* or *radiogram* is sent according to a specific format. The first part of the radiogram is the *preamble.* It contains a message number, a *precedence* (see below), handling instructions (if any apply), the callsign of the originating station, the *check* (number of words or groups in the message), and the date and time (optional).

The second part of a radiogram is the *address* of the person to whom the message is to be delivered. The third part of the message is the body or *text.* The fourth and final part of a radiogram is the *signature* (the name and/or callsign of the originating person).

Precedence of Message *Precedence* is the degree of importance attached to a message. There are three levels of precedence. A *routine* message, abbreviated *R* on CW, is by far the most common. It is general greeting type material, such as "Happy new year." It has the lowest priority.

A *priority* message, abbreviated *P* on CW, is a message that must be delivered within a certain time frame, and/or that has significance over and above routine. Traffic in an emergency situation, but not itself of great urgency, falls into this category. An example is "We all survived the hurricane just fine." Priority messages supersede routine messages.

An *emergency* message, always spelled out fully on CW, is any message that is a matter of life or death to a person or persons. An example is "Send more medication immediately." Emergency messages always supersede priority or routine messages.

For Further Information The best way to become familiar with traffic handling is to join a traffic net and check in regularly. There's no substitute for doing; that's what ham radio is all about!

Books are available that explain the mechanics of traffic handling in great detail. One of the best is *The ARRL Operating Manual*, published by the American Radio Relay League. The very name of this organization comes from its original function, which was as a nationwide message-relaying network in the early days of ham radio. *See also* AMERICAN RADIO RELAY LEAGUE, NET, and NATIONAL TRAFFIC SYSTEM.

TRANSCEIVER

A *transceiver* is a combination *trans*mitter and re*ceiver* in a single cabinet, usually controlled by a single variable-frequency oscillator (VFO). *See* RECEIVER, and TRANSMITTER. The block diagram shows the basic design for a transceiver.

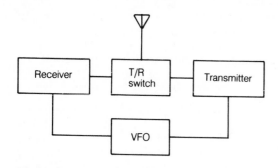

TRANSCEIVER: A block diagram of a transceiver.

Transceivers are almost exclusively used at very high frequencies (VHF) and above. In recent years, the availability of external VFOs, and of features such as dual VFOs and receiver incremental tuning (RIT) has made transceivers almost as versatile as a completely separate receiver and transmitter on the high-frequency (HF) bands. Some hams still prefer to have separate units on HF, but most use transceivers.

Types of Transceivers Transceivers can be categorized as fixed-station, mobile and portable. Fixed-station units work from 117-V or 234-V household utility mains. Mobile units generally need 12 V to 14 V dc, such as is supplied by an automotive battery. Portable units are almost always handheld "handy-talkies" (HTs). These use rechargeable batteries, usually of the nickel-cadmium (NICAD) type, supplying 3 V to 12 V. *See* MOBILE EQUIPMENT, and PORTABLE EQUIPMENT.

Another way to categorize transceivers is by frequency. Numerous models are available to cover the "low bands" or HF bands from 160 through 10 meters, or from 80 through 10 meters. The photograph shows the front panel of a typical ham transceiver that covers 1.8 through 29.7 MHz. Many HF transceivers have receive capability continuously throughout the HF range, and some even work down to 100 kHz or lower. The transmit capability is usually limited to ham bands only. "Lowband" rigs usually work all popular modes.

Transceivers at VHF and UHF might cover one band, such as 2 meters, or two bands, such as 2 meters and ¾ meters. Some cover several bands. Some work only on frequency modulation (FM); some can work all popular modes, including code (CW), frequency-shift keying (FSK), single sideband (SSB), slow-scan television (SSTV) and FM. "Multiband" or "multimode" VHF/UHF radios cost a little more than the simpler rigs.

TRANSCEIVER: Photograph of the front panel of a sophisticated amateur radio transceiver. TEN-TEC, INC., SEVIERVILLE, TN

Typical Features The following are short descriptions of some of the features that can be found in modern ham transceivers.

Receiver incremental tuning (RIT). This allows the receiver frequency to be adjusted to a certain extent, usually a few kilohertz, up or down from the transmitter frequency, which remains constant as determined by the main VFO. This is popular in HF rigs. *See* RECEIVER INCREMENTAL TUNING.

Microcomputer control. Sophisticated convenience functions are handled by a microcomputer. First, it became popular with VHF rigs around 1980; now it is common at HF as well as VHF and UHF. *See* MICROCOMPUTER CONTROL.

Programmable frequencies. This allows certain frequencies to be stored in a memory, and called up at the push of a button, without the need for changing the band switch or VFO setting. This is almost universal in VHF/UHF transceivers.

External VFO. This lets the receiver frequency be tuned independently of the transmitter frequency within the same band. It is an optional feature in HF rigs that is largely being replaced by the dual-VFO standard feature.

Dual VFOs. This feature allows the receiver and transmitter frequencies to be determined independently within a given band, or, in some units, in crossband mode (transmit and receive frequencies in different amateur bands). This is seen more often in HF rigs than at VHF/UHF.

Semi break-in operation. The transmitter is keyed automatically when the key is closed; receiver is disabled with a delay.

See SEMI BREAK-IN OPERATION. This is common in multi-mode rigs at all frequencies.

Break-in operation, also known as *full break-in* or *full QSK*. The operator can actually hear the band in between dits and dahs of code, and in some units, even between words or syllables of speech. This is popular with traffic handlers and contesters, especially at HF. *See* BREAK-IN OPERATION.

Continuous receive coverage. The receiver works all the way from the lowest transceiver frequency to the highest, without any gaps. Some HF rigs have this feature so that the transceiver can be used for shortwave listening.

Out-of-band receive capability. Sometimes this is seen in VHF/UHF radios. It allows reception of Civil Air Patrol (CAP), Military Affiliated Radio Service (MARS) and other nonham communications.

No-tune (broadband) function. There's no need to tune the transmitter final amplifier. This is almost universal at VHF and UHF. It's popular with many hams at HF.

Digital and analog displays. Digital displays are almost universal at all frequency ranges nowadays; some HF operators like analog readouts in addition. *See* ANALOG CONTROL, and DIGITAL CONTROL.

Phase-locked loop. This allows precision frequency control. It's almost universal at VHF/UHF, and is increasingly common at HF. *See* PHASE-LOCKED LOOP.

Precision readout. The frequency is displayed to the tenth of a kilohertz, and sometimes down to 0.01 kHz (100 Hz), from 160 through 10 meters. This degree of resolution is not often seen at VHF/UHF; it isn't usually necessary there.

MF, LF and VLF coverage. The receiver works below 160 meters, all the way down to 100 kHz or even 10 kHz. This is an extension of continuous-receive capability for HF rigs.

Other features. Some units have built-in keyers, built-in transmatches, adjustable automatic gain control, digital speech processing, squelch, speech clipping, speech compression, noise blankers and/or limiters, adjustable passband, adjustable selectivity, notch filtering, audio filters and scanning functions. *See* AUDIO FILTERING, AUTOMATIC GAIN CONTROL, DIGITAL SIGNAL PROCESSING, KEYER, NOISE LIMITER, NOTCH FILTER, SELECTIVITY, SPEECH CLIPPING, SPEECH COMPRESSION, SQUELCH, and TRANSMATCH.

The best way to get acquainted with all the features available in modern ham transceivers is to go to a hamfest and look at radios, or to get brochures from the various manufacturers. New features are being devised every day.

TRANSCONDUCTANCE

Transconductance is a parameter that is sometimes used as an expression of the performance of a vacuum tube or transistor. The symbol for transconductance is g_m, and the unit is the mho or siemens (*see* CONDUCTANCE, MHO, and SIEMENS).

Transconductance is defined as the ratio of the change in plate, collector or drain current to the change in grid, base or gate voltage over a defined, arbitrarily small interval on the plate-current-versus-grid-voltage, collector-current-versus-base-voltage, or drain-current-versus-gate-voltage characteristic curve. If ΔI represents a change in plate, collector or drain current caused by a change in the grid, base or gate voltage ΔE, then the transconductance is approximately:

$$g_m = \Delta I / \Delta E$$

The precise value at a given point on the characteristic curve is given by:

$$g_m = dI / dE$$

and corresponds to the slope of the tangent line to the curve at that point. *See also* CHARACTERISTIC CURVE.

TRANSDUCER

A *transducer* is a device that converts one form of energy or disturbance into another. In electronics, transducers convert alternating or direct electric current into sound, light, heat, radio waves, or other forms. Transducers also convert sound, light, heat, radio waves, or other energy forms into alternating or direct electric current.

Common examples of electrical and electronic transducers include buzzers, speakers, microphones, piezoelectric crystals, light-emitting and infrared-emitting diodes, photocells, radio antennas, and many other devices. For specific information concerning a particular kind of electrical or electronic transducer, please refer to the article, according to the name of the device.

TRANSEQUATORIAL PROPAGATION

Transequatorial (TE) propagation is a form of ionospheric refraction of radio waves. It occurs along paths running north and south, when one station is in the Northern Hemisphere and the other is in the Southern Hemisphere.

During periods of maximum sunspot activity, the F layer reaches its highest density levels. The great circle equidistant from the magnetic poles is called the *geomagnetic equator*. In the ionosphere above this great circle, the F-layer can become so densely ionized that propagation takes place at frequencies much higher than usual (*see* F LAYER, and IONOSPHERE).

In TE propagation, the stations should be roughly equidistant from the geomagnetic equator, and the path should run nearly perpendicular to the geomagnetic equator, or parallel to the earth's magnetic lines of flux. Under ideal conditions, F-layer contacts have been made at frequencies as high as the 222-MHz band over paths of several thousand miles. *See also* PROPAGATION CHARACTERISTICS.

TRANSFER IMPEDANCE

In a direct-current or alternating-current circuit, a voltage applied at one point can result in a current at another point. The ratio E/I of the applied voltage to the resultant current is known as the *transfer impedance*.

Consider the example of a transformer with a stepdown ratio of 5:1, an input voltage E of 120 volts, and a load impedance Z of 12 ohms. The output voltage, E_o, is:

$$E_o = 120/5 = 24 \text{ V}$$

and the load current I is therefore:

$$I = E_o/Z = 24/12 = 2 \text{ A}$$

therefore, the transfer impedance, Z_t, is given by:

$$Z_t = E/I = 120/2 = 60 \text{ ohms}$$

This example assumes that the efficiency of the transformer is 100 percent. *See also* IMPEDANCE.

TRANSFORMER

A *transformer* is a device that is used to change the voltage of an alternating current, or the impedance of an alternating-current circuit path. Transformers operate by means of inductive coupling. The windings of a transformer can have a ferromagnetic core or a dielectric core.

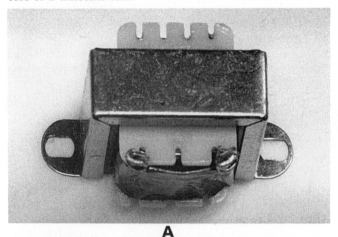

A

B

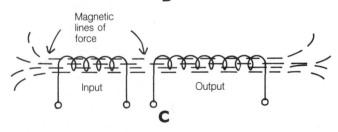

C

TRANSFORMER: At A, a small audio unit. At B, a utility power unit. At C, the basic concept of a transformer.

Transformers, such as the one shown in illustration A, are extensively used in electronic power supplies and audio circuits. The unit shown is an audio-frequency transformer, intended for matching the output impedance of a transistorized amplifier to the 8-ohm impedance of a typical speaker. This transformer is about the size of a walnut. Larger transformers are used in high-current power supplies. Still larger ones provide the 117-V utility power to which we are accustomed in our homes and businesses (see B). The largest transformers handle many thousands of volts and thousands of amperes, and are used at power-transmission stations.

At radio frequencies, transformers are used for impedance matching between amplifying stages, or between a power amplifier and an antenna.

A transformer operates according to the principle of magnetic induction, as illustrated at C. Transformers work only with alternating currents. The input winding is called the *primary*, and the output winding is called the *secondary* (*see* TRANSFORMER PRIMARY, and TRANSFORMER SECONDARY). If a transformer has a primary winding with n_1 turns and a secondary winding with n_2 turns, then the primary-to-secondary turns ratio, t, is defined as:

$$T = n_1/n_2$$

This ratio determines the input-to-output impedance and voltage ratios.

For a transformer with primary-to-secondary turns ratio T and an efficiency of 100 percent (that is, no power loss), the output voltage E_o is related to the input voltage E_i by the equation:

$$E_i = TE_o$$

The output, or load, impedance Z_o in such a transformer is related to the input impedance according to the equation:

$$Z_i = T^2Z_o$$

See also IMPEDANCE TRANSFORMER, and TRANSFORMER EFFICIENCY.

TRANSFORMER COUPLING

In an audio-frequency or radio-frequency system, the signal can be transferred from stage to stage in various ways. *Transformer coupling* was once a commmonly used method.

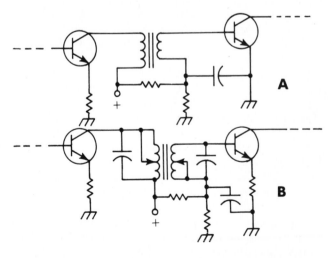

TRANSFORMER COUPLING: For audio (A) and RF (B) circuits.

The illustration at A illustrates transformer coupling between two audio-frequency amplifier stages. The transformer is chosen so that the output impedance of the first stage is matched to the input impedance of the second stage. B shows transformer coupling in the intermediate-frequency chain of a radio receiver. In this arrangement, the windings are tapped for optimum impedance matching. Capacitors are connected across the transformer windings to provide resonance at the intermediate frequency.

An important advantage of transformer coupling is that it minimizes the capacitance between stages. This is especially desirable in radio-frequency circuits, because stray capacitance reduces the selectivity. In transmitter power amplifiers, the output is almost always coupled to the antenna by means of a tuned transformer. This minimizes unwanted harmonic radiation. An electrostatic shield provides additional attenuation of harmonics by practically eliminating stray capacitance between the primary and secondary windings (see ELECTROSTATIC SHIELDING).

The main disadvantage of transformer coupling is cost: It is more expensive than simpler coupling methods. Transformer coupling is not often seen in modern circuitry. See also CAPACITIVE COUPLING.

TRANSFORMER EFFICIENCY

In any transformer, some power is lost in the coil windings. If the transformer has a ferromagnetic core, some power is lost in the core. Conductor losses occur because of ohmic resistance and sometimes skin effect (see SKIN EFFECT). Core losses occur because of eddy currents and hysteresis (see EDDY-CURRENT LOSS, and HYSTERESIS LOSS). The efficiency of a transformer is therefore always less than 100 percent.

Let E_p and I_p represent the primary-winding voltage and current in a hypothetical transformer, and let E_s and I_s be the secondary-winding voltage and current. In a perfect transformer, the product E_pI_p would be equal to the product E_sI_s. However, in a real transformer, E_pI_p is always greater than E_sI_s. The efficiency, in percent, of a transformer is:

$$Eff = 100E_sI_s/(E_pI_p)$$

The power P that is lost in the transformer windings and core is equal to the difference between E_pI_p and E_sI_s:

$$P = E_pI_p - E_sI_s$$

The efficiency of a transformer varies depending on the load connected to the secondary winding. If the current drain is excessive, the efficiency is reduced. Transformers are generally rated according to the maximum amount of power they can deliver without serious degradation in efficiency.

In audio and radio circuits, the frequency affects the efficiency of a transformer. In transformers with air cores, the efficiency gradually decreases as the frequency becomes higher. This occurs because of skin effect in the windings. In a ferromagnetic-core transformer, the efficiency gradually decreases as the frequency increases, until a certain critical point is reached. As the frequency rises beyond this point, the efficiency drops rapidly because of hysteresis loss in the core material.

Transformer efficiency is maximized by the use of the proper type of transformer in a given application. See also TRANSFORMER.

TRANSFORMER LOSS

See EDDY-CURRENT LOSS, HYSTERESIS LOSS, SKIN EFFECT, and TRANSFORMER EFFICIENCY.

TRANSFORMER PRIMARY

The *primary winding* of a transformer is the winding to which a voltage or signal is applied from an external source. In a stepdown transformer, the primary winding has more turns than the secondary. In a step-up transformer, the primary has fewer turns than the secondary (see STEP-DOWN TRANSFORMER, STEP-UP TRANSFORMER, and TRANSFORMER SECONDARY).

Some transformers have center-tapped primary windings. This type of transformer is used in the output of push-pull circuits (see PUSH-PULL CONFIGURATION) and in various amplifiers and oscillators. A center-tapped primary can be used for obtaining a balanced input for an audio-frequency or radio-frequency amplifier.

The size of the wire used for a primary winding depends on the amount of current it must carry. For a given amount of power, a step-up transformer must use larger primary-winding wire than a step-down transformer requires. See also TRANSFORMER.

TRANSFORMER SECONDARY

The *secondary winding* of a transformer is the winding from which the output is taken. In a step-up transformer, the secondary winding has more turns than the primary winding; in a step-down transformer, the secondary has fewer turns (see STEP-DOWN TRANSFORMER, STEP-UP TRANSFORMER, and TRANSFORMER PRIMARY).

The secondary winding of a transformer can have one or more taps for obtaining different voltages or impedances. A center-tapped secondary winding is used in full-wave rectifier circuits, and in oscillators or amplifiers having balanced outputs (see BALANCED OUTPUT, and FULL-WAVE RECTIFIER). A tapped secondary is used in the input transformer of a balanced modulator, push-pull amplifier, or push-push circuit (see BALANCED MODULATOR, PUSH-PULL CONFIGURATION, and PUSH-PUSH CONFIGURATION).

The size of the wire in the secondary winding of a transformer depends on the turns ratio. A step-up transformer can have smaller secondary-winding wire than a step-down transformer for a given amount of power. See also TRANSFORMER.

TRANSIENT

A *transient* is a sudden, very brief surge of high voltage on a utility power line. Normally, the peak voltage at a 117-V root-mean-square (RMS) household utility outlet is about 165 V. However, lightning or arcing can result in momentary surges of several hundred volts (see illustration) because of the electromagnetic pulse generated by such phenomena.

Transients, or "spikes," do not normally affect ordinary appliances, but they can cause damage to some types of electronic devices that are not properly protected. Transients can also blow fuses and trip circuit breakers for no apparent reason. Transients are quite common, and all solid-state electronic equipment should have some sort of protection circuit to reduce the chances of damage.

A transient is not the same as a surge. A surge results from a sudden demand for current. *See also* ELECTROMAGNETIC PULSE, SURGE, and TRANSIENT PROTECTION.

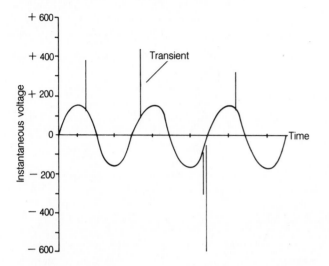

TRANSIENT: Peaks can be many times the RMS ac voltage.

TRANSIENT PROTECTION

Certain types of electronic apparatus, especially solid-state equipment, can be damaged by transients (*see* TRANSIENT). Some form of protection should be provided for such devices.

The simplest method of transient protection is the connection of a moderate-sized capacitor (approximately 0.01 μF) in parallel with the ac power line. A more effective method is the installation of moderate-sized chokes (approximately 10 mH) in series with the lines, in addition to the capacitor. The most effective transient protection is afforded by a thyrector (*see* THYRECTOR).

No form of transient protection is 100-percent reliable. A nearby lightning strike can result in a transient of such severity that the protection device, as well as the equipment connected to the power line, can be destroyed. Fortunately, transients of such magnitude do not occur very often.

TRANSISTOR

A *transistor* is a semiconductor device consisting of three or four sections, or layers, of N-type and P-type material (*see* N-TYPE SEMICONDUCTOR, and P-TYPE SEMICONDUCTOR). There are many types of transistors, manufactured for various applications. Transistors are used as oscillators, amplifiers, and switches at frequencies ranging from direct-current to ultrahigh.

Transistors were invented by scientists at the Bell Laboratories in the late 1940s. Their discovery has revolutionized electronics. Ultimately, it led to miniaturization, making possible the computers and handheld calculators that we have today.

Types of Transistors There are two main types of transistors: the bipolar device and the field-effect device. Bipolar transistors are classified as npn or pnp (*see* NPN TRANSISTOR, and PNP TRANSISTOR). Field-effect transistors can be either N-channel or P-channel (*see* FIELD-EFFECT TRANSISTOR).

Bipolar transistors are made from germanium or silicon; field-effect transistors can be fabricated from germanium, silicon, or metal oxides (*see* METAL-OXIDE-SEMICONDUCTOR FIELD-EFFECT TRANSISTOR). Methods of construction vary considerably, depending on the application for which the device is intended.

The photograph illustrates two small transistors that are commonly used in miniaturized equipment (a dime is placed next to the transistors for size comparison). The device with the metal case and four leads is a dual-gate, metal-oxide-semiconductor field-effect transistor (MOSFET). It can be used as a radio-frequency amplifier or mixer in receiving circuits. The device with the black case is an npn bipolar transistor, suitable for digital switching or amplification at low to medium frequencies. These are just two examples among many different kinds of field-effect or bipolar transistors.

Some transistors are massive, and are capable of handling hundreds or even thousands of watts at audio and radio frequencies. These transistors are of the bipolar type, usually npn, and are called *power transistors* (*see* POWER TRANSISTOR). Other devices are far smaller than the ones shown, consisting of microminiature silicon chips that are soldered, welded, or etched onto an integrated circuit (*see* INTEGRATED CIRCUIT, and THICK-FILM AND THIN-FILM INTEGRATED CIRCUITS).

Biasing and Operation Bipolar transistors consist essentially of two diodes, connected in reverse series, and having a common center element. This results in variable resistance through the device. An electrode is attached to each of the three semiconductor wafers. The electrode at the center is called the *base,* and the outer two electrodes are called the *emitter* and *collector.* A source of direct-current voltage, generally between 3 and 30 V, is applied to the emitter and collector. In the npn transistor, the collector is positive, with respect to the emitter; in the pnp device, the emitter is positive, with respect to the collector. The base is supplied with a voltage somewhere in between the emitter and collector potentials in most applications. Sometimes the base is made negative with respect to the emitter in the npn transistor, or positive in the pnp device. The base-emitter voltage depends on the class of amplification desired (*see* CLASS-A AMPLIFIER, CLASS-AB AMPLIFIER, CLASS-B AMPLIFIER, and CLASS-C AMPLIFIER).

The signal is applied either in series with the emitter, or between the emitter and the base. The resulting fluctuations in the emitter-base current cause large changes in the collector current. This is how a transistor amplifies: it produces large signal changes as a result of small ones. The base draws some current, and the input impedance of a bipolar transistor is therefore usually low to moderate.

The field-effect transistor is constructed somewhat differently. A single layer of semiconductor material, called the *channel,* is surrounded by sections of the opposite type semiconductor. The outer sections are called the *gates.* An electrode is attached to each end of the channel; one of these is the *drain* and the other is the *source.* The gates can be tied together, in which case they are attached to a single electrode; or they can be separated and connected to different electrodes. If the channel is fabricated from N-type material, the device is called an *N-channel field-effect transistor.* If the channel is made from P-type material, the device is a *P-channel field-effect transistor.*

In the N-channel device, the drain is biased positively, with respect to the source; in the P-channel field-effect transistor, the polarity is reversed. The gate is biased at some intermediate potential, or can be negative in the N-channel device or positive in the P-channel device. The gate bias depends on the class of amplification for which the field-effect transistor is to be used.

The signal is applied either in series with the source, or between the gate and the source. Small changes in the input signal cause large fluctuations in the current through the channel (the principle of operation is similar to that of a water faucet). The gate draws almost no current. Therefore, the input impedance is high. In metal-oxide-semiconductor devices, the input impedance is extremely large. This makes the field-effect transistor especially useful as a weak-signal amplifier.

Characteristics Important characteristics of bipolar transistors include the alpha, the alpha-cutoff frequency, the beta, and the beta-cutoff frequency. In the field-effect transistor, the principal characteristic is the transconductance. All of these parameters are related to the gain. *See* ALPHA, ALPHA-CUTOFF FREQUENCY, BETA, and TRANSCONDUCTANCE.

For More Information Various types of bipolar and field-effect transistors and their construction, operation, and uses are discussed in more detail in other articles of this book. In addition to the subjects referenced above, related articles in this book include CHARACTERISTIC CURVE, COMMON CATHODE/EMITTER/SOURCE, COMMON GRID/BASE/GATE, COMMON PLATE/COLLECTOR/DRAIN, COMPLEMENTARY METAL-OXIDE SEMICONDUCTOR, CUTOFF, CUTOFF VOLTAGE, DOPING, EPITAXIAL TRANSISTOR, GALLIUM-ARSENIDE FIELD-EFFECT TRANSISTOR, IMPURITY, PINCHOFF, PINCHOFF VOLTAGE, P-N JUNCTION, SEMICONDUCTOR, SOLID-STATE ELECTRONICS, SWITCHING TRANSISTOR, and VERTICAL METAL-OXIDE-SEMICONDUCTOR FIELD-EFFECT TRANSISTOR.

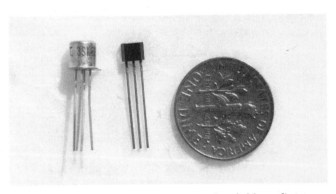

TRANSISTOR: Some transistors are dwarfed by a dime.

TRANSISTOR TESTER

A *transistor tester* is a device that quantitatively determines whether or not a transistor is functioning properly, and if so, gives a quantitative indication of the operating characteristics.

Sophisticated transistor testers are commercially available for checking the performance of bipolar and field-effect transistors. Such devices measure the alpha and beta of a bipolar device and the transconductance of a field-effect device (*see* ALPHA, BETA, TRANSCONDUCTANCE, and TRANSISTOR).

If it is suspected that a bipolar transistor has been destroyed, an ohmmeter can be used for a quick check. All ohmmeters produce some voltage at the test terminals. The ohmmeter should be switched to the lowest resistance ($R \times 1$) scale, and the po-

larity of the test leads determined by means of an external voltmeter. Designate P+ as the lead at which a positive voltage appears, and P— the lead at which negative voltage appears. (Note: in a volt-ohm-milliammeter, the ohmmeter-mode polarity might not correspond to the red/black lead colors used for measurement of voltage and current!) To test an npn transistor:

- Connect P+ to the collector and P— to the emitter. The resistance should appear infinite.
- Connect P+ to the emitter and P— to the collector. The resistance should again appear infinite.
- Connect P+ to the emitter and P— to the base. The resistance should appear infinite.
- Connect P+ to the base and P— to the emitter. The resistance should appear finite and measurable.
- Connect P+ to the collector and P— to the base. The resistance should appear infinite.
- Connect P— to the collector and P+ to the base. The resistance should appear finite and measurable.

If any of the above conditions is not met, the transistor is suspect.

To check a pnp transistor, the procedure is just the same as above, except that every P+ should be changed to P—, and every P— should be changed to P+. Again, if any of the above conditions is not met, the transistor is suspect.

TRANSISTOR-TRANSISTOR LOGIC

Transistor-transistor logic (TTL) is a bipolar logic design in which transistors act on direct-current pulses. Transistor-transistor logic is similar to diode-transistor logic, except that a multiple-emitter transistor replaces several diodes.

Transistor-transistor logic is characterized by high speed and good noise immunity. For these reasons, TTL is widely used in digital applications. Many TTL logic gates can be fabricated into a single integrated-circuit package. Transistor-transistor logic can be used for either negative or positive logic circuits. *See also* DIODE-TRANSISTOR LOGIC, DIRECT-COUPLED TRANSISTOR LOGIC, EMITTER-COUPLED LOGIC, HIGH-THRESHOLD LOGIC, INTEGRATED INJECTION LOGIC, METAL-OXIDE-SEMICONDUCTOR LOGIC FAMILIES, NEGATIVE LOGIC, POSITIVE LOGIC, RESISTOR-CAPACITOR-TRANSISTOR LOGIC, RESISTOR-TRANSISTOR LOGIC, and TRIPLE-DIFFUSED EMITTER-FOLLOWER LOGIC.

TRANSITION ZONE

The electromagnetic field surrounding a transmitting antenna can be divided into three distinct zones. They are known as the *far field*, the *near field*, and the *transition zone*. The far field exists at a great distance from the antenna, and is sometimes called the *Fraunhofer region*. The near field occurs very close to the transmitting antenna; it is sometimes called the *Fresnel zone*. The region between the far field and the near field is known as the *transition zone*.

The far field, near field, and transition zone are not precisely defined. They exist in qualitative form. In general, the transition zone occurs at distances where the electric and magnetic lines are somewhat curved. The lines of flux are decidedly curved in the near field, but they are practically straight in the far field. In communications practice, the far field is the only important consideration; the near field and transition zone fall within a few wavelengths of the radiating element. *See also* FAR FIELD, and NEAR FIELD.

TRANSIT TIME

Transit time is the time required for a charge carrier to get from one place to another. Transit time is of significance in the operation of vacuum tubes and semiconductor devices.

In a tube, the electrons travel from the cathode to the plate through a vacuum or near vacuum. Although the electrons move very fast, they require a small amount of transit time to cross the gap. The transit time depends on the type and size of the tube, the plate voltage, and the current drawn.

In a semiconductor diode or bipolar transistor, electrons and holes move and diffuse through the wafers. The transit time is the effective time required for a charge carrier to get to the anode from the cathode in a diode, or from the emitter-base junction to the collector-base junction in a bipolar transistor. In a field-effect transistor, the transit time is the time required for the electrons or holes to pass through the channel. Transit time in a semiconductor device is related to the carrier mobility, which depends on the type of semiconductor material and the extent of doping. *See also* CARRIER MOBILITY, FIELD-EFFECT TRANSISTOR, TRANSISTOR, and TUBE.

TRANSLATION

The conversion of data from one language into another is called *translation*. A translation is a definable, constant mathematical function between two sets of data. Although the two sets can differ greatly in format and speed, the translation preserves all of the information without altering the content or meaning.

Encoding and decoding are forms of translation; most often, however, the term is used to denote the changing of languages in, or by, a computer. The assembler, for example, is responsible for translating assembly language into machine language. A modem translates between printed symbols and audio-frequency tones. An analog-to-digital or digital-to-analog converter is a form of translator.

Computers are extensively used for translating data to achieve compatibility between or among systems. This allows one computer to "talk to" another, even if they use different languages. The computer that performs the data conversion is called the *translator* or the *interface computer*. The operation of the interface computer is governed by a special program called the *translator program*. *See also* ANALOG-TO-DIGITAL CONVERTER, DATA CONVERSION, DECODING, DIGITAL-TO-ANALOG CONVERTER, ENCODING, HIGHER-ORDER LANGUAGE, INTERFACE, MACHINE LANGUAGE, and MODEM.

TRANSMATCH

The output impedance of a radio transmitter, or the input impedance of a receiver, should ideally be matched to the antenna-system impedance. This requires that the antenna system present a nonreactive load of a certain value, usually 50 or 75 ohms. Very few antenna systems meet this requirement, but almost any load impedance can be matched to a transmitter or receiver by means of a device called a *transmatch*.

A transmatch consists of one or more inductors and capacitors arranged so that they cancel any existing reactance in the antenna system, and convert the remaining resistance to the appropriate value. Transmatches vary in complexity from simple L or pi networks to sophisticated adjustable circuits.

The illustration shows four transmatch circuits. The arrangements at A and B are for use with unbalanced loads; the circuits at C and D are for use with balanced loads. Most transmatch circuits incorporate reflectometers in the input circuit to facilitate adjustment.

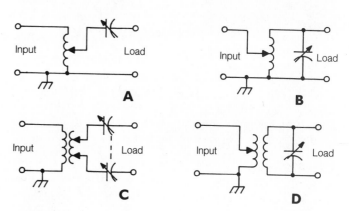

TRANSMATCH: At A, circuit for matching low-Z unbalanced loads. At B, for high-Z unbalanced loads; at C, for low-Z balanced loads; at D, for high-Z balanced loads.

When a transmatch is used with a radio transmitter and receiver, adjustments should be made using an impedance bridge if possible. This eliminates the need for transmitting signals over the airwaves. If an impedance bridge is not available, transmatch adjustments should be performed according to the following procedure:

- Set the receiver and transmitter to the same frequency.
- Adjust the transmatch until the signals and/or noise are maximized in the receiver.
- Preset the transmitter tuning and loading controls to the approximate positions for the frequency in use.
- Switch the transmitter on, and apply a low-power signal to the transmatch.
- Adjust the transmatch for minimum standing-wave ratio (SWR) as indicated on a reflectometer placed between the transmitter and the transmatch.
- Retune the transmitter for normal operating output power.
- Identify your station.
- Record the positions of the transmatch controls for future reference.

Transmatches are extremely useful for portable station operation. A random-wire antenna can be strung in any fashion and tuned to resonance by using a transmatch. Generally, such antennas must measure at least ¼ electrical wavelength if good results are to be obtained, although some transmatches can provide an impedance match with much shorter wires.

A well-designed and properly operated transmatch (see photo on pg. 510) acts as a bandpass filter. This provides additional front-end selectivity for a receiver, improving the image response and reducing the chances of front-end overload (See FRONT END, and IMAGE REJECTION). The transmatch also attenuates spurious outputs from a transmitter, including the harmonics (*see* HARMONIC SUPPRESSION, and SPURIOUS EMISSIONS).

TRANSMATCH: Interior view of a high-efficiency transmatch. Ten-Tec, Inc., Sevierville, TN

TRANSMISSION CONTROL PROTOCOL

All packet-radio network protocols have some means of maintaining accuracy, efficiency and reliability. *Transmission control protocol (TCP)* is favored by some hams because it has unique features.

TCP puts packets together for transmission, and also takes apart received packets so they can be displayed as data on a terminal. TCP checks constantly for errors. It maintains accurate, efficient and reliable circuit connections for individual packets.

In addition, TCP acts as a "network manager," keeping track of all packets in the network. This optimizes the operation of the whole system.

The software for TCP was developed by Phil Karn, KA9Q, for use at the *transport layer* of the *open systems interconnection reference model (OSI-RM)*. This protocol is part of a networking protocol known as TCP/IP. TCP/IP consists of a synergistic combination of features from protocols that existed previously, with a few new twists.

Details of TCP/IP and other networking protocols can be found in books about packet radio. The ham magazines should be consulted for the most up-to-date volumes on this rapidly evolving field of amateur radio. *See also* OPEN SYSTEMS IN-TERCONNECTION REFERENCE MODEL, PACKET RADIO, and PROTOCOL.

TRANSMISSION LINE

A *transmission line* is a medium by which energy is transferred from one place to another. In ham radio, this refers to an antenna feed line (*see* FEED LINE).

Transmission lines can be categorized as unbalanced or balanced. Unbalanced lines include the single-wire line (not often used nowadays), coaxial cables and hard lines, and the waveguide. Balanced lines include open-wire and twin-lead line. Twin-lead line is also sometimes called *ribbon line*. Less often, a four-wire line is used.

Sections of transmission line are sometimes used as matching stubs or as tuned circuits. *See* QUARTER-WAVE TRANS-MISSION LINE, and STUB. Any transmission line exhibits a property called *characteristic impedance*. This is the ratio of voltage to current in the line if it has a 1:1 standing-wave ratio (SWR). *See* CHARACTERISTIC IMPEDANCE, and STAND-ING-WAVE RATIO. Coaxial lines typically have a characteristic impedance, or Z_o, from 50 to 100 ohms. Twin-lead is available in 75-ohm and 300-ohm Z_o values. Open-wire line has Z_o ranging from about 300 to 600 ohms, depending on the spacing between the conductors and the type of dielectric.

All transmission lines have some loss; they do not transfer energy with perfect efficiency. The loss occurs because of the ohmic resistance of the conductors and skin effect in the conductors (*see* SKIN EFFECT), and because of losses in the dielectric (*see* DIELECTRIC LOSS). This loss is measured in decibels per unit length, such as db/100 ft. It increases with increasing frequency. It also increases as the *SWR* increases. *See* STANDING-WAVE-RATIO LOSS.

Transmission lines are rated according to the power they can handle, assuming a 1:1 SWR. If the SWR is more than 1:1, the power-handling ability of a transmission line decreases. This can be significant, especially if the SWR is very high or if the line is used at or near its maximum power rating.

The choice of a transmission line for a given application will depend on several factors. These include whether the antenna is balanced or unbalanced, the impedance of the antenna, the operating frequency, the transmitter output power, whether or not a transmatch is used, whether the SWR on the line will be high or low, and whether the transmitter or transmatch output is balanced or unbalanced. Space does not allow a complete discussion of all these factors and considerations here. For further information, a book about ham radio antennas is recommended. There are several good books on this subject; check the ham magazines for the latest offerings.

TRANSMIT-RECEIVE SWITCH

When a transmitter and a receiver are used with a common antenna, some method must be devised to switch the antenna between the two units. It is especially important that the receiver be disconnected while the transmitter is operating. A *transmit-receive (T-R) switch* accomplishes this.

The simplest form of transmit-receive switch is a relay. When the transmitter is keyed, the relay connects the antenna to the transmitter. When the transmitter is unkeyed, the relay connects the antenna to the receiver. This type of arrangement is used in many transceivers.

A more sophisticated type of T-R switch consists of a radio-frequency-actuated electronic circuit. The transmitter is always connected to the antenna, even while receiving. But when the transmitter is keyed, the receiver is disconnected.

The T-R switch allows full break-in operation if A1 emission (continuous-wave morse code) is used. *See also* BREAK-IN OPERATION.

TRANSMITTANCE

Transmittance is an expression of the extent to which a substance or circuit passes energy. Transmittance is defined as a ratio between 0 and 1, or between 0 percent and 100 percent. In radio-frequency circuits, transmittance is sometimes called *transmittivity*.

If e_1 represents the incident energy and e_2 represents the energy transmitted through a substance or circuit, the transmittance T is given in percent by:

$$T = 100e_2/e_1$$

A transmittance value of 0 indicates that a material is perfectly opaque, or that all of the available power is absorbed by a circuit. A transmittance figure of 100 percent indicates that a substance is perfectly transparent, or that none of the power is absorbed by a circuit.

TRANSMITTER

A *transmitter* converts data into electromagnetic waves. A transmitter is an information *transducer*. Transmitters for amateur service are generally manufactured to work only within the ham bands. A few ham transmitters are capable of working on frequencies outside the ham bands. Hams must always observe the limits of their bands when transmitting. In addition, hams must use only those frequencies and modes that their license classes allow.

High-frequency (HF) ham transmitters cover the ham bands from 1.8 MHz through 29.7 MHz. Some transmitters work one or more very-high and ultra-high-frequency (VHF and UHF) bands. Most transmitters work in several different modes. See EMISSION CLASS. At HF, transmitters commonly work code (CW) and single-sideband (SSB). Frequency-shift keying (FSK) is done by inputting audio tones at the microphone jack in SSB mode. The illustration is a block diagram of a typical CW/SSB transmitter. Each stage shown is discussed briefly below.

Local Oscillator   The *local oscillator (L.O.)* is where the signal is generated. This signal is tailored or keyed by other stages, for ultimate conversion to the transmitting frequency, followed by amplification to the final power output level.

The L.O. is crystal-controlled for optimum stability. Its frequency might be switchable, facilitating band changes when mixed with the VFO output. Or the L.O. might always have the same frequency, if the VFO tunes over a very wide range. *See* CRYSTAL OSCILLATOR, and LOCAL OSCILLATOR.

Keyer   Some transmitters have internal electronic *keyers*. Others simply have a key jack that turns the L.O. on and off, or blocks its output. *See* KEYER.

Microphone Amplifier   The audio from the microphone must be amplified for use by the modulator. This is done by a simple, low-distortion *microphone amplifier*. This amplifier usu-

ally has a tailored frequency response, attenuating audio of less than about 300 Hz or more than about 3 kHz. The range 300 Hz to 3 kHz is all that is needed to convey intelligible voice signals.

Modulator   The *modulator* in a single-sideband transmitter is of the balanced variety. This cancels out the carrier from the local oscillator, leaving only the sidebands. *See* BALANCED MODULATOR. Also included (but not shown in the diagram) is a filtering circuit that gets rid of either the lower sideband or the upper sideband, producing SSB. This filter also provides further attenuation of sidebands from audio outside the range 300 Hz to 3 kHz. *See* LOWER SIDEBAND, SINGLE SIDE-BAND, and UPPER SIDEBAND.

If frequency modulation (FM) is used, the modulator works by varying the frequency or phase of the L.O. *See* FREQUENCY MODULATION, PHASE MODULATION, and REACTANCE MODULATOR. Amplitude modulation (AM) is rarely used today by radio amateurs. It wastes power and spectrum space. It is mainly of historical interest. *See* AMPLITUDE MODULATION.

Frequency-shift keying (FSK) can be done by varying the reactance in the L.O. But more often, especially in HF transceivers, it is done indirectly via SSB mode. Two audio sine-wave tones are fed into the microphone input, producing two different sideband frequencies that come out as carrier waves. *See* FREQUENCY-SHIFT KEYING.

Variable-frequency Oscillator (VFO)   The transmitter VFO is either of two types: a tuned inductance-capacitance (LC) oscillator, or a *frequency synthesizer*. The LC type of VFO tunes over a span of frequencies, usually 500 kHz or 1 MHz. It is carefully designed and sealed so that it maintains its frequency despite mechanical shock, temperature changes and other variables in the environment. *Permeability tuning* is the preferred method of adjusting the frequency in this type of VFO. *See* PERMEABILITY TUNING.

The frequency synthesizer is becoming standard nowadays because it is more stable than any LC-tuned oscillator. A synthesizer can be tuned over a wide range, doing away with the need for bandswitching at the L.O. Or it can cover a 500-kHz or 1-MHz range, like the older LC-tuned VFOs. *See* FREQUENCY SYNTHESIZER, and PHASE-LOCKED LOOP.

Mixer   The *mixer* combines the L.O. and VFO signals to obtain the actual transmitting frequency. This is a nonlinear circuit that produces sum and difference signals, either of which can be selected for amplification by later stages. *See* MIXER.

RF Amplifiers   From one to several stages of radio-frequency (RF) amplification follow the mixer. The last of these stages, prior to the final amplifier, is called the *driver* because it produces the drive for the final amplifier. In a typical ham transmitter with 20 W to 200 W output, the driver produces approximately 100 mW to 5 W. All of the amplifying stages must be linear to ensure that SSB is not distorted.

Final Amplifier   The *final amplifier* boosts the signal power to the level to be transmitted. Most ham transmitters have outputs from 20 W to 200 W or so. The final amplifier is always linear. Sometimes "outboard" linear amplifiers are used to obtain the legal maximum power. Some transmitters have "no-tune"

Ant.

L.O. → Modulator → Mixer → rf amps. → Power amp.

Keyer → L.O.
Audio amp. → Modulator
VFO → Mixer

Audio Input

TRANSMITTER: The general block diagram for a CW/SSB transmitter.

final amplifiers; others require adjustment of the output circuitry for resonance and/or coupling to the antenna system. *See* FINAL AMPLIFIER, and LINEAR AMPLIFIER.

TRANSMITTING TUBE

A vacuum tube, designed for operation in a radio-frequency power amplifier, is sometimes called a *transmitting tube*. Transmitting tubes are characterized by their ability to dissipate large amounts of power.

Although tubes have been largely replaced by transistors in recent years, tubes are still used in some transmitters. This is especially true of high-power circuits operating at microwave frequencies. Tubes are also fairly tolerant of momentary excessive power dissipation. *See also* TUBE.

TRANSPARENT MODE

When using packet radio in the normal mode, called *converse mode*, certain character sequences are used as commands for the *terminal node controller (TNC)*. The TNC performs the functions, but does not send the characters themselves.

This is a problem if you want the destination station to get such a character sequence in text. In converse mode, you can't get the character combination across in text, because the TNC will think you're giving it a command.

You can overcome this problem by using *transparent mode*. This mode is enabled by sending TRANS followed by a carriage return. When in transparent mode, the TNC interprets only one set of characters as a command: the letters CONV followed by a carriage return, telling it to go back to the converse mode. *See also* PACKET RADIO, and TERMINAL NODE CONTROLLER.

TRANSPONDER

In an active communications satellite, there is a broadband repeater that receives the signals sent up from the earth, converts them to another frequency, and retransmits them back to the earth. This device is a *transponder*. Some satellites have two or more transponders, each working on different sets of frequency bands. *See* ACTIVE COMMUNICATIONS SATELLITE.

Transponders vs. Repeaters   A *transponder* is something like a repeater, because it retransmits the signals at the same time they are received. But a transponder converts a whole band of frequencies at once, whereas a repeater works with only one input frequency and one output frequency. *See* REPEATER.

A transponder can carry many contacts (QSOs) at a time. Also, full duplex is possible using a transponder. *See* DUPLEX OPERATION. You can even listen to your own downlink signal. This is commonly done in satellite communications, because when you call CQ, other stations will most likely come back at or near your own downlink frequency.

Two Types of Transponders   The illustration is a block diagram of a transponder. The mixer can convert a band of frequencies, containing many signals, to another band. Depending on whether the sum or difference output of the mixer is used for retransmitting, the signals in the transponder output might be right-side-up (the same as the way they came in) or upside-down (inverted in frequency from the way they came in). The first kind of transponder is called a *noninverting device*. The second kind is called an *inverting device*. If a satellite has an inverting transponder, code (CW) will not be affected. But the mark and space in frequency-shift keying (FSK) will be reversed, and a single-sideband (SSB) signal will be transformed to the opposite sideband. This does not cause a loss of signal information, but it must be taken into account if the transponder is of the inverting type.

In a noninverting transponder, the output, or *downlink*, band comes out in the same sense as the input, or *uplink*, band. That is, the lowest-frequency signal going in is also the lowest-frequency signal coming out. In an inverting transponder, however, the entire band is flipped upside-down so that the lowest-frequency signal going in becomes the highest-frequency signal coming out.

Transponders in Ham Satellites   The transponders in ham satellites have bandwidths on the order of a few hundred kilohertz. This can accommodate dozens of signals simultaneously. A transponder can only produce a certain maximum power output, distributed among all the signals it sends back to earth. Therefore, when a transponder is heavily used, the individual signals tend to be weaker than they are when the transponder is not dealing with many signals.

A very strong uplink signal can cause a transponder to use up almost all of its output power to retransmit that signal. If there are many QSOs in progress via a satellite, and an excessively powerful uplink signal is received, all the other signals will be attenuated. This is something like the desensing that

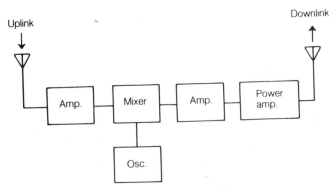

TRANSPONDER: Simplified block diagram of a satellite transponder.

takes place when the front end of a receiver is overloaded. But it does not really take a fantastic amount of uplink power to cause this with a satellite transponder. It is important, therefore, when using a ham satellite, not to use any more power than is necessary to have a decent QSO.

See also OSCAR, and SATELLITE TRANSPONDER MODES.

TRANSPORT LAYER
See OPEN SYSTEMS INTERCONNECTION REFERENCE MODEL.

TRANSVERSE WAVE

A *transverse wave* is a disturbance in which the displacement is perpendicular to the direction of travel. All electromagnetic waves are transverse waves because the electric and magnetic lines of flux are perpendicular to the direction of propagation. In some substances, sound propagates as transverse waves.

The waves on a lake or pond illustrate the principle of transverse disturbances. When a pebble is dropped into a mirror-smooth pond, waves form and are propagated outward. The effects of the disturbance travel at a defined and constant speed. The waves have measurable length and amplitude. The actual water molecules, however, move up and down, at right angles to the wave motion. Some waves are propagated by movement of particles in directions parallel to the wave motion. This kind of disturbance is called a *longitudinal wave*.

TRANSVERTER

A *transverter* is a device that allows operation of a transceiver on a frequency much different from the design frequency. A transverter consists of a transmitting converter and a receiving converter in a single package (*see* CONVERTER).

The illustration is a block diagram of a hypothetical transverter. The transceiver operates in the band 28.0 to 30.0 MHz. The actual operating frequency band is 144.0 to 146.0 MHz. A common local oscillator at 116.0 MHz provides the conversion.

The receiving converter heterodynes the incoming frequency fr_1 MHz to a new frequency fr_2 MHz according to the relation:

$$fr_2 = fr_1 - 116.0$$

The transmitting converter heterodynes the transceiver output frequency ft_1 to a new frequency ft_2 according to:

$$ft_2 = 116.0 + ft_1$$

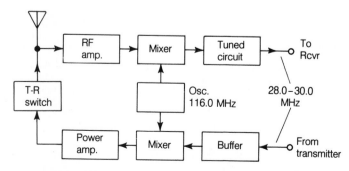

TRANSVERTER: This block diagram shows a transverter for using a 10-meter radio on 2 meters.

Transverters are commercially manufactured for use with most kinds of transceivers. *See also* MIXER, RECEIVER, TRANSCEIVER, and TRANSMITTER.

TRAP

A *trap* is a form of band-rejection filter, designed for blocking energy at one frequency while allowing energy to pass at all other frequencies. A trap consists of a parallel-resonant circuit in series with the signal path (see illustration A) or a series-resonant circuit in parallel with the signal path, as at B. The configuration at A is the more common.

Traps can be constructed from discrete inductors and capacitors, as shown, or they can be fabricated from sections of transmission line. In the latter case, the trap is usually called a *stub* (*see* STUB). Traps are used extensively in the design and construction of multiband antenna systems. *See also* TRAP ANTENNA.

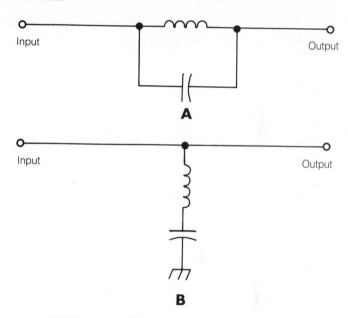

TRAP: At A, parallel-resonant; at B, series-resonant.

TRAP ANTENNA

A *trap antenna* is a form of multiband antenna in which traps are used to obtain resonance on two or more different frequencies (*see* TRAP). The traps are placed at certain points along the radiating element and parasitic elements (if any), as shown in the illustration.

The number of frequencies at which a trap antenna is resonant is usually one greater than the number of traps in the radiating element. For example, an antenna with three traps will be resonant on four frequencies. This is the case in the example shown.

The design of a trap antenna is rather complicated, because several interacting parameters are involved: the spacing and positions of the traps, the size of the inductor in each trap, and the number of traps. The greater the number of traps, the more complicated the antenna design.

Antenna traps are designed so that they have the least possible amount of loss, and the highest possible Q factor. This involves the use of large wire for the coil of a trap, and a low-loss, high-voltage capacitor. The capacitor can be placed inside or outside the coil.

Trap antennas, while commonly used at high frequencies, are not the only means of obtaining multiband operation. *See also* MULTIBAND ANTENNA.

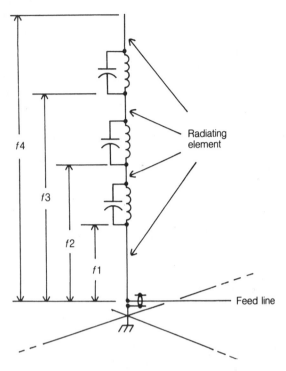

TRAP ANTENNA: In this example, traps isolate portions of the antenna for use on frequencies near f_1, f_2, f_3, and f_4.

TRAPEZOIDAL PATTERN

The percentage of modulation of an amplitude-modulated signal can be determined using an oscilloscope, connected in such a way that a trapezoid-shaped pattern appears on the screen. The radio-frequency signal is applied to the vertical deflection plates of the oscilloscope, and the modulating audio signal is applied to the horizontal deflection plates.

The modulation percentage is determined according to how much the shape of the trapezoid differs from a perfect rectangle. If there is no modulation, the pattern appears as a perfect rectangle. As the modulation percentage increases, the pattern becomes more and more distorted until, with 100-percent modulation, it appears as a triangle.

As the modulation percentage is increased past 100 percent, the triangle shrinks in the horizontal dimension, and a line appears at the right. The line indicates clipping of the negative peaks of the modulated signal. Positive peak clipping shows up as flattening of the left-hand end of the triangle.

The modulation percentage can be determined mathematically from the lengths of the vertical sides of the trapezoid. If the long side measures L graduations on the oscilloscope screen and the short side measures S divisions, then the modulation percentage, m, is given by the formula:

$$m = 100(L - S)/(L + S)$$

within the limits of 0 to 100 percent. The formula does not hold for modulation percentages over 100. *See also* AMPLITUDE MODULATION, and MODULATION PERCENTAGE.

TRAPEZOIDAL WAVE

A *trapezoidal wave* results from the combination of a square wave and a sawtooth wave, having identical frequencies, but perhaps different amplitudes. The trapezoidal wave gets its name from the fact that its shape somewhat resembles a trapezoid.

The period of a trapezoidal wave is the length of time from any point on the waveform to the same point on the next pulse in the train. This period is divided into 360 degrees of phase, with the zero point usually corresponding to the moment at which the amplitude is zero and rapidly changing. The frequency of the trapezoidal wave, in hertz, is the reciprocal of the period in seconds.

In a cathode-ray tube, the voltage applied to the deflecting coils always has a trapezoidal waveshape. This provides the necessary sawtooth-wave variation in the currents through the coils, and thus in the magnetic fields, ensuring a linear forward sweep and a fast return sweep. *See also* CATHODE-RAY TUBE, and SAWTOOTH WAVE.

TRAVELING-WAVE TUBE

A *traveling-wave tube* is a specialized vacuum tube that is used in oscillators and amplifiers at ultra-high and microwave frequencies. There are various different types of traveling-wave tubes.

The electron gun produces a high-intensity beam of electrons that travels in a straight line to the anode. A helical conductor is wound around the tube. The distributed inductance and capacitance of this winding result in a very low velocity factor (*see* VELOCITY FACTOR), typically from 10 to 20 percent, approximately matching the speed of the electrons in the beam inside the tube. A signal is applied at one end of the helix and is taken from the other end.

When the helical winding is energized, the electron beam inside the tube is phase-modulated. Some of the electrons travel in synchronization with the wave in the helix, because of the low velocity factor of the winding. This produces waves in the electron beam. The waves can travel in the same direction as the electrons (forward-wave mode) or in the opposite direction (backward-wave mode). In either case, energy from the electrons is transferred to the signal in the winding, producing gain when the beam voltage is within a certain range. Gain figures of 15 to 20 dB are common, and some traveling-wave tubes can produce more than 50 dB of gain.

A traveling-wave tube can be used to produce energy at ultra-high and microwave frequencies by coupling some of the output back into the input. This type of oscillator is called a *backward-wave oscillator* because the feedback is applied opposite to the direction of movement of the electrons inside the tube. The backward-wave oscillator produces about 20 to 100 mW of radio-frequency power at frequencies up to several gigahertz. The backward-wave arrangement can also be used for amplification, but the most common traveling-wave amplifier configuration is the parametric amplifier, which uses the forward-wave mode. *See also* PARAMETRIC AMPLIFIER.

TRIAC

A *triac* is a semiconductor device that is used for controlling ac voltage. The triac is constructed in a manner similar to a pnp transistor, except that no distinction is made between the emitter and collector. A triac is identical to a diac, except that a control electrode is connected to one of the anodes via an N-type section of semiconductor. The triac is used as a light dimmer and motor-speed control in common household circuits. *See also* DIAC.

TRIANGULAR WAVE

A *triangular wave* is a form of sawtooth wave in which the rise and decay times are identical. The triangle wave is therefore symmetrical, with respect to the amplitude peaks. *See* SAWTOOTH WAVE.

TRIANGULATION

Sometimes it is necessary to locate the position of a transmitter. Some hams enjoy transmitter hunting just for fun; one ham hides a transmitter and the others look for it using direction-finding (DF) equipment (*see* DIRECTION FINDER). This is called a *foxhunt*.

To pinpoint the position of a transmitter, a technique called *triangulation* is used. Generally, three direction-finding receivers are used. The receivers are mobile. As bearings are taken, lines are drawn on a map, in the directions of the nulls in the DF loops. When the receivers are located so that the three lines converge from points more or less equally spaced around the compass, that point is marked as the approximate location of the transmitter (see the illustration).

The receivers are moved close to the point indicated on the map as the approximate location of the transmitter, and bearings are taken again, resulting in a new point that more closely approximates the transmitter position. This process is repeated as many times as necessary.

Once a receiver is extremely close to the hidden transmitter, the signal will get stronger and weaker with walking motion. Then, nondirectional antennas can be used, and the foxhunt becomes a matter of walking around to find the strongest signal point.

Transmitter-hiding hams sometimes play rough. One trick is to hide the transmitter high in a tree. Then, the "hunters" will have to send somebody up the tree to get the "fox."

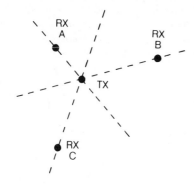

TRIANGULATION: Bearings are taken from three receiving points RX A, RX B, and RX C, to zero in on the transmitter location, TX.

TRICKLE CHARGE

A rechargeable battery can be charged at a low current level or at a high current level. In general, a rechargeable cell or battery having a capacity of x ampere hours can be fully charged, starting at zero charge, with I amperes of current for t hours, such that $x = tI$. When I is very small, t must be large. Low-current, long-term charging is known as *trickle charging*.

Trickle charging is the best method for maintaining the energy in rechargeable cells and batteries. The nickel-cadmium memory backup batteries in some microcomputer-controlled devices are trickle-charged whenever power is on. The nickel-cadmium battery packs in some handheld transceivers and small calculators are trickle-charged with plug-in transformer/rectifier devices.

When a battery is initially in a state of nearly complete discharge, there is not always time for trickle charging. Then, quick charging is used. *See also* CHARGING, and QUICK CHARGE.

TRIFILAR WINDING

A set of three coil windings, usually oriented in the same sense and having the same number of turns on a common core, is called a *trifilar winding*. Trifilar windings are sometimes used in transformers at audio or radio frequencies.

Trifilar windings on a toroidal core can be connected to obtain various different impedance-transfer ratios. This is useful in radio-frequency impedance-matching applications. *See also* TRANSFORMER.

TRIGGER

A *trigger* is a digital circuit that changes state or initiates conduction when a pulse is received. Trigger circuits can diodes, transistors, resistors, capacitors, and/or other components. *See also* FLIP-FLOP.

TRIGGERING

Triggering is a term with a specific meaning in digital electronics. The term is also used in oscilloscope operation.

Triggering in Digital Circuits The inputs of a digital circuit can be either edge-triggered or level-triggered. These terms are also known as *dynamic* and *static* triggering respectively; they work either from a change of state or from the existence of a given state.

In edge or dynamic triggering, the inputs are sampled every time the clock state is altered. If the sampling occurs during a transition from low to high, it is *positive edge triggering*; if it occurs during a transition from high to low, it is *negative edge triggering*.

Level or static triggering works from the logic state itself, sampling the inputs during each logic high or each logic low. *See also* LOGIC.

Triggering in Oscilloscopes In order to keep the display on a scope stable, or to have it show up a certain way, the sweep rate must be synchronized with some standard frequency. This usually is done by having the sweep *triggered*.

If the sweep frequency is triggered every 10th cycle of the input signal, for example, there will be 10 complete waveforms displayed on the screen. If it occurs every cycle of the signal, there will be one complete waveform shown on the scope screen. Triggering can be made to begin at any point on the waveform, such as the positive peak, the negative peak, or the positive-going zero point.

Oscilloscope triggering can be done either externally or automatically. Most scopes have an auto-trigger feature that supplies a trigger signal from whatever waveform is being analyzed. *See also* OSCILLOSCOPE.

TRIGONOMETRIC FUNCTION

A *trigonometric function*, also called a *circular function*, is a mathematical function that relates angular measure to the set of real numbers. There are six trigonometric functions. The three most common ones are the cosine, sine, and tangent. Less well known are the cosecant, cotangent, and secant. The trigonometric functions are abbreviated in equations and formulas.

Although trigonometric values can be expressed for negative angles and angles measuring 360 degrees or more, the domains of the trigonometric functions are usually restricted to angles θ, in degrees, such that $0 < \theta < 360$, or in radians such that $0 < \theta < 2\pi$. When the domain of a trigonometric function is deliberately restricted in this way, the first letter of the abbreviation is sometimes capitalized (this semantic distinction is important to mathematicians, but not as important to engineers).

All of the trigonometric functions can be expressed in either of two ways: the *unit-circle model* and the *triangle model*.

The Unit-Circle Model

Mathematicians usually express the trigonometric functions in terms of values on a circle in the Cartesian plane. The circle has a radius of 1 unit and is centered at the origin. The equation of the circle is:

$$x^2 + y^2 = 1$$

If θ represents an angle, measured counterclockwise from the positive x axis, then the trigonometric-function values of this angle can be determined from the values of x and y at the point (x,y) on the circle that corresponds to the angle θ. The functions are:

$$\cos \theta = x$$
$$\sin \theta = y$$
$$\tan \theta = y/x$$
$$\csc \theta = 1/y$$
$$\cot \theta = x/y$$
$$\sec \theta = 1/x$$

A special category of trigonometric functions exists, based on the unit hyperbola having the equation:

$$x^2 - y^2 = 1$$

These are known as *hyperbolic trigonometric functions*.

The Triangle Model

A right triangle can be used to express trigonometric-function values. The angle θ corresponds to one of the angles not measuring 90 degrees. If the sides of the triangle have lengths a, b, and c, and the angle θ is opposite the side of length b, then:

$$\cos \theta = a/c$$
$$\sin \theta = b/c$$
$$\tan \theta = b/a$$
$$\csc \theta = c/b$$
$$\cot \theta = a/b$$
$$\sec \theta = c/a$$

These formulas are valid no matter how large or small the right triangle can be, in terms of the magnitudes of the numbers a, b, and c. The triangle model is normally used only for angles θ measuring between 0 and 90 degrees.

Inverse Trigonometric Functions

When the domain of a trigonometric function is properly restricted, an inverse function exists that maps a real number into a unique angular value. This type of function is called an *inverse trigonometric function*. Each of the six trigonometric functions has an inverse. The inverse of a given function can be denoted by the prefix "arc-" or by an exponent -1. The inverse trigonometric functions are defined only when the angles are restricted to certain values, some of which are negative:

If $\cos \theta = z$, then $\arccos z = \theta$ for $0 < \theta < 180$

If $\sin \theta = z$, then $\arcsin z = \theta$ for $-90 < \theta < 90$

If $\tan \theta = z$, then $\arctan z = \theta$ for $-90 < \theta < 90$

If $\csc \theta = z$, then $\text{arccsc } z = \theta$ for
$$-90 < \theta < 0 \text{ and } 0 < \theta < 90$$

If $\cot \theta = z$, then $\text{arccot } z = \theta$ for $0 < \theta < 180$

If $\sec \theta = z$, then $\text{arcsec } z = \theta$ for
$$0 < \theta < 90 \text{ and } 90 < \theta < 180$$

Significance and Uses of Trigonometric Functions

Trigonometric Functions, especially the sine and cosine functions, are manifested in many different natural phenomena. These functions are representations of circular motion, which occurs frequently in the physical universe.

The most common example of a trigonometric function in electricity and electronics is the classical alternating-current wave. A wave disturbance, containing energy at only one frequency, can be perfectly represented by a sine or cosine function. The wave cycle is divided into 360 degrees of phase, beginning at the point where the amplitude is zero and increasing in a positive direction (*see* PHASE ANGLE, and SINE WAVE). Trigonometric functions arise in certain antenna and feed-line calculations, frequency and phase modulation, and many other instances. Trigonometric functions are important in polar and spherical coordinate systems. *See also* COORDINATE SYSTEMS.

TRIGONOMETRY

Trigonometry is a branch of mathematics that is concerned with angles and straight lines in a plane or in space. *See* COSINE, SINE, TANGENT, and TRIGONOMETRIC FUNCTION.

TRIMMER CAPACITOR

A *trimmer capacitor* is a small, variable capacitor used for the purpose of adjusting the characteristics of a tuned circuit. A trimmer capacitor can, for example, be placed in parallel with the fixed or variable capacitor in the tank circuit of a radio-frequency amplifier or oscillator. Trimmer capacitors are also used in filters and in other networks containing capacitance.

Trimmer capacitors are often used in the same manner as padder capacitors (*see* PADDER CAPACITOR). A trimmer capacitor contains two sets of plates separated by thin layers of dielectric material, usually mica. The plates are pressed together more or less tightly by means of an adjustable screw (see illustration). The whole assembly is mounted on a porcelain base.

The capacitance depends on how the screw is set. When the screw is turned clockwise, pressing the plates more tightly together, the capacitance increases. When the screw is turned counterclockwise, the plates move apart and the capacitance decreases.

Trimmer capacitors are available in various sizes, ranging in maximum value from about 3 pF to 150 pF. Typical trimmer capacitors can handle around 100 V, although some can tolerate higher voltages. The frequency and loss characteristics are similar to those of silver-mica capacitors. *See also* CAPACITOR, and SILVER-MICA CAPACITOR.

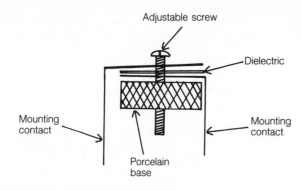

TRIMMER CAPACITOR: The screw adjusts plate spacing.

TRIODE TUBE

A *triode tube* is a vacuum tube having three elements. Early in the twentieth century, engineer and inventor Lee deForest discovered that the conductance of a diode tube could be controlled by putting an electrode, called the *grid*, between the cathode and the plate. A signal applied to the grid caused large changes in the plate current, resulting in amplification; if some of the output was coupled back to the input and shifted 180 degrees in phase, oscillation occurred.

The control grid in a triode is usually biased at a voltage somewhat negative with respect to the cathode. The bias can be supplied by a resistance-capacitance network (*see* SELF BIAS), or by means of a separate power supply called the *C supply*.

TRIPLE-DIFFUSED EMITTER-FOLLOWER LOGIC

A bipolar logic family, similar to emitter-coupled logic (but using a simple triple-diffusion process in manufacture), is known as *triple-diffused emitter-follower logic (3DEFL)*. A 3DEFL gate operates below saturation, allowing fast operating speed.

The main advantage of 3DEFL is that it allows many switching gates to be impressed on a chip. Manufacturing is simple and relatively inexpensive. *See also* DIODE-TRANSISTOR LOGIC, DIRECT-COUPLED TRANSISTOR LOGIC, EMITTER-COUPLED LOGIC, HIGH-THRESHOLD LOGIC, INTEGRATED INJECTION LOGIC, RESISTOR-CAPACITOR-TRANSISTOR LOGIC, RESISTOR-TRANSISTOR LOGIC, and TRANSISTOR-TRANSISTOR LOGIC.

TRIPLER

A *tripler* is a radio-frequency circuit that produces an output signal at a frequency three times that of the input signal. There are various different methods to obtain frequency multiplication by 3.

Triplers are commonly used with radio transmitters to obtain operation at a higher frequency. Two triplers can be connected in cascade to obtain multiplication by 9. *See also* FREQUENCY MULTIPLIER, HARMONIC, and PUSH-PULL CONFIGURATION.

TROPOSPHERIC PROPAGATION

The lower part of the earth's atmosphere has an effect on electromagnetic-field propagation at certain frequencies. At wavelengths shorter than about 15 m (or a frequency of 20 MHz), refraction and reflection occur within and between air masses of different density. The air also produces some scattering of electromagnetic energy at wavelengths shorter than 3 m (or a frequency of 100 MHz). All of these effects are generally known as *tropospheric propagation.*

Tropospheric propagation can result in communication over distances of hundreds of miles. The most common type of tropospheric propagation occurs when radio waves are refracted in the lower atmosphere. This occurs to a certain extent all the time, but is most dramatic in the vicinity of a weather front, where warm, relatively light air lies above cool, denser air. The cooler air has a higher index of refraction than the warm air, causing the electromagnetic fields to be bent downward at a considerable distance from the transmitter (see illustration A). This phenomenon is called *tropospheric bending.*

If the boundary between a cold air mass and a warm air mass is extremely well-defined, and an electromagnetic field strikes the boundary at a near-grazing angle of incidence from within the cold air mass, total internal reflection occurs. This is known as *tropospheric reflection*, as at B. If a cold air mass is sandwiched in between warm air masses, the energy can be

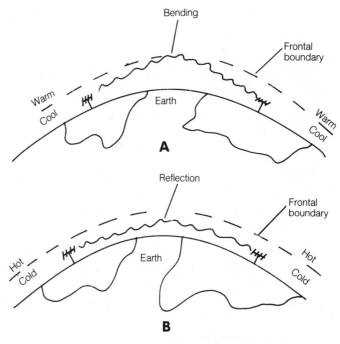

TROPOSPHERIC PROPAGATION: At A, bending; at B, reflection.

propagated for long distances because of repeated total internal reflection. This is called the *tropospheric ducting* or *duct effect* (*see* DUCT EFFECT).

Tropospheric propagation is often responsible for anomalies in reception of television and frequency-modulation broadcast signals. A television station hundreds of miles away can suddenly appear on a channel that is normally dead. Unfamiliar stations can be received on a hi-fi radio tuner. Sometimes two or more distant stations come in on a single channel, causing interference. Sporadic-E propagation has similar effects, and sometimes it is difficult to tell whether tropospheric or sporadic-E propagation is responsible for long-distance reception or communication (*see* SPORADIC-E PROPAGATION). Usually, sporadic-E events result in propagation over longer distances than tropospheric effects, but this is not always the case.

At very high, ultra-high, and microwave frequencies, the atmosphere scatters electromagnetic energy in much the same way as it scatters visible light. This can result in propagation over the horizon. It is called *tropospheric scatter* or *troposcatter*. *See also* PROPAGATION CHARACTERISTICS, and TROPOSPHERIC-SCATTER PROPAGATION.

TROPOSPHERIC-SCATTER PROPAGATION

At frequencies above about 150 MHz, the atmosphere has a scattering effect on electromagnetic fields. The scattering allows over-the-horizon communication at very-high, ultra-high, and microwave frequencies. This mode of propagation is called *tropospheric scatter*, or *troposcatter* for short.

Dust and clouds in the air increase the scattering effect, but some troposcatter occurs regardless of the weather. Troposcatter takes place mostly at low altitudes, but some effects occur at altitudes up to about 10 miles. Troposcatter propagation can provide reliable communication over distances of several hundred miles when the appropriate equipment is used.

Communication via troposcatter requires the use of high-gain antennas. Fortunately, the size of a high-gain antenna is manageable at ultra-high and microwave frequencies. The transmitting and receiving antennas are aimed at a common parcel of air, ideally located midway between the two stations and at as low an altitude as possible (see illustration). The maximum obtainable range depends not only on the gain of the antennas used for transmitting and receiving, but on their height above the ground: the higher the antennas, the greater the range. The terrain also affects the range; flat terrain is best, while mountains seriously impede troposcatter propagation.

In order to realize communication via troposcatter over long distances, sensitive receivers and high-power transmitters must be used, because the path loss is high.

Tropospheric scatter, while occurring for different reasons than bending, reflection, or ducting, is often observed along with other modes of tropospheric propagation. While communicating via troposcatter, a sudden improvement in conditions can be caused by bending, reflection, or duct effect. *See also* DUCT EFFECT, TROPOSPHERIC PROPAGATION, and PROPAGATION CHARACTERISTICS.

TROUBLESHOOTING

Troubleshooting is the process of finding the faulty component or components in a malfunctioning piece of equipment. In electronics, troubleshooting saves time and expense in the repair of equipment. Faulty components can be located without "shotgun-style" removal and replacement when the proper troubleshooting procedures are followed.

The Basic Concept   Most complicated electronic devices have service manuals. A service manual contains instructions on how to troubleshoot for specific problems. The most common, and most effective, troubleshooting instructions are given in the form of flowcharts for various potential malfunctions (*see* FLOWCHART). Troubleshooting data can also be written in tabular form.

Some digital devices, especially computers, can be used to find their own problems. The two most common methods are the *diagnostic program* and the service *read-only-memory (ROM)*. These methods are used in conjunction with a comprehensive service manual. Troubleshooting of faulty computer programs and equipment prototypes is called *debugging* (*see* DEBUGGING).

In any troubleshooting situation, the most effective procedure is to work from output to input, from general systems to specific circuits and components, or from the most obvious to the least obvious potential problems. For example, if a radio transmitter does not work, the troubleshooting procedure would be as follows, assuming a transmitter design as shown in the accompanying block diagram.

- Check to be sure equipment is operated within the quoted specifications
- Check power supply
- Check output at antenna terminals (point X1)
- Check output of driver stage (point X2)
- Check output of modulator stage (point X3)
- Check modulating-signal output (point X4)
- Check oscillator output (point X5)

When the faulty stage is found, the bad component(s) can be located using standard test equipment.

Intermittent Problems   Electronic equipment does not always malfunction catastrophically; sometimes the problem occurs intermittently. Intermittent problems are well known to the experienced technician. In order to find the source of a problem, the equipment must be evaluated while the malfunc-

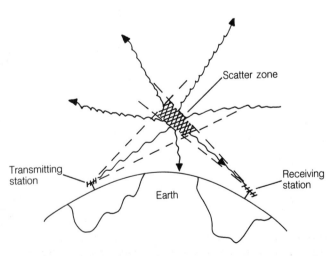

TROPOSPHERIC-SCATTER PROPAGATION: Random scattering allows over-the-horizon communication.

tion can be observed. The process is sometimes time-consuming, tedious, and frustrating.

Intermittents usually occur either because of a faulty connection, or as a result of temperature sensitivity. Some intermittents, however, occur for no apparent reason. A bad connection can often be located visually, or by wiggling the leads in a suspected area. Thermal intermittents require manual temperature control. Small heaters, along with "freeze mist," can be used to subject individual components to thermal shock until the source of a thermal intermittent is found. In some stubborn cases, a technician must use intuition or perhaps even guesswork to solve an intermittent problem!

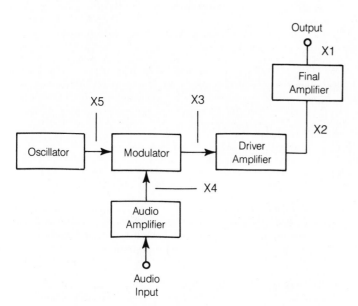

TROUBLESHOOTING: In this example, points are checked in order from X1 to X5.

T-R SWITCH

See TRANSMIT-RECEIVE SWITCH.

TRUE POWER

In a direct-current circuit, or in an alternating-current circuit containing resistance but not reactance, the determination of power is relatively simple. Given a root-mean-square (RMS) voltage E across a load of resistance R, and an RMS current I through the load, the dissipated or radiated power P_t is given by the formulas:

$$P_t = EI = E^2/R = I^2R$$

In this case, the value P_t is called the *true power* because it represents power that is actually manifested as dissipation or radiation. True power is often called *real power* because it is represented by a real number.

In an ac circuit containing reactance, the true power cannot be found by the above formulas. If we try to determine the power according to the above formulas, we get an inflated (artificially large) value called the *apparent power*. The greater the reactance in proportion to the resistance, the greater the difference between the true power and the apparent power. In the extreme, if a load is a pure capacitive or inductive reactance, the true power is zero. A pure reactance will not radiate or dissipate power, but can only store power. The difference between the apparent power and the true power is called *reactive* or *imaginary power*. It is represented in the equations by an imaginary number.

If the apparent power is represented by P_a, the reactive power by P_r, and the true power by P_t, then:

$$P_t^2 + P_r^2 = P_a^2$$

See also APPARENT POWER, POWER, POWER FACTOR, and REACTIVE POWER.

TRUE POWER GAIN

The *true power gain* of an amplifier is an expression of the extent to which the amplifier can increase the realizable power of a signal. True power gain is generally given in decibels.

Suppose the true power input to an amplifier is P_{ti} watts. This means that, if the amplifier input circuit is replaced by a dissipating or radiating load of the same impedance, the load will consume P_{ti} watts. Let the true power output of the amplifier, in watts, be denoted P_{to}. Then, the true power gain, in decibels, is:

$$G = 10 \log_{10} (P_{to}/P_{ti})$$

The true power gain of an amplifier is always maximized when the input and output impedances are perfectly matched. This means there must be no reactance, and the resistance must be of the proper value. Any impedance mismatch results in a reduction in the true power output, and therefore in the true power gain. *See also* DECIBEL, GAIN, and TRUE POWER.

TRUTH TABLE

A *truth table* is an expression of a logical, or Boolean, function. Truth tables provide a useful means of showing logical equivalences, and for analyzing the meanings of complicated logical expressions. The table is an example of a truth table. *See also* BOOLEAN ALGEBRA.

TRUTH TABLE: AN EXAMPLE OF A TRUTH TABLE, DENOTING A LOGICAL FUNCTION $(X + Y)' + XZ$.

X	Y	Z	X + Y	(X + Y)'	XZ	(X + Y)' + XZ
0	0	0	0	1	0	1
0	0	1	0	1	0	1
0	1	0	1	0	0	0
0	1	1	1	0	0	0
1	0	0	1	0	0	0
1	0	1	1	0	1	1
1	1	0	1	0	0	0
1	1	1	1	0	1	1

TTL

See TRANSISTOR-TRANSISTOR LOGIC.

TUBE

A *tube* is an electronic component that is used for amplification, rectification, and as the active device in some forms of oscillators. Tubes are also used as voltage regulators and display indicators. There are two basic types of tubes: the *vacuum tube* and the *gas-filled tube*.

Vacuum Tubes A *vacuum tube* consists of two or more metal electrodes, enclosed in a glass envelope that has been evacuated to allow electrons to flow freely. The electrode at the center is charged negatively and emits electrons. It is called the *cathode*. A cylindrical electrode surrounds the cathode, and is charged positively; it is called the *plate*. There can be one or more intermediate electrodes, charged variously to control the flow of electrons from the cathode to the plate; these electrodes are called *grids*.

Vacuum tubes can be used for rectification because electrons readily flow from the cathode to the plate but not in the other direction. However, semiconductor diodes are much more commonly used today for rectification.

When an alternating-current signal is applied to the control grid of a vacuum tube, large plate-current fluctuations occur (see TRIODE TUBE). This makes amplification possible. Amplification is enhanced by the addition of screen and suppressor grids (see TETRODE TUBE, and PENTODE TUBE).

Vacuum tubes can be used for weak-signal amplification in receivers, or for power amplification in transmitters at all radio frequencies.

There are other types of vacuum tubes that are used for diverse purposes such as television and oscilloscope displays, oscillators and amplifiers at very short wavelengths, and the generation of X rays. *See also* CATHODE-RAY TUBE, ICONOSCOPE, IMAGE ORTHICON, MAGNETRON, TRAVELING-WAVE TUBE, and VIDICON.

Gas-filled Tubes *Gas-filled tubes* generally have no grids, but can have several cathodes or anodes. Gas-filled tubes are used for voltage regulation and the generation of visible light. All gas-filled tubes operate on the principle of ionization of elements in vapor form.

A gas-filled tube normally exhibits a high resistance at voltages below a certain level. When the anode or anodes are sufficiently positive with respect to the cathode or cathodes, conduction occurs because the rarefied gas ionizes. The electrons then flow easily among the atoms, from the cathode to the anode, but not vice-versa.

A common form of gas-filled tube, the mercury-vapor rectifier, is still used in some high-voltage power supplies today. Gas-filled tubes are also used for regulation in high-voltage supplies. Semiconductor devices have taken the place of rectifier and regulator tubes, however, in power supplies designed for low and moderate voltages.

Various elements emit colored light when they are heated and ionized. A common example of this is the neon tube, which produces pink-orange light. Mercury-vapor and sodium-vapor lamps, as well as common fluorescent lamps, are examples of gas-filled tubes.

TUBULAR CAPACITOR

A capacitor can be made by rolling layers of metal foil between layers of flexible dielectric material. This results in a cylindrical or tube-shaped package, and for this reason, such a capacitor is called a *tubular capacitor*.

Tubular capacitors are less common today than they were a few years ago. Paper is the most common dielectric material used in tubular capacitors. *See* PAPER CAPACITOR.

TUNABLE CAVITY RESONATOR

Metal enclosures are often used as tuned circuits at frequencies above about 200 MHz. Such enclosures exhibit excellent selectivity characteristics. Such devices are known as *cavity resonators*.

A cavity resonator can be either fixed or tunable. In a tunable resonator, the physical length is mechanically adjustable. The wavelength at resonance is directly proportional to the physical length of the cavity. Therefore, to increase the frequency, the cavity is shortened, and to lower the frequency, the cavity is lengthened.

Tunable cavity resonators can be used in power amplifiers at very high, ultra-high, and microwave frequencies. Tunable resonators are used in some types of frequency meters. An adjustable cavity can also be used to provide selectivity at the front end of a receiver. *See also* CAVITY FREQUENCY METER, and CAVITY RESONATOR.

TUNED CIRCUIT

Any circuit that displays resonance at one or more frequencies is called a *tuned circuit*. A tuned circuit can be either fixed or adjustable.

Physical Forms Tuned circuits are found in many different physical forms, such as inductor-capacitor combinations, sections of transmission lines, or metal enclosures. The most familiar tuned circuit consists of a combination of discrete inductors and capacitors. A resonant section of transmission line is called a *stub* (see STUB). A resonant metal enclosure is called a *cavity resonator* (see CAVITY RESONATOR). An antenna is a form of tuned circuit.

Electrical Forms All tuned circuits operate according to the same principle: the interaction of capacitive reactance and inductive reactance at different frequencies (see IMPEDANCE, REACTANCE, and RESONANCE).

There are four electrical types of tuned circuit. Some act as bandpass or band-rejection filters, and are classified as either series-resonant or parallel-resonant. Other tuned circuits can exhibit highpass or lowpass properties (see BANDPASS FILTER, BAND-REJECTION FILTER, HIGHPASS FILTER, and LOWPASS FILTER).

A series-resonant tuned circuit acts as a short circuit at the resonant frequency or frequencies. A parallel-resonant tuned circuit acts as an open circuit at resonance. At nonresonant frequencies, a series- or parallel-resonant tuned circuit behaves as a pure reactance, the value of which depends on the frequency. A parallel-resonant tuned circuit is sometimes called a *tank circuit* (see PARALLEL RESONANCE, SERIES RESONANCE, and TANK CIRCUIT).

TUNED FEEDERS

An antenna feed line is usually operated with a standing-wave ratio that is as low as possible. This involves operating the antenna at resonance, and matching the characteristic impedance of the feeder to the radiation resistance of the antenna (see CHARACTERISTIC IMPEDANCE, RADIATION RESISTANCE, RESONANCE, and STANDING-WAVE RATIO). This type of feed system is called *untuned*, because the length of the feed line does not affect the antenna-system impedance at the transmitter/receiver.

An antenna system having a low-loss feed line can be operated with a high standing-wave ratio. This is sometimes done with an open-wire line. The electrical feed-line length is important, and therefore the line is said to be *tuned.* This type of feed system was popular in the early days of radio, when open-wire line was almost universally used, and is still used today by some radio amateurs.

A tuned feeder is "pruned" to such a length that the input impedance is a pure resistance. This necessitates that the system have the following properties:

- Each side of the whole system must measure an integral multiple of ¼ electrical wavelength.
- The lengths of the two sides of the system must either be identical, or differ by an integral multiple of ½ electrical wavelength.

Ideally, both halves of the system should measure the same.

The illustration shows four examples of antennas with tuned feeders. The system at A is symmetrical, with each side measuring 2 wavelengths. The system at B incorporates off-center feed, with one side measuring 1¾ wavelength and the other side measuring 2¼ wavelength. The system at C uses off-center feed, with sides measuring 1½ and 2 wavelengths. The antenna system at D uses end feed, in which the feed line measures 2 wavelengths and the radiating element measures 1 wavelength.

The impedance at the input of a tuned feed system can range from a few ohms to several hundred ohms in practice. It is, therefore, usually necessary to use an impedance-matching transformer between a transmitter and a tuned feeder system (*see* IMPEDANCE TRANSFORMER).

If a transmatch, capable of tuning out reactance, is used at the input of a tuned-feeder system, the feed line length can be varied at will. This simplifies the installation procedure by eliminating the need for "pruning" the feed line. The two halves of the antenna must, however, still differ in length by some integral multiple of ½ wavelength (*see* TRANSMATCH). A center-fed radiator, fed with open-wire line and tuned with a transmatch, can be operated over a wide range of frequencies.

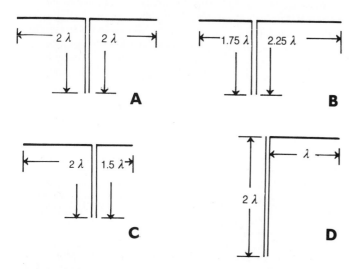

TUNED FEEDERS: Examples of center feed (A), off-center feed (B and C), and end feed (D).

TUNED-INPUT/TUNED-OUTPUT OSCILLATOR

A radio-frequency amplifier, having tuned input and output circuits resonant at the same frequency, often breaks into oscillation. This can be avoided by neutralization (*see* NEUTRALIZATION). However, it can also be used to advantage if oscillation is wanted.

TUNING

The process of adjusting the resonant frequency of an inductance-capacitance circuit, a stub, or a cavity resonator for the purpose of obtaining certain operating characteristics, is called *tuning.* In an oscillator, tuning is used to control the frequency. An amplifier is tuned for maximum efficiency, power output, or gain. An antenna system is tuned for resonance at a particular frequency or set of frequencies.

Tuning of an inductance-capacitance circuit can be accomplished by the use of variable inductors or capacitors. Inductors can be tapped in various places and different portions switched in; some inductors have continuously variable taps (*see* TAP). Inductors with ferromagnetic cores can be permeability-tuned (*see* PERMEABILITY-TUNING). If the capacitance in a tuned circuit is less than 0.001 μF (1000 pF), air-variable capacitors can be used to obtain continuous tuning (*see* AIR-VARIABLE CAPACITOR).

Tuning of a stub involves the use of a movable shorting bar. The longer the stub, the lower the resonant frequency or frequencies (*see* STUB). Tuning of a cavity resonator is accomplished by a movable barrier inside the enclosure (*see* CAVITY RESONATOR, and TUNABLE CAVITY RESONATOR). An antenna is tuned either by cutting the radiator to a certain length, or by adding reactances in series. Antenna tuning is sometimes called *loading* (*see* CAPACITIVE LOADING, and INDUCTIVE LOADING).

To facilitate adjustment of a tuned circuit, some mechanical means must be devised that is convenient for operators of the equipment. In tuned circuits having variable inductors or capacitors, this usually involves the use of a calibrated scale, in conjunction with a gear or belt drive. *See also* TUNING DIAL.

TUNING DIAL

The adjustment of a tuned circuit can be accomplished in various different ways. A tuning dial is a mechanical device that facilitates convenient and simple adjustment of an inductance-capacitance circuit. A good tuning dial has the following characteristics:

- A calibrated scale with sufficient accuracy for the intended application, or a digital display
- Ease of adjustment and readability
- Ease of resettability
- Minimal backlash
- Immunity to accidental misadjustment because of vibration

A tuning dial can operate continuously, or in defined increments. The display can be analog or digital. Analog tuning mechanisms are usually mechanical, and operate via a gear drive or belt drive. Even if the tuning control is an analog device, a digital display can be used. Some tuning dials are entirely electronic, using keyboard entry and a digital display. *See also* ANALOG CONTROL, DIAL SYSTEM, and DIGITAL CONTROL.

TUNNEL DIODE

A *tunnel diode* is a special form of semiconductor device that can be used as an oscillator and amplifier at ultra-high and microwave frequencies. A tunnel diode does not rectify, as does an ordinary semiconductor P-N junction. This is because the semiconductor materials contain large amounts of impurities. The current-versus-voltage curve has a negative-resistance region that makes oscillation and amplification possible when a suitable bias is applied.

In recent years, Gunn diodes have largely replaced tunnel diodes as oscillators and amplifiers at ultra-high and microwave frequencies. Gunn diodes are more efficient. *See also* DIODE OSCILLATOR, and GUNN DIODE.

TURNAROUND TIME

In error-correcting digital modes such as AMTOR and packet radio, it is often necessary to have a radio that switches rapidly from receive to transmit, and vice-versa. This is because some error-correction features require constant back-and-forth checking between stations; transmissions occur in short bursts with listening periods interspersed.

The time delay for a radio to effectively begin receiving after a transmission ends is called *transmit-to-receive (TX-RX) turnaround time*. The time delay for a radio to effectively start transmitting is called *receive-to-transmit (RX-TX) turnaround time*.

There are several factors that affect turnaround time. When you are shopping for a transceiver, the manufacturer or dealer should be able to give you information about the turnaround time. This is important if you plan to use AMTOR or packet radio. *See also* AMTOR, and PACKET RADIO.

TURNS RATIO

In a transformer, the *turns ratio* is defined as the number of turns in the primary winding divided by the number of turns in the secondary winding. In a step-up transformer, the turns ratio is between 0 and 1; in a step-down transformer, the turns ratio is greater than 1.

The primary-to-secondary turns ratio of a transformer is usually denoted by two integers, separated by a colon to indicate the ratio. The ratio is reduced to lowest terms. A transformer with 500 turns in the primary and 200 turns in the secondary, for example, has a turns ratio of 5:2. *See also* STEP-DOWN TRANSFORMER, STEP-UP TRANSFORMER, TRANSFORMER, TRANSFORMER PRIMARY, and TRANSFORMER SECONDARY.

TURNSTILE ANTENNA

A *turnstile antenna* is a pair of horizontal half-wave dipoles, oriented at right angles to each other and fed 90 degrees out of phase. This results in radiation in all possible directions. The polarization is linear in the horizontal plane, and elliptical in directions above the horizon. The polarization is circular toward the zenith (*see* CIRCULAR POLARIZATION, and ELLIPTICAL POLARIZATION). The illustration is a pictorial diagram of a turnstile antenna.

Turnstile antennas are commonly used in satellite communications when high gain is not needed. They are also used at VHF television transmitting stations. A turnstile antenna presents a reasonable match to 50-ohm coaxial cable. *See also* DIPOLE ANTENNA.

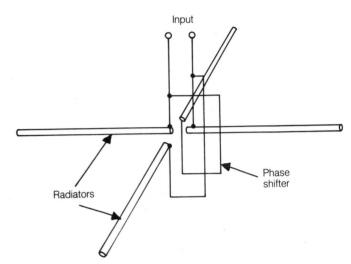

TURNSTILE ANTENNA: Two horizontal dipoles are fed 90 degrees out of phase.

TV

See TELEVISION.

TVI

See ELECTROMAGNETIC INTERFERENCE.

TWIN-LEAD

Twin-lead is a prefabricated, parallel-wire, balanced transmission line, molded in polyethylene and extensively used for television reception. Most twin-lead lines have a characteristic impedance of 300 ohms. A few have a characteristic impedance of 75 ohms. Twin-lead is sometimes called *ribbon line*.

Twin-lead is available in a wide variety of different forms. The illustration at A is a cross-sectional illustration of the most common form of twin-lead. The conductors are stranded copper, of size AWG (American Wire Gauge) #18 or #20. Illustration B shows a form of twin-lead with a foamed-polyethylene dielectric for reduced loss. Illustration C is a cross-sectional view of tubular twin-lead, with even lower loss.

Twin-lead can be used in low-power transmitting applications in place of open-wire line. However, the loss is somewhat greater in twin-lead than in a conventional open-wire line. This is attributable to the higher dielectric loss of polyethylene as compared with air.

TWO-TONE TEST

The distortion in a single-sideband transmitter can be evaluated using a procedure called a *two-tone test*. The intermodulation distortion of a receiver can be evaluated by means of a similar procedure, also called a *two-tone test*.

Testing a Single-sideband Transmitter   The single-sideband (SSB) transmitter two-tone test consists of the appli-

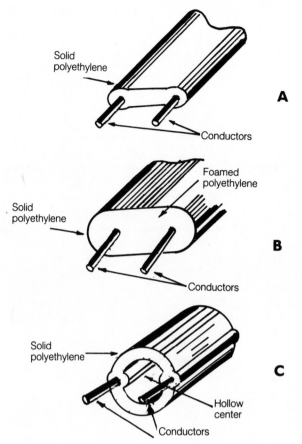

Solid
polyethylene

A

Conductors

Foamed
polyethylene

Solid
polyethylene

B

Conductors

Solid
polyethylene

C

Hollow
center

Conductors

TWIN-LEAD: At A, ribbon; at B, foamed ribbon; at C, tubular.

cation of a pair of audio tones, both within the audio passband, to the microphone terminals. Any two audio frequencies between about 300 Hz and 3 kHz can be used; the most common

are 1 kHz and 2 kHz. The output of the transmitter is analyzed with an oscilloscope. *See also* SINGLE SIDEBAND.

Testing a Receiver   For testing the intermodulation distortion of a receiver, two unmodulated signals are applied at the antenna terminals. These signals are usually close together in frequency; the spacing can vary between a few kilohertz and about 100 kHz.

If the two applied frequencies are f_1 and f_2, then the intermodulation products will be most noticeable at frequencies f_3 and f_4, such that:

$$f_3 = 2f_1 - f_2$$
$$f_4 = 2f_2 - f_1$$

The lower the levels of signals at frequencies f_3 and f_4, the better the intermodulation-distortion performance of the receiver. *See also* INTERMODULATION.

TWO-WIRE SYSTEM

A *two-wire system* is a scheme for transmitting electric power, either in direct-current form or in alternating-current form at any frequency. Two-wire systems were originally used in house wiring. Most modern house wiring uses the three-wire arrangement (*see* THREE-WIRE-SYSTEM).

A two-wire electrical system can be either unbalanced or balanced. It usually has one grounded wire and one live wire that carries the voltage, commonly 117 V at 60 Hz. Some two-wire systems operate with both conductors floating (that is, above ground potential), but carrying 117 V at 60 Hz, relative to each other.

A two-wire system for transmitting radio-frequency energy is called a *parallel-wire* or *balanced line*. Parallel-wire lines include the open-wire and the twin-lead types of feed line.

UART

See UNIVERSAL ASYNCHRONOUS RECEIVER/TRANS-MITTER.

UHF

See ULTRA-HIGH FREQUENCY.

UHF CONNECTOR

The *UHF connector* is a form of connector that is widely used with coaxial cables. The term *UHF*, which actually means *ultra-high-frequency*, is somewhat imprecise, because this type of connector is normally used at low, medium, and high frequencies. At very-high and ultra-high frequencies, the N-type connector is preferred (*see* N-TYPE CONNECTOR).

A

B

UHF CONNECTOR: At A, male; at B, female.

A UHF connector has a central pin or receptacle and an outer screw-on shell, separated by hard plastic dielectric. The dimensions are designed to present a characteristic impedance of 50 ohms, so that there will be no impedance "bump" when 50-ohm coaxial cable is used. This scheme works very well for most applications up to 150 or 200 MHz. A UHF connector can handle from a few hundred watts to several kilowatts of radio-frequency power, depending on the particular type of connector and the standing-wave ratio on the feed line.

Illustration A shows a male UHF connector on the end of a length of RG-58/U coaxial cable. The male connector is sometimes called a PL-259, which is the catalog number of a specific male UHF connector manufactured by Amphenol (the connector at A is not a true PL-259, but instead is a solderless male UHF connector that can be interchanged with a PL-259 for use with small-diameter cable). Illustration B shows a chassis-mounted female UHF connector. It is sometimes called an *SO-239*, which is the catalog number of a specific chassis-mounted female UHF connector manufactured by Amphenol. Not all female UHF connectors are true SO-239 types. There are various other sorts of UHF connectors, such as male-to-male, female-to-female, right-angle, and tee configurations.

All UHF connectors are suitable for use indoors. When a UHF connector is used outdoors for splicing cable or for connecting a cable to an antenna, some means must be devised to keep water from entering the cable through the connector. Electrical tape can be used to wrap a connection. Various special tapes, for sealing splices in radio-frequency cable, are available from commercial sources.

ULSI

See ULTRA-LARGE-SCALE INTEGRATION.

ULTRA-HIGH FREQUENCY

The *ultra-high-frequency (UHF) range* of the radio spectrum is the band extending from 300 MHz to 3 GHz. The wavelengths corresponding to these limit frequencies are 1 m and 10 cm. Ultra-high-frequency waves are sometimes called *decimetric waves* because the wavelength is on the order of tenths of a meter. Channels and bands at ultra-high frequencies are allocated by the International Telecommunication Union, headquartered in Geneva, Switzerland.

At UHF, electromagnetic fields are unaffected by the ionosphere of the earth. Signals in this range pass through the ionized layers without being bent or reflected in any way. The UHF waves behave very much like waves at higher frequencies. In the upper portion of the UHF band, waves can be focused by moderate-sized dish antennas for high gain and directivity. Because of their tendency to propagate through the ionosphere, UHF waves are extensively used in satellite communications. Because the frequency of a decimetric signal is so high, wide-band modulation is practical.

Signals in the lower portion of the UHF band are sometimes bent or reflected within or between air masses of different temperatures. A certain amount of scattering also takes place in the atmosphere. *See* also DUCT EFFECT, ELECTROMAGNETIC SPECTRUM, FREQUENCY ALLOCATIONS, TROPOSPHERIC PROPAGATION, and TROPOSPHERIC-SCATTER PROPAGATION.

ULTRA-LARGE-SCALE INTEGRATION

Miniaturization of digital electronic circuits has led some engineers to speculate that we will someday have the capability to put well in excess of 1,000 gates on an individual chip. A chip having from 1,000 to 10,000 logic gates would be referred to as an *ultra-large-scale integration (ULSI)* chip.

Many modern integrated circuits contain from 100 to 1,000 gates; this level of miniaturization is called very-large-scale integration. *See* VERY-LARGE-SCALE INTEGRATION.

UMBRELLA ANTENNA

An *umbrella antenna* is a wire antenna that is used mostly at low and very low frequencies. An umbrella antenna consists of several wire conductors and a metal mast. The mast and the wires all radiate, and serve to support each other. The conductors extend from the top of the mast to points at ground level (see illustration). The wires and the mast are inductively loaded so that they are $\frac{1}{4}$ electrical wavelength at the operating frequency. The feed point is at the base of the vertical mast. The umbrella antenna produces a vertically polarized signal.

The advantage of the umbrella antenna is that it increases the radiation resistance, as compared with a single radiator. At very low frequencies, even a fairly large tower has a radiation resistance so small that efficiency is severely impaired by ground loss (*see* ANTENNA EFFICIENCY, and RADIATION RESISTANCE). When conductors are connected in parallel in a configuration such as that shown, the radiation resistance increases in proportion to the square of the number of conductors. If the radiation resistance of a single conductor is x ohms, the radiation resistance of n conductors in parallel is n^2x ohms.

Consider the example of an antenna for transmission at a frequency of 12 kHz. A single radiator, 500 feet high, would have a radiation resistance of approximately 0.01 ohm at this frequency. If an umbrella antenna with 10 radiators (including a 500-foot support mast) is used, the radiation resistance will increase to $0.01 \times 10^2 = 1$ ohm; if 70 conductors are used, the radiation resistance will theoretically be $0.01 \times 70^2 = 49$ ohms (a good match for 50-ohm cable). In practice, the radiation resistance does not increase quite this much, because of ground-current effects. However, the umbrella is the antenna of choice at frequencies below 300 kHz.

UNBALANCED LOAD

An *unbalanced load* is a load with one side or terminal at ground potential and the other side above ground potential. An unbalanced load is required at the termination at an unbalanced transmission line, such as coaxial cable, to ensure that unwanted currents do not flow on the ground side of the transmission line (*see* UNBALANCED TRANSMISSION LINE).

A good example of an unbalanced load is an end-fed or asymmetrical antenna. Ground-plane and vertical antennas are the most common types of unbalanced antennas (*see* GROUND-PLANE ANTENNA, and VERTICAL ANTENNA).

UNBALANCED OUTPUT

An *unbalanced-output circuit* is designed to be used with an unbalanced load and an unbalanced line. There is only one output terminal in an unbalanced circuit. The output is usually shielded.

A balanced output can be used as an unbalanced output by simply grounding one of the terminals. However, if an unbalanced output is to be used with a balanced load or line, a special transformer, called a *balun*, is required. *See also* BALANCED OUTPUT, BALUN, UNBALANCED LINE, and UNBALANCED LOAD.

UNBALANCED SYSTEM

An *unbalanced system* is any circuit in which one side is at ground potential and the other side is at some ac or dc voltage different from ground potential.

Unbalanced systems are common in electronics. Many types of audio-frequency and radio-frequency oscillators, amplifiers, and transmission lines are unbalanced. Unbalanced systems are somewhat more stable than balanced systems, since a condition of near-perfect balance is difficult to maintain. *See also* BALANCED LINE, and BALANCED OUTPUT.

UNBALANCED TRANSMISSION LINE

An *unbalanced transmission line* is a form of unbalanced line used in radio-frequency antenna transmitting and receiving applications. Such a line is usually a coaxial cable. In some instances, an unbalanced transmission line consists of a single wire, or a parallel-wire line in which one conductor is at ground potential. The illustration on pg. 526 shows the principle of operation of a coaxial unbalanced transmission line.

The current in the center conductor flows back and forth at a certain frequency (heavy arrow). This current sets up an electric, or *E*, field, shown by the dotted lines, and a magnetic, or *M*, field, shown by the solid circles. The *E* and *M* fields are mutually orthogonal. They are prevented by the shield from leaking outside the line.

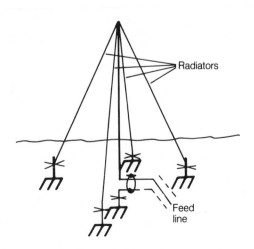

UMBRELLA ANTENNA: Multiple radiators reduce ground losses.

Because the *E* and *M* fields are perpendicular within the coaxial line, an electromagnetic field exists there. This field propagates in a direction perpendicular to both the *E* and *M* lines of flux. This direction is along the length of the transmission line. The speed of propagation depends on the dielectric material between the center conductor and the shield. If the dielectric is air, the speed of propagation is about 95 percent of the speed of light. If the dielectric is polyethylene, the speed varies between 66 and 80 percent of the speed of light (*see* VELOCITY FACTOR).

With an unbalanced transmission line such as that shown, it is important that the shielding be as complete as possible. This keeps electromagnetic fields from being radiated or picked up by the center conductor. In the case of an unshielded, unbalanced transmission line, radiation is inevitable, and is accepted as a characteristic of the line. Radiation from, or reception of, signals by, a transmission line degrades the directional characteristics of the antenna with which the line is used. This might or might not be important, depending on the type of antenna.

Unbalanced transmission lines usually have more loss than balanced transmission lines. This is because the electromagnetic field in an unbalanced transmission line must pass through more dielectric material than it does in a balanced transmission line. Shielded, unbalanced lines are easier to install, however, than are balanced transmission lines, and this is their main advantage.

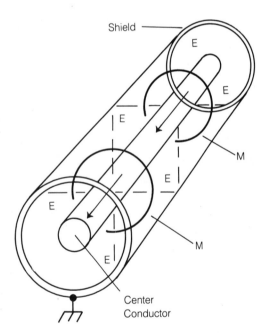

UNBALANCED TRANSMISSION LINE: Electric flux (*E*) is radial; magnetic flux (*M*) is in circles around the center conductor.

UNCONNECTED PACKET

When it is necessary to call CQ (meaning "calling anybody") on packet radio, a one-way transmission is made. This transmission is configured so that stations who receive it will know it is a CQ. An example is:

CQ CQ CQ DE W1GV K K K

A CQ packet is called an *unconnected packet*, because there is no connection, or contact, in progress. See *also* PACKET RADIO.

UNIDIRECTIONAL ANTENNA

Any antenna that exhibits maximum radiation or sensitivity in one direction, with less radiation or sensitivity in all other directions, is called a *unidirectional antenna*. Many types of antennas have this property (*see* UNIDIRECTIONAL PATTERN).

At high frequencies, the most common unidirectional antennas are the log-periodic, terminated longwire and rhombic, quad, and Yagi configurations. Various types of phased arrays operate as unidirectional antennas. These antennas will also work well at very-high and ultra-high frequencies (*see* LOG-PERIODIC ANTENNA, LONGWIRE ANTENNA, PHASED ARRAY, QUAD ANTENNA, RHOMBIC ANTENNA, and YAGI ANTENNA).

As the wavelength gets shorter, other antenna arrangements become reasonable in size, and can provide excellent unidirectional operation. These include the billboard, Chireix, helical, horn, parabolic, and Sterba antennas. Some of these antennas are bidirectional unless a reflector is added. *See* BILLBOARD ANTENNA, CHIREIX ANTENNA, HELICAL ANTENNA, HORN ANTENNA, PARABOLOID ANTENNA, and STERBA ANTENNA.

UNIDIRECTIONAL PATTERN

Any transducer that performs well in one direction, but poorly in all other directions, is said to exhibit a *unidirectional pattern*. Such a transducer can be an antenna, microphone, speaker, or other device designed for converting one form of energy to another.

Devices with unidirectional radiation or response are quite common. A unidirectional pattern is usually associated with gain in the favored direction. For example, a Yagi antenna has gain compared with a dipole antenna.

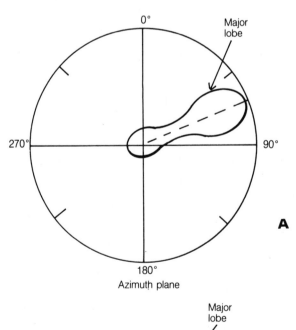

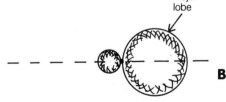

UNIDIRECTIONAL PATTERN: At A, an example of unidirectional pattern in horizontal plane. At B, a three-dimensional example.

For antennas at low, medium, and high frequencies, only the azimuth plane is generally used to define unidirectionality. Both the azimuth and elevation become increasingly significant as the wavelength gets shorter. *See also* UNIDIRECTIONAL ANTENNA.

UNIJUNCTION TRANSISTOR

A *unijunction transistor (UJT)* is a semiconductor device similar to a silicon-controlled rectifier. The UJT is sometimes called a *double-base diode.* The UJT is a three-terminal device.

Two metal contacts, called the *bases*, are attached to either end of a silicon rod. A third contact, called the *emitter*, is attached to the side of the rod. When a voltage is applied at the base terminals, the emitter-substrate junction becomes reverse-biased. The current through the UJT can be controlled by varying the voltage at the emitter.

The schematic symbol for the UJT resembles the symbol for an N-channel field-effect transistor. The two devices are used in much different situations. Typical applications of the UJT include relaxation oscillators, multivibrators, and voltage-control devices.

UNIVERSAL ASYNCHRONOUS RECEIVER/TRANSMITTER

In a serial input/output (SIO) interface, a *universal asynchronous receiver/transmitter (UART)* is often used. It is available as an integrated circuit (IC). The heart of the UART is a *shift register.*

Most digital equipment uses *parallel data transfer.* If *serial data transfer* is desired, conversion is required. The UART is a serial-to-parallel converter and vice-versa. The transmitter takes parallel data and converts it to asynchronous serial form (hence the "asynchronous" in the name of the device). The receiver converts serial data to parallel form.

The UART works in full duplex; it can handle data going both ways at once, performing the serial-to-parallel and parallel-to-serial conversions at the same time. *See also* ASYNCHRONOUS DATA, PARALLEL DATA TRANSFER, SERIAL DATA TRANSFER, SHIFT REGISTER, UNIVERSAL SYNCHRONOUS/ASYNCHRONOUS RECEIVER/TRANSMITTER, and UNIVERSAL SYNCHRONOUS RECEIVER/TRANSMITTER.

UNIVERSAL SYNCHRONOUS/ ASYNCHRONOUS RECEIVER/ TRANSMITTER

A *universal synchronous/asynchronous receiver/transmitter (USART)* is a combination of a *universal asynchronous receiver/transmitter (UART)* and a *universal synchronous receiver/transmitter (USRT)* in a single integrated-circuit chip. The USART is popular because of its versatility; it can work with either asynchronous or synchronous data. *See* UNIVERSAL ASYNCHRONOUS RECEIVER/TRANSMITTER, and UNIVERSAL SYNCHRONOUS RECEIVER/TRANSMITTER.

UNIVERSAL SYNCHRONOUS RECEIVER/TRANSMITTER

In an interface that uses *serial data transfer*, some means is generally needed to convert the serial data to parallel data, and

vice-versa. This is because most digital equipment uses *parallel data transfer.* The conversion can be done by means of a *universal synchronous receiver/transmitter (USRT)* when the serial data is synchronous.

The USRT works in basically the same way as a UART. *See also* PARALLEL DATA TRANSFER, SERIAL DATA TRANSFER, SYNCHRONOUS DATA, UNIVERSAL ASYNCHRONOUS RECEIVER/TRANSMITTER, and UNIVERSAL SYNCHRONOUS/ASYNCHRONOUS RECEIVER/TRANSMITTER.

UP CONVERSION

Up conversion refers to the heterodyning of an input signal with the output of a local oscillator, resulting in an intermediate frequency that is higher than the frequency of the incoming signal. Technically, any situation in which this occurs is up conversion. But usually, the term is used in reference to converters designed for reception of low-frequency and very-low frequency signals, using receivers designed for higher frequencies.

Up conversion is accomplished using heterodyne techniques common to all mixers and converters. *See also* DOWN CONVERSION, and MIXER.

UPLINK

The term *uplink* is used to refer to the band on which an active communications satellite receives its signals from earth-based stations. The uplink frequency is much different from the downlink frequency, on which the satellite transmits signals back to the earth. The different uplink and downlink frequencies allow the satellite to receive and retransmit signals at the same time, so it can function as a transponder.

At a ground-based satellite-communications station, the uplink antenna can be relatively nondirectional if high transmitter power is used. But for more distant satellites, and especially for satellites in geostationary orbits, directional antennas are preferable. *See also* ACTIVE COMMUNICATIONS SATELLITE, DOWNLINK, and TRANSPONDER.

UPPER SIDEBAND

An amplitude-modulated signal carries the information in the form of sidebands, or energy at frequencies just above and below the carrier frequency (*see* AMPLITUDE MODULATION, and SIDEBAND). The upper sideband is the group of frequencies immediately above the carrier frequency. Upper sideband is often abbreviated USB.

The USB frequencies result from additive mixing between the carrier signal and the modulating signal. For a typical voice amplitude-modulated signal, the upper sideband occupies approximately 3 kHz of spectrum space (see illustration on pg. 528). For reproduction of television video information, the sideband may be several megahertz wide.

The upper sideband contains all of the modulating-signal intelligence. For this reason, the rest of the signal (carrier and lower sideband) can be eliminated. This results in a single-sideband (SSB) signal on the upper sideband. *See also* LOWER SIDEBAND, and SINGLE SIDEBAND.

UPWARD MODULATION

If the average power of an amplitude-modulated transmitter increases when the operator speaks into the microphone, the modulation is said to be *upward*. Ordinary amplitude modulation is of the upward type. The average amplitude is minimum when there is no modulation.

Upward modulation is not the only means by which a carrier wave can be amplitude-modulated. Information can also be impressed on a carrier by decreasing the amplitude. *See also* AMPLITUDE MODULATION, and DOWNWARD MODULATION.

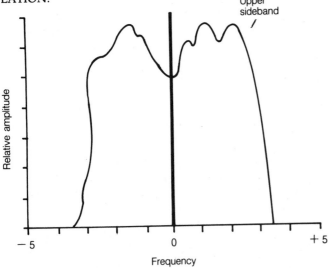

UPPER SIDEBAND: Energy is above the carrier frequency.

USART
See UNIVERSAL SYNCHRONOUS/ASYNCHRONOUS RECEIVER/TRANSMITTER.

USB
See UPPER SIDEBAND.

USRT
See UNIVERSAL SYNCHRONOUS RECEIVER/TRANSMITTER.

UTILIZATION FACTOR
The extent to which an electrical or electronic system is loaded, compared with its maximum capability, is called *utilization factor*. The utilization factor is generally expressed as a percentage.

Suppose, for example, that a generator can deliver a maximum continuous power of 5 kW, and appliances are connected that draw 3 kW. Then the utilization factor is $3/5$, or 60 percent. As another example, consider a microcomputer memory that has a maximum capacity of 16 kilobytes. If 12 kilobytes are filled with data, then the utilization factor is $12/16$, or 75 percent.

VACUUM TUBE

See TUBE.

VACUUM-TUBE VOLTMETER

A *vacuum-tube voltmeter (VTM)* is a device that allows measurement of voltages in electronic circuits without drawing significant current. The input impedance is theoretically infinite, although in practice it is finite, but very large. The vacuum-tube voltmeter gets its name from the fact that it uses a tube in a class-A configuration to obtain the high input impedance.

Some vacuum-tube voltmeters have provision for measuring current and resistance. The tube is not required for these modes; the operation is identical to that of the volt-ohm-milliameter (*see* VOLT-OHM-MILLIAMMETER). A VTVM with the capability to measure current and resistance is sometimes called a *vacuum-tube volt-ohm-milliammeter (VTVOM)*.

In recent years, field-effect transistors have replaced vacuum tubes for measurement of low and medium voltage when it is essential that the current drain be small. The VTVM is still sometimes used, however, for measuring high voltages. *See also* FET VOLTMETER.

VACUUM-TUBE VOLT-OHM-MILLIAMMETER

See VACUUM-TUBE VOLTMETER.

VACUUM-VARIABLE CAPACITOR

A *vacuum-variable capacitor* is a variable capacitor that is enclosed in an evacuated chamber. The construction of the vacuum variable is similar to that of an air variable capacitor.

Vacuum-variable capacitors have extremely low loss and high voltage ratings, because a vacuum cannot become ionized. This makes the vacuum variable ideal for high-power radio-frequency transmitting applications. *See also* AIR-VARIABLE CAPACITOR, and VARIABLE CAPACITOR.

VALENCE BAND

The electrons in an atom orbit the nucleus at defined distances, following average paths that lie in discrete spheres called *shells*. The more energy an electron possesses, the farther its orbit will be from the nucleus. If an electron gets enough energy, it will escape the nucleus altogether and move to another atom. This is how conduction occurs, and it is also responsible for reactions between or among various elements.

An electron is most likely to escape from a nucleus when the electron is in the outermost shell, or conduction band, of an atom. Electrons are stripped easily from nuclei in good conductors, such as copper and silver, and less readily in semiconductors, such as germanium and silicon. In insulators, the nuclei hold onto the electrons very tightly. The willingness with which a nucleus will give up its outermost electrons depends on how "full" the outer shell is. The shell just below the outer shell of an atom is called the *valence shell*. The range of energy states of electrons in this shell is known as the *valence band. See also* ELECTRON ORBIT.

VALENCE ELECTRON

In an atom, any electron that orbits in the partially filled outermost shell, without escaping from the nucleus, is called a *valence electron*. In an atom having a "full" outer shell (*see* ELECTRON ORBIT), there are, by definition, no valence electrons.

Some atoms have valence shells that are occupied by just one or two electrons, and thus the electrons are given up easily. Other atoms have nearly full valence shells; such elements tend to attract electrons. The various elements have properties that depend, to some extent, on the state of the outer shell. *See also* VALENCE NUMBER.

VALENCE NUMBER

The *valence number* of an atom is an expression of the conditions in the outer (valence) shell of that element in its natural state (*see* VALENCE ELECTRON). The valence numbers of various elements indicate how their atoms conduct an electric current, and how they behave in chemical reactions and mixtures. In some cases, relative valence numbers are important for elements combined as mixtures.

Definition An element with a filled valence shell has a valence number of 0. If the outer shell contains either an excess or a shortage of electrons, the valence number is the number of electrons that must be added or taken away to result in an outer shell that is completely filled.

Valence numbers are sometimes expressed as positive or negative quantities, indicating an excess or deficiency of electrons compared to the most stable possible state. These numbers are called *oxidation numbers*. If the outer shell is less than half full, containing n electrons, the oxidation number is positive, and is defined as $+n$. If the outer shell is more than half full, containing n electrons (but having a capacity to hold k electrons), then the oxidation number is negative, and is defined as $-(k - n)$. Substances with positive oxidation numbers tend to give up electrons; substances with negative oxidation numbers tend to accept electrons from outside the atom.

Sometimes both the oxidation values $+n$ and $-(k - n)$ are given, especially if the shell is approximately half filled. Such an element has two effective valence numbers. A few elements behave in such a way that they can be considered to have several different oxidation numbers.

Conduction and Chemical Reactions The oxidation number of an element determines how readily it will conduct an electric current. Metallic substances have low positive oxidation numbers, indicating that the outer shell has just a few electrons. Semiconductors have outer shells that are approximately half full. Poor conductors have valence shells that are almost completely filled, and thus their oxidation numbers are small and negative. Those elements that have totally filled outer shells are the worst conductors, but the best dielectric substances.

The oxidation number of an element also indicates how readily it will react with certain other elements. When the non-zero oxidation numbers of two or more atoms add up to 0, the elements will react very easily when brought together. For example, two atoms of hydrogen (each with oxidation number +1) will combine readily with one atom of oxygen (oxidation number −2) because the three values add up to 0.

When an element has two or more oxidation numbers, it can react with other atoms according to any of the values. An example of such a substance is hydrogen, which can be considered to have an outer shell that is either half full or half empty, having oxidation value −1 in some cases and +1 in other cases. Hydrogen thus reacts with many other elements; it can either accept or give up an electron.

Relative Valence in Semiconductors   In a semiconductor material, impurity substances are added to modify the conducting characteristics. The valence of the impurity, relative to that of the original semiconductor, determines the major and minor charge carriers.

If an impurity has more electrons than the semiconductor, it is called a *donor*. If the impurity has fewer electrons than the semiconductor, it is called an *acceptor*. The addition of a donor impurity causes conduction mostly via electrons, and the resulting material is called *N-type*. The addition of an acceptor impurity causes conduction mostly in the form of electron-deficient atoms called *holes*, and the material is called *P-type*. *See also* ELECTRON, HOLE, IMPURITY, N-TYPE SEMICONDUCTOR, and P-TYPE SEMICONDUCTOR.

VALVE
See TUBE.

VANE ATTENUATOR
A piece of resistive material, placed in a waveguide to absorb some of the electromagnetic field, is called a *vane attenuator*. The device can be moved back and forth to provide varying degrees of attenuation, from zero to maximum. The resistivity of the vane is chosen so that it does not upset the characteristic impedance of the waveguide.

Vane attenuators can provide exact resettability by means of a calibrated scale marked on the sliding element. Vane attenuators are extensively used at ultra-high and microwave frequencies in various applications. *See also* ATTENUATOR, and WAVEGUIDE.

VAR
The unit of reactive power is called the *reactive voltampere (VAR)*. Reactive power does not represent power actually dissipated, although it has an effect on power measurement. *See* REACTIVE POWER.

VARACTOR DIODE
All diodes exhibit a certain amount of capacitance when they are reverse-biased. This capacitance limits the frequency at which a diode can be used as a rectifier or detector. But the reverse-bias capacitance properties of a semiconductor diode can be used to advantage in certain situations. A *varactor diode* is deliberately designed to provide an electronically variable capacitance. A varactor diode is sometimes called a *variable-capacitance diode (varicap)*, for this reason.

The illustration shows the principle of a varactor diode. At A, the reverse-bias voltage is very small, and the depletion region is therefore narrow, resulting in a fairly large capacitance. At B, the reverse-bias voltage is greater, and the depletion region is thus larger. The result is a smaller value of capacitance. Varactor diodes exhibit widely varying reverse-bias-versus-capacitance characteristics, depending on the method of manufacture. Typical minimum capacitance values range from less than 1 pF to 20 or 30 pF; maximum values can be as much as 200 pF.

Varactor diodes can be used to provide frequency modulation of an oscillator (*see* REACTANCE MODULATOR). Varactor diodes can also be used to tune radio-frequency amplifiers at practically all wavelengths. Varactor diodes operate efficiently as frequency multipliers. *See also* DIODE CAPACITANCE, P-N JUNCTION, and REVERSE BIAS.

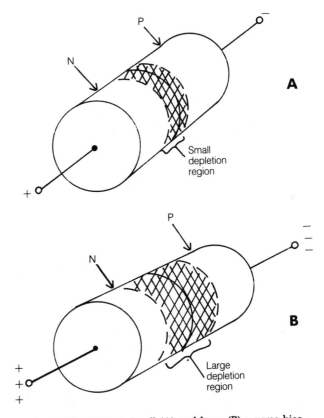

VARACTOR DIODE: Small (A) and large (B) reverse bias.

VARIABLE BANDWIDTH TUNING
In a receiver, different modes require different intermediate-frequency (IF) bandwidths. A morse code (CW) receiver works well with an IF bandwidth of 500 Hz; in single-sideband (SSB) mode, 2.4 kHz is typical.

In high-noise or crowded band conditions, narrower IF bandwidth improves sensitivity. A DXer might want a 200-Hz CW filter and a 1.8-kHz SSB filter. It is expensive to equip a receiver with several IF filters. But there is a way to achieve adjustable IF bandwidth with only two filters. This is *variable bandwidth tuning (VBT)*.

VBT is obtained by cascading two IF filters with independently adjustable passbands. When the filters' center frequencies match, the bandwidth is widest. By offsetting the center frequencies, the overlap between filter responses can be reduced, narrowing the bandwidth.

If both filters have 2.4-kHz bandwidth and very steep skirts, then the IF selectivity can be varied continuously from 2.4 kHz down to perhaps 100 or 200 Hz.

VARIABLE CAPACITOR

A *variable capacitor* is any device that exhibits a capacitance that can be adjusted over a continuous range. Variable capacitors are extensively used for tuning oscillators and amplifiers, and for adjusting transmatches and other impedance-matching networks.

Variable capacitors can be found in a wide variety of voltage and capacitance ratings for various applications. Most variable capacitors can be adjusted within a range of values having a maximum-to-minimum ratio of between 10:1 and 40:1. Very few variable capacitors have maximum values of more than a few hundred picofarads.

A special form of variable capacitor, known as a *varactor diode*, allows electronic adjustment of capacitance. The varactor is unique because it facilitates rapid variation of capacitance — in some cases, at ultra-high or microwave frequencies. *See also* AIR-VARIABLE CAPACITOR, PADDER CAPACITOR, TRIMMER CAPACITOR, VACUUM-VARIABLE CAPACITOR, and VARACTOR DIODE.

VARIABLE CRYSTAL OSCILLATOR

The oscillating frequency of a piezoelectric crystal is normally fixed. However, it can be varied to a certain extent by the addition of a small capacitor or inductor. The components are connected in parallel or in series with the crystal. This is called *pulling* (*see* PULLING). A variable crystal oscillator (VXO) is a crystal oscillator in which the frequency can be adjusted over a small range by means of added reactances.

Variable crystal oscillators are sometimes used in radio transmitters or transceivers to obtain operation over a small part of a band. Any crystal oscillator can be made into a VXO by adding the proper parallel or series reactances. Two examples of VXO circuits are shown in the illustration.

The main advantage of the VXO over an ordinary variable-frequency oscillator is excellent frequency stability. The main disadvantage of the VXO is its limited frequency coverage. Usually, the frequency of the VXO can be varied up or down by a maximum of approximately 0.1 percent of the operating frequency without crystal failure or loss of stability. *See also* CRYSTAL OSCILLATOR, OSCILLATOR, PIERCE OSCILLATOR, and VARIABLE-FREQUENCY OSCILLATOR.

VARIABLE FREQUENCY

A *variable-frequency device* is an oscillator, receiver, or transmitter that is designed to operate on more than one frequency.

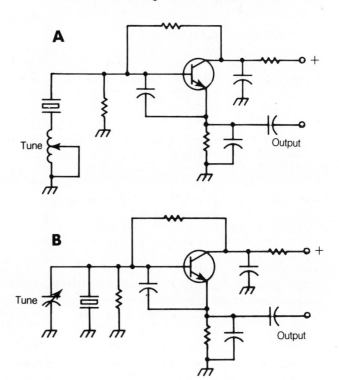

VARIABLE CRYSTAL OSCILLATOR: Inductive (A) and capacitive (B) tuning.

Variable-frequency devices are usually tuned by means of continuously adjustable inductance-capacitance circuits, or with synthesizers. Variable-frequency operation can be conducted on discrete channels, or over a continuous range of frequencies.

The main advantage of variable-frequency communication is that interference from other stations can be avoided by choosing an unoccupied channel or frequency. The primary disadvantage of variable-frequency operation is that a scheduled contact is not feasible unless the operators have previously agreed on the frequency. *See also* FIXED FREQUENCY, VARIABLE CRYSTAL OSCILLATOR, and VARIABLE-FREQUENCY OSCILLATOR.

VARIABLE-FREQUENCY OSCILLATOR

Any oscillator in which the frequency can be adjusted, either in discrete channels or over a continuous range, is called a *variable-frequency oscillator (VFO)*. Many different types of oscillators can be used for variable-frequency operation.

Some VFO circuits use inductance-capacitance tuning to obtain oscillation over a band of frequencies. The Colpitts and Hartley configurations are probably the most common. The tuned-input/tuned-output circuit is also sometimes used. The frequency of a crystal-controlled oscillator can be varied to a certain extent (*see* COLPITTS OSCILLATOR, HARTLEY OSCILLATOR, and VARIABLE CRYSTAL OSCILLATOR).

In recent years, another type of VFO has evolved: the *frequency synthesizer*. This device contains a reference oscillator, usually crystal-controlled, various multiplier and divider circuits, in conjunction with a phase-locked loop, allowing operation at hundreds, thousands, or millions of discrete frequencies over a wide band. *See also* FREQUENCY SYNTHESIZER, and PHASE-LOCKED LOOP.

VARIABLE INDUCTOR

A *variable inductor* is a device having an inductance that can be adjusted over a continuous range. The inductance of a coil can be varied in several ways. Basically, any variable inductor can be considered to fall into one of three categories: adjustable-turns, adjustable-core, or adjustable-phase.

The number of turns in a coil can be varied by means of a roller device. Roller inductors are sometimes used in trans-match circuits (*see* INDUCTOR TUNING, and TRANS-MATCH). The permeability of the core can be controlled by moving the slug in and out of the coil (*see* PERMEABILITY TUNING). Two inductors can be connected in series and the relative phase controlled by mechanical means, resulting in variable net inductance (*see* VARIOMETER).

VARIAC

A *Variac* is a variable transformer that is sometimes used to control the line voltage in a common household utility circuit. Any variable transformer can be loosely termed a ''variac.'' However, a true Variac is a special type of adjustable autotransformer consisting of a toroidal winding and a rotary contact. The name *Variac* was originally used by the General Radio Company for their adjustable transformer.

A Variac provides continuous adjustment of the root-mean-square output voltage from 0 to 134 V in a 117-V line. The output voltage is essentially independent of the current drawn. A Variac can be used at the input to a high-voltage power supply to obtain adjustable direct-current output. This scheme is sometimes used in tube type radio-frequency power amplifiers.

VARICAP

See VARACTOR DIODE.

VARIOMETER

A *variometer* is a type of adjustable inductor that was popular at one time in transmatches and in the output circuits of radio transmitters. Today, the variometer is not often seen.

The variometer has two windings, one inside the other. The inner winding is mounted on a bearing, and can be turned. The outer winding is not movable. The mutual inductance is maximum when the axes of the two coils coincide, and is minimum when the axes are at right angles. Continuous adjustment is possible by turning the inner coil through 90 degrees. This facilitates variable coupling between two circuits, such as a radio-frequency power amplifier and an antenna system.

The two coils of a variometer can be connected in series to obtain adjustable inductance. The inductance is maximum when the currents in the two coils are in phase, and is minimum when they are out of phase. The phase is controlled by turning the inner coil through 180 degrees. *See also* VARIABLE INDUCTOR.

VAR METER

The reactive power in a radio-frequency circuit can be determined by a device called a *VAR meter*. A simple VAR meter consists of a radio-frequency ammeter in series and a radio-frequency voltmeter in parallel with a circuit. If the true power P_t, is known, the reactive power P_r, in VAR, is given by:

$$P_r{}^2 = E^2 I^2 - P_t{}^2$$

where E and I are the radio-frequency voltage and current, as indicated by the meters, in volts and amperes, respectively.

A reflectometer can be used as a VAR meter if the scales are calibrated in watts. The reactive power in VAR is simply the wattmeter reading on the reflected-power scale, or with the function switch set for reflected power. *See also* POWER, REACTIVE POWER, REFLECTED POWER, REFLEC-TOMETER, TRUE POWER, and VAR.

V BEAM

A *V beam* is a form of longwire antenna that exhibits gain in one or two fixed directions. V-beam antennas are used mostly at the high frequencies (3 MHz to 30 MHz), although they are sometimes used at medium frequencies (300 kHz to 3 MHz). The V beam is essentially half of a rhombic antenna (*see* RHOMBIC ANTENNA). The illustration is a diagram of the configuration.

The gain and directional characteristics of a V beam depend on the physical lengths of the wires. The longer the wires in terms of the wavelength, the greater the gain and the sharper the directional pattern. The gain and directional characteristics of the V beam are similar to those of the rhombic.

The V beam operates as two single longwire antennas in parallel. As the wires are made longer, the lobe of maximum radiation becomes more nearly in line with the wires (*see* LONG-WIRE ANTENNA). The apex angle is chosen so that the major lobes from the wires coincide; thus, long V-beam antennas have small apex angles and short V beams have larger apex angles.

The V beam shown is fed at the apex with open-wire line. A bidirectional pattern results, as shown by the double arrow. The feed-point impedance varies with the frequency, but this is of no concern if low-loss line is used in conjunction with a transmatch. Efficient operation can be expected at all frequencies at which the lengths of the legs are at least ½ wavelength. Optimum performance will be at the frequency for which the angle of radiation, with respect to either wire, is exactly half the apex angle.

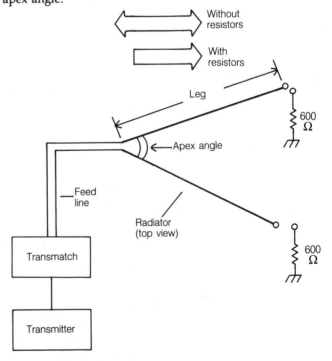

V BEAM: This is a balanced, directional system.

To obtain unidirectional operation with a V beam, a 600-ohm, noninductive, high-power resistor can be connected between the far end of each wire and ground. This eliminates half of the pattern, as shown by the single arrow. The addition of the terminating resistors results in a nearly constant feed-point impedance of 300 to 450 ohms at all frequencies for which the wires measure ½ wavelength or longer.

The main advantage of the V beam is its power gain and its simplicity. The main disadvantages are that it cannot be rotated, and that it requires a large amount of space. In recent years, the quad and Yagi antennas have become more popular than the V beam for high-frequency operation. *See also* QUAD ANTENNA, and YAGI ANTENNA.

VECTOR

A *vector* is a quantity that has magnitude and direction. Examples of vectors include the force of gravity, the movement of electrons in a wire, and the propagation of an electromagnetic field through space. If direction is not specified, a quantity is called a *scalar*. A vector for which just two directions can exist is sometimes called a *directed number*.

A vector is usually represented in equations by a capital, such as X. Vectors can also be represented in equations by capital letters in boldface or italics, or both. Vectors are geometrically represented by arrows whose length indicate magnitude and whose orientation indicate direction.

Sometimes it is necessary to consider only the length of a vector. The magnitude alone is denoted by placing vertical lines on either side of the designator, such as $|X|$, $|Y|$, and $|Z|$. Alternatively, the length of a vector can be denoted by removing the arrow above the capital letter, or by eliminating the boldface or italic notation.

Vectors are important in applied mathematics. In electronics, vectors simplify the definitions and concepts of such abstract things as impedance and the behavior of fields. Vector quantities can be added, subtracted, and multiplied, although the procedures differ from the familiar arithmetic operations used with scalars.

VELOCITY FACTOR

The speed of electromagnetic progagation through a vacuum is approximately 186,282 miles (299,792 kilometers) per second. In material substances, however, the speed is reduced somewhat. The velocity factor, v, is expressed as a percentage by:

$$v = 100c'/c$$

where c' represents the speed of propagation in a particular substance, and c is the speed of light in a perfect vacuum. Both c' and c must be specified in the same units.

The velocity factor affects the wavelength of an electromagnetic field at a fixed frequency. For a disturbance having a frequency f in megahertz, the wavelength λ in a substance with velocity factor v is given in feet by:

$$\lambda = 984v/f$$

and in meters by:

$$\lambda = 300v/f$$

Velocity factor is an important consideration in the construction of antennas. An electromagnetic field travels along a single wire in air at a speed of about $0.95c$. An electrical wavelength is therefore only 95 percent as long as in free space. As the con-

ductor diameter is made larger, the velocity factor decreases a bit, perhaps to 92 or 93 percent for large-diameter metal tubing at very-high frequencies. The velocity factor must be taken into account when cutting an antenna to a resonant length for a particular frequency. The radiating and parasitic elements must be cut to length vm, if m represents the length determined in terms of a free-space wavelength.

Velocity factor is also important when it is necessary to cut a transmission line to a certain length. To make a half-wave-length stub, for example, from a transmission line with velocity factor v, you must multiply the free-space half wavelength by v to obtain the proper length for the line. You must do the same thing if tuned feeders are used with an antenna (*see* STUB, and TUNED FEEDERS).

In an open-wire transmission line having few spacers, the velocity factor is near 95 percent. But when dielectric material is present, for example in a twin-lead line, the velocity factor is reduced. In a coaxial line, the velocity factor is still smaller. The velocity factor of a transmission line depends on:

- The type of dielectric material used in construction
- The amount of the electromagnetic field that is contained within the dielectric material
- The presence of objects near a parallel-wire line
- The extent of contamination of the dielectric because of aging in a coaxial cable

The most common dielectric materials used in fabrication of feed lines are solid polyethylene, with a velocity factor of 66 percent, and foamed polyethylene, with a velocity factor of 75 to 85 percent.

The table lists the velocity factors of some common types of transmission lines, assuming ideal operating conditions. These values are approximate. For precise velocity-factor ratings in critical transmission-line applications, the manufacturer should be consulted.

VELOCITY FACTOR: FACTORS FOR VARIOUS TYPES OF TRANSMISSION LINES.

Line type or manufacturer's no.	Velocity factor, percent
Coaxial cable, RG-58/U, solid dielectric	66
Coaxial cable, RG-59/U, solid dielectric	66
Coaxial cable, RG-8/U, solid dielectric	66
Coaxial cable, RG-58/U, foam dielectric	75–85
Coaxial cable, RG-59/U, foam dielectric	75–85
Coaxial cable, RG-8/U, foam dielectric	75–85
Twin-lead, 75-ohm, solid dielectric	70–75
Twin-lead, 300-ohm, solid dielectric	80–85
Twin-lead, 300-ohm, foam dielectric	85–90
Open-wire with plastic spacers, 300-ohm	90–95
Open-wire with plastic spacers, 450-ohm	90–95
Open-wire, homemade, 600-ohm	95

VELOCITY MICROPHONE

A *velocity microphone* is a transducer that converts sound waves into electric currents by means of the interaction of a moving conductor with a magnetic field. The principle of operation of the velocity microphone is similar to that of the dynamic microphone (*see* DYNAMIC MICROPHONE), but the construction is different. A velocity microphone is sometimes called a *ribbon microphone*.

A thin ribbon, made from a nonmagnetic metal (such as aluminum or copper), is suspended in a strong magnetic field. The magnetic lines of flux are parallel with the plane of the ribbon. When an acoustic disturbance is present, the ribbon moves back and forth, cutting across the magnetic lines in the ribbon. The current has the same waveshape as the sound that moves the ribbon. The output terminals are connected to the ends of the ribbon.

VELOCITY MODULATION

A beam of electrons can be modulated by varying the speed at which the electrons move. This is the principle of operation of certain types of vacuum tubes. When the electrons are accelerated, their energy is increased, and when they are decelerated, their energy is reduced. Although the number of electrons per unit volume fluctuates, the actual current remains constant or nearly constant.

When an electron beam is velocity-modulated, a pattern of waves is produced. The density of the beam is inversely proportional to the acceleration of the electrons, and 90 degrees out of phase with their instantaneous velocity. The pattern of waves can stand still along the length of the beam, move toward the anode, or move toward the cathode. If the pattern of waves moves toward the anode, the beam is *forward-modulated*; if it moves toward the cathode, the beam is *backward-modulated*.

Velocity modulation is used in traveling-wave tubes to obtain amplification and oscillation at ultra-high and microwave frequencies. *See also* PARAMETRIC AMPLIFIER, and TRAVELING-WAVE TUBE.

VERIFIER

After information has been recorded on magnetic tape, disks, paper tape, cards, or any other storage medium, it is desirable to ascertain that the data is free from errors before running it through a computer. A *verifier* is a device that checks the recording to ensure that the data is properly stored. This saves expensive computer time if errors are found.

A verifier operates by comparing the recorded data, bit by bit, with the original data. The verifier identifies any discrepancies according to location in the list or program.

VERNIER

Vernier is a special form of gear-drive and scale that is used for precision adjustment of variable capacitors, inductors, potentiometers, and other controls. The vernier drive is a simple gear mechanism. The control knob must be rotated several times for the shaft to turn through its complete range. The gear ratio determines the ratio of the knob and shaft rotation rates.

The vernier scale is designed to aid an operator in interpolating readings between divisions. An auxiliary scale, located above the main scale, is marked in divisions that are 90 percent of the size of the main-scale divisions. The auxiliary-scale lines are numbered 0 through 9, representing tenths of a division on the main scale.

When the dial is set to a point between two divisions on the main scale, one mark on the auxiliary scale will line up with some mark on the main scale. The operator looks for this coincidence, and notes the number of the auxiliary-scale mark at which the alignment occurs. This is the number of tenths of the way that the control is set between the two main-scale divisions.

VERTICAL ANTENNA

A *vertical antenna* is any antenna in which the radiating element is perpendicular to the average terrain. There are many types of vertical antennas, and they are used at all frequencies from the very low to the ultra-high.

Quarter-wave Verticals The simplest form of vertical antenna is a *quarter-wave radiator* mounted at ground level. The radiator is fed with a coaxial cable. The center conductor is connected to the base of the radiator, and the shield is connected to a ground system. The feed-point impedance of a ground-mounted vertical is a pure resistance, and is equal to 37 ohms plus the ground-loss resistance. The ground-loss resistance affects the efficiency of this type of antenna system. A set of grounded radials helps to minimize the loss (*see* ANTENNA EFFICIENCY, ANTENNA GROUND SYSTEM, EARTH CONDUCTIVITY, and RADIAL).

The base of a vertical antenna can be elevated at least a quarter wavelength above the ground, and a system of radials used to reduce the loss. This is called a *ground-plane antenna* (*see* GROUND-PLANE ANTENNA). The impedance can be varied by adjusting the angle of the radials, with respect to the horizontal (*see* DROOPING RADIAL).

At some frequencies, the height of a quarter-wave vertical is unmanageable unless inductive loading is used to reduce the physical length of the radiator. This technique is used mostly at very low, low, and medium frequencies, and to some extent at high frequencies (*see* INDUCTIVE LOADING).

A vertical antenna can be made resonant on several frequencies by the use of multiple loading coils, or by inserting traps at specific points along the radiator (*see* TRAP ANTENNA).

Half-wave Verticals A half-wave radiator can be fed at the base by using an impedance transformer. This provides more efficient operation for a ground-mounted antenna, as compared with a quarter-wavelength radiator, because the radiation resistance is much higher. A quarter-wave matching transformer can be attached to a half-wave vertical to form a J antenna (*see* J ANTENNA). A half-wave radiator can be fed at the center with coaxial cable to form a vertical dipole (*see* VERTICAL DIPOLE ANTENNA). If the feed line is run through one element of a vertical dipole, the configuration is called a *coaxial antenna*. A half-wave element can even be fed at the top with open-wire line (*see* TOP-FED VERTICAL ANTENNA).

Multiple Verticals Two or more vertical radiators can be combined in various ways to obtain gain and/or directivity. Parasitic elements can be placed near vertical radiators to achieve the same results (*see* PARASITIC ARRAY, PARASITIC ELEMENT, and PHASED ARRAY). Two or more vertical antennas can be stacked in collinear fashion to obtain an omnidirectional gain pattern. A Yagi can be oriented so that all of its elements are vertical (*see* YAGI ANTENNA).

Advantages and Disadvangates of Vertical Antennas

Vertical antennas radiate, and respond to, electromagnetic fields having vertical polarization (*see* VERTICAL POLARIZATION). This is an asset at very low, low, and medium frequencies, because it facilitates efficient surface-wave propagation (*see* SURFACE WAVE). Vertical antennas provide good low-angle radiation at high frequencies, an attribute for long-distance ionospheric communication. At medium and high frequencies, vertical antennas are often more convenient to install than horizontal antennas because of limited available space. Self-supporting vertical antennas require essentially no real estate. At very high and ultra high frequencies, some communications systems use vertical polarization; a vertical antenna is obviously preferable to a horizontal antenna in such cases.

One of the chief disadvantages of a vertical antenna, especially at high frequencies, is its susceptibility to reception of man-made noise; most artificial noise sources emit vertically polarized electromagnetic fields. Ground-loss problems present another disadvantage of vertical antennas at high frequencies. A ground-mounted vertical antenna usually requires an extensive radial system if the efficiency is to be reasonable.

VERTICAL DIPOLE ANTENNA

A half-wave vertical radiator, fed at the center, is called a *vertical dipole antenna*. A vertical dipole exhibits excellent low-angle radiation characteristics, good efficiency without the need for a radial system, and relative simplicity.

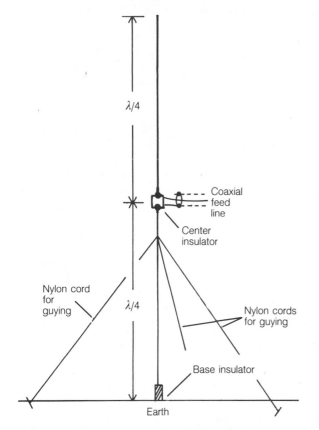

VERTICAL DIPOLE ANTENNA: The total height is ½ wavelength.

The illustration shows a vertical dipole for use at high frequencies. The design is very practical at frequencies of about 10 MHz or more, and is conceivable at frequencies considerably lower than that, provided adequate guying is used. The radiating element is constructed from aluminum tubing. The feed line should be run away from the radiator at a right angle for a distance of at least a quarter wavelength. The feed-point impedance is approximately 73 ohms at the resonant frequency of the antenna.

A vertical dipole can be fed with coaxial line through one of the halves of the radiating element. This type of arrangement is called a *coaxial antenna*. *See also* DIPOLE ANTENNA.

VERTICAL LINEARITY

In a television video system, *vertical linearity* refers to the faithful reproduction of the image along the vertical axis. Ideally, all of the scanning lines in the raster are equally spaced. If this is not the case, the image will appear distorted.

Most television receivers have a vertical-linearity control, usually placed on the chassis inside the cabinet to prevent accidental tampering. *See also* RASTER.

VERTICAL METAL-OXIDE SEMICONDUCTOR FIELD-EFFECT TRANSISTOR

The most common way to make a power metal-oxide-semiconductor (MOS) field-effect transistor (FET) is to diffuse the semiconductor material vertically. This was popular during the 70s and early 80s, and was known as *vertical MOS* or *VMOS* technology. More recently, the manufacturing process has been modified, and it is now known as *double-diffused MOS* or *DMOS. See* DOUBLE-DIFFUSED MOSFET.

VERTICAL POLARIZATION

Vertical polarization is a condition in which the electric lines of flux of an electromagnetic wave are vertical, or perpendicular to the surface of the earth. In communications, vertical polarization has certain advantages and disadvantages, depending on the application and the wavelength.

At low and very-low frequencies, vertical polarization is ideal, because surface-wave propagation, the major mode of propagation at these wavelengths, requires a vertically polarized field. Surface-wave propagation is effective in the standard amplitude-modulation (AM) broadcast band as well; most AM broadcast antennas are vertical.

Vertical polarization is often used at high frequencies, mainly because vertical antennas can be erected in a very small physical space (*see* VERTICAL ANTENNA). At very high and ultra high frequencies, vertical polarization is used mainly for mobile communications, and also in repeater communications.

The main disadvantage of vertical polarization is that most man-made noise tends to be vertically polarized. Thus, a vertical antenna picks up more of this interference than would a horizontal antenna in most cases. At very high and ultra high frequencies, vertical polarization results in more "flutter" in mobile communications, as compared with horizontal polarization. *See also* CIRCULAR POLARIZATION, HORIZONTAL POLARIZATION, and POLARIZATION.

VERTICAL SYNCHRONIZATION

In television communications, the picture signals must be synchronized at the transmitter and receiver. The electron beam in the television picture tube scans from left to right and top to bottom, just as you read the pages of a book. At any given instant of time, in a properly operating television system, the electron beam in the receiver picture tube is in exactly the same relative position as the scanning beam in the camera tube. This requires synchronization of the vertical and the horizontal positions of the beams.

When the *vertical synchronization* of a television system is lost, the picture appears split along a horizontal line. The picture can appear to "roll" upward or downward continually, much like the way a movie picture "rolls" when the film slips in the projector. Even a small error in synchronization results in a severely disturbed picture at the receiver.

The transmitted television picture signal contains vertical-synchronization pulses at the end of every frame (complete picture). This tells the receiver to move the electron beam from the end of one frame to the beginning of the next.

Both vertical and horizontal synchronization are necessary for a picture signal to be transmitted and received without distortion. *See also* HORIZONTAL SYNCHRONIZATION, PICTURE SIGNAL, and TELEVISION.

VERY HIGH FREQUENCY

The *very-high-frequency (VHF) range* of the radio spectrum is the band extending from 30 MHz to 300 MHz. The wavelengths corresponding to these limit frequencies are 10 m and 1 m. Very-high-frequency waves are sometimes called *metric waves* because the wavelength is on the order of several meters. Channels and bands at very high frequencies are allocated by the International Telecommunication Union, headquartered in Geneva, Switzerland.

At VHF, electromagnetic fields are somewhat affected by the ionosphere and the troposphere. Propagation can occur via the E and F layers of the ionosphere (*see* E LAYER, F LAYER, and PROPAGATION CHARACTERISTICS) in the lower part of the VHF spectrum. Tropospheric bending, ducting, reflection, and scattering take place throughout the VHF band (*see* DUCT EFFECT, TROPOSPHERIC PROPAGATION, and TROPOSPHERIC-SCATTER PROPAGATION). Exotic modes of signal propagation, such as auroral, meteor-scatter, and moonbounce, are also feasible at VHF (*see* AURORAL PROPAGATION, METEOR SCATTER, and MOONBOUNCE).

The VHF band is very popular for mobile communications and for repeater operation. Some satellite communications is also done at VHF. Wideband modulation is used to some extent; the most common example is television broadcasting. *See also* ELECTROMAGNETIC SPECTRUM, and FREQUENCY ALLOCATIONS.

VERY-LARGE-SCALE INTEGRATION

A common kind of integrated-circuit chip incorporates from 100 to 1,000 individual logic gates. Such a chip is called a *very-large-scale-integration (VLSI) chip*. VLSI technology is widely used in the manufacture of microcomputers and peripheral circuits.

VERY LOW FREQUENCY

The range of electromagnetic frequencies extending from 10 kHz to 30 kHz is known as the *very-low-frequency (VLF) band*. The wavelengths corresponding to these frequencies are 30 km and 10 km. Some engineers consider the lower limit of the VLF band to be 3 kHz (a wavelength of 100 km).

The ionosphere returns all vlf signals to the earth, even if they are sent straight upward. If a VLF signal arrives from space, you will not hear it at the surface because the ionosphere will completely block it.

Electromagnetic fields at VLF propagate very well along the surface of the earth if they are vertically polarized. Very-low-frequency signals might someday be used by space travelers on planets having no ionosphere. Even on the moon, which has no ionosphere to return radio waves back to ground, VLF might be useful for over-the-horizon communication via the surface wave. Another mode of signal travel, known as *waveguide propagation*, occurs in the VLF band on our planet. *See* PROPAGATION CHARACTERISTICS, SURFACE WAVE, and WAVEGUIDE PROPAGATION.

VESTIGIAL SIDEBAND

Vestigial-sideband transmission is a form of amplitude modulation (AM) in which one of the sidebands has been largely eliminated. The carrier wave and the other sideband are unaffected. Vestigial-sideband transmission differs from single sideband in that the carrier is not suppressed (*see* SIDEBAND, and SINGLE SIDEBAND).

In television broadcasting, vestigial-sideband transmission is used to optimize the efficiency with which the channel is utilized. This mode can also be used for ordinary communications. A spectral illustration of a typical vestigial-sideband signal is shown in the illustration. *See also* AMPLITUDE MODULATION.

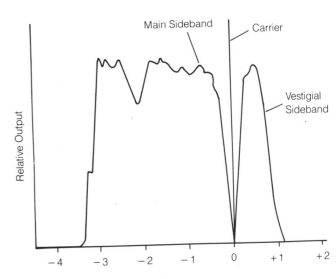

VESTIGIAL SIDEBAND: Most of the energy is in one sideband.

VFO

See VARIABLE-FREQUENCY OSCILLATOR.

VHF
See VERY HIGH FREQUENCY.

VIBRATOR
A *vibrator* is a device that interrupts a direct current, producing square-wave pulses. A vibrator operates on the same principle as a solid-state chopper. The terms *chopper* and *vibrator* are often used interchangeably. Vibrators are employed in certain direct-current power supplies.

VIBRATOR POWER SUPPLY
A *vibrator power supply* is a circuit that produces a high direct-current voltage from a lower direct-current voltage. Such devices are often used in the operation of mobile or portable equipment containing vacuum tubes. Some vibrator power supplies can provide 117 V of alternating current, allowing certain types of household appliances to be used with low-voltage, direct-current sources of power.

The vibrator consists of a relay connected in a relaxation-oscillator configuration. The relay contacts chatter, interrupting the current from the source. The resulting direct-current pulses are converted into true alternating current, as well as boosted in voltage, by a transformer. If the output voltage is 117 V root-mean-square (RMS) with a frequency near 60 Hz, the vibrator supply will provide satisfactory power for simple appliances. Devices that require a good sine wave, or a frequency of precisely 60 Hz, cannot generally be used with vibrator supplies.

VIDEO DISPLAY TERMINAL
A *video display terminal* is a device that facilitates transmission and reception of digital and graphic information. A typical video terminal contains a cathode-ray tube, which shows the data to the operator as it is sent and received, and a keyboard for sending or storing the information. Video display terminals are used in conjunction with modems or terminal units for various communications purposes (*see* MODEM). A printer can be used to obtain hard copies of displayed data.

VIDEO SIGNAL
See COMPOSITE VIDEO SIGNAL, and PICTURE SIGNAL.

VIDICON
A *vidicon* is a small, relatively simple television camera tube. The vidicon is preferred over other types of camera tubes in portable and mobile applications because it is less bulky and heavy. The vidicon is extensively used in closed-circuit television surveillance systems in business and industry.

The image is focused, by means of a lens, onto a layer of photoconductive material. The electron gun and grids generate an electron beam that scans the photoconductor via the action of external deflecting coils. This beam causes the photoconductive material to become electrically charged. The rate of discharge is roughly proportional to the intensity of the light in a particular part of the photoconductor. The result is different conductivity for different portions of the visual image.

The vidicon produces very little noise because it operates at a low level of current. Thus, the sensitivity is very high. Because of capacitive effects in the photoconductor, the vidicon has some lag. The image orthicon is preferable if a rapid image response is needed. *See also* CAMERA TUBE, ICONOSCOPE, and IMAGE ORTHICON.

VIRTUAL CIRCUIT
In packet radio, messages are often left for a destination station when the operator of that station is not immediately able to retrieve it. This is called *time-shifting communications* and is a major feature of packet radio.

Sometimes, real-time conversations (QSOs) are carried out via packet radio. When this is done, the path seems as if it is a direct connection between the two stations, even though one or more nodes might intervene. Such a path is called a *virtual circuit* because it appears to work just like a much simpler, direct link. *See also* PACKET RADIO.

VIRTUAL CIRCUIT PROTOCOL
See CONNECTION PROTOCOL.

VIRTUAL HEIGHT
Virtual height is an expression of the altitude of the ionosphere in terms of radio-frequency energy sent directly upward. The virtual height H_v of an ionized layer is always greater than the actual height H because electromagnetic energy requires some time to be turned around by interaction with the ionized molecules. The D, E, and F layers have different virtual heights, as well as different actual altitudes (*see* D LAYER, E LAYER, and F LAYER).

The virtual height of the ionosphere is determined from the delay in the return of a pulse sent straight upward. If the pulse returns after t milliseconds, then the virtual height is given in miles by:

$$H_v = 186t$$

and in kilometers by:

$$H_v = 300t$$

The virtual height of an ionized layer depends to some extent on the frequency. As the frequency rises, the virtual height increases very slightly up to a certain point. As the frequency increases further, the virtual height rapidly grows greater. At a certain frequency, called the *critical frequency*, the ionosphere no longer returns signals sent directly upward. *See also* CRITICAL FREQUENCY, and IONOSPHERE.

VLSI
See VERY-LARGE-SCALE INTEGRATION.

VMOSFET
See VERTICAL METAL-OXIDE SEMICONDUCTOR FIELD-EFFECT TRANSISTOR.

VOCODER
A *vocoder* is a device that greatly reduces the occupied bandwidth of a voice signal. The unprocessed voice signal needs 2 to 3 kHz of spectrum space for intelligible transmission. Using a vocoder, a voice can be compressed into a much smaller space,

with some sacrifice in inflection. The vocoder was originally invented by W. H. Dudley of the Bell Laboratories.

There are different types of vocoders, and a detailed discussion of the technology involved is beyond the scope of this book. The basic principle of the vocoder is the replacement of certain voice signals with electronically synthesized impulses. A typical vocoder reduces the signal bandwidth to approximately 1 kHz with essentially no degradation of intelligibility. Greater compression ratios are possible with some sacrifice in signal quality. *See also* VODER, and VOICE FREQUENCY CHARACTERISTICS.

VODER

A *voder* is a speech synthesizer that was originally designed by H. W. Dudley of the Bell Laboratories. The intended purpose was the transmission of recognizable voice signals over a channel having a narrow bandwidth. The voder is similar to the vocoder, except that the voder produces entirely synthesized signals.

A voder generates the various voice sounds by means of a keyboard, in somewhat the same way a Moog synthesizer operates. In theory, the voder could reduce the occupied bandwidth of a voice signal by a factor of several hundred. This might result in a bandwidth of as little as a few hertz. But in practice, the quality of a synthesized voice, compressed to such a tiny part of the spectrum, is degraded. Virtually all inflection, and therefore the emotional content of the voice, is lost in the process. Therefore, the realizable compression factor is smaller than that predicted by theory.

The voder experiments by Dudley eventually led to the development of the vocoder, which partially synthesizes speech and provides a significant reduction in occupied bandwidth. *See also* VOCODER, and VOICE FREQUENCY CHARACTERISTICS.

VOICE FREQUENCY CHARACTERISTICS

We can easily recognize, and mentally decode, the sounds of speech, but the electrical characteristics of a voice wave are complicated. A voice signal transmits information, not only as characters and words, but via emotional content as a result of subtle inflections.

A voice signal contains energy in three distinct bands of frequencies. These bands are called *formants*. The lowest frequency band is called the *first formant*, and occupies the range from a few hertz to near 800 Hz. The second formant ranges in frequency from approximately 1.5 to 2.0 kHz. The third formant ranges from near 2.2 to 3.0 kHz and above. The exact frequencies differ slightly, but not greatly, from person to person.

It has been discovered that speech information can be adequately conveyed by restricting the audio response to two ranges, approximately 300 to 600 Hz and 1.5 to 3.0 kHz. A unique method of doing this was first put forth by R. W. Harris and J. C. Gorski in the December, 1977 issue of *QST* magazine, published by the American Radio Relay League, Inc. (*see* NARROW-BAND VOICE MODULATION).

There is another way to look at speech. There are five basic types of sounds made by humans. These include the vowels, the semivowels, the plosives, the fricatives, and the nasal consonants. The lowest audio frequencies are generated by the utterance of vowels and semivowels: A, E, I, L, O, R, U, V, W, and

Y. Medium frequencies are generated by nasal sounds: M, N, and NG. The highest frequencies are generated by the utterance of fricatives and plosives: B, C, D, F, G, H, J, K, P, PL, Q, SH, T, TH, X, and Z (there are a few irregularities because of language anomalies).

The analysis of speech characteristics is important in communications, especially for the purpose of minimizing the occupied bandwidth of voice signals. Devices, such as the vocoder and the voder, have been used to synthesize some voice sounds, resulting in improved communications efficiency. When speech synthesis is used, however, some degradation occurs in voice inflection. Thus, although the bandwidth can be reduced, the subtle meanings, conveyed by emotion, are sacrificed or even lost. This might or might not be important, depending on the intended communications applications. *See also* VOCODER, and VODER.

VOICE-OPERATED TRANSMISSION

See VOX.

VOICE TRANSMITTER

A *voice transmitter* is any radio transmitter intended for the purpose of conveying information by means of the human voice. The most common methods of voice transmission are amplitude modulation, frequency modulation, and single sideband. Less often used are phase modulation, polarization modulation, pulse modulation, and other methods.

A voice transmitter consists of an oscillator, a modulator, a filter (if needed), and a chain of amplifiers. In variable-frequency voice transmitters, mixing circuits are often used for multiband operation. *See also* AMPLITUDE MODULATION, FREQUENCY MODULATION, PHASE MODULATION, PULSE MODULATION, and SINGLE SIDEBAND.

VOLT

The *volt* is the unit of electric potential. A potential difference of one volt across a resistance of one ohm will result in a current of one ampere, or one coulomb of electrons per second (*see* AMPERE, COULOMB, OHM, and OHM'S LAW).

Various units smaller than the volt are often used to measure electric potential. This is especially true for weak radio signals, which can have root-mean-square magnitudes of less than a millionth of a volt (*see* ROOT-MEAN-SQUARE). A millivolt (mV) is one thousandth of a volt. A microvolt (μV) is one millionth of a volt. A nanovolt (nV) is a billionth of a volt. It is not likely that you will ever encounter a voltage smaller than 1 nV and be able to detect it.

Some voltages commonly encountered in electricity and electronics are as follows: electrochemical cell, 1.0 to 1.8 V; automobile battery, 12 to 14 V; household utility outlet, 117 or 234 V (alternating-current, root mean square).

VOLTAGE

Voltage is the existence of a potential (charge) difference between two objects or points in a circuit. Normally, a potential difference is manifested as an excess of electrons at one point, and/or a deficiency of electrons at another point. Sometimes, other charge carriers, such as holes or protons, can be responsible for a voltage between two objects or points in a circuit (*see* ELECTRON, and HOLE).

Electric voltage is measured in units called *volts.* A potential difference of one volt causes a current of one ampere to flow through a resistance of one ohm (*see* AMPERE, COULOMB, OHM, and OHM'S LAW). Voltage can be either alternating or direct. In the case of alternating voltage, the value can be expressed in any of three ways: peak, peak-to-peak, or root-mean-square (*see* PEAK-TO-PEAK VALUE, PEAK VALUE, PEAK VOLTAGE, and ROOT MEAN SQUARE). Voltage is symbolized by the letter E or V in most equations involving electrical quantities.

If a given point P has more electrons than another point Q, the point P is said to be negative with respect to Q, and Q is said to be positive with respect to P. This is relative voltage. Relative voltage is not the same as absolute voltage, which is determined, with respect to a point at ground (neutral) potential.

When two points having different voltage are connected together, physicists consider the current to flow away from the point having the more positive voltage, although the movement of electrons is actually toward the positive pole. *See also* CURRENT.

VOLTAGE AMPLIFICATION

Voltage amplification is the increase in the magnitude of a voltage between the input and the output of a circuit. It is also sometimes called *voltage gain.* Some circuits are designed specifically for the purpose of amplifying a direct or alternating voltage. Other circuits are intended for amplification of current, and still others are designed to amplify power.

In theory, a voltage amplifier can operate with zero driving power and an infinite input impedance. In practice, the input impedance can be made extremely high, and the driving power is therefore very small, but not zero. A class-A amplifier, using a field-effect transistor, is the most common circuit used for voltage amplification (*see* CLASS-A AMPLIFIER). Bipolars can also be used for this purpose.

Voltage amplification is measured in decibels. Mathematically, if E_{in} is the input voltage and E_{out} is the output voltage, then:

$$\text{Voltage gain (dB)} = 20 \log_{10} (E_{out}/E_{in})$$

See also DECIBEL, and GAIN.

VOLTAGE-CONTROLLED OSCILLATOR

Varactor diodes can be used to adjust the frequency of an oscillator circuit. This is commonly done to provide frequency modulation (*see* FREQUENCY MODULATION, REACTANCE MODULATOR, and VARACTOR DIODE). A direct-current voltage, supplied to the varactor diode or pair of diodes, facilitates variable-frequency operation in a circuit called a *voltage-controlled oscillator* (VCO).

In a VCO, one or two varactor diodes are connected in parallel with, or in place of, the tank capacitor of the oscillator. A pair of diodes can be connected in reverse series, replacing the tuning capacitor, as shown in the illustration. Frequency adjustment is accomplished by means of a potentiometer. The tuning range depends on the ratio of inductance to capacitance in the tank circuit. The maximum-to-minimum oscillator frequency ratio can be made as high as 2 : 1 without difficulty; considerably greater ratios are possible.

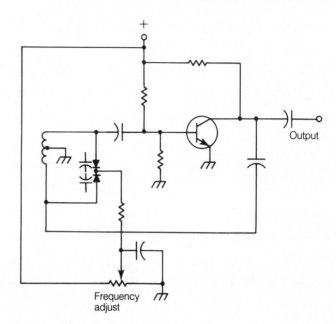

VOLTAGE-CONTROLLED OSCILLATOR: Uses varactor tuning.

VOLTAGE-DEPENDENT RESISTOR

Some resistors exhibit variable ohmic value, depending on the voltage across them. Such a resistor is said to be *voltage-dependent.* The current through a voltage-dependent resistor is a nonlinear function of the voltage.

Voltage-dependent resistors are used in certain current-regulating circuits. *See also* CURRENT REGULATION.

VOLTAGE DIVIDER

A *voltage divider* is a network of passive resistors, inductors, or capacitors that is used for the purpose of obtaining different voltages for various purposes.

A resistive voltage divider is commonly used for providing the direct-current bias in an amplifier or oscillator circuit. The desired bias is obtained by selecting resistors having the proper ratio of values. If the supply voltage is E and the resistors have values R_1 and R_2, then the bias voltage is:

$$E_b = E \times R_1/(R_1 + R_2)$$

Although there exists an infinite number of values R_1 and R_2 that will provide a given bias voltage E_b, the actual values are chosen on the basis of the circuit impedance.

VOLTAGE DOUBLER

For obtaining high voltages needed in vacuum-tube circuits, a *voltage-doubler power supply* is often used. The voltage doubler makes it possible to use a transformer having a lower step-up ratio than would be needed if an ordinary full-wave supply was used. Voltage doublers are sometimes used in radio-frequency-actuated circuits to obtain the control voltage. Voltage-doubler circuits are not generally used when excellent regulation is needed, or when the current drain is high.

VOLTAGE DROP

Voltage drop is the potential difference that appears across a current-carrying impedance. For direct currents and resistances, the voltage drop E across a resistance of R ohms that carries a current of I amperes is given by Ohm's law as:

$$E = IR$$

For alternating currents and reactances or complex impedances, the root-mean-square voltage drop E is determined from the root-mean-square current I and the impedance $R \pm jX$ as:

$$E = I\sqrt{R^2 + X^2}$$

See also CURRENT, IMPEDANCE, OHM'S LAW, REACTANCE, RESISTANCE, and VOLTAGE.

VOLTAGE FEED

Voltage feed is a method of connecting a radio-frequency feed line to an antenna at a point on the radiator where the voltage is maximum. Such a point is called a *voltage loop* or a *current node* (*see* CURRENT NODE, and VOLTAGE LOOP). In a half-wavelength radiator, the voltage maxima occur at the ends. Therefore, voltage feed for a half-wavelength antenna can only be accomplished by connecting the line to either end of the radiating element (see illustration A). In an antenna longer than ½ wavelength, voltage maxima exist at multiples of ½ wavelength from either end. There can be several different places on a resonant antenna that are suitable for voltage feed, as at B.

The feed-point impedance of a voltage-fed antenna is high. It is a pure resistance of 200 ohms or more, depending on the amount of end-effect capacitance and the diameter of the radiating element. In wire antennas, the feed-point impedance is difficult to predict with certainty; it can range from several hundred ohms to 2000 ohms or more. Because of the uncertainty, low-loss open-wire line is generally used in voltage-fed antennas. Voltage feed results in good electrical balance for a two-wire line, provided the current and voltage distribution are reasonably symmetrical in the antenna. *See also* CURRENT FEED.

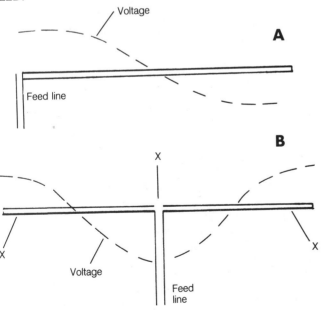

VOLTAGE FEED: At A, end feed; at B, a 1-wavelength radiator can be voltage-fed at points marked X.

VOLTAGE GAIN
See VOLTAGE AMPLIFICATION.

VOLTAGE GRADIENT

Whenever a potential difference exists between two points, such as at either end of a long wire, resistor, or semiconductor wafer, there are points at which the voltage is at some intermediate value. The rate of voltage change per unit length is called the *voltage gradient*.

The voltage gradient in a given circuit or conductor depends on the potential difference between the two end points, and on the distance between these points. In general, if the end points have voltages E_1 and E_2, and the length of the conductor or component is d meters, then the voltage gradient, in volts per meter, is:

$$G = (E_2 - E_1)/d$$

Voltage gradient can also be specified in volts per foot, centimeter, millimeter, or micron.

The voltage gradient is sometimes specified for semiconductor materials. This is especially true for field-effect transistors, in which a voltage gradient exists along the channel (*see* FIELD-EFFECT TRANSISTOR).

VOLTAGE HOGGING

Voltage hogging is a phenomenon that sometimes occurs in resistive components connected in series. Normally, if several components having identical resistances are connected in series, the voltage is divided equally among them. But if the components have slightly different resistances and a positive temperature coefficient, the component with the highest initial resistance will "hog" most of the voltage. This can cause improper operation of a circuit and, in some cases, component failure.

Voltage hogging sometimes occurs with semiconductor devices, especially diodes. In a high-voltage supply, several semiconductor diodes can be connected in series to increase the peak-inverse-voltage rating of the combination. Ideally, if n diodes, each with a peak-inverse-voltage rating of E volts, are connected in series, the combination will tolerate En volts. However, voltage hogging can substantially reduce the peak-inverse-voltage rating of the combination.

Voltage hogging is prevented by placing large-value resistors across components connected in series. All of the resistors have the same value, and they act as a voltage divider to maintain a constant voltage across each component. *See also* VOLTAGE DIVIDER.

VOLTAGE LIMITING

Voltage limiting is a method of preventing a direct-current or alternating-current voltage from exceeding a certain value. In the case of alternating current, voltage limiting is usually called *clipping*.

Voltage limiting is sometimes used as a means of regulation in a direct-current power supply. A Zener diode, connected across the source of voltage, serves this purpose. *See also* VOLTAGE REGULATION, and ZENER DIODE.

VOLTAGE LOOP

In an antenna radiating element, the voltage in the conductor depends on the location. At any free end, the voltage is maximum. At a distance of ½ wavelength from a free end, or any multiple of ½ wavelength from a free end, the voltage is also at a maximum. The points of maximum voltage are called *voltage loops.*

A ½-wavelength radiator has two voltage loops, one at each end. A full-wavelength radiator has three voltage loops: one at each end and one at the center. In general, the number of voltage loops in an antenna radiator is 1 plus the number of half wavelengths.

Voltage loops will occur along the length of a feed line not terminated in an impedance identical to its characteristic impedance. These loops occur at multiples of ½ wavelength from the resonant antenna feed point when the antenna impedance is greater than the characteristic impedance of the line. The loops exist at odd multiples of ¼ wavelength from the feed point when the resonant antenna impedance is less than the characteristic impedance of the line. Ideally, the voltage on a transmission line should be the same everywhere, and equal to the product of the characteristic impedance and the line current. *See also* VOLTAGE NODE, and STANDING WAVE.

VOLTAGE MULTIPLIER

A *voltage multiplier* is a form of power supply designed to deliver a very high voltage using a transformer with a secondary that supplies only a moderate voltage. The most common type of voltage multiplier is the voltage doubler (*see* VOLTAGE DOUBLER). However, some supplies have multiplication factors of 3, 4, or more.

In a voltage multiplier, capacitors and diodes are connected in a lattice configuration. For a multiplication factor of n, the circuit requires $2n$ capacitors and $2n$ diodes. The direct-current output voltage is n times the peak input voltage, or about $1.4n$ times the root-mean-square input voltage.

In practice, there is a limit to the extent to which a voltage can be multiplied. The larger the multiplication factor n, the worse the voltage regulation becomes. For this reason, high-order voltage multipliers are not often used except in situations where the current drain is extremely small.

VOLTAGE NODE

A *voltage node* is a voltage minimum in an antenna radiator or transmission line. The voltage on an antenna depends, to some extent, on the location of the radiator. In general, voltage nodes occur at odd multiples of ¼ wavelength from the nearest free end of a radiating element. The number of voltage nodes is equal to the number of half wavelengths.

Voltage nodes can occur along a transmission line that is not terminated in an impedance identical to its characteristic impedance. These nodes occur at multiples of ½ wavelength from the resonant antenna feed point when the antenna impedance is smaller than the feed-line characteristic impedance. Voltage nodes exist at odd multiples of ¼ wavelength from the resonant antenna feed point when the antenna impedance is larger than the characteristic impedance of the line.

Voltage nodes are always spaced at intervals of ¼ wavelength from voltage loops. Ideally, the voltage on a transmission line is the same everywhere, being equal to the product of the characteristic impedance and the line current. *See also* VOLTAGE LOOP, and STANDING WAVE.

VOLTAGE REGULATION

Voltage regulation is the process of maintaining the voltage at a constant value across a load. This is done by means of a regu-

lated, or constant-voltage, power supply. In some applications, good voltage regulation is very important.

Voltage regulation is expressed as a percentage (*see* REGULATION). In most power supplies, the output voltage falls slightly as the load resistance decreases to a certain minimum tolerable load resistance. As the load resistance continues to drop past the critical point, the power-supply output voltage decreases rapidly. The regulation percentage indicates the extent to which the voltage drops while the load resistance remains within the tolerable range.

In low-voltage power supplies, good voltage regulation can be realized in various different ways. The simplest method is the shunt connection of a Zener diode across the output terminals. The Zener voltage is about half the output voltage of the supply without the diode (*see* ZENER DIODE). A more sophisticated means of obtaining good regulation is the use of a series transistor that inserts more or less resistance into the circuit as the load resistance changes.

In high-voltage power supplies, a vacuum tube is most often used for regulation. Various gas-filled tubes are available for use as voltage regulators in such supplies. Voltage regulation can be obtained manually in any power supply by means of a small transformer in series with the alternating-current supply input. *See also* REGULATING TRANSFORMER, SERIES REGULATOR, SHUNT REGULATOR, and STEP-VOLTAGE REGULATOR.

VOLTAGE TRANSFORMER

A *voltage transformer* is a two-winding device, intended especially for stepping an alternating-current voltage up or down. Voltage transformers are used in almost all power supplies designed for operation from the 117-V utility mains. Voltage transformers are also extensively employed in power transmission.

In general, the input-to-output voltage ratio of a voltage transformer is identical to the primary-to-secondary turns ratio. *See also* STEP-DOWN TRANSFORMER, STEP-UP TRANSFORMER, and TRANSFORMER.

VOLT-AMPERE

In any electrical or electronic circuit, the *volt-ampere* is an expression of the apparent power. If the voltage between two points is E and the current in amperes is I, then the apparent power in volt-amperes is:

$$P_a = EI$$

In direct-current circuits, the volt-ampere is identical to the watt because the apparent power is the same as the true power. However, in alternating-current circuits, the voltampere might not be a true indication of the dissipated power. *See also* APPARENT POWER, POWER, REACTIVE POWER, TRUE POWER, VAR, and WATT.

VOLT-AMPERE-HOUR METER

A *volt-ampere-hour meter* is a device that determines the product of the voltage, current, and time in an electrical circuit. The common utility meter operates as a voltampere-hour meter. Such a device integrates the current in amperes over a period of time, obtaining the reading in watt hours or kilowatt hours, under the assumption that the voltage is constant (234 V in most cases). *See* WATT-HOUR METER.

VOLT BOX

A *volt box* is a precision voltage-divider circuit, used in the calibration of meters and other test instruments. A typical volt box is constructed using series-connected, fixed resistors. There is one pair of input terminals and several output terminals. The various voltages are obtained from the appropriate output terminals, provided the load resistance is significantly higher than that of the series combination in the box. This is always the case for voltmeters and oscilloscopes. *See also* VOLTAGE DIVIDER.

VOLTMETER

A *voltmeter* is a device that is used for measuring the voltage between two points. There are many types of voltmeters, used for various purposes in equipment design, testing, operation, and troubleshooting.

The simplest type of voltmeter consists of a microammeter in series with a large-value resistor. The size of the resistor determines the voltage required to produce a full-scale indication on the meter. Most voltmeters contain several resistors that can be switched in to obtain the desired meter range.

The illustration is a schematic diagram of a typical direct-current voltmeter. The meter is a 100-uA device. Full-scale ranges of 1 V, 10V, 100V, and 1 kV are obtained using resistors of 10 kΩ, 100 kΩ, 1 MΩ, and 10 MΩ respectively (kΩ = 1,000 ohms; MΩ = 1,000,000 ohms). The device can be used to measure ac voltage by switching in a diode and capacitor to rectify and filter the incoming voltage. The meter scale must be calibrated differently for alternating current.

In some applications, it is necessary that the internal resistance of a voltmeter be extremely high. The internal resistance of a voltmeter is expressed as a figure called *ohms per volt* (*see* OHMS PER VOLT). The meter illustrated has an internal resistance of 10 kΩ ohms per volt—a low enough value to cause problems in high-impedance circuits. More sophisticated voltmeters are available for use in such cases. *See also* FET VOLTMETER, and VACUUM-TUBE VOLTMETER.

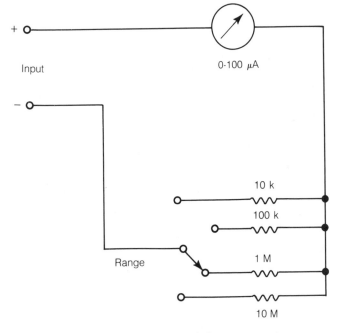

VOLTMETER: A voltmeter with four range settings.

VOLT-OHM-MILLIAMMETER

A common type of test meter consists of a combination voltmeter, ohmmeter, and milliammeter, and is known as a *volt-ohm-milliammeter (VOM)* or *multimeter*.

A multimeter contains a microammeter, a source of direct current, and a switchable network of resistors. In the voltmeter mode, series resistances are selected to determine the meter range (*see* VOLTMETER). In the ohmmeter mode, the meter is connected in series with the test leads and the power source (*see* OHMMETER). In the milliammeter mode, shunt resistances are placed across the meter to obtain the desired full-scale range (*see* AMMETER, and SHUNT RESISTOR).

The photograph shows a typical VOM. This device can measure voltages up to 1 kV (either dc or ac), direct currents to 10 A, and resistances to several megohms. A range-doubler switch allows for optimal meter reading.

Although the instrument shown is adequate for most testing and service purposes, the ohms-per-volt rating might not be high enough in some cases. Then, a special voltmeter must be used. *See also* FET VOLTMETER, and VACUUM-TUBE VOLTMETER.

VOLT-OHM-MILLIAMMETER: This is a common lab instrument.

VOLUME

Volume is an expression of the loudness of the sound produced by an electronic device. In particular, volume refers to the level of the sound coming from the speakers of a radio, tape recorder, or record player. This is a function of the audio-frequency current and voltage supplied to the speakers.

Volume can be expressed in decibels, relative to the threshold of hearing. Volume is also specified in terms of electrical decibels, relative to a power level of 2.51 mW in a load of 600 ohms resistive impedance. These units are called *volume units* (*see* DECIBEL, VOLUME UNIT, and VOLUME-UNIT METER).

VOLUME COMPRESSION

Volume compression is a technique that is sometimes used for improving the signal-to-noise ratio in a communications, high-fidelity, or public-address system. The basic principle is to enhance low-level sounds at some or all frequencies in the transmitting or recording process. There are several volume-compression schemes, used for various purposes. *See* AUTOMATIC GAIN CONTROL, AUTOMATIC LEVEL CONTROL, COMPRESSION, COMPRESSION CIRCUIT, and SPEECH COMPRESSION.

VOLUME CONTROL

A *volume control* is a component, usually a potentiometer or set of potentiometers, that is used for adjusting the volume in an audio circuit. There are two basic types of volume control: input adjustment and output adjustment. Either method will provide satisfactory results. In high-power audio circuits, input adjustment is more commonly used.

VOLUME UNIT

In high-fidelity applications, the amplitude of a music signal is generally expressed in terms of *volume units (VU)*. The VU is a relative indicator of the root-mean-square power output from an audio amplifier.

A level of 0 VU corresponds to +4 dBm = 2.51 mW, across a purely resistive load of 600 ohms impedance. This is a root-mean-square voltage of 1.23 V and a root-mean-square current of 2.05 mA. In general, a level of x VU corresponds to a signal x dB louder than the reference level. Thus, for a power level of P mW, the VU level is:

$$x = 10 \log_{10} (P/2.51)$$

Volume units are measured by means of a special meter, called a *volume-unit meter (VU meter)* at the output of an audio amplifier. *See also* DECIBEL, VOLUME, and VOLUME-UNIT METER.

VOLUME-UNIT METER

A *volume-unit (VU) meter* is a device that is used for measuring the root-mean-square volume level in an audio amplifier. The VU meter is calibrated in decibels relative to +4 dBm (*see* VOLUME UNIT).

Most VU meters are fast-acting devices, with just enough damping to allow easy reading. In a sophisticated stereo high-fidelity amplifier, each channel has a VU meter. The scale is marked off in black and red numerals, with a black and red reference line. The amplifier gain should normally be set so that the meter needle never enters the red range. If the needle moves into the red range, it indicates that distortion is likely to occur on audio peaks.

VOM

See VOLT-OHM-MILLIAMMETER.

VOX

VOX is a means of actuating a circuit, such as a radio transmitter, using the electrical voice impulses from a microphone or audio amplifier. Such a voice-actuated system can use a relay or an electronic switch.

VOX circuits are commonly used in communications transceivers. This is especially true of single-sideband units. VOX eliminates the need for pressing a lever or switch to change from receive mode to transmit mode. This feature is especially useful in mobile operation, or when both hands are needed for such things as taking notes. The VOX circuit can be disabled and push-to-talk switching used instead if the operator desires.

In VOX operation, the transmitter is actuated within a few milliseconds after the operator speaks into the microphone. The transmitter remains actuated for a short time after the operator stops speaking; this prevents unwanted "tripping out" during short pauses.

The schematic diagram shows a VOX circuit. The sensitivity and delay are adjustable. An anti-VOX circuit prevents the received signals from actuating the transmitter if a speaker is used for listening.

VOX is a form of semi break-in operation (*see* SEMI BREAK-IN OPERATION). If full break-in operation is wanted with single-sideband communications, a more sophisticated circuit is needed. *See also* BREAK-IN OPERATION.

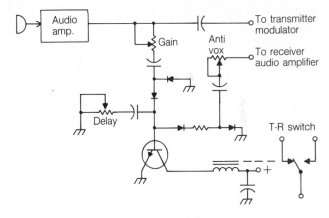

VOX: A circuit for voice-actuated transmission.

VSWR

See STANDING-WAVE RATIO.

VU METER

See VOLUME-UNIT METER.

VXO

See VARIABLE CRYSTAL OSCILLATOR.

W1AW

The headquarters of the American Radio Relay League (ARRL) maintains an official code-practice and bulletin station at 225 Main Street, Newington, CT 06111. This station transmits information of interest to all ham radio operators on a daily basis, on all of the major ham bands.

The station bears the original call letters of Hiram Percy Maxim, W1AW, one of the founders of the League. Hams often visit ARRL headquarters and W1AW. Visiting hams may operate the station during certain hours. Schedules and transmitting frequencies for W1AW can be found in *QST*, the official magazine of the ARRL, or can be obtained by writing a request letter to ARRL headquarters, enclosing a business-sized, self-addressed, stamped envelope. *See also* AMERICAN RADIO RELAY LEAGUE.

WAFER

See CHIP, and INTEGRATED CIRCUIT.

WAFER SWITCH

A *wafer switch* is a type of rotary switch, having one or more poles and one or more throw positions for each pole. The shaft passes through the center of one or more parallel, roughly circular wafers of plastic porcelain, or other dielectric material. The terminals are attached around each wafer.

As the shaft is turned, a rotating terminal makes contact with each of the fixed terminals, one after the other in a circle. If the center shaft breaks contact with a terminal before making contact with the next, the switch is called *nonshorting*. If the shaft maintains contact with a terminal until after contact has been established with the next, the switch is called *shorting*. Either type has certain advantages and disadvantages, depending on the application.

Wafer switches are commercially manufactured in various sizes and configurations. Larger wafer switches can be used at radio frequencies. Wafer switches are commonly used for such purposes as bandswitching in receivers, transmitters, and tuning networks. Wafer switches are also used in some types of test equipment.

WATT

The *watt (W)* is the unit of true, or real, power. In a dc circuit, or in an ac circuit containing no reactance, the power *P*, in watts, can be found from any of the formulas:

$$P = EI = I^2R = E^2/R$$

where E is the RMS voltage in volts, I is the RMS current in amperes, and R is the resistance in ohms. In an alternating-current circuit containing reactance as well as resistance, the true power in watts must be determined in a different way (*see* POWER, and TRUE POWER).

A power level of 1 W represents the expenditure of 1 joule of energy per second. A flashlight or small lantern bulb dissipates approximately 1 W of power. The audio output from a typical small portable radio is about 1 W when the volume is turned to maximum. An average fluorescent light bulb consumes 15 to 50 W; an incandescent bulb, as much as 500 W; a heavy appliance, as much as 2,000 W.

Fractions of the watt are sometimes used to express low power. The milliwatt (mW) is a thousandth (0.001) watt; the microwatt (μW) is a millionth (0.000001) watt. For larger power levels, the kilowatt (kW) or megawatt (MW) are used: 1 kW = 1,000 W and 1 MW = 1,000,000 W.

The radio-frequency output of a transmitter can be expressed in watts. This is the amount of power that would be dissipated in a pure resistor having a value equal to the matched output impedance of the transmitter—usually 50 ohms. *See also* APPARENT POWER, JOULE, REACTIVE POWER, TRUE POWER, VAR, VOLT-AMPERE, and WATTMETER.

WATT HOUR

The *watt hour* is a unit of energy equivalent to 3,600 joules (watt seconds). The watt hour is sometimes used to express the energy consumed by an electrical device over a period of time. The watt hour may also be used to express the storage capacity of a battery. Energy in watt hours is the product of the average power drawn, in watts, and the time, in hours. The abbreviation for watt hour is Wh.

Suppose that a 60-W bulb burns for one hour. Then the energy consumed by the bulb is 60 watt hours. In two hours the same bulb consumes 120 watt hours; in three hours it uses up 180 watt hours, and so on. A 12-V battery having 50 ampere hours of storage capacity can deliver 12 × 50 = 600 watt hours of energy. This might constitute 1 watt for 600 hours, 2 watts for 300 hours, or any of a number of other situations.

In utility applications, the kilowatt hour is more often specified than the watt hour. This is equivalent to 1,000 watt hours. The abbreviation for kilowatt hour is kWh. *See also* ENERGY, JOULE, KILOWATT HOUR, and WATT-HOUR METER.

WATT-HOUR METER

A *watt-hour meter* is an instrument that is used to measure consumed energy. Your electric meter is a form of watthour meter, although it actually registers kilowatt hours (*see* KILOWATT HOUR, and WATT HOUR). The abbreviation for watt hour is Wh, and the abbreviation for kilowatt hour is kWh.

Most watt-hour meters actually register the integral of the root-mean-square current in a circuit over a period of time. Thus, such a meter is suitable for operation only at a single, rated voltage, such as 234 V. A motor runs at a speed that is directly proportional to the current drawn at any given time. The motor is connected to a set of gears, in turn attached to pointers or drum indicators that numerically indicate the energy used up to a given time.

WATTMETER

A *wattmeter* is an instrument that measures power. There are various types of wattmeters; some measure true power, and others measure only the apparent power (*see* APPARENT POWER, POWER, REACTIVE POWER, and TRUE POWER).

The simplest wattmeter consists of an ammeter in series with a circuit, and a voltmeter in parallel with the component or set of components for which the power is to be determined. The power (in watts) is given by the product of the voltage (in volts) and the current (in amperes). This method is satisfactory for direct-current circuits and for alternating-current circuits in which no reactance is present. However, if reactance exists, the product of current and voltage is artificially large, and does not represent true power.

In a common household circuit, there is usually no reactance. Therefore, a root-mean-square ammeter can be used to measure the power consumed by a device at a particular voltage. The ammeter is simply connected in series with the appliance. The consumed power, P, is given (in watts) in terms of the current reading, I (in amperes) and the line voltage, E (in volts) as:

$$P = EI$$

In radio-frequency circuits, directional wattmeters are generally used to measure true power. Such wattmeters, such as the device shown in the illustration, can measure forward or reflected power. The true power is the difference between the forward and reflected readings. The greater the amount of reactance in the antenna system, the larger the reflected reading will be, in proportion to the forward reading. *See also* REFLECTED POWER, and REFLECTOMETER.

WAVE

A *wave* is any form of disturbance that exhibits a periodic (repeating) pattern. Examples of waves include acoustic disturbances, pulsating or alternating currents or voltages, and electromagnetic fields.

All waves have a definable period T (usually given in seconds), a specific frequency f, a propagation speed c (usually expressed in units per second), and a wavelength A. These quantities are related according to the formulas:

$$c = f\lambda = \lambda/T$$

Some disturbances exhibit changes in amplitude that have no identifiable period. These disturbances are not true waves, although they can produce noticeable effects (such as acoustic or electromagnetic noise) over a wide band of frequencies.

A wave does not, in itself, necessarily appear "wavelike." The wave nature of some disturbances, such as ripples on a pond or swells in the ocean, is obvious. But in most cases, some means of artificial display is required to see the wave nature of a disturbance. *See also* ELECTROMAGNETIC FIELD, TRANSVERSE WAVE, and WAVEFORM.

WAVEFORM

Waveform is an expression used to describe the "shape" of a wave disturbance, either as seen directly or as observed on a display instrument, such as an oscilloscope. A wave disturbance can have any of an infinite number of forms.

The simplest waveform occurs when a disturbance has only one frequency. The resulting waveform is sinusoidal, and is known as a *sine wave* because its shape is identical to the graph

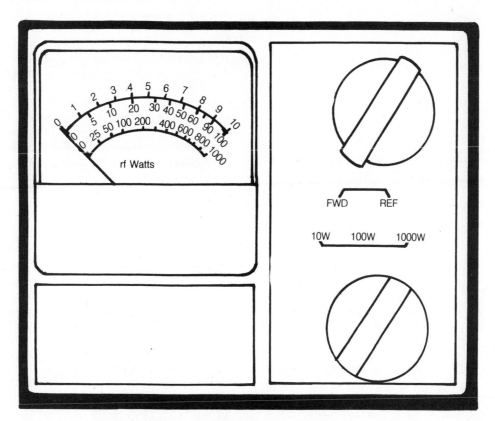

WATTMETER: A typical RF wattmeter. GOLD LINE CONNECTOR

of the sine function (*see* SINE, and SINE WAVE). More complex waveforms result when energy is concentrated at frequencies that are integral multiples of a lowest, fundamental frequency (*see* FUNDAMENTAL FREQUENCY, and HARMONIC). Examples include the sawtooth, square, trapezoidal, and triangular waves (*see* SAWTOOTH WAVE, SQUARE WAVE, TRAPEZOIDAL WAVE, and TRIANGULAR WAVE).

When energy is concentrated at many different frequencies, waveforms become exceedingly complicated. Various musical instruments produce waveforms ranging from a near-perfect sine wave to a complex-jumble.

All periodic waves, no matter how complicated they appear, have one common characteristic: they repeat themselves at defined intervals. Thus, it is possible to predict the instantaneous amplitude of the waveform for any specified moment in the future. However, some waveforms do not have this property. Natural electromagnetic noise, for example, has a waveform that is random. Although such disturbances are not true waves, they nevertheless can be observed on display instruments such as the oscilloscope. *See also* WAVE.

WAVEGUIDE

A *waveguide* is a feed line that is often used at ultra-high and microwave frequencies. A waveguide consists of a hollow metal pipe, usually having a rectangular or circular cross section (see illustration). The electromagnetic field travels down the pipe, provided that the wavelength is short enough. A waveguide provides excellent shielding and low loss.

Cross-Sectional Size Requirements In order to efficiently propagate an electromagnetic field, a rectangular waveguide must have sides measuring at least 0.5 wavelength, and preferably more than 0.7 wavelength. A circular waveguide should be at least 0.6 wavelength in diameter, and preferably 0.7 wavelength or more.

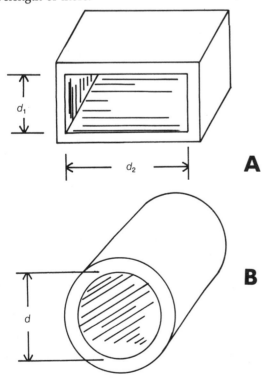

WAVEGUIDE: At A, rectangular with dimensions d_1 and d_2. At B, circular with inside diameter d.

Consider the example of a waveguide operating at a frequency of 3 GHz. The wavelength is 3.9 in (10 cm). Thus, a rectangular waveguide must be at least 2.0 in (5.1 cm) wide, and preferably 2.7 in (7.0 cm). A circular waveguide should be at least 2.3 in (6.0 cm) in diameter, but preferably 2.7 in (7.0 cm). The frequency at which the length of the side, or the diameter, is ½ wavelength, is known as the *cutoff frequency*. The waveguide exhibits a highpass response (*see* HIGHPASS RESPONSE).

Propagation An electromagnetic field can travel down a waveguide in various ways. If all of the electric lines of flux are perpendicular to the axis of the waveguide, the propagation mode is called *transverse-electric (TE)*. If all of the magnetic lines of flux are perpendicular to the axis, the mode is called *transverse-magnetic (TM)*. The electromagnetic field can be coupled into the waveguide via either the electric or the magnetic components.

In a waveguide, electromagnetic fields tend to circulate in eddies. Depending on the frequency of the applied energy, there will be one or more eddies. In general, as the frequency becomes higher, the number of eddies in the cross section increases.

Operation When a waveguide is used, it is important that the impedance of the antenna be purely resistive, and that the resistance be matched to the characteristic impedance of the waveguide (*see* CHARACTERISTIC IMPEDANCE). Otherwise, there will be standing waves in the line, and the loss will be increased compared with the perfectly matched condition (*see* STANDING WAVE, STANDING-WAVE RATIO, and STANDING-WAVE-RATIO LOSS). The waveguide behaves exactly like other types of feed lines in this respect.

The characteristic impedance of a waveguide varies with frequency. In this sense, it differs from coaxial or parallel-wire lines. In most cases, matching transformers are needed to achieve a low standing-wave ratio on a waveguide because few antennas present impedances that are favorable to a waveguide if direct coupling is used. A quarter-wave section of waveguide, coaxial cable, or parallel wires or rods, can be used for this purpose (*see* QUARTER-WAVE TRANSMISSION LINE).

Because of the highpass characteristics of waveguides, they can be used as filters and attenuators. They are very effective for this purpose because of their high efficiency. It is important that the interior of a waveguide be kept clean and free from condensation. Even a small obstruction can seriously degrade the performance of a waveguide.

For Further Information The theory of waveguides cannot be thoroughly treated here. For more details, a volume on communications technology or microwave theory and practice is recommended.

WAVEGUIDE PROPAGATION

Waveguide propagation is an effect that is observed at long wavelengths. The ionosphere acts as a total reflector at very low and low frequencies. If the wavelength is great enough, the earth and ionosphere act like the surfaces of a waveguide, and propagation occurs in a manner similar to that in a waveguide transmission line at ultra-high and microwave frequencies (*see* WAVEGUIDE).

In the waveguide mode, the transmitting antenna, which must be vertical, couples energy into the space between the earth and ionized upper atmosphere. The receiving antenna picks up the energy in a manner similar to a probe in a waveguide feed line. The earth-ionosphere waveguide has a cutoff (lower limit) of about 10 kHz. Below this frequency, propagation is very poor because of severe attenuation.

WAVELENGTH

All wave disturbances have a certain physical length in space. This length depends on two factors: the frequency of the disturbance, and the speed at which it is propagated (*see* WAVE). The *wavelength* is inversely proportional to the frequency, and directly proportional to the speed of propagation. Wavelength is denoted in equations by the lowercase Greek letter lambda (λ).

In a periodic disturbance, the wavelength is defined as the distance between identical points of two adjacent waves. In electromagnetic fields, this distance can be millions of meters or a tiny fraction of a millimeter, or anything in between. The wavelength determines many aspects of the behavior of a wave disturbance.

For electromagnetic waves in free space, the wavelength, in feet, is given in terms of the frequency by the formula:

$$\lambda = 9.84 \times 10^8/f$$

where f is in hertz. The wavelength in meters is:

$$\lambda = 3.00 \times 10^8/f$$

For sound waves in air at sea level, wavelength is given in feet by:

$$\lambda = 1.10 \times 10^3/f$$

where f is in hertz; the wavelength in meters is:

$$\lambda = 3.35 \times 10^2/f$$

Wavelength is an important consideration in the design of antenna systems for radio frequencies. Wavelength is also important to the designers of optical apparatus and various acoustic devices. *See also* ELECTROMAGNETIC FIELD, ELECTROMAGNETIC SPECTRUM, and VELOCITY FACTOR.

WAVEMETER

A *wavemeter* is a device that makes use of the properties of resonant circuits to determine the frequency or wavelength of a radio signal. The most common type of wavemeter is called the absorption wavemeter (*see* ABSORPTION WAVEMETER). However, there are other methods of using resonant circuits to measure frequency. *See also* CAVITY FREQUENCY METER, COAXIAL WAVEMETER, LECHER WIRES, and WAVE-TRAP FREQUENCY METER.

WAVE POLARIZATION

See POLARIZATION.

WAVE-TRAP FREQUENCY METER

A trap can be used to measure the frequency of a radio signal by means of a circuit called a *wave trap* or *wave-trap frequency meter*. The device consists of a tunable trap (parallel-resonant circuit) in series with the signal path, and an indicating device such as a meter (see illustration).

The tunable trap is adjusted until a dip is observed in the meter reading. This condition occurs when the trap is set for resonance at the frequency of the applied signal. The frequency is then read from a calibrated scale.

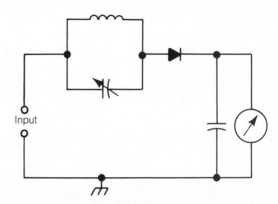

WAVE-TRAP FREQUENCY METER: Tune for dip in meter reading.

WEBER

The *weber* is a unit of magnetic flux, representing one line of flux in the meter-kilogram-second (MKS) or standard international (SI) system of units. The weber is equivalent to a volt second.

Magnetic flux is sometimes expressed in units called maxwells. The maxwell is equal to 10^{-8} weber. *See also* MAGNETIC FIELD, and MAXWELL.

WEBER-FECHNER LAW

Our senses do not perceive things in direct proportion to their actual intensity. Instead, we sense effects according to the logarithm of their actual intensity. This rule is called the *Weber-Fechner Law*. The Weber-Fechner Law is of special significance for audible and visible effects. *See also* DECIBEL.

WEIGHT

In digital transmission, the duration ratio of mark to space is sometimes called *weight*. This expression is especially used in morse-code transmission. A string of dots has a weight ratio of 1:1 if the duration of a dot is identical to the duration of the spaces between the dots. If the dots last longer than the spaces, the weight ratio is greater than 1:1, and is said to be *heavy*. If the dots are briefer than spaces, the ratio is less than 1:1, and is called *light*. Normally, a weight ratio of 1:1 is used, although some telegraphers prefer a slightly heavier weight at slow speeds and a slightly lighter weight at very high speeds.

Some manufacturers define weight according to a different formula than that described above; the dot-to-dash ratio is sometimes specified. Normally it is 1:3. Small ratios such as 1:4 represent lighter weight, and larger values such as 1:1.2 represent heavier weight. *See also* MARK/SPACE, MORSE CODE, and WEIGHT CONTROL.

WEIGHT CONTROL

Many keying devices, especially for morse-code transmission, have *weight controls*. A weight control is used to compensate for the varying tastes of receiving operators, and for the effects of shaping circuits in code transmitters (*see* SHAPING).

A typical weight control allows adjustment of the dot-to-space ratio between limits of about 1 : 2 and 2 : 1. Some weight controls have larger ranges. In certain digital terminals, the weight is adjustable in discrete steps; in many electronic keyers, the weight can be varied continuously. *See* MORSE CODE, and WEIGHT.

WHEATSTONE BRIDGE

A *Wheatstone bridge* is a device for measuring unknown resistances. The Wheatstone bridge operates by comparing the ratio of the unknown resistance to three other known resistances. The Wheatstone bridge is similar to the ratio-arm bridge. *See also* RATIO-ARM BRIDGE, and RESISTANCE MEASUREMENT.

WHIP ANTENNA

A *whip antenna* is a short radiator, usually loaded at the base and measuring ¼ physical wavelength or less. Whip antennas are often used in mobile communications, especially at high frequencies.

The construction of a typical whip antenna is illustrated in the drawing. A tapered rod of stainless steel forms the radiator. A tapped coil at the base facilitates adjustment of the resonant frequency. A spring allows for wind resistance. A magnetic mounting can be used for small whips. For antennas longer than about 3 feet, a ball mounting is generally used.

The efficiency of a whip antenna depends on the size of the vehicle on which it is mounted. The larger the vehicle, the better the grounding system, and the better the antenna performance. In general, efficiency improves as the frequency is increased, and worsens as the frequency is decreased. *See also* ANTENNA EFFICIENCY, and VERTICAL ANTENNA.

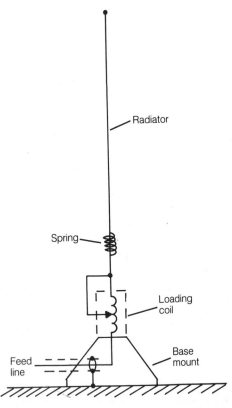

WHIP ANTENNA: In this example, inductive base loading is used.

WHITE NOISE

Broadband noise is sometimes called *white noise*. It gets its name from the fact that white light contains energy at all visible wavelengths, and thus is wideband electromagnetic energy; wideband noise is therefore, in a sense, "white."

White noise at audio frequencies consists of a more or less uniform distribution of energy at wavelengths from 20 Hz to 20 kHz. If energy is wideband in nature, but concentrated toward one end or the other, the noise is called *pink* or *violet*. White noise sounds like a rough hiss and roar, similar to the sound of ocean breakers on a beach. White noise is pleasing to the ear, and is even used in some cases to induce sleep.

At radio frequencies, the term *white noise* is quite general. If a given band of frequencies contains noise of essentially equal amplitude at all points, the noise can be called *white*. *See also* NOISE.

WHITE TRANSMISSION

White transmission is a form of amplitude-modulated facsimile signal (*see* FACSIMILE) in which the greatest copy density, or darkest shade, corresponds to the minimum amplitude of the signal. White transmission is the opposite of black transmission, in which the darkest shade corresponds to the maximum signal amplitude (*see* BLACK TRANSMISSION). White transmission can be considered, in a sense, right-side-up. In a frequency-modulated facsimile system, *white transmission* means that the darkest copy corresponds to the highest transmitted frequency.

WIEN BRIDGE

A *Wien bridge* is a resistance-capacitance circuit that can be used for measurement of unknown values of capacitance. The Wien bridge can also be used as a resonant circuit. The Wien bridge requires an audio-frequency generator, and an indicating device, such as a meter or headset.

A Wien bridge can be connected in the feedback circuit of an oscillator, resulting in a nearly pure sine wave at audio frequencies. This type of oscillator is a two-stage circuit, and is known as a *Wien-bridge oscillator*.

In order for the Wien-bridge oscillator to produce a sine wave, the components must have values such that the Wien bridge is balanced.

WIND LOADING

Wind loading is an expression of the maximum wind speed, in miles per hour or meters per second, that an antenna can withstand without damage. In general, the larger antennas exhibit more wind resistance than smaller ones. Therefore, larger antennas require more rugged physical construction than small antennas for a given wind-loading figure.

Wind loading is affected by the lengths and diameters of conductors used in an antenna. In some cases, wind loading depends on the direction of the wind with respect to the antenna. If an antenna becomes laden with ice or wet snow, the wind loading is greatly reduced (*see* ICE LOADING).

Wind-loading figures are sometimes given for towers. The wind speed that a tower can withstand is affected by the sizes and types of antennas it supports. The ability of a tower to resist high winds can be enhanced by guying (*see* GUYING).

WINDOM ANTENNA

A *Windom antenna* is a multiband wire antenna that uses a single-wire feed line. The antenna gets its name from the radio amateur who devised it. The Windom is a half-wave horizontal antenna, fed slightly off center (see illustration A).

For a given fundamental frequency f in megahertz, the length d of a windom, in feet, is:

$$d = 468/f$$

If d is expressed in meters, then:

$$d = 143/f$$

The feed line is attached at a point 36 percent of the way from one end of the radiator to the other end. For a given frequency f in megahertz, the distance r, in feet, from the near end of the radiator to the line is:

$$r = 168/f$$

If r is expressed in meters:

$$r = 51/f$$

The Windom, thus constructed, will operate satisfactorily at all of the even harmonics, as well as the fundamental frequency. An antenna designed for a frequency f can be used at $2f$, $4f$, $6f$, and so on. A transmatch can be used to match the antenna to a transmitter output circuit. Some radiation occurs from the feed line because it is unbalanced and not shielded.

A parallel-wire line can be used with off-center feed in an antenna that is sometimes called a *Windom*. Actually, this antenna is not a true Windom; it could be called a *quasi-Windom*.

The design is illustrated at B. The radiator measures ½ electrical wavelength at the fundamental frequency. The length in feet or meters is found by the same formula as given earlier. The feed line is attached ⅓ of the way from one end of the radiator to the other. For a fundamental frequency f, the distance r, in feet is:

$$r = 156/f$$

and in meters:

$$r = 48/f$$

This configuration results in a feed-point impedance of about 300 ohms at frequencies f, $2f$, $3f$, $4f$, $5f$, and so on. Some radiation occurs from the line because the system is not perfectly balanced. A 4:1 balun or a transmatch can be used at the transmitter end of the feed line to obtain a more or less purely resistive impedance of approximately 75 ohms.

WIND POWER

The wind can be used to generate electrical or mechanical power. Farmers have been harnessing the wind for centuries; in recent years, some attention has been given to the possibility of using wind as an alternative source of energy, as fossil fuels have become more and more expensive.

A simple wind-power "plant" can be constructed by attaching an ordinary electric generator to the shaft of a windmill. As the wind turns the blades of the mill, the shaft turns the coils or magnets in the generator (*see* GENERATOR). The resulting ac power is proportional to the angular speed of the shaft. The frequency, too, is a direct function of the angular speed of the shaft. The principle is illustrated in the drawing (A). Commercially manufactured windmill generators are available (B).

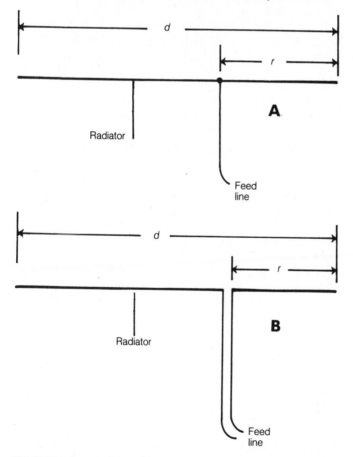

WINDOM ANTENNA: At A, single-wire feed; at B, parallel-wire feed. Dimensions are covered in the text.

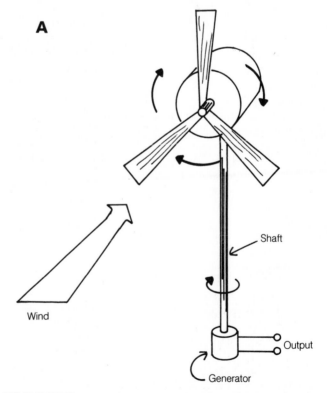

WIND POWER: At A, a functional drawing of a wind-power generator. At B, a wind-power generator on the roof of a house.

B

WIND POWER Continued.

The main advantage of wind power is that a generating system is simple and easy to maintain. Rectifiers and choppers can be used to convert the variable-frequency current to 117 V at 60 Hz. The main problem with a wind-power generator is that the wind does not always blow! Nevertheless, a wind-power generator can provide considerable energy savings if it is interconnected with the utility mains so that power is available when the weather is calm.

WIRE

Wire is the most common and universal way of getting electrical energy from one place to another. Wires are also used for physical reinforcement in a variety of situations. Some types of wire are used to make lamps, resistors, and heating elements.

The most common type of wire is a cylindrical length of metal, such as iron, steel, copper, or aluminum. Some wires consist of just one conductor, and others can have several. A single-conductor is called *solid*, and a multiconductor wire is called *stranded* (*see* STRANDED WIRE). Some wires have square or rectangular cross sections. Two or more wires can be combined to form a cable, cord, or transmission line (*see* CABLE, and TRANSMISSION LINE).

Wires are sometimes bare, or are sometimes coated with a layer of enamel, rubber, plastic, or other insulating material. Insulated wires are used when a short circuit must be avoided (*see* INSULATED CONDUCTOR). Uninsulated wires are less expensive, and are preferable in situations where a short circuit is unlikely or of no consequence.

For wire having a circular cross section, the size is expressed in various ways. The simplest method is to simply state the diameter in inches, millimeters, or some other units. The cross-sectional area might also be specified. The most common way of expressing wire size is by means of standard gauges (*see* AMERICAN WIRE GAUGE, BIRMINGHAM WIRE GAUGE, and BRITISH STANDARD WIRE GAUGE).

For electrical purposes, the current that a wire can handle is a function of the material used and the diameter of the conductor. In general, larger diameters allow for more current. Copper is the most efficient conductor for reasonable cost, and has the highest carrying capacity of the common types of wire (*see* CARRYING CAPACITY, and COPPER). The resistance of a wire, per unit length, is also a function of diameter (*see* RESISTANCE). The tensile strength of a wire is also related to material

and size, and to the method of manufacture (*see* TENSILE STRENGTH).

WIRE ANTENNA

At very low, low, medium, and high frequencies, antennas are often constructed of wire. Any type of antenna can, theoretically, be made from wire. The dipole, longwire, rhombic, V beam, and Windom are the most common examples of antennas that are usually constructed using wire (*see* DIPOLE ANTENNA, LONGWIRE ANTENNA, RHOMBIC ANTENNA, V BEAM, and WINDOM ANTENNA).

When erecting a wire antenna that must be a certain length, it is best to use wire that will not stretch significantly. Hard-drawn copper or copper-clad steel wire are best for such applications. These types of wire also have excellent tensile strength, and are preferable for use in long spans.

If the antenna length is not critical, or if the span is relatively short, soft-drawn or annealed copper wire can be used. It is not necessary to use insulated wire unless short circuits with other objects must be avoided. The ends of the antenna should be electrically isolated, however, by means of suitable insulators (*see* INSULATOR).

Any convenient supports can be used to anchor the ends of a wire antenna: trees, towers, and the sides of buildings are all excellent for this purpose. Utility poles should not be used; this is illegal in some municipalities, and can present an electrocution hazard. A wire antenna should never be run over or under a utility wire; if the antenna or the wire falls, the antenna can become live with hundreds or thousands of volts.

Wire antennas are especially popular among shortwave listeners and radio amateurs in the medium-frequency and high-frequency bands. At very high frequencies and above, antennas are small enough to make metal tubing practical for construction.

WIRE GAUGE

See AMERICAN WIRE GAUGE, BIRMINGHAM WIRE GAUGE, and BRITISH STANDARD WIRE GAUGE.

WIRELESS MICROPHONE

A small radio transmitter can be installed in a microphone enclosure, and a receiver attached to the audio input of the device with which the microphone is used. This eliminates the necessity for using a cord for short distances. This type of device is called a *wireless microphone*. Wireless microphones are sometimes used in newscasting and entertainment, when personnel must move around a studio or stage.

Most wireless microphones operate at very-high or ultra-high frequencies. This is primarily because an efficient short antenna is practical at these wavelengths. Frequency modulation is usually used for optimum immunity to noise.

In the United States and some other countries, the use of wireless microphones is regulated by government agencies. The power level of the transmitter, and the frequency at which it operates, are such that there is little chance of interference to other services.

WIRE SPLICING

In wiring of electrical circuits and antennas, it is often necessary to splice two lengths of wire. There are various splicing methods; the most common are described here.

Twist Splice The simplest way to splice two wires is the method illustrated at A, B, and C. The wires are brought parallel with each other and twisted together. This method can be used with solid wire or stranded wire. About six to eight twists should be made. If the wires are of unequal diameter, the smaller wire is twisted around the larger wire. Electrical tape can be put over the connection if insulation is important. This type of splice has very poor mechanical strength unless solder is used. Even if a twist splice is soldered, the connection cannot withstand much strain.

The twist splice is used in most household electrical systems with AWG #12 or #14 solid copper wire. Special caps are placed over the splices for insulation purposes.

Western Union Splice When it is desired that a splice have high breaking strength, the method shown at D, E, and F is preferable. This type of splice is called a *Western Union splice*.

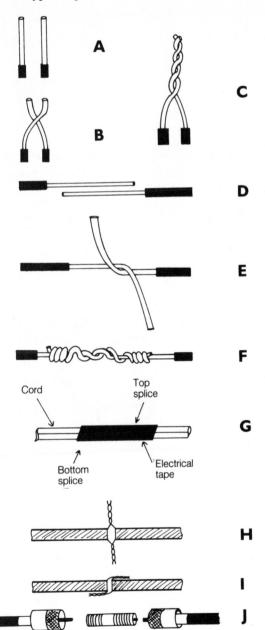

WIRE SPLICING: A twist splice (A, B, C); a Western Union splice (D, E, F); a two-wire cord splice (G); a twist splice for twin-lead (H, I); a preferred method of coaxial splice (J).

The wires are brought together end-to-end, overlapping a couple of inches. They are hooked around as shown at E, and twisted several times as shown at F. For guy-wire splicing, each end should be twisted around 10 to 12 times. Any protruding ends are removed using a diagonal cutter. The splice is then soldered if a good electrical bond is needed, and a layer of electrical tape or other insulation is applied. For large-diameter wires, the ends can be tinned with solder before the splice is made to facilitate the best possible electrical bond.

This splicing method can be used with solid or stranded wire of any gauge, but it is rather difficult to splice wires having diameters that differ by more than 4 in the AWG system (*see* AMERICAN WIRE GAUGE). The Western Union splice is also somewhat awkward when one wire is solid and the other is stranded. For maximum strength, both wires should be the same size and the same type.

Cord and Cable Splicing When splicing cords or lengths of open-wire transmission line, the Western Union splice is preferable because the wires remain in line. Insulation is very important when splicing electrical cords; both splices should be thoroughly wrapped with electrical tape, and the combination wrapped afterwards (G). For additional insulation, the splices can be made at slightly different points along the cord.

Two lengths of twin-lead transmission line can be effectively and conveniently twist-spliced, as illustrated at H and I. After the twists have been soldered, the splices are trimmed to about 1/4 inch and folded back. The whole area is then carefully wrapped with electrical tape.

Multiconductor cables can be spliced wire by wire, using Western Union splices in each instance. All of the splices must be individually wrapped with insulating material to prevent short circuiting.

Coaxial cables, and other cables in which a constant characteristic impedance must be maintained, are generally spliced by using special connectors. The N-type or UHF connectors are the most common for coaxial cable. A male connector is soldered to each of the two ends to be spliced; a female-to-female adaptor is used between them. This is illustrated at J. The whole joint must be thoroughly wrapped with insulating tape. *See also* N-TYPE CONNECTOR, and UHF CONNECTOR.

WIREWOUND RESISTOR

A length of resistive wire can be wrapped on a cylindrical core to obtain a specific ohmic value. Such a device is called a *wirewound resistor*. Wirewound resistors are commonly used in high-current applications because of their ability to dissipate large amounts of power: some wirewound resistors can handle several hundred watts on a continuous basis.

The value of a wirewound resistor depends on the type and size of wire used, and also on the length of the winding. The power-dissipating capability depends on the size of the wire, the diameter of the winding form, and the material from which the winding form is made. Ceramic forms are common.

Wirewound resistors are not generally suitable for use at radio frequencies because they exhibit inductive reactance, as well as resistance, for high-frequency alternating currents. *See also* RESISTOR.

WIRE WRAPPING

Wire wrapping is a method of circuit wiring in which solder is not used. Components are mounted on lugs that are inserted into a perforated board. The lugs protrude underneath the board. Small-gauge, solid wire is used for interconnection among the lugs. A connection is made by neatly wrapping the wire several times around the lug (see illustration). A special tool is used for wrapping the wire. The lugs are long enough so that three or more connections can be made on top of each other.

Wire wrapping is extremely convenient, and makes it possible to assemble complicated circuits in a small amount of physical space. Wire wrapping is extensively used in conjunction with computer systems and large switching networks in controlled environments. Wire wrapping is also ideal for construction of experimental circuits because the wires are easy to remove.

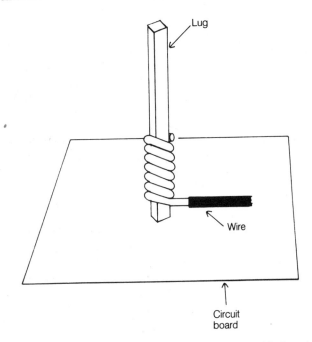

WIRE WRAPPING: A square lug enhances contact with the wire.

WIRING DIAGRAM

Wiring diagrams are designed to show point-to-point connections or printed circuit patterns from one point to another on system or circuit boards, including cabling connections between boards. They are completely separate from schematics and have no electrical symbols indicated.

Schematic diagrams do not locate parts, but are the electrical signal flow and operating voltage charts giving all component values, waveforms, test conditions, and I/O points as important references. Whenever engineers and technicians troubleshoot defective electronic equipment, they always refer to schematics first for service information and electrical measurements. Afterwards, PC board or wiring diagrams can be consulted for specific cabling or physical parts locations on the equipment's plug-in boards or chassis. *See also* SCHEMATIC DIAGRAM.

WOODPECKER

In the 1970s, a previously nonexistent signal began to be heard on the high-frequency radio bands. Because of its constant pulse rate and the way the pulses sound, it has become known as the *woodpecker*. It has also been called the *buzz saw*.

At times, the woodpecker signal is very strong and causes interference to radio communications over a wide band of wavelengths. A well-designed noise blanker can provide some reduction in the interference. Switching off the automatic gain control (AGC) and using an audio clipper can also help to alleviate the interference.

The woodpecker is believed to have originated in Russia as an over-the-horizon radar system (*see* RADAR). Some engineers have theorized that the woodpecker might be used to modify the ionosphere for the purpose of gaining some control over the propagation conditions at shortwave frequencies. The signal consists of sharp, regular pulses at a rate near 10 Hz. The bandwidth is usually several hundred kilohertz. The transmitter power is on the order of several megawatts. The center frequency changes, apparently following the maximum usable frequency (MUF) to take advantage of optimum long-distance ionospheric propagation (*see* MAXIMUM USABLE FREQUENCY).

WORD

In digital and computer practice, a group of bits or characters, having a specified length or number of units, is called a *word*. A word conveys a certain amount of information, and occupies a certain amount of space in memory. Words can be used as data, instructions, designators, or numerical values. *See also* BYTE, and CHARACTER.

In teletype and telegraph communications, a word consists of five characters plus one space. To obtain the total word count in a message, the number of characters, symbols, and spaces is counted. In teletype, symbols and punctuation marks count as single characters, but in morse code, they count as two characters each. The result is divided by 6. Speed of transmission can then be expressed in terms of the number of words sent per minute. *See also* WORDS PER MINUTE.

WORD PROCESSING

Computers and microcomputers can be used as an aid in writing and editing letters and manuscripts. This is known as *word processing*. Some computers are designed specifically for this purpose. These computers are called *word processors*. They are often completely self-contained units, including a video terminal, keyboard, disk drive, and printer.

Some Common Word-processing Capabilities Different writing styles require different word-processing functions. For example, letters require a different set of functions than magazine articles require. Tables and charts require still another set of functions. Any of several specialized software packages can be used with a single word processor for various kinds of writing. A few of the most common word-processing functions are listed below.

- Left-margin adjustment and justification: all rows are aligned along the left side of the page with a margin width the operator can select.
- Right-margin adjustment and justification: if desired, all rows can be aligned along the right side of the page. The margin width is adjustable.
- Insertion and deletion of characters, words, lines, and paragraphs: text can be added or removed and the continuity of the manuscript will be readjusted.

- Exchange of characters, words, lines, and paragraphs: text can be altered as necessary.
- Row and column alignment: used for making tables and charts.
- Word wrap around: automatically moves a word to next line in the text if the word runs past the right-hand margin limit.
- Double spacing: inserts an extra space between each line of text.

Data Storage and Printing Almost all word processors allow storage of information on magnetic disks. The floppy disk is by far the most commonly used (see FLOPPY DISK). Information is easily erased or rewritten onto the disks.

Every word processor uses a printer. There are various types of printers available for use with word processors. The dot-matrix printer operates at high speed, and is useful when a lot of text must be printed in a short time. Ball-type and daisy-wheel printers give crisper copy, but run at a somewhat slower speed than dot-matrix printers.

An important feature of a word processor is the memory capacity for each file of text. All computers have a certain maximum memory limit. In general, each page of double-spaced text, containing 220 words, has 1400 to 1500 bytes. The number of such pages that can be stored in a file is thus about 70 percent of the number of kilobytes memory. Larger computers, of course, contain more of memory space than small computers. Small files are adequate for letters and magazine articles. For maximum flexibility in book writing, larger files are desirable.

As our society becomes progressively more information-oriented, the need for word processors in everyday life will increase. Many moderately priced word processors and software packages are now available. See also KEYBOARD, and VIDEO DISPLAY TERMINAL.

WORDS PER MINUTE

The speed of transmission of a digital code is sometimes measured in words per minute. A word generally consists of five characters plus one space (see WORD). The abbreviation for words per minute is WPM.

For teleprinter codes, speeds range from 60 WPM to hundreds or even thousands of words per minute. The original Baudot speed was 60 WPM, and this speed is still in widespread use (see BAUDOT). The standard ASCII speed is 110 WPM, and the highest commonly used ASCII speed is 19,200 WPM (see ASCII). Sometimes the speed is given in bauds, rather than words per minute. For Baudot transmissions, the speeds in words per minute is about 33 percent greater than the baud rate. For ASCII transmissions, the speed in words per minute is the same as the baud rate (see BAUD RATE).

The speed of an International Morse Code transmission, in words per minute, is about 1.2 times the baud rate, where one baud is the length of a single dot or the space between dots and dashes in a character. If a string of dots is sent, the speed in words per minute is equal to 2.4 times the number of dots in one second (see INTERNATIONAL MORSE CODE).

WRITING GUN

A *writing gun* is an electron gun that is used in a storage cathode-ray tube. The writing gun is basically an ordinary electron gun that emits a narrow beam of electrons that strikes the phosphor surface at a certain point, resulting in a bright spot (see CATHODE-RAY TUBE, and ELECTRON-BEAM GENERATOR).

In a storage oscilloscope, the writing-gun beam has constant intensity, and scans from left to right at a constant speed. The vertical position is varied by the incoming signal (see OSCILLOSCOPE, and STORAGE OSCILLOSCOPE). The intensity of the writing-gun beam can be modulated, and the beam moved through a standard raster, for storage of television pictures. The display is observed using another electron gun, known as the *flood gun* (see FLOOD GUN).

WWV/WWVH/WWVB

The National Bureau of Standards maintains time/frequency stations on 2.5, 5, 10, 15, and 20 MHz. Amplitude modulation (AM) is used. The transmitter power is 10 kW at 5, 10, and 15 MHz, and 2.5 kW at 2.5 and 20 MHz. The frequencies are exact for practical amateur purposes. The time signals are precise to within a tiny fraction of a second. Cesium-based reference oscillators are used as the standard for time and frequency.

There is also a station, WWVB, in the low-frequency (LF) band at 60 kHz. This station transmits digital information only. Transmissions are also made via the Geostationary Operational Environmental Satellites (GOES).

Stations WWV and WWVB are located in Colorado, and WWVH is located on the Hawaiian island of Kauai. Time is announced every minute on WWV and WWVH by computerized voices. A man's simulated voice is used at WWV, and a woman's voice is used at WWVH. Standard audio tones are sent during the first 45 seconds for most minutes. No tone is sent when voice bulletins are given. No voices or audio tones are sent on WWVB.

Propagation forecasts, weather bulletins and geophysical data are transmitted periodically on these stations. There are numerous other features of these broadcasts; *Special Publication 432* details them. This can be obtained free of charge by writing to:

Special Publication 432
National Institute of Standards and Technology
Mail Station 847
325 Broadway
Boulder, CO 80303.

WYE CONNECTION

In a three-phase alternating-current system, a *wye connection* is a method of interconnecting the windings of a transformer to a common point. The configuration gets its name from the fact that, in a schematic diagram, it appears like a capital letter Y.

The phases are 120 electrical degrees ($1/3$ cycle) out of phase with respect to each other (see THREE-PHASE ALTERNATING CURRENT). The center point is neutral. Wye configurations are often used in polyphase rectifiers (see POLYPHASE RECTIFIER).

X BAND

In the electromagnetic spectrum, either of two wavelength ranges can be called the *X band*. The range of frequencies from 5.2 to 11 GHz, corresponding to wavelengths of 5.8 and 2.7 cm, is known as the *X band*. This band is used for radar, line-of-sight communications, and satellite communications. The X band falls in the superhigh range, which is considered to be in the microwave part of the electromagnetic spectrum (*see* SUPERHIGH FREQUENCY).

At much shorter wavelengths, ranging from 4 nm (nanometers) down to about 0.01 nm, electromagnetic fields have extreme energy and penetrating properties. These are *X rays*. This band is occasionally called the *X band*.

YAGI ANTENNA

Two Japanese engineers, Yagi and Uda, researched how parasitic elements affect the directional properties of an antenna. *See* PARASITIC ARRAY, and PARASITIC ELEMENT. An array of parallel, straight antenna elements, one or more driven and one or more parasitic, was originally called a *Yagi-Uda* array. In recent years this has been shortened to *Yagi*. Sometimes an antenna of this kind is called a *beam*.

The Yagi is the most popular directional array on the amateur bands from 14 MHz through VHF. It is sometimes seen on 30 and 40 meters also. Most hams use rotators with their Yagi arrays so that the antenna can be pointed in any desired horizontal direction.

The Driven Element

The driven element or elements is/ are connected to the feed line. In amateur Yagi antennas, there is usually just one driven element. This element is sometimes physically shortened by inductive loading. It might contain traps so that it resonates on more than one band. It is half-wave resonant, is center-fed, and by itself would be a dipole antenna. *See* DIPOLE ANTENNA, and DRIVEN ELEMENT.

Gain and Front-to-back Ratio

The *forward gain*, or *gain*, of a Yagi is measured relative to a half-wave dipole (dBd) in most cases. The *front-to-back (F/B) ratio* is specified in decibels, and is the ratio of signal strength in the favored direction to the strength 180 degrees opposite the favored direction. *See* FRONT-TO-BACK RATIO. Gain figures for Yagi antennas are typically 3 dBd to 10 dBd at high frequencies (HF). F/B ratios at HF are from 15 to 25 dB. At very-high and ultra-high frequencies, where Yagis can have 15 to 20 elements, the gain can be close to 20 dBd, and the F/B ratio can be 40 dB or more.

When a Yagi is optimized for forward gain, there is some compromise in F/B ratio. Conversely, when it is optimized for the best possible F/B ratio, there is some compromise in the forward gain.

Two-element Yagi

A *two-element Yagi* can be formed by adding either a *director* or a *reflector* alongside the driven element. *See* DIRECTOR, and REFLECTOR. The optimum spacing for a driven-element/director Yagi is about 0.1 to 0.2 wavelength, with the director tuned to a frequency 5 to 10 percent higher than the resonant frequency of the driven element. The optimum spacing for a driven-element/reflector Yagi is about 0.15 to 0.2 wavelength, with the reflector tuned to a frequency 5 to 10 percent lower than the resonant frequency of the driven element.

The gain of an optimally built, full-size two-element Yagi is about 5 dBd using either a director or a reflector. If the elements are shortened by inductive loading or with traps, the gain is somewhat compromised.

Three-element Yagi

A Yagi with one director and one reflector, along with the driven element, increases the gain and F/B ratio, compared with a two-element beam. An optimally designed three-element Yagi has about 7 dBd gain. Figure A is a diagram of a three-element Yagi that can be scaled for universal construction purposes. Optimization must be done after the antenna is built. Matching schemes and construction details are not included.

Long Yagis

The gain and F/B ratio of a Yagi increases as more elements are added. This is usually done by placing extra directors in front of a three-element Yagi. Some Yagis have 10, 15, or even 20 elements. The extra directors tend to have optimum spacings that increase as the number of elements increases. Thus, to make a 10-element Yagi into an 11-element Yagi, a director is added about 0.4 wavelength in front of the 10-element Yagi for optimum performance. The directors also tend to get shorter and shorter, converging to a minimum limit. It is rare to see a Yagi of more than 20 elements. An ideal 20-element Yagi can theoretically give 19 dBd forward gain. If the gain needed in an antenna requires more than 20 elements on a single boom, two or more Yagis can be fed together in phase,

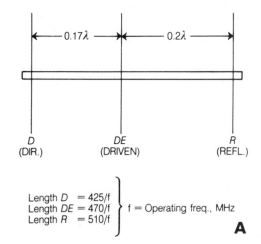

Length D = 425/f
Length DE = 470/f f = Operating freq., MHz
Length R = 510/f

A

YAGI ANTENNA: A. The general design for a three-element Yagi.

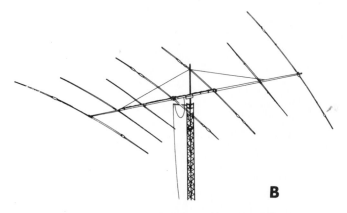

B

YAGI ANTENNA: B. A multiband Yagi. TELEX/HY-GAIN

and placed end-to-end (*collinear*) or *stacked*, or both. Other antennas, such as the dish, become easier to construct and tune than the Yagi at high UHF and microwave frequencies, when large gain and F/B ratio are wanted.

Multiband Yagis Traps can be placed in the elements of a Yagi, allowing operation on more than one band. Extra elements for some bands can be placed in between the elements for other bands, because of the differences in optimum element spacing. A multiband Yagi often takes on a complex appearance, as shown in Fig. B. The most common multiband Yagis work the bands from 14 MHz through 29.7 MHz (20 through 10 meters). A few also work other combinations, such as 40/30/20 meters or 30 through 10 meters.

Variations on the Yagi The driven element of a Yagi can be a full-wave loop, fed so that its polarization is parallel to that of the other elements. This antenna is called a *quagi* because it is a hybrid between a Yagi and a *quad*. Sometimes the reflector is also formed into a loop to make a quagi. *See* QUAD ANTENNA, and QUAGI ANTENNA.

A pair of Yagis can be oriented at right angles on a single boom, and fed 90 degrees out of phase to get *circular polarization*. *See* CIRCULAR POLARIZATION. This is favored when working through ham satellites.

YOKE

In a cathode-ray tube, the electron beam can be deflected by means of charged plates, or by means of coils. A complete set of magnetic deflection coils is called the *yoke* of the tube (*see* CATHODE-RAY TUBE, and PICTURE TUBE).

In an electric motor or generator, a ring-shaped piece of ferromagnetic material is used to hold the pole pieces together, and also to provide magnetic coupling between the poles. This assembly is known as a *yoke* (*see* GENERATOR, and MOTOR).

ZENER DIODE

All semiconductor diodes conduct poorly in the reverse-bias condition as long as the voltage remains below a certain critical value called the *avalanche voltage* (*see* AVALANCHE VOLTAGE). If the reverse bias exceeds the avalanche voltage, the resistance of the diode decreases greatly. The avalanche voltage of a particular diode depends on the way in which it is constructed.

A *zener diode* is manufactured specifically to have a constant, predictable avalanche voltage. This makes it possible to use reverse-biased diodes as voltage regulators, voltage dividers, reference-voltage sources, threshold detectors, transient-protection devices, and in other situations where a controlled avalanche response is needed.

Zener diodes are rated according to their avalanche voltage, and also according to the amount of power they can dissipate on a continuous basis. Zener diodes are available at ratings from less than 3 V to more than 150 V, and at power-dissipation ratings of up to 50-60 W. The high-voltage, high-power zener diodes can be used in place of regulator tubes in most types of medium-voltage power supplies.

A typical zener-diode characteristic curve is illustrated. In the forward-biased condition, the zener diode behaves very much like an ordinary silicon rectifier. In the reverse-biased condition, the impedance is practically infinite up to the critical, or zener, voltage (in this case, 14 V). Then, the impedance drops to practically zero. The reverse-bias voltage drop across this particular zener diode is a constant 14 V. This diode might be connected in parallel with the output of a power supply to obtain voltage regulation.

ZEPPELIN ANTENNA

A *zeppelin antenna* is a half-wavelength radiator, fed at one end with a quarter-wavelength section of open-wire line. The antenna gets its name from the fact that it was originally used for radio communications aboard zeppelins. The antenna, also called a *zepp*, was dangled in flight (see illustration A).

The zeppelin antenna can be thought of as a current-fed, full-wavelength radiator, part of which has been folded up to form the transmission line. The impedance at the feed point is extremely high, and at the transmitter end of the line, the impedance is practically zero. Any desired resistive impedance can be obtained by short-circuiting the end of the line opposite the feed point, and connecting the transmitter/receiver at some intermediate point, as at B. This configuration is sometimes used to feed a vertical half-wavelength radiator (*see* J ANTENNA). A zeppelin antenna will operate well at all odd harmonics of the design frequency.

If a transmatch is available, a half-wavelength radiator can be end fed with an open-wire line of any length (C). This arrangement is popular among many radio amateurs. The antenna will operate well at all harmonics of the design frequency. Thus, if the antenna radiator is cut for 3.5 MHz, the antenna will also work (as a longwire) at 7, 14, 21, and 28 MHz.

The primary advantage of the zeppelin antenna, especially the configuration at C, is ease of installation. The main difficulty with the zepp is that the feed line radiates to some extent

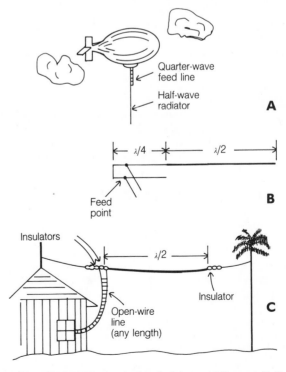

ZEPPELIN ANTENNA: At A, dangled from a blimp; at B, J-pole matching; at C, a typical ham installation.

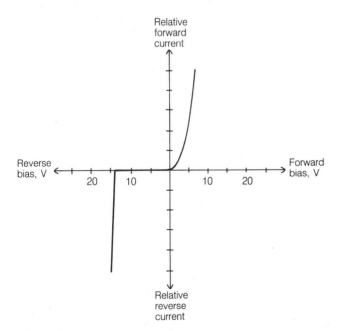

ZENER DIODE: The characteristic curve for a 14-V Zener diode.

because the system is not perfectly balanced. Feed-line radiation can be kept to a minimum by carefully cutting the radiator to ½ wavelength at the design frequency, and by using the antenna only at this frequency or one of its harmonics. For a wire radiator, properly insulated at the ends and placed well away from obstructions, the length L, in feet, at the design frequency f, is approximately:

$$L = 468/f$$

The length is meters is:

$$L = 143/f$$

Some ''pruning'' will probably be necessary to minimize radiation from the line. A field-strength meter can be used to observe the line radiation while adjusting the length of the antenna for optimum performance at the design frequency.

The radiation pattern from a zepp antenna is the same as that from a dipole at the design frequency, provided that the radiation from the line is not excessive. At harmonic frequencies, the radiation pattern is identical with that of an end-fed longwire of the same length. *See also* DIPOLE ANTENNA, and LONGWIRE ANTENNA.

ZERO BEAT

When two signals are mixed, beat notes, or heterodynes, are generated at the sum and difference frequencies (*see* BEAT, and HETERODYNE). As the frequencies of two signals are brought closer and closer together, the difference frequency decreases. When the two signals have the same frequency, the difference signal disappears altogether. This condition is called *zero beat*.

In a receiver having a product detector, a carrier produces an audible beat note (*see* PRODUCT DETECTOR). As the receiver is tuned, the frequency of the audio note varies. When the receiver is tuned closer to the carrier frequency, the beat note gets low-pitched and then disappears, indicating zero beat between the carrier and the local oscillator.

In communications practice, two stations often *zero beat* each other. This means that they both transmit on exactly the same frequency, reducing the amount of spectrum space used.

ZERO BIAS

When the control grid of a vacuum tube is at the same potential as the cathode, the tube is said to be operating at *zero bias*. In a field-effect transistor, zero bias is the condition in which the gate and the source are provided with the same voltage. Similarly, in a bipolar transistor, zero bias means that the base is at the same potential as the emitter.

In vacuum tubes and depletion-mode field-effect transistors, zero bias usually results in fairly large plate or drain current. In enhancement-mode field-effect transistors and in bipolar transistors, little or no drain or collector current normally flows with zero bias. There are a few exceptions, however.

Some devices are designed to operate at zero bias in certain applications. Some class-AB and class-B vacuum-tube power amplifiers operate in this way. Bipolar transistors at zero bias are often used as class-C amplifiers. *See also* BIAS, CHARACTERISTIC CURVE, CLASS-AB AMPLIFIER, CLASS-B AMPLIFIER, CLASS-C AMPLIFIER, CUTOFF VOLTAGE, DEPLETION MODE, ENHANCEMENT MODE, FIELD-EFFECT TRANSISTOR, GRID BIAS, NPN TRANSISTOR, PINCHOFF VOLTAGE, PNP TRANSISTOR, and TUBE.

ZERO BIT INSERTION
See BIT STUFFING.

ZERO-CENTER METER

Some meters are capable of indicating both positive and negative values, with the zero reading at the center. Such a meter is called a *zero-center meter*. Zero-center meters are used to measure negative and positive voltage or current, frequency centering, plus-or-minus errors, and other parameters that may fluctuate to either side of a zero or balanced condition. Zero-center meters are universally employed to indicate balance in bridge circuits.

The illustration shows an example of a zero-center meter for determining the carrier frequency of a radio transmitter. If the carrier is on the correct frequency, the meter needle will be at center scale, indicating zero error. If the transmitter frequency is too high, the needle position is to the right of center; if the transmitter frequency is low, the needle position is left of center. The meter shows that the transmitter frequency is approximately 2.5 kHz below the center of the channel. The most widely used type of zero-center meter is the galvanometer. *See also* GALVANOMETER.

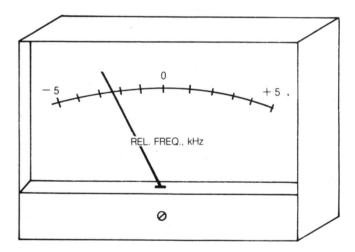

ZERO-CENTER METER: Indicates negative and positive values.

ZINC

Zinc is an element with atomic number 30 and atomic weight 65. In its pure form, zinc appears as a grayish metal of low luster. Zinc is a fairly good conductor of electric current.

Zinc is extensively used to coat iron and steel structures. This reduces the tendency for such materials to rust, since zinc corrodes less easily then many other metals. Zinc is also used in the manufacture of various types of electrochemical cells (*see* ZINC CELL).

ZINC CELL

Many different types of electrochemical cells use zinc as the negative electrode. Such cells are sometimes called *zinc cells*.

The most well-known type of zinc cell is the zinc-carbon dry cell (the common flashlight cell).

The zinc-carbon cell consists of a zinc case, usually cylindrical in shape, filled with an electrolyte paste. The zinc case forms the negative electrode of the cell. A carbon rod, immersed in the paste, forms the positive electrode. This type of cell provides approximately 1.5 V of direct current. Several such cells are often connected in series in a common case, forming a battery. *See also* BATTERY, CELL, and DRY CELL.

ZONE OF SILENCE
See SKIP ZONE.

Appendix 1
CW abbreviations

This list contains alphabetic/numeric abbreviations commonly used by hams when communicating in morse code (CW). See APPENDIX 2 for acronyms and slang. *See also* Q SIGNAL, and RST SYSTEM.

Abbreviation	Meaning
aa	all after
ab	all before
abt	about
adr	address
agn	again
ant	antenna
ar	end of transmission (letters run together)
bk	break; back
bn	been; between
bt	dash; pause (letters run together)
b4	before
c	yes
cfm	confirm
ck	check
cl	going off the air
clg	calling
cq	calling any station for a contact
cud	could
dr	dear
dx	long-distance communication; foreign station
es	and
fb	fine business
fist	quality of sending
ga	go ahead
gb	goodbye
ge	good evening
gg	going
gm	good morning
gn	good night
gud	good
hi	laughter
hr	here; hear; hour
hv	have
hw	how
ie	Is the frequency in use?
k	go ahead and transmit
kb	keyboard
lid	inept or discourteous operator
msg	message
n	no; numeral nine
ncs	net control station
nil	nothing
nr	number
nw	now
ob	old boy
oc	old chap
om	old man
op	operator
ot	old timer
pls	please (less common)
pse	please (more common)
pwr	power
px	publicity; press
qrq	high-speed code
r	roger (message received); decimal point
rcvr	receiver
rig	station hardware
rpt	repeat; report
rx	receiver; receive
rst	readability/strength/tone signal report
sed	said
sig	signal
sk	end of contact (letters run together)
sked	regular schedule for contacts
sri	sorry
svc	service
t	numeral zero
tfc	traffic
tmw	tomorrow
tks	thanks (less common)
tnx	thanks (more common)
tt	that
tu	thank you; terminal unit
tx	transmitter; transmit
txt	text
ur	your; you're
urs	yours
vy	very
wa	word after
wb	word before
wd	word
wkd	worked
wl	well; will
wpm	words per minute
wud	would
wx	weather
xcvr	transceiver
xmtr	transmitter
xtal	crystal
xyl	wife
yl	young lady ham
73	best regards
88	love and kisses

Appendix 2
Ham radio lingo

This list will help you identify slang used in voice modes. On the air, it's best to avoid excessive use of slang. For technical expressions, refer to article titles in the main body of this book, or to the index. For CW abbreviations, see APPENDIX 1. *See also* the articles Q SIGNAL, and RST SYSTEM.

Abbreviation	Meaning
afterburner	linear amplifier
amp	ampere; linear amplifier
barefoot	operation without a linear amplifier
beam	Yagi antenna
bird	satellite
birdie	spurious response in a receiver
boat anchor	old, antiquated, bulky equipment
brass pounder	ham who handles much traffic; straight key
breaker	person trying to enter a QSO in progress
bug	semiautomatic key
call	callsign
clone	a piece of equipment that is identical to another; a "copy" of a computer
copy	receive; readability; written messages
desensing	receiver overloading by strong signal
download	receive packets from another station or from a bulletin board
DXpedition	trip to foreign land to "be DX"
eyeball	meeting in person
feedback	comments; corrections
flat	SWR of 1:1
foxhunt	hidden-transmitter search
full quieting	strong, clear FM reception
gateway	means for stations to communicate when on different frequencies and/or modes
handle	name
handy-talky; HT	handheld transceiver
header	start of a packet frame, containing no information
hound	enthusiast (e.g., DX hound, CW hound)
intermod	intermodulation distortion
junk box	collection of spare parts
kerchunk	to actuate a repeater without saying anything
key; key up	to actuate a transmitter
landline	telephone
league, the	American Radio Relay League
lid	inept or discourteous operator
log off	sign off from a packet bulletin board
log on	sign onto a packet bulletin board
machine	repeater; digipeater; satellite
mike	microphone
monkey chatter	sound of SSB signal improperly tuned in
over	reference to dB over s-9
overhead	part of a packet that does not contain message information
paddle	key for electronic keyer
patch	connect(ion); interconnect(ion)
phone	voice communications
pileup	many stations calling a single station
rag chewing	long conversations on the radio
rig	station hardware
ritty	RTTY (radioteletype)
roundtable	contact involving three or more stations
rubber duck	shortened, flexible VHF antenna
scanner	receiver that scans for open or vacant channels in a band
shack	station
skyhook	antenna
solid copy	easy-to-receive signal
splatter	spurious sidebands
swisher	person who transmits while tuning VFO
s-1 through s-9	expression of signal strength. See the article RST SYSTEM
test	contest
top band	160 meters
traffic	messages sent and received, especially on behalf of third parties
trailer	end of a packet frame, containing no information
tuner	antenna tuner; operator tuning up on frequency
upload	send packet data to another station or to a bulletin board
vox	voice-actuated operation
wallpaper	collection of QSL cards
woodpecker	over-the-horizon HF radar signals
wormhole	packet link via satellite
XYL	wife
YL	female ham
zulu	Coordinated Universal Time
73	best regards
88	love and kisses

Appendix 3
Physical conversions

Electrical Properties

Unit	To Change Multiplied by	Yields	Unit	To Change Back Multiplied by	Yields
Power					
ft-lb/min	0.0000303	hp	hp	33,000.0	ft-lb/min
ft-lb/s	0.001818	mp	hp	550.0	ft-lb/min
ft-lb/min	0.0226	W	W	44.25	ft-lbs/min
ft-lb/s	1.356	W	W	0.7373	ft-lb/s
W	0.001341	hp	hp	746.0	W
Btu/hr	0.000393	hp	hp	2,545.0	Btu/hr

1 hp = 746 W	1 kW = 1.000 J/s
0.746 kW	1.34 hp
33.000 ft lbs/min	44,250 ft-lb/min
550 ft-lbs/s	737.3 ft-lb/s
2,545 Btus/hr	3,412 Btus/hr
0.175 lb carbon oxidized/hr	0.227 lb carbon oxidized/hr
17 lbs water/hr heated from 62 to 212 °F	22.75 lbs water/hr heated from 62 to 212 °F
2.64 lbs water/hr evaporated from and at 212 °F	3.53 lbs water/hr evaporated from and at 212 °F

Energy

Unit	To Change Multiplied by	Yields	Unit	To Change Back Multiplied by	Yields
ergs	0.0000001	J	J	10,000,000.0	ergs
J	0.2388	g-cal	g-cal	4.186	J
J	0.10198	kg-m	kg-m	9.8117	J
J	0.375	ft-lb	ft-lb	1.356	J
ft-lb	0.1383	kg-m	kg-m	7.233	ft-lb
g-cal	0.003968	Btu	Btu	252.0	g-cal
J	0.000947	Btu	Btu	1,055.0	J
ft-lb	0.001285	Btu	Btu	778.0	ft-lb
Btu	0.293	W-hr	W-hr	3,416.0	Btu

Miscellaneous

Unit	Multiplied by	Yields	Unit	Multiplied by	Yields
ohms/ft	0.3048	ohms/meter	ohms/m	3.2808	ohms/ft
ohms/km	0.3048	ohms/1000 ft	ohms/1000 ft	3.2808	ohms/km
ohms/km	0.9144	ohms/1000 yds	ohms/1000 yds	1.0936	ohms/km

Comparison of Electric and Magnetic Circuits

Property	Electric circuit	Magnetic circuit
force	volts (V or E) or emf	gilberts (F)
flow	amperes (I)	flux (Φ) in maxwells
opposition	ohms (R)	reluctance ($\Re$) or rels
law	Ohm's law ($I = E/R$)	Rowland's law ($\Phi = F/R$)
intensity of force	volts per cm of length	$H = (1.2571N)/l$ (in gilberts per cm of length)
density	current density (i.e., amperes per cm^2)	flux density (i.e., lines per cm^2, or gausses)

Conversion between Electrical Systems

Property	mks	System cgs electromagnetic	cgs electrostatic
Capacitance	1 farad	10^{-9} abfarad	9×10^{11} statfarad
	10^9 F	1 abF	9×10^{20} statF
	$10^{-11}/9$ F	$10^{-20}/9$ abF	1 statF
Charge	1 coulomb	0.1 abcoulomb	3×10^9 statC
	10 C	1 abc	3×10^{10} statC
	$10^{-9}/3$ C	$10^{-10}/3aC$	1 statC
Charge density	1 coulomb/m^3	10^{-7} abcoulomb/cm^3	3×10^3 statcoulomb/cm^3
	10^7 C/m^3	1 abC/cm^3	3×10^{10} statCcm^3
	$10^{-3}/3$ C/m^3	$10^{-10}/3$ aC/cm^3	1 statC/cm^3
Conductivity	1 siemens/m	10^{-11} absiemens/cm	9×10^9 statSiemens/cm
	10^{11} S/m	1 abS/cm	9×10^{20} statS/cm
	$10^{-9}/9$ S/m	$10^{-20}/9$ abS/cm	1 statS/cm
Current	1 ampere	10^{-1} abampere	3×10^9 statampere
	10 a	1 abA	3×10^{10} statA
	$10^{-9}/3$ a	$10^{-10}/3$ abA	1 statA
Current density	1 ampere/m^2	10^{-5} abampere/cm^2	3×10^5 statampere/cm^2
	10^5 A/m^2	1 abA/cm^2	3×10^{10} statA/cm^2
	$10^{-3}/3$ A/m^2	$10^{-10}/3$ aA/cm^2	1 statA/cm^2
Electric field intensity	1 volt/m	10^6 abvolt/cm	$10^{-4}/3$ statvolt/cm
	10^{-6} V/m	1 abV/cm	$10^{-10}/3$ statV/cm
	3×10^4 V/m	3×10^{10} aV/cm	1 stat V/cm
Electric potential	1 volt	10^8 abvolt	$10^{-2}/3$ statvolt
	10^{-8} V	1 abV	$10^{-10}/3$ statV
	3×10^2 V	3×10^{10} aV	1 statV
Electric dipole moment	1 coulomb-m	10 abC-cm	3×10^{11} statC-cm
	0.1 C-m	1 abC-cm	3×10^{10} statC-cm
	$10^{-11}/3$ C-m	$10^{-10}/3$ abC-cm	1statC-cm
Energy	1 joule	10^7 erg	10^7 erg
	10^{-7} J	1 e	1 e
	10^{-7} J	1 e	1 e
Force	1 newton	10^5 dyne	10^5 dyne
	10^{-5} N	1 d	1 d
	10^{-5} N	1 d	1 d
Flux density	1 Weber/m^2	10^4 gauss (or abtesia)	$10^{-6}/3$ electrostatic unit
	10^{-4} Wb/m^2	1 G	$10^{-10}/3$ esu
	3×10^6 Wb/m^2	3×10^{10} G	1 esu
Inductance	1 henry	10^9 abhenry	$10^{-11}/9$ stathenry
	10^{-9} H	1 abH	$10^{-20}/9$ statH
	9×10^{11} H	9×10^{20} abH	1 statH
Inductive capacity	1 farad/m	10^{-11} abfarad/cm	9×10^9 statfarad/cm
	10^{11} F/m	1 abF/cm	9×10^{20} statF/cm
	$10^{-9}/9$ F/m	$10^{-20}/9$ abF/cm	1 statF/cm

Conversion between Electrical Systems

Property	mks	System cgs electromagnetic	cgs electrostatic
Magnetic flux	1 weber	10^8 Maxwell	$10^{-2}/3$ electrostatic unit
	10^{-8} W	1 Mx	$10^{-10}/3$ esu
	3×10^2 W	3×10^{10} Mx	1 esu
Magnetic dipole moment	1 ampere-m^2	10^3 abampere-cm^2	3×10^{13} statampere-cm^2
	10^{-3} A-m^2	1 abA-cm^2	3×10^{10} statA-cm^2
	$10^{-13}/3$ A-m^2	$10^{-10}/3$ abA-cm^2	1 statA-cm^2
Permeability	1 henry/m	10^7 abhenry/cm	$10^{-13}/9$ stathenry/cm
	10^{-7} H/m	1 abH/cm	$10^{-20}/9$ statH/cm
	9×10^{13} H/m	9×10^{20} abH/cm	1 statH/cm
Power	1 watt	10^7 erg/s	10^7 erg/s
	10^{-7} W	1 e/s	1e/s
	10^{-7} W	1 e/s	1e/s
Resistance	1 ohm	10^9 abohm	$10^{-11}/9$ statohm
	10^{-9} ohm	1 abohm	$10^{-20}/9$ statohm
	9×10^{11} ohm	9×10^{20} abohm	1 statohm

Appendix 4
Schematic symbols

ammeter	
amplifier (operational)	or
AND gate	
antenna (balanced, dipole)	
antenna (general)	or or
antenna (loop, shielded)	
antenna (loop, unshielded)	
antenna (unbalanced)	or
antenna (whip, or telescopic)	
attenuator (or resistor, fixed)	
attenuator (or resistor, variable)	
battery	
capacitor (feedthrough)	or
capacitor (fixed, nonpolarized)	
capacitor (fixed, polarized)	
capacitor (ganged, variable)	
capacitor (single, variable)	
capacitor (split-rotor, variable)	
capacitor (split-stator, variable)	
cathode (directly heated)	
cathode (indirectly heated)	or
cathode (cold)	
cavity resonator	
cell	

circuit breaker	
coaxial cable	or
coaxial cable (grounded shield)	or
crystal (piezoelectric)	
delay line	or
diode (field-effect)	
diode (general)	
diode (Gunn)	
diode (light-emitting)	
diode (photosensitive)	
diode (photovoltaic)	
diode (pin)	or
diode (Schottky)	
diode (tunnel)	or
diode (varactor)	
diode (zener)	
directional coupler (or wattmeter)	or
exclusive-OR gate	
female contact (general)	
ferrite bead	or
fuse	or or
galvanometer	or
ground (chassis)	or
ground (earth)	
handset	
headphone (double)	
headphone (single)	
headphone (stereo)	
inductor (air-core)	

inductor (bifilar)

inductor (iron-core)

inductor (tapped)

inductor (variable)   or

integrated circuit

inverter or inverting amplifier

jack (coaxial or phono)

jack (phone, two-conductor)

jack (phone, two-conductor interrupting)

jack (phone, three-conductor)

jack (phono)

key (telegraph)

lamp (incandescent)

lamp (neon)   or

male contact (general)

meter (general)

microammeter

microphone   or

microphone (directional)

milliammeter

NAND gate

negative voltage connection

NOR gate

operational amplifier   or

OR gate

outlet (nonpolarized)

outlet (polarized)

outlet (utility, 117 V, nonpolarized)

outlet (utility, 234 V)

photocell (tube)

plug (nonpolarized)

plug (polarized)

plug (phone, two-conductor)

plug (phone, three-conductor)

plug (phono)

plug (utility, 117 V)

plug (utility, 234 V)

positive-voltage connection

potentiometer (variable resistor, or rheostat)   or

probe (radio-frequency)   or

rectifier (semiconductor)

rectifier (silicon-controlled)

rectifier (tube-type)

rectifier (tube-type, gas-filled)

relay (DPDT)

relay (DPST)

relay (SPDT)

relay (SPST)

resistor (fixed)

resistor (preset)

resistor (tapped)

resonator

rheostat (variable resistor, or potentiometer)   or

saturable reactor

shielding

signal generator

solar cell

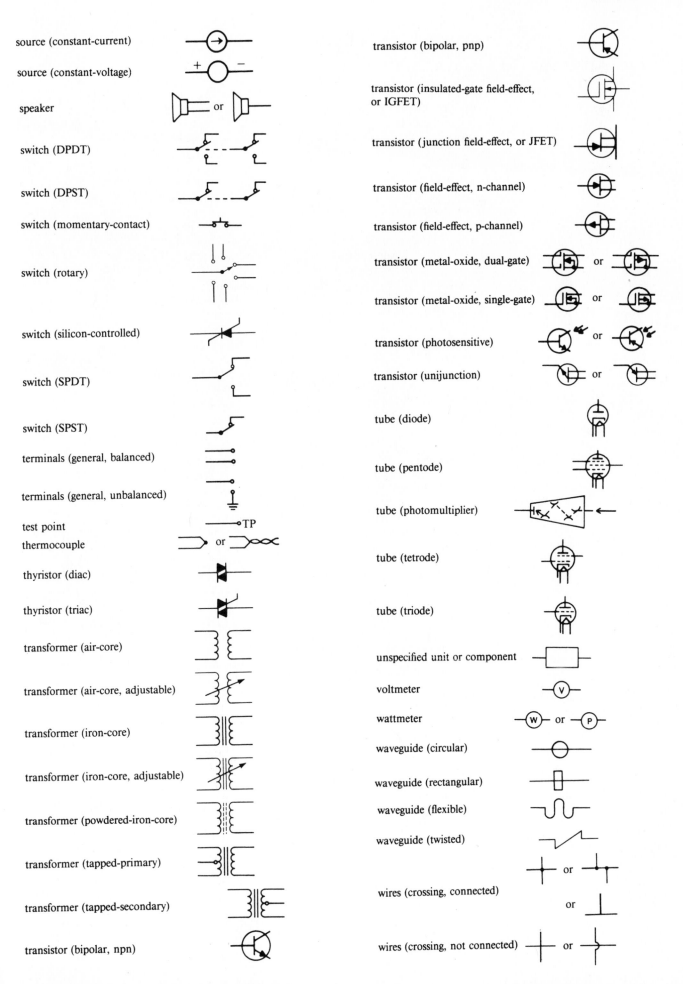

source (constant-current)

source (constant-voltage)

speaker   or

switch (DPDT)

switch (DPST)

switch (momentary-contact)

switch (rotary)

switch (silicon-controlled)

switch (SPDT)

switch (SPST)

terminals (general, balanced)

terminals (general, unbalanced)

test point   TP

thermocouple   or

thyristor (diac)

thyristor (triac)

transformer (air-core)

transformer (air-core, adjustable)

transformer (iron-core)

transformer (iron-core, adjustable)

transformer (powdered-iron-core)

transformer (tapped-primary)

transformer (tapped-secondary)

transistor (bipolar, npn)

transistor (bipolar, pnp)

transistor (insulated-gate field-effect, or IGFET)

transistor (junction field-effect, or JFET)

transistor (field-effect, n-channel)

transistor (field-effect, p-channel)

transistor (metal-oxide, dual-gate)   or

transistor (metal-oxide, single-gate)   or

transistor (photosensitive)   or

transistor (unijunction)   or

tube (diode)

tube (pentode)

tube (photomultiplier)

tube (tetrode)

tube (triode)

unspecified unit or component

voltmeter   V

wattmeter   W   or   P

waveguide (circular)

waveguide (rectangular)

waveguide (flexible)

waveguide (twisted)

wires (crossing, connected)   or   or

wires (crossing, not connected)   or

Index

*Boldface page numbers refer to art

About the author

Stan Gibilisco was born in 1953 and is the son of Dr. Joseph A. Gibilisco, who served for more than 30 years as a staff physician at the Mayo Clinic in Rochester, Minnesota. A mathematician educated at the University of Minnesota, Stan has been a radio ham since 1966, and has held the Extra Class license since 1973. His callsign, W1GV, is known to many hams from dozens of technical articles in amateur-radio magazines.

Between 1977 and 1982, Stan served as Assistant Technical Editor for *QST* Magazine of the American Radio Relay League, Inc., in Newington, Connecticut, and as Vice President of Engineering for International Electronic Systems, Inc., in Miami, Florida. In 1982, Stan began writing about science and electronics full-time.

Stan first attracted attention with his book *Understanding Einstein's Theories of Relativity* (TAB Books, 1983) and as Editor-in-Chief of *Encyclopedia of Electronics* (TAB Professional and Reference Books, 1985). The *Encyclopedia* was annotated by the American Library Association as one of the Best Reference Books of the 1980s. To date, Stan has authored and co-authored more than 20 books. His books about astronomy and mathematical subjects enjoy a growing audience in Japan.

Stan plans to continue writing books and articles about science and electronics. In addition, he aspires to write science books for children, as well as fiction stories and novels.